# LIST OF THE ELEMENTS WITH THEIR SYMBOLS AND ATOMIC WEIGHTS

| Element | Symbol | Atomic number | Atomic weight | Element | Symbol | Atomic number | Atomic weight | Element | Symbol | Atomic number | Atomic weight |
|---|---|---|---|---|---|---|---|---|---|---|---|
| Actinium | Ac | 89 | 227.0278 | Hahnium [b] | Ha | 105 | (262) | Protactinium | Pa | 91 | 231.0359 |
| Aluminum | Al | 13 | 26.98154 | Helium | He | 2 | 4.00260 | Radium | Ra | 88 | 226.0254 |
| Americium | Am | 95 | (243)[a] | Holmium | Ho | 67 | 164.9304 | Radon | Rn | 86 | (222) |
| Antimony | Sb | 51 | 121.75 | Hydrogen | H | 1 | 1.0079 | Rhenium | Re | 75 | 186.207 |
| Argon | Ar | 18 | 39.948 | Indium | In | 49 | 114.82 | Rhodium | Rh | 45 | 102.9055 |
| Arsenic | As | 33 | 74.9216 | Iodine | I | 53 | 126.9045 | Rubidium | Rb | 37 | 85.4678 |
| Astatine | At | 85 | (210) | Iridium | Ir | 77 | 192.22 | Ruthenium | Ru | 44 | 101.07 |
| Barium | Ba | 56 | 137.33 | Iron | Fe | 26 | 55.847 | Rutherfordium [b] | Rf | 104 | (261) |
| Berkelium | Bk | 97 | (247) | Krypton | Kr | 36 | 83.80 | Samarium | Sm | 62 | 150.36 |
| Beryllium | Be | 4 | 9.01218 | Lanthanum | La | 57 | 138.9055 | Scandium | Sc | 21 | 44.9554 |
| Bismuth | Bi | 83 | 208.9804 | Lawrencium | Lr | 103 | (260) | Selenium | Se | 34 | 78.96 |
| Boron | B | 5 | 10.81 | Lead | Pb | 82 | 207.2 | Silicon | Si | 14 | 28.0855 |
| Bromine | Br | 35 | 79.904 | Lithium | Li | 3 | 6.941 | Silver | Ag | 47 | 107.8682 |
| Cadmium | Cd | 48 | 112.41 | Lutetium | Lu | 71 | 174.967 | Sodium | Na | 11 | 22.98977 |
| Calcium | Ca | 20 | 40.08 | Magnesium | Mg | 12 | 24.305 | Strontium | Sr | 38 | 87.62 |
| Californium | Cf | 98 | (251) | Manganese | Mn | 25 | 54.9380 | Sulfur | S | 16 | 32.06 |
| Carbon | C | 6 | 12.011 | Mendelevium | Md | 101 | (258) | Tantalum | Ta | 73 | 180.9479 |
| Cerium | Ce | 58 | 140.12 | Mercury | Hg | 80 | 200.59 | Technetium | Tc | 43 | (98) |
| Cesium | Cs | 55 | 132.9054 | Molybdenum | Mo | 42 | 95.94 | Tellurium | Te | 52 | 127.60 |
| Chlorine | Cl | 17 | 35.453 | Neodymium | Nd | 60 | 144.24 | Terbium | Tb | 65 | 158.9254 |
| Chromium | Cr | 24 | 51.996 | Neon | Ne | 10 | 20.179 | Thallium | Tl | 81 | 204.383 |
| Cobalt | Co | 27 | 58.9332 | Neptunium | Np | 93 | 237.0482 | Thorium | Th | 90 | 232.0381 |
| Copper | Cu | 29 | 63.546 | Nickel | Ni | 28 | 58.69 | Thulium | Tm | 69 | 168.9342 |
| Curium | Cm | 96 | (247) | Niobium | Nb | 41 | 92.9064 | Tin | Sn | 50 | 118.69 |
| Dysprosium | Dy | 66 | 162.50 | Nitrogen | N | 7 | 14.0067 | Titanium | Ti | 22 | 47.88 |
| Einsteinium | Es | 99 | (254) | Nobelium | No | 102 | (259) | Tungsten | W | 74 | 183.85 |
| Erbium | Er | 68 | 167.26 | Osmium | Os | 76 | 190.2 | Unnilennium [b] | Une | 109 | (266) |
| Europium | Eu | 63 | 151.96 | Oxygen | O | 8 | 15.9994 | Unnilhexium [b] | Unh | 106 | (263) |
| Fermium | Fm | 100 | (257) | Palladium | Pd | 46 | 106.4 | Unnilseptium [b] | Uns | 107 | (262) |
| Fluorine | F | 9 | 18.998403 | Phosphorus | P | 15 | 30.97376 | Uranium | U | 92 | 238.0289 |
| Francium | Fr | 87 | (223) | Platinum | Pt | 78 | 195.08 | Vanadium | V | 23 | 50.9415 |
| Gadolinium | Gd | 64 | 157.25 | Plutonium | Pu | 94 | (244) | Xenon | Xe | 54 | 131.29 |
| Gallium | Ga | 31 | 69.72 | Polonium | Po | 84 | (209) | Ytterbium | Yb | 70 | 173.04 |
| Germanium | Ge | 32 | 72.59 | Potassium | K | 19 | 39.0983 | Yttrium | Y | 39 | 88.9059 |
| Gold | Au | 79 | 196.9665 | Praseodymium | Pr | 59 | 140.9077 | Zinc | Zn | 30 | 65.38 |
| Hafnium | Hf | 72 | 178.49 | Promethium | Pm | 61 | (145) | Zirconium | Zr | 40 | 91.22 |

[a] Approximate values for radioactive elements are listed in parentheses.
[b] The official name and symbol have not been agreed to. The names for elements 106, 107, and 109 represent their atomic numbers, as in un (1) nil (0) hex (6) = unnilhexium (Unh) for element 106. Element 109 was reported in 1982; 108 is not known.

# CHEMISTRY
## The Central Science

**4**TH EDITION

# CHEMISTRY
## The Central Science

## THEODORE L. BROWN
*University of Illinois*

## H. EUGENE LeMAY, JR.
*University of Nevada*

PRENTICE HALL, Englewood Cliffs, New Jersey 07632

*Library of Congress Cataloging-in-Publication Data*

Brown, Theodore L.
    Chemistry: the central science/Theodore L. Brown, H. Eugene
LeMay, Jr.—4th ed.
        p.    cm.
    Includes index.
    ISBN 0-13-130246-9
    1. Chemistry.    I. LeMay, H. Eugene (Harold Eugene), 1940–
II. Title.
QD31.2.B78 1987                                                    87-26358
540—dc19                                                          CIP

Development editor: Stephen Deitmer
Editorial/production supervision: Fay Ahuja
Interior design and cover design: Janet Schmid
Manufacturing buyer: Paula Benevento
Page layout: Maureen Eide
Photo research: Tobi Zausner
Photo editor: Lorinda Morris-Nantz
Cover photomicrograph courtesy of McCrone Research Institute, Chicago

Cover: Crystals of picric acid viewed under polarized light. Picric acid, or
2,4,6-trinitrophenol, melts at 122°C. It is a rather strong acid, which reacts with
active metals to form metal picrate salts. These salts are highly sensitive
explosives. Picric acid is itself an explosive; like TNT (trinitrotoluene), it
requires a detonator. Picric acid has also been used as a yellow dye, as an
antiseptic, and in the synthesis of certain insecticides.

© 1988, 1985, 1981, 1977 by Prentice-Hall, Inc.
A Division of Simon & Schuster
Englewood Cliffs, New Jersey 07632

Printed in the United States of America

10  9  8  7  6  5  4


ISBN 0-13-130246-9


Prentice-Hall International (UK) Limited, *London*
Prentice-Hall of Australia Pty. Limited, *Sydney*
Prentice-Hall Canada Inc., *Toronto*
Prentice-Hall Hispanoamericana, S.A., *Mexico*
Prentice-Hall of India Private Limited, *New Delhi*
Prentice-Hall of Japan, Inc., *Tokyo*
Simon & Schuster Asia Pte. Ltd., *Singapore*
Editora Prentice-Hall do Brasil, Ltda., *Rio de Janeiro*

To our students, whose enthusiasm and
curiosity have often inspired us,
and whose questions and suggestions have
sometimes taught us.

# Brief Contents

# Contents

**CHAPTER 9**    **Geometries of Molecules; Molecular Orbitals**       **260**

**CHAPTER 10**    **Gases**       **300**

---

**CHAPTER 21 Nuclear Chemistry** 688

---

**CHAPTER 22 Chemistry of Hydrogen, Oxygen, Nitrogen, and Carbon** 722

# Preface

*Chemistry: The Central Science*, fourth edition, has been written to introduce you to modern chemistry. During the many years we have been practicing chemists, we have found chemistry to be an exciting intellectual challenge and an extraordinarily rich and varied part of our cultural heritage. We hope that as you advance in your study of chemistry, you will share with us some of that enthusiasm and appreciation. As authors, we have, in effect, been engaged by your instructor to help you learn chemistry. Based on the comments of students and instructors who have used this book in its previous editions, we believe that we have done that job well. Of course, we expect the text to continue to evolve through future editions. We invite you to write to us to tell us what you like about the book, so that we will know where we have helped you most. Also, we would like to learn of any shortcomings, so that we might further improve it in subsequent editions.

**Features of the Text**

As you page through any chapter of this text, you will see sections entitled *Sample Exercise*. Each exercise illustrates the use of a key concept or skill. The accompanying solution provides the reasoning required to answer the exercise. Paired to every Sample Exercise is a *Practice Exercise*, which addresses the same concept. Test your understanding by working the Practice Exercise and comparing your answer with the one that is given.

*Full-color illustrations*, which include photographs, charts, graphs, and diagrams, help you to visualize abstract ideas. Don't neglect to study the figures and their captions as you read the text.

Three kinds of brief *supplemental sections* appear throughout the text as enrichment material. The *Chemistry at Work* sections introduce interesting applications of the concepts under consideration in the text. Those supplements entitled *A Closer Look* enrich the discussion by delving more deeply into a topic. Those entitled *A Historical Perspective* expand your knowledge of what led to a particularly important development or discovery in chemistry.

Each chapter closes with a mini-study guide—entitled *For Review*—to help you master key concepts. The *Summary* points out the chapter highlights. The *Learning Goals* outline the skills you are expected to have learned in studying the chapter. The *Key Terms* help you build your vocabulary.

The *Exercises* at the end of each chapter test your understanding of the material. Many exercises are grouped according to topic, and these are also arranged in matched pairs. Each exercise in a pair deals with the same principle or procedure, so if you have difficulty with a particular exercise, its companion will provide you with further practice. The answer to the odd-numbered exercise of each pair is given in a section following the appendices. *Additional exercises* appear at the end of each chapter's exercise set. The additional exercises test your ability to solve problems that are not identified by topic. These exercises often combine ideas from more than one part of the chapter. The additional exercises and the matched-pair exercises whose answers appear near the end of the book are numbered in color. The more challenging exercises are marked with brackets.

Finally, you should note that there are several *appendices* near the back of the book, as well as useful tables in the front and back inside covers. You should find the pull-out periodic table at the front inside cover to be particularly useful.

## Advice for Studying Chemistry

*Keep up with your studying day to day.* New material builds on the old. It is important not to fall behind; if you do, you will find it much harder to follow the lectures and discussions on current topics. Trying to "cram" just before an exam is generally a very ineffective way to study chemistry.

*Focus your study.* The amount of information you will receive in your chemistry course can sometimes seem overwhelming. Certainly, there are more facts and details than any student can hope to assimilate in a first course. It is important to focus on the key concepts and skills. Listen intently for what your lecturer and discussion leader emphasize. Pay attention to the skills stressed in the sample exercises and homework assignments. Notice the italicized statements in the text, and study the concepts presented in the chapter summaries.

*Keep good lecture notes,* so that you have a clear and concise record of the required material. You will find it easier to take useful notes if you *skim topics in the text before they are covered in lecture*. To skim a chapter, first read the introduction and summary. Then quickly read through the chapter, skipping Sample Exercises and supplemental sections. Pay attention to section heads and subheads, which give you a feeling for the scope of topics. Avoid the compulsion to learn and understand everything right away.

After lecture, *carefully read the topics covered in class*. Remember, though, that you'll need to read assigned material more than once to master it. As you read, pay particular attention to the Sample Exercises. Once you think you understand a Sample Exercise, test your understanding by working the accompanying Practice Exercise.

Finally, *attempt all of the assigned end-of-chapter exercises*. If you get stuck on a problem, get help from your instructor, tutor, or a fellow student. It is rarely effective to spend more than 20 minutes on a single exercise.

# TO THE INSTRUCTOR

## Philosophy

Throughout the evolution of this text, certain goals have guided our writing efforts. The first is that a text should endeavor to show students the usefulness of chemistry in their major areas of study as well as in the world around them. It has been our experience that as students become aware of the importance of chemistry to their own goals and interests, they become more enthusiastic about learning the subject. With this in mind, we have attempted, as much as space and our imaginations permit, to bring in interesting and significant applications as an integral part of the subject matter development. We attempt to show that chemistry is indeed the *central science*. At the same time, of course, we seek to provide students with the background in modern chemistry that they will need for their specialized studies, including more advanced chemistry courses.

Second, we want to show not only that chemistry provides the basis for much of what goes on in our world, but that it is a vital, continually developing science. We have tried to keep the book up-to-date in terms of new concepts and applications and to convey some of the excitement of the field.

Third, we feel that any text should be written to the students and not just to their instructors. We have sought to keep our writing clear and interesting and the book attractive and well illustrated. Furthermore, we have provided numerous in-text study helps for students. A more subtle aspect of this student orientation is the care we have taken to describe problem-solving strategies.

## Organization

In the present edition, the first four chapters give a largely macroscopic, phenomenological view of chemistry. They introduce many basic concepts (including nomenclature and stoichiometry) that provide the background required for many of the laboratory experiments usually performed in general chemistry. Chapter 5, also macroscopic in breadth, gives an introduction to some of the common elements and to the use of the periodic table. It serves as a bridge to the next four chapters, which deal with electronic structure and bonding (Chapters 6–9). After the treatment of bonding, the orientation changes to a focus on the states of matter (Chapters 10 and 11) and solutions (Chapters 12 and 13). Chapter 14 is an optional chapter in which the concepts developed in the preceding chapters are applied to a discussion of the atmosphere and hydrosphere. The next several chapters examine the factors that determine the speed and extent of chemical reactions: kinetics (Chapter 15), equilibria (Chapters 16–18), thermodynamics (Chapter 19), and electrochemistry (Chapter 20). After a discussion of nuclear chemistry (Chapter 21), the final block of material surveys the chemistry of nonmetals, metals, organic chemistry, and biochemistry (Chapters 22–28).

Although this is a fairly standard organization, not everyone teaches all of these topics in exactly this order. For example, many instructors prefer to introduce gases after stoichiometry or after thermochemistry rather than with states of matter. Some instructors may prefer an earlier introduction to redox reactions, and others may choose to discuss equilibria before kinetics. We have structured our writing so that changes in teaching sequence may be made with no loss in student comprehension.

We have introduced students to the properties of elements and their compounds in several ways throughout the text. Most obvious are those chapters that emphasize descriptive chemistry (especially Chapters 5, 13, 14, and 22–28). In addition, material on the properties of elements and compounds is woven throughout the text to illustrate principles and applications. Furthermore, such descriptive chemistry is also used in end-of-chapter exercises.

**Changes in this Edition**

The most obvious change in this edition is the new full-color design and the inclusion of a large number of *full-color photographs and illustrations*. There were 454 figures in the third edition; the fourth edition contains over 600, most of them in full color. The greatly increased use of color photographs allows us to illustrate chemical reactions and convey the beauty and excitement of chemistry. The functional use of color throughout the text helps call the students' attention to aspects of chemistry that we wish to emphasize and helps make the text attractive and inviting.

Many sections have been extensively rewritten to improve clarity and to incorporate full-color graphics and photos. For example, material on elements and compounds has been moved to the beginning of Chapter 1 so the book begins with a discussion of chemical concepts. Treatment of units, significant figures, and dimensional analysis remains in Chapter 1, however.

More applications and *descriptive chemistry* have been added to the first half of the text. The most significant of these changes is the new Chapter 5, Introduction to the Elements and the Periodic Table. This chapter discusses the discovery and use of the periodic table from an empirical perspective and introduces the chemistry of many of the common representative elements. It provides a background for discussing electronic structure and bonding in subsequent chapters. The chapter is short and can be skipped, if so desired, with no adverse effect on students' grasp of subsequent material.

Chapter 13, Reactions in Aqueous Solution, is another new chapter. It brings together a variety of topics dealing with aqueous solution chemistry: net ionic equations, solubility rules and precipitation reactions, Brønsted acids and bases, and oxidation-reduction reactions. In earlier editions, most of this material was discussed in various later chapters.

A *Practice Exercise* has been added to each Sample Exercise. Practice Exercises provide an opportunity for students to test and reinforce their understanding of the concepts or skills treated in the Sample Exercises.

The *end-of-chapter exercises* that are arranged by topic are now also arranged in matched pairs. The answer to the odd-numbered member of each pair and the answer to selected additional exercises are given in a section following the appendices. Each exercise that is answered is labeled in red. As with earlier editions, we use real-life situations as often as possible as the context for exercises. Furthermore, the large variety of exercises allows the instructors to select those that are most appropriate for their course.

**AVAILABLE SUPPLEMENTS**

Several supplements accompany *Chemistry: The Central Science*, fourth edition:

An *Instructor's Guide* to the text includes suggested readings, alternate arrangement of chapters, and a general commentary on each chapter.

*Student's Guide* has been written by Professor James C. Hill of California State University, Sacramento. This soft-cover book has been carefully structured to help students in their study of chemistry. It is filled with helpful ideas, problem-solving techniques, and fresh insights into the materials presented in the text. Each chapter of the text has a corresponding chapter in the student's guide that gives an overview of the chapter, summarizes the major topics and concepts, offers sample problems, and features an end-of-chapter test. These chapter tests include true-false, multiple-choice, and essay problems. Solutions are also given to each question.

*Laboratory Experiments for Chemistry: The Central Science*, written by Professors John H. Nelson and Kenneth C. Kemp of the University of Nevada, Reno, includes 40 experiments sequenced to follow the topical development of the text. *The instructor's manual* that accompanies *Laboratory Experiments* includes sample data and calculations for all answer sheets, as well as directions for preparation of all solutions.

A combination text and laboratory manual on *Qualitative Inorganic Analysis* has been prepared by the authors of this text. This supplement contains chapters on laboratory techniques, a review of aqueous solution equilibria, and discussions of the chemistry of the qualitative analysis groups, including complete laboratory procedures for identifying both cations and anions.

*Solution manuals* are available that provide completely worked-out solutions either to all of the end-of-chapter exercises (*Solutions to Exercises in Chemistry: The Central Science*) or to only those exercises whose short answers are provided in the text (*Solutions to Selected Exercises in Chemistry: The Central Science*).

Several other supplements are also available. We offer a new *test-bank file*, in hard copy or on IBM-compatible or Apple-compatible disks. Using our *telephone testing service*, with a toll-free number, the instructor may select questions from the test bank to be prepared by Prentice-Hall for the instructor at no charge. The supplements package also includes *transparencies* for key illustrations and tables in the text. Also, a text on problem solving, *Chemical Problem Solving Using Dimensional Analysis*, second edition, by Robert Nakon, focuses on strategies useful in working problems. Several interactive *software packages* are available free to the instructor upon adoption of the text. Contact your sales representative for further details.

## ACKNOWLEDGMENTS

This book owes its final shape and form to the assistance and hard work of many people. Several colleagues reviewed the manuscript and helped us immensely by sharing their insights and criticizing our initial writing efforts. We would like especially to thank the following:

Wayne Anderson
*Bloomsburg State College*

James Birk
*Arizona State*

Susan Brennan
*University of North Florida*

Marcia Davies
*Creighton University*

Jim Davis
*University of Pennsylvania*

Charles Fosha
*University of Colorado*

Wade Freeman
*University of Illinois*

Roy Garvey
*North Dakota State University*

L. Peter Gold
*Pennsylvania State University*

Tom Greenbowe
*Southeastern Massachussetts University*

Roger Hoburg
*University of Nebraska at Omaha*

Colin D. Hubbard
*University of New Hampshire*

Harold Hunt
*Georgia Institute of Technology*

Donald Jicha
*University of North Carolina, Chapel Hill*

Stephen Kozub
*University of Pennsylvania*

Robert M. Kren
*University of Michigan, Flint*

Russell D. Larsen
*Texas Tech University*

Harvey Moody
*United States Air Force, Colorado*

Leon Morgan
*University of Texas*

W. D. Perry
*Auburn University*

Helen Place
*Washington State University*

Curtis T. Sears, Jr.
*Georgia State University*

Robert Sprague
*Lehigh University*

Mable Ruth Stephanic
*Oklahoma State University*

Jack Stocker
*University of New Orleans*

Charles Trapp
*University of Louisville*

We deeply appreciate the assistance of the following members of Prentice-Hall's College Division: Dan Joraanstad, our chemistry editor, who provided enthusiastic support and encouragement throughout the project; Stephen Deitmer, our developmental editor, who provided excellent counsel and guidance; Fay Ahuja, our production editor, who did an outstanding job in overseeing the photo research and production of the book; Janet Schmid, the designer, for the excellent design of the text; and Beth Stoll, our marketing manager, for her fine work in publicizing our efforts. We also continue to owe a debt of thanks to Raymond Mullaney, editor in chief, whose impact on this book from its earlier editions continues to be felt. Other co-workers and friends deserve a special thanks: Kenneth Kemp (University of Nevada) and James Hill (California State University, Sacramento) for their helpful advice; Robyn Green (University of Nevada) for her help in proofreading the text; Roxy Wilson (University of Illinois) for her assistance with the photography program and for performing the difficult job of working end-of-chapter exercises and checking our calculations.

THEODORE L. BROWN
*School of Chemical Sciences,*
*University of Illinois, Urbana*
61801
H. EUGENE LeMAY, Jr.
*Department of Chemistry,*
*University of Nevada, Reno*
89557

THEODORE L. BROWN          H. EUGENE LeMAY, JR.

# ABOUT THE AUTHORS

Theodore L. Brown received his Ph.D. from Michigan State University in 1956. Since that time he has been a member of the faculty of the University of Illinois, Urbana, where he is Professor of Chemistry. He has served as Vice-Chancellor for Research and as Dean, The Graduate College. In March 1987 he became Director of the Beckman Institute.

Professor Brown has been an Alfred P. Sloan Research Fellow and has received the American Chemical Society Award for Research in Inorganic Chemistry. He has also been awarded a Guggenheim Fellowship and has held several offices with the American Chemical Society. He is a member of the Board of Governors of Argonne National Laboratory and Chairman of the Scientific and Technical Advisory Committee of the Laboratory.

H. Eugene LeMay, Jr., received his Ph.D. in 1966 from the University of Illinois. He then joined the faculty of the Chemistry Department at the University of Nevada, Reno, where he has served as Vice-Chairman, Director of Freshman Chemistry, and Interim Chairman. He is presently Associate Chairman and Professor of Chemistry. He has enjoyed Visiting Professorships at the University of North Carolina at Chapel Hill and at the University College of Wales in Great Britain.

Professor LeMay has received university awards for outstanding teaching at the undergraduate and graduate levels. He is the author of thirty research publications and review articles, mainly in the area of solid-phase reactions. In addition to *Chemistry: The Central Science*, Professors Brown and LeMay are coauthors of *Qualitative Inorganic Analysis*.

# CHEMISTRY
## The Central Science

## CHAPTER 1

# Introduction:
# Some Basic Concepts

Have you ever walked down a path through the forest in autumn and wondered why leaves change their colors? Perhaps you have marveled at the vivid colors of a fireworks display or sat around a campfire at night, captivated by the bright flames. These colorful phenomena (Figure 1.1) are the results of chemical changes. Chemistry does not happen only in laboratories. It occurs all around us, right before our eyes.

Chemistry deals with all the multitude of materials and changes we see around us that make our world so diverse, beautiful, and at times mysterious. It explains the rusting of nails, the melting of ice, the digestion of a candy bar, the colors of your vacation slides, the baking of bread, and the aging of a person.

(a)

**FIGURE 1.1** Changes in materials occur all around us. For example, trees change color in autumn (a), snow melts (b), and iron rusts (c). Such changes have long fascinated people and have prompted them to look more closely at nature's working in hopes of better understanding themselves and their environment. Chemistry is the science that is concerned primarily with matter and the changes it undergoes. (© Townsend P. Dickinson, 1985; James R. Fisher, 1984; H. E. LeMay, Jr.)

(b) (c)

Fireworks display over New York City. The red, blue, and yellow colors are produced by salts of strontium, copper, and sodium, respectively. Magnesium and aluminum metals are largely responsible for the white flames. © Wesley Bocxe/Photo Researchers

1

Most of you are studying chemistry because it has been declared an essential part of your curriculum. Your major may be agriculture, dental hygiene, electrical engineering, geology, biology, or one of many other related areas of study. Why do so many diverse subjects share an essential tie to chemistry? The answer is that chemistry, by its very nature, is the *central* science. In any area of human activity that deals with some aspect of the material world, concern invariably arises about the fundamental nature of the materials involved: their composition and endurance, how they interact with other materials and with their environment, and how they undergo change. This interest is present whether the material involved is a polymer used to coat electronics components, the pigments used by a Renaissance painter (Figure 1.2), or the blood cells of a child born with sickle-cell anemia. It is very likely that chemistry will play an important role in your future profession. You will be a more versatile and creative person if you understand the chemical concepts relevant to your work and can apply these concepts as needed.

The relationship of chemistry to professional goals, however, is not the sole reason to study the subject. Because chemistry is central to our lives and intimately tied to almost every aspect of our contact with the material world, this science is an integral part of our culture. The role of chemistry in our lives goes much deeper than the obvious products of chemical research that we use: items like plastic bags, batteries, photographic films, and computer chips. Thousands of chemical products make up the foods we eat, the cars we drive, and the medical care we receive (Figure 1.3). We have become increasingly aware that the widespread use of some chemicals has had a profound effect on our environment. To be a responsible citizen you will need to be informed on many complex issues of chemistry and the use of chemicals. You can more fully appreciate and analyze the complex issues put before you if you keep the relevant chemical principles in mind as you read and study current events.

This text introduces you to basic chemical facts and theories, not as ends in themselves but as the means to help you to understand the world

**FIGURE 1.2** (*a*) A microscopic view of a computer logic chip. (*b*) A portrait of Francesco Sassetti with his grandson by D. Ghirlandaio from approximately 1490. (John Walsh/Photo Researchers, The Granger Collection)

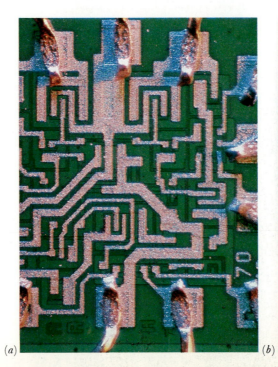

(*a*)

(*b*)

around you and to recognize the constraints and opportunities that the world provides. We begin our studies by considering some basic ideas about matter and its properties and by discussing some important background information dealing with scientific measurements and calculations: the metric system, uncertainty in measurement, and problem solving in chemistry.

Our present understanding of the changes we see around us—like the melting of ice and the burning of wood—is intimately tied to our understanding of the nature and composition of matter. **Matter** is the physical material of the universe; it is anything that occupies space and has mass.

Matter exists in three physical *states*: **gas** (also known as vapor), liquid, and solid (Figure 1.4). A gas has no fixed volume or shape. It takes the volume and shape of its container. A gas can be compressed to fit a small container, and it will expand to occupy a large one. A **liquid** has a definite volume but no specific shape. It assumes the shape of the portion of the container that it occupies. A **solid** is rigid. It has both a fixed volume and a fixed shape. Neither liquids nor solids are compressible to any appreciable extent.

## 1.1 STATES AND PROPERTIES OF MATTER

*Mass — quantity of matter*

**FIGURE 1.4** The three physical states of water are common and familiar to us: water vapor, liquid water, and ice. We cannot see water vapor. What we see when we look at steam is tiny droplets of liquid water dispersed through the water vapor. In this photo we see both the liquid and solid states of water. This iceberg is about fifteen feet high and of glacial origin. It has probably been in the water for over a year and shows the effects of weathering by wind and waves. (Robert W. Hernandez)

No substance in nature is ever totally pure, even when we take great care to separate it from other substances. One substance may be the major component of a given sample, but other substances, called *impurities*, are always present to some extent. A substance may be described as 99.9 percent pure if it contains only 0.1 percent impurities by mass. For many purposes we may describe such a sample as "pure." Sometimes, though, much greater purity may be required. For example, the silicon used to make computer chips has to be 99.99999 percent pure. At times we may need to know not only the percent purity of a substance but also the nature of the impurities. Labels of chemical substances used in laboratories often list the major impurities and their amounts (Figure 1.5).

**FIGURE 1.5** Label for bottle of potassium bromide indicating impurities. This potassium bromide is about 99.8% pure. It contains only slightly more than 0.2% total impurities, which are listed on the label. (EM Science)

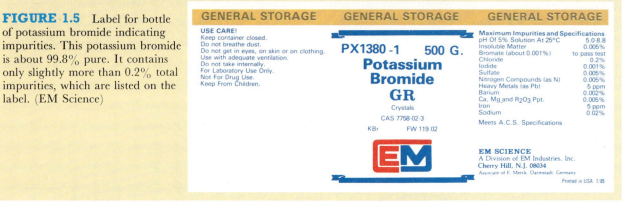

We can resolve, or separate, various kinds of matter into different pure substances. A **pure substance** (from here on referred to simply as a *substance*) is a kind of matter that has a fixed composition and distinct properties. For example, seawater is composed of several different substances, the most abundant being water and ordinary table salt (sodium chloride).

## Chemical and Physical Properties and Changes

Every substance has a unique set of **properties**, or characteristics, that allow us to recognize it and to distinguish it from other substances. Properties of matter can be grouped into two categories: physical and chemical. **Physical properties** are those properties that we can measure without changing the basic identity of the substance. Physical properties include color, hardness, density, melting point, and boiling point. **Chemical properties** describe the way a substance may change or "react" to form other substances. A common chemical property of iron is its ability to combine with oxygen in the presence of water to form a red-brown substance that we call rust (an oxide of iron).

Matter can undergo changes of two basic types: physical changes and chemical changes. **Physical changes** are processes in which a material changes its physical appearance but not its basic identity. The evaporation of water is a physical change. When water evaporates, it changes from the liquid state to the gas state, but it is still water; it has not changed into any other substance. All changes of state are physical changes. In **chemical changes**, also called **chemical reactions**, substances change not only in physical appearance but also in basic identity; that is, one substance is converted into another. For example, hydrogen burning in air undergoes a chemical change in which it is converted to water.

The following account describes the first experiences a young man had with chemical reactions. The writer is Ira Ramsen, who was later to become the author of a popular chemistry text published in 1901. The chemical reaction that he set out to observe is shown in Figure 1.6.

> While reading a textbook of chemistry, I came upon the statement "nitric acid acts upon copper," and I determined to see what this meant. Having located some nitric acid, I had only to learn what the words "act upon" meant. In the interest of knowledge I was even willing to sacrifice one of the few copper cents then in my possession. I put one of them on the table, opened a bottle labeled "nitric acid," poured some of the liquid on the copper, and prepared to make an observation. But what was this wonderful thing which I beheld? The cent was already changed, and it was no small change either. A greenish-blue liquid foamed and fumed over the cent and over the table. The air became colored dark red. How could I stop this? I tried by picking the cent up and throwing it out the window. I learned another fact: nitric acid acts upon fingers. The pain led to another unpremeditated experiment. I drew my fingers across my trousers and discovered nitric acid acts upon trousers. That was the most impressive experiment I have ever performed. I tell of it even now with interest. It was a revelation to me. Plainly the only way to learn about such remarkable kinds of action is to see the results, to experiment, to work in the laboratory.

**FIGURE 1.6**   The chemical reaction between a copper penny and nitric acid. The dissolved copper produces the blue-green solution; the reddish brown gas produced is nitrogen dioxide. (Donald Clegg and Roxy Wilson)

## 1.2 ELEMENTS, COMPOUNDS, AND MIXTURES

Most matter around us—the air we breathe, the milk we drink, and the sidewalks on which we walk—consists of mixtures. **Mixtures** are combinations of two or more substances in which each substance retains its own identity. In some mixtures, such as soil, rocks, and wood, the components are readily distinguished. Such mixtures are **heterogeneous** [Figure 1.7(a)]. Other mixtures are **homogeneous**—that is, uniform throughout. For example, salt, sugar, and many other substances dissolve in water, forming homogeneous mixtures [Figure 1.7(b)]. Homogeneous mixtures are known as **solutions**.

The compositions of mixtures can vary widely. However, each component in a mixture retains its identity and its properties. Thus we can separate a mixture by taking advantage of the different physical properties of its components. For example, water has a much lower boiling point than salt. If we boil a mixture of salt and water, the water evaporates, leaving the salt behind.

(a)                                                                         (b)

**FIGURE 1.7**   Many common materials, including rocks, are heterogeneous. (a) This photo shows a copper-containing mineral called malachite. Homogeneous mixtures are called solutions. (b) Many substances, including the blue solid shown in this photo (copper sulfate), dissolve in water to form solutions. (Runk, Schoenberger/Grant Heilman; Richard Megna/Fundamental Photographs)

---

### A CLOSER LOOK                          Separation of Mixtures

Chemists have developed many procedures to separate (or partly separate) mixtures into their component substances. One of the most common separation procedures is *filtration*. Filtration separates solids from liquids as illustrated in Figure 1.8. Examples of separations based on filtration abound outside the lab: air and oil filters in automobiles, air filters on furnaces in our homes, filters in coffee makers, and so forth.

Another familiar separation procedure is *distillation*. This procedure is based on differences in the volatilities of substances (that is, differences in the ease with which substances form gases). Distillation

is the procedure by which the "moonshiner" obtains whiskey using a "still," and by which petrochemical plants achieve the separation of crude petroleum into gasoline, diesel fuel, lubricating oil, and so forth. Figure 1.9 shows a simple laboratory distillation apparatus. Imagine that the solution in the distilling flask is a salt–water mixture. Water, of course, vaporizes at a much lower temperature than salt, so the water boils off, leaving the salt in the distilling flask. The water is condensed by cooling elsewhere in the system and collected. The liquid obtained by condensation in a distillation is known as the distillate.

A number of separation procedures are based

**FIGURE 1.8** Separation of a liquid and solid by filtration. (Donald Clegg and Roxy Wilson)

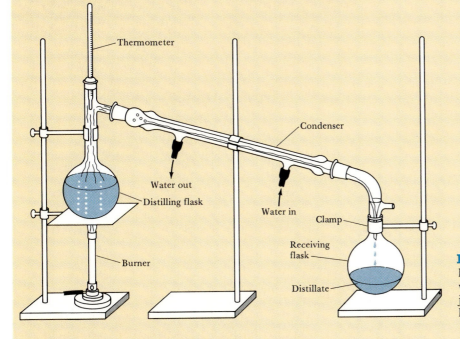

Thermometer

Condenser

Water out

Distilling flask

Water in

Clamp

Receiving flask

Burner

Distillate

**FIGURE 1.9** Simple laboratory distillation setup. Cool water circulating through the jacket of the condenser causes the liquid to condense.

on differences in the degree to which various substances are adsorbed onto the surface of an inert material. (An inert material is one that does not undergo a chemical change.) The difference between *adsorption* and *absorption* should be noted. *Adsorption* means adherence to a surface. *Absorption* denotes passage into the interior; water is absorbed by a sponge.

Separations that use adsorption differences are known as *chromatographic procedures*. The word *chromatography* means "graphing of colors" and arose from its use in separating pigments. For example, a solution containing the colored pigments of a leaf may be washed through a column packed with alumina (aluminum oxide). The various components will move through the column at different speeds due to differences in the degree to which they are adsorbed. This type of separation, illustrated in Figure 1.10, is known as *column chromatography*.

When the adsorbent material is paper and the solution containing the mixture moves upward through the paper, the technique is called *paper chromatography*. As the liquid moves upward, it carries the mixture along. The components that are adsorbed most strongly to the paper move most slowly. The technique is illustrated in Figure 1.11.

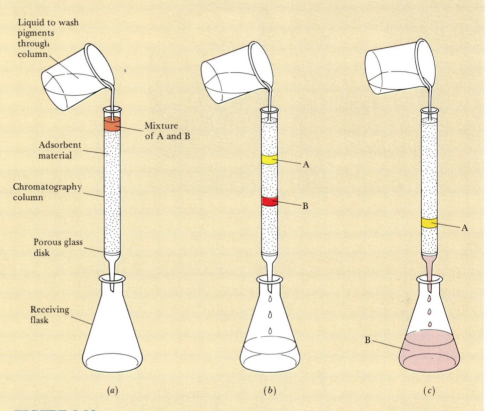

**FIGURE 1.10** (*a*) A chromatography column separates a mixture of A and B, seen here at the top of the column. (*b*) Since B has a higher affinity than A does for the mobile phase, the band of B moves more quickly down the column. (*c*) Finally, B is washed off the column into the receiving flask, while A remains on the column.

**FIGURE 1.11** Separation of ink into components by paper chromatography. (*a*) Water begins to move up the paper. (*b*) Water moves past the ink spot, lifting different components of the ink at different rates. (*c*) Water has separated the ink into three different components.

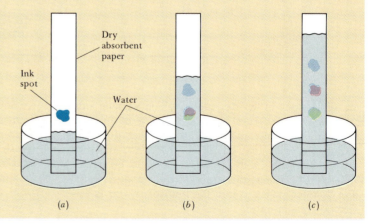

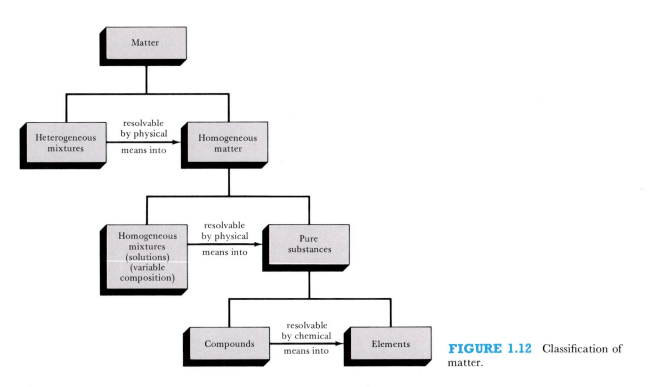

**FIGURE 1.12** Classification of matter.

Pure substances are homogeneous. In addition, they have a constant, invariable composition. There are two classes of substances: elements and compounds. **Elements** are substances that cannot be decomposed into simpler substances. **Compounds**, however, can be decomposed (by chemical means) into two or more elements. Figure 1.12 summarizes the classification of matter into mixtures, compounds, and elements.

Elements are the basic substances out of which all matter is composed. In light of the seemingly endless variety in our world, it is perhaps surprising that there are only 108 known elements. Not all of these are of equal importance or abundance (see Figure 1.13). Over 90 percent, by

**Elements**

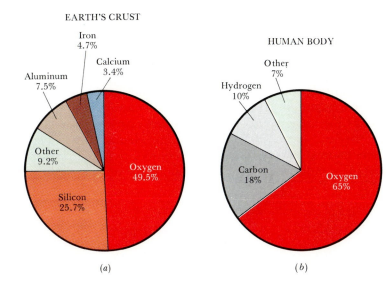

**FIGURE 1.13** Elements, in percent by mass, in (*a*) the earth's crust (including oceans and atmosphere) and (*b*) the human body.

**TABLE 1.1**  Some common elements and their symbols

| | | |
|---|---|---|
| Carbon (C) | Aluminum (Al) | Copper (Cu, from *cuprum*) |
| Fluorine (F) | Barium (Ba) | Iron (Fe, from *ferrum*) |
| Hydrogen (H) | Calcium (Ca) | Lead (Pb, from *plumbum*) |
| Iodine (I) | Chlorine (Cl) | Mercury (Hg, from *hydrargyrum*) |
| Nitrogen (N) | Helium (He) | Potassium (K, from *kalium*) |
| Oxygen (O) | Magnesium (Mg) | Silver (Ag, from *argentum*) |
| Phosphorus (P) | Platinum (Pt) | Sodium (Na, from *natrium*) |
| Sulfur (S) | Silicon (Si) | Tin (Sn, from *stannum*) |

weight, of the portion of the earth to which we have access for raw materials is composed of only five elements: oxygen, silicon, aluminum, iron, and calcium. Over 90 percent of the human body is composed of just three elements: oxygen, carbon, and hydrogen.

Some of the more familiar elements are listed in Table 1.1, together with the chemical symbols used to denote them. All of the known elements are listed on the front inside cover of this text. The abbreviation— or symbol—for an element consists of one or two letters, with the first letter capitalized.* These symbols are often derived from the English name (first and second columns of Table 1.1), but sometimes they are derived instead from a foreign name (third column). You will need to know these symbols and to learn others as we encounter them in the text.

## Compounds

Compounds are substances composed of two or more elements united chemically in definite proportions by mass. We can gain a clearer understanding of the distinctions among elements, compounds, and mixtures by examining a common substance, water. With the discovery of methods of generating electricity, chemists found that water could be decomposed into the elements hydrogen and oxygen, as shown in Figure 1.14. This decomposition clearly indicates that water is not an element. However, it is not merely a mixture of hydrogen and oxygen either. The properties of water are clearly unique and much different from those of its constituent elements, as seen in Table 1.2. Furthermore, the composition of water is not variable. Pure water consists of 11 percent hydrogen and 89 percent oxygen by mass, regardless of its source.

The observation that the elemental composition of a pure compound is always the same is known both as the **law of constant composition** and the **law of definite proportions**. Although this law has been known for over 150 years, the general belief persists among some people that a fundamental difference exists between compounds prepared in the laboratory and the corresponding compounds found in nature. However, a pure compound has the same composition and properties regardless of source. Both chemists and nature must use the same elements and operate under the same natural laws. Differences in composition and properties between substances indicate that the compounds are not the same or that at least one is impure.

* Three-letter abbreviations have been proposed for elements beyond 103.

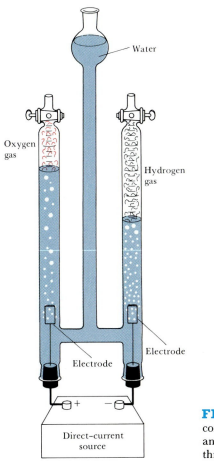

**FIGURE 1.14**  Decomposition of the compound water into the elements hydrogen and oxygen by passing a direct electrical current through it.

Water — Water

Oxygen gas

Hydrogen gas

Electrode

Electrode

Direct–current source

**TABLE 1.2**  Comparison of water, hydrogen, and oxygen

|                          | **Water**   | **Hydrogen** | **Oxygen** |
| ------------------------ | ----------- | ------------ | ---------- |
| Physical state[a]        | Liquid      | Gas          | Gas        |
| Normal boiling point     | 100°C       | − 253°C      | − 183°C    |
| Density[a]               | 1.00 g/mL   | 0.090 g/L    | 1.43 g/L   |
| Combustible?             | No          | Yes          | No         |

[a] At room temperature and atmospheric pressure.

## 1.3 UNITS OF MEASUREMENT

A great many properties of matter, like density and boiling point, are quantitative; that is, they are associated with numbers. When a number represents a measurement, the units of that measurement must always be indicated. To say that the length of a paper clip is 3.2 would be completely meaningless. To say that it is 3.2 cm gives proper meaning. The units used for measurements in science are those of the **metric system**.

The metric system is used as the system of measurement in most societies throughout the world. In recent years the use of metric units has grown in the United States as well. For example, the contents of

**FIGURE 1.15** Road sign along an interstate highway in California, showing distance in metric and English-system units. (Tom McHugh/Photo Researchers)

most canned products in grocery stores are now given in grams as well as ounces, some highway signs show distance in both miles and kilometers (Figure 1.15), and baseball stadiums give their dimensions in meters and feet.

According to international agreement reached in 1960, certain basic metric units and units derived from them are preferred in scientific use. The preferred units are known as **SI units** after the French *Système International d'Unités*. The seven basic units of the SI system are given in Table 1.3. All other SI units of measure are derived from these base units. For example, speed, which is the ratio of distance to elapsed time, has units of meters per second, or m/s.

Adoption of SI units is an attempt to further systematize the metric system. Non-SI metric units are being phased out. However, until SI units are fully adopted by practicing scientists, we must understand both

**TABLE 1.3**   Basic SI units

| Physical quantity | Name of unit | Abbreviation |
|---|---|---|
| Mass | Kilogram | kg |
| Length | Meter | m |
| Time | Second | s[a] |
| Electric current | Ampere | A |
| Temperature | Kelvin | K |
| Luminous intensity | Candela | cd |
| Amount of substance | Mole | mol |

[a] The abbreviation "sec" is frequently used.

**TABLE 1.4**  Selected prefixes used in the SI system

| Prefix | Abbreviation | Meaning | Example |
|--------|-------------|---------|---------|
| Mega- | M | $10^6$ | 1 megameter (Mm) = $1 \times 10^6$ m |
| Kilo- | k | $10^3$ | 1 kilometer (km) = $1 \times 10^3$ m |
| Deci- | d | $10^{-1}$ | 1 decimeter (dm) = 0.1 m |
| Centi- | c | $10^{-2}$ | 1 centimeter (cm) = 0.01 m |
| Milli- | m | $10^{-3}$ | 1 millimeter (mm) = 0.001 m |
| Micro- | $\mu$[a] | $10^{-6}$ | 1 micrometer ($\mu$m) = $1 \times 10^{-6}$ m |
| Nano- | n | $10^{-9}$ | 1 nanometer (nm) = $1 \times 10^{-9}$ m |
| Pico- | p | $10^{-12}$ | 1 picometer (pm) = $1 \times 10^{-12}$ m |

[a] This is the Greek letter mu (pronounced "mew").

SI units and the non-SI units that are still in use. Whenever we first encounter a non-SI unit in the text, the proper SI unit will also be given.

The SI system employs a series of prefixes to indicate decimal fractions or multiples of various units. For example, the prefix milli- represents a $10^{-3}$ fraction of a unit: a milligram is $10^{-3}$ gram, a millimeter is $10^{-3}$ meter, and so forth. Table 1.4 presents the prefixes most commonly encountered in chemistry. In using the SI system and in working problems throughout this text, it is important to have a comfortable familiarity with exponential notation. If you are unfamiliar with exponential notation or want to review it, refer to Appendix A.1.

**Length**

The basic SI unit of length is the meter (m). From the comparison shown in Figure 1.16, we see that the meter is only slightly longer than a yard. We will consider interconversion of English and metric system measures in Section 1.5. For the moment it is more important to understand clearly the use of the prefixes given in Table 1.4. (The relations between the English and metric system units that we will use most frequently in this text appear on the back inside cover.)

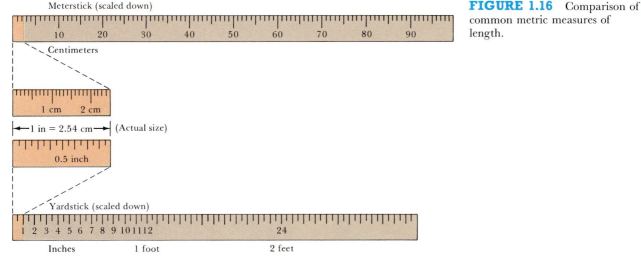

**FIGURE 1.16**  Comparison of common metric measures of length.

**Volume**

The measure for volume is a derived unit based on the fundamental SI unit of length cubed, $m^3$. The cubic meter, $m^3$, is the volume of a cube that is 1 m on each edge. Related units, like the cubic centimeter, $cm^3$ (sometimes written cc), or cubic decimeter, $dm^3$, are also used. Another common measure of volume is the liter (L), a volume roughly the size of a quart. A liter is the volume occupied by 1 cubic decimeter, $dm^3$. There are 1000 mL in a liter, and each milliliter is the same volume as a cubic centimeter (Figure 1.17). Thus the terms milliliter and cubic centimeter are used interchangeably in expressing volume. The liter is the first metric unit that we have encountered that is not an SI unit.

The devices most frequently used in chemistry to measure volume are illustrated in Figure 1.18. Pipets and burets allow delivery of liquids with more accurately known volumes than do graduated cylinders. Volumetric flasks are used to prepare accurately a designated volume of solution.

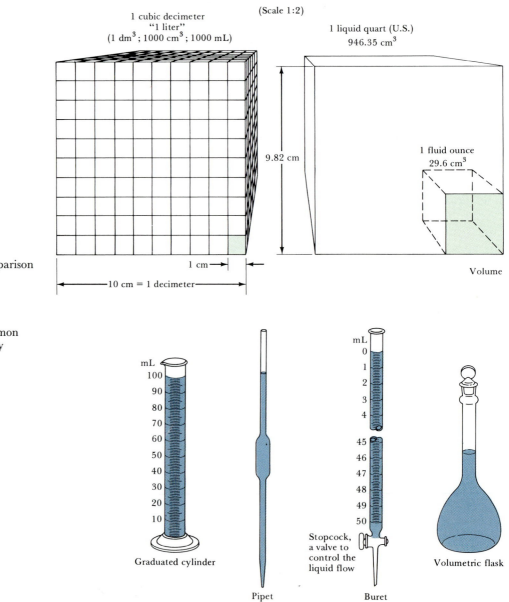

**FIGURE 1.17** Comparison of common measures of volume.

**FIGURE 1.18** Common devices used in chemistry laboratories to measure volume.

The basic SI unit of **mass*** is the kilogram (kg). This base unit is un-usual because it uses a prefix, kilo-, instead of the word gram alone. As shown in Figure 1.19, a kilogram is equal to about 2.2 lb. We obtain other SI units for mass by adding prefixes to the word gram.

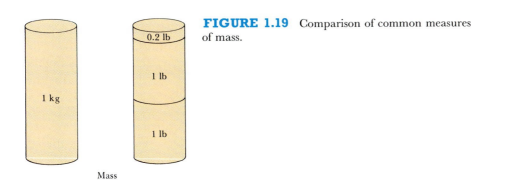

**FIGURE 1.19**  Comparison of common measures of mass.

---

## SAMPLE EXERCISE 1.1

What is the name given to the unit that equals **(a)** $10^{-9}$ gram; **(b)** $10^{-6}$ liter; **(c)** $10^{-3}$ meter?

**Solution:**   In each case we can refer to Table 1.4, finding the prefix related to each of the decimal frac-

tions: **(a)** nanogram, ng; **(b)** microliter, $\mu$L; **(c)** milli-meter, mm.

### PRACTICE EXERCISE
What fraction of a second is a picosecond, ps?
***Answer:***   $10^{-12}$ second

---

**Density** is a quantity widely employed by chemists to identify substances. It is defined as the amount of mass in a unit volume of the substance:

$$\text{Density} = \frac{\text{mass}}{\text{volume}} \qquad [1.1]$$

Density is commonly expressed in units of grams per cubic centimeter ($g/cm^3$ or $g\ cm^{-3}$). The densities of some common substances are listed in Table 1.5.

**Density**

**TABLE 1.5**   Densities of some selected substances

| Substance | Density ($g/cm^3$) |
|---|---|
| Air | 0.001 |
| Balsa wood | 0.16 |
| Water | 1.00 |
| Table salt | 2.16 |
| Iron | 7.9 |
| Gold | 19.32 |

---

## SAMPLE EXERCISE 1.2

**(a)** Calculate the density of mercury if $1.00 \times 10^2$ g occupies a volume of 7.36 cm$^3$. **(b)** Calculate the mass of 65.0 cm$^3$ of mercury.

**Solution:**

**(a)** Density $= \dfrac{\text{mass}}{\text{volume}} = \dfrac{1.00 \times 10^2 \text{ g}}{7.36 \text{ cm}^3} = 13.6 \text{ g/cm}^3$

$$\frac{1.00 \times 100}{7.36}$$

---

*** Mass and weight are often incorrectly thought to be the same. Mass is a measure of the amount of material in an object; the weight of that object, however, depends not only on its mass but also on the attractive force of gravity. In outer space, where gravitational forces are very weak, an astronaut may be weightless, but he is not massless. In fact, he has the *same* mass as he has on earth. Nevertheless, it is common practice to use the terms mass and weight interchangeably.

**(b)** Solving Equation 1.1 for mass gives

Mass = volume × density

Using the density of mercury that we calculated in part (a) gives

Mass = $(65.0 \text{ cm}^3)(13.6 \text{ g/cm}^3) = 884$ g

**PRACTICE EXERCISE** ————————

A student needs 15.0 g of ethyl alcohol for an experiment. If the density of the alcohol is 0.789 g/mL, how many mL of alcohol are needed?

***Answer:*** 19.0 mL

The terms *density* and *weight* are sometimes confused. A person who says that iron weighs more than air generally means that iron has a higher density than air; 1 kg of air has the same mass as 1 kg of iron, but the iron is confined to a smaller volume, thereby giving it a higher density.

## Temperature

The temperature scales commonly employed in scientific studies are the Celsius (also called centigrade) and Kelvin scales. The kelvin is the SI unit of temperature. The Celsius scale is based on assignment of 0°C to the freezing point of water and 100°C to its boiling point at sea level. The corresponding temperatures in the Fahrenheit scale are 32°F and 212°F. There are 100° between the freezing point and boiling point of water in the Celsius scale, and there are 180° between these points on the Fahrenheit scale. Consequently, the Celsius and Fahrenheit scales are related as follows:

$$°\text{C} = \frac{100}{180} \, (°\text{F} - 32°) = \frac{5}{9} \, (°\text{F} - 32°) \qquad [1.2]$$

The Kelvin scale is based on the properties of gases, and its origins will be considered more fully in Chapter 10. Zero on this scale corresponds to −273.15°C, and the size of a kelvin is the same as a degree Celsius. The Kelvin and Celsius scales are therefore related by .

$$\text{K} = °\text{C} + 273.15 \qquad [1.3]$$

**FIGURE 1.20** Comparison of the Kelvin, Celsius, and Fahrenheit temperature scales.

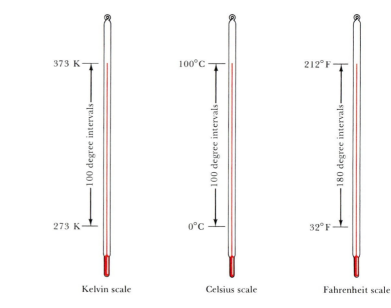

For most purposes we will use 273 instead of 273.15 to convert temperature. [According to the SI convention, a degree sign (°) is not used with the Kelvin scale. Thus we write 273 K and not 273°K.] Further comparisons between the Celsius, Kelvin, and Fahrenheit scales are made in Figure 1.20 and Table 1.6.

**TABLE 1.6**  Some comparisons of Fahrenheit, Celsius, and Kelvin temperatures

| | | | |
|---|---|---|---|
| Absolute zero | $-460°F$ | $-273°C$ | $0$ K |
| Freezing point of water | $32°F$ | $0°C$ | $273$ K |
| Average room temperature | $68°F$ | $20°C$ | $293$ K |
| Normal body temperature | $98.6°F$ | $37°C$ | $310$ K |
| Boiling point of water | $212°F$ | $100°C$ | $373$ K |

**SAMPLE EXERCISE 1.3**

If a weather forecaster predicts that the temperature for the day will reach 30°C, what is the predicted temperature **(a)** in K; **(b)** in °F?

**Solution:**

**(a)**   $K = 30 + 273 = 303$ K

**(b)**
$$°C = \tfrac{5}{9}(°F - 32°)$$
$$30° = \tfrac{5}{9}(°F - 32°)$$
$$\left(\tfrac{9}{5}\right)(30°) = °F - 32°$$
$$54° + 32° = °F$$
$$86° = °F$$

**PRACTICE EXERCISE**

Ethylene glycol, the major ingredient in antifreeze, boils at 199°C. What is the boiling point in **(a)** K; **(b)** °F?

*Answers:*   **(a)** 472 K; **(b)** 390°F

We sense temperature as a measure of the hotness or coldness of an object. Indeed, temperature determines the direction of heat flow. Heat always flows spontaneously from a substance at higher temperature to one at lower temperature. Thus we feel the influx of energy when we touch a hot stove, and we know that the stove is at a higher temperature than our hand.

Temperature is an **intensive property**, meaning that its value does not depend on the amount of material chosen. Thus two samples of a liquid may have the same temperature though one has a volume of one cup and the other sample fills a bathtub. Density is another example of an intensive property. The density of mercury, calculated in Sample Exercise 1.2, is 13.6 g/cm$^3$ whether one has 100 g or 1000 g. By contrast, both volume and mass are **extensive properties**. They depend on the amount of material. Similarly, the *heat content* of a sample is an extensive property. In some types of solar heating systems for homes, air is heated by passage through solar panels exposed to the sun. The heated air is then passed through a large bed of stones. Because the heat content of the stones is large, the heat stored in them during the day is sufficient to heat the house during the night. Note the important distinction between heat content, an extensive property, and temperature, an intensive property.

In the following description, label each property or characteristic as intensive or extensive: "The yellow sample is solid at 25°C. It weighs 6.0 g and has a density of 2.3 g/cm³."

**Solution:** Mass is an extensive property; color, phy- sical state (that is, solid), temperature, and density are intensive properties.

**PRACTICE EXERCISE** _____

Is the heat required to evaporate a gram of liquid water an intensive or extensive property?

***Answer:*** intensive

# 1.4 UNCERTAINTY IN MEASUREMENT

In scientific work we recognize two kinds of numbers: **exact numbers** (those whose values are known exactly) and **inexact numbers** (those whose values have some uncertainty). Exact numbers result from counting or have defined values. For example, there are exactly 3 ft in a yard, exactly four people in my immediate family, exactly 1000 g in a kilo- gram, and exactly 12 eggs in a dozen. The number 1 in any conversion factor between units, as in 1 m = 1.0936 yd, is also an exact number.

Numbers obtained from measurements are always *inexact*. For example, suppose you and some friends each weigh the same dime on different balances. You will likely find that your measurements vary slightly. The differences may result from how well calibrated each balance is or from your judgment in reading the scale. Remember this important fact: *un- certainties always exist in measured quantities.*

In discussing the uncertainties in measured quantities, two terms are frequently used: precision and accuracy. **Precision** refers to how closely individual measurements agree with each other. **Accuracy** refers to how closely measurements agree with the correct or standard value. Highly precise measurements are usually accurate. However, the same error may possibly enter each of several measurements, making them inaccurate even though they are precise. This sort of error could arise, for example, from a faulty balance.

Suppose you weigh a dime on a balance capable of measuring to the nearest 0.0001 g. You could report the mass as 2.2405 ± 0.0001 g. The ± notation (read "plus or minus 0.0001") is a useful way to express the uncertainty of a measurement. In much scientific work we drop the ± notation with the understanding that an uncertainty of at least one unit exists in the last digit of the measured quantity. That is, measured quantities are generally reported in such a way that only the last digit is uncertain. All the digits, including the uncertain one, are called **signif- icant figures**. The number 2.2405 has five significant figures. The num- ber of significant figures indicates the preciseness of a measurement.

## SAMPLE EXERCISE 1.5

What is the difference between 4.0 g and 4.00 g?

**Solution:** Many people would say there is no dif- ference, but a scientist would note the difference in the number of significant figures between the two measurements. The number 4.0 has two significant figures; 4.00 has three. This implies that the second measurement has been made more precisely. A mass of 4.0 g indicates that the mass of the sample is closer to 4.0 g than to 3.9 g or to 4.1 g; the range of un- certainty is within 4.0 ± 0.1. A mass of 4.00 g means that the mass of the sample must be closer to 4.00 g than to 3.99 g or to 4.01 g; the range of uncertainty is within 4.00 ± 0.01. We see that 4.00 is more precise because it has more significant figures.

**PRACTICE EXERCISE** _____
How many significant figures are contained in the number 6402.13?
*Answer:* 6

The following rules apply to determining the number of significant figures in a measured quantity:

1. All nonzero digits are significant—457 cm (three significant figures); 0.25 g (two significant figures).
2. Zeros between nonzero digits are significant—1005 kg (four significant figures); 1.03 cm (three significant figures).
3. Zeros to the left of the first nonzero digit in a number are not significant; they merely indicate the position of the decimal point—0.02 g (one significant figure); 0.0026 cm (two significant figures).
4. Zeros that fall both at the end of a number and to the right of the decimal point are significant—0.0200 g (three significant figures); 3.0 cm (two significant figures).
5. When a number ends in zeros that are not to the right of a decimal point, the zeros are not necessarily significant—130 cm (two or three significant figures); 10,300 g (three, four, or five significant figures). We describe how to remove this ambiguity below.

Use of standard exponential notation (Appendix A) avoids the potential ambiguity of whether the zeros at the end of a number are significant (rule 5). For example, a mass of 10,300 g can be written in exponential notation showing three, four, or five significant figures:

$$1.03 \times 10^4 \text{ g} \qquad \text{(three significant figures)}$$
$$1.030 \times 10^4 \text{ g} \qquad \text{(four significant figures)}$$
$$1.0300 \times 10^4 \text{ g} \qquad \text{(five significant figures)}$$

In these numbers all the zeros to the right of the decimal point are significant (rules 2 and 4). (All significant figures come before the exponent; the exponential term does not add to the number of significant figures.)

In carrying measured quantities through calculations, observe this point: The precision of the result is limited by the least precise measurement. *In multiplication and division the result must be reported as having no more significant figures than the measurement with the fewest significant figures.* When the result contains more than the correct number of significant figures it must be rounded off.

**Combining Measured Quantities in Calculations**

For example, the area of a rectangle whose edge lengths are 6.221 cm and 5.2 cm should be reported as 32 cm$^2$:

$$\text{Area} = (6.221 \text{ cm})(5.2 \text{ cm}) = 32.3492 \text{ cm}^2 \longrightarrow \text{round off to } 32 \text{ cm}^2$$

We round off to two significant figures because 5.2 cm has only two significant figures.

In rounding off numbers, the following rules are followed (each example is rounded to two digits):

1. If the leftmost digit to be removed is more than 5, the preceding number is increased by 1; 2.376 rounds to 2.4.
2. If the leftmost digit to be removed is less than 5, the preceding number is left unchanged; 7.248 rounds to 7.2.
3. If the leftmost digit to be removed is 5, the preceding number is not changed if it is even and is increased by 1 if it is odd; 2.25 rounds to 2.2; 4.35 rounds to 4.4.

The rule used to determine the number of significant figures in multiplication and division cannot be used for *addition and subtraction*. For these operations, *the result should be reported to the same number of decimal places as that of the term with the least number of decimal places*. In the following example the uncertain digits appear in color:

| | | |
|---|---|---|
| This number limits | 20.4 | ← one decimal place |
| the number of significant | 1.322 | ← three decimal places |
| figures in the result ——→ | 83 | ← zero decimal places |
| | 104.722 ——→ | round off to 105 |
| | | (one uncertain digit) |

---

## SAMPLE EXERCISE 1.6

How many significant figures are there in each of the following numbers (assume that each number is a measured quantity): (**a**) 4.003; (**b**) $6.023 \times 10^{23}$; (**c**) 5000; (**d**) the sum $8.7 + 1.966$; (**e**) the product $(16)(5.7793)$?

**Solution:** (**a**) Four; the zeros are significant figures. (**b**) Four; the exponential term does not add to the number of significant figures. (**c**) One, two, three, or four. In this case, the ambiguity could have been avoided by using standard exponential notation. Thus $5 \times 10^3$ has only one significant figure; $5.00 \times 10^3$ has three. (**d**) Three; the sum, expressed to the proper three significant figures, is 10.7. The first uncertain digit comes in the tenths column. (**e**) Two; the product, expressed to the proper two significant figures, is 92.

### PRACTICE EXERCISE
Express the result of the following calculations to the proper number of significant figures: (**a**) $0.335 + 1.774 + 10.82$; (**b**) $6.447 \times 1.30/0.9258$.
***Answers:*** (**a**) 12.93; (**b**) 9.05

---

## SAMPLE EXERCISE 1.7

A gas at 25°C exactly fills a container previously determined to have a volume of $1.05 \times 10^3$ cm³. The container plus gas are weighed and found to have a mass of 837.6 g. The container, when emptied of all gas, has a mass of 836.2 g. What is the density of the gas at 25°C?

**Solution:** The mass of the gas is just the difference in the two masses: $(837.6 - 836.2)$ g $= 1.4$ g. Notice that 1.4 g has only two significant figures, even though the masses from which it is obtained have four.

From the definition of density we have

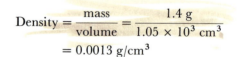

$$\text{Density} = \frac{\text{mass}}{\text{volume}} = \frac{1.4 \text{ g}}{1.05 \times 10^3 \text{ cm}^3}$$
$$= 0.0013 \text{ g/cm}^3$$

There are two significant figures in this quantity, corresponding to the smaller number of significant figures in the two numbers that form the ratio.

### PRACTICE EXERCISE
To how many significant figures should the container be weighed (with and without the gas) in Sample Exercise 1.7 in order for the density to be calculated to three significant figures?
***Answer:*** 5

---

It is important to have a feeling for significant figures when you use a calculator, because calculators ordinarily display more digits than are significant. For example, a typical calculator would give $1.3333333 \times 10^{-3}$ as the answer to the calculation in Sample Exercise 1.7. This result must be rounded off because of the uncertainties in the measured quantities used in the calculation.

When a calculation involves two or more steps, retain at least one additional digit—past the number of significant figures— for intermediate answers. This procedure ensures that small errors from rounding at each step do not combine to appreciably affect the final result. In using a calculator, you may enter the numbers one after another, rounding only the final answer. Accumulated round-off errors may often be the origin of small differences between results you obtain and answers given in the text for numerical problems.

Before we go on, perhaps a word of caution is in order. Sometimes students have little difficulty reading their chemistry text or following the lecture and yet have trouble on exams. In some instances the problem is lack of familiarity with terms. Often the problem lies in students' having a passive but not an active understanding of the material. They can see how someone else has worked a problem, but they are unable to work any problem on their own. An active understanding means being able to use the material in new situations, especially in working problems that are not identical to the sample exercises in the text. It is important to use the practice exercises that accompany each sample exercise. End-of-chapter exercises provide additional questions to help you to determine how well you are able to use the material in the chapter. Colored numbers indicate exercises whose answers can be found at the back of the book. Bracketed numbers indicate exercises of above-average difficulty. If you need a review of basic mathematics as we proceed, refer to Appendix A.

Throughout the text we use what is called **dimensional analysis** as part of our problem-solving approach. In this approach we carry units through all calculations. Units are multiplied together, divided into each other, or canceled like algebraic quantities. The units help to guide us through calculations. We illustrate dimensional analysis in the examples and sample exercises that follow.

Consider the conversion of mass from pounds to kilograms. If a man weighs 175 lb, what is his mass in kilograms? From the table on the back inside cover we have the following relationship: 1 kg = 2.205 lb. From this equality we can write two **conversion factors**:

$$\frac{1\,\text{kg}}{2.205\,\text{lb}}; \quad \frac{2.205\,\text{lb}}{1\,\text{kg}}$$

A conversion factor is a fraction obtained from a valid relationship between equivalent quantities. Its numerator and denominator are equivalent quantities. The conversion factors shown above can be read "1 kg per 2.205 lb" and "2.205 lb per 1 kg." Multiplication of a quantity by a conversion factor changes the units in which the quantity is expressed but not its value. To convert pounds to kilograms we choose the conversion factor that cancels pounds:

$$? \text{ kg} = (175 \text{ lb})\left(\frac{1 \text{ kg}}{2.205 \text{ lb}}\right) = 79.4 \text{ kg}$$

If we were converting kilograms to pounds, we would use the inverse conversion factor. For example, the mass, in pounds, of a 5.00-kg object is given by the following calculation:

$$? \text{ lb} = (5.00 \text{ kg})\left(\frac{2.205 \text{ lb}}{1 \text{ kg}}\right) = 11.0 \text{ lb}$$

Now consider a slightly more complex conversion of units, the calculation of the number of inches in 3.00 km. We can begin by writing the equality we are working toward:

$$? \text{ in.} = 3.00 \text{ km}$$

From the relations shown in the table on the back inside cover and from our basic knowledge of the metric and the English systems we can write the following equalities:

$$1 \text{ km} = 1000 \text{ m}; \qquad 1 \text{ m} = 1.094 \text{ yd}; \qquad 1 \text{ yd} = 36 \text{ in.}$$

From these we obtain the following conversion factors:

$$\frac{1000 \text{ m}}{1 \text{ km}}; \qquad \frac{1.094 \text{ yd}}{1 \text{ m}}; \qquad \frac{36 \text{ in.}}{1 \text{ yd}}$$

These conversion factors permit us to convert km to m, then m to yd, and finally yd to in. (If the table provided a single relationship giving the number of inches in a kilometer, the whole problem would be simpler because only one conversion factor would be needed.) If we multiply 3.00 km by the conversion factors given above, we have

$$? \text{ in.} = (3.00 \text{ km})\left(\frac{1000 \text{ m}}{1 \text{ km}}\right)\left(\frac{1.094 \text{ yd}}{1 \text{ m}}\right)\left(\frac{36 \text{ in.}}{1 \text{ yd}}\right)$$
$$= 1.18 \times 10^5 \text{ in.}$$

Note that each conversion factor is applied so as to cancel the units of the preceding factor. This process converts kilometers successively to meters to yards to inches. Because we are left at the end with the appropriate units, we know that the problem has been set up correctly. (There are are exactly 1000 m in a kilometer and exactly 36 in. in a yard. The number of significant figures in the result, in this case three, is thus determined by the number of significant figures in the quantity being converted.) If we were to apply an incorrect conversion factor, a meaningless mixture of units would result. For example, suppose we applied the conversion factor 1 m/1.094 yd instead of its reciprocal. We would then obtain:

$$? \text{ in.} = (3.00 \text{ km})\left(\frac{1000 \text{ m}}{1 \text{ km}}\right)\left(\frac{1 \text{ m}}{1.094 \text{ yd}}\right)\left(\frac{36 \text{ in.}}{1 \text{ yd}}\right) = 9.88 \times 10^4 \, \frac{\text{m}^2\text{-in.}}{\text{yd}^2}$$

Clearly, the units do not cancel to give the desired units, indicating that the calculation is incorrect.

Conversion factors are ratios, or fractions, obtained from a valid relationship between quantities. We may always write a conversion factor in two ways, each form being the reciprocal of the other. The correct form of the factor will cancel the units that we wish to change. The calculation may be summarized as follows:

Summary of Dimensional Analysis

$$\text{Desired units} = \text{given units} \times \text{conversion factor(s)} \qquad [1.4]$$

You should always carry units thoughout all your calculations, making sure that they cancel properly. Whenever you finish a calculation, look at both the units and magnitude of your answer and ask yourself whether your answer makes any sense.

---

### SAMPLE EXERCISE 1.8

You are approaching a city and see a sign indicating a speed limit of 40 km/hr. What is the corresponding speed in miles per hour?

$$? \, \frac{\text{mi}}{\text{hr}} = \left(40 \, \frac{\text{km}}{\text{hr}}\right)\left(\frac{1 \text{ mi}}{1.609 \text{ km}}\right) = 25 \, \frac{\text{mi}}{\text{hr}}$$

### PRACTICE EXERCISE

One quart is equivalent to 0.946 L. Calculate the number of liters in 10.0 U.S. gallons of gasoline. (1 gal = 4 qt.)

**Solution:** From the table on the back inside cover we have 1 mi = 1.609 km. Thus

*Answer:* 37.8 L

---

### SAMPLE EXERCISE 1.9

You have to pour 2.0 cubic yards ($\text{yd}^3$) of concrete for a patio. What is this volume in cubic meters ($\text{m}^3$)?

$$= (2.0 \text{ yd}^3)\left(\frac{1 \text{ m}^3}{1.309 \text{ yd}^3}\right)$$

$$= 1.5 \text{ m}^3$$

**Solution:** Because 1 m = 1.094 yd and $(1 \text{ m})^3 = (1.094 \text{ yd})^3$, then $1 \text{ m}^3 = 1.309 \text{ yd}^3$.

### PRACTICE EXERCISE

Convert 5.00 $\text{in.}^2$ to $\text{cm}^2$.

$$? \, \text{m}^3 = (2.0 \text{ yd}^3)\left(\frac{1 \text{ m}}{1.094 \text{ yd}}\right)^3$$

*Answer:* 32.3 $\text{cm}^2$

---

### SAMPLE EXERCISE 1.10

The density of mercury is 13.6 g/$\text{cm}^3$. Convert this to the basic SI units of kg/$\text{m}^3$.

$$= 1.36 \times 10^4 \text{ kg/m}^3$$

Note that the first unit conversion factor is cubed to provide the correct dimension of volume.

**Solution:** The relations 1 kg = 1000 g and 1 m = 100 cm provide the conversion factors necessary to change the units in both the numerator and denominator:

### PRACTICE EXERCISE

An airplane flying at the speed of sound is traveling at 1087 ft/s. What is the speed in km/hr?

*Answer:* 1193 km/hr

$$\text{Density} = \left(\frac{13.6 \text{ g}}{1 \text{ cm}^3}\right)\left(\frac{100 \text{ cm}}{1 \text{ m}}\right)^3\left(\frac{1 \text{ kg}}{1000 \text{ g}}\right)$$

## SAMPLE EXERCISE 1.11

The acid in an automobile battery (a solution of sulfuric acid) has a density of 1.2 g/mL. What is the mass (in grams) of 2.00 L of this acid?

**Solution:**

$$\text{Density} = \text{mass/volume}$$

Thus

$$\text{Mass} = \text{volume} \times \text{density}$$

However, we cannot merely multiply 2.00 and 1.2 and get the correct answer. We must pay attention to units. Because 1 L = 1000 mL we have

$$? \text{ g} = (2.00 \text{ L})\left(\frac{1000 \text{ mL}}{1 \text{ L}}\right)\left(1.2 \frac{\text{g}}{\text{mL}}\right)$$
$$= 2.4 \times 10^3 \text{ g}$$

Notice that density can be thought of as a unit conversion factor for converting volume to mass, and vice versa.

**PRACTICE EXERCISE**

What is the volume, in liters, of 2.00 kg of aluminum? Its density is 2.70 g/cm³.
*Answer:* 0.741 L

# FOR REVIEW

## SUMMARY

Chemistry is the study of the properties, composition, and changes of **matter**. Matter exists in three states: **gas**, **liquid**, and **solid**. Most matter consists of a **mixture** of substances. Mixtures can be either **homogeneous** or **heterogeneous**; homogeneous mixtures are called **solutions**. Mixtures can be separated into two types of **pure substances**: **elements** and **compounds**. Each substance has a unique set of **physical** and **chemical properties** that can be used to identify it. Matter can undergo **physical changes** and **chemical changes** (**chemical reactions**).

Measurements in chemistry are made using the **metric system**. Special emphasis is placed on a particular set of metric units called **SI** units, which are based on the meter, kilogram, and second as the basic units of length, mass, and time, respectively. The metric system employs a set of prefixes to indicate decimal fractions or multiples of the base units.

All measured quantities are inexact to some extent. The number of **significant figures** indicates the exactness of the measurement. Certain rules must be followed so that a calculation involving measured quantities is reported to the proper number of significant figures.

In the dimensional analysis approach to problem solving we keep track of units as we carry measurements through calculations. The units are multiplied together, divided into each other, or canceled like algebraic quantities. Obtaining the proper units for the final result is an important means of checking the method of calculation. In converting units, and in several other types of problems, **conversion factors** can be used. These factors are ratios constructed from valid relations between equivalent quantities.

## LEARNING GOALS

Having read and studied this chapter, you should be able to:

1. Distinguish between physical and chemical properties and also between physical and chemical changes.

2. Differentiate among the three states of matter.

3. Distinguish among elements, compounds, and mixtures.

4. Give the symbols for the elements discussed in this chapter.

5. List the basic SI units and the common metric prefixes and their meanings.

6. Perform calculations involving density.

7. Convert temperatures among the Fahrenheit, Celsius, and Kelvin scales.

8. Determine the number of significant figures in a measured quantity.

9. Express the result of a calculation with the proper number of significant figures.

10. Interconvert units using dimensional analysis.

# KEY TERMS

Among the more important terms and expressions used for the first time in this chapter are the following:

A **chemical change** (Section 1.1) is a process in which one or more substances are converted into other substances.

**Chemical properties** (Section 1.1) describe a substance's composition and its reactivity—how the substance reacts, or changes into other substances.

A **compound** (Section 1.2) is a substance composed of two or more elements united chemically in a definite proportion.

**Density** (Section 1.3) is the ratio of an object's mass to its volume.

An **element** (Section 1.2) is a substance that cannot be separated into simpler substances.

An **intensive property** (Section 1.3) is independent of the amount of material considered; an **extensive property** depends on the amount of material considered.

**Mass** (Section 1.3) is a measure of material in an object. It measures the resistance of a stationary object to being moved. In SI units, mass is measured in kilograms.

**Matter** (Section 1.1) is the physical material of the universe; it is anything that occupies space and has mass.

A **physical change** (Section 1.1) is a change (such as a phase change) that occurs with no change in chemical composition.

**Physical properties** (Section 1.1) are those properties that can be measured without changing the composition of a substance. (Color and freezing point are examples.)

**Significant figures** (Section 1.4) are the digits that indicate the precision with which a measurement is made; all digits of a measured quantity are significant, including the last digit, which is uncertain.

# EXERCISES

## Properties of Matter; Substances

**1.1** Identify each of the following substances as a gas, a liquid, or a solid under ordinary conditions: (**a**) mercury; (**b**) copper; (**c**) hydrogen; (**d**) isopropyl alcohol (used as rubbing alcohol); (**e**) sodium bicarbonate (baking soda).

**1.2** Give the state of matter (gas, liquid, or solid) for each of the following under normal conditions: (**a**) oxygen; (**b**) iron; (**c**) ethanol (the alcohol in alcoholic beverages); (**d**) carbon dioxide; (**e**) sodium chloride (table salt).

**1.3** Which of the following are chemical processes and which are physical processes: (**a**) rusting of iron; (**b**) melting of aluminum; (**c**) souring of milk; (**d**) forming of gold foil from a bar of gold; (**e**) burning of carbon; (**f**) digesting of sugar?

**1.4** Characterize the following as chemical changes or physical changes: (**a**) evaporation of water; (**b**) pulverizing of rocks; (**c**) cooking of a steak; (**d**) explosion of a firecracker; (**e**) corrosion of magnesium.

**1.5** Consider the following description of the element sodium: "Sodium is silver-white and soft. It is a good conductor of electricity. It can be prepared by passing electricity through molten sodium chloride. The metal boils at 883°C; the vapor is violet-colored. Sodium metal tarnishes rapidly in air. It burns on heating in air or in an atmosphere of bromine vapor." Indicate which of the properties in this description are physical and which are chemical.

**1.6** Consider the following description of the element bromine: "Bromine is a reddish-brown liquid at room temperature. It vaporizes readily to form a red vapor. It boils at 58.8°C and freezes at −7.2°C. The density of the vapor is 7.59 g/L, and the density of the liquid is 3.12 g/mL (at 20°C). Bromine can be prepared by bubbling chlorine gas through solutions that contain bromide compounds. Sodium metal will burn in an atmosphere of bromine vapor." Indicate which of the properties in this description are physical and which are chemical.

**1.7** Indicate whether each of the following is a pure substance or a mixture; if a mixture, indicate whether it is homogeneous or heterogeneous: (**a**) salt; (**b**) a cube of sugar; (**c**) sawdust; (**d**) orange juice; (**e**) ice; (**f**) vodka.

**1.8** Classify each of the following as an element, compound, or mixture; if the material is a mixture, indicate whether it is homogeneous or heterogeneous: (**a**) ink; (**b**) mercury metal; (**c**) hydrogen peroxide; (**d**) battery acid; (**e**) air; (**f**) a Bufferin or Anacin tablet.

**1.9** Give the chemical symbol for each of the following elements: (**a**) carbon; (**b**) sodium; (**c**) oxygen; (**d**) iron; (**e**) magnesium; (**f**) bromine; (**g**) copper; (**h**) nitrogen.

**1.10** What chemical elements are represented by the following chemical symbols: (**a**) Mg; (**b**) K; (**c**) S; (**d**) Pb; (**e**) Si; (**f**) P; (**g**) Zn; (**h**) Al?

**1.11** In 1807 Humphry Davy passed an electric current through molten potassium hydroxide and isolated a bright, shiny, very reactive substance. He claimed the discovery of a new element, which he named potassium. In those days, before the advent of modern instruments, what was the basis on which one could claim that a substance was an element?

**1.12** A solid white substance A is heated strongly in the absence of air. It decomposes to form a new white substance B and a gas C. The gas has exactly the same properties as the product obtained when a carbon rod is burned in an excess of oxygen. What can we say about whether solids A and B and the gas C are elements or compounds?

## Metric System; SI Units

**1.13** What are the *basic* SI units appropriate to express the following: **(a)** the mass of a gold bar; **(b)** the volume of a gas storage tank; **(c)** the length of a garden hose; **(d)** the area of a desk top; **(e)** the time required to run 800 m?

**1.14** Indicate whether the following are measurements of length, area, volume, mass, density, time, or temperature: **(a)** 5 ns; **(b)** 3.2 kg/L; **(c)** 0.88 pm; **(d)** 540 km$^2$; **(e)** 173 K; **(f)** 15 mL; **(g)** 2.9 g/cm$^3$; **(h)** 2 mm$^3$; **(i)** 23°C.

**1.15** What word prefixes are used in the metric system to indicate the following multipliers: **(a)** $1 \times 10^{-3}$; **(b)** $1 \times 10^{-9}$; **(c)** 0.01; **(d)** $1 \times 10^{-6}$?

**1.16** What word prefixes are used in the metric system to indicate the decimal fraction represented in each of the following measurements: **(a)** $3.2 \times 10^{-12}$ g; **(b)** $8.8 \times 10^{-6}$ s; **(c)** $5.7 \times 10^{-3}$ L; **(d)** 0.04 m?

**1.17** **(a)** How many grams is 4.53 pg? **(b)** How many microseconds are there in 35 ms? **(c)** 1327 cm is equivalent to how many meters?

**1.18** Convert **(a)** 326 g to kg; **(b)** 2.5 ns to ps; **(c)** 0.125 L to mL; **(d)** $3.88 \times 10^5$ m to km.

**1.19** **(a)** A sample of chloroform, a liquid once used as an anesthetic, has a mass of 37.25 g and a volume of 25.0 mL. What is its density? **(b)** The density of platinum is 23.4 g/cm$^3$. Calculate the mass of 50.0 cm$^3$ of platinum. **(c)** The density of liquid bromine is 3.12 g/mL. What is the volume occupied by 5.00 g of bromine?

**1.20** **(a)** A gem-quality ruby has a mass of 5.2 g and a volume of 1.3 cm$^3$. What is its density? **(b)** The density of magnesium is 1.74 g/cm$^3$. What is the mass of a piece of magnesium having a volume of 175 cm$^3$? **(c)** The density of a piece of ebony wood is 1.20 g/cm$^3$. What is its volume if its mass is 874 g?

**1.21** Make the following temperature conversions: **(a)** 68°F to °C; **(b)** 105°C to K; **(c)** −15°C to °F; **(d)** 305 K to °C.

**1.22** Make the following conversions: **(a)** 180°F to °C; **(b)** 225°C to K; **(c)** 58°C to °F; **(d)** 188 K to °C.

## Significant Figures

**1.23** Indicate which of the following are exact numbers: **(a)** the number of donuts in a dozen; **(b)** the area of a postage stamp; **(c)** the number of players on a football team; **(d)** the number of days in September; **(e)** the number of grams in a pound; **(f)** the number of ounces in a pound.

**1.24** Which of the following are exact numbers: **(a)** the volume of a milk carton; **(b)** the distance from your home to your school; **(c)** the number of seconds in an hour; **(d)** the number of pages in this book; **(e)** the mass of a dime; **(f)** the number of inches in a meter?

**1.25** Indicate the number of significant figures in each of the following measured quantities: **(a)** 122 g; **(b)** $5.0 \times 10^{-5}$ m; **(c)** 0.0002796 s; **(d)** 8.007 mm; **(e)** 273 K.

**1.26** How many significant figures are there in each of the following numbers: **(a)** 6.300; **(b)** $3.55 \times 10^3$; **(c)** 0.0709; **(d)** 1200; **(e)** 27.8995.

**1.27** Round off each of the following numbers to four significant figures: **(a)** 4,567,985; **(b)** $6.3375 \times 10^3$; **(c)** 0.00238866; **(d)** 0.98758; **(e)** $0.322589 \times 10^{-3}$.

**1.28** Round off each of the following numbers to three significant figures: **(a)** 0.033390; **(b)** 1.5538; **(c)** $6.022045 \times 10^{23}$; **(d)** 10,537; **(e)** 0.95938.

**1.29** Carry out the following operations and round off the answers to the appropriate number of significant figures, assuming that each number is inexact: **(a)** $0.166 \times 48.3557$; **(b)** $38.5 - 7.376$; **(c)** $(8.55 + 1.6933)/1.478$; **(d)** $0.8775(6.02 \times 10^{23})/20.745$.

**1.30** Carry out the following mathematical operations and round to the appropriate significant figures, assuming that each number is inexact: **(a)** $3.4 \times 2.668/1012$; **(b)** $15.67 + 0.8896 + 2.0 + 1.2 \times 10^{-2}$; **(c)** $(4.43 \times 1.254) + 0.87$; **(d)** $(16.788 - 15.990)/118.9$.

## Conversion of Units; Dimensional Analysis

**1.31** Perform the following conversions: **(a)** 36 in. to cm; **(b)** 4.45 qt to mL; **(c)** 0.885 ft$^2$ to cm$^2$; **(d)** 13.4 m/hr to m/s; **(e)** 55 mi/hr to m/s; **(f)** 5.0 gal/mi to L/km.

**1.32** Perform the following conversions: **(a)** 154 cm to in.; **(b)** $1.25 \times 10^{-8}$ m to pm; **(c)** 5.00 days to seconds; **(d)** $1.99 per pound to dollars per gram; **(e)** 177 ft$^3$ to m$^3$; **(f)** 1.54 lb/ft$^3$ to kg/m$^3$.

**1.33** **(a)** If an athlete has a height of 216 cm, what is the athlete's height in feet and inches? **(b)** If the gasoline tank of a compact car has a capacity of 12 U.S. gal, what is its capacity in liters? **(c)** If a person weighs 135 lb, what is the person's weight in kilograms? **(d)** If a bee flies at an average speed of 3.4 m/s, what is its average speed in mi/hr? **(e)** If an automobile is able to cover 22.0 mi on a gallon of gasoline, what is the gas mileage in km/L? (There are 1.61 km in a mile.) **(f)** What is the engine piston displacement in liters of an engine whose displacement is listed at 320 in.$^3$?

**1.34** **(a)** How many liters of wine can be held in a wine barrel whose capacity is 31 gal? **(b)** The deepest mine is the Western Deep Levels Gold Mine in South Africa, which is 11,647 ft deep. What is this distance in meters? **(c)** The 1928 Indianapolis 500-mi race was won by Lou Meyer with an average speed of 99.482 mph. What is this speed in km/hr? **(d)** Gasoline has a density of about 0.65 g/mL. What is the mass, in kg, of 16 gal of gasoline? **(e)** A small metal block measures 3.3 in. by 4.0 in. by 2.1 in. What is its volume in $cm^3$? **(f)** The recommended adult dose of Elixophyllin, a drug used to treat asthma, is 6 mg/kg of body weight. Calculate the dose, in milligrams, for a 170-lb person.

**1.35** The maximum allowable concentration of carbon monoxide in urban air is 10 $mg/m^3$ over an 8-hr period. At this level, what mass of carbon monoxide in grams is present in a room measuring $2.5 \times 15 \times 40$ m?

**1.36** The density of air at ordinary atmospheric pressure and 25°C is 1.19 g/L. What is the mass, in kg, of the air in a room that measures $8.2 \times 13.5 \times 2.75$ m?

## Additional Exercises

**1.37** Which of the following properties are intensive: **(a)** mass; **(b)** density; **(c)** temperature; **(d)** area; **(e)** color; **(f)** volume?

**1.38** It is sometimes said that aluminum is light and lead is heavy. How might this comparison by stated in a scientifically more precise fashion?

**1.39** Determine the number of **(a)** centimeters in 1 km; **(b)** picoseconds in 1 ms; **(c)** micrograms in 34.2 mg; **(d)** kilograms in $3.05 \times 10^5$ g.

**1.40** A cube composed of one of the substances listed in Table 1.5 measures 2.00 cm on each edge and has a mass of 63 g. **(a)** What is its density? **(b)** What is the identity of the substance?

**1.41** Gallium metal has one of the largest liquid ranges of any substance. It melts at 30°C and boils at 1983°C. What are its melting and boiling points in °F?

**1.42** **(a)** What is the temperature in °C of an animal whose body has a temperature of 100.6°F? **(b)** On August 24, 1960, the temperature at the Vostok Station in Antarctica was recorded at −127°F. What is this temperature in °C? **(c)** Helium has the lowest boiling point of any liquid, 4 K. What is the boiling point in °C? in °F?

**1.43** Express the following numbers in appropriate exponential notation: **(a)** 1245; **(b)** 65,000 (to show three significant figures); **(c)** 59,750 (to show four significant figures); **(d)** 0.00456.

**1.44** Is the use of significant figures in each of the following statements appropriate? Why or why not? **(a)** The 1976 circulation of *Reader's Digest* magazine was 17,887,299. **(b)** In the United States, more than 1.4 million persons have the surname Brown. **(c)** The average annual rainfall in San Diego, California, is 20.54 in. **(d)** The population of East Lansing, Michigan, in 1979 was 51,237.

**1.45** Convert the measures in the following statements into the indicated units. **(a)** A furlong, a measure of distance used in horse racing, is defined as 201.168 m. What is the distance in km of the Kentucky Derby, which is a 10-furlong race? **(b)** A fathom, used as a measure of water depths, is defined as 1.8288 m, exactly. How deep, in feet, is Lake Superior, which is 216.9 fathoms at maximum depth? **(c)** If Jules Verne expressed the title of his famous book *Twenty Thousand Leaques Under the Sea* in meters, what would the title be? (1 league = 3.45 mi; 1 mi = 1609 m.) **(d)** In 1978 the world's record for the 100-m dash was 9.9 s. To cover 100 m in 9.9 s, what must the average speed be in miles per hour?

**1.46** A cylindrical container of radius $r$ and height $h$ has a volume of $\pi r^2 h$. **(a)** Calculate the volume in cubic centimeters of a cylinder with radius 6.5 cm and a height of 28.6 cm. **(b)** Calculate the volume in cubic meters of a cylinder that is 8.0 ft high and 20.0 in. in diameter. **(c)** Calculate the mass in kilograms of water required to fill the cylinder in part (b) if the dimensions listed are the inside dimensions. The density of water is 1.00 $g/cm^3$.

**1.47** A cylindrical glass tube 15.0 cm in length is filled with mercury. The mass of mercury needed to fill the tube is found to be 110.5 g. Calculate the inner diameter, in cm, of the tube. The density of mercury is 13.6 $g/cm^3$.

**1.48** What is the radius, in centimeters, of a metal sphere whose mass is $2.00 \times 10^2$ g if the metal is **(a)** iron, density = 7.86 $g/cm^3$; **(b)** aluminum, density = 2.70 $g/cm^3$; **(c)** lead, density = 11.3 $g/cm^3$. The volume of a sphere is given by the equation $V = 4\pi r^3/3$.

**1.49** A graduated cylinder on a balance has a mass of 57.832 g. An organic liquid, toluene, with a density of 0.866 g/mL, is added until the combined mass reads 87.127 g. What is the volume of the liquid in the graduated cylinder?

**1.50** A graduated cylinder contains 20.0 mL of water. An irregularly shaped object is placed in the cylinder, and the water level rises to the 31.2-mL mark. If the object has a mass of 49.7 g, what is its density?

**1.51** A few years ago a cartoon pictured a thief making his getaway, gun in one hand and a bucket of gold dust in the other. If the bucket had a volume of 8 qt and was full of gold—with a density of 19.3 $g/cm^3$—what was the mass? Comment on the thief's strength.

**[1.52]** Suggest a method of separating each of the following mixtures: **(a)** sugar and water; **(b)** sugar and powdered glass; **(c)** alcohol and water; **(d)** the color pigments in a leaf; **(e)** the components of air.

**[1.53]** Suppose you were given a sample of a homogeneous liquid. What would you do to determine whether it is a solution or a pure substance?

**[1.54]** Using the *Handbook of Chemistry and Physics* or a similar source of data, determine **(a)** the densest known solid element; **(b)** the solid element with the highest known melting point; **(c)** the element with the lowest known boiling point; **(d)** the only two elements that are liquids at room temperature (about 20°C).

**[1.55]** Using the *Handbook of Chemistry and Physics* or a similar source of data, list as many properties as possible that would allow you to distinguish between **(a)** silver and aluminum; **(b)** water, $H_2O$, and ethyl alcohol, $C_2H_5OH$.

## CHAPTER 2

# Atoms, Molecules, and Ions

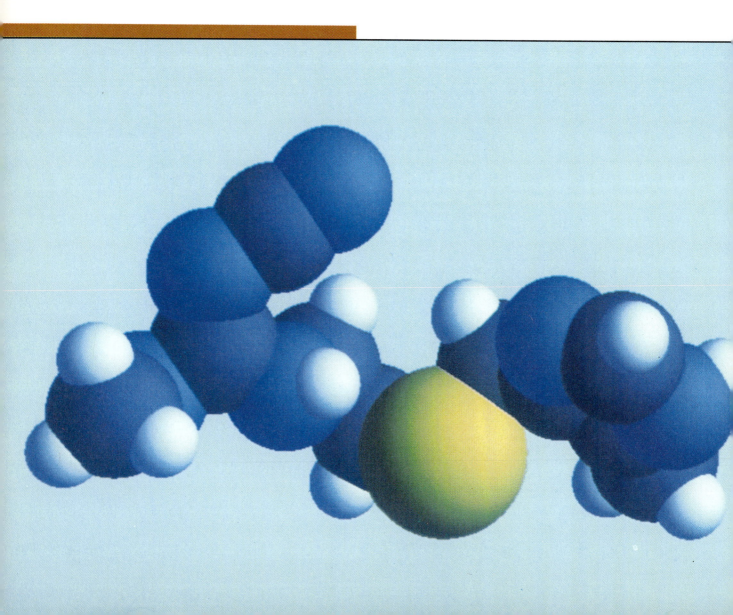

In 1661, Robert Boyle (1627–1691) published a book entitled *The Sceptical Chymist*. In this book he proposed that substances that cannot be decomposed into simpler substances be called elements. The classification of substances as compounds and elements continues to be a central idea of modern chemistry. It helps us to systematize many chemical facts. However, it also raises a number of questions. Why is one element different from another? Why is a compound different from a mixture? Why do elements combine to form compounds? In the years following Boyle's work, such questions often occupied the thoughts of chemists. The atomic theory of matter, first published by John Dalton (1766–1844) over the period 1803 to 1807, formed the basis for answering such questions.

The atomic theory is fundamental to chemistry. When chemists seek to understand why a particular material is hard, or sweet, or explosive, or toxic, their experiments are directed by the concept that matter consists of atoms. In this chapter we introduce the atomic view of matter, examining some basic concepts of atomic structure and considering briefly the formation of molecules and ions. We then build on these concepts by discussing the naming of compounds.

## CONTENTS

## 2.1 THE ATOMIC THEORY

The seeds of the atomic theory go back at least to the time of the ancient Greeks. The Greeks pondered a seemingly abstract question: Can matter be divided endlessly into smaller and smaller pieces, or is it composed of some ultimate particle that cannot be further divided? Plato (427–347 B.C.), Aristotle (384–322 B.C.), and most other Greek philosophers believed that matter is continuous. However, some Greek philosophers, notably Democritus (460–370 B.C.), argued that matter is composed of small indivisible particles called *atomos*, meaning "indivisible."

The erroneous idea that matter is continuous was widely held from the time of the Greeks until the early nineteenth century. Although the atomic idea was occasionally revived, its proponents relied largely on intuitive arguments to support their views. During this long period, however, there was a thin, intermittent stream of experimental work. Much of it was prompted by incorrect notions, such as the alchemical belief that common metals such as lead might be transformed into precious metals. Nevertheless, experience of how substances react with one another accumulated, and more quantitative methods of studying chemical reactions were developed. The way was prepared for the reemergence of the notion of atoms.

A meaningful statement of an atomic theory was finally published in the early nineteenth century by John Dalton, an English schoolteacher (Figure 2.1). Dalton's atomic theory was strongly tied to experimental

A computer-generated model of an organic molecule which is used as an anti-ulcer drug. (Courtesy of DuPont)

**FIGURE 2.1** John Dalton (1766–1844) was the son of a poor English weaver. Dalton began teaching at the age of 12; he spent most of his years in Manchester, where he taught both grammar school and college. His lifelong interest in meteorology led him to study gases and hence to chemistry and eventually to the atomic theory. It was perhaps because Dalton's training was not in chemistry that he was able to approach problems with a viewpoint different from the chemists of the time. (Library of Congress)

observation. His efforts were so successful that his theory has had to undergo little revision.

The basic postulates of Dalton's theory were as follows:

1. Each element is composed of extremely small particles called atoms.
2. All atoms of a given element are identical.
3. Atoms of different elements have different properties (including different masses).
4. Atoms of an element are not changed into different types of atoms by chemical reactions; atoms are neither created nor destroyed in chemical reactions.
5. Compounds are formed when atoms of more than one element combine.
6. In a given compound, the relative number and kind of atoms are constant.

Dalton's theory provides us with a mental picture of matter. As represented schematically in Figure 2.2, we visualize an element as being composed of tiny particles called atoms (Figure 2.3). **Atoms** are the basic building blocks of matter. They are the smallest units of an element that can combine with other elements (that is, take part in a chemical

**FIGURE 2.2** Difference between elements, compounds, and mixtures as visualized through Dalton's atomic theory. (*a*) Elements are composed of small particles called atoms. All atoms of a given element are identical; atoms of different elements are different. (*b*) Compounds involve atoms of two or more elements combined in definite arrangements. (*c*) Mixtures have variable compositions. There is no restriction on the relative number of atoms of elements 1 and 2.

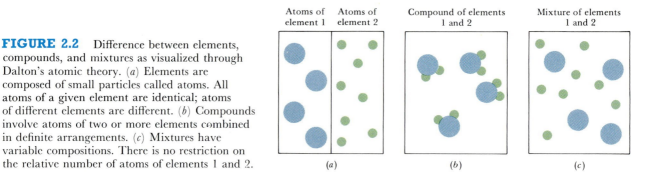

reaction). In compounds the atoms of two or more elements combine in definite arrangements. Mixtures do not involve the intimate interactions between atoms that are found in compounds.

Dalton's theory embodies several simple laws of chemical combination that were known at the time. Because atoms are neither created nor destroyed in the course of chemical reactions (postulate 4), it is readily evident that matter is neither created nor destroyed in such reactions. Thus we have the **law of conservation of matter** (also known as the **law of conservation of mass**): The total mass of materials present after a chemical reaction is the same as the total mass before the reaction. This law provides the basis for much of what we discuss in Chapter 3. Dalton's postulate 6 explains the **law of constant composition** (which was cited in Section 1.2): In a given compound the relative number and kind of atoms are constant. A third law discovered by Dalton and consistent with his theory is the **law of multiple proportions**. Consider two elements that form more than one compound. For a fixed mass of one element the different masses of the second element in the different compounds are related to each other by small whole numbers. For example, the substances water and hydrogen peroxide both consist of the elements hydrogen and oxygen. In water there are 8.0 g of oxygen for each gram of hydrogen. In hydrogen peroxide there are 16.0 g of oxygen for each gram of hydrogen. The masses of oxygen that combine with a gram of hydrogen in these compounds are in the ratio of the small whole number two: Hydrogen peroxide has twice as much oxygen per unit mass of hydrogen as water does. Using the atomic theory, we understand this to mean that hydrogen peroxide contains twice as many oxygen atoms per hydrogen atom as does water. We now know that water contains one oxygen atom for each two hydrogen atoms and that hydrogen peroxide contains two oxygen atoms for each two hydrogen atoms.

Thus we see that the atomic theory ties together many observations and helps us explain them. To answer our earlier question of what makes one element different from another, we can now answer that they have different types of atoms. However, this explanation only begs a further question: How do the atoms of the various elements differ from each other? We need to consider the structure of the atom to answer this question. Sections 2.2 and 2.3 take up this topic. We will see that as we begin to understand the structure of the atom, we will begin to understand many more aspects of matter.

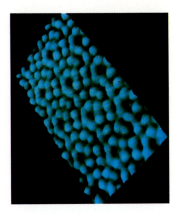

**FIGURE 2.3**  The image of the surface of a silicon crystal as obtained by a relatively new technique called tunneling electron microscopy. (The 1986 Nobel Prize in physics was awarded for development of this technique.) The blue spheres are silicon atoms situated on the surface of the crystal. (IBM Research)

---

Scientists presently have a large arsenal of sophisticated equipment with which to measure the properties of individual atoms in great detail. Consequently, we now know a great deal about atoms and their structures. However, only 150 years ago very little was known about atoms beyond what was contained in Dalton's atomic theory. Dalton and his contemporaries viewed the atom as an indivisible object, like a tiny indestructible and unchangeable ball. By 1850, data had begun to accumulate that indicated that the atom is composed of smaller particles. Before we summarize our current model of atomic structure, we will consider a few of the most important experiments that led to that model. In order to understand these experiments we need to keep in mind a basic rule

## 2.2
## THE DISCOVERY OF ATOMIC STRUCTURE

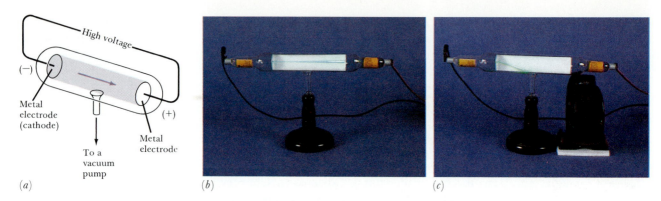

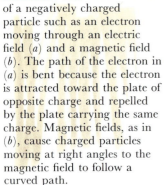

(a)

(b)

(c)

**FIGURE 2.4** (a) In a cathode-ray tube electrons move from the negative electrode (cathode) to the positive electrode (anode). (b) A photo of a cathode-ray tube containing a fluorescent screen to show the path of the cathode rays. (c) The path of the cathode rays is deflected by the presence of a magnet. (Donald Clegg and Roxy Wilson)

regarding the behavior of electrically charged particles: *Like charges repel each other; unlike charges attract.*

## Cathode Rays and Electrons

In the mid-1800s a number of investigators began to study electrical discharge through partially evacuated tubes, such as that shown in Figure 2.4. A high voltage produces radiation within the tube. This radiation became known as **cathode rays** because it emanated from the negative electrode, or cathode. Although the rays themselves could not be seen, their movement could be determined because the rays cause certain materials, including glass, to give off light, or fluoresce. (Television picture tubes are cathode-ray tubes; the television picture results from fluorescence from the television screen.) In the absence of magnetic or electric fields, cathode rays travel in straight lines. However, magnetic and electric fields deflect the rays in the manner expected for negatively charged particles, as shown in Figure 2.5. These facts as well as additional observations suggested that the radiation consists of a stream of negatively charged particles that were named **electrons**. The rays (streams of electrons) were found to be independent of the nature of the cathode material. Thus scientists deduced that electrons are a basic component of all matter.

**FIGURE 2.5** The behavior of a negatively charged particle such as an electron moving through an electric field (a) and a magnetic field (b). The path of the electron in (a) is bent because the electron is attracted toward the plate of opposite charge and repelled by the plate carrying the same charge. Magnetic fields, as in (b), cause charged particles moving at right angles to the magnetic field to follow a curved path.

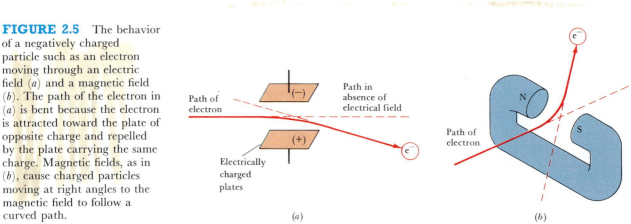

(a)

(b)

In 1897 the British physicist J. J. Thomson (Figure 2.6) measured the ratio of the electrical charge to the mass of the electron using a cathode-ray tube such as that shown schematically in Figure 2.7. When only the magnetic field is turned on, the electron strikes point *A* of the tube. When the magnetic field is off and the electric field is on, the electron strikes point *C*. When both the magnetic and electric fields are off or when they are balanced so as to cancel each other's effects, the electron strikes point *B*. By carefully and quantitatively determining the effects of magnetic and electric fields on the motion of the cathode rays, Thomson was able to determine the charge-to-mass ratio of $1.76 \times 10^8$ coulombs per gram.*

In 1909 Robert Millikan of the University of Chicago determined the charge on the electron by measuring the effect of an electric field on the rate at which charged oil droplets fall under the influence of gravity. The

* The coulomb (C) is the SI unit for electrical charge.

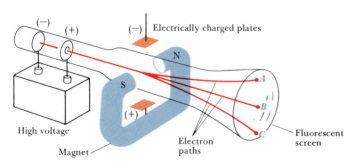

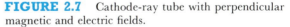

**FIGURE 2.7** Cathode-ray tube with perpendicular magnetic and electric fields.

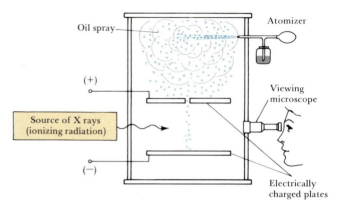

**FIGURE 2.8**  Schematic representation of Millikan's apparatus for studying the rate of fall of oil droplets.

apparatus that he used is shown schematically in Figure 2.8. The rate at which the droplets fall in air is determined by their size and mass. By watching a particular droplet, Millikan could measure its rate of fall and calculate from this its mass. The experiment was arranged so that a source of radioactivity was near the droplets. This caused the oil drops to acquire an electrical charge. When an electrical potential was applied to the plates, the electric field would act on the charged oil droplets in the region between the plates. Their fall could be accelerated, retarded, or even reversed, depending on the charge on the droplet and the polarity of the voltage applied to the plates. By carefully measuring the effects of the electrical field on the movements of many droplets, Millikan found that the charge on the oil drops was always an integral multiple of $1.60 \times 10^{-19}$ C, which he deduced was the charge of the electron. The mass of the electron was then calculated by combining Millikan's value of the charge with Thomson's charge-to-mass ratio:

$$\text{Mass} = \frac{1.60 \times 10^{-19}\,\text{C}}{1.76 \times 10^8\,\text{C/g}} = 9.10 \times 10^{-28}\,\text{g}$$

Using slightly more accurate values for the charge and the charge-to-mass ratio, we obtain the presently accepted value for the mass of the electron, $9.10939 \times 10^{-28}$ g.

**Radioactivity**

The discovery of radioactivity by the French scientist Henri Becquerel in 1896 provided additional evidence for the complexity of the atom. Becquerel's imagination had been captured by W. C. Roentgen's discovery of X rays, which had been reported in January 1896. Roentgen had been quick to grasp the practical importance of his discovery, and within a short time X rays had been used in medicine. Members of the international scientific community also sensed that this was something big. Becquerel was well aware that certain substances upon exposure to sunlight become luminous, a phenomenon referred to as fluorescence. He sought to determine whether such fluorescent substances gave off X rays. In his initial experiments, Becquerel chose to work with a fluorescent uranium mineral. He placed this in the sunlight over a photographic plate that had been carefully wrapped to protect it from the direct radiation of the sun. When the plate was developed, he found the image of the mineral on the plate. Toward the end of February 1896, Becquerel incorrectly reported that penetrating rays, presumably X rays, could be induced by sunlight and emitted as part of fluorescence. However, the weather turned bad, and Becquerel had to postpone further studies. While the sun stayed behind the clouds, Becquerel kept the mineral and the wrapped photographic plate in a desk drawer. On March 1, 1896, he decided to develop the plate, not expecting to find any images. He was surprised to find very intense silhouettes. Becquerel concluded correctly this time that the mineral was producing a spontaneous radiation and referred to this phenomenon as **radioactivity.** At Becquerel's suggestion, Marie Sklodowska Curie (Figure 2.9) and her husband, Pierre, began their famous experiments to isolate the radioactive components of the mineral, called pitchblende.

Further study of the nature of radioactivity, principally by the British scientist Ernest Rutherford (1871–1937), revealed three types of radiation—alpha ($\alpha$), beta ($\beta$), and gamma ($\gamma$) radiation. Each type differs in electrical behavior and penetrating ability. The behavior of these three types of radiation in an electric field is shown in Figure 2.10.

Rutherford showed that the $\beta$ radiation was identical to a stream of high-speed electrons. The particles of this radiation were called $\beta$ particles. In units of the charge of the electron, each $\beta$ particle has a charge of $1-$. The $\alpha$ rays were found to consist of particles with a charge of $2+$. The $\alpha$ particles are comparatively much more massive than the $\beta$ particles. Rutherford was able to show that the $\alpha$ particles combined with electrons to form atoms of helium. He thus concluded that the $\alpha$ rays consist of the positively charged core of the helium atom. The $\gamma$ rays are high-energy radiation like X rays; they do not consist of particles. The $\alpha$ rays have low penetrating power; they are stopped by paper. The $\beta$

**FIGURE 2.9** Marie Sklodowska Curie (1867–1934). When M. Curie presented her doctoral thesis, it was described as the greatest single contribution of any doctoral thesis in the history of science. Among other things, two new elements, polonium and radium, had been discovered. In 1903, Becquerel, M. Curie, and her husband, Pierre, were jointly awarded the Nobel Prize in physics. In 1911 M. Curie won a second Nobel Prize, this time in chemistry. Irene Curie, daughter of Marie and Pierre Curie, was also a scientist. She and her husband, Frederic Joliot, shared the 1935 Nobel Prize in chemistry for their work in artificial production of radioactive substances by bombarding certain elements with particles. (The Granger Collection)

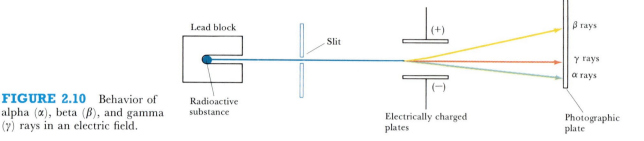

**FIGURE 2.10** Behavior of alpha ($\alpha$), beta ($\beta$), and gamma ($\gamma$) rays in an electric field.

particles have about 100 times greater penetrating ability, and the $\gamma$ rays have about 1000 times greater penetrating ability than $\alpha$ rays. The comparative properties of the three types of radiation are summarized in Table 2.1.

**TABLE 2.1**  Summary of the properties of alpha, beta, and gamma rays

| | Type of radiation | | |
|---|---|---|---|
| | $\alpha$ | $\beta$ | $\gamma$ |
| Charge | 2+ | 1− | 0 |
| Mass | $6.64 \times 10^{-24}$ g | $9.11 \times 10^{-28}$ g | 0 |
| Relative penetrating power | 1 | 100 | 1000 |
| Identity | $^{4}_{2}\text{He}^{a}$ nuclei | Electrons | High-energy radiation |

$^{a}$ This notation will be explained in the next section.

## The Nuclear Atom

By 1909 Rutherford had firmly established that $\alpha$ rays consisted of particles with a 2+ charge. He then began to study the ways these $\alpha$ particles interact with matter. By this time it was well accepted that the atom was electrical in nature and contained electrons. The prevalent model of the atom, as developed by J. J. Thomson, pictured the atom as a cloud of positive charge in which negatively charged electrons were embedded like seeds in a watermelon.

In 1910 Rutherford and his co-workers performed an experiment that led to the downfall of Thomson's model. Rutherford was studying the manner of scattering of a narrow beam of $\alpha$ particles as they passed through a thin gold foil. He had found slight scattering, on the order of 1 degree, which was consistent with Thomson's model. One day Hans Geiger, an associate of Rutherford's, proposed that Ernest Marsden, a 20-year-old undergraduate working in their laboratory, get some experience in conducting such experiments. Rutherford suggested that Marsden see if $\alpha$ particles were scattered through large angles. In Rutherford's own words:

> I may tell you in confidence that I did not believe they would be since we knew that the $\alpha$ particle was a very massive particle with a great deal of energy. . . . Then I remember two or three days later Geiger coming to me in great excitement and saying, "We have been able to get some $\alpha$ particles coming backwards." . . . It was quite the most incredible event

that has ever happened to me in my life. It was almost as if you fired a 15-inch shell into a piece of tissue paper and it came back and hit you.

What Rutherford and his co-workers had observed was that the vast majority of α particles passed directly through the foil without deflection. However, a few did undergo deflection, some even bouncing back in the direction from which they had come, as shown in Figure 2.11.

By 1911 Rutherford was able to explain these observations by postulating that all of the positive charge and most of the mass of the atom reside in a very small, extremely dense region, which he called the **nucleus**. Most of the total volume of the atom is empty space in which electrons move around the nucleus. In the α-scattering experiment, most α particles pass directly through the foil because they do not encounter the minute nucleus; they merely pass through the empty space of the atom. Occasionally an α particle, however, comes into the close vicinity of a nucleus. The repulsion between the highly charged gold nucleus and an α particle is strong enough to deflect the less massive α particle, as depicted in Figure 2.12.

**FIGURE 2.11**
Rutherford's experiment on the scattering of α particles.

Beam of α particles

Scattered α particles

Source of α particles

Circular fluorescent screen

Thin metal foil

Most particles are undeflected

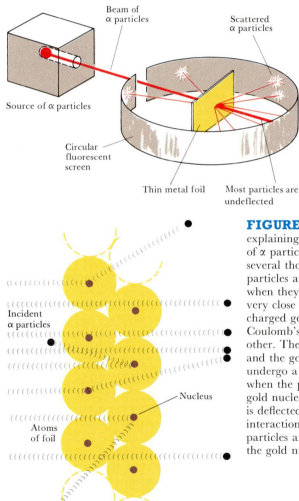

Incident α particles

Nucleus

Atoms of foil

**FIGURE 2.12** Rutherford's model explaining his experiment with scattering of α particles. The gold foil is actually several thousand atoms thick. The α particles are deflected backward only when they collide directly with, or pass very close to, the much heavier, positively charged gold nuclei. According to Coulomb's law, like charges repel each other. The α particles, with a 2+ charge, and the gold nuclei, with 79+ charge, undergo a strongly repulsive interaction when the particle closely approaches a gold nucleus. The less massive α particle is deflected from its path by the repulsive interaction. Only a small fraction of the particles are strongly deflected, because the gold nucleus has a small volume.

## 2.3 THE MODERN VIEW OF ATOMIC STRUCTURE

Since the time of Rutherford, physicists have learned much about the detailed structure of atomic nuclei. In the course of these discoveries the list of subnuclear particles has grown long and continues to increase. As chemists, we can take a very simple view of the atom, because only three subatomic particles—the **proton**, **neutron**, and **electron**—have a bearing on chemical behavior.

The charge of a proton is $+1.602 \times 10^{-19}$ C, and that of an electron is $-1.602 \times 10^{-19}$ C. The quantity $1.602 \times 10^{-19}$ C is called the electronic charge. The charges of atomic and subatomic particles are usually expressed as multiples of this charge rather than in coulombs. Thus the charge of the proton is $1+$, and that of the electron is $1-$. Neutrons are uncharged. Atoms have no net electrical charge because they have an equal number of electrons and protons.

Protons and neutrons reside together in the very small volume within the atom known as the **nucleus**. Most of the rest of the atom is space in which the electrons move. The electrons are attracted to the protons in the nucleus and kept from flying off completely free in space by the attraction that exists between particles of unlike electrical charge.

Atoms have extremely small masses. For example, the mass of the heaviest known atom is on the order of $4 \times 10^{-22}$ g. It would be cumbersome to have to continually express such small masses in grams. Therefore, we use a unit called the *atomic mass unit*, or *amu.** An amu equals

* The SI abbreviation for the atomic mass unit is merely u. We will use the more common abbreviation amu.

---

## A CLOSER LOOK                                                    Basic Forces

There are four basic forces, or interactions, known in nature: gravity, electromagnetism, and the strong and the weak nuclear forces. *Gravitational forces* act between all objects in proportion to their masses. Gravitational forces between atoms or subatomic particles are so small that they are of no chemical significance.

*Electromagnetic forces* act between electrically charged or magnetic objects. Electric and magnetic forces are intimately related. Electric forces are of fundamental importance in understanding the chemical behavior of atoms. The magnitude of the electrical force between two charged particles is given by **Coulomb's law:** $F = kQ_1Q_2/d^2$, where $Q_1$ and $Q_2$ are the magnitudes of the charges on the two particles, $d$ is the distance between their centers, and $k$ is a constant determined by the units for $Q$ and $d$. A negative value for the force indicates attraction, and a positive value indicates repulsion.

All nuclei except those of hydrogen atoms contain two or more protons. Since like charges repel, the electrical repulsion would cause the protons to fly apart if a stronger force did not keep them together in the nucleus. This stronger force is called the *strong nuclear force*. It acts between subatomic particles that are extremely close together, as they are in the nucleus. At this small distance this force is stronger than the electrical force, so the nucleus holds together. The *weak nuclear force* is weaker than the electric force but stronger than gravity. We are aware of its existence only because it shows itself in certain types of radioactivity. The implications of nuclear forces in our daily lives are much less evident than the effects of electromagnetic forces, and we will not consider them further.

All forces that we experience around us are derived from these four basic interactions. For example, the force that your fingers exert on a pencil is the result of electrical repulsion between the outer electrons of the atoms of your fingers and those of the pencil. We will often refer to electrical interactions as we discuss chemical behavior in the coming chapters.

**TABLE 2.2** Comparison of the proton, neutron, and electron

| Particle | Charge | Mass (amu) |
|----------|--------|------------|
| Proton | Positive $(1+)$ | 1.0073 |
| Neutron | None (neutral) | 1.0087 |
| Electron | Negative $(1-)$ | $5.486 \times 10^{-4}$ |

$1.66053 \times 10^{-24}$ g. On this scale a proton has a mass of 1.0073 amu, a neutron 1.0087 amu, and an electron $5.486 \times 10^{-4}$ amu. The masses of the proton and neutron are very nearly equal, and both are much greater than that of an electron. In fact, it would take 1836 electrons to equal the mass of one proton. Thus the nucleus carries most of the mass of an atom. Table 2.2 summarizes the charges and masses of the subatomic particles. We will have much more to say about atomic masses in Section 3.3.

Atoms have diameters on the order of 1–5 Å. The angstrom unit (Å), which is $10^{-10}$ m, is widely used to indicate dimensions on the atomic scale. For example, the diameter of a chlorine atom is 2.0 Å. In time the angstrom may be replaced by a more acceptable SI unit such as the picometer or nanometer. In these units the diameter of a chlorine atom is 200 pm or 0.20 nm. For the most part, we will employ angstroms throughout the text to indicate atomic and molecular dimensions.

---

**SAMPLE EXERCISE 2.1**

The diameter of a U.S. penny is 19 mm. How many chlorine atoms would fit side by side along this diameter?

**Solution:** As mentioned in the text, the diameter of a single chlorine atom is 2.0 Å. Starting with the diameter of the penny, we can convert its dimensions to Å and then, using the diameter of the Cl atom, calculate the number of Cl atoms:

Cl atoms =

$$(19 \text{ mm})\left(\frac{1 \text{ m}}{10^3 \text{ mm}}\right)\left(\frac{10^{10} \text{ Å}}{1 \text{ m}}\right)\left(\frac{1 \text{ Cl atom}}{2.0 \text{ Å}}\right)$$
$$= 9.5 \times 10^7 \text{ Cl atoms}$$

This exercise helps to illustrate how very small atoms are compared to more familiar dimensions.

**PRACTICE EXERCISE**
The diameter of a carbon atom is 1.5 Å. **(a)** Express this diameter in picometers. **(b)** How many carbon atoms could be aligned side by side in a straight line across the width of a pencil line that is 0.10 mm wide?
*Answers:* **(a)** 150 pm; **(b)** $6.7 \times 10^5$ C atoms

---

The diameters of atomic nuclei are on the order of $10^{-4}$ Å, only a small fraction of the diameter of the atom as a whole. If the atom were scaled upward in size so that the nucleus were 2 cm in diameter (about the diameter of a penny), the atom would have a diameter of 200 m (about twice the length of a football field). Because the tiny nucleus carries most of the mass of the atom in such a small volume, it has an incredible density—on the order of $10^{13}$–$10^{14}$ g/cm³. A matchbox full of material of such density would weigh over $2\frac{1}{2}$ billion tons! Astrophysicists have suggested that the interior of a collapsed star may reach approximately this density.

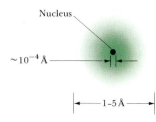

**FIGURE 2.13** Schematic cross-sectional view through the center of an atom. The nucleus, which contains positive protons and neutral neutrons, is the location of virtually all the mass of the atom. The rest of the atom is mainly space in which the light, negatively charged electrons move.

An illustration of the atom that incorporates the features we have just discussed is shown in Figure 2.13. The electrons, which take up most of the volume of the atom, play the major role in chemical reactions. The significance of representing the region containing the electrons as an indistinct cloud will become clear in later chapters when we consider the energies and spatial arrangements of the electrons.

## Isotopes, Atomic Numbers, and Mass Numbers

The identity of an element depends on the number of protons in the nucleus of an atom of that element. In fact we may define an element as a substance whose atoms all have the same number of protons. Because an atom has no net electronic charge, there must be one electron for each proton. For example, all atoms of the element carbon have six protons and six electrons. Most carbon atoms also have six neutrons, although some have more and some have less. Atoms of a given element that differ in number of neutrons, and consequently in mass, are called **isotopes**. The symbol $^{12}_{6}C$ or simply $^{12}C$ (read "carbon twelve," carbon-12) represents the carbon atom with six protons and six neutrons. The number of protons, which is called the **atomic number**, is shown by the subscript. Since all atoms of a given element have the same atomic number, this subscript is redundant and hence often omitted. The superscript is called the **mass number** and is the total number of protons plus neutrons in the atom. Some carbon atoms contain six protons and eight neutrons and are consequently represented as $^{14}C$ (read "carbon fourteen").

Subscripts and superscripts are generally used with the symbol for an element only when reference is made to a particular isotope of that element. Three isotopes of oxygen and their chemical symbols are shown schematically in Figure 2.14. The term **nuclide** is applied in a general way to a nucleus with a specified number of protons and neutrons. For example, the nucleus of $^{16}_{8}O$ is referred to as the $^{16}_{8}O$ nuclide. (We will have more to say about the isotopic compositions of the elements in Section 3.3, when we examine atomic masses.)

**FIGURE 2.14** Distribution of subatomic particles in three isotopes of oxygen ($p^+$ = proton, n = neutron, $c^-$ = electron). The shaded circle represents the nucleus; the eight electrons are in space surrounding the nucleus.

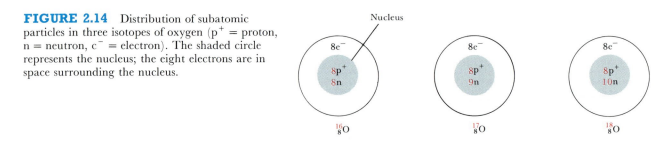

## SAMPLE EXERCISE 2.2

How many protons, neutrons, and electrons are there in an atom of $^{197}$Au?

**Solution:** According to the list of elements given in the front inside cover of this text, gold has an atomic number of 79. Consequently, an $^{197}$Au atom has 79 protons, 79 electrons, and $197 - 79 = 118$ neutrons.

**PRACTICE EXERCISE** _____

How many protons, neutrons, and electrons are there in a $^{39}$K atom?

**Answer:** 19 protons, 19 electrons, and 20 neutrons

## SAMPLE EXERCISE 2.3

Write the nuclear isotope symbols for the three isotopes of hydrogen, with mass numbers of 1, 2, and 3.

**Solution:** Because all three of the hydrogen isotopes must have the same number of protons, 1, the three symbols are: $^1_1$H; $^2_1$H; $^3_1$H.

**PRACTICE EXERCISE** _____

Give the complete chemical symbol for the nuclide with 15 protons, 15 electrons, and 16 neutrons.

**Answer:** $^{31}_{15}$P

On the atomic level gold, oxygen, and carbon differ in terms of the number of protons, neutrons, and electrons their respective atoms contain. These subatomic particles, however, are common to all substances. We can therefore state that an atom is the smallest representative sample of an element, because breaking the atom into subatomic particles destroys its identity.

In order to change a base or common metal like lead, atomic number 82, to gold, atomic number 79, requires removal of three protons from the nucleus of the lead atom. Because the nucleus is extremely small and buried in the heart of the atom, and because of the very strong binding forces between particles in the nucleus, this removal is exceedingly difficult. The energies required to cause changes in the nucleus are enormously greater than the energies associated with even the most vigorous chemical reactions. Thus, we still agree with Dalton that atoms of an element are not changed into different types of atoms by chemical reactions. Therein lies the futility of the alchemists' attempts to change base metals to gold.

## 2.4 THE PERIODIC TABLE

Dalton's atomic theory, and the various empirical laws that it helped to explain (Section 2.1), set the stage for a vigorous growth in chemical experimentation during the early part of the nineteenth century. As the body of chemical observations grew and the list of known elements expanded, attempts were made to find regularities in chemical behavior. These efforts culminated in the development of the periodic table in 1869. We will have much to say about the periodic table in later chapters, but it is so important and useful that you should become acquainted with it now.

Many elements show very strong similarities to each other. For example, lithium (Li), sodium (Na), and potassium (K) are all soft, very reactive metals. The elements helium (He), neon (Ne), and argon (Ar) are very nonreactive gases. If the elements are arranged in order of increasing

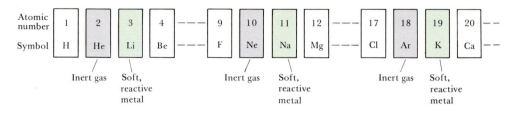

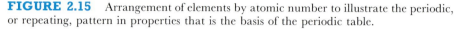

**FIGURE 2.15** Arrangement of elements by atomic number to illustrate the periodic, or repeating, pattern in properties that is the basis of the periodic table.

atomic number, their chemical and physical properties are found to show a repeating, or periodic, pattern. For example, each of the soft, reactive metals—lithium, sodium, and potassium—comes immediately after one of the nonreactive gases—helium, neon, and argon—as shown in Figure 2.15. The arrangement of elements in order of increasing atomic number, with elements having similar properties placed in vertical columns, is known as the **periodic table**. The periodic table is shown in Figure 2.16 and is also given on the front inside cover of your text for easy reference. In most chemistry classrooms large periodic tables are hung on the walls—a testimony to their usefulness. You may notice slight variations in periodic tables from one book to another, or between the lecture hall and the text. These are matters of style or the particular information included; there are no fundamental differences.

The elements in a column of the periodic table are known as a **family** or **group**. They are identified as group 1A, 2A, and so forth, as shown at the top of the periodic table.* For example, three familiar elements that have similar properties are copper (Cu), silver (Ag), and gold (Au), which occur together in group 1B. Some groups are also described by a family name. The members of group 1A—lithium, sodium, potassium, rubidium (Rb), cesium (Cs), and francium (Fr)—are known as the **alkali metals**. The members of group 2A—beryllium (Be), magnesium (Mg), calcium (Ca), strontium (Sr), barium (Ba), and radium (Ra)—are known as the **alkaline earth metals**. The members of group 7A—fluorine (F), chlorine (Cl), bromide (Br), iodine (I), and astatine (At)—are known as the **halogens**. The members of group 8A—helium (He), neon (Ne), argon (Ar), krypton (Kr), xenon (Xe), and radon (Rn)—are known as the **noble gases**, **inert gases**, or **rare gases**.

We will learn in Chapters 6 and 7 that the elements in a family of the periodic table have similar properties because they have the same type of arrangement of electrons at the periphery of their atoms. However, we need not wait until then to make good use of the periodic table; after all, the table in pretty much its modern form was invented by chemists who knew nothing of the electronic structures of atoms! We can use the table, as they intended, to correlate the behaviors of elements and to aid in

---

* The labeling of the families is basically arbitrary, and three different labeling schemes are presently in use: (1) the scheme shown in Figure 2.16, which is presently most common; (2) the scheme that numbers the columns from 1A through 8A and then from 1B through 8B, thus giving the label 8B to the noble gases; and (3) the scheme recently proposed that numbers the columns from 1 (for the alkali metals) through 18 (for the noble gases), with no A or B designations. When using the first two schemes, Roman numerals, rather than Arabic ones, are sometimes employed. Thus the halogens are often labeled VIIA.

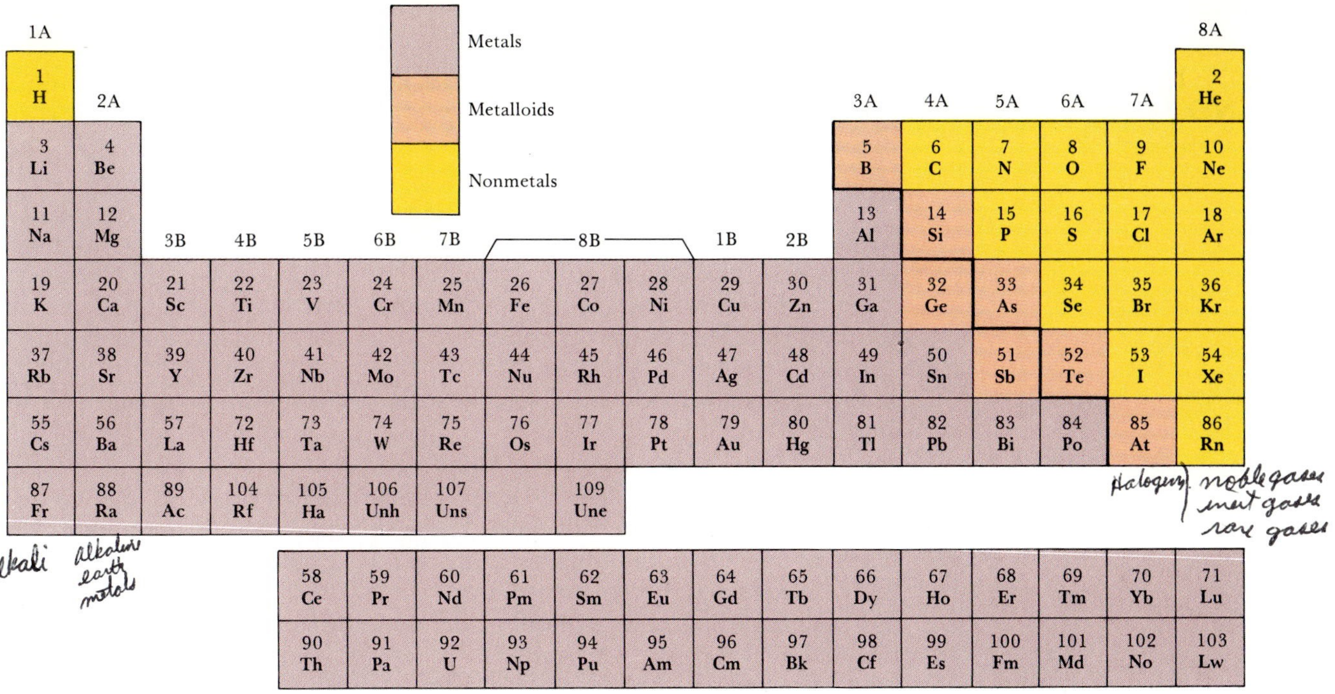

| 1A | | | | | | | | | | | | 3A | 4A | 5A | 6A | 7A | 8A |
|---|---|---|---|---|---|---|---|---|---|---|---|---|---|---|---|---|---|
| 1 H | 2A | | | | | | | | | | | | | | | | 2 He |
| 3 Li | 4 Be | 3B | 4B | 5B | 6B | 7B | 8B | | 1B | 2B | | 5 B | 6 C | 7 N | 8 O | 9 F | 10 Ne |
| 11 Na | 12 Mg | | | | | | | | | | | 13 Al | 14 Si | 15 P | 16 S | 17 Cl | 18 Ar |
| 19 K | 20 Ca | 21 Sc | 22 Ti | 23 V | 24 Cr | 25 Mn | 26 Fe | 27 Co | 28 Ni | 29 Cu | 30 Zn | 31 Ga | 32 Ge | 33 As | 34 Se | 35 Br | 36 Kr |
| 37 Rb | 38 Sr | 39 Y | 40 Zr | 41 Nb | 42 Mo | 43 Tc | 44 Nu | 45 Rh | 46 Pd | 47 Ag | 48 Cd | 49 In | 50 Sn | 51 Sb | 52 Te | 53 I | 54 Xe |
| 55 Cs | 56 Ba | 57 La | 72 Hf | 73 Ta | 74 W | 75 Re | 76 Os | 77 Ir | 78 Pt | 79 Au | 80 Hg | 81 Tl | 82 Pb | 83 Bi | 84 Po | 85 At | 86 Rn |
| 87 Fr | 88 Ra | 89 Ac | 104 Rf | 105 Ha | 106 Unh | 107 Uns | 109 Une | | | | | | | | | | |

| 58 Ce | 59 Pr | 60 Nd | 61 Pm | 62 Sm | 63 Eu | 64 Gd | 65 Tb | 66 Dy | 67 Ho | 68 Er | 69 Tm | 70 Yb | 71 Lu |
|---|---|---|---|---|---|---|---|---|---|---|---|---|---|
| 90 Th | 91 Pa | 92 U | 93 Np | 94 Pu | 95 Am | 96 Cm | 97 Bk | 98 Cf | 99 Es | 100 Fm | 101 Md | 102 No | 103 Lw |

Legend: Metals · Metalloids · Nonmetals

*(handwritten annotations: "alkali", "alkaline earth metals", "Halogens", "noble gases / inert gases / rare gases")*

remembering many facts. You will find it helpful to refer to the periodic table frequently in studying the remainder of this chapter.

One pattern that is evident when elements are arranged in the periodic table is the grouping together of the **metallic elements**. These elements, which are grouped together on the left side of the periodic table, share many characteristic properties, such as luster and high electrical and heat conductivity. The metallic elements are separated from the **nonmetallic elements** by the diagonal steplike line that runs across the right side of the periodic table from boron (B) to astatine (At). Hydrogen, although on the left of this line, is a nonmetal. Note that the majority of the elements are metallic. The nonmetals, which occupy the upper right side of the periodic table, are gases, liquids, or crystalline solids. They lack those physical characteristics that distinguish the metallic elements (Figure 2.17). Many of the elements that lie along the line that separates metals from nonmetals, such as antimony (Sb), possess properties intermediate between those of metals and nonmetals. These elements are often referred to as **semimetals** or **metalloids**.

**FIGURE 2.16** Periodic table of the elements, showing the division of elements into metals, metalloids, and nonmetals.

**FIGURE 2.17** Some familiar examples of metals and nonmetals. The metals pictured are in the form of silver dollars, an aluminum whisk, brass keys, copper pennies, and gold jewelry. The nonmetals are in the form of tincture of iodine, graphite and diamonds (both forms of carbon), and sulfur. (Fundamental Photographs)

## SAMPLE EXERCISE 2.4

Which of the following elements would you expect to show the greatest similarity in chemical and physical properties: Li, Be, F, S, Cl?

**Solution:** The elements F and Cl should be most alike because they are in the same family (group 7A, the halogen family).

## PRACTICE EXERCISE

Locate P (phosphorus) and K (potassium) on the periodic table. Give the atomic number of each, and label each a metal, nonmetal, or semimetal.

**Answer:** P, atomic number 15, is a nonmetal; K, atomic number 19, is a metal.

## 2.5 MOLECULES AND IONS

We have seen that the atom is the smallest representative sample of an element. However, only the noble gas elements are normally found in nature as isolated atoms. Most matter is composed of molecules or ions, both of which are formed from atoms.

### Molecules

**Molecules** consist of combinations of tightly bound atoms. The resultant assembly or package of atoms behaves in many ways as a single object, just as a television set composed of many parts can be recognized as a single object. The nature of the forces (bonds) that bind the atoms together will be examined in Chapter 8.

When elements exist in molecular form, they contain only one type of atom. For instance, the element oxygen, as it is normally found in air, consists of molecules composed of pairs of oxygen atoms. This molecular form of oxygen is represented by the **chemical formula** $O_2$ ("oh two") and is sometimes called dioxygen. The subscript in this formula indicates that two oxygen atoms are present in each molecule. The molecule is said to be **diatomic**. Oxygen also exists in a form known as ozone, the molecules of which consist of three bound oxygen atoms. Correspondingly, the chemical formula for ozone is $O_3$. Ozone and "normal" oxygen exhibit quite different chemical and physical properties. For example, $O_2$ sustains life, and $O_3$ is toxic; $O_2$ is odorless, whereas $O_3$ has a sharp, pungent smell. You have probably smelled ozone, which is formed in small concentrations in the sparks of faulty electrical motors and is a common component of smog.

The elements that normally occur as diatomic molecules are hydrogen, oxygen, nitrogen, and the halogens. Their location in the periodic table is shown in Figure 2.18. When we speak of the substance hydrogen, we mean $H_2$ unless we indicate explicitly otherwise. Likewise, when we speak of oxygen, nitrogen, or any of the halogens, we are referring to $O_2$, $N_2$, $F_2$, $Cl_2$, $Br_2$, or $I_2$. Thus the properties of oxygen and hydrogen listed in Table 1.2 are those of $O_2$ and $H_2$. In other forms, these elements behave much differently.

Molecules of compounds contain more than one type of atom. For example, a molecule of water consists of two hydrogen atoms and one oxygen atom. It is therefore represented by the chemical formula $H_2O$ (read "aitch two oh"). Lack of a subscript on O implies one atom of O per molecule. Another compound composed of these same elements is

**FIGURE 2.18** Elements that exist as diatomic molecules at room temperature.

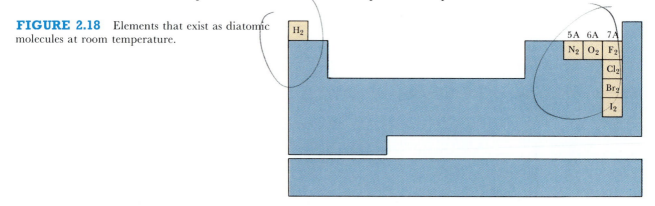

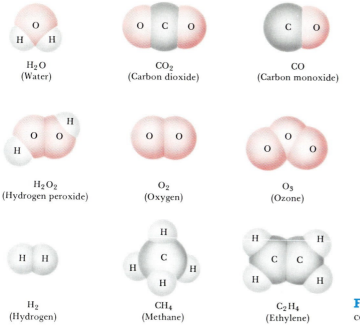

$H_2O$
(Water)

$CO_2$
(Carbon dioxide)

CO
(Carbon monoxide)

$H_2O_2$
(Hydrogen peroxide)

$O_2$
(Oxygen)

$O_3$
(Ozone)

$H_2$
(Hydrogen)

$CH_4$
(Methane)

$C_2H_4$
(Ethylene)

**FIGURE 2.19**  Representation of some simple, common molecules.

hydrogen peroxide, $H_2O_2$. These two compounds have quite different properties. A number of common molecules are shown in Figure 2.19. Pay close attention to how the chemical formula of each molecule reflects its composition. Note also that these substances are composed only of non-metallic elements. Most molecular substances that we will encounter contain only nonmetals.

**SAMPLE EXERCISE 2.5**

Which of the molecules shown in Figure 2.19 are compounds and which are elements?

**Solution:**  The compounds are $H_2O$, $CO_2$, CO, $H_2O_2$, $CH_4$, and $C_2H_4$. The elements are $O_2$, $O_3$, and $H_2$ (one type of atom).

**PRACTICE EXERCISE** _____

Which of the following formulas represent molecules of elemental substances: $P_4$, HCl, $NH_3$?

*Answer:*  $P_4$

Chemical formulas that indicate the *actual* numbers and types of atoms in a molecule are called **molecular formulas**. (The formulas in Figure 2.19 are molecular formulas.) Chemical formulas that give only the *relative* number of atoms of each type in a molecule are called **empirical** or **simplest formulas**. The subscripts in an empirical formula are always smallest whole-number ratios; conversely, the subscripts in a molecular formula are always integral multiples (that is $1\times$, $2\times$, $3\times$, and so forth) of the subscripts in the empirical formula of the substance. For example, the empirical formula for hydrogen peroxide is HO; its molecular formula is $H_2O_2$. The empirical formula for ethylene is $CH_2$; its molecular formula is $C_2H_4$. For some substances the empirical formula and molecular formula are identical, as in the case of water, $H_2O$.

## SAMPLE EXERCISE 2.6

Write the empirical formulas for the following molecules: **(a)** glucose, a substance also known as blood sugar and as dextrose, whose molecular formula is $C_6H_{12}O_6$; **(b)** nitrous oxide, a substance used as an anesthetic and commonly called laughing gas, whose molecular formula is $N_2O$.

**Solution:** **(a)** The empirical formula has subscripts that are smallest whole-number ratios. The smallest ratios are obtained by dividing each subscript by the largest common factor, in this case 6. The resultant empirical formula is $CH_2O$. **(b)** Because the subscripts in $N_2O$ are already the lowest integral numbers, the empirical formula for nitrous oxide is the same as its molecular formula, $N_2O$.

### PRACTICE EXERCISE

Give the empirical formula for the substance whose molecular formula is $Si_2H_6$.

***Answer:*** $SiH_3$

Molecular formulas are preferred over empirical formulas because they provide more information. Only empirical formulas, however, can be written for substances that exist as three-dimensional structures in which there are no discrete molecules. (As we shall see shortly, this situation includes ionic substances.) For example, the element carbon normally exists in extended structures rather than as isolated atoms or molecules. Because the empirical formula for any element is simply the symbol for that element, carbon is represented by its symbol, C. As a further example, boron nitride consists of a three-dimensional array of boron and nitrogen atoms; it is represented by its empirical formula BN.

Often the formula of a molecule is written to show how its atoms are joined together. For example, the formulas for water and hydrogen peroxide can be written as follows:

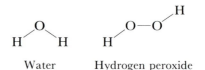

Water          Hydrogen peroxide

Such formulas are known as **structural formulas**. The lines between the symbols for the elements represent the bonds that hold the respective atoms together. These formulas indicate which atoms are attached to which; however, they do not necessarily tell anything about the shapes of molecules (that is, about the actual angles at which the atoms are joined together).

**Ions**

The nucleus of an atom is unchanged by ordinary chemical processes, but atoms readily gain or lose electrons. If electrons are removed or added to a neutral atom, a charged particle called an **ion** is formed. For example, the sodium atom, which has 11 protons and 11 electrons, can lose an electron. The resulting ion has 11 protons and 10 electrons; its net charge is consequently $1+$. This ion is symbolically represented as $Na^+$. The net charge on an ion is represented by a superscript; $+$, $2+$, and $3+$ mean a net charge resulting from the loss of one, two, or three electrons, respectively. The superscripts $-$, $2-$, and $3-$ represent net charges resulting from gain of one, two, or three electrons, respectively. The formation of the $Na^+$ ion from a Na atom is shown schematically below:

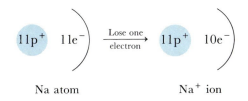

Na atom         $Na^+$ ion

Chlorine, with 17 protons and 17 electrons, can gain an electron in chemical reactions, producing the $Cl^-$ ion:

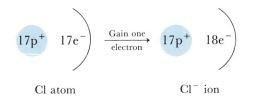

Cl atom         $Cl^-$ ion

*In general, metal atoms lose electrons most readily, and nonmetal atoms tend to gain electrons.*

**SAMPLE EXERCISE 2.7**

Give the chemical symbol for the ion with 26 protons and 24 electrons.

**Solution:** The element whose atoms have 26 protons (atomic number 26) is Fe (iron). If the ion has two more protons than electrons, it has a net charge of $2+$; thus the symbol for the ion is $Fe^{2+}$.

**PRACTICE EXERCISE**
How many protons and electrons does the $Se^{2-}$ ion possess?
*Answer:* 34 protons and 36 electrons

In addition to simple ions such as $Na^+$ and $Cl^-$, there exist **polyatomic ions** such as $NO_3^-$ (nitrate ion) and $SO_4^{2-}$ (sulfate ion). These ions consist of atoms joined together as in a molecule, but they have a net positive or negative charge. We will consider further examples of polyatomic ions in Section 2.6.

The chemical properties of ions are greatly different from those of the atoms from which they are derived. The change of an atom or molecule to an ion is like that from Dr. Jekyll to Mr. Hyde: Although the body may be essentially the same (plus or minus a few electrons), the behavior is much different.

Many atoms gain or lose electrons so as to end up with the same number of electrons as the noble gas closest to it in the periodic table. The members of the noble gas family are chemically very nonreactive and form very few compounds. We might deduce that this is because they have very stable electron arrangements. Nearby elements can obtain these same stable arrangements by losing or gaining electrons. For example, loss of one electron from an atom of sodium leaves it with the same number of electrons as the neutral neon atom (atomic number 10). Similarly, when chlorine gains an electron it ends up with 18, the same as argon (atomic number 18). We will content ourselves with this simple observation in explaining the formation of ions until later chapters in which we consider chemical bonding.

## SAMPLE EXERCISE 2.8

Predict the charges expected for the most stable ions of barium and oxygen.

**Solution:** Refer to the periodic table. Barium has atomic number 56. The nearest noble gas is xenon, atomic number 54. Barium can obtain the stable arrangement of 54 electrons by losing two of its electrons, thereby forming the $Ba^{2+}$ ion.

Oxygen has atomic number 8. The nearest noble gas is neon, atomic number 10. Oxygen can obtain this stable electron arrangement by gaining two electrons, thereby forming an ion of $2-$ charge, $O^{2-}$.

## PRACTICE EXERCISE

Predict the charge of the most stable ion of aluminum.
**Answer:** $3+$

Gains and losses of electrons that result in the formation of ions occur in transactions between different kinds of atoms. The compounds that result from these electron transfers are composed of positively charged and negatively charged ions. The positively charged ions are usually metal ions; the negatively charged ones are usually nonmetal ions. Consequently, *ionic compounds tend to be composed of metals combined with nonmetals.* For example, NaCl, ordinary table salt, is such a compound; it consists of equal numbers of $Na^+$ and $Cl^-$ ions. *Molecular compounds tend to be composed of nonmetals combined with other nonmetals.*

## SAMPLE EXERCISE 2.9

Which of the following compounds would you expect to be ionic: $N_2O$, $Na_2O$, $CaCl_2$, $SF_4$?

**Solution:** We would predict that the ionic compounds are $Na_2O$ and $CaCl_2$ because they are composed of a metal combined with a nonmetal. The other two compounds, which are composed entirely of nonmetals, are molecular.

## PRACTICE EXERCISE

Which of the following compounds are molecular: $CI_4$, $FeS$, $P_4O_6$, $PbF_2$?
**Answer:** $CI_4$ and $P_4O_6$

The ions in ionic compounds are arranged in three-dimensional structures. The arrangement of $Na^+$ and $Cl^-$ ions in NaCl is shown in Figure 2.20. Because there is no discrete molecule of NaCl, we are able to write only an empirical formula for this substance. Indeed, only empirical formulas can be written for most ionic compounds.

It is a simple matter to write the empirical formula for an ionic compound if we know the charges of the ions of which it is composed. Chemical compounds are always electrically neutral. Consequently, the ions in an ionic compound always occur in such a ratio that the total positive charge is equal to the total negative charge. Thus there is one $Na^+$ to one $Cl^-$ giving NaCl, one $Ba^{2+}$ to two $Cl^-$ giving $BaCl_2$, and so forth.

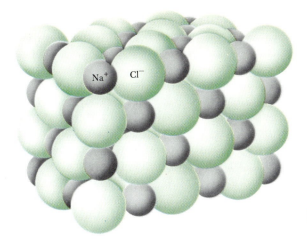

**FIGURE 2.20** Arrangement of ions in sodium chloride (NaCl).

## SAMPLE EXERCISE 2.10

What are the empirical formulas of the compounds formed by **(a)** $Al^{3+}$ and $Cl^-$ ions; **(b)** $Al^{3+}$ and $O^{2-}$ ions; **(c)** $Mg^{2+}$ and $NO_3^-$ ions?

**Solution:** **(a)** It requires three $Cl^-$ ions to balance the charge of one $Al^{3+}$ ion. Thus the formula is $AlCl_3$. **(b)** It requires two $Al^{3+}$ ions to balance the charge of three $O^{2-}$ ions (that is, the total positive charge is $6+$ and the total negative charge is $6-$). Thus the formula is $Al_2O_3$. **(c)** Two $NO_3^-$ ions are needed to balance the charge of one $Mg^{2+}$. Thus the

formula is $Mg(NO_3)_2$. In this case the formula for the entire negative ion must be enclosed in parentheses so that it is clear that the subscript 2 applies to all the atoms of that ion.

## PRACTICE EXERCISE

Write the empirical formulas for the compounds formed by the following ions: **(a)** $Na^+$ and $PO_4^{3-}$; **(b)** $Zn^{2+}$ and $SO_4^{2-}$; **(c)** $Fe^{3+}$ and $CO_3^{2-}$.
*Answers:* **(a)** $Na_3PO_4$; **(b)** $ZnSO_4$; **(c)** $Fe_2(CO_3)_3$

As you proceed in your study of this text and in chemistry laboratory work, you will need to refer to specific chemical substances by name. We present here some of the basic rules for naming simple compounds. You may not have immediate use for some of the rules, but they are gathered here in one place for your convenience to use whenever a question of **nomenclature**, or naming of substances, arises.

There are now about 7 million known chemical substances. Naming them all would be a hopelessly complicated task if each had a special name independent of all the others. Many important substances that have been known for a long time, such as water ($H_2O$) and ammonia ($NH_3$), do have individual, traditional names. For most substances, however, we rely upon a set of rules that lead to an informative, systematic name for each substance.

One of the earliest classification schemes in chemistry was the distinction between inorganic and organic compounds. Organic compounds contain carbon, usually in combination with hydrogen, oxygen, nitrogen, or sulfur. Organic compounds were first associated only with plants and animals. However, a great number of organic compounds have now been prepared that do not occur in nature. We will discuss the chemistry and naming of organic compounds in Chapter 27. However, we will have many occasions throughout the text to illustrate chemical principles with organic compounds as examples. In this section we will consider the basic rules for naming inorganic compounds, those that the early chemists

## 2.6 NAMING INORGANIC COMPOUNDS

associated with the nonliving portion of our world. Let us first consider the naming of ionic compounds.

**Ionic Compounds**

The names of ionic compounds are based on the names of the ions of which they are composed. For example, NaCl is called sodium chloride after the $Na^+$ or sodium ion, and the $Cl^-$ or chloride ion. The positive ion is always named first and listed first in writing the formula for the compound. The negative ion is named and written last. To see how the names of these ions arise, consider first the naming of positive ions, also called **cations** (pronounced CAT-ion). Ions may be monatomic (composed of a single atom) or polyatomic (formed from two or more atoms). Monatomic cations are most commonly formed from metallic elements. These ions take the name of the element itself:

$$Na^+ \quad \text{sodium ion} \qquad Zn^{2+} \quad \text{zinc ion} \qquad Al^{3+} \quad \text{aluminum ion}$$

If an element can form more than one positive ion, the positive charge of the ion is indicated by a Roman numeral in parentheses following the name of the metal:

$$Fe^{2+} \quad \text{iron(II) ion} \qquad Cu^+ \quad \text{copper(I) ion}$$
$$Fe^{3+} \quad \text{iron(III) ion} \qquad Cu^{2+} \quad \text{copper(II) ion}$$

At this stage you have no way of knowing which elements commonly exist in more than one charge state. This need not be a source of difficulty. If there is any doubt in your mind, use the Roman numeral designation of charge as part of the name. It is never wrong to use this form of charge designation, even though it may sometimes be unnecessary.

An older method still widely used for distinguishing between two differently charged ions of a metal is to apply the endings *-ous* or *-ic*. These endings represent the lower and higher charged ions, respectively. They are added to the root of the Latin name of the element:

$$Fe^{2+} \quad \text{ferrous ion} \qquad Cu^+ \quad \text{cuprous ion}$$
$$Fe^{3+} \quad \text{ferric ion} \qquad Cu^{2+} \quad \text{cupric ion}$$

The only common polyatomic cations are those given below:

$$NH_4^+ \quad \text{ammonium ion} \qquad Hg_2^{2+} \quad \text{mercury(I) or mercurous ion}$$

The name mercury(I) ion is given to $Hg_2^{2+}$ because it can be considered to consist of two $Hg^+$ ions. Mercury also occurs as the monatomic $Hg^{2+}$ ion, which is known as the mercury(II) or mercuric ion.

Negative ions are called **anions** (pronounced AN -ion). Monatomic anions (those derived from a single atom) are most commonly formed from atoms of the nonmetallic elements. They are named by dropping the ending of the name of the element and adding the ending *-ide*:

$$H^- \quad \text{hydride ion} \qquad O^{2-} \quad \text{oxide ion} \qquad N^{3-} \quad \text{nitride ion}$$
$$F^- \quad \text{fluoride ion} \qquad S^{2-} \quad \text{sulfide ion} \qquad P^{3-} \quad \text{phosphide ion}$$

Only a few common polyatomic ions end in *-ide*:

$OH^-$   hydroxide ion     $CN^-$   cyanide ion     $O_2^{2-}$   peroxide ion

Table 2.3 lists the most common cations and anions. Notice that many polyatomic anions contain oxygen. Anions of this kind are referred to as **oxyanions**. A particular element such as sulfur may form more than one oxyanion. When this occurs, there are rules for indicating the relative numbers of oxygen atoms in the anion. When an element forms only two oxyanions, the name of the one that contains more oxygen ends in *-ate*; the name of the one with less oxygen ends in *-ite*:

$NO_2^-$   nitrite ion     $SO_3^{2-}$   sulfite ion

$NO_3^-$   nitrate ion     $SO_4^{2-}$   sulfate ion

When the series of anions of a given element extends to three or four members, as with the oxyanions of the halogens, prefixes are also employed. The prefix *hypo-* indicates less oxygen, and the prefix *per-* indicates more oxygen:

$ClO^-$      hypochlorite ion (less oxygen than chlorite)

$ClO_2^-$     chlorite ion

$ClO_3^-$     chlorate ion

$ClO_4^-$     perchlorate ion (more oxygen than chlorate)

**TABLE 2.3**   Common ions

| Positive ions (cations) | Negative ions (anions) |
|---|---|
| **1+ (IA)** | **1−** |
| Ammonium ($NH_4^+$) | Acetate ($C_2H_3O_2^-$) |
| Hydrogen ($H^+$) | Bromide ($Br^-$) |
| Potassium ($K^+$) | Chlorate ($ClO_3^-$) |
| Silver ($Ag^+$) | Chloride ($Cl^-$) |
| Sodium ($Na^+$) | Cyanide ($CN^-$) |
| | Fluoride ($F^-$) |
| **2+ (2A)** | Hydride ($H^-$) |
| Barium ($Ba^{2+}$) | Hydrogen carbonate ($HCO_3^-$) |
| Calcium ($Ca^{2+}$) |   or bicarbonate |
| Cobalt(II) or cobaltous ($Co^{2+}$) | Hydrogen sulfate ($HSO_4^-$) |
| Copper(II) or cupric ($Cu^{2+}$) |   or bisulfate |
| Iron(II) or ferrous ($Fe^{2+}$) | Hydroxide ($OH^-$) |
| Lead(II) or plumbous ($Pb^{2+}$) | Iodide ($I^-$) |
| Magnesium ($Mg^{2+}$) | Nitrate ($NO_3^-$) |
| Manganese(II) or manganous ($Mn^{2+}$) | Perchlorate ($ClO_4^-$) |
| Mercury(II) or mercuric ($Hg^{2+}$) | Permanganate ($MnO_4^-$) |
| Tin(II) or stannous ($Sn^{2+}$) | |
| Zinc ($Zn^{2+}$) | **2−** |
| | Carbonate ($CO_3^{2-}$) |
| **3+ (3A)** | Chromate ($CrO_4^{2-}$) |
| Aluminum ($Al^{3+}$) | Oxide ($O^{2-}$) |
| Chromium(III) or chromic ($Cr^{3+}$) | Peroxide ($O_2^{2-}$) |
| Iron(III) or ferric ($Fe^{3+}$) | Sulfate ($SO_4^{2-}$) |
| | Sulfide ($S^{2-}$) |
| | Sulfite ($SO_3^{2-}$) |
| | |
| | **3−** |
| | Arsenate ($AsO_4^{3-}$) |
| | Phosphate ($PO_4^{3-}$) |

Notice that if you memorize the rules just indicated, you need know only the name for one oxyanion in a series to deduce the names for the other members.

---

**SAMPLE EXERCISE 2.11**

The formula for the selenate ion is $SeO_4^{2-}$. Write the formula for the selenite ion.

**Solution:** The selenite ion should have one less oxygen than the selenate ion; hence, $SeO_3^{2-}$.

**PRACTICE EXERCISE**

The formula for the bromate ion is $BrO_3^-$. Write the formula for the hypobromite ion.

*Answer:* $BrO^-$

---

Because many names of ions predate the establishment of systematic rules, there are many exceptions to the rules. For example, the permanganate ion is $MnO_4^-$; we thus expect that the manganate ion should be $MnO_3^-$, but this ion is unknown. The name *manganate* is given to the species $MnO_4^{2-}$.

Many polyatomic anions that have high charges readily add one or more hydrogen ions $(H^+)$ to form anions of lower charge. These ions are named by prefixing the word *hydrogen* or *dihydrogen*, as appropriate, to the name of the hydrogen-free anion. An older method, which is still used, is to use the prefix *bi*-:

$HCO_3^-$    hydrogen carbonate (or bicarbonate) ion

$HSO_4^-$    hydrogen sulfate (or bisulfate) ion

$H_2PO_4^-$    dihydrogen phosphate ion

We are now in a position to combine the names of cations and anions to name and write the formulas for ionic compounds. The following examples illustrate the relationship between formula and name:

| | | | |
|---|---|---|---|
| barium bromide | $BaBr_2$ | mercurous chloride | $Hg_2Cl_2$ |
| copper(II) nitrate | $Cu(NO_3)_2$ | aluminum oxide | $Al_2O_3$ |

---

**SAMPLE EXERCISE 2.12**

Name the following compounds: (**a**) $K_2SO_4$; (**b**) $Ba(OH)_2$; (**c**) $FeCl_3$.

**Solution:** (**a**) This compound is composed of $K^+$ and $SO_4^{2-}$ ions. Because $K^+$ is called the potassium ion and $SO_4^{2-}$ is called the sulfate ion, the name of the compound is potassium sulfate. (**b**) This compound is composed of $Ba^{2+}$ and $OH^-$ ions. $Ba^{2+}$ is the barium ion; $OH^-$ is the hydroxide ion. Thus the compound is called barium hydroxide. (**c**) This com-

pound is composed of $Fe^{3+}$, which is called iron(III) or ferric ion, and chloride ions, $Cl^-$. The compound is iron(III) chloride or ferric chloride.

**PRACTICE EXERCISE**

Name the following compounds: (**a**) $NH_4Cl$; (**b**) $Cr_2O_3$; (**c**) $Co(NO_3)_2$.

*Answers:* (**a**) ammonium chloride; (**b**) chromium(III) oxide; (**c**) cobalt(II) nitrate

---

## SAMPLE EXERCISE 2.13

Write the chemical formulas for the following compounds: **(a)** calcium carbonate; **(b)** stannous fluoride; **(c)** iron(II) perchlorate.

**Solution:** **(a)** The calcium ion is $Ca^{2+}$; the carbonate ion is $CO_3^{2-}$. Because of the charges of each ion, there will be one $Ca^{2+}$ ion for each $CO_3^{2-}$ in the compound, giving the empirical formula $CaCO_3$. **(b)** The stannous ion, also known as the tin(II) ion, is $Sn^{2+}$. The fluoride ion is $F^-$. Two $F^-$ ions are needed to balance the positive charge of $Sn^{2+}$, giving the formula $SnF_2$. This compound, incidentally, is a common ingredient in many toothpastes. **(c)** The iron(II) ion is $Fe^{2+}$; the perchlorate ion is $ClO_4^-$. Two $ClO_4^-$ ions are required to balance the charge on one $Fe^{2+}$, giving $Fe(ClO_4)_2$.

These examples should indicate the importance of remembering the charges of the common ions. Indeed, you may find it necessary to memorize many of the entries in Table 2.3.

### PRACTICE EXERCISE

Give the chemical formula for **(a)** magnesium sulfate; **(b)** silver sulfide; **(c)** lead(II) nitrate.
*Answers:* **(a)** $MgSO_4$; **(b)** $Ag_2S$; **(c)** $Pb(NO_3)_2$

## Acids

The important class of compounds known as acids are named in a special way. These compounds will be discussed further in Section 3.2 and then extensively considered in Chapter 15. An acid may be defined as a substance whose molecules each yield one or more hydrogen ions ($H^+$) when dissolved in water. The formula for an acid is formed by adding sufficient $H^+$ ions to balance the anion's charge, as seen in the examples in the table below. The name of the acid is related to the name of the anion. Anions whose names end in *-ide* have associated acids that have the *hydro-* prefix and an *-ic* ending, as in these examples:

| Anion | Corresponding acid |
|---|---|
| $Cl^-$ (chlor*ide*) | HCl (*hydro*chlor*ic* acid) |
| $S^{2-}$ (sul*fide*) | $H_2S$ (*hydro*sulfur*ic* acid) |

Many of the most important acids are derived from oxyanions. If the anion has an *-ate* ending, the corresponding acid is given an *-ic* ending. Anions whose names end in *-ite* have associated acids whose names end in *-ous*. Prefixes in the name of the anion are retained in the name of the acid. These rules are illustrated by the oxyacids of chlorine:

| Anion | Corresponding acid |
|---|---|
| $ClO^-$ (*hypo*chlor*ite*) | HClO (*hypo*chlor*ous* acid) |
| $ClO_2^-$ (chlor*ite*) | $HClO_2$ (chlor*ous* acid) |
| $ClO_3^-$ (chlor*ate*) | $HClO_3$ (chlor*ic* acid) |
| $ClO_4^-$ (*per*chlor*ate*) | $HClO_4$ (*per*chlor*ic* acid) |

## SAMPLE EXERCISE 2.14

Name the following acids: **(a)** HCN; **(b)** $HNO_3$; **(c)** $H_2SO_4$; **(d)** $H_2SO_3$.

**Solution:** **(a)** The anion from which this acid is derived is $CN^-$, the cyanide ion. Because this ion has an *-ide* ending, the acid is given a *hydro-* prefix and an *-ic* ending: hydrocyanic acid. Incidentally, only water solutions of HCN are referred to as hydrocyanic acid; the pure compound is called hydrogen cyanide. **(b)** Because $NO_3^-$ is the nitrate ion, $HNO_3$ is called nitric acid (the *-ate* ending of the anion is replaced with an *-ic* ending in naming the acid). **(c)** Because $SO_4^{2-}$ is the sulfate ion, $H_2SO_4$ is called sulfuric acid. **(d)** Because $SO_3^{2-}$ is the sulfite ion, $H_2SO_3$ is sulfurous acid (the *-ite* ending of the anion is replaced with an *-ous* ending).

**PRACTICE EXERCISE** _____
Give the chemical formulas for **(a)** hydrobromic acid; **(b)** phosphoric acid.
*Answers:* **(a)** HBr; **(b)** $H_3PO_4$

## Molecular Compounds

The procedures for naming binary (two element) molecular compounds are similar to those for naming ionic compounds. In these molecular compounds it is possible to associate a more positive nature with one element in the molecule and a more negative nature with the other element. (In Chapter 8 we will consider how the more positive atom is selected.) The element with the more positive nature is named first and also appears first in the chemical formula. The second element is named with an *-ide* ending. For example, the name for HCl is hydrogen chloride. (This is the name used when referring to the pure compound; water solutions of HCl are referred to as hydrochloric acid; see Figure 2.21.)

Often a pair of elements can form several different molecular compounds. For example, carbon and oxygen form CO and $CO_2$. To distinguish these compounds from one another, the prefixes given in Table 2.4 are used to denote the numbers of atoms of each element present. Thus CO is called carbon monoxide, and $CO_2$ is called carbon dioxide. When the prefix ends in *a* and the name of the anion begins with a vowel (such as *oxide*), the *a* is often dropped. A few examples follow:

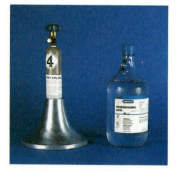

**FIGURE 2.21** Hydrogen chloride, which is a gas at room temperature, is shown in a compressed-gas cylinder (left). Hydrochloric acid, which is a water solution of hydrogen chloride, is shown in the glass bottle (right). (Donald Clegg and Roxy Wilson)

| | |
|---|---|
| $NF_3$ | nitrogen trifluoride |
| $N_2O_4$ | dinitrogen tetroxide |
| $SO_3$ | sulfur trioxide |

**TABLE 2.4** Prefixes used in naming binary compounds formed between nonmetals

| Prefix | Meaning |
|--------|---------|
| Mono- | 1 |
| Di- | 2 |
| Tri- | 3 |
| Tetra- | 4 |
| Penta- | 5 |
| Hexa- | 6 |
| Hepta- | 7 |
| Octa- | 8 |

**SAMPLE EXERCISE 2.15**

Name the following compounds: **(a)** $PCl_5$; **(b)** $N_2O_3$.

**Solution:** Because the compounds consist entirely of nonmetals, you should expect that they are molecular rather than ionic. Using the prefixes in Table 2.4, we have **(a)** phosphorus pentachloride, and

**(b)** dinitrogen trioxide.

**PRACTICE EXERCISE** _____
Give the chemical formula for **(a)** silicon tetrabromide; **(b)** disulfur dichloride.
*Answers:* **(a)** $SiBr_4$; **(b)** $S_2Cl_2$

# FOR REVIEW

## SUMMARY

**Atoms** are the basic building blocks of matter; they are the smallest units of an element that can combine with other elements. Atoms are composed of a **nucleus** (containing **protons** and **neutrons**) and **electrons** that move around the nucleus. We considered some of the historically significant experiments that led to this model of the atom: Thomson's experiments on the behavior of cathode rays (a stream of electrons) in magnetic and electric fields; Millikan's oil-drop experiment; Becquerel's and Rutherford's studies of radioactivity; and Rutherford's studies of the scattering of α particles by thin metal foils.

Elements can be classified by **atomic number**, the number of protons in the nucleus of an atom. All atoms of a given element have the same atomic number. The **mass number** of an atom is the sum of the number of protons and neutrons. Atoms of the same element that differ in mass number are known as **isotopes**.

The **periodic table** is an arrangement of elements in order of increasing atomic number with elements with similar properties placed in vertical columns. The elements in a vertical column are known as a **periodic family** or **group**. The **metallic elements**, which comprise the majority of the elements, are on the left side of the table; the **nonmetallic**

**elements** are located on the upper right side.

Atoms can combine to form **molecules**. Atoms can also either gain or lose electrons, thereby forming charged particles called **ions**. Metals tend to lose electrons, becoming positively charged ions (**cations**). Nonmetals tend to gain electrons, forming negatively charged ions (**anions**). Because ionic compounds are electrically neutral, containing both cations and anions, they usually contain both metallic and nonmetallic elements. Compounds composed of molecules (molecular compounds) usually contain only nonmetallic elements.

The chemical formulas used for ionic compounds are **simplest formulas**. The simplest formula of an ionic compound can be written readily if the charges of the ions are known. Although simplest formulas can also be written for molecular substances, the **molecular formula** is preferred because it gives the actual numbers of each type of atom in a molecule of the substance. **Structural formulas** show the order in which the atoms in a molecule are connected to each other.

A set of systematic rules has been developed for naming inorganic compounds. We considered ionic compounds, acids, and binary molecular compounds.

## LEARNING GOALS

Having read and studied this chapter, you should be able to:

1. Describe the composition of an atom in terms of protons, neutrons, and electrons.

2. Give the approximate size, relative mass, and charge of an atom, proton, neutron, and electron.

3. Write the chemical symbol for an element (for example $^{12}_{6}C$), having been given its mass number and atomic number, and perform the reverse operation.

4. Write the symbol and charge for an atom or ion, having been given the number of protons, neutrons, and electrons, and perform the reverse operation.

5. Describe the properties of the electron as seen in cathode rays. Describe the means by which J. J. Thomson determined the ratio $e/m$ for the electron.

6. Describe Millikan's oil-drop experiment, and indicate what property of the electron he was able to measure.

7. Cite the evidence from studies of radioactivity for the existence of subatomic particles.

8. Describe the experimental evidence for the nuclear nature of the atom.

9. Use the periodic table to predict whether an element is a metal or nonmetal.

10. Use the periodic table to predict the charges of monatomic ions.

**11.** Distinguish between simplest formulas, molecular formulas, and structural formulas.

**12.** Write the simplest formula of a compound, having been given the charges of the ions of which it is composed.

**13.** Write the name of a simple inorganic compound, having been given its chemical formula, and perform the reverse operation.

## KEY TERMS

Among the more important terms and expressions used for the first time in this chapter are the following:

**Alpha (α) particles** (Section 2.2) are helium nuclei (consisting of two protons and two neutrons) emitted from the nuclei of certain radioactive atoms.

An **anion** (Section 2.6) is a negatively charged ion.

The **atomic number** (Section 2.3) of an element is the number of protons in the nucleus of one of its atoms.

**Beta (β) particles** (Section 2.2) are electrons emitted from the nuclei of certain radioactive atoms.

A **cation** (Section 2.6) is a positively charged ion.

An **electron** (Sections 2.2, 2.3) is a negatively charged subatomic particle found outside the atomic nucleus. It forms a part of all atoms. It has a mass 1/1836 times that of a proton.

An **empirical formula** or **simplest formula** (Section 2.5) is a chemical formula that shows the kinds of atoms and their relative numbers in a substance.

**Gamma (γ) radiation** (Section 2.2), or **gamma rays**, is a form of radiation similar to X rays that is emitted by radioactive atoms.

**Ions** (Section 2.5) are electrically charged atoms or groups of atoms (polyatomic ions). Ions can be positively or negatively charged, depending on whether electrons are lost (positive) or gained (negative) from the atoms.

**Isotopes** (Section 2.3) are atoms of the same element containing different numbers of neutrons and therefore having different masses.

The **mass number** (Section 2.3) is the sum of the number of protons and neutrons in the nucleus of a particular atom.

A **molecular formula** (Section 2.5) is a chemical formula that indicates the actual number of atoms of each element in one molecule of a substance.

A **molecule** (Section 2.5) is a chemical combination of two or more atoms.

A **neutron** (Section 2.3) is a neutral particle found in the nucleus of an atom; it has approximately the same mass as a proton.

A **proton** (Section 2.3) is a positively charged subatomic particle found in the nucleus of any atom.

**Radioactivity** (Section 2.2) is the spontaneous disintegration of an unstable atomic nucleus with accompanying emission of alpha, beta, or gamma radiation.

A **structural formula** (Section 2.5) is a formula that shows not only the number and kind of atoms in a molecule, but also the arrangement of the atoms.

## EXERCISES

### Dalton's Atomic Theory; Atomic Structure

**2.1** When the elements hydrogen and bromine are caused to react, a gas is formed. The composition of this product gas is the same, no matter what the relative amounts of hydrogen and bromine are used to form it. How is this observation explained in terms of Dalton's atomic theory? What "law" does it illustrate?

**2.2** Ascorbic acid, vitamin C, can be isolated from the juices of citrus fruits. It can also be synthesized in the laboratory in a series of reactions beginning with sorbose, a readily obtained sugar. In what respects should the two types of vitamin C differ in their nutritional value? Bearing in mind the material of Section 2.1, if the two materials did differ, in what respects would they be different?

**2.3** A chemistry student prepared a series of compounds containing only nitrogen and oxygen:

| Compound | Mass of nitrogen | Mass of oxygen |
|----------|------------------|----------------|
| A | 16.8 | 19.2 |
| B | 17.1 | 39.0 |
| C | 33.6 | 57.3 |

**(a)** Calculate the mass of oxygen combined with 1 g of nitrogen in each compound. **(b)** Show how the numbers calculated in part (a) illustrate the law of multiple proportions.

**2.4** Imagine that some of Dalton's contemporaries decomposed samples of three different iron–sulfur compounds and obtained the following results:

| Compound | Mass of iron (g) | Mass of sulfur (g) |
|----------|------------------|--------------------|
| A | 4.654 | 2.672 |
| B | 5.150 | 4.434 |
| C | 3.019 | 3.466 |

Show how these data obey the law of multiple proportions.

**2.5** When an iron object rusts, its mass increases, but the mass of a match decreases when it burns. Do these observations violate the law of conservation of matter? Explain.

**2.6** When the magnesium foil within a flashbulb burns, the flashbulb undergoes no change in mass. When magnesium ribbon burns in open air, the magnesium gains mass. What is the difference between the two experiments? Does the second observation violate the law of conservation of mass? Explain.

**2.7** Static electrical charges are caused by the transfer of electrons. In Millikan's oil-drop experiment, a radioactive source was used to produce a static charge in oil drops. How many excess electrons are present in an oil drop whose static charge is $-1.922 \times 10^{-18}$ C?

**2.8** In a physics laboratory a student carried out the Millikan oil-drop experiment, using several oil droplets for her measurements, and calculated the charges on the drops. She obtained the following data:

| Droplet | Calculated charge (C) |
|---------|----------------------|
| A | $1.60 \times 10^{-19}$ |
| B | $3.15 \times 10^{-19}$ |
| C | $4.81 \times 10^{-19}$ |
| D | $6.31 \times 10^{-19}$ |

What is the significance of the fact that the droplets carried different charges? What conclusion can the student draw from these data regarding the charge of the electron? What value (and to how many significant figures) should she report for the electronic charge?

**2.9** What differences would you expect if magnesium foil is used instead of gold foil in the $\alpha$ scattering experiment depicted in Figures 2.11 and 2.12?

**2.10** The tracks of energetic particles such as $\alpha$ and $\beta$ rays through gases can be followed in a device called the Wilson cloud chamber. The tracks appear as streaks. The tracks of a few $\alpha$ particles through air are shown in Figure 2.22. Explain why the paths of the $\alpha$ particles are mostly long and account for the sudden change in direction of one particle, as shown.

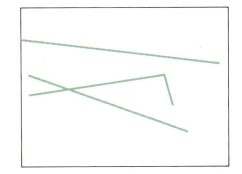

**FIGURE 2.22**

**2.11** The diameter of the iridium (Ir) atom is estimated to be 2.7 Å. **(a)** Express this distance in picometers (pm); in nanometers (nm). **(b)** What number of Ir atoms would form a line 1.0 mm long?

**2.12** If $1.0 \times 10^9$ sodium (Na) atoms could be arranged side by side, they would form a straight line measuring 3.1 cm. What is the diameter of an Na atom?

**2.13** Indicate the number of protons, neutrons, and electrons present in a neutral atom of each of the following: **(a)** $^{80}$Br; **(b)** $^{109}$Ag; **(c)** $^{48}$Ti; **(d)** $^{37}$Cl; **(e)** $^{40}$Ar; **(f)** $^{137}$Ba.

**2.14** Each of the following nuclides is used in medicine. Indicate the number of protons and neutrons in each nuclide: **(a)** cobalt-60; **(b)** iodine-131; **(c)** technetium-99; **(d)** phosphorus-32 **(e)** chromium-51; **(f)** iron-59.

**2.15** Write the correct symbol, with both superscript and subscript, for each of the following (use the list of elements on the front inside cover): **(a)** the nuclide of potassium with mass 39; **(b)** the nuclide of chlorine with mass 35; **(c)** the nuclide of silicon with mass 29; **(d)** two isotopes of sulfur, one with 16 neutrons, the other with 18 neutrons.

**2.16** Write the full symbol (with the appropriate superscript and subscript for mass number and atomic number, respectively) for a neutral atom **(a)** with atomic number 27 and mass number 32; **(b)** with 50 protons and 69 neutrons; **(c)** of zinc having 35 neutrons; **(d)** with 38 electrons and 51 neutrons.

**2.17** Fill in the gaps in the following table:

| Symbol | $^{79}$Se | | | | $^{198}$Au |
|--------|-----------|---|---|---|-----------|
| Protons | | | 56 | | |
| Neutrons | | | | 15 | |
| Electrons | | 35 | | 16 | |
| Mass number | | 80 | 137 | | |

**2.18** Fill in the gaps in the following table:

| Symbol | $^{19}F$ | $^{52}Cr$ | | | |
|---|---|---|---|---|---|
| Atomic number | | | 34 | | |
| Mass number | | | 79 | 27 | |
| Electrons | | | | 13 | |
| Protons | | | | | 55 |
| Neutrons | | | | | 78 |

## The Periodic Table; Molecules and Ions

**2.19**  Using the table on the front inside cover as a guide, write the symbol for each of the following elements; locate the element in the periodic table and indicate whether it is a metal, metalloid, or nonmetal: **(a)** manganese; **(b)** bromine; **(c)** chromium; **(d)** germanium; **(e)** hafnium; **(f)** selenium; **(g)** argon.

**2.20**  Locate each of the following elements on the periodic table, indicate whether it is a metal, metalloid, or nonmetal, and give the name of the element: **(a)** K; **(b)** Kr; **(c)** S; **(d)** Sb; **(e)** C; **(f)** Ca; **(g)** H.

**2.21**  Which pair in each of the following groups of elements would you expect to be the most similar in chemical and physical properties: **(a)** Ca, Si, I, P, Sr, Sc; **(b)** Mg, Al, S, I, Sb, Ga? Explain.

**2.22**  Describe what is meant by the term "family" in speaking of the elements and the periodic table.

**2.23**  Write the empirical formulas corresponding to each of the following molecular formulas: **(a)** $C_4H_8$; **(b)** $C_3H_8$; **(c)** $P_4O_{10}$; **(d)** $N_2O_4$; **(e)** $C_6H_6$; **(f)** $C_4H_2O_4$.

**2.24**  What are the empirical formulas of the following substances: **(a)** $C_6H_6$; **(b)** $B_2H_6$; **(c)** $K_2Cr_2O_7$; **(d)** $S_8$; **(e)** $C_6H_8O_6$?

**2.25**  Each of the following elements is capable of forming an ion in chemical reactions. Refer to the periodic table, and predict the charge found on the most stable ion formed by each: **(a)** K; **(b)** F; **(c)** Ba; **(d)** S; **(e)** Al; **(f)** Sc.

**2.26**  What charge would you expect on a stable ion of each of the following elements: **(a)** Br; **(b)** Li; **(c)** Se; **(d)** Mg; **(e)** N?

**2.27**  Predict the formula for the ionic compound formed between the following ions: **(a)** $Ca^{2+}$ and $Br^-$; **(b)** $NH_4^+$ and $SO_4^{2-}$; **(c)** $Mg^{2+}$ and $NO_3^-$; **(d)** $Na^+$ and $S^{2-}$; **(e)** $K^+$ and $OH^-$.

**2.28**  Predict the formula for the ionic compound formed between each of the following pairs of elements: **(a)** Li, O; **(b)** Sr, S; **(c)** Cs, Br; **(d)** Cl, Mg; **(e)** Sc, F.

**2.29**  Based on their compositions, predict whether each of the following compounds is molecular or ionic: **(a)** $N_2O$; **(b)** $Na_2O$; **(c)** $CaO$; **(d)** $CO$; **(e)** $ScF_3$; **(f)** $NF_3$; **(g)** $KBr$.

**2.30**  Which of the following are ionic substances and which are molecular: **(a)** $Cl_2$; **(b)** $CrCl_3$; **(c)** $HNO_3$; **(d)** $Na_2SO_4$; **(e)** $C_3H_8$?

## Naming Inorganic Compounds

**2.31**  Name the following compounds: **(a)** $ZnCl_2$; **(b)** $PbCrO_4$; **(c)** $Hg(NO_3)_2$; **(d)** $Ca(CN)_2$; **(e)** $FeF_3$; **(f)** $Na_2CO_3$; **(g)** $KClO_4$; **(h)** $Cu(OH)_2$; **(i)** $H_2S$; **(j)** $(NH_4)_2SO_4$; **(k)** $Cr_2O_3$; **(l)** $Ag_3PO_4$.

**2.32**  Write the chemical formula for each of the following compounds: **(a)** aluminum hydroxide; **(b)** potassium permanganate; **(c)** silver nitrate; **(d)** calcium hydride; **(e)** copper(I) iodide; **(f)** magnesium phosphate; **(g)** lead(II) acetate; **(h)** barium peroxide; **(i)** sodium sulfite; **(j)** ammonium carbonate; **(k)** iron(II) sulfide; **(l)** zinc hydrogen sulfate.

**2.33**  Give the name or chemical formula, as appropriate, for each of the following acids: **(a)** HI; **(b)** $H_2CO_3$; **(c)** $HNO_2$; **(d)** chromic acid; **(e)** hydrofluoric acid; **(f)** hypochlorous acid.

**2.34**  Give the name or chemical formula, as appropriate, for each of the following acids: **(a)** $H_3PO_4$; **(b)** $HIO_4$ (named like the corresponding acid of chlorine); **(c)** $H_2S$; **(d)** nitric acid; **(e)** hydrocyanic acid; **(f)** acetic acid.

**2.35**  Give the name or chemical formula, as appropriate, for each of the following binary molecular compounds: **(a)** $SeO_2$; **(b)** $CCl_4$; **(c)** $P_4O_6$; **(d)** iodine pentafluoride; **(e)** hydrogen cyanide; **(f)** carbon disulfide.

**2.36**  Give the name or chemical formula, as appropriate, for each of the following binary molecular compounds: **(a)** $XeO_3$; **(b)** $SiF_4$; **(c)** $Cl_2O_3$; **(d)** hydrogen selenide; **(e)** sulfur dichloride; **(f)** phosphorus pentafluoride.

**2.37**  Write the chemical formula for each substance mentioned in the following word descriptions (use the front inside cover to find the symbols for the elements you don't know). **(a)** Zinc carbonate can be heated to form zinc oxide and carbon dioxide. **(b)** On treatment with hydrofluoric acid, silicon dioxide forms silicon tetrafluoride and water. **(c)** Sulfur dioxide reacts with water to form sulfurous acid. **(d)** The substance hydrogen phosphide is commonly called phosphine. **(e)** Perchloric acid reacts with cadmium to form cadmium(II) perchlorate. **(f)** Vanadium(III) bromide is a colored solid.

**2.38**  Assume that you encounter the following phrases in your reading. What is the chemical formula for each compound mentioned? **(a)** Potassium chlorate is used as a laboratory source of oxygen. **(b)** Sodium hypochlorite is used as a household bleach. (**c**) Ammonia is important in the synthesis of fertilizers such as ammonium nitrate. (**d**) Hydrofluoric acid is used to etch glass. (**e**) The smell of rotten eggs is due to hydrogen sulfide. (**f**) When hydrochloric acid is added to sodium bicarbonate (baking powder), carbon dioxide gas forms.

## Additional Exercises

**2.39** Describe the contributions to atomic theory made by the following persons: (**a**) Dalton; (**b**) Rutherford; (**c**) Thomson; (**d**) Millikan; (**e**) Becquerel.

**2.40** (**a**) Explain why $^{18}O$ and $^{16}O$ have essentially identical chemical properties. (**b**) Explain why the symbols for an alpha particle and a beta particle could be given as $^{4}_{2}\alpha$ and $^{0}_{-1}\beta$.

**2.41** Fill in the gaps in the following table:

| Symbol | $^{12}_{6}C$ | $^{17}_{8}O^{2-}$ | | | |
|---|---|---|---|---|---|
| Protons | 6 | | 12 | | 8 |
| Neutrons | | | 13 | 12 | 10 |
| Electrons | | 10 | | 10 | 10 |
| Net charge | 0 | 2− | 0 | 1+ | |

**2.42** Fill in the gaps in the following table:

| Symbol | $^{37}Cl^{-}$ | $^{9}Be^{2+}$ | | | |
|---|---|---|---|---|---|
| Protons | | | 21 | | 28 |
| Neutrons | | | 23 | 76 | 31 |
| Electrons | | | | 54 | 26 |
| Net charge | | | 3+ | 2− | |

**2.43** Identify each of the following: (**a**) an element of group 2A that has 20 electrons in the neutral atom; (**b**) an unreactive element with 36 protons in the nucleus; (**c**) an element with atomic mass between 70 and 90 that tends to form ions of 1− charge; (**d**) a metallic element that forms 2+ ions having 28 electrons.

**2.44** From the following list of elements—Ar, H, Ga, Al, Ca, Br, Ge, K, O—pick the one that best fits each description; use each element only once: (**a**) an alkali metal; (**b**) an alkaline earth metal; (**c**) a noble gas; (**d**) a halogen; (**e**) a metalloid; (**f**) a nonmetal listed in family 1A; (**g**) a metal that forms a 3+ ion; (**h**) a nonmetal that forms a 2− ion; (**i**) an element that resembles aluminum.

**2.45** The formula of the compound formed between zinc and oxygen is ZnO. Predict the formula of the compound formed between (**a**) cadmium and oxygen; (**b**) zinc and tellurium; (**c**) cadmium and sulfur; (**d**) zinc and chlorine; (**e**) cadmium and bromine.

**2.46** Distinguish between the members of the following pairs: (**a**) atomic number and mass number; (**b**) chemical and physical properties; (**c**) Ca and $Ca^{2+}$; (**d**) hydrochloric acid and chloric acid; (**e**) iron(II) and iron(III); (**f**) sodium carbonate and sodium bicarbonate; (**g**) $H_2O$ and $H_2O_2$; (**h**) metal and nonmetal; (**i**) H and He; (**j**) chloride and chlorate.

**2.47** Distinguish between the members of the following pairs: (**a**) alpha and beta particles; (**b**) proton and neutron; (**c**) ion and atom; (**d**) molecular formula and empirical formula; (**e**) O and $O^{2-}$; (**f**) C and Ca; (**g**) sodium sulfate and sodium sulfite; (**h**) sodium hydrogen phosphate and sodium dihydrogen phosphate; (**i**) cuprous chloride and cupric chloride; (**j**) $O_2$ and $O_3$.

**2.48** Many ions and compounds have very similar names, and there is great potential of confusing them. Write the correct chemical formulas to distinguish between (**a**) calcium sulfide and calcium hydrogen sulfide; (**b**) hydrobromic acid and bromic acid; (**c**) aluminum nitride and aluminum nitrite; (**d**) iron(II) oxide and iron(III) oxide; (**e**) ammonia and ammonium ion; (**f**) potassium sulfite and potassium bisulfite; (**g**) mercurous chloride and mercuric chloride; (**h**) chloric acid and perchloric acid.

**2.49** Many familiar substances have common, unsystematic names. In each of the following cases, give the correct systematic name: (**a**) saltpeter ($KNO_3$); (**b**) soda ash ($Na_2CO_3$); (**c**) lime (CaO); (**d**) baking soda ($NaHCO_3$); (**e**) lye (NaOH); (**f**) muriatic acid (HCl); (**g**) milk of magnesia ($Mg(OH)_2$); (**h**) dry ice ($CO_2$); (**i**) ammonia ($NH_3$).

**2.50** Give the chemical name for each of the following compounds: (**a**) $SF_6$; (**b**) $H_2O_2$; (**c**) $BaSO_4$; (**d**) $HNO_3$; (**e**) $KHCO_3$; (**f**) $NO_2$; (**g**) $CoBr_2$; (**h**) $Fe(ClO_4)_2$.

**2.51** Write the chemical formula for each of the following: (**a**) zinc sulfide; (**b**) xenon tetrafluoride; (**c**) chlorous acid; (**d**) iron(III) chromate; (**e**) cupric nitrate; (**f**) magnesium sulfate; (**g**) mercurous chloride; (**h**) sodium oxide.

**[2.52]** The two principal isotopes of lithium, $^6Li$ and $^7Li$, have masses of 6.01513 amu and 7.01601 amu, respectively. The average mass of a lithium atom found in nature is 6.941 amu. (**a**) Which isotope is more abundant? (**b**) Calculate the percent abundance of each isotope.

**[2.53]** Using a suitable reference such as the *Handbook of Chemistry and Physics*, look up the following information for sulfur: (**a**) the number of known isotopes; (**b**) the natural abundance of the four most abundant isotopes.

**[2.54]** Using the *Handbook of Chemistry and Physics*, find the density, melting point, and boiling point for $PF_3$.

## CHAPTER 3

# Stoichiometry

In the late 1700s a wealthy French nobleman named Antoine Lavoisier (Figure 3.1) began a significant series of careful, quantitative studies of chemical reactions. He observed what countless subsequent chemists have verified: The total mass of all substances present after a chemical reaction is the same as the total mass before the reaction. This important fact is known as the **law of conservation of mass**, one of the fundamental laws of chemistry. In 1789, Lavoisier published a textbook of chemistry in which he stated:

> We may lay it down as an incontestable axiom that, in all the operations of art and nature, nothing is created; an equal quantity of matter exists both before and after the experiment.*

Today we state the law in a somewhat simpler and more cautious fashion: There is no *detectable* change in mass in any chemical reaction.† From

---

\* A. L. Lavoisier, *Traité élémentaire de chimie* (Paris: 1789); English translation in *Great Books*, vol. 45 (Chicago: Encyclopedia Britannica, 1952), pp. 9–133.

† In Chapter 21, we will discuss the relationship between mass and energy summarized by the equation $E = mc^2$ ($E$ is energy, $m$ is mass, and $c$ is the speed of light). We will find that whenever an object loses energy it loses mass, and whenever it gains energy it gains mass. These changes in mass are too small to detect in chemical reactions. However, for nuclear reactions, such as those involved in a nuclear reactor or in a hydrogen bomb, the energy changes are enormously larger; in these reactions there are detectable changes in mass.

**FIGURE 3.1** Antoine Lavoisier (1734–1794), as pictured in a nineteenth century French engraving. Lavoisier conducted many important studies on combustion reactions. Here he is shown experimenting to determine the composition of water by igniting a mixture of hydrogen and water with an electric spark. Unfortunately, Lavoisier's career was cut short by the French Revolution. He was not only a member of the French nobility but also a tax collector. He was guillotined in 1794 during the final months of the Reign of Terror. He is now generally considered to be the father of modern chemistry because of his reliance on carefully controlled experiments and his use of quantitative measurements. The tribunal that sentenced him to death, however, declared that France had no need for scientists. The great mathematician Lagrange, then living in Paris, remarked: "It took but a moment to cut off that head, though a hundred years perhaps will be required to produce another like it." (The Granger Collection)

Combustion reactions, discussed in this chapter, are widely used both to generate electricity and to dispose of combustible substances. This photo shows a natural gas flare at the Abadan Refinery in Iran. The natural gas is burned off where facilities are unavailable for its collection. (Frederick Ayer/Photo Researchers)

atomic theory we understand this law to exist because *atoms are neither created nor destroyed during a chemical reaction.* Although matter is conserved, it can be changed from one form to another by rearranging atoms.

The law of conservation of mass, a foundation of this chapter, serves as a basis for treating many practical problems in quantitative relationships between substances undergoing chemical changes. The study of these quantitative relationships is known as **stoichiometry** (pronounced stoy-key-AHM-uh-tree), a word derived from the Greek words *stoicheion* ("element") and *metron* ("measure").

## 3.1 CHEMICAL EQUATIONS

Chemical reactions can be represented in a concise way using **chemical equations**. For example, when hydrogen, $H_2$, burns it reacts with molecular oxygen, $O_2$, in the air to form water, $H_2O$. We write the chemical equation for this reaction as follows:

$$2H_2 + O_2 \longrightarrow 2H_2O \qquad [3.1]$$

We read the + sign to mean "reacts with" and the arrow as "produces." The chemical formulas on the left of the arrow represent the starting substances, called **reactants**. The substances produced in a reaction, called **products**, are shown to the right of the arrow. The numbers in front of the formulas are called **coefficients**. (The numeral "1" is usually not written.)

Because atoms are neither created nor destroyed in any reaction, an equation must have an equal number of atoms of each element on each side of the arrow. When this condition is met, the equation is said to be **balanced**. The number of atoms of an element is the product of the coefficient in front of a formula and the subscript for that element in the formula. Thus $2H_2O$ contains $2 \times 2 = 4$ hydrogen atoms and $2 \times 1 = 2$ oxygen atoms. Count the number of hydrogen and oxygen atoms on each side of Equation 3.1 to verify that the equation is balanced.

Before we can write the chemical equation for a reaction, we must determine by experiment those substances that are reactants and those that are products. Once we know the chemical formulas of these substances, we can write the chemical equation. Let's consider how we go about balancing an equation given the reactants and products. This process amounts to finding coefficients that provide equal numbers of each type of atom on each side of the equation.

Consider the reaction that occurs when methane ($CH_4$), the principal component of natural gas, burns in air to produce carbon dioxide ($CO_2$) and water ($H_2O$). Combustion in air is "supported by" oxygen ($O_2$), meaning that oxygen is a reactant. The unbalanced equation is

$$CH_4 + O_2 \longrightarrow CO_2 + H_2O \qquad \text{(unbalanced)} \qquad [3.2]$$

It is usually best to balance first the elements that occur in only one substance on each side of the equation. In our example, both C and H appear in only one reactant and, separately, in one product each, so we begin by focusing attention on $CH_4$. We consider first carbon and then hydrogen.

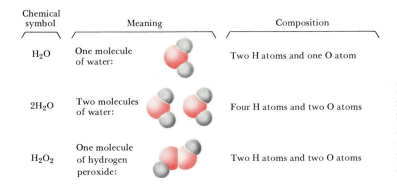

| Chemical symbol | Meaning | | Composition |
|---|---|---|---|
| $H_2O$ | One molecule of water: | | Two H atoms and one O atom |
| $2H_2O$ | Two molecules of water: | | Four H atoms and two O atoms |
| $H_2O_2$ | One molecule of hydrogen peroxide: | | Two H atoms and two O atoms |

**FIGURE 3.2** Illustration of the difference in meaning between a subscript in a chemical formula and a coefficient in front of the formula. Notice that the number of atoms of each type (listed under composition) is obtained by multiplying the coefficient and the subscript associated with each element in the formula.

Notice that the reactants and products both contain one carbon atom. Thus a coefficient 1 (understood) in front of $CO_2$ is appropriate. However, the $CH_4$ contains more hydrogen atoms (four) than the products (two). If we place a coefficient 2 in front of $H_2O$, there will be four hydrogens on each side of the equation:

$$CH_4 + O_2 \longrightarrow CO_2 + 2H_2O \quad \text{(unbalanced)} \qquad [3.3]$$

Before we continue to balance this equation, let's make sure that we clearly understand the distinction between a coefficient in front of a formula and a subscript in a formula. Refer to Figure 3.2. Notice that changing a subscript in a formula—from $H_2O$ to $H_2O_2$, for example—changes the identity of the chemical. The substance $H_2O_2$, hydrogen peroxide, is quite different from water. *Subscripts should never be changed in balancing an equation.* Placing a coefficient in front of a formula changes only the *amount* and not the *identity* of the substance; $2H_2O$ means two molecules of water, $3H_2O$ means three molecules of water, and so forth. Now let's continue balancing the example equation. Equation 3.3 has equal numbers of carbon and hydrogen atoms on both sides. However, the products have more oxygen atoms (four) than the reactants have (two). If we place a coefficient 2 in front of $O_2$, we make equal the number of oxygen atoms on both sides of the equation:

$$CH_4 + 2O_2 \longrightarrow CO_2 + 2H_2O \quad \text{(balanced)} \qquad [3.4]$$

The balanced equation is shown schematically in Figure 3.3. For most

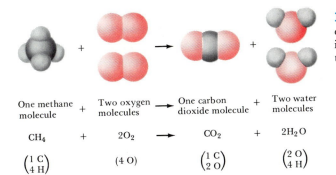

| One methane molecule | + | Two oxygen molecules | | One carbon dioxide molecule | + | Two water molecules |
|---|---|---|---|---|---|---|
| $CH_4$ | + | $2O_2$ | | $CO_2$ | + | $2H_2O$ |
| $\begin{pmatrix} 1\ C \\ 4\ H \end{pmatrix}$ | | $(4\ O)$ | | $\begin{pmatrix} 1\ C \\ 2\ O \end{pmatrix}$ | | $\begin{pmatrix} 2\ O \\ 4\ H \end{pmatrix}$ |

**FIGURE 3.3** Balanced chemical equation for the combustion of $CH_4$. The drawings of the molecules involved call attention to the conservation of atoms through the reaction.

purposes, a balanced equation should contain the smallest whole-number coefficients, as in this example. This approach to balancing equations is largely trial and error. Verifying that an equation is balanced is easier than actually balancing one, so practice in balancing equations is essential.

The physical state of each chemical in a chemical equation is often indicated parenthetically. We use the symbols $(g)$, $(l)$, $(s)$, and $(aq)$ for gas, liquid, solid, and aqueous (water) solution, respectively. Thus the balanced equation above can be written

$$CH_4(g) + 2O_2(g) \longrightarrow CO_2(g) + 2H_2O(l) \qquad [3.5]$$

Sometimes an upward arrow ($\uparrow$) is used to indicate the escape of a gaseous product, and a downward arrow ($\downarrow$) to indicate a precipitating solid (that is, a solid that separates from solution during the reaction). Often the conditions under which the reaction proceeds appear above the arrow between the two sides of the equation. For example, the temperature or pressure at which the reaction occurs could be so indicated. The symbol $\Delta$ is often placed above the arrow to indicate the addition of heat.

## SAMPLE EXERCISE 3.1

Balance the following equation:

$$Na(s) + H_2O(l) \longrightarrow NaOH(aq) + H_2(g)$$

**Solution:** A quick inventory of atoms reveals an equal numbers of Na and O atoms on both sides of the equation, but that there are two H atoms among reactants and three H atoms among products. To increase the number of H atoms among reactants, we might place a coefficient 2 in front of $H_2O$:

$$Na(s) + 2H_2O(l) \longrightarrow NaOH(aq) + H_2(g)$$

Now we have four H atoms among reactants but only three H atoms among the products. The H atoms can be balanced with a coefficient 2 in front of NaOH:

$$Na(s) + 2H_2O(l) \longrightarrow 2NaOH(aq) + H_2(g)$$

If we again inventory the atoms on each side of the equation, we find that the H atoms and O atoms are balanced but not the Na atoms. However, a coefficient 2 in front of Na gives two Na atoms on each side of the equation:

$$2Na(s) + 2H_2O(l) \longrightarrow 2NaOH(aq) + H_2(g)$$

If the atoms are counted once more, we find two Na atoms, four H atoms, and two O atoms on each side of the equation. The equation is therefore balanced.

## PRACTICE EXERCISE
Balance the following equations by providing the missing coefficients:

**(a)** __$C_2H_4$ + __$O_2$ $\longrightarrow$ __$CO_2$ + __$H_2O$
**(b)** __Al + __HCl $\longrightarrow$ __$AlCl_3$ + __$H_2$

*Answers:* (a) 1, 3, 2, 2; (b) 2, 6, 2, 3

## 3.2 SOME COMMON CHEMICAL REACTIONS

Our discussion in Section 3.1 focused on how to balance chemical equations given the reactants and products for the reactions. You were not asked to say anything about the type of reaction or to predict the products. Students often ask how the products are determined. Discussing this matter is a bit of a diversion from the basic theme of this chapter. However, some discussion now will help you to feel more at ease with the reactions used to illustrate concepts in the chapter, and it will certainly help you in the lab.

We may identify reactants and products experimentally by their properties. However, we can often predict what will happen if we have seen

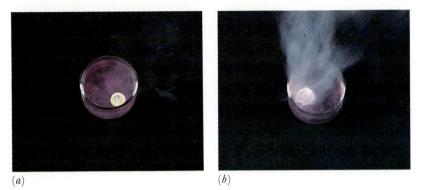

(a)                                    (b)

**FIGURE 3.4** The reaction of sodium with water (left) and of potassium with water (right). An acid-base indicator (phenolpthalein) has been added to both solutions; the pink color indicates the presence of hydroxide ions in the solution. Although the reactions of the two metals with water are very similar, potassium reacts more vigorously. (Donald Clegg and Roxy Wilson)

the specific reaction or a similar reaction before. Naturally, recognizing a general pattern of reactivity for a class of substances gives you a broader understanding than merely memorizing a large number of unrelated reactions. The periodic table can be a powerful ally in this regard. For example, knowing that sodium (Na) reacts with water ($H_2O$) to form NaOH and $H_2$ (see Sample Exercise 3.1), we can predict what happens when potassium (K) is placed in water. Both sodium and potassium are in the same family of the periodic table (the alkali metal family, family 1A). We would expect them to behave similarly, producing the same kinds of products. Indeed, this prediction is correct:

$$2K(s) + 2H_2O(l) \longrightarrow 2KOH(aq) + H_2(g) \qquad [3.6]$$

In fact, all alkali metals behave in this general fashion. If we let **M** represent any alkali metal, we can write the general reaction as follows:

$$2M(s) + 2H_2O(l) \longrightarrow 2MOH(aq) + H_2(g) \qquad [3.7]$$

Figure 3.4 shows photographs of the reactions of sodium and of potassium with water. We now examine a few common types of reactions that you will be encountering in your laboratory work and in the chapters ahead.

Combustion reactions are rapid reactions that produce a flame. Most of the combustions we observe involve $O_2$ as a reactant. Equation 3.5 and Practice Exercise 3.1(a) illustrate a general class of reactions involving the combustion of compounds called hydrocarbons. (Hydrocarbons are compounds that contain only carbon and hydrogen, like $C_2H_4$.) In these reactions the hydrocarbon reacts with $O_2$ to form $CO_2$ and $H_2O$.* For example, propane ($C_3H_8$), a gas used for cooking and home heating, burns in air as described by the following equation:

$$C_3H_8(g) + 5O_2(g) \longrightarrow 3CO_2(g) + 4H_2O(l) \qquad [3.8]$$

Combustion of compounds containing oxygen atoms as well as carbon and hydrogen (for example, $CH_3OH$) also produce $CO_2$ and $H_2O$.

**Combustion in Oxygen**

* When there is an insufficient quantity of $O_2$ present, carbon monoxide, CO, will be produced. Even more severe restriction of $O_2$ will cause the production of fine particles of carbon that we call soot. *Complete* combustion produces $CO_2$.

**TABLE 3.1**  Some common types of reactions

| Combination reactions | |
|---|---|
| $A + B \longrightarrow AB$ <br> $C(s) + O_2(g) \longrightarrow CO_2(g)$ <br> $N_2(g) + 3H_2(g) \longrightarrow 2NH_3(g)$ <br> $CaO(s) + H_2O(l) \longrightarrow Ca(OH)_2(s)$ | Two reactants combine to form a single product. Many elements react with one another in this fashion to form compounds. |
| **Decomposition reactions** | |
| $AB \longrightarrow A + B$ <br> $2KClO_3(s) \longrightarrow 2KCl(s) + 3O_2(g)$ <br> $CaCO_3(s) \longrightarrow CaO(s) + CO_2(g)$ | A single reactant breaks apart to form two or more substances. Many compounds behave in this fashion when heated. |
| **Single displacement reactions** | |
| $A + BX \longrightarrow AX + B$ <br> $Fe(s) + 2HCl(aq) \longrightarrow FeCl_2(aq) + H_2(g)$ <br> $Zn(s) + CuSO_4(aq) \longrightarrow ZnSO_4(aq) + Cu(s)$ <br> $2Na(s) + 2H_2O(l) \longrightarrow 2NaOH(aq) + H_2(g)$ | One element replaces another in a compound. (The elements are often hydrogen or a metal.) |
| **Double displacement (metathesis) reactions** | |
| $AX + BY \longrightarrow AY + BX$ <br> $BaBr_2(aq) + K_2SO_4(aq) \longrightarrow 2KBr(aq) + BaSO_4(s)$ <br> $Ca(OH)_2(aq) + 2HCl(aq) \longrightarrow CaCl_2(aq) + 2H_2O(l)$ | Atoms or ions exchange partners. Two kinds of reactions of this type, precipitation and acid-base reactions, are discussed in the text. |

**Other Common Reaction Types**

Table 3.1 summarizes several additional types of reactions. A large number of reactions fit these patterns, or classes. These classes show how atoms or groups of atoms often rearrange during chemical reactions. However, they do not allow us to predict products, given just reactants. In order to make such predictions, we need to look more closely at chemical behavior, as we have done in considering the combustion of hydrocarbons.

Rather than study each reaction type in greater detail at this time, we will examine only double displacement reactions. In double displacement reactions, positive ions (cations) and negative ions (anions) exchange partners. These reactions are also known as **metathesis reactions** (muh-TATH-uh-sis, which is Greek for "to transpose"). Two kinds of metathesis reactions will be considered: acid-base reactions and precipitation reactions.

**Acids, Bases, and Neutralization**

**Acids** are substances that increase the $H^+$ ion concentration in aqueous solution. For example, hydrochloric acid, which we often represent as $HCl(aq)$, exists in water as $H^+(aq)$ and $Cl^-(aq)$ ions. Thus the process of dissolving hydrogen chloride in water to form hydrochloric acid can be represented as follows:

$$HCl(g) \xrightarrow{\text{H}_2\text{O}} HCl(aq) \qquad [3.9]$$

or

$$HCl(g) \xrightarrow{\text{H}_2\text{O}} H^+(aq) + Cl^-(aq) \qquad [3.10]$$

The $H_2O$ given above the arrows in these equations reminds us that the reaction medium is water. Pure sulfuric acid is a liquid. When it dissolves in water it releases $H^+$ ions in two successive steps:

$$H_2SO_4(l) \xrightarrow{H_2O} H^+(aq) + HSO_4^-(aq) \qquad [3.11]$$

$$HSO_4^-(aq) \xrightarrow{H_2O} H^+(aq) + SO_4^{2-}(aq) \qquad [3.12]$$

Thus, although we frequently represent aqueous solutions of sulfuric acid as $H_2SO_4(aq)$, these solutions actually contain a mixture of $H^+(aq)$, $HSO_4^-(aq)$, and $SO_4^{2-}(aq)$.

**Bases** are compounds that increase the hydroxide ion, $OH^-$, concentration in aqueous solution. The base sodium hydroxide does this because it is an ionic substance composed of $Na^+$ and $OH^-$ ions. When NaOH dissolves in water, the cations and anions simply separate in the solution:

$$NaOH(s) \xrightarrow{H_2O} Na^+(aq) + OH^-(aq) \qquad [3.13]$$

Thus, although aqueous solutions of sodium hydroxide might be written as $NaOH(aq)$, sodium hydroxide exists as $Na^+(aq)$ and $OH^-(aq)$ ions. Many other bases such as $Ca(OH)_2$ are also ionic hydroxide compounds. However, $NH_3$ (ammonia) is a base although it is not a compound of this sort.

It may seem odd at first glance that ammonia is a base, because it contains no hydroxide ions. However, we must remember that the definition of a base is that it *increases* the concentration of $OH^-$ ions in water. Ammonia does this by a reaction with water. We can represent the dissolving of ammonia gas in water as follows:

$$NH_3(g) + H_2O(l) \longrightarrow NH_4^+(aq) + OH^-(aq) \qquad [3.14]$$

Solutions of ammonia in water are often labeled ammonium hydroxide, $NH_4OH$, to remind us that ammonia solutions are basic. (Ammonia is referred to as a weak base, which means that not all the $NH_3$ that dissolves in water goes on to form $NH_4^+$ and $OH^-$ ions. That is a matter for Chapter 17 and need not concern us here.)

Acids and bases are among the most important compounds in industry and in the chemical laboratory. Table 3.2 lists several acids and bases and the amount of each compound produced in the United States each year. You can see that these substances are produced in enormous quantities.

**TABLE 3.2**  U.S. production of some acids and bases, 1986.

| Compound | Formula | Annual production (kg) |
|---|---|---|
| Acids: | | |
| Sulfuric | $H_2SO_4$ | $3.4 \times 10^9$ |
| Phosphoric | $H_3PO_4$ | $8.4 \times 10^9$ |
| Nitric | $HNO_3$ | $6.0 \times 10^9$ |
| Hydrochloric | HCl | $2.7 \times 10^9$ |
| Bases: | | |
| Sodium hydroxide | NaOH | $1.0 \times 10^{10}$ |
| Calcium hydroxide | $Ca(OH)_2$ | $1.4 \times 10^{10}$ |
| Ammonia | $NH_3$ | $1.3 \times 10^{10}$ |

Solutions of acids and bases have very different properties. Acids have a sour taste, whereas bases have a bitter taste.* Acids can change the colors of certain dyes in a specific way that differs from the effect of a base. For example, the dye known as litmus is changed from blue to red by an acid, and from red to blue by a base (Figure 3.5). In addition, acidic and basic solutions differ in chemical properties in several important ways. When a solution of an acid is mixed with a solution of a base, a **neutralization reaction** occurs. The products of the reaction have none of the characteristic properties of either the acid or base. For example, when a solution of hydrochloric acid is mixed with precisely the correct quantity of a sodium hydroxide solution, the result is a solution of sodium chloride, a simple ionic compound possessing neither acidic nor basic properties. (In general, such ionic products are referred to as **salts**.) The neutralization reaction can be written as follows:

$$\text{HCl}(aq) + \text{NaOH}(aq) \longrightarrow \text{H}_2\text{O}(l) + \text{NaCl}(aq) \qquad [3.15]$$

$$\text{(acid)} \qquad \text{(base)} \qquad \text{(water)} \qquad \text{(salt)}$$

Keep in mind that the substances shown as (aq) are present in the form of the separated ions, as discussed above. Notice that the acid and base in Equation 3.15 have combined to form water as a product. The general description of an acid-base neutralization reaction in aqueous solution, then, is that *an acid and base react to form a salt and water.* Using this general description we can predict the products formed in any acid-base neutralization reaction.

* Tasting chemical solutions is, of course, not a good practice. However, we have all had acids such as ascorbic acid (vitamin C), acetylsalicylic acid (aspirin), and citric acid (in citrus fruits) in our mouths, and we are familiar with the characteristic sour taste. It differs from the taste of soaps, which are mostly basic.

**FIGURE 3.5** The acid-base indicator litmus changes from blue to red upon addition of an acid (left), and from red to blue upon addition of a base (right). (Dr. E.R. Degginger)

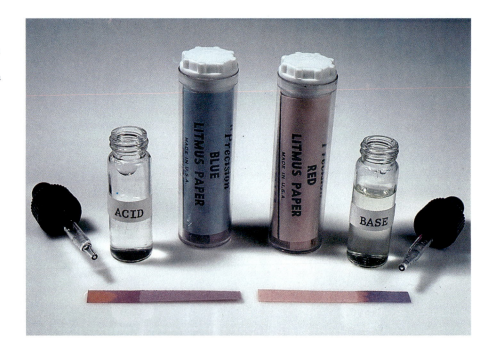

## SAMPLE EXERCISE 3.2

Write a balanced equation for the reaction of hydrobromic acid, HBr, with barium hydroxide, $Ba(OH)_2$.

**Solution:** The products of any acid-base reaction are a salt and water. The salt is that formed from the cation of the base, $Ba(OH)_2$, and the anion of the acid, HBr. The charge on the barium ion is $2+$ (see Table 2.3), and that on the bromide ion is $1-$. Therefore, to maintain electrical neutrality, the formula for the salt must be $BaBr_2$. The unbalanced equation for the neutralization reaction is therefore

$$HBr(aq) + Ba(OH)_2(aq) \longrightarrow H_2O(l) + BaBr_2(aq)$$

To balance the equation we must provide two molecules of HBr to furnish the two $Br^-$ ions and to supply the two $H^+$ ions needed to combine with the two $OH^-$ ions of the base. The balanced equation is thus

$$2HBr(aq) + Ba(OH)_2(aq) \longrightarrow 2H_2O(l) + BaBr_2(aq)$$

### PRACTICE EXERCISE

Write a balanced equation for the reaction between phosphoric acid, $H_3PO_4$, and potassium hydroxide, KOH.

*Answer:*   $H_3PO_4 + 3KOH \longrightarrow 3H_2O + K_3PO_4$

---

In a **precipitation reaction** a solid product, called a **precipitate**, separates from solution. As an example, consider the reaction between a solution of potassium iodide, KI, and a solution of lead nitrate, $Pb(NO_3)_2$. When the two solutions are mixed, a finely divided yellow solid forms, as shown in Figure 3.6. Upon analysis, this solid proves to be lead iodide, $PbI_2$, a salt that has a very low solubility in water.* The following equation presents the reaction:

**Precipitation Reactions**

$$2KI(aq) + Pb(NO_3)_2(aq) \longrightarrow PbI_2(s) + 2KNO_3(aq) \qquad [3.16]$$

The following equations provide further examples of precipitation reactions:

$$HCl(aq) + AgNO_3(aq) \longrightarrow AgCl(s) + HNO_3(aq) \qquad [3.17]$$

$$CuCl_2(aq) + 2NaOH(aq) \longrightarrow Cu(OH)_2(s) + 2NaCl(aq) \qquad [3.18]$$

* Solubility will be considered in some detail in Chapter 13. It is a measure of the amount of substance that can be dissolved in a given quantity of solvent (see Section 3.10).

(*a*)        (*b*)        (*c*)

**FIGURE 3.6** The addition of a solution of the colorless potassium iodide, KI, to one of the colorless lead nitrate, $Pb(NO_3)_2$, produces a yellow precipitate of lead iodide, $PbI_2$. (Lawrence Migdale/ Photo Researchers)

## SAMPLE EXERCISE 3.3

When solutions of sodium phosphate and barium nitrate are mixed, a precipitate of barium phosphate forms. Write a balanced equation to describe the reaction.

**Solution:**  Our first task is to determine the formulas of the reactants. The sodium ion is $Na^+$ and the phosphate ion is $PO_4^{3-}$; thus sodium phosphate is $Na_3PO_4$. The barium ion is $Ba^{2+}$ and the nitrate ion is $NO_3^-$; thus barium nitrate is $Ba(NO_3)_2$. The $Ba^{2+}$ and $PO_4^{3-}$ ions combine to form the barium phosphate precipitate, $Ba_3(PO_4)_2$. The other ions, $Na^+$ and $NO_3^-$, remain in solution and are represented as $NaNO_3(aq)$. The *unbalanced* equation for the reaction is thus

$$Na_3PO_4(aq) + Ba(NO_3)_2(aq) \longrightarrow$$
$$Ba_3(PO_4)_2(s) + NaNO_3(aq)$$

Because the $NO_3^-$ and $PO_4^{3-}$ ions maintain their identity through the reaction, we can treat them as units in balancing the equation. There are two $(PO_4)$ units on the right, so we place a coefficient 2 in front of $Na_3PO_4$. This then gives six Na atoms on the left, necessitating a coefficient of 6 in front of $NaNO_3$. Finally, the presence of six $(NO_3)$ units on the right requires a coefficient of 3 in front of $Ba(NO_3)_2$:

$$2Na_3PO_4(aq) + 3Ba(NO_3)_2(aq) \longrightarrow$$
$$Ba_3(PO_4)_2(s) + 6NaNO_3(aq)$$

### PRACTICE EXERCISE

When aqueous solutions of NaOH and $Mg(NO_3)_2$ are mixed, $Mg(OH)_2$ precipitates from solution. Write a balanced equation for this reaction.
***Answer:***  $2NaOH(aq) + Mg(NO_3)_2(aq) \longrightarrow$
$$Mg(OH)_2(s) + 2NaNO_3(aq)$$

## 3.3 ATOMIC AND MOLECULAR WEIGHTS

Chemical formulas and chemical equations have a quantitative significance. The formula $H_2O$ is not merely an abbreviation for the word "water"; it indicates that a molecule of this substance contains exactly two atoms of hydrogen and one atom of oxygen. Similarly, the chemical equation for the combustion of propane—$C_3H_8(g) + 5O_2(g) \longrightarrow 3CO_2(g) + 4H_2O(l)$, shown in Equation 3.8—indicates more than the qualitative idea that propane reacts with $O_2$ to form $CO_2$ and $H_2O$. The equation is a quantitative statement that the combustion of one molecule of $C_3H_8$ requires five molecules of $O_2$ and produces exactly three molecules of $CO_2$ and four of $H_2O$. Although we can not directly count atoms or molecules, we can indirectly determine their numbers if we know the masses of atoms. Therefore, before we can pursue the quantitative aspects of chemical formulas or equations further, we must explore the concept of atomic and molecular masses.

### The Atomic Mass Scale

Scientists of the nineteenth century knew nothing about subatomic particles, but they were aware that atoms of different elements have different masses. They found, for example, that each 100 g of water contains 11.1 g of hydrogen and 88.9 g of oxygen. Thus water contains $88.9/11.1 = 8$ times as much oxygen, by mass, as hydrogen. Once it became clear that water contains two hydrogen atoms for each oxygen, they concluded that an oxygen atom must weigh $2 \times 8 = 16$ times as much as a hydrogen atom. Hydrogen, the lightest atom, was arbitrarily assigned a relative mass of 1 (no units), and atomic masses of other elements were determined relative to this value. Thus oxygen was assigned an atomic mass of 16.

Today we can measure the masses of individual atoms with a high degree of accuracy. For example, we know that the hydrogen-1 atom has a mass of $1.6736 \times 10^{-24}$ g and the oxygen-16 atom has a mass of $2.6561 \times 10^{-23}$ g. Rather than express atomic masses in grams, however,

At the time of Dalton, the problem of establishing even a relative atomic weight scale for the elements was formidable. Since atoms and molecules cannot be seen, scientists of the nineteenth century had no simple way to be sure about the relative numbers of atoms in any compound. Dalton thought that the formula for water was HO. However, the French scientist Gay-Lussac showed in a brilliant set of measurements that it required two volumes of hydrogen gas to react with one volume of oxygen to form two volumes of water vapor. This observation was inconsistent with Dalton's formula for water. Furthermore, if oxygen were assumed to be a monatomic gas, as Dalton did, one could obtain two volumes of water vapor only by splitting the oxygen atoms in half, which of course violates the concept of the atom as indivisible in chemical reactions.

The Italian physicist Amedeo Avogadro suggested in 1811 that Gay-Lussac's results could be explained if it were assumed that both hydrogen and oxygen exist in the gas phase as diatomic molecules $H_2$ and $O_2$. Avogadro also proposed that equal volumes of gases, if measured at the same temperature and pressure, contain equal numbers of molecules. (When we talk about gases in Chapter 10 we will see the basis for Avogadro's hypothesis, which is entirely correct.) The way in which Avogadro's hypothesis explains Gay-Lussac's observations is illustrated in Figure 3.7. The key here is in assuming that the reacting gases are diatomic. Thus it is possible to obtain two volumes of water vapor from one volume of oxygen gas. Because the combining volume of hydrogen is twice that of oxygen, the formula for water must be $H_2O$, not HO as Dalton had suggested.

Not until about 1860 did Avogadro's ideas gain acceptance. In the meantime chemists such as Berzelius (Figure 3.8) had been making painstaking measurements of the masses of the elements that combined with one another and had established the atomic weights of many elements with quite good accuracy.

**FIGURE 3.8** Jons Jakob Berzelius (1779–1848), Swedish chemist. Berzelius is one of the truly great figures in nineteenth-century science. He developed the modern system of symbols and formulas in chemistry and discovered several elements. Berzelius was an exceptionally gifted experimentalist. His determinations of the atomic weights of several elements were the most accurate known for many years. (The Granger Collection)

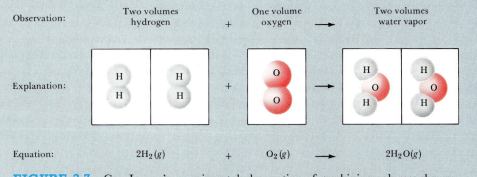

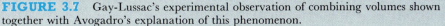

**FIGURE 3.7** Gay-Lussac's experimental observation of combining volumes shown together with Avogadro's explanation of this phenomenon.

it is more convenient to use a unit called the **atomic mass unit** (amu):

$$1 \text{ amu} = 1.66056 \times 10^{-24} \text{ g}$$

The amu is defined by assigning a mass of exactly 12 amu to the $^{12}C$ isotope of carbon. In these units, the mass of the hydrogen-1 atom is 1.008 amu and that of the oxygen-16 atom is 15.995 amu.

## Average Atomic Masses

We can determine the average atomic mass of each element by using the masses of the various isotopes of an element and their relative abundances. For example, naturally occurring carbon contains three isotopes: $^{12}C$ (98.892 percent), $^{13}C$ (1.108 percent), and $^{14}C$ ($1 \times 10^{-10}$ percent). The masses of these nuclides are 12 amu (exactly), 13.00335 amu, and 14.00317 amu, respectively. The computation of the average atomic mass of carbon follows:

$$(0.98892)(12 \text{ amu}) + (0.01108)(13.00335 \text{ amu}) + (1 \times 10^{-12})(14.00317 \text{ amu}) = 12.011 \text{ amu}$$

The average atomic mass of each element (expressed in amu) is also known as its **atomic weight**. Although the term *average atomic mass* is more proper, the term *atomic weight* has become so commonly used that there is little point in attempting to change. The atomic weights of the elements are listed below the symbol for the element on the periodic table found on the front inside cover of this text. They are also listed in the table of the elements on the front inside cover.

---

### SAMPLE EXERCISE 3.4

Naturally occurring chlorine is 75.53 percent $^{35}Cl$, which has an atomic mass of 34.969 amu, and 24.47 percent $^{37}Cl$, which has a mass of 36.966 amu. Calculate the average atomic mass (that is, the atomic weight) of chlorine.

**Solution:**

$(75.53\%)(34.969 \text{ amu}) + (24.47\%)(36.966 \text{ amu})$

$= (0.7553)(34.969 \text{ amu}) + (0.2447)(36.966 \text{ amu})$

$= 26.41 \text{ amu} + 9.05 \text{ amu}$

$= 35.46 \text{ amu}$

### PRACTICE EXERCISE

Three isotopes of silicon occur in nature: $^{28}Si$ (92.21 percent), which has a mass of 27.97693 amu; $^{29}Si$ (4.70 percent), which has a mass of 28.97649 amu; and $^{30}Si$ (3.09 percent), which has a mass of 29.97376 amu. Calculate the atomic weight of silicon.
*Answer:* 28.09 amu

---

## Formula and Molecular Weights

The **formula weight** of a substance is merely the sum of the atomic weights of each atom in its chemical formula. For example, $H_2SO_4$, sulfuric acid, has a formula weight of 98.0 amu:*

$$FW = 2(\text{AW of H}) + \text{AW of S} + 4(\text{AW of O})$$
$$= 2(1.0 \text{ amu}) + 32.0 \text{ amu} + 4(16.0 \text{ amu})$$
$$= 98.0 \text{ amu}$$

* The abbreviation AW is used for atomic weight, MW for molecular weight, and FW for formula weight. The SI symbol for amu is simply u, as in 35.46 u. We will not be using this symbol.

Here we have rounded off the atomic weights so that our result has three significant figures. We will round off the atomic weights in this way for most problems.

If the chemical formula of a substance is its molecular formula, then the formula weight is often called the **molecular weight**. For example, the molecular formula for glucose (the sugar transported by the blood to body tissues to satisfy energy requirements) is $C_6H_{12}O_6$. The molecular weight of glucose is therefore

$$6(12.0 \text{ amu}) + 12(1.0 \text{ amu}) + 6(16.0 \text{ amu}) = 180.0 \text{ amu}$$

With ionic substances such as NaCl, which exist as three-dimensional arrays of ions (Figure 2.17), it is really inappropriate to speak of molecules. Thus we cannot write molecular formulas and molecular weights for such substances. The formula weight of NaCl is

$$23.0 \text{ amu} + 35.5 \text{ amu} = 58.5 \text{ amu}$$

---

### SAMPLE EXERCISE 3.5

Calculate the formula weight of (a) sucrose, $C_{12}H_{22}O_{11}$ (table sugar); (b) calcium nitrate, $Ca(NO_3)_2$.

**Solution:** (a) By adding the weights of the atoms in sucrose we find it to have a formula weight of 342.0 amu:

$$
\begin{aligned}
12\text{C atoms} &= 12(12.0 \text{ amu}) = 144.0 \text{ amu} \\
22\text{H atoms} &= 22(1.0 \text{ amu}) = 22.0 \text{ amu} \\
11\text{O atoms} &= 11(16.0 \text{ amu}) = \underline{176.0 \text{ amu}} \\
& \phantom{= 11(16.0 \text{ amu}) = } 342.0 \text{ amu}
\end{aligned}
$$

(b) If a chemical formula has parentheses, the sub-

script outside the parentheses is a multiplier for all atoms inside. Thus for $Ca(NO_3)_2$ we have

$$
\begin{aligned}
1\text{Ca atom} &= 1(40.1 \text{ amu}) = 40.1 \text{ amu} \\
2\text{N atoms} &= 2(14.0 \text{ amu}) = 28.0 \text{ amu} \\
6\text{O atoms} &= 6(16.0 \text{ amu}) = \underline{96.0 \text{ amu}} \\
& \phantom{= 6(16.0 \text{ amu}) = } 164.1 \text{ amu}
\end{aligned}
$$

### PRACTICE EXERCISE
Calculate the formula weight of (a) $Al(OH)_3$; (b) $CH_3OH$.
*Answers:* (a) 78.0 amu; (b) 32.0 amu

---

Occasionally we must calculate the percentage composition of a compound (that is, the percentage by mass or weight contributed by each element in the substance). For example, we may wish to compare the calculated composition of a substance with that found experimentally in order to verify the purity of the compound. Calculating percentage composition is a straightforward matter if the chemical formula is known. The calculation of such percentages is illustrated in Sample Exercise 3.6.

**Percentage Composition from Formulas**

---

### SAMPLE EXERCISE 3.6

Calculate the percentage composition of $C_{12}H_{22}O_{11}$.

**Solution:** In general, the percentage of a given element in a compound is given by

$$\frac{(\text{Atoms of element})(\text{AW})}{\text{FW of compound}} \times 100$$

The formula weight of $C_{12}H_{22}O_{11}$ is 342 amu (Sample Exercise 3.5). Therefore, the percentage composition is

$$\%C = \frac{12(12.0 \text{ amu})}{342 \text{ amu}} \times 100 = 42.1\%$$

$$\%H = \frac{22(1.0 \text{ amu})}{342 \text{ amu}} \times 100 = 6.4\%$$

$$\%O = \frac{11(16.0 \text{ amu})}{342 \text{ amu}} \times 100 = 51.5\%$$

The same elemental composition is obtained from the empirical formula of a substance as from its molecular formula. In Section 3.7 we shall see how the experimentally determined percentage composition of a compound can be used to calculate the empirical formula of the compound.

### PRACTICE EXERCISE

Calculate the percentage of nitrogen, by mass, in $Ca(NO_3)_2$.
***Answer:*** 17.1%

## 3.4  THE MASS SPECTROMETER

The most direct and accurate means for determining atomic and molecular weights is provided by the **mass spectrometer**. This instrument is illustrated schematically in Figure 3.9. A gaseous sample is introduced into the instrument at $A$ and then bombarded by a stream of high-energy electrons at $B$. Collisions between the electrons and the atoms or molecules of the gas produce positive ions. These ions (of mass $m$ and charge $e$) are accelerated toward a negatively charged wire grid ($C$). After they pass through the grid, they encounter two slits, which produce a narrow beam of the ions. This beam of ions is then passed between the poles of a magnet, which forces the ions into a circular path whose radius depends on the charge-to-mass ratio of the ions, $e/m$. Ions of smaller $e/m$ ratio follow a curved path of larger radius than those having a larger $e/m$ ratio. Consequently, ions with equal charges but different masses are separated from each other. By continuously changing either the strength of the magnetic field or the accelerating voltage on the negatively charged grid, ions of varying $e/m$ ratios can be caused to enter the detector at the end of the instrument. Because the ions are principally singly charged, the separation is primarily on the basis of mass. A graph of the intensity of signal from the detector versus the mass of the ion is called a **mass spectrum**.

The mass spectrometer provided the first unambiguous evidence for the existence of isotopes. Suppose we allow a stream of mercury vapor to enter the mass spectrometer. If all the atoms of mercury were in fact identical, all of them upon ionization would possess the same $e/m$ ratio and would therefore appear at the same place in the mass spectrum. But

**FIGURE 3.9**  Schematic diagram of a modern mass spectrometer.

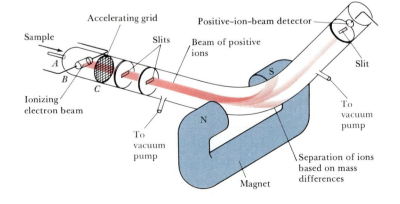

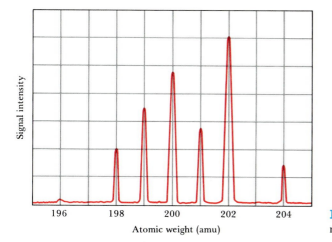

**FIGURE 3.10**   Mass spectrum of $Hg^+$ ions in mercury vapor.

**TABLE 3.3**   Mass spectrum of mercury

| Mass number | Atomic weight (amu) | Fractional abundance |
|---|---|---|
| 196 | 195.965 | 0.0014 |
| 198 | 197.967 | 0.10039 |
| 199 | 198.967 | 0.1683 |
| 200 | 199.968 | 0.2312 |
| 201 | 200.970 | 0.1323 |
| 202 | 201.970 | 0.2979 |
| 204 | 203.973 | 0.0685 |

the mass spectrum actually observed for mercury is as shown in Figure 3.10. It is clear that there are several kinds of atoms of mercury, differing slightly in their masses.

Using the mass spectrometer, it is possible to measure relative values of the ratio $e/m$ for ions with great accuracy. It is also possible to measure with high accuracy the relative numbers of the different isotopes of an element. As we have noted previously, the atomic weight is the weighted average of the masses of all the isotopes of that element. For example, from the mass spectrum of mercury (Figure 3.10), we obtain the data shown in Table 3.3, from which the atomic weight of mercury, 200.59 amu, can be calculated.

# 3.5 THE CHEMICAL MOLE

We indicated earlier that the concept of atomic weights is important because it permits us to count atoms indirectly, by weighing samples. A convenient number of atoms, molecules, or formula units is that number whose mass in grams is equal numerically to the atomic weight, molecular weight, or formula weight. This quantity is called the **mole**, abbreviated mol.* Thus 1 mol of $^{12}C$ atoms is the number of $^{12}C$ atoms in exactly 12 g of $^{12}C$. It has been determined experimentally that the number of atoms in this quantity of $^{12}C$ is $6.022 \times 10^{23}$. This number is given a special name: **Avogadro's number**. For most purposes, we will use $6.02 \times 10^{23}$ for Avogadro's number throughout the text; this number should be committed to memory.

A mole of ions, molecules, or anything else contains Avogadro's number of these objects:

$$1 \text{ mol } ^{12}C \text{ atoms} = 6.02 \times 10^{23} \text{ } ^{12}C \text{ atoms}$$

$$1 \text{ mol } H_2O \text{ molecules} = 6.02 \times 10^{23} \text{ } H_2O \text{ molecules}$$

$$1 \text{ mol } NO_3^- \text{ ions} = 6.02 \times 10^{23} \text{ } NO_3^- \text{ ions}$$

The concept of a mole as being $6.02 \times 10^{23}$ of something is analogous to

* The term *mole* comes from the Latin word *moles* meaning "a mass." The term *molecule* is the diminutive form of this word and means "a small mass."

the concept of a dozen as 12 of something or a gross as 144 of something. Because a $^{24}$Mg atom has twice the mass of a $^{12}$C atom, 1 mol of $^{24}$Mg has twice the mass of 1 mol of $^{12}$C. A $^{12}$C atom has a mass of exactly 12 amu, whereas a $^{24}$Mg atom has a mass of 24.0 amu. Thus 1 mol of $^{12}$C atoms weighs 12 g (by definition), and 1 mol of $^{24}$Mg atoms weighs 24.0 g. In fact, a mole of atoms of any element has a mass in grams numerically equal to the atomic weight of that element:

One $^{12}$C atom has a mass of 12 amu; 1 mol $^{12}$C weighs 12 g (exactly).
One $^{24}$Mg atom has a mass of 24.0 amu; 1 mol $^{24}$Mg weighs 24.0 g.
One Au atom has a mass of 197 amu; 1 mol Au weighs 197 g.

We can generalize this idea to include molecules and ions: The mass of a mole of formula units of any substance (that is, $6.02 \times 10^{23}$ of them) is always equal to the formula weight expressed in grams:

One $H_2O$ molecule has a mass of 18.0 amu; 1 mol $H_2O$ weighs 18.0 g.
One $NO_3^-$ ion has a mass of 62.0 amu; 1 mol $NO_3^-$ weighs 62.0 g.
One NaCl unit has a mass of 58.5 amu; 1 mol NaCl weighs 58.5.

Further examples of mole relationships are shown in Table 3.4.

The first entries in Table 3.4, those for N and $N_2$, point out the importance of stating the chemical form of a substance exactly when we use the mole concept. Suppose you read that 1 mol of nitrogen is produced in a particular reaction. You might interpret this statement to mean 1 mol of nitrogen atoms (14.0 g). Unless otherwise stated, what was probably meant is 1 mol of nitrogen molecules, $N_2$ (28.0 g), because $N_2$ is the usual chemical form of the element. However, to avoid ambiguity it is always best to state explicitly the chemical form being discussed. Using the chemical formula ($N_2$) or referring to the substance as "dinitrogen" avoids ambiguity.

**TABLE 3.4**  Mole relationships

| Name | Formula | Formula weight | Mass of 1 mol of formula units (g) | Number and kind of particles in 1 mol |
|------|---------|----------------|-------------------------------------|----------------------------------------|
| Atomic nitrogen | N | 14.0 | 14.0 | $6.02 \times 10^{23}$ N atoms |
| Molecular nitrogen | $N_2$ | 28.0 | 28.0 | $\begin{cases} 6.02 \times 10^{23} \text{ N}_2 \text{ molecules} \\ 2(6.02 \times 10^{23}) \text{ N atoms} \end{cases}$ |
| Silver | Ag | 107.9 | 107.9 | $6.02 \times 10^{23}$ Ag atoms |
| Silver ions | $Ag^+$ | 107.9[a] | 107.9 | $6.02 \times 10^{23}$ $Ag^+$ ions |
| Barium chloride | $BaCl_2$ | 208.2 | 208.2 | $\begin{cases} 6.02 \times 10^{23} \text{ BaCl}_2 \text{ units} \\ 6.02 \times 10^{23} \text{ Ba}^{2+} \text{ ions} \\ 2(6.02 \times 10^{23}) \text{ Cl}^- \text{ ions} \end{cases}$ |

[a] Recall that the electron has negligible mass; thus ions and atoms have essentially the same mass.

**SAMPLE EXERCISE 3.7**

What is the mass of 1 mol of glucose, $C_6H_{12}O_6$?

**Solution:**  By adding the weights of the atoms in glucose, we find it to have a formula weight of 180 amu:

$$
\begin{aligned}
6\text{C atoms} &= 6(12.0 \text{ amu}) = 72.0 \text{ amu} \\
12\text{H atoms} &= 12(1.0 \text{ amu}) = 12.0 \text{ amu} \\
6\text{O atoms} &= 6(16.0 \text{ amu}) = \underline{96.0 \text{ amu}} \\
& \qquad\qquad\qquad\qquad\quad 180.0 \text{ amu}
\end{aligned}
$$

Hence 1 mol of $C_6H_{12}O_6$ weighs 180 g. Glucose is sometimes called dextrose. Also known as blood sugar, glucose is found widely in nature, occurring, for example, in honey and fruits. Other types of sugars used as food must be converted into glucose in the stomach or liver before they are used by the body as energy sources. Because glucose requires no conversion, it is often given intravenously to patients who need immediate nourishment.

**PRACTICE EXERCISE** —————————
Calculate the mass of 1 mol of $Ca(NO_3)_2$.
*Answer:* 164.1 g

## SAMPLE EXERCISE 3.8

How many C atoms are in 1 mol of $C_6H_{12}O_6$?

$$\text{C atoms} = (1 \text{ mol } C_6H_{12}O_6)$$
$$\times \left( \frac{6.02 \times 10^{23} \text{ molecules}}{1 \text{ mol}} \right) \left( \frac{6C \text{ atoms}}{1 \text{ molecule}} \right)$$
$$= 3.61 \times 10^{24} \text{ C atoms}$$

**Solution:** There are $6.02 \times 10^{23}$ $C_6H_{12}O_6$ molecules in 1 mol. Each molecule contains 6C atoms; hence there are $6(6.02 \times 10^{23})$C atoms:

**PRACTICE EXERCISE** —————————
How many nitrogen atoms are in 1 mol of $Ca(NO_3)_2$?
*Answer:* $1.204 \times 10^{24}$

To illustrate how the mole concept and Avogadro's number allow us to interconvert masses and numbers of particles, let's calculate the number of copper atoms in a traditional copper penny. Such a penny weighs 3 g, and we'll assume that it is 100 percent copper:

$$\text{Cu atoms} = (3 \text{ g Cu}) \left( \frac{1 \text{ mol Cu}}{63.5 \text{ g Cu}} \right) \left( \frac{6.02 \times 10^{23} \text{ Cu atoms}}{1 \text{ mol Cu}} \right)$$
$$= 3 \times 10^{22} \text{ Cu atoms}$$

Notice how we were able to use dimensional analysis (Section 1.5) in a straightforward manner to go from grams to numbers of atoms; the conversion sequence is grams ⟶ moles ⟶ atoms.

We might reflect momentarily on the number of copper atoms in a penny, $3 \times 10^{22}$. This is a tremendously large number. It becomes more impressive when we realize that the entire United States could be covered to a depth of about 30 mi with this number of ice cubes, each 1 in. on an edge. Keep in mind that Avogadro's number is even larger.

## SAMPLE EXERCISE 3.9

How many moles of glucose, $C_6H_{12}O_6$, are in (a) 538 g and (b) 1.00 g of this substance?

**Solution:** (a) One mol of $C_6H_{12}O_6$ weighs 180 g (Sample Exercise 3.7). Therefore, there must be more than 1 mol in 538 g.

Moles $C_6H_{12}O_6$

$$= (538 \text{ g } C_6H_{12}O_6) \left( \frac{1 \text{ mol } C_6H_{12}O_6}{180 \text{ g } C_6H_{12}O_6} \right)$$
$$= 2.99 \text{ mol}$$

(b) In this case there must be less than 1 mol.

Moles $C_6H_{12}O_6$
$$= (1.00 \text{ g } C_6H_{12}O_6) \left( \frac{1 \text{ mol } C_6H_{12}O_6}{180 \text{ g } C_6H_{12}O_6} \right)$$
$$= 5.56 \times 10^{-3} \text{ mol}$$

The conversion of mass to moles and of moles to mass is frequently encountered in calculations using the mole concept. Notice that the number of moles is always the mass divided by the mass of 1 mol (the formula weight expressed in grams).

**PRACTICE EXERCISE** —————————
How many moles of $NaHCO_3$ are present in 5.08 g of this substance?
*Answer:* 0.0605 mol

## SAMPLE EXERCISE 3.10

What is the mass, in grams, of 0.433 mol of $C_6H_{12}O_6$?

**Solution:**  Because this is less than 1 mol, the mass will be less than 180 g, the mass of 1 mol.

Grams $C_6H_{12}O_6$

$$= (0.433 \text{ mol } C_6H_{12}O_6)\left(\frac{180 \text{ g } C_6H_{12}O_6}{1 \text{ mol } C_6H_{12}O_6}\right)$$

$$= 77.9 \text{ g}$$

Notice that the mass of a certain number of moles of a substance is always the number of moles times the mass of 1 mol.

**PRACTICE EXERCISE** _____
What is the mass, in grams, of 6.33 mol of $NaHCO_3$?
*Answer:*  532 g

## SAMPLE EXERCISE 3.11

How many glucose molecules are in 5.23 g of $C_6H_{12}O_6$?

**Solution:**  In this case we need to carry out unit conversions in the order grams ⟶ moles ⟶ molecules. Because the mass we begin with corresponds to less than a mole, there should be fewer than $6.02 \times 10^{23}$ molecules.

Molecules $C_6H_{12}O_6 =$

$$(5.23 \text{ g } C_6H_{12}O_6)\left(\frac{1 \text{ mol } C_6H_{12}O_6}{180 \text{ g } C_6H_{12}O_6}\right)$$

$$\times \left(\frac{6.02 \times 10^{23} \text{ } C_6H_{12}O_6 \text{ molecules}}{1 \text{ mol } C_6H_{12}O_6}\right)$$

$$= 1.75 \times 10^{22} \text{ molecules}$$

**PRACTICE EXERCISE** _____
How many oxygen atoms are present in 4.20 g of $NaHCO_3$?
*Answer:*  $9.03 \times 10^{22}$

## 3.6 EMPIRICAL FORMULAS FROM ANALYSES

The empirical formula for a substance tells us the relative number of atoms of each element it contains. Thus the formula $H_2O$ indicates that water contains two H atoms for each O atom. The ratio of the number of atoms of each element in a substance is also the ratio of the number of moles of these elements. Thus $H_2O$ contains 2 mol of H atoms for each mole of O atoms. Conversely, the ratio of the number of moles of each element in a compound gives the subscripts in a compound's empirical formula. Thus the mole concept provides a way of calculating the empirical formulas of chemical substances, as shown in the following examples.

Mercury forms a compound with chlorine that is 73.9 percent mercury and 26.1 percent chlorine by mass. This means that if we had a 100-g sample of the solid, the sample would contain 73.9 g of mercury (Hg) and 26.1 g of chlorine (Cl). We divide each of these weights by the appropriate atomic weight to obtain the number of moles of each element in 100 g:

$$73.9 \text{ g Hg}\left(\frac{1 \text{ mol Hg}}{200.6 \text{ g Hg}}\right) = 0.368 \text{ mol Hg}$$

$$26.1 \text{ g Cl}\left(\frac{1 \text{ mol Cl}}{35.5 \text{ g Cl}}\right) = 0.735 \text{ mol Cl}$$

We then divide the larger number of moles by the smaller to obtain the ratio 1.99 mole of Cl/mole of Hg. Because of experimental errors, the results of an analysis may not lead to exact integers for the ratios of moles. The ratio obtained in this case is very close to 2; the formula for the compound is thus $HgCl_2$. This is the simplest, or empirical, formula because it uses as subscripts the smallest set of integers that express the correct ratios of atoms present (Section 2.5).

---

**SAMPLE EXERCISE 3.12**

Phosgene, a poison gas used during World War I, contains 12.1 percent C, 16.2 percent O, and 71.7 percent Cl. What is the empirical formula of phosgene?

**Solution:** For simplicity, we may assume that we have 100 g of material (although any number can be used). The number of moles of each element is then

$$\text{Moles C} = (12.1 \text{ g})\left(\frac{1 \text{ mol C}}{12.0 \text{ g}}\right) = 1.01 \text{ mol C}$$

$$\text{Moles O} = (16.2 \text{ g})\left(\frac{1 \text{ mol O}}{16.0 \text{ g}}\right) = 1.01 \text{ mol O}$$

$$\text{Moles Cl} = (71.7 \text{ g})\left(\frac{1 \text{ mol Cl}}{35.5 \text{ g}}\right) = 2.02 \text{ mol Cl}$$

The simplest ratio, found by dividing each number by the smallest, 1.01, is $C:O:Cl = 1:1:2$ and the empirical formula is $COCl_2$. Because other experiments show that the molecular weight of the phosgene molecule is 99 amu, $COCl_2$ is also the molecular formula.

**PRACTICE EXERCISE** _____

A 5.325-g sample of methyl benzoate, a compound used in the manufacture of perfumes, is found to contain 3.758 g of carbon, 0.316 g of hydrogen, and 1.251 g of oxygen. What is the empirical formula of this substance?
*Answer:* $C_4H_4O$

---

Remember that the formula determined from percentage compositions is always the empirical formula. To obtain the molecular formula from the empirical formula we must know the molecular weight of the compound. In the practice exercise above, the empirical formula of the substance is $C_4H_4O$, which has a formula weight of 68.0 amu. The experimentally determined molecular weight of the compound is 136.0 amu. Because the molecular weight is twice the formula weight, the subscripts in the molecular formula must be twice those in the empirical formula: $C_8H_8O_2$.

When a compound containing carbon and hydrogen is combusted in an apparatus such as that shown in Figure 3.11, the carbon of the compound is converted to $CO_2$, and all the hydrogen to $H_2O$. The amount of $CO_2$ produced can be measured by determining the mass increase in the $CO_2$ absorber. Similarly, the amount of $H_2O$ formed is determined from the

**Combustion Analysis**

**FIGURE 3.11** Apparatus for determining the percentages of carbon and hydrogen in a compound. Copper oxide serves to oxidize traces of carbon and carbon monoxide to carbon dioxide, and to oxidize hydrogen to water. Magnesium perchlorate, $Mg(ClO_4)_2$, is used to absorb water, whereas sodium hydroxide, NaOH, absorbs carbon dioxide.

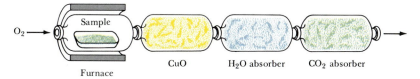

O₂ →  | Sample | CuO | H₂O absorber | CO₂ absorber | →
Furnace

increase in mass of the water absorption tube. As an example, let's consider an analysis of a sample of ascorbic acid (vitamin C). Combustion of 1.000 g of ascorbic acid produces 1.500 g of $CO_2$ and 0.405 g of $H_2O$. From these two bits of experimental information we must calculate the quantities of C and H in the 1.000-g sample of ascorbic acid:

$$(1.500 \text{ g CO}_2)\left(\frac{1 \text{ mol CO}_2}{44.00 \text{ g CO}_2}\right)\left(\frac{1 \text{ mol C}}{1 \text{ mol CO}_2}\right)\left(\frac{12.01 \text{ g C}}{1 \text{ mol C}}\right) = 0.409 \text{ g C}$$

$$(0.405 \text{ g H}_2\text{O})\left(\frac{1 \text{ mol H}_2\text{O}}{18.0 \text{ g H}_2\text{O}}\right)\left(\frac{2 \text{ mol H}}{1 \text{ mol H}_2\text{O}}\right)\left(\frac{1.01 \text{ g H}}{1 \text{ mol H}}\right) = 0.045 \text{ g H}$$

Other experiments establish that ascorbic acid contains only C, H, and O. Thus the amount of oxygen in the compound must be

$$1.000 \text{ g} - (0.409 \text{ g} + 0.045 \text{ g}) = 0.546 \text{ g}$$

From these data, we can now proceed to calculate the number of moles of each element present in 1 g of ascorbic acid:

$$\text{Moles C} = (0.409 \text{ g C})\left(\frac{1 \text{ mol C}}{12.0 \text{ g C}}\right) = 0.0341 \text{ mol C}$$

$$\text{Moles H} = (0.045 \text{ g H})\left(\frac{1 \text{ mol H}}{1.01 \text{ g H}}\right) = 0.045 \text{ mol H}$$

$$\text{Moles O} = (0.546 \text{ g O})\left(\frac{1 \text{ mol O}}{16.0 \text{ g O}}\right) = 0.0341 \text{ mol O}$$

The relative number of moles of each element can be found by dividing each number by the smallest number, 0.0341. The ratio of C:H:O so obtained is 1:1.32:1. The second number is too far from 1 to attribute the difference to experimental error; in fact, it is quite close to $1\frac{1}{3}$. The fraction can be cleared by multiplying by 3. Using this factor on each member of the ratio gives 3:3.96:3, which we may take to be 3:4:3. Thus the empirical formula is $C_3H_4O_3$.

The experimentally determined molecular weight of ascorbic acid is 176 amu. Using the molecular weight and the empirical formula, we can determine the molecular formula. The subscripts in the molecular formula of a substance are always a whole-number multiple of those in its empirical formula (Section 2.5). The formula weight associated with the empirical formula of ascorbic acid, $C_3H_4O_3$, is 3(12.0 amu) + 4(1.0 amu) + 3(16.0 amu) = 88.0 amu. The molecular weight is twice this value: 176/88.0 = 2.00. Thus the subscripts in the empirical formula must be multiplied by 2 to obtain the molecular formula, $C_6H_8O_6$.

## 3.7 QUANTITATIVE INFORMATION FROM BALANCED EQUATIONS

The mole concept provides a key to placing the quantitative information available in a balanced chemical equation on a practical, macroscopic level. Consider the following balanced equation:

$$2H_2(g) + O_2(g) \longrightarrow 2H_2O(l) \qquad \text{[3.19]}$$

The coefficients tell us that two molecules of $H_2$ react with each mole-

**TABLE 3.5**   Interpretations of equations

| Ratio | 2H$_2$ | + | O$_2$ | $\longrightarrow$ | 2H$_2$O |
|---|---|---|---|---|---|
| Molecular | 2 molecules | React with | 1 molecule | To form | 2 molecules |
|  | 2(6.02 × 10$^{23}$) | React with | 6.02 × 10$^{23}$ | To form | 2(6.02 × 10$^{23}$) |
|  | molecules |  | molecules |  | molecules |
| Mole | 2 mol | React with | 1 mol | To form | 2 mol |
| Weight | 2(2.02) g = 4.04 g | React with | 32.00 g | To form | 2(18.02) g = 36.04 g |

cule of O$_2$ to form two molecules of H$_2$O. Therefore, 2(6.02 × 10$^{23}$) molecules of H$_2$ will react with 6.02 × 10$^{23}$ molecules of O$_2$ to form 2(6.02 × 10$^{23}$) molecules of H$_2$O. This is the same as saying that 2 mol of H$_2$ react with 1 mol of O$_2$ to form 2 mol of H$_2$O. The point is that the coefficients in a balanced equation can be interpreted *both* as the *relative numbers of molecules* (or formula units) involved in a reaction *and* as the *relative number of moles*. These interpretations are summarized in Table 3.5. Notice that 4.04 g of H$_2$ will react with each 32.00 g of O$_2$ to form 36.04 g of H$_2$O, because these are the masses of 2 mol of H$_2$, 1 mol of O$_2$, and 2 mol of H$_2$O, respectively. Notice also that the sum of the masses of the reactants equals the mass of the product as it must in any chemical reaction according to the law of conservation of mass. The quantities 2 mol of H$_2$, 1 mol of O$_2$, and 2 mol of H$_2$O, which are related by Equation 3.19, are called stoichiometrically equivalent quantities. The relationship between these quantities can be represented as follows:

$$2 \text{ mol H}_2 \simeq 1 \text{ mol O}_2 \simeq 2 \text{ mol H}_2\text{O}$$

where the symbol $\simeq$ is taken to mean "stoichiometrically equivalent to." These stoichiometric relations can be used to give conversion factors to relate quantities of reactants and products in a chemical reaction. For example, the number of moles of H$_2$O produced from 1.57 mol of O$_2$ can be calculated as follows:

$$\text{Moles H}_2\text{O} = (1.57 \text{ mol O}_2)\left(\frac{2 \text{ mol H}_2\text{O}}{1 \text{ mol O}_2}\right)$$
$$= 3.14 \text{ mol H}_2\text{O}$$

As an additional example, consider the following reaction:

$$2\text{CuFeS}_2(s) + 5\text{O}_2(g) \longrightarrow 2\text{Cu}(s) + 2\text{FeO}(s) + 4\text{SO}_2(g) \quad \text{[3.20]}$$

This equation describes a process in the smelting of copper using chalcopyrite (CuFeS$_2$) as the mineral source of the copper. Using the mole concept we can calculate the mass of Cu that can be produced from 1.00 g of chalcopyrite. From the coefficients in Equation 3.20 we can write the following stoichiometric relationship: 2 mol CuFeS$_2 \simeq$ 2 mol Cu. In order to use this relationship, however, the quantity of CuFeS$_2$ must be converted to moles. Because 1 mol CuFeS$_2$ = 183 g CuFeS$_2$, we have

$$\text{Moles CuFeS}_2 = (1.00 \text{ g CuFeS}_2)\left(\frac{1 \text{ mol CuFeS}_2}{183 \text{ g CuFeS}_2}\right)$$
$$= 5.46 \times 10^{-3} \text{ mol CuFeS}_2$$

The stoichiometric factor from the balanced equation, 2 mol $CuFeS_2 \rightleftharpoons$ 2 mol Cu, can then be used to calculate moles of Cu:

$$\text{Moles Cu} = (5.46 \times 10^{-3} \text{ mol } CuFeS_2)\left(\frac{2 \text{ mol Cu}}{2 \text{ mol } CuFeS_2}\right)$$
$$= 5.46 \times 10^{-3} \text{ mol Cu}$$

Finally, the mass of the Cu, in grams, can be calculated using the relationship 1 mol Cu = 63.5 g Cu:

$$\text{Grams Cu} = (5.46 \times 10^{-3} \text{ mol Cu})\left(\frac{63.5 \text{ g Cu}}{1 \text{ mol Cu}}\right)$$
$$= 0.347 \text{ g Cu}$$

These steps, of course, can be combined in a single sequence of factors:

$$\text{Grams Cu} = (1.00 \text{ g } CuFeS_2)\left(\frac{1 \text{ mol } CuFeS_2}{183 \text{ g } CuFeS_2}\right)\left(\frac{2 \text{ mol Cu}}{2 \text{ mol } CuFeS_2}\right)\left(\frac{63.5 \text{ g Cu}}{1 \text{ mol Cu}}\right)$$
$$= 0.347 \text{ g Cu}$$

We can similarly calculate the amount of $SO_2$ produced in the production of this quantity of copper using the relationship 2 mol $CuFeS_2 \rightleftharpoons$ 4 mol $SO_2$, which also comes from the coefficients in Equation 3.20:

$$\text{Grams } SO_2 = (1.00 \text{ g } CuFeS_2)\left(\frac{1 \text{ mol } CuFeS_2}{183 \text{ g } CuFeS_2}\right)\left(\frac{4 \text{ mol } SO_2}{2 \text{ mol } CuFeS_2}\right)\left(\frac{64.1 \text{ g } SO_2}{1 \text{ mol } SO_2}\right)$$
$$= 0.701 \text{ g } SO_2$$

It is interesting to note that the mass of $SO_2$ produced in this reaction is approximately twice the mass of the copper. Consequently, considerable air pollution from sulfur dioxide is generated in the vicinity of copper smelters (Figure 3.12).

**FIGURE 3.12** Extended exposure to high concentrations of $SO_2$ and other pollutants can cause extensive damage to plants, animals, and structural materials. This photograph, taken at Queenstown, Tasmania, Australia, shows the hills denuded of vegetation by pollution produced by an ore-smelting operation. (Bill Bachman/Photo Researchers)

## SAMPLE EXERCISE 3.13

How much water is produced in the combustion of 1.00 g of glucose, $C_6H_{12}O_6$:

$$C_6H_{12}O_6(s) + 6O_2(g) \longrightarrow 6CO_2(g) + 6H_2O(l)$$

**Solution:** The chemical equation indicates how much $H_2O$ is produced from $C_6H_{12}O_6$: 1 mol $C_6H_{12}O_6 \simeq 6$ mol $H_2O$. To use this information, we must convert the amount of $C_6H_{12}O_6$ from grams to moles. Thus the problem can be solved by a stepwise conversion of grams of $C_6H_{12}O_6$ to moles of $C_6H_{12}O_6$ to moles of $H_2O$ to grams of $H_2O$:

$$\text{Grams } H_2O = (1.00 \text{ g } C_6H_{12}O_6)\left(\frac{1 \text{ mol } C_6H_{12}O_6}{180 \text{ g } C_6H_{12}O_6}\right)$$
$$\times \left(\frac{6 \text{ mol } H_2O}{1 \text{ mol } C_6H_{12}O_6}\right)\left(\frac{18.0 \text{ g } H_2O}{1 \text{ mol } H_2O}\right)$$
$$= 0.600 \text{ g } H_2O$$

We may note that an average person ingests 2 L of water daily and eliminates 2.4 L. The difference is produced in metabolism of foodstuffs. (Metabolism is a general term used to describe all the chemical processes of a living animal or plant.) The desert rat (kangaroo rat) is able to take great advantage of its metabolic water to help it survive in the dry desert. In fact, it apparently never drinks water.

### PRACTICE EXERCISE

A common laboratory method for preparing small amounts of $O_2$ involves the decomposition of $KClO_3$: $2KClO_3(s) \longrightarrow 2KCl(s) + 3O_2(g)$. How many grams of $O_2$ can be prepared from 4.50 g of $KClO_3$?
*Answer:* 1.77 g

## SAMPLE EXERCISE 3.14

Solid lithium hydroxide is used in space vehicles to remove exhaled carbon dioxide. The lithium hydroxide reacts with gaseous carbon dioxide to form solid lithium carbonate and liquid water. How many grams of carbon dioxide can be absorbed by each 1.00 g of lithium hydroxide?

**Solution:** Using the verbal description of the reaction we can write the unbalanced chemical equation:

$$LiOH(s) + CO_2(g) \longrightarrow Li_2CO_3(s) + H_2O(l)$$

The balanced equation is

$$2LiOH(s) + CO_2(g) \longrightarrow Li_2CO_3(s) + H_2O(l)$$

The problem can be solved by a stepwise conversion of grams of LiOH to moles of LiOH to moles of $CO_2$ to grams of $CO_2$. The conversion from grams of LiOH to moles of LiOH requires this substance's

formula weight $(6.94 + 16.00 + 1.01 = 23.95)$. The conversion of moles of LiOH to moles of $CO_2$ is based on the balanced chemical equation: 2 mol LiOH $\simeq$ 1 mol $CO_2$. To convert the number of moles of $CO_2$ to grams, we must use the formula weight of $CO_2$: $12.01 + 2(16.00) = 44.01$:

$$(1.00 \text{ g LiOH})\left(\frac{1 \text{ mole LiOH}}{23.95 \text{ g LiOH}}\right)\left(\frac{1 \text{ mol } CO_2}{2 \text{ mol LiOH}}\right)$$
$$\times \left(\frac{44.01 \text{ g } CO_2}{1 \text{ mol } CO_2}\right) = 0.919 \text{ g } CO_2$$

### PRACTICE EXERCISE

Propane, $C_3H_8$, is a common fuel used for cooking and home heating. What mass of $O_2$ is consumed in the combustion of 1.00 g of propane?
*Answer:* 3.64 g

In many situations an excess of one or more substances is available for chemical reaction. Some will therefore be left over when the reaction is complete; the reaction stops as soon as one of the reactants is totally consumed. Before we examine a chemical example, consider a simple analogy: Suppose that you have 20 slices of bread and 15 pieces of bologna. You wish to make as many sandwiches as possible using only one slice of bologna and two slices of bread per sandwich. You will be able

## 3.8 LIMITING REACTANT

The quantity of product that is calculated to form when all of the limiting reagent reacts is called the *theoretical yield*. The amount of product actually obtained in a reaction is called the *actual yield*. The actual yield is almost always less than the theoretical yield. There are many reasons for this difference. For example, part of the reactants may not react or they may react in a way different from that desired (side reactions). In addition, it is not always possible to recover all of the reaction product from the reaction mixture. The *percent yield* of a reaction relates the actual yield to the theoretical (calculated) yield:

$$\text{Percent yield} = \frac{\text{actual yield}}{\text{theoretical yield}} \times 100$$

For example, in the experiment described in Sample Exercise 3.16, we calculate that 4.92 g of $Ba_3(PO_4)_2$ should form when 3.50 g of $Na_3PO_4$ is mixed with 6.40 g of $Ba(NO_3)_2$. This is the theoretical yield of $Ba_3(PO_4)_2$ in the reaction. If the actual yield turned out to be 4.70 g, the percent yield would be

$$\frac{4.70 \text{ g}}{4.92 \text{ g}} \times 100 = 95.5\%.$$

to make only 10 sandwiches, at which point you run out of bread; at that point you have 5 slices of bologna left over.

Now consider the combustion of hydrogen:

$$2H_2(g) + O_2(g) \longrightarrow 2H_2O(l)$$

Suppose this reaction is initiated by passing a spark through a reaction vessel containing 2 mol of $H_2$ and 2 mol of $O_2$. The balanced equation tells us that only 1 mol of $O_2$ is required to completely consume 2 mol of $H_2$, thereby forming 2 mol of $H_2O$. Once all of the $H_2$ is consumed, the reaction stops. Thus 1 mol of $O_2$ is left over at the end of the reaction. This situation is represented schematically in Figure 3.13. The substance that is completely consumed in a reaction is called the **limiting reactant** or **limiting reagent**, because it determines, or limits, the amount of product formed. In our present example, $H_2$ is the limiting reactant.

## SAMPLE EXERCISE 3.15

Part of the $SO_2$ that is introduced into the atmosphere ends up being converted to sulfuric acid, $H_2SO_4$. The net reation is

$$2SO_2(g) + O_2(g) + 2H_2O(l) \longrightarrow 2H_2SO_4(l)$$

How much $H_2SO_4$ can be prepared from 5.0 mol of $SO_2$, 1.0 mol of $O_2$, and an unlimited quantity of $H_2O$?

**Solution:** The number of moles of $O_2$ needed for complete consumption of 5.0 mol of $SO_2$ is

$$\text{Moles } O_2 = (5.0 \text{ mol } SO_2)\left(\frac{1 \text{ mol } O_2}{2 \text{ mol } SO_2}\right)$$

$$= 2.5 \text{ mol } O_2$$

This quantity of $O_2$ is not available; therefore, all of the $SO_2$ cannot be consumed; $O_2$ must be the

limiting reactant. We use the quantity of the limiting reagent, $O_2$, to calculate the quantity of $H_2SO_4$ prepared:

$$\text{Moles } H_2SO_4 = (1.0 \text{ mol } O_2)\left(\frac{2 \text{ mol } H_2SO_4}{1 \text{ mol } O_2}\right)$$

$$= 2.0 \text{ mol } H_2SO_4$$

We might note that in forming 2.0 mol of $H_2SO_4$, 2.0 mol of $SO_2$ are required. Therefore, 3.0 mol of $SO_2$ is left over.

## PRACTICE EXERCISE

Consider the reaction $2Al(s) + 3Cl_2(g) \longrightarrow Al_2Cl_6(s)$. 1.5 mol of Al and 3.0 mol of $Cl_2$ are caused to react. (a) What is the limiting reactant? (b) How many moles of $Al_2Cl_6$ are formed?
*Answers:* (a) Al; (b) 0.75 mol

Before reaction

After reaction

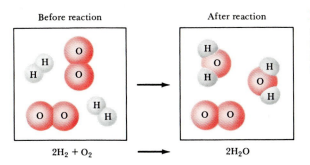

$2H_2 + O_2 \longrightarrow 2H_2O$

**FIGURE 3.13**  Diagram showing the complete utilization of a limiting reagent in a reaction. Because the $H_2$ is completely consumed, it is the limiting reagent in this case. Because there is a stoichiometric excess of $O_2$, some is left over at the end of the reaction.

Another approach to the problem of the limiting reactant is to calculate the amount of product that could be formed from each of the given amounts of reactants, assuming they were all completely consumed. The reagent that leads to the smallest amount of product is the limiting reactant.

### SAMPLE EXERCISE 3.16

Consider the precipitation reaction described in Sample Exercise 3.3. Suppose that a solution containing 3.50 g of $Na_3PO_4$ were mixed with a solution containing 6.40 g of $Ba(NO_3)_2$. How many grams of $Ba_3(PO_4)_2$ can be formed?

**Solution:**  The balanced equation for the reaction is

$$2Na_3PO_4(aq) + 3Ba(NO_3)_2(aq) \longrightarrow$$
$$Ba_3(PO_4)_2(aq) + 6NaNO_3(aq)$$

From this equation we have the following stoichiometric relations:

$$2 \text{ mol } Na_3PO_4 \simeq 3 \text{ mol } Ba(NO_3)_2$$
$$\simeq 1 \text{ mol } Ba_3(PO_4)_2$$

The mass of 1 mol of each substance can be found by determining the formula weight for each substance. The results are as follows:

$$1 \text{ mol } Na_3PO_4 = 164 \text{ g } Na_3PO_4$$
$$1 \text{ mol } Ba(NO_3)_2 = 261 \text{ g } Ba(NO_3)_2$$
$$1 \text{ mol } Ba_3(PO_4)_2 = 602 \text{ g } Ba_3(PO_4)_2$$

Let us now calculate the amount of product that could be formed from each of our given amounts of reactants, assuming that each in turn is the limiting reactant. Assuming first that $Na_3PO_4$ is completely consumed gives:

Grams $Ba_3(PO_4)_2$

$$= (3.50 \text{ g } Na_3PO_4)\left(\frac{1 \text{ mol } Na_3PO_4}{164 \text{ g } Na_3PO_4}\right)$$
$$\times \left(\frac{1 \text{ mol } Ba_3(PO_4)_2}{2 \text{ mol } Na_3PO_4}\right)\left(\frac{602 \text{ g } Ba_3(PO_4)_2}{1 \text{ mol } Ba_3(PO_4)_2}\right)$$
$$= 6.42 \text{ g } Ba_3(PO_4)_2$$

Assuming that $Ba(NO_3)_2$ is completely consumed gives:

Grams $Ba_3(PO_4)_2$

$$= (6.40 \text{ g } Ba(NO_3)_2)\left(\frac{1 \text{ mol } Ba(NO_3)_2}{261 \text{ g } Ba(NO_3)_2}\right)$$
$$\times \left(\frac{1 \text{ mol } Ba_3(PO_4)_2}{3 \text{ mol } Ba(NO_3)_2}\right)\left(\frac{602 \text{ g } Ba_3(PO_4)_2}{1 \text{ mol } Ba_3(PO_4)_2}\right)$$
$$= 4.92 \text{ g } Ba_3(PO_4)_2$$

The lesser quantity, 4.92 g, is the amount of $Ba_3(PO_4)_2$ that will form, indicating that $Ba(NO_3)_2$ is the limiting reactant.

### PRACTICE EXERCISE

A strip of zinc metal weighing 2.00 g is placed in an aqueous solution containing 2.50 g of silver nitrate, causing the following reaction to occur:

$$Zn(s) + 2AgNO_3(aq) \longrightarrow 2Ag(s) + Zn(NO_3)_2(aq)$$

How many grams of Ag will form?
***Answer:***  1.59 g

## 3.9 MOLARITY AND SOLUTION STOICHIOMETRY

We have seen how to use the concept of the mole to determine the quantities of substances that react with one another or to determine the quantity of product that can be obtained from a given mass of reactant. However, often chemicals are employed in the form of solutions, particularly aqueous solutions. For example, the reaction described in Sample Exercise 3.16 involves aqueous solutions. It is convenient to measure out quantities of dissolved substances by measuring out volumes of the solutions, rather than by weighing out solids or liquids each time and then dissolving these in water.

In discussing solutions we often call one component the **solvent** and the others **solutes**. The component of a solution whose physical state is preserved when the solution is formed is known as the solvent. For example, when sodium chloride (a solid) is mixed with water, the resultant solution is a liquid. Consequently, water is referred to as the solvent and sodium chloride as the solute. If all components of a solution are in the same state, the one present in greatest amount is called the solvent.

The term **concentration** is used to designate the amount of solute dissolved in a given quantity of solvent or solution. The method for expressing concentration that is most useful for discussing solution stoichiometry is **molarity**. The molarity (symbol $M$) of a solution is defined as the number of moles of solute in a liter of solution (soln):

$$\text{Molarity} = \frac{\text{moles solute}}{\text{volume of soln in liters}} \qquad [3.21]$$

A 1.00 molar solution (written 1.00 $M$) contains 1.00 mol of solute in every liter of solution. Figure 3.14 shows the preparation of 250 mL of a 1.00 $M$ solution of $CuSO_4$ in a 250 mL volumetric flask. First 0.250 mol of $CuSO_4$ (39.9 g) is weighed out and placed in the volumetric flask. Water is added to dissolve the salt, and the resultant solution is diluted to a total volume of 250 mL.

**FIGURE 3.14**  Procedure for preparation of 0.250 L of 1.00 $M$ solution of $CuSO_4$. (*a*) Weigh out 0.250 mol (39.9 g) of $CuSO_4$ (formula weight = 159.6 amu). (*b*) Put the $CuSO_4$ (solute) into a 250-mL volumetric flask, and add a small quantity of water. (*c*) Dissolve the solute by swirling the stoppered flask. (*d*) Add more water until the solution just reaches the calibration mark etched on the neck of the flask. Shake the stoppered flask to assure complete mixing (Donald Clegg and Roxy Wilson)

(*a*)          (*b*)          (*c*)          (*d*)

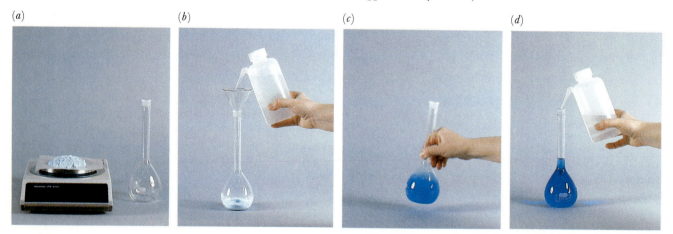

## SAMPLE EXERCISE 3.17

Calculate the molarity of a solution made by dissolving 23.4 g of sodium sulfate ($Na_2SO_4$) in enough water to form 125 mL of solution.

**Solution:**

$$\text{Molarity} = \frac{\text{moles } Na_2SO_4}{\text{liters soln}}$$

$$\text{Moles } Na_2SO_4 = (23.4 \text{ g } Na_2SO_4)\left(\frac{1 \text{ mol } Na_2SO_4}{142 \text{ g } Na_2SO_4}\right)$$

$$= 0.165 \text{ mol } Na_2SO_4$$

$$\text{Liters soln} = (125 \text{ mL})\left(\frac{1 \text{ L}}{1000 \text{ mL}}\right) = 0.125 \text{ L}$$

$$\text{Molarity} = \frac{0.165 \text{ mol } Na_2SO_4}{0.125 \text{ L soln}}$$

$$= 1.32 \frac{\text{mol } Na_2SO_4}{\text{L soln}}$$

$$= 1.32 \ M$$

**PRACTICE EXERCISE**

Calculate the molarity of a solution made by dissolving 5.00 g of $C_6H_{12}O_6$ (MW = 180 amu) in sufficient water to form 100 mL of solution.

***Answer:*** 0.278 *M*

The expression for concentration in molarity allows us to measure out a solution volume of known concentration and readily calculate the number of moles of solute dispensed. Molarity can be used to interconvert volume and moles just as density can be used to interconvert volume and mass (Sample Exercise 1.11). Calculation of the number of moles of $HNO_3$ in 2.0 L of 0.200 *M* $HNO_3$ solution illustrates conversion of volume to moles:

$$\text{Moles } HNO_3 = (2.0 \text{ L soln})\left(\frac{0.200 \text{ mol } HNO_3}{1 \text{ L soln}}\right)$$

$$= 0.40 \text{ mol } HNO_3$$

Notice how dimensional analysis can be used in this conversion if we express molarity as moles $HNO_3$/liters soln. Notice also that to obtain moles we multiplied liters and molarity: moles = liters × *M*. This same expression for moles can be obtained directly by algebraic rearrangement of Equation 3.21. The use of molarity to convert moles to volume can be illustrated by calculating the volume of 0.30 *M* $HNO_3$ solution required to supply 2.0 mol of $HNO_3$:

$$\text{Liters soln} = (2.0 \text{ mol } HNO_3)\left(\frac{1 \text{ L soln}}{0.30 \text{ mol } HNO_3}\right)$$

$$= 6.7 \text{ L soln}$$

In this case we needed to apply the reciprocal of molarity to convert moles to volume: liters = moles × 1/*M*.

## SAMPLE EXERCISE 3.18

How many grams of $Na_2SO_4$ are required to make 350 mL of 0.50 *M* $Na_2SO_4$?

**Solution:**

$$M \ Na_2SO_4 = \frac{\text{moles } Na_2SO_4}{\text{liters soln}}$$

thus

$$\text{moles } Na_2SO_4 = \text{liters soln} \times M \ Na_2SO_4$$

$$= (0.350 \text{ L soln})\left(\frac{0.50 \text{ mol } Na_2SO_4}{\text{L soln}}\right)$$

$$= 0.175 \text{ mol } Na_2SO_4$$

Because each mole of $Na_2SO_4$ weighs 142 g, the required number of grams of $Na_2SO_4$ is

$$(0.175 \text{ mol } Na_2SO_4)\left(\frac{142 \text{ g } Na_2SO_4}{1 \text{ mol } Na_2SO_4}\right)$$
$$= 24.8 \text{ g } Na_2SO_4$$

**PRACTICE EXERCISE**

(a) How many grams of $Na_2SO_4$ are there in 15 mL of 0.50 $M$ $Na_2SO_4$? (b) How many mL of 0.50 $M$ $Na_2SO_4$ solution is required to supply 0.035 mol of this salt?
*Answers:* (a) 1.1 g; (b) 70 mL

## Dilution

A common method for preparing solutions of specific concentration is to start with an available solution of higher concentration and to add solvent to dilute it.* For example, you could prepare a 0.10 $M$ $CuSO_4$ solution by diluting some 1.0 $M$ $CuSO_4$.

When solvent is added to dilute a solution, the number of moles of solute remains unchanged:

Moles solute before dilution = moles solute after dilution     [3.22]

Because number of moles = $M$ × liters, we can write

(Initial molarity)(initial volume) = (final molarity)(final volume)
$$M_{\text{initial}}V_{\text{initial}} = M_{\text{final}}V_{\text{final}}$$     [3.23]

* In diluting an acid or base, the acid or base should be added to water, then further diluted by addition of more water. Adding water directly to concentrated acid or base can cause spattering because of the intense heat generated.

**FIGURE 3.15**     Procedure for preparation of 250 mL of 0.10 $M$ $CuSO_4$ by dilution of 1.0 $M$ $CuSO_4$. (a) Draw 25 mL of the 1.0 $M$ solution into a pipet. (b) Add this amount to a 250-mL volumetric flask. (c) Add water to dilute the solution to a total volume of 250 mL (Donald Clegg and Roxy Wilson)

(a)                    (b)                    (c)

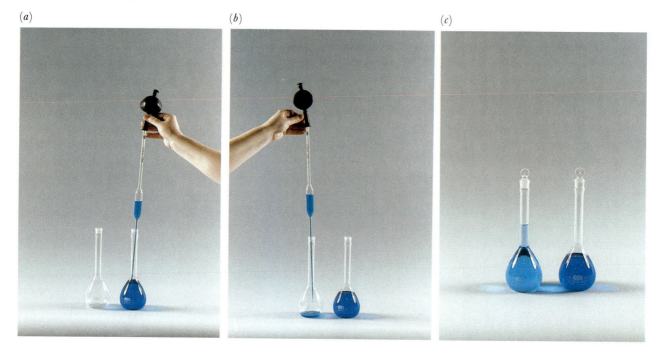

Suppose you had to prepare 250 mL of 0.10 $M$ $CuSO_4$ solution by diluting some 1.0 $M$ $CuSO_4$. In this case:

$$V_{initial} = \frac{M_{final}V_{final}}{M_{initial}} = \frac{(0.10\ M)(250\ mL)}{1.0\ M} = 25\ mL$$

Thus this dilution is achieved by withdrawing 25 mL of the 1.0 $M$ solution using a pipet, adding it to a 250-mL volumetric flask, and then diluting it to 250 mL, as shown in Figure 3.15.

## SAMPLE EXERCISE 3.19

How much 3.0 $M$ $H_2SO_4$ would be required to make 500 mL of 0.10 $M$ $H_2SO_4$?

**Solution:**  Using Equation 3.23, $M_{initial}V_{initial} = M_{final}V_{final}$, we can write

$$V_{initial} = \frac{M_{final}V_{final}}{M_{initial}}$$
$$= \frac{(0.10\ M)(500\ mL)}{3.0\ M} = 17\ mL$$

We see that if we start with 17 mL of 3.0 $M$ $H_2SO_4$ and dilute it to a total volume of 500 mL, the desired 0.10 $M$ solution will be obtained.

## PRACTICE EXERCISE

How many mL of 5.0 $M$ $K_2Cr_2O_7$ solution must be diluted in order to prepare 250 mL of 0.10 $M$ solution?
*Answer:*  5.0 mL

We are often faced with situations in which we wish to know the concentration of a solution. If we have a second solution of known concentration that undergoes chemical reaction with the first, we can use that reaction as a means of determining the unknown concentration. This procedure is known as a **titration**. As an example, suppose we have an HCl solution of unknown concentration and an NaOH solution that we know to be 0.100 $M$. The solution of known concentration is referred to as a **standard solution**. To determine the concentration of the HCl solution, we would take a specific volume of that solution, say 20.00 mL. We would then slowly add the standard NaOH solution to it until the reaction between the HCl and NaOH is complete. The point at which stoichiometrically equivalent quantities of substances have been brought together is known as the **equivalence point** of the titration.

In order to titrate an unknown with a standard solution, there must be some way to determine when the equivalence point of the titration has been reached. In acid-base titrations, organic dyes known as acid-base **indicators** are used for this purpose. For example, the dye known as phenolphthalein is colorless in acidic solution but is red in basic solution. If phenolphthalein is added to an unknown solution of acid, the solution will be colorless. Standard base can then be added from the buret until the solution barely turns from colorless to red. This indicates that the acid has been neutralized, and the drop of base that caused the solution

**Titration**

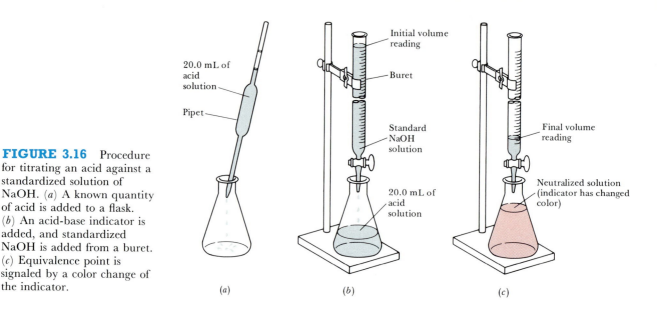

**FIGURE 3.16** Procedure for titrating an acid against a standardized solution of NaOH. (*a*) A known quantity of acid is added to a flask. (*b*) An acid-base indicator is added, and standardized NaOH is added from a buret. (*c*) Equivalence point is signaled by a color change of the indicator.

*(a)*   *(b)*   *(c)*

Labels on figure:
- 20.0 mL of acid solution
- Pipet
- Initial volume reading
- Buret
- Standard NaOH solution
- 20.0 mL of acid solution
- Final volume reading
- Neutralized solution (indicator has changed color)

to become colored has no acid to react with. The solution therefore becomes basic, and the dye turns red. The experimental procedure for an acid-base titration is summarized in Figure 3.16.

The following sample exercises illustrate stoichiometric calculations involving solution reactions such as those encountered in titrations. Recall that the coefficients in a balanced chemical equation give the relative numbers of moles of reactants and products in a reaction (Section 3.7). To use this information we have to convert the amounts of substances involved in a reaction into moles. If we are dealing with grams of a substance, we accomplish this conversion by using the formula weight of the substance. When we deal with substances in solution, we use the relationship moles $= M \times$ L.

**SAMPLE EXERCISE 3.20**

What volume of 0.500 $M$ NaCl is required to react completely with 0.200 mol of $Pb(NO_3)_2$? The chemical equation for this reaction is

$$2NaCl(aq) + Pb(NO_3)_2(aq) \longrightarrow 2NaNO_3(aq) + PbCl_2(s)$$

**Solution:**   According to the chemical equation

$$2 \text{ mol NaCl} \approx 1 \text{ mol Pb(NO}_3)_2$$

Therefore,

Moles NaCl

$$= (0.200 \text{ mol Pb(NO}_3)_2)\left(\frac{2 \text{ mol NaCl}}{1 \text{ mol Pb(NO}_3)_2}\right)$$

$$= 0.400 \text{ mol NaCl}$$

Because mol NaCl $= M$ NaCl soln $\times$ liters NaCl soln,

$$\text{Liters NaCl soln} = \frac{\text{mol NaCl}}{M \text{ NaCl soln}}$$

$$\text{Liters NaCl soln} = \frac{0.400 \text{ mol NaCl}}{0.500 \text{ mol NaCl/L soln}}$$

$$= 0.800 \text{ L}$$

Therefore, 800 mL of the 0.500 $M$ NaCl solution could be measured out and added to the 0.200 mol of $Pb(NO_3)_2$. This problem can also be solved by direct conversion of moles $Pb(NO_3)_2$ to moles NaCl to volume of NaCl solution:

Liters NaCl soln $= (0.200 \text{ mol Pb(NO}_3)_2)$

$$\times \left(\frac{2 \text{ mol NaCl}}{1 \text{ mol Pb(NO}_3)_2}\right)\left(\frac{1 \text{ L NaCl soln}}{0.500 \text{ mol}}\right)$$

$$= 0.800 \text{ L}$$

## SAMPLE EXERCISE 3.21

One method used commercially to peel potatoes is to soak them in a solution of NaOH for a short time, remove them from the NaOH, and spray off the peel. The concentration of NaOH is normally in the range 3 to 6 $M$. The NaOH is analyzed periodically. In one such analysis, 45.7 mL of 0.500 $M$ $H_2SO_4$ is required to react completely with a 20.0-mL sample of NaOH solution:

$$H_2SO_4(aq) + 2NaOH(aq) \longrightarrow$$
$$2H_2O(l) + Na_2SO_4(aq)$$

What is the concentration of the NaOH solution?

**Solution:**

Moles $H_2SO_4 = (45.7 \text{ mL soln})\left(\dfrac{1 \text{ L soln}}{1000 \text{ mL soln}}\right)$

$$\times \left(0.500 \frac{\text{mol } H_2SO_4}{\text{L soln}}\right)$$

$$= 2.28 \times 10^{-2} \text{ mol } H_2SO_4$$

According to the balanced equation, 1 mol $H_2SO_4 \rightleftharpoons$ 2 mol NaOH. Therefore,

## PRACTICE EXERCISE

What volume of 1.05 $M$ NaOH is needed to completely neutralize 0.0135 mol of $H_2SO_4$:

$$2NaOH(aq) + H_2SO_4(aq) \longrightarrow$$
$$2H_2O(l) + Na_2SO_4(aq)?$$

*Answer:* 25.7 mL

Moles NaOH

$$= (2.28 \times 10^{-2} \text{ mol } H_2SO_4)\left(\frac{2 \text{ mol NaOH}}{1 \text{ mol } H_2SO_4}\right)$$

$$= 4.56 \times 10^{-2} \text{ mol NaOH}$$

Knowing the number of moles of NaOH present in 20.0 mL of solution allows us to calculate the concentration of this solution:

$M$ NaOH

$$= \frac{\text{mol NaOH}}{\text{L soln}}$$

$$= \left(\frac{4.56 \times 10^{-2} \text{ mol NaOH}}{20.0 \text{ mL soln}}\right)\left(\frac{1000 \text{ mL soln}}{1 \text{ L soln}}\right)$$

$$= 2.28 \frac{\text{mol NaOH}}{\text{L soln}} = 2.28 \text{ } M$$

## PRACTICE EXERCISE

What is the molarity of a NaOH solution if 48.0 mL is needed to neutralize 35.0 mL of 0.144 $M$ $H_2SO_4$? The equation for the chemical reaction is given in Sample Exercise 3.21.
*Answer:* 0.210 $M$

## SAMPLE EXERCISE 3.22

The quantity of $Cl^-$ in a water supply is determined by titrating the sample with $AgNO_3$:

$$AgNO_3(aq) + Cl^-(aq) \longrightarrow AgCl(s) + NO_3^-(aq)$$

What mass of chloride ion is present in a 10.0-g sample of the water if 20.2 mL of 0.100 $M$ $AgNO_3$ is required to react with all the chloride in the sample?

**Solution:** We must first determine the number of moles of $AgNO_3$ required in the titration:

$$(20.2 \text{ mL soln})\left(\frac{1 \text{ L soln}}{1000 \text{ mL soln}}\right)\left(0.100 \frac{\text{mol } AgNO_3}{\text{L soln}}\right)$$
$$= 2.02 \times 10^{-3} \text{ mol } AgNO_3$$

From the balanced equation we see that 1 mol $AgNO_3 \rightleftharpoons$ 1 mol $Cl^-$. Therefore, the sample must contain $2.02 \times 10^{-3}$ mol of $Cl^-$:

Moles $Cl^-$

$$= (2.02 \times 10^{-3} \text{ mol } AgNO_3)\left(\frac{1 \text{ mol } Cl^-}{1 \text{ mol } AgNO_3}\right)$$

$$= 2.02 \times 10^{-3} \text{ mol } Cl^-$$

The number of moles of $Cl^-$ can then be converted to grams:

Grams $Cl^- = (2.02 \times 10^{-3} \text{ mol } Cl^-)\left(\frac{35.5 \text{ g } Cl^-}{1 \text{ mol } Cl^-}\right)$

$$= 7.17 \times 10^{-2} \text{ g } Cl^-$$

We might note that the weight percentage of $Cl^-$ in the water is

$$\% \text{ } Cl^- = \frac{7.17 \times 10^{-2} \text{ g } Cl^-}{10.0 \text{ g soln}} \times 100 = 0.717\%$$

Chloride ion is one of the major ions in water and sewage. Ocean water contains 1.92 percent $Cl^-$. Whether or not water containing $Cl^-$ exhibits a salty taste depends on the other ions present. If the accompanying ions are $Na^+$ ions, a salty taste may be detected with as little as 0.03 percent $Cl^-$.

# FOR REVIEW

## SUMMARY

The **law of conservation of mass** states that there are no detectable changes in mass in any chemical reaction. This observation indicates that the same number of atoms of each type are present before and after a chemical reaction. A **balanced equation** shows equal numbers of atoms of each element on each side of the equation, consistent with the law of conservation of mass. Equations are balanced by placing coefficients in front of the chemical formulas for the **reactants** and **products** of a reaction.

You can predict the products of simple reactions by noting the analogy to known reaction types and by using the periodic table. Among the reaction types seen in this chapter are the following: (1) **combustion** in oxygen, in which a hydrocarbon reacts with $O_2$ forming $CO_2$ and $H_2O$; (2) **neutralization,** in which an **acid** and **base** react to form water and a salt; (3) **precipitation reaction**, in which an insoluble product forms and separates from the solutions. Both neutralization and precipitation reactions are double displacement reactions, also called **metathesis reactions**.

Much quantitative information can be determined from chemical formulas and equations by using atomic weights. The atomic weight (average atomic mass) of an element can be calculated from the relative abundances and masses of that element's isotopes. Atomic weights are conveniently expressed in **atomic mass** units (amu). The amu is defined by assigning a mass of exactly 12 amu to the $^{12}C$ atom. The **formula weight** of a compound equals the sum of the atomic weights of the atoms in its formula. If the formula is a molecular formula, the formula weight is also called the **molecular weight**. The **mass spectrometer** provides the most direct and accurate means of experimentally determining atomic and molecular weights.

A **mole** of any substance is **Avogadro's number** ($6.022 \times 10^{23}$) of formula units of that substance. The mass of a mole of atoms, molecules, or ions is the formula weight of that material expressed in grams. For example, a single molecule of $H_2O$ weighs 18 amu; a mole of $H_2O$ weighs 18 g. The mole concept is very useful in chemistry, and interconverting grams and moles for any substance is important.

The **empirical formula** of any substance can be determined from its percent composition by calculating the relative number of moles of each atom in 100 g of the substance. If the substance is molecular in nature, its **molecular formula** can be determined from the empirical formula if the molecular weight is known.

The mole concept can be used to calculate the relative quantities of reactants and products involved in chemical reactions. The coefficients in a balanced equation give the relative numbers of moles of these substances. A **limiting reactant** is the reactant that is completely consumed in a reaction. When it is gone, the reaction stops, thus limiting the quantities of products formed.

The concentration of a solution can be expressed in terms of its **molarity**, the number of moles of solute per liter of solution. Molarity serves as a conversion factor for interconverting solution volume and number of moles of solute. Solutions of known concentration can be formed either by weighing out the solute and diluting it to a known volume or by dilution of a more concentrated solution of known concentration. In the process called **titration** we bring a solution of known concentration—a **standard solution**—into reaction with a solution of unknown concentration in order to determine the unknown concentration or the quantity of solute in the unknown.

# LEARNING GOALS

Having read and studied this chapter, you should be able to:

1. Balance chemical equations.
2. Predict the products of a chemical reaction, having seen a suitable analogy.
3. Predict the products of the combustion reactions of hydrocarbons and other simple organic compounds, and predict the products of precipitation and neutralization reactions.
4. Calculate the atomic weight of an element given the abundances and masses of its isotopes.
5. Calculate the formula weight of any substance.
6. Interconvert number of moles, mass in grams, and number of atoms, ions, or molecules.
7. Calculate the empirical formula of a compound, having been given appropriate analytical data such as elemental percentages or the quantity of $CO_2$ and $H_2O$ produced by combustion.
8. Calculate the molecular formula, having been given the empirical formula and molecular weight.
9. Calculate the mass of a particular substance produced or used in a chemical reaction (mass-mass problems).
10. Determine the limiting reactant in a reaction.
11. Define molarity.
12. Solve problems involving interconversions among molarity, solution volume, and number of moles of solute.
13. Explain what is meant by the term *titration*.
14. Calculate concentration or mass of solute in a sample from titration data.

# KEY TERMS

Among the more important terms and expressions used for the first time in this chapter are the following:

An **acid** (Section 3.2) is a substance that produces an excess of $H^+$ ions when it dissolves in water.

The **atomic mass unit** (amu) (Section 3.3) is based on the value of exactly 12 for the isotope of carbon with six protons and six neutrons in the nucleus.

**Atomic weight** (Section 3.3) or average atomic mass is the average mass of the atoms of an element in atomic mass units. It is numerically equal to the mass in grams of one mole of the element.

**Avogadro's number** (Section 3.5) is the number of $^{12}C$ atoms in exactly 12 g of $^{12}C$, $6.022 \times 10^{23}$.

A **balanced chemical equation** (Section 3.1) is one that satisfies the law of conservation of mass; it contains equal numbers of atoms of each element on both sides of the equation.

A **base** (Section 3.2) is a substance that produces an excess of $OH^-$ ions when it dissolves in water.

A **combustion reaction** (Section 3.2) is one that proceeds with evolution of heat and usually also a flame. Most combustion involves reaction with oxygen, as in the burning of a match.

Solution **concentration** (Section 3.9) denotes the number of moles of a solute present in a given quantity of solvent.

The **equivalence point** (Section 3.9) is the point in a titration at which the added solute just completely reacts with the solute present in solution.

**Formula weight** (Section 3.3) is the weight of the collection of atoms represented by a chemical formula. For example, the formula weight of $NO_2$ (46.0 amu) is the sum of the weights of one nitrogen atom and two oxygen atoms. If the formula is the molecular formula of the substance, the formula weight is the **molecular weight** of the substance.

An **indicator** (Section 3.9) is a substance added to a solution to indicate by a color change the point at which the added solute has just reacted with all the solute present in solution.

The **law of conservation of mass** (introduction) states that atoms are neither created nor destroyed during a chemical reaction.

A **limiting reactant** (Section 3.8) is the reactant present in smallest stoichiometric quantity in a mixture of reactants. The amount of product that can form is limited by the complete consumption of the limiting reactant.

The **mass spectrometer** (Section 3.4) is an instrument used to measure the precise masses and relative amounts of atomic and molecular ions.

A **metathesis reaction** (Section 3.2) is one in which two substances react to form two new substances through an exchange of the component parts of each reacting substance. For example, $AX + BY \rightarrow AY + BX$. Precipitation and neutralization reactions are examples of metathesis reactions.

**Molarity** (Section 3.9) is the concentration of a solution expressed as moles of solute per liter of solution; abbreviated $M$.

A **mole** (Section 3.5) is a collection of **Avogadro's number** ($6.022 \times 10^{23}$) of objects; for example, a mole of $H_2O$ is $6.022 \times 10^{23}$ $H_2O$ molecules.

A **neutralization reaction** (Section 3.2) is one in which an acid and a base react in stoichiometrically equivalent amounts. The product of a neutralization reaction is water and a salt.

A **precipitation reaction** (Section 3.2) is one occurring between substances in solution in which one of the products is insoluble.

A **salt** (Section 3.2) is an ionic compound that can be formed by an acid-base neutralization reaction.

A **titration** (Section 3.9) is the process of reacting a solution of unknown concentration with one of known concentration (a **standard solution**). The procedure is commonly used to determine the amounts of solute or the concentrations of solutions of unknown concentrations.

# EXERCISES

## Balancing Chemical Equations

**3.1** Which of the following equations, as written, are consistent with the law of conservation of mass?
**(a)** $H_2SO_4(aq) + Ca(OH)_2(aq) \longrightarrow$
$$H_2O(l) + CaSO_4(aq)$$
**(b)** $2AgNO_3(aq) + Na_2SO_4(aq) \longrightarrow$
$$2Ag_2SO_4(s) + 2NaNO_3(aq)$$
**(c)** $2C_6H_6(l) + 15O_2(g) \longrightarrow 12CO_2(g) + 6H_2O(l)$
**(d)** $CHF_3(g) + O(g) \longrightarrow OH(g) + CF_3(g)$

**3.2** Which of the following equations are not balanced?
**(a)** $CCl_4(g) + O_2(g) \longrightarrow COCl_2(g) + Cl_2(g)$
**(b)** $Al(OH)_3(s) + 3HCl(aq) \longrightarrow$
$$AlCl_3(aq) + 3H_2O(l)$$
**(c)** $C_2H_5OH(l) + 3O_2(g) \longrightarrow 2CO_2(g) + 3H_2O(l)$
**(d)** $C_3H_5NO(g) + 6O_2(g) \longrightarrow$
$$3CO_2(g) + NO_2(g) + 5H_2O(l)$$

**3.3** Balance the following equations by providing the missing coefficients:
**(a)** $\_N_2O_5(g) + \_H_2O(l) \longrightarrow \_HNO_3(aq)$
**(b)** $\_HClO_4(aq) + \_Ca(OH)_2(aq) \longrightarrow$
$$\_Ca(ClO_4)_2(aq) + \_H_2O(l)$$
**(c)** $\_Mg_2C_3(s) + \_H_2O(l) \longrightarrow$
$$\_Mg(OH)_2(s) + \_C_3H_4(g)$$
**(d)** $\_Cu(s) + \_AgNO_3(aq) \longrightarrow$
$$\_Cu(NO_3)_2(aq) + \_Ag(s)$$
**(e)** $\_PCl_5(l) + \_H_2O(l) \longrightarrow$
$$\_H_3PO_4(aq) + \_HCl(aq)$$

**3.4** Balance the following equations:
**(a)** $La_2O_3(s) + H_2O(l) \longrightarrow La(OH)_3(s)$
**(b)** $C_4H_{10}(g) + O_2(g) \longrightarrow CO_2(g) + H_2O(g)$
**(c)** $Pb(NO_3)_2(aq) + H_3AsO_4(aq) \longrightarrow$
$$PbHAsO_4(s) + HNO_3(aq)$$
**(d)** $Au_2S_3(s) + H_2(g) \longrightarrow H_2S(g) + Au(s)$
**(e)** $NO_2(g) + H_2O(l) \longrightarrow HNO_3(aq) + NO(g)$

**3.5** Write a balanced chemical equation to correspond to each of the following word descriptions. **(a)** Phosphine, $PH_3(g)$, is combusted in air to form gaseous water and solid diphosphorus pentoxide. **(b)** Barium metal reacts with methyl alcohol, $CH_3OH(l)$, to form hydrogen gas and dissolved barium methoxide, $Ba(OCH_3)_2$. **(c)** Boron sulfide, $B_2S_3(s)$, reacts violently with water to form dissolved boric acid, $H_3BO_3$, and hydrogen sulfide gas, $H_2S$.

**(d)** Copper metal reacts with hot concentrated sulfuric acid to form aqueous copper(II) sulfate, sulfur dioxide gas, and liquid water. **(e)** When ammonia gas, $NH_3$, is passed over hot liquid sodium metal, hydrogen gas, $H_2$, is released and sodium amide, $NaNH_2$, is formed as a solid product.

**3.6** Write balanced chemical equations to correspond to each of the following verbal descriptions. **(a)** Cyanic acid, HNCO, is quite unstable. The gas reacts with water to form ammonia and gaseous carbon dioxide. **(b)** When solid mercury(II) nitrate is heated, it decomposes to form solid mercury(II) oxide and gaseous nitrogen dioxide and oxygen. **(c)** When liquid phosphorus trichloride is added to water, it reacts violently to form aqueous phosphorous acid and aqueous hydrogen chloride. **(d)** When solid potassium nitrate is heated, it decomposes to solid potassium nitrite, and oxygen gas is evolved. **(e)** When hydrogen sulfide ($H_2S$) gas is passed over solid hot iron(III) hydroxide, it reacts to form solid iron(III) sulfide and gaseous water.

## Predicting Reaction Products

**3.7** Balance the following equations and classify each reaction into one of the four categories given in Table 3.1:
**(a)** $Zn(s) + H_2SO_4(aq) \longrightarrow H_2(g) + ZnSO_4(aq)$
**(b)** $H_2O_2(aq) \longrightarrow H_2O(l) + O_2(g)$
**(c)** $Fe(s) + Cl_2(g) \longrightarrow FeCl_3(s)$
**(d)** $Na_2CO_3(aq) + Ca(OH)_2(aq) \longrightarrow$
$$CaCO_3(s) + NaOH(aq)$$

**3.8** Balance the following equations and classify each reaction as one of the following types: combustion, neutralization, or precipitation:
**(a)** $Ba(OH)_2(aq) + HCl(aq) \longrightarrow$
$$BaCl_2(aq) + H_2O(l)$$
**(b)** $CO(g) + O_2(g) \longrightarrow CO_2(g)$
**(c)** $CaCl_2(aq) + Na_2CO_3(aq) \longrightarrow$
$$CaCO_3(s) + NaCl(aq)$$
**(d)** $CH_3SH(g) + O_2(g) \longrightarrow$
$$CO_2(g) + H_2O(l) + SO_2(g)$$
**(e)** $Ni(NO_3)_2(aq) + KOH(aq) \longrightarrow$
$$Ni(OH)_2(s) + KNO_3(aq)$$

**3.9** Complete and balance each of the following equations:
**(a)** $C_4H_{10}(g) + O_2(g) \longrightarrow \_(g) + \_(l)$
**(b)** $C_2H_5OH(l) + O_2(g) \longrightarrow \_(g) + \_(l)$
**(c)** $Al(OH)_3(s) + HNO_3(aq) \longrightarrow \_(aq) + \_(l)$

**(d)**  $Cu(OH)_2(s) + HCl(aq) \longrightarrow \underline{\phantom{x}}(aq) + \underline{\phantom{x}}(l)$
**(e)**  $AgNO_3(aq) + H_2SO_4(aq) \longrightarrow Ag_2SO_4(s) + \underline{\phantom{x}}(aq)$
**(f)**  $CaCl_2(aq) + Na_3PO_4(aq) \longrightarrow$
$$Ca_3(PO_4)_2(s) + \underline{\phantom{x}}(aq)$$

**3.10**  Complete and balance each of the following equations (the states of the reactants and products need not be given):
**(a)**  $C_6H_{12} + O_2 \longrightarrow$
**(b)**  $Fe(OH)_3 + H_2SO_4 \longrightarrow$
**(c)**  $ZnCl_2 + Na_2S \longrightarrow$
**(d)**  $Li + H_2O \longrightarrow$
**(e)**  $HC_2H_3O_2 + Cu(OH)_2 \longrightarrow$
**(f)**  $CH_3OC_2H_5 + O_2 \longrightarrow$

**3.11**  Drawing upon the following list of substances as reactants—$BaCl_2$, $C_2H_5OH$, $Na_2CrO_4$, $O_2$, $H_2O$, $KOH$, $H_2SO_4$, $Ca(OH)_2$, $Pb(NO_3)_2$, $HNO_3$, and $K$—write balanced equations for each of the following: **(a)** a combustion reaction; **(b)** formation of insoluble $PbCrO_4$ from solution; **(c)** formation of hydrogen gas; **(d)** neutralization of sulfuric acid; **(e)** solid $Ca(OH)_2$, which has a low solubility in water, is dissolved by reaction with another reagent; **(f)** formation of insoluble $BaSO_4$ from solution.

**[3.12]**  We can sometimes extend our ability to write balanced equations by recognizing the analogies between compounds. From the hints provided, write complete balanced equations for the following. **(a)** Combustion of nitromethane, $CH_3NO_2(g)$, leads to $NO_2(g)$ as one of the products. **(b)** Reaction of potassium with liquid ammonia is very much like reaction of this metal with water. **(c)** Fluorine, like oxygen, can support combustion; for example, methane, $CH_4(g)$, can be made to "burn" in an atmosphere of $F_2$. **(d)** Reaction of a metal with an acid solution is like reaction of active metals with water, except that a salt of the metal rather than the hydroxide is the product. For example, Zn reacts with dilute HCl solution.

## Atomic Weights and the Mass Spectrometer

**3.13**  The element magnesium consists of three isotopes with masses 23.9924, 24.9938, and 25.9898 amu. These three isotopes are present in nature to the extent of 78.6, 10.1 and 11.3 percent by mass, respectively. From these data calculate the average atomic mass of magnesium.

**3.14**  The element neon consists of three isotopes with masses 19.99, 20.99, and 21.99 amu. These three isotopes are present in nature to the extent of 90.92, 0.25, and 8.83 percent, respectively. From these data, calculate the atomic weight of neon.

**[3.15]**  The element silver consists in nature of two isotopes, $^{107}Ag$ with atomic mass 106.905 amu, and $^{109}Ag$ with atomic mass 108.905 amu. The accepted atomic weight of Ag is 107.870. From this calculate the relative amounts of $^{107}Ag$ and $^{109}Ag$ in nature.

**[3.16]**  Gallium consists of two naturally occurring isotopes with masses of 68.926 and 70.926 amu. **(a)** How many protons and neutrons are in the nucleus of each isotope? Write the complete atomic symbol for each, showing the atomic number and mass number. **(b)** Calculate the abundance of each isotope.

**3.17**  The mass spectrum of a sample of lead oxide contains ions of the formula $PbO^+$. The lead oxide has been prepared from isotopically pure $^{16}O$. The ion masses seen and their relative intensities are listed as follows:

| $PbO^+$ ion mass | Fractional intensity |
| --- | --- |
| 220.002 | 0.0137 |
| 222.056 | 0.2630 |
| 223.050 | 0.2080 |
| 224.055 | 0.5153 |

The mass of $^{16}O$ is 15.9948. Calculate the average atomic weight of lead in this sample.

**[3.18]**  The element bromine consists of two isotopes. The mass spectrum of molecular bromine ($Br_2$) consists of three peaks:

| Mass (amu) | Relative size |
| --- | --- |
| 157.84 | 0.2534 |
| 159.84 | 0.5000 |
| 161.84 | 0.2466 |

**(a)** What is the origin of each peak (that is, of what isotopes does each consist)? **(b)** What is the mass of each isotope? **(c)** Determine the average molecular mass of a $Br_2$ molecule. **(d)** Determine the average atomic mass of a bromine atom. **(e)** Calculate the abundances of the two isotopes.

**3.19**  We know that chlorine and fluorine are both diatomic gases. One volume of chlorine gas reacts with three volumes of fluorine gas to yield two volumes of product, with all gases measured at the same temperature and pressure. What is the formula for the product?

**3.20**  Gay-Lussac found that one volume of hydrogen gas reacts with one volume of chlorine gas to form *two* volumes of hydrogen chloride as product. Is this observation consistent with an assumption that hydrogen and chlorine gas are monatomic? Explain. What does this experiment tell us directly about the formula for hydrogen chloride? Explain.

**3.21**  In his determination of the atomic weight of the element zinc, Berzelius determined in 1818 that the weight ratio of Zn to O in the oxide of zinc was 4.032. He thought that the formula of the compound was $ZnO_2$. Assuming an atomic weight of 16.00 for oxygen, what value does this give for the atomic weight of zinc? By comparing this with the presently accepted value, what can you say about Berzelius's assumption? If it was in error, what should it have been?

**[3.22]**  Stas reported in 1865 that he had reacted a weighed amount of pure silver with nitric acid and had recovered all the silver as pure silver nitrate, $AgNO_3$. The weight ratio of Ag to $AgNO_3$ was found to be 0.634985. Using only this ratio and the presently accepted values of the atomic weights of silver and oxygen, calculate the atomic weight of nitrogen. Compare this calculated atomic weight with the currently accepted value.

## Formula Weights; Percent Composition

**3.23** Calculate the formula weight of each of the following: **(a)** $SO_2$; **(b)** $C_3H_8$; **(c)** $K_2Cr_2O_7$; **(d)** $(NH_4)_2SO_4$; **(e)** $Ca(OH)_2$; **(f)** $CH_3OH$; **(g)** $CH_3OC_2H_5$.

**3.24** Determine the formula weight of each of the following compounds: **(a)** acetylene, $C_2H_2$; **(b)** barium sulfate, $BaSO_4$; **(c)** thallium(I) oxalate, $Tl_2(C_2O_4)$; **(d)** ascorbic acid (vitamin C), $C_6H_8O_6$; **(e)** xenon tetrafluoride, $XeF_4$; **(f)** zinc nitrate, $Zn(NO_3)_2$.

**3.25** Calculate the weight percentage of each element in the following compounds: **(a)** $NO_2$; **(b)** $SiF_4$; **(c)** $NH_4Cl$; **(d)** $C_2H_5OH$; **(e)** $Ca(HCO_3)_2$; **(f)** $MgNH_4PO_4$.

**3.26** Calculate the weight percentage of the indicated element in each of the following compounds: **(a)** nitrogen in ammonia, $NH_3$; **(b)** sulfur in thionyl chloride, $SOCl_2$; **(c)** carbon in butane, $C_4H_{10}$; **(d)** sodium in sodium bicarbonate (baking soda), $NaHCO_3$; **(e)** hydrogen in morphine, $C_{17}H_{19}NO_3$; **(f)** copper in copper(II) nitrate, $Cu(NO_3)_2$.

## The Chemical Mole

**3.27** **(a)** What is the mass of 1 mol of calcium nitrate, $Ca(NO_3)_2$? **(b)** How many moles of $Ca(NO_3)_2$ are there in 2.50 g of this substance? **(c)** What is the mass, in grams, of 0.325 mol of $Ca(NO_3)_2$? **(d)** How many $Ca^{2+}$ ions are there in 8.73 g of $Ca(NO_3)_2$?

**3.28** Aspartamine, the artificial sweetener marketed by G. D. Searle as Nutra-Sweet, has a molecular formula $C_{14}H_{18}N_2O_5$. **(a)** What is the mass of 1.00 mol of aspartamine? **(b)** How many moles of aspartamine are present in 5.88 g of this substance? **(c)** What is the mass, in grams, of 0.376 mol of aspartamine? **(d)** How many molecules are present in 8.22 mg of aspartamine?

**3.29** Calculate the mass, in grams, of each of the following: **(a)** 3.00 mol of $CO_2$; **(b)** 0.00850 mol of ethylene, $C_2H_4$; **(c)** $3.58 \times 10^{22}$ atoms of Ar; **(d)** $1.50 \times 10^{22}$ molecules of caffeine, $C_8H_{10}N_4O_2$.

**3.30** Calculate the mass, in grams, of each of the following: **(a)** 2.49 mol of sodium cyanide, $NaCN$; **(b)** 0.0590 mol of aspirin, $C_9H_8O_4$; **(c)** $8.25 \times 10^{20}$ molecules of ozone, $O_3$; **(d)** $1.83 \times 10^{24}$ molecules of HCl.

**3.31** Calculate the number of moles present in the following samples: **(a)** 125 g of $Fe_2O_3$; **(b)** 5.35 mg of $Ca(OH)_2$; **(c)** $1.40 \times 10^{20}$ molecules of $H_2O$; **(d)** a 1.15-carat diamond, which contains $1.38 \times 10^{23}$ carbon atoms.

**3.32** Calculate the number of molecules present in the following samples: **(a)** 0.500 mol of CO; **(b)** 10.8 g of $H_3PO_4$; **(c)** a 500-mg tablet of vitamin C, $C_6H_8O_6$; **(d)** an average snowflake containing $5.0 \times 10^{-5}$ g of $H_2O$.

**3.33** It requires about 25 $\mu$g minimum of tetrahydrocannabinol, THC, the active ingredient in marijuana, to produce intoxication from smoking substances containing THC. The molecular formula of THC is $C_{21}H_{30}O_2$. How many molecules of THC does this 25 $\mu$g represent? How many moles?

**3.34** The allowable concentration level of vinyl chloride, $C_2H_3Cl$, in the atmosphere in a chemical plant is $2.05 \times 10^{-6}$ g/L. How many moles of vinyl chloride in each liter does this represent? How many molecules per liter?

## Empirical Formulas

**3.35** **(a)** A sample of a compound contains 0.36 mol of hydrogen and 0.090 mol of carbon. What is the empirical formula of this compound? **(b)** A sample of a compound contains 11.66 g of iron and 5.01 g of oxygen. What is the empirical formula of this compound? **(c)** What is the empirical formula of hydrazine, which contains 87.5% N and 12.5% H?

**3.36** Give the empirical formula for each of the following compounds if a sample contains: **(a)** 0.014 mol of sulfur and 0.042 mol of oxygen; **(b)** 5.28 g of tin and 3.37 g of fluorine; **(c)** 26.4 percent sodium, 36.8 percent sulfur, and 36.8 percent oxygen.

**3.37** What are the empirical formulas of compounds with the following compositions: **(a)** 40.0 percent C, 6.7 percent H, 53.3 percent O; **(b)** 10.4 percent C, 27.8 percent S, 61.7 percent Cl; **(c)** 32.79 percent Na, 13.02 percent Al, 54.19 percent F; **(d)** 83.0 percent I, 7.85 percent C, 9.15 percent N?

**3.38** Determine the empirical formulas of the compounds with the following compositions: **(a)** 1.6 percent hydrogen, 22.2 percent nitrogen, and 76.2 percent oxygen; **(b)** 62.1 percent carbon, 5.21 percent hydrogen, 12.1 percent nitrogen, and 20.7 percent oxygen; **(c)** 21.7 percent carbon, 9.6 percent oxygen, and 68.7 percent fluorine; **(d)** 21.2 percent nitrogen, 6.1 percent hydrogen, 24.3 percent sulfur, and 48.4 percent oxygen.

**3.39** In a laboratory, 1.55 g of an organic compound containing carbon, hydrogen, and oxygen is combusted for analysis, as illustrated in Figure 3.11. Combustion resulted in 1.45 g of $CO_2$ and 0.890 g of $H_2O$. What is the empirical formula?

**3.40** Cyclopropane, a substance used with oxygen as a general anesthetic, contains only two elements, carbon and hydrogen. When 1.00 g of this substance is completely combusted, 3.14 g of $CO_2$ and 1.29 g of $H_2O$ are produced. What is the empirical formula of cyclopropane?

**3.41** The elemental analysis of acetylsalicylic acid (aspirin) is 60.0 percent carbon, 4.48 percent hydrogen, and 35.5 percent oxygen. If the molecular mass of this substance is 180.2 amu, what is its molecular formula?

**3.42** The characteristic odor of pineapple is due to ethyl butyrate, a compound containing carbon, hydrogen, and oxygen. Combustion of 2.78 mg of ethyl butyrate leads to formation of 6.32 mg of $CO_2$ and 2.58 mg of $H_2O$. What is the empirical formula? The properties of the compound suggest that the molecular weight should be between 100 and 150. What is the likely molecular formula?

**3.43** Strontium hydroxide is isolated as a hydrate, which means that a certain number of water molecules are included in the solid structure. The formula of the hydrate can be written as $Sr(OH)_2 \cdot xH_2O$, where $x$ indicates the number of moles of water in the solid per mole of $Sr(OH)_2$. When 6.85 g of the hydrate is dried in an oven, 3.13 g of anhydrous $Sr(OH)_2$ is formed. What is the value for $x$?

**3.44** The compound $MgI_2 \cdot xH_2O$ is analyzed to determine the value of $x$. A 1.557-g sample of the compound is heated to remove all the water; 1.0254 g of $MgI_2$ remains after heating. What is the value of $x$?

**[3.45]** Fungal laccase, a blue protein found in wood-rotting fungi, is approximately 0.39 percent copper. If a laccase molecule contains four copper atoms, what is its approximate molecular weight?

**[3.46]** Hemoglobin, the oxygen-carrying protein in red blood cells, has four iron atoms per molecule and contains 0.340 percent iron by mass. Calculate the molecular mass of hemoglobin.

## Chemical Calculations Involving Equations

**3.47** **(a)** What weight of $NH_3$ is formed when 5.38 g of $Li_3N$ reacts with water according to the equation

$$Li_3N(s) + 3H_2O(l) \longrightarrow 3LiOH(s) + NH_3(g)$$

**(b)** What mass of $CaCO_3(s)$ is required to produce 2.87 g $CO_2(g)$ according to the reaction

$$CaCO_3(s) \longrightarrow CaO(s) + CO_2(g)$$

**3.48** **(a)** What mass of $CO_2$ is formed when 9.53 g of $C_2H_5OH$ is produced in the fermentation of sugar according to the reaction

$$C_6H_{12}O_6(aq) \longrightarrow 2C_2H_5OH(aq) + 2CO_2(g)$$

**(b)** What mass of Mg is required to react with excess $CuSO_4$ to form 1.89 g $Cu_2O$ according to the reaction

$$2Mg(s) + 2CuSO_4(aq) + H_2O(l) \longrightarrow$$
$$2MgSO_4(aq) + Cu_2O(s) + H_2(g)$$

**3.49** The reusable booster rockets of the U.S. space shuttle use a mixture of aluminum and ammonium perchlorate for fuel. The reaction between these substances is as follows:

$$3Al(s) + 3NH_4ClO_4(s) \longrightarrow$$
$$Al_2O_3(s) + AlCl_3(s) + 3NO(g) + 6H_2O(g)$$

What mass of ammonium perchlorate should be used in the fuel mixture for each kilogram of aluminum?

**3.50** The fizz produced when an Alka-Seltzer tablet is dissolved in water is due to the reaction between sodium bicarbonate and citric acid:

$$3NaHCO_3(aq) + H_3C_6H_5O_7(aq) \longrightarrow$$
$$3CO_2(g) + 3H_2O(l) + Na_3C_6H_5O_7(aq)$$

**(a)** How many grams of citric acid should be used for each 1.00 g of sodium bicarbonate? **(b)** How many grams of carbon dioxide are produced for each 1.00 g of sodium bicarbonate?

**3.51** A sample of aluminum carbide, $Al_4C_3(s)$, is added to a dilute acid solution. The following reaction goes to completion:

$$Al_4C_3(s) + HCl(aq) \longrightarrow AlCl_3(aq) + CH_4(g)$$

Balance this equation. In one particular experiment, the methane produced was collected and found to have a mass of 1.754 g. What mass of $Al_4C_3(s)$ was added to the acid solution? What mass of $AlCl_3$ is formed in solution as the other product?

**3.52** Magnesium oxide reacts with gaseous phosphorus pentachloride to form solid magnesium chloride and solid diphosphorus pentoxide as products. Write a balanced equation for the reaction. Assuming an excess of $MgO(s)$ and complete reaction, what mass of $PCl_5$ is required to produce 4.50 kg of diphosphorus pentoxide?

## Limiting Reactants; Theoretical Yields

**3.53** **(a)** In the reaction

$$Mg_2Si(s) + 4H_2O(l) \longrightarrow 2Mg(OH)_2(s) + SiH_4(g)$$

how many moles of $SiH_4$ are formed by complete reaction of 3.95 g of $Mg_2Si$? **(b)** Calculate the weight of water necessary to react with all the $Mg_2Si$. **(c)** If 42.5 mg of $Mg_2Si$ is reacted with 27.0 mg of water, what mass of $SiH_4$ is formed, assuming that the limiting reactant is completely reacted?

**3.54** **(a)** In the reaction

$$H_2S(g) + 2NaOH(aq) \longrightarrow Na_2S(aq) + 2H_2O(l)$$

how many moles of $Na_2S$ are formed by complete reaction of 5.00 g of $H_2S$? **(b)** Calculate the mass of NaOH necessary to react with this quantity of $H_2S$. **(c)** If 3.05 g of $H_2S$ is bubbled into a solution containing 1.84 g of NaOH, what mass of $Na_2S$ is produced, assuming that the limiting reactant is completely consumed?

**3.55** What mass of $AgBr(s)$ is formed when a solution containing 3.45 g of KBr is mixed with a solution containing 7.28 g of $AgNO_3$?

**3.56** Ethylene, $C_2H_4$, burns in air according to the following equation:

$$C_2H_4(g) + 3O_2(g) \longrightarrow 2CO_2(g) + 2H_2O(l)$$

How many grams of $CO_2$ will be formed when a mixture of 2.93 g of $C_2H_4$ and 4.29 g of $O_2$ is ignited assuming that the reaction above is the only one to occur?

**3.57** A student reacts benzene, $C_6H_6$, with bromine, $Br_2$, in an attempt to prepare bromobenzene, $C_6H_5Br$:

$$C_6H_6 + Br_2 \longrightarrow C_6H_5Br + HBr$$

**(a)** What is the theoretical yield of bromobenzene in this reaction when 30.0 g of benzene reacts with 65.0 g $Br_2$? **(b)** Dibromobenzene, $C_6H_4Br_2$, is produced as a by-product in the synthesis of bromobenzene. If the actual yield of bromobenzene was 56.7 g, what was the percentage yield?

**3.58** Azobenzene ($C_{12}H_{10}N_2$) is an important intermediate in the manufacture of dyes. It can be prepared by the reaction between nitrobenzene ($C_6H_5NO_2$) and triethylene glycol ($C_6H_{14}O_6$) in the presence of zinc and potassium hydroxide:

$$2C_6H_5NO_2 + 4C_6H_{14}O_4 \xrightarrow[\text{KOH}]{\text{Zn}}$$
$$C_{12}H_{10}N_2 + 4C_6H_{12}O_4 + 4H_2O$$

(a) What is the theoretical yield of azobenzene when 115 g of nitrobenzene and 327 g of triethylene glycol are allowed to react? (b) If the reaction yields 55 g of azobenzene, what is the percent yield of azobenzene?

## Molarity: Solution Stoichiometry

**3.59** (a) Calculate the molarity of a solution containing 5.44 g of $Na_2CrO_4$ in 0.250 L. (b) How many moles of HCl are present in 25.0 mL of a 12.0 $M$ solution of hydrochloric acid? (c) How many mL of 2.00 $M$ NaOH solution is needed to obtain 0.100 mol of NaOH?

**3.60** (a) Calculate the molarity of a solution made by dissolving 5.63 g of $NaHCO_3$ in enough water to form 250 mL of solution. (b) How many moles of $K_2Cr_2O_7$ are present in 50.0 mL of a 0.105 $M$ solution? (c) How many mL of 9.0 $M$ $H_2SO_4$ solution is required to obtain 0.050 mol of $H_2SO_4$?

**3.61** (a) What mass of $Na_2SO_4$ is required to prepare 250.0 mL of 0.100 $M$ $Na_2SO_4$? (b) What volume of 1.000 $M$ $KNO_3$ must be diluted with water to prepare 500.0 mL of 0.250 $M$ $KNO_3$?

**3.62** Describe how you would prepare 100.0 mL of 0.1000 $M$ glucose solution starting with (a) solid glucose, $C_6H_{12}O_6$; (b) 1.000 L of 2.000 $M$ glucose solution.

**3.63** (a) What volume of 0.105 $M$ $HClO_4$ solution is required to react completely with 50.0 mL of 0.0875 $M$ NaOH? (b) What volume of 0.158 $M$ HCl is required to neutralize 2.87 g of $Mg(OH)_2$? (c) If it requires 25.8 mL of $AgNO_3$ to precipitate all of the chloride ion in an 895-mg sample of KCl (forming AgCl), what is the molarity of the $AgNO_3$ solution? (d) If it requires 35.8 mL of 0.117 $M$ HCl solution to neutralize a solution of KOH, how many grams of KOH must be present in the solution?

**3.64** (a) How many mL of 0.210 $M$ HCl is needed to completely neutralize 35.0 mL of 0.101 $M$ $Ba(OH)_2$ solution? (b) How many mL of 3.50 $M$ $H_2SO_4$ is needed to neutralize 75.0 g of NaOH? (c) If 45.2 mL of $BaCl_2$ solution is needed to precipitate all of the sulfate in a 544-mg sample of $Na_2SO_4$ (forming $BaSO_4$), what is the molarity of the solution? (d) If 42.7 mL of 0.250 $M$ HCl solution is needed to neutralize a solution of $Ca(OH)_2$, how many grams of $Ca(OH)_2$ must be present in the solution?

**3.65** In the laboratory, 6.67 g of $Sr(NO_3)_2$ is dissolved in enough water to form 0.750 L. A 0.100-L sample is withdrawn from this stock solution and titrated with a 0.0460 $M$ solution of $Na_2CrO_4$. What volume of $Na_2CrO_4$ solution is required to precipitate all the $SrCrO_4$?

**3.66** A sample of solid $Ca(OH)_2$ is allowed to stand in contact with water at 30°C for a long time, until the solution contains as much dissolved $Ca(OH)_2$ as it can hold. A 100-mL sample of this solution is withdrawn and titrated with $5.00 \times 10^{-2}$ $M$ HBr. It requires 48.8 mL of the acid solution for neutralization. What is the concentration of the $Ca(OH)_2$ solution? What is the solubility of $Ca(OH)_2$ in water, at 30°C, in grams of $Ca(OH)_2$ per 100 mL of solution?

## Additional Exercises

**3.67** Balance the following equations:
(a) $Li_3N(s) + H_2O(l) \longrightarrow NH_3(g) + LiOH(aq)$
(b) $C_3H_7OH(l) + O_2(g) \longrightarrow CO_2(g) + H_2O(g)$
(c) $PBr_3(l) + H_2O(l) \longrightarrow H_3PO_3(aq) + HBr(aq)$
(d) $Mg_3B_2(s) + H_2O(l) \longrightarrow Mg(OH)_2(aq) + B_2H_6(g)$
(e) $CCl_4(g) + O_2(g) \longrightarrow CCl_2O(g) + Cl_2(g)$
(f) $La(NO_3)_3(aq) + Ba(OH)_2(aq) \longrightarrow$ $La(OH)_3(s) + Ba(NO_3)_2(aq)$

**3.68** Given the following reactions:

$$2HCl(aq) + CaCO_3(s) \longrightarrow$$
$$CaCl_2(aq) + CO_2(g) + H_2O(l)$$

$$Zn(s) + 2HCl(aq) \longrightarrow H_2(g) + ZnCl_2(aq)$$

$$Na_2O(s) + H_2O(l) \longrightarrow 2NaOH(aq)$$

predict what will happen in the following cases and write a balanced equation for each reaction. (a) Hydrochloric acid is added to solid $BaCO_3$. (b) Nitric acid is added to solid $CaCO_3$. (c) Solid potassium oxide, $K_2O$, is added to water. (d) Zinc metal is added to hydrobromic acid, HBr. (e) Barium oxide, $BaO(s)$, is added to water.

**3.69** Write complete balanced chemical equations for each of the following reactions: (a) the complete combustion of butyric acid, $C_4H_8O_2$, a compound produced when butter becomes rancid; (b) the neutralization of battery acid, $H_2SO_4$, by $Al(OH)_3$; (c) the precipitation of $Fe(OH)_3$ by the addition of NaOH to a solution of $Fe(SO_4)_3$.

**3.70** The element copper consists of two stable, naturally occurring isotopes: $^{63}Cu$ (mass = 62.9298 amu; abundance = 69.09 percent), $^{65}Cu$ (mass = 64.9278 amu; abundance = 30.91 percent). Calculate the average atomic mass (the atomic weight) of copper.

**3.71** Calculate the percent by mass of the indicated element in each of the following compounds: (a) the percent samarium in $Co_5Sm$, an alloy used to form permanent magnets in very lightweight headsets such as in the Sony Walkman and similar units; (b) the percentage of nitrogen in $N_2O$, a gas used as an anesthetic by dentists (known as laughing gas); (c) the percentage of carbon in dopamine, $C_{18}H_{11}O_2N$, a neurotransmitter; (d) the percentage of platinum in cisplatin, $Pt(NH_3)_2Cl_2$, a compound used as an antitumor agent.

**3.72** The molecule pyridine, $C_5H_5N$, was found to adsorb on the surfaces of certain metal oxides. A 5.0-g sample of finely divided zinc oxide, ZnO, was found to adsorb 0.068 g of pyridine. How many pyridine molecules are adsorbed? What is the ratio of pyridine molecules to formula units of zinc oxide? If the surface area of the oxide is 48 $m^2/g$, what is the average area of surface per adsorbed pyridine molecule?

**[3.73]** One of the earliest accurate formula weight measurements involved measurement of the weight ratio of $KClO_3$ to KCl, based on decomposition of $KClO_3$: $2KClO_3(s) \longrightarrow 2KCl(s) + 3O_2(g)$. In 1911 Stähler and Meyer measured this ratio and found it to be 1.64382. Using only this ratio and the presently accepted atomic weight of oxygen, calculate the formula weight for KCl. Compare this calculated formula weight with the presently accepted value.

**3.74** Calculate the empirical formula and molecular formula of each of the following substances: **(a)** epinephrine (adrenaline), a hormone secreted in the bloodstream in times of danger or stress: 59.0 percent C, 7.1 percent H, 26.2 percent O, and 7.7 percent N; MW about 183 amu; **(b)** nicotine, a component of tobacco; 74.1 percent C, 8.6 percent H, and 17.3 percent N; MW = $160 \pm 5$ amu; **(c)** ethylene glycol, the substance used as the primary component of most antifreeze solutions: 38.7 percent C, 9.7 percent H, and 51.6 percent O; MW = 62.1 amu; **(d)** caffeine, a stimulant found in coffee; 49.5 percent C, 5.15 percent H, 28.9 percent N, and 16.5 percent O; MW about 195 amu.

**3.75** The koala bear dines exclusively on eucalyptus leaves. Its digestive system detoxifies the eucalyptus oil, a poison to other animals. The chief constituent in eucalyptus oil is a substance called eucalyptol, which contains 77.87 percent C, 11.76 percent H, and the remainder O. **(a)** What is the empirical formula of this substance? **(b)** What is its molecular formula if the substance has a molecular mass of 154.2 amu?

**3.76** Butadiene, a substance used to manufacture certain types of synthetic rubber, contains only two elements, carbon and hydrogen. When 1.00 g of this substance is completely combusted, 3.26 g of $CO_2$ and 0.998 g of $H_2O$ are produced. What is the simplest formula of butadiene?

**[3.77]** An oxybromate compound, $KBrO_x$, where $x$ is unknown, is analyzed and found to contain 52.92 percent Br. What is the value for $x$?

**3.78** Many "antacids" contain $Al(OH)_3$. **(a)** Write the equation for the neutralization of HCl in stomach acid by $Al(OH)_3$. **(b)** How many moles of HCl are neutralized by 5.0 g of $Al(OH)_3$?

**3.79** What mass of ZnO is obtained by heating 75 kg of $ZnSO_3(s)$ according to the following equation:

$$ZnSO_3(s) \longrightarrow ZnO(s) + SO_2(g)$$

How many grams of $ZnSO_3$ are required to form 2.0 kg of $SO_2$ in this reaction?

**[3.80]** Under a certain set of laboratory conditions, 2.1 g of Na reacts with water to form 1.14 L of $H_2$ gas. When 3.4 g of another alkali metal is reacted with water under the same conditions, 497 mL of hydrogen gas is evolved. Which of the other alkali metals was reacted?

**3.81** A mixture of 3.50 g of $H_2$ and 26.0 g of $O_2$ is caused to react to form $H_2O$. How much $H_2$, $O_2$, and $H_2O$ remain after reaction is complete?

**[3.82]** Chloromycetin is an antibiotic with the formula $C_{11}H_{12}O_5N_2Cl_2$. A 1.03-g sample of an ophthalmic ointment containing chloromycetin was chemically treated to convert its chlorine into $Cl^-$ ions. The $Cl^-$ was precipitated as AgCl. If the AgCl weighed 0.0129 g, calculate the weight percentage of chloromycetin in the sample.

**3.83** A particular coal contains 2.8 percent sulfur by weight. When this coal is burned, the sulfur appears as $SO_2(g)$. This $SO_2$ is reacted with CaO to form $CaSO_3(s)$. If the coal is burned in a power plant that uses 2000 tons of coal per day, what is the daily production of $CaSO_3$?

**3.84** Adipic acid, $C_6H_{10}O_4$, is a raw material used for the production of nylon. Adipic acid is made industrially by oxidation of cyclohexane, $C_6H_{12}$:

$$5O_2 + 2C_6H_{12} \longrightarrow 2C_6H_{10}O_4 + 2H_2O$$

**(a)** If $2.00 \times 10^2$ kg of $C_6H_{12}$ reacts with an unlimited supply of $O_2$, what is the theoretical yield of adipic acid? **(b)** If the reaction produces $2.95 \times 10^2$ kg of adipic acid, what is the percent yield for the reaction?

**3.85** Aspirin, $C_9H_8O_4$, is produced from salicylic acid, $C_7H_6O_3$, and acetic anhydride, $C_4H_6O_3$:

$$C_7H_6O_3 + C_4H_6O_3 \longrightarrow C_9H_8O_4 + HC_2H_3O_2$$

**(a)** How much salicylic acid is required to produce $1.5 \times 10^2$ kg of aspirin, assuming that all of the salicylic acid is converted to aspirin? **(b)** How much salicylic acid would be required if only 80 percent of the salicylic acid is converted to aspirin? **(c)** What is the theoretical yield of aspirin if 185 kg of salicylic acid is allowed to react with 125 kg of acetic anhydride? **(d)** If the situation described in part (c) produces 182 kg of aspirin, what is the percentage yield?

**3.86** Calculate molarity of the following solutions: **(a)** 35.0 g of $H_2SO_4$ in 600 mL of solution; **(b)** 42.0 mL of 0.550 $M$ $HNO_3$ solution diluted to a volume of 0.500 L; **(c)** a solution formed by dissolving 2.56 g of NaBr in water to form 65.0 mL of solution; **(d)** a solution formed by mixing 35 mL of 0.50 $M$ KBr solution with 65 mL of 0.36 $M$ KBr solution (assume a total volume of 100 mL).

**3.87** Describe how you would prepare each of the following solutions: **(a)** 0.300 L of 0.35 $M$ KOH solution, beginning with solid KOH; **(b)** 1.00 L of a 0.500 $M$ $AgNO_3$ solution, starting with solid $AgNO_3$; **(c)** 1 liter of a 0.200 $M$ solution of $H_2SO_4$, starting with a concentrated $H_2SO_4$ solution of unknown concentration and a quantity of 0.150 $M$ NaOH solution.

**3.88** Calculate the molarity of the solution produced by mixing **(a)** 50.0 mL of 0.200 $M$ NaCl and 100.0 mL of 0.100 $M$ NaCl; **(b)** 24.5 mL of 1.50 $M$ NaOH and 20.5 mL of 0.850 $M$ NaOH. (Assume that the volumes are additive.)

**3.89** Using modern analytical techniques it is possible to detect sodium ions in concentrations as low as 50 pg/mL. What is this detection limit expressed in **(a)** molarity of $Na^+$; **(b)** $Na^+$ ions per cubic centimeter?

**3.90** What volume of 0.20 $M$ HBr is required to react with 18.0 g of zinc according to the equation

$$2HBr(aq) + Zn(s) \longrightarrow ZnBr_2(aq) + H_2(g)$$

**[3.91]** The arsenic in a 1.22-g sample of a pesticide was converted to $AsO_4^{3-}$ by suitable chemical treatment. It was then titrated using $Ag^+$ to form $Ag_3AsO_4$ as a precipitate. If it took 25.0 mL of 0.102 $M$ $Ag^+$ to reach the equivalence point in this titration, what is the percentage of arsenic in the pesticide?

**[3.92]** A solid sample of $Zn(OH)_2$ is added to 0.400 L of a 0.550 $M$ solution of HBr. The solution that remains is still acidic. It is then titrated with 0.500 $M$ NaOH solution, and 165 mL of the NaOH solution is required to reach the equivalence point. What was the mass of $Zn(OH)_2$ added to the HBr solution?

# Thermochemistry

When you eat an orange, the sugar it contains reacts in your body with oxygen to ultimately form $CO_2$ and $H_2O$. But more occurs than merely this conversion of chemicals—energy is released. The food you eat is the fuel that your body uses to operate your muscles and to maintain proper body temperature. This example illustrates a general point: Chemical reactions involve changes in energy. Some reactions, like the oxidation of sugar, produce energy. Others, like the electrolysis of water, use energy. At the present time, over 90 percent of the energy produced in our society comes from chemical reactions, principally from the combustion of coal, petroleum products, and natural gas.

The study of energy and its transformations is known as **thermodynamics** (Greek: *therme*, "heat"; *dynamis*, "power"). This area of study began during the Industrial Revolution as the relationships among heat, work, and the energy content of fuels were studied in an effort to maximize the performance of steam engines. Thermodynamics is important not only to chemistry but to other areas of science and to engineering as well. It touches our daily lives as we use energy for manufacturing, travel, and communications. Thermodynamics relates to such diverse topics as the metabolism of foods, the operation of batteries, and the design of engines.

In this chapter we will take a look at the relationships between chemical reactions and energy changes. This aspect of thermodynamics is called **thermochemistry**. We will discuss additional concepts of thermodynamics later, in Chapter 19.

## CONTENTS

## 4.1 THE NATURE OF ENERGY

Because energy is not tangible, as are material objects, it is a little difficult to define it completely. The accomplishment of work and the appearance of heat are two ways in which we can tell that an energy change has occurred. In fact, energy is defined as the capacity to do work or to transfer heat. Heat and work are the two ways we experience energy changes.

The most familiar kind of work, **mechanical work** ($w$), is defined and measured by the product of the net force ($F$) operating on an object and the distance ($d$) through which that force moves:

$$w = F \times d \qquad [4.1]$$

That is, work is the movement of an object against some force, and energy is required to accomplish this work. For example, energy is required to lift a book against the force of gravity and to separate a positively charged ion from a negatively charged one, overcoming the attractive force that

Explosions, such as that of a Titan rocket shown in this photo, are dramatic demonstrations that energy changes accompany chemical reactions. (Hanan Isachar/Sygma)

exists between them. **Heat** is energy that is transferred from one object to another because of a difference of temperature. For example, as a hot dish fresh from the oven cools, it transfers energy as heat to its surroundings.

## Forms of Energy

Energy can be classified into two principal types: kinetic energy and potential energy. **Kinetic energy** is the energy of motion. The magnitude of the kinetic energy of an object depends on its mass ($m$) and velocity ($v$):

$$E_k = \tfrac{1}{2}mv^2 \qquad\qquad [4.2]$$

This equation expresses quantitatively what our experience teaches—that both the mass and speed of an object determine how much work it can accomplish. For example, the heavier a football player is and the faster he moves, the more work he can accomplish against opposing players.

An object can also possess energy by virtue of its position. This stored energy is called **potential energy**. A heavy brick held high in the air has potential energy because of its position above the ground and because of the force of gravity acting on it. If it falls, its potential energy will be converted to kinetic energy, and the brick will do work on whatever object it strikes. Likewise, an electron has potential energy when located near a proton because of the electrostatic force between them.

Besides the kinetic and potential energies of objects, other forms of energy exist. These forms include electrical energy, nuclear energy, radiant energy (light), and chemical energy. However, with the advent of atomic theory, these other forms of energy have come to be considered as kinetic or potential energy at the atomic or molecular level. For example, the chemical energy of gasoline can be regarded as potential energy stored in the electrons and nuclei because of their positions within the molecules that comprise the gasoline.

## Energy Units

The SI unit of energy is the **joule** (J); $1\ J = 1\ kg\text{-}m^2/s^2$. A mass of 2 kg moving at a velocity of 1 m/s possesses a kinetic energy of 1 J:

$$E = \tfrac{1}{2}(2\ kg)(1\ m/s)^2 = 1\ kg\text{-}m^2/s^2 = 1\ J$$

To further appreciate the size of a joule, we might note that a 100-watt (W) light bulb produces 100 J of energy each second it operates.

Traditionally, energy changes accompanying chemical reactions have been expressed in units of calories. A **calorie** (cal) is the amount of energy required to raise the temperature of 1 g of water by 1°C, from 14.5°C to 15.5°C.* A kilocalorie (1000 cal) is the same as the Calorie (capitalized), used in expressing the energy values of foods. One calorie is the same amount of energy as 4.184 J:

$$1\ cal = 4.184\ J$$

---

* The temperature interval 14.5 to 15.5°C is specified because the energy required to raise the temperature of water by 1°C is slightly different for different temperatures. However, it is constant to three significant figures, 1.00 cal, over the entire range of liquid water, from 0 to 100°C.

In much of engineering work it is common to use the British thermal unit (Btu); 1 Btu = 1.05 kJ. A Btu is the amount of heat required to raise the temperature of 1 lb of water by 1°F. In most places in this text, the kilojoule will be used as the unit of energy.

---

**SAMPLE EXERCISE 4.1**

(a) A 145-g baseball is thrown with a speed of 25 m/s. Calculate the kinetic energy of the ball in joules. (b) What is the kinetic energy of the ball in calories?

**Solution:** (a) To calculate the kinetic energy, we use Equation 4.2. Recall that a joule is a kg-m$^2$/s$^2$. Thus mass needs to be expressed in kilograms: 145 g = 0.145 kg.

$$E_k = \tfrac{1}{2}mv^2 = \tfrac{1}{2}(0.145 \text{ kg})(25 \text{ m/s})^2 = 45 \text{ J}$$

(b) To obtain the kinetic energy in calories, we use the conversion factor 1 cal = 4.184 J:

$$(45 \text{ J})\left(\frac{1 \text{ cal}}{4.184 \text{ J}}\right) = 11 \text{ cal}$$

**PRACTICE EXERCISE**

(a) Calculate the kinetic energy, in joules, of a 6.0-kg object moving at the speed of 5.0 m/s. (b) What is its kinetic energy in calories?
*Answers:* (a) 75 J; (b) 18 cal

---

**Systems and Surroundings**

To study energy changes we focus our attention on a limited and well-defined part of the universe. The portion we single out for study is called the **system**; everything else is called the **surroundings**. The systems we can most readily study are those that exchange energy but not matter with their surroundings. As an example, consider a mixture of hydrogen gas, $H_2$, and oxygen gas, $O_2$, in a cylinder, as illustrated in Figure 4.1. The system in this case is just the hydrogen and oxygen; the cylinder, piston, and everything beyond them are their surroundings. If the hydrogen and oxygen react to form water, energy is produced:

$$2H_2(g) + O_2(g) \longrightarrow 2H_2O(l) + \text{energy}$$

Although the chemical form of the hydrogen and oxygen atoms in the system is changed by this reaction, the system has not lost or gained mass; it undergoes no exchange of matter with its surroundings. However, it does exchange energy with its surroundings in the form of heat and work. These are quantities that we can measure.

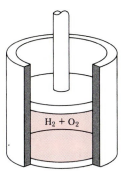

**FIGURE 4.1** Hydrogen and oxygen gases in a cylinder. If we are interested in studying the properties of the gases only, the gases are the system and the cylinder and piston are considered part of the surroundings.

---

**4.2 THE FIRST LAW OF THERMODYNAMICS**

A very important fact is that *energy is conserved*. Although energy can be converted from one form to another, it is neither created nor destroyed. The energy lost by a system equals the energy gained by its surroundings. Likewise, the energy gained by a system equals that lost by its surroundings. This important observation is known as the **law of conservation of energy**. Because it is the most elementary of thermodynamic concepts, it is also known as the **first law of thermodynamics**.

**Internal Energy Changes**

The total energy of a system is the sum of all the kinetic and potential energies of its component parts. For the system shown in Figure 4.1, total energy includes not only the motions and interactions of the $H_2$ and $O_2$

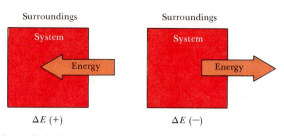

**FIGURE 4.2** If the system gains energy (left), $\Delta E$ will have a positive sign. If the system loses energy (right), $\Delta E$ will be negative.

$$\Delta E \, (+) \qquad\qquad \Delta E \, (-)$$

molecules themselves, but also of their component nuclei and electrons. This total energy is called the **internal energy** of the system. Because there are so many types of motions and interactions, it is impossible to determine the exact internal energy of any system of practical interest. We can, however, measure the *changes* in internal energy that accompany chemical and physical processes.

We define the change in internal energy, which we represent as $\Delta E$ (read "delta E"),* as the difference between the internal energy of the system at the completion of a process and that at the beginning:

$$\Delta E = E_{final} - E_{initial} \qquad\qquad [4.3]$$

Thermodynamic quantities like $\Delta E$ have two parts: a number, giving the magnitude of the change, and a sign, giving the direction. A *positive* $\Delta E$ results when $E_{final} > E_{initial}$, indicating that the system has gained energy from its surroundings. A negative $\Delta E$ is obtained when $E_{final} < E_{initial}$, indicating that the system has lost energy to its surroundings. These two situations are compared in Figure 4.2.

In a chemical reaction, of course, the initial state of the system refers to the reactants and the final state refers to the products. When hydrogen and oxygen react to form water, the system loses energy; $\Delta E$ for this process is negative. This means that the internal energy of the hydrogen and oxygen is greater than that of the water. This fact is represented by the energy diagram shown in Figure 4.3. Systems of higher energy tend to lose energy. Higher-energy systems therefore are generally less stable and more likely to undergo change than systems possessing lower energy.

**FIGURE 4.3** A system composed of $H_2(g)$ and $O_2(g)$ has a greater energy content than one composed of $H_2O(l)$. The system loses energy (negative $\Delta E$) when $H_2$ and $O_2$ are converted to $H_2O$. It gains energy (positive $\Delta E$) when $H_2O$ is decomposed into $H_2$ and $O_2$.

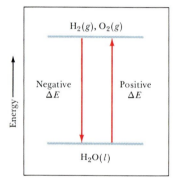

### Relating $\Delta E$ to Heat and Work

Any system can exchange energy with its surroundings in two general ways: as heat or as work. The internal energy of a system changes in magnitude as heat is added to or removed from the system or as work is done on it or by it. We can use these ideas to write a very useful algebraic expression of the first law of thermodynamics. When a system undergoes any chemical or physical change, the accompanying change in its internal energy, $\Delta E$, is given by the heat added to the system ($q$) plus the work done on the system ($w$):

$$\Delta E = q + w \qquad\qquad [4.4]$$

The *heat added to the system is assigned a positive sign*. Likewise, *work done to the system is positive*. Both the heat added to the system and the work done on it increase the internal energy. On the other hand, both the heat lost

---

* The symbol $\Delta$ is commonly used to denote *change*. For example, a change in volume can be represented by $\Delta V$.

by the system and the work done by the system on its surroundings are negative; they reduce the internal energy. For example, if the system absorbs 50 J of heat and does 10 J of work on its surroundings, $q = 50$ J and $w = -10$ J. Thus $\Delta E = 50$ J $+ (-10$ J$) = 40$ J.*

A reaction that results in the evolution of heat is **exothermic** (*exo-* is a prefix meaning "out of"); that is, energy flows out of the system and into its surroundings. Combustion reactions are exothermic. Reactions that absorb heat from their surroundings are **endothermic** (*endo-* is a prefix meaning "into"). The melting of ice is an endothermic process.

## SAMPLE EXERCISE 4.2

The hydrogen and oxygen gases in the cylinder illustrated in Figure 4.1 are ignited. As the reaction occurs, the system loses 550 J of heat to its surroundings (the reaction is exothermic). The reaction also causes the piston to move upward as the gases expand. It is determined that the expanding gas does 240 J of work on the surroundings. What is the change in the internal energy of the system?

**Solution:**  From the sign conventions for $q$ and $w$, we have $q = -550$ J and $w = -240$ J (that is, energy

flows from the system both as heat and as work). Using Equation 4.4, we have

$$\Delta E = q + w = (-550\text{ J}) + (-240\text{ J}) = -790\text{ J}$$

### PRACTICE EXERCISE

Calculate $\Delta E$ for an endothermic process in which the system absorbs 65 J of heat and also receives 12 J of work from its surroundings.
*Answer:*  77 J

Although we usually have no way of knowing the precise value of the internal energy of a system, we do know that it has a fixed value for a given set of conditions. The conditions that influence this energy include the temperature and pressure. We also know that the total internal energy of a system is proportional to the total quantity of matter in the system; energy is an extensive property (Section 1.3).

Suppose we define our system as 50 g of water at 25°C, as in Figure 4.4. Our system could have arrived at that state by our cooling 50 g of water from 100°C or by our melting 50 g of ice and subsequently warming the

**State Functions**

---

* Equation 4.4 is sometimes written $\Delta E = q - w$. When written this way, work done *by* the system is defined as positive. This convention has merit in the many engineering applications that focus interest on a machine that does work on its surroundings.

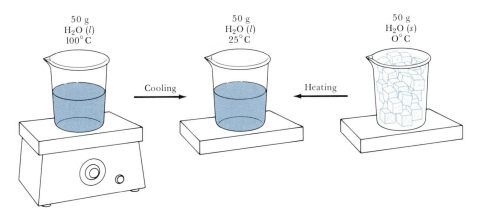

50 g
$H_2O$ (*l*)
100°C

Cooling →

50 g
$H_2O$ (*l*)
25°C

← Heating

50 g
$H_2O$ (*s*)
0°C

**FIGURE 4.4**  Internal energy, a state function, depends only on the present state of the system and not on the path by which it arrived at that state. The internal energy of 50 g of water at 25°C is the same whether the water is cooled from a higher temperature to 25°C or is obtained by melting 50 g of ice, and then warming to 25°C.

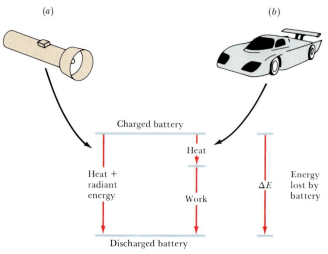

**FIGURE 4.5** When a battery is discharged in lighting a flashlight, all the energy of the battery appears as radiant energy and heat; no work is done. When the battery is used in the toy car, work is done in moving the car from place to place. Thus the work done by the system (the battery) is not a state function, because its magnitude depends on the particular path by which the system gets from its initial to its final state.

water to 25°C. The internal energy of the water is the same in either case. The internal energy of a system is a **state function**, a property of a system that is determined by specifying its condition, or its state (in terms of temperature, pressure, location, and so forth). *The value of a state function does not depend on the particular previous history of the sample, only on its present condition.* Thus $\Delta E$ depends only on the initial and final states of the system and not on how the change occurs.

You may well ask: What is an example of something that is *not* a state function? The work done by a system in a given process is not a state function. Rather, it depends on the manner in which the process is carried out. As an example, let's consider a flashlight battery as our system, and let the change in the system be the complete discharge of the battery at constant temperature. If the battery is discharged in a flashlight [Figure 4.5(a)], no mechanical work is accomplished. All of the energy lost from the battery appears as radiant energy and heat. If the battery is used in a mechanical toy [Figure 4.5(b)], the same change in state of the battery produces mechanical work and heat. The change in state of the system, and thus the change in $\Delta E$, is the same in both cases. However, the amount of work done in the two cases is different. We see from this example that although $\Delta E$ is always the same for a given change in the system, the way in which the change is performed will determine the relative contributions of $q$ and $w$, the means by which energy is transferred.

## 4.3 *P–V* WORK AND ENTHALPY

There are many varieties of work, but in chemistry we are interested mainly in electrical work and in the mechanical work done by expanding gases. In this chapter we restrict our discussion to the latter type of work. Expansion work, which results from a change in the volume of the system, is usually called pressure–volume or *P–V* work. Expanding gases in the cylinder of an automobile engine do *P–V* work on the piston, and this work eventually turns the wheels. Expanding gases from an open reaction vessel do *P–V* work in pushing back the atmosphere, but this work is lost to us; it accomplishes nothing in a practical sense. However, all kinds of work, whether useful or not, must be considered if we desire to keep track of all the energy changes of a system.

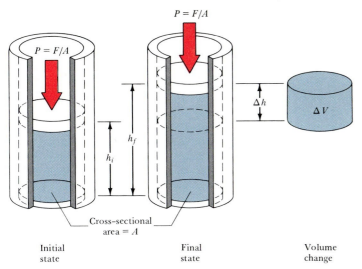

$P = F/A$

$\Delta h$

$\Delta V$

$P = F/A$

$h_f$

$h_i$

Cross-sectional
area = $A$

Initial
state

Final
state

Volume
change

**FIGURE 4.6** A piston moving a distance $\Delta h$ against a pressure $P$ does work on the surroundings. The amount of work done is given by the product of the pressure and the change in volume of the gas: $w = -P\Delta V$.

To understand $P-V$ work better, consider a gas confined to a cylinder with a movable piston, as shown in Figure 4.6. A downward force, $F$, acts on the piston. The pressure on the gas is defined as the ratio of this force to the area of the piston, $A$: $P = F/A$. We will assume that the piston itself is weightless and that the only force acting on it ($F$) is the weight of the earth's atmosphere. This pressure is called the atmospheric pressure. We will have more to say about atmospheric pressure and about the properties of gases in Chapter 10; at this point we need only be aware that the atmosphere exerts a pressure. Now assume that the gas in the cylinder expands against this constant external pressure. This expansion could be caused by heating the gas or by a reaction that increases the number of moles of gas in the system. The magnitude of the work accomplished by the expanding gas is given by the product of the pressure and the volume change of the piston, $\Delta V$; because the expanding gas does work on its surroundings in pushing back the atmosphere, the work has a negative sign:

$$w = -P\Delta V \qquad [4.5]$$

For the case where only $P-V$ work is done, Equation 4.5 can be substituted into Equation 4.4 to give

**Enthalpy**

$$\Delta E = q + w = q - P\Delta V \qquad [4.6]$$

When a reaction is carried out in a constant-volume container ($\Delta V = 0$), the heat flow equals the change in internal energy:

$$q_V = \Delta E \qquad \text{(constant volume)} \qquad [4.7]$$

(We put a subscript $v$ on $q$ to remind ourselves that we are considering a special case where volume is constant.) This equation tells us that if the volume of the system is constant, any heat added to or lost from the system equals the change in internal energy. In contrast, if the volume

Equation 4.5 is readily derived from the definitions of work and pressure. The magnitude of the work accomplished by a moving piston, such as that shown in Figure 4.6, is given by the product of the force, $F$, and the distance the piston moves, $\Delta h$:

$$\text{Work} = \text{force} \times \text{distance} = F \times \Delta h$$

Rearranging the equation that defines pressure, $P = F/A$, gives $F = P \times A$. Thus

$$\text{Work} = F \times \Delta h = P \times A \times \Delta h$$

The product $A \times \Delta h$ is the volume change, $\Delta V$, resulting from the piston moving. Thus

$$\text{Work} = P \times A \times \Delta h = P \times \Delta V$$

But in our derivation we have ignored the sign of this work. Because an expanding gas must do work on its surroundings, the sign of the work must be negative. Thus we have Equation 4.5: $w = -P\Delta V$. This equation applies equally well to the case of a contracting gas. When a gas is compressed, $\Delta V$ is negative. In that case the product $-P\Delta V$ will be positive, as it should be when work is done *on* the system by the surroundings.

is allowed to change and the pressure is constant, we have

$$q_P = \Delta E + P\Delta V \qquad \text{(constant pressure)} \qquad [4.8]$$

Thus the heat needed to bring about any change at constant pressure is the sum of the internal energy change plus the $P-V$ work.

The two conditions of constant volume and constant pressure are frequently encountered. The constant-pressure case is especially important in chemistry, because most chemical reactions, including those in living systems, take place under the essentially constant pressure of the earth's atmosphere.

If we write Equation 4.8 in full we have

$$
\begin{aligned}
q_P &= (E_{\text{final}} - E_{\text{initial}}) + (PV_{\text{final}} - PV_{\text{initial}}) \\
&= (E_{\text{final}} + PV_{\text{final}}) - (E_{\text{initial}} + PV_{\text{initial}})
\end{aligned}
$$

Thus $q_P$ is the difference between two terms, each having the form $E + PV$. Because of the importance of constant-pressure processes, the quantity $E + PV$ is given a special symbol, $H$, and a special name, **enthalpy** (from the Greek word *enthalpein*, which means "to warm"). Thus we can write

$$q_P = \Delta E + P\Delta V = \Delta H \qquad [4.9]$$

*Enthalpy*, like internal energy, *is a state function.*

Equation 4.9 tells us that at constant pressure, the change in enthalpy ($\Delta H$) of a system is equal to the energy flow as heat. That is, if a reaction is studied at constant pressure (as it would be in an open container), the flow of heat equals the change of enthalpy of the system. For this reason the enthalpy change for a reaction is commonly called the heat of reaction.

Enthalpy changes are extremely important in thermochemistry. In fact, chemists tend to think in terms of $\Delta H$ rather than $\Delta E$. The volume change accompanying many reactions is close to zero, so $\Delta H$ is close to $\Delta E$. Even when the volume change is not near zero, the difference in magnitude between $\Delta H$ and $\Delta E$ is generally small.

## 4.4 ENTHALPIES OF REACTION

The enthalpy change for a chemical reaction is given by the enthalpy of the products minus that of the reactants:

$$\Delta H = H(\text{products}) - H(\text{reactants}) \qquad [4.10]$$

If the products of the reaction have a greater enthalpy than the reactants, $\Delta H$ will be *positive*. In this case the system absorbs heat, and the reaction is *endothermic*. An example of an endothermic process is the formation of nitric oxide, NO, from nitrogen, $N_2$, and oxygen, $O_2$:

$$\text{Heat} + N_2(g) + O_2(g) \longrightarrow 2NO(g)$$

This reaction accompanies high-temperature combustion reactions in air and contributes greatly to smog. The enthalpy content of the $2NO(g)$ is greater than that of the $N_2(g) + O_2(g)$, as represented graphically in Figure 4.7.

If the enthalpy content of the products of a reaction is less than that of the reactants (an *exothermic* reaction), $\Delta H$ will be *negative*. This situation arises, for example, when hydrogen is combusted in the presence of oxygen:

$$2H_2(g) + O_2(g) \longrightarrow 2H_2O(g) + \text{heat}$$

This reaction produced the disastrous explosion of the space shuttle *Challenger* in 1986. Figure 4.8 shows the explosion of a balloon filled with $H_2$ as it is ignited by a burning candle. The change in the enthalpy of the system during this reaction is diagrammed in Figure 4.8c.

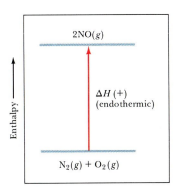

**Figure 4.7** The enthalpy change for the combustion of $N_2(g) + O_2(g)$ to form $2NO(g)$. The system absorbs heat from its surroundings, increasing in enthalpy. The process is endothermic, and $\Delta H$ is positive.

**FIGURE 4.8** The enthalpy change for the combustion of $H_2(g)$ in $O_2(g)$ to form $H_2O(l)$. The system gives off heat to the surroundings, decreasing in enthalpy. The reaction is exothermic, and $\Delta H$ is negative, as shown in (*c*). [(*a, b*) Donald Clegg and Roxy Wilson]

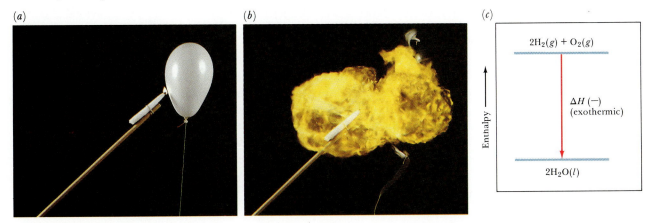

Before exploring further how we might use the concept of enthalpy, let's consider three of its important characteristics.

1. *Enthalpy is an extensive property.* This fact means that the magnitude of $\Delta H$ is directly proportional to the amount of reactant consumed in the process. Consider the combustion of methane to form carbon dioxide and water. It is found experimentally that 802 kJ of heat is produced when 1 mol of $CH_4$ is burned in a constant-pressure system. We can express this fact as follows:

$$CH_4(g) + 2O_2(g) \longrightarrow CO_2(g) + 2H_2O(g) \qquad \Delta H = -802 \text{ kJ} \qquad [4.11]$$

The negative sign for $\Delta H$ tells us that this reaction is exothermic. Notice that $\Delta H$ is reported at the end of a balanced equation, with no explicit mention of the amounts of chemicals involved. In such cases it is understood that the coefficients in the balanced equation represent the number of moles of reactants producing the associated enthalpy change. Thus combustion of 1 mol of $CH_4$ with 2 mol of $O_2$ produces 802 kJ of heat. The combustion of 2 mol of $CH_4$ with 4 mol of $O_2$ produces 1604 kJ.

---

**SAMPLE EXERCISE 4.3**

How much heat is produced when 4.50 g of methane gas is burned in a constant-pressure system? (Use the information given in Equation 4.11.)

**Solution:** According to Equation 4.11, 802 kJ is produced when a mole of $CH_4$ is burned ($\Delta H = -802$ kJ/mol). A mole of $CH_4$ has a mass of 16.0 g.

$$\text{Heat} = (4.50 \text{ g } CH_4)\left(\frac{1 \text{ mol } CH_4}{16.0 \text{ g } CH_4}\right)$$
$$\times \left(-802 \frac{\text{kJ}}{\text{mol } CH_4}\right) = -226 \text{ kJ}$$

**PRACTICE EXERCISE**

Ammonium nitrate can decompose explosively by the following reaction:

$$NH_4NO_3(s) \longrightarrow N_2O(g) + 2H_2O(g)$$
$$\Delta H = -37.0 \text{ kJ}$$

Calculate the quantity of heat produced when 2.50 g of $NH_4NO_3$ decomposes at constant pressure.

*Answer:* $-1.16$ kJ

---

2. *The enthalpy change for a reaction is equal in magnitude but opposite in sign to $\Delta H$ for the reverse reaction.* For example, when Equation 4.11 is reversed, $\Delta H$ for the process is $+802$ kJ:

$$CO_2(g) + 2H_2O(g) \longrightarrow CH_4(g) + 2O_2(g) \qquad \Delta H = 802 \text{ kJ} \qquad [4.12]$$

If more energy were produced by combustion of $CH_4$ than was required for the reverse reaction, it would be possible to use these processes to create an unlimited supply of energy. Some $CH_4$ could be combusted, and that portion of the energy necessary to reform $CH_4$ could be saved. The rest could be used to do work of some type. After $CH_4$ is reformed, it could again be combusted, and so forth, continually supplying energy. This, of course, is clearly contrary to our experience—such behavior does not obey the law of conservation of energy. The observed situation is shown in Figure 4.9.

**3.** *The enthalpy change for a reaction depends on the state of the reactants and products.* If the product in the combustion of methane (Equation 4.11) were liquid $H_2O$ instead of gaseous $H_2O$, $\Delta H$ would be $-890$ kJ instead of $-802$ kJ. More heat is available for transfer to the surroundings because 88 kJ is released when 2 mol of gaseous water is condensed to the liquid state:

$$2H_2O(g) \longrightarrow 2H_2O(l) \qquad \Delta H = -88 \text{ kJ} \qquad [4.13]$$

Therefore, the states of the reactants and products must be specified. In addition, we will generally assume that the reactants and products are both at the same temperature: 25°C, unless otherwise indicated.

The enthalpy change associated with a given chemical process is often of great significance. In the sections that follow we will consider some ways in which we can evaluate this important quantity. As we shall see, $\Delta H$ for a reaction can be either directly determined by experiment or calculated from a knowledge of the enthalpy changes associated with other reactions by invoking the first law of thermodynamics. Let's begin by discussing how we make use of the concept of a state function to carry out calculations of enthalpy changes.

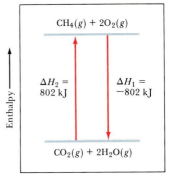

**FIGURE 4.9** If a substance is combusted and then reformed from its combustion products, the total enthalpy change in the system must be zero $(\Delta H_1 + \Delta H_2 = 0)$. In general, the net energy change in the system for a series of reactions that eventually regenerate the original substances in their original states is zero.

---

## 4.5  HESS'S LAW

Because enthalpy is a state function, the enthalpy change, $\Delta H$, associated with any chemical process depends only on the amount of matter that undergoes change and on the nature of the initial state of the reactants and the final state of the products. This means that if a particular reaction can be carried out in one step or in a series of steps, the sum of the enthalpy changes associated with the individual steps must be the same as the enthalpy change associated with the one-step process. For example, the enthalpy change for the combustion of methane to form carbon dioxide and liquid water can be calculated from $\Delta H$ for the condensation of water vapor and $\Delta H$ for combustion to methane to form gaseous water:

$$CH_4(g) + 2O_2(g) \longrightarrow CO_2(g) + 2H_2O(g) \qquad \Delta H = -802 \text{ kJ}$$

(Add) 
$$\underline{2H_2O(g) \longrightarrow 2H_2O(l) \qquad\qquad\qquad\qquad \Delta H = \phantom{-}-88 \text{ kJ}}$$

$$CH_4(g) + 2O_2(g) + 2H_2O(g) \longrightarrow CO_2(g) + 2H_2O(l) + 2H_2O(g) \qquad \Delta H = -890 \text{ kJ}$$

Net equation:

$$CH_4(g) + 2O_2(g) \longrightarrow CO_2(g) + 2H_2O(l) \qquad \Delta H = -890 \text{ kJ}$$

To obtain the net equation, the sum of the reactants of the two equations are placed on one side of the arrow, and the sum of the products on the other side. Because $2H_2O(g)$ occurs on both sides of the arrow, it can be canceled like an algebraic quantity that appears on both sides of an equal sign.

**Hess's law** states that *if a reaction is carried out in a series of steps, $\Delta H$ for the reaction will be equal to the sum of the enthalpy changes for each step.* The overall enthalpy change for the process is independent of the number of steps or the particular nature of the path by which the reaction is carried out. Thus we can calculate $\Delta H$ for any process, as long as we find a route for which $\Delta H$ is known for each step. This important fact permits us to use a relatively small number of experimental measurements to calculate $\Delta H$ for a vast number of different reactions.

Hess's law provides a useful means of calculating energy changes that are difficult to measure directly. For instance, it is not possible to measure directly the heat of combustion of carbon to form carbon monoxide. Combustion of 1 mol of carbon with $\frac{1}{2}$ mol of $O_2$ produces not only CO but also $CO_2$, leaving some carbon unreacted. However, the heat of the reaction forming CO can be calculated as shown in Sample Exercise 4.4.

## SAMPLE EXERCISE 4.4

The heat of combustion of C to $CO_2$ is $-393.5$ kJ/mol C and the heat of combustion of CO to $CO_2$ is $-283.0$ kJ/mol CO:

(1) $C(s) + O_2(g) \longrightarrow CO_2(g)$
$$\Delta H = -393.5 \text{ kJ}$$

(2) $CO(g) + \frac{1}{2}O_2(g) \longrightarrow CO_2(g)$
$$\Delta H = -283.0 \text{ kJ}$$

Using these data, calculate the heat of combustion of C to CO:

(3) $C(s) + \frac{1}{2}O_2(g) \longrightarrow CO(g)$

**Solution:** In order to use equations (1) and (2), we need to arrange them so that $C(s)$ is on the reactant side and $CO(g)$ is on the product side of the arrow, as we see in the target reaction, equation (3). Because equation (1) has $C(s)$ as a reactant, we can use that equation just as it is. Notice that the target reaction has CO as a product. Thus we need to turn equation (2) around so that $CO(g)$ is a product. Remember that when reactions are turned around, the sign of $\Delta H$ is reversed. We arrange the two equations so that they can be added to give the target equation:

$$
\begin{array}{lll}
C(s) + O_2(g) \longrightarrow & CO_2(g) & \Delta H = -393.5 \text{ kJ} \\
CO_2(g) \longrightarrow & CO(g) + \frac{1}{2}O_2(g) \\
& & \Delta H = \quad 283.0 \text{ kJ} \\
\hline
C(s) + \frac{1}{2}O_2(g) \longrightarrow & CO(g) & \Delta H = -110.5 \text{ kJ}
\end{array}
$$

When we add the two equations, $CO_2(g)$ appears on both sides of the arrow and therefore cancels out. Likewise $\frac{1}{2}O_2(g)$ is eliminated from each side when the two equations are summed.

## PRACTICE EXERCISE

Carbon occurs in two forms, graphite and diamond. The heat of combustion of graphite is $-393.5$ kJ/mol, and that of diamond is $-395.4$ kJ/mol:

$$C(\text{graphite}) + O_2(g) \longrightarrow CO_2(g)$$
$$\Delta H = -393.5 \text{ kJ}$$

$$C(\text{diamond}) + O_2(g) \longrightarrow CO_2(g)$$
$$\Delta H = -395.4 \text{ kJ}$$

Calculate $\Delta H$ for the conversion of graphite to diamond:

$$C(\text{graphite}) \longrightarrow C(\text{diamond})$$

*Answer:* $+1.9$ kJ

## SAMPLE EXERCISE 4.5

Calculate $\Delta H$ for the reaction

$$2C(s) + H_2(g) \longrightarrow C_2H_2(g)$$

given the following reactions and their respective enthalpy changes:

$$C_2H_2(g) + \frac{5}{2}O_2(g) \longrightarrow 2CO_2(g) + H_2O(l)$$
$$\Delta H = -1299.6 \text{ kJ}$$

$$C(s) + O_2(g) \longrightarrow CO_2(g)$$
$$\Delta H = -393.5 \text{ kJ}$$

$$H_2(g) + \frac{1}{2}O_2(g) \longrightarrow H_2O(l)$$
$$\Delta H = -285.9 \text{ kJ}$$

**Solution:** Because the target equation has $C_2H_2$ as a product, we turn the first equation around; the sign of $\Delta H$ is therefore changed. Because the target equation has $2C(s)$ as a reactant, we multiply the second equation and its $\Delta H$ by 2. Because the target equation has $H_2$ as a reactant, we keep the third equation as it is. We then add the three equations and their enthalpy changes in accordance with Hess's law:

$$2CO_2(g) + H_2O(l) \longrightarrow C_2H_2(g) + \frac{5}{2}O_2(g)$$
$$\Delta H = \quad 1299.6 \text{ kJ}$$

$$2C(s) + 2O_2(g) \longrightarrow 2CO_2(g)$$
$$\Delta H = -787.0 \text{ kJ}$$

$$H_2(g) + \tfrac{1}{2}O_2(g) \longrightarrow H_2O(l)$$
$$\Delta H = -285.9 \text{ kJ}$$

$$2C(s) + H_2(g) \longrightarrow C_2H_2(g)$$
$$\Delta H = 226.7 \text{ kJ}$$

When the equations are added, there are $2CO_2, \tfrac{5}{2}O_2$, and $H_2O$ on both sides of the arrow. These are canceled in writing the net equation.

**PRACTICE EXERCISE** _____

Calculate $\Delta H$ for the reaction

$$NO(g) + O(g) \longrightarrow NO_2(g)$$

given the following information:

$$NO(g) + O_3(g) \longrightarrow NO_2(g) + O_2(g)$$
$$\Delta H = -198.9 \text{ kJ}$$

$$O_3(g) \longrightarrow \tfrac{3}{2}O_2(g) \quad \Delta H = -142.3 \text{ kJ}$$

$$O_2(g) \longrightarrow 2O(g) \quad \Delta H = 495.0 \text{ kJ}$$

*Answer:* $-304.1$ kJ

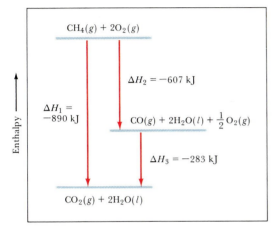

**FIGURE 4.10** The quantity of heat generated by combustion of $CH_4$ is independent of whether the reaction takes place in one or more steps ($\Delta H_1 = \Delta H_2 + \Delta H_3$).

The first law of thermodynamics, in the form of Hess's law, teaches us that we can never expect to obtain more (or less) energy from a chemical reaction by changing the method of carrying out the reaction. For example, for the reaction of methane, $CH_4$, and oxygen, $O_2$, to form $CO_2$ and $H_2O$, we may envision the reaction to occur either directly or with the initial formation of CO, which is then subsequently combusted. This set of choices is illustrated in Figure 4.10. Because $\Delta H$ is a state function, either path produces the same change in the enthalpy content of the system. That is, $\Delta H_1 = \Delta H_2 + \Delta H_3$. Again, we may note that if this were not so, it would be possible to create energy continuously, in conflict with the first law of thermodynamics.

**4.6 HEATS OF FORMATION**

By using the methods we have just discussed, we can calculate the enthalpy changes for a great many reactions from a few tabulated values. Many experimental data are tabulated according to the type of process. For example, extensive tables exist of heats of vaporization ($\Delta H$ for converting liquids to gas), heats of fusion ($\Delta H$ for melting solids), heats of combustion ($\Delta H$ for combusting a substance in oxygen), and so forth. A particularly important process used for tabulating thermodynamic data is the formation of a compound from its constituent elements. The enthalpy change associated with this process is called the **heat of formation**, labeled $\Delta H_f$.

**TABLE 4.1** Standard heats of formation, $\Delta H_f^\circ$, at 25°C

| Substance | Formula | $\Delta H_f^\circ$ (kJ/mol) | Substance | Formula | $\Delta H_f^\circ$ (kJ/mol) |
|---|---|---|---|---|---|
| Acetylene | $C_2H_2(g)$ | 226.7 | Hydrogen chloride | $HCl(g)$ | −92.30 |
| Ammonia | $NH_3(g)$ | −46.19 | Hydrogen fluoride | $HF(g)$ | −268.6 |
| Benzene | $C_6H_6(l)$ | 49.04 | Hydrogen iodide | $HI(g)$ | 25.9 |
| Calcium carbonate | $CaCO_3(s)$ | −1207.1 | Methane | $CH_4(g)$ | −74.85 |
| Calcium oxide | $CaO(s)$ | −635.5 | Methanol | $CH_3OH(l)$ | −238.6 |
| Carbon dioxide | $CO_2(g)$ | −393.5 | Silver chloride | $AgCl(s)$ | −127.0 |
| Carbon monoxide | $CO(g)$ | −110.5 | Sodium bicarbonate | $NaHCO_3(s)$ | −947.7 |
| Diamond | $C(s)$ | 1.88 | Sodium carbonate | $Na_2CO_3(s)$ | −1130.9 |
| Ethane | $C_2H_6(g)$ | −84.68 | Sodium chloride | $NaCl(s)$ | −411.0 |
| Ethanol | $C_2H_5OH(l)$ | −277.7 | Sucrose | $C_{12}H_{22}O_{11}(s)$ | −2221 |
| Ethylene | $C_2H_4(g)$ | 52.30 | Water | $H_2O(l)$ | −285.8 |
| Glucose | $C_6H_{12}O_6(s)$ | −1260 | Water vapor | $H_2O(g)$ | −241.8 |
| Hydrogen bromide | $HBr(g)$ | −36.23 | | | |

The magnitude of any enthalpy change depends on the conditions of temperature, pressure, and state (gas, liquid, solid, crystalline form) of the reactants and products. A standard enthalpy change, $\Delta H^\circ$, is one that takes place with all reactants and products in their **standard states**. That is, all substances are in the forms most stable at the particular temperature of interest and at standard atmospheric pressure (1 atm; see Section 10.2). The temperature usually chosen for purposes of tabulating data is 298 K (25°C). Thus the **standard heat of formation** of a compound, $\Delta H_f^\circ$, is the change in enthalpy that accompanies the formation of one mole of that substance from its elements, with all substances in their standard states. For example, the standard heat of formation for ethanol—$C_2H_5OH$—is the enthalpy change for the following reaction:

$$2C(\text{graphite}) + 3H_2(g) + \tfrac{1}{2}O_2(g) \longrightarrow C_2H_5OH(l) \qquad \Delta H_f^\circ = -277.7 \text{ kJ} \qquad [4.14]$$

The elemental source of oxygen is $O_2$, not O or $O_3$, because $O_2$ is the stable form of oxygen at 25°C and standard atmospheric pressure. Similarly, the elemental source of carbon is graphite and not diamond, because the former is the stable (lower energy) form at 25°C and standard atmospheric pressure. The conversion of graphite to diamond requires the addition of energy:

$$C(\text{graphite}) \longrightarrow C(\text{diamond}) \qquad \Delta H^\circ = 1.88 \text{ kJ} \qquad [4.15]$$

As shown in Equation 4.14, the stable chemical source of hydrogen is $H_2$.

A few standard heats of formation are given in Table 4.1. A more complete table is provided in Appendix D. *By convention, the standard heat of formation of the stable form of any element is zero.* For example, $\Delta H_f^\circ$ for C(graphite) is zero.

## Calculating Heats of Reactions from Heats of Formation

The standard enthalpy change for any reaction, $\Delta H_{\text{rxn}}$, can be found by summing the heats of formation of all reaction products, taking care to multiply each molar heat of formation by the coefficient of that substance in the balanced equation, then subtracting a similar sum for all the heats of formation of all the reactants:

$$\Delta H^{\circ}_{\text{rxn}} = \sum n\, \Delta H^{\circ}_{f}(\text{products}) - \sum m\, \Delta H^{\circ}_{f}(\text{reactants}) \qquad [4.16]$$

The symbol $\sum$ (sigma) means "the sum of," and $n$ and $m$ are the stoichiometric coefficients of the chemical equation. For example, $\Delta H^{\circ}$ for the combustion of glucose,

$$C_6H_{12}O_6(s) + 6O_2(g) \longrightarrow 6CO_2(g) + 6H_2O(l) \qquad [4.17]$$

$$\Delta H^{\circ}_{\text{rxn}} = [6\,\Delta H^{\circ}_{f}(\text{CO}_2) + 6\,\Delta H^{\circ}_{f}(\text{H}_2\text{O})] - [\Delta H^{\circ}_{f}(\text{C}_6\text{H}_{12}\text{O}_6) + 6\,\Delta H^{\circ}_{f}(\text{O}_2)] \qquad [4.18]$$

Using the heats of formation recorded in Table 4.1, this process gives

$$\Delta H^{\circ}_{\text{rxn}} = \left[ (6 \text{ mol CO}_2)\left( -393.5\,\frac{\text{kJ}}{\text{mol CO}_2} \right)\right.$$
$$\left. + (6 \text{ mol H}_2\text{O})\left( -285.8\,\frac{\text{kJ}}{\text{mol H}_2\text{O}} \right)\right]$$
$$- \left[ (1 \text{ mol C}_6\text{H}_{12}\text{O}_6)\left( -1260\,\frac{\text{kJ}}{\text{mol C}_6\text{H}_{12}\text{O}_6} \right)\right.$$
$$\left. + (6 \text{ mol O}_2)\left( 0\,\frac{\text{kJ}}{\text{mol O}_2} \right)\right]$$
$$= -2816 \text{ kJ}$$

The general relationship expressed in Equation 4.16 is just an application of Hess's law. It follows directly from the fact that $\Delta H$ is a state function. This fact allows calculation of the enthalpy change for any reaction from the energies required to convert the initial reactants into elements and then combine the elements into the desired products. This reaction pathway is shown in Figure 4.11 for the combustion of glucose.

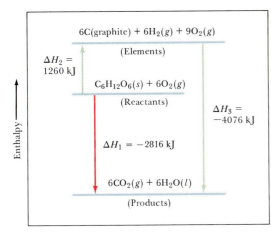

**FIGURE 4.11** Because $\Delta H$ is a state function, $\Delta H_1 = \Delta H_2 + \Delta H_3$. Note that $\Delta H_2$ is $-\Delta H^{\circ}_{f}(\text{C}_6\text{H}_{12}\text{O}_6)$, while $\Delta H_3$ is $6\Delta H^{\circ}_{f}(\text{H}_2\text{O}) + 6\Delta H^{\circ}_{f}(\text{CO}_2)$. That is, the enthalpy change for combustion of $C_6H_{12}O_6$ equals the sum of the enthalpy changes for converting the reactants into the standard states of the elements and then forming the products from the elements. This same result is given by Equation 4.18, because $\Delta H^{\circ}_{f}(\text{O}_2)$ is zero.

## SAMPLE EXERCISE 4.6

Compare the quantity of heat produced by combustion of 1.00 g of glucose ($C_6H_{12}O_6$) with that produced by 1.00 g of sucrose ($C_{12}H_{22}O_{11}$).

**Solution:** The example worked in the text gave $\Delta H° = -2816$ kJ for the combustion of a mole of glucose. The molecular weight of glucose is 180 amu. Therefore, the heat produced per gram is

$$\left(-2816 \frac{kJ}{mol}\right)\left(\frac{1\ mol}{180\ g}\right) = -15.6 \frac{kJ}{g}$$

For sucrose:

$$C_{12}H_{22}O_{11}(s) + 12O_2(g) \longrightarrow$$
$$12CO_2(g) + 11H_2O(l)$$
$$\Delta H°_{rxn} = [12\,\Delta H°_f(CO_2) + 11\,\Delta H°_f(H_2O)]$$
$$- [\Delta H°_f(C_{12}H_{22}O_{11}) + 12\,\Delta H°_f(O_2)]$$
$$= 12(-393.5\ kJ) + 11(-285.8\ kJ)$$
$$- (-2221\ kJ)$$
$$= (-4722 - 3144 + 2221)\ kJ$$
$$= -5645\ kJ/mol\ C_{12}H_{22}O_{11}$$

The molecular weight of sucrose is 342 amu. Therefore, the heat produced per gram of sucrose is

$$\left(-5645 \frac{kJ}{mol}\right)\left(\frac{1\ mol}{342\ g}\right) = -16.5 \frac{kJ}{g}$$

Both sucrose and glucose are carbohydrates. As a rule of thumb the energy obtained from the combustion of a gram of carbohydrate is 17 kJ (4 kcal or 4 Cal).

### PRACTICE EXERCISE
Using the standard heats of formation listed in Table 4.1, calculate the enthalpy change for the combustion of ethanol:

$$C_2H_5OH(l) + 3O_2(g) \longrightarrow 2CO_2(g) + 3H_2O(l)$$

(Remember that $O_2$ is an element in its standard state and so its standard heat of formation is zero.)

*Answer:* $-1366.7$ kJ

## SAMPLE EXERCISE 4.7

The standard enthalpy change for the reaction

$$CaCO_3(s) \longrightarrow CaO(s) + CO_2(g)$$

is 178.1 kJ. From the values for the standard heats of formation given in Table 4.1, calculate the standard heat of formation of $CaCO_3(s)$.

**Solution:** The standard enthalpy change in the reaction is

$$\Delta H°_{rxn} = [\Delta H°_f(CaO) + \Delta H°_f(CO_2)]$$
$$- \Delta H°_f(CaCO_3)$$

Inserting the known values, we have

$$178.1\ kJ = -635.5\ kJ - 393.5\ kJ$$
$$- \Delta H°_f(CaCO_3)$$

Solving for $\Delta H°_f(CaCO_3)$, we have

$$\Delta H°_f(CaCO_3) = -1207.1\ kJ/mol$$

### PRACTICE EXERCISE
Given the following standard heat of reaction, use the standard heats of formation in Table 4.1 to calculate the standard heat of formation of CuO(s):

$$CuO(s) + H_2(g) \longrightarrow Cu(s) + H_2O(l)$$
$$\Delta H° = -130.6\ kJ$$

*Answer:* $-155.2$ kJ

## 4.7 CALORIMETRY

$\Delta H$ can be determined experimentally by measuring the heat flow accompanying a reaction at constant pressure. When heat flows into or out of a substance, its temperature changes. Experimentally, we can determine the heat flow associated with a chemical reaction by measuring the temperature change it produces. The measurement of heat flow is called **calorimetry**; an apparatus that measures heat flow is called a **calorimeter**.

The temperature change experienced by a body when it absorbs a certain amount of heat is determined by its **heat capacity**—the energy required to raise the temperature of the body by 1°C. An object of large heat capacity absorbs more heat when it undergoes a certain temperature change than does an object of smaller heat capacity. For example, it requires considerably more heat to raise the temperature of water in an outdoor swimming pool from 15°C to 25°C than it does to produce the same temperature change in a fish tank. The heat capacity, $C$, of an object can be expressed as the ratio $q/\Delta T$, where $q$ is the total heat flow into or out of an object and $\Delta T$ is the temperature change produced. Thus the heat flow equals the heat capacity times the temperature change:

$$q = C \times \Delta T \qquad [4.19]$$

The heat capacity of an object depends on its mass and its composition. The heat capacity of one gram of a substance is called the **specific heat**. That is, specific heat is the amount of heat required to produce a change of 1°C in exactly 1 g of substance. Thus the quantity of heat transferred in any process is determined by the product of the mass, $m$, specific heat, $s$, and temperature change, $\Delta T$, of the system:

$$q = m \times s \times \Delta T \qquad [4.20]$$

The specific heats of several substances are listed in Table 4.2. Liquid water has one of the highest specific heats known, 1 cal/g-°C = 4.184 J/g-°C. This specific heat, for example, is about five times as great as that of aluminum metal. Thus 4.184 J of heat will increase the temperature of 1 g of aluminum by nearly 5°C, but it raises the temperature of water by only 1°C. The high specific heat of water makes our body's important task of maintaining a constant body temperature, 37°C, much easier. The adult body is about 60 percent water by mass and consequently has the ability to absorb or release considerable energy with little effect on its temperature.

**TABLE 4.2**  Specific heats of some selected substances

| Compound | Temperature (°C) | Specific heat (J/°C-g) |
|---|---|---|
| $H_2O(l)$ | 15 | 4.184 |
| $H_2O(s)$ | −11 | 2.03 |
| $Al(s)$ | 20 | 0.89 |
| $C(s)$ | 20 | 0.71 |
| $Fe(s)$ | 20 | 0.45 |
| $Hg(l)$ | 20 | 0.14 |
| $CaCO_3(s)$ | 0 | 0.85 |
| $MgO(s)$ | 0 | 0.87 |
| $HgS(s)$ | 0 | 0.21 |

## SAMPLE EXERCISE 4.8

The specific heat of iron(III) oxide, $Fe_2O_3$, is 0.75 J/°C-g. **(a)** What is the heat capacity of a 2.00-kg brick of $Fe_2O_3$? **(b)** What quantity of heat is required to increase the temperature of 1.75 g of $Fe_2O_3$ from 25°C to 380°C?

**Solution:**   **(a)** To obtain the total heat capacity of the brick we simply multiply its mass by its specific heat. Because 2.00 kg = $2.00 \times 10^3$ g, we have

$$C = (2.00 \times 10^3 \text{ g})(0.75 \text{ J/°C-g}) = 1.5 \times 10^3 \text{ J/°C}$$

**(b)** Using Equation 4.20, we have

$$q = m \times s \times \Delta T = (1.75 \text{ g})(0.75 \text{ J/°C-g})(355°C)$$
$$= 4.7 \times 10^2 \text{ J}$$

## PRACTICE EXERCISE

How much heat, in kJ, is required to increase the temperature of 150 g of water from 25°C to 42°C? The specific heat of water is 4.18 J/°C-g.
*Answer:*   11 kJ

## Constant-Pressure Calorimetry

The techniques and equipment employed in calorimetry depend on the nature of the process being studied. For many reactions, such as those occurring in solution, it is a simple matter to control pressure so that $\Delta H$ can be measured directly. (Recall that $\Delta H = q_P$, Equation 4.9.) Although the calorimeters used for highly accurate work are precision instruments, a very simple "coffee cup" calorimeter, such as that shown in Figure 4.12, is often used in freshman chemistry labs to illustrate the principles of calorimetry. Because the calorimeter is not sealed, the reaction occurs under the essentially constant pressure of the atmosphere. The heat of a reaction is determined from the temperature change of a known quantity of solution in the calorimeter, as shown in Sample Exercise 4.9.

## SAMPLE EXERCISE 4.9

When 50 mL of 1.0 $M$ HCl and 50 mL of 1.0 $M$ NaOH are mixed in a "coffee cup" calorimeter, the temperature of the resultant solution increases from 21.0°C to 27.5°C. Assuming that the calorimeter absorbs only a negligible quantity of heat, that the total volume of the solution is 100 mL, that its density is 1.0 g/mL, and that its specific heat is 4.18 J/°C-g, calculate the enthalpy change for the reaction

$$HCl(aq) + NaOH(aq) \longrightarrow H_2O(l) + NaCl(aq)$$

**Solution:** Because the total volume of the solution is 100 mL, the mass of the solution is

$$(100 \text{ mL})(1.0 \text{ g/mL}) = 100 \text{ g}$$

The temperature change is $27.5°C - 21.0°C = 6.5°C$. Because the temperature increases, the reaction must be exothermic. The heat flow is given by the product of the mass, specific heat, and temperature change (Equation 4.20):

$$q_P = -\Delta H = (100 \text{ g})(4.18 \text{ J/°C-g})(6.5°C)$$
$$= 2700 \text{ J} = 2.7 \text{ kJ}$$

To put the enthalpy change on a molar basis, we use the fact that the quantity of HCl and of NaOH is given by the product of the respective solution volumes (50 mL = 0.050 L) and concentrations:

$$(0.050 \text{ L})(1.0 \text{ mol/L}) = 0.050 \text{ mol}$$

Thus the heat evolved per mole is

$$-2.7 \text{ kJ/0.050 mol} = -54 \text{ kJ/mol}$$

### PRACTICE EXERCISE

When 50.0 mL of 0.100 $M$ AgNO$_3$ and 50.0 mL of 0.100 $M$ HCl are mixed in a constant-pressure calorimeter, the temperature of the mixture increases from 22.30°C to 23.11°C. The temperature increase is caused by the following reaction:

$$AgNO_3(aq) + HCl(aq) \longrightarrow AgCl(s) + NaCl(aq)$$

Calculate $\Delta H$ for this reaction, assuming that the combined solution has a mass of 100.0 g and a specific heat of 4.18 J/°C-g.
*Answer:* $-68000 \text{ J/mol} = -68 \text{ kJ/mol}$

## Bomb Calorimetry

One of the most important types of reaction studied by means of calorimetry is combustion (Section 3.2). A compound, usually an organic compound, is allowed to react completely with excess oxygen. Equation 4.11 is an example of such a reaction. Combustion reactions are most conveniently studied by means of a **bomb calorimeter**, a device shown schematically in Figure 4.13. The substance to be studied is placed in a small cup within a sealed vessel called a bomb. The bomb, which is designed to withstand high pressures, has an inlet valve for adding oxygen and also has electrical contacts to initiate the combustion reaction. After the sample has been placed in the bomb, the bomb is sealed and pressurized with oxygen. It is then placed in the calorimeter, which is essentially an insulated container, and covered with an accurately measured quantity of water. When all of the components within the calorimeter have come to the same temperature, the combustion reaction is initiated by passing an electrical current through a fine wire that is in contact with the sample. The sample ignites when the wire gets hot.

Heat is evolved when combustion occurs. This heat is absorbed by the calorimeter contents, causing a rise in the temperature of the water. The temperature of the water is very carefully measured before reaction, and then again after reaction when the contents of the calorimeter have again arrived at a common temperature. In keeping with the first law of thermodynamics, the heat evolved in the combustion of the sample is absorbed by its surroundings, that is, the calorimeter contents.

To calculate the heat of combustion from the measured temperature increase in the calorimeter, it is necessary to know the heat capacity of the calorimeter. This is normally ascertained by combusting a sample that gives off a known quantity of heat. For example, it is known that combustion of exactly 1 g of benzoic acid, $C_7H_6O_2$, in a bomb calorimeter produces 26.38 kJ of heat. Suppose that 1 g of benzoic acid is combusted in our calorimeter, and it causes a temperature increase of 5.022°C. The heat capacity of the calorimeter is then given by 26.38 kJ/5.022°C = 5.253 kJ/°C. Once we know the value of the heat capacity of the calorimeter, we can measure temperature changes produced by other reactions, and from these we can calculate the heat, $q$, evolved in the reaction:

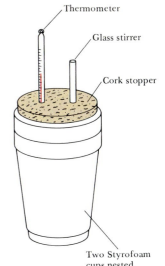

**FIGURE 4.12** "Coffee cup" calorimeter, in which reactions occur at constant pressure.

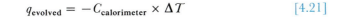

$$q_{\text{evolved}} = -C_{\text{calorimeter}} \times \Delta T \qquad [4.21]$$

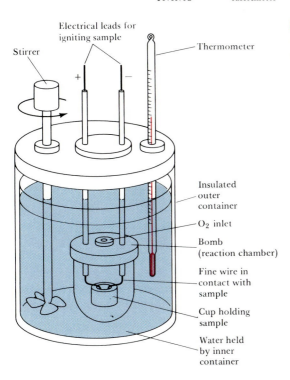

**FIGURE 4.13** Bomb calorimeter, in which reactions occur at constant volume.

## SAMPLE EXERCISE 4.10

The temperature of a bomb calorimeter is raised 3.51°C when 1.00 g of hydrazine, $N_2H_4$, is burned in it. If the calorimeter has a heat capacity of 5.510 kJ/°C, what is the quantity of heat evolved? What is the heat evolved upon combustion of a mole of $N_2H_4$?

**Solution:** The quantity of heat evolved is just the product of the temperature change times the heat capacity (Equation 4.21):

$$-\left(\frac{5.510 \text{ kJ}}{1°C}\right) 3.51°C = -19.34 \text{ kJ}$$

Since this is the amount of heat that results from com-

bustion of 1.00 g of hydrazine, the amount released by combustion of 1 mol of $N_2H_4$ is

$$\left(\frac{-19.3 \text{ kJ}}{1 \text{ g } N_2H_4}\right)\left(\frac{32.0 \text{ g } N_2H_4}{1 \text{ mol } N_2H_4}\right) = -619 \text{ kJ/mol } N_2H_4$$

### PRACTICE EXERCISE

A 0.5865-g sample of lactic acid $(HC_3H_5O_3)$ is burned in a calorimeter whose heat capacity is 4.812 kJ/°C. The temperature increases from 23.10°C to 24.95°C. Calculate the heat of combustion of lactic acid per gram and then per mole.
*Answer:* −15.2 kJ/g; −1370 kJ/mol

Because the reactions in a bomb calorimeter occur at constant volume rather than constant pressure, the heats evolved correspond to $\Delta E$ rather than to $\Delta H$. The measured $\Delta E$ values can readily be converted to $\Delta H$ values, using Equation 4.9, $\Delta H = \Delta E + P\Delta V$. However, even when the formation or consumption of gases in a reaction leads to large pressure changes, the correction factor $P\Delta V$ is quite small.* For example, the combustion of 1 mol of octane, $C_8H_{18}$, gives $\Delta E = -5461$ kJ and $\Delta H = -5450$ kJ. Thus we will not concern ourselves with the correction procedure. The important point is that $\Delta H$ can be calculated from measurements made at constant volume.

## 4.8 FOODS AND FUELS

Most chemical reactions used to produce heat are combustion reactions. The energy released when 1 g of a material is combusted is often called its **fuel value**. Because all heats of combustion are exothermic, it is common to report fuel values without their associated negative sign. The fuel value of any food or fuel can be measured by calorimetry.

### Foods

Most of the energy our bodies need comes from carbohydrates and fats. Carbohydrates are decomposed in the stomach into glucose, $C_6H_{12}O_6$. Glucose is soluble in blood and is known as blood sugar. It is transported by the blood to cells, where it reacts with $O_2$ in a series of steps, eventually producing $CO_2(g)$, $H_2O(l)$, and energy:

$$C_6H_{12}O_6(s) + 6O_2(g) \longrightarrow 6CO_2(g) + 6H_2O(l) \qquad \Delta H° = -2816 \text{ kJ}$$

The breakdown of carbohydrates is rapid, so their energy is quickly supplied to the body. However, the body stores a very small amount of carbohydrates. The average fuel value of carbohydrates is 17 kJ/g (4 kcal/g). (See Sample Exercise 4.6.)

---

\* To calculate $P\Delta V$, you need to know about the pressure-volume properties of gases, a subject discussed in Chapter 10. It works out that the $P\Delta V$ term is equal to $RT\Delta n$, where $\Delta n$ is the total number of moles of gas in the products minus the total number of moles of gas in the reactants, $R$ is the molar gas constant (8.314 J/K-mol), and $T$ is the temperature in degrees kelvin.

Like carbohydrates, fats produce $CO_2$ and $H_2O$ in both their metabolism and their combustion in a bomb calorimeter. The reaction of tristearin, $C_{57}H_{110}O_6$, a typical fat, is as follows:

$$2C_{57}H_{110}O_6(s) + 163O_2(g) \longrightarrow 114CO_2(g) + 110H_2O(l) \qquad \Delta H^\circ = -75{,}520 \text{ kJ}$$

The body puts the chemical energy from foods to different uses: to maintain body temperature, to drive muscles, and to construct and repair tissues. Any excess energy is stored as fats. Fats are well suited to serve as the body's energy reserve for at least two reasons: (1) they are insoluble in water, which permits their storage in the body; and (2) they produce more energy per gram than either proteins or carbohydrates, which makes them efficient energy sources on a weight basis. The average fuel value of fats is 38 kJ/g (9 kcal/g).

In the case of proteins, metabolism in the body produces less energy than combustion in a calorimeter does because the products are different. Proteins contain nitrogen, which is released in the bomb calorimeter as $N_2$. In the body this nitrogen ends up mainly as urea, $(NH_2)_2CO$. Proteins are used by the body mainly as building materials for organ walls, skin, hair, muscle, and so forth. On the average, the metabolism of proteins produces 17 kJ/g (4 kcal/g).

The fuel values for a variety of common foods are shown in Table 4.3. The amount of energy the body requires varies considerably depending on such factors as body weight, age, and muscular activity. When a person is doing average work, his or her daily energy requirement is about 10,000 to 13,000 kJ (2500 to 3000 kcal). This is about the same amount of energy as a 100-W light bulb consumes in operating for a 24-hr period.

**TABLE 4.3**  Fuel values and compositions of some common foods

| | Approximate composition (%) | | | Fuel value | |
|---|---|---|---|---|---|
| | Protein | Fat | Carbohydrate | kJ/g | kcal/g |
| Apples (raw) | 0.4 | 0.5 | 13 | 2.5 | 0.59 |
| Beer[a] | 0.3 | 0 | 1.2 | 1.8 | 0.42 |
| Bread (white, enriched) | 9 | 3 | 52 | 12 | 2.8 |
| Cheese (cheddar) | 28 | 37 | 4 | 20 | 4.7 |
| Eggs | 13 | 10 | 0.7 | 6 | 1.4 |
| Fudge | 2 | 11 | 81 | 18 | 4.4 |
| Green beans (frozen) | 1.9 | — | 7.0 | 1.5 | 0.38 |
| Hamburger | 22 | 30 | — | 15 | 3.6 |
| Milk | 3.3 | 4.0 | 5.0 | 3.0 | 0.74 |
| Peanuts | 26 | 39 | 22 | 23 | 5.5 |

[a] Beers typically contain 3.5 percent ethanol, which has fuel value.

## SAMPLE EXERCISE 4.11

Scientists estimate that a person of average weight requires the consumption of 100 Cal/mi when running or jogging. What weight of hamburger provides the fuel value requirements for running 3 mi?

**Solution:**  Recall that the dietary Calorie is equivalent to 1 kcal. The running requires 300 Cal, or 300 kcal. Using data from Table 4.3:

Hamburger required

$$= 300 \text{ kcal} \times \left( \frac{1 \text{ g hamburger}}{3.6 \text{ kcal}} \right)$$

$$= 83 \text{ g hamburger}$$

Thus the calories consumed in running 3 mi are replaced by less hamburger than a "quarter-pounder" offers.

**PRACTICE EXERCISE**

Dry red beans contain 62 percent carbohydrate, 22 percent protein, 1.5 percent fat, and the remainder water. Estimate the fuel value of these beans.
*Answer:* 15 kJ/g

**Fuels**

Several common fuels are compared in Table 4.4. Notice that an increase in the percentage of carbon or hydrogen in the fuel increases the fuel value. For example, the fuel value of bituminous coal is greater than that of wood because of its greater carbon content.

Coal, oil, and natural gas, which are presently our major sources of energy, are known as **fossil fuels**. All are thought to have formed over millions of years from the decomposition of plants and animals. All are presently being depleted far more rapidly than they are being formed. **Natural gas** consists of gaseous hydrocarbons, compounds of hydrogen and carbon. It varies in composition but contains primarily methane ($CH_4$), with small amounts of ethane ($C_2H_6$), propane ($C_3H_8$), and butane ($C_4H_{10}$). **Oil**, which is also known as petroleum, is a liquid composed of hundreds of compounds. Most of these compounds are

**TABLE 4.4**  Fuel values and compositions of some common fuels

|  | Approximate elemental composition (%) | | | |
|---|---|---|---|---|
|  | **C** | **H** | **O** | **Fuel value (kJ/g)** |
| Wood (pine) | 50 | 6 | 44 | 18 |
| Anthracite coal (Pennsylvania) | 82 | 1 | 2 | 31 |
| Bituminous coal (Pennsylvania) | 77 | 5 | 7 | 32 |
| Charcoal | 100 | 0 | 0 | 34 |
| Crude oil (Texas) | 85 | 12 | 0 | 45 |
| Gasoline | 85 | 15 | 0 | 48 |
| Natural gas | 70 | 23 | 0 | 49 |
| Hydrogen | 0 | 100 | 0 | 142 |

## CHEMISTRY AT WORK

### Hydrogen as a Fuel

Hydrogen ($H_2$) is a very attractive fuel because of its high fuel value and because its combustion produces water, a "clean" chemical that produces no negative environmental effects. However, hydrogen cannot be used as a primary energy source because there is so little $H_2$ in nature. Most hydrogen is produced by the decomposition of water or hydrocarbons. This decomposition requires energy; in fact, because of heat losses, more energy must be used to generate hydrogen than can be reclaimed when the hydrogen is subsequently used as a fuel. However, should large, cheap sources of energy become available because of technical advances, perhaps in areas such as nuclear or solar power generation, a portion of this energy could be used to generate hydrogen. The hydrogen could then serve as a convenient energy carrier. It would be cheaper to transport hydrogen using existing gas pipelines than to transport electrical energy; hydrogen is both portable and storable. Because present industrial technology is based on combustible fuels, hydrogen could replace oil and natural gas as these fuels become scarcer and more expensive.

hydrocarbons, with the remainder being mainly organic compounds containing sulfur, nitrogen, or oxygen. **Coal**, which is solid, contains hydrocarbons of high molecular weight as well as compounds containing sulfur, oxygen, and nitrogen. The sulfur in oil and coal is important from the standpoint of air pollution, as we shall discuss in Chapter 14.

## 4.9 ENERGY USAGE: TRENDS AND PROSPECTS

The average energy consumption per person in the United States each day amounts to about $8.8 \times 10^5$ kJ. This amount of energy is about 100 times greater than the individual food-energy requirement. We are a very energy-intensive society.

During the past century the fossil fuels—first coal, later oil and gas—have become the major energy sources in our society. Together they account for more than 80 percent of national energy consumption. Eventually, such fossil fuel energy sources must decline in importance as we exhaust the readily available supplies. Current projections are that 80 percent of all the known and estimated oil reserves will have been consumed before the children born during the 1980s will have lived out their lives. Clearly we must develop alternative energy sources to replace those fossil fuels if the 8 billion people expected as the world's population by 2025 are to have a decent standard of living.

Of the various sources of energy that could serve as alternatives to fossil fuels, only nuclear and solar energies are potentially capable of furnishing sufficient quantities of energy to satisfy the world's needs. Nuclear power supplied about 4 percent of the gross energy consumed in the United States in 1985. Aspects of the use of nuclear reactors for energy production are discussed in Chapter 21. In the paragraphs that follow we'll consider some aspects of coal and solar radiation as energy sources.

Coal is the most abundant fossil fuel; it constitutes 80 percent of the fossil fuel reserves of the United States and 90 percent of those of the world. However, the use of coal presents a number of problems. Coal is a complex mixture of substances, and it contains components that produce air pollution. Because it is a solid, recovery from its underground deposits is expensive and often dangerous. Furthermore, coal deposits are not always close to the locations where energy use is high, so there are often substantial shipping costs to add to the total cost of coal use.

Some experts feel that coal could be used most effectively if it were converted to a gaseous form, often called **syngas** (for "*syn*thetic *gas*"). In such a conversion the sulfur is removed, thereby decreasing air pollution when the syngas is burned. Syngas could be easily transported in pipelines and could supplement our diminishing supplies of natural gas. Gasification of coal requires addition of hydrogen to coal. Typically, the coal is pulverized and treated with superheated steam. The product contains a mixture of $CO$, $H_2$, and $CH_4$, all of which can be used as fuels. However, conditions are maintained to maximize production of $CH_4$. A simplified schematic showing some of the reactions that occur is given in Figure 4.14. The syngas produced in coal gasification can also be used to make chemicals. Methyl alcohol, $CH_3OH$, and acetic anhydride, $(CH_3CO)_2O$, are examples of chemicals produced in large quantities from synthesis gas (Figure 4.15).

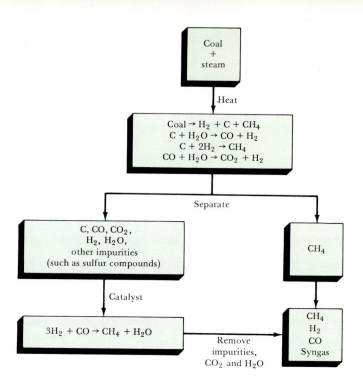

$$\text{Coal} \rightarrow H_2 + C + CH_4$$
$$C + H_2O \rightarrow CO + H_2$$
$$C + 2H_2 \rightarrow CH_4$$
$$CO + H_2O \rightarrow CO_2 + H_2$$

Separate

C, CO, CO$_2$, H$_2$, H$_2$O, other impurities (such as sulfur compounds)

CH$_4$

Catalyst

$$3H_2 + CO \rightarrow CH_4 + H_2O$$

Remove impurities, CO$_2$ and H$_2$O

CH$_4$
H$_2$
CO
Syngas

**FIGURE 4.14** Basic processes involved in the gasification of coal to form synthetic gas (syngas). A catalyst is a substance that increases the speed of a reaction without itself being consumed in the reaction.

**FIGURE 4.15** View of a large, new plant for conversion of coal to acetic anhydride, $(CH_3CO)_2O$. This plant is designed to process 900 tons of high-sulfur-content Appalachian coal daily, for both feedstock and to furnish plant power. Acetic anhydride is used in the manufacture of photographic film, synthetic fibers, cigarette filters, and coatings. (Eastman Kodak)

Solar energy is the world's largest energy source. The solar energy that falls on only 0.1 percent of the land area of the United States is equivalent to all the energy that this nation currently uses. The problem with the use of solar energy is that it is dilute (because it is distributed over a wide area) and fluctuates with time and weather conditions. The effective use of solar energy will depend on the development of some means of storing the energy collected for use at a later time. Any practical means for doing this will almost certainly involve use of an endothermic chemical process that can be reversed at a later time to release heat. One such reaction is the following:

$$CH_4(g) + H_2O(g) + heat \rightleftharpoons CO(g) + 3H_2(g) \qquad [4.22]$$

This reaction can be made to proceed in the forward direction at high temperatures, which can be obtained directly in a solar furnace. The CO and $H_2$ formed in the reaction could then be reacted later, with the resulting evolution of heat put to useful work.

Solar energy can be converted directly into electricity by use of photovoltaic devices, sometimes called solar cells. The efficiencies of solar energy conversion by use of such devices have increased dramatically during the past few years as a result of intensive research efforts. Photovoltaics are vital to the generation of power for satellites, as illustrated in Figure 4.16. However, for large-scale generation of useful energy at the earth's surface, they are not yet practical because of high unit cost. Even if the costs are reduced, some means must be found to store the energy produced by the solar cells, because the sun shines only intermittently and during only part of the day at any place. Once again, the solution to this problem will almost certainly be to use the energy to run a chemical reaction in the direction in which it is endothermic.

**FIGURE 4.16**  The NASA space telescope depicted in this artist's drawing will orbit the earth at 310 miles. This telescope will permit observation of stars and galaxies 1/50th as bright as can now be observed by ground-based telescopes. Photovoltaic cells supply power needed for the operation of the telescope and for transmission of information to earth. (NASA)

# FOR REVIEW

## SUMMARY

This chapter has been about **energy** and the **first law of thermodynamics**. **Thermodynamics** is the study of heat, work, and energy and the rules that govern their interconversions. To study the thermodynamic properties of matter, we define some specific amount of matter as our **system** and study the interactions between this system and its **surroundings**. The system may exchange heat with its surroundings, or have work done on it by the surroundings, or do work on the surroundings. **Work** comes in many forms. Mechanical work ($w$) is measured by the product of a force ($F$) acting through a distance ($d$): $w = F \times d$. The term **heat** refers to the flow of energy between two bodies at different temperatures when they are placed in thermal contact.

The **joule** (J) is the SI unit of energy. $1\,\text{J} = 1\,\text{kg-m}^2/\text{s}^2$. Another common energy unit is the **calorie** (cal). A calorie is the quantity of energy necessary to increase the temperature of 1 g of water by $1°C$; $1\,\text{cal} = 4.184\,\text{J}$.

The **internal energy**, $E$, of a system is all the energy possessed by the atomic, molecular, or ionic units of which the system is composed. We do not in general know the magnitude of this internal energy; we are concerned primarily with *changes* in the internal energy, $\Delta E$, that accompany changes in the system. The first law of thermodynamics tells us that the change in internal energy is given by the heat added to the system during the change, $q$, plus the work done on the system by the surroundings: $\Delta E = q + w$. The internal energy is a **state function**; that is, the value of the internal energy for the system depends only on the condition of the system, as described by variables such as temperature, pressure, and so forth. It does not depend on the details of how the system came to be in that state.

The **enthalpy**, $H$, of a system, is a state function defined as $E + PV$, where $P$ is the pressure acting on the system and $V$ is its volume. Enthalpy is a most useful function for dealing with changes that occur at constant pressure; then $\Delta H = \Delta E + P\Delta V$. The enthalpy change for any process is the heat flow into or out of a system maintained at constant pressure, so $q_P = \Delta H$. For **endothermic** processes (those that absorb heat) $\Delta H$ is positive. For **exothermic** processes (those that transfer heat from the system to the surroundings) $\Delta H$ is negative.

Enthalpy is an extensive property; the magnitude of the enthalpy change for any process is proportional to the quantity of reactants. The magnitude of $\Delta H$ also depends on the physical states of the reactants and products. If a reaction is reversed, the sign of $\Delta H$ is also reversed.

Because enthalpy is a state function, the enthalpy change for a given chemical process is the same whether the process is carried out in one step or in a series of steps. **Hess's law** states this fact in a very useful form: If a reaction is carried out in a series of steps, $\Delta H$ for the reaction will be equal to the sum of the enthalpy changes for each step. This means that we can calculate the enthalpy changes for any reaction as long as we can find a route for which $\Delta H$ is known for each step. Tabulated values for the **standard heat of formation**, $\Delta H_f^\circ$, are particularly helpful for such calculations. The standard heat of formation of a substance is defined as the enthalpy change for the formation of a mole of that substance from the elements, with all reactants and products in their standard states. The **standard state** refers to the conditions of 1 atmosphere pressure, with each reactant and product in its stable form (gas, liquid, or solid) at the temperature in question, usually 298 K. By convention, the enthalpies of formation of elements in their standard states are zero. The enthalpy change of any reaction can be calculated from the heats of formation of the reactants and of products in the reaction:

$$\Delta H_{\text{rxn}}^\circ = \sum n\,\Delta H_f^\circ(\text{products}) - \sum m\,\Delta H_f^\circ(\text{reactants})$$

The quantity of heat absorbed or evolved in chemical and physical processes is measured experimentally by **calorimetry**. A **calorimeter** measures the temperature changes accompanying the process being investigated. The magnitude of the temperature change that an object undergoes when it absorbs a certain quantity of heat depends on its **heat capacity**. The magnitude of the heat flow is given by the product of the heat capacity and the temperature change: $q = C\Delta T$. The **specific heat** of any substance is the heat required to raise the temperature of 1 g of the substance by $1°C$. Thus the quantity of heat transferred to a substance is given by the product of its mass, specific heat, and temperature change: $q = m \times s \times \Delta T$. If the system is at constant pressure, the heat flow equals $\Delta H$ for the process; if the system is at constant volume, the heat flow equals $\Delta E$. A **bomb calorimeter** is a constant-volume device used to measure the heat evolved when a substance is combusted in oxygen.

The **fuel value** of any substance or mixture is defined as the heat energy released when a gram of that material is combusted. We considered briefly the fuel values of some common foods and fuels, including various forms of fossil fuels. Finally, we have considered current and projected rates of fuel consumption and the prospects for various sources of energy in the future.

# LEARNING GOALS

Having read and studied this chapter, you should be able to:

1. Give examples of different forms of energy.
2. List the important units in which energy is expressed, and convert from one unit to another.
3. Define the first law of thermodynamics both verbally and by means of an equation.
4. Describe how the change in the internal energy of a system is related to the exchanges of heat and work between the system and its surroundings.
5. Define the term *state function*, and describe its importance in thermochemistry.
6. Define enthalpy, and relate the enthalpy change in a process occurring at constant pressure to the change in internal energy and the change in volume taking place during the process.
7. Sketch an energy diagram like that shown in Figure 4.7, given the enthalpy changes in the processes involved, and know that the sign of $\Delta H$ is positive in an endothermic reaction and negative in an exothermic reaction.
8. Calculate the quantity of heat involved in a reaction at constant pressure, given the quantity of reactants and the enthalpy change on a mole basis for the reaction.
9. State Hess's law, and apply it in calculating the enthalpy change in a process, given the enthalpy changes in other processes that could be combined to yield the reaction of interest.
10. Define and illustrate what is meant by the term *standard state*, and identify the standard states for common elements.
11. Define the term *standard heat of formation*, and identify the type of chemical reaction with which it is associated.
12. Calculate the enthalpy change in a reaction occurring at constant pressure, given the standard enthalpies of formation of each reactant and product.
13. Define the terms *heat capacity* and *specific heat*.
14. Calculate any one of the following quantities given the other three: heat, quantity of material, temperature change, and specific heat.
15. Calculate the heat capacity of a calorimeter, given the temperature change and quantity of heat involved; also calculate the heat evolved or absorbed in a process from a knowledge of the heat capacity of the system and its temperature change.
16. Define the term *fuel value*; calculate the fuel value of a substance given its heat of combustion, and estimate the fuel value of a material given its composition.
17. List the major sources of energy on which humankind must depend, and discuss the likely availability of these sources in the foreseeable future.
18. Describe how coal is converted to syngas; indicate some of the uses that might be made of syngas.

# KEY TERMS

Among the more important terms and expressions used for the first time in this chapter are the following:

A **bomb calorimeter** (Section 4.7) is a device for measurements of heat evolved in combustion of a substance under constant-volume conditions.

**Calorimetry** (Section 4.7) refers to the experimental measurements of heat effects in chemical and physical processes.

An **endothermic process** (Section 4.2) is one in which a system absorbs heat from its surroundings.

**Energy** (Section 4.1) is the ability to do work or to transfer heat.

The **enthalpy**, $H$ (Section 4.3) is defined by the relationship $H = E + PV$. The enthalpy change, $\Delta H$, for a reaction that occurs at constant pressure is the heat evolved or absorbed in the reaction, $q_P$.

An **exothermic process** (Section 4.2) is one in which a system loses heat to its surroundings.

The **first law of thermodynamics** (Section 4.2) is a statement of our experience that energy is conserved in any process. We can express the first law in many ways. One of the more useful expressions is that the change in internal energy, $\Delta E$, of a system in any process is equal to the heat, $q$, added *to* the system, plus the work, $w$, done *on* the system by its surroundings: $\Delta E = q + w$.

**Heat** (Section 4.1) refers to the flow of energy from a body at higher temperature to one at lower temperature when they are placed in thermal contact.

The **heat capacity** (Section 4.7) is defined as the quantity of heat required to cause a 1°C change in temperature. The **specific heat** is the heat required to raise the temperature of 1 g of a substance by 1°C.

The **heat of formation** (Section 4.6) of a substance is defined as the heat evolved when a substance is formed from the elements in their standard states.

**Hess's law** (Section 4.5) states that the heat evolved in a given process can be expressed as the sum of the heats of several processes that, when added, yield the process of interest.

**Internal energy**, $E$ (Section 4.2), is the total energy possessed by a system. When a system undergoes a change, the change in internal energy, $\Delta E$, is defined as the heat, $q$, added to the system, plus the work, $w$, done on the system by its surroundings: $\Delta E = q + w$.

The **joule**, J (Section 4.1), is the SI unit of energy, 1 kg-m$^2$/s$^2$. A related unit is the **calorie**; 4.184 J = 1 cal.

**Kinetic energy** (Section 4.1) is the energy that an object possesses by virtue of its motion.

**Mechanical work**, $w$ (Section 4.1), is defined and measured by the product of the net force, $F$, and distance, $d$, through which that force acts: $w = F \times d$.

**Potential energy** (Section 4.1) is the energy that an object possesses as a result of its position with respect to another object, or by virtue of its composition.

A **standard state** (Section 4.6) for a substance is defined as the pure material in the physical state (gas, liquid, or solid) most stable at the temperature in question (most usually 25°C), under 1 atm pressure.

A **state function** (Section 4.2) is a property of a system that is determined by the state or condition of the system and not by how it got to that state. Its value is fixed when temperature, pressure, composition, and physical form are specified. $P$, $V$, $T$, $E$, and $H$ are state functions.

The **surroundings** (Section 4.1) are everything that lies outside the system that we study.

**Syngas** (Section 4.9) is a mixture of gases, mainly $H_2$ and CO, that results from heating coal in the presence of steam. The mixture can be used as a fuel or for synthesis of other substances.

A **system** (Section 4.1) is a portion of the universe that we single out for study. We must be careful to state exactly what the system contains and what transfers of energy it may have with its surroundings.

## EXERCISES

### Nature of Energy; the First Law

**4.1** (a) Calculate the kinetic energy, in joules, of a 28-g bullet moving with a speed of $3.9 \times 10^4$ cm/s. (b) Convert this energy to calories. (c) What happens to this energy when the bullet slams into an impenetrable steel plate?

**4.2** (a) Calculate the kinetic energy, in kilojoules, of a car weighing $4.8 \times 10^3$ lb if it is traveling at a speed of 45 mi/hr. (Use the necessary conversion factors from the back inside cover of the text.) (b) Convert this energy to calories. (c) What happens to this energy when the car stops for a red light?

**4.3** (a) Verbally state the first law of thermodynamics. (b) Define the term "system." (c) How is the energy change of a system related to that of its surroundings?

**4.4** State the first law of thermodynamics in equation form. Explain this equation in your own words.

**4.5** Calculate the change in internal energy of a system in each of the following processes. (a) A gas expands very rapidly, so that there is no heat exchange with the surroundings; in the expansion it does 450 J of work on the surroundings. (b) 200 g of water is heated from 30°C to 40°C, a process that requires approximately 8360 J of heat. (c) A gas contracts as it is cooled; it has 300 J of work done on it and loses 146 J of heat to its surroundings.

**4.6**  Calculate $\Delta E$ for the following cases:
**(a)**  $q = 42$ J and $w = -18$ J
**(b)**  $q = -87$ J and $w = -22$ J
**(c)**  $q = 1.1$ kJ and $w = 580$ J
**(d)**  $q = -128$ J and $w = 367$ J
In each case indicate whether the system absorbs or releases heat and whether the system does work or has work done on it.

**4.7**  Indicate which of the following is independent of the path by which a change occurs: **(a)** the change in potential energy when a book is transferred from table to shelf; **(b)** the heat evolved when a cube of sugar is oxidized to $CO_2(g)$ and $H_2O(g)$; **(c)** the work accomplished in burning a gallon of gasoline.

**4.8**  Which of the following are state functions: **(a)** the internal energy of a mole of ice at a particular temperature and pressure; **(b)** the kinetic energy of a golf ball; **(c)** the distance traveled in going from New York to Miami; **(d)** the difference in altitude between Denver and Chicago?

## Enthalpy

**4.9**  What is the special name given to the heat exchanged between a system and its surroundings under constant pressure? What is the symbol used to represent this quantity?

**4.10**  During a process, the volume of a system remains constant. What is the value of the work for the process? Will the heat exchanged between the system and its surroundings be equal to $\Delta H$ or $\Delta E$?

**4.11**  A gas is confined to a cylinder under constant atmospheric pressure, as illustrated in Figure 4.1. When the temperature of the gas is increased, 600 J of heat is added, and the gas expands, doing 140 J of work on the surroundings. What are the values of $\Delta H$ and $\Delta E$ in this process?

**4.12**  A gas is confined to a cylinder under constant atmospheric pressure, as illustrated in Figure 4.1. When the gas undergoes a particular chemical reaction, it loses 850 kJ of heat to its surroundings and does 35 kJ of $P$–$V$ work on its surroundings. What are the values of $\Delta H$ and $\Delta E$ for this reaction?

**4.13**  Consider the following reaction:

$$2Al(s) + 3Cl_2(g) \longrightarrow 2AlCl_3(s) \qquad \Delta H = -1390.8 \text{ kJ}$$

**(a)** Is the reaction exothermic or endothermic? **(b)** Draw a diagram similar to Figure 4.7 for this reaction. **(c)** Calculate the quantity of heat produced when 10.0 g of $AlCl_3(s)$ forms by this reaction at constant pressure. **(d)** How many grams of aluminum must react to produce an enthalpy change of 1.00 kJ? **(e)** How many kilojoules of heat are required to decompose 15.0 g of $AlCl_3(s)$ into $Al(s)$ and $Cl_2(g)$ at constant pressure?

**4.14**  Consider the following reaction:

$$2CuO(s) \longrightarrow 2Cu(s) + O_2(g) \qquad \Delta H = +310.4 \text{ kJ}$$

**(a)** Is the reaction exothermic or endothermic? **(b)** Draw a diagram similar to Figure 4.7 for this reaction. **(c)** Calculate the quantity of heat produced when 12.0 g of $CuO(s)$ reacts according to this reaction at constant pressure. **(d)** How many grams of copper metal correspond to absorption of 1.00 kJ of heat at constant pressure? **(e)** How many kilojoules of heat are produced when 25.0 g of $CuO(s)$ is produced from $Cu(s)$ and $O_2(g)$ at constant pressure?

**4.15**  When solutions containing silver ions and chloride ions are mixed, silver chloride precipitates:

$$Ag^+(aq) + Cl^-(aq) \longrightarrow AgCl(s) \qquad \Delta H = -65.5 \text{ kJ}$$

**(a)** Calculate $\Delta H$ for formation of 0.200 mol of AgCl by this reaction. **(b)** Calculate $\Delta H$ for formation of 2.50 g of AgCl. **(c)** Calculate $\Delta H$ when 0.350 mol of AgCl dissolves in water.

**4.16**  The common means of forming small quantities of oxygen gas in the laboratory is to heat $KClO_3$:

$$2KClO_3(s) \longrightarrow 2KCl(s) + 3O_2(g) \qquad \Delta H = -89.4 \text{ kJ}$$

For this reaction, calculate $\Delta H$ for the formation of **(a)** 0.345 mol of $O_2$; **(b)** 7.85 g of KCl; **(c)** 9.22 g of $KClO_3$ from KCl and $O_2$.

**4.17**  Which of the following has the highest enthalpy at a given temperature and pressure: $H_2O(s)$, $H_2O(l)$, or $H_2O(g)$? Which has the lowest enthalpy?

**4.18**  How do the enthalpies of any substance change as the substance is converted from a solid to a liquid? from a liquid to a gas?

## Hess's Law

**4.19**  State Hess's law. Why is it important to thermochemistry?

**4.20**  What connection is there between the fact that $\Delta H$ is a state function and Hess's law?

**4.21**  From the following heats of reaction:

$$N_2(g) + 2O_2(g) \longrightarrow 2NO_2(g) \qquad \Delta H = +67.6 \text{ kJ}$$
$$NO(g) + \tfrac{1}{2}O_2(g) \longrightarrow NO_2(g) \qquad \Delta H = -56.6 \text{ kJ}$$

calculate the heat of the reaction

$$N_2(g) + O_2(g) \longrightarrow 2NO(g)$$

**4.22**  Given:

$$2NH_3(g) + 3N_2O(g) \longrightarrow 4N_2(g) + 3H_2O(l) \\ \Delta H = -1010 \text{ kJ}$$
$$4NH_3(g) + 3O_2(g) \longrightarrow 2N_2(g) + 6H_2O(l) \\ \Delta H = -1531 \text{ kJ}$$

calculate the enthalpy change for the following reaction:

$$N_2(g) + \tfrac{1}{2}O_2(g) \longrightarrow N_2O(g)$$

**4.23** Given the following data:

$$2C_2H_6(g) + 7O_2(g) \longrightarrow 4CO_2(g) + 6H_2O(l)$$
$$\Delta H = -3120 \text{ kJ}$$

$$C(s) + O_2(g) \longrightarrow CO_2(g)$$
$$\Delta H = -394 \text{ kJ}$$

$$2H_2(g) + O_2(g) \longrightarrow 2H_2O(l)$$
$$\Delta H = -572 \text{ kJ}$$

use Hess's law to calculate $\Delta H$ for the reaction

$$2C(s) + 3H_2(g) \longrightarrow C_2H_6(g)$$

**4.24** Given the following reactions and their associated enthalpy changes:

$$H_2(g) + Br_2(g) \longrightarrow 2HBr(g) \qquad \Delta H = -72 \text{ kJ}$$
$$H_2(g) \longrightarrow 2H(g) \qquad \Delta H = 436 \text{ kJ}$$
$$Br_2(g) \longrightarrow 2Br(g) \qquad \Delta H = 224 \text{ kJ}$$

calculate $\Delta H$ for the reaction

$$H(g) + Br(g) \longrightarrow HBr(g)$$

## Heats of Formation

**4.25** Write the balanced equations that describe the formation of the following compounds from their elements in their standard states (as was done for $C_2H_5OH$ in Equation 4.14; use Appendix D to obtain the appropriate heat of formation of each compound): **(a)** $CaCO_3(s)$; $-1207.1$ **(b)** $Fe_2O_3(s)$; **(c)** $NH_4NO_3(s)$. $-822.16$ $-365.6$

**4.26** Write the balanced equations that describe the formation of the following compounds from their elements in their standard states (use Appendix D to obtain the heat of formation of each compound): **(a)** $NaOH(s)$; **(b)** $CH_3OH(l)$; **(c)** $Al_2O_3(s)$.

**4.27** Given the following information:

$$2Ga_2O_3(s) \longrightarrow 4Ga(s) + 3O_2(g) \quad \Delta H° = +2158 \text{ kJ}$$

what is the standard heat of formation of $Ga_2O_3(s)$?

**4.28** Given the following information:

$$2AsH_3(g) \longrightarrow 2As(s) + 3H_2(g) \qquad \Delta H° = +773.4 \text{ kJ}$$

what is the standard heat of formation of $AsH_3(g)$?

**4.29** The following reaction is known as the thermite reaction:

$$2Al(s) + Fe_2O_3(s) \longrightarrow Al_2O_3(s) + 2Fe(s)$$

This highly exothermic reaction is used for welding massive units such as propellers for large ships. Using heats of formation from Appendix D, calculate $\Delta H°$ for this reaction.

**4.30** Nitric acid is obtained when nitrogen dioxide dissolves in water:

$$3NO_2(g) + H_2O(l) \longrightarrow 2HNO_3(aq) + NO(g)$$

Using heats of formation from Appendix D, calculate $\Delta H°$ for this reaction.

**4.31** Using Appendix D, calculate the standard enthalpy change for each of the following reactions:
**(a)** $3H_2(g) + N_2(g) \longrightarrow 2NH_3(g)$
**(b)** $2Na(s) + 2H_2O(l) \longrightarrow 2NaOH(aq) + H_2(g)$
**(c)** $NO(g) + O_3(g) \longrightarrow NO_2(g) + O_2(g)$
**(d)** $4NH_3(g) + 5O_2(g) \longrightarrow 4NO(g) + 6H_2O(g)$

**4.32** Using Appendix D, calculate the standard enthalpy change for each of the following reactions:
**(a)** $3H_2(g) + P_2(g) \longrightarrow 2PH_3(g)$
**(b)** $2NOCl(g) \longrightarrow 2NO(g) + Cl_2(g)$
**(c)** $MgO(s) + 2HCl(g) \longrightarrow MgCl_2(s) + H_2O(g)$
**(d)** $BaCO_3(s) \longrightarrow BaO(s) + CO_2(g)$

**4.33** Calcium carbide, $CaC_2$, reacts with water to form acetylene, $C_2H_2$, and $Ca(OH)_2$. From the following heat of reaction data and the data in Appendix D, calculate $\Delta H_f°$ for $CaC_2(s)$:

$$CaC_2(s) + 2H_2O(l) \longrightarrow Ca(OH)_2(s) + C_2H_2(g)$$
$$\Delta H° = -127.2 \text{ kJ}$$

**4.34** Complete combustion of acetone, $C_3H_6O$, results in the liberation of 1790 kJ:

$$C_3H_6O(l) + 4O_2(g) \longrightarrow 3CO_2(g) + 3H_2O(l)$$
$$\Delta H° = -1790 \text{ kJ}$$

Using this information together with the data in Appendix D, calculate the heat of formation of acetone.

## Calorimetry

**4.35** **(a)** What is the heat capacity of 855 g of water? **(b)** How many kJ of heat is needed to raise the temperature of 1.74 kg of water from 23.00°C to 68.10°C?

**4.36** **(a)** What mass of water has a heat capacity of 325 J/°C? **(b)** How many kJ of heat is needed to raise the temperature of 2.66 kg of water from 22.00°C to 58.20°C?

**4.37** The specific heat of ethanol is 2.46 J/g-°C. How many joules of heat are required to heat 193 g of ethanol from 19.00°C to 35.00°C?

**4.38** The specific heat of lead is 0.129 J/g-°C. How many joules of heat are required to raise the temperature of 382 g of lead from 22.50°C to 37.20°C?

**4.39** When a 5.00-g sample of $CaCl_2$ dissolves in 100.0 g of water in a "coffee cup" calorimeter, the temperature rises from 21.3°C to 30.2°C. Calculate $\Delta H$ (in kJ/mol $CaCl_2$) for the solution process

$$CaCl_2(s) \longrightarrow Ca^{2+}(aq) + 2Cl^-(aq)$$

Assume that the specific heat of the solution is the same as that of pure water.

**4.40** When a 4.50-g sample of $NH_4NO_3$ dissolves in 80.0 g of water in a "coffee cup" calorimeter, the temperature drops from 22.0°C to 17.7°C. Calculate $\Delta H$ (in kJ/mol $NH_4NO_3$) for the solution process

$$NH_4NO_3(s) \longrightarrow NH_4^+(aq) + NO_3^-(aq)$$

Assume that the specific heat of the solution is the same as that of pure water.

**4.41** A 2.20-g sample of quinone, $C_6H_4O_2$, was burned in a bomb calorimeter whose total heat capacity is 7.854 kJ/°C. The temperature of the calorimeter and its contents increased from 23.44°C to 30.57°C. What quantity of heat would be liberated by 1 mol of quinone?

**4.42** A 1.45-g sample of caffeine, $C_8H_{10}O_2N_2$, is burned in a bomb calorimeter whose total heat capacity is 14.38 kJ/°C. The temperature of the calorimeter increases from 20.66°C to 23.24°C. What is the heat of combustion of caffeine per mole?

**4.43** Under constant-volume conditions, the heat of combustion of benzoic acid, $HC_7H_5O_2$, is 26.38 kJ/g. A 1.200-g sample of benzoic acid is burned in a bomb calorimeter. The temperature of the calorimeter increased from 22.45°C to 26.10°C. **(a)** What is the total heat capacity of the calorimeter? **(b)** If the calorimeter contained 1.500 kg of water, what is the heat capacity of the calorimeter when it contains no water? **(c)** What temperature increase would be expected in this calorimeter if the 1.200-g sample of benzoic acid were combusted when the calorimeter contained 1.000 kg of water?

**4.44** Under constant-volume conditions, the heat of combustion of glucose is 15.57 kJ/g. A 2.500-g sample of glucose is burned in a bomb calorimeter. The temperature of the calorimeter increased from 20.55°C to 23.25°C. **(a)** What is the total heat capacity of the calorimeter? **(b)** If the calorimeter contained 2.700 kg of water, what is the heat capacity of the dry calorimeter? **(c)** What temperature increase would be expected in this calorimeter if the glucose sample had been combusted when the calorimeter contained 2.000 kg of water?

## Foods, Fuels, and Sources of Energy

**4.45** The heat of combustion of ethanol, $C_2H_5OH(l)$, is $-1371$ kJ/mol. A 12-oz (355-mL) bottle of beer contains 3.7 percent ethanol by mass. Assuming the density of the beer to be 1.0 g/mL, what caloric content does the alcohol in a bottle of beer have?

**4.46** The heat of combustion of fructose, $C_6H_{12}O_6$, is $-2812$ kJ/mol. If a freshly harvested yellow delicious apple weighing 4.23 oz (120 g) contains 16.0 g of fructose, what caloric content does the fructose contribute to the apple?

**4.47** A 1-oz serving of a popular "high protein" breakfast cereal contains 6 g of protein, 19 g of carbohydrate, and 1 g of fat. What is the fuel value in kJ and in kcal for this cereal?

**4.48** The total caloric value of 100 g of fresh pineapple is 52 kcal, due mainly to carbohydrate. However, the pineapple also contains 0.4 g of protein and 0.4 g of fat. Given these data, calculate the percentage of the pineapple that is water.

**4.49** The standard enthalpies of formation of gaseous propyne, $C_3H_4$, propylene, $C_3H_6$, and propane, $C_3H_8$, are $+185.4$, $+20.4$, and $-103.8$ kJ/mol, respectively. Cal-culate the heat evolved per mole on combustion of each substance to yield $CO_2(g)$ and $H_2O(g)$. Also calculate the heat evolved on combustion of 1 kg of each substance. Which is the most efficient fuel in terms of heat evolved per unit mass?

**4.50** From the following data for three prospective fuels, calculate which could provide the most energy per unit volume.

| Fuel | Density (g/cm³) at 20°C | Molar heat of combustion (kJ/mol) |
|---|---|---|
| Nitroethane, $C_2H_5NO_2(l)$ | 1.052 | $-1348$ |
| Ethanol, $C_2H_5OH(l)$ | 0.789 | $-1371$ |
| Diethyl ether, $(C_2H_5)_2O(l)$ | 0.714 | $-2727$ |

**4.51** In 1977, the "proved" U.S. reserves of natural gas were estimated to be $2.09 \times 10^{11}$ ft³. Assuming the fuel value of the gas to be 980 kJ/ft³, calculate the total fuel value of this quantity of natural gas. What weight of anthracite coal has an equivalent total fuel value?

**4.52** The United States uses about $2.0 \times 10^{16}$ kJ of natural gas annually. Using the fuel value of the natural gas given in Table 4.4, and assuming its density to be 0.70 g/L, estimate the number of cubic meters of natural gas burned each year in the United States.

## Additional Exercises

**4.53** Which has the greater kinetic energy, a 1.0-kg object moving at 2.0 m/s or a 2.0-kg object moving at 1.0 m/s? Calculate the kinetic energy of each in joules.

**4.54** Occasionally, people try to patent machines that are claimed to perform work without a corresponding consumption of energy or heat flow from another system. How do you evaluate these claims in light of the first law of thermodynamics? Explain.

**4.55** When a mole of dry ice, $CO_2(s)$, is converted to $CO_2(g)$ at atmospheric pressure and $-78°C$, the heat absorbed by the system exceeds the increase in internal energy of the $CO_2$. Why is this so? What happens to the remaining energy?

**4.56** An expanding gas absorbs 1.55 kJ of heat. If its internal energy increases by 1.32 kJ, does the system do work on its surroundings or have work done on it? What quantity of work is involved?

**4.57** Limestone stalactites are formed in caves by the following reaction:

$$Ca^{2+}(aq) + 2HCO_3^-(aq) \longrightarrow$$
$$CaCO_3(s) + CO_2(g) + H_2O(l)$$

If 1 mol of $CaCO_3$ forms at 298 K under 1 atm pressure, the reaction performs 2.47 kJ of expansion work, pushing back the atmosphere as the gaseous $CO_2$ forms. At the same time, 38.95 kJ of heat are absorbed from the environment. What are the values of $\Delta H$ and of $\Delta E$ for this reaction?

**4.58** When fruits and grains are fermented, glucose is converted to ethyl alcohol. Given the following data:

$$C_6H_{12}O_6(s) \longrightarrow 2C_2H_5OH(l) + 2CO_2(g)$$
$$\Delta H = 67.0 \text{ kJ}$$

**(a)** Is this reaction exothermic or endothermic? **(b)** Which has higher enthalpy, the reactants or products of this reaction? **(c)** Calculate $\Delta H$ for the formation of 5.00 g of $C_2H_5OH$. **(d)** What quantity of heat is liberated when 95.0 g of $C_2H_5OH$ is formed at constant pressure?

**4.59** Given

$$OF_2(g) + H_2O(l) \longrightarrow O_2(g) + 2HF(g)$$
$$\Delta H = -276.6 \text{ kJ}$$

$$SF_4(g) + 2H_2O(l) \longrightarrow SO_2(g) + 4HF(g)$$
$$\Delta H = -827.5 \text{ kJ}$$

$$S(s) + O_2(g) \longrightarrow SO_2(g) \quad \Delta H = -296.9 \text{ kJ}$$

calculate $\Delta H$ for the reaction

$$2S(s) + 2OF_2(g) \longrightarrow SO_2(g) + SF_4(g)$$

**4.60** The heat of formation of $C_2H_5OH(l)$ is $-277.7$ kJ/mol. The heat of vaporization of this substance at 25°C is 40.5 kJ/mol. Write the equations for these processes, and calculate the heat of formation of $C_2H_5OH(g)$.

**4.61** Which is the more negative quantity: the standard heat of formation of $H_2O(l)$ or of $H_2O(g)$? Explain.

**4.62** Calculate the enthalpy change for each of the following reactions:
**(a)** $3Fe(s) + 2O_2(g) \longrightarrow Fe_3O_4(s)$
**(b)** $2HF(g) + Br_2(l) \longrightarrow 2HBr(g) + F_2(g)$
**(c)** $N_2H_4(g) + 2H_2O_2(g) \longrightarrow N_2(g) + 4H_2O(g)$
**(d)** $2C_6H_6(l) + 15O_2(g) \longrightarrow 12CO_2(g) + 6H_2O(l)$

**4.63** A house is being designed to have passive solar energy features. Brickwork is to be incorporated into the interior of the house to act as a heat absorber. Each brick weighs approximately 1.8 kg. The specific heat of the brick is 0.85 kJ/°C-g. How many bricks will need to be incorporated into the interior of the house to provide the same total heat capacity as 1000 gal of water?

**4.64** When 50.0 mL of 1.00 $M$ $CuSO_4$ and 50.0 mL of 2.00 $M$ KOH are mixed in a constant-pressure calorimeter, the temperature of the mixture rises from 21.5°C to 27.7°C. From these data, calculate $\Delta H$ for the process

$$CuSO_4(aq; 1\ M) + 2KOH(aq; 2\ M) \longrightarrow$$
$$Cu(OH)_2(s) + K_2SO_4(aq; 0.5\ M)$$

Assume that the calorimeter absorbs only a negligible quantity of heat, that the total volume of the solution is 100.0 mL, and that the specific heat and density of the solution following mixing is the same as that of pure water.

**4.65** A 1.85-g sample of acetic acid, $HC_2H_3O_2$, was burned in excess oxygen in a bomb calorimeter. The calorimeter, which alone has a heat capacity of 2.76 kJ/°C, contained 0.850 kg of water. The temperature of the calorimeter and its contents increased from 23.32°C to 27.58°C. What quantity of heat would be liberated by combustion of 1 mol of acetic acid?

**[4.66]** A dry cell battery is connected to a heater that is immersed in an insulated water bath. As the dry cell discharges, the heater delivers 1300 J of energy, raising the temperature. The dry cell undergoes a slight temperature increase during the time of discharge. Consider the dry cell and the water plus heater as two separate systems. What kinds of changes have taken place in the internal energy and enthalpy of each of these systems, neglecting any exchanges of heat that each may have with its surroundings?

**4.67** By looking in a table of thermodynamic data such as Appendix D, how can you tell when an element is not in its standard state? Find two examples in Appendix D of thermodynamic data for elements that are not in their standard states at 25°C.

**[4.68]** Ammonia, $NH_3$, boils at $-33°C$; at this temperature, it has a density of 0.81 g/cm³. The heat of formation of $NH_3(g)$ is $-46.2$ kJ/mol, and the heat of vaporization of $NH_3(l)$ is 4.6 kJ/mol. Calculate the heat evolved when 1 L of liquid $NH_3$ is burned in air to give $N_2(g)$ and $H_2O(g)$. How does this compare with the heat evolved upon complete combustion of a liter of liquid methyl alcohol, $CH_3OH$ (density at 25°C = 0.792 g/cm³, and heat of formation $\Delta H_f^\circ(CH_3OH(l)) = -239$ kJ/mol)?

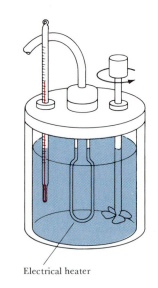

Electrical heater

**FIGURE 4.17**

[4.69]  (a) When a 0.235-g sample of benzoic acid is combusted in a bomb calorimeter, a 1.642°C rise in temperature is observed. When a 0.265-g sample of caffeine, $C_8H_{10}O_2N_4$, is burned, a 1.525°C rise in temperature is measured. Using the value 26.38 kJ/g for the heat of combustion of benzoic acid, calculate the heat of combustion per mole of caffeine at constant volume. (b) Assuming that there is an uncertainty of 0.002°C in each temperature reading and that the masses of samples are measured to 0.001 g, what is the estimated uncertainty in the value calculated for the heat of combustion per mole of caffeine?

[4.70]  When 4.00 g of KCl is added to a calorimeter of the type illustrated in Figure 4.17, the temperature is observed to change from 25.818°C to 24.886°C. The heater in the calorimeter has a resistance of 6.00 Ω. A current of 4.00 A is passed through the heater for a period of 20.0 s. As a result, the temperature in the calorimeter is observed to increase by 2.290°C. The heat dissipated in the elec-

trical heater is given by the formula $q = i^2 Rt$, where $i$ is the current in amperes, $R$ the resistance in ohms, $t$ the time the current flows in seconds, and $q$ is given in joules. From these data calculate the heat of solution per mole of KCl in water at approximately 25°C.

[4.71]  Aspirin is produced commercially from salicylic acid, $C_7O_3H_6$. A large shipment of salicylic acid is contaminated with boric oxide, which, like salicylic acid, is a white powder. The heat of combustion of salicylic acid at constant volume is known to be $-3.00 \times 10^3$ kJ/mol. Boric oxide, because it is fully oxidized, does not burn. When a 3.556-g sample of the contaminated salicylic acid is burned in a bomb calorimeter, the temperature increases 2.556°C. From previous measurements, the heat capacity of the calorimeter is known to be 13.62 kJ/°C. What is the amount of boric oxide in the sample, in terms of percent by weight?

## CHAPTER 5

# Introduction to the Elements and the Periodic Table

Chemists of the nineteenth century must have been bewildered at times by the great variety of behavior exhibited by the elements and their compounds. They sought general patterns of behavior and rules to organize the facts that were so quickly accumulating in their laboratories. The most important product of these attempts is the periodic table, which was first formulated in 1869.

In this chapter we briefly examine the periodic table and use it to become acquainted with some chemical properties of the elements. As we proceed, we may find ourselves asking the kinds of questions that puzzled chemists in the early 1900s. Why does the periodic table have its particular form? Why do the elements in a periodic family chemically resemble each other? Why are some elements more reactive than others? In the chapters ahead we will address such questions. Meanwhile, we will use the periodic table as a tool in examining a few of the more common elements. As we do so, we will also study the ways that chemical and physical properties vary according to an element's location in the periodic table.

## CONTENTS

During the early nineteenth century, many new elements were discovered in a short period of time. Only 31 elements were known in 1800. By 1830 the number had climbed to 54, and by 1865 it had reached 63. As the number increased, chemists began to investigate the possibilities of classifying elements in useful ways. The classification process in chemistry in those years is common to the development of all science. As the quantity and variety of data increase, scientists seek ways to organize and manage the information. When a workable classification system is found, it often stimulates new research as the system is used and tested. It may also form the basis for a theory that accounts for the observed regularities.

In 1869, Dmitri Mendeleev in Russia (Figure 5.1) and Lothar Meyer in Germany published nearly identical schemes for classifying elements. Both scientists noted that similar chemical and physical properties recur periodically when the elements are arranged according to their atomic weights. Using this observation, Mendeleev constructed a table in which he listed the elements in order of increasing atomic weight, placing those with similar properties in vertical columns. His table was the forerunner of the modern periodic table.

Although Meyer and Mendeleev came to essentially the same conclusion about the periodicity of the properties of the elements, Mendeleev is given credit for more vigorously advancing his ideas and stimulating much new work in chemistry. By adhering to his notion that elements

## 5.1 HISTORICAL DEVELOPMENT OF THE PERIODIC TABLE

**Predicting the Existence of New Elements**

An iron object covered with rust. Rusting is due to a reaction of iron with oxygen and water to produce a hydrated form of $Fe_2O_3$. (Barry L. Runk from Grant Heilman)

**FIGURE 5.1** Dmitri Mendeleev. Mendeleev rose from very poor beginnings to a position of great eminence in nineteenth-century science. He was born in Siberia, the youngest child in a family of at least 14. His mother endured great personal sacrifice to make it possible for him to enroll in a university in Saint Petersburg. Mendeleev proved to be a brilliant student in sciences and mathematics and eventually was able to study in France and Germany. He spent most of his career as a professor of chemistry in the University of Saint Petersburg. Despite his eminence as a scientist, he was often in trouble because of his liberal, unorthodox opinions. (Library of Congress)

| B 10.81 | C 12.01 | N 14.01 |
|---------|---------|---------|
| Al 26.98 | Si 28.09 | P 30.97 |
| Zn 65.37 | ? ? | ? ? | As 74.92 |
| Cd 112.41 | In 114.82 | Sn 118.69 | Sb 121.75 |

**FIGURE 5.2** Portion of Mendeleev's periodic table showing the symbols for elements and the modern values for their atomic masses.

with similar characteristics must be listed in the same groups, he was forced to leave several blank spaces in his table. For example, at the time Mendeleev was developing his periodic table, arsenic (As) was the element of next highest atomic weight after zinc (Zn). However, putting arsenic immediately after zinc in the table would have situated arsenic under aluminum (Al). Because arsenic resembles phosphorus (P) much more than it resembles aluminum, Mendeleev placed As under P. In doing so, he left two blank spaces in his table, as shown in Figure 5.2. He boldly predicted that these spaces belonged to two undiscovered elements. He gave these elements the names eka-aluminum and eka-silicon and suggested that they should occur in nature with other members of their respective families. For example, eka-aluminum might be found in ores containing aluminum.

By noting that the properties of elements within a vertical column vary in a regular way with increasing atomic weight, Mendeleev was able to predict the properties of these unknown elements. He predicted the properties of eka-aluminum in 1870. The element was discovered five years later in France and named gallium (Ga) after Gallia, the Latin name for France. Eka-silicon, whose properties he predicted in 1871, was discovered in Germany in 1886 and named germanium (Ge). The properties of both of these elements closely matched those predicted by Mendeleev. Table 5.1 compares some of Mendeleev's predicted properties of eka-silicon with the properties measured for germanium. Notice that the properties of germanium are intermediate between those of silicon (Si) and tin (Sn), the elements above and below Ge in the periodic table. The excellent agreement between the properties of gallium and germanium and those predicted for these elements by Mendeleev did much to establish the acceptance of the periodic table.

## Organization by Atomic Number

Mendeleev's periodic table had a serious flaw. When arranged by increasing atomic weight, some elements are out of order. For example, the atomic weight of iodine (I, 126.9) is less than that of tellurium (Te, 127.6). Nevertheless, their properties indicate that tellurium should be placed

**TABLE 5.1**  Comparison of the properties of eka-silicon predicted by Mendeleev with the known properties of germanium

| Property | Observed for silicon | Predicted for eka-silicon | Observed for tin | Found for germanium |
|---|---|---|---|---|
| Atomic weight | 28 | 72 | 118 | 72.56 |
| Density (g/cm³) | 2.33 | 5.5 | 7.28 | 5.35 |
| Specific heat (J/g-K) | 0.70 | 0.31 | 0.21 | 0.32 |
| Formula of oxide | $SiO_2$ | $XO_2$ | $SnO_2$ | $GeO_2$ |
| Density of oxide (g/cm³) | 2.66 | 4.7 | 6.95 | 4.70 |
| Formula of chloride | $SiCl_4$ | $XCl_4$ | $SnCl_4$ | $GeCl_4$ |
| Density of chloride (g/cm³) | 1.48 | 1.9 | 2.23 | 1.84 |

before iodine. Mendeleev believed so strongly in his periodic law that he suggested that the atomic weights of these elements were wrong. Indeed, some atomic weights were wrong, but not those of iodine and tellurium. The problem was not resolved until 1913, when scientists came to understand more fully the structure of the atom.

In 1913, just two years after Rutherford had proposed the nuclear model of the atom, Henry Moseley (Figure 5.3) discovered the concept of atomic numbers. Moseley bombarded different elements with energetic electrons and studied the X rays they emitted. He observed that the frequency of the X rays was different for each element. He was able to arrange these frequencies in order by assigning each element a unique, integral (whole) number, which he called the atomic number of the ele-

**FIGURE 5.3**  Henry G. J. Moseley (1887—1915). A brilliant young British physicist, Moseley studied under Rutherford at the University of Manchester after graduating from Oxford in 1910. He completed the work that led to the discovery of atomic numbers in 1913. He was killed in action at age 27 during World War I on a hillside at Gallipoli, Turkey. His early death deprived the world of further fruits of his genius. (Courtesy of the Edgar Fahs Smith Collection, ACS Center for History of Chemistry, University of Pennsylvania)

ment. The atomic number is the correct order for the elements in the periodic table. Moseley offered a simple explanation: "There is every reason to suppose that the integer that controls the X-ray spectrum is the charge on the nucleus." Today we know the atomic number to be the number of protons in the nucleus of an element (Section 2.3). The ordering of the elements in the periodic table by atomic number resolved the difficulties presented by certain elements, like tellurium and iodine. Indeed, tellurium has the atomic number 52 and so properly precedes iodine, which has the atomic number 53.

The atomic number equals not only the number of protons in the nucleus of an atom, but also the number of electrons in that atom. As we shall see in Chapters 6 and 7, it is the arrangement of electrons that gives rise to the chemical behavior of an element. The elements in a vertical column of the periodic table have similar chemical properties because their electron arrangements, or configurations, are similar.

## 5.2 THE MODERN PERIODIC TABLE

In the modern periodic table the elements are arranged in order of increasing atomic number. In this way, elements with similar physical and chemical properties occur at regular intervals. Elements that bear close chemical resemblances are placed in vertical columns referred to as **groups** or **families**. For example, the elements Li, Na, K, Rb, and Cs (the alkali metals) are closely related and are aligned vertically in group 1A. Although each member of a group has a distinct chemical personality, a family resemblance can always be recognized. One of the family characteristics of the alkali metals is their high chemical reactivity. They also form compounds having the same general chemical formulas. For example, all alkali metals react vigorously with chlorine to form ionic chlorides of the same stoichiometry: LiCl, NaCl, KCl, RbCl, and CsCl.

The horizontal rows of the table are called **rows** or **periods**. The first row consists of only two members, H and He. The next two rows, beginning with Li and Na, contain eight members each. The following rows, beginning with K, Rb, and Cs, contain 18, 18, and 32 members, respectively. Gradual changes in properties occur as we move from one element to another across a period. For example, in moving from Na to Mg to Al, reactivity decreases. It is important to consider not only how properties change within a group, but also within a period.

The major divisions of the periodic table are shown in Figure 5.4. Notice the three distinct regions:

1. The **representative elements** are comprised of groups 1A through 8A on the main portion of the table. The representative elements in groups 1A and 2A are often called the **active metals**, because they are chemically more reactive than most other metals.
2. The **transition metals** are the elements in the ten columns near the center of the table, in groups 3B through 2B.
3. The **inner transition metals** are the elements in the two 14-member rows that are placed below the main portion of the table. The elements in the first row are called the **lanthanides**, and those in the second are called the **actinides**.

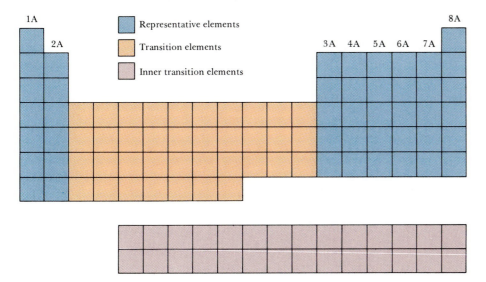

**FIGURE 5.4** Diagram of the periodic table showing the representative elements, transition elements, and inner transition elements.

The members of groups 1A, 2A, 6A, 7A, and 8A—at the left and right sides of the table—show particularly strong family resemblances and are given distinct family names, as summarized in Table 5.2. Groups 1A and 2A are known as the **alkali metals** and the **alkaline earth metals**, respectively. The elements of group 6A are sometimes called the **chalcogens**, a name less commonly used than the other group names. The elements in groups 7A and 8A are called the **halogens** and the **noble gases**, respectively.

Why do certain elements possess chemical similarities, and why does the periodic table have its particular form, with rows containing 2, 8, 8, 18, 18, and 32 members? The answers to both questions will become evident in Chapters 6 and 7. For the present we are more concerned with how to use the periodic table to organize information. We want to associate common chemical properties with the position of an element in the periodic table. We also want to develop some understanding of how properties change as we move from one part of the table to another.

The periodic table is a powerful tool for learning chemical facts. It will be useful to you, however, only if you take time to study it and work with it. Refer to the table often in reading the material that follows to be sure you know the locations of the elements being discussed.

**TABLE 5.2** Named families of elements

| Group | Name | Members |
|-------|------|---------|
| Group 1A | Alkali metals | Li, Na, K, Rb, Cs, Fr |
| Group 2A | Alkaline earth metals | Be, Mg, Ca, Sr, Ba, Ra |
| Group 6A | Chalcogens | O, S, Se, Te, Po |
| Group 7A | Halogens | F, Cl, Br, I, At |
| Group 8A | Noble gases | He, Ne, Ar, Kr, Xe, Rn |

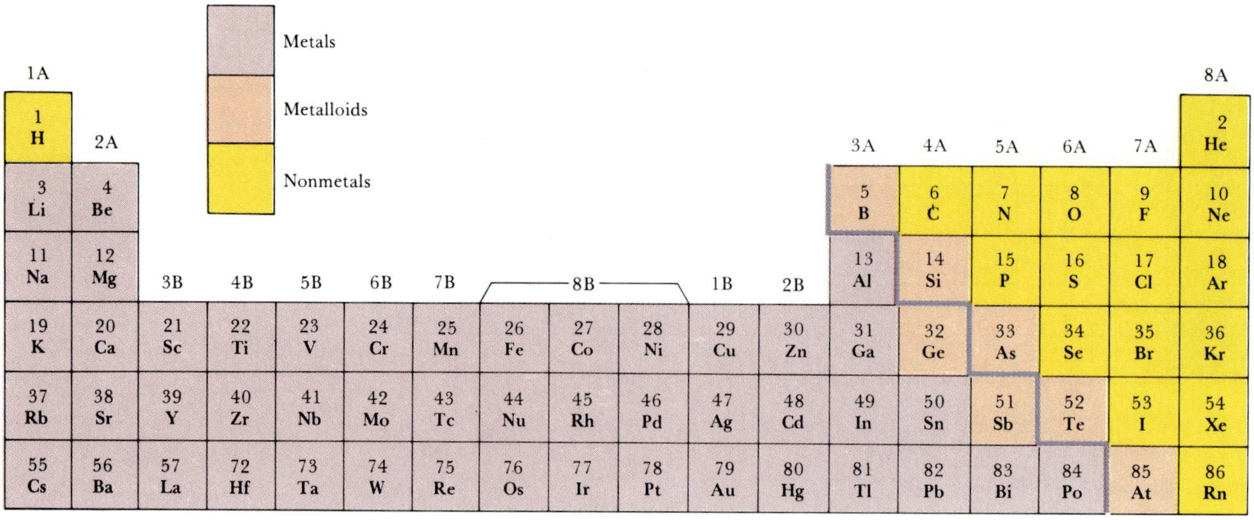

**FIGURE 5.5** Portion of the periodic table showing metals, metalloids (semimetals), and nonmetals.

## 5.3 METALS, NONMETALS, AND METALLOIDS

Elements can be divided into three broad categories: metals, nonmetals, and semimetals, also called metalloids. The metals, which comprise roughly 70 percent of the known elements, are situated on the left side of the table, as shown in Figure 5.5. The nonmetals are located in the upper right-hand corner, and the metalloids lie between the metals and nonmetals. Notice, however, that hydrogen, at the top-left corner, is a nonmetal. Some of the distinguishing properties of metals and nonmetals are summarized in Table 5.3.

**TABLE 5.3** Characteristic properties of metallic and nonmetallic elements

| Metallic elements | Nonmetallic elements |
| --- | --- |
| Distinguishing luster | Nonlustrous; various colors |
| Malleable and ductile as solids | Solids are usually brittle, may be hard or soft |
| Good thermal and electrical conductivity | Poor conductors of heat and electricity |
| Most metallic oxides are basic, ionic solids | Most nonmetallic oxides are molecular, acidic compounds |
| Exist in aqueous solution mainly as cations | Exist in aqueous solution mainly as anions or oxyanions |

**Metals**

Most metals look much alike, their appearance dominated by metallic luster (Figure 5.6). Metals exhibit good electrical and thermal conductivity. They are malleable (can be flattened), ductile (can be drawn into wire), and can be formed into various shapes (Figure 5.7). All are solids at room temperature except mercury (melting point = $-39°C$), which is a liquid. Two metals melt when slightly above room temperature: cesium at $28.4°C$ and gallium at $29.8°C$. At the other extreme, many metals melt at very high temperatures. For example, chromium melts at $1900°C$. Figure 5.8 shows how the melting points of the metals vary through the fourth, fifth, and sixth periods of the periodic table. Notice that the melting

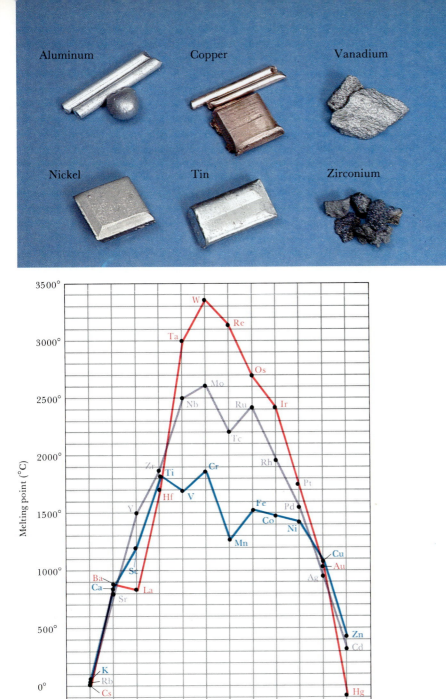

Aluminum  Copper  Vanadium

Nickel  Tin  Zirconium

**FIGURE 5.6** Metals are readily recognized by their characteristic luster. (Dr. E. R. Degginger)

**FIGURE 5.7**
Characteristic properties of metals include their luster and the ease with which they can be formed into various shapes. The object pictured here is a pre-Columbian gold frog from Central America. (Tom McHugh/Photo Researchers)

**FIGURE 5.8** Melting points of some metals as a function of their location in the periodic table.

points are periodic in character. In each period they tend to increase with increasing atomic number up to group 6B, near the center of the transition metals.

Metals tend to lose electrons when they undergo chemical reactions; that is, they transfer electrons to other substances and become **cations** (positively charged ions). For example, the reaction between magnesium metal and oxygen produces magnesium oxide, an ionic compound containing $Mg^{2+}$ and $O^{2-}$ ions:

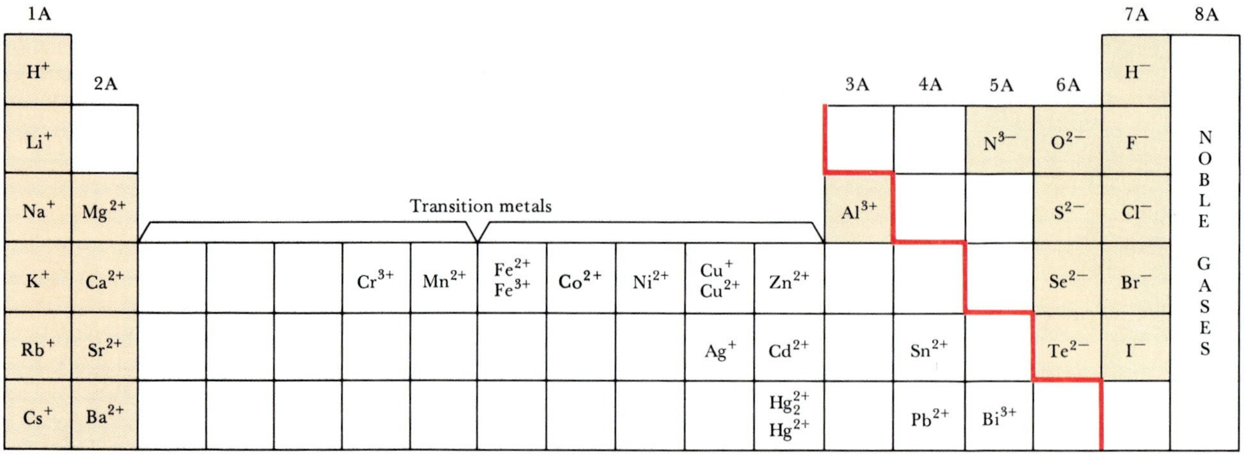

**FIGURE 5.9** Charges of some common ions found in ionic compounds. Notice that the steplike line that divides metals from nonmetals also separates cations from anions. Ions shown in color have the same number of electrons as the nearest noble gas atom. For example, $S^{2-}$, $Cl^-$, $K^+$, and $Ca^{2+}$ all have 18 electrons, the number found in the argon atom.

$$2Mg(s) + O_2(g) \longrightarrow 2MgO(s) \qquad [5.1]$$

Figure 5.9 shows the charges of some common ions. Notice that the charges of alkali metals are always $1+$ and that the charges of the alkaline earth metals are always $2+$ in their compounds. The charges on the transition metal ions do not follow an obvious pattern. Many transition metal ions have $2+$ charges, but $1+$ and $3+$ are also encountered. One of the characteristic features of the transition metals is their ability to form more than one positive ion. For example, iron may be $2+$ in some compounds and $3+$ in others. Figure 5.9 shows other examples.

Compounds of metals with nonmetals tend to be ionic substances with relatively high melting points. For example, most oxides, halides, and hydrides of metals are ionic solids. The oxides are particularly important because of the great abundance of oxygen in our environment.

Most metal oxides are basic; those that dissolve in water react to form metal hydroxides, as in the following examples:

Metal oxide + water $\longrightarrow$ base (*metal hydroxides*)

$$Na_2O(s) + H_2O(l) \longrightarrow 2NaOH(aq) \qquad [5.2]$$

$$CaO(s) + H_2O(l) \longrightarrow Ca(OH)_2(aq) \qquad [5.3]$$

Metal oxides also react with acids to form salts and water:

Metal oxide + acid $\longrightarrow$ salt + water

$$MgO(s) + 2HCl(aq) \longrightarrow MgCl_2(aq) + H_2O(l) \qquad [5.4]$$

$$NiO(s) + H_2SO_4(aq) \longrightarrow NiSO_4(aq) + H_2O(l) \qquad [5.5]$$

## SAMPLE EXERCISE 5.1

(a) Write the chemical formula for aluminum oxide (b) Would you expect this substance to be a solid, liquid, or gas at room temperature? (c) Write the balanced chemical equation for the reaction of aluminum oxide with nitric acid.

**Solution:** (a) In all its compounds, aluminum has a 3+ charge, $Al^{3+}$; the oxide ion is $O^{2-}$. Consequently, the chemical formula of aluminum oxide is $Al_2O_3$.

(b) Because aluminum oxide is the oxide of a metal, we would expect it to be a solid. Indeed it is, and it happens to have a very high melting point, 2072°C.

(c) Metal oxides generally react with acids to form salts and water. In this case the salt is aluminum nitrate, $Al(NO_3)_3$. The balanced equation is

$$Al_2O_3(s) + 6HNO_3(aq) \longrightarrow 2Al(NO_3)_3(aq) + 3H_2O(l)$$

## PRACTICE EXERCISE

Write the balanced chemical equation for the reaction between copper(II) oxide and sulfuric acid.
**Answer:** $CuO(s) + H_2SO_4(aq) \longrightarrow CuSO_4(aq) + H_2O(l)$

---

**Nonmetals**

Nonmetals vary greatly in appearance (Figure 5.10). They are not lustrous and are generally poor conductors of heat and electricity. Although diamond, the most stable form of carbon, has a high melting point (3570°C), the melting points of nonmetals are generally lower than those of metals (see Figure 5.11). Seven of the nonmetals exist under ordinary conditions as diatomic molecules. Included in this list are gases ($H_2$, $N_2$, $O_2$, $F_2$, and $Cl_2$), one liquid ($Br_2$), and one volatile solid ($I_2$). The remaining nonmetals are solids that may be hard like diamond or soft like sulfur.

**FIGURE 5.10**    Nonmetals are very diverse in their appearances. Shown here are, (left to right) sulfur, white phosphorus, bromine, and carbon. (Dr. E. R. Degginger)

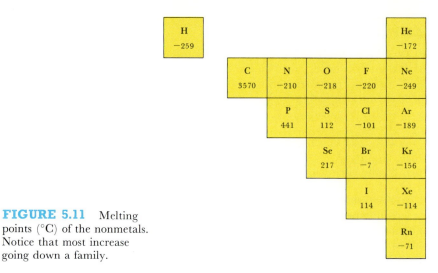

**FIGURE 5.11**  Melting points (°C) of the nonmetals. Notice that most increase going down a family.

Nonmetals, in reacting with metals, tend to gain or accept electrons and become **anions** (negatively charged ions). For example, the reaction of aluminum with bromine produces aluminum bromide, an ionic compound containing the bromide ion, $Br^-$, and the aluminum ion, $Al^{3+}$:

Metal + nonmetal $\longrightarrow$ salt

$$2Al(s) + 3Br_2(l) \longrightarrow 2AlBr_3(s) \qquad [5.6]$$

---

## SAMPLE EXERCISE 5.2

Predict the formulas of compounds formed between **(a)** Ba and Te; **(b)** Ga and Br.

**Solution:**  **(a)** Notice the location of these elements in the periodic table. The charge on a barium ion is $2+$; the charge on a tellurium ion (telluride) is $2-$, like that of oxides and sulfides. Thus the formula of the compound is BaTe. **(b)** Gallium is in group 3A.

Like the more familiar aluminum ion, gallium forms a $3+$ ion. Bromine forms a $1-$ ion (bromide). Thus the formula is $GaBr_3$.

**PRACTICE EXERCISE**
Predict the formula of the compound formed by Rb and Se.
***Answer:***  $Rb_2Se$

---

Compounds of nonmetals with each other are molecular substances. For example, the oxides, halides, and hydrides of the nonmetals are molecular substances that tend to be gases, liquids, or low-melting solids.

Most nonmetal oxides are acidic; those that dissolve in water react to form acids, as in the following examples:

Nonmetal oxide + water $\longrightarrow$ acid

$$CO_2(g) + H_2O(l) \longrightarrow H_2CO_3(aq) \qquad [5.7]$$

$$P_4O_{10}(s) + 6H_2O(l) \longrightarrow 4H_3PO_4(aq) \qquad [5.8]$$

The reaction of carbon dioxide with water (Figure 5.12) accounts for the acidity of carbonated water. Nonmetal oxides also dissolve in basic solutions to form salts, as in the following examples:

(a)                                    (b)

Nonmetal oxide + base $\longrightarrow$ salt + water

$$CO_2(g) + 2NaOH(aq) \longrightarrow Na_2CO_3(aq) + H_2O(l) \qquad [5.9]$$

$$SO_3(g) + 2KOH(aq) \longrightarrow K_2SO_4(aq) + H_2O(l) \qquad [5.10]$$

**SAMPLE EXERCISE 5.3**

Write the balanced chemical equations for the reactions of solid selenium dioxide with **(a)** water; **(b)** sodium hydroxide.

**Solution:**   **(a)** Selenium dioxide is $SeO_2$. Its reaction with water is like that of carbon dioxide (Equation 5.7):

$$SeO_2(s) + H_2O(l) \longrightarrow H_2SeO_3(aq)$$

(It doesn't matter that $SeO_2$ is a solid and $CO_2$ is a gas; the point is that both are nonmetal oxides.)

**(b)** The reaction with sodium hydroxide is like the reaction summarized by Equation 5.9:

$$SeO_2(s) + 2NaOH(aq) \longrightarrow$$
$$Na_2SeO_3(aq) + H_2O(l)$$

**PRACTICE EXERCISE** _____

Write the balanced chemical equation for the reaction of tetraphosphorus hexaoxide with water.
**Answer:**   $P_4O_6(s) + 6H_2O(l) \longrightarrow 4H_3PO_3(aq)$

Metalloids tend to have properties intermediate between those of metals and nonmetals. They may have *some* characteristic metallic properties but lack others. For example, antimony *looks* like a metal, but it is brittle rather than malleable and is a poor conductor of heat and electricity. Several of the metalloids, including silicon and germanium, are electrical semiconductors.

**Metalloids**

The more fully an element exhibits the physical and chemical properties characteristic of metals, the greater its **metallic character**. Similarly, we can speak of the nonmetallic character of an element. Metallic and non-metallic character exhibit some important periodic trends, summarized in Figure 5.13 and in the following list:

## 5.4   TRENDS IN METALLIC AND NONMETALLIC CHARACTER

**1.**   *Metallic character is strongest for the elements on the leftmost part of the periodic table and tends to decrease as we move to the right in any period.* For example, as

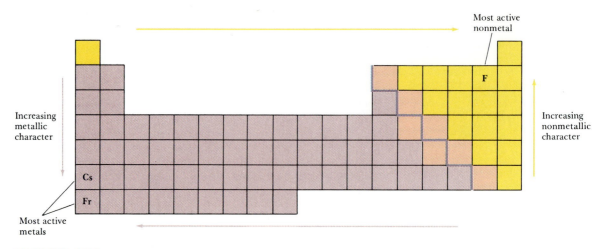

**FIGURE 5.13** Periodic trends in metallic and nonmetallic character. Notice the positions of the most active metals and the most active nonmetal.

we move across the fourth period, from K to Kr, there is a decrease in the ease with which each element loses electrons. K displays a greater chemical reactivity than Co does towards substances that readily accept electrons, like oxygen.

2. *Within any group of representative elements, the metallic character increases progressively from top to bottom.* For example, Ba loses electrons more readily than Mg. This rule is generally not followed among the transition metals. The first-row transition metals are generally more active than those in the second or third rows. For example, Fe is a more active metal than Ru or Os.

3. *Nonmetallic character is greatest for the elements in the upper right of the table and increases moving from left to right across a period.* For example, Br gains electrons more readily than As and thus displays greater chemical reactivity toward metals and other substances that readily lose electrons.

4. *Nonmetallic character decreases moving down a family.* The general shift from nonmetallic to metallic character as we go down a family of representative elements is most evident in groups 3A, 4A, and 5A. For example, carbon, the first element in group 4A, is clearly a nonmetal. Below carbon are the metalloids silicon and germanium. Below them are the metals tin and lead.

## 5.5 OXIDATION OF METALS AND THE ACTIVITY SERIES

Many metals react directly with oxygen in air to form metal oxides. In these reactions the metal loses electrons to oxygen, forming an ionic compound of the metal ion and oxide ion. For example, when calcium metal is exposed to the air, the bright metallic surface of the metal tarnishes as CaO forms:

$$2Ca(s) + O_2(g) \longrightarrow 2CaO(s) \qquad [5.11]$$

The reaction of iron with oxygen is responsible for the formation of rust, an oxide of iron.

In Equation 5.11, Ca (no net charge) reacts to form the $Ca^{2+}$ ion in CaO. When a substance has become more positively charged (that is, when it has lost electrons), it is said to be oxidized; the loss of electrons by a substance is called **oxidation**. As Ca is oxidized in Equation 5.11, oxygen goes from $O_2$ (no net charge) to $O^{2-}$. When a substance has become more negatively charged (gained electrons), it is said to be reduced; the gain of electrons by a substance is called **reduction**. Sodium and calcium are examples of metals that oxidize readily; silver and gold are examples of metals that are quite resistant to oxidation.

Many metals react with acids to form salts and hydrogen gas. For example, magnesium metal reacts with hydrochloric acid to form magnesium chloride and hydrogen gas:

**Oxidation of Metals by Acids**

$$\text{Metal} + \text{acid} \longrightarrow \text{salt} + \text{hydrogen}$$
$$Mg(s) + 2HCl(aq) \longrightarrow MgCl_2(aq) + H_2(g) \qquad [5.12]$$

Similarly, iron reacts with sulfuric acid (see Figure 5.14):

$$Fe(s) + H_2SO_4(aq) \longrightarrow FeSO_4(aq) + H_2(g) \qquad [5.13]$$

In each instance, the metal is oxidized by the acid to form the metal cation; the $H^+$ ion of the acid is reduced to form $H_2$.

The oxidation of iron by acids occurs when acidic foods are cooked in cast-iron cookware, providing a useful dietary source of iron. Half a cup of spaghetti sauce contains only 3 mg of iron. However, spaghetti sauce simmered several hours in an iron pot increases in iron content to at least 50 mg. Another example of the reaction of a metal with an acid is the corrosion of clamps at the terminals of an automobile battery (Figure 5.15). In this instance, the metal clamps react with battery acid, $H_2SO_4$.

**FIGURE 5.14**  Many metals such as the iron in the nail shown here react with acids such as sulfuric acid, to form hydrogen gas. The bubbles are due to the hydrogen gas. (Dr. E. R. Degginger)

**FIGURE 5.15**  Corrosion at the terminals of a battery caused by attack of the metal by sulfuric acid from the battery. (Dr. E. R. Degginger)

## SAMPLE EXERCISE 5.4

Write the balanced chemical equation for the reaction of aluminum with hydrobromic acid.

**Solution:** Aluminum reacts with hydrobromic acid (HBr) to form $H_2$ and the salt of $Al^{3+}$ and $Br^-$:

$$2Al(s) + 6HBr(aq) \longrightarrow 2AlBr_3(aq) + 3H_2(g)$$

### PRACTICE EXERCISE

Write the balanced equation for the reaction between sulfuric acid and lead.

*Answer:* $Pb(s) + H_2SO_4(aq) \longrightarrow PbSO_4(s) + H_2(g)$

### The Activity Series

The reaction of a metal with an acid to form hydrogen gas is characteristic of a much broader class of reactions called **single displacement reactions**. These reactions conform to the following general equation:

$$A + BX \longrightarrow AX + B \qquad [5.14]$$

Element A reacts with compound BX to form compound AX and the element B, as in the following examples:

$$Metal + salt \longrightarrow salt + metal$$
$$A + BX \longrightarrow AX + B$$
$$Fe(s) + Ni(NO_3)_2(aq) \longrightarrow Fe(NO_3)_2(aq) + Ni(s) \qquad [5.15]$$
$$Zn(s) + CuSO_4(aq) \longrightarrow ZnSO_4(aq) + Cu(s) \qquad [5.16]$$

Notice the similarity between the reaction of zinc and copper(II) sulfate (Equation 5.16) and that between zinc and sulfuric acid:

$$Zn(s) + H_2SO_4(aq) \longrightarrow ZnSO_4(aq) + H_2(g) \qquad [5.17]$$

In both cases the zinc metal is oxidized, going from Zn to the $Zn^{2+}$ ion in $ZnSO_4$.

Comparing single displacement reactions between metals allows us to rank metals according to the ease with which they can undergo oxidation in solution. For example, although zinc reacts with aqueous solutions of copper(II) sulfate, silver metal does not. We conclude that zinc loses electrons more readily than does silver. By comparing the abilities of metals to react, we may rank metals in order of decreasing ease of oxidation.

A list of metals arranged in order of decreasing ease of oxidation is called an **activity series**. Table 5.4 gives the activity series for many of the most common metals. Hydrogen is also included in the table. Those metals at the top of the table are most easily oxidized; that is, they most readily react to form compounds. Notice that the alkali metals and alkaline earth metals are at the top. The metals at the bottom of the activity series are very stable and form compounds less readily. Notice that the transition elements from groups 8B and 1B are at the bottom of the list. These metals, which are used in making coins and jewelry, are called noble metals because of their low reactivity.

The activity series can be used to predict the outcome of single displacement reactions between metals and either metal salts or acids. *Any*

**TABLE 5.4** Activity series of metals

| Metal | Oxidation reaction |
|-------|---------------------|
| Lithium | Li $\longrightarrow$ Li$^+$ + e$^-$ |
| Potassium | K $\longrightarrow$ K$^+$ + e$^-$ |
| Barium | Ba $\longrightarrow$ Ba$^{2+}$ + 2e$^-$ |
| Calcium | Ca $\longrightarrow$ Ca$^{2+}$ + 2e$^-$ |
| Sodium | Na $\longrightarrow$ Na$^+$ + e$^-$ |
| Magnesium | Mg $\longrightarrow$ Mg$^{2+}$ + 2e$^-$ |
| Aluminum | Al $\longrightarrow$ Al$^{3+}$ + 3e$^-$ |
| Manganese | Mn $\longrightarrow$ Mn$^{2+}$ + 2e$^-$ |
| Zinc | Zn $\longrightarrow$ Zn$^{2+}$ + 2e$^-$ |
| Chromium | Cr $\longrightarrow$ Cr$^{3+}$ + 3e$^-$ |
| Iron | Fe $\longrightarrow$ Fe$^{2+}$ + 2e$^-$ |
| Cobalt | Co $\longrightarrow$ Co$^{2+}$ + 2e$^-$ |
| Nickel | Ni $\longrightarrow$ Ni$^{2+}$ + 2e$^-$ |
| Tin | Sn $\longrightarrow$ Sn$^{2+}$ + 2e$^-$ |
| Lead | Pb $\longrightarrow$ Pb$^{2+}$ + 2e$^-$ |
| Hydrogen | H$_2$ $\longrightarrow$ 2H$^+$ + 2e$^-$ |
| Copper | Cu $\longrightarrow$ Cu$^{2+}$ + 2e$^-$ |
| Silver | Ag $\longrightarrow$ Ag$^+$ + e$^-$ |
| Mercury | Hg $\longrightarrow$ Hg$^{2+}$ + 2e$^-$ |
| Platinum | Pt $\longrightarrow$ Pt$^{2+}$ + 2e$^-$ |
| Gold | Au $\longrightarrow$ Au$^{3+}$ + 3e$^-$ |

Ease of oxidation decreases

metal on the list is able to displace the elements below it from their compounds. For example, copper is above silver in the series. Thus copper metal will displace silver from aqueous solutions of its compounds, as in the following example, pictured in Figure 5.16:

$$Cu(s) + 2AgNO_3(aq) \longrightarrow Cu(NO_3)_2(aq) + 2Ag(s) \qquad [5.18]$$

Similarly, because aluminum is above iron in the series, aluminum metal will displace iron from aqueous solutions of its compounds. For example,

$$2Al(s) + 3FeSO_4(aq) \longrightarrow Al_2(SO_4)_3(aq) + 3Fe(s) \qquad [5.19]$$

However, because copper is below iron, copper metal will not displace iron. Thus no reaction occurs when copper metal is added to a solution of iron(II) sulfate.

**FIGURE 5.16** The displacement reaction between copper metal and a solution of silver nitrate. The product, silver metal, is evident on the surface of the copper wires. The other product, copper(II) nitrate, produces the blue color in the solution. (© Fundamental Photographs)

## SAMPLE EXERCISE 5.5

Will a single displacement reaction occur between magnesium metal and iron(II) chloride? If so, write the balanced chemical equation for the reaction.

**Solution:** We see in Table 5.4 that Mg is above Fe in the activity series, which indicates that a displacement reaction will occur. To write the equations for displacement reactions, we must remember the charges of common ions. Magnesium is always present in compounds as the Mg$^{2+}$ ion; the chloride ion

is Cl$^-$. Thus the compound formed is MgCl$_2$. The balanced equation is

$$Mg(s) + FeCl_2(aq) \longrightarrow MgCl_2(aq) + Fe(s)$$

### PRACTICE EXERCISE
Which of the following metals will displace lead from Pb(NO$_3$)$_2$: Zn, Cu, Fe?
**Answer:** Zn and Fe

TABLE 5.5 Hydrogen
displacement by metals

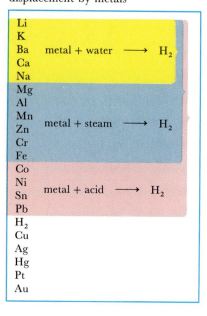

As shown in Table 5.5, only those metals above hydrogen in the activity series are able to react with acids to form $H_2$. Thus Mn reacts with HCl to form $H_2$:

$$Mn(s) + 2HCl(aq) \longrightarrow MnCl_2(aq) + H_2(g) \qquad [5.20]$$

Those metals from iron and above react with steam to form $H_2$ and insoluble oxides. For example,

$$Fe(s) + H_2O(g) \longrightarrow FeO(s) + H_2(g) \qquad [5.21]$$

The most active metals (from Na on up) are also able to displace hydrogen from cold water, forming $H_2$ and the metal hydroxide. For example,

$$2Na(s) + 2H_2O(l) \longrightarrow 2NaOH(aq) + H_2(g) \qquad [5.22]$$

Because elements below hydrogen in the activity series do not displace hydrogen from acids, Cu does not react with HCl. Interestingly, copper does react with nitric acid, as shown earlier in Figure 1.6. However, this reaction is not a single displacement reaction. Instead, the nitrate ion of the acid, rather than the hydrogen ion, removes electrons from the copper. The reaction between copper and concentrated nitric acid proceeds as follows:

$$Cu(s) + 4HNO_3(aq) \longrightarrow Cu(NO_3)_2(aq) + 2H_2O(l) + 2NO_2(g) \qquad [5.23]$$

## 5.6 A SURVEY OF SELECTED REPRESENTATIVE METALS

### Group 1A: The Alkali Metals

The group 1A elements consist of lithium (Li), sodium (Na), potassium (K), rubidium (Rb), cesium (Cs), and francium (Fr). All have characteristic metallic properties such as a silvery, metallic luster and high thermal and electrical conductivities. The name "alkali" comes from an Arabic word meaning "ashes." Many compounds of sodium and potassium, the two most abundant alkali metals, were isolated from wood ashes by early chemists. The names "soda ash" and "potash" are still sometimes used for $Na_2CO_3$ and $K_2CO_3$, respectively.

Owing to their high reactivities, the alkali metals exist in nature only as compounds. The metals can be obtained by passing an electric current through a molten salt. For example, sodium is prepared commercially by electrolysis of molten NaCl. The electrical energy is used to remove an electron from $Cl^-$ and to force it onto $Na^+$:

$$Cl^- \longrightarrow Cl + e^- \qquad [5.24]$$
$$Na^+ + e^- \longrightarrow Na \qquad [5.25]$$

Unlike most metals, the alkali metals are soft enough to be cut with a knife (Figure 5.17), and their softness increases going down the family. Some of their physical properties are given in Table 5.6. Notice that the elements have low densities and melting points and that these properties vary in a fairly regular way with increasing atomic number.

The alkali metals, extremely reactive elements, combine directly with

**FIGURE 5.17** Sodium and the other alkali metals are soft. Here we see sodium being cut with a knife. The shiny metallic surface quickly tarnishes as the metal reacts with oxygen in the air. (H. E. LeMay, Jr.)

**TABLE 5.6** Some physical and chemical properties of the alkali metals

| Element | Symbol | Atomic number | Atomic weight | Melting point (°C) | Density (g/cm³) | Formula of hydroxide | Formula of chloride |
|---------|--------|---------------|---------------|--------------------|------------------|----------------------|---------------------|
| Lithium | Li | 3 | 6.939 | 181 | 0.53 | LiOH | LiCl |
| Sodium | Na | 11 | 22.9898 | 98 | 0.97 | NaOH | NaCl |
| Potassium | K | 19 | 39.102 | 63 | 0.86 | KOH | KCl |
| Rubidium | Rb | 37 | 85.47 | 39 | 1.53 | RbOH | RbCl |
| Cesium | Cs | 55 | 132.905 | 29 | 1.87 | CsOH | CsCl |

most nonmetals. For example, they react with hydrogen to form solid hydrides and with chlorine to form solid chlorides, as shown in the following reactions of potassium:

$$2K(s) + H_2(g) \longrightarrow 2KH(s) \qquad [5.26]$$

$$2K(s) + Cl_2(g) \longrightarrow 2KCl(s) \qquad [5.27]$$

In KH and other hydrides of the alkali metals (such as LiH), hydrogen is present as $H^-$, called the hydride ion. The hydride ion is to be distinguished from the hydrogen ion, $H^+$, formed when hydrogen loses an electron.

It is particularly noteworthy that the alkali metals react readily with oxygen and water, two very common substances in our environment. The reaction with oxygen causes the metals to tarnish quickly when exposed to air. For example, lithium reacts readily with $O_2$ to form $Li_2O$:

$$4Li(s) + O_2(g) \longrightarrow 2Li_2O(s) \qquad [5.28]$$

The reactions of the alkali metals with water are quite vigorous, producing hydrogen gas and solutions of the alkali-metal hydroxides. For example,

$$2Li(s) + 2H_2O(l) \longrightarrow 2LiOH(aq) + H_2(g) \qquad \Delta H° = -445.1 \text{ kJ} \qquad [5.29]$$

In many cases, enough heat is generated to ignite the $H_2$, producing a fire or explosion. Consequently, alkali metals are usually stored under a hydrocarbon, such as kerosene, to protect them from $O_2$ and $H_2O$. As expected from the general trends in metallic character, the reactivity of the alkali metals increases going down the family.

## SAMPLE EXERCISE 5.6

Write balanced chemical equations for the reactions of cesium with (a) $H_2(g)$; (b) $Cl_2(g)$; (c) $H_2O(l)$.

**Solution:** By analogy to the equations above, we have

$$2Cs(s) + H_2(g) \longrightarrow 2CsH(s)$$
$$2Cs(s) + Cl_2(g) \longrightarrow 2CsCl(s)$$
$$2Cs(s) + 2H_2O(l) \longrightarrow 2CsOH(aq) + H_2(g)$$

In each case, cesium forms a $1+$ ion in its compounds, $Cs^+$. The hydride ion, chloride ion, and hydroxide ion all have $1-$ charges: $H^-$, $Cl^-$, and $OH^-$.

## PRACTICE EXERCISE

Solid sodium hydride reacts with water to form an aqueous solution of sodium hydroxide and hydrogen gas. Write a balanced chemical equation for this reaction.

*Answer:* $NaH(s) + H_2O(l) \longrightarrow NaOH(aq) + H_2(g)$

## Group 2A: The Alkaline Earth Metals

The group 2A metals consist of beryllium (Be), magnesium (Mg), calcium (Ca), strontium (Sr), barium (Ba), and radium (Ra). Because of their reactivities, these metals are not found in nature in the uncombined state. Be is the least reactive of the 2A elements.

Like the alkali metals, the group 2A elements are all solids with typical metallic properties, some of which are listed in Table 5.7. Compared with the alkali metals, the alkaline earth metals are harder, more dense, and melt at higher temperatures.

Beryllium is relatively unreactive even at elevated temperatures. Magnesium reacts much more readily with most nonmetals. For example, Mg reacts at room temperature with $Cl_2$ to form $MgCl_2$, and it burns with dazzling brilliance in air to give MgO (Figure 5.18):

$$Mg(s) + Cl_2(g) \longrightarrow MgCl_2(s) \qquad [5.30]$$
$$2Mg(s) + O_2(g) \longrightarrow 2MgO(s) \qquad [5.31]$$

The metal is protected from many chemicals by a thin, compact surface coat of water-insoluble MgO. Thus even though it is high on the activity series, magnesium can be used as a structural metal. The heavier alkaline earth metals (Ca, Sr, and Ba) react even more readily with nonmetals.

2A

| 4 |
|---|
| Be |

| 12 |
|---|
| Mg |

| 20 |
|---|
| Ca |

| 38 |
|---|
| Sr |

| 56 |
|---|
| Ba |

| 88 |
|---|
| Ra |

**TABLE 5.7** Some physical and chemical properties of the alkaline earth metals

| Element | Symbol | Atomic number | Atomic weight | Melting point (°C) | Density (g/cm³) | Formula of hydroxide | Formula of chloride |
|---------|--------|---------------|---------------|--------------------|-----------------|----------------------|---------------------|
| Beryllium | Be | 4 | 9.01 | 1283 | 1.85 | $Be(OH)_2$ | $BeCl_2$ |
| Magnesium | Mg | 12 | 24.30 | 650 | 1.74 | $Mg(OH)_2$ | $MgCl_2$ |
| Calcium | Ca | 20 | 40.08 | 851 | 1.54 | $Ca(OH)_2$ | $CaCl_2$ |
| Strontium | Sr | 38 | 87.62 | 757 | 2.58 | $Sr(OH)_2$ | $SrCl_2$ |
| Barium | Ba | 56 | 137.33 | 704 | 3.65 | $Ba(OH)_2$ | $BaCl_2$ |

**FIGURE 5.18**
Magnesium metal burns with a brilliant flame. (Dr. E. R. Degginger)

**FIGURE 5.19**   Calcium metal reacts with water to form hydrogen gas and calcium hydroxide, $Ca(OH)_2$. (Dr. E. R. Degginger)

Like the alkali metals, the heavier alkaline earth metals also react with water at room temperature, as illustrated by the following reaction, shown in Figure 5.19:

$$Ca(s) + 2H_2O(l) \longrightarrow Ca(OH)_2(aq) + H_2(g) \qquad [5.32]$$

## Aluminum

Among the other important representative metals is aluminum, the second member of group 3A. Aluminum is the third most abundant element, after oxygen and silicon, in the earth's crust. Aluminum metal is a relatively soft, light metal of density $2.7 \text{ g/cm}^3$. It is widely used as a structural metal where light weight is an important consideration. As shown in Table 5.8, aluminum ranks second among metals after iron in annual production.

From the position of the metal in the activity series, we should expect aluminum to be very reactive. However, it is essentially free from corrosion in air because of the formation of a thin protective oxide coat that adheres tenaciously and prevents further oxidation. When the protective oxide coat is destroyed—for example, by seawater—the metal corrodes rapidly.

Aluminum is situated in the periodic table just to the left of the line dividing metals and nonmetals. It is less metallic (and consequently more nonmetallic) than sodium and magnesium, its neighbors to the left. The acid-base behavior of aluminum oxide reflects this difference. We have seen that the oxides of metals tend to be basic, whereas those of nonmetals

**TABLE 5.8**   Estimated annual worldwide production of metals

| Metal | Production |
|---|---|
| Iron | 700 million tons ($6.4 \times 10^{11}$ kg) |
| Aluminum | 16 million tons ($1.5 \times 10^{10}$ kg) |
| Manganese | 10 million tons ($9 \times 10^9$ kg) |
| Copper | 8 million tons ($7 \times 10^9$ kg) |
| Zinc | 6 million tons ($5 \times 10^9$ kg) |
| Lead | 4 million tons ($4 \times 10^9$ kg) |

The calcium ion is an essential component of our diet, required for clotting blood and maintaining heartbeat rhythm. However, most calcium is used in forming and maintaining bones and teeth. Bone is composed of protein fibers called collagen, water, and inorganic salts (or "minerals"). The primary mineral is known as hydroxyapatite, $Ca_5(PO_4)_3OH$.

The recommended daily allowance for calcium for adults is 1000 mg a day. Most of our dietary calcium is provided by dairy products, such as milk and cheese. However, many weight-conscious people shy away from these foods because they are high in calories. Thus many people fail to meet the daily requirement of calcium in their diet.

One of the major health problems associated with calcium metabolism is osteoporosis (Greek *osteon*, meaning "bone," and *poros*, meaning "a passage"), the increased porosity or softening of bone (see Figure 5.20). This condition can lead to severe cur-

vature of the spine and to increased likelihood of hip fractures. Osteoporosis is more a problem for women than for men because calcium metabolism is affected by levels of the female sex hormone, estrogen. Reduced estrogen production, which occurs after menopause, changes the normal reactions by which the removal of calcium from bones is balanced by new deposits. Calcium leaves faster than it can be replaced, and the bones become thin. Another problem for older women is reduced vitamin D production, which in turn lessens the efficiency with which dietary calcium is absorbed from the intestine.

Many experts now recommend the use of calcium supplements, especially for women with a family history of osteoporosis. Supplements must be taken with caution, though, because an excess of calcium can lead to kidney stones and other medical problems.

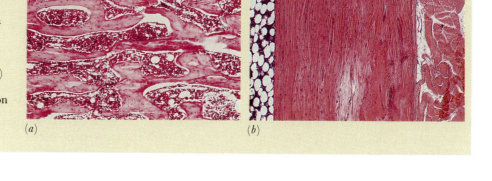

**FIGURE 5.20** (*a*) Normal, cancellous (spongy) bone, composed of a network of bony material separated by a labyrinth of interconnecting spaces containing bone marrow. (© Biophoto Associates/Photo Researchers) (*b*) Decalcified, cancellous bone, which shows a reduction in the white areas composed of bony material. (Patrick Lynch/Photo Researchers)

(*a*)                    (*b*)

are acidic (Section 5.3). The oxides of sodium and magnesium are typical basic oxides; they react with acids but not with bases. In contrast, aluminum oxide reacts with both acids and bases:

$$Al_2O_3(s) + 6HCl(aq) \longrightarrow 2AlCl_3(aq) + 3H_2O(l) \qquad [5.33]$$

$$Al_2O_3(s) + 6NaOH(aq) \longrightarrow 2Na_3AlO_3(aq) + 3H_2O(l) \qquad [5.34]$$

The metal itself also reacts with both acids and bases:

$$2Al(s) + 6HCl(aq) \longrightarrow 2AlCl_3(aq) + 3H_2(g) \qquad [5.35]$$

$$2Al(s) + 6NaOH(aq) \longrightarrow 2Na_3AlO_3(aq) + 3H_2(g) \qquad [5.36]$$

The popular drain cleaner Drāno contains pieces of aluminum metal mixed with crystals of sodium hydroxide (lye). When added to water, the mixture reacts exothermally to produce bubbles of hydrogen gas. The bubbles and the heat help to dislodge or dissolve hair, grease, and other materials clogging the drain. The reaction of bases with aluminum also explains why many oven cleaners contain warnings not to use them on aluminum utensils. These oven cleaners contain sodium hydroxide, which would react with the aluminum.

## 5.7 A SURVEY OF SELECTED NONMETALS

### Hydrogen

The first element in the periodic table, hydrogen, is usually placed above the alkali metals. However, it is a unique element and does not truly belong to any family. Hydrogen is a nonmetal that occurs as a colorless, diatomic gas, $H_2(g)$, under most conditions. Whereas the chlorides and oxides of metals tend to be solids at room temperature, HCl is a gas and $H_2O$ a liquid.

Small quantities of hydrogen are usually prepared in the laboratory by reaction between zinc and an acid, such as hydrochloric acid:

$$Zn(s) + 2HCl(aq) \longrightarrow ZnCl_2(aq) + H_2(g) \qquad [5.37]$$

Many metals, such as iron, can be used in place of zinc.

Hydrogen burns readily to form water:

$$2H_2(g) + O_2(g) \longrightarrow 2H_2O(l) \qquad \Delta H° = -571.7 \text{ kJ} \qquad [5.38]$$

A simple test for $H_2$ often performed in general chemistry lab consists of bringing a glowing splint of wood to the open end of a test tube containing a gas. If the test tube holds only hydrogen, the gas burns with a colorless flame. If it contains oxygen together with hydrogen, a characteristic "pop" is heard as $H_2$ and $O_2$ combine rapidly to form $H_2O$.

### Group 6A: The Oxygen Family

Group 6A consists of oxygen (O), sulfur (S), selenium (Se), tellurium (Te), and polonium (Po). Oxygen is a colorless gas at room temperature; all of the others are solids. Oxygen, sulfur, and selenium are typical nonmetals. Tellurium has some metallic properties and is classified as a semimetal. Polonium is a metallic element; it is very radioactive and quite rare. Some of the physical properties of the group 6A elements are given in Table 5.9.

Oxygen is encountered in two molecular forms, $O_2$ and $O_3$. The $O_2$ form is the common one. People generally mean $O_2$ when they say "oxygen," although the name *dioxygen* is more descriptive. $O_3$ is called ozone. The two forms of oxygen are examples of **allotropes**. Allotropes are different forms of the same element in the same state (in this case, both forms are gases).

About 21 percent of dry air consists of $O_2$ molecules. Because $O_2$ is so plentiful in our environment, its reactions are of great practical interest. All those reactions in which a substance combines with oxygen to form oxides are examples of oxidation reactions. Those oxidations that proceed with evolution of heat and light are known as combustion reactions. Combustion reactions were among the first reactions to be carefully studied by chemists. These important reactions have been encountered previously, in Section 3.2.

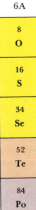

**TABLE 5.9**  Some physical and chemical properties of the group 6A elements

| Element | Symbol | Atomic number | Atomic weight | Melting point (°C) | Density (g/cm³) | Formula of hydride |
|---------|--------|---------------|---------------|--------------------|-----------------|--------------------|
| Oxygen | O | 8 | 16.00 | −218 | 1.43 g/L | $H_2O$ |
| Sulfur | S | 16 | 32.06 | 112 | 2.07 | $H_2S$ |
| Selenium | Se | 34 | 78.96 | 217 | 4.79 | $H_2Se$ |
| Tellurium | Te | 52 | 127.60 | 450 | 6.24 | $H_2Te$ |
| Polonium | Po | 84 | (210)[a] | 254 | 9.32 | $H_2Po$ |

[a] Most stable isotope.

Oxygen is separated from air on a vast scale (about 100 million tons per year worldwide). A large portion of this oxygen is used in making steel (Section 24.2). Very small quantities of $O_2$ are often generated in the laboratory by moderate heating of potassium chlorate ($KClO_3$) in the presence of a small quantity of manganese dioxide ($MnO_2$), which serves to catalyze, or speed, the reaction:

$$2KClO_3(s) \xrightarrow[\Delta]{MnO_2} 2KCl(s) + 3O_2(g) \qquad [5.39]$$

Ozone is present in very small amounts in the upper atmosphere and in polluted air. It is also formed near high-voltage generators and where significant electrical discharges occur, such as in lightning storms:

$$3O_2(g) \longrightarrow 2O_3(g) \qquad \Delta H° = 264.6 \text{ kJ} \qquad [5.40]$$

Ozone has a characteristic, pungent odor. This allotrope is less stable and more reactive than $O_2$.

Another important member of group 6A is sulfur, which also exists in several allotropic forms. The most stable and common allotrope at room temperature is the yellow solid with molecular formula $S_8$. This molecule consists of an eight-membered ring of sulfur atoms, as shown in Figure 5.21. Most sulfur in nature is present in metal-sulfide compounds. Sulfur is also a constituent of most animal and plant proteins. Both coal and petroleum contain sulfur because of their biological origin (Section 4.8). When sulfur or sulfur-containing compounds are burned, sulfur dioxide is formed:

$$S(s) + O_2(g) \longrightarrow SO_2(g) \qquad [5.41]$$

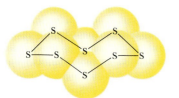

**FIGURE 5.21**  Structure of $S_8$ molecules as found in the most common allotropic form of sulfur at room temperature.

Sulfur dioxide is one of the principal air pollutants and is the major cause of acid rain (Section 14.4).

---

**SAMPLE EXERCISE 5.7**

Both magnesium and graphite burn in air. **(a)** Write balanced chemical equations for these reactions, assuming that only $O_2$ reacts with these elements and that complete combustion occurs. **(b)** Using heats of formation (Appendix D), determine which produces more heat when it burns, a gram of magnesium or a gram of graphite.

**Solution:** **(a)** Oxidation of magnesium produces MgO (containing $Mg^{2+}$ and $O^{2-}$ ions). Combustion of carbon (graphite) produces $CO_2$:

$$2Mg(s) + O_2(g) \longrightarrow 2MgO(s)$$
$$C(s) + O_2(g) \longrightarrow CO_2(g)$$

Note that the metal oxide, MgO, is solid. The non-metal oxide, $CO_2$, is gaseous.

**(b)** From Appendix D we note that the heat of formation of MgO is $-143.84$ kJ/mol and that of $CO_2$ is $-393.5$ kJ/mol. Proceeding as in Section 4.4, we have

$$(1.00 \text{ g Mg})\left(\frac{1 \text{ mol Mg}}{24.3 \text{ g Mg}}\right)\left(\frac{-143.84 \text{ kJ}}{1 \text{ mol Mg}}\right) = -5.92 \text{ kJ}$$

$$(1.00 \text{ g C})\left(\frac{1 \text{ mol C}}{12.0 \text{ g C}}\right)\left(\frac{-393.5 \text{ kJ}}{1 \text{ mol C}}\right) = -32.8 \text{ kJ}$$

Thus a gram of carbon produces more heat than a gram of magnesium.

**PRACTICE EXERCISE**

Write the balanced equation for the combustion of hydrogen sulfide, and calculate $\Delta H°$ for this process.

*Answer:* $2H_2S(g) + 3O_2(g) \longrightarrow 2H_2O(l) + 2SO_2(g)$
$\Delta H° = -1125.2$ kJ

**TABLE 5.10** Some physical and chemical properties of the halogens

| Element | Symbol | Atomic number | Atomic weight | Melting point (°C) | Boiling point (°C) | Density (g/cm³) | Formula of hydride |
|---------|--------|--------------|---------------|-------------------|-------------------|-----------------|--------------------|
| Fluorine | F | 9 | 19.00 | $-220$ | $-188$ | 1.81 g/L | HF |
| Chlorine | Cl | 17 | 35.45 | $-101$ | $-34$ | 3.21 g/L | HCl |
| Bromine | Br | 35 | 79.91 | $-7.3$ | 58.8 | 3.12 | HBr |
| Iodine | I | 53 | 126.90 | 114 | 134 | 4.94 | HI |

**Group 7A: The Halogens**

The elements of group 7A consist of fluorine (F), chlorine (Cl), bromine (Br), iodine (I), and astatine (At). These elements are known as the halogens, after the Greek words *halos* and *gennao*, meaning "salt formers." Some of the properties of these elements are given in Table 5.10. Astatine is omitted because it is both extremely rare and radioactive.

All the halogens are typical nonmetals. Their melting and boiling points increase going down the family. As a consequence, fluorine and chlorine are gases at room temperature, while bromine is a liquid and iodine is a solid. Fluorine gas is pale yellow; chlorine has a yellow-green color; bromine is reddish brown and readily forms a reddish-brown vapor; solid iodine is grayish black and readily forms a violet vapor (Figure 5.22). Each element consists of diatomic molecules: $F_2$, $Cl_2$, $Br_2$, and $I_2$.

The halogens react with hydrogen to form gaseous compounds: HF, HCl, HBr, and HI. For example,

$$H_2(g) + F_2(g) \longrightarrow 2HF(g) \qquad [5.42]$$

These compounds are all very soluble in water and dissolve to form important acid solutions. The halogens react with metals to form halide salts. For example, the reaction between magnesium and chlorine produces magnesium chloride:

$$Mg(s) + Cl_2(g) \longrightarrow MgCl_2(s) \qquad [5.43]$$

7A

| 9 F |
|---|
| 17 Cl |
| 35 Br |
| 53 I |
| 85 At |

**FIGURE 5.22** Chlorine, $Cl_2$, is a greenish-yellow gas at room temperature and at ordinary pressure; bromine, $Br_2$, is a reddish-brown liquid that vaporizes readily to form a vapor of the same color; iodine, $I_2$, is a grayish-black solid that readily vaporizes (sublimes) to form a violet vapor. (Dr. E. R. Degginger)

## A HISTORICAL PERSPECTIVE

## Discovery of the Noble Gases

None of the noble gases were known when Mendeleev proposed his periodic table. Their discovery created considerable scientific controversy but proved invaluable in understanding chemical behavior. The first significant hint of their existence was obtained during the early 1890s in the laboratories of the British physicist Lord Rayleigh. Rayleigh was at that time involved in a program of making exact measurements of the densities of simple gases. During his studies he discovered a discrepancy between the densities of nitrogen obtained from air and nitrogen obtained from chemical reactions, as in the decomposition of ammonia:

$$2NH_3(g) \longrightarrow N_2(g) + 3H_2(g)$$

When all the other gases then known were removed from air, the remaining nitrogen was found to have a density of 1.2572 g/L at 0°C and standard atmospheric pressure. However, the density of nitrogen from chemical sources was found to be 1.2506 g/L under the same conditions. The difference was slight but reproducible. In April 1894, Rayleigh published a paper devoted entirely to the matter of the densities of nitrogen from the two sources. This report excited the interest of the British chemist Sir William Ramsay (Figure 5.23), who began to study the atmospheric nitrogen.

**FIGURE 5.23** Sir William Ramsay (1852–1916), from the *Vanity Fair* series of caricatures. He was one of the principals in the discovery and study of the noble gases. He was awarded the Nobel Prize in chemistry in 1904 for this work. (The Granger Collection)

Rayleigh and Ramsay deduced that air must contain a previously unidentified component. They were able to isolate small quantities of this substance and measure some of its properties. The gas was composed of a new element with an atomic weight of 39.95, which presented some problems in placing the element in the periodic table. Based on its atomic weight, the element would fall between potassium (K) and calcium (Ca), but this placement makes no chemical sense. The scientists met with a second surprise: The element had no chemical reactivity. Try as they may, they were unable to form any compounds of the element. They consequently named the element argon, which means "the lazy one" in Greek.

Rayleigh and Ramsay publicly announced the discovery of argon in August 1894. Chemists found it hard to believe the existence of this curious element, and many ingenious but erroneous alternative proposals were put forth for its identity. In January 1895, Ramsay presented a paper on the discovery of argon before the Royal Society. The largest crowd in that scientific society's history (over 800 people) assembled to hear him.

Later that same year, Ramsay isolated helium (He), the lightest of the noble gases, from uranium ores. The helium in these ores forms when alpha particles produced in radioactive decay (Section 2.2) pick up electrons. During the spring and summer of 1898, Ramsay and his coworkers isolated three additional noble gases from air: neon (Ne), krypton (Kr), and xenon (Xe). The discovery of these elements contributed to a partial resolution of the problem of argon's location in the periodic table. Clearly, there is an entire family of elements whose most obvious characteristic is their lack of chemical reactivity. The atomic weights of family members other than argon placed them after the halogens, a position that made chemical sense. The apparent problem of argon's position in the periodic table was finally resolved by Moseley's work (Section 5.1), which led to arranging elements by their atomic numbers.

The elements of group 8A consist of helium (He), neon (Ne), argon (Ar), krypton (Kr), xenon (Xe), and radon (Rn). All these elements are gases at room temperature and are nonmetals. They are all monoatomic (that is, they consist of single atoms and are not combined to form molecules). Once commonly called inert gases, they have very low chemical reactivities. The first noble-gas compound was not characterized until 1962.

The heavier members of the family—Kr, Xe, and Rn—react directly with fluorine to form fluorides. For example, $XeF_2$, $XeF_4$, and $XeF_6$ are formed by direct reaction between $Xe(g)$ and $F_2(g)$. Other compounds of xenon can be formed starting with these three xenon fluorides. For example, $XeOF_4$, $XeO_2F_2$, and $XeO_3$ can be obtained by reacting $XeF_6$ with $H_2O$. Most of the chemistry of xenon involves Xe–F and Xe–O bonds. Only a single stable compound of krypton is known, $KrF_2$. The high radioactivity of radon has inhibited study of its chemistry.

### Group 8A: The Noble Gases

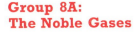

8A

| 2 He |
| 10 Ne |
| 18 Ar |
| 36 Kr |
| 54 Xe |
| 86 Rn |

*FOR REVIEW*

## SUMMARY

The periodic table developed in an entirely empirical way. Chemists had begun to amass a great many facts about the elements, and a need arose for some basis on which to organize them. Mendeleev found that when elements are arranged in order of increasing atomic weight, similar properties recur at regular, repeating intervals. He placed elements that have similar chemical and physical properties in columns and left several blanks in the table to accommodate undiscovered elements, whose properties he predicted. Many years later, Moseley discovered the concept of atomic numbers. In the modern periodic table, elements are arranged in order of increasing atomic number. The elements are placed in rows called

**periods** and columns called **families** or **groups**. Groups 1A through 8A are called the **representative elements**. Families 3B through 2B, near the center of the table, are called the **transition metals**.

Most elements are metals. Metals are located on the left side of the periodic table. Nonmetals appear in the upper-right section. Metalloids occupy a narrow band between the metals and nonmetals.

Metals have a characteristic luster. They are good electrical and thermal conductors. Metals tend to transfer electrons to other substances to form **cations** when they react. The compounds of metals with nonmetals, such as the oxides, hydrides, and chlorides, are mostly ionic solids. Metal oxides are basic; they react with acids to form salts and water.

Nonmetals lack metallic luster. Several are gases at room temperature. When nonmetals react with metals, the nonmetal generally gains electrons from the metal to become an **anion**. Compounds of nonmetals with each other are molecular; thus the oxides hydrides, and chlorides of the nonmetals are usually gases, liquids, or low-melting solids. Nonmetal oxides are acidic; they tend to react with bases to form salts and water.

**Metallic character** increases as we move down a family of representative elements; it also increases moving to the left in any period. Metallic character is strongest for elements at the bottom left of the periodic table; nonmetallic character is greatest for elements in the upper right.

The loss of electrons by a substance is called **oxidation**; the gain of electrons is called **reduction**. Many metals are oxidized by molecular oxygen, $O_2$, and by acids. The reaction of a metal and an acid to form a salt and $H_2$ is characteristic of a broad class of reactions called **single displacement reactions**.

This class also includes the reactions of metals with metal salts. Comparing single displacement reactions between metals allows us to rank metals according to their ease of oxidation. A list of metals arranged in order of decreasing ease of oxidation is called an **activity series**. Any metal on the list is able to displace the metals below it from their compounds; that is, metals on the top of the list are able to reduce those below them.

The most active metals are those in group 1A, the **alkali metals**. These reactive metals are soft, melt at low temperatures, and have low density; they react readily with air and water. The metals in family 2A, the **alkaline earth metals**, are generally harder, more dense, and melt at higher temperatures. The heavier alkaline earths also react readily with oxygen and water. Other important representative metals include aluminum, tin, and lead. Aluminum metal and $Al_2O_3$ dissolve in both acids and bases.

Hydrogen is a nonmetal that readily reacts with oxygen. The element can be prepared by the reaction of a metal like zinc with an acid like hydrochloric acid. Oxygen and sulfur are the most important members of family 6A. About 21 percent of air consists of $O_2$. $O_2$ has a strong attraction for electrons and is able to oxidize many substances. Ozone, $O_3$, is an important **allotrope** of oxygen.

The elements of group 7A are called the **halogens**. These nonmetals react readily with hydrogen to form gaseous compounds (HF, HCl, HBr, and HI) that dissolve easily in water to form important acid solutions. The members of family 8A are called the **noble gases**. They have very low chemical reactivities, and only the heavier members of the family—Kr, Xe, and Rn—form compounds.

## LEARNING GOALS

Having read and studied this chapter, you should be able to:

1. Describe the roles of Mendeleev and Moseley in the historical development of the periodic table.

2. Identify an element as a metal, nonmetal, or metalloid, based on its position in the periodic table.

3. Use the periodic table to predict the formulas of simple compounds of the representative elements, such as their oxides, hydrides, and chlorides.

4. List the characteristic physical and chemical properties of metals (see Table 5.3).

5. List the characteristic physical and chemical properties of nonmetals (see Table 5.3).

6. Identify a compound as molecular or ionic based on its composition. (Compounds of metals with nonmetals are generally ionic; those of nonmetals with nonmetals are molecular.)

7. Write balanced chemical equations for the reaction of a metal oxide with water or an acid.

8. Write balanced chemical equations for the reaction of a nonmetal oxide with water or a base.

9. Describe the periodic trends in metallic and nonmetallic character.

10. Define oxidation and reduction.

11. Use the activity series to predict whether or not

a single-displacement reaction will occur when a metal is added to an aqueous solution of either a metal salt or an acid; write the balanced chemical equation for the reaction.

12. Describe the general physical and chemical behavior of the alkali metals and alkaline earth metals, and write balanced chemical equations for their reactions with $H_2$, $Cl_2$, $O_2$, and $H_2O$.

13. Write balanced chemical equations for the reactions of Al and $Al_2O_3$ with either acids or bases.

14. Write a balanced chemical equation for the preparation of hydrogen gas using a metal and an acid.

15. Describe the allotropy of oxygen.

16. Write a balanced chemical equation for the preparation of oxygen from potassium chlorate.

17. Describe the physical states and colors of the halogens.

## KEY TERMS

Among the more important terms and expressions used for the first time in this chapter are the following:

The **actinides** (Section 5.2) are those elements in the second row of the inner transition metals, atomic numbers 90–103.

The **active metals** (Section 5.2) are the members of the alkali metal and alkaline earth metal groups.

The **activity series** (Section 5.5) is a list of metals in order of decreasing ease of oxidation.

The **alkali metals** (Section 5.6) are the members of group 1A in the periodic table.

The **alkaline earth metals** (Section 5.6) are the members of group 2A in the periodic table.

**Allotropes** (Section 5.7) are different chemical forms of the same element existing in the same physical state.

The **halogens** (Section 5.7) are the members of group 7A in the periodic table.

The **inner transition metals** (Section 5.2) are the elements usually listed in two long rows below the main body of the periodic table (atomic numbers 58–71 and 90–103).

The **lanthanides** (Section 5.2) are those elements in the first row of the inner transition metals, atomic numbers 58–71.

The **noble gases** (Section 5.7) are the members of group 8A in the periodic table.

**Oxidation** (Section 5.5) is the process by which a substance loses one or more electrons.

**Ozone** (Section 5.7) is the name given to $O_3$, an allotrope of oxygen.

**Reduction** (Section 5.5) is the process by which a substance gains one or more electrons.

**Representative elements** (Section 5.2) are the elements in the A groups of the periodic table.

**Single displacement reactions** (Section 5.5) are reactions that conform to the general pattern $A + BX \longrightarrow AX + B$.

The **transition metals** (Section 5.2) are the elements in the middle of the periodic table, between groups 2A and 3A.

## EXERCISES

### The Periodic Table

**5.1** In what fundamental way did Mendeleev's periodic table differ from the modern periodic table?

**5.2** By examining the modern periodic table, find as many examples as you can of violations of Mendeleev's periodic law that the chemical and physical properties of the elements are periodic functions of their atomic weights.

**5.3** Classify each of the following elements—Rb, Cu, Al, U, Fe, Pt, Pb—as (a) a representative element, (b) a transition metal, or (c) an inner transition metal.

**5.4** Which of the following—Mo, Ag, S, Ce, Th, Ba, P, He, H—are (a) representative elements; (b) transition elements; (c) inner transition elements?

**5.5** Which of the following—Ca, As, Ce, Xe, Li, Ar, Se— are (a) active metals; (b) noble gases; (c) lanthanides?

**5.6** Which of the following—F, Si, Na, Th, Mg, Kr—are (a) alkali metals; (b) alkaline earth metals; (c) halogens; (d) noble gases?

### Properties of Metals and Nonmetals

**5.7** Consult the periodic table and arrange the following pure solid elements in order of increasing electrical conductivity at room temperature: Ca, Sn, S, and Si. Explain the reason for the order you chose.

**5.8** Consult the periodic table and arrange the following elements in order of increasing melting points: Al, F, Na, Mg. Explain the reason for the order you chose.

**5.9** In each pair, choose the element expected to have the more metallic character: **(a)** Li and Be; **(b)** Li and Na; **(c)** Sn and P; **(d)** B and Al.

**5.10** **(a)** Arrange the following series of elements in order of increasing metallic character: Si, Sn, C, Ge, Pb. **(b)** Arrange the following series in order of increasing nonmetallic character: As, P, Bi, Sb, N.

**5.11** Write the chemical formula of the compound formed by **(a)** magnesium and iodine; **(b)** gallium and oxygen; **(c)** potassium and sulfur.

**5.12** Write the chemical formula of the compound formed by **(a)** aluminum and fluorine; **(b)** lithium and selenium; **(c)** barium and bromine.

**5.13** Which of the following compounds are ionic and which are molecular: $N_2O$, $Na_2O$, $CaO$, $CO$, $P_2O_5$, $Cl_2O_7$, $Fe_2O_3$? Explain the reason for your choices.

**5.14** Which of the following substances are solids at room temperature, and which are gases: NO, $Na_2S$, BaO, $CO_2$, $PbCl_2$, $OF_2$, $MgF_2$? Explain the reason for your choices.

**5.15** In light of the trends in metallic character, which oxide would you expect to be more basic: **(a)** MgO or BaO; **(b)** MgO or $Al_2O_3$?

**5.16** In light of the trends in nonmetallic character, which oxide would you expect to be more acidic: **(a)** $CO_2$ or $SiO_2$; **(b)** $P_2O_5$ or $SO_3$?

**5.17** Write balanced chemical equations for the following reactions; **(a)** sodium oxide with water; **(b)** copper(II) oxide with nitric acid; **(c)** sulfur trioxide with water; **(d)** selenium dioxide with sodium hydroxide.

**5.18** Write balanced chemical equations for the following reactions: **(a)** dichlorine heptaoxide with water; **(b)** iron(II) oxide with hydrochloric acid; **(c)** barium oxide with water; **(d)** carbon dioxide with potassium hydroxide.

**5.19** Magnesium metal reacts directly with oxygen gas to form magnesium oxide, with nitrogen gas to form magnesium nitride, with hydrogen gas to form magnesium hydride, and with chlorine gas to form magnesium chloride. Write balanced chemical equations for each of these reactions.

**5.20** Lithium metal reacts directly with oxygen gas to form lithium oxide, with nitrogen gas to form lithium nitride, with hydrogen gas to form lithium hydride, and with chlorine gas to form lithium chloride. Write balanced chemical equations for each of these reactions.

## Oxidation of Metals and the Activity Series

**5.21** Where, in general, do the most easily oxidized metals occur in the periodic table? Where do the least easily oxidized metals occur in the periodic table?

**5.22** **(a)** Why are platinum and gold called noble metals? **(b)** Why are the alkali metals and alkaline earth metals called the active metals?

**5.23** Write balanced chemical equations for the reactions of **(a)** hydrochloric acid with nickel; **(b)** sulfuric acid with iron; **(c)** hydrobromic acid with zinc; **(d)** acetic acid, $HC_2H_3O_2$, with magnesium.

**5.24** Write balanced chemical equations for the reactions of **(a)** manganese with sulfuric acid; **(b)** chromium with hydrobromic acid; **(c)** tin with hydrochloric acid; **(d)** aluminum with formic acid, $HCHO_2$.

**5.25** Based on the activity series (Tables 5.4 and 5.5), what is the outcome of each of the following reactions?

**(a)** $Al(s) + NiCl_2(aq) \longrightarrow$
**(b)** $Ag(s) + Pb(NO_3)_2(aq) \longrightarrow$
**(c)** $Cr(s) + NiSO_4(aq) \longrightarrow$
**(d)** $Mn(s) + HBr(aq) \longrightarrow$
**(e)** $H_2(g) + CuCl_2(aq) \longrightarrow$
**(f)** $Ba(s) + H_2O(l) \longrightarrow$

**5.26** Using the activity series (Tables 5.4 and 5.5) write balanced chemical equations for the following reactions. If no reaction occurs, simply write NR. **(a)** Zinc metal is added to a solution of silver nitrate; **(b)** iron metal is added to a solution of aluminum sulfate; **(c)** hydrochloric acid is added to cobalt metal; **(d)** hydrogen gas is passed over a sample of copper(II) oxide; **(e)** lithium metal is added to water.

**5.27** Use the following equations to prepare an activity series for the hypothetical elements A, B, C, and D: $A + DX \longrightarrow AX + D$; $B + DX$ gives no reaction; $C + AX \longrightarrow CX + A$.

**5.28** **(a)** Use the following reactions to prepare an activity series for the halogens: $Br_2(aq) + 2NaI(aq) \longrightarrow 2NaBr(aq) + I_2(aq)$; $Cl_2(aq) + 2NaBr(aq) \longrightarrow 2NaCl(aq) + Br_2(aq)$. **(b)** Relate the positions of the halogens in the periodic table with their locations in this activity series. **(c)** Predict whether the following reactions will occur: $Cl_2(aq) + 2KI(aq) \longrightarrow$; $Br_2(aq) + LiCl(aq)) \longrightarrow$.

**5.29** Why is HCl called a nonoxidizing acid and $HNO_3$ an oxidizing acid?

**5.30** Nitric acid dissolves silver, but hydrochloric acid does not. Explain.

## The Representative Elements

**5.31** What is the charge of each of the following ions: **(a)** hydride; **(b)** nitride; **(c)** beryllium; **(d)** gallium?

**5.32** What is the charge of each of the following ions: **(a)** selenide; **(b)** astinide; **(c)** phosphide; **(d)** cesium?

**5.33** From the following elements—hydrogen, copper, bromine, potassium, nitrogen, argon, aluminum—pick one that best fits each of the following statements: **(a)** a dark red liquid; **(b)** an inert gas; **(c)** a metal that readily reacts with water; (d) a metal that forms a 3+ ion; (e) a gas that burns in air.

**5.34** From the following elements—oxygen, sulfur, lithium, silver, iodine, germanium, aluminum—pick one that best fits each of the following descriptions; (a) an active metal that forms a cation with a 1+ charge; (b) a grayish-black solid that readily forms a purple vapor; (c) a metal that dissolves in NaOH solution; (d) a metalloid; (e) a yellow nonmetal.

**5.35** A student had three tubes containing colorless gases. One tube held $H_2$, one $O_2$, and one $CO_2$. A glowing splint was inserted into each tube. In tube A, the splint was extinguished; in tube B, a sharp pop was produced; in tube C, the splint burst into flames. Which tube contained which gas?

**5.36** A student had three metals, Ag, Ca, and Mg. Metal A reacts with water at room temperature, whereas metals B and C do not. Metal B reacts with hydrochloric acid, metal C does not. Identify the metals.

## Additional Exercises

**5.37** In what portion of the periodic table do we find the most active metals? The most active nonmetals?

**5.38** List the elements that occur under ordinary conditions as diatomic molecules.

**5.39** A sample of an element is broken into small pieces with a hammer, then treated with chlorine gas to produce a liquid chloride compound. Is the element copper, silicon, or krypton? Explain briefly.

**5.40** From the following list of substances, identify the three compounds that are gases at room temperature: $PH_3$, NaH, $N_2O$, NiO, $O_3$, $GaCl_3$, FeS. What is the general reason for your choices?

**5.41** The label falls off a flask containing a colorless gas. The gas does not react with oxygen, but it does dissolve in water to form an acidic solution. Which of the following substances could the substance be: $H_2$, $Na_2SO_4$, $SO_2$, $Cl_2$? Explain your answer.

**5.42** Relying on the periodic table, predict which of the substances in the following pairs will (a) oxidize Si: $F_2$ or Al; (b) reduce $CuCl_2$: Fe or $Br_2$; (c) be a solid: $CrCl_3$ or $NCl_3$; (d) react with an acid: NiO or $NO_2$.

**5.43** Write balanced chemical equations for the reactions of (a) calcium metal and molecular iodine, $I_2$; (b) tin(II) oxide and perchloric acid; (c) sulfur trioxide and calcium hydroxide; (d) barium metal and water; (e) zinc and nickel nitrate; (f) aluminum and potassium hydroxide.

**[5.44]** Using the *Handbook of Chemistry and Physics*, determine the five metallic elements with the lowest melting points; highest melting points; greatest density; lowest density.

**[5.45]** Some properties of sulfur and tellurium follow:

|                          | S      | Te      |
|--------------------------|--------|---------|
| Atomic weight            | 32.06  | 127.6   |
| Melting point (°C)       | 112.8  | 449.5   |
| Boiling point (°C)       | 445    | 990     |
| Density (g/cm³)          | 2.07   | 6.25    |
| Formula of oxide         | $SO_2$ | $TeO_2$ |
| Formula of hydride       | $H_2S$ | $H_2Te$ |

Use this information to estimate the same six properties for selenium. Find these properties of selenium in the *Handbook of Chemistry and Physics*, and compare them with your predictions.

## CHAPTER 6

# Electronic Structures of Atoms: Basic Concepts

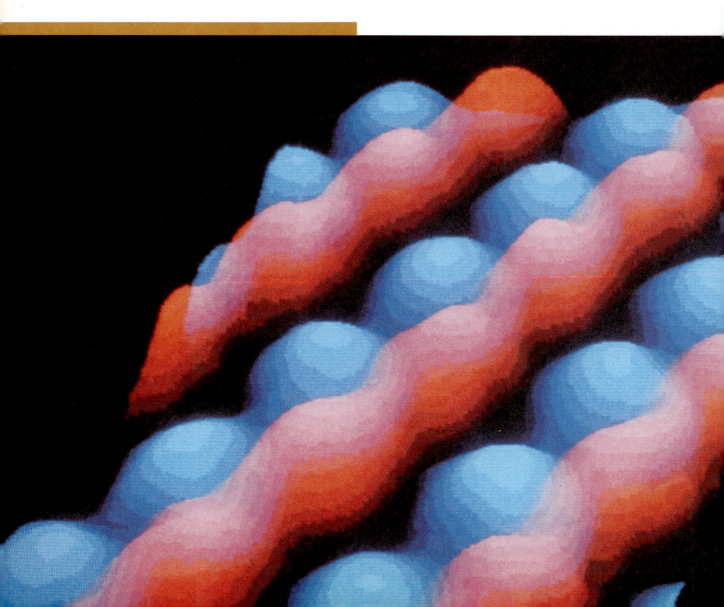

In earlier chapters we did not address the question of why atoms form molecules or ions. Why, for example, do hydrogen and oxygen atoms "stick" together to form the familiar water molecule, $H_2O$? Before we can answer this question and many others like it, we must know more about atoms themselves. It is especially important to know how electrons are arranged in atoms. We refer to the electron arrangement of an atom as its electronic structure. The goal of this chapter is to give you an understanding of the electronic structure of the simplest atom, the hydrogen atom. In subsequent chapters we will apply these ideas to other atoms and then to the bonding between atoms. A great deal of our understanding of electronic structure has come from studies of the properties of light or radiant energy. We begin our study, then, by considering the characteristics of radiant energy.

## 6.1 RADIANT ENERGY

Different kinds of radiant energy—such as the warmth from a glowing fireplace, the light reflected off snow in the mountains, and the X rays used by a dentist—*seem* very different from one another. Yet they share certain fundamental characteristics. All types of radiant energy, also called **electromagnetic radiation**, move through a vacuum at a speed of $2.9979250 \times 10^8$ m/s, the "speed of light." The speed of light is one of the most accurately known physical constants. For our purposes, however, it will be sufficient to use $3.00 \times 10^8$ m/s.

All radiant energy has wavelike characteristics analogous to those of waves that move through water. Water waves are the result of energy imparted to the water, perhaps by the dropping of a stone, the movement of a boat, or the force of wind on the water surface. This energy is expressed as the up-and-down movement of the water. A cork bobbing on water as waves pass by is not swept along with the wave. Rather, it moves up and down with the wave motion.

If we look at a cross section of a water wave (Figure 6.1), we see that it is periodic in character. That is, the wave form repeats itself at regular intervals. The distance between identical points on successive waves is called the **wavelength**. The number of times per second that one complete wavelength passes a given point is the **frequency** of the wave. We can measure this frequency by counting the number of times per second that the cork moves through one complete cycle of upward and downward motion.

In a similar way, radiant energy has a characteristic frequency and wavelength associated with it. These are illustrated schematically in Figure 6.2. The frequency of the radiation is the number of cycles that occur

Scanning tunneling microscope image of the semiconductor GaAs (gallium arsenide), magnified about one hundred million times. Gallium atoms are shown in blue, arsenic in red. (Courtesy of IBM, Thomas J. Watson Research Center)

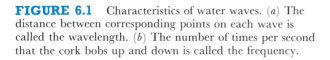

**FIGURE 6.1** Characteristics of water waves. (*a*) The distance between corresponding points on each wave is called the wavelength. (*b*) The number of times per second that the cork bobs up and down is called the frequency.

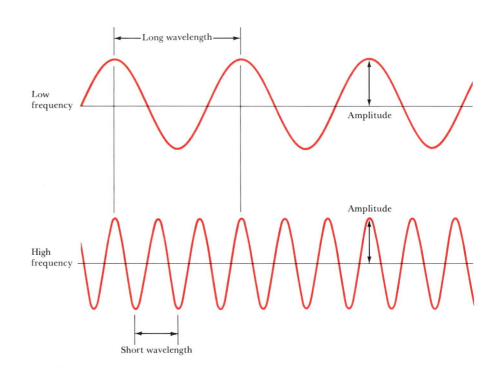

**FIGURE 6.2** Wave characteristics of radiant energy. Unlike water waves, radiant energy is not associated with the movement of matter. Instead, it is associated with periodic changes in the electric and magnetic fields. That is why it is referred to as electromagnetic radiation. Radiant-energy waves move through space with the speed of light. The frequency of the high-frequency radiation in this illustration is three times that of the low-frequency radiation. In the time that the low-frequency, long-wavelength radiation goes through one cycle, the high-frequency, short-wavelength radiation goes through three cycles. The *amplitude* of the wave relates to the intensity of the radiation. Here both waves have the same amplitude.

in 1 s as the radiation flows past a given point. Short wavelengths correspond to high frequency, and long wavelengths to low frequency. Wavelength and frequency are related. The product of frequency, $v$ (nu), and wavelength, $\lambda$ (lambda), equals the speed of light, $c$:

$$v\lambda = c \qquad\qquad [6.1]$$

Wavelength is expressed in units of length per cycle. The unit of length chosen depends on the type of radiation, as shown in Table 6.1. Figure

6.3 shows the ranges of wavelength that characterize the various types of radiant energy. Frequency is sometimes given in units of hertz (Hz), which is the number of cycles per second. For example, the frequency of an AM radio station might be written as 810 kilohertz (kHz) or as 810,000 cycles/s. Because it is understood that cycles are involved, the units are normally given simply as /s or $s^{-1}$.

**TABLE 6.1**  Wavelength units for electromagnetic radiation

| Unit | Symbol | Length (m) | Type of radiation |
|------|--------|-----------|-------------------|
| Angstrom | Å | $10^{-10}$ | X ray |
| Nanometer | nm | $10^{-9}$ | Ultraviolet, |
| (or millimicron)[a] | (m$\mu$) | | visible |
| Micrometer | $\mu$ | $10^{-6}$ | Infrared |
| (or micron) | | | |
| Millimeter | mm | $10^{-3}$ | Infrared |
| Centimeter | cm | $10^{-2}$ | Microwaves |
| Meter | m | 1 | TV, radio |

[a] The millimicron is the same size unit as the nanometer; use of the latter is preferred.

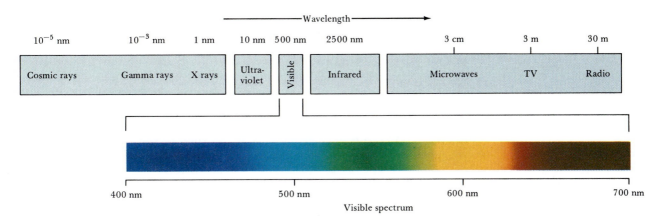

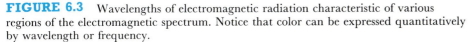

**FIGURE 6.3**  Wavelengths of electromagnetic radiation characteristic of various regions of the electromagnetic spectrum. Notice that color can be expressed quantitatively by wavelength or frequency.

## SAMPLE EXERCISE 6.1

The yellow light given off by a sodium lamp has a wavelength of 589 nm. What is the frequency of this radiation?

**Solution:**  We can rearrange Equation 6.1 to give $v = c/\lambda$. We insert the value for $c$ and $\lambda$, then convert nm to m. This gives us

$$v = \left(\frac{3.00 \times 10^8 \text{ m/s}}{589 \text{ nm}}\right)\left(\frac{10^9 \text{ nm}}{1 \text{ m}}\right)$$

$$= 5.09 \times 10^{14}/s$$

Note that frequency has units of /s or $s^{-1}$.

## PRACTICE EXERCISE

A laser used to weld detached retinas produces radiation with a frequency of $4.69 \times 10^{14}$/s. What is the wavelength of this radiation?
***Answer:***  $6.40 \times 10^{-7}$ m = 640 nm

**FIGURE 6.4** Max Planck (1858–1947), physicist. Born in Kiel, Germany, Planck was the son of a law professor at the University of Kiel. When he announced his intention to study physics, Planck was warned that all the major discoveries had already been made in this field. Nevertheless, Planck became a physicist, and in 1892 was named professor of physics at the University of Berlin. In 1900 he presented a paper before the Berlin Physical Society that launched the greatest intellectual revolution in the history of science. (Library of Congress)

Radiations of different wavelengths affect matter differently. For example, overexposure of a part of your body to infrared radiation may cause a burn, overexposure to visible and near-ultraviolet light may cause sunburn and suntan, and overexposure to X radiation may cause tissue damage, possibly even cancer. These diverse effects are due to differences in the energy of the radiation. Radiation of high frequency (and thus short wavelength) is more energetic than radiation of lower frequency (and thus longer wavelength). The quantitative relation between frequency and energy was developed at the turn of the century in the far-reaching and revolutionary *quantum theory* of Max Planck (Figure 6.4).

## 6.2 THE QUANTUM THEORY

The quantum theory is concerned with the rules that govern the gain or loss of energy from an object. Planck's contribution was to see that when we deal with gain or loss of energy from objects in the atomic or subatomic size range, the rules *seem* to be different from those that apply when we are dealing with energy gain or loss from objects of ordinary dimensions. A very crude and fanciful analogy might best illustrate the point. Imagine a large dump truck loaded with perhaps 20 tons of fine-grained sand. Let's say that the amount of sand on the truck is measured by driving it onto a supersensitive scale that measures the weight of a 30-ton object to the nearest pound. With this as our measuring device, the gain or loss of a few grains of sand from the truck would be too small to be measured; a spoonful of sand might be added or deleted with no change in scale reading.

Now imagine, if you will, a tiny truck, operated by people the size of tiny mites. For this little truck a full load would consist of perhaps a dozen grains of sand of the same size as those carried by the large truck. In this microscopic world, the load on the truck can be added to or decreased only by rolling on or off one or more full grains of sand. On a scale that weighs this tiny truck, even one grain of sand, the smallest piece attainable, represents a substantial fraction of the full load and is easily measurable.

In our analogy the sand represents energy. An object of ordinary, or macroscopic, dimensions, like the trucks we see on highways, contains energy in so many tiny pieces that the gain or loss of individual pieces is completely unnoticed. However, an object of atomic dimensions, such as our imaginary little truck, contains such a small amount of energy that the gain or loss of even the smallest possible piece makes a substantial difference. The essence of Planck's quantum theory is that there *is* such a thing as a smallest allowable gain or loss of energy. Even though the amount of energy gained or lost at one time may be very tiny, there is a limit to how small it may be. Planck termed the smallest allowed increment of energy gained or lost a **quantum**. In our analogy, a single grain of sand represents a quantum of sand.

You should keep in mind that the rules regarding the gain or loss of energy are always the same, whether we are concerned with objects on the size scale of our ordinary experience or with microscopic objects. However, it is only when dealing with matter at the atomic level of size that the impact of the quantum restriction is evident. Humans, being creatures of macroscopic dimensions, had no reason to suppose that the quantum restriction existed until they devised the means of observing the behavior of matter at the atomic level. The method for doing this at the time of Planck's work was by observing the radiant energy absorbed or emitted by matter. An object can gain or lose energy by absorbing or emitting radiant energy. Planck assumed that the amount of energy gained or lost at the atomic level by absorption or emission of radiation had to be a whole-number multiple of a constant times the frequency ($v$) of the radiant energy. Let us call this amount of energy gained or lost $\Delta E$. Then, according to Planck's theory,

$$\Delta E = hv, 2hv, 3hv, \text{ and so on} \qquad [6.2]$$

The constant $h$ is known as Planck's constant. It has the value $6.63 \times 10^{-34}$ joule second, or J-s. The smallest increment of energy at a given frequency, $hv$, is called a *quantum* of energy.

## SAMPLE EXERCISE 6.2

Calculate the smallest increment of energy (that is, the quantum of energy) that an object can absorb from yellow light whose wavelength is 589 nm.

**Solution:** We obtain the magnitude of a quantum of energy from Equation 6.2, $\Delta E = hv$. The value of Planck's constant is given both in the text above and in the table of physical constants on the back inside cover of the text: $h = 6.63 \times 10^{-34}$ J-s. The frequency, $v$, is calculated from the given wavelength, as shown in Sample Exercise 6.1: $v = c/\lambda = 5.09 \times 10^{14}$/s. Thus we have

$$\Delta E = hv = (6.63 \times 10^{-34} \text{ J-s})(5.09 \times 10^{14}/\text{s})$$
$$= 3.37 \times 10^{-19} \text{ J}$$

Planck's theory tells us that an atom or molecule emitting or absorbing radiation whose wavelength is 589 nm cannot lose or gain energy by radiation except in multiples of $3.37 \times 10^{-19}$ J. It cannot, for example, gain $5.00 \times 10^{-19}$ J from this radiation because this amount is not a multiple of $3.37 \times 10^{-19}$ J.

If one quantum of radiant energy supplies $3.37 \times 10^{-19}$ J, then one mole of these quanta will supply $(6.02 \times 10^{23}$ quanta$)(3.37 \times 10^{-19}$ J/quantum$)$, which equals $2.03 \times 10^5$ J.

This is the magnitude of heats of reactions (Section 4.4). Indeed, radiation can cause chemical bonds to break, producing what are called photochemical reactions.

**PRACTICE EXERCISE** ————————————
A laser that emits light energy in pulses of short duration has a frequency of $4.69 \times 10^{14}$/s and deposits $1.3 \times 10^{-2}$ J of energy during each pulse. How many quanta of energy does each pulse deposit? (Hint: First calculate the energy of one quantum of this frequency.)

***Answer:*** $4.2 \times 10^{16}$ photons

At this point you may be wondering about the practical applications of Planck's quantum theory. A few years after Planck presented his theory, scientists began to see its applicability to a great many experimental observations. It soon became apparent that Planck's theory had within it the seeds of a revolution in the way the physical world is viewed. Let's consider a few applications that are of special importance for chemistry.

**The Photoelectric Effect**

Light shining on a clean metallic surface can cause the surface to emit electrons. This phenomenon, known as the photoelectric effect, can be demonstrated as shown in Figure 6.5. For each metal there is a minimum frequency of light below which no electrons are emitted, regardless of how intense the beam of light. In 1905, Albert Einstein (1879–1955) used the quantum theory to explain the photoelectric effect. He assumed that the radiant energy striking the metal surface is a stream of tiny energy packets. Each energy packet, called a **photon**, is a quantum of energy $h\nu$. Thus radiant energy itself is considered to be quantized. *Photons of higher-frequency radiation have higher energies, and photons of lower-frequency radiation have lower energy:*

$$E_{\text{photon}} = h\nu \qquad [6.3]$$

When the photons are absorbed by the metal, their energy is transferred to an electron in the metal. A certain amount of energy is required for the electron to overcome the attractive forces that hold it within the

**FIGURE 6.5** The photoelectric effect. When photons of sufficiently high energy strike the metal surface in the tube, electrons are emitted from the metal. The electrons are drawn toward the other electrode, which is a positive terminal. As a result, current flows in the circuit. If the energy of the photon, $h\nu$, is too small, no electrons are freed and no current flows in the circuit. Photocells in automatic doors use the photoelectric effect to generate the electrons that operate the door-opening circuits.

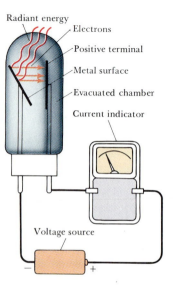

metal. If the photons of the radiation have less energy than this energy threshold, the electron cannot escape from the metal surface, even if the light beam is quite intense. If a photon does have sufficient energy, the electron is emitted. If a photon has more than the minimum energy required to free an electron, the excess appears as the kinetic energy of the emitted electron. Thus the kinetic energy of the emitted electron, $E_k$, equals the energy supplied by the photon, $hv$, minus the binding energy holding the electron in the metal, $E_b$:

$$E_k = hv - E_b \qquad [6.4]$$

A particular source of radiant energy may emit a single wavelength, as in the light from a laser (Figure 6.6). Radiation composed of a single wavelength is said to be **monochromatic**. However, most common radiation sources, including light bulbs and stars, produce radiation containing many different wavelengths. When radiation from such sources is separated into its different wavelength components, a **spectrum** is produced. Figure 6.7 shows how light from a light bulb can be dispersed by a prism. The spectrum so produced consists of a continuous range of

## Continuous and Line Spectra

**FIGURE 6.6**
Monochromatic light being emitted from the discharge tube of an argon-gas laser. (Laser Analytics Division, Spectra Physics)

**FIGURE 6.7**  Production of a continuous visible spectrum by passing a narrow beam of white light through a prism. The white light could be sunlight or light from an incandescent lamp.

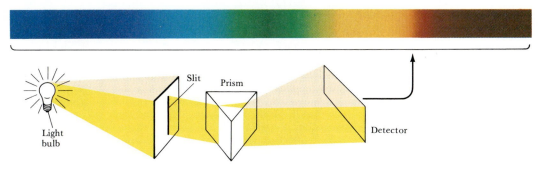

colors; violet merges into blue, blue into green, and so forth with no blank spots. This rainbow of colors, containing light of all wavelengths, is called a **continuous spectrum**. The most familiar example of a continuous spectrum is the rainbow, produced by the dispersal of sunlight by raindrops or mist.

Not all radiation sources produce a continuous spectrum. For example, when hydrogen is placed under reduced pressure in a tube such as that depicted in Figure 6.8 and a high voltage is applied, light is emitted. (If, instead of hydrogen, neon gas were placed in the tube, the familiar red-orange glow of neon lights would be produced; sodium vapor would produce the yellow radiation of many modern streetlights.) When the light coming from such tubes is dispersed by a prism, only certain colors or wavelengths of light are found to be present. A spectrum containing

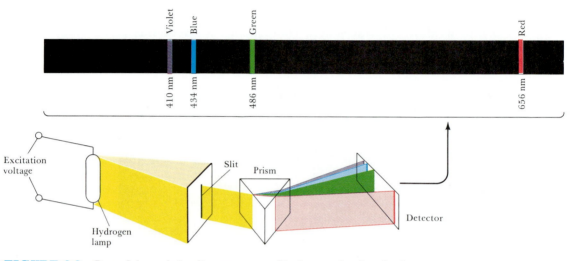

**FIGURE 6.8**  Part of the emission line spectrum of hydrogen showing the four most intense lines. This spectrum is obtained by passing a narrow beam of the light emitted by a hydrogen lamp through a prism. The lamp emits light when hydrogen atoms are excited in an electrical discharge.

**FIGURE 6.9**  Niels Bohr (1885–1962), a Danish physicist, born and educated in Copenhagen. From 1911 to 1913 he studied in England, working first with J. J. Thomson at Cambridge University and then with Ernest Rutherford at the University of Manchester. He published his quantum theory of the atom in 1914 and was awarded the Nobel Prize in physics in 1922. (AIP Niels Bohr Library, Uhlenbeck Collection)

radiation of only specific wavelengths is called a **line spectrum** (see Figure 6.8).

Although it was not realized at the time by the scientists who gathered the data, line spectra provide beautiful applications of the quantum theory. In 1914, Niels Bohr (Figure 6.9) incorporated Plank's theory to explain the line spectrum of hydrogen. Bohr introduced the basic idea that the absorptions and emissions of light by hydrogen atoms correspond to energy changes of electrons within the atoms. The fact that only certain frequencies are absorbed or emitted by an atom tells us that only certain energy changes are possible. We say that the energy changes within an atom are *quantized*. Further details of Bohr's model appear in the next section.

The spectrum of the radiation emitted by a substance is called an *emission spectrum*. The line spectrum of hydrogen shown in Figure 6.8 is an emission spectrum. Substances also exhibit *absorption spectra*. When continuous electromagnetic radiation, such as that from a light bulb, passes through a substance, certain wavelengths may be absorbed. The spectrum of the radiation that passes through is called an absorption spectrum. The absorption spectrum of hydrogen consists of what looks like a continuous spectrum interrupted by black lines at 656.3 nm, 486.3 nm, and the other wavelengths found in the emission spectrum. The absorption spectrum of hydrogen is a line spectrum that is complementary to its emission spectrum.

Each element has a characteristic line spectrum that can be used to identify it. Figure 6.10 shows the spectrum of the light emitted by the sun. Notice that there are several dark bands in the spectrum. Because of its high temperature, the core of the sun emits a continuous spectrum of radiation; however, elements that are present in the outer regions of the sun, where the temperatures are not so high, absorb radiation at characteristic wavelengths. Absorptions due to hydrogen, helium, iron, and other atoms give rise to the dark bands evident in Figure 6.10.

Helium was first discovered in 1868 from a similar spectrum of our sun. Some of the absorption lines of the sun's spectrum could not be matched with those of any element then known. It was concluded that the sun contained an element previously unknown on earth. This element was named helium after *helios*, the Greek word for the sun. Helium was subsequently isolated and characterized in the laboratory in 1895.

**FIGURE 6.10** Spectrum of the sun's radiation. Notice the presence of dark lines, called Fraunhofer lines, due to absorption of radiation by hydrogen and other atoms at the outer limits of the sun. The lower spectrum is that of an incandescent source of the same temperature as the sun's surface. (Courtesy Bausch & Lomb)

## 6.3  BOHR'S MODEL OF HYDROGEN

Bohr's model of the hydrogen atom took into account two important developments that were relatively new at that time. The first was Rutherford's experiments, which established the nuclear nature of the atom (Section 2.2). The other was Einstein's work, which showed that radiant energy could be thought of as a stream of discrete bundles of energy called photons. We are not concerned here with the details of the Bohr treatment, because it is not strictly correct. However, it does introduce an important concept: the quantization of the energy of electrons in atoms.

Bohr's model consists of a series of postulates that we may summarize as follows:

1. The electron of a hydrogen atom moves about the central proton in a circular orbit. However, an electron in an atom cannot have just any energy; only orbits of certain radii, corresponding to certain energies, are permitted. An electron in a permitted orbit is said to be in an "allowed" energy state.

2. In the absence of radiant energy, an electron in an atom remains indefinitely in one of the allowed energy states. When radiant energy is present, however, the atom may absorb energy. When this happens, the electron changes from one allowed energy state to another. The frequency of the radiant energy absorbed ($\nu$) corresponds exactly to the energy difference ($\Delta E$) between two of the allowed energies: $\Delta E = h\nu$.

We need not concern ourselves with the details of how Bohr used quantum theory to calculate the energies of the electron. The main point is that he was able to calculate a set of allowed energies. Each of these allowed energies corresponds to a circular path of different radius. In Bohr's model, each allowed orbit was assigned an integer $n$, known as the **principal quantum number**, that may have values from 1 to infinity. The radius of the electron orbit in a particular energy state varies as $n^2$:

$$\text{Radius} = n^2(5.30 \times 10^{-11} \text{ m}) \qquad [6.5]$$

Thus the larger the value of $n$, the farther the electron from the nucleus.

The energy of the electron depends on the orbit it occupies:

$$E_n = -R_H\left(\frac{1}{n^2}\right) \qquad [6.6]$$

The constant $R_H$ in Equation 6.6 is called the Rydberg constant; it has the value of $2.18 \times 10^{-18}$ J. From Equation 6.6 we have that the energy of the electron is $-2.18 \times 10^{-18}$ J when the electron is in the orbit closest to the nucleus, $n = 1$. When it is in the second orbit, $n = 2$, its energy is

$$E_2 = (-2.18 \times 10^{-18} \text{ J})\left(\frac{1}{2^2}\right) = -5.45 \times 10^{-19} \text{ J}$$

Figure 6.11 illustrates the orbit radii and energies for $n = 1$, $n = 2$, and $n = 3$. A more common way of representing the allowed energies is shown in Figure 6.12.

Energy (J)

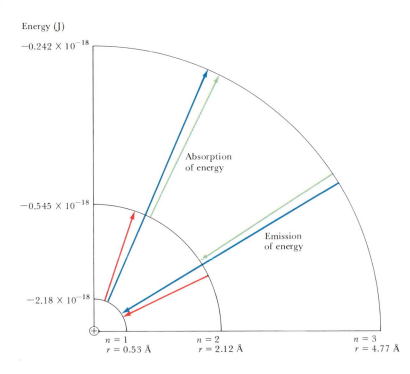

−0.242 × 10⁻¹⁸

−0.545 × 10⁻¹⁸

Absorption
of energy

Emission
of energy

−2.18 × 10⁻¹⁸

$n = 1$
$r = 0.53$ Å

$n = 2$
$r = 2.12$ Å

$n = 3$
$r = 4.77$ Å

**FIGURE 6.11** Radii and energies of the three lowest-energy orbits in the Bohr model of hydrogen. The arrows refer to transitions of the electron from one allowed energy state to another. When the transition takes the electron from a lower- to a higher-energy state, absorption occurs. When the transition is from a higher- to a lower-energy state, emission occurs.

**FIGURE 6.12** Energy levels in the hydrogen atom from the Bohr model. The arrows refer to transitions of the electron from one allowed energy state to another, as described in Figure 6.11. Only the lowest six energy levels are shown.

The negative sign in Equation 6.6 denotes stability relative to some reference state. In other words, the more negative the value for energy, the more stable the system is. (One way to remember this convention is to think of a ball rolling around on a surface with many hills and valleys. The ball will naturally come to rest in a valley; it is more stable when its potential energy is lowest.) The reference, or zero-energy, state for the electron in hydrogen is the state in which the electron is completely separated from the nucleus. This corresponds to an infinitely large value for the principal quantum number, $n$:

$$E_\infty = (-2.18 \times 10^{-18}\ \text{J})\left(\frac{1}{\infty^2}\right) = 0$$

The energy of the electron in any other orbit is then negative relative to this reference state. The lowest-energy, or most stable, state, with $n = 1$, is known as the **ground state**. When the electron is in a higher energy orbit—that is, $n = 2$ or higher—the atom is said to be in an electronically **excited state**.

Bohr's model of the hydrogen atom quantitatively explained the observed line spectra of that substance. The absorptions or emissions in the line spectra correspond to transitions of the electron from one orbit to another. Radiant energy is absorbed when the electron moves from one orbit to another having a larger radius; it requires energy to pull the electron away from the nucleus. Conversely, energy is emitted when the electron moves from a larger orbit to another having a smaller radius. The changes in energy, $\Delta E$, are given by the difference between the energy of the final state of the electron, $E_f$, and the initial state, $E_i$:

$$\Delta E = E_f - E_i$$

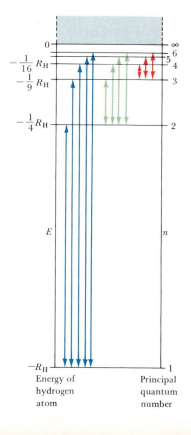

0
∞
6
5
4

$-\frac{1}{16} R_\text{H}$

$-\frac{1}{9} R_\text{H}$  3

$-\frac{1}{4} R_\text{H}$  2

$E$  $n$

$-R_\text{H}$  1

Energy of
hydrogen
atom

Principal
quantum
number

Substituting the expression for the energy of the electron, Equation 6.6, gives

$$\Delta E = \left(\frac{-R_H}{n_f^2}\right) - \left(\frac{-R_H}{n_i^2}\right) = R_H\left(\frac{1}{n_i^2} - \frac{1}{n_f^2}\right)$$

Because $\Delta E = h\nu$, we have

$$\Delta E = h\nu = R_H\left(\frac{1}{n_i^2} - \frac{1}{n_f^2}\right) \qquad [6.7]$$

In this expression, $n_i$ and $n_f$ represent the quantum numbers for the initial and final states, respectively. Notice that when the final state quantum number $(n_f)$ is larger than the initial state quantum number $(n_i)$, the term in parentheses is positive, and $\Delta E$ is positive. This means that the system has absorbed a photon and thus increased in energy (an endothermic process). When $n_i$ is larger than $n_f$, as happens in emission, energy is given off; $\Delta E$ is negative.

Complete removal of the electron from a hydrogen atom, corresponding to a transition from the $n = 1$ (ground) state to the $n = \infty$ state, is known as ionization. This is represented as

$$H(g) \longrightarrow H^+(g) + e^-$$

The energy required for ionization from the ground state is called the **ionization energy**.

---

## SAMPLE EXERCISE 6.3

Calculate the frequency of the hydrogen line that corresponds to the transition of the electron from the $n = 4$ to the $n = 2$ states.

**Solution:** We employ Equation 6.7, substituting $n_i = 4$ and $n_f = 2$ because these are the quantum numbers for the initial and final orbits, respectively:

$$\Delta E = h\nu = R_H\left(\frac{1}{n_i^2} - \frac{1}{n_f^2}\right)$$

$$\nu = \frac{R_H}{h}\left(\frac{1}{n_i^2} - \frac{1}{n_f^2}\right)$$

$$= \frac{2.18 \times 10^{-18}\,\text{J}}{6.63 \times 10^{-34}\,\text{J-s}}\left(\frac{1}{4^2} - \frac{1}{2^2}\right)$$

$$= \frac{2.18 \times 10^{-18}\,\text{J}}{6.63 \times 10^{-34}\,\text{J-s}}\left(\frac{1}{16} - \frac{1}{4}\right)$$

$$= \frac{2.18 \times 10^{-18}\,\text{J}}{6.63 \times 10^{-34}\,\text{J-s}}\left(-\frac{3}{16}\right)$$

$$= -6.17 \times 10^{14}/\text{s}$$

The negative sign simply indicates that the light is emitted. A negative frequency or wavelength is physically meaningless, so the sign is ignored. This value for $\nu$ can be used to calculate the wavelength of the radiation, so we can compare it with the hydrogen spectrum shown in Figure 6.8:

$$\lambda = \frac{c}{\nu} = \frac{3.00 \times 10^8\,\text{m/s}}{6.17 \times 10^{14}/\text{s}}$$

$$= 4.86 \times 10^{-7}\,\text{m} = 486\,\text{nm}$$

All of the lines shown in Figure 6.8 correspond to transitions of an electron from higher orbits to the $n = 2$ orbit.

## PRACTICE EXERCISE

Calculate the frequency of the hydrogen line that corresponds to the transition of the electron from the $n = 4$ to the $n = 1$ states.
***Answer:*** $3.08 \times 10^{15}/\text{s}$

## SAMPLE EXERCISE 6.4

Calculate the energy required for ionization of an electron from the ground state of the hydrogen atom.

**Solution:** The ionization energy may be written as the difference between the final and initial state energies. We have $n_f = \infty$, $n_i = 1$.

$$\Delta E = E_f - E_i = R_H\left(\frac{1}{1^2} - \frac{1}{\infty^2}\right)$$
$$= R_H(1 - 0)$$

This is just equal to $R_H$, $2.18 \times 10^{-18}$ J.

It is often useful to express this energy on a molar basis. To do this we simply multiply by Avogadro's number:

$$\left(2.18 \times 10^{-18}\,\frac{J}{\text{atom}}\right)\left(6.02 \times 10^{23}\,\frac{\text{atoms}}{\text{mol}}\right)$$
$$\times \left(\frac{1\text{ kJ}}{1000\text{ J}}\right) = 1.31 \times 10^3\text{ kJ/mol}$$

### PRACTICE EXERCISE

Calculate the energy, in kJ/mol, required for ionization of an electron from the $n = 2$ state of the hydrogen atom.
*Answer:* 328 kJ/mol

Bohr's model was very important because it introduced the idea of quantized energy states for electrons in atoms. This feature is incorporated into our current model of the atom. However, Bohr's model was adequate for explaining only atoms and ions with one electron, such as H, $He^+$, or $Li^{2+}$. It was not adequate to account for the atomic spectra of other atoms or ions, except in a rather crude way. Consequently, Bohr's model was eventually replaced by a new way of viewing atoms that is called **quantum mechanics** or **wave mechanics**. This newer model maintains the concept of quantized energy states, but additional applications of Planck's quantum theory enter the picture.

## 6.4 MATTER WAVES

In the years following Bohr's development of a model for the hydrogen atom, the dual nature of radiant energy had become a familiar concept. Depending on the experimental circumstances, radiation appears to have either a wavelike or a particlelike (photon) character. Louis de Broglie, a young man working on his Ph.D. thesis in physics at the Sorbonne, in Paris, made a rather daring, intuitive extension of this idea. If radiant energy could under appropriate conditions behave as though it were a stream of particles, could not matter under appropriate conditions possibly show the properties of a wave? Suppose that the electron in orbit around the nucleus of a hydrogen atom could be thought of as a wave, with a characteristic wavelength. De Broglie suggested that the electron in its circular path about the nucleus has associated with it a particular wavelength. He went on to propose that the characteristic wavelength of the electron or of any other particle depends on its mass, $m$, and velocity, $v$:

$$\lambda = \frac{h}{mv} \qquad [6.8]$$

($h$ is Planck's constant). The quantity $mv$ for any object is called its **momentum**. De Broglie used the term **matter waves** to describe the wave characteristics of material particles.

Because de Broglie's hypothesis is perfectly general, any object of mass $m$ and velocity $v$ would give rise to a characteristic matter wave. However, it is easy to see from Equation 6.8 that the wavelength associated with an object of ordinary size, such as a golf ball, is so tiny as to be completely out of the range of any possible observation. This is not so for electrons, because their mass is so small.

## SAMPLE EXERCISE 6.5

What is the characteristic wavelength of an electron with a velocity of $5.97 \times 10^6$ m/s? (The mass of the electron is $9.11 \times 10^{-28}$ g.)

**Solution:** The value of Planck's constant, $h$, is $6.63 \times 10^{-34}$ J-s ($1 \text{ J} = 1 \text{ kg-m}^2/\text{s}^2$).

$$\lambda = \frac{h}{mv}$$

$$= \frac{6.63 \times 10^{-34} \text{ J-s}}{(9.11 \times 10^{-28} \text{ g})(5.97 \times 10^6 \text{ m/s})}$$

$$\times \left(\frac{1 \text{ kg-m}^2/\text{s}^2}{1 \text{ J}}\right)\left(\frac{10^3 \text{ g}}{1 \text{ kg}}\right)$$

$$= 1.22 \times 10^{-10} \text{ m} = 0.122 \text{ nm}$$

By comparing this value with the wavelengths of electromagnetic radiations shown in Figure 6.3, we see that the characteristic wavelength is about the same as that of X rays.

### PRACTICE EXERCISE

At what velocity must a neutron be moving in order for it to exhibit a wavelength of 500 pm? The mass of a neutron is given in the table on the back inside cover of the text.

*Answer:* $7.92 \times 10^2$ m/s

Within a few years after de Broglie published his theory, the wave properties of the electron were demonstrated experimentally. Electrons were diffracted by crystals, just as X rays, which are definitely radiant energy, are diffracted. (We shall have more to say about the X-ray diffraction experiment in Chapter 11.)

The technique of electron diffraction has been highly developed. In the electron microscope, the wave characteristics of electrons are used to obtain electron diffraction pictures of tiny objects. The electron microscope is an important tool for studying surface phenomena at the very highest magnifications. Figure 6.13 is an example of an electron microscope picture. Such pictures are powerful demonstrations that tiny particles of matter can indeed behave as waves.

**FIGURE 6.13** The surface of silicon, magnified about ten million times. The image was obtained using a scanning tunneling electron microscope. (Courtesy of IBM, Thomas J. Watson Research Center)

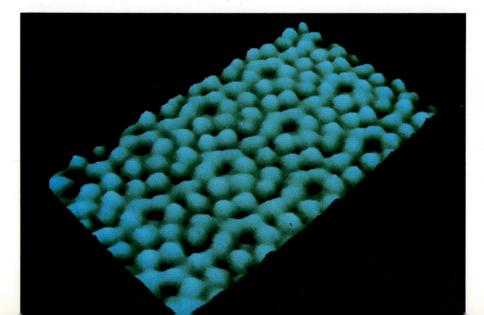

Discovery of the wave properties of matter raised new and interesting questions. If a subatomic particle can exhibit the properties of a wave, is it possible to say precisely just where that particle is located? One can hardly speak of the precise location of a wave. The amplitude, or intensity, of a wave can be defined at a certain point, (as illustrated in Figure 6.2), but the wave as a whole extends in space. Its location is therefore not defined precisely, at least not in the same sense that one can define the location of a particle. We might logically expect to be able to measure not only a particle's location but also its direction and speed of motion.

However, the German physicist Werner Heisenberg (1901–1976) concluded that there is a fundamental limitation on how precisely we can know both the location and the momentum of a particle. The limitation becomes important only when we deal with matter at the subatomic level, that is, with masses as small as that of an electron. Heisenberg's principle is called the **uncertainty principle**. When applied to the electrons in an atom, this principle states that it is inherently impossible for us to know simultaneously both the exact momentum of the electron and its

## A CLOSER LOOK — Measurement and the Uncertainty Principle

Whenever any measurement is made, some uncertainty exists. Our experience with objects of ordinary dimensions, like balls or trains or laboratory equipment, indicates that the uncertainty of a measurement can be decreased by using more precise instruments. In fact, we might expect that the uncertainty in a measurement can be made indefinitely small. However, the uncertainty principle states that there is an actual limit to the accuracy of measurements. This limit is not a restriction on how well instruments can be made; rather, it is inherent in nature. This limit has no practical consequences when we are dealing with ordinary-sized objects, but its implications are enormous when we are dealing with subatomic particles, such as electrons.

To measure an object, we must disturb it, at least a little, with our measuring device. Imagine that you use a flashlight to locate a large rubber ball in a dark room. You see the ball when the light from the flashlight bounces off the ball and strikes your eyes. When a beam of photons strikes an object of this size, it does not alter its position or momentum to any practical extent. Imagine, however, that you wish to locate an electron by similarly bouncing light off it into some detector. Objects can be located to an accuracy no greater than the wavelength of the radiation used. Thus if we want an accurate position measurement for an electron, we must use a short

wavelength. This means that photons of high energy must be employed. The more energy the photons have, the more momentum they impart to the electron when they strike it, which changes the electron's motion in an unpredictable way. The attempt to accurately measure the electron's position introduces considerable uncertainty in its momentum; the act of measuring the electron's position at one moment makes our knowledge of its future position inaccurate.

Suppose, then, that we use photons of longer wavelength. Because these photons have lower energy, the momentum of the electron is not so appreciably changed during measurement, but its position will be correspondingly less accurately known. This is the essence of the uncertainty principle: *There is an uncertainty in either the position or the momentum of the electron that cannot be reduced beyond a certain minimum level.* The more accurately one is known, the less accurately the other is known. Although we can never know with certainty the exact position and motion of the electron, we can talk about the probability of the electron being at certain locations in space. In the next section we introduce a model of the atom that provides the probability of finding electrons of specific energies at certain positions in atoms.

exact location in space. Thus it is not appropriate to imagine the electrons as moving in well-defined circular orbits about the nucleus.

De Broglie's hypothesis and Heisenberg's uncertainty principle set the stage for a new and more broadly applicable theory of atomic structure. In this new approach, any attempt to define precisely the instantaneous location and momentum of the electrons is abandoned. The wave nature of the electron is recognized, and its behavior is described in terms appropriate to waves.

## 6.5 THE QUANTUM-MECHANICAL DESCRIPTION OF THE ATOM

The mathematics employed to determine electron energy levels in even so simple a system as the hydrogen atom by quantum-mechanical methods are quite advanced. There is no point in presenting any mathematical details here. We can, however, understand the idea of quantum mechanics with the aid of a qualitative description.

Quantum mechanics describes mathematically the wave properties of the electron. This approach also takes into account the attraction of the electron for the nucleus and the kinetic energy of the electron. Certain conditions are imposed on the possible solutions to the resultant mathematical equation to make them physically reasonable. The result is a series of solutions that describe the allowed energy states of the electron. These solutions, usually represented by the symbol $\psi$ (the Greek lowercase letter *psi*), are called **wave functions**.*

A wave function provides information about an electron's location in space when it is in a certain allowed energy state. The allowed energy states are the same as those predicted by the Bohr model. However, the Bohr model suggests that the electron is in a circular orbit of some particular radius about the nucleus. In the quantum-mechanical model for hydrogen, it is not so simple to describe the electron's location. The uncertainty principle suggests that if we know the momentum of the electron with high accuracy, our knowledge of its location is very uncertain. Thus, for an individual electron, we cannot hope to specify its location around the nucleus. Rather we must be content with a kind of statistical knowledge. In the quantum-mechanical model we therefore speak of the *probability* that the electron will be in a certain region of space at a given instant. As it turns out, the square of the wave function $(\psi^2)$ at a given point in space represents the probability that the electron will be found at that location.

One way of representing the probability of finding the electron in various regions of an atom is shown in Figure 6.14. In this figure the density of the dots represents the probability of finding the electron. The regions with a high density of dots correspond to relatively large values for $\psi^2$. **Electron density** is another way of expressing probability; regions where there is a high probability of finding the electron are said to be regions of high electron density. In Section 6.6 we will say more about the ways in which we can represent electron density.

**FIGURE 6.14** Electron density distribution in the ground state of the hydrogen atom.

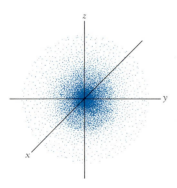

* In mathematics, the term *function* is used to describe a relationship between two variables—for example, time and distance. Thus the distance $(d)$ covered by an auto moving at a constant speed $(s)$ is a *function* of the amount of time $(t)$ it has been in motion: $d = st$.

**TABLE 6.2** The azimuthal quantum numbers and corresponding labels for atomic orbitals

| Value of $l$ | 0 | 1 | 2 | 3 | 4 |
|---|---|---|---|---|---|
| Letter used | $s$ | $p$ | $d$ | $f$ | $g$ |

The complete quantum-mechanical solution to the hydrogen atom problem yields a set of wave functions and a corresponding set of energies. Each wave function represents an allowed solution. The term **orbital** denotes an allowed energy state for the electron. It also refers to the probability function that defines the distribution of electron density in space. Thus an orbital has both a characteristic energy and a characteristic shape. For example, the lowest energy orbital in the hydrogen atom has an energy of $-2.18 \times 10^{-18}$ J and the shape illustrated in Figure 6.14. Note that an *orbital* (quantum-mechanical model) is not the same as an *orbit* (Bohr model).

The Bohr model introduced a single quantum number ($n$) to describe an orbit. The quantum-mechanical model uses three quantum numbers ($n$, $l$, and $m_l$) to describe an orbital. We will consider what information we obtain from each of these and how they are interrelated.

1. The *principal quantum number* ($n$) can have integral values of 1, 2, 3, and so forth. As $n$ increases, the orbital becomes larger and the electron spends more time farther from the nucleus. An increase in $n$ also means that the electron has a higher energy and is therefore less tightly bound to the nucleus. For hydrogen, $E_n = -(2.18 \times 10^{-18} \text{ J})(1/n^2)$, as in the Bohr model.

2. The second quantum number—the *azimuthal quantum number* ($l$)—can have integral values from 0 to $n - 1$ for each value of $n$. This quantum number defines the shape of the orbital. (We will consider these shapes in Section 6.6.) The value of $l$ for a particular orbital is generally designated by the letters $s$, $p$, $d$, $f$, and $g$,* corresponding to $l$ values of 0, 1, 2, 3, and 4, respectively, as summarized in Table 6.2.

3. The *magnetic quantum number* ($m_l$) can have integral values between $l$ and $-l$, including zero. This quantum number describes the orientation of the orbital in space. (The orientations will be considered in Section 6.6.)

A collection of orbitals with the same value of $n$ is called an **electron shell**. For example, all the orbitals with $n = 3$ are said to be in the third shell. One or more orbitals with the same set of $n$ and $l$ values is called a **subshell**. Each subshell is designated by a number (the value of $n$) and a letter ($s$, $p$, $d$, $f$, or $g$, corresponding to the value of $l$). For example, all of the orbitals of an atom with $n = 3$ and $l = 1$ are collectively referred to as $3p$ orbitals and are said to be in the $3p$ subshell. The possible values of the three quantum numbers through $n = 4$ (the fourth shell) are summarized in Table 6.3.

The restrictions in the possible values of the quantum numbers give rise to a pattern that is very important:

* The letters $s$, $p$, $d$, and $f$ came from the words *sharp*, *principal*, *diffuse*, and *fundamental*. These words were used to describe certain features of spectra before quantum mechanics was developed.

**TABLE 6.3**  Relationship among values of $n$, $l$, and $m_l$ through $n = 4$

| $n$ | $l$ | Subshell designation | $m_l$ | Number of orbitals in subshell |
|---|---|---|---|---|
| 1 | 0 | $1s$ | 0 | 1 |
| 2 | 0 | $2s$ | 0 | 1 |
|  | 1 | $2p$ | 1, 0, −1 | 3 |
| 3 | 0 | $3s$ | 0 | 1 |
|  | 1 | $3p$ | 1, 0, −1 | 3 |
|  | 2 | $3d$ | 2, 1, 0, −1, −2 | 5 |
| 4 | 0 | $4s$ | 0 | 1 |
|  | 1 | $4p$ | 1, 0, −1 | 3 |
|  | 2 | $4d$ | 2, 1, 0, −1, −2 | 5 |
|  | 3 | $4f$ | 3, 1, 2, 0, −1, −2, −3 | 7 |

*(Handwritten margin notes)*
M-level
S-orbital = 0
P  '' = +1, 0, −1
D  '' = 2, 1, 0, −1, −2
F  '' = 3, 2, 1, 0, −1, −2, −3
G  '' = 4, 3, 2, 1, 0, −1, −2, −3, −4
l-level

1. Each shell is divided into the number of subshells equal to the principal quantum number, $n$, for that shell. Thus the first shell consists of only the $1s$ subshell; the second shell consists of two subshells, $2s$ and $2p$; the third shell consists of three subshells, $3s$, $3p$, and $3d$. Thus every shell has an $s$ subshell; every shell beginning with the second has a $p$ subshell; every shell beginning with the third has a $d$ subshell, and so forth.

2. Each subshell is divided into orbitals. Each $s$ subshell consists of one orbital; each $p$ subshell consists of three orbitals; each $d$ subshell consists of five orbitals; each $f$ subshell consists of seven orbitals. (Notice that these are the odd numbers, 1, 3, 5, 7.)

**FIGURE 6.15**  Orbital energy levels in the hydrogen atom and in hydrogen-like ions (those containing just one electron). Note that all orbitals with the same value for the principal quantum number, $n$, have the same energy. This is true only in one-electron systems.

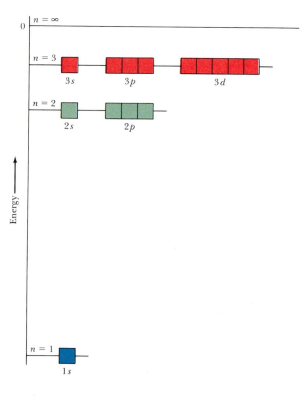

Figure 6.15 shows the number and relative energies of all hydrogen atom orbitals through $n = 3$. Each box represents an orbital; orbitals of the same subshell, such as the $2p$, are grouped together. When the electron is in the lowest energy orbital (the $1s$ orbital), the hydrogen atom is said to be in its *ground state*. When the electron is in any other orbital, the atom is in an *excited state*. At ordinary temperatures essentially all hydrogen atoms are in their ground states. The electron may be promoted to an excited-state orbital by absorption of a photon of appropriate energy.

**SAMPLE EXERCISE 6.6**

**(a)** Without referring to Table 6.3, predict the number of subshells in the fourth shell (that is, for $n = 4$). **(b)** Give the label for each of these subshells. **(c)** How many orbitals are in each of these subshells?

**Solution:** **(a)** There are four subshells in the fourth shell, corresponding to the four possible values of $l$ (0, 1, 2, and 3).
   **(b)** These subshells are labeled $4s$, $4p$, $4d$, and $4f$. The number given in the designation of a subshell is the principal quantum number, $n$; the following letter designates the value of the azimuthal quantum number, $l$, as given in Table 6.2.
   **(c)** There is one $4s$ orbital (when $l = 0$, there is only one possible value of $m_l$: 0). There are three $4p$ orbitals (when $l = 1$, there are three possible values of $m_l$: 1, 0, and $-1$). There are five $4d$ orbitals (when $l = 2$, there are five allowed values of $m_l$: 2, 1, 0, $-1$, $-2$). There are seven $4f$ orbitals (when $l = 3$, there are seven permitted values of $m_l$: 3, 2, 1, 0, $-1$, $-2$, $-3$).

**PRACTICE EXERCISE**
**(a)** What is the designation for the subshell with $n = 5$ and $l = 1$? **(b)** How many orbitals are in this subshell? **(c)** Indicate the values of $m_l$ for each of these orbitals.
*Answers:* **(a)** $5p$; **(b)** 3; **(c)** 1, 0, $-1$

In our discussion of orbitals we have so far emphasized their energies. But the wave function also provides information about the electron's location in space when it is in a particular allowed energy state. We need to examine the ways that we can picture the orbitals.

The lowest-energy (most stable) orbital, the $1s$ orbital, is spherically symmetric, as shown in Figure 6.14. Figures of this type, showing electron density, are one of the several ways we have to help us visualize orbitals. This figure indicates that the probability of finding the electron around the nucleus decreases as we move away from the nucleus in any direction. When the probability function ($\psi^2$) for the $1s$ orbital is graphed as a function of the distance from the nucleus ($r$), it rapidly approaches zero, as shown in Figure 6.16($a$). This effect indicates that the electron, which is drawn toward the nucleus by electrostatic attraction, is unlikely ever to get very far from the nucleus.
   If we similarly consider the $2s$ and $3s$ orbitals of hydrogen, we find that they are also spherically symmetric. Indeed, *all* $s$ orbitals are spherically symmetric. The manner in which the probability function ($\psi^2$) varies with $r$ for the $2s$ and $3s$ orbitals is shown in Figure 6.16($b$) and ($c$). Notice that for the $2s$ orbital, $\psi^2$ goes to zero and then increases again in value before finally approaching zero at a larger value of $r$. The intermediate regions where $\psi^2$ goes to zero are called **nodal surfaces** or simply **nodes**. The number of nodes increases with increasing value for the principal quantum number ($n$). The $3s$ orbital possesses two nodes,

## 6.6 REPRESENTATIONS OF ORBITALS

### The s Orbitals

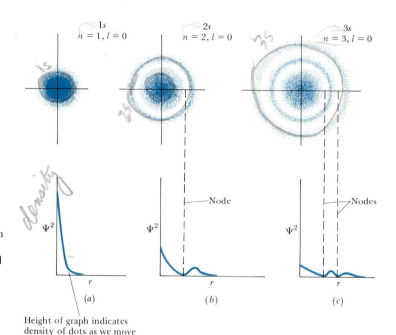

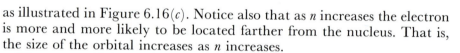

**FIGURE 6.16** Electron-density distribution in $1s$, $2s$, and $3s$ orbitals. The lower part of the figure shows how the electron density, represented by $\psi^2$, varies as a function of distance from the nucleus. In the $2s$ and $3s$ orbitals, the electron-density function drops to zero at certain distances from the nucleus. The spherical surfaces around the nucleus at which $\psi^2$ is zero are called *nodes*.

Height of graph indicates density of dots as we move from origin

as illustrated in Figure 6.16(*c*). Notice also that as *n* increases the electron is more and more likely to be located farther from the nucleus. That is, the size of the orbital increases as *n* increases.

The most widely used method of representing orbitals is to display a boundary surface that encloses some substantial fraction, say 90 percent, of the total electron density for the orbital. For the *s* orbitals these contour representations are merely spheres. The contour or boundary surface representations of the $1s$, $2s$, and $3s$ orbitals are shown in Figure 6.17. They have the same shape, but they differ in size. The fact that there are nodes within the $2s$ and $3s$ surfaces is lost in these representations. This is not a serious disadvantage; it turns out that for most qualitative

**FIGURE 6.17** Contour representations of the $1s$, $2s$, and $3s$ orbitals. The spherical surfaces connect points of equal value of $\psi^2$. The surface encloses 90 percent of total $\psi^2$ for each orbital.

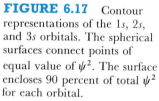

**FIGURE 6.18** Electron density distribution in a $2p$ orbital.

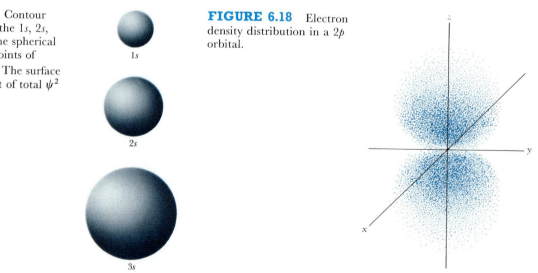

discussions the most important features of orbitals are their size and shape. These features are adequately represented by the contour diagrams.

The distribution of electron density for a 2*p* orbital is shown in Figure 6.18. As we can see from this figure, the electron density is not distributed in a spherically symmetric fashion as in an *s* orbital. Instead, the electron density is concentrated on two sides of the nucleus, separated by a node at the nucleus; we often say that this orbital has two lobes. It is useful to recall that we are making no statement of how the electron is moving within the orbital; Figure 6.18 portrays the *averaged* distribution of the 2*p* electron in space.

Each shell beginning with $n = 2$ has three *p* orbitals. For example, there are three 2*p* orbitals, three 3*p* orbitals, and so forth. The orbitals of a given principal quantum number have the same size and shape but differ from each other in orientation. The contour surfaces of the three 2*p* orbitals are shown in Figure 6.19. It is convenient to label these as the $2p_x$, $2p_y$, and $2p_z$ orbitals. The letter subscript indicates the axis along which the orbital is oriented. As it turns out, there is no necessary connection between one of these subscripts and a particular value of $m_l$. To explain why this is so would require discussion of material beyond the scope of an introductory text.

The distance from the nucleus to the center of the electron density moves outward as we go from the 2*p* to 3*p* to 4*p* orbitals. In other words, orbital size increases with increase in the principal quantum number ($n$). In accurate contour representations of the 3*p* and higher *p* orbitals, small regions of electron density separate the major lobes. These details of the

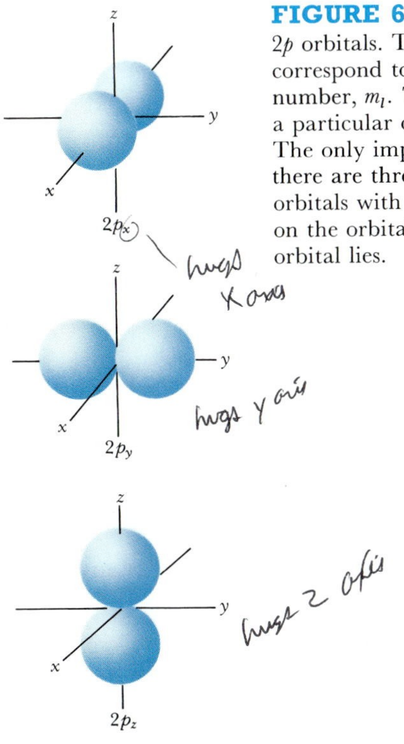

**FIGURE 6.19**   Contour representations of the three 2*p* orbitals. The three orbitals with differing orientations correspond to different values of the magnetic quantum number, $m_l$. There is no necessary connection between a particular orientation and a particular value of $m_l$. The only important point to remember is that because there are three possible values for $m_l$, there are three *p* orbitals with differing orientation. Note that the subscript on the orbital label indicates the axis along which the orbital lies.

shapes of the *p* orbitals are not of major chemical importance. The general overall shape of the *p* orbitals is more significant. We shall therefore always represent the *p* orbitals as shown in Figure 6.19, regardless of the value for the principal quantum number ($n$).

## The *d* and *f* Orbitals

When $l = 2$, the orbital is a *d* orbital. No *d* orbitals may exist with *n* lower than 3 because *l* can never be larger than $n - 1$. There are five equivalent 3*d* orbitals corresponding to the five possible values for $m_l$: 2, 1, 0, −1, and −2. Similarly, there are five 4*d* orbitals, and so forth. Just as in the case of the *p* orbitals, the differing values of $m_l$ correspond to different orientations of orbitals in space. The most useful representations of the 3*d* orbitals are shown in Figure 6.20. Notice that four of the orbitals are of the same shape but have differing orientations. The fifth orbital, labeled the $d_{z^2}$, has a different shape. It is not possible to represent the five *d* orbitals as having the same shape in an ordinary *x*, *y*, *z* axis system. Although the fifth *d* orbital looks different, it has the same energy in an atom as any of the other four orbitals.

The representations of higher *d* orbitals are very much like those for the 3*d*. The contour representations shown in Figure 6.20 are commonly employed for all *d* orbitals, regardless of major quantum number.

There are seven equivalent *f* orbitals (for which $l = 3$) for each value of *n* of 4 or greater. The *f* orbitals are difficult to represent in three-dimensional contour diagrams. We shall have no need to concern ourselves with orbitals having values for *l* greater than 3.

As we shall see in Chapter 9, an understanding of the number and shapes of atomic orbitals is important to a proper understanding of the

**FIGURE 6.20**  Contour representations of the five 3*d* orbitals. Notice that four of these orbitals have lobes centered in the plane indicated by the subscript on the orbital label. Both the $d_{xy}$ and $d_{x^2-y^2}$ orbitals are centered in the *xy* plane; however, the lobes of the $d_{x^2-y^2}$ orbital lie along the *x* and *y* axes, whereas the lobes of the $d_{xy}$ orbital lie between the axes. The lobes of the $d_{xz}$ and $d_{yz}$ orbitals also lie between the respective axes. The one orbital of unique shape is the $d_{z^2}$, which has two lobes along the *z* axis and a "donut" centered in the *xy* plane.

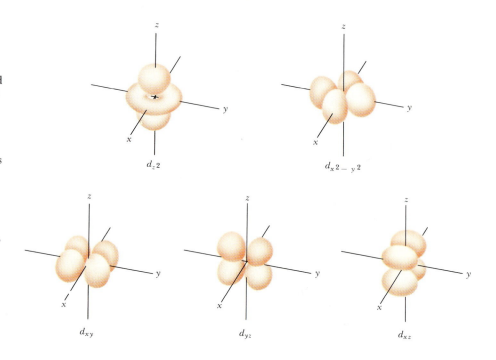

molecules formed by combining atoms. *You should commit to memory the orbital representations shown in Figures 6.17, 6.19, and 6.20.*

The orbitals we have been describing are those of a hydrogen atom. In the atoms of other elements, multiple electrons move about a single nucleus. These electrons are attracted by the nucleus and at the same time are repelled by one another. The resulting distribution of electron density and the energies of the allowed energy states of the electrons are thus the products of extremely complex forces. Nevertheless, as we shall see in the following chapter, the electronic structures of atoms having many electrons can be built up by the progressive addition of electrons to orbitals that are very much like those of the hydrogen atom.

 *FOR REVIEW*

## SUMMARY

**Radiant energy** moves through a vacuum at the "speed of light," $c = 3.00 \times 10^8$ m/s. It has wavelike characteristics that allow it to be described in terms of **wavelength** ($\lambda$) and **frequency** ($\nu$), which are interrelated: $c = \lambda \nu$. The dispersion of radiation into its component wavelengths produces a **spectrum**. If all wavelengths are present, the spectrum is said to be **continuous**; if only certain wavelengths are present, it is called a **line spectrum**.

The quantum theory describes the minimum amount of radiant energy that an object can gain or lose, $E = h\nu$; this smallest quantity is called a **quantum**. A quantum of radiant energy is called a **photon**. The quantum theory was used to explain the **photoelectric effect** and the line spectrum of the hydrogen atom. The absorptions or emissions of light by an atom, which produce its line spectrum, correspond to energy changes of electrons within the atom; the energy of the electron in an atom is quantized.

Electrons exhibit wave properties and can be described by a wavelength, $\lambda = h/mv$. Discovery of the wave properties of the electron led to the **uncertainty principle**, which indicates that the position and momentum of an electron can be determined simultaneously with only limited accuracy.

The ideas described above culminated in our current model of the electronic structures of atoms, in which we speak of the probability of the electron being found at a particular point in space. Although the positions of the electrons are defined in this averaged sense, their energies are precisely known. Each allowed state of an electron in an atom corresponds to a particular set of values for three **quantum numbers**. Each such allowed energy state is termed an **orbital**. An orbital is described by a combination of an integer and letters, corresponding to the three values for the quantum numbers. The **principal quantum number** ($n$) is indicated by the integers 1, 2, 3, .... This quantum number relates most directly to the size and energy of an orbital. The **azimuthal quantum number** ($l$) is indicated by the letters $s$, $p$, $d$, $f$, and so on, corresponding to values of $l$ of 0, 1, 2, 3, .... The $l$ quantum number defines the shape of the orbital. The **magnetic quantum number** ($m_l$) describes the orientation of the orbital in space. For example, the three $3p$ orbitals are designated $3p_x$, $3p_y$, and $3p_z$, the subscript letters indicating the axis along which the orbital is oriented.

Restrictions in the values of the three quantum numbers gives rise to the following allowed **subshells**:

$1s$

$2s, 2p$

$3s, 3p, 3d$

$4s, 4p, 4d, 4f$

$\vdots$

There is one orbital in an $s$ subshell, three in a $p$ subshell, five in a $d$ subshell, and seven in an $f$ subshell. The **contour representations** are the most generally useful way to visualize the spatial characteristics of the orbitals.

# LEARNING GOALS

Having read and studied this chapter, you should be abe to:

1. Describe the wave properties and characteristic speed of propagation of radiant energy (electromagnetic radiation).

2. Use the relationship $\lambda v = c$, which relates the wavelength ($\lambda$) and frequency ($v$) of radiant energy to its speed ($c$).

3. Describe how the wavelength and frequency differ in the various parts of the electromagnetic spectrum, such as the infrared, visible, and ultraviolet.

4. Explain the essential feature of Planck's quantum theory, namely, that the smallest increment, or quantum, of radiant energy of frequency, $v$, that can be emitted or absorbed is $hv$, where $h$ is Planck's constant.

5. Explain how Einstein accounted for the photoelectric effect by considering the radiant energy to be a stream of particlelike photons striking a metal surface. In other words, you should be able to explain all the observations about the photoelectric effect using Einstein's model.

6. Explain what is meant by the term *spectrum* and by the expression *line spectrum* in referring to the light emitted or absorbed by an atom.

7. List the assumptions made by Bohr in his model of the hydrogen atom. Most important, you should be able to explain how Bohr's model relates to Planck's quantum theory.

8. Explain the concept of an allowed energy state and how this concept is related to the quantum theory.

9. Calculate the energy differences between any two allowed energy states of the electron in hydrogen.

10. Explain the concept of ionization energy.

11. Calculate the characteristic wavelength of a particle from a knowledge of its mass and velocity.

12. Describe the uncertainty principle and explain the limitations it places on our ability to define simultaneously the location and momentum of a subatomic particle, particularly an electron.

13. Explain the concepts of *orbital, electron density,* and *probability* as used in the quantum-mechanical model of the atom; explain the physical significance of $\psi^2$.

14. Describe the three quantum numbers used to define an orbital in an atom, and list the limitations placed on the values each may have.

15. Describe the correspondence between letter designations and values for the azimuthal quantum number ($l$).

16. Describe the shapes of $s$, $p$, and $d$ orbitals.

# KEY TERMS

Among the more important terms and expressions used for the first time in this chapter are the following:

The **azimuthal quantum number**, $l$ (Section 6.5), is one of the quantum numbers that specifies an atomic orbital. The value for $l$ defines the shape of the orbital. The values allowed for $l$ are restricted by the value for the principal quantum number ($n$); $l$ may take on integral values from 0 to $n - 1$.

The **electron density** (Section 6.5) at a particular point in space in an atom is the probability that the electron will be found in the region immediately around that point. This probability is expressed by $\psi^2$, the square of the wave function.

The **ionization energy** (Section 6.3) for the hydrogen atom is the energy required to move an electron from its lowest-energy, or most stable, state to a point infinitely far from the nucleus, in effect removing the electron from the atom.

A **line spectrum** (Section 6.2) contains radiation only of certain specific wavelengths.

The **magnetic quantum number**, $m_l$ (Section 6.5), describes the orientation of an orbital in space. This quantum number may have integral values ranging from $l$ to $-l$. For each value of $l$ there are thus $2l + 1$ values for $m_l$.

**Matter wave** (Section 6.4) is the term used to describe the wave characteristics of a subatomic particle.

A **node** (Section 6.6), as applied to electron density in atoms, is the locus of points at which the electron density is zero. For example, the node in a $2s$ orbital (Figure 6.16) is a spherical surface.

An **orbit** (Section 6.3) in the Bohr theory of the hydrogen atom is any one of the allowed energy states of the electron. Each orbit corresponds to a different value for the principal quantum number ($n$)

An **orbital** (Section 6.5) represents an allowed

energy state of an electron in the modern, quantum-mechanical model of the atom. The term *orbital* is also used to describe the spatial distribution of the electron. An orbital is defined by specifying the values of three quantum numbers: $n$, $l$, and $m_l$.

A **photon** (Section 6.2) is a quantum, or the smallest increment, of radiant energy, $h\nu$.

The **principal quantum number**, $n$ (Section 6.5), is the quantum number that relates most directly to the size and energy of an atomic orbital. It may take on integer values of 1, 2, 3, . . . .

A **quantum** (Section 6.2) is the smallest increment of radiant energy that may be absorbed or emitted. The magnitude of the quantum of radiant energy is $h\nu$.

**Radiant energy**, or **electromagnetic radiation** (Section 6.1), is a form of energy that has wave characteristics and that propagates through space with the characteristic speed, $3.00 \times 10^8$ m/s.

The term **spectrum** (Section 6.2) refers to the distribution among various wavelengths of the light emitted or absorbed by an object. A continuous spectrum contains radiation distributed over all wavelengths.

A **subshell** (Section 6.5) is one or more orbitals with the same set of quantum numbers $n$ and $l$. For example, we speak of the $2p$ subshell ($n = 2$, $l = 1$), which is composed of three orbitals ($2p_x$, $2p_y$, and $2p_z$).

The **uncertainty principle** (Section 6.4) states that there is an inherent uncertainty in the precision with which we can simultaneously specify the location and momentum of a particle. It is of importance only for the lightest particles, such as the electron.

A **wave function** (Section 6.5) is a mathematical description of an allowed energy state, or orbital, for an electron in the quantum-mechanical model for an atom. It is usually symbolized by the Greek letter $\psi$.

## EXERCISES

### Radiant Energy

**6.1** List the following types of electromagnetic radiation in order of increasing wavelength: **(a)** radiation from a room heater; **(b)** radiation from an FM station; **(c)** the green light from a traffic signal; **(d)** cosmic radiation from outer space; **(e)** X rays used in medical diagnosis.

**6.2** List the following types of electromagnetic radiation in order of increasing wavelength: **(a)** radiation from a microwave oven; **(b)** the red light from a traffic signal; **(c)** the gamma rays produced in the radioactive decay of a cobalt-60 nucleus; **(d)** the light from a UV lamp; **(e)** the yellow light from a sodium-vapor light.

**6.3** **(a)** What is the wavelength of radiation whose frequency is $6.24 \times 10^{13}$/s? **(b)** What is the frequency of radiation whose wavelength is $2.20 \times 10^{-6}$ m? **(c)** What distance does light travel in 1.00 h?

**6.4** **(a)** What is the wavelength of radiation whose frequency is $5.00 \times 10^{12}$/s? **(b)** What is the frequency of light whose wavelength is 525 nm? **(c)** What distance can electromagnetic radiation travel in 1.00 min?

**6.5** Excited barium atoms emit radiation of 455-nm wavelength. What is the frequency of this radiation? Using Figure 6.3, indicate the color of the radiation.

**6.6** A neon light strongly emits radiation whose wavelength is 616 nm. What is the frequency of this radiation? Using Figure 6.3, suggest the color associated with this wavelength.

### Quantum Theory

**6.7** **(a)** Calculate the smallest increment of energy (a quantum) that can be emitted or absorbed at a wavelength of 667 nm. **(b)** Calculate the energy of a photon of frequency $4.5 \times 10^{12}$/s. **(c)** What frequency of radiation has photons of energy $2.15 \times 10^{-18}$ J?

**6.8** **(a)** Calculate the smallest increment of energy (a quantum) that can be emitted or absorbed at a wavelength of 455 nm. **(b)** Calculate the energy of a photon of frequency $6.00 \times 10^{13}$/s. **(c)** What wavelength of radiation has photons of energy $1.30 \times 10^{-19}$ J?

**6.9** Calculate and compare the energy of an ultraviolet-light photon of wavelength 106 nm with that of an infrared photon of wavelength 44 $\mu$m.

**6.10** Calculate and compare the energy of an X-ray photon of wavelength 0.15 nm with that of a microwave photon whose wavelength is 50 cm.

**6.11** Excited chromium atoms strongly emit radiation of 427 nm. What color is this radiation? What is the energy, in kilojoules per photon, of radiation of this wavelength? What is the energy in kilojoules per mole?

**6.12** The strong red line of the lithium spectrum occurs at 670.8 nm. What is the energy, in kilojoules per photon, of radiation of this wavelength? What is the energy in kilojoules per mole?

**6.13** In astronomy it is often necessary to be able to detect just a few photons, because the light signals from distant stars are so weak. The photon detector receives a signal of a total energy of $4.05 \times 10^{-18}$ J from radiation of 540-nm wavelength. How many photons have been detected?

**6.14** A high-powered laser is pulsed for a period of 100 ns. During that time it emits a signal of a total energy of 945 J. If the wavelength of the signal is 1.05 $\mu$m, how many photons have been emitted?

**6.15** Radiation that impinges on chemical substances may cause rupture of a chemical bond. A minimum energy of 332 kJ/mol is required to break a carbon-chlorine bond in a plastic material. What is the longest wavelength of radiation that possesses the necessary energy?

**6.16** Nitrogen dioxide is one of the components of photochemical smog. The energy required to dissociate this molecule into NO and O atoms is 305 kJ/mol. What is the maximum wavelength that can cause this dissociation? What type of radiation is it?

**6.17** Potassium metal must absorb radiation with a wavelength shorter than 540 nm before it can emit an electron from its surface via the photoelectric effect. **(a)** What is the minimum energy required to produce this effect? **(b)** What is the frequency of radiation of the minimum required energy? **(c)** If potassium is irradiated with light of 440-nm wavelength, what is the maximum possible kinetic energy of the emitted electron?

**6.18** Cesium metal must absorb radiation with a minimum frequency of $4.60 \times 10^{14}$/s before it can emit an electron from its surface via the photoelectric effect. **(a)** What is the minimum energy required to produce this effect? **(b)** What wavelength radiation will provide a photon of this energy? **(c)** If cesium is irradiated with light whose wavelength is 540 nm, what is the kinetic energy of the emitted electron?

## Bohr Model; Matter Waves

**6.19** Indicate whether energy is emitted or absorbed when the following electronic transitions occur in hydrogen: **(a)** $n = 3$ to $n = 5$; **(b)** from an orbit with radius 8.48 Å to one with radius 2.12 Å; **(c)** ionization of an electron from the $n = 3$ state.

**6.20** Indicate whether energy is emitted or absorbed when the following electronic transitions occur in hydrogen: **(a)** $n = 2$ to $n = 4$; **(b)** from an orbit with radius 53 pm to one of radius 477 pm; **(c)** ionization of an electron from the $n = 2$ state.

**6.21** What wavelength of light is absorbed when an electron moves from the $n = 2$ to $n = 5$ state in hydrogen? Using Figure 6.8, select the line in the spectrum to which this transition belongs.

**6.22** What wavelength of light is emitted when an electron moves from the $n = 5$ to $n = 1$ states in hydrogen? Using Figure 6.3, indicate the region of the spectrum characteristic of the radiation emitted.

**6.23** **(a)** Calculate the radius of the electron orbit in hydrogen when $n = 3$. **(b)** The radius of the electron orbit for one-electron ions of nuclear charge $Z$ is given by the equation $r = n^2 (5.30 \times 10^{-11} \text{ m})/Z$. Why is the radius inversely proportional to the nuclear charge?

**6.24** **(a)** Calculate the radius of the electron orbit in hydrogen when $n = 2$. **(b)** The radius of the electron orbit for one-electron ions of nuclear charge (atomic number) $Z$ is given by the equation $r = n^2 (5.30 \times 10^{-11} \text{ m})/Z$. Calculate the radius of the electron orbit for $n = 2$ in the $He^+$ ion.

**6.25** Major league pitchers can throw a baseball at speeds near 95 miles per hour. What is the wavelength of a 5.0-oz ball thrown at this speed?

**6.26** What is the wavelength of a 1.6-oz golf ball if it is traveling at a speed of 200 ft/s?

**6.27** Neutron diffraction has become an important technique for determining the structures of molecules. Calculate the velocity of a neutron that has a characteristic wavelength of 0.880 Å. (Refer to the back inside cover for the mass of the neutron.)

**6.28** The electron microscope has been widely used to obtain highly magnified images of biological and other types of materials. Calculate the velocity of an electron in such a microscope if it has a characteristic wavelength of 400 pm. (Refer to the back inside cover for the mass of the electron.)

## Wave Functions; Orbitals; Quantum Numbers

**6.29** **(a)** For $n = 5$, what are the possible values of $l$? **(b)** For $l = 3$, what are the possible values of $m_l$?

**6.30** **(a)** For $n = 4$, what are the possible values of $l$? **(b)** For $l = 4$, what are the possible values of $m_l$?

**6.31** Give the values for $n$, $l$, and $m_l$ **(a)** for each orbital in the $4d$ subshell; **(b)** for each orbital in the $n = 3$ shell.

**6.32** Give the values for $n$, $l$, and $m_l$ **(a)** for each orbital in the $4f$ subshell; **(b)** for each orbital in the $n = 2$ shell.

**6.33** Which of the following are incorrect (forbidden) designations for atomic orbitals: $3f$, $3d$, $2p$, $4s$, $4f$, $2d$?

**6.34** Which of the following are incorrect (forbidden) designations for atomic orbitals: $6s$, $5f$, $1p$, $2s$, $2f$, $3p$?

**6.35** Sketch the contour representation for the following types of orbitals: **(a)** $s$; **(b)** $p_x$; **(c)** $d_{x^2-y^2}$.

**6.36** Sketch the contour representation for the following types of orbitals: **(a)** $p_y$; **(b)** $d_{xy}$; **(c)** $d_{z^2}$.

**6.37** What is the physical significance of the square of the wave function?

**6.38** What characteristics of an orbital are determined by **(a)** the principal quantum number, $n$; **(b)** the azimuthal quantum number, $l$; **(c)** the magnetic quantum number, $m_l$?

## Additional Exercises

**6.39** What are the wavelengths (in meters) of the following radio signals: **(a)** an FM radio station that broadcasts at 98.6 MHz; **(b)** an AM station whose signal is at 740 kHz?

**6.40** Radar signals are electromagnetic radiation in the microwave region of the spectrum. A typical radar signal has a wavelength of 3.19 cm. What is its frequency?

**6.41** The planetary space probe *Voyager 2* should pass close to Neptune in 1989. At that time it will be $2.82 \times 10^9$ mi from Earth. How long will it take for the pictures transmitted from *Voyager 2* to reach Earth?

**6.42** The laser in a compact audio disc player uses light whose wavelength is 780 nm. **(a)** What is the frequency of this radiation? **(b)** What is the energy (in joules) of a single photon of this wavelength?

**6.43** The strong blue line in the mercury spectrum occurs at $4.358 \times 10^{-7}$ m. **(a)** What is the energy of a photon of this wavelength? **(b)** What is the energy of this radiation in kJ/mol?

**6.44** Chorophyll absorbs blue light, $\lambda = 460$ nm, and emits red light, $\lambda = 660$ nm. Calculate the net energy change in the chlorophyll system when a single photon of 460-nm wavelength is absorbed and a photon of 660-nm wavelength is emitted.

**6.45** In the photoelectric effect, the minimum energy required to eject an electron from a metal is called the work function. Rubidium metal will emit electrons when the wavelength of radiation is 574 nm or less. **(a)** Calculate the work function for rubidium. **(b)** If rubidium is irradiated with 420-nm light, what is the maximum kinetic energy of the emitted electrons?

**[6.46]** Using Appendix D, calculate the energy change for the process $O_2(g) \longrightarrow 2O(g)$. This process, called photodissociation, occurs in the earth's upper atmosphere upon absorption of solar radiation. Calculate the maximum wavelength that a photon may have if it is to possess sufficient energy to cause this dissociation.

**6.47** The work function is the energy required to remove an electron from an atom on the surface of a metal. How does this definition differ from that for ionization energy?

**6.48** When the light from a neon light is dispersed in a prism to form a spectrum, the spectrum is not continuous; rather, it consists of several sharp lines, each of a specific frequency. Explain in general terms why a line spectrum is produced.

**6.49** In which of the following states does the electron in a hydrogen atom have the higher energy: $n = 1$ or $n = 4$? Which is the ground state, and which is an excited state?

**6.50** Calculate the wavelength of light emitted or absorbed in each of the following spectral transitions in the hydrogen atom: **(a)** $n = 3 \longrightarrow n = 2$; **(b)** $n = 2 \longrightarrow n = 5$; **(c)** $n = 4 \longrightarrow n = 3$.

**6.51** Make a figure like Figure 6.12, showing each of the transitions indicated in Exercise 6.50.

**6.52** A hydrogen emission line in the infrared region, at 1875.6 nm, corresponds to a transition from a higher $n$ level to the $n = 3$ level. What is the value of $n$ for the higher-energy level?

**6.53** What is the maximum wavelength of light capable of removing an electron from a hydrogen atom in the energy state characterized by **(a)** $n = 1$; **(b)** $n = 2$; **(c)** $n = 3$?

**6.54** For one-electron ions, the energy of the electron is given by the equation $E_n = -Z^2 R_H(1/n^2)$, where $Z$ is the atomic number of the nucleus. What is the energy required to remove the remaining electron from the $He^+$ ion? Refer to Sample Exercise 6.4, and compare this value with that for the hydrogen atom. Explain the reason for the difference.

**6.55** Calculate the wavelength for each of the following: **(a)** an electron moving at 20 percent the speed of light; **(b)** a neutron moving at 15 percent the speed of light.

**[6.56]** The Heisenberg uncertainty principle can be stated in the following form: The product of the uncertainty in the position of an object, $\Delta x$, and the uncertainty in its momentum, $\Delta p$, can be no less than $h/2\pi$ [that is, $(\Delta p)(\Delta x) \geq h/2\pi$]. Using this relationship, calculate the uncertainty of position for each of the following: **(a)** an electron whose uncertainty of velocity, $\Delta v$, is 0.1 m/s (assume that the uncertainty in the momentum is due entirely to the uncertainty in velocity); **(b)** a 12-g bullet whose uncertainty of velocity is 0.1 m/s. **(c)** Compare the two results; how does each compare with the size of the object itself?

**6.57** Discuss the differences between Bohr's theory of the atom and the quantum-mechanical theory. Explain why Bohr's theory of the atom is not compatible with quantum mechanics.

**6.58** What is the subshell designation for each of the following cases: **(a)** $n = 2$, $l = 0$; **(b)** $n = 4$, $l = 2$; **(c)** $n = 5$, $l = 1$; **(d)** $n = 3$, $l = 2$; **(e)** $n = 4$, $l = 3$?

**6.59** **(a)** How many subshells are present in the fourth shell? **(b)** How many orbitals are present in the $4d$ subshell?

**6.60** What are the possible values of $m_l$ for **(a)** $l = 2$; **(b)** $l = 1$; **(c)** $n = 2$ (all sublevels)?

**6.61** Indicate the number of orbitals that can have each of the following designations: **(a)** $n = 5$; **(b)** $3p$; **(c)** $3p_x$; **(d)** $3f$; **(e)** $4s$; **(f)** $4d$; **(g)** $4d_{xz}$.

**6.62** How do the three $3p$ orbitals differ from one another? How do the $2p$ and the $3p$ orbitals differ?

**6.63** Draw the contour representations for all of the orbitals with quantum numbers $n = 3$, $l = 2$.

**6.64** Indicate whether each of the following statements is true or false. Correct those that are false. **(a)** The energy of an electron in hydrogen depends only on the principal quantum number, $n$. **(b)** The energies of electrons in H and $He^+$ are the same when the principal quantum number, $n$, is the same. **(c)** The number of orbitals in a subshell of azimuthal quantum number $l$ is the same regardless of the value of the principal quantum number, $n$. **(d)** The set $n = 4$, $l = 3$, and $m_l = 3$ is a permissible set of quantum numbers for an electron in hydrogen. **(e)** The contour representation of the $3p_z$ orbital looks much like that for the $3d_{z^2}$ orbital.

CHAPTER **7**

# Electronic Structure:
# Periodic Relationships

In Chapter 6, we saw how the work of physicists led to our present understanding of atomic structure, at least as far as the hydrogen atom is concerned. The application of the concepts outlined there to atoms containing more than one electron proved to be possible in principle, but difficult in practice. While physicists were wrestling with this problem, chemists were making significant progress in understanding atomic structure from another point of view. In brief, they asked what sorts of inferences about the arrangements of electrons in atoms they might be able to make by using the periodic table and the known chemical behavior of the elements. One of the chief contributors to this development was the American chemist Gilbert N. Lewis (1875–1946). Even before Bohr had proposed his theory of the hydrogen atom, Lewis had suggested that electrons in atoms are arranged in shells.

In this chapter we will see why there is a close connection between the periodic table and the ways in which electrons are arranged in atoms. Our first goal will be to use the concepts developed in Chapter 6 to understand the electronic structures of many-electron atoms. We will then relate these electronic structures to the periodic table and to the chemical behavior of the elements.

## CONTENTS

The electronic structures of atoms with two or more electrons can be described in terms of orbitals like those describing hydrogen (Section 6.6). Thus we can continue to use orbital designations such as $1s$, $2p_x$, and so forth. Furthermore, these orbitals have the same shape as the corresponding hydrogen orbitals.

However, because of electron-electron repulsions, the different subshells in a many-electron atom are at different energies, as shown in Figure 7.1. For example, the $2s$ subshell is at a lower energy than the $2p$ subshell. To understand why, we need to consider the forces that operate between the electrons and the way these forces are influenced by the shapes of the orbitals.

If an atom or ion has only one electron, the energy of that electron can be calculated exactly. That energy depends on the principal quantum number, $n$, and the charge of the nucleus, $Z$:

$$E_n = -R_H\left(\frac{Z^2}{n^2}\right) = -(1312 \text{ kJ/mol})\left(\frac{Z^2}{n^2}\right) \qquad [7.1]$$

Notice that Equation 6.6, which applies to hydrogen, is just a special case of this equation in which $Z = 1$.

## 7.1 ORBITALS IN MANY-ELECTRON ATOMS

**Effective Nuclear Charge**

Neon signs, such as those seen in this photo of Hong Kong, owe their brilliant colors to the light emitted when high voltage electricity is passed through glass tubes filled with various gases. Neon produces the red color in such signs. (Alain Evrard/Photo Researchers)

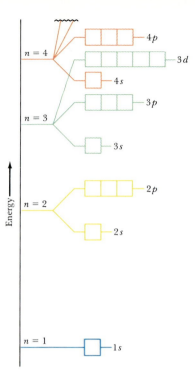

**FIGURE 7.1** Ordering of orbital energy levels in many-electron atoms. As in Figure 6.14, which shows the orbital energy levels for the hydrogen atom, each box represents an orbital. Note that orbitals with the same value for the principal quantum number, $n$, but with differing azimuthal quantum numbers, $l$ differ in energy. (That is, orbitals in different subshells differ in energy.)

For many-electron atoms, the energy of an electron depends not only on $n$ and $Z$, but also on the azimuthal quantum number, $l$; that is, the energy differs for the different subshells. Because there are so many electron-electron interactions, the situation is too complex to be analyzed in an exact way. However, we may estimate the energy of each electron by considering how it interacts with the *average* environment created by the nucleus and all the other electrons in the atom.

Any electron density between the nucleus and the electron of interest will reduce the nuclear charge acting on that electron. The net positive charge that the electron experiences is called the **effective nuclear charge**. The effective nuclear charge, $Z_{eff}$, equals the number of protons in the nucleus, $Z$, minus the average number of electrons, $S$, that are between the nucleus and the electron in question:

$$Z_{eff} = Z - S \qquad [7.2]$$

Thus the positive charge experienced by outer-shell electrons is always less than the full nuclear charge, because the inner-shell electrons partly offset the positive charge of the nucleus. The inner electrons are said to shield or screen the outer electron from the full charge of the nucleus. This effect, which is called the **screening effect**, is illustrated in Figure 7.2.

**Energies of Orbitals**

The manner in which the electron distribution varies as we move outward from the nucleus differs for orbitals in different subshells. Consider the orbitals for which $n = 3$. The 3s electron distribution extends closer to the nucleus than does the 3p; the 3p, in turn, extends closer than does

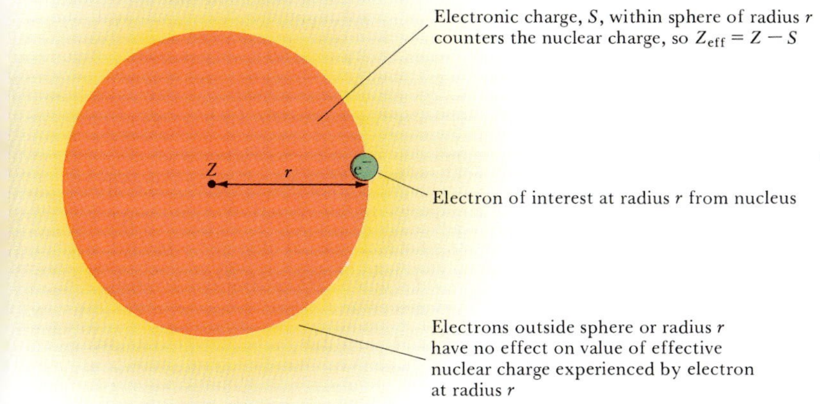

Electronic charge, $S$, within sphere of radius $r$
counters the nuclear charge, so $Z_{eff} = Z - S$

Electron of interest at radius $r$ from nucleus

Electrons outside sphere or radius $r$
have no effect on value of effective
nuclear charge experienced by electron
at radius $r$

**FIGURE 7.2** Shielding of the nuclear charge from an electron by other electrons in an atom. As an example, if the nuclear charge were 5 and the sphere of radius $r$ contained three electrons, the effective nuclear charge at the radius of the sphere would be $5 - 3 = 2$.

the $3d$. As a result, the other electrons of the atom (that is, the $1s$, $2s$, and $2p$) shield the $3s$ electrons from the nucleus less effectively than they shield the $3p$ or $3d$. Thus the $3s$ electrons experience a larger $Z_{eff}$ than do the $3p$ electrons, and these in turn experience a larger $Z_{eff}$ than do the $3d$ electrons.

The energy of an electron depends on the effective nuclear charge, $Z_{eff}$. Because $Z_{eff}$ is larger for the $3s$ electrons, they have a lower energy (that is, they are more stable) than the $3p$, which in turn are lower in energy than the $3d$. As a result, the energy-level diagram for the orbitals of many-electron atoms is like that shown in Figure 7.1. Keep in mind that this is a *qualitative* energy-level diagram; the exact energies and their spacings differ from one atomic species to another. In all cases, however, the relative energies of the orbitals through $n = 3$ are as shown. Notice that all orbitals of a given subshell, such as the $3d$ orbitals, have the same energy. Orbitals that have the same energy are said to be **degenerate**.

## SAMPLE EXERCISE 7.1

Based on the energy-level diagram of Figure 7.1, would you expect the average distance from the nucleus of a $3d$ electron to be greater or less than that of a $2p$? Explain.

**Solution:** The energy of the $2p$ orbitals is considerably lower than for a $3d$. This indicates that the average attractive interaction of the $2p$ electron with the nucleus is much greater than for an electron in a $3d$ orbital. The increased attractive interaction is due to a smaller average distance of the $2p$ electron from the nucleus.

## PRACTICE EXERCISE

The sodium atom has 11 electrons. Two of these occupy a $1s$ orbital, two occupy a $2s$ orbital, and one occupies a $3s$ orbital. Which of these electrons experience the lowest effective nuclear charge?
*Answer:* the electron in the $3s$ orbital

## 7.2 ELECTRON SPIN AND THE PAULI EXCLUSION PRINCIPLE

An important property of the electron that we have not yet discussed is **electron spin**; the electron behaves as though it were spinning on its own axis. The concept of electron spin was developed to account for certain features of the line spectra of atoms. Lines that were at first thought to be single lines were found under high resolution to be closely spaced pairs. The only way to account for these extra splittings was to introduce a new quantum number in addition to the three quantum numbers that

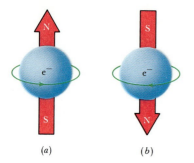

(a)                    (b)

**FIGURE 7.3** The electron behaves as if it were spinning about an axis through its center. A spinning charge produces a magnetic field. The orientation of the magnetic field depends upon the direction in which the particle spins. The two directions of spin correspond to the two possible values for the spin quantum number, $m_s$; in (a) $m_s = \frac{1}{2}$ and in (b) $m_s = -\frac{1}{2}$.

we have already encountered. This new quantum number, **the electron spin quantum number** $(m_s)$, can have only one of two values: $+\frac{1}{2}$ or $-\frac{1}{2}$. We interpret this to mean that the electron can spin in one of two opposite directions. A spinning charge produces a magnetic field. The two opposing spins produce oppositely directed magnetic fields, as shown in Figure 7.3.

The presence of electron spin turns out to be important in determining the electronic structures of atoms. The first to recognize this was the Austrian physicist Wolfgang Pauli (1900–1958). In 1924, he spelled out what has become known as the **Pauli exclusion principle**, which declares that no two electrons in an atom can have the same set of four quantum numbers ($n$, $l$, $m_l$, and $m_s$). Electrons in the same orbital have the same values of the first three quantum numbers, leaving two possible values for the spin quantum number. Thus the Pauli exclusion principle indicates that *an orbital can hold a maximum of two electrons, and they must have opposite spins*. This restriction provides the key to what had been one of the great problems of chemistry: explaining the structure of the periodic table of the elements.

## A CLOSER LOOK                    Experimental Evidence for Electron Spin

In 1921, Otto Stern and Walter Gerlach succeeded in separating a beam of atoms into two groups according to the orientation of the electron spin. Their experiment is diagrammed in Figure 7.4. Let us assume that the beam of atoms is hydrogen. We saw in Section 2.2 that charged particles are deflected upon moving through a magnetic field. In the Stern-Gerlach experiment, the atoms moving through the magnetic field are electrically neutral. Thus any de-

flection of the beam cannot be due to the charge on the atoms. However, the magnetic field arising from the electron's spin causes the atom to interact with the magnet's field, deflecting the atom from its straight-line path. The direction in which the atom is deflected depends on the orientation of the spin of the electron. We expect that there will be equal numbers of electrons with each of the two possible orientations, as is found to be the case.

**FIGURE 7.4** Diagrammatic illustration of the Stern-Gerlach experiment. A beam of hydrogen atoms is allowed to pass through an inhomogeneous magnetic field. Atoms in which the electron-spin quantum number $m_s$ is $+\frac{1}{2}$ are deflected in one direction, whereas those in which $m_s$ is $-\frac{1}{2}$ are deflected in the other.

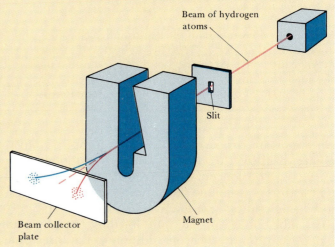

Beam of hydrogen atoms

Slit

Beam collector plate

Magnet

Elements are arranged in the periodic table according to atomic number. The atomic number corresponds not only to the number of protons in the nucleus of the neutral atom, but also to the number of electrons in that atom. With Rutherford's discovery of the nuclear atom in 1911, scientists began to seek relationships between atomic structure and chemical behavior. Gilbert N. Lewis, who was mentioned in the introduction, reasoned that if the chemical properties of the elements are repeated when the elements are arranged by atomic number, then their electron configurations must be repeating in some way. He suggested that electrons in atoms are arranged in shells and that once a shell of electrons is filled, additional electrons must go into a new shell. He further reasoned that the noble gases, which are chemically very nonreactive, have closed or filled shells of electrons. Chemical behavior might then be understood in terms of the tendency of an atom to achieve a closed shell by gaining, losing, or sharing electrons. For example, we saw in Section 2.5 that sodium, atomic number 11, loses one electron in chemical reactions to form the sodium ion, $Na^+$. This ion has the stable, closed-shell electron arrangement of the noble gas neon, atomic number 10.

These ideas of Lewis generate additional questions. Why do certain electron configurations occur periodically? Why are electrons arranged in shells? To answer these questions and to account in more detail for the way in which certain atomic properties vary with atomic number, we must apply the ideas about orbitals that we have learned about in this chapter and Chapter 6.

The arrangement of electrons in the orbitals of an atom is called the **electron configuration**. The most stable, or ground, state of an atom is that in which all the electrons are in the lowest possible energy states. If there were no restrictions on the possible values for the quantum numbers of the electrons, all the electrons would crowd into the 1s orbital, because this is the lowest in energy (Figure 7.1). The Pauli exclusion principle, however, tells us that there can be at most two electrons in any single orbital. Thus the orbitals are filled in order of increasing energy, with no more than two electrons per orbital. For example, consider the lithium atom, which has three electrons. (Recall that the number of electrons in a neutral atom is equal to its atomic number, $Z$.) In the lowest-energy state of lithium, two electrons are in the 1s orbital and one electron is in the 2s orbital.

We can summarize any electron configuration by writing the symbol for the occupied subshell and adding a superscript to indicate the number of electrons in that subshell. Thus for lithium we write $1s^2 2s^1$. We can also show the arrangement of electrons in more detail in the following way:

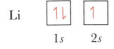

In this kind of diagram each orbital is represented by a box, and each electron by a half-arrow. A half-arrow pointing upward (↑) represents an electron spinning in one direction ($m_s = +\frac{1}{2}$), and a downward half-arrow (↓) represents an electron spinning in the opposite direction

**The Electron Configurations of the Elements**

$(m_s = -\frac{1}{2})$. We shall refer to representations of this type as **orbital diagrams**. Electrons having opposite spins are said to be *paired* when they are in the same orbital. An *unpaired electron* is not accompanied by a partner of opposite spin. In the lithium atom the two electrons in the 1*s* orbital are paired, and the electron in the 2*s* orbital is unpaired.

It is informative to consider how the electron configurations of the elements change as we move from element to element across the periodic table. Hydrogen has one electron, which occupies the 1*s* orbital in its ground state:

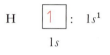

The next element, helium, has two electrons. Because two electrons with opposite spins can occupy an orbital, both of helium's electrons are in the 1*s* orbital:

The two electrons present in helium complete the filling of the first shell. This arrangement represents a very stable configuration, as evidenced by the chemical inertness of helium.

The electron configurations of lithium and several elements that follow it in the periodic table are shown in Table 7.1. Because a maximum of two electrons can be placed in each orbital, the third electron of lithium cannot enter the 1*s* orbital. Instead, it must be placed in the next-most-stable orbital, the 2*s* (refer to Figure 7.1). The change in principal quantum number for the third electron represents a large jump in energy and a corresponding jump in the average distance of the electron from the nucleus. We may say that it represents the start of a new shell of electrons. As you can see by examining the periodic table, lithium represents the start of a new row of the periodic table. It is the first member of the alkali metals family (group 1A).

**TABLE 7.1**   Electron configurations of several lighter elements

| Element | Total electrons | Orbital diagram | | | | Electron configuration |
|---------|-----------------|-----------------|--|--|--|------------------------|
| Li | 3 | ↑↓ | ↑ | | | $1s^2 2s^1$ |
| Be | 4 | ↑↓ | ↑↓ | | | $1s^2 2s^2$ |
| B | 5 | ↑↓ | ↑↓ | ↑ | | $1s^2 2s^2 2p^1$ |
| C | 6 | ↑↓ | ↑↓ | ↑ ↑ | | $1s^2 2s^2 2p^2$ |
| Ne | 10 | ↑↓ | ↑↓ | ↑↓ ↑↓ ↑↓ | | $1s^2 2s^2 2p^6$ |
| Na | 11 | ↑↓ | ↑↓ | ↑↓ ↑↓ ↑↓ | ↑ | $1s^2 2s^2 2p^6 3s^1$ |
| | | 1*s* | 2*s* | 2*p* | 3*s* | |

The element that follows lithium is beryllium; its electron configuration is $1s^2 2s^2$ (Table 7.1). Boron, atomic number 5, has the electron configuration $1s^2 2s^2 2p^1$. The fifth electron must be placed in a $2p$ orbital, because the $2s$ orbital is filled. Because each of the three $2p$ orbitals are of equal energy, it doesn't matter which $2p$ orbital is occupied.

With the next element, carbon, we come to a new situation. We know that the sixth electron must go into a $2p$ orbital, where there is already one electron. However, does this new electron go into the $2p$ orbital that already has one electron, or into one of the others? This question is answered by **Hund's rule**, which states that *electrons occupy degenerate orbitals singly to the maximum extent possible, and with their spins parallel.* In the case of carbon, then, the sixth electron goes into one of the other $2p$ orbitals, with its spin in the same orientation as the other $2p$ electron. Hund's rule is based on the fact that electrons repel one another because they have the same electrical charge. By occupying different orbitals, the electrons remain as far as possible from one another in space, thus minimizing electron-electron repulsions. When electrons must occupy the same orbital, the repulsive interaction between the paired electrons is greater than between electrons in different, equivalent orbitals.

Neon, the last member of the second row, has ten electrons. Two electrons fill the $1s$ orbital, two electrons fill the $2s$ orbital, and the remaining six electrons fill the $2p$ orbitals. The electron configuration is thus $1s^2 2s^2 2p^6$ (Table 7.1). In neon, all of the orbitals with $n = 2$ are filled. The filling of the $2s$ and $2p$ orbitals by the eight electrons that they can hold represents a very stable configuration. As a result, neon is chemically quite inert. Lewis, in his model for the electron configurations of elements, noted that the octet of electrons (eight) in the outermost shell of an atom or ion represents an especially stable arrangement.

Sodium, atomic number 11, marks the beginning of a new row of the periodic table. Sodium has a single $3s$ electron beyond the stable configuration of neon. We can abbreviate the electron configuration of sodium as follows:

$$\text{Na} \qquad [\text{Ne}]3s^1$$

The symbol [Ne] represents the electron configuration of the ten electrons of neon, $1s^2 2s^2 2p^6$. Writing the electron configuration in this manner helps us focus attention on the outermost electrons of the atom. The outer electrons are the ones largely responsible for the chemical behavior of an element. For example, we can write the electron configuration of lithium as follows:

$$\text{Li} \qquad [\text{He}]2s^1$$

By comparing this with the electron configuration for sodium, it is easy to appreciate why lithium and sodium are so similar chemically: They have the same type of outer-shell electron configuration. All the members of the alkali metal family (group 1A) have a single $s$ electron beyond an inner-core noble-gas configuration. The outer-shell electrons are often referred to as **valence electrons**. The electrons in the inner shells are called the **core electrons**.

## SAMPLE EXERCISE 7.2

Draw the orbital diagram representation for the electron configuration of oxygen, atomic number 8.

**Solution:** The ordering of orbitals is shown in Figure 7.1. Two electrons each go into the $1s$ and $2s$ orbitals. This leaves four electrons for the three $2p$ orbitals. Following Hund's rule, we put one electron into each $2p$ orbital until all three have one each. The fourth electron must then be paired up with one of the three electrons already in a $2p$ orbital, so that the correct representation is

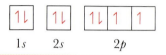

The corresponding electron configuration is written $1s^2 2s^2 2p^4$ or $[He]2s^2 2p^4$. The $1s^2$ or $[He]$ electrons are the inner-shell or core electrons of the oxygen atom. The $2s^2 2p^4$ electrons are the outer-shell or valence electrons.

### PRACTICE EXERCISE

Write the electron configuration of phosphorus, element 15.

***Answer:*** $1s^2 2s^2 2p^6 3s^2 3p^3 = [Ne]3s^2 3p^3$

## SAMPLE EXERCISE 7.3

What is the characteristic outer-shell electron configuration of the group 7A elements, the halogens?

**Solution:** The first member of the halogen family is fluorine, atomic number 9. The abbreviated form of the electronic configuration for fluorine is

$$F \qquad [He]2s^2 2p^5$$

Similarly, the abbreviated form of the electron configuration for chlorine, the second halogen, is

$$Cl \qquad [Ne]3s^2 3p^5$$

From these two examples we see that the characteristic outer-shell electron configuration of a halogen is $ns^2 np^5$, where $n$ ranges from 2 in the case of fluorine to 5 in the case of iodine.

### PRACTICE EXERCISE

What family of elements is characterized by having an $ns^2$ electron configuration?

***Answer:*** the alkaline earth metals, family 2A

The rare-gas element argon marks the end of the row started by sodium. The configuration for argon is $1s^2 2s^2 2p^6 3s^2 3p^6$. The element following argon in the periodic table is potassium (K), atomic number 19. In all its chemical properties, potassium is very obviously a member of the alkali metal family. The experimental facts about the properties of potassium leave no doubt that the outermost electron of this element occupies an $s$ orbital. But this means that the highest-energy electron has *not* gone into a $3d$ orbital, which we might naïvely have expected it to do. In this case the ordering of energy levels is such that the $4s$ orbital is lower in energy than the $3d$ (see Figure 7.1).

Following complete filling of the $4s$ orbital (this occurs in the calcium atom), the next set of equivalent orbitals to be filled is the $3d$. (You will find it helpful as we go along to refer often to the periodic table on the front inside cover.) Beginning with scandium and extending through zinc, electrons are added to the five $3d$ orbitals until they are completely filled. Thus the fourth row of the periodic table is ten elements wider than the previous rows. These ten elements are known as the **transition metals**. Note the position of these elements in the periodic table (front inside cover).

In accordance with Hund's rule, electrons are added to the $3d$ orbitals singly until all five orbitals have one electron each. Additional electrons are then placed in the $3d$ orbitals with spin pairing until the shell is completely filled. The orbital diagram representations and electron configurations of two transition elements are as follows:

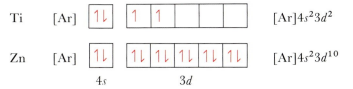

The $3d$ and $4s$ orbital energies are very close. Occasionally an electron we would expect in a $4s$ orbital actually exists in a $3d$ orbital in the atom's most stable state. For example, we might expect chromium to have the outer electron configuration $4s^2 3d^4$, but it is actually $4s^1 3d^5$. This anomalous behavior is partly due to the special stability associated with precisely half-filled sets of degenerate orbitals. Apparently, there is just enough gain in stability in arriving at this arrangement to cause the electron to move from a $4s$ to a $3d$ orbital. A similar anomaly occurs at copper (Cu), whose electron configuration is $[Ar]4s^1 3d^{10}$. In this case the stability is associated with completing the $3d$ subshell.

Upon completion of the $3d$ transition series, the $4p$ orbitals begin to be occupied, until the completed octet of outer electrons ($4s^2 4p^6$) is reached with krypton (Kr), atomic number 36. Krypton is another of the rare gases. Rubidium (Rb) marks the beginning of the fifth row of the periodic table. This row is in every respect like the preceding one, except that the value for $n$ is one greater. The sixth row of the table begins similarly to the preceding one: one electron in the $6s$ orbital of cesium (Cs) and two electrons in the $6s$ orbital of barium (Ba). The next element, lanthanum (La), represents the start of the third series of transition elements. But with cerium (Ce), element 58, a new set of orbitals, the $4f$, enter the picture. The energies of the $5d$ and $4f$ orbitals are very close. For lanthanum itself, the $5d$ orbital energy is just a little lower than the $4f$. However, for the elements immediately following lanthanum, the $4f$ orbital energies are a little lower, so that the $4f$ orbitals fill before the $5d$ orbitals.

There are seven equivalent $4f$ orbitals, corresponding to the seven allowed values of $m_l$, ranging from 3 to $-3$. Thus it requires 14 electrons to fill the $4f$ orbitals completely. The 14 elements corresponding to the filling of the $4f$ orbitals are elements 58 to 71, known as the **rare-earth**, or **lanthanide**, elements. In order not to make the periodic table unduly wide, the rare-earth elements are set together below the other elements. The properties of the rare-earth elements are all quite similar, and they occur together in nature. For many years it was virtually impossible to separate them from one another.

After the rare-earth series, the third transition element series is completed, followed by the filling of the $6p$ orbitals. This brings us to radon (Rn), heaviest of the rare-gas elements. The final row of the periodic table begins as the one before it. The **actinide** elements are built up by the completion of the $5f$ electron orbitals. This series consists mainly of elements not found in nature but synthesized in nuclear reactions.

## 7.4 USING THE PERIODIC TABLE TO WRITE ELECTRON CONFIGURATIONS

Our rather brief survey of electron configurations of the elements has taken us through the periodic table. We have seen that the electron configurations of elements are related to their locations in the periodic table. The periodic table is structured so that elements with the same type of outer-shell electron configuration are arranged in columns. For example, for groups 2A and 3A we have

| Group 2A | | Group 3A | |
|---|---|---|---|
| Be | $[He]2s^2$ | B | $[He]2s^22p^1$ |
| Mg | $[Ne]3s^2$ | Al | $[Ne]3s^23p^1$ |
| Ca | $[Ar]4s^2$ | Ga | $[Ar]3d^{10}4s^24p^1$ |
| Sr | $[Kr]5s^2$ | In | $[Kr]4d^{10}5s^25p^1$ |
| Ba | $[Xe]6s^2$ | Tl | $[Xe]5d^{10}6s^26p^1$ |
| Ra | $[Rn]7s^2$ | | |

If you understand how the periodic table is organized, it is not necessary to memorize the order in which orbitals fill. You can write the electron configuration of an element based on its location in the periodic table. The pattern is summarized in Figure 7.5. Notice that the elements can be grouped in terms of the *type* of orbital into which the electrons are placed. On the left are *two* columns of elements. These elements, known as the **active metals**, are those in which the outer-shell *s* orbitals are being filled. On the right is a block of *six* columns. These are the elements in which the outermost *p* orbitals are being filled. The *s* block and the *p* block of the periodic table contain the **representative elements**. In the middle of the table is a block of *ten* columns that contains the **transition metals**. These are the elements in which the *d* orbitals are being filled. Below the main portion of the table are two rows that contain *fourteen* columns. These elements, sometimes called the **inner-transition metals**, are the ones in which the *f* orbitals are being filled. Recall that the numbers 2, 6, 10, and 14 are precisely the number of electrons that can fill the *s*, *p*, *d*, and *f* subshells, respectively. Recall also that the 1*s* subshell is the first *s* subshell, the 2*p* is the first *p* subshell, the 3*d* is the first *d* subshell, and the 4*f* is the first *f* subshell.

**FIGURE 7.5** Block diagram of the periodic table showing the groupings of the elements according to the type of orbital being filled with electrons.

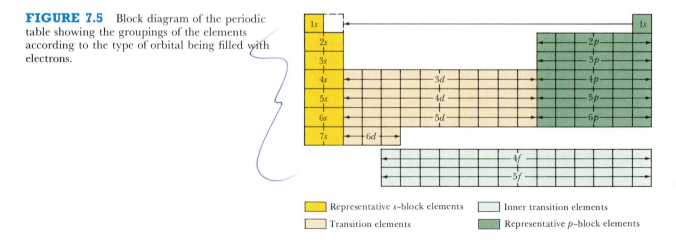

Representative *s*-block elements

Inner transition elements

Transition elements

Representative *p*-block elements

There are a few instances in which minor shifts of an electron or two from one orbital to another occur. We have given two examples in the preceding section: Chromium possesses the configuration $[Ar]4s^13d^5$, rather than $[Ar]4s^23d^4$ as we might have expected, and the configuration of copper is $[Ar]4s^13d^{10}$ instead of the expected $[Ar]4s^23d^9$. There are a few similar instances among the transition metals and the inner-transition metals. Although these minor departures from the expected are interesting, they are not of great chemical significance.

## SAMPLE EXERCISE 7.4

Write the electron configuration for the element bismuth, atomic number 83.

**Solution:** We can do this by simply moving across the periodic table one row at a time and writing the occupancies of the orbitals corresponding to each row (refer to Figure 7.5),

First row     $1s^2$
Second row     $2s^22p^6$
Third row     $3s^23p^6$
Fourth row     $4s^23d^{10}4p^6$
Fifth row     $5s^24d^{10}5p^6$
Sixth row     $6s^24f^{14}5d^{10}6p^3$
Total:     $1s^22s^22p^63s^23p^63d^{10}4s^24p^64d^{10}4f^{14}$
                 $5s^25p^65d^{10}6s^26p^3$

Note that 3 is the lowest possible value that $n$ may have for a $d$ orbital, and that 4 is the lowest possible value of $n$ for an $f$ orbital.

The total of the superscripted numbers should equal the atomic number of bismuth, 83. It does not matter a great deal precisely in which order the orbitals are listed. They may be listed, as shown above, in the order of increasing major quantum number. However, it is also possible to list them in the sequence read from the periodic table: $1s^22s^22p^63s^23p^6$ $4s^23d^{10}4p^65s^24d^{10}5p^66s^24f^{14}5d^{10}6p^3$.

It is a simple matter to write the abbreviated electron configuration of an element using the periodic table. First locate the element of interest (in this case element 83) and then move backward until the first noble gas is encountered (in this case Xe, element 54). Thus the inner core is [Xe]. The outer electrons are then read from the periodic table as before. Moving from Xe to Cs, element 55, we find ourselves in the sixth row. Moving across this row to Bi gives us the outer electrons. The complete electron configuration is thus $[Xe]6s^24f^{14}5d^{10}6p^3$.

## PRACTICE EXERCISE

Using the periodic table, write the electron configurations for the following atoms by giving the appropriate noble-gas inner core plus the electrons beyond it: **(a)** Co (atomic number 27); **(b)** Te (atomic number 52).
*Answers:* **(a)** $[Ar]4s^23d^7$; **(b)** $[Kr]5s^24d^{10}5p^4$

## SAMPLE EXERCISE 7.5

Draw the orbital diagram representation for zirconium, atomic number 40; show only those electrons beyond the krypton inner core.

**Solution:** Zirconium has four electrons beyond the nearest noble gas, krypton, atomic number 36. Examining the periodic table, we see that zirconium is a transition element from the fifth row of the table. This means that its outermost electrons are in $5s$ and $4d$ orbitals. Two electrons occupy the $5s$ orbital; two must be placed in the five $4d$ orbitals. As indicated by Hund's rule, the $4d$ electrons occupy separate orbitals. Thus we have

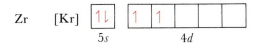

## PRACTICE EXERCISE

Using orbital diagrams for the outer electrons, determine the number of unpaired electrons in **(a)** Ni (atomic number 28); **(b)** Br (atomic number 35)
*Answers:* **(a)** 2; **(b)** 1

A complete list of the electron configurations of the elements is contained in Table 7.2. You can use this table to check your answers as you practice writing electron configurations. We have written these configurations as they would be read off the periodic table. However, they are sometimes written with orbitals of a given principal quantum number gathered together. Thus we might write the electron configuration of arsenic (atomic number 33) as $[Ar]3d^{10}4s^24p^3$ instead of $[Ar]4s^23d^{10}4p^3$ as shown in Table 7.2.

The configurations of many of the heavier elements are not known for certain. The configurations must be deduced by analysis of atomic spectra. These methods are extremely complicated, especially for the heavier elements, in which the energy levels are closely spaced.

## 7.5 ELECTRON SHELLS IN ATOMS

We have seen how the electron configuration of an atom may be built up by adding electrons to orbitals of successively higher energy, in accordance with the Pauli exclusion principle and Hund's rule. How do our results compare with Lewis's idea of electron shells? Consider the noble gases helium, neon, and argon, whose electron configurations follow:

He $\quad 1s^2$
Ne $\quad 1s^22s^22p^6$
Ar $\quad 1s^22s^22p^63s^23p^6$

Very accurate calculations of the total electronic charge distribution in these atoms can be made using large computers. These distributions are shown in Figure 7.6. The quantity plotted on the vertical axis is called the radial electron density. It corresponds to the probability of the electron being located at a particular distance from the nucleus. As Figure 7.6 shows, the radial electron density does not fall off continuously as we move away from the nucleus. Rather it shows maxima corresponding to distances at which there are higher probabilities of finding electrons.

**FIGURE 7.6** Radial electron-density graphs for the first three rare-gas elements, He, Ne, and Ar. The maxima that occur in the radial electron density correspond to electrons with the same value of the principal quantum number, $n$.

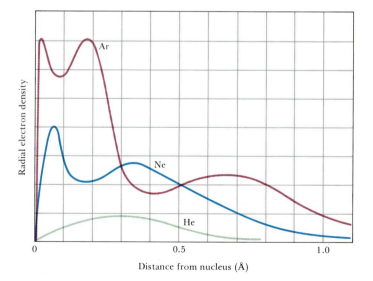

**TABLE 7.2**  The electron configurations of the elements

| Atomic number | Symbol | Electron configuration | Atomic number | Symbol | Electron configuration |
|---|---|---|---|---|---|
| 1 | H | $1s^1$ | 55 | Cs | $[Xe]6s^1$ |
| 2 | He | $1s^2$ | 56 | Ba | $[Xe]6s^2$ |
| 3 | Li | $[He]2s^1$ | 57 | La | $[Xe]6s^25d^1$ |
| 4 | Be | $[He]2s^2$ | 58 | Ce | $[Xe]6s^24f^{15}d^1$ |
| 5 | B | $[He]2s^22p^1$ | 59 | Pr | $[Xe]6s^24f^3$ |
| 6 | C | $[He]2s^22p^2$ | 60 | Nd | $[Xe]6s^24f^4$ |
| 7 | N | $[He]2s^22p^3$ | 61 | Pm | $[Xe]6s^24f^5$ |
| 8 | O | $[He]2s^22p^4$ | 62 | Sm | $[Xe]6s^24f^6$ |
| 9 | F | $[He]2s^22p^5$ | 63 | Eu | $[Xe]6s^24f^7$ |
| 10 | Ne | $[He]2s^22p^6$ | 64 | Gd | $[Xe]6s^24f^75d^1$ |
| 11 | Na | $[Ne]3s^1$ | 65 | Tb | $[Xe]6s^24f^9$ |
| 12 | Mg | $[Ne]3s^2$ | 66 | Dy | $[Xe]6s^24f^{10}$ |
| 13 | Al | $[Ne]3s^23p^1$ | 67 | Ho | $[Xe]6s^24f^{11}$ |
| 14 | Si | $[Ne]3s^23p^2$ | 68 | Er | $[Xe]6s^24f^{12}$ |
| 15 | P | $[Ne]3s^23p^3$ | 69 | Tm | $[Xe]6s^24f^{13}$ |
| 16 | S | $[Ne]3s^23p^4$ | 70 | Yb | $[Xe]6s^24f^{14}$ |
| 17 | Cl | $[Ne]3s^23p^5$ | 71 | Lu | $[Xe]6s^24f^{14}5d^1$ |
| 18 | Ar | $[Ne]3s^23p^6$ | 72 | Hf | $[Xe]6s^24f^{14}5d^2$ |
| 19 | K | $[Ar]4s^1$ | 73 | Ta | $[Xe]6s^24f^{14}5d^3$ |
| 20 | Ca | $[Ar]4s^2$ | 74 | W | $[Xe]6s^24f^{14}5d^4$ |
| 21 | Sc | $[Ar]4s^23d^1$ | 75 | Re | $[Xe]6s^24f^{14}5d^5$ |
| 22 | Ti | $[Ar]4s^23d^2$ | 76 | Os | $[Xe]6s^24f^{14}5d^6$ |
| 23 | V | $[Ar]4s^23d^3$ | 77 | Ir | $[Xe]6s^24f^{14}5d^7$ |
| 24 | Cr | $[Ar]4s^13d^5$ | 78 | Pt | $[Xe]6s^14f^{14}5d^9$ |
| 25 | Mn | $[Ar]4s^23d^5$ | 79 | Au | $[Xe]6s^14f^{14}5d^{10}$ |
| 26 | Fe | $[Ar]4s^23d^6$ | 80 | Hg | $[Xe]6s^24f^{14}5d^{10}$ |
| 27 | Co | $[Ar]4s^23d^7$ | 81 | Tl | $[Xe]6s^24f^{14}5d^{10}6p^1$ |
| 28 | Ni | $[Ar]4s^23d^8$ | 82 | Pb | $[Xe]6s^24f^{14}5d^{10}6p^2$ |
| 29 | Cu | $[Ar]4s^13d^{10}$ | 83 | Bi | $[Xe]6s^24f^{14}5d^{10}6p^3$ |
| 30 | Zn | $[Ar]4s^23d^{10}$ | 84 | Po | $[Xe]6s^24f^{14}5d^{10}6p^4$ |
| 31 | Ga | $[Ar]4s^23d^{10}4p^1$ | 85 | At | $[Xe]6s^24f^{14}5d^{10}6p^5$ |
| 32 | Ge | $[Ar]4s^23d^{10}4p^2$ | 86 | Rn | $[Xe]6s^24f^{14}5d^{10}6p^6$ |
| 33 | As | $[Ar]4s^23d^{10}4p^3$ | 87 | Fr | $[Rn]7s^1$ |
| 34 | Se | $[Ar]4s^23d^{10}4p^4$ | 88 | Ra | $[Rn]7s^2$ |
| 35 | Br | $[Ar]4s^23d^{10}4p^5$ | 89 | Ac | $[Rn]7s^26d^1$ |
| 36 | Kr | $[Ar]4s^23d^{10}4p^6$ | 90 | Th | $[Rn]7s^26d^2$ |
| 37 | Rb | $[Kr]5s^1$ | 91 | Pa | $[Rn]7s^25f^26d^1$ |
| 38 | Sr | $[Kr]5s^2$ | 92 | U | $[Rn]7s^25f^36d^1$ |
| 39 | Y | $[Kr]5s^24d^1$ | 93 | Np | $[Rn]7s^25f^46d^1$ |
| 40 | Zr | $[Kr]5s^24d^2$ | 94 | Pu | $[Rn]7s^25f^6$ |
| 41 | Nb | $[Kr]5s^14d^4$ | 95 | Am | $[Rn]7s^25f^7$ |
| 42 | Mo | $[Kr]5s^14d^5$ | 96 | Cm | $[Rn]7s^25f^76d^1$ |
| 43 | Tc | $[Kr]5s^24d^5$ | 97 | Bk | $[Rn]7s^25f^9$ |
| 44 | Ru | $[Kr]5s^14d^7$ | 98 | Cf | $[Rn]7s^25f^{10}$ |
| 45 | Rh | $[Kr]5s^14d^8$ | 99 | Es | $[Rn]7s^25f^{11}$ |
| 46 | Pd | $[Kr]4d^{10}$ | 100 | Fm | $[Rn]7s^27s^25f^{12}$ |
| 47 | Ag | $[Kr]5s^14d^{10}$ | 101 | Md | $[Rn]7s^25f^{13}$ |
| 48 | Cd | $[Kr]5s^24d^{10}$ | 102 | No | $[Rn]7s^25f^{14}$ |
| 49 | In | $[Kr]5s^24d^{10}5p^1$ | 103 | Lr | $[Rn]7s^25f^{14}6d^1$ |
| 50 | Sn | $[Kr]5s^24d^{10}5p^2$ | 104 | Rf | $[Rn]7s^25f^{14}6d^2$ |
| 51 | Sb | $[Kr]5s^24d^{10}5p^3$ | 105 | Ha | $[Rn]7s^25f^{14}6d^3$ |
| 52 | Te | $[Kr]5s^24d^{10}5p^4$ | 106 | Unh | $[Rn]7s^25f^{14}6d^4$ |
| 53 | I | $[Kr]5s^24d^{10}5p^5$ | 107 | Uns | $[Rn]7s^25f^{14}6d^5$ |
| 54 | Xe | $[Kr]5s^24d^{10}5p^6$ | 109 | Une | $[Rn]7s^25f^{14}6d^7$ |

These maxima correspond to the traditional idea of shells of electrons; however, these shells are diffuse and overlap considerably.

Helium shows a single shell, neon two, and argon three. Each of these maxima is due mainly to electrons in the atom that have the same value for the principal quantum number $n$. Thus for helium the $1s$ electrons possess a maximum in radial electron density at about 0.3 Å. In argon, the maximum in the $1s$ radial electron density occurs at only 0.05 Å. The second maximum, which occurs at larger radial distance, is due to both the $2s$ and $2p$ electrons. The third maximum is due to $3s$ and $3p$ electrons.

The reason for the smaller radial distance of the orbital of the $1s$ electrons in the heavier atom is clear when we recall that the nuclear charge of helium is only 2, whereas that for argon is 18. The $1s$ electrons are the innermost electrons of the atom. The electrons of quantum number $n = 2$ and greater, present in elements beyond helium, therefore do not do much to shield the $1s$ electrons from the increasing nuclear charge. As a result, the size of the $1s$ orbital shrinks steadily as nuclear charge increases. Thus the calculations show that in many-electron atoms the inner electrons are pulled with ever-increasing force into the region around the nucleus as the nuclear charge increases.

Now that we have some understanding of the electronic structure of atoms, we can examine some properties of atoms that are strongly dependent on electron configuration. We will consider three properties that provide important insights into chemical behavior: atomic size, ionization energy, and electron affinity.

## 7.6 SIZES OF ATOMS

One conclusion we can draw from the quantum-mechanical model is that an atom does not have a sharply defined boundary that determines its size. This is evident in Figure 7.6. The electron-density distributions illustrated do not end sharply; rather, they simply drop off with increasing distance from the nucleus, approaching zero at large distance. Given this state of affairs we might well ask whether it makes sense to speak of an atomic radius for an atom. Certainly such a concept would be difficult to define for an isolated atom. Suppose, however, that two atoms form a chemical bond between them, as in $Br_2$. The distance between the centers of the two bromine atoms in $Br_2$ can be thought of as the sum of two bromine atomic radii. The Br—Br distance in $Br_2$ is 2.286 Å; we might then say that, to the nearest 0.01 Å, the radius of the bromine atom is 1.14 Å. Similarly, in compounds containing carbon-carbon bonds, the C—C distance is found to be very close to 1.54 Å. We can thus assign an atomic radius of 0.77 Å to carbon. If our concept of an atomic radius is going to be very useful, these atomic radii should remain pretty much the same when the atom is bound to another element. Thus the distance between carbon and bromine in a C—Br bond should be 1.14 + 0.77 Å, or 1.91 Å. It turns out that carbon-bromine bonds in various compounds are about this length. Atomic radii are not so easy to evaluate for metallic elements, but a list of atomic radii has been assembled for many elements, based on a large body of experimental data. These atomic radii are graphed in Figure 7.7 as a function of atomic number.

Provided we keep in mind that there are uncertainties in these values because of the means by which they are obtained, we can discern some

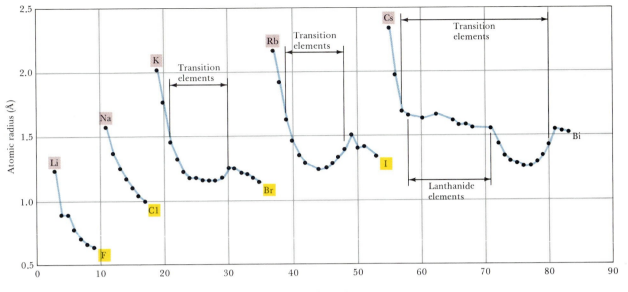

**FIGURE 7.7** Atomic radii versus atomic number. The noble gases are not included in this graph because there is no simple way of relating their radii to those of the other elements on the basis of solid-state structure determinations. Gaps in the graph are due to lack of experimental data.

interesting trends in the data:

1. Within each group, the atomic radius tends to increase going from top to bottom.
2. Within each period, the atomic radius tends to decrease moving left to right.

These general trends, which are summarized in Figure 7.8, can be understood in terms of two factors that determine the size of the outermost orbital: its principal quantum number and the effective nuclear charge acting on its electrons. Increasing the principal quantum number increases the size of the orbital; increasing the effective nuclear charge reduces the size.

Proceeding across any horizontal row of the table, the number of core electrons remains the same while the nuclear charge increases. The electrons that are added to counterbalance the increasing nuclear charge are

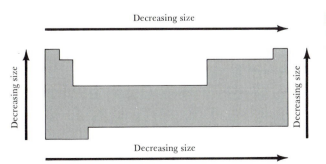

**FIGURE 7.8**
General periodic trends in atomic size.

very ineffective in shielding each other. Thus the effective nuclear charge increases steadily while the principal quantum number remains constant. For example, the inner $1s^2$ electrons of lithium $(1s^2 2s^1)$ shield the outer $2s$ electron from the 3+ charge nucleus. Consequently, the outer electron experiences an effective nuclear charge of about 1+. For beryllium $(1s^2 2s^2)$, the effective nuclear charge experienced by each outer $2s$ electron is larger; in this case the inner $1s^2$ electrons are shielding a 4+ nucleus, and each $2s$ electron only partially shields the other from the nucleus. As the effective nuclear charge increases, the electrons are drawn closer to the nucleus. Thus the radius of the atom decreases.

In going down a column of elements, the effective nuclear charge remains essentially constant, while the principal quantum number increases. Thus the size of the orbital and consequently of the atomic radius increases.

## SAMPLE EXERCISE 7.6

Referring to a periodic table, arrange the following atoms in order of increasing size: O, S, F.

**Solution:** Notice that F is to the right of O in the same period; thus we expect that F is smaller than O. Now notice that S is below O in group 6A: hence S is larger than O. The resultant order of increasing radii is F < O < S.

## PRACTICE EXERCISE

Arrange the following atoms in order of increasing atomic radius: Na, Be, Mg.

*Answer:* Be < Mg < Na

## 7.7 IONIZATION ENERGY

The ionization energy $(I)$ is the energy required to remove an electron from a gaseous atom or ion. The first ionization energy for an element, $I_1$, is that required for the process shown in Equation 7.3:

$$M(g) \longrightarrow M(g)^+ + e^- \qquad [7.3]$$

where M is a gaseous, neutral atom. The second ionization energy, $I_2$, is then the energy for the removal of the second electron, Equation 7.4:

$$M(g)^+ \longrightarrow M(g)^{2+} + e^- \qquad [7.4]$$

Successive ionization energies are defined in a similar manner. The values of successive ionization energies are known for many elements. Values for the elements sodium through argon are listed in Table 7.3. Remember that a larger value of $I$ corresponds to tighter binding of the electron to the atom or ion.

As we might expect, each successive removal of an electron requires more energy. The reason for this is that the positive nuclear charge that provides the attractive force remains the same, whereas the number of electrons, which produce repulsive interaction, steadily decreases. For example, the electronic configuration for silicon (Si) is $1s^2 2s^2 2p^6 3s^2 3p^2$. If we look at the successive ionization energies for silicon given in Table 7.3, we see a steady increase from 780 kJ/mol to 4350 kJ/mol for the four values of $I$ that correspond to loss of the four valence-shell electrons with

**TABLE 7.3**  Successive values of ionization energies ($I$) for the elements sodium through argon (kJ/mol)[a]

| Element | $I_1$ | $I_2$ | $I_3$ | $I_4$ | $I_5$ | $I_6$ | $I_7$ |
|---------|-------|-------|-------|-------|-------|-------|-------|
| Na | 490 | 4560 | (Inner-shell electrons) | | | | |
| Mg | 735 | 1445 | 7730 | | | | |
| Al | 580 | 1815 | 2740 | 11,600 | | | |
| Si | 780 | 1575 | 3220 | 4350 | 16,100 | | |
| P | 1060 | 1890 | 2905 | 4950 | 6270 | 21,200 | |
| S | 1005 | 2260 | 3375 | 4565 | 6950 | 8490 | 27,000 |
| Cl | 1255 | 2295 | 3850 | 5160 | 6560 | 9360 | 11,000 |
| Ar | 1525 | 2665 | 3945 | 5770 | 7230 | 8780 | 12,000 |

[a] Although the ionization energies are given here in units of kJ/mol, they are also often given in units of electron volts; 1 electron volt is equal to 96.49 kJ/mol.

principal quantum number $n = 3$. The fifth electron, however, requires considerably more energy for removal: 16,100 kJ/mol. This sharp increase in ionization energy occurs because the fifth electron is an inner-shell electron. This electron is in a $2p$ orbital and consequently penetrates closer to the nucleus than do the $3s$ and $3p$ electrons. The $2p$ electron in silicon not only has a smaller average distance from the nucleus, but also experiences a larger effective nuclear charge because it penetrates the charge distribution of the other electrons.

The effect of change in the principal quantum number can be seen in other comparisons as well. For example, if we compare Mg and Al, we see from Table 7.3 that the energies required to remove first one and then two electrons from the two metals are not so very different. Yet the energy required to remove the third electron from Mg, 7730 kJ/mol, is much greater than the energy required to remove a third electron from Al, 2740 kJ/mol. The difference in energies arises predominately from the fact that a third electron removed from Mg is an inner-shell $2p$ electron; in contrast, the third electron removed from Al is an outer-shell $3s$ electron.

These and similar ionization energy data thus support the idea that only the outermost electrons, those beyond the noble-gas core, are involved in the sharing and transfer of electrons that give rise to chemical change. The inner electrons are too tightly bound to the nucleus to be lost from the atom or even shared with another atom.

## SAMPLE EXERCISE 7.7

As can be seen in Table 7.3, the energy required to remove an electron from $P^{4+}$ is 6270 kJ/mol, as compared with 16,100 kJ/mol for removal of an electron from $Si^{4+}$. Account for the large difference.

**Solution:** The outer electron configuration of phosphorus is $3s^2 3p^3$. After removal of four of these electrons, the highest-energy electron remaining is a $3s$. The outer electron configuration of silicon is $3s^2 3p^2$. After removal of these four electrons the highest-

energy electron remaining is a $2p$. It requires considerably less energy to remove the $3s$ electron, which lies largely outside the $1s^2 2s^2 2p^6$ core of electrons, than to remove an electron from the $2p$ level.

## PRACTICE EXERCISE

Which atom should have a larger second ionization energy, lithium or beryllium?
*Answer:*  lithium

**Periodic Trends in Ionization Energies**

Let's observe how the first ionization energies, $I_1$, vary according to the position of an element in the periodic table. Figure 7.9 shows a graph of $I_1$ versus atomic number, and an overall periodicity in $I_1$ is evident. Overlooking for the moment the slight irregularities, there is a gradual increase in $I_1$ with increasing atomic number in any horizontal row. Thus the alkali metals show the lowest ionization energy in each row and the noble gases the highest. There is also a general decrease in ionization energy with increasing atomic number in any column or family of elements. For example, it requires more energy to remove an electron from a lithium atom than from a potassium atom. These general periodic trends are summarized in Figure 7.9.

A few simple considerations help to explain these trends. The energy needed to remove an electron from the outer shell depends on both the effective nuclear charge and the average distance of the electron from the nucleus. Either increasing the effective nuclear charge or decreasing the distance from the nucleus increases the attraction between the electron and the nucleus. As this attraction increases, it becomes harder to remove the electron, and thus the ionization energy increases. As we move across a period, there is both an increase in effective nuclear charge and a decrease in atomic radius, causing the ionization energy to increase. On the other hand, as we move down a column, the atomic radius increases while the effective nuclear charge remains essentially constant. Thus the attraction between the nucleus and the electron decreases in this direction, causing the ionization energy to decrease.

The irregularities within a given period are somewhat more subtle but are readily explained. For example, the decrease in ionization in going from beryllium ($[\text{He}]2s^2$) to boron ($[\text{He}]2s^2 2p^1$) arises because the

**FIGURE 7.9**  Ionization energy versus atomic number.

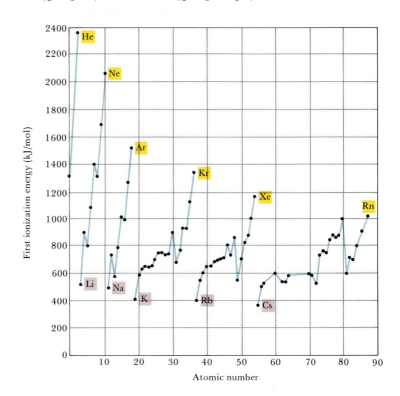

electrons in the filled $2s$ orbital provide some shielding for the electrons in the $2p$ subshell. The decrease in ionization in going from nitrogen ([He]$2s^22p^3$) to oxygen ([He]$2s^22p^4$) is due to repulsion of paired electrons in the $p^4$ configuration.

---

## SAMPLE EXERCISE 7.8

Referring to the periodic table, select the atom from the following list that has the greatest first ionization energy: S, Cl, Se, Br.

**Solution:** Because ionization energy tends to increase as we move from the bottom of any family to the top, S and Cl should have greater ionization energies than Se and Br. Because the ionization energy increases as we move left to right in any period, Cl

should have a greater ionization energy than S. Thus Cl has the largest ionization energy of these four elements.

## PRACTICE EXERCISE
Which of the following atoms—B, Al, C, and Si— has the lowest ionization energy?
*Answer:* Al

---

Atoms not only lose electrons to form positively charged ions but also gain them to form negatively charged ones. The ionization energy measures the energy changes associated with removing electrons from gaseous atoms. The energy change that occurs when an electron is added to a gaseous atom or ion is called the **electron affinity**, $E$.* The process may be represented for a neutral atom as

$$\mathrm{M}(g) + e^- \longrightarrow \mathrm{M}^-(g) \qquad [7.5]$$

For most neutral atoms and for all cations, energy is evolved when an electron is added; $E$ is thus negative in sign. The greater the attraction between the species and the added electron, the more negative the electron affinity. On the other hand, the electron affinities of anions and some neutral atoms are positive, meaning that work must be done to force the electron onto the species.

The electron affinities of neutral atoms are quite difficult to measure, and accurate values are now known for only about 50 elements. Figure 7.10 shows the variation of electron affinities for the first 20 elements in the periodic table. Notice that the electron affinities of the elements in group 2A (Be, Mg, and Ca) and group 8A (He, Ne, and Ar) are positive. These elements have filled subshells, so adding an electron requires energy. The outer $s$ subshells are filled for the 2A elements; both the $s$ and $p$ subshells are filled for the noble gases. Notice also that the halogens (F and Cl) have the most negative electron affinities. By picking up an electron, they form very stable negative ions with noble-gas electron configurations.

Now notice how the electron affinities change as we move from Li to F and from Na to Cl. In general, electron affinities become more negative

## 7.8 ELECTRON AFFINITIES

---

* We have defined electron affinity in such a way that a negative $E$ is associated with an exothermic process. Thus the more negative the value for $E$, the greater the attraction for electrons. However, $E$ is sometimes defined as the energy given off when an electron is added to a gaseous atom or ion. In references that define $E$ in this second way, the more positive the $E$ value, the greater the attraction for electrons.

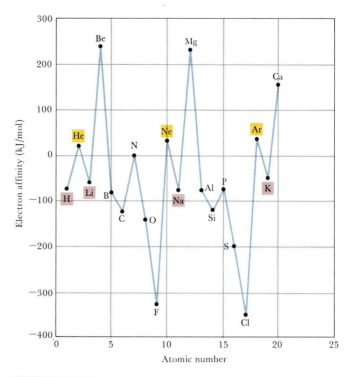

**FIGURE 7.10**   Electron affinity versus atomic number.

as we move across a period. The group 5A elements (N and P) have electron affinities that are higher than those of the preceding group 4A elements (C and Si) because of half-filled subshells. For example, nitrogen has a half-filled $2p$ subshell $(1s^2 2s^2 2p^3)$; thus an added electron must be placed into an orbital that is already occupied by an electron. The resultant electron-electron repulsion causes the nitrogen to have less attraction for an extra electron than does carbon. In general, atoms with filled or half-filled subshells have more positive electron affinities than do elements on either side of them in the periodic table.

In going down a group, the electron affinities do not change greatly. For example, consider the electron affinities of halogens, which are listed in Table 7.4. As we proceed from fluorine to iodine the added electron is going into a $p$ orbital of increasing major quantum number. The average distance of the electron from the nucleus steadily increases, and electron-nuclear attraction should thus steadily decrease. If this were all that is involved, fluorine would have the highest electron affinity. But the orbitals that hold the outermost electrons of the halogen are increasingly spread out as we proceed from fluorine to iodine. The electron-electron repulsions between these electrons and the added one therefore decrease with increasing atomic number of the halogen. A lower electron-nuclear attraction is thus counterbalanced by lower electron-electron repulsion. The overall result is that the electron affinities differ very little among the halogens. Other families likewise show little variation in electron affinities among their members.

**TABLE 7.4** Electron affinities of the halogens

| Element | Ion formed | $E$ (kJ/mol) |
|---------|-----------|--------------|
| H | $H^-$ | $-73$ |
| F | $F^-$ | $-332$ |
| Cl | $Cl^-$ | $-349$ |
| Br | $Br^-$ | $-325$ |
| I | $I^-$ | $-295$ |

## 7.9 GROUP TRENDS: THE ACTIVE METALS

Our discussion of atomic size, ionization energy, and electron affinity gives some indication of the way the periodic table can be used to organize and remember facts. Not only do elements in a family possess general similarities, but there are also trends in behavior as we move through a family or from one family to another. In this section we want to use the periodic table and our knowledge of electron configurations to examine the chemistry of the alkali metals (group 1A) and the alkaline earths (group 2A), and to compare these elements briefly with those of groups 1B and 2B. We discussed some of the chemistry of the alkali metals and alkaline earths earlier, in Section 5.6.

### The Alkali Metals

The alkali metals are soft, metallic solids of relatively low density and melting point (Section 5.6). They are all very reactive, readily losing one electron to form ions with a 1+ charge:

$$M \longrightarrow M^+ + e^- \qquad [7.6]$$

(The symbol M in this equation and others in this section represent any one of the alkali metals.) This behavior corresponds to their $ns^1$ outer electron configurations and their low ionization energies. As we have noted in Section 7.7, ionization energies decrease as we move to the left across any period. Thus the alkali metals have the lowest ionization energies in each period. Furthermore, the ease with which elements lose electrons increases as we move down a family. Consequently, Cs is especially reactive. Of course, Fr is even more reactive, but it is an extremely rare, radioactive element; Cs is therefore the heaviest alkali metal generally encountered in the laboratory.

The alkali metals combine directly with water, generating hydrogen gas and forming hydroxides:

$$2M(s) + 2H_2O(l) \longrightarrow 2MOH(aq) + H_2(g) \qquad [7.7]$$

This reaction is most violent in the case of the heavier members of the family, in keeping with their weaker hold on the single outer-shell electron.

The reactivity of the alkali metals permits them to react directly with most elements. The following reactions illustrate this behavior:

$$2M(s) + H_2(g) \longrightarrow 2MH(s) \qquad [7.8]$$

$$2M(s) + S(s) \longrightarrow M_2S(s) \qquad [7.9]$$

$$2M(s) + Cl_2(g) \longrightarrow 2MCl(s) \qquad [7.10]$$

The reactions between the alkali metals and oxygen are more complex, because three types of compounds may possibly be formed: oxides, peroxides, and superoxides, which contain $O^{2-}$, $O_2^{2-}$, and $O_2^-$ ions, respectively. Lithium forms only the oxide:

$$4Li(s) + O_2(g) \longrightarrow 2Li_2O(s) \qquad [7.11]$$

The other metals all form peroxides. For example,

$$2Rb(s) + O_2(g) \longrightarrow Rb_2O_2(s) \qquad [7.12]$$

Potassium, rubidium, and cesium also form superoxides:

$$Cs(s) + O_2(g) \longrightarrow CsO_2(s) \qquad [7.13]$$

The small size of lithium confers some special properties on the metal and its compounds. For example, Li is the only alkali metal that reacts directly with $N_2$ to form a nitride ($Li_3N$). Thus when lithium is burned in air, it forms not only lithium oxide, $Li_2O$, but lithium nitride as well.

Alkali metal salts and their aqueous solutions are colorless unless they contain a colored anion like the yellow $CrO_4^{2-}$. Color is produced when an electron in an atom is excited from one energy level to another by visible radiation. Alkali metal ions, having lost their outermost electrons, have no electrons that can be excited by visible radiation.

When alkali metal compounds are placed in a flame, they emit characteristic colors, as shown in Figure 7.11. The alkali metal ions are reduced to gaseous metal atoms in the lower, central region of the flame. The atoms are electronically excited by the high temperature of the flame; they then emit energy in the form of visible light as they return to the ground state. For example, sodium gives a yellow flame due to emission at 589.2 nm. This wavelength is produced by transition of the excited valence electron from the $3p$ to the $3s$ subshell.

**FIGURE 7.11**  Flame test for (a) Li (crimson-red), (b) Na (yellow), and (c) K (lilac). (H. E. LeMay, Jr.)

(a)

(b)

(c)

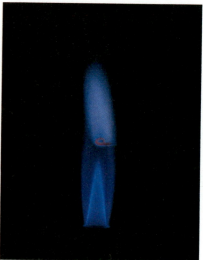

The alkaline earth metals, like the alkali metals, are reactive elements. However, they lose their electrons less readily than do their alkali metal neighbors. As we have noted in Section 7.7, the ease with which the elements lose electrons decreases as we move across the periodic table from left to right, and the ease with which these elements lose electrons increases as we move down the family. Thus beryllium and magnesium, the lightest members of the family, are least reactive.

The trend of increasing reactivity within the family is illustrated by the behavior of the elements toward water. Beryllium does not react with water or steam, even when heated red hot. Magnesium does not react with water in the liquid phase but does react with steam:

$$Mg(s) + H_2O(g) \longrightarrow MgO(s) + H_2(g) \qquad [7.14]$$

Calcium and the elements below it react readily with water at room temperature (although more slowly than the alkali metals adjacent to them in the periodic table):

$$Ca(s) + 2H_2O(l) \longrightarrow Ca(OH)_2(aq) + H_2(g) \qquad [7.15]$$

Because of their $ns^2$ outer electron configurations, the alkali metals are present in compounds only as 2+ ions, as in MgO and $Ca(OH)_2$.

Like the alkali metals, the alkaline earths form colorless or white compounds unless they contain a colored anion. The heavier alkaline earths, Ca, Sr, and Ba, have characteristic flames. The calcium flame is brick red, strontium is crimson red, and barium is green. The strontium flame produces the familiar red color of flares and fireworks (Figure 7.12).

The label attached to each family in the periodic table has either an A or B designation.* Elements of the same group number but different letter labels may share certain characteristics, but in general there is not a close connection. Let's consider the distinction between A and B elements in terms of electron configuration. As an example, locate the group 1A and 1B elements on the periodic table. In both families of group 1 elements the outer electron configuration has a single $s$ electron. In the case of the alkali metals (group 1A), this $s$ electron is outside a stable noble-gas core of electrons. For example, the single $4s$ electron in potassium is outside a set of eight electrons associated with filled $3s$ and $3p$ subshells, $3s^23p^6$. In the corresponding group 1B element, copper, the single electron is outside filled $3s$, $3p$, and $3d$ subshells, $3s^23p^63d^{10}$. The presence of the additional ten $d$ electrons in copper, and of course the increase of ten in nuclear charge that goes along with them, has a profound effect on the chemical behavior of the single $s$ electron. Because these $d$ electrons only partially shield the nucleus, the $s$ electron of the 1B group element experiences a larger effective nuclear charge than does the $s$ electron of the 1A group. As a result, the 1B elements have higher ionization energies

## The Alkaline Earth Metals

**FIGURE 7.12** The red color of the fireworks and flares is produced by strontium salts. (Peter Sarula/Photo Researchers)

## Comparison of A and B Groups

---

* The designation of groups as "A" or "B" is purely arbitrary. In fact, some periodic tables use a different system from that used in this text; however, the system we use is presently the most common one for designating groups. (See the footnote in Section 2.4, p. 42.)

**TABLE 7.5** Comparative values of ionization energies for group 1A, 1B, 2A, and 2B elements (kJ/mol)

| 1A | $I_1$ | 1B | $I_1$ | 2A | $I_1$ | $I_2$ | 2B | $I_1$ | $I_2$ |
|----|-------|----|-------|----|-------|-------|----|-------|-------|
| K  | 418   | Cu | 859   | Ca | 589   | 1145  | Zn | 907   | 1733  |
| Rb | 403   | Ag | 730   | Sr | 549   | 1064  | Cd | 867   | 1630  |
| Cs | 375   | Au | 890   | Ba | 503   | 965   | Hg | 994   | 1805  |

than do the corresponding 1A elements. Table 7.5 lists the ionization energies for the group 1 and 2 elements. Compare the values for the A and B groups.

Because the 1B members have a much stronger hold on their outer-shell electron, these metals are much less reactive than are the 1A metals. The 1B metals, often referred to as the *coinage metals*, can be found in elemental form in nature and are widely used in jewelry and coins (Figure 7.13). They are among the most chemically unreactive of the metals.

**FIGURE 7.13** Some of the silver and gold coins pictured here date from before the time of Christ. Their condition attests to the low chemical reactivity of these elements. (Paolo Koch/Photo Researchers)

# *FOR REVIEW*

## SUMMARY

Our major concern in this chapter has been the relationship between electron configurations and the properties of atoms, especially as organized by the periodic table. We saw that the electron configurations of many-electron atoms can be written by placing electrons into orbitals in the following order:

$$1s, 2s, 2p, 3s, 3p, 4s, 3d, 4p, \ldots$$

Subshells with a given principal quantum number, such as the $3s$, $3p$, and $3d$ subshells, do not have the same energies. This fact can be understood in terms of the **effective nuclear charge** and the average distance of an electron from the nucleus in each of these subshells.

The **Pauli exclusion principle** places a limit of two on the number of electrons that may occupy any one atomic orbital. These two electrons differ in their **electron-spin quantum number**, $m_s$. As the electrons populate orbitals of equal energy, they do not pair up until each orbital contains one electron; this observation is called **Hund's rule**. Using the relative energies of the orbitals, the Pauli exclusion principle, and Hund's rule, it is possible to write the electron configuration of any atom. When we do so, we see that the elements in any given family in the periodic table have the same type of electron arrangements in their outermost shells. For example, the electron configurations of fluorine and chlorine, which are both members of the halogen family, are $[He]2s^22p^5$ and $[Ne]3s^23p^5$, respectively. This periodicity in electron configurations, summarized in Figure 7.5, permits us to write the electron configuration of an element from its position in the periodic table.

Many properties of atoms that have chemical significance exhibit periodic character. Among the most important of these are **atomic radii, ionization energy**, and **electron affinity**. Electron configurations and the periodic table help us to understand trends in these properties. We also illustrated how electron configurations help us to organize and understand some of the chemistry of the alkali and alkaline earth metals.

# LEARNING GOALS

Having read and studied this chapter, you should be able to:

1. List the factors that determine the energy of an electron in a many-electron atom.

2. Explain the fact that electrons with the same value of principal quantum number ($n$) but differing values of the azimuthal quantum number ($l$) possess different energies.

3. Explain the concepts of effective nuclear charge and the screening effect as they relate to the energies of electrons in atoms.

4. State the Pauli exclusion principle and Hund's rule, and illustrate how they are used in writing the electronic structures for the elements.

5. Define the term "group" or "family" in terms of electron configuration.

6. Describe the various blocks of elements in the periodic table in terms of the type of orbital being occupied by electrons in that block ($s$, $p$, $d$, and $f$ blocks).

7. List the names and give the locations in the periodic table for the active metals ($s$-block), representative elements ($s$- and $p$-blocks), transition metals ($d$-block), and inner transition metals ($f$-block).

8. Write the electron configuration for any element once you know its place in the periodic table.

9. Write the orbital diagram representation for electron configurations of atoms.

10. Explain the effect of increasing nuclear charge on the radial density function in many-electron atoms.

11. Explain the variations in atomic radii among the elements, as shown in Figures 7.7 and 7.8; predict the relative sizes of atoms based on their positions in the periodic table.

12. Explain the general variations in first ionization energies among the elements, as shown in Figure 7.9, and relate these variations to variations in atomic radii.

13. Explain the observed changes in values of the successive ionization energies for a given atom.

14. Explain the variations in electron affinities among the elements.

15. Relate the properties of the alkali metals and alkaline earths to their electron configurations.

16. Explain the differences in chemical activity between the group 1A and 2A elements and the group 1B and 2B elements based on their electron configurations.

# KEY TERMS

Among the more important terms and expressions used for the first time in this chapter are the following:

The **active metals** (Section 7.4), groups 1A and 2A, are those in which electrons occupy only the *s* orbitals of the valence shell.

The **effective nuclear charge** (Section 7.1) is the net positive charge experienced by an electron in a many-electron atom. This charge is not the full nuclear charge, because there is some shielding of the nuclear charge by other electrons in the atom. How much shielding occurs depends on the average distance of the electron from the nucleus, compared with the other electrons in the atom.

The **electron affinity** (Section 7.8) is the energy change that occurs when an electron is added to a gaseous atom or ion.

An **electron configuration** (Section 7.3) is a particular arrangement of electrons in the orbitals of an atom.

**Electron spin** (Section 7.2) is a property of the electron that makes it behave as though it were a tiny magnet. Associated with the electron spin is a **spin quantum number** ($m_s$), which may have values of $+\frac{1}{2}$ or $-\frac{1}{2}$.

**Hund's rule** (Section 7.3) states that electrons must occupy degenerate orbitals one at a time and with their spins parallel. All orbitals must have at least one electron before pairing of electrons in the orbitals occurs. Note carefully that the rule applies only to orbitals that are **degenerate**, which means that they have the same energy.

The **inner transition elements** (Section 7.4), or **lanthanides** and **actinides**, are those in which the 4*f* or 5*f* orbitals are partially occupied.

The **Pauli exclusion principle** (Section 7.2) states that no two electrons in an atom may have the same four quantum numbers—$n$, $l$, $m_l$, and $m_s$. As a consequence of this principle, there can be no more than two electrons in any one atomic orbital.

The **representative elements** (Section 7.4) are those in which the *s* and *p* orbitals are partially occupied.

The **screening effect** (Section 7.1) is the effect of inner electrons in decreasing the nuclear charge experienced by outer electrons.

**Transition elements** (Section 7.4) are those in which the *d* orbitals are partially occupied.

**Valence-shell electrons** (Section 7.3) are the electrons in the outermost shell of an atom.

# EXERCISES

## Energies of Orbitals

**7.1** What quantum numbers must be the same in order that the orbitals be degenerate (have the same energy) **(a)** in a hydrogen atom and **(b)** in a many-electron atom?

**7.2** Within a given shell, how do the energies of the *s*, *p*, *d*, and *f* subshells compare for a many-electron atom? How do the energies of the orbitals of a given subshell compare?

**7.3** Explain why the effective nuclear charge experienced by a 3*s* electron in magnesium is larger than that experienced by a 3*s* electron in sodium.

**7.4** Suppose that we could follow the forces acting on an individual electron in its motion in a many-electron atom. What changes occurs in the effective nuclear charge as an electron moves closer to the nucleus? Explain.

**7.5** Which of the electrons in each of the following sets experiences the larger effective nuclear charge in a many-electron atom: **(a)** 2*s*, 2*p*; **(b)** 3*d*, 4*d*; **(c)** 3*d*, 2*p*?

**7.6** Which of the electrons in each of the following sets experiences the larger effective nuclear charge: **(a)** a 3*s* and a 3*p* electron of Cl; **(b)** a 3*p* electron of Cl or a 3*p* electron of S?

## Electron Spin; the Pauli Principle

**7.7** What is the maximum number of electrons that can occupy each of the following subshells: **(a)** 4*d*; **(b)** 4*f*; **(c)** 5*f*; **(d)** 2*p*?

**7.8** What is the maximum number of electrons in an atom that can have the following quantum numbers: **(a)** $n = 3$; **(b)** $n = 4$, $l = 2$; **(c)** $n = 4$, $l = 3$, $m_l = 2$; **(d)** $n = 2$, $l = 1$, $m_l = 0$, $m_s = \frac{1}{2}$?

**7.9** List the possible values of the four quantum numbers for each electron in the lithium atom.

**7.10** List the possible values of the four quantum numbers for each electron in the beryllium atom.

**7.11** Indicate whether each of the following statements is true or false. If the statement is false, correct it. **(a)** The $m_l$ value for an electron in a $5d$ orbital can have any value less than 5. **(b)** As many as nine electrons with the principal quantum number $n = 3$ can have the $m_s$ value of $+\frac{1}{2}$.

**7.12** Indicate whether each of the following statements is true or false. If the statement is false, correct it. **(a)** The set of quantum numbers $n = 3$, $l = 0$, $m_l = -2$, $m_s = -\frac{1}{2}$ is allowable. **(b)** As a general rule, the maximum number of electrons that can occupy all the orbitals of major quantum number $n$ is $2n^2$.

## Electron Configurations of the Elements

**7.13** Write the electron configurations for the following atoms using the appropriate noble-gas inner core for abbreviation: **(a)** Ca; **(b)** Ge; **(c)** Br; **(d)** Co; **(e)** Eu; **(f)** Ti.

**7.14** Write the electron configurations for the following atoms using the appropriate noble-gas inner core for abbreviation: **(a)** Fe; **(b)** Cs; **(c)** Ga; **(d)** Cr; **(e)** Pb; **(f)** Lu.

**7.15** Draw the orbital diagrams for the electrons beyond the appropriate noble-gas inner core for each of the following elements: **(a)** As; **(b)** Mn; **(c)** Sn; **(d)** Lu; **(e)** Te.

**7.16** Using orbital diagrams, determine the number of unpaired electrons in each of the following atoms: **(a)** Ge; **(b)** Ni; **(c)** B; **(d)** Kr; **(e)** In.

**7.17** In which family or group of the periodic table are the elements with the following electron configurations found: **(a)** $1s^2 2s^2$; **(b)** $[\text{Ne}]3s^2 3p^5$; **(c)** $[\text{Ar}]4s^1$; **(d)** $[\text{Ar}]3d^{10}4s^1$?

**7.18** Identify the specific element or group of elements that can have the following electron configurations: **(a)** $1s^2 2s^2 2p^4$; **(b)** $[\text{Ar}]4s^1 3d^{10}$; **(c)** [noble gas]$ns^2(n-1)d^{10}np^3$; **(d)** $[\text{Kr}]5s^2 4d^2$; **(e)** $1s^2 2s^2 2p^6$.

**7.19** On the basis of electron configurations, indicate the number of **(a)** unpaired electrons on an atom of P; **(b)** valence-shell $p$ electrons on an atom of Cl; **(c)** $3d$ electrons on an atom of Fe; **(d)** total valence-shell electrons on an atom of N.

**7.20** Give the symbol for the first element in the periodic table that has **(a)** a completed $3d$ subshell; **(b)** a completed second shell; **(c)** two $p$ electrons; **(d)** two $3d$ electrons; **(e)** three unpaired electrons; **(f)** a completed $4f$ subshell.

## Periodicity and Atomic Properties

**7.21** How do the following properties vary as we move from left to right across the periodic table: **(a)** effective nuclear charge; **(b)** atomic size; **(c)** ionization energy; **(d)** electron affinity?

**7.22** How do the following properties vary as we move down a group in the periodic table: **(a)** principal quantum number of valence electrons; **(b)** atomic size; **(c)** ionization energy; **(d)** electron affinity?

**7.23** List the following elements in order of increasing atomic radius: Mg, C, Kr, S, K, Cl, Co.

**7.24** Arrange the members of each of the following groups of atoms in order of increasing atomic radius: **(a)** Mg, Ca, Sr; **(b)** Al, Si, P; **(c)** O, F, S; **(d)** Ne, Na, Mg.

**7.25** Based on their positions in the periodic table, select the atom with the larger ionization energy from each of the following pairs: **(a)** B, Cl; **(b)** N, P; **(c)** Hf, Cs; **(d)** O, N; **(e)** Ga, Ge.

**7.26** In each of the following pairs, indicate which element has the larger ionization energy: **(a)** P, Cl; **(b)** Al, Ga; **(c)** Cs, La; **(d)** La, Hf. In each case provide an explanation in terms of electron configuration and effective nuclear charge.

**7.27** Although $SiO_2$—which can be thought of as containing $Si^{4+}$ ions—is common, $AlO_2$ is not a known compound. Explain this fact in terms of the data in Table 7.3.

**7.28** Compare the effective nuclear charges operating on the $3s$ electrons in $Mg^+$ and $Al^{2+}$. What evidence for your response can be found in Table 7.3?

**7.29** The electron affinity of chlorine is strongly negative; that is, addition of an electron to Cl is an exothermic process. On the other hand, addition of an electron to Ar is an endothermic process. Account for the difference in terms of the electron configurations of the two elements.

**7.30** Addition of an electron to $Na(g)$ is a slightly exothermic process, whereas addition of an electron to $Mg(g)$ is strongly endothermic. Explain this difference in terms of the ground-state electron configurations of the two elements.

## Group 1A and 2A Elements

**7.31** Compare the elements potassium and calcium with respect to the following properties: **(a)** electron configuration; **(b)** typical ionic charge; **(c)** first ionization energy; **(d)** formula of hydroxide; **(e)** melting point (see Tables 5.6 and 5.7); **(f)** atomic radius; **(g)** electron affinity. Account for the differences in the two elements.

**7.32** Compare the elements rubidium and silver with respect to the following properties: **(a)** electron configuration; **(b)** typical ionic charge; **(c)** first ionization energy; **(d)** formula of chloride; **(e)** melting point (see Figure 5.8); **(f)** atomic radius. Account for the differences in the two elements.

**7.33** Write a balanced chemical equation for the reaction that occurs in each of the following cases. **(a)** Potassium is added to water. (**b**) Barium is added to water. (**c**) Lithium is heated in nitrogen. (**d**) Sodium vapor reacts with bromine vapor.

**7.34** Write a balanced chemical equation for the reaction that occurs in each of the following cases. (**a**) Potassium burns in air. (**b**) Hydrogen gas is bubbled through molten sodium. (**c**) Strontium oxide is added to water. (**d**) Rubidium is added to water.

## Additional Exercises

**7.35** How does the average distance from the nucleus of an electron in a $2s$ orbital of neon compare with that for a $2p$ electron? Explain.

**7.36** In a phosphorus atom, which electrons experience the largest effective nuclear charge? Which experience the lowest?

**7.37** Equation [7.1] can be modified as follows for a many-electron atom: $E_n = -(Z_{eff}^2/n^2)(1312 \text{ kJ/mol})$. Using this equation together with first ionization energies from Table 7.3, calculate the effective nuclear charges experienced by $3s$ electrons in sodium and magnesium atoms. Explain the difference. Are the values reasonable?

**7.38** The quantum numbers listed below are for four different electrons in an atom. Arrange them in order of increasing energy. If any two are of the same energy, so indicate. (**a**) $n = 4, l = 0, m_l = 0, m_s = \frac{1}{2}$; (**b**) $n = 3, l = 2, m_l = 1, m_s = \frac{1}{2}$; (**c**) $n = 3, l = 2, m_l = -2, m_s = -\frac{1}{2}$; (**d**) $n = 3, l = 1, m_1 = 1, m_s = -\frac{1}{2}$.

**7.39** Which of the following electron configurations are not allowed by the Pauli exclusion principle? Explain why. (**a**) $1s^2 2s^2 2p^8$; (**b**) $[Ne]3s^2 4s^1$; (**c**) $[Ar]3d^{10} 4s^3$.

**7.40** The electron configurations that we have written in this chapter are ground-state configurations. An atom can absorb energy to promote one or more electrons to higher-energy levels (excited states). The following configurations are excited states of neutral atoms. In each case identify the element, write its ground-state electron configuration, and indicate the electron that has been excited. (**a**) $1s^2 2s^1 3s^1$; (**b**) $[Ne]3s^2 3p^3 4p^1$; (**c**) $[Ar]3d^4 4s^2 5s^1$; (**d**) $[Kr]4f^1 6s^2$.

**7.41** Write the complete electron configuration for the following atoms without referring to any data other than the periodic table: (**a**) In; (**b**) Se; (**c**) La; (**d**) Ag; (**e**) Cl. Do not use the noble-gas inner core abbreviation, and list the orbitals in the order in which they are filled.

**7.42** Without looking at Table 7.2, write the electron configurations for the following elements: (**a**) Ne; (**b**) As; (**c**) Mo; (**d**) Be; (**e**) Pd. When you have completed writing the electron configurations, compare your results with those in Table 7.2. Any differences should be readily explained in terms of the near-degeneracy of different subshells.

**7.43** Name the element (or elements) that fits each of the following descriptions: (**a**) alkaline earth metal with two

$5s$ valence-shell electrons; (**b**) element with the most negative electron affinity; (**c**) element with a half-filled $4p$ subshell; (**d**) element with the lowest ionization energy; (**e**) all the elements that possess two unpaired $3p$ electrons.

**7.44** Explain why the radii of atoms do not increase uniformly as the atomic number of the atom increases.

**7.45** Are there any atoms for which the second ionization energy ($I_2$) is smaller than the first? Explain.

**7.46** Although the first ionization energy for Na is lower than that of Mg, the second ionization energy of Mg is much lower than that of Na. Explain this in terms of the electron configuration for each element.

**7.47** Arrange the following elements in terms of (**a**) increasing ionization energy and (**b**) increasing atomic radius: O, Se, C, Si, F.

**7.48** The successive ionization energies for boron are 0.801, 2.427, 3.670, 25.026, and 32.827 MJ/mol. (**a**) Why do the values increase with successive ionization? (**b**) Why is there such a large change in ionization energy in moving from the third to the fourth ionization energy?

**7.49** Arrange O, F, and Ne in order of increasing effective nuclear charge.

**7.50** Atomic radii normally increase going down a group in the periodic table. Suggest a reason why hafnium breaks this rule, as shown in the following data:

| | *Atomic* | *radii* (Å) | |
|---|---|---|---|
| Sc | 1.57 | Ti | 1.477 |
| Y | 1.693 | Zr | 1.593 |
| La | 1.915 | Hf | 1.476 |

**7.51** Going down a group in the periodic table, the change in electron affinity is not nearly as great as the change in first ionization energy. Explain.

**7.52** The electron affinities of F and of the $O^-$ ion are given below:

$$F(g) + e^- \longrightarrow F^-(g) \qquad \Delta H = -332 \text{ kJ/mol}$$
$$O^-(g) + e^- \longrightarrow O^{2-}(g) \qquad \Delta H = +710 \text{ kJ/mol}$$

What is the essential difference in these two processes, and how does it account for the difference in the two energy changes?

**7.53** Explain, in terms of electron configurations, why hydrogen exhibits properties similar to both Li and F.

**7.54** The electron affinity of copper is $-123$ kJ/mol. Compare this number with that for potassium, and account for the difference. The electron affinity for zinc is not known. Predict roughly what value it should have.

**[7.55]** What is the lowest value of the principal quantum number $n$ for which there can be a $g$ subshell? Are there any elements known whose ground-state electron configuration contain electrons in the $g$ subshell?

**[7.56]** Where should you go to look up the electron affinities of the singly charged positive ions of the elements? Explain.

[7.57] Make a graph of all the second ionization energies, $I_2$, listed in Table 7.3, as a function of atomic number. Account for the general trend in the series from Mg through Ar. Account for the exceptionally large value for Na. Suggest why the value for Al is larger than one might expect from the general trend.

[7.58] The element technetium, atomic number 43, is not observed in nature because it is radioactive. Assuming that it can be synthesized in a nuclear reactor in quantity, predict some of its characteristic properties. These should include electron configuration, density, melting point, and the formula of oxides. Consult a handbook of chemistry for information on related elements.

[7.59] There are certain similarities in properties that exist between the first member of any periodic family and the element located below it and to the right in the periodic table. For example, in some ways Li resembles Mg, Be resembles Al, and so forth. This observation is called the diagonal relationship. In terms of what we have learned in this chapter, offer a possible explanation for this relationship.

## CHAPTER 8

# Basic Concepts
# of Chemical Bonding

Deep within the earth below the city of Detroit and beneath the rolling plains of Kansas lie enormous deposits of the white mineral halite (see Figure 8.1). This substance, also known as sodium chloride, NaCl, was deposited in these and other places millions of years ago, when extensive primordial seas dried upon the changing surface of the earth. Sodium chloride is the most abundant dissolved substance present in seawater and is found in human body tissues in large quantities. However, it is most familiar to us as ordinary table salt. This substance consists of sodium ions and chloride ions, $Na^+$ and $Cl^-$.

Water, $H_2O$, is another very abundant substance. We drink it, swim in it, and use it as a cooling agent. It is essential to life as we know it. This substance is composed of molecules.

Why are some substances composed of ions while others are composed of molecules? The key to this question is found in the electronic structures of the atoms involved and in the nature of the chemical forces within the compounds. In this chapter and the next we shall examine the relationships between electronic structure, chemical bonding forces, and the properties of substances. As we do this, we shall find it useful to classify chemical forces into three broad groups: (1) ionic bonds, (2) covalent

**FIGURE 8.1**  One million tons of salt awaiting shipment near a saltworks. (© Georg Gerster/Photo Researchers)

The *Voyager* is a testimony to the contributions that chemistry has made to modern materials science. Most of the components of this aircraft were unavailable just a decade or so ago. These new materials owe their special properties to the covalent bonds that define the shapes and stabilities of their constituent molecules. (© Mark Greenberg/Visions, 1986)

(a)        (b)        (c)

**FIGURE 8.2**   Examples of substances in which (*a*) ionic, (*b*) covalent, and (*c*) metallic bonds are found. (Donald Clegg and Roxy Wilson)

bonds, and (3) metallic bonds. Figure 8.2 shows examples of substances in which we find these types of bonds.

The term **ionic bond** refers to the electrostatic forces that exist between particles of opposite charge. As we shall see, ions may be formed from atoms by transfer of one or more electrons from one atom to another. Ionic substances generally result from the interaction of metals from the far left side of the periodic table with the nonmetallic elements from the far right side (excluding the rare gases, group 8A).

The **covalent bond** results from a sharing of electrons between two atoms. The familiar examples of covalent bonding are seen in the interactions of nonmetallic elements with one another.

**Metallic bonds** are found in solid metals such as copper, iron, and aluminum. In the metals, each metal atom is bonded to several neighboring atoms. The bonding electrons are relatively free to move throughout the three-dimensional structure. Metallic bonds give rise to such typical metallic properties as high electrical conductivity and luster. We will postpone further discussion of metallic bonding until Chapter 24.

## 8.1  LEWIS SYMBOLS AND THE OCTET RULE

The term **valence** is commonly used in discussions of both ionic and covalent bonding. The valence of an element is a measure of its capacity to form chemical bonds. Originally, it was determined by the number of hydrogen atoms with which an element combined. Thus the valence of oxygen is 2 in $H_2O$; in $CH_4$ the valence of carbon is 4.

We speak of **valence electrons** when referring to the electrons that take part in chemical bonding. These electrons are the ones residing in the outermost electron shell of the atom, the valence shell. **Electron-dot symbols** (also known as **Lewis symbols**, after G. N. Lewis) are a simple and convenient way of showing the valence electrons of atoms and keeping track of them in the course of bond formation. The electron-dot symbol for an element consists of the chemical symbol for the element plus a dot for each valence electron. For example, sulfur has the electron configuration $[Ne]3s^23p^4$; its electron-dot symbol therefore shows six valence electrons:

$$\cdot\ddot{\underset{\cdot\cdot}{S}}\cdot$$

Other examples are shown in Table 8.1.

The number of valence electrons of any representative element is the same as the column number of the element in the periodic table. Thus the electron-dot symbols for both oxygen and sulfur, members of family 6A, show six dots.

The way atoms gain, lose, or share electrons can often be viewed as an attempt on the part of the atoms to achieve the same number of electrons as the noble gas closest to them in the periodic table. The noble gases, you will recall, have very stable electron arrangements, as evidenced by their high ionization energies, low affinity for additional electrons, and general lack of chemical reactivity. Because all noble gases (except He) have eight valence electrons, many atoms undergoing reactions also end up with eight valence electrons. This observation has led to what is known as the **octet rule**. Of course, because He has only two electrons, atoms near it in the periodic table, such as H, will generally tend to obtain an arrangement of two electrons. As we shall see, there are many exceptions to the octet rule. Nevertheless, it provides a useful framework for introducing many important concepts of bonding.

**TABLE 8.1**  Electron-dot symbols

| Element | Electron configuration | Electron-dot symbol |
|---------|------------------------|---------------------|
| Li | $[He]2s^1$ | Li$\cdot$ |
| Be | $[He]2s^2$ | $\cdot$Be$\cdot$ |
| B | $[He]2s^22p^1$ | $\cdot\dot{B}\cdot$ |
| C | $[He]2s^22p^2$ | $\cdot\dot{\underset{\cdot}{C}}\cdot$ |
| N | $[He]2s^22p^3$ | $\cdot\dot{\underset{\cdot}{N}}:$ |
| O | $[He]2s^22p^4$ | $:\dot{\underset{\cdot}{O}}:$ |
| F | $[He]2s^22p^5$ | $\cdot\ddot{\underset{\cdot\cdot}{F}}:$ |
| Ne | $[He]2s^22p^6$ | $:\ddot{\underset{\cdot\cdot}{Ne}}:$ |

*know how to draw them*

## 8.2 IONIC BONDING

When sodium metal is brought into contact with chlorine gas, $Cl_2$, a violent reaction ensues (see Figure 8.3). The product of that reaction is sodium chloride, NaCl, a substance composed of $Na^+$ and $Cl^-$ ions:

$$2Na(s) + Cl_2(g) \longrightarrow 2NaCl(s)$$

These ions are arranged throughout the solid NaCl in a regular three-dimensional array, as shown in Figure 8.4.

The formation of $Na^+$ from Na and of $Cl^-$ from $Cl_2$ indicates that an

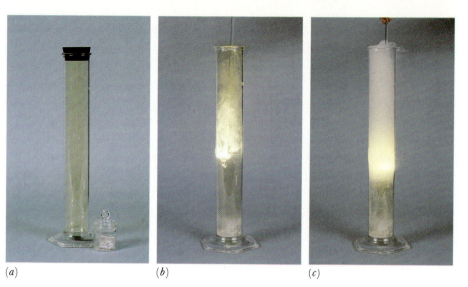

**FIGURE 8.3**  The reaction between sodium metal and chlorine gas to form sodium chloride. (*a*) A container of chlorine gas (left) and sodium metal (right). (*b*) Formation of NaCl begins as sodium is added to the chlorine. (*c*) The reaction a few minutes later. (Donald Clegg and Roxy Wilson)

(*a*)　　　　　(*b*)　　　　　(*c*)

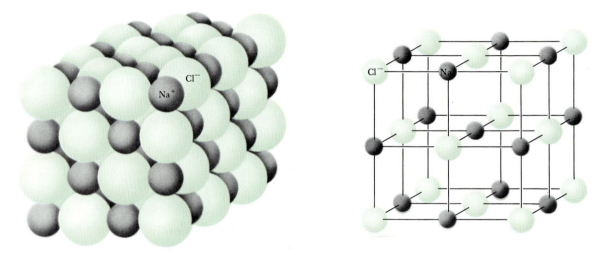

**FIGURE 8.4**  Two ways of representing the crystal structure of sodium chloride. The structure on the left shows the ions in their correct sizes relative to the distances between them. The larger spheres represent the chloride ions. The structure on the right emphasizes the symmetry of the structure, whereas the one on the left better illustrates the way the ions are packed together in the solid.

electron has been lost by a sodium atom and gained by a chlorine atom. Such electron transfer to form oppositely charged ions occurs when the atoms involved differ greatly in their attraction for electrons. Our example of NaCl is rather typical for ionic ompounds; it involves a metal of low ionization energy and a nonmetal with a high affinity for electrons. Using electron-dot symbols (and showing a chlorine atom rather than the $Cl_2$ molecule) we can represent this reaction as follows:

$$Na\cdot + \cdot \ddot{\underset{\cdot\cdot}{Cl}}: \longrightarrow Na^+ + [:\ddot{\underset{\cdot\cdot}{Cl}}:]^-$$

Each ion has an octet of electrons, the octet on $Na^+$ being the $2s^2 2p^6$ electrons that lie below the single $3s$ valence electron of the Na atom.

The formation of ionic compounds is not merely the result of the low ionization energies of metals and the high affinities of nonmetals for electrons, although these factors are very important. The formation of an ionic compound from the elements is always an exothermic process; the compound forms because it is more stable (lower in energy) than its elements. Much of the stability of NaCl results from the packing of the oppositely charged $Na^+$ and $Cl^-$ ions together, as shown in Figure 8.4. A measure of just how much stabilization results from this packing is given by the **lattice energy**. This quantity is the energy required for 1 mol of the solid ionic substance to be separated completely into ions far removed from one another. To get a picture of this process for NaCl, imagine that the lattice shown in Figure 8.4 expands from within, so that the spaces between the ions grow larger and larger, until the ions are very far apart. We can write the process as

$$NaCl(s) \longrightarrow Na^+(g) + Cl^-(g) \qquad [8.1]$$

The energy that would be required for this to occur for a lattice containing 1 mol $Na^+$ and 1 mol $Cl^-$ of NaCl is the lattice energy.

The lattice energy for NaCl(s) amounts to 785 kJ/mol. This is a very large amount of energy; it accounts for the fact that sodium chloride is a stable, solid substance with a high melting point. We see from the structure of NaCl (Figure 8.4) that each sodium ion is surrounded by six nearest-neighbor chloride ions of opposite charge. Similarly, each chloride ion is surrounded by six sodium ions. The attractive force between each ion and its nearest neighbors of opposite charge provides much of the stabilizing lattice energy. Furthermore, each ion also experiences repulsive interactions with ions of like charge in the lattice and is attracted to other ions of opposite charge in addition to its nearest neighbors. The lattice energy is the result of all such electrostatic interactions, taken over the entire lattice.

Ionic compounds may have arrangements of ions that differ from that shown in Figure 8.4. Two other examples are shown in Figure 8.5. The overall structure of an ionic compound depends on the charges of the ions and their relative sizes. However, in every case the forces between ions lead to hard, brittle solids with high melting points.

The charges and sizes of ions are the principal factors that determine the magnitude of the lattice energy of an ionic solid. The potential energy of two interacting charges is given by

$$E = k \frac{Q_1 Q_2}{d} \qquad [8.2]$$

$Q_1$ and $Q_2$ are the magnitudes of the charges on the particles in coulombs and $d$ is the distance between their centers in meters. The constant $k$ has the value $8.99 \times 10^9$ J-m/C$^2$. As Equation 8.2 indicates, the attractive interaction between two oppositely charged ions increases as the magnitudes of their charges increase and as the distance between their centers decreases. Thus for a given arrangement of ions, the lattice energy increases as the charges on the ions increase and as their radii decrease. (The decrease in radii allows ions to approach more closely to one another.)

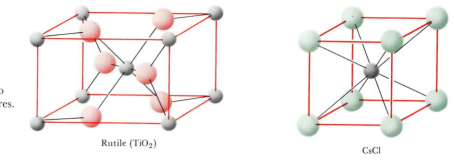

**FIGURE 8.5** Examples of two important types of crystal structures. In these illustrations, the color spheres represent the anions, the gray spheres the cations.

Rutile (TiO₂)

CsCl

---

**SAMPLE EXERCISE 8.1**

Which substance would have the higher lattice energy, NaF or MgO? Explain.

**Solution:** Magnesium oxide, MgO, would have the higher lattice energy. The electrostatic attraction between oppositely charged ions increases with the charge on the ion. For this reason, it would require a greater amount of energy to separate a mole of $Mg^{2+}$ ions and a mole of $O^{2-}$ ions to infinite separation, as compared with separating a mole of $Na^+$ and a mole of $F^-$. In this case, the effect of larger charge is much more important than any slight differences in ionic radii in the two compounds.

**PRACTICE EXERCISE**

Which substance would you expect to have the greater lattice energy, FeO or $Fe_2O_3$?
***Answer:*** $Fe_2O_3$

---

Now if lattice energy increases as the charges of the ions increase, why doesn't sodium lose two electrons to form $Na^{2+}$? The second electron would have to come from the inner shell of the sodium atom. We can see from the large value for the second ionization energy of Na (Table 7.3) that it would require too much energy to form the $Na^{2+}$ ion in chemical compounds. Thus sodium and the other group 1A metals are found in ionic substances only as 1+ ions. Similarly, addition of a second electron to a chloride ion to form a hypothetical $Cl^{2-}$ is never observed. The other group 7A elements (the halogens) are also found in ionic compounds only as the 1− ions $F^-$, $Br^-$, or $I^-$, which possess noble-gas configurations.

Magnesium, an element of group 2A, also forms an ionic compound with chlorine, $MgCl_2$. In this instance, the metal achieves a noble-gas configuration by loss of two electrons, forming $Mg^{2+}$. It requires much less energy to remove two electrons from Mg (Table 7.3) than from Na because both Mg electrons are in its valence shell. Of course, removing two electrons requires more energy than removing just one. This energy is more than recovered, however, in the increased lattice energy of $MgCl_2$, which comes from the higher charge on the metal ion. Similarly, the group 6A elements form $O^{2-}$, $S^{2-}$, and so forth, in which the ion possesses a noble-gas configuration. In the formation of MgO, both magnesium and oxygen attain the noble-gas configuration by the transfer of two electrons:

$$Mg + :\ddot{O}: \longrightarrow Mg^{2+} + \left[:\ddot{\underset{..}{O}}:\right]^{2-}$$

Thus the balance between ionization energies and lattice energies determines the magnitude of the positive charge on a metal ion. Similarly, the balance between electron affinities and lattice energies determines the magnitude of the negative charge on nonmetal ions.

## SAMPLE EXERCISE 8.2

Predict the formula of the compound formed between aluminum and fluorine; between aluminum and oxygen.

**Solution:** Aluminum, with atomic number 13, has three electrons beyond the inert-gas configuration. We expect it to lose three electrons to form the $Al^{3+}$ ions. Each fluorine atom attains the noble-gas configuration by accepting one electron to form $F^-$. As we saw in Chapter 2, the formula of an ionic compound depends on the charges on the ions. The formula may be derived by the principle of electroneutrality: The total charge of the cations must balance that of the anions. Three $F^-$ ions are required to balance the charge of one $Al^{3+}$ ion; the expected formula is thus $AlF_3$.

In forming a compound with oxygen, each aluminum atom again loses three electrons to form $Al^{3+}$. Each oxygen atom accepts two electrons to form $O^{2-}$, thereby achieving the noble-gas configuration (an octet of valence-shell electrons). Two $Al^{3+}$ balance the charge of three $O^{2-}$; the formula for aluminum oxide is therefore $Al_2O_3$.

### PRACTICE EXERCISE
Give the formula and name of the compound formed between barium and bromine.
***Answer:*** $BaBr_2$; barium bromide

Ionic bond theory correctly predicts the charges found on many simple ions, based on the notion that attainment of a noble-gas configuration leads to maximum stability. For some elements, however, the rule must be modified, and for others it is not applicable at all. For example, metals of group 1B (Cu, Ag, Au) are observed to occur often as the 1+ ions (as in CuBr and AgCl). Silver possesses a $4d^{10}5s^1$ outer electron configuration. In forming $Ag^+$ the 5s electron is lost. This leaves a completely filled shell of 18 electrons in the $n = 4$ level. Because it is a completed shell, it is somewhat like a noble-gas arrangement. Similarly, the group 2B elements most commonly are seen as the 2+ ions ($Zn^{2+}$, $Cd^{2+}$, $Hg^{2+}$) in ionic compounds. The valence-shell s electrons are lost in forming the ions, leaving an electronic arrangement consisting of 18 electrons in the highest occupied level.

For most of the transition metals, the attainment of a noble-gas configuration by loss of electrons is not feasible; that would require the loss of too many electrons. The outer electron configurations of these elements are either $(n - 1)d^x ns^2$ or $(n - 1)d^x ns^1$, where n is 4, 5, or 6, and x may vary from 1 to 10. *In forming ions the transition metals lose the valence-shell s electrons first, then as many d electrons as are required to form an ion of particular charge.* Most of the transition metals are found in more than one charge state. For example, the element chromium is found in compounds as $Cr^{2+}$ or $Cr^{3+}$. No simple rules tell which charge state of a transition-metal ion will exist in a particular case.

**Transition-Metal Ions**

## SAMPLE EXERCISE 8.3

Write the electron configuration for the $Co^{2+}$ ion and for the $Co^{3+}$ ion.
**Solution:** Cobalt (atomic number 27) has an electron configuration $[Ar]4s^2 3d^7$. To form a 2+ ion, two electrons must be removed. As discussed in the text above, the 4s electrons are removed before the

$3d$. Consequently, the $Co^{2+}$ ion has an electron configuration of $[Ar]3d^7$. To form $Co^{3+}$ requires the removal of an additional electron; the electron configuration for this ion is $[Ar]3d^6$.

This is a good point at which to review Table 2.3, which lists common ions. Recall that positively charged ions are called **cations** and negatively charged ones are called **anions**. Notice that some ions are polyatomic. Examples of polyatomic cations include the vanadyl ion, $VO^{2+}$, and the familiar ammonium ion, $NH_4^+$. However, most polyatomic ions are anions. Examples include the carbonate ion, $CO_3^{2-}$, found in many mineral deposits, and the brightly colored yellow chromate ion, $CrO_4^{2-}$.

In polyatomic ions, two or more atoms are bound together by predominantly covalent bonds. They form a stable grouping that carries a charge, either positive or negative. We will examine the covalent bonding forces in these ions in Chapter 9. For now, you must realize only that the group of atoms as a whole acts as a charged species in forming an ionic compound with an ion of opposite charge.

**SAMPLE EXERCISE 8.4**

The dichromate ion, $Cr_2O_7^{2-}$, is readily obtained in the form of its ammonium salt. Write the formula for ammonium dichromate.

**Solution:**   The charge on the dichromate ion is $2-$; that on the ammonium ion is $1+$, $NH_4^+$. We therefore require two ammonium ions to balance the charge of the dichromate ion. The formula for the salt is thus $(NH_4)_2Cr_2O_7$.

## 8.3 SIZES OF IONS

Ionic size plays an important role in determining the structure and stability of ionic solids and the properties of ions in solution. For example, the sizes of ions are important in determining the lattice energy in an ionic solid. Often only a small difference in ionic size is sufficient for one metal ion to be biologically active and another not to be.

The size of an ion depends on its nuclear charge, the number of electrons it possesses, and the orbitals in which the outer-shell electrons reside. Consider first the relative sizes of an ion and its parent atom. Positive ions are formed by removing one or more electrons from the outermost

**TABLE 8.2**   Radii (Å) of ions with rare-gas electron configurations

| Group 1A | | Group 2A | | Group 3A, 3B | | Group 6A | | Group 7A | |
|---|---|---|---|---|---|---|---|---|---|
| $Li^+$ | 0.68 | $Be^{2+}$ | 0.30 | | | $O^{2-}$ | 1.45 | $F^-$ | 1.33 |
| $Na^+$ | 0.98 | $Mg^{2+}$ | 0.65 | $Al^{3+}$ | 0.45 | $S^{2-}$ | 1.90 | $Cl^-$ | 1.81 |
| $K^+$ | 1.33 | $Ca^{2+}$ | 0.94 | $Sc^{3+}$ | 0.68 | $Se^{2-}$ | 2.02 | $Br^-$ | 1.96 |
| $Rb^+$ | 1.48 | $Sr^{2+}$ | 1.10 | $Y^{3+}$ | 0.90 | $Te^{2-}$ | 2.22 | $I^-$ | 2.19 |
| $Cs^+$ | 1.67 | $Ba^{2+}$ | 1.31 | | | | | | |

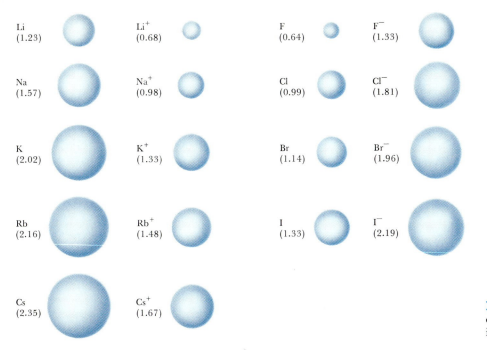

Li
(1.23)

Li⁺
(0.68)

F
(0.64)

F⁻
(1.33)

Na
(1.57)

Na⁺
(0.98)

Cl
(0.99)

Cl⁻
(1.81)

K
(2.02)

K⁺
(1.33)

Br
(1.14)

Br⁻
(1.96)

Rb
(2.16)

Rb⁺
(1.48)

I
(1.33)

I⁻
(2.19)

Cs
(2.35)

Cs⁺
(1.67)

**FIGURE 8.6** Relative sizes of atoms and ions. The values in parentheses are radii (Å).

region of the atom. This not only vacates the most spatially extended orbitals but also decreases the total electron-electron repulsions. As a consequence, *cations are smaller than their parent atoms*, as illustrated in Figure 8.6. The opposite is true of negative ions. When electrons are added to form an anion, the increased electron-electron repulsions cause the electrons to spread out more in space. Thus *anions are larger than their parent atoms*.

It is also important to note that *for ions of the same charge, size increases as we go down a family in the periodic table*. This trend is seen in Figure 8.6 and also in Table 8.2, which gives the radii for several ions with noble-gas electron configurations. As the principal quantum number of the outer occupied orbital of an ion increases, the size of both the ion and its parent atom increases.

## SAMPLE EXERCISE 8.5

Arrange the following atoms and ions in order of decreasing size: $Mg^{2+}$, $Ca^{2+}$, and Ca.

**Solution:** Cations are smaller than their parent atoms; thus $Ca^{2+}$ is smaller than the Ca atom. Ca is below Mg in group 2A of the periodic table; hence $Ca^{2+}$ is larger than $Mg^{2+}$. These observations lead

to the following order: $Ca > Ca^{2+} > Mg^{2+}$.

## PRACTICE EXERCISE
Which of the following atoms and ions is largest: $S^{2-}$, S, $O^{2-}$.
*Answer:* $S^{2-}$

The effect of varying nuclear charge on ionic radii is seen in the variation in radius in an **isoelectronic series** of ions. The term "isoelectronic" means that the ions possess the same number and arrangement of electrons. For example, each ion in the series $O^{2-}$, $F^-$, $Na^+$, $Mg^{2+}$, and

$Al^{3+}$ has ten electrons arranged in the neon electron configuration about its nucleus. The nuclear charge in this series increases steadily in the order listed. With the number of electrons remaining constant, the radius of the ion decreases as the nuclear charge increases, attracting the electrons more strongly toward the nucleus:

$$\xrightarrow{\text{Increasing nuclear charge}}$$

| $O^{2-}$ | $F^-$ | $Na^+$ | $Mg^{2+}$ | $Al^{3+}$ |
|---|---|---|---|---|
| 1.45 Å | 1.33 Å | 0.98 Å | 0.65 Å | 0.45 Å |

The charge on the nucleus of an atom or a monatomic ion, of course, is given by the atomic number of the element. Notice the positions of these particular elements in the periodic table and also their atomic numbers. The nonmetal anions precede the noble gas Ne in the table. The metal cations follow Ne. Oxygen, the largest ion in this isoelectronic series, has the lowest atomic number, 8. Aluminum, the smallest of these ions, has the highest atomic number, 13.

**SAMPLE EXERCISE 8.6**

Arrange the ions $S^{2-}$, $Cl^-$, $K^+$, and $Ca^{2+}$ in order of decreasing size.

**Solution:** This is an isoelectronic series of ions, with all ions having 18 electrons. In an isoelectronic series, size decreases as the nuclear charge (atomic number) of the ion increases. The atomic numbers of the ions are S (16), Cl (17), K (19), and Ca (20). Thus the ions decrease in size in the following order: $S^{2-} > Cl^- > K^+ > Ca^{2+}$.

**PRACTICE EXERCISE**
Which of the following ions is largest: $Rb^+$, $Sr^{2+}$, $Y^{3+}$?

*Answer:* $Rb^+$

One further comparison worth keeping in mind is the relative sizes of ions from the A and B subgroups. You may recall from the discussion in Section 7.6 that the outermost $s$ electron of the group 1B elements experiences a higher effective nuclear charge. This happens because the ten $d$ electrons added in going from a group 1A element to a group 1B element in the same row (for example, in going from K to Cu) do not completely shield the valence $s$ electron from the nucleus. They also do not completely shield one another from the nucleus. As a result, not only is the atom of the group 1B element smaller, the ion formed by removal of the valence $s$ electron is smaller also. For example, the radius of $Cu^+$, 0.96 Å, is less than that for the corresponding group 1A ion, $K^+$, radius 1.33 Å.

## 8.4 COVALENT BONDING

We have seen that ionic substances possess several characteristic properties. They are usually brittle substances with high melting points. They are usually also crystalline, meaning that the solids have flat surfaces that make characteristic angles with one another. Ionic crystals can often be cleaved; that is, they break apart along smooth, flat surfaces. The characteristics of ionic substances result from the ionic forces that maintain the ions in a rigid, well-defined, three-dimensional arrangement such as one of those illustrated in Figures 8.4 and 8.5.

The vast majority of chemical substances do not have the characteristics of ionic materials; water, gasoline, banana peelings, hair, antifreeze, and plastic bags are examples. Most of the substances with which we come in daily contact tend to be gases, liquids, or solids with low melting points; many vaporize readily—for example, mothball crystals. Many in their solid forms are plastic rather than rigidly crystalline—for example, paraffin and plastic bags.

For the very large class of substances that do not behave like ionic substances, we need a different model for the bonding between atoms. G. N. Lewis reasoned that an atom might acquire a noble-gas electron configuration by sharing electrons with other atoms. A chemical bond formed by sharing a pair of electrons is called a **covalent bond**.

The hydrogen molecule, $H_2$, furnishes the simplest possible example of a covalent bond. Using electron-dot symbols, formation of the $H_2$ molecule by combination of two hydrogen atoms can be represented as

$$H\cdot + \cdot H \longrightarrow \left(H \overset{\cdot}{\underset{\cdot}{}} H\right)$$

The shared pair of electrons provides each hydrogen atom with two electrons in its valence shell (the $1s$ orbital), so that in a sense it has the electron configuration of the rare gas helium (the shared electrons are counted with both atoms). Similarly, when two chlorine atoms combine to form the $Cl_2$ molecule, we have:

$$:\overset{\cdot\cdot}{\underset{\cdot\cdot}{Cl}}\cdot + \cdot\overset{\cdot\cdot}{\underset{\cdot\cdot}{Cl}}: \longrightarrow \left(:\overset{\cdot\cdot}{\underset{\cdot\cdot}{Cl}} \overset{\cdot}{\underset{\cdot}{}} \overset{\cdot\cdot}{\underset{\cdot\cdot}{Cl}}:\right)$$

Each chlorine atom, by sharing the bonding electron pair, acquires eight electrons (an octet) in its valence shell. It thus achieves the noble-gas electron configuration of argon. The structures shown above for $H_2$ and $Cl_2$ are called **Lewis structures**. In writing Lewis structures, we usually show each electron pair shared between atoms as a line and the unshared electron pairs as dots. Thus the Lewis structures for $H_2$ and $Cl_2$ are shown as follows:

$$H\!-\!H \qquad :\overset{\cdot\cdot}{\underset{\cdot\cdot}{Cl}}\!-\!\overset{\cdot\cdot}{\underset{\cdot\cdot}{Cl}}:$$

Of course, the shared pairs of electrons are not located in fixed positions between nuclei. Figure 8.7 shows the distribution of electron density in the $H_2$ molecule. Notice that electron density is concentrated between nuclei. The two atoms are bound into the $H_2$ molecule principally because of the electrostatic attractions of the two positive nuclei for the concentration of negative charge between them. In Chapter 9 we shall look more closely at the spatial distribution of electron density within molecules. At that time we shall treat covalent bonds in terms of orbitals. Meanwhile, our discussions shall rely primarily on Lewis structures.

In the Lewis model, the valence of an element is the number of electron pairs shared to complete the octet of electrons. Because the number of valence electrons is the same as the group number for the nonmetals, one might predict that the 7A elements, such as F, would form one covalent bond to achieve an octet; 6A elements, such as O, would form two covalent

**FIGURE 8.7** Electron distribution in the $H_2$ molecule.

bonds; 5A elements, such as N, would form three covalent bonds; and 4A elements, such as C, would form four covalent bonds. These predictions are borne out in many compounds. For example, consider the simple hydrides of the nonmetals of the second row (period) of the periodic table:

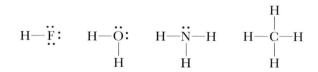

Thus the Lewis model succeeds in accounting for the compositions of compounds of the nonmetals, in which covalent bonding predominates.

## SAMPLE EXERCISE 8.7

Using the Lewis theory and the theory of ionic bonding described earlier, explain the formulas of the following hydrides: NaH, $MgH_2$; $AlH_3$; $SiH_4$; $PH_3$; $H_2S$; HCl.

**Solution:**   The hydrides of the metallic elements are ionic compounds consisting of metallic cations and the hydride ion, $H^-$. These ionic substances are formed by transfer of one or more electrons from the metal to hydrogen atoms. Each hydrogen atom accepts one electron to form $H^-$. One hydride ion is required for each electron removed from the metal to form the noble-gas configuration. The first three compounds thus correspond to the compositions $Na^+H^-$; $Mg^{2+}2H^-$; $Al^{3+}3H^-$. The remaining compounds are best formulated as covalent, in which an electron pair is shared between the central atom and each hydrogen. Si, an element of group 4A, requires four electrons to attain the noble-gas configuration of eight valence-shell electrons. Phosphorus requires three, sulfur two, and chlorine one. The formulas of the hydrides are in accord with the number of electrons needed.

## PRACTICE EXERCISE

How many H atoms must bond to selenium (atomic number 34) to give the selenium atom an octet of valence-shell electrons?
*Answer:*   2

**Multiple Bonds**

The sharing of a pair of electrons constitutes a single covalent bond, generally referred to simply as a **single bond**. In many molecules, atoms attain complete octets by sharing more than one pair of electrons between them. When two electron pairs are shared, two lines are drawn, representing a **double bond**. A **triple bond** corresponds to the sharing of three pairs of electrons. Such **multiple bonding** is found, for example, in the $N_2$ molecule:

$$:\overset{\displaystyle .}{N}\cdot + \cdot\overset{\displaystyle .}{N}: \longrightarrow :N:::N: \quad (\text{or } :N\equiv N:)$$

Because each nitrogen atom possesses five electrons in its valence shell, the sharing of three electron pairs is required to achieve the octet configuration. The properties of $N_2$ are in complete accord with this Lewis structure. Nitrogen gas is a diatomic gas with exceptionally low reactivity that results from the very stable nitrogen-nitrogen bond. Study of the structure of $N_2$ reveals that the nitrogen atoms are separated by only 1.10 Å. The short N≡N bond distance is a result of the triple bond between the atoms. From structure studies of many different substances in which nitrogen atoms share one or two electron pairs, it has been learned that the average distance between bonded nitrogen atoms varies with the number of shared electron pairs:

$$N—N \qquad N=N \qquad N≡N$$
$$1.47 \text{ Å} \qquad 1.24 \text{ Å} \qquad 1.10 \text{ Å}$$

As a general rule, the distance between bonded atoms decreases as the number of shared electron pairs increases.

Carbon dioxide, $CO_2$, provides a further example of a molecule containing multiple bonds:

$$:\ddot{O}: + \cdot\dot{C}\cdot + :\ddot{O}: \longrightarrow \ddot{O}::C::\ddot{O} \qquad (or \ \ddot{O}=C=\ddot{O})$$

We have seen in the foregoing section some simple examples of Lewis structures. It is actually fairly easy to draw the Lewis structures for most compounds and ions formed from nonmetallic elements. Doing so is a skill that is important in mastering the material in this chapter and the next.

In drawing Lewis structures it is a good idea to follow a regular procedure. We'll first outline the procedure, then go through several examples to show its application.

## 8.5 DRAWING LEWIS STRUCTURES

1. *Sum the valence electrons from all atoms.* (Use the periodic table as necessary to help you determine the number of valence electrons in each atom.) If the species is an ion, add an electron for each negative charge or subtract an electron for each positive charge. Do not worry about keeping track of which electrons come from which atoms. Only their total number is important.

2. *Write the symbols for the atoms involved so as to show which atoms are connected to which.* Often atoms are written in the order in which they are connected in the molecule or ion, as in HCN. When a central atom has a group of other atoms bonded to it, we usually write the central atom first, as in $CO_3{}^{2-}$ and $CCl_4$. In other cases you may need more information before you can draw the Lewis structure.

3. *Draw a single bond between each pair of atoms bonded together.*

4. *Complete the octets of the atoms bonded to the central atom.* (Remember, however, that hydrogen needs only two electrons.)

5. *Place any leftover electrons on the central atom.*

6. *If there are not enough electrons to give the central atom an octet, try multiple bonds.* Use one or more of the unshared pairs of electrons on the atoms bonded to the central atom to form double or triple bonds.

These rules are illustrated in the following sample exercise.

## SAMPLE EXERCISE 8.8

Draw Lewis structures for the following molecules: (a) $PCl_3$; (b) HCN; (c) $ClO_3{}^-$.

**Solution:** (a) Phosphorus (group 5A) has five valence electrons and each chlorine (group 7A) has seven. The total number of valence-shell electrons is

therefore $5 + (3 \times 7) = 26$. There are various ways the atoms might be arranged. However, in binary (two element) compounds the first element listed in the chemical formula is generally surrounded by the remaining atoms. Thus we begin with a skeleton structure that shows single bonds between phosphorus and each chlorine:

$$Cl—P—Cl$$
$$|$$
$$Cl$$

(It is not important that we place the atoms in exactly this arrangement; Lewis structures are not drawn to show geometry. However, it is important to show correctly which atoms are bonded to which.) The octets around each Cl are completed, accounting for 24 electrons. The remaining two electrons are placed on P, completing the octet around that atom as well:

$$:\ddot{C}l—\ddot{P}—\ddot{C}l:$$
$$|$$
$$:\ddot{C}l:$$

(Remember that the bonding electrons are counted for both atoms.)

**(b)** Hydrogen has one valence-shell electron, carbon (group 4A) has four, and nitrogen (group 5A) has five. The total number of valence-shell electrons is therefore $1 + 4 + 5 = 10$. Again there are various ways we might choose to arrange the atoms. Because hydrogen can accommodate only one electron pair, it always has only one single bond associated with it in any compound. This fact causes us to reject C—H—N as a possible arrangement. The remaining two possibilities are H—C—N and H—N—C. The first is the arrangement found experimentally. You might have guessed this to be the atomic arrangement because the formula is written with the atoms in this order. Thus we begin with a skeleton structure that shows single bonds between hydrogen, carbon, and nitrogen:

$$H—C—N$$

These two bonds account for four electrons. If we then place the remaining six electrons around N to give it an octet, we do not achieve an octet on C:

$$H—C—\ddot{N}:$$

We therefore try a double bond between C and N, using an unshared pair of electrons that we had placed on N. Again there are fewer than eight electrons on C, so we try a triple bond. This structure gives an octet around both C and N:

$$H—C≡N:$$

**(c)** Chlorine has seven valence electrons, and oxygen (group 6A) has six. An extra electron is added to account for the ion having a 1− charge. The total number of valence-shell electrons is therefore $7 + (3 \times 6) + 1 = 26$. After putting in the single bonds and distributing the unshared electron pairs, we have

$$\left[:\ddot{O}—\ddot{C}l—\ddot{O}:\right]^{-}$$
$$|$$
$$:\ddot{O}:$$

(For oxyanions—$ClO_3^-$, $SO_4^{2-}$, $NO_3^-$, $CO_3^{2-}$, and so forth—the oxygen atoms surround the central nonmetal atom.)

**PRACTICE EXERCISE**

Draw the Lewis electron dot structure for the $NO^+$ ion.

***Answer:***   $:N≡O:^{+}$

## 8.6 RESONANCE FORMS

We sometimes encounter substances in which the known arrangement of atoms is not adequately described by a single Lewis structure. The structural chemistry of nonmetallic elements affords several examples. Consider ozone, $O_3$, about which we shall have much to say in Chapter 14. This fascinating substance consists of bent molecules with both O—O distances the same, as shown in Figure 8.8. Because each oxygen atom contributes 6 valence-shell electrons, the ozone molecule has 18 valence-shell electrons. In writing the Lewis structure, we find that we must have one double bond to attain an octet of electrons about each atom:

But this structure cannot by itself be correct, because it requires that one O—O bond be different from the other, contrary to the observed struc-

**FIGURE 8.8**  Molecular structure of ozone, $O_3$.

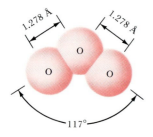

The concept of **formal charge** is sometimes used as an aid in deciding between alternative Lewis structures. The formal charge is largely a means of "bookkeeping" for the valence-shell electrons. To establish the formal charge on any atom in a molecule or ion, we assign electrons to the atom as follows:

1. All bonding electrons are divided equally between the atoms which form bonds.

2. All nonbonding electrons are assigned entirely to the atom on which they are found.

The formal charge is defined as *the number of valence-shell electrons in the isolated atom, minus the number of electrons assigned to the atom in the Lewis structure*. Let's illustrate these rules by calculating the formal charge on the central atom in the second-row nonmetal hydrides:

| | Hydride | | | |
|---|---|---|---|---|
| | $H-\overset{..}{\underset{..}{F}}:$ | $H-\overset{..}{\underset{\underset{H}{\vert}}{O}}:$ | $H-\overset{..}{\underset{\underset{H}{\vert}}{N}}-H$ | $H-\underset{\underset{H}{\vert}}{\overset{\overset{H}{\vert}}{C}}-H$ |
| Bonding electrons assigned | 1 | 2 | 3 | 4 |
| Nonbonding electrons assigned | 6 | 4 | 2 | 0 |
| Total electrons assigned | 7 | 6 | 5 | 4 |
| Electrons in isolated atom | 7 | 6 | 5 | 4 |
| Formal charge | 0 | 0 | 0 | 0 |

Note that the formal charge on the central atom is zero in all cases.

To see how the idea of formal charge can help in making a distinction between alternative Lewis structures, consider the cyanate ion, $NCO^-$. There are three possible orders for the atoms in this ion. For each we can write a Lewis structure that yields an octet about each atom. For each structure we can then calculate the formal charge on each atom. The results are as follows:

$$[\overset{..}{N}=C=\overset{..}{O}]^- \quad [\overset{..}{\underset{..}{C}}=O=\overset{..}{N}]^- \quad [\overset{..}{\underset{..}{O}}=N=\overset{..}{\underset{..}{C}}]^-$$

Formal charge    $-1$   $0$   $0$     $-2$ $+2$ $-1$     $0$ $+1$ $-2$

Because the ion as a whole has a charge of $1-$, the formal charges of all atoms must sum to $1-$. As a general rule the most stable Lewis structure will be that in which the atoms bear the smallest formal charges. The Lewis structure on the left is clearly superior to the other two in producing the smallest variations in formal charge among the atoms. This suggests that the arrangement shown at the left is the preferred structure for the ion; indeed, it is the observed structure.

Although the concept of formal charge is useful in helping to decide between alternative Lewis structures, you must keep in mind that *the formal charges do not represent real charges on the atoms*. Other factors that we will be discussing in the material ahead contribute to determine the actual net charges on atoms in molecules and ions.

ture. However, in drawing the Lewis structure we could just as easily have put the O=O bond on the left:

The two alternative Lewis structures for ozone are equivalent except for the placement of electrons. Equivalent Lewis structures of this sort are called **resonance forms**. To properly describe the structure of ozone, we write both Lewis structures and indicate that the real molecule is described by an average of the structures suggested by the two resonance forms:

The double-headed arrow indicates that the structures shown are resonance forms.

The fact that we must write more than one Lewis structure to describe a molecule or ion does not imply anything especially different about these species. You must not suppose that the molecule really exists in two or more different forms and oscillates rapidly between them. There is only one form of the molecule, that which is observed experimentally. We need to write two or more different resonance forms only because Lewis structures are limited in describing the electron distributions in molecules.

One rule that must be followed in writing resonance forms is that the arrangement of the nuclei must be the same in each structure. That is, the same atoms must be bonded to one another in all the structures, so that the only differences are in the arrangements of electrons.

As an additional example of resonance forms, let us draw the Lewis structure for the nitrate ion, $NO_3^-$, one of the most commonly encountered anions. We find that three equivalent Lewis structures are required in this instance:

Note that the arrangement of nuclei is the same in each structure; only the placement of electrons differs. All three Lewis structures taken together adequately describe the nitrate ion, which is observed to be planar (all atoms in the same plane), with all three N—O distances equal.

In the examples of resonance structures we have seen so far, all the resonance forms have the same importance in contributing to the overall description of the molecule or ion. In the example in Sample Exercise 8.9, the contributing resonance structures are not equally important.

---

**SAMPLE EXERCISE 8.9**

The molecule chlorine dioxide, $ClO_2$, is bent, with a central chlorine atom bound to two oxygen atoms. The two Cl—O distances are observed to be equal. Describe the $ClO_2$ molecule in terms of three possible resonance forms.

**Solution:**  The chlorine atom has 7 valence-shell electrons, and oxygen has 6. We therefore have a total

of 19 electrons $[(2 \times 6) + 7]$ to place in this molecule. Note that $ClO_2$ has an odd number of electrons. This is an unusual situation; it means that one electron must remain unpaired.

We draw the arrangement of atoms in accord with the experimental facts and put one single bond between chlorine and each oxygen.

This leaves us with 15 electrons to place. We put electron pairs on the atoms to achieve an octet on each atom, or to come as close to it as we can get. Because the molecule possesses an odd number of electrons, we can be sure that at least one atom will not possess an octet of electrons. The odd electron must go on one or the other of the atoms, so that three possible structures result:

The first two of these are equivalent, but the third is different; the odd electron is on chlorine rather than oxygen. To find out how much weight to attach to this resonance structure in comparison with the first two, we would have to have experimental information on how the odd electron is distributed in the real $ClO_2$ molecule.

**PRACTICE EXERCISE**

Draw two equivalent resonance structures for the $NO_2^-$ ion.

*Answer:* $[\ddot{O}{=}\ddot{N}{-}\ddot{O}:]^- \longleftrightarrow [:\ddot{O}{-}\ddot{N}{=}\ddot{O}]^-$

## 8.7 EXCEPTIONS TO THE OCTET RULE

The octet rule is so simple and useful in introducing the basic concepts of bonding that one might get the impression that it is always obeyed. However, in many situations the octet rule fails. These exceptions are of three main types:

1. Molecules with an odd number of electrons
2. Molecules in which an atom has less than an octet
3. Molecules in which an atom has more than an octet

**Odd Number of Electrons**

In the vast majority of molecules the number of electrons is even, and complete pairing of electron spins occurs. However, in a few molecules, such as $ClO_2$, NO, and $NO_2$, the number of electrons is odd. For example, NO contains $5 + 6 = 11$ valence electrons. Obviously, complete pairing of these electrons is impossible, and an octet around each atom cannot be achieved.

**Less Than an Octet**

A second type of exception occurs when there are fewer than eight electrons around an atom in a molecule or ion. This is also a relatively rare situation and is most often encountered in compounds of boron and beryllium. One example is boron trifluoride, $BF_3$. A Lewis structure with a double bond between B and F can be drawn that satisfies the octet rule. However, the properties of $BF_3$ are more consistent with a Lewis structure in which there are single bonds between B and each F, as shown in Figure 8.9. In this Lewis structure there are only six electrons around boron. The chemical behavior of the molecule reflects the lack of an octet. $BF_3$ reacts very energetically with molecules that have an unshared pair of electrons that can be used to form a bond with boron. For example, it reacts with ammonia, $NH_3$, to form the compound $NH_3BF_3$:

**FIGURE 8.9** Molecular structure and Lewis structure of boron trifluoride, $BF_3$.

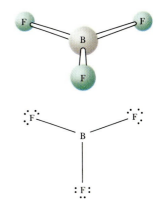

In this stable compound boron has an octet of electrons.

**More Than an Octet**    The third and largest class of exceptions consists of molecules or ions in which there are more than eight electrons in the valence shell of an atom. As an example, consider $PCl_5$. When we draw the Lewis structure for this molecule, we are forced to "expand" the valence shell and place ten electrons around the central phosphorus atom:

Among the other molecules and ions with "expanded" valence shells are $SF_4$, $AsF_6^-$, and $ICl_4^-$. Let's take a look at why expanded valence shells are observed only for elements in period 3 and beyond in the periodic table.

The octet rule works as well as it does because the representative elements usually employ only an $ns$ and three $np$ valence-shell orbitals in bonding, and these orbitals hold eight electrons. Because elements of the second period have only $2s$ and $2p$ orbitals available for bonding, they can never have more than an octet of electrons in their valence shells. However, from the third period on, the elements have unfilled $nd$ orbitals that can be used in bonding. For example, the orbital diagram for the valence shell of a phosphorus atom is as follows:

Although third-period elements like phosphorus often satisfy the octet rule, they also often exceed it by using their empty $d$ orbitals to accommodate electrons.

Size also plays an important role in determining whether an atom can accommodate more than eight electrons. The larger the central atom, the larger the number of atoms that can surround it. Thus the occurrences of expanded valence shells increase with increasing size of the central atom. The size of the surrounding atoms is also important. Expanded valence shells occur most often when the central atom is bonded to the smallest and most strongly electron-attracting atoms, such as F, Cl, and O.

---

**SAMPLE EXERCISE 8.10**

Draw the Lewis structure for $ICl_4^-$.

**Solution:** Iodine (group 7A) has seven valence electrons; each chlorine (group 7A) also has seven; an extra electron is added to account for the 1− charge of the ion. Thus the total number of valence electrons is $7 + 4(7) + 1 = 36$. The I atom is the central atom in the ion. Putting eight electrons around each Cl atom (including a pair of electrons between I and each Cl to represent the single bonds between these atoms) requires $8 \times 4 = 32$ electrons. Thus we are left with $36 - 32 = 4$ electrons to be placed on the larger iodine:

Thus iodine has 12 electrons around it, exceeding the common octet of electrons.

**PRACTICE EXERCISE**

Which of the following atoms is never found with more than an octet of electrons around it: S, C, P, Br?
*Answer:*  C

The stability of a molecule can be related to the strengths of the co-valent bonds it contains. The strength of a covalent bond between two atoms is determined by the energy required to break that bond. The **bond-dissociation energy**, also called the **bond energy**, is the enthalpy change, $\Delta H$, required to break a particular bond in a mole of gaseous substance. For example, the dissociation energy for the bond between chlorine atoms in the $Cl_2$ molecule is the energy required to dissociate a mole of $Cl_2$ into chlorine atoms:

$$:\ddot{C}l\!-\!\ddot{C}l\!:(g) \longrightarrow 2:\ddot{C}l\cdot(g) \qquad \Delta H = D(Cl\!-\!Cl) = 242 \text{ kJ}$$

We use the designation $D$(bond type) in this equation and elsewhere to represent bond-dissociation energies.

It is a relatively easy matter to assign bond energies to bonds in dia-tomic molecules. As we have seen, the bond energy is just the energy required to break the diatomic molecule into its component atoms. How-ever, for bonds that occur only in polyatomic molecules (such as the C—H bond) we must often utilize average bond energies. For example, the enthalpy change for the process shown below (called "atomization") can be used to define an average bond strength for the C—H bond:

$$\begin{array}{c} H \\ | \\ H\!-\!\underset{|}{\overset{|}{C}}\!-\!H(g) \\ H \end{array} \longrightarrow \cdot\dot{C}\cdot(g) + 4H\cdot(g) \qquad \Delta H = 1660 \text{ kJ}$$

Since there are four equivalent C—H bonds in methane, the heat of atomization is equal to the total bond energies of the four C—H bonds. Thus the average C—H bond energy is $D(C\!-\!H) = (1660/4)$ kJ/mol = 415 kJ/mol.

The bond energy for a given set of atoms, say C—H, depends on the rest of the molecule of which it is a part. However, the variation from one molecule to another is generally small. This supports the idea that the bonding electron pairs are localized between atoms. If we consider C—H bond strengths in many different compounds, we find that the average strength is 413 kJ/mol, which compares closely with the 415 kJ/mol value calculated from $CH_4$.

Table 8.3 lists several average bond energies. Notice that the bond energy is always a positive quantity; energy is always required to break chemical bonds. Conversely, energy is given off when a bond forms be-tween two gaseous atoms or molecular fragments. Of course, the greater the bond energy, the stronger the bond.

Referring to Table 8.3, compare the bond energies for the C—C, C=C, and C≡C bonds. Notice that the bond strength increases as the number of pairs of electrons shared between the atom increases. We noted in Section 8.4 that the distance between atoms decreases as we move from a single to a double to a triple bond. For carbon-carbon bonds we have the following average bond lengths and bond strengths:

| C—C | C=C | C≡C |
|---|---|---|
| 1.54 Å | 1.34 Å | 1.20 Å |
| 348 kJ/mol | 614 kJ/mol | 839 kJ/mol |

**TABLE 8.3** Average bond energies (kJ/mol)

**Single bonds**

| | | | | | | | |
|---|---|---|---|---|---|---|---|
| C—H | 413 | N—H | 391 | O—H | 463 | F—F | 155 |
| C—C | 348 | N—N | 163 | O—O | 146 | | |
| C—N | 293 | N—O | 201 | O—F | 190 | Cl—F | 253 |
| C—O | 358 | N—F | 272 | O—Cl | 203 | Cl—Cl | 242 |
| C—F | 485 | N—Cl | 200 | O—I | 234 | | |
| C—Cl | 328 | N—Br | 243 | | | Br—F | 237 |
| C—Br | 276 | | | | | Br—Cl | 218 |
| C—I | 240 | | | S—H | 339 | Br—Br | 193 |
| C—S | 259 | H—H | 436 | S—F | 327 | | |
| | | H—F | 567 | S—Cl | 253 | I—Cl | 208 |
| Si—H | 323 | H—Cl | 431 | S—Br | 218 | I—Br | 175 |
| Si—Si | 226 | H—Br | 366 | S—S | 266 | I—I | 151 |
| Si—C | 301 | H—I | 299 | | | | |
| Si—O | 368 | | | | | | |

**Multiple bonds**

| | | | | | |
|---|---|---|---|---|---|
| C=C | 614 | N=N | 418 | O₂ | 495 |
| C≡C | 839 | N≡N | 941 | | |
| C=N | 615 | | | | |
| C≡N | 891 | | | S=O | 323 |
| C=O | 799 | | | S=S | 418 |
| C≡O | 1072 | | | | |

In general, longer bonds have lower bond energies. Consequently, stronger bonds are generally associated with smaller atoms.

A molecule with strong chemical bonds generally has less tendency to undergo chemical change than does one with weak bonds. This relationship between strong bonding and chemical stability helps explain the chemical form in which many elements are found in nature. For example, Si—O bonds are among the strongest ones that silicon forms. It is not surprising therefore that $SiO_2$ and other substances containing Si—O bonds (silicates) are so common; it is estimated that over 90 percent of the earth's crust is composed of $SiO_2$ and silicates. We will have more to say about these compounds in Chapter 23.

**Bond Energies and Chemical Reactions**

A knowledge of bond energies is helpful in understanding why some reactions are exothermic (negative $\Delta H$) while others are endothermic (positive $\Delta H$). An exothermic reaction results when the bonds in the product molecules are stronger than those in the reactants. Consider the following reaction:

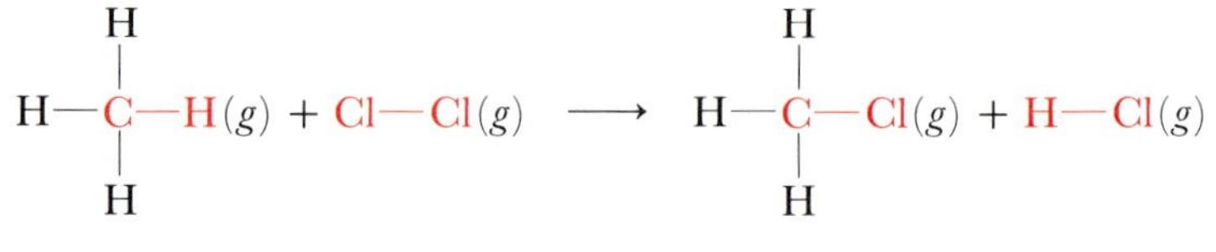

$$H\!-\!\underset{\underset{H}{|}}{\overset{\overset{H}{|}}{C}}\!-\!H(g) + Cl\!-\!Cl(g) \longrightarrow H\!-\!\underset{\underset{H}{|}}{\overset{\overset{H}{|}}{C}}\!-\!Cl(g) + H\!-\!Cl(g)$$

In the course of this reaction, 1 mol of C—H bonds and 1 mol of Cl—Cl bonds must be broken (reactants). Using average bond energies, we can estimate that this requires 655 kJ:

$$\Delta H(\text{bond breakage}) = D(\text{C—H}) + D(\text{Cl—Cl}) = (413 + 242) \text{ kJ} = 665 \text{ kJ}$$

However, 1 mol of C—Cl bonds and 1 mol of H—Cl bonds are formed in the products; this produces 759 kJ:

$$\Delta H \text{(bond formation)} = -D(\text{C—Cl}) - D(\text{H—Cl})$$
$$= (-328 - 431) \text{ kJ} = -759 \text{ kJ}$$

The overall energy change for a reaction is the sum of the energy required to break the pertinent bonds in the reactants (always a positive quantity) and the energy given up in forming the new bonds in the products (always a negative quantity). In the present example we have

$$\Delta H_{rxn} = \Delta H \text{(bond breakage)} + \Delta H \text{(bond formation)}$$
$$= 655 \text{ kJ} - 759 \text{ kJ} = -104 \text{ kJ} \qquad [8.3]$$

We see that the reaction is exothermic because the bonds in the products (especially the H—Cl bond) are stronger than those in the reactants (especially the Cl—Cl bond).

This illustrates how bond energies can be used to estimate heats of reactions using Equation 8.3. In practice, we seldom calculate $\Delta H$ in this way, because it can be determined more accurately from heats of formation (Section 4.6). However, we might use bond energies to estimate $\Delta H$ for a reaction if the heat of formation of some reactant or product is not known. Whenever one uses bond energies in this fashion, it is important to remember that they refer to gaseous molecules and that they are often just averaged values.

---

### SAMPLE EXERCISE 8.11

Using Table 8.3, estimate $\Delta H$ for the following reaction (where we show explicitly the bonds involved in the reactants and products):

H—C(H)(H)—C(H)(H)—H$(g)$ + $\frac{7}{2}O_2(g) \longrightarrow$

$2O{=}C{=}O(g) + 3H—O—H(g)$

**Solution:** Among the reactants, we must break six C—H bonds and a C—C bond in $C_2H_6$; we also break $\frac{7}{2}O_2$ bonds. Among the products, we form four C=O bonds (two in each $CO_2$) and six O—H bonds (two in each $H_2O$). Using Equation 8.3 and data from Table 8.3, we have

$$\Delta H = 6D(\text{C—H}) + D(\text{C—C}) + \tfrac{7}{2}D(O_2)$$
$$- 4D(\text{C=O}) - 6D(\text{O—H})$$

$$= 6(413 \text{ kJ}) + 348 \text{ kJ} + \tfrac{7}{2}(495 \text{ kJ}) - 4(799 \text{ kJ})$$
$$- 6(463 \text{ kJ})$$
$$= 4558 \text{ kJ} - 5974 \text{ kJ} = -1416 \text{ kJ}$$

This estimate can be compared with the value of $-1428$ kJ calculated from more precise thermochemical data; the agreement is excellent.

### PRACTICE EXERCISE

Using Table 8.3, estimate $\Delta H$ for the following reaction:

H—N(H)—N(H)—H$(g) \longrightarrow$ N$\equiv$N$(g)$ + 2H—H$(g)$

*Answer:* $-86$ kJ

---

The electron pairs shared between two different atoms are not necessarily shared equally. We can visualize two extreme cases in the degree to which electron pairs are shared. On the one hand, we have bonding between two identical atoms, as in $Cl_2$ or $N_2$, where the electron pairs must be equally shared. At the other extreme, illustrated by NaCl, there will

## 8.9 BOND POLARITY AND ELECTRO-NEGATIVITIES

be essentially no sharing of electrons. We know that in this case the compound is best described as composed of $Na^+$ and $Cl^-$ ions. The $3s$ electron of the Na atom is, in effect, transferred completely to chlorine. The bonds occurring in most covalent substances fall somewhere between these extremes. In practice we describe bonds as either ionic or covalent depending on which extreme the bond more closely resembles.

The concept of **bond polarity** is useful in describing the sharing of electrons between atoms. A **nonpolar bond** is one in which the electrons are shared equally between two atoms. In a **polar covalent bond**, one of the atoms exerts a greater attraction for the electrons than the other. If the difference in relative abilities to attract electrons is large enough an ionic bond is formed.

**Electronegativity**

The ability of an atom to attract electrons to itself in a chemical bond is referred to as **electronegativity**. Electronegativity can be related to electron affinity and ionization energy, properties that describe the tendency of an isolated, gaseous atom to gain or lose an electron. In practice, numerical estimates of electronegativity are based on a variety of other properties. For example, Linus Pauling (1901–   ), who first developed the concept of electronegativity, based his scale on bond-energy relationships. We will not be concerned in detail with how the electronegativity values are obtained, but rather with using the concept in discussing chemical bonding.

Figure 8.10 shows electronegativity values for many of the elements. Notice that the most electronegative element is fluorine, with an electronegativity of 4.0. The least electronegative element, cesium, has an electronegativity of 0.79. The values for all other elements lie between these extremes. The values listed for the transition elements are those for the 2+ state. When the element is in a higher charge state, its electronegativity is higher.

**FIGURE 8.10**  Electronegativities of the elements. The values for the transition elements are those for the 2+ state.

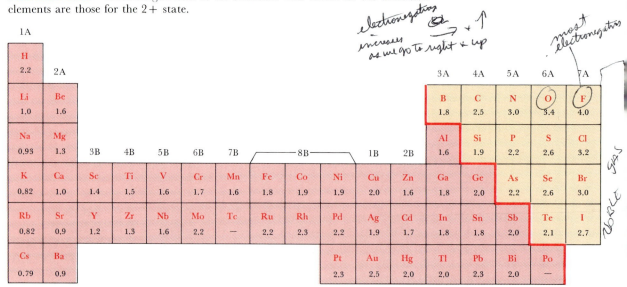

Note that in a horizontal row of the table there is a more or less steady increase in electronegativity in moving from left to right, that is, from the most metallic to the most nonmetallic elements. Notice also that, with a few exceptions, there is an overall decrease in electronegativity with increasing atomic number in any one group of the periodic table. This is what we might expect, because we know that ionization energies tend to decrease with increasing atomic number in a group, and electron affinities don't change very much. You do not need to memorize numerical values for electronegativity. However, you should know the periodic trends so you can predict which of two elements is more electronegative.

Keep in mind that electronegativities are *approximate* measures of the *relative* tendencies of these elements to attract electrons to themselves in a chemical bond. The electronegativity varies with the type of chemical environment in which an element is situated. We have noted, for example, that the electronegativities of transition metals vary with their valence. Similarly, the electronegativity of chlorine in its bonding to phosphorus in $PCl_3$ is likely to be different from its value in the chlorate ion, $ClO_3^-$, in which it is in a much different bonding situation. Variations of this sort are not so large as to render the concept of electronegativity useless, but we must avoid placing too much reliance on precise values for electronegativities. As long as we remain aware of their limitations, electronegativity values provide a useful guide to polarities in chemical bonds.

The electronegativity difference between two atoms is a measure of the polarity of the bond between them; the greater the difference in electronegativity, the more polar the bond. The shared electron pair has a greater probability of being located on the more electronegative of the two atoms. For example, the electronegativity difference between F and H is $4.0 - 2.2 = 1.8$. Consequently, the sharing of electrons is unequal (the bond is polar), with the more electronegative fluorine attracting electron density off the hydrogen. We represent this situation in the following two ways:

$$\overset{\delta+ \quad \delta-}{H\text{—}F} \quad \text{or} \quad \overset{\longmapsto}{H\text{—}F}$$

The $\delta+$ and $\delta-$ are meant to represent partial positive and negative charges, respectively. (The symbol $\delta$ is the Greek lowercase letter delta.) The arrow represents the pull of electron density off the hydrogen by the fluorine, leaving the hydrogen with a partial positive charge; the head of the arrow points in the direction in which the electrons are attracted, toward the more electronegative atom. We will consider the molecular consequences of bond polarities in Chapter 9 after we have discussed molecular shapes.

## Electronegativity Differences and Bond Polarity

*If δ− is > 1.8 is considered ionic*

---

### SAMPLE EXERCISE 8.12

Which of the following bonds is more polar: (a) B—Cl or C—Cl; (b) P—F or P—Cl? Indicate in each case which atom has the partial negative charge.

**Solution:** (a) The difference in the electronegativities of boron and chlorine is $3.2 - 1.8 = 1.4$; the difference between carbon and chlorine is $3.2 - 2.5 = 0.7$. Consequently, the B—Cl bond is the more polar; the chlorine atom carries the partial negative charge because it has a higher electronegativity. We should be able to reach this same conclusion without using a table of electronegativities; instead, we can rely on periodic trends. Because boron is to the left of carbon in the periodic table, we would predict that it has a lower attraction for electrons. Chlorine, being on the right side of the table, has a strong attraction for electrons. The most polar bond will be the one between the atoms having the lowest attraction for electrons (boron) and the highest attraction (chlorine).

**(b)** Because fluorine is above chlorine in the periodic table, we would predict it to be more electronegative. Consequently, the P—F bond will be more polar than the P—Cl bond. You should compare the electronegativity differences for the two bonds to verify this prediction. The fluorine atom carries the partial negative charge.

**PRACTICE EXERCISE**
Which of the following bonds is most polar: S—Cl, S—Br, Se—Cl, or Se—Br?
*Answer:* Se—Cl

## 8.10 OXIDATION NUMBERS

In light of our discussion of polar covalent bonds, it is useful to examine the similarities between reactions in which electrons are completely transferred and those in which only partial shifts of electron density occur. In the reaction between sodium and chlorine atoms to form NaCl, an electron transfers from sodium to chlorine. We can imagine the overall reaction as the result of two separate processes:

$$Na\cdot \longrightarrow Na^+ + e^- \qquad [8.4]$$

$$:\ddot{Cl}\cdot + e^- \longrightarrow [:\ddot{Cl}:]^- \qquad [8.5]$$

Each of these processes is called a **half-reaction**. Adding these two half-reactions together gives the overall formation of the ionic species from the neutral atoms:

$$Na + Cl \longrightarrow Na^+ + Cl^-$$

A process in which an electron is lost (Equation 8.4, for example) is called an **oxidation**. A process in which an electron is gained (Equation 8.5, for example) is called a **reduction**. A substance that has lost an electron is said to be oxidized; one that has gained an electron is said to be reduced.

The definitions of oxidation and reduction in terms of the loss and gain of electrons apply readily to the formation of ionic compounds, because electrons are transferred completely. As we shall see, the concept of oxidation and reduction is also useful in treating the reactions of molecular compounds.

Consider the reaction of hydrogen atoms and chlorine atoms to form HCl, a gaseous substance with a normal boiling point of −84°C. Since ionic compounds are solids at room temperature, we see that HCl is not ionic. The bonding in HCl is best described as polar covalent. Because chlorine is more electronegative than hydrogen, the electrons in the H—Cl bond are displaced toward chlorine:

$$H\cdot + \cdot\ddot{Cl}: \longrightarrow H:\ddot{Cl}:$$

The H atom therefore carries a somewhat positive charge and the Cl atom a somewhat negative one.

In keeping track of electrons, it is a reasonable simplification to assign the shared electrons to the more electronegative Cl atom:

$$H \left[ \:\overset{\displaystyle ..}{\underset{\displaystyle ..}{Cl}}\: \right.$$

This procedure gives Cl eight valence-shell electrons, one more than the neutral atom. We have in effect given a $1-$ charge to the chlorine. Hydrogen, stripped of its electron, is assigned a charge of $1+$.

Charges assigned in this fashion are called **oxidation numbers** or **oxidation states**. The oxidation number of an atom is the charge that results when the electrons in a covalent bond are assigned to the more electronegative atom; it is the charge an atom would possess *if* the bonding were ionic. In HCl the oxidation number of H is $+1$ and that of Cl is $-1$. (In writing oxidation numbers we will write the sign before the number to distinguish them from actual electronic charges, which we write with the number first.)

Although we can determine oxidation numbers for atoms using Lewis structures and electronegativities as we have done for HCl, we seldom use this procedure. It is generally easier to determine oxidation numbers using the following set of rules.

1. *The oxidation number of an element in its elemental form is zero.* For an isolated atom—like an Na atom, where there is no bonding and no net charge—the oxidation state must be 0. It is also zero for any elemental substance in which there is bonding. For example, in $N_2$, $Cl_2$, and $P_4$ the bonding electrons are shared equally between identical atoms, giving each atom an oxidation state of 0.

2. *The oxidation number of a monatomic ion is the same as its charge.* For example, the oxidation number of sodium in $Na^+$ is $+1$ and that of sulfur in $S^{2-}$ is $-2$.

3. *In binary compounds (those with two different elements), the element with greater electronegativity is assigned a negative oxidation number equal to its charge in simple ionic compounds of the element.* For example, consider the oxidation state of Cl in $PCl_3$. Cl is more electronegative than P. In its simple ionic compounds, chlorine appears as the chloride ion, $Cl^-$. Thus is $PCl_3$, Cl is assigned an oxidation number of $-1$.

4. *The sum of the oxidation numbers equals zero for an electrically neutral compound and equals the overall charge for an ionic species.* For example, $PCl_3$ is a neutral molecule. Thus the sum of the oxidation numbers of the P and Cl atoms must equal zero. Because the oxidation number of each Cl in this compound is $-1$ (rule 3), the oxidation number of P must be $+3$. In like manner, the sum of the oxidation number of C and O in $CO_3^{2-}$ must equal $-2$. The oxidation number of O in this ion is $-2$, because O is more electronegative than C and $-2$ is the charge on the oxide ion (rule 3). Thus the oxidation number on C must be $+4$, because $+4 + 3(-2) = -2$.

The periodic table provides us with many additional guidelines for assigning oxidation numbers. As shown in Figure 8.11, oxidation numbers exhibit periodic trends. Some observations are particularly helpful. The alkali metals (group 1A) exhibit only the oxidation state of $+1$ in their compounds. The alkaline earth metals (group 2A) are always found in compounds in the $+2$ oxidation state. The most commonly encountered element in group 3A, Al, is always found in the $+3$ oxidation state.

The most electronegative element, F, is always found in compounds in the $-1$ oxidation state. Oxygen in compounds is nearly always in the

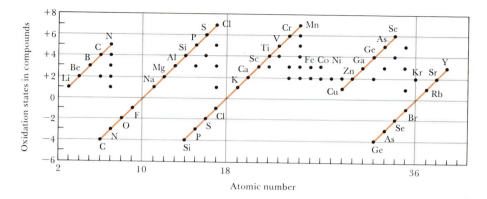

**FIGURE 8.11** Common oxidation numbers for elements with atomic numbers 3–39. Notice that the maximum and minimum oxidation states (through which lines have been drawn for emphasis) are a periodic function of atomic number.

−2 oxidation state. The only common exception to this general rule occurs in peroxides. In the peroxide ion, $O_2^{2-}$, and in molecular peroxides, such as $H_2O_2$, oxygen has an oxidation state of −1. Hydrogen is +1 when it is bonded to a more electronegative element (most nonmetals) and −1 when bonded to a less electronegative element (most metals).

## SAMPLE EXERCISE 8.13

Determine the oxidation state of sulfur in each of the following molecules or ions: **(a)** $H_2S$; **(b)** $S_8$; **(c)** $SCl_2$; **(d)** $Na_2SO_3$; **(e)** $SO_4^{2-}$.

**Solution:** **(a)** This is a binary (two-element) compound. Because S is more electronegative than H, S must have a negative oxidation number equal to its ionic charge (rule 3). Thus S has an oxidation number of −2. This is consistent with the oxidation number of +1 expected for H.

**(b)** Because this is an elemental form of sulfur, the oxidation state of S is 0 (rule 1).

**(c)** Cl is more electronegative than S. Thus the Cl in this binary compound must have a negative oxidation number equal to its ionic charge (rule 3), which is −1. The sum of the oxidation numbers must equal zero (rule 4). Letting $x$ equal the oxidation number of S, we have $x + 2(-1) = 0$. Consequently, the oxidation state of S, $x$, must be +2.

**(d)** Na, an alkali metal, is always found in compounds in the +1 oxidation state. O has a common oxidation state of −2. If we let $x$ equal the oxidation number of S, we have $2(+1) + x + 3(-2) = 0$. Thus

the oxidation number of S in this compound is +4.

**(e)** The oxidation state of O is −2. The sum of the oxidation numbers equals −2, the net charge of the $SO_4^{2-}$ ion. Letting $x$ equal the oxidation number of S, we have $x + 4(-2) = -2$. From this relation we conclude that the oxidation state of S is +6.

These examples illustrate the range of oxidation states exhibited by sulfur, from −2 to +6. In general, the most negative oxidation state of a nonmetal corresponds to the number of electrons that must be added to the atom to give it an octet. In this case S, which belongs to periodic family 6A, has six valence-shell electrons. Thus two are needed to give it an octet, as in the $S^{2-}$ ion. The most positive oxidation state, in this case +6, corresponds to loss of all valence-shell electrons.

## PRACTICE EXERCISE

What is the oxidation state of the underlined element in each of the following: **(a)** $\underline{P}_2O_5$; **(b)** $Na\underline{H}$; **(c)** $\underline{Cr}_2O_7^{2-}$; **(d)** $\underline{Sn}Br_4$; **(e)** $Ba\underline{O}_2$?
*Answers:* **(a)** +5; **(b)** −1; **(c)** +6; **(d)** +4; **(e)** −1

It is important to keep in mind that the oxidation numbers, or oxidation states, do not correspond to real charges on the atoms, except in the special case of simple ionic substances. Nevertheless, they furnish a useful means of organizing chemical facts, especially in the case of metallic elements. Their most frequent use is in naming compounds and in balancing chemical equations for reactions in which changes in oxidation numbers occur.

The rules for naming simple inorganic compounds were discussed in Chapter 2. This is a good point at which to review those rules. Now that the concept of oxidation number has been explained, we can add another widely used method for naming binary compounds. You learned in Section 2.6 that the name of a binary compound consists of the name of the less electronegative element first, followed by the name of the more electronegative element, modified to have an -ide ending. The following examples are illustrative:

Oxidation Numbers and Nomenclature

| | |
|---|---|
| BaSe | barium selenide |
| ZnS | zinc sulfide |
| $MgH_2$ | magnesium hydride |

When one or the other of the elements involved has more than one possible oxidation state, the number of atoms may be included in the name as a Greek prefix (1 = *mono*-, 2 = *di*-, 3 = *tri*-, 4 = *tetra*-, 5 = *penta*-, 6 = *hexa*-, 7 = *hepta*-), *or* the oxidation state is indicated using a roman numeral, as in these examples:

| | | |
|---|---|---|
| $MnO_2$ | manganese dioxide | *or* manganese(IV) oxide |
| $Mn_2O_3$ | dimanganese trioxide | *or* manganese(III) oxide |
| $P_2O_5$ | diphosphorus pentoxide | *or* phosphorus(V) oxide |
| $P_2O_3$ | diphosphorus trioxide | *or* phosphorus(III) oxide |
| $SnCl_4$ | tin tetrachloride | *or* tin(IV) chloride |
| $SnCl_2$ | tin dichloride | *or* tin(II) chloride |

The use of roman numerals to indicate the oxidation state of the element has the disadvantage that it gives only the simplest formula for the compound, and not the molecular formula. For example, the compounds $NO_2$ and $N_2O_4$ are both nitrogen(IV) oxide and are not distinguished by this name. However, the names nitrogen dioxide and dinitrogen tetroxide do distinguish them. This shortcoming is often unimportant because many compounds are referred to only by their simplest formula. As an example, phosphorus(V) oxide normally exists as $P_4O_{10}$ molecules, but it is frequently represented by its simplest formula, $P_2O_5$.

## SAMPLE EXERCISE 8.14

(a) Give the name of $TiO_2$. (b) What is the chemical formula of chromium(III) sulfide?

**Solution:** (a) One way to name the compound is to use the Greek prefix *di*- to indicate the presence of two O atoms: titanium dioxide. The compound can also be named using oxidation numbers. Because O has an oxidation number of $-2$, the oxidation number of Ti must be $+4$ [that is, when $x + 2(-2) = 0$, $x$ must be $+4$]. Thus the compound can be called titanium(IV) oxide, the roman numeral indicating the oxidation state of Ti.

(b) Chromium(III) indicates that Cr has an oxidation state of $+3$. The sulfide ion is $S^{2-}$. It takes two $Cr^{3+}$ ions to balance the charge of three $S^{2-}$ ions to give a neutral compound, $Cr_2S_3$.

## PRACTICE EXERCISE

(a) Name the chemical substance, $Cl_2O_7$. (b) Give the chemical formula for sulfur(IV) fluoride.
*Answers:* (a) dichlorine heptaoxide or chlorine(VII) oxide; (b) $SF_4$

Many chemical reactions involve atoms that undergo changes in oxidation numbers. Atoms that lose electrons (increase in oxidation number) are said to be oxidized. Atoms that gain electrons (decrease in oxidation number) are said to be reduced. Oxidation and reduction are always discussed together because whenever one substance loses electrons another must gain them. We will take up the topic of balancing oxidation-reduction (or redox) equations in Chapter 13.

---

### SAMPLE EXERCISE 8.15

Is the following reaction an oxidation-reduction reaction? If so, indicate which atom is oxidized and which is reduced:

$$Zn(s) + 2HCl(aq) \longrightarrow ZnCl_2(aq) + H_2(g)$$

**Solution:** To decide whether this is an oxidation-reduction reaction, we must determine whether any atom undergoes a change in oxidation state. Zinc starts with an oxidation state of 0 and ends up as +2 in $ZnCl_2$. Consequently, the reaction is an oxidation-reduction reaction in which zinc is oxidized. But what is reduced? Hydrogen starts with an oxi-

dation state of +1 in HCl and ends up as zero in $H_2$. It has gained electrons and has therefore been reduced. The chlorine maintains an oxidation state of −1 throughout the reaction; it does not undergo either oxidation or reduction.

### PRACTICE EXERCISE

Which element is oxidized in the following reaction: $PbS(s) + 4H_2O_2(aq) \longrightarrow PbSO_4(s) + 4H_2O(l)$? What are the oxidation states of this element before and after reaction?
***Answer:*** S; −2 and +6

---

## 8.11 BINARY OXIDES

With even such simple ideas about bonds as those discussed in this chapter, we can organize and understand many chemical facts. Before taking a closer look at covalent bonding, let's apply some of what we have learned to a discussion of the binary compounds of oxygen.

Oxygen is a very important element that is abundant in air and is found in numerous compounds in the environment. The chemistry of the element is dominated by its high ionization energy and strong attraction for electrons. The term "oxidation," which we have defined as an increase in oxidation number, was originally applied only to reactions of oxygen, $O_2$. Because of its high electronegativity, oxygen almost invariably gains electrons in its reactions. It thereby oxidizes (removes electrons) from the substance with which it reacts. For example, $O_2$ reacts with magnesium to produce magnesium oxide, MgO (Mg in a +2 oxidation state), as shown in Figure 8.12:

$$2Mg(s) + O_2(g) \longrightarrow 2MgO(s)$$

With the exception of the noble gases He, Ne, Ar, and Kr, oxygen forms compounds with every element. Many elements form more than one oxide. For example, iron forms FeO, $Fe_2O_3$, and $Fe_3O_4$, and chlorine forms $Cl_2O$, $Cl_2O_3$, $ClO_2$, $Cl_2O_6$, and $Cl_2O_7$. The oxides of the metals are generally ionic substances, and those of the nonmetals are generally covalent.

**FIGURE 8.12** The combustion of a piece of magnesium ribbon; slower oxidation occurs at lower temperatures. (Lawrence Migdale/Science Source)

Because of their low ionization energies, metals generally combine with oxygen to form ionic solids. Metal oxides tend to have high lattice energies because metal cations pack efficiently in the solid in the interstices (vacant spaces) between the larger oxide ions. The melting points of ionic oxides are typically high; some examples are listed in Table 8.4. Many metal oxides find uses as high-temperature materials, such as firebrick and ceramics. Many other uses of oxides, such as the use of $Y_2O_3$ in fluorescent lights (Figure 8.13), also depend on their thermal stability.

The solubilities of metals oxides in water vary over a wide range, but most are rather insoluble. The solubilities are greatest when the lattice energies are lowest. Thus the solubilities tend to be greatest when the charge on the metal ion is low. The oxides of the alkali metals, in which the metal has a $+1$ oxidation state, are quite soluble. The oxides of the alkaline earth metals, in which the metal ion has a $+2$ oxidation state are less soluble, but their solubilities increases as the size of the metal ion increases: BeO and MgO have extremely low solubilities.

The oxides of metals in low oxidation states ($+1$, $+2$) tend to be basic. These oxides dissolve in water to form hydroxide solutions and in acids to form salt solutions (see Section 5.3). When a metal oxide dissolves in water, the $H_2O$ molecule transfers $H^+$ (hydrogen without an electron) to the $O^{2-}$ ion to form the $OH^-$ ions:

$$:\!\overset{..}{\underset{..}{O}}\!:^{2-} + H\!:\!\overset{..}{\underset{..}{O}}\!:\!H \longrightarrow :\!\overset{..}{\underset{..}{O}}\!:\!H^- + :\!\overset{..}{\underset{..}{O}}\!:\!H^-$$

In acids, $H_2O$ is formed along with the metal salt.

## Covalent Oxides

Nonmetals have sufficiently high electronegativities to keep oxygen from removing electrons to form $O^{2-}$ ions. Instead, these elements form molecules containing polar covalent bonds with oxygen. These molecular substances are typically gases, liquids, or solids with low melting points. A notable exception is silicon dioxide, $SiO_2$, which we shall consider in more detail in Section 11.5. This substance, also known as silica, is relatively hard and has a high melting point (approximately 1600°C).

Whereas metal oxides are basic in character, nonmetal oxides—especially in high oxidation states—are acidic (Section 5.3). For example, chlorine(VII) oxide, $Cl_2O_7$, dissolves in water, reacting to form perchloric acid, $HClO_4$:

$$Cl_2O_7(l) + H_2O(l) \longrightarrow 2HClO_4(aq)$$

No oxidation-reduction occurs in this reaction. Notice that Cl has an oxidation state of $+7$ in both $Cl_2O_7$ and $HClO_4$.

Metallic oxides involving metals in high oxidation states are also better described as polar covalent rather than as ionic. These oxides are also acidic. For example, $Mn_2O_7$ forms a strongly acidic solution of the acid $HMnO_4$ when it dissolves in water.

In comparing the acid-base character of oxides, two simple rules are useful:

## Ionic Oxides

**TABLE 8.4** Melting points of some metal oxides

| Oxide | Melting point (°C) |
|---|---|
| $Na_2O$ | 1132 |
| $TiO_2$ | 1857 |
| $Al_2O_3$ | 2045 |
| BeO | 2530 |
| CaO | 2610 |
| MgO | 2826 |

**FIGURE 8.13** Yttrium oxide, $Y_2O_3$, is a white solid. Here we see the red fluorescence of a sample of yttrium oxide caused by the UV lamp seen in the background. This oxide material is used to coat the walls of fluorescent lamps and TV picture tubes to enhance the red light they produce. This use of $Y_2O_3$ relies on its thermal stability at high operating temperature. (Courtesy of GTE Products Corporation)

1. *For a given element, the acidity of the oxide increases with increasing oxidation state of the element.* For example, as we go through the series MnO, $Mn_2O_3$, $MnO_2$, and $Mn_2O_7$, in which the oxidation state of Mn is increasing from $+2$ to $+7$, the acidity of the oxide increases. Thus MnO is insoluble in water but dissolves readily in acid solution. Conversely, $Mn_2O_7$ dissolves readily in water to form a strongly acidic solution.

2. *For a given oxidation state, the acidity of an oxide increases with increasing electronegativity of the element.* For example, in both $SeO_2$ and $SO_2$ the oxidation state of the central atom is $+4$. However, the electronegativity of S is greater than that of Se. Correspondingly, $SO_2$ is the more acidic oxide of the two.

**Amphoteric Oxides**

Certain oxides are on the borderline of being acidic or basic. These oxides, which tend to be virtually insoluble in water, are soluble in both acids and bases. They are said to be **amphoteric** (from the Greek word *amphoteros*, which means "each of two"). As shown in Figure 8.14, most of the representative elements that form amphoteric oxides lie near the diagonal line in the periodic table that divides metals from nonmetals. For example, aluminum oxide, $Al_2O_3$, is amphoteric and reacts with both acidic and basic solutions:

$$Al_2O_3(s) + 6HCl(aq) \longrightarrow 2AlCl_3(aq) + 3H_2O(l)$$

$$Al_2O_3(s) + 6NaOH(aq) \longrightarrow 2Na_3AlO_3(aq) + 3H_2O(l)$$

Chromium(III), which has the same charge as the aluminum ion and has a virtually identical ionic radius, also forms an amphoteric oxide, $Cr_2O_3$. Those metals with amphoteric oxides also form amphoteric hydroxides. Thus $Al(OH)_3$ and $Cr(OH)_3$ are water insoluble, but they dissolve in both acid and base solutions. We will consider amphoterism again, in Chapter 18, when we consider acids and bases in more detail.

**FIGURE 8.14** Simplest formulas of the oxides of the representative elements in their maximum oxidation states. Those that are basic are shown with a light blue shading, those that are amphoteric with violet, and those that are acidic with red.

Increasing acidic character →

Increasing base character ↓

| 1A | 2A | 3A | 4A | 5A | 6A | 7A |
|---|---|---|---|---|---|---|
| $Li_2O$ | BeO | $B_2O_3$ | $CO_2$ | $N_2O_5$ | | $F_2O$ |
| $Na_2O$ | MgO | $Al_2O_3$ | $SiO_2$ | $P_2O_5$ | $SO_3$ | $Cl_2O_7$ |
| $K_2O$ | CaO | $Ga_2O_3$ | $GeO_2$ | $As_2O_5$ | $SeO_3$ | $Br_2O_7$ |
| $Rb_2O$ | SrO | $In_2O_3$ | $SnO_2$ | $Sb_2O_5$ | $TeO_3$ | $I_2O_7$ |
| $Cs_2O$ | BaO | $Tl_2O_3$ | $PbO_2$ | $Bi_2O_5$ | $PoO_3$ | $At_2O_7$ |

# FOR REVIEW

## SUMMARY

In this chapter we have dealt with the interactions that lead to the formation of chemical bonds. The tendencies of atoms to gain, lose, or share their valence electrons to form bonds can often be viewed in terms of attempts to achieve a noble-gas electron configuration (the **octet rule**).

**Ionic bonding** results from the complete transfer of electrons from one atom to another, with formation of a three-dimensional lattice of charged particles. The stabilities of ionic substances result from the powerful electrostatic attractive forces between an ion and all the surrounding ions of opposite charge. These interactions are measured by the **lattice energy**. The magnitude of the lattice energy depends primarily on the charges and sizes of the ions. In general, lattice energies increase as the charges of the ions increase and as their sizes decrease.

Cations are smaller than their parent atom; anions are larger than their parent atom. For ions of the same charge, size increases going down a family. For an **isoelectronic series**, size decreases with increasing nuclear charge (atomic number). Not all ions have noble-gas configurations. Many transition metal ions do not. In forming transition-metal ions, the atom first loses its outer *s* electrons.

**Covalent bonding** results from the sharing of electrons. The octet rule is useful in describing this sharing. We can represent shared electron-pair structures of molecules by means of **Lewis structures**, which show the sharing of electron pairs between atoms. The sharing of one pair of electrons produces a **single bond**; the sharing of two or three pairs of electrons between atoms produces **double** and **triple bonds**, respectively.

Sometimes a single Lewis structure is inadequate to represent a particular molecule, but an average of two or more Lewis structures does form a satisfactory representation. In these cases, the Lewis structures are called **resonance forms**. Sometimes the octet rule is not obeyed; this situation occurs mainly when a large atom is surrounded by small, electronegative atoms like F, O, or Cl. In such instances the large atom has unfilled *d* orbitals in its valence shell to accommodate more than an octet of electrons. Thus expanded octets are observed for atoms in the third period and beyond in the periodic table.

The strength of a covalent bond is measured by its **bond** energy. The strengths of covalent bonds increase with the number of electron pairs shared between two atoms. In single bonds, the bond strengths are generally higher between atoms of smaller size.

It is important to recognize that even in covalent bonding, electrons may not be shared equally between two atoms. **Electronegativity** is a measure of the ability of an atom to compete with other atoms for the electrons shared between them. Highly electronegative elements strongly attract electrons. The electronegativities of the elements, which show a regular periodic relationship, are an important guide to chemical behavior. We shall be using the concept of electronegativity often throughout the text. The difference in electronegativities of bonded atoms is used to determine the polarity of a bond.

Another application of electronegativity is in the assignment of **oxidation numbers**, formal whole-number charges assigned to atoms in molecules and ions. Although the oxidation numbers do not represent the real charges on atoms except in simple ionic substances, they are of great value in helping us to organize chemical facts, to balance equations, and to name compounds.

**Oxidation** may be defined as a process in which an atom undergoes an increase in oxidation number. **Reduction** is a process in which an element undergoes a decrease in oxidation number. In an **oxidation-reduction reaction**, both oxidation and reduction occur in such a manner as to balance the total increases and decreases in oxidation numbers.

Several ideas presented in this chapter were used in discussing the chemistry of binary oxides. Metal oxides tend to be ionic substances that are either soluble in water to give basic solutions or that can be dissolved by acids; they are called basic oxides. Nonmetal oxides, by contrast, are covalent substance. They tend to be acidic oxides, which dissolve in water to produce acidic solutions or which dissolve in bases. Some oxides are borderline; they are called **amphoteric oxides**.

# LEARNING GOALS

Having read and studied this chapter, you should be able to:

1. Determine the number of valence electrons for any atom and write its Lewis symbol.
2. Describe the origin of the energy terms that lead to stabilization of ionic lattices.
3. Describe the NaCl lattice, as illustrated in Figure 8.4.
4. Predict on the basis of the periodic table the probable formulas of ionic substances formed between common metals and nonmetals.
5. Write the electron configurations of ions.
6. Describe how the gain or loss of an electron affects atomic radii.
7. Explain the concept of an isoelectronic series and the origin of changes in ionic radius within such a series.
8. Describe the basis of the Lewis theory, and predict the valence of common nonmetallic elements from their position in the periodic table.
9. Write the Lewis structures for molecules and ions containing covalent bonds (including those with expanded octets), using the periodic table.
10. Write resonance forms for molecules or polyatomic ions that are not adequately described by a single Lewis structure.

11. Explain the significance of electronegativity, and in a general way relate the electronegativity of an element to its position in the periodic table.
12. Predict the relative polarities of bonds using either the periodic table or electronegativity values.
13. Relate bond energies to bond strengths, and use bond energies to estimate $\Delta H$ for reactions.
14. Assign oxidation numbers to atoms in molecules and ions.
15. Give the meaning of the terms *oxidation*, *reduction*, and *oxidation-reduction reactions*.
16. Determine whether oxidation-reduction has occurred in a reaction; if it has, be able to identify the substance that is oxidized and the one that is reduced.
17. Assign acceptable names to simple inorganic compounds and ions.
18. Describe the general differences in physical properties between substances with ionic bonds and those with covalent bonds.
19. Describe how the water solubility of a metal oxide is related to cation size and charge.
20. Write balanced chemical equations for the reactions of oxides with water, metallic oxides with acids, and nonmetal oxides with bases.

# KEY TERMS

Among the more important terms and expressions used for the first time in this chapter are the following:

An **amphoteric oxide** (Section 8.11) is a water-insoluble oxide that dissolves in either an acidic or basic solution.

**Bond energy** (Section 8.8) is the enthalpy change, $\Delta H$, required to break a chemical bond when a substance is in the gas phase.

**Bond polarity** (Section 8.9) is a measure of the difference in ability of the two atoms in a chemical bond to attract electrons.

A **covalent bond** (Section 8.4) is a bond formed between two or more atoms by a sharing of electrons.

A **double bond** (Section 8.4) is a covalent bond involving two electron pairs.

**Electronegativity** (Section 8.9) is a measure of the ability of an atom that is bonded to another atom to attract electrons to itself.

A **half-reaction** (Section 8.10) is half of an overall oxidation-reduction reaction (either the oxidation half or the reduction half).

An **ionic bond** (Section 8.2) is a bond formed on the basis of the electrostatic forces that exist between oppositely charged species in solid lattices made up of ions. The ions are formed from atoms by transfer of one or more electrons.

An **isoelectronic series** (Section 8.3) is a series of atoms, ions, or molecules having the same number of electrons.

**Lattice energy** (Section 8.2) is the energy required to separate completely the ions in an ionic solid.

A **Lewis structure** (Section 8.4) is a representation of covalent bonding in a molecule that is drawn using Lewis symbols. Covalently shared electron pairs are shown as lines, and unshared electron pairs are shown as a pair of dots. Only the valence-shell electrons are shown.

The **octet rule** (Section 8.1) states that bonded atoms tend to possess or share a total of eight valence-shell electrons.

**Oxidation** (Section 8.10) is the half of an oxidation-reduction process that corresponds to an increase in oxidation number.

**Oxidation number** (Section 8.10) or oxidation state is a positive or negative whole number assigned to an element in a molecule or ion on the basis of a set of formal rules; to some degree it reflects the positive or negative character of that atom.

A **polar covalent bond** (Section 8.9) is a covalent bond in which the electrons are not shared equally.

**Reduction** (Section 8.10) is the half of an oxidation-reduction process that corresponds to a decrease in oxidation number.

**Resonance forms** (Section 8.6) are individual Lewis structures in cases where two or more Lewis structures are equally good descriptions of a single molecule. The resonance structures in such an instance are "averaged" to give a correct description of the real molecule.

A **triple bond** (Section 8.4) is a covalent bond involving three electron pairs.

**Valence** (Section 8.1) may be defined as the capacity of an atom for entering into chemical combination with other atoms. Ionic valence is equal to the number of electrons gained or lost in forming the ionic species. Covalence is equal to the number of electrons from an atom that are involved in shared electron-pair bonds with other atoms.

**Valence electrons** (Section 8.1) are the outer-shell electrons of an atom; these are the ones the atom uses in bonding.

# EXERCISES

## Valence Electrons, Lewis Symbols, Ionic Bonding

**8.1**  Write the Lewis symbol for each of the following elements: (**a**) sulfur, S; (**b**) silicon, Si; (**c**) aluminum, Al; (**d**) argon, Ar.

**8.2**  Write the Lewis symbol for each of the following atoms or ions: (**a**) P; (**b**) Br; (**c**) $S^{2-}$; (**d**) $Ca^{2+}$.

**8.3**  Using Lewis symbols, diagram the reaction that occurs between sodium and oxygen atoms to give $Na^+$ and $O^{2-}$ ions.

**8.4**  Using Lewis symbols, diagram the reaction that occurs between Al atoms and F atoms.

**8.5**  Predict the chemical formula of the ionic compound formed between the following pairs of elements: (**a**) Sc, O; (**b**) Mg, Br; (**c**) Ba, S; (**d**) Ti, Cl.

**8.6**  Indicate whether each of the following formulas is likely to represent a stable compound, and give an explanation for your answer: (**a**) $Rb_2O$; (**b**) BaCl; (**c**) $MgF_3$; (**d**) $ScBr_3$; (**e**) $Na_3N$.

**8.7**  Write the electron configuration for each of the following ions, and state which possess noble-gas configurations: (**a**) $Ba^{2+}$; (**b**) $Br^-$; (**c**) $Mn^{2+}$; (**d**) $Ti^{4+}$; (**e**) $Zn^{2+}$; (**f**) $In^+$.

**8.8**  Write the electron configuration for each of the following ions, and state which possess noble-gas configurations: (**a**) $Mn^{3+}$; (**b**) $Sr^{2+}$; (**c**) $Mo^{3+}$; (**d**) $Te^{2-}$; (**e**) $Tl^+$; (**f**) $Cu^{2+}$.

**8.9**  It requires energy to remove two electrons from Ca to form $Ca^{2+}$. It also requires energy to add two electrons to O to form $O^{2-}$. Why, then, is CaO stable relative to the free elements?

**8.10**  The energy required to evaporate solid argon is very low; on the other hand, the energy required to evaporate RbCl, formed from the two elements on each side of Ar, is quite high. What accounts for the difference?

**8.11**  The lattice energy of LiH is 858 kJ/mol, whereas that for $MgH_2$ is 2790 kJ/mol. Account for the large difference in these two quantities. *(don it calculate)*

**8.12**  Explain the following trends in lattice energies: (**a**) CaS > KCl; (**b**) LiF > CsBr; (**c**) MgO > MgS; (**d**) MgO > BaO.

## Sizes of Ions

**8.13**  Explain the following variations in atomic or ionic radii: (**a**) $I^- > I > I^+$; (**b**) $Ca^{2+} > Mg^{2+} > Be^{2+}$; (**c**) $Br^- > Kr > Rb^+$; (**d**) $N^{3-} > O^{2-} > F^-$.

**8.14**  Arrange the members of each of the following sets in order of increasing size: (**a**) $Li^+$, $Rb^+$, $K^+$; (**b**) $Br^-$, $Na^+$, $Mg^{2+}$; (**c**) Ar, $Cl^-$, $S^{2-}$, $K^+$; (**d**) Cl, $Cl^-$, Ar.

**8.15**  Based on the data of Figure 8.6, how would you compare the effective nuclear charge experienced by a $3p$ electron in K with that of a $3p$ electron in $K^+$? Explain.

**8.16** Compare the effective nuclear charge experienced by a 3$p$ electron in $K^+$ with that experienced by a 3$p$ electron in $Cl^-$. Explain.

**8.17** Identify the neutral atom that is isoelectronic with each of the following: **(a)** $Br^-$; **(b)** $I^+$; **(c)** $Sr^{2+}$; **(d)** $Ga^{3+}$.

**8.18** Select the ions or atoms from each of the following sets that are isoelectronic with each other: **(a)** $K^+$, $Rb^+$, $Ca^{2+}$; **(b)** $Cu^{2+}$, $Ca^{2+}$, $Sc^{3+}$; **(c)** $S^{2-}$, $Se^{2-}$, Ar; **(d)** $F^-$, Ne, $Na^+$.

## Lewis Structures; Resonance Forms

**8.19** Draw the Lewis structures for **(a)** $SiH_4$; **(b)** $ClO_2^-$; **(c)** $CO_2$; **(d)** $HBrO_3$ (H is bonded to O); **(e)** $TeCl_2$.

**8.20** Write the Lewis structures for **(a)** $SO_4^{2-}$; **(b)** CO; **(c)** $BH_4^-$; **(d)** ClOH; **(e)** ONCl.

**8.21** Draw the Lewis structures for each of the following compounds. Identify those that do not obey the octet rule, and indicate the nature of the departure from the octet rule. **(a)** $ClO_2$; **(b)** $GeF_4$; **(c)** $TeF_4$; **(d)** $BCl_3$; **(e)** $XeF_4$.

**8.22** Draw the Lewis structures for each of the following ions. Identify those that do not obey the octet rule, and indicate the nature of the departure from the octet rule. **(a)** $ClO_3^-$; **(b)** $NO^+$; **(c)** $I_3^-$; **(d)** $AsF_6^-$; **(e)** $O_2^-$.

**8.23** Draw resonance structures for each of the following: **(a)** $SO_3$; **(b)** $C_2O_4^{2-}$ (each C is bonded to two O atoms and another C); **(c)** $HNO_3$ (H is bonded to O); **(d)** $ClO_3$.

**8.24** Draw the resonance forms for the following: **(a)** $SeO_2$; **(b)** $CO_3^{2-}$; **(c)** $SCN^-$; **(d)** $HCO_2^-$ (H and each O bonded to C).

**8.25** Based on their Lewis structures, predict the relative N—O bond lengths in NO, $NO_2^-$, and $NO_3^-$.

**8.26** Based on their Lewis structures, predict the relative S—O bond lengths in $SO_2$, $SO_3$, and $SO_4^{2-}$.

## Bond Energies

**8.27** Using the bond energies tabulated in Table 8.3, estimate $\Delta H$ for each of the following gas-phase reactions:

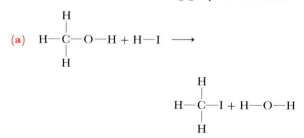

**8.28** Using bond energies, estimate $\Delta H$ for the following gas-phase reactions:

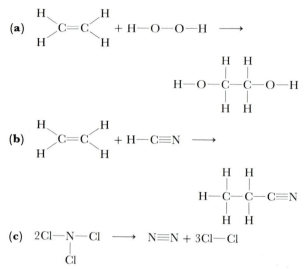

**8.29** Using the bond energies in Table 8.3, estimate the enthalpy change for each of the following gas-phase reactions:
**(a)** $HCN + 3H_2 \longrightarrow CH_4 + NH_3$
**(b)** $HBr + 2F_2 \longrightarrow BrF_3 + HF$

**8.30** Using bond energies (Table 8.3), estimate $\Delta H$ for the following gas-phase reactions:
**(a)** $CO + 2H_2 \longrightarrow CH_3OH$
**(b)** $CH_2{=}CH_2 + F_2 \longrightarrow CH_2F{-}CH_2F$

**8.31** Given the following bond dissociation energies, calculate the average bond energy for the Ti—Cl bond.

|  | $\Delta H$ **(kJ/mol)** |
|---|---|
| $TiCl_4(g) \longrightarrow TiCl_3(g) + Cl(g)$ | 335 |
| $TiCl_3(g) \longrightarrow TiCl_2(g) + Cl(g)$ | 423 |
| $TiCl_2(g) \longrightarrow TiCl(g) + Cl(g)$ | 444 |
| $TiCl(g) \longrightarrow Ti(g) + Cl(g)$ | 519 |

**8.32** Based on data given in this chapter, what is the general relationship between bond order (the number of bonding electron pairs in the bond) and the bond energy?

## Bond Polarities; Electronegativities

**8.33** What is the general rule regarding the variation in electronegativity in moving from left to right in a horizontal row of the periodic table? Explain the origin of the trend.

**8.34** In a given family of the periodic table, what is the general relationship between electronegativity and size?

**8.35** Without looking at the table of electronegativities (refer to a periodic table instead), arrange the members of each of the following sets in order of increasing electronegativity: **(a)** O, P, S; **(b)** Mg, Al, Si; **(c)** S, Cl, Br; **(d)** C, Si, N.

**8.36** Using only a periodic table, select the most electronegative atom in each of the following sets: **(a)** Be, B, C, Si; **(b)** Cl, S, Br, Se; **(c)** Si, Ge, Al, Ga; **(d)** K, Ca, As, Se.

**8.37** Which of the following bonds is polar: **(a)** B—Cl; **(b)** Cl—Cl; **(c)** Hg—Sb; **(d)** As—F; **(e)** Co—C? Indicate the more electronegative atom in each polar bond.

**8.38** Arrange the bonds in each of the following sets in order of increasing polarity: **(a)** C—S, B—F, N—O; **(b)** Pb—Cl, Pb—Pb, Pb—C; **(c)** H—F, O—F, Be—F.

## Oxidation Numbers; Oxidation-Reduction

**8.39** Determine the oxidation state of the underlined element in each of the following: **(a)** $K\underline{Mn}O_4$, **(b)** $\underline{Fe}_2O_3$; **(c)** $\underline{Xe}OF_4$; **(d)** $\underline{Sn}Cl_4$; **(e)** $K_2\underline{O}_2$; **(f)** $\underline{N}O_2^-$; **(g)** $\underline{U}O_2^{2+}$; **(h)** $\underline{Cu}Cl_4^{2-}$.

**8.40** Determine the oxidation state of the underlined element in each of the following: **(a)** $\underline{P}_4$; **(b)** $\underline{Hg}_2Cl_2$; **(c)** $Mg_2\underline{P}_2O_7$; **(d)** $\underline{S}F_4$; **(e)** $\underline{Cr}_2O_7^{2-}$; **(f)** $\underline{N}H_4^+$; **(g)** $\underline{N}O^+$; **(h)** $H\underline{As}O_4^{2-}$.

**8.41** What are the maximum and minimum oxidation states exhibited by each of the following elements: **(a)** Br; **(b)** As; **(c)** Ba; **(d)** Cr?

**8.42** What are the maximum and minimum oxidation states exhibited by each of the following elements: **(a)** Se; **(b)** Ge; **(c)** Al; **(d)** Mn?

**8.43** Give the name or chemical formula, as appropriate, for each of the following substances: **(a)** iron(III) fluoride; **(b)** molybdenum(VI) oxide; **(c)** arsenic(V) bromide; **(d)** $V_2O_3$; **(e)** $CoF_3$; **(f)** MnS.

**8.44** Give the name or chemical formula, as appropriate, for each of the following substances: **(a)** manganese(IV) oxide; **(b)** gallium(III) sulfide; **(c)** selenium(VI) fluoride; **(d)** $Cu_2O$; **(e)** $ClF_3$; **(f)** $TeO_3$.

**8.45** Indicate the oxidation states of the elements that undergo a change in oxidation state in each of the following reactions:

**(a)** $S(s) + 2F_2(g) \longrightarrow SF_4(g)$

**(b)** $2CuSO_4(aq) + 4KI(aq) \longrightarrow$ $2CuI(s) + 2K_2SO_4(aq) + I_2(s)$

**(c)** $NH_3(g) + 3Cl_2(g) \longrightarrow NCl_3(g) + 3HCl(g)$

**(d)** $I_2(aq) + SO_2(aq) + 2H_2O(l) \longrightarrow$ $2HI(aq) + H_2SO_4$

**(e)** $2PbS(s) + 3O_2(g) \longrightarrow 2PbO(s) + 2SO_2(g)$

**8.46** Indicate which of the following are oxidation-reduction reactions. In each oxidation-reduction reaction identify the element that is oxidized and the element that is reduced:

**(a)** $Na_2CO_3(s) + 2HCl(aq) \longrightarrow$ $2NaCl(aq) + H_2O(l) + CO_2(g)$

**(b)** $2KI(aq) + F_2(g) \longrightarrow 2KF(aq) + I_2(s)$

**(c)** $MnO_2(s) + 4HCl(aq) \longrightarrow$ $MnCl_2(aq) + Cl_2 + 2H_2O(l)$

**(d)** $CaS(s) + 2HCl(aq) \longrightarrow CaCl_2(aq) + H_2S(g)$

**(e)** $2PbO_2(s) \longrightarrow 2PbO(s) + O_2$

**(f)** $O_2(g) + 4HCl(g) \longrightarrow 2Cl_2(g) + 2H_2O(g)$

**[8.47]** Calculate the formal charge on the indicated atom in each of the following molecules or ions: **(a)** sulfur in $SO_2$; **(b)** fluorine in $BF_3$; **(c)** phosphorus in $OPCl_3$; **(d)** phosphorus in $PO_4^{3-}$; **(e)** iodine in $ICl_3$; **(f)** boron in $BH_4^-$.

**[8.48]** Use the concept of formal charge to choose the more likely skeleton structure in each of the following cases: **(a)** NNO or NON; **(b)** HCN or HNC; **(c)** NOBr or ONBr.

## Binary Oxides

**8.49** Write the chemical formulas of the oxides of the following elements in their highest oxidation state: **(a)** Mg; **(b)** Si; **(c)** V; **(d)** Cl.

**8.50** Write the chemical formulas of the oxides of the following elements in its highest oxidation state: **(a)** K; **(b)** Ga; **(c)** Ti; **(d)** P.

**8.51** Label each of the following as acidic, basic, or amphoteric: **(a)** CoO; **(b)** $NO_2$; **(c)** BaO; **(d)** $As_2O_5$; **(e)** $Cr_2O_3$; **(f)** $Fe_2O_3$.

**8.52** Each of the following compounds reacts with water. Write a balanced chemical equation for each reaction: **(a)** CaO; **(b)** $N_2O_5$; **(c)** $K_2O$; **(d)** $SO_3$.

**8.53** Complete and balance each of the following chemical equations. (In each case the products are water-soluble.)

**(a)** $ZnO(s) + HCl(aq) \longrightarrow$

**(b)** $P_2O_5(s) + NaOH(aq) \longrightarrow$

**(c)** $Cr_2O_3(s) + H_2SO_4(aq) \longrightarrow$

**8.54** Complete and balance each of the following equations. (In each case the products are water soluble.)

**(a)** $Li_2O(s) + H_2O(l) \longrightarrow$

**(b)** $NiO(s) + H_3PO_4(aq) \longrightarrow$
**(c)** $SeO_2(s) + KOH(aq) \longrightarrow$

**8.55** Select the more acidic oxide in each of the following pairs **(a)** $SO_2$, $SO_3$; **(b)** $CO_2$, $NO_2$; **(c)** $Cr_2O_3$, $CrO_3$.

**8.56** Select the more basic oxide in each of the following pairs **(a)** $Al_2O_3$, $SiO_2$; **(b)** $TiO$, $TiO_2$; **(c)** $CaO$, $ZnO$.

**8.57** Which compound in each of the following pairs would you expect to be less soluble in water: **(a)** $Al_2O_3$ or $CaO$; **(b)** $CaO$ or $SrO$? Explain your answer in each case.

**8.58** Which compound in each of the following pairs would you expect to be less soluble in water: **(a)** $MgO$ or $CaO$; **(b)** $Na_2O$ or $NiO$? Explain your answer in each case.

## Additional Exercises

**8.59** In each of the following examples of a Lewis symbol, indicate the group in the periodic table in which the element X belongs: **(a)** $\cdot\overset{\cdot}{X}\cdot$; **(b)** $\cdot X\cdot$; **(c)** $:\overset{\cdot}{X}\cdot$.

**8.60** What change must occur in the electron configuration of each of the following elements if it is to achieve a configuration that obeys the octet rule: **(a)** Cl; **(b)** Mg; **(c)** N; **(d)** Rb?

**8.61** Which of the following contain metal ions that do not have noble-gas electron configurations: **(a)** $CuCl$; **(b)** $CdO$; **(c)** $TiO_2$; **(d)** $ScF_3$.

**8.62** Write the electron configuration for each of the following ions: **(a)** $Cu^{2+}$; **(b)** $Pb^{2+}$; **(c)** $Co^{3+}$; **(d)** $Sc^{3+}$.

**8.63** The $+2$ oxidation state is common for transition-metal ions. Explain why this might be expected.

**8.64** Describe electron configurations other than a completed octet in the valence shell that are relatively stable arrangements often found in ions.

**8.65** Explain the following trend in lattice energies: LiH, 858 kJ/mol; NaH, 782 kJ/mol; KH, 699 kJ/mol; RbH, 674 kJ/mol.

**8.66** From the ionic radii given in Table 8.2, calculate the potential energy of each of the following ion pairs, assuming that they are separated by the sum of their ionic radii (the magnitude of the electronic charge is given on the back inside cover): **(a)** $Mg^{2+}$, $O^{2-}$; **(b)** $Na^+$, $Br^-$.

**[8.67]** From the ionic radii given in Figure 8.6, calculate the potential energy of a $Na^+$ and $Cl^-$ ion pair that are just touching (the magnitude of the electronic charge is given on the back inside cover). Calculate the energy of a mole of such pairs. How does this value compare with the lattice energy of NaCl? Explain the difference.

**8.68** In each of the following pairs, select the species that is smaller in size: **(a)** Cl, $Cl^-$; **(b)** Ca, $Ca^{2+}$; **(c)** $Cl^-$, $Ca^{2+}$; **(d)** $Ca^{2+}$, $Ba^{2+}$.

**8.69** What type of bond (ionic, covalent, or metallic)

would you expect to find in each of the following substances: **(a)** $K_2S$; **(b)** $SCl_2$; **(c)** $MnF_2$; **(d)** $B_2H_6$; **(e)** Cr; **(f)** $P_4S_{10}$?

**8.70** Use the octet rule to predict the formula of the simplest compound formed between **(a)** hydrogen and silicon; **(b)** sulfur and fluorine; **(c)** phosphorus and chlorine.

**8.71** Use Lewis symbols and Lewis structures to represent the formation of $H_2O$ from hydrogen and oxygen atoms.

**8.72** Why does hydrogen never form more than one bond? Why do period-2 elements never form more than four covalent bonds? Why are period-3 elements able to exceed an octet when they form bonds?

**8.73** Write the Lewis structure for each of the following substances: **(a)** nitrous oxide, $N_2O$, known as laughing gas and used as an inhalation anesthetic in dental surgery; **(b)** ethyl alcohol, $CH_3CH_2OH$, the alcohol of alcoholic beverages (contains a C—C bond); **(c)** formaldehyde, $H_2CO$, used to preserve biological samples; **(d)** urea, $H_2NCONH_2$, which is excreted in the urine of mammals (contains two N—C bonds).

**8.74** Although $I_3^-$ is known, $F_3^-$ is not. Using Lewis structures, explain why $F_3^-$ does not form.

**8.75** Using bond energies, estimate the enthalpy change for the following reactions:
**(a)** $Cl_2(g) + F_2(g) \longrightarrow 2ClF(g)$
**(b)** $N_2(g) + 3H_2(g) \longrightarrow 2NH_3(g)$

**8.76** Use bond energies (Table 8.3), electron affinities (Table 7.4), and the ionization energy of hydrogen (1312 kJ/mol) to estimate $\Delta H$ for each of the following reactions:
**(a)** $HF(g) \longrightarrow H^+(g) + F^-(g)$
**(b)** $HCl(g) \longrightarrow H^+(g) + Cl^-(g)$

**[8.77]** The enthalpy of formation of $NH(g)$ at 25°C is 360 kJ/mol. From this value, and using the data in Table 8.3, calculate the N—H bond dissociation energy in NH.

**8.78** Based on the positions of the elements in the periodic table, select the most polar and least polar bond from the following list: P—N, P—O, P—P, P—S.

**8.79** Assign oxidation states to all atoms in each of the following compounds: **(a)** $N_2O$; **(b)** $KBiO_3$; **(c)** $ClF_3$; **(d)** $PBr_3$; **(e)** $HAsO_2$; **(f)** $N_2H_4$; **(g)** $Na_2S_2O_3$; **(h)** $(NH_4)_2SO_4$.

**8.80** Give the chemical formula for each of the following compounds: **(a)** sodium hypchlorite, used in household bleaches; **(b)** calcium dihydrogen phosphate, used in soft drinks to add tartness; **(c)** calcium nitrite, used as a protective additive in cement formulations; **(d)** molybdenum(IV) sulfide, used as a lubricant.

**8.81** Write balanced chemical equations for each of the following reactions. **(a)** Barium metal reacts with oxygen to form barium peroxide. **(b)** Gaseous dichlorine heptoxide dissolves in water to form an acidic solution. **(c)** Gaseous sulfur dioxide dissolves in aqueous sodium hydroxide solution. **(d)** Solid chromium(III) oxide dissolves in hydrochloric acid solution. **(e)** Solid strontium oxide is added to water. **(f)** Solid diphosphorus pentoxide is added to water.

**8.82** A white solid, melting at 1115°C, insoluble in water but slightly soluble in aqueous NaOH, is likely to be which of the following: SrO; $GeO_2$; $SeO_2$; $N_2O_3$?

**8.83** Write a set of reactions for the formation of each of the following compounds, starting with the materials indicated: **(a)** $CaSO_3(s)$ beginning with $CaO(s)$ and elemental sulfur; **(b)** $H_3AsO_4$, beginning with $As_2O_5(s)$;

**(c)** $BaSO_4(s)$, beginning with $BaO(s)$ and $SO_3(g)$.

**[8.84]** Using the bond energies in Table 8.3 and the following bond distances, construct a graph of bond energies versus bond distances. What conclusions can you reach from your graph? Bond distances are as follows: C—C (1.54 Å), Si—Si (2.35 Å), N—N (1.45 Å), O—O (1.48 Å), S—S (2.05 Å), F—F (1.42 Å), Cl—Cl (1.99 Å), Br—Br (2.28 Å), I—I (2.67 Å), C—F (1.35 Å), C—O (1.43 Å), and S—Br (2.27 Å).

## CHAPTER 9

# Geometries
# of Molecules;
# Molecular Orbitals

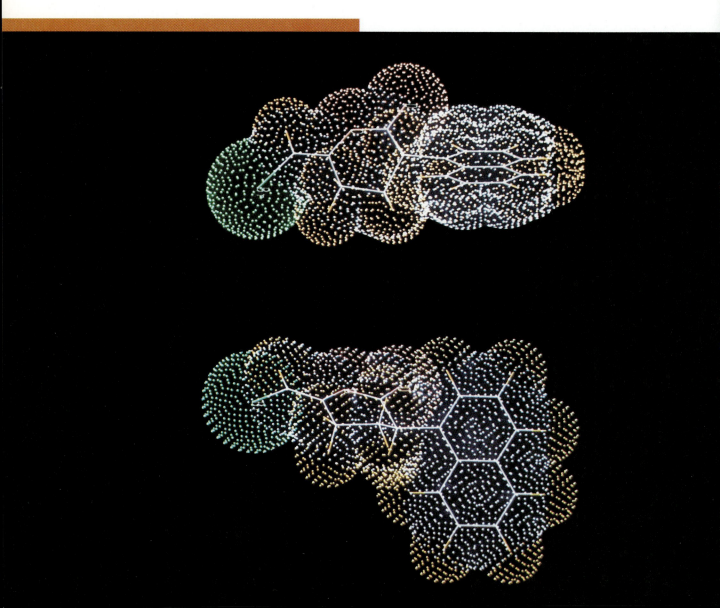

When you smell the aroma of freshly baked bread, it is because molecules from the bread interact in some way with receptor sites in the oral cavity of your nose. It is believed that the shape and size of molecules determine their ability to fit properly on these receptor sites. The interactions between the molecules and the receptors sensitize nerve endings that transmit impulses to the brain. The brain then identifies these impulses as a particular smell. Two molecules may produce different sensations of odor even when their structures differ as subtly as your right hand differs from your left.

The shape and size of a molecule, together with the strength and polarity of its bonds, largely determine the physical and chemical properties of that substance. Some of the most dramatic examples of the importance of shape and size are seen in biochemical reactions. For example, a small change in the shape or size of a drug molecule may enhance its effectiveness or reduce its side effects.

In this chapter we will consider how the shapes of simple molecules can be predicted and described. We will also examine why molecules have the shapes they do and how atoms employ their orbitals to form bonds.

In Chapter 8 we employed the simple Lewis model to account for the formulas of covalent compounds. Lewis structures do not represent the shapes of molecules. They simply describe the bonding connections between atoms in a two-dimensional representation. For example, the Lewis structure of carbon tetrachloride is intended to tell us only that four Cl atoms are bonded to a central C by four single covalent bonds:

## 9.1 MOLECULAR GEOMETRIES

The Lewis structure is drawn with the atoms in the same plane, but the actual structure is tetrahedral, as shown in Figure 9.1. Notice not only the shape of the molecule, but also the different ways used to depict the molecular geometry.

The overall shape of a molecule is determined by its **bond angles**, the angles made by the lines (the internuclear axes) joining the nuclei of the atoms in the molecule. The size of a molecule is determined by its **bond distances**, the distances between the nuclei of bonded atoms. Figure 9.1 shows the bond angles and bond distances that define the

Two views of a computer-generated molecular model of an organic molecule that functions as an inhibitor of chymotrypsin, an enzyme important in digestion. (Tripos Associates, St. Louis, SYBYL molecular modeling software)

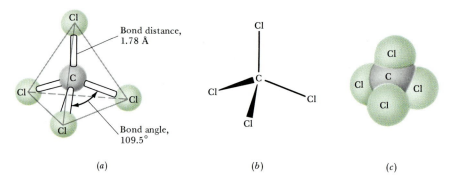

(a)　　　　　　　　　(b)　　　　　　　　　(c)

**FIGURE 9.1** The geometrical structure of the $CCl_4$ molecule, showing bond distance and bond angle. Each C—Cl bond in the molecule points toward a corner of a regular tetrahedron. A tetrahedron is a symmetrical figure consisting of four corners and four faces. Each face is an equilateral triangle. The positions of all Cl atoms in the molecule are equivalent, and all Cl—C—Cl bond angles are the same. The drawing in (a) is called a "ball-and-stick model." A simpler perspective drawing of the molecule is given in (b). The model in (c) is called a "space-filling model." It shows the relative sizes of the atoms, but the geometry of the molecule is a bit harder to see.

shape and size of $CCl_4$. It is possible to determine bond angles and bond lengths from various experimental techniques. We will not discuss these techniques here. Our concern is with how molecular geometries can be described and explained.

Although an enormous number of different molecules exist, the number of different ways that bonds can be arranged in space is rather limited. For example, three atoms can be arranged in only two different ways: in a linear fashion or in a bent fashion. Figure 9.2 shows these and several

**FIGURE 9.2** The molecular geometries of some simple molecules.

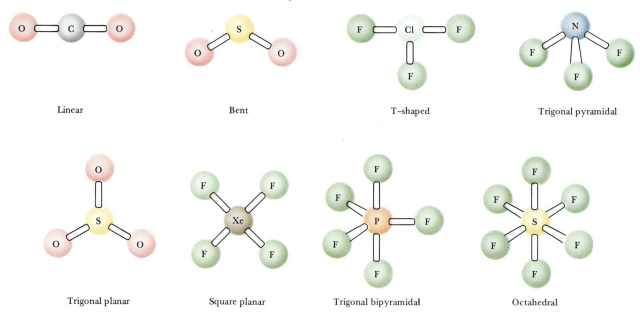

other common bonding arrangements. These arrangements, as well as the tetrahedral arrangement shown in Figure 9.1, can be predicted using the **valence-shell electron-pair repulsion (VSEPR) model**. Although the name is rather imposing, the model is quite simple.

**The Valence-Shell Electron-Pair Repulsion (VSEPR) Model**

We have seen that atoms are bonded to each other in molecules by the sharing of pairs of valence-shell electrons. Each electron pair in the valence shell of an atom is repelled by the other electron pairs about the atom. As a result, electron pairs remain as far apart from one another as possible. Thus *electron pairs orient themselves so as to make the angles between themselves as large as possible*. This simple idea is the basis for the VSEPR model.

To visualize the possible arrangements of electron pairs in the valence shell, imagine the electron pairs situated on the surface of a sphere with the nucleus of the atom at the center. Figure 9.3 shows how different numbers of electron pairs are arranged so as to minimize electron-pair repulsions. Table 9.1 summarizes the relationships between the number of electron pairs, the angles between them, and the name associated with each arrangement. You should familiarize yourself with these arrangements.

To use the VSEPR model, we must consider two types of valence-shell electron pairs: **bonding pairs**, which are shared by atoms in covalent bonds, and **nonbonding pairs** (also called lone pairs or unshared pairs). For example, the Lewis structure for ammonia reveals three bonding pairs and one nonbonding pair around the central nitrogen atom:

$$H—\overset{\displaystyle ..}{N}—H$$
$$|$$
$$H$$

The electron-pair repulsions about the nitrogen atom are minimized when these four electron pairs are arranged tetrahedrally (Table 9.1). Thus the **electron-pair geometry** of $NH_3$ is tetrahedral.

When we experimentally determine the structure of a molecule, we locate the atoms and not the electron pairs. The **molecular geometry**

**FIGURE 9.3**   Valence-shell electron pairs (red spheres) are arranged on the surface of an imaginary sphere centered at the atom's nucleus. The arrangements shown are those that minimize the mutual repulsion between the electron pairs.

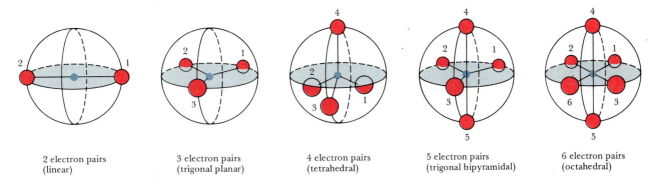

| 2 electron pairs (linear) | 3 electron pairs (trigonal planar) | 4 electron pairs (tetrahedral) | 5 electron pairs (trigonal bipyramidal) | 6 electron pairs (octahedral) |

**TABLE 9.1**  Electron-pair geometries as a function of the number of electron pairs

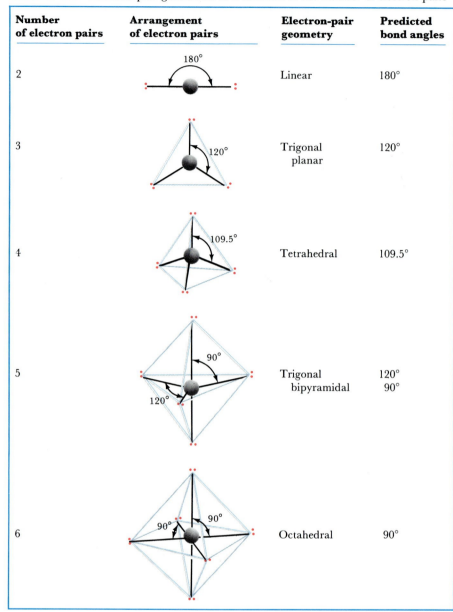

| Number of electron pairs | Arrangement of electron pairs | Electron-pair geometry | Predicted bond angles |
|---|---|---|---|
| 2 | 180° | Linear | 180° |
| 3 | 120° | Trigonal planar | 120° |
| 4 | 109.5° | Tetrahedral | 109.5° |
| 5 | 90° 120° | Trigonal bipyramidal | 120° 90° |
| 6 | 90° 90° | Octahedral | 90° |

of a molecule or ion is described in terms of how the atoms are arranged in space. However, the locations of atoms bonded to a central atom are determined by the locations of the bonding pairs, because the bonding electrons lie along the lines between nuclei (the internuclear axes). Thus we can derive the molecular geometry from the electron-pair geometry. *The name given to the shape of a molecule is the word that best describes the relative positions of the atoms.* That name is not always the same as that associated with the location of the electron pairs. For example, the $NH_3$ molecule, shown in Figure 9.4, has a tetrahedral electron-pair geometry, but the molecular geometry is described as pyramidal or trigonal pyramidal.

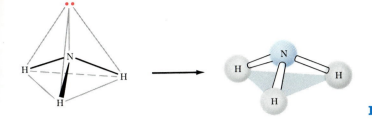

FIGURE 9.4    The molecular geometry of $NH_3$.

The steps in predicting molecular geometries using the VSEPR model follow:

1.  Sketch the Lewis dot structure of the molecule or ion.
2.  Count the total number of electron pairs around the central atom, and arrange them in the way that minimizes electron-pair repulsions (see Table 9.1).
3.  Describe the molecular geometry in terms of the angular arrangement of the *bonding* pairs. (The angular arrangement of the bonding pairs corresponds to the angular arrangement of the bonded atoms.)

In the following discussions, we apply the VSEPR model first to molecules and ions that obey the octet rule and then to those that have expanded octets.

The molecular geometries that can arise when a central atom has four or fewer valence-shell electron pairs are summarized in Table 9.2. These geometries are important because they include all of the commonly occurring structural types found for molecules or ions that obey the octet rule. Application of the VSEPR model to molecules containing double or triple bonds reveals one further rule: *Multiple bonds are considered the same as single bonds in determining molecular geometry*. That is, a double or triple bond has essentially the same effect on bond angles as a single bond and should therefore be counted as one bonding pair when predicting geometry.

**Four or Fewer Valence-Shell Electron Pairs Around a Central Atom**

---

## SAMPLE EXERCISE 9.1

Using the VSEPR model, predict the molecular geometries of the following: (**a**) $SnCl_3^-$; (**b**) $SO_2$.

**Solution:**   (**a**) The Lewis structure for the $SnCl_3^-$ ion is as follows:

$$\left[ :\overset{..}{\underset{..}{Cl}}-\overset{..}{\underset{|}{Sn}}-\overset{..}{\underset{..}{Cl}}: \right]^-$$
$$:\overset{..}{\underset{..}{Cl}}:$$

The central Sn atom is surrounded by one nonbonding electron pair and three single bonds. Thus the electron-pair geometry is tetrahedral. That is, the four electron pairs are disposed at the corners of a tetrahedron. Three of the corners are occupied by the bonding pairs of electrons. The molecular geometry is thus pyramidal:

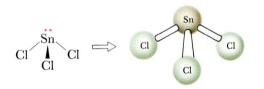

(**b**) Two resonance structures for the $SO_2$ molecule can be drawn:

$$:\overset{..}{\underset{..}{O}}-\overset{..}{S}=\overset{..}{\underset{..}{O} } \longleftrightarrow \overset{..}{\underset{..}{O}}=\overset{..}{S}-\overset{..}{\underset{..}{O}}:$$

Both resonance structures show one nonbonding electron pair, one single bond, and one double bond. When we predict geometry, a double bond is counted as one electron pair. Thus the arrangement of valence-shell electrons is trigonal planar. Two of these positions are occupied by O atoms, so the molecule has a bent shape:

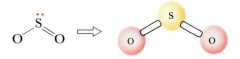

As this example illustrates, when a molecule exhibits resonance, any one of the resonance structures can be used to predict the geometry.

### PRACTICE EXERCISE

Predict the electron-pair geometry and the molecular geometry for **(a)** $H_2S$; **(b)** $CO_3^{2-}$.

***Answers:*** **(a)** tetrahedral, bent; **(b)** trigonal planar, trigonal planar

**TABLE 9.2** Electron-pair geometries and molecular shapes for molecules with two, three, and four electron pairs about the central atom

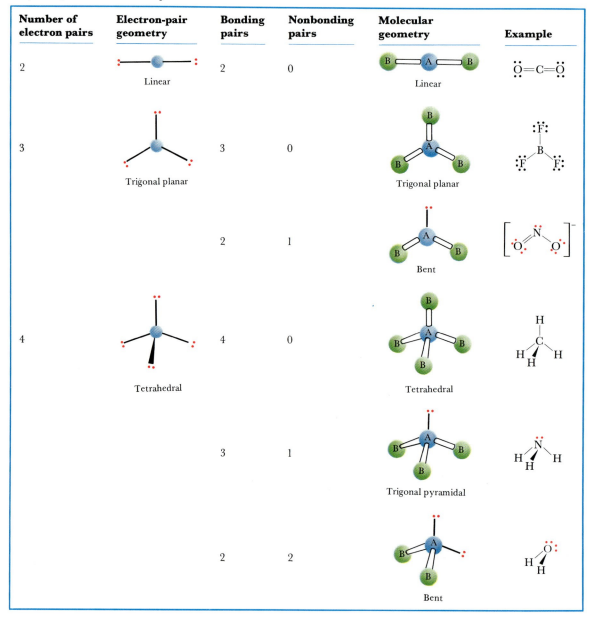

We may extend the VSEPR model to predict and explain slight distortions of molecules from the ideal geometries summarized in Table 9.2. For example, consider methane ($CH_4$), ammonia ($NH_3$), and water ($H_2O$). All three have tetrahedral electron-pair geometries, but their bond angles differ slightly:

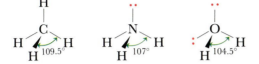

Notice that the bond angles decrease as the number of nonbonding electron pairs increases. Bonding pairs are relatively confined by the attractive influence of the two nuclei of the bonded atoms. By contrast, nonbonding electrons move under the attractive influence of only one nucleus and thus spread out more in space. As a result, *nonbonding electron pairs exert greater repulsive forces on adjacent electron pairs and thus tend to compress the angles between the bonding pairs.*

Because multiple bonds contain a higher electronic charge density than do single bonds, multiple bonds also affect bond angles. Notice how the bond angles in the formaldehyde molecule, $H_2CO$, differ from the ideal $120°$ angles:

The double bond seems to act much like a nonbonding pair of electrons, reducing the H—C—H bond angle from $120°$ to $116°$.

When a molecule contains a central atom from the third period of the periodic table and beyond, that atom may have more than four electron pairs around it (Section 8.7). There are numerous molecules with five or six electron pairs around a central atom. The molecular geometries that have been observed are summarized in Table 9.3.

The most stable electron-pair geometry for five electron pairs is the trigonal bipyramid. In the trigonal bipyramid, there are two geometrically distinct types of electron pairs, labeled the axial and radial (or equatorial) pairs, as shown in Figure 9.5. In an axial position, an electron pair is situated at $90°$ from the three radial electron pairs. In a radial position, an electron pair is situated $120°$ from the other two radial pairs and $90°$ from the two axial pairs. Because a radial electron pair has fewer neighbors at $90°$, it experiences less repulsion from nearby electron pairs than does an electron pair in an axial position. Because nonbonding electron pairs exert larger repulsions than bonding pairs, the nonbonding electron pairs always occupy the radial positions in the trigonal bipyramid, as shown in Table 9.3.

**TABLE 9.3** Electron-pair geometries and molecular shapes for molecules with five and six electron pairs about the central atom

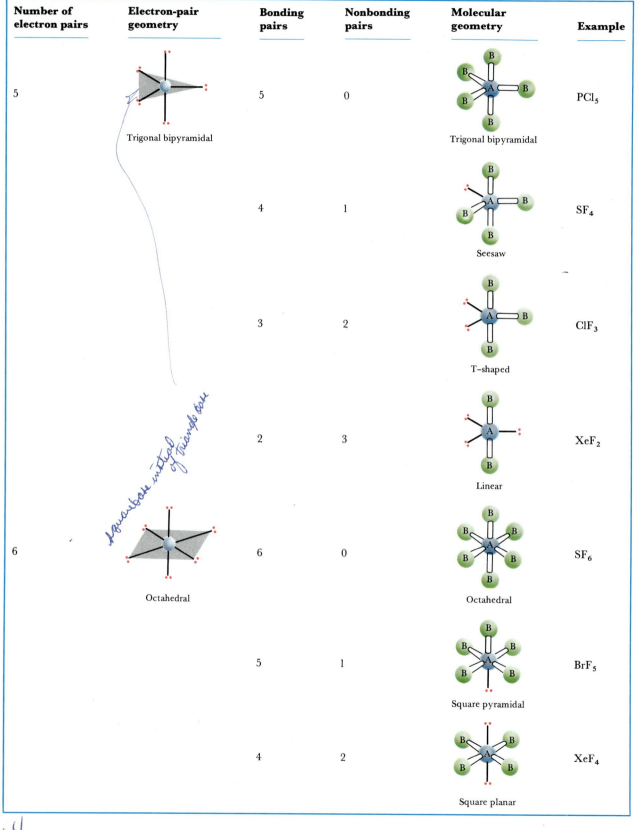

| Number of electron pairs | Electron-pair geometry | Bonding pairs | Nonbonding pairs | Molecular geometry | Example |
|---|---|---|---|---|---|
| 5 | Trigonal bipyramidal | 5 | 0 | Trigonal bipyramidal | $PCl_5$ |
| | | 4 | 1 | Seesaw | $SF_4$ |
| | | 3 | 2 | T-shaped | $ClF_3$ |
| | | 2 | 3 | Linear | $XeF_2$ |
| 6 | Octahedral | 6 | 0 | Octahedral | $SF_6$ |
| | | 5 | 1 | Square pyramidal | $BrF_5$ |
| | | 4 | 2 | Square planar | $XeF_4$ |

square base instead of triangle base

p268

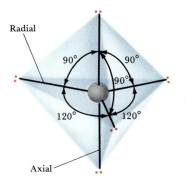

Radial

90° 90°

90°

120° 120°

Axial

**FIGURE 9.5** Trigonal bipyramidal arrangement of five electron pairs about a central atom. There are two geometrically distinct types of electron pairs, the two axial and three radial pairs. Unshared electron pairs occupy the radial positions.

The most stable electron-pair geometry for six electron pairs is the octahedron. All the angles in the octahedron are 90°, and there is no distinction between axial and radial positions (that is, all the positions are the same). Thus if a molecule has five bonding pairs of electrons and one nonbonding pair, it makes no difference where we place them on the octahedron. However, when there are two nonbonding electron pairs, their repulsions are minimized by placing them on opposite sides of the octahedron, as shown in Table 9.3.

## SAMPLE EXERCISE 9.2

Using the VSEPR model, predict the molecular geometry of **(a)** $SF_4$; **(b)** $IF_5$.

**Solution:** **(a)** The Lewis dot structure for $SF_4$ is

$$:\ddot{F}: \quad \ddot{F}:$$
$$:\ddot{F}-\ddot{S}.$$
$$:\ddot{F}:$$

The sulfur has five valence-shell electron pairs around it. Each pair points toward a corner of a trigonal bipyramid. The one nonbonding pair occupies a radial position. The four bonding electron pairs occupy the remaining four positions:

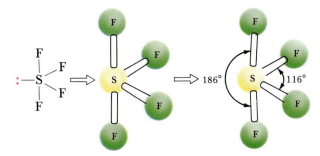

The experimentally observed structure is shown on the right. From this structure we can infer that the unshared electron pair resides in a radial position, as

predicted. Note that the axial S—F bonds are slightly bent back, suggesting that they are pushed by the nonbonding electron pair, with its greater repulsive effect.

**(b)** The Lewis structure of $IF_5$ (nonbonding pairs excluded from F) is

The iodine has six pairs of valence-shell electrons around it, five of them bonding pairs. The electron pairs should point toward the corners of an octahedron. The geometrical arrangement of atoms is therefore square pyramidal (Table 9.3).

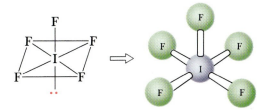

## PRACTICE EXERCISE

Predict the molecular geometries of **(a)** $ClF_3$; **(b)** $ICl_4^-$.

*Answers:* **(a)** T-shaped; **(b)** square planar

## Molecules with No Single Central Atom

The molecules and ions whose structures we have thus far considered contain only a single central atom. The VSEPR model is readily extended to more complex molecules. Consider the acetic acid molecule, whose Lewis structure is shown below:

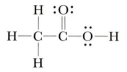

We can predict, in turn, the geometry around the leftmost C atom, the central C atom, and finally the rightmost O atom.

Notice that the leftmost C has four electron pairs around it, all of them bonding pairs. Thus the geometry around that atom is tetrahedral. The central C has effectively three bonding pairs around it (counting the double bond as if it were one pair). Thus the geometry around that atom is trigonal planar. The O atom has four electron pairs, giving it a tetrahedral electron pair geometry. However, only two of these pairs are bonding pairs, so the molecular geometry around O is bent. The molecular geometry is shown in Figure 9.6.

---

### SAMPLE EXERCISE 9.3

Predict the approximate values for the H—O—C and O—C—O bond angles in oxalic acid:

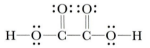

**Solution:** To predict the H—O—C bond angle, consider the number of electron pairs around the central O of this angle. Because there are four electron pairs (two single bonds and two nonbonding electron pairs), the electron-pair geometry is tetrahedral, and thus the bond angle is approximately 109°.

In the case of the O—C—O bond angle, the central C atom is surrounded by a double bond and two single bonds. To predict their arrangement, we count the double bond as a single bond, so we have three electron pairs. Thus the electron-pair geometry is trigonal planar, and the bond angle is approximately 120°.

### PRACTICE EXERCISE

Predict the approximate C—C—C bond angle in the following molecule:

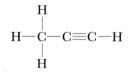

**_Answer:_** 180°

---

**FIGURE 9.6** The molecular structure of acetic acid, $HC_2H_3O_2$.

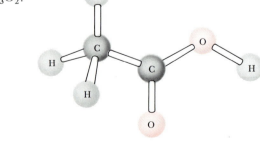

The shape of a molecule and the polarity of its bonds together determine the charge distribution in the molecule. A molecule is said to be **polar** if its centers of negative and positive charge do not coincide. A polar molecule has one end with a slightly negative charge and the other with a slight positive one.

Any diatomic molecule with a polar bond (Section 8.9) is a polar molecule. For example, the HF molecule is polar, having a concentration of negative charge on the more electronegative F atom, leaving the less electronegative H atom as the positive end:

$$\overset{\longmapsto}{\text{H—F}} \qquad \text{polar}$$

Polar molecules align themselves in an electric field, as shown in Figure 9.7. They also align themselves with respect to each other and with respect to ions. The negative end of one polar molecule and the positive side of another attract each other. Polar molecules are likewise attracted to ions. The negative end of a polar molecule is attracted to a positive ion; the positive end is attracted to a negative ion. These interactions are extremely important in explaining the properties of liquids, solids, and solutions, as you will see in Chapters 11 and 12.

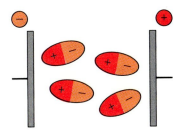

**FIGURE 9.7** Polar molecules align themselves in an electric field, with their negative sides pointing toward the positive plate.

The degree of polarity of a molecule is measured by its **dipole moment**. The dipole moment, $\mu$, is defined as the product of the charge at either end of the dipole, $Q$, times the distance, $r$, between the charges: $\mu = Qr$. Thus the dipole moment increases as the quantity of charge that is separated increases and/or as the distance between the positive and negative centers increases.

Dipole moments are generally reported in units of debye, D; a debye is $3.33 \times 10^{-30}$ C-m (C-m stands for coulomb-meter). Let's consider what the dipole moment value tells us about the separation of charge in a polar molecule. As an example, the dipole moment of HCl is 1.03 D. The H—Cl bond distance in this molecule is 1.36 Å. From these data, we can calculate the value of $Q$ in the formula for the dipole moment, assuming that the charges are centered on the H and Cl atoms. We have

$$\mu = Qr$$

$$(1.03\ \text{D})\left(3.33 \times 10^{-30}\ \frac{\text{C-m}}{\text{D}}\right) = Q \times (1.36\ \text{Å})\left(10^{-10}\ \frac{\text{m}}{\text{Å}}\right)$$

$$Q = \frac{3.43 \times 10^{-30}\ \text{C-m}}{1.36 \times 10^{-10}\ \text{m}}$$

$$= 2.52 \times 10^{-20}\ \text{C}$$

Note from the back inside cover that the charge of the electron is $1.60 \times 10^{-19}$ C. Thus, as a percentage of the electronic charge, $Q$ is $100 \times (2.52 \times 10^{-20})/(1.60 \times 10^{-19}) = 15.8\%$. If the H—Cl were ionic, there would be a full + charge on H and a full − charge on Cl. Thus, $Q$ would be 100 percent of the electronic charge. In reality, $Q$ is less than this because the H—Cl bond is not completely ionic but rather polar covalent. The dipole moments of the hydrogen halides are listed in Table

**TABLE 9.4**   Some properties of hydrogen halides

| Compound | Electronegativity difference | Dipole moment (D) |
|---|---|---|
| HF | 1.8 | · 1.91 |
| HCl | 1.0 | 1.03 |
| HBr | 0.8 | 0.79 |
| HI | 0.5 | 0.38 |

9.4. Notice that the dipole moment decreases as the electronegativity difference decreases.

To predict when a simple molecule containing more than two atoms will be polar, we need to consider if the molecule has polar bonds and how these bonds are positioned relative to each other. Consider the linear $CO_2$ molecule:

O=C=O    nonpolar

Each C—O bond is polar, but their orientation does not give the molecule a positive end. Rather, both ends of the molecule carry negative charge.

In the $H_2O$ molecule, the polar bonds are not positioned opposite each other, as in $CO_2$. Rather, both bonds pull electron density toward the O end of the molecule:

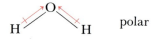
polar

As a result, the $H_2O$ molecule is polar, with the negative side being the O atom.

Figure 9.8 shows several further examples of both polar and nonpolar molecules, all of which have polar bonds. Notice that the molecules in which the central atom is symmetrically surrounded by identical atoms ($BF_3$ and $CCl_4$) are nonpolar. Although our examples do not show it, nonbonding valence-shell electrons as well as polar bonds can contribute to molecular polarity.

---

**SAMPLE EXERCISE 9.4**

Predict whether the following molecules are polar or nonpolar: **(a)** ICl; **(b)** $SO_2$; **(c)** $SF_6$.

**Solution:**   **(a)** Chlorine is a more electronegative element than iodine. Consequently, ICl will be polar with chlorine as the negative end:

I—Cl

All diatomic molecules with polar bonds are polar molecules.

**(b)** Oxygen is more electronegative than sulfur; the molecule therefore has the polar bonds necessary

for the molecule to be polar. The molecule has the following resonance forms:

$$ :\ddot{O}-\ddot{S}=\ddot{O} \quad \longleftrightarrow \quad \ddot{O}=\ddot{S}-\ddot{O}: $$

The nonlinear shape of the molecule results from the trigonal planar arrangement of electron pairs around sulfur (Table 9.2), with one unshared pair of electrons on the sulfur atom. Because of the molecular shape, the bond polarities do not cancel and the molecule is polar:

(It is unlikely that the unshared electron pair on sulfur completely cancels the effect of the polar S—O bonds, though it certainly does so to some extent.)

**(c)** Fluorine is more electronegative than sulfur. The bond dipoles therefore point toward fluorine. The six S—F bonds are arranged in a symmetric octahedral fashion around the central sulfur (there are no unshared valence-shell electron pairs on sulfur):

The bond dipoles cancel each other; the molecule does not have a negative and positive side. It is nonpolar, with an overall dipole moment of zero.

### PRACTICE EXERCISE
Are the following molecules polar or nonpolar: **(a)** $NF_3$; **(b)** $BCl_3$?
*Answers:* **(a)** polar; **(b)** nonpolar

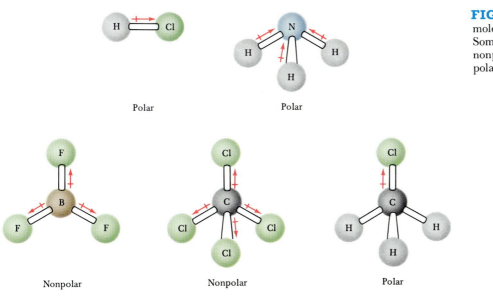

**FIGURE 9.8** Examples of molecules with polar bonds. Some of these molecules are nonpolar because their bond polarities cancel each other.

The VSEPR theory provides a simple model for predicting the shapes of molecules. However, it does not explain why bonds exist between atoms. The shared and unshared pairs are taken as given; the model is used simply to deduce the shape of the molecule or ion. In developing a theory of covalent bonding, chemists have also approached the problem from another direction. Suppose we take the formula and geometrical structure of the molecule as given. How can we account for the observed geometries in terms of the atomic orbitals used by the atoms in forming bonds to one another?

In the Lewis theory, covalent bonding occurs when atoms share electrons. Such sharing concentrates electron density between nuclei. This buildup of electron density occurs when a valence atomic orbital of one atom merges with that of another atom. The orbitals are then said to share a region of space, or to **overlap**. The overlap of two $1s$ orbitals from two H

## 9.3 HYBRID ORBITALS AND MOLECULAR SHAPE

**Orbital Overlap**

Atoms approach each other

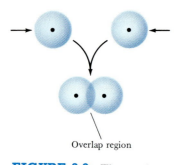

Overlap region

**FIGURE 9.9**  The overlap
of two 1*s* orbitals from two
H atoms to form H$_2$.

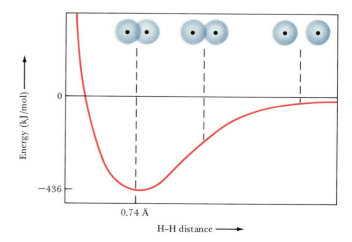

**FIGURE 9.10**  The total potential-energy change during the formation of the
stable H$_2$ molecule by the overlap of two hydrogen 1*s* orbitals. The minimum in
the energy, at 0.74 Å, represents the equilibrium bond distance. The energy at that
point, $-436$ kJ, corresponds to the energy change for formation of the H—H bond.

**FIGURE 9.11**  Formation
of $\sigma$ bonds by overlap (*a*) of
an *s* orbital with a *p* orbital,
and (*b*) of two *p* orbitals.

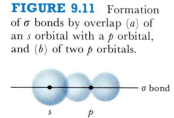

**FIGURE 9.12**  Formation
of a $\pi$ bond by overlap of two
*p* orbitals. The two regions of
overlap constitute *one* $\pi$ bond.

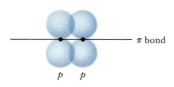

atoms to form H$_2$ is shown in Figure 9.9. Due to the orbital overlap, the
pair of electrons in H$_2$ is found within the region influenced by both
nuclei. Indeed, both electrons are simultaneously attracted to both nuclei,
and this attraction leads to bonding.

There is always an optimum distance between the two bonded nuclei
in any covalent bond. This fact is illustrated in Figure 9.10, which shows
the coming together of two H atoms to form the H$_2$ molecule. As the
atoms come closer together, the overlap between their 1*s* orbitals increases. Because of the resultant increase of electron density between the
nuclei, the potential energy of the system decreases. That is, the strength
of the bond increases, as shown by the decrease in the energy on the curve.
However, the figure also shows that as the atoms come very close together,
the energy increases rapidly. This rapid increase is due to the electrostatic
repulsion between nuclei, which becomes significant at short internuclear
distances. The internuclear distance at the minimum in the potential
energy curve corresponds to the observed bond distance. Thus the observed bond distance is a compromise between increased overlap of the
atomic orbitals, which draws the atoms together, and the nuclear-nuclear
repulsions, which force them apart.

The bond that results from the overlap of two *s* orbitals, one from
each atom, is called a **sigma ($\sigma$) bond**. A $\sigma$ bond is a covalent bond in
which the electron density is concentrated symmetrically along the internuclear axis. That is, in a $\sigma$ bond the line joining the two nuclei passes
through the middle of the overlap region. The overlap of an *s* orbital
with a *p* orbital and the overlap of two *p* orbitals also produces $\sigma$ bonds,
as shown in Figure 9.11.

A different kind of bond results from the overlap between two *p* orbitals
oriented perpendicular to the line connecting the nuclei (Figure 9.12).
This sideways overlap of *p* orbitals produces a **pi ($\pi$) bond**. A $\pi$ bond is
a covalent bond in which the overlap regions lie above and below the
internuclear axis. Because the total overlap in a $\pi$ bond tends to be less
than that in a $\sigma$ bond, $\pi$ bonds tend to be weaker than $\sigma$ bonds.

A single bond is always of the $\sigma$ type. A double bond consists of a $\sigma$ bond and a $\pi$ bond, and a triple bond consists of a $\sigma$ bond and two $\pi$ bonds:

$$H\!-\!H \qquad \ddot{O}\!=\!C\!=\!\ddot{O} \qquad :C\!\equiv\!O:$$

one $\sigma$ bond $\qquad$ one $\sigma$ bond $\qquad$ one $\sigma$ bond
$\qquad\qquad\qquad\quad$ + one $\pi$ bond $\qquad$ + two $\pi$ bonds

Double bonds (and hence $\pi$ bonds) are more common in molecules with small atoms, especially O, N, and C. Larger atoms (for example, S, P, and Si) form $\pi$ bonds less readily. We will take a closer look at $\pi$ bonds in Section 9.4. Meanwhile, let's consider how $\sigma$ bonding and molecular structure are related.

## Hybrid Orbitals

The idea of the overlap of atomic orbitals allows us to understand why covalent bonds form. However it is not so obvious how to relate molecular geometries with the directional characteristics of atomic orbitals. Consider $CH_4$, whose Lewis structure and molecular geometry are shown in Figure 9.13. This molecule is tetrahedral with four equivalent $\sigma$ bonds between C and the four surrounding H atoms. In order to have the overlap regions along the bond axes, C must have orbitals that point toward the H atoms. If carbon has four equivalent orbitals that point toward the corners of a tetrahedron, overlap of these orbitals with the $1s$ atomic orbital of each H will form bonds having the correct directions, as shown in Figure 9.13(c).

What kinds of orbitals point toward the corners of a tetrahedron? The valence-shell $2s$ and $2p$ orbitals of C do not have these orientations. However, the electron arrangements characteristic of an isolated atom can be perturbed or changed by the influence of nearby atoms. We may imagine that as four H atoms approach the C atom, the $s$ and $p$ orbitals mix, losing their identities and forming a new set of four equivalent orbitals, as shown in Figure 9.14. Each of these equivalent orbitals has electron density pointed toward a corner of a tetrahedron. This orientation increases electron density between atoms, thereby increasing orbital overlap and producing bonds that are more stable. The mixing of one $s$ and three $p$ orbitals leads to four **hybrid orbitals**. Because these four hybrid orbitals are formed by mixing, or hybridizing, an $s$ orbital and

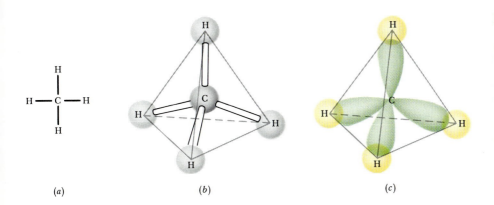

(a) $\qquad\qquad$ (b) $\qquad\qquad$ (c)

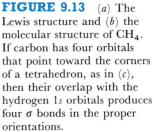

**FIGURE 9.13** (a) The Lewis structure and (b) the molecular structure of $CH_4$. If carbon has four orbitals that point toward the corners of a tetrahedron, as in (c), then their overlap with the hydrogen $1s$ orbitals produces four $\sigma$ bonds in the proper orientations.

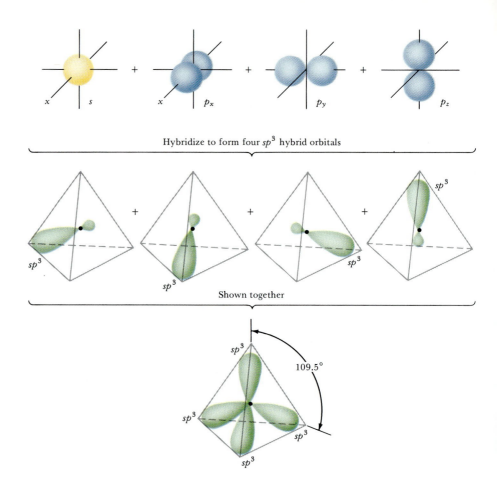

**FIGURE 9.14** Formation of four $sp^3$ hybrid orbitals from a set of one $s$ orbital and three $p$ orbitals.

three $p$ orbitals, they are called $sp^3$ hybrids. (Notice that the superscript 3 denotes the relative proportions of $s$ and $p$ orbitals, not the number of electrons they contain. Likewise, hybrid orbitals formed from an $s$ and two $p$ orbitals are called $sp^2$ hybrids.)

The hybridization process can be represented in an orbital energy diagram, as shown in Figure 9.15. Each of the $sp^3$ hybrid orbitals has the same shape and the same energy. Notice also that four atomic orbitals produce four hybrid orbitals. In general, the number of hybrid orbitals formed always equals the number of atomic orbitals used.

The idea of $sp^3$ hybridization can also be used to describe the bonding in other molecules, such as $NH_3$ and $H_2O$, where the electron-pair geometry around the central atom is approximately tetrahedral. In $NH_3$, one of the $sp^3$ hybrid orbitals contains the nonbonding pair of electrons, and the other three contain the bonding pairs, as shown in Figure 9.16(*a*). In $H_2O$, two of the hybrid orbitals contain nonbonding pairs of electrons, while the other two are used in forming $\sigma$ bonds with hydrogen atoms, as shown in Figure 9.16(*b*).

Various combinations of atomic orbital sets can be mixed, or hybridized, to obtain different geometries of orbitals about a central atom. Table 9.5 shows several of the more important hybrid orbital combina-

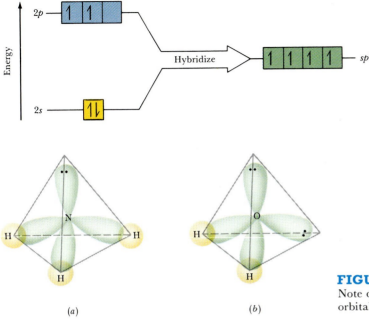

FIGURE 9.15 An orbital energy-level diagram showing formation of $sp^3$ hybrid orbitals.

**FIGURE 9.16** (*a*) Ammonia, $NH_3$, and (*b*) water. Note overlap of hydrogen $1s$ orbitals with the $sp^3$ hybrid orbitals of the central atom in both molecules.

(*a*)

(*b*)

tions and the geometries to which they correspond. The hybrid orbital sets are illustrated in Figure 9.17. Note that the expanded valence shells about atoms—that is, those in which there are more than eight electrons in the valence shell—can be accommodated by mixing in *d* orbitals along with *s* and *p*.

The purpose in formulating hybrid orbitals is to provide a convenient model in which we can imagine the electrons to be localized in the region between two atoms. The picture of hybrid orbitals has limited predictive value; that is, we cannot say in advance that in $NH_3$ the nitrogen uses essentially $sp^3$ hybrid orbitals. Once given the molecular geometry, however, we can employ the concept of hybridization to describe the atomic orbitals employed by the central atom in bonding.

In order to predict the hybrid orbitals used by an atom in bonding, we use the following three steps:

**TABLE 9.5** Geometrical arrangements characteristic of hybrid orbitals

| Atomic orbital set | Hybrid orbital set | Geometrical arrangement | Examples |
|---|---|---|---|
| $s,p$ | $sp$ | Linear (180° angle) | $Be(CH_3)_2$, $HgCl_2$ |
| $s,p,p$ | $sp^2$ | Trigonal planar (120° angles) | $BF_3$ |
| $s,p,p,p$ | $sp^3$ | Tetrahedral (109.5° angles) | $CH_4$, $AsCl_4^+$, $TiCl_4$ |
| $d,s,p,p^a$ | $dsp^2$ | Square planar (90° angles) | $PdBr_4^{2-}$ |
| $d,s,p,p,p^a$ | $dsp^3$ | Trigonal bipyramidal (120° and 90° angles) | $PF_5$ |
| $d,d,s,p,p,p^a$ | $d^2sp^3$ | Octahedral (90° angles) | $SF_6$, $SbCl_6^-$ |

[a] Depending on the particular element, the *d* orbital that mixes with the *s* and *p* may be the same quantum number as the *s* and *p* or one number lower.

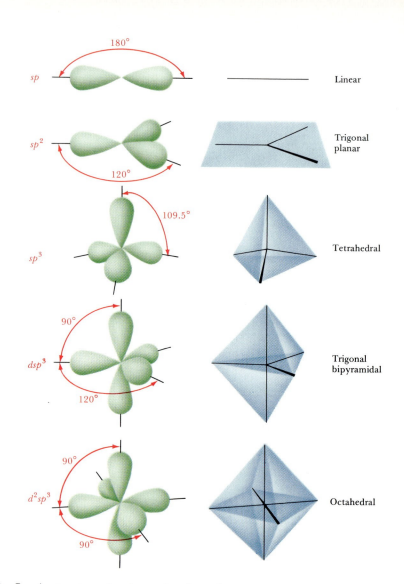

**FIGURE 9.17** Geometrical arrangements characteristic of hybrid orbital sets.

1. Draw the Lewis structure for the molecule or ion.
2. Determine the arrangement of electron pairs (the electron-pair geometry) using the VSEPR model.
3. Specify the hybrid orbitals needed to accommodate the electron pairs based on their geometrical arrangement (Table 9.5).

## SAMPLE EXERCISE 9.5

The molecule $BeH_2$ is known to be linear. Account for the bonding in $BeH_2$ in terms of the hybrid orbitals employed by Be in bonding to the two hydrogen atoms.

**Solution:** The orbital diagram for the Be atom is as follows:

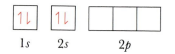

The Be atom in its ground state is incapable of forming bonds with other atoms, because all the electrons are paired. However, suppose one of the electrons is promoted to the $2p$ orbital:

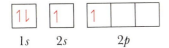

1s    2s        2p

and the 2s and one of the 2p orbitals are mixed to form two sp hybrid orbitals:

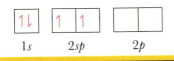

1s    2sp     2p

## SAMPLE EXERCISE 9.6

Indicate the hybridization of orbitals employed by the central atom in each of the following: **(a)** $NH_2^-$; **(b)** $SF_4$ (see Sample Exercise 9.2).

**Solution:** **(a)** The Lewis structure for $NH_2^-$ is

$$[H:\ddot{N}:H]^-$$

From the VSEPR model we conclude that the four electron pairs around N should be arranged in a tetrahedral fashion. Such a tetrahedral arrangement is characteristic of $sp^3$ hybridization (Figure 9.17); two of the hybrid orbitals contain unshared electron pairs, the other two contain pairs shared with hydrogen.

The electrons in the two sp hybrid orbitals can form shared electron pair bonds with two hydrogen atoms. The sp hybrid orbitals are directed at 180° angles from one another (see Figure 9.17) so the $BeH_2$ molecule is linear.

### PRACTICE EXERCISE
Predict the electron-pair geometry and the hybridization of boron in $BF_3$.
***Answer:*** trigonal planar; $sp^2$

**(b)** As shown in Sample Exercise 9.2, there are 10 valence-shell electrons around sulfur in $SF_4$. With an expanded octet of ten electrons, the use of a d orbital on the sulfur is indicated. The trigonal-bipyramidal arrangement of valence-shell electron pairs shown in Sample Exercise 9.2 corresponds to $dsp^3$ hybridization (Figure 9.17). One of the hybrid orbitals contains an unshared electron pair; the other four are bonded to fluorine.

### PRACTICE EXERCISE
Predict the electron-pair geometry and the hybridization of the central atom in **(a)** $SO_3^{2-}$; **(b)** $SF_6$.
***Answers:*** **(a)** tetrahedral, $sp^3$; **(b)** octahedral, $d^2sp^3$

---

The concept of hybridization may be applied also to molecules containing multiple bonds. For example, consider ethylene, $C_2H_4$, which possesses a C=C double bond. This molecule is planar with approximately 120° bond angles, as shown in Figure 9.18. The planar arrangement and 120° bond angles suggest that each carbon uses $sp^2$ hybrid orbitals (Table 9.5) to bond to the other carbon and to the two hydrogens. Because the valence orbitals on carbon consist of a 2s and *three* 2p orbitals, one 2p orbital remains unused after forming the $sp^2$ hybrid set, as shown in Figure 9.19.

Figure 9.20 shows how the C—H bonds are formed by overlap of $sp^2$ hybrid orbitals on C with 1s orbitals on H. The resultant electron density is concentrated symmetrically between the nuclei along the lines joining them. Bonds of this type are $\sigma$ bonds (Section 9.3). A $\sigma$ bond is also formed between C atoms as a result of overlap of $sp^2$ hybrid orbitals.

Figure 9.20 also shows that each carbon atom has a p orbital that is perpendicular to the plane of the molecule. These p orbitals can overlap with one another in a sideways fashion, as shown in Figure 9.21. The resultant electron density is concentrated above and below the C—C bond axis. Bonds of this type are $\pi$ bonds (Section 9.3). Thus the C=C double bond in ethylene consists of one $\sigma$ bond and one $\pi$ bond.

## 9.4 HYBRIDIZATION IN MOLECULES CONTAINING MULTIPLE BONDS

**FIGURE 9.18** The molecular geometry of ethylene, $C_2H_4$.

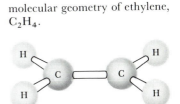

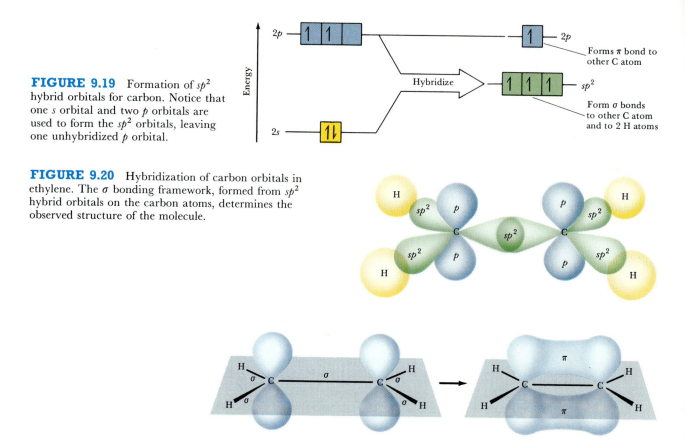

**FIGURE 9.19** Formation of $sp^2$ hybrid orbitals for carbon. Notice that one s orbital and two p orbitals are used to form the $sp^2$ orbitals, leaving one unhybridized p orbital.

**FIGURE 9.20** Hybridization of carbon orbitals in ethylene. The $\sigma$ bonding framework, formed from $sp^2$ hybrid orbitals on the carbon atoms, determines the observed structure of the molecule.

**FIGURE 9.21** Formation of the $\pi$ bond in ethylene by overlap of the 2p orbitals on each carbon atom. Note that the centers of charge density in the $\pi$ bond are above and below the bond axis, whereas in the $\sigma$ bonds the centers of charge density lie on the bond axes. The two lobes constitute *one* $\pi$ bond.

A $\pi$ bond does not appreciably influence molecular geometry. That is why we can treat a double or a triple bond as a single bond when we use the VSEPR model to predict molecular shapes.

The only kind of $\pi$ bond that we consider is formed by the overlap of p orbitals, one on each of two atoms. This $\pi$ bond can form only if unhybridized p orbitals are present on the bonded atoms. Therefore, only atoms having $sp$ or $sp^2$ hybridization can be involved in such $\pi$ bonding.

Consider acetylene, $C_2H_2$, a linear molecule containing a triple bond: H—C≡C—H. Each carbon may be visualized as using $sp$ hybrid orbitals in forming $\sigma$ bonds with the other carbon and with a hydrogen. Each

**FIGURE 9.22** Formation of two $\pi$ bonds in acetylene, $C_2H_2$, from the overlap of two sets of carbon 2p orbitals.

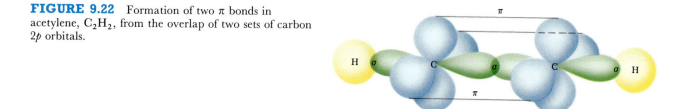

carbon then has two remaining valence $p$ orbitals at right angles to the axis of the $sp$ hybrid set (Figure 9.22). These overlap to form a pair of $\pi$ bonds. Thus the triple bond can be thought of as formed from one $\sigma$ and two $\pi$ bonds.

### SAMPLE EXERCISE 9.7

Formaldehyde, which is a planar molecule, has the following Lewis structure:

Describe the bonding in formaldehyde in terms of an appropriate set of hybrid orbitals at the carbon atom.

**Solution:** Using the VSEPR model, we would predict the bond angles around C to be about 120° (trigonal-planar geometry). The 120° bond angles about the central atom suggests $sp^2$ hybrid orbitals for the $\sigma$ bonds (Table 9.5). There remains a $2p$ orbital on carbon, perpendicular to the plane of the three $\sigma$ bonds.

   Using the VSEPR model, we predict that oxygen will have two unshared electron pairs and that the O—C $\sigma$ bond will be in a trigonal plane, with approximately 120° angles between them. In this case also, there remains a $2p$ orbital on oxygen, perpendicular to the plane of the three $\sigma$ bonds. This orbital overlaps with the similarly oriented $2p$ orbital on carbon to form a $\pi$ bond between carbon and oxygen, as illustrated in Figure 9.23.

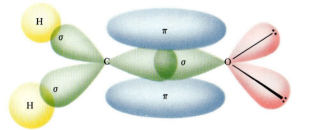

**FIGURE 9.23**   Formation of $\sigma$ and $\pi$ bonds in formaldehyde, $H_2CO$.

### PRACTICE EXERCISE

Consider the acetonitrile molecule:

$$H-\overset{\displaystyle \underset{|}{\overset{|}{H}}}{\underset{\displaystyle \underset{|}{H}}{C}}-C\equiv N\colon$$

**(a)** Predict the bond angles around each carbon; **(b)** give the hybridizations on both carbon atoms; **(c)** determine the total number of $\sigma$ and $\pi$ bonds in the molecule.
*Answers:*   **(a)** 109° around the leftmost C and 180° on the rightmost C; **(b)** $sp^3$, $sp$; **(c)** 5 $\sigma$ bonds and 2 $\pi$ bonds

In each of the molecules that we've discussed in this chapter, the bonding electrons are *localized*; by this we mean that the $\sigma$ and $\pi$ electrons are associated totally with the two atoms forming the bond. In some molecules, however, the $\pi$ electrons are free to move over several atoms; we say that such electrons are *delocalized* over those atoms. Benzene, $C_6H_6$, is an example of such a molecule. This molecule has the following resonance forms:

**Delocalized Orbitals**

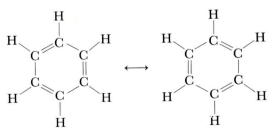

All C—C bonds in benzene are of equal length; the C—C distance is 1.395 Å, intermediate between the values for C—C single bonds (1.54 Å)

and C=C double bonds (1.34 Å). The bond angles around each carbon are 120°.

To describe benzene in terms of hybridization of carbon orbitals, we follow the procedure of setting up a hybrid orbital set consistent with the skeletal structure for the molecule. Because each carbon is surrounded by three atoms at 120° angles in a plane, the appropriate hybrid set (Table 9.5) is $sp^2$, as shown in Figure 9.24(a). This leaves a $p$ orbital on each carbon perpendicular to the plane of the benzene ring. The situation is very much like that in ethylene, except that now we have six carbon $2p$ orbitals, in a cyclic arrangement [Figure 9.24(b)]. The six carbon $2p$ atomic orbitals interact with one another to form $\pi$ orbitals. Each of the $2p$ orbitals overlaps with two others, one on each adjacent carbon atom, to form a kind of doughnut of electron density above and below the plane of the benzene molecule, as illustrated in Figure 9.24(c).

Because there is one electron from each $2p$ orbital, there is a total of three electron pairs in the $\pi$ orbitals formed in this manner. The electrons in the $\pi$ orbitals of benzene are said to be *delocalized* in the sense that any one electron is free to move around the entire circle of carbon atoms. This delocalization of the $\pi$ electrons gives benzene a special stability. For example, this substance does not react readily with bromine, as does ethylene.

It is common to represent benzene and related molecules by leaving off the hydrogens attached to the carbon and showing only the carbon-carbon framework. The presence of $\pi$ electrons may be shown using one of the Lewis structures or by placing a circle in the center of the carbon ring. Thus we can represent benzene as

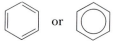

**FIGURE 9.24**   The $\sigma$ and $\pi$ bond networks in benzene, $C_6H_6$. (a) The C—C sigma bonds all lie in the molecular plane and are formed by overlap of the carbon $sp^2$ hybrid orbitals. (b) Each carbon atom has a $2p$ orbital that lies perpendicular to the molecular plane. (c) The $\pi$-bond network is formed from overlap of the $2p$ orbital on each carbon atom with the $2p$ orbitals of each of its neighbors. Six $\pi$ molecular orbitals result. The lowest-energy bonding $\pi$ molecular orbital is illustrated here.

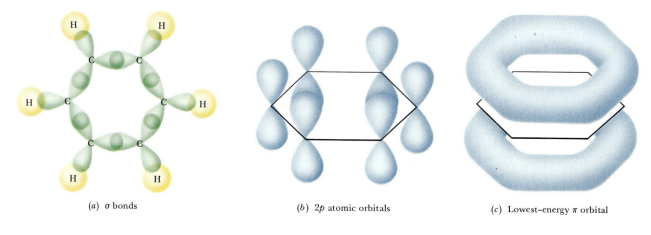

(a) $\sigma$ bonds            (b) $2p$ atomic orbitals            (c) Lowest-energy $\pi$ orbital

On the basis of all the examples we've seen, we can formulate a few general conclusions that are helpful in using the concept of hybrid orbitals to discuss molecular structures.

1. Every pair of bonded atoms shares one or more pairs of electrons. In every bond at least one pair of electrons is localized in the space between the atoms, in a $\sigma$ bond. The appropriate set of hybrid orbitals used to form the $\sigma$ bonds between an atom and its neighbors is determined by the observed geometry of the molecule. The relationship between hybrid orbital set and geometry about an atom is given in Table 9.5.

2. The electrons in $\sigma$ bonds are localized in the region between two bonded atoms and do not make a significant contribution to the bonding between any other two atoms.

3. When atoms share more than one pair of electrons, the additional pairs are in $\pi$ bonds. The centers of charge density in a $\pi$ bond lie above and below the bond axis.

4. $\pi$ bonds may extend over more than two bonded atoms. Electrons in $\pi$ bonds that extend over more than two atoms are said to be delocalized.

## 9.5 MOLECULAR ORBITALS

The models of covalent bonding and molecular geometries discussed in this and the preceding chapter are very useful. They provide a nice way of relating the formulas and structures of molecules to the electron configurations of their atoms. For example, we can understand why methane has the formula $CH_4$, and why the arrangement of C—H bonds about the central carbon is tetrahedral. But in all of this discussion, we have only briefly touched on a rather important question: Why do atoms combine to form covalent bonds in the first place? The answer to this question has to be expressed in terms of energy.

We have seen that electrons in atoms exist in allowed energy states called atomic orbitals. The quantum theory tells us that, in a similar way, electrons exist in molecules in allowed energy states that are called **molecular orbitals**.

We can think of a molecular orbital as forming from a combination of atomic orbitals. As atoms approach each other and their atomic orbitals overlap, molecular orbitals are formed. Molecular orbitals have many of the same characteristics as atomic orbitals. For example, they can hold a maximum of two electrons with opposite spins, and their electron-density distributions can be visualized by using contour representations, as we have done in discussing atomic orbitals. Furthermore, we shall see that when two atomic orbitals interact, two molecular orbitals form.

### The H$_2$ Molecule

The hydrogen molecule, $H_2$, is the simplest example of molecular bonding. As two hydrogen atoms approach each other to form $H_2$, their $1s$ orbitals can interact in two ways, as shown in Figure 9.25. The orbitals can reinforce each other so that the resultant electron density is concentrated between the nuclei. This interaction produces a molecular orbital that is more stable (lower energy) than the separate atomic orbitals. This molecular orbital is called a **bonding molecular orbital**, because the concen-

**FIGURE 9.25** Contour diagrams for the $1s$ atomic orbitals of two H atoms and for the $\sigma_{1s}$ and $\sigma_{1s}^*$ molecular orbitals that result from their interaction. In $\sigma_{1s}$, constructive interference of the $1s$ orbitals builds electron density between the nuclei. In $\sigma_{1s}^*$, destructive interference removes electron density between nuclei. Notice the node in the $\sigma_{1s}^*$ orbital, where electron density goes to zero, between the nuclei.

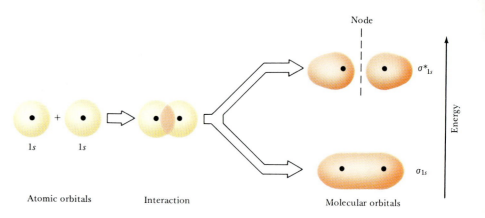

Atomic orbitals     Interaction     Molecular orbitals

tration of electron density between the nuclei holds the atoms together in a covalent bond.

The $1s$ orbitals also can interact so that the electron density cancels out where they overlap (Figure 9.25). In this case the resultant molecular orbital has a node between nuclei where the electron density is zero (see Section 6.6); in fact, the greatest electron density is on opposite sides of the nuclei. This molecular orbital is called an **antibonding molecular orbital**. It is less stable—and so at a higher energy—than the separated atomic orbitals from which it forms.

The electron density in both the bonding and the antibonding molecular orbitals of $H_2$ is centered along an imaginary line passing through the two nuclei (see Figure 9.25). Molecular orbitals of this type are called **sigma ($\sigma$) molecular orbitals**. The bonding sigma molecular orbital of $H_2$ is labeled $\sigma_{1s}$, the subscript indicating that the molecular orbital is formed from two $1s$ orbitals. The antibonding sigma orbital of $H_2$ is labeled $\sigma_{1s}^*$, the asterisk denoting that the orbital is antibonding.

The interactions between two $1s$ orbitals to form $\sigma_{1s}$ and $\sigma_{1s}^*$ molecular orbitals can be represented by an orbital **energy-level diagram**, shown in Figure 9.26. Because each H atom contains one electron, there are two electrons in $H_2$. These electrons occupy the lowest-energy molecular orbital ($\sigma_{1s}$), with their spins paired. Because the $\sigma_{1s}$ orbital is lower in energy than the isolated $1s$ orbitals, the $H_2$ molecule is more stable than the two separate H atoms.

When two helium atoms come together, each atom has two electrons in a $1s$ orbital. The $1s$ orbitals interact, forming $\sigma_{1s}$ and $\sigma_{1s}^*$ molecular

**FIGURE 9.26** Energy-level diagram for the molecular orbitals in the $H_2$ molecule.

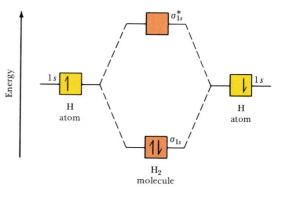

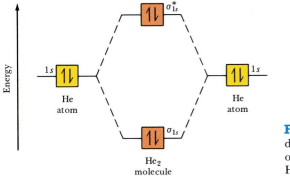

**FIGURE 9.27** Energy-level diagram for the molecular orbitals in the hypothetical $He_2$ molecule.

orbitals, just as described for hydrogen. As seen in Figure 9.27, two of the electrons are placed in the $\sigma_{1s}$ orbital; the other two must be placed in the $\sigma_{1s}^*$ orbital. The energy decrease resulting from the bonding electrons is offset by the energy increase resulting from the antibonding electrons. Hence $He_2$ is not a stable molecule. Molecular-orbital theory correctly predicts that hydrogen forms diatomic molecules but that helium does not.

The stability of a covalent bond is related to its **bond order**, which is defined as follows:

**Bond Order**

$$\text{bond order} = \tfrac{1}{2}(\text{number of bonding electrons} - \text{number of antibonding electrons}) \qquad [9.1]$$

That is, the bond order is half the difference between the number of bonding electrons and the number of antibonding electrons. We take half the difference because we are used to thinking of bonds in terms of pairs of electrons. A bond order of 1 represents a single bond, a bond order of 2 represents a double bond, and a bond order of 3 represents a triple bond. Because molecular-orbital theory also treats molecules with an odd number of electrons, bond orders of $\tfrac{1}{2}$, $\tfrac{3}{2}$, or $\tfrac{5}{2}$ are possible.

Because $H_2$ has two bonding electrons and no antibonding ones (Figure 9.26), it has a bond order of $\tfrac{1}{2}(2 - 0) = 1$. Because $He_2$ has two bonding electrons and two antibonding ones (Figure 9.27), it has a bond energy of $\tfrac{1}{2}(2 - 2) = 0$. A bond energy of 0 means that the molecule has no stability.

**SAMPLE EXERCISE 9.8**

What is the bond order of the $He_2^+$ ion? Would you expect this ion to be stable relative to the separated He atom and $He^+$ ion?

**Solution:** The energy-level diagram for this system is shown in Figure 9.28. The $He_2^+$ ion has a total of three electrons. Two are placed in the bonding orbital, the third in the antibonding orbital. Thus the bond order is

$$\text{bond order} = \tfrac{1}{2}(2 - 1) = \tfrac{1}{2}$$

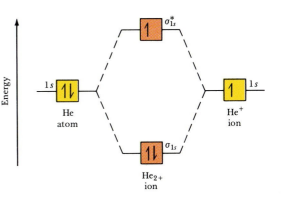

**FIGURE 9.28** Energy-level diagram for the $He_2^+$ ion.

Because the bond order is above 0, the $He_2{}^+$ molecular ion is predicted to be stable relative to the separated He and $He^+$. Formation of $He_2{}^+$ in the gas phase has been demonstrated in laboratory experiments.

## 9.6 MOLECULAR ORBITAL CONFIGURATIONS FOR DIATOMIC MOLECULES

Just as we treated the electron configuration of $H_2$ in terms of molecular-orbital theory, we can also consider the electron configurations of other diatomic molecules. In this section we will restrict our considerations to homonuclear diatomic molecules (those composed of two identical atoms). As we shall see, the procedure for determining the distribution of electrons in these molecules is analogous to that for $H_2$.

### Rules Governing Molecular Electron Configurations

There are various rules and restrictions governing how atomic orbitals can be combined to form molecular orbitals and how these orbitals are populated by electrons. Some of the most important of these follow:

1.  The number of molecular orbitals formed equals the number of atomic orbitals combined.
2.  Atomic orbitals combine most effectively with other atomic orbitals of similar energy.
3.  Each molecular orbital can accommodate up to two electrons with opposite spins (Pauli exclusion principle).
4.  When molecular orbitals have the same energy, one electron enters each orbital (with parallel spins) before pairing occurs (Hund's rule).

### Homonuclear Diatomic Molecules of the Second-Period Elements

Among the most familiar homonuclear diatomic molecules are $N_2$, $O_2$, and $F_2$, all of which are stable at ordinary temperatures. Other elements in the second period of the periodic table also form diatomic molecules under appropriate conditions. For example, when lithium metal is heated above its boiling point, 1347°C, $Li_2$ molecules are found in the vapor phase.

To apply the molecular-orbital theory to the elements of the second period requires that we consider the molecular orbitals that form from atomic orbitals in the second electron shell. (The $1s$ orbitals are essentially buried beneath the valence-shell orbitals and are therefore not involved to any appreciable extent in bonding in these species.) There are four atomic orbitals in the valence shell on each atom ($2s$, $2p_x$, $2p_y$, and $2p_z$). The interactions among these orbitals produces eight molecular orbitals, four bonding and four antibonding.

Because orbitals of comparable energy interact most effectively (rule 2), the $2s$ orbital on one atom interacts primarily with a $2s$ orbital on another atom. Similarly, the $2p$ orbitals interact mainly with one another.

The combination of any pair of $s$ orbitals produces a $\sigma$ and a $\sigma^*$ molecular orbital, like those shown earlier in Figure 9.25. Thus two $2s$ orbitals interact to form a $\sigma_{2s}$ molecular orbital and a $\sigma_{2s}^*$ molecular orbital.

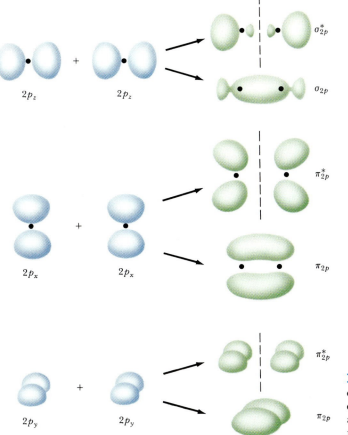

$\sigma_{2p}^*$

$\sigma_{2p}$

$\pi_{2p}^*$

$\pi_{2p}$

$\pi_{2p}^*$

$\pi_{2p}$

$2p_z$      $2p_z$

$2p_x$      $2p_x$

$2p_y$      $2p_y$

**FIGURE 9.29** Contour diagrams for molecular orbitals formed by overlap of $p$ orbitals. These diagrams illustrate the general shapes of bonding and antibonding orbitals and are not accurate representations.

The interactions between $p$ orbitals are shown in Figure 9.29. We have arbitrarily chosen the $z$ axis as the internuclear axis. Notice that the $2p_z$ orbitals overlap in a head-to-head fashion. The resultant molecular orbitals have electron density concentrated along the line through the nuclei. Thus they are $\sigma$ orbitals: $\sigma_{2p}$ and $\sigma_{2p}^*$. The other $2p$ orbitals overlap in a sideways fashion. The resultant molecular orbitals have electron density concentrated above and below the line through the nuclei. Molecular orbitals of this type are called **pi ($\pi$) molecular orbitals**. These interactions produce two $\pi$ and two $\pi^*$ orbitals.

Figure 9.30 shows the relative energies of the molecular orbitals obtained from the $2s$ and $2p$ orbitals.[†] Using this energy-level diagram, we can readily deduce the electron configuration of any of the second-row diatomic molecules. The filling of the molecular orbitals in these elements is given in Table 9.6.

[†] There is actually some slight interaction of $s$ orbitals with $p$ orbitals. For $B_2$, $C_2$, and $N_2$, this interaction causes the $\sigma_{2p}$ orbital to be of higher energy than the $\pi_{2p}$ orbital, as shown in Figure 9.30. For $O_2$ and $F_2$, $\sigma_{2p}$ is lower in energy than $\pi_{2p}$. Nevertheless, the energy-level diagram given in Figure 9.30 leads to the correct bond order and correctly predicts the magnetic properties of these molecules. Thus we will discuss only this one energy-level diagram.

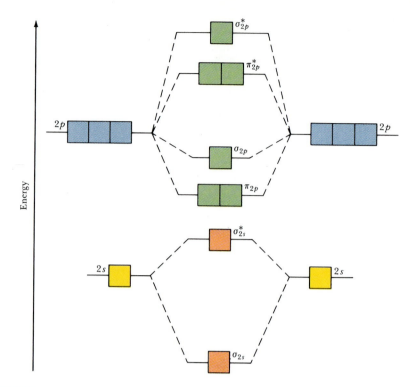

**FIGURE 9.30** General energy-level diagram for molecular orbitals of second-row diatomic molecules.

**TABLE 9.6** Electronic configuration and some experimental data for several second-row diatomic molecules

| | $B_2$ | $C_2$ | $N_2$ | $O_2$ | $F_2$ |
|---|---|---|---|---|---|
| $\sigma_{2p}^*$ | ☐ | ☐ | ☐ | ☐ | ☐ |
| $\pi_{2p}^*$ | ☐☐ | ☐☐ | ☐☐ | ↑ ↑ | ↑↓ ↑↓ |
| $\sigma_{2p}$ | ☐ | ☐ | ↑↓ | ↑↓ | ↑↓ |
| $\pi_{2p}$ | ↑ ↑ | ↑↓ ↑↓ | ↑↓ ↑↓ | ↑↓ ↑↓ | ↑↓ ↑↓ |
| $\sigma_{2s}^*$ | ↑↓ | ↑↓ | ↑↓ | ↑↓ | ↑↓ |
| $\sigma_{2s}$ | ↑↓ | ↑↓ | ↑↓ | ↑↓ | ↑↓ |
| Bond order | One | Two | Three | Two | One |
| Bond-dissociation energy (kJ/mol) | 290 | 620 | 941 | 495 | 155 |
| Bond distance (Å) | 1.59 | 1.31 | 1.10 | 1.21 | 1.43 |
| Ionization energy (kJ/mol) | — | 1150 | 1495 | 1205 | 1700 |
| Magnetic behavior | Paramagnetic | Diamagnetic | Diamagnetic | Paramagnetic | Diamagnetic |

It is interesting to compare the electronic structures of these diatomic molecules with their observable properties. The behavior of a substance in a magnetic field provides an important insight into arrangements of electrons. Molecules with one or more unpaired electrons are attracted into a magnetic field. The more unpaired electrons in a species, the stronger the force of attraction. This type of magnetic behavior is called **paramagnetism**.

Substances with no unpaired electrons are weakly repelled from a magnetic field. This property is called **diamagnetism**. A straightforward method for measuring the magnetic properties of a substance, illustrated in Figure 9.31, involves weighing the substance in the presence and absence of a magnetic field. If the substance is paramagnetic, it appears to weigh more in the magnetic field; if it is diamagnetic, it appears to weigh less. The magnetic behaviors observed for the diatomic molecules of the second-period elements agree with the electron configurations shown in Table 9.6.

The electron configurations can also be related to the bond distances and bond-dissociation energies of the molecules. As bond orders increase, bond distances decrease and bond-dissociation energies increase. Notice the short bond distance and high bond-dissociation energy of $N_2$, whose bond order is 3. The $N_2$ molecule does not react readily with other substances to form nitrogen compounds. The high bond order of the molecule helps explain its exceptional stability. We should also note that molecules with the same bond orders do not have the same bond distances and bond-dissociation energies. Bond order is only one factor influencing these properties. Other factors, including the nuclear charges and the extent of orbital overlap, also contribute.

The bonding in the dioxygen molecule, $O_2$, is especially interesting. The Lewis structure of this molecule shows a double bond and complete pairing of electrons:

$$\ddot{O}=\ddot{O}$$

**FIGURE 9.31** Experiment for determining the magnetic properties of a sample. The sample is first weighed in the absence of a magnetic field (*a*). When a field is applied, a diamagnetic sample tends to move out of the field (*b*), and thus appears to have a lower mass. A paramagnetic sample is drawn into the field (*c*), and thus appears to gain mass.

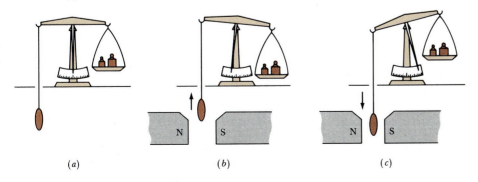

(*a*)          (*b*)          (*c*)

**FIGURE 9.32** Liquid $O_2$ being poured between the poles of a magnet. Because the $O_2$ is attracted into the magnetic field, it forms a bridge between the magnetic poles. (Donald Clegg and Roxy Wilson)

The short O—O bond distance (1.21 Å) and the relatively high bond-dissociation energy (495 kJ/mol) of the molecule are in agreement with the presence of a double bond. However, the molecule is found to contain two unpaired electrons. The paramagnetism of $O_2$ is demonstrated in Figure 9.32. Although the Lewis structure fails to account for the paramagnetism of $O_2$, molecular-orbital theory predicts that there are two unpaired electrons in the $\pi_{2p}^*$ orbital of the molecule (Table 9.6). The molecular-orbital description also correctly indicates a bond order of 2.

---

**SAMPLE EXERCISE 9.9**

Predict the following properties of $O_2{}^+$: **(a)** number of unpaired electrons; **(b)** bond order; **(c)** bond-dissociation energy and bond length.

**Solution:** **(a)** The $O_2{}^+$ ion has one electron less than $O_2$. The electron removed from $O_2$ to form $O_2{}^+$ is one of the two unpaired $\pi^*$ electrons (see Table 9.6). $O_2{}^+$ should therefore have just one unpaired electron left.

**(b)** The molecule has eight bonding electrons (the same number as $O_2$) and three antibonding ones (one less than $O_2$). Thus its bond order is

$$\text{bond order} = \tfrac{1}{2}(8 - 3) = 2\tfrac{1}{2}$$

**(c)** The bond-dissociation energy and bond length should be about midway between that for $O_2$ and $N_2$, say 720 kJ/mol and 1.15 Å, respectively. The observed bond-dissociation energy and bond length of the ion are 625 kJ/mol and 1.123 Å.

**PRACTICE EXERCISE** ⎯⎯⎯⎯⎯⎯⎯⎯

Predict the magnetic properties and bond order of **(a)** the peroxide ion, $O_2{}^{2-}$; **(b)** the acetylide ion, $C_2{}^{2-}$.

*Answers:* **(a)** diamagnetic, 1; **(b)** diamagnetic, 3

---

## 9.7 STRUCTURES OF SELECTED NONMETALLIC ELEMENTS

Our discussions in this chapter have provided the background for understanding many of the properties of molecular substances in terms of their molecular geometries and bonding. It is instructive to close the chapter by applying some of these ideas in examining four common nonmetallic elements: oxygen, sulfur, carbon, and silicon.

### Oxygen and Sulfur

Oxygen, the first member of family 6A, occurs as a diatomic molecule, $O_2$. This substance is familiar as an important component of the atmosphere. The short O—O bond distance (1.21 Å) and the relatively high bond-dissociation energy (495 kJ/mol) of $O_2$ suggest that it possesses a double bond, as correctly predicted by molecular-orbital theory (Section 9.6).

Oxygen also occurs as a gaseous triatomic molecule, $O_3$, known as ozone. This substance is responsible for the pungent odor often detected around electric motors. In contrast to $O_2$, which is essential for life, $O_3$ is toxic. Ozone and $O_2$ are allotropes of oxygen. **Allotropes** are different forms of the same element in the same state (see Section 5.7). Ozone is less stable and more reactive than $O_2$. In particular, it is a much stronger oxidizing agent. Although $O_2$ reacts directly with most elements, high temperatures are often required; $O_3$ is more likely to react readily at lower temperatures.

The ozone molecule may be represented by the following resonance forms:

$$\ddot{O}\diagdown\quad\longleftrightarrow\quad\diagup\ddot{O}$$

The molecule is bent, as we would predict using the VSEPR model; the bond angle is found experimentally to be 116.8° and the O—O bond distance is 1.278 Å. The $\sigma$ bonds between the oxygen atoms involve the $sp^2$ hybridization of the valence-shell electrons on the central oxygen atom. A $\pi$ bond that is delocalized over the three atoms is formed by sideways overlap of $p$ orbitals on each atom (Figure 9.33). The bond-dissociation energy of the molecule, corresponding to the reaction $O_3(g) \longrightarrow O_2(g) + O(g)$, is 107 kJ/mol. This is much lower than the bond energy of $O_2$. The lower bond energy in $O_3$ explains why $O_3$ is more reactive than $O_2$.

Sulfur, the second member of family 6A, exists in several allotropic forms. The most stable and common allotrope at room temperature is a yellow solid with molecular formula $S_8$. The $S_8$ molecule consists of an eight-membered ring of sulfur atoms, as shown in Figure 9.34. The S—S—S bond angle is 107.8°, close to the tetrahedral angle associated with $sp^3$ hybridization. Each sulfur atom in this structure achieves an octet of electrons by bonding to two other sulfur atoms by single $\sigma$-type bonds. By contrast, oxygen satisfies its bonding capacity by forming $\pi$ bonds in the $O_2$ and $O_3$ molecules. $\pi$ bonds are generally more common between smaller atoms like O, N, and C than between larger ones like S, P, and Si.

**FIGURE 9.34** Structure of $S_8$ molecules as found in the most common allotropic form of sulfur at room temperature.

**FIGURE 9.33** Delocalized $\pi$ bond in ozone formed by overlap of $2p$ orbitals on each of the O atoms.

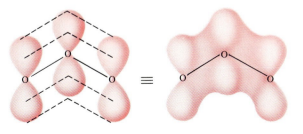

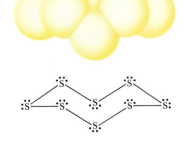

## Carbon and Silicon

Carbon, the lightest of the group 4A elements, exists in two major allotropic forms, graphite and diamond. The structure of diamond is shown in Figure 9.35. You can see that each carbon atom is surrounded by four other carbon atoms arranged at the corners of a tetrahedron. Each carbon can therefore be described as using $sp^3$ hybrid orbitals in forming $\sigma$ bonds with each of its four neighboring carbon atoms. The stability of the diamond lattice arises because the carbon atoms are interconnected in a three-dimensional array of strong carbon-carbon single bonds. Diamond is a very hard and brittle material. In fact, industrial-grade diamonds are employed in the blades of saws for the most demanding cutting jobs. The melting point of diamond, above 3500°C, is higher than that for any other element. It is exceptionally inert chemically. However, when heated to about 1800°C in the absence of air, diamond converts to graphite, the other allotropic form of the element. Diamonds also burn at about 900°C when heated in air or in oxygen. (Naturally, there is not much interest in carrying out experiments of this kind.)

The structure of graphite is shown in Figure 9.36. The carbon atoms are arranged in layers; each carbon atom is surrounded by three other carbon atoms, all at the same distance of separation. The distance between adjacent carbon atoms in the plane is 1.42 Å, very close to the C—C distance in benzene (1.395 Å). The distance between adjacent layers is 3.41 Å, too great a distance for a covalent bond to exist. Graphite occurs in gray, shiny plates and is usually found in masses of thin, easily separated sheets (see Figure 9.37). The layers readily slide past one another when rubbed, giving the substance a greasy feel. Graphite has been used as a lubricant and in making the "lead" in pencils.

The properties of graphite are due to the nature of the bonding between carbon atoms. It is because the individual layers in graphite are not directly bonded together that they slide past one another easily and are readily separated. We can understand the bonding within each layer by supposing that each carbon is bonded to its three neighbors by $sp^2$ hybrid $\sigma$ bonds. The remaining $2p$ orbital is employed in $\pi$ bonding to the same three atoms, as illustrated in Figure 9.38. But because those

**FIGURE 9.35** Structure of diamond, a major allotropic form of carbon.

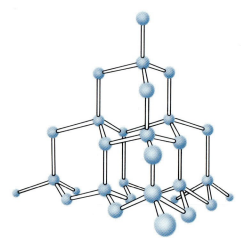

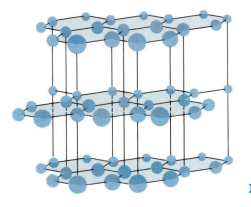

**FIGURE 9.36**   Structure of graphite.

**FIGURE 9.37**   Piece of graphite photographed with an electron microscope (magnified about 15 million times). The bright bands are layers of carbon atoms that are only 3.41 Å apart. (P. A. Marsch and A. Voet. J. M. Huber Corp.)

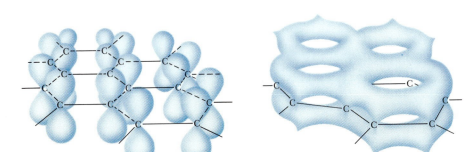

**FIGURE 9.38**   Formation of the $\pi$ bonds in graphite. Only a portion of a plane of carbon atoms is shown.

three neighboring atoms are also involved in $\pi$ bonding to two other carbons, the network of $\pi$ bonds extends over essentially the entire plane. The electrons that occupy the $\pi$ bonds are free to move from one bond to the next. Thus graphite represents an extension to an entire plane of the kind of delocalization we saw in benzene. In benzene the $\pi$ electrons are free to move about the circumference of the ring. In graphite they are free to move over the entire plane. Because of this freedom of motion, graphite is a good conductor of heat and electricity in directions along the planes of carbon atoms. (If you have ever taken apart a flashlight battery, you know that the central electrode in the battery is made of graphite.) However, graphite is an insulator in the direction normal to (perpendicular to) the planes. This is so because there is no means by which electrons can move easily from one plane to the other.

The structure of elemental silicon is the same as that for diamond. No graphitelike allotrope of this element is known. In graphite, atoms are bonded to each other by both $\sigma$ and $\pi$ bonds; in diamond, only $\sigma$ bonds are used. The absence of a graphite allotrope of silicon is consistent with our expectation that $\pi$ bonds between silicon atoms are of lower strength than $\pi$ bonds between the smaller carbon atoms. Thus silicon fills its octet by forming only $\sigma$ bonds, with each Si atom bonded to four neighboring Si atoms.

# FOR REVIEW

## SUMMARY

In this chapter, we've applied the basic principles of chemical bonding to several important areas of chemical structure and behavior. The three-dimensional structures of molecules are determined by the distances between bonded atoms and by the directions of chemical bonds with respect to one another around a particular atom. The **valence-shell electron-pair repulsion (VSEPR) model** explains these relative directions in terms of the repulsions that exist between electron pairs. According to this model, electron pairs around an atom orient themselves so as to minimize electrostatic repulsions; that is, they remain as far apart as possible. By recognizing that unshared electron pairs take up more space (exert greater repulsive forces) than shared electron pairs, it is possible to account for the departures of bond angles from the idealized values and to explain many other aspects of molecular structure. The shape of a molecule and the bond polarities determine whether or not a molecule will be polar. The degree of polarity of a molecule is measured by its **dipole moment**.

The Lewis model for covalent bonding introduced in Chapter 8 can be extended to account very nicely for the geometrical properties of molecules. We can imagine that the atoms in a molecule are bonded to one another by electron pairs that occupy pairs of overlapping atomic orbitals. The extent to which the atomic orbitals share the same region of space, called **overlap**, is important in determining the amount of stability that results from bond formation. The bonds directed along the internuclear axes are called $\sigma$ **bonds**. It is possible to formulate orbitals on an atom that are directed toward each of the other atoms surrounding it by forming **hybrid orbitals**. These orbitals are made up of mixtures of the familiar $s$, $p$, and $d$ atomic orbitals. Depending on the particular number of other atoms bonded to an atom and their arrangement in space, a particular set of hybrid orbitals can be formulated that has the necessary directional characteristics. For example, $sp^3$ hybrid orbitals are directed toward the corners of a tetrahedron.

In addition to the $\sigma$ **bonds**, which determine the geometry of the bonding around a particular atom, there may be also $\pi$ **bonds** constructed from remaining, unhybridized atomic orbitals. Thus a double bond, consisting of a $\sigma$ and a $\pi$ bond, or a triple bond, consisting of a $\sigma$ and two $\pi$ bonds, may be formed. In some molecules the $\pi$ bonds may extend, or be delocalized, over several atoms. Delocalization of the $\pi$ electrons in a cyclic structure, such as in benzene, or throughout a plane, as in graphite, leads to a special stability.

The coming together of atoms to form molecules may be viewed also as the coming together of atomic orbitals to form **molecular orbitals**. Atomic orbitals may combine with one another in various ways. The rules for combining atomic orbitals on atoms to form molecular orbitals allow us to account very well for the observed properties of the diatomic molecules formed by the first several elements of the periodic table. The molecular-orbital model is particularly impressive in explaining why the $O_2$ molecule contains two unpaired electrons. Many of the ideas presented in this chapter are important in understanding the structures of the nonmetallic elements.

## LEARNING GOALS

Having read and studied this chapter, you should be able to:

1. Relate the number of electron pairs in the valence shell of an atom in a molecule to their geometrical arrangement around that atom.

2. Explain why nonbonding electron pairs exert a greater repulsive interaction on other pairs than do bonding electron pairs.

3. Predict the geometrical structure of a molecule or ion from its Lewis structure.

4. Predict from the molecular shape and the electronegativities of the atoms involved whether a molecule has a dipole moment.

5. Explain the concept of hybridization and its relationship to geometrical structure.

6. Assign a hybridization to the valence orbitals of an atom in a molecule, knowing the number and geometrical arrangement of the atoms to which it is bonded.

7. Formulate the bonding in a molecule, in terms of $\sigma$ bonds and $\pi$ bonds, from its Lewis structure.

8. Explain the concept of delocalization in $\pi$ bonds.

9. Explain the concept of orbital overlap.

10. Describe how molecular orbitals are formed by overlap of atomic orbitals.

11. Explain the relationship between bonding and antibonding molecular orbitals.

12. Construct the molecular-orbital energy-level diagram for a diatomic molecule or ion built from elements of the first or second row, and predict the bond order and number of unpaired electrons.

13. Describe the structures of elemental oxygen, sulfur, carbon, and silicon.

# KEY TERMS

Among the more important terms and expressions used for the first time in this chapter are the following:

An **antibonding molecular orbital** (Section 9.5) is a molecular orbital in which electron density is concentrated outside the region between the two nuclei of bonded atoms. Such orbitals, designed as $\sigma^*$ or $\pi^*$, are less stable (of higher energy) than bonding molecular orbitals.

A **bonding molecular orbital** (Section 9.5) is one in which the electron density is concentrated in the internuclear region. The energy of a bonding molecular orbital is lower than the energy of the separate atomic orbitals from which it forms.

**Bond order** (Section 9.5) is expressed as the number of bonding electron pairs shared between two atoms, less the number of antibonding electron pairs: bond order = $\frac{1}{2}$(number of bonding electrons − number of antibonding electrons).

**Delocalized electrons** (Section 9.4) are electrons that are spread over a number of atoms in a molecule rather than localized between a pair of atoms.

**Diamagnetism** (Section 9.6) is a type of magnetism that causes a substance with no unpaired electrons to be weakly repelled from a magnetic field.

The **dipole moment** (Section 9.2) is a measure of the separation between centers of positive and negative charges in polar molecules.

**Hybridization** (Section 9.3) refers to the mixing of different types of atomic orbitals to produce a set of equivalent hybrid orbitals.

A **molecular orbital** (Section 9.5) is an allowed state for an electron in a molecule. A molecular orbital is entirely analogous to an atomic orbital, which is an allowed state for an electron in an atom. A molecular orbital may be classified as $\sigma$ or $\pi$, depending on the disposition of electron density with respect to the internuclear axis.

The term **overlap** (Section 9.3) refers to the extent to which atomic orbitals on different atoms share the same region of space to form a molecular orbital. When overlap is large, a strong bond may be formed.

**Paramagnetism** (Section 9.6) is a property that a substance possesses if it contains one or more unpaired electrons. A paramagnetic substance is drawn into a magnetic field.

A **pi ($\pi$) bond** (Section 9.3) is a covalent bond in which electron density is concentrated above and below the line joining the bonding atoms.

A **sigma ($\sigma$) bond** (Section 9.3) is a covalent bond in which electron density is concentrated along the internuclear axis.

The **valence-shell electron-pair repulsion (VSEPR) model** (Section 9.1) accounts for the geometrical arrangements of shared and unshared electron pairs around a central atom in terms of the repulsions between electron pairs.

# EXERCISES

## The VSEPR Model

**9.1** Describe the electron-pair geometry characteristic of each of the following numbers of electron pairs about a central atom: (a) 3; (b) 4; (c) 5; (d) 6.

**9.2** Indicate the number of electron pairs about a central atom, given the following angles between them: (a) 120°; (b) 180°; (c) 109°; (d) 90°.

**9.3** What is the difference between the electron-pair geometry and the molecular geometry of a molecule? Use the water molecule as an example in your discussion.

**9.4** Are the following statements true or false? If a statement is false, correct it. (a) Only bonding electron pairs around an atom are considered in determining molecular geometry. (b) The electron-pair geometry and the molecular geometry of a molecule are always the same.

**9.5** Name the electron-pair geometry and the molecular geometry for each of the following molecules and ions: (a) $ClO_2^-$; (b) $ClO_3^-$; (c) $CO_3^{2-}$; (d) $CF_4$; (e) $NF_3$; (f) $ICl_2^-$.

**9.6** Name the electron-pair geometry and the molecular geometry for each of the following molecules and ions: (a) $BH_4^-$; (b) $SO_3^{2-}$; (c) $SO_3$; (d) $SCl_2$; (e) $Cl_2SO$; (f) $ICl_3$.

**9.7** The molecules $NF_3$, $BF_3$, and $ClF_3$ all have molecular formulas of the type $XF_3$, but the molecules have different molecular geometries. Predict the shape of each molecule, and explain the origin of the differing shapes.

**9.8** The molecules $SiF_4$, $SF_4$, and $XeF_4$ all have molecular formulas of the type $XF_4$, but the molecules have different molecular geometries. Predict the shape of each molecule, and explain the origin of the differing shapes.

**9.9** The three species $NO_2^+$, $NO_2$, and $NO_2^-$ all have a central N atom. The ONO bond angles in the three species are 180°, 134°, and 115°, respectively. Explain this variation in bond angles.

**9.10** The three species $NH_2^-$, $NH_3$, and $NH_4^+$ have H—N—H bond angles of 105°, 107°, and 109°, respectively. Explain this variation in bond angles.

**9.11** Give approximate values for the indicated bond angles in the following molecules:

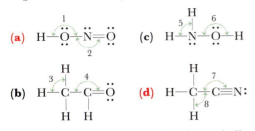

**(a)** H—Ö—N=Ö

**(c)** H—N—Ö—H

**(b)** H—C—C=Ö

**(d)** H—C—C≡N:

**9.12** Give the approximate values for the indicated bond angles in the following molecules:

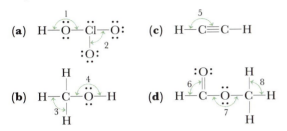

**(a)** H—Ö—Cl—Ö:

**(c)** H—C≡C—H

**(b)** H—C—Ö—H

**(d)** H—C—Ö—C—H

## Dipole Moments

**9.13** Indicate whether each of the following molecules possesses a dipole moment: **(a)** $CCl_4$; **(b)** $NF_3$; **(c)** $SO_3$; **(d)** $CS_2$; **(e)** $SCl_2$; **(f)** $N_2O$.

**9.14** Predict whether the following molecules are polar or nonpolar: **(a)** $PH_3$; **(b)** $SiCl_4$; **(c)** $BF_3$; **(d)** IF; **(e)** $C_2H_4$; **(f)** $SO_2$.

**9.15** Despite the larger electronegativity difference between the bonded atoms, $BeCl_2(g)$ has no dipole moment, whereas $SCl_2(g)$ does possess one. What is the origin of this difference in properties?

**9.16** The molecule $C_2H_2Cl_2$ can exist in the two forms shown below:

Does either of these have a dipole moment? Explain.

**9.17** The bond length in HBr is 1.41 Å. From the dipole moment (Table 9.4), calculate the percentage ionic character in the H—Br bond, assuming that the partial charges are centered on H and Br.

**9.18** The bond length in HF is 91.7 pm. From the dipole moment (Table 9.4), calculate the percentage ionic character in the H—F bond, assuming that the partial charges are centered on H and F.

## Hybrid Orbitals

**9.19** Without referring to tables in the chapter, indicate the designation for the hybrid orbitals formed from each of the following combinations of atomic orbitals: **(a)** one *s* and two *p*; **(b)** one *d*, one *s*, and three *p*; **(c)** two *d*, one *s*, and three *p*. What bond angles are associated with each?

**9.20** Without referring to tables or figures in the chapter, indicate the hybridization and bond angles associated with each of the following electron-pair geometries: **(a)** linear; **(b)** tetrahedral; **(c)** trigonal planar; **(d)** octahedral; **(e)** trigonal bipyramidal.

**9.21** Draw the Lewis structure of $H_2S$. What is its electron-pair geometry and molecular shape? Describe the bonding of the molecule, using hybrid orbitals.

**9.22** Draw the Lewis structure of $BF_3$. What is its electron-pair geometry and molecular shape? Describe the bonding of the molecule, using hybrid orbitals.

**9.23** Indicate the hybrid orbital set employed by the central atom in each of the following molecules or ions: **(a)** $CH_3^+$; **(b)** $PF_5$; **(c)** $AlCl_4^-$; **(d)** $TiCl_4$; **(e)** $OSCl_2$; **(f)** $BrF_3$; **(g)** $ClO_2^-$; **(h)** $XeF_4$; **(g)** $SO_2$.

**9.24** For each of the following molecules, predict the molecular geometry (including approximate bond angles), and indicate the hybrid orbitals on the central atoms. **(a)** $OF_2$; **(b)** $SiF_4$; **(c)** $SeF_6$; **(d)** $CS_2$; **(e)** $NO_2^-$; **(f)** $ClF_5$.

**9.25** If an atom is *sp* hybridized, how many *p* orbitals in the same valence shell remain on the atom? How many π bonds can the atom form?

**9.26** Indicate the number of *p* orbitals on a particular atom available for π bonding when the following hybrid orbital sets are used to define the molecular framework: **(a)** $sp^2$; **(b)** $sp^3$; **(c)** $dsp^3$; **(d)** $d^2sp^3$.

**9.27** Consider the Lewis structure for glycine, the simplest amino acid:

$$H_2N—C—C—O—H$$

**(a)** What are the approximate bond angles about each of the two carbon atoms, and what are the hybridizations of the orbitals on each of them? **(b)** What are the hybridizations of the orbitals on the two oxygens and the nitrogen atom, and what are the approximate bond angles at the nitrogen? **(c)** What is the total number of σ bonds in the entire molecule, and the total number of π bonds?

**9.28** The compound whose Lewis structure is shown below is acetylsalicylic acid, better known as aspirin:

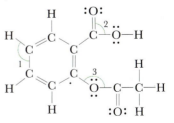

**(a)** What are the approximate values of the bond angles marked 1, 2, and 3? **(b)** What hybrid orbitals are used about the central atom of each of these angles? **(c)** How many σ bonds are there in the molecule?

**9.29** The geometrical structure of the nitrate ion, $NO_3^-$, can be accounted for either in terms of Lewis structures, using resonance forms, or in terms of delocalized $\pi$ bonding. Explain the structure of $NO_3^-$ in both of these terms.

**9.30** (a) Draw two resonance structures for $SO_2$. (b) What is the hybridization of orbitals around S in this molecule? (c) Describe the $\pi$ bonding in the molecule. (d) Label both the $\sigma$ and $\pi$ bonds as either localized or delocalized.

## Molecular Orbitals

**9.31** How do bonding and antibonding molecular orbitals differ with respect to (a) energies; (b) the spatial distribution of electron density?

**9.32** How do $\sigma$ and $\pi$ molecular orbitals differ with respect to the spatial distribution of electron density? Sketch the shapes of the $\sigma_{2p}$ and $\pi_{2p}$ molecular orbitals.

**9.33** What is meant by the following terms: (a) bond order; (b) paramagnetism; (c) antibonding molecular orbital?

**9.34** How are bond order, bond length, and bond-dissociation energy related?

**9.35** Assuming that the homonuclear molecular-orbital energy-level diagram in Figure 9.30 can be applied to heteronuclear diatomic molecules and ions, predict the bond order and magnetic behavior of the following: (a) NO; (b) $NO^+$; (c) $CN^-$; (d) OF.

**9.36** Using Figure 9.30 as a guide, sketch the molecular-orbital energy-level diagram for each of the following: (a) $B_2^+$; (b) $Li_2^+$; (c) $C_2^+$; (d) $Ne_2^{2+}$. In each case indicate whether the stability of the species increases or decreases upon addition of one electron.

**9.37** The ions $O_2^-$, $O_2^{2-}$, and $O_2^+$ occur in several compounds. Compare these three ions with $O_2$ by listing the four in order of increasing bond length.

**9.38** List the members of the following series in order of increasing bond length: $N_2^+$, $N_2$, $N_2^-$.

**9.39** Which of the following molecules should have the lowest ionization energy: NO, $N_2$, $O_2^+$? Explain your answer.

**9.40** Predict the nitrogen-oxygen bond orders in the series $NO^-$, NO, $NO^+$. Predict the variation in N—O bond distances in this same series.

## Structures of Nonmetallic Elements

**9.41** Among the second-row nonmetallic elements the internuclear bond distances vary as follows: carbon (diamond), 1.54 Å; nitrogen ($N_2$), 1.09 Å; oxygen ($O_2$), 1.208 Å; fluorine ($F_2$), 1.417 Å. Account for these variations in terms of the concepts of bond order and effective nuclear charge.

**9.42** The following table provides data regarding O—O bond distances and dissociation energies:

| Compound | O—O distance (Å) | Bond-dissociation energy (kJ/mol) |
|---|---|---|
| $H_2O_2$ | 1.48 | 213 |
| $O_2$ | 1.21 | 495 |
| $O_3$ | 1.28 | 107 (for $O_3 \longrightarrow O_2 + O$) |

(a) Explain in terms of the structures of the molecules involved why the O—O distance in $O_3$ is intermediate between that for $O_2$ and $H_2O_2$. (b) Explain why the energy requirement for rupture of the O—O bond in $O_3$ is lower than for either $H_2O_2$ or $O_2$.

**9.43** Account for the following observations. (a) There is no silicon analogue of graphite. (b) Diamond is one of the hardest known substances.

**9.44** Account for the following observations. (a) Under ordinary conditions, the stable form of elemental oxygen is $O_2$, but $S_2$ is seen only under special conditions. (b) Graphite is a good conductor of electricity in the direction parallel to the graphite sheets.

## Additional Exercises

**9.45** Using the VSEPR model, predict the molecular geometry of each of the following: (a) $PO_4^{3-}$; (b) $AsF_3$; (c) $OCN^-$; (d) $H_2CO$; (e) $ICl_4^-$; (f) $I_3^-$.

**9.46** The H—P—H bond angle in $PH_3$ is 93°; in $PH_4^+$ it is 109.5°. Account for this difference.

**9.47** When applying the VSEPR model, we count double and triple bonds as a single pair of electrons. Present an argument as to why this is justified.

**9.48** From their Lewis structures, determine the number of $\sigma$ and $\pi$ bonds in each of the following molecules or ions: (a) $H_2CO$; (b) $CN^-$; (c) $SO_2$; (d) $SO_3^{2-}$.

**9.49** Which of the following molecules will have a dipole moment: (a) HI; (b) $SCl_2$; (c) $BCl_3$; (d) HCN; (e) $HC\equiv CH$; (f) $Cl_2C=CH_2$?

**[9.50]** Dichlorobenzene, $Cl_2C_6H_4$, can have one of the following three structures:

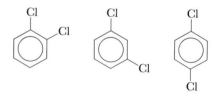

Assume (1) that the value for a C—Cl bond dipole moment is 1.70 D, (2) that the value for the C—H bond dipole moment is zero, and (3) that there are no direct influences of the bond dipoles on one another. Calculate the approximate dipole moments to be expected for the three structures. (Such substances, which have the same molecular formulas but different atomic arrangements, are called *isomers*.) The three isomers are found experimentally to have dipole moments of 0, 1.72, and 2.50 D. Assign the dipole moments to the three isomers.

**[9.51]** The Lewis structure for allene is

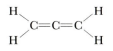

Make a sketch of the structure of this molecule that is analogous to Figure 9.22. In addition, answer the following two questions: **(a)** Is the molecule planar? **(b)** Does it have a dipole moment?

**9.52** Describe the molecular shape and bonding in the unstable molecule diimine, $HN{=}NH$, by answering the following questions. **(a)** Draw the Lewis structure. **(b)** Use the VSEPR model to predict the molecular geometry. **(c)** Indicate the type of hybrid orbitals employed by nitrogen. **(d)** Sketch the structure using $\sigma$ and $\pi$ bond orbitals, and label them. **(e)** Experimentally, it is found that the dipole moment of the molecule is zero. Is this consistent with your structure? Is there a form of the molecule, with the same Lewis structure, that could have a nonzero dipole moment?

**9.53** Indicate the hybrid orbital set employed by the underlined atom in each of the following structures: **(a)** $\underline{S}Cl_2$; **(b)** $F_2\underline{C}{=}CH_2$; **(c)** $\underline{Br}F_4^+$; **(d)** $\underline{Tl}Cl_3$; **(e)** $\underline{Se}O_3^{2-}$; **(f)** $\underline{Sn}Cl_6^{2-}$.

**9.54** Cumene hydroperoxide, for which the structure is

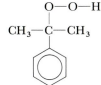

is an intermediate in the formation of phenol, $C_6H_5OH$, an important industrial chemical. Indicate the hybrid orbital set employed by all the carbon and oxygen atoms of cumene hydroperoxide.

**9.55** What change in the hybridization of orbitals at the central atom occurs in each of the following reactions?
**(a)** $GaCl_4^- \longrightarrow GaCl_3 + Cl^-$
**(b)** $PCl_5 \longrightarrow PCl_3 + Cl_2$
**(c)** $SF_6 \longrightarrow SF_4 + F_2$

**9.56** Draw the Lewis structure and describe the state of hybridization of the nitrogen atoms in the azide ion, $N_3^-$. Then N—N bond distances in this ion are both 1.15 Å. By comparison with the data for representative N—N bond distances presented in Section 8.4, is this distance consistent with your Lewis structure?

**9.57** The structure of indium triiodide is shown in Figure 9.39. It is evident from this structure that the molecular formula is $In_2I_6$. Draw the Lewis structure for this mol-

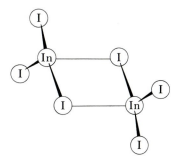

**FIGURE 9.39** Structure of $In_2I_6$.

ecule. What is the hybridization about In? The $In_2I_6$ molecules dissociate in the gas phase into $InI_3$ molecules. Draw the Lewis structure for this molecule and indicate the hybridization about In? What characteristic of the electronic structure of iodine in this compound is of importance in forming $In_2I_6$? What characteristic of In in $InI_3$ is of importance?

**[9.58]** Telluric acid, $Te(OH)_6$, is formed by reaction of $TeO_2$ with hydrogen peroxide in aqueous solution. Write a balanced chemical equation for this process. Indicate the oxidation numbers of all species that undergo oxidation or reduction in this reaction. Draw the Lewis structure for telluric acid. What hybridization is employed by Te in this compound? Fusion of $Te(OH)_6$ with NaOH gives $Na_6TeO_6$. What is the structure of the $TeO_6^{6-}$ ion?

**9.59** The nitrogen-nitrogen distances in $N_2H_4$, $N_2F_2$, and $N_2$ are 1.45, 1.25, and 1.10 Å, respectively. Account for this variation in bond distances.

**9.60** In which of the following molecules would you expect to find delocalized orbitals: $C_2H_4$; $NO_2^-$; $H_2CO$? Explain.

**9.61** Indicate the expected order of bond-dissociation energies in the series $H_2$, $Ca_2$, $K_2$, $Li_2$. Explain your reasoning.

**9.62** The $S_2$ molecule, seen in the gas phase under special conditions, has a dissociation energy of 425 kJ/mol. **(a)** Compare this with the value of $O_2$, and sketch a possible energy-level diagram for the $S_2$ molecule. **(b)** Why doesn't sulfur exist in this form at ordinary temperature and pressures?

**[9.63]** In each of the following pairs of molecules, one is stable and well known, the other is unstable or unknown. Identify the unstable member in each pair, and explain why it is the less stable compound: **(a)** $CH_4$, $CH_5$; **(b)** HI, $NI_3$; **(c)** $OF_6$, $SF_6$; **(d)** $HC{\equiv}CH$, $HSi{\equiv}SiH$; **(e)** $PO_3$, $SO_3$.

# Gases

In the past several chapters we have learned about the electronic structures of atoms and about how atoms come together to form molecules or ionic substances. We commonly observe matter, however, not on the atomic and molecular level, but as a solid, liquid, or gas. In the next few chapters we will be considering some of the important characteristics of these states of matter. We will be interested in learning why substances are found in one or the other state; what forces operate between atoms, ions, or molecules in these states; what transitions may occur between states; and about some of the characteristic properties of matter in each state.

## 10.1 CHARACTERISTICS OF GASES

Many familiar substances exist at ordinary temperature and pressure as gases. These include several elements ($H_2$, $N_2$, $O_2$, $F_2$, $Cl_2$, and the noble gases) and a great variety of compounds. Table 10.1 lists a few of the more common gaseous compounds. Under appropriate conditions, substances that are ordinarily liquids or solids can also exist in the gaseous state, where they are often referred to as *vapors*. The substance $H_2O$, for example, is familiar to us as liquid water, ice, or water vapor. Frequently, a substance exists in all three separate states of matter, or phases, at the same time. A Thermos flask containing a mixture of ice and water at $0°C$ has a certain pressure of water vapor in the gas phase over the liquid and solid phases.

Normally, the three states of matter differ very obviously from one another. Gases differ dramatically from solids and liquids in several respects. A gas expands to fill its container. Consequently, the volume of a gas is given simply by specifying the volume of the container in which it is held. Volumes of solids and liquids, on the other hand, are not determined by the container. The corollary of this is that gases are highly compressible. When pressure is applied to a gas, its volume readily contracts. Liquids and solids, on the other hand, are not very compressible at all. Great pressures must be applied to cause the volume of a liquid or solid to diminish by even as little as 5 percent.

Two or more gases form homogeneous mixtures in all proportions, regardless of how different the gases may be. Liquids, on the other hand, often do not form homogeneous mixtures. For example, when water and gasoline are poured into a bottle, the water vapor and gasoline vapors above the liquids form a homogeneous gas mixture. The two liquids, by

---

A tornado is one of the most destructive manifestations of the gaseous state of matter. Tornados form when moist, warm air at lower levels converges with cooler dry air aloft. The resultant air flows lead to winds that can attain speeds of 200 to 300 miles per hour, with updraft speeds in the range of 200 miles per hour. (Dr. E. R. Degginger)

**TABLE 10.1** Some common compounds that are gases

| Formula | Name | Characteristics |
|---------|------|-----------------|
| HCN | Hydrogen cyanide | Very toxic, slight odor of bitter almonds |
| HCl | Hydrogen chloride | Toxic, corrosive, choking odor |
| $H_2S$ | Hydrogen sulfide | Very toxic, odor of rotten eggs |
| CO | Carbon monoxide | Toxic, colorless, odorless |
| $CO_2$ | Carbon dioxide | Colorless, odorless |
| $CH_4$ | Methane | Colorless, odorless, flammable |
| $N_2O$ | Nitrous oxide | Colorless, sweet odor, "laughing gas" |
| $NO_2$ | Nitrogen dioxide | Red-brown, irritating odor |
| $NH_3$ | Ammonia | Colorless, pungent odor |
| $SO_2$ | Sulfur dioxide | Colorless, irritating odor |

contrast, remain largely separate; each dissolves in the other to only a slight extent.

The characteristic properties of gases arise because the individual molecules are relatively far apart. For example, in the air we breathe, the molecules take up only about 0.1 percent of the volume. The average distance between molecules in air is about 10 times as great as the sizes of the molecules themselves. Thus each molecule behaves largely as though the others weren't present; different gases behave similarly, even though they are made up of different molecules. By contrast, the individual molecules in a liquid are close together and occupy perhaps 70 percent of the total space. The attractive forces among the molecules keep the liquid together. Because these forces differ from one substance to another, different liquids behave differently.

## 10.2 PRESSURE

Among the most readily measured properties of a gas are its temperature, volume, and pressure. It is not surprising, therefore, that many early studies of gases focused on relationships among these properties. We have already discussed volume and temperature (Section 1.3). Let us now consider the concept of pressure.

In general terms, **pressure** carries with it the idea of a force, something that tends to move something else in a given direction. Pressure is, in fact, the force that acts on a given area $(P = F/A)$. Gases exert a pressure on any surface with which they are in contact. For example, the gas in an inflated balloon exerts a pressure on the inside surface of the balloon.

To understand better the concept of pressure and the units in which it is expressed, consider the aluminum cylinder illustrated in Figure 10.1. Because of the gravitational force, this cylinder exerts a downward force upon the surface on which it rests. According to Newton's second law of motion, the force exerted by an object is the product of its mass, $m$, times its acceleration, $a$: $F = ma$. The acceleration due to the gravitational force of earth is 9.81 m/s$^2$. The mass of the cylinder is 1.06 kg. Thus the force with which the earth attracts it is

$$(1.06 \text{ kg})(9.81 \text{ m/s}^2) = 10.4 \text{ kg-m/s}^2 = 10.4 \text{ N}$$

A kg-m/s$^2$ is the SI unit for force; it is called the **newton**, abbreviated N: 1 N = 1 kg-m/s$^2$. The cylinder has a cross-sectional area of 7.85 ×

$10^{-3}$ m$^2$; thus the pressure exerted by the cylinder is

$$P = \frac{F}{A} = \frac{10.4 \text{ N}}{7.85 \times 10^{-3} \text{ m}^2} = 1.32 \times 10^3 \text{ N/m}^2$$

A N/m$^2$ is the standard unit of pressure in SI units. It is given the name **pascal** (abbreviated Pa) after Blaise Pascal (1623–1662), a French mathematician and scientist: 1 Pa = 1 N/m$^2$.

Like the aluminum cylinder in our example above, the earth's atmosphere is also attracted toward earth by gravitational attraction. A column of air 1 m$^2$ in cross section extending through the atmosphere has a mass of roughly 10,000 kg and produces a resultant pressure of about 100 kPa (Figure 10.2):

$$P = \frac{F}{A} = \frac{(10,000 \text{ kg})(9.81 \text{ m/s}^2)}{1 \text{ m}^2} = 1 \times 10^5 \text{ Pa} = 1 \times 10^2 \text{ kPa}$$

Of course, the actual atmospheric pressure at any location depends on altitude and weather conditions.

Atmospheric pressure can be measured by use of a mercury **barometer**, like that illustrated in Figure 10.3. Such a barometer is formed by filling a glass tube more than 76 cm long, which is closed on one end, with mercury and inverting it in a dish of mercury. Care must be taken that no air gets into the tube. When the tube is inverted in this manner, some of the mercury runs out, but a column remains.

The mercury surface outside the tube experiences the full force of the earth's atmosphere over each unit area. However, the atmosphere is not in contact with the mercury surface within the tube. The atmosphere pushes the mercury up the tube until the pressure due to the mass of the mercury column balances the atmospheric pressure. Thus the height of the mercury column fluctuates as the atmospheric pressure fluctuates. The **standard atmospheric pressure**, which corresponds to the typical pressure at sea level, is defined as the pressure sufficient to support

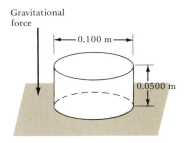

**FIGURE 10.1**  Aluminum cylinder resting on a flat surface. The diameter of the cylinder is 0.100 m and its height is 0.0500 m. The area of its base is thus $A = \pi r^2 = \pi(0.0500 \text{ m})^2 = 7.85 \times 10^{-3}$ m$^2$. Its volume is $V = \pi r^2 h = \pi(0.0500 \text{ m})^2(0.0500 \text{ m}) = 3.93 \times 10^{-4}$ m$^3$. The density of aluminum is 2.70 g/cm$^3$ = $2.70 \times 10^3$ kg/m$^3$. The mass of the cylinder is then $M = d \times V = (2.70 \times 10^3$ kg/m$^3) \times (3.93 \times 10^{-4}$ m$^3) = 1.06$ kg. Calculation of the pressure exerted by this cylinder on the surface upon which it rests is described in the text.

**FIGURE 10.3**  Mercury barometer invented by Torricelli. The space in the tube above mercury is nearly a vacuum; a negligible amount of mercury vapor occupies it.

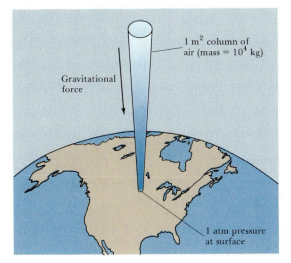

**FIGURE 10.2**  Illustration of the manner in which earth's atmosphere exerts pressure at the surface of the planet. The mass of a column of atmosphere 1 m$^2$ in cross-sectional area and extending to the top of the atmosphere exerts a force of $1.01 \times 10^5$ N. Thus the pressure is 101 kPa, corresponding to 760 mm Hg.

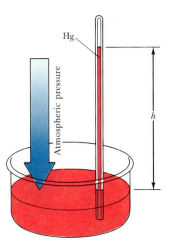

a column of mercury 760 mm in height. This pressure, which corresponds to $1.01325 \times 10^5$ Pa, is used to define another unit in common use, the atmosphere (abbreviated **atm**):

$$1 \text{ atm} = 760 \text{ mm Hg} = 1.01325 \times 10^5 \text{ Pa} = 101.325 \text{ kPa}$$

One mm Hg pressure is also referred to as a **torr**, after the Italian scientist Evangelista Torricelli (1608–1647), who invented the barometer: 1 mm Hg = 1 torr.

In this text we will ordinarily express gas pressure in units of atm or mm Hg. However, you should be able to convert gas pressures from one set of units to another.

---

### SAMPLE EXERCISE 10.1

**(a)** Convert 0.605 atm to millimeters of mercury (mm Hg). **(b)** Convert $3.5 \times 10^{-4}$ mm Hg to atmospheres.

**Solution:** **(a)** Because 1 atm = 760 mm Hg, conversion of atm to mm Hg is made by multiplying the number of atm by the factor 760 mm Hg/1 atm:

$$(0.605 \text{ atm})\left(\frac{760 \text{ mm Hg}}{1 \text{ atm}}\right) = 460 \text{ mm Hg}$$

Notice that the units cancel in the required manner.

**(b)** To convert from mm Hg to atm, we must multiply by the conversion factor 1 atm/760 mm Hg:

$$(3.5 \times 10^{-4} \text{ mm Hg})\left(\frac{1 \text{ atm}}{760 \text{ mm Hg}}\right)$$
$$= 4.6 \times 10^{-7} \text{ atm}$$

**PRACTICE EXERCISE**

What is the pressure, in atmospheres, exerted by a mercury column 1340 cm in height?
**Answer:** 17.6 atm

---

### SAMPLE EXERCISE 10.2

In countries that use the metric system—for example, Canada—atmospheric pressure is expressed in weather reports in units of kPa. Convert a pressure of 735 mm Hg to kPa.

**Solution:** From the material discussed earlier we know that 1 atm = 101.3 kPa = 760 mm Hg. Thus the conversion factor we want is of the form 101.3 kPa/760 mm Hg. We use this conversion factor

to convert the pressure given:

$$(735 \text{ mm Hg})\left(\frac{101.3 \text{ kPa}}{760 \text{ mm Hg}}\right) = 98.0 \text{ kPa}$$

**PRACTICE EXERCISE**

A pressure of 1.0 kPa corresponds to how many mm Hg?
**Answer:** 7.5 mm Hg

---

We can use a device called a **manometer**, whose principle of operation is similar to that of a barometer, to measure the pressures of enclosed gases. Figure 10.4(*a*) shows a closed-tube manometer, a device normally used to measure pressures below atmospheric pressure. The pressure is just the difference in the heights of the mercury levels in the two arms.

An open-tube manometer, like that pictured in Figure 10.4(*b*) and (*c*), is often employed to measure gas pressures that are near atmospheric pressure. The difference in the heights of the mercury levels in the two arms of the manometer relates the gas pressure to atmospheric pressure. If the pressure of the enclosed gas is the same as atmospheric pressure, the levels in the two arms are equal. If the pressure of the enclosed gas is

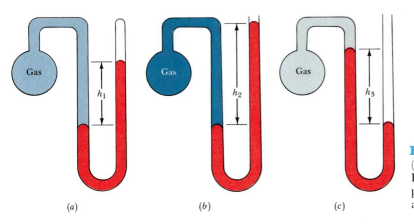

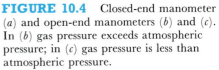

**FIGURE 10.4** Closed-end manometer (*a*) and open-end manometers (*b*) and (*c*). In (*b*) gas pressure exceeds atmospheric pressure; in (*c*) gas pressure is less than atmospheric pressure.

greater than atmospheric pressure, mercury is forced higher in the arm exposed to the atmosphere, as in Figure 10.4(*b*). Conversely, if atmospheric pressure exceeds the gas pressure, the mercury is higher in the arm exposed to the gas, as in Figure 10.4(*c*).

Although mercury is most often used as the liquid in a manometer, other liquids can be employed. For a given pressure difference, the difference in heights of the liquid levels in the two arms of the manometer is inversely proportional to the density of the liquid. That is, the less dense the liquid, the greater the difference in column heights.

**SAMPLE EXERCISE 10.3**

Consider a container of gas with an attached open-tube manometer. The manometer is not filled with mercury but rather with another nonvolatile liquid, dibutylphthalate. The density of mercury is 13.6 g/mL; that of dibutylphthalate is 1.05 g/mL. If the conditions are as shown in Figure 10.4(*b*) with $h = 12.2$ cm when atmospheric pressure is 0.964 atm, what is the pressure of the enclosed gas in mm Hg?

**Solution:** Converting atmospheric pressure to mm Hg, we have

$$(0.964 \text{ atm})\left(\frac{760 \text{ mm Hg}}{1 \text{ atm}}\right) = 733 \text{ mm Hg}$$

The pressure associated with a 12.2-cm column of dibutylphthalate is equivalent to a mercury column of

$$(12.2 \text{ cm})\left(\frac{1.05 \text{ g/mL}}{13.6 \text{ g/mL}}\right) = 0.94 \text{ cm} = 9.4 \text{ mm}$$

If the situation is like that in Figure 10.4(*b*), the pressure of the enclosed gas exceeds atmospheric pressure by this amount:

$$P = 733 \text{ mm Hg} + 9 \text{ mm Hg} = 742 \text{ mm Hg}$$

**PRACTICE EXERCISE**
What height of a dibutylphthalate column would be supported by 1 atm pressure?
***Answer:*** $9.84 \times 10^3$ mm

**10.3**
**THE GAS LAWS**

Experiments with a large number of gases reveal that the four variables temperature ($T$), pressure ($P$), volume ($V$), and quantity of gas in moles ($n$) are sufficient to define the state, or condition, of many gaseous substances. The first relationship between these variables was found in 1662 by Robert Boyle (1627–1691). **Boyle's law** states that *the volume of a fixed quantity of gas maintained at constant temperature is inversely proportional to the gas pressure.*

**FIGURE 10.5** An illustration of Boyle's experiment. In (*a*) the volume of the gas trapped in the J-tube is 60 mL when the gas pressure is 760 mm Hg, corresponding to a balance in the two columns of mercury between the trapped gas and the atmospheric pressure. When additional mercury is added, as shown in (*b*), the trapped gas is compressed. The volume is 30 mL when its total pressure is 1520 mm Hg, corresponding to atmospheric pressure plus the pressure exerted by the 760-mm column of mercury.

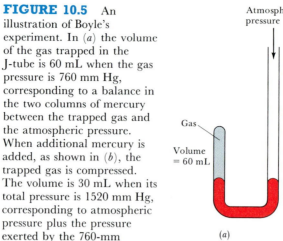
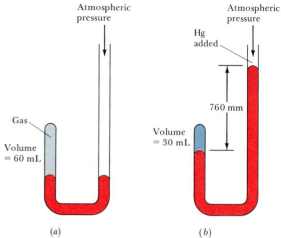

Atmospheric pressure

Atmospheric pressure

Hg added

760 mm

Gas

Volume = 60 mL

Volume = 30 mL

(*a*)

(*b*)

**FIGURE 10.6** Graphs based on Boyle's law:
(*a*) pressure versus volume;
(*b*) volume versus 1/P.

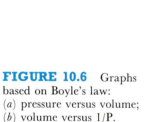

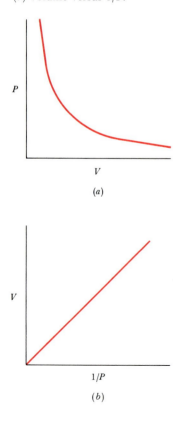

$P$

$V$

(*a*)

$V$

$1/P$

(*b*)

Boyle was interested in determining how the volume of a fixed quantity of gas varies with pressure. To perform his experiments he used a J-shaped tube in which a quantity of gas was trapped behind a column of mercury (see Figure 10.5). Boyle measured changes in the gas volume as he added mercury to the column to increase the pressure. He found that the volume of gas decreased as pressure increased, as shown in Figure 10.5. Doubling the pressure results in the gas volume decreasing to one half its original value. The product of pressure times volume is constant for a given quantity of gas: $PV = c$. The value for $c$ depends on the temperature and amount of gas in the sample. A graph of Boyle's pressure versus volume data is shown in Figure 10.6(*a*).

Boyle's relationship can be rearranged to yield $V = c/P$. This is the equation for a straight line with slope $c$ and zero intercept (Appendix A.4). Figure 10.6(*b*) shows a graph of $V$ versus $1/P$ for Boyle's data. Notice that a linear relationship is obtained.

The relationship between gas volume and temperature was discovered in 1787 by Jacques Charles (1746–1823), a French scientist. Charles found that the volume of a fixed quantity of gas at constant pressure increases in a linear fashion with temperature. Some typical data are shown in Figure 10.7. Notice that the extrapolated (extended) line (which is dashed) through the data points passes through −273.15°C. Note also that the gas is predicted to have zero volume at this temperature. Of course, this condition is never fulfilled, because all gases liquefy or solidify before reaching this temperature. In 1848, William Thomson (1824–1907), a British physicist whose title was Lord Kelvin, proposed the idea of an absolute temperature scale, now known as the Kelvin scale, with −273°C = 0 K. In terms of this scale, **Charles's law** can be stated as follows: *The volume of a fixed amount of gas maintained at constant pressure is directly proportional to absolute temperature*. This relationship can be expressed as $V \propto T$ or as $V = cT$, where the proportionality constant, $c$, depends on the pressure and amount of gas. Thus doubling absolute temperature, say from 200 K to 400 K, causes the gas volume to double.

The relationship between gas volume and the quantity of gas follows from the work of the French scientist Joseph Louis Gay-Lussac (1778–

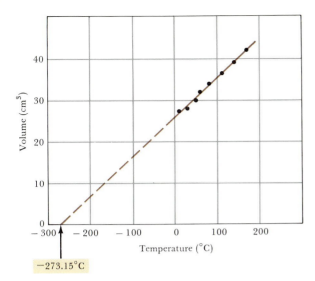

**FIGURE 10.7** Volume of an enclosed gas as a function of temperature at constant pressure.

−273.15°C

1850) and the Italian scientist Amedeo Avogadro (1776–1856). Gay-Lussac is one of those extraordinary figures in the early history of modern science who can truly be called an adventurer. He was interested in lighter-than-air balloons and in 1804 made an ascent to 23,000 ft (Figure 10.8). This exploit set the altitude record for several decades, but Gay-Lussac had other reasons for making the flight: He tested the variation of the earth's magnetic field and sampled the composition of the atmosphere as a function of elevation.

To control lighter-than-air balloons properly, Gay-Lussac needed to know more about the properties of gases. He therefore carried out several

**FIGURE 10.8** A representation of Gay-Lussac's 1804 balloon ascent, taken from *Appleton's Beginner's Handbook of Chemistry*, a popular nineteenth-century chemistry text. (Culver Pictures)

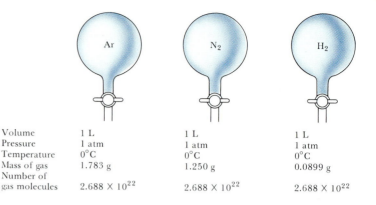

**FIGURE 10.9** Comparison illustrating Avogadro's hypothesis. Note that argon gas consists of argon atoms; we can regard these as one-atom molecules. Each gas has the same volume, temperature, and pressure and thus contains the same number of molecules. Because a molecule of one substance differs in mass from a molecule of another, the masses of gas in the three containers differ.

| | Ar | N₂ | H₂ |
|---|---|---|---|
| Volume | 1 L | 1 L | 1 L |
| Pressure | 1 atm | 1 atm | 1 atm |
| Temperature | 0°C | 0°C | 0°C |
| Mass of gas | 1.783 g | 1.250 g | 0.0899 g |
| Number of gas molecules | $2.688 \times 10^{22}$ | $2.688 \times 10^{22}$ | $2.688 \times 10^{22}$ |

experiments. The most important of these led to his discovery in 1808 of the law of combining volumes (Section 3.3). Recall that this law states that the volumes of gases that react with one another at the same pressure and temperature are in the ratios of small whole numbers.

Gay-Lussac's work led Avogadro in 1811 to propose what is known as **Avogadro's hypothesis:** *Equal volumes of gases at the same temperature and pressure contain equal numbers of molecules.* We saw in Chapter 3 the importance of both Gay-Lussac's and Avogadro's work in setting the stage for a correct appreciation of atomic weights. Let us now consider how their results can help us understand the nature of the gaseous state. Suppose that we have three 1-L bulbs containing Ar, $N_2$, and $H_2$, respectively (Figure 10.9), and that each gas is at the same pressure and temperature. According to Avogadro's hypothesis, those bulbs contain *equal numbers of gaseous particles*, although the masses of the substances in the bulbs differ greatly.

**Avogadro's law** follows from Avogadro's hypothesis: *The volume of a gas maintained at constant temperature and pressure is directly proportional to the quantity of gas:* $V \propto n$ or $V = cn$. Thus doubling the number of moles of gas will cause the volume to double if $T$ and $P$ remain constant.

## 10.4 THE IDEAL-GAS EQUATION

In the preceding section, we examined three historically important gas laws:

Boyle's law: $\quad V \propto \dfrac{1}{P} \quad$ (constant $n$, $T$)

Charles's law: $\quad V \propto T \quad$ (constant $n$, $P$)

Avogadro's law: $\quad V \propto n \quad$ (constant $P$, $T$)

We can combine these relationships to make a more general gas law:

$$V \propto \frac{nT}{P}$$

If we call the proportionality constant $R$, we have

$$V = R\left(\frac{nT}{P}\right)$$

Rearranging, we have this relationship in its more familiar form:

$$PV = nRT \qquad \text{[10.1]}$$

This equation is known as the **ideal-gas equation**. The term $R$ is called the **gas constant**. The numerical value of $R$ depends on the units chosen for the four variables in the equation. Temperature, $T$, must be expressed in an absolute temperature scale, normally the Kelvin scale. The quantity of gas, $n$, is normally expressed in moles. The units chosen for pressure, $P$, and volume, $V$, are most often atm and liters, respectively. However, other units could be used. Table 10.2 shows the numerical values for $R$ in a few of the more important units. The last two values given, which include calories and joules among the units, arise because the product $P \times V$ has energy units (Section 4.3).

The ideal-gas equation is called an *equation of state* for a gas because it contains all the variables needed to describe completely the condition, or state, of any gas sample. Knowledge of any three of the variables (temperature, pressure, volume, and number of moles of gas) is enough to specify the system completely, because the fourth variable can be calculated from the other three using the ideal-gas equation.

**TABLE 10.2** Numerical values of the gas constant, $R$, in various units

| Units | Numerical value |
|---|---|
| Liter-atm/K-mol | 0.08206 |
| Calories/K-mol | 1.987 |
| Joules/K-mol[a] | 8.314 |

[a] SI units.

## SAMPLE EXERCISE 10.4

Using the ideal-gas equation, calculate the volume of exactly 1 mol of gas at 0°C (273.15 K) and exactly 1 atm pressure.

**Solution:** Rearranging Equation 10.1 to solve for $V$ gives

$$V = \frac{nRT}{P}$$

Inserting the numerical values for each term gives

$$V = \frac{(1 \text{ mol})(0.08206 \text{ L-atm/mol-K})(273.15 \text{ K})}{1 \text{ atm}}$$

$$= 22.41 \text{ L}$$

**PRACTICE EXERCISE**

What is the pressure exerted by 2.0 mol of an ideal gas when it occupies a volume of 12.0 L at 100°C (373 K)?
*Answer:* 5.1 atm

The conditions 0°C and 1 atm are referred to as the **standard temperature and pressure** (abbreviated **STP**). Many properties of gases are tabulated for these conditions. The volume calculated in Sample Exercise 10.4, 22.41 L, is known as the *molar volume* of an ideal gas at STP.

## SAMPLE EXERCISE 10.5

A flashbulb of volume 2.6 cm³ contains $O_2$ gas at a pressure of 2.3 atm and a temperature of 26°C. How many moles of $O_2$ does the flashbulb contain?

**Solution:** Because we know volume, temperature, and pressure, the only unknown quantity in the ideal-gas equation (Equation 10.1) is the number of moles, $n$. Solving Equation 10.1 for $n$ gives

$$n = \frac{PV}{RT}$$

The values of all quantities involved are tabulated and changed to units consistent with those for $R$:

$$P = 2.3 \text{ atm}$$
$$V = 2.6 \text{ cm}^3 = 2.6 \times 10^{-3} \text{ L}$$

$n = ?$

$R = 0.0821$ L-atm/mol-K

$T = 26°C = 299$ K

Thus we have

$$n = \frac{(2.3 \text{ atm})(2.6 \times 10^{-3} \text{ L})}{(0.0821 \text{ L-atm/mol-K})(299 \text{ K})}$$

$$= 2.4 \times 10^{-4} \text{ mol}$$

**PRACTICE EXERCISE**

We may make an exotic—but quite accurate—thermometer by measuring the volume of a known quantity of gas at a known pressure. If 0.200 mol of helium, He, occupies a volume of 64.0 L at a pressure of 0.150 atm, what is the temperature of the gas?

*Answer:* 585 K

The fact that Equation 10.1 is called the *ideal*-gas equation correctly suggests that there may be conditions where gases don't exactly obey this equation. For example, we might calculate the quantity of a gas, $n$, for given conditions of $P$, $V$, and $T$ and find it to differ somewhat from the measured quantity under these conditions. Ordinarily, however, the difference between ideal and real behavior is so small that we may ignore it. We will examine deviations from ideal behavior later, in Section 10.9.

**Relationship between the Ideal-Gas Equation and the Gas Laws**

The simple gas laws, such as Boyle's law, which we discussed in Section 10.3, are special cases of the ideal-gas equation. For example, when the temperature and quantity of gas are held constant, $n$ and $T$ have fixed values. Thus the product $nRT$ is the product of three constants and must itself be a constant:

$$PV = nRT = \text{constant} = c \qquad \text{or} \qquad PV = c \qquad [10.2]$$

Thus we have Boyle's law. Equation 10.2 expresses the fact that even though the individual values of $P$ and $V$ can change, their product $PV$ remains constant.

Similarly, when $n$ and $P$ are constant, the ideal-gas equation gives the following relationship:

$$V = \left(\frac{nR}{P}\right)T = \text{constant} \times T \qquad \text{or} \qquad \frac{V}{T} = c \qquad [10.3]$$

As we have noted earlier, this relationship was first observed by Charles and is known as Charles's law.

**SAMPLE EXERCISE 10.6**

The pressure of nitrogen gas in a 12.0-L tank at 27°C is 2300 lb/in.² What volume would the gas in this tank have at 1 atm pressure (14.7 lb/in.²) if the temperature remains unchanged?

**Solution:** Let us begin by making a table that lists the initial and final values for the pressure, temperature, and volume of the gas. (Always convert temperature to the Kelvin scale.)

|         | Temperature (K) | Pressure (lb/in.²) | Volume (L) |
| ------- | --------------- | ------------------ | ---------- |
| Initial | 300             | 2300               | 12.0       |
| Final   | 300             | 14.7               | ?          |

If the quantity of gas $(n)$ and the temperature $(T)$ do not change, the product $PV$ must remain constant (see Equation 10.2). Thus, if we have two different sets of conditions for the same quantity of gas at constant temperature, we can write

$$P_1V_1 = P_2V_2$$

In our example, $P_1$ is 2300 lb/in.$^2$, $P_2$ is 14.7 lb/in.$^2$, $V_1$ is 12 L, and $V_2$ is unknown. Inserting all the known quantities and solving for $V_2$, we obtain

$$V_2 = \frac{P_1V_1}{P_2} = \frac{(2300 \text{ lb/in.}^2)(12.0 \text{ L})}{14.7 \text{ lb/in.}^2} = 1880 \text{ L}$$

You can see from this expression that we have multiplied the initial volume $(V_1)$ by a ratio of pressures $(P_1/P_2)$. You can check that the result makes sense. If the pressure is to decrease in going from the initial state to the final state, volume should increase, as it does.

### PRACTICE EXERCISE

A large natural-gas storage tank is arranged so that the pressure is maintained at 2.2 atm. On a cold day in December, when the temperature is $-15°C$ (4°F), the volume of gas in the tank is 28,500 ft$^3$. What is the volume of the same quantity of gas on a warm July day when the temperature is 31°C (88°F)?
*Answer:*   33,600 ft$^3$

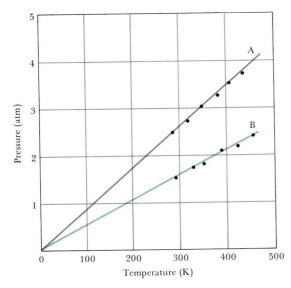

**FIGURE 10.10**   Variation of gas pressure with temperature under constant-volume conditions.

We know that when a confined gas is heated at constant volume, the pressure increases. For example, a popcorn kernel bursts open under the pressure of steam that forms within the kernel when it is heated in oil. We could make quantitative measurements of the change in pressure of a confined gas by placing the gas in a steel container fitted with a pressure gauge and then varying the temperature. We would find that the pressure increases linearly with absolute temperature, perhaps as shown by the sample data labeled A in Figure 10.10. If the experiment were repeated with a different-sized sample of the same gas, we might obtain the results labeled B in the figure. Note that in both cases the extrapolated pressure at 0 K is zero.

If both $n$ and V in Equation 10.1 are fixed, the pressure varies with temperature as expressed in the equation

$$P = \left(\frac{nR}{V}\right)T = \text{constant} \times T \qquad [10.4]$$

Thus the ideal-gas equation predicts a linear relationship between pressure and absolute temperature, extrapolating to zero pressure at 0 K. Again, we must remind ourselves that real gases lose their gaseous properties before absolute zero is reached.

## SAMPLE EXERCISE 10.7

Why do the two samples of gas for which data are shown in Figure 10.10 show different linear relationships?

**Solution:** Inspection of Equation 10.4 shows us that the linear relationship between $P$ and $T$ passes through the origin and has slope $nR/V$. (You may wish to review linear equations, Appendix A.4.) In our example, both $R$ and $V$ are the same for the two samples, but the number of moles of gas, $n$, is different. The slope of the pressure-versus-temperature relationship for a given volume is proportional to the amount of gas that is confined. There is more gas in sample A.

**PRACTICE EXERCISE**
The slope of the line in Figure 10.10 for gas A is 1.64 times that for gas B. If there are 0.30 mol of gas in sample A, how many moles are there in sample B?
*Answer:* 0.18 mol

## SAMPLE EXERCISE 10.8

The gas pressure in an aerosol can is 1.5 atm at 25°C. Assuming that the gas inside obeys the ideal-gas equation, what would the pressure be if the can were heated to 450°C?

**Solution:** Let us proceed, as in Sample Exercise 10.6, by writing down the initial and final conditions of temperature, pressure, and volume that the problem gives us. (Remember that we must convert temperatures to kelvin.)

|         | Volume | Pressure (atm) | Temperature (K) |
|---------|--------|----------------|-----------------|
| Initial | $V_1$  | 1.5            | 298             |
| Final   | $V_1$  | $P_2$          | 723             |

From Equation 10.1 we can see that $P/T = nR/V$ In this problem the quantity of gas ($n$) and volume are constant. Thus, if we have two different sets of conditions for the same quantity of gas at constant volume, we have

$$\frac{P_1}{T_1} = \frac{P_2}{T_2}$$

Rearranging gives us

$$P_2 = \left(\frac{T_2}{T_1}\right)P_1$$

$$P_2 = (1.5 \text{ atm})\left(\frac{723 \text{ K}}{298 \text{ K}}\right) = 3.6 \text{ atm}$$

It is evident from this example why aerosol cans carry the warning not to incinerate.

**PRACTICE EXERCISE**
An inflatable raft is filled with gas at a pressure of 800 mm Hg at 16°C. When the raft is left in the sun, the gas heats up to 44°C. Assuming no volume change, what is the gas pressure in the raft under these conditions?
*Answer:* 878 mm Hg

## Combined Gas Law

The three variables $P$, $V$, and $T$ may all change for a given sample of gas. Under these circumstances we have, from the ideal gas law,

$$\frac{PV}{T} = nR = \text{constant}$$

Thus as long as the total quantity of gas, $n$, is constant, $PV/T$ is a constant. If we represent the initial and final conditions of pressure, temperature, and volume by subscripts 1 and 2, respectively, we can write the following expression:

$$\frac{P_1 V_1}{T_1} = \frac{P_2 V_2}{T_2} \qquad [10.5]$$

## SAMPLE EXERCISE 10.9

A quantity of helium gas occupies a volume of 16.5 L at 78°C and 45.6 atm. What is its volume at STP?

**Solution:** It is best to begin problems of this sort by writing down all we know of the initial and final values of temperature, pressure, and volume. (Remember that you must always convert temperatures to the absolute temperature scale, K.)

|         | Pressure (atm) | Volume (L) | Temperature (K) |
|---------|----------------|------------|-----------------|
| Initial | 45.6           | 16.5       | 351             |
| Final   | 1 (exactly)    | $V_2$      | 273             |

Putting the quantities into Equation 10.5, we have

$$\frac{(45.6 \text{ atm})(16.5 \text{ L})}{351 \text{ K}} = \frac{(1 \text{ atm})(V_2)}{273 \text{ K}}$$

$$V_2 = \left(\frac{45.6 \text{ atm}}{1 \text{ atm}}\right)\left(\frac{273 \text{ K}}{351 \text{ K}}\right)(16.5 \text{ L}) = 585 \text{ L}$$

Does this answer make sense? Notice that the initial volume, 16.5 L, is multiplied by a ratio of temperatures and by a ratio of pressures. The temperature of the gas decreases; this should cause a decrease in volume. Thus the temperature ratio should be less than 1. The decrease in pressure from 45.6 atm to 1 atm will cause expansion of the gas. Thus the pressure ratio should be greater than 1. Our intuitive sense of what should happen is in accord with the expression we have used. It is always a good idea in checking your work to ask whether the answer makes sense.

### PRACTICE EXERCISE
A pocket of gas is discovered in a deep drilling operation. The gas has a temperature of 480°C and is at a pressure of 12.8 atm. Assume ideal behavior. What volume of the gas is required to provide 18.0 L at the surface at 1.00 atm and 22°C?

*Answer:* 3.59 L

## 10.5 DALTON'S LAW OF PARTIAL PRESSURES

The pressure of a gas under conditions of constant volume and temperature is directly proportional to the number of moles of gas:

$$P = \left(\frac{RT}{V}\right)n = \text{constant} \times n \qquad [10.6]$$

Suppose that the gas with which we are concerned is not made up of a single kind of gas particle but is rather a mixture of two or more different substances. We expect that the total pressure exerted by the gas mixture is the sum of pressures due to the individual components. Each of the individual components, if present alone under the same temperature and volume conditions as the mixture, would exert a pressure that we term the **partial pressure**. John Dalton was the first to observe that the *total pressure of a mixture of gases is just the sum of the pressures that each gas would exert if it were present alone:*

$$P_t = P_1 + P_2 + P_3 + \cdots \qquad [10.7]$$

Each of the gases obeys the ideal-gas equation. Thus we can write

$$P_1 = n_1\left(\frac{RT}{V}\right), \qquad P_2 = n_2\left(\frac{RT}{V}\right), \qquad P_3 = n_3\left(\frac{RT}{V}\right), \qquad \text{etc.}$$

All of the gases experience the same temperature and volume. Therefore, by substituting into Equation 10.7, we obtain

$$P_t = \frac{RT}{V}(n_1 + n_2 + n_3 + \cdots) \qquad\qquad [10.8]$$

That is, the total pressure at constant temperature and volume is determined by the total number of moles of gas present, whether that total represents just one substance or a mixture.

---

## SAMPLE EXERCISE 10.10

From data gathered by *Voyager 1*, scientists estimate that methane, $CH_4$, constitutes 6.0 mol percent of the atmosphere of Titan, Saturn's largest moon. The total pressure at the surface of Titan is 1.6 earth atmospheres. Calculate the partial pressure of methane, in earth units.

**Solution:** Let's first calculate the total pressure on Titan in units of mm Hg:

$$P_t = (1.6 \text{ atm})\left(\frac{760 \text{ mm Hg}}{1 \text{ atm}}\right) = 1216 \text{ mm Hg}$$

The total pressure of the atmosphere on Titan is the sum of the pressures exerted by each of the components. The pressure exerted by each component is proportional to its mol fraction, $f$, in the gas mixture. We can see this by using Equation 10.8 to set up the following ratio:

$$\frac{P_1}{P_t} = \frac{(RT/V)n_1}{(RT/V)(n_1 + n_2 + n_3 + \cdots)}$$

$$= \frac{n_1}{n_1 + n_2 + n_3 + \cdots} = f_1$$

Thus $P_1 = P_t f_1$ and $P_t = P_t f_1 + P_t f_2 + P_t f_3 + \cdots$

The sum of the mol fractions must, of course, equal 1.

$$f_1 + f_2 + f_3 + \cdots = 1$$

Mol percent is just mol fraction times 100. Thus for $CH_4$ we have $f = 0.060$.

$$P_{CH_4} = 1216 \times 0.060 = 73 \text{ mm Hg}$$
$$= 0.096 \text{ earth atm}$$

### PRACTICE EXERCISE

If a 0.20-L sample of $O_2$ at 0°C and 1.0 atm pressure and a 0.10-L sample of $N_2$ at 0°C and 2.0 atm pressure are both placed in a 0.40-L container at 0°C, what is the total pressure in the container?
*Answer:* 1.0 atm

---

## SAMPLE EXERCISE 10.11

What pressure, in atm, is exerted by a mixture of 2.00 g of $H_2$ and 8.00 g of $N_2$ at 273 K in a 10.0-L vessel?

**Solution:** The pressure depends on the total moles of gas (Equation 10.8). Calculating the moles of $H_2$ and $N_2$, we have

$$n_{H_2} = (2.00 \text{ g } H_2)\left(\frac{1 \text{ mol } H_2}{2.02 \text{ g } H_2}\right) = 0.990 \text{ mol } H_2$$

$$n_{N_2} = (8.00 \text{ g } N_2)\left(\frac{1 \text{ mol } N_2}{28.0 \text{ g } N_2}\right) = 0.286 \text{ mol } N_2$$

Using Equation 10.8 gives us

$$P_t = \frac{RT}{V}(n_{H_2} + n_{N_2})$$

$$= \frac{(0.0821 \text{ L-atm/mol-K})(273 \text{ K})}{10.0 \text{ L}}$$

$$\times (0.990 \text{ mol} + 0.286 \text{ mol})$$

$$= 2.86 \text{ atm}$$

This total pressure is the sum of the partial pressures of $H_2$ and $N_2$.

A study of the effects of certain gases on plant growth requires a synthetic atmosphere in a 120-L space at 745 mm Hg and 295 K. This atmosphere is to be composed of 1.5 mol percent $CO_2$, 18.0 mol percent $O_2$, and 80.5 mol percent Ar. Calculate the partial pressure of each gas in the mixture, and compute the volume of each gas, measured at atmospheric pressure and 295 K, required to form the 120 L of mixture.

***Answers:*** $CO_2$, 11.2 mm Hg, 1.76 L; $O_2$, 134 mm Hg, 21.2 L; Ar, 600 mm Hg, 94.7 L

## 10.6 MOLECULAR WEIGHTS AND GAS DENSITIES

We can make many applications of the ideal-gas equation in measuring and calculating gas density. Density has the units of mass per unit volume. We can arrange the gas equation to obtain

$$\frac{n}{V} = \frac{P}{RT}$$

Now $n/V$ has the units of moles per liter. Suppose that we multiply both sides of this equation by molecular weight ($\mathcal{M}$), which is the number of grams in 1 mol of a substance:

$$\frac{n\mathcal{M}}{V} = \frac{P\mathcal{M}}{RT} \qquad [10.9]$$

But the product of the quantities $n/V$ and $\mathcal{M}$ equals density, because the units multiply as follows:

$$\frac{\text{Moles}}{\text{Liter}} \times \frac{\text{grams}}{\text{mole}} = \frac{\text{grams}}{\text{liter}}$$

Thus the density of the gas is given by the expression on the right in Equation 10.9:

$$d = \frac{P\mathcal{M}}{RT} \qquad [10.10]$$

or, rearranging, we obtain

$$\mathcal{M} = \frac{dRT}{P} \qquad [10.11]$$

**SAMPLE EXERCISE 10.12**

What is the density of carbon dioxide gas at 745 mm Hg and 65°C?

**Solution:** The molecular weight of carbon dioxide, $CO_2$, is $12.0 + (2)(16.0) = 44.0$ g/mol. If we are to use 0.0821 L-atm/K-mol for $R$, we must convert pressure to atmospheres. Using Equation 10.10, we have then

$$d = \frac{(\frac{745}{760} \text{ atm})(44.0 \text{ g/mol})}{(0.0821 \text{ L-atm/K-mol})(338 \text{ K})}$$

$$= 1.55 \text{ g/L}$$

The problem can be turned around a bit to determine the molecular weight of a gas from its density, as shown in Sample Exercise 10.13.

The mean molecular weight of the atmosphere of Titan at the surface is 28.6 amu. The surface temperature is 95 K, and the pressure is 1.6 earth atm.

Assuming ideal behavior, calculate the density of Titan's atmosphere.

*Answer:* 5.9 g/L

## SAMPLE EXERCISE 10.13

A large flask fitted with a stopcock is evacuated and weighed; its mass is found to be 134.567 g. It is then filled to a pressure of 735 mm Hg at 31°C with a gas of unknown molecular weight and then reweighed; its mass is 137.456 g. The flask is then filled with water and again weighed; its mass is now 1067.9 g. Assuming that the ideal-gas equation applies, what is the molecular weight of the unknown gas? (The density of water at 31°C is 0.997 g/cm³.)

**Solution:** First we must determine the volume of the flask. This is given by the difference in weights of the empty flask and the flask filled with water, divided by the density of water at 31°C, which is 0.997 g/cm³:

$$V = \frac{1067.9 \text{ g} - 134.6 \text{ g}}{0.997 \text{ g/cm}^3} = 936 \text{ cm}^3$$

Because the mass of the gas is 137.456 g − 134.567 g =

2.889 g, its density is 2.889 g/0.936 L = 3.087 g/L. Using Equation 10.11, we have

$$\mathcal{M} = \frac{dRT}{P}$$

$$= \frac{(3.087 \text{ g/L})(0.0821 \text{ L-atm/mol-K})(304 \text{ K})}{(735/760) \text{ atm}}$$

$$= 79.7 \text{ g/mol}$$

**PRACTICE EXERCISE** ⎯⎯⎯⎯⎯⎯⎯

One method for accurately determining the molecular weight of a gas is to measure its density as a function of pressure. The graph of the quantity $d/P$ against pressure is extrapolated to zero pressure to obtain a limiting value. In one set of experiments a certain gas was shown to have a limiting value $d/P$ of 2.86 at 0°C. Calculate the molecular weight.

*Answer:* 64.1 g/mol

## 10.7 QUANTITIES OF GASES INVOLVED IN CHEMICAL REACTIONS

A knowledge of the properties of gases is important for chemists because gases are so often reactants or products in chemical reactions. We are thus often faced with calculating the volumes of gases required as reactants or yielded as products, or with calculating pressure changes in reaction vessels of fixed volume. In this context the gas laws are a part of chemical stoichiometry, a subject discussed in Chapter 3.

An experiment that often comes up in the course of laboratory work is the determination of the number of moles of gas collected from a chemical reaction. Sometimes this gas is collected over water. For example, solid potassium chlorate ($KClO_3$) may be decomposed by heating it in a

## FIGURE 10.11

(*a*) Collection of gas over water. (*b*) When the gas has been collected, the bottle is raised or lowered to equalize pressures inside and outside before measuring the volume of the gas collected.

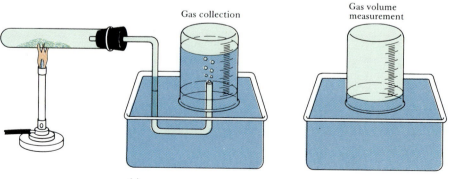

Gas collection

Gas volume measurement

(*a*)

(*b*)

test tube in an arrangement shown in Figure 10.11. The balanced equation for the reaction is

$$2KClO_3(s) \longrightarrow 2KCl(s) + 3O_2(g)$$

The oxygen gas is collected in a bottle that is initially filled with water and inverted in a water pan.

The volume of gas collected is measured by raising or lowering the bottle as necessary until the water levels inside and outside the bottle are the same. When this condition is met, the pressure inside the bottle is equal to the atmospheric pressure outside. The total pressure inside is the sum of the pressure of gas collected and the pressure of water vapor in equilibrium with liquid water:

$$P_{total} = P_{gas} + P_{H_2O} \qquad\qquad [10.12]$$

The pressure exerted by water vapor, $P_{H_2O}$, at various temperatures is shown in Appendix C.

---

### SAMPLE EXERCISE 10.14

Suppose that 0.200 L of oxygen gas is collected over water, as shown in Figure 10.10. The temperature of the water and gas is 26°C, and the atmospheric pressure is 750 mm Hg. **(a)** How many moles of $O_2$ are collected? **(b)** What volume would the $O_2$ gas collected occupy when dry, at the same temperature and pressure?

**Solution:** **(a)** The pressure of $O_2$ gas in the vessel is the difference between the total pressure, 750 mm Hg, and the vapor pressure of water at 26°C, 25 mm (Appendix C):

$$P_{O_2} = 750 - 25 = 725 \text{ mm Hg}$$

Solving the ideal-gas equation for $n$, we have

$$n = \frac{PV}{RT} = \frac{\left(\frac{725}{760}\text{ atm}\right)(0.200 \text{ L})}{(0.0821 \text{ L-atm/K-mol})(299 \text{ K})}$$
$$= 7.77 \times 10^{-3} \text{ mol}$$

**(b)** Suppose that we dried the water from the gas sample while maintaining the same total pressure. The $O_2$ pressure after drying would be 750 mm Hg. The corrected $O_2$ volume would thus be less for the dry gas, because its partial pressure is greater than for the wet gas:

$$V_{O_2} = (0.200 \text{ L})\left(\frac{725 \text{ mm Hg}}{750 \text{ mm Hg}}\right) = 0.193 \text{ L}$$

### PRACTICE EXERCISE

Assume that 260 mL of dry nitrogen at 20°C and 760 mm Hg are bubbled through water and stored in an inverted beaker, as shown in Figure 10.10(b). What is the volume of the stored gas at a pressure of 740 mm Hg and a temperature of 26°C?

*Answer:* 282 mL

---

We often collect a gas over water in an experiment to determine the amount of gaseous product. To compute the number of moles of gas collected, a correction must be applied for the partial pressure of water vapor in the collection bottle. We shall go through a complete analysis of such an experiment in Sample Exercise 10.15 to show how the stoichiometric relationships come together.

---

### SAMPLE EXERCISE 10.15

A 2.55-g sample of ammonium nitrite ($NH_4NO_2$) is heated in a test tube, as shown in Figure 10.11. The ammonium nitrite is expected to decompose according to the equation

$$NH_4NO_2(s) \longrightarrow N_2(g) + 2H_2O(g)$$

If it does decompose in this way, what volume of $N_2$ will be collected in the flask? The water and gas temperature are 26°C, and the barometric pressure is 745 mm Hg.

**Solution:** We begin by calculating the number of moles of $N_2$ gas formed:

$$2.55 \text{ g NH}_4\text{NO}_2 \left( \frac{1 \text{ mol NH}_4\text{NO}_2}{64.0 \text{ g NH}_4\text{NO}_2} \right)$$
$$\times \left( \frac{1 \text{ mol N}_2}{1 \text{ mol NH}_4\text{NO}_2} \right) = 0.0398 \text{ mol N}_2$$

To predict the volume of $N_2$ gas that is to be collected, we might be tempted simply to calculate the volume that would be occupied by 0.0398 mol of $N_2$ at 745 mm Hg. But we must also take into account the partial pressure of water vapor at 26°C, because the gas within the bottle is saturated with water vapor. From a table of water-vapor pressure versus temperature (see Appendix C), we can determine that the vapor pressure of water at 26°C is 25 mm Hg. The pressure of nitrogen gas in the flask when the water levels inside and out have been equalized is thus $745 - 25 = 720$ mm Hg. We must calculate the predicted volume of gas using this pressure (720/760 = 0.947 atm). (Pressure must be expressed in atm when we employ the value for the gas constant, $R$, expressed in L-atm/mol-K.) Rearranging Equation 10.2, we obtain

$$V = \frac{nRT}{P}$$

Inserting all known quantities in correct units, we obtain

$$V = \frac{(0.0398 \text{ mol})(0.0821 \text{ L-atm/K-mol})(299 \text{ K})}{0.947 \text{ atm}}$$
$$= 1.03 \text{ L}$$

### PRACTICE EXERCISE

A 1.60-g sample of $KClO_3$ is heated to produce $O_2$ according to the equation

$$2KClO_3(s) \longrightarrow 2KCl(s) + 3O_2(g)$$

Assume complete decomposition and ideal-gas behavior. What volume of $O_2$ collects over water at 26°C, 740 mm Hg pressure?
*Answer:* 511 mL

## 10.8 KINETIC-MOLECULAR THEORY

The ideal-gas equation describes *how* gases behave, but it doesn't explain *why* they behave as they do. For example, why does a gas expand when heated at constant pressure, or why does its pressure increase when the gas is compressed at constant temperature? To understand the physical properties of gases, we need a model that helps us picture what happens to gas particles as experimental conditions such as pressure or temperature change. Such a model, known as the **kinetic-molecular theory**, was developed over a period of about 100 years, culminating in 1857 when Rudolf Clausius (1822–1888) published a complete and satisfactory form of the theory.

The kinetic-molecular theory is summarized by the following statements:

1. Gases consist of large numbers of molecules that are in continuous, random motion. (The word "molecule" is used here to designate the smallest particle of any gas; some gases, such as the noble gases, consist of uncombined atoms.)

2. The volume of all the molecules of the gas is negligible compared to the total volume in which the gas is contained.

3. Attractive and repulsive forces between gas molecules are negligible.

4. Energy can be transferred between molecules during collisions, but the *average* kinetic energy of the molecules does not change with time, as long as the temperature of the gas remains constant. In other words, the collisions are perfectly elastic.

5. The average kinetic energy of the molecules is proportional to absolute temperature. At any given temperature the molecules of all gases have the same average kinetic energy.

The kinetic-molecular theory gives us an understanding of both pressure and temperature at the molecular level. The pressure of a gas is caused by collisions of the molecules with the walls of the container; it is determined both by the frequency of collisions per unit area and by the impulse imparted per collision (that is, by how frequently and by how "hard" the molecules strike the walls). The absolute temperature of a substance is a measure of the average kinetic energy of its molecules; absolute zero is the temperature at which the average kinetic energy of the molecules would be zero.

The idea that average kinetic energy and temperature are proportional provides a particularly important insight into matter. Let's consider this idea further. The molecules of a gas move at varying speeds. At one instant some of them are moving rapidly, others slowly. Figure 10.11 illustrates the distribution of molecular speeds within nitrogen gas at 0°C (red line) and at 100°C (blue line). Notice that at higher temperatures the distribution curve has shifted toward higher speeds.

Figure 10.12 also shows the value of the **root-mean-square** (rms) **speed**, $u$, of the molecules at each temperature. This quantity is the square root of the average squared speeds of the molecules. The rms speed is not the same as the average speed. The difference between the two, however, is so small that for most purposes they can be considered equal.* The rms speed is important because the average kinetic energy of the gas molecules, $\epsilon$, is related directly to $u^2$:

$$\epsilon = \tfrac{1}{2}mu^2 \qquad\qquad [10.13]$$

where $m$ is the mass of the molecule. When the temperature of a gas increases, the average kinetic energy of the gas molecules increases

---

* To illustrate the difference between rms speed and average speed, suppose that we have four objects with speeds of 4.0, 6.0, 10.0, and 12.0 m/s. Their average speed is $\tfrac{1}{4}(4.0 + 6.0 + 10.0 + 12.0) = 8.0$ m/s. However, the rms speed, $u$, is

$$\sqrt{\tfrac{1}{4}(4.0^2 + 6.0^2 + 10.0^2 + 12.0^2)} = \sqrt{74.0} = 8.6 \text{ m/s}$$

In general, the average speed equals $0.921 \times u$. Thus the average speed is directly proportional to the rms speed, and the two are in fact nearly equal.

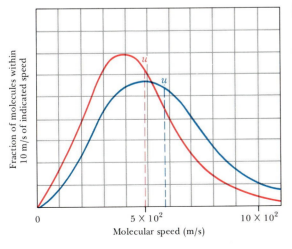

**FIGURE 10.12**
Distribution of molecular speeds for nitrogen at 0°C (red line) and 100°C (blue line).

proportionally. The relationship between average kinetic energy and temperature that is obtained from kinetic-molecular theory is given in Equation 10.14, where $R$ is the ideal gas constant, $\mathcal{N}$ is Avogadro's number, and $T$ is the absolute temperature:

$$\epsilon = \frac{1}{2}mu^2 = \frac{3RT}{2\mathcal{N}} \qquad [10.14]$$

## SAMPLE EXERCISE 10.16

What is the average kinetic energy of a nitrogen molecule at 27°C?

**Solution:** Using Equation 10.14, we have

$$\epsilon = \frac{3RT}{2\mathcal{N}} = \frac{3(8.314\ \text{J/K-mol})(300\ \text{K})}{2(6.022 \times 10^{23}/\text{mol})}$$

$$= 6.21 \times 10^{-21}\ \text{J}$$

The average kinetic energy of a gas molecule at 27°C will have this same value for any other gaseous substance.

### PRACTICE EXERCISE

Without repeating the foregoing calculation, give the average kinetic energy of gaseous argon atoms at 600 K.
***Answer:*** $1.24 \times 10^{-20}$ J

The empirical observations of gas properties as expressed in the various gas laws are readily understood in terms of the kinetic-molecular theory. The following examples are illustrative.

1.  *Effect of a volume increase at constant temperature:* The fact that temperature remains constant means that the average kinetic energy of the gas molecules remains unchanged. This in turn means that the rms speed of the molecules, $u$, is unchanged. However, if the volume is increased, the molecules must move a longer distance between collisions. Consequently, there are fewer collisions per unit time with the container walls, and pressure decreases. Thus the model accounts in a simple way for Boyle's law.

2.  *Effect of a temperature increase at constant volume:* An increase in temperature means an increase in the average kinetic energy of the molecules, and thus an increase in $u$. If there is no change in volume, there will be more collisions with the walls per unit time. Furthermore, the change in momentum in each collision increases (the molecules strike the walls harder). Hence the model explains the observed pressure increase.

## SAMPLE EXERCISE 10.17

A sample of $O_2$ gas initially at STP is transferred from a 2-L container to a 1-L container at constant temperature. What effect does this change have on (**a**) the average kinetic energy of $O_2$ molecules; (**b**) the average speed of the $O_2$ molecules; (**c**) the total number of collisions of $O_2$ molecules with the container walls in a unit time; (**d**) the number of collisions of $O_2$ molecules with a unit area of container wall in a unit time?

**Solution:** (**a**) The average kinetic energy of $O_2$ molecules is determined only by temperature. The

average kinetic energy is not changed by the compression of $O_2$ from 2 L to 1 L at constant temperature. (**b**) If the average kinetic energy of the $O_2$ molecules doesn't change, neither does $u$ (see Equation 10.14). Both the average and rms speeds remain constant. (**c**) The total number of collisions with the container walls in a unit time must increase, because the molecules are moving within a smaller volume but with the same average speed as before. Under these conditions they must encounter a wall more frequently. (**d**) The number of collisions with a unit area of wall increases, because the total number of collisions with the walls is higher and the area of wall is smaller than before.

**PRACTICE EXERCISE**

How is the rms speed of $N_2$ molecules in a gas sample
changed by (**a**) an increase in temperature; (**b**) an increase in volume of sample; (**c**) mixing with an Ar sample at the same temperature?

***Answers:*** (**a**) increases; (**b**) no effect; (**c**) no effect

---

Beginning with the postulates of the kinetic-molecular theory it is possible to derive the ideal-gas equation. Rather than proceed through a derivation, let's consider in somewhat qualitative terms how the ideal-gas equation might follow. As we have seen (Section 10.2), pressure is force per unit area. The total force of the molecular collisions on the walls, and hence the pressure produced by these collisions, depend both on how strongly the molecules strike the walls (impulse imparted per collision) and on the rate at which these collisions occur:

**The Ideal-Gas Equation**

$$P \propto \text{impulse imparted per collision} \times \text{rate of collisions} \qquad [10.15]$$

The impulse imparted by a collision of a molecule with a wall depends on the momentum of the molecule; that is, it depends on the product of its mass and speed, $mu$. The rate of collisions is proportional to both the number of molecules per unit volume, $n/V$, and their speed, $u$. If there are more molecules in a container, there will be more frequent collisions with the container walls. As the molecular speed increases or the volume of the container decreases, the time required for molecules to traverse the distance from one wall to another is reduced, and the molecules collide more frequently with the walls. Thus we have

$$P \propto mu \times \frac{n}{V} \times u \propto \frac{nmu^2}{V} \qquad [10.16]$$

From Equation 10.14 we have that $mu^2 \propto T$. (This proportionality also follows from the basic idea that the average kinetic energy is proportional to temperature.) Making this substitution into Equation 10.16, we have

$$P \propto \frac{n(mu^2)}{V} \propto \frac{nT}{V} \qquad [10.17]$$

Let us now convert the proportionality sign to an equal sign by expressing $n$ as the number of moles of gas; we then insert a proportionality constant—$R$, the molar gas constant:

$$P = \frac{nRT}{V} \qquad [10.18]$$

This expression, of course, is the familiar ideal-gas equation.

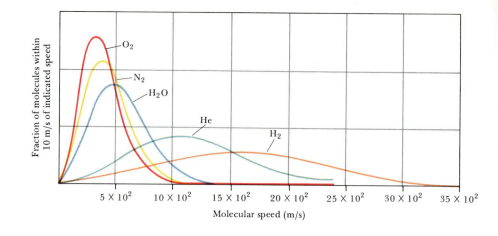

**FIGURE 10.13**
Distribution of molecular speeds for different gases at 25°C.

We have already made reference to the fact that the molecules of a gas do not all move at the same speed. Instead, the molecules are distributed over a range of speeds, as shown for nitrogen at two different temperatures in Figure 10.12. The distribution of molecular speeds depends on the mass of the gas molecules and on temperature. The average kinetic energy of the molecules in any gas is determined only by temperature; thus the quantity $\frac{1}{2}mu^2$, which is kinetic energy, must have the same value for two gases at the same temperature, though their masses may differ. This in turn means that molecules of larger mass must have smaller average speeds. From the kinetic-molecular theory it can be shown that the rms speed, $u$, is given by Equation 10.19. Note that $u$ is proportional to the square root of the absolute temperature and *inversely* proportional to the square root of the molecular weight, $\mathscr{M}$:

$$u = \sqrt{\frac{3RT}{\mathscr{M}}}$$

[10.19]

This relationship tells us that at any given temperature lighter molecules have higher rms speeds. In fact, the entire distribution of molecular speeds is skewed to higher values for gases of lower molecular weights, as shown for several gases in Figure 10.13.

The dependence of molecular speeds on mass has several interesting consequences. For example, the rate at which a gas is able to escape through a tiny hole, as when a gas escapes through a hole in a balloon, depends on the molecular mass of the gas. This process is known as **effusion**. Effusion is related to, but is not quite the same as, **diffusion**. The latter term refers to the spread of one substance throughout a space, or throughout a second substance. For example, the molecules of a perfume diffuse through a room.

**Graham's Law of Effusion**

In about 1830, Thomas Graham discovered that the effusion rates of gases are inversely related to the square roots of their molecular weights. Assume that we have two gases at the same initial pressure contained in identical containers, each with an identical pinhole in one wall. Let the rate of effusion be called $r$. **Graham's law** states that

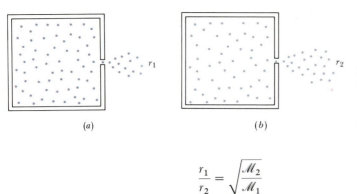

**FIGURE 10.14** Effusion of gases through a pinhole leak. In this illustration the molecular weight of gas molecules in (*a*) is higher than in (*b*). The rate of effusion $r_1$, from box (*a*), is slower than $r_2$, from box (*b*), in accordance with Equation 10.20.

$$\frac{r_1}{r_2} = \sqrt{\frac{\mathcal{M}_2}{\mathcal{M}_1}} \qquad [10.20]$$

Equation 10.20 compares the *rates* of effusion of two different gases; the lighter gas effuses more rapidly (see Figure 10.14).

Graham's law follows from our previous discussion if we assume that the rate of effusion is proportional to the rms speed of the molecules. Because $R$ and $T$ are constant, we have from Equation 10.19:

$$\frac{r_1}{r_2} = \frac{u_1}{u_2} = \sqrt{\frac{3RT/\mathcal{M}_1}{3RT/\mathcal{M}_2}} = \sqrt{\frac{\mathcal{M}_2}{\mathcal{M}_1}} \qquad [10.21]$$

## SAMPLE EXERCISE 10.18

**(a)** Calculate the rms speed, in m/s, of an $O_2$ molecule at 27°C. **(b)** If an unknown gas effuses at a rate that is only 0.468 times that of $O_2$ at the same temperature, what is the molecular weight of the unknown gas?

**Solution:** **(a)** Using Equation 10.19, we have

$$u = \sqrt{\frac{3RT}{\mathcal{M}}}$$

$$= \sqrt{\frac{(3)(8.314 \text{ J/K-mol})(300 \text{ K})}{32.0 \text{ g/mol}} \times \frac{10^3 \text{ g}}{1 \text{ kg}}}$$

$$= 484 \text{ m/s}$$

The factor $10^3$ g/kg is needed to convert mass to SI units consistent with the rest of the units in the problem. Recall (Section 4.1) that $1 \text{ J} = 1 \text{ kg-m}^2/\text{s}^2$.

**(b)** Using Equation 10.20, we have

$$\frac{r_x}{r_{O_2}} = \sqrt{\frac{\mathcal{M}_{O_2}}{\mathcal{M}_x}}$$

Thus

$$\frac{r_x}{r_{O_2}} = 0.468 = \sqrt{\frac{32.0 \text{ g/mol}}{x}}$$

Solving for $x$ yields

$$\frac{32.0 \text{ g/mol}}{x} = (0.468)^2 = 0.219$$

$$x = \frac{32.0 \text{ g/mol}}{0.219} = 146 \text{ g/mol}$$

**PRACTICE EXERCISE**

Calculate the ratio of effusion rates of $N_2$ and $O_2$, $r_{N_2}/r_{O_2}$.
**Answer:** $r_{N_2}/r_{O_2} = 1.07$

The explanation for diffusive flow is more complicated than that for effusion, because diffusion, unlike effusion, involves the effects of molecular collisions. Nevertheless, the dependence of diffusion rate on molecular mass is also given by Equation 10.20: The ratio of the rates of diffusion of two gases under identical experimental conditions is inversely proportional to the square root of the ratio of molecular masses.

That both effusion and diffusion rates are greater for lighter gases has many interesting applications. One popular lecture demonstration,

**FIGURE 10.15** Hydrogen-fountain demonstration of the greater rate of diffusion of hydrogen compared with diffusion of air. A container filled with $H_2$ gas is placed over the porous cup containing air. Hydrogen diffuses into the cup more rapidly than the molecules of air diffuse outward. As a result the pressure inside the vessel increases and water is pushed out of the tube, which is open to the outside. (Donald Clegg and Roxy Wilson)

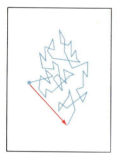

**FIGURE 10.16** Schematic illustration of the diffusion of a gas molecule. For the sake of clarity all the other gas molecules in the container are not shown. The path of the molecule of interest begins at the dot. Each short segment of line represents travel between collisions. The path traveled by the molecule is often described as a "random walk." The red arrow indicates the net distance traveled by the molecule.

## Mean Free Path and Thermal Conductivity

the hydrogen fountain, is illustrated in Figure 10.15. The diffusion of hydrogen through the walls of the porous cup is faster than diffusion of atmospheric gases out of the cup through the wall. Excess pressure builds up in the enclosure, forcing water through the tube.

The effort during World War II to develop the atomic bomb necessitated separating the relatively low-abundance uranium isotope $^{235}U$ (0.7 percent) from the much more abundant $^{238}U$ (99.3 percent). This was done by converting the uranium into a volatile compound, $UF_6$, which boils at 56°C. The gaseous $UF_6$ was allowed to diffuse from one chamber into a second through a porous barrier. Because of the slight difference in molecular weights, the relative rates of passage through the barrier for $^{235}UF_6$ and $^{238}UF_6$ are not exactly the same. The ratio of diffusion rates is given by the square root of the ratio of molecular weights, Equation 10.20:

$$\frac{r_{235}}{r_{238}} = \sqrt{\frac{352.04}{349.03}} = 1.0043$$

Thus the gas initially appearing on the opposite side of the barrier would be very slightly enriched in the lighter molecule. The diffusion process was repeated thousands of times, leading to a nearly complete separation of the two nuclides of uranium.

We can see from the horizontal scale of Figure 10.13 that the speeds of molecules are quite high. The average speed of $N_2$ at room temperature, 515 m/s, corresponds to 1850 km/hr, or 1150 mi/hr. Yet we know that if a vial of perfume is opened at one end of a room, some time elapses, perhaps a few minutes, before the odor is detected at the other end. The diffusion of gases is much slower than molecular speeds because of molecular collisions. These collisions occur quite frequently for a gas at atmospheric pressure—about $10^{10}$ times per second for each molecule. The paths of the gas molecules are therefore interrupted very often. Diffusion of gas molecules, depicted in Figure 10.16, thus consists of a random mo-

The rate of diffusion of a gas through a porous medium is not always determined solely by the molecular mass of the gas molecules. Even weak interactions between the molecules of gas and the molecules of the porous medium will affect the rate. Attractive intermolecular interactions, which we will discuss in more detail in the next chapter, slow the rate at which a gas molecule passes through the narrow passages of the porous medium.

The medium's molecules may attract one substance in a mixture more than another. Thus the rates of diffusion of a mixture's components through a medium are not uniform. Dow Chemical Company has recently developed a method of separating gaseous mixtures into their components by using the difference in the rates at which molecules pass through a porous membrane. Figure 10.17 shows Dow's tiny hollow fiber made of polyolefin, an organic polymer. The fiber is smaller than a human hair. Its walls are permeable to the passage of $N_2$ and $O_2$. However, $O_2$ and water vapor diffuse through the fiber walls considerably more readily than does $N_2$. To take advantage of this property, thousands of the fibers are formed into bundles.

Compressed air is introduced around the outside of the bundles, as shown schematically in Figure 10.18. The gas stream taken from inside the fibers is selectively enriched in $O_2$. The gas stream on the outside of the fibers is correspondingly enriched in $N_2$.

The Dow process, called the Generon system, makes available a stream of gas containing $O_2$ up to 35 percent by volume (air is 20 percent $O_2$ by volume) at substantially less cost than previous methods of $O_2$ enrichment. The oxygen-rich gas stream can be used both to speed up many chemical oxidation processes and to replace other, more expensive chemicals—such as hydrogen peroxide—that might otherwise be used to bring about oxidation.

**FIGURE 10.18** A diagrammatic illustration of gas separation that takes advantage of the difference in ease of diffusion of gases through polyolefin fiber walls. The gas that enters under pressure (feed air) passes from the distribution core all around the outsides of the tiny hollow fibers. Oxygen and water vapor preferentially pass through the fiber walls, are collected at the ends of the fibers, and pass out as oxygen-enriched air. The gas that has not passed through the fibers exits the assembly as nitrogen-enriched gas. (Dow Chemical Company)

**FIGURE 10.17** Polyolefin fibers used to carry out gas separations. The solid cylinder is human hair. Gases such as nitrogen and oxygen differ significantly in the rate at which they pass through the wall of the hollow fiber. This difference in permeability provides the basis for gas separations, as shown in Figure 10.18. (Dow Chemical Company)

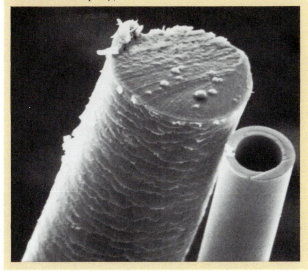

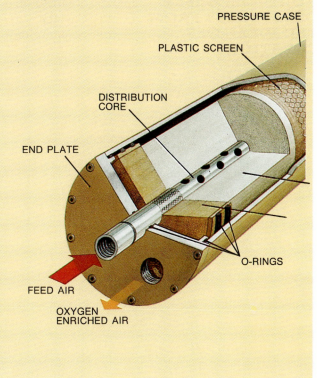

**TABLE 10.3** Mean free paths for several gases at 0°C, 1 atm

| Gas | Mean free path (nm) |
|---|---|
| Carbon dioxide | 39.7 |
| Carbon monoxide | 58.4 |
| Nitrogen | 60.0 |
| Argon | 63.5 |
| Oxygen | 64.7 |
| Hydrogen | 112.3 |
| Helium | 179.8 |

tion, first in one direction, then in another, at one instant at high speed, the next at low speed.

The average distance traveled by a molecule between collisions is called the **mean free path**. This distance depends on the effective radius of the molecules, because larger molecules are more likely to undergo a collision. It depends also on the number of molecules in a unit volume— the larger the number of molecules per unit volume, the more likely is collision. Table 10.3 lists some values of experimentally determined mean free paths for several gases at STP. As you can see from these values, the molecules of a gas at STP do not travel very far before undergoing collision. By contrast, at an elevation of about 100 km in the earth's atmosphere, the mean free paths of nitrogen and oxygen molecules are on the order of 10 cm, more than a million times longer than at the earth's surface.

The **thermal conductivity** of a gas—the rate at which heat energy can be transferred through the gas—depends on the average speed of its molecules and on their mean free path. Those gases whose molecules move fastest and with the largest mean free path have the highest thermal conductivities. Among the gases in Table 10.3, carbon dioxide has the highest molecular weight and thus has the lowest average speed per molecule for a given temperature. It also has the shortest mean free path of the gases listed. We can conclude that carbon dioxide has the lowest thermal conductivity of any of the gases in the table. Helium has the highest thermal conductivity.

## 10.10 NONIDEAL GASES: DEPARTURES FROM THE IDEAL-GAS EQUATION

Although the ideal-gas equation is a very useful description of gases, all real gases fail to obey the relationship to a greater or lesser degree. The extent to which a real gas departs from ideal behavior may be seen by slightly rearranging the ideal-gas equation:

$$\frac{PV}{RT} = n \qquad [10.22]$$

**FIGURE 10.19** $PV/RT$ versus pressure for several gases at 300 K. The data for $CO_2$ pertain to a temperature of 313 K, because $CO_2$ liquefies under high pressure at 300 K.

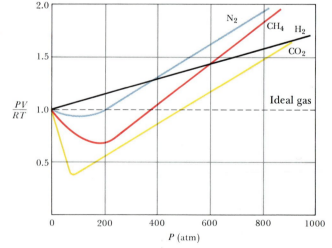

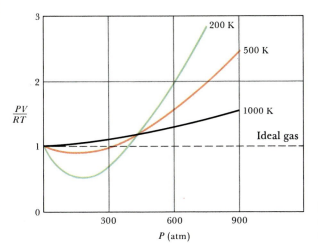

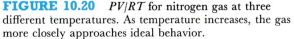

**FIGURE 10.20** $PV/RT$ for nitrogen gas at three different temperatures. As temperature increases, the gas more closely approaches ideal behavior.

For a mole of ideal gas ($n = 1$) the quantity $PV/RT$ equals 1 at all pressures. Figure 10.19 shows the quantity $PV/RT$ plotted as a function of pressure for a few gaseous substances, compared with the expected behavior of an ideal gas. Clearly, real gases are simply not ideal. However, the pressures shown are very high; at more ordinary pressures, below 10 atm, the deviations from ideal behavior are not so large, and the ideal-gas equation can be used without serious error.

Thus we see that deviations from ideal behavior tend to be larger at higher pressures than at lower ones. Temperature also has an effect.

Figure 10.20 shows graphs of $PV/RT$ for $N_2$ at three different temperatures. As temperature increases, the properties of the gas more nearly approach the ideal gas behavior. In general, gases tend to show significant deviations from ideal behavior at temperatures near their liquefaction points; that is, the deviations increase as temperature decreases, becoming significant near the temperature at which the gas is converted into a liquid.

We can understand these pressure and temperature effects on non-ideality by considering two factors that are considered negligible in the kinetic-molecular theory: (1) the molecules of a gas possess finite volumes, and (2) at short distances of approach they exert attractive forces upon one another.

When gases are contained at relatively low pressure, say 1 atm, the volume of the space they occupy is very large in comparison with the volumes of the gas molecules themselves. At increasingly high pressures, however, the volume taken up by the molecules becomes a larger fraction of the total. This effect is illustrated in Figure 10.21. Thus, at pressures of several hundred atmospheres, the free volume in which the gas molecules can move is considerably smaller than the volume of the container. As a result, the value of $V$ that *should* be used in the product $PV$ is smaller than the volume of the container. Hence, when we use the container volume in the ideal-gas equation we obtain a product $PV$ that is larger than it should be. The data for $H_2$ in Figure 10.19 illustrate this situation. Notice that $PV/RT$ increases steadily with increasing pressure. This is due entirely to the finite volume of the $H_2$ molecules.

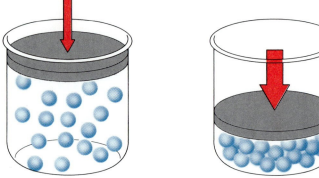

(a)                                (b)

**FIGURE 10.21** Illustration of the effect of the finite volume of gas molecules on the properties of a real gas at high pressure. In (a), at low pressure, the volume of the gas molecules is small compared with the container volume. In (b), at high pressure, the volume of the gas molecules themselves is a large fraction of the total space available.

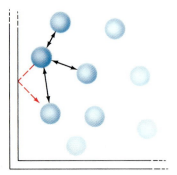

**FIGURE 10.22** Effect of attractive intermolecular forces on the pressure exerted by a gas on its container walls. The pressure is determined by the change in momentum when molecules strike the container wall. The molecule that is about to strike the wall in the figure experiences attractive forces from nearby molecules, and its impact on the wall is thereby lessened. The attractive forces become noticeable only under high-pressure conditions, when the average distance between molecules is small.

The attractive forces between molecules come into play at short distances, when the molecules undergo collisions or pass very close to one another. Because of these attractive forces, the impact of a given molecule with the wall of the container is lessened. If we could stop the action in a gas, we might see something like that illustrated in Figure 10.22. The molecule about to make contact with the wall experiences the attractive forces of nearby adjacent molecules, and thus the change of momentum when it hits the wall is lessened. As a result the quantity $PV/RT$ is less than we would expect on the basis of the ideal-gas equation. The effect of such intermolecular attractive forces becomes more significant as the pressure of the gas increases and the average distance between gas molecules correspondingly decreases.

From the data illustrated in Figure 10.19 we can guess that the attractive forces between molecules are greatest for $CO_2$ among the sample gases. The product $PV$ shows a substantial negative departure from the ideal-gas relationship over a wide range of pressure. The attractive forces are less important for $CH_4$, and less important still for $N_2$. Because there is very little attractive interaction between $H_2$ molecules, the quantity $PV/RT$ for this gas is continuously larger than that expected for an ideal gas.

Temperature determines how effective attractive forces between gas molecules are. As the gas is cooled, the motional energies decrease while intermolecular attractions remain constant. In a sense, cooling a gas deprives molecules of the energy they need to overcome their mutual attractive influence.

The effects of temperature shown in Figure 10.20 illustrate this point very well. Notice that as temperature increases, the negative departure of $PV/RT$ from the ideal-gas behavior disappears. What difference remains at high temperature stems mainly from the effect of volume.

## The van der Waals Equation

Engineers and scientists who work with gases at high pressures often cannot use the ideal-gas equation to predict the pressure-volume properties of gas, because departures from the ideal-gas behavior are too large. Various equations of state have been developed to predict more realistically the pressure-volume behavior of real gases. These equations, though more realistic, are also considerably more complicated than the simple

**TABLE 10.4** Van der Waals constants for gas molecules

| Substance | $a$ (L²-atm/mol²) | $b$ (L/mol) |
|---|---|---|
| He | 0.0341 | 0.02370 |
| Ne | 0.211 | 0.0171 |
| Ar | 1.34 | 0.0322 |
| Kr | 2.32 | 0.0398 |
| Xe | 4.19 | 0.0510 |
| $H_2$ | 0.244 | 0.0266 |
| $N_2$ | 1.39 | 0.0391 |
| $O_2$ | 1.36 | 0.0318 |
| $Cl_2$ | 6.49 | 0.0562 |
| $CO_2$ | 3.59 | 0.0427 |
| $CH_4$ | 2.25 | 0.0428 |
| $CCl_4$ | 20.4 | 0.1383 |

ideal-gas equation (Equation 10.1). Equation 10.23 shows the **van der Waals equation**, named after Johannes van der Waals, who presented it in 1873.

$$\left(P + \frac{an^2}{V^2}\right)(V - nb) = nRT \qquad [10.23]$$

This equation differs from the ideal-gas equation by the presence of two correction terms; one corrects the volume, the other modifies the pressure. The term $nb$ in the expression $(V - nb)$ is a correction for the finite volume of the gas molecules; the van der Waals constant $b$, different for each gas, has units of liters/mole. It is a measure of the actual volume occupied by the gas molecules. Values of $b$ for several gases are listed in Table 10.4. Note that $b$ increases with an increase in mass of the molecule or in the complexity of its structure.

The correction to the pressure takes account of the intermolecular attractions between molecules. Notice that it consists of the constant $a$, different for each gas, times the quantity $(n/V)^2$. The units of $n/V$ are moles/liter. This quantity is squared because the number of molecular-molecular interactions, as illustrated in Figure 10.22, is proportional to the square of the number of molecules per unit volume. Values of the van der Waals constant $a$ are listed in Table 10.4 for several gases. Notice that $a$, like $b$, increases with an increase in molecular weight and with an increase in complexity of molecular structure.

To get some feeling for the magnitudes of the departures from ideal behavior, let's calculate these departures for $CO_2$ at STP.

## SAMPLE EXERCISE 10.19

Calculate the correction terms to pressure and volume for $CO_2$ at STP, using the data in Table 10.4. Compare these terms with the ideal-gas values for $P$ and $V$.

**Solution:** From Equation 10.23 we see that the volume correction term is given by $nb$. Since $n = 1$, $nb$ equals 0.0427 L, which is to be compared with

22.4 L, the molar volume of an ideal gas at STP. The correction to volume is thus

$$\frac{0.0427}{22.4} \times 100 = 0.191\% \simeq 0.2\%$$

The correction to pressure is given by $an^2/V^2$. Inserting the value of $a$ from Table 10.4, $n = 1$, and $V = 22.4$ L, we obtain

$$\frac{an^2}{V^2} = \frac{\left(\dfrac{3.59 \text{ L}^2\text{-atm}}{\text{mol}^2}\right)1 \text{ mol}^2}{(22.4 \text{ L})^2} = 0.007 \text{ atm}$$

The required correction to pressure is thus $0.007 \times 100 = 0.7\%$. We conclude that the ideal-gas law is obeyed by $CO_2$ at STP conditions to within 1 percent.

**PRACTICE EXERCISE**

Which of the following molecules would you expect to have the largest values for $a$ and $b$: $CO_2$, Ar, $SO_2$?

*Answer:* $SO_2$

# FOR REVIEW

## SUMMARY

Many substances are capable of existing in any one of the three states of matter—solid, liquid, or gas. This chapter has been concerned with the gaseous state. To describe the state or condition of a gas, we must specify four variables: pressure, temperature, volume, and quantity of gas. Volume is usually measured in liters (L), and temperature in the Kelvin scale. Pressure is defined as the force per unit area. It is expressed in SI units as pascals, Pa (1 Pa = 1 N/m$^2$ = 1 kg/m-s$^2$), or more commonly in millimeters of mercury (mm Hg) or in atmospheres (atm). One **standard atmosphere** pressure equals 101.325 kPa, or 760 mm Hg. A **barometer** is often used to measure the atmospheric pressure. A **manometer** can be used to measure the pressure of enclosed gases.

The **ideal-gas equation**, $PV = nRT$, is the equation of state for an ideal gas. Most gases at pressures of about 1 atm and temperatures of 300 K and above obey the ideal-gas equation reasonably well. We can use the ideal-gas equation to calculate variations in one variable when one or more of the others are changed. For example, for a constant quantity of gas at constant temperature, the volume of the gas is inversely proportional to the pressure (**Boyle's law**). Similarly, for a constant quantity of gas at constant pressure, the volume of a gas is directly proportional to temperature (**Charles's law**). **Avogadro's law** states that at constant temperature and pressure the volume of a gas is directly proportional to the quantity of gas, that is, to the number of gas molecules. In gas mixtures, the total pressure is the sum of the partial pressures that each gas would exert if it were present alone under the same conditions (**Dalton's law of partial pressures**). In all applications of the ideal-gas equation we must remember to convert temperatures to the absolute temperature scale, Kelvin.

We use the ideal-gas equation to solve problems involving gases as reactants or products in chemical reactions. From the gas density, $d$, under given conditions of pressure and temperature, we may calculate the molecular weight of the gas: $\mathcal{M} = dRT/P$. In calculating the quantity of gas collected over water, correction must be made for the partial pressure of water vapor in the container.

The **kinetic-molecular theory** accounts for the properties of an ideal gas in terms of a set of assumptions about the nature of gases. Briefly these assumptions are that molecules are in ceaseless, chaotic motion; that the volume of gas molecules is negligible in relation to the volume of their container; that the gas molecules have no attractive forces for one another; that the average kinetic energy does not change; and that the average kinetic energy of the gas molecules is proportional to absolute temperature.

The molecules of a gas do not all have the same kinetic energy at a given instant. Their speeds are distributed over a wide range; the distribution varies with the molecular weight of the gas and with temperature. The root-mean-square (rms) speed, $u$, varies in proportion to the square root of absolute temperature and inversely with the square root of molecular weight: $u = (3RT/\mathcal{M})^{1/2}$. It follows that the rate at which a gas escapes (effuses) through a tiny hole is inversely proportional to the square root of its molecular weight (**Graham's law**). Molecules in a real gas possess finite volume and thus undergo frequent collisions with one another. Because of these collisions, the mean free path—the mean distance traveled between collisions—is short. Collisions between molecules limit the rate at which a gas molecule can diffuse through the space occupied by other gas molecules and influence the thermal conductivity of a gas.

Departures from ideal behavior increase in magnitude as pressure increases and as temperature decreases. The extent of nonideality of a real gas can be seen by examining the quantity $PV/RT$ for 1 mol of the gas as a function of pressure; for an ideal gas this quantity is exactly 1 at all pressures. Real gases depart from the ideal behavior because the molecules possess finite volume (leads to $PV/RT > 1$), or because the molecules experience attractive forces for one another upon collision (leads to $PV/RT < 1$). The van der Waals equation is an equation of state for gases that modifies the ideal-gas equation to more faithfully represent the pressure and volume behavior of real gases.

## LEARNING GOALS

Having read and studied this chapter, you should be able to:

1. Describe the general characteristics of gases as compared with other states of matter, and list the ways in which gases are distinctly different.

2. List the variables that are required to define the state of a gas.

3. Define atmosphere, millimeters of mercury, and kilopascals, the most important units in which pressure is expressed. You should also understand the principle of operation of a barometer and manometer.

4. Explain the way in which pressure, volume, and temperature are related in the ideal-gas equation. That is, you should remember $PV = nRT$ and be able to solve for one unknown given the other quantities.

5. Solve problems involving changes in the condition or state of a gas. You should be able to explain how one variable is affected by a change in another, when the other variables are maintained constant.

6. Calculate the quantity of a gas under a given set of conditions that is required as a reactant or formed as product in a chemical reaction.

7. Correct for the effects of water vapor pressure in calculating the quantity of a gas collected over water.

8. Explain the concept of gas density and describe how it is related to temperature, pressure, and molecular weight.

9. Calculate molecular weight, given gas density under defined conditions of temperature and pressure. You should also be able to calculate gas density under stated conditions, knowing molecular weight.

10. List and explain the assumptions on which the kinetic theory of gases is based.

11. Describe graphically how gas molecules are distributed over a range of speeds and how that distribution changes with temperature.

12. Describe how the relative rates of diffusion or effusion of two gases depend on their relative molecular weights (Graham's law).

13. Explain the concept of mean free path and how it relates to the rates of diffusion of molecules in the gas state and to thermal conductivity.

14. Explain the origin of deviations shown by real gases from the relationship $PV/RT = 1$ for a mole of ideal gas.

15. List the two major factors responsible for deviations of gases from ideal behavior.

16. Explain the origins of the correction terms to $P$ and $V$ that appear in the van der Waals equation of state for a gas.

## KEY TERMS

Among the more important terms and expressions used for the first time in this chapter are the following:

According to **Avogadro's hypothesis** (Section 10.3), equal volumes of gases at the same temperature and pressure contain equal numbers of molecules. Avogadro's law follows from this: At constant temperature and pressure, the volume of a gas is directly proportional to the quantity of gas.

A **barometer** (Section 10.2) is a device for measuring atmospheric pressure in terms of the height of a liquid column sustained by that pressure.

According to **Boyle's law** (Section 10.3), at constant temperature, the product of the volume and pressure of a given amount of gas is a constant.

According to **Charles's law** (Section 10.3), at constant pressure, the volume of a given quantity of gas is proportional to absolute temperature.

**Dalton's law of partial pressures** (Section 10.5) states that the total pressure of a mixture of gases is just the sum of the pressures that each gas would exert if it were present alone.

**Diffusion** (Section 10.9) refers to the rate at which a substance speeds into and throughout a space. Thus a gas might diffuse throughout a room, or atmospheric oxygen might diffuse through the waters of a lake. **Effusion** refers to the rate at which a gas escapes through an orifice or hole.

The **gas constant**, $R$ (Section 10.4), is the constant of proportionality in the ideal-gas equation.

**Graham's law** (Section 10.9) states that the relative rate of effusion of two gases is inversely proportional to the square root of the ratio of their molecular weights.

The **ideal-gas equation** (Section 10.4) is an equation of state for gases that embodies Boyle's law, Charles's law, and Avogadro's hypothesis in the form $PV = nRT$.

The **kinetic-molecular theory of gases** (Section 10.8) consists of a set of assumptions about the nature of gases. These assumptions, when translated into mathematical form, yield the ideal-gas equation.

A **manometer** (Section 10.2) is a device for measuring gas pressure by measuring the difference in heights of the mercury columns in a U-tube.

The **mean free path** (Section 10.9) in a gas sample is the average distance traveled by a gas molecule between collisions.

**Pressure** (Section 10.2) is a measure of the force exerted on a unit area. In work with gases, pressure is most commonly expressed in units of atmospheres (atm) or of millimeters of mercury (mm Hg)—760 mm Hg = 1 atm; in SI units, pressure is expressed in pascal (Pa).

The **root-mean-square** (rms) **speed**, $u$ (Section 10.8), of a gas is given by the square root of the average of the squared speeds of the gas molecules.

The **standard atmospheric pressure** (Section 10.2) is defined as 760 mm Hg or, in SI units, 101.325 kPa.

**Standard temperature and pressure** (STP) (Section 10.4), 0°C and 1 atm pressure, are frequently used reference conditions for a gas.

The **thermal conductivity** of a gas (Section 10.9) is a measure of the rate at which heat energy can be transferred through it.

The **torr** (Section 10.2) is a unit of pressure (1 torr = 1 mm Hg).

The **van der Waals equation** (Section 10.10) is an equation of state for real gases containing terms that correct for the existence of attractive forces between molecules and for their finite volumes.

# EXERCISES

## Introduction; Pressure

**10.1** Compressibility, which is the change in volume of a substance in response to a change in pressure, is much greater for gases than for liquids or solids. Explain.

**10.2** The mass of Venus is 0.82 times that of the earth; the atmospheric pressure at the surface of Venus is about 90 earth atmospheres. The atmosphere is 96 percent $CO_2$, and the surface temperature is over 700 K. Given these facts, explain how it is possible for the atmospheric pressure on Venus to be so much greater than that on the earth.

**10.3** Perform the following conversions: (**a**) 0.910 atm to kPa; (**b**) 730 mm Hg to kPa; (**c**) 29.86 in Hg to atm; (**d**) 1.06 atm to mm Hg.

**10.4** Perform the following conversions: (**a**) 607 mm Hg to atm; (**b**) 636 kPa to atm; (**c**) 615 mm Hg to torr; (**d**) 0.393 atm to mm Hg.

**10.5** (**a**) If the weather report for Montreal states that the temperature is 11°C and the barometric pressure is 98.8 kPa, what is the temperature in degrees Fahrenheit and pressure in inches of mercury? (**b**) If the weather in Chicago is reported to be 89°F with a barometric pressure of 29.32 in. Hg, what is the temperature in degrees Celsius and pressure in kilopascals?

**10.6** (**a**) On Titan, the largest moon of Saturn, atmospheric pressure is 1.6 earth atmospheres. What is the Titan atmosphere in kPa? (**b**) On Venus the surface atmospheric pressure is about 90 earth atmospheres. What is the Venusian atmosphere in kPa?

**10.7** (**a**) A plastic bucket has a flat bottom of area 4.6 × $10^3$ $cm^2$. When filled with water it weighs 12.8 kg. What pressure, in pascals, does the bucket exert on a flat surface on which it rests? (**b**) A phonograph stylus resting on a record has an effective mass of 2.5 g. The area of the stylus making contact with the record averages 2.7 × $10^{-4}$ $mm^2$. What pressure, in pascals, does the stylus exert on the record?

**10.8** (**a**) Calculate the pressure in pascals exerted on a tabletop by an iron cube (density 7.87 g/$cm^3$) measuring 8.0 cm on an edge. (**b**) Suppose that a woman weighing 130 lb and wearing high-heeled shoes momentarily places all her weight on the heel of one foot. If the area of the heel is $0.50$ in.$^2$, calculate the pressure exerted on the underlying surface, in atm and in kPa.

**10.9** An open-tube manometer containing mercury is connected to a container of gas. What is the pressure of the enclosed gas, in mm Hg, in each of the following situations? **(a)** The mercury in the arm attached to the gas is 6.5 cm higher than the one open to the atmosphere; atmospheric pressure is 0.960 atm. **(b)** The mercury in the arm attached to the gas is 4.8 cm lower than the one open to the atmosphere; atmospheric pressure is 1.02 atm.

**10.10** Suppose that the mercury in the barometer of Figure 10.2 were replaced by a liquid metal alloy with a density of $6.28$ g/cm$^3$. The density of mercury is $13.6$ g/cm$^3$. What height of column of liquid alloy would be supported by 1.00 atm pressure?

## The Gas Laws

**10.11** What are the variables that make up the equation of state for an ideal gas? How many of these variables need to be specified to completely determine the state of an ideal gas?

**10.12** Which of the variables in the ideal-gas equation is *not* a variable in **(a)** Boyle's law; **(b)** Charles's law; **(c)** Avogadro's law.

**10.13** Write an equation or proportionality expression that expresses each of the following statements: **(a)** for a given quantity of gas at constant temperature, the product of pressure times volume is constant; **(b)** for a given temperature and pressure, the volume of a gas is proportional to the number of moles of gas present; **(c)** for a given volume and quantity of gas, the pressure is proportional to the absolute temperature; **(d)** for a given quantity of gas, the product of pressure and volume is proportional to absolute temperature.

**10.14** Which of the following statements, if any, are false? Correct any false statements. **(a)** At constant temperature and volume, pressure is inversely proportional to the number of moles of gas. **(b)** At constant volume the pressure of a given amount of gas increases in proportion to the absolute temperature. **(c)** An increase in volume of a given quantity of gas at constant pressure arises from an increase in absolute temperature. **(d)** At constant volume the pressure of a gas is inversely proportional to temperature.

**10.15** Assuming a gas to behave ideally, calculate **(a)** its pressure if $7.25 \times 10^{-2}$ mol occupies 184 mL at $-14$°C; **(b)** its volume if 3.28 mol has a pressure of 0.688 atm and a temperature of 27°C; **(c)** the quantity of gas, in moles, if 2.00 L at $-25$°C has a pressure of 2.48 atm; **(d)** the temperature, in kelvin, at which $9.87 \times 10^{-2}$ mol occupies 164 mL at 722 mm Hg.

**10.16** For an ideal gas, calculate **(a)** the pressure of the gas if 1.34 mol occupies 3.28 L at 28°C; **(b)** the volume occupied by 0.150 mol at $-14$°C and 60.0 atm; **(c)** the number of moles in 1.50 L at 37°C and 725 mm Hg;

**(d)** the temperature at which 0.270 mol occupies 15.0 L at 2.54 atm.

**10.17** Many gases are shipped in high-pressure containers. If a steel tank whose volume is 42.0 L contains $O_2$ gas at a total pressure of 18,000 kPa at 23°C, what mass of oxygen does it contain? What volume would the gas occupy at STP?

**10.18** Fluorine gas, which is dangerously reactive, is shipped in steel cylinders of 30.0 L capacity, at a pressure of 160 lb/in.$^2$ at 26°C. What mass of $F_2$ is contained in such a tank?

## The Ideal-Gas Equation

**10.19** Starting with the ideal-gas equation, derive Avogadro's law.

**10.20** Starting with the ideal-gas equation, show that at constant volume the pressure of a given quantity of gas is proportional to absolute temperature.

**10.21** **(a)** A fixed quantity of gas is compressed at constant temperature from a volume of 368 mL to 108 mL. If the initial pressure was 522 mm Hg, what is the final pressure? **(b)** A fixed quantity of gas is allowed to expand at constant temperature from 2.45 L to 3.40 L. If the initial pressure was 1.05 atm, what is the final pressure? **(c)** What is the final volume of a gas if a 1.20-L sample is heated from 32°C to 450°C at constant pressure?

**10.22** **(a)** A gas originally at 15°C and having a volume of 182 mL is reduced in volume to 82.0 mL while its pressure is held constant. What is its final temperature? **(b)** A gas exerts a pressure of 187 kPa at 27°C. The temperature is increased to 108°C with no volume change. What is the gas pressure at the higher temperature? **(c)** The helium in a 0.75-L tank at 105 atm, 27°C, is expanded to 54.5 L, $-10$°C. What is the final pressure of the gas?

**10.23** At 36°C and 1.00 atm pressure, a gas occupies a volume of 0.600 L. How many liters will it occupy **(a)** at 0°C and 0.205 atm; **(b)** at STP?

**10.24** Chlorine is widely used to purify municipal water supplies and to treat swimming pool waters. Suppose that the volume of a particular sample of $Cl_2$ is 6.18 L at 740 torr and 33°C. **(a)** What volume will the $Cl_2$ occupy at 107°C and 680 torr? **(b)** What volume will the $Cl_2$ occupy at STP? **(c)** At what temperature will the volume be 3.00 L if the pressure is $8.00 \times 10^2$ mm Hg? **(d)** At what pressure will the volume be 5.00 L if the temperature is 67°C?

**10.25** In an experiment recently reported in the scientific literature, male cockroaches were made to run at different speeds on a miniature treadmill while their oxygen consumption was measured. The average cockroach running at 0.08 km/hr consumed 0.8 mL of $O_2$ at 1 atm pressure and 24°C per gram of insect weight per hour. **(a)** On this basis, how many moles of $O_2$ would be consumed in 1 hr by a 5.2-g cockroach moving at this speed? **(b)** This same cockroach is caught by a child and placed in a 1-qt fruit jar with a tight lid. Assuming the same level of continuous activity as in the research, will the cockroach consume more than 20 percent of the available $O_2$ in a 48-hr period? (Air is 21 mol percent $O_2$.)

**10.26** After the large eruption of Mount St. Helens in 1980, gas samples from the volcano were taken by sampling the downwind gas plume. The unfiltered gas samples were passed over a gold-coated wire coil to absorb mercury, Hg, present in the gas. The mercury was recovered from the coil by heating it, then analyzed. In one particular set of experiments, scientists found a mercury vapor level of 1800 ng of Hg per cubic meter in the plume, at a gas temperature of 10°C. Calculate **(a)** the partial pressure of Hg vapor in the plume; **(b)** the number of Hg atoms per $m^3$ in the gas; **(c)** the total mass of Hg emitted per day by the volcano if the daily plume volume is 1600 $km^3$.

### Dalton's Law of Partial Pressures

**10.27** Consider a mixture composed of 2.10 g of $H_2$, 65.2 g of $O_2$, and 12.5 g of Ar confined to a volume of 4.10 $m^3$ at 88°C. Calculate **(a)** the partial pressure of $H_2$; **(b)** the total pressure in the vessel.

**10.28** A mixture of cyclopropane gas, $C_3H_6$, and oxygen, $O_2$, is widely used as an anesthetic. **(a)** How many moles of each gas are present in a 16.5-L container at 23°C if the partial pressure of cyclopropane is 140 mm Hg and that of oxygen is 605 mm Hg? **(b)** What is the partial pressure of $C_3H_6$ in a mixture of 3.00 g of $C_3H_6$ and 20.0 g of $O_2$ if the total pressure in the vessel is 850 mm Hg?

**10.29** Consider the arrangement of bulbs shown in Figure 10.23. Each of the bulbs contains a gas at the pressure shown. What is the pressure of the system when all the stopcocks are opened, assuming that the temperature remains constant? (We can neglect the volume of the capillary tubing connecting the bulbs.)

### FIGURE 10.23

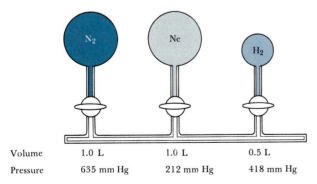

| Volume | 1.0 L | 1.0 L | 0.5 L |
|---|---|---|---|
| Pressure | 635 mm Hg | 212 mm Hg | 418 mm Hg |

**10.30** A quantity of $N_2$ gas originally held at 4.60 atm pressure in a 1.00-L container at 26°C is transferred to a 10.0-L container at 20°C. A quantity of $O_2$ gas orig-

inally at 3.50 atm and 26°C in a 5.00-L container is transferred to this same container. What is the total pressure in the new container?

### Density and Molecular Weight

**10.31** **(a)** Calculate the density of argon gas at STP. **(b)** Calculate the density of a gas at 37°C and 700 mm Hg if the gas has a molecular weight of 64.1 g/mol.

**10.32** **(a)** Calculate the molecular weight of a gas if 0.835 g occupies 800 mL at 400 mm Hg and 34°C. **(b)** Calculate the molecular weight of a gas that has a density of 2.18 g/L at 66°C and 720 mm Hg.

**10.33** **(a)** Cyanogen is 46.2 percent carbon and 53.8 percent nitrogen by mass. At 25°C and 750 mm Hg, 1.05 g of cyanogen occupies 0.500 L. What is the molecular formula of cyanogen? **(b)** Benzene is 92.3 percent carbon and 7.7 percent hydrogen by mass. At 120°C and 698 mm Hg, 0.555 g of benzene occupies 0.250 L. What is the molecular formula of benzene?

**10.34** The atmosphere of Titan is believed to consist of 82 mol percent $N_2$, 12 mol percent Ar, and 6.0 mol percent $CH_4$, as its principal constituents. Calculate the average molecular weight. At the surface the temperature is 95 K, and the pressure is 1.6 earth atmospheres. Calculate the density of Titan's atmosphere at the surface, assuming ideal-gas behavior.

**10.35** In the Dumas bulb technique for determining the molecular weight of an unknown liquid, one vaporizes the sample of a liquid that boils below 100°C in a boiling-water bath and determines the mass of vapor required to just fill the bulb (see Figure 10.24). From the following data, calculate the molecular weight of the unknown liquid; mass of unknown vapor, 1.012 g; volume of bulb, 354 $cm^3$; pressure, 742 mm Hg; temperature, 99°C.

### FIGURE 10.24

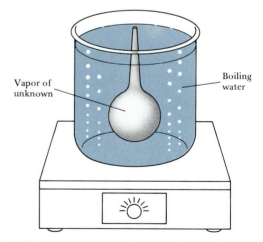

Vapor of unknown

Boiling water

**10.36** The molecular weight of a volatile substance was determined by the Dumas bulb method described in Exercise 10.35. The unknown vapor had a mass of 0.846 g; the volume of the bulb was 354 $cm^3$, pressure 752 mm Hg, temperature 100°C. Calculate the molecular weight of the unknown vapor.

[10.37]  A gaseous mixture of He and $N_2$ has a density of 0.550 g/L at 25°C and 715 mm Hg. What is the percent He (by mass) in the mixture?

[10.38]  A gaseous mixture of $O_2$ and Kr has a density of 1.104 g/L at 435 mm Hg and 300 K. What is the mole percent $O_2$ in the mixture?

## Quantities of Gases in Chemical Reactions

10.39  Magnesium can be used as a "getter" in evacuated enclosures, to react with the last traces of oxygen. (The magnesium is usually heated by passing an electric current through a wire or ribbon of the metal.) If an enclosure of 0.382 L has a partial pressure of $O_2$ of 3.5 × $10^{-6}$ mm Hg at 27°C, what mass of magnesium will react according to the following equation?

$$2Mg(s) + O_2(g) \longrightarrow 2MgO(s)$$

10.40  A 2.16-g sample of lead nitrate, $Pb(NO_3)_2$, is heated in an evacuated cylinder with a volume of 1.18 L. The salt decomposes when heated:

$$2Pb(NO_3)_2(s) \longrightarrow 2PbO(s) + 4NO_2(g) + O_2(g)$$

Assuming complete decomposition, what is the pressure in the cylinder after decomposition and cooling to a final temperature of 30°C?

10.41  A piece of solid magnesium is reacted with dilute hydrochloric acid to form hydrogen gas:

$$Mg(s) + 2HCl(aq) \longrightarrow MgCl_2(aq) + H_2(g)$$

What volume of $H_2$ is collected over water at 28°C (see Appendix C) by reaction of 1.25 g of Mg with 50.0 mL of 0.10 M HCl? The barometer records an atmospheric pressure of 748 mm Hg.

10.42  Ammonium nitrate can be decomposed thermally to form nitrous oxide, $N_2O$:

$$NH_4NO_3(s) \xrightarrow{\Delta} N_2O(g) + 2H_2O(l)$$

In one experiment 860 cm³ of $N_2O$ was collected over mineral oil, which has no appreciable vapor pressure, at 22°C, and 740 mm Hg pressure, using the experimental arrangement shown in Figure 10.10. (a) Assuming complete decomposition, how much $NH_4NO_3$ was decomposed? (b) What reason can you give for why mineral oil rather than water would be used in this case?

10.43  Recall from Chapter 4 that heats of combustion are often measured in a bomb calorimeter (Figure 4.13). Let's assume that we want to measure the heat of combustion of naphthalene, $C_{10}H_8$:

$$C_{10}H_8(s) + 12O_2(g) \longrightarrow 10CO_2(g) + 4H_2O(g)$$

The bomb is loaded with 0.202 g of $C_{10}H_8$. It has a gas volume of 132 cm³. The bomb is pressurized with completely with the naphthalene? If yes, what pressure of $O_2$ remains at the end of the reaction, when the temperature has returned to 25°C?

10.44  Assume that a single cylinder of an automobile engine has a volume of 600 cm³. (a) If the cylinder is full of air at 80°C and 0.980 atm, how many moles of $O_2$ are present? (The mole fraction of $O_2$ in dry air is 0.2095.) (b) How many grams of $C_8H_{18}$ could be combusted by this quantity of $O_2$ assuming complete combustion with formation of CO and $H_2O$?

## Kinetic-Molecular Theory; Graham's Law

10.45  Suppose that we have two 1-L flasks, one containing $N_2$ at STP, the other containing $SF_6$ at STP. How do these systems differ with respect to (a) the average kinetic energies of the molecules; (b) the total number of collisions occurring per unit time with the container walls; (c) the shapes of the distribution curves of molecular speeds; (d) the relative rates of effusion through a pinhole leak?

10.46  Vessel A contains $H_2$ gas at 0°C and 1 atm. Vessel B contains $O_2$ gas at 20°C and 0.5 atm. The two vessels have the same volume. (a) Which vessel contains more molecules? (b) Which contains more mass? (c) In which vessel is the average kinetic energy of molecules higher? (d) In which vessel is the average speed of molecules higher?

10.47  Indicate which of the following statements regarding the kinetic-molecular theory of gases are correct. For those that are false, formulate a correct version of the statement. (a) The average kinetic energy of a collection of gas molecules at a given temperature is proportional to $M^{1/2}$. (b) The gas molecules are assumed to exert no forces on each other. (c) All the molecules of a gas at a given temperature have the same kinetic energy. (d) The volume of the gas molecules is negligible in comparison to the total volume in which the gas is contained.

10.48  What change or changes in the state of a gas bring about each of the following effects? (a) The number of impacts per unit time on a given container wall increases. (b) The average energy of impact of molecules with the wall of the container decreases. (c) The average distance between gas molecules increases. (d) The average speed of molecules in the gas mixture is increased.

10.49  Calculate the rms speed, in m/s, for each of the following: (a) $H_2$ at 400 K; (b) $Cl_2$ at 400 K; (c) Ar at 100 K.

10.50  (a) What is the ratio of the average kinetic energies of $H_2$ and $N_2$ molecules at 300 K? (b) What is the ratio of the rms speeds of the molecules in the two gases at 300 K?

10.51  Which gas will effuse faster, $NH_3$ or CO? What are their relative rates of effusion?

10.52  Calculate the ratio of rates of effusion of (a) $H_2$ and Ne; (b) CO and $CO_2$; (c) $XeF_4$ and $XeF_6$.

10.53  A gas of unknown molecular mass is allowed to effuse through a small opening under constant pressure conditions. It required 72 s for 1 L of the gas to effuse. Under identical experimental conditions it required 28 s for 1 L of $O_2$ gas to effuse. Calculate the molecular weight of the unknown gas. (Remember that the faster the rate of effusion, the shorter the time required for effusion of 1 L; that is, rate and time are inversely proportional.)

**10.54** Arsenic(III) sulfide sublimes readily, even below its melting point of 320°C. The molecules of the vapor phase are found to effuse through a tiny hole at 0.28 times the rate of effusion of Ar atoms under the same conditions of temperature and pressure. What is the molecular formula of arsenic(III) sulfide in the gas phase?

## Departures from Ideal-Gas Behavior

**10.55** Describe two respects in which the ideal-gas law is not obeyed by real gases. How do these departures from ideal behavior show up in the behavior of real gases?

**10.56** The planet Jupiter has a mass of 318 earth masses, and its surface temperature is 140 K. Mercury has a mass of 0.05 earth mass, and its surface temperature is between 600 and 700 K. On which planet is the atmosphere more likely to obey the ideal-gas law? Explain.

**10.57** Which of the following gases would you expect to show the largest negative departure from the $PV/RT$ relationship expected for an ideal gas: $SO_2$, Ar, $H_2$, $CH_4$? For which of these gases should the correction for finite volume be largest?

**10.58** For each of the following pairs of gases, indicate which you would expect to deviate more from the $PV/RT$ relationship expected for an ideal gas: **(a)** $N_2$ and $SF_6$; **(b)** $BF_3$ and $SiCl_4$; **(c)** Ar and HCl. In each case indicate the reason for your choice.

**10.59** It turns out that the van der Waals constant $b$ is equal to four times the total volume actually occupied by the molecules of a mole of gas. Using this figure, calculate the fraction of the volume in a container actually occupied by Ar atoms **(a)** at STP; **(b)** at 100 atm pressure and 0°C. (Assume for simplicity that the ideal-gas equation still holds.)

**10.60** Calculate the pressure that $CCl_4$ will exert at 40°C if 1.00 mol occupies 28.0 L, assuming that **(a)** $CCl_4$ obeys the ideal-gas equation; **(b)** $CCl_4$ obeys the van der Waals equation. (Values for the van der Waals constants are given in Table 10.4.)

## Additional Exercises

**10.61** Atmospheric pressure is usually given in weather reports in units of inches of mercury. What is 1 atm pressure in inches of mercury?

**10.62** At 25°C, the density of mercury is 13.6 g/ml and that of water is 1.00 g/mL. On a day when tne atmospheric pressure is 735 mm Hg, how high would a column in a water-filled barometer be? **(a)** Ignore the vapor pressure of water. **(b)** Include the vapor pressure of water.

**10.63** Suppose that the manometer illustrated in Figure 10.4(c) contains mineral oil (density 0.752 $g/cm^3$) as fluid in place of mercury, which has a density of 13.6 $g/cm^3$.

The oil has no significant partial pressure of its own. When $h$ is 36.5 cm and the atmospheric pressure is 742 mm Hg, what is the pressure of the enclosed gas?

**10.64** A large lighter-than-air craft similar to the Goodyear blimp has a gas capacity of $5.0 \times 10^7$ L. Assume that this volume is filled with He at 1.0 atm pressure, 20°C. What lifting capacity does the craft possess? That is, what is the maximum mass that it can carry, including the gas bag itself, and still be lighter than air? (Air has a mean molecular mass of 28.7 g/mol.)

**10.65** A bubble of helium gas is trapped in a cavity of 0.132 $cm^3$ volume inside a mineral sample at a pressure of 75 atm at 18°C. What pressure does the helium exert after release from the mineral and storage in a bulb of 90-$cm^3$ volume at 23°C?

**10.66** The atmosphere of Venus at the surface consists of 96 mol percent $CO_2$, 3 mol percent $N_2$, 1 mol percent $SO_2$, and traces of other gases. **(a)** Calculate the average molecular weight of the Venusian atmosphere. **(b)** The surface atmospheric pressure on Venus is 90 earth atmospheres. The temperature is about 750 K. Calculate the density of the Venusian atmosphere at the surface.

**10.67** Under British law the maximum level of tetraethyllead, $Pb(C_2H_5)_4$, allowable in the work environment is 0.10 $mg/m^3$. To what partial pressure of $Pb(C_2H_5)_4$ does this correspond?

**10.68** Some of Robert Boyle's data from his experiments in 1662 with air trapped in a J-tube follow. (These were pre-SI-unit days; pressure was measured in inches of mercury. The gas volume units are arbitrary.)

| Gas volume | 48 | 40 | 32 | 24 | 20 | 16 | 12 |
|---|---|---|---|---|---|---|---|
| Difference in Hg levels (in.) | 0 | 6.2 | 15.1 | 29.7 | 41.6 | 58.1 | 88.4 |

Atmospheric pressure on that day was recorded at 29.1 in. Show whether the hypothesis that pressure is inversely related to gas volume is supported by Boyle's data.

**[10.69]** A mixture of methane, $CH_4$, and acetylene, $C_2H_2$, occupies a certain volume at a total pressure of 70.5 mm Hg. The sample is burned, forming $CO_2$ and $H_2O$. The $H_2O$ is removed and the remaining $CO_2$ found to have a pressure of 96.4 mm Hg at the same volume and temperature as the original mixture. What fraction of the gas was acetylene?

**10.70** Anhydrous copper(II) nitrate, $Cu(NO_3)_2$, is a sublimable solid. The vapor at a temperature of 182°C and a pressure of 0.32 mm Hg has a density of 2.11 mg/L. Is $Cu(NO_3)_2$ monomeric or dimeric in the gas phase?

**10.71** Boron forms a gaseous compound with hydrogen with empirical formula $BH_3$. A 1.05-L sample of this compound at a pressure of 88 mm Hg at 22°C has a mass of 0.138 g. What is the molecular formula of the compound?

**10.72** Calcium hydride, $CaH_2$, reacts with water to produce $H_2$ gas:

$$CaH_2(s) + 2H_2O(l) \longrightarrow$$
$$2H_2(g) + Ca^{2+}(aq) + 2OH^-(aq)$$

This reaction is used to generate $H_2$ to inflate life rafts and for similar uses where a simple compact means of $H_2$ generation is desired. Assuming complete reaction with water, how many grams of $CaH_2$ are required to fill a balloon to a total pressure of 1.08 atm at 15°C if its volume is 6.16 L?

**10.73** Nickel carbonyl, $Ni(CO)_4$, is one of the most toxic substances known. The present maximum allowable concentration in laboratory air during an 8-hr workday is 1 part in $10^9$. Assume 24°C at 1 atm pressure. What mass of $Ni(CO)_4$ is allowable in a laboratory that is 110 $m^2$ in area, with a ceiling height of 2.7 m?

**[10.74]** A glass vessel fitted with a stopcock has a mass of 337.428 g when evacuated. When filled with Ar it has a mass of 339.712 g. When evacuated and refilled with a mixture of Ne and Ar, under the same conditions of temperature and pressure, it weighs 339.218 g. What is the mole percentage of Ne in the gas mixture?

**10.75** Derive Equation 10.19, $u = \sqrt{3RT/M}$, starting with Equation 10.14.

**10.76** A balloon made of rubber permeable to small molecules is filled with helium. This balloon is then placed in a box that contains pure hydrogen, $H_2$. Will the balloon expand or contract? Explain.

**[10.77]** It is very difficult to obtain uniform gas samples from a tank containing a mixture of gases. After the tank has been in place for a time, the gas that initially flows when the tank is first opened is enriched in the lowest-molecular-weight components. Is this result predicted by the kinetic-molecular theory of gases? Suggest an explanation for the behavior.

**10.78** Suppose that a relatively rare isotope of carbon, $^{13}C$, could be separated from the more abundant $^{12}C$ by using a diffusion process similar to that described in the text for separating uranium isotopes. The uranium diffusion requires $UF_6$. The carbon diffusion requires CO or $CO_2$. Calculate the relative rates of diffusion for $^{12}CO$ and $^{13}CO$ and similarly for $^{12}CO_2$ and $^{13}CO_2$. Which substance would give the greater degree of separation?

**[10.79]** Scandium metal reacts with excess hydrochloric acid to produce $H_2$ gas. When 2.25 g of scandium is treated in this way, it is found that 2.41 L of $H_2$ measured at 100°C and 722 mm Hg pressure is liberated. Write the balanced chemical equation for the reaction that occurred.

**[10.80]** Consider the experiment illustrated in Figure 10.25. A gas is confined in the leftmost cylinder under 1 atm pressure, and the other cylinder is evacuated. When the stopcock is opened, the gas expands to fill both cylinders. Only a very small temperature change is noted when this expansion occurs. Explain how this observation relates to assumption 4 of the kinetic-molecular theory (Section 10.8.)

**FIGURE 10.25**

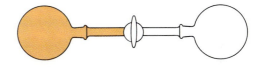

**[10.81]** The density of a gas of unknown molecular mass was measured as a function of pressure at 0°C:

| Pressure (atm) | 1.00 | 0.666 | 0.500 | 0.333 | 0.250 |
|---|---|---|---|---|---|
| Density (g/L) | 2.3074 | 1.5263 | 1.1401 | 0.7571 | 0.5660 |

**(a)** Determine a precise molecular weight for the gas. (Hint: Graph $d/P$ versus $P$.) **(b)** Why is $d/P$ not a constant as a function of pressure?

## CHAPTER 11

# Liquids
# and Solids

The water vapor in air (which we recognize as humidity), the water in a lake, and the ice in a glacier are all forms of the same chemical substance, $H_2O$. All have the same chemical properties; however, their physical properties differ greatly. The physical properties of any substance depend on the physical state of the substance. Some properties characteristic of each state of matter are listed in Table 11.1. In Chapter 10 we discussed the gaseous state in some detail. We now turn our attention to the physical properties of liquids and solids. You will remember that we could explain the most important physical properties of gases in terms of the kinetic-molecular theory. This same theory, with some modifications, can also help us to understand the characteristics of liquids and solids. In this chapter we will also examine those properties of the molecules themselves that are important in determining whether a substance is a gas, liquid, or solid under a given set of conditions. A key concept in comparing the properties of substances is that of intermolecular forces, the attractive forces responsible for keeping molecules together in liquids and solids. In this chapter we will be exploring the relationships among structure, intermolecular forces, and physical properties.

## CONTENTS

**TABLE 11.1**   Some characteristic properties of the states of matter

| Gas | 1. Assumes both the volume and shape of container. |
| | 2. Is compressible. |
| | 3. Diffusion within a gas occurs rapidly. |
| | 4. Flows readily. |
| Liquid | 1. Assumes the shape of the portion of the container it occupies. |
| | 2. Does not expand to fill container. |
| | 3. Is virtually incompressible. |
| | 4. Diffusion within a liquid occurs slowly. |
| | 5. Flows readily. |
| Solid | 1. Retains its own shape and volume. |
| | 2. Is virtually incompressible. |
| | 3. Diffusion within a solid occurs extremely slowly. |
| | 4. Does not flow. |

## 11.1 THE KINETIC-MOLECULAR DESCRIPTION OF LIQUIDS AND SOLIDS

In the kinetic-molecular theory a gas is viewed as a collection of widely separated molecules in constant, chaotic motion. The average kinetic energy of the molecules is much larger than the energy of the attractive forces between them. Indeed, in the derivation of the ideal-gas equation, attractive forces are ignored altogether.

In liquids the attractive energies between molecules are much larger; they are comparable to the kinetic energies of the molecules. The attrac-

A sample of quartz, one of the commonest of all rock-forming minerals and one of the most important constituents of the earth's crust. The sample of quartz shown here is nearly pure $SiO_2$. It occurs in a hexagonal form that appears as six-sided prisms. Samples this large and pure are rare, but quartz is present in every handful of beach sand. (Runk, Schoenberger/Grant Heilman Photography)

tive forces are thus able to hold the molecules in close proximity. However, the intermolecular attractions are not sufficiently strong (relative to the kinetic energies of molecules) to keep the molecules from moving in a more or less chaotic fashion. Thus liquids flow. The molecules in a liquid are close together, so liquids have much larger densities than gases. Because there is so little free space between molecules, liquids are much less compressible than gases. They have definite volumes, independent of the shape or size of their container. Because the molecules are free to move relative to each other, liquids do not have definite shape. They can be poured, and they flow to assume the shape of their container.

In solids, the intermolecular attractions are sufficiently strong relative to kinetic energies to virtually lock the molecules in place. Each molecule takes up a certain position relative to its neighbors, often in a highly regular pattern that extends through the solid. Solids that possess highly ordered structures are said to be crystalline. The transition from a liquid to a crystalline solid is rather like the change that occurs on a military parade ground when the troops are called to formation. Solids, like liquids, are not very compressible because the molecules are close together, with little free space available between them. Because the particles of a solid are not free to undergo long-range movement (translational motion), solids are rigid. Although translational motion is restricted, the molecules within a solid undergo vibrational motion; that is, they move back and forth periodically about the positions they occupy.

Figure 11.1 illustrates schematically the comparisons among the three states of matter based on the kinetic-molecular model. The particles that compose the substance can be individual atoms, as in Ar; molecules, as in $H_2O$; or ions, as in NaCl. For convenience, we will use the term "molecule" to refer to all these possibilities. The key factor that determines the physical state of a substance is the average kinetic energy of the molecules relative to the average energy of the attractive forces between them. Substances that exist as gases at room temperature possess weaker intermolecular attractions than do liquid substances; solid substances possess stronger intermolecular attractions. Conversions from one state to another occur by heating or cooling, which changes the average kinetic energy of the molecules. For example, NaCl, which is a solid at room temperature, melts at 804°C and boils at 1465°C when heated under

**FIGURE 11.1** Molecular-level comparison of gases, liquids, and solids. The density of particles in the gas phase is exaggerated as compared with most real situations.

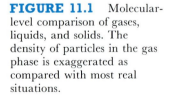

Gas

Total disorder; much empty space; particles have complete freedom of motion owing to separation and ineffective attractive forces.

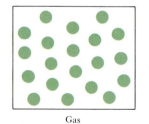

Cool and compress →

← Heat and reduce pressure

Liquid

Disorder; particles or clusters of particles are free to move relative to each other; particles close together.

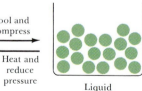

Cool →

← Heat

Crystalline solid

Ordered arrangement; particles can vibrate, but are in fixed positions; particles close together.

1 atm pressure. Ar, which is a gas at room temperature, can be liquefied at −186°C and solidified at −189°C when the pressure is 1 atm. Before we examine liquids, solids, or intermolecular forces in more detail, let's consider the interconversions that can occur between the states of matter.

Imagine an experiment in which we begin with a quantity of ethyl alcohol, also called ethanol, in a closed container, with a large space above the liquid. Suppose that we could somehow begin the experiment with all the ethanol in liquid form. The pressure in the container under these conditions would be zero, as illustrated in Figure 11.2(a). We would find that the pressure in the container quickly rises. After a short time the pressure would attain a constant value, termed the **vapor pressure** of liquid ethanol [Figure 11.2(b)].

We can account for these observations by using the kinetic-molecular theory. We know that the molecules of liquid ethanol are in constant motion. At any given temperature some of the molecules possess more than the average thermal energy, some less. That is, there is a distribution of molecular energies in the liquid similar to the distribution in a gas, illustrated in Figure 10.11.

At any instant some of the molecules at the surface of the liquid possess sufficient energy to escape from the attractive forces of their neighbors. Transfer of molecules into the gas phase is called **vaporization** or **evaporation**. The movement of molecules from the liquid to the gas phase goes on continuously. However, as the number of gas-phase molecules increases, the probability increases that a molecule in the gas phase will strike the liquid surface and stick there. We call this process **condensation**. Eventually, the two opposing processes occur at an equal rate, as illustrated in Figure 11.3. The pressure in the gas phase at this point becomes constant.

The system we have just discussed provides a simple example of **dynamic equilibrium**, in which two opposing processes take place at equal rates. To the observer it may appear that nothing is taking place. In our example of ethyl alcohol, the pressure remains constant at some value determined by the temperature of the liquid. In fact, a great deal is happening; molecules continuously pass from the liquid state to the gas state, and from the gas state to the liquid state. All equilibria between matter in different phases possess this dynamic character.

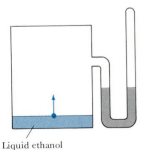

Liquid ethanol

(a) Initial

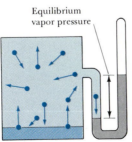

Equilibrium vapor pressure

(b) At equilibrium

**FIGURE 11.2** Illustration of the equilibrium vapor pressure over liquid ethanol. In (a) we imagine that no ethanol molecules exist in the gas phase; there is zero pressure in the cell. In (b) the rate at which molecules of ethanol leave the surface equals the rate at which gas molecules pass into the liquid phase. Thus the rates of condensation and of vaporization are equal. This produces a stable vapor pressure that does not change with time, as long as temperature remains constant.

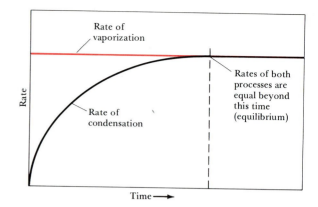

**FIGURE 11.3** Comparison of the rates of vaporization and condensation of a liquid. Time begins with the situation shown in Figure 11.2(*a*). The dotted vertical line marks equilibrium, which we see in Figure 11.2(*b*).

Although equilibria can exist simultaneously between all three states of matter, we usually encounter equilibria occurring between just two states, as in our example of the equilibrium between liquid and gaseous ethanol. Figure 11.4 summarizes the possible two-phase equilibria and the terms associated with them. Some terms are more familiar than others, but a little thought will bring to mind several examples of each possible kind of phase equilibrium.

## SAMPLE EXERCISE 11.1

Using water as the substance, give examples of all the possible equilibria between two different phases.

**Solution:** *Vaporization* of liquid water occurs from lakes or other water bodies in the summer months. *Condensation* of water vapor occurs on a cold window surface or on the lawn in the early morning hours. Ice *melts* in the spring, and lakes *freeze* in winter. When air moves over snow or ice at temperatures below freezing, the solid *sublimes*. The formation of snow crystals in clouds in the winter provides an example of *deposition*.

### PRACTICE EXERCISE

Name the phase change described in each of the following: (**a**) Beads of liquid mercury form on a cold surface when mercury vapor passes over it. (**b**) Crystalline naphthalene, sometimes used to ward off moths, disappears from a clothes closet. (**c**) Liquid benzene disappears from an open beaker in the fume hood.

*Answers:* (**a**) condensation; (**b**) sublimation; (**c**) vaporization

The change of matter from one state to another is called a **phase change**. Each of the phase changes diagramed in Figure 11.4, when occurring at constant temperature, has an associated enthalpy change. Energy is required in vaporization to overcome the attractive forces between molecules in the liquid phase. Similarly, sublimation requires energy; the attractive forces that bind the molecules to one another in the solid must be overcome to produce a gas of widely separated particles. Melting also requires energy; the regularity of the solid state, in which molecules are packed to attain the strongest intermolecular interactions, must be disrupted to attain the liquid state. Thus vaporization, sublimation, and melting are all endothermic processes. To proceed they require an input of heat. For example, an input of 2.26 kJ is required to cause vaporization of 1 g of water at 100°C.

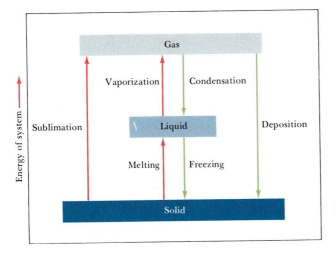

**FIGURE 11.4** Energy changes accompanying phase changes between the three states of matter and the names associated with them.

Condensation is just the reverse of the process of vaporization. Similarly, deposition is the reverse of sublimation, and freezing is the reverse of melting. Thus the enthalpies of condensation, deposition, and freezing are the same magnitude but opposite in sign from the enthalpies of vaporization, sublimation, and melting, respectively. The melting process is also referred to as fusion; thus the enthalpy of melting is frequently called the enthalpy of fusion, or heat of fusion. The **normal melting point** is the temperature at which melting occurs under 1 atm pressure. Unlike the boiling point, which we will discuss shortly, the melting point is not highly sensitive to external pressure. Incidentally, the **freezing point** of a substance is identical to its **melting point**; the two differ only in the temperature direction from which the phase change is approached.

Table 11.2 lists the molar heat capacities for water in all three states and the molar enthalpies of fusion and vaporization. Using these values,

**ΔH for Phase Changes**

**TABLE 11.2** Enthalpy properties of $H_2O$

| Molar heat capacities[a] | |
| --- | --- |
| Ice: | 37.6 J/mol-°C |
| Water: | 75.2 J/mol-°C |
| Water vapor: | 33.1 J/mol-°C |
| Molar enthalpies of phase changes | |
| Fusion: | 6.01 kJ/mol |
| Vaporization: | 40.67 kJ/mol |

[a] The enthalpy change of any substance as temperature changes depends on the molar heat capacity of the phase in which the substance happens to be. The molar heat capacity at constant pressure, $C_p$, is the product of the specific heat and the molar mass. For example, for liquid water, $C_p =$ (4.18 J/g-°C) × (18.0 g/mol) = 75.2 J/mol-°C.

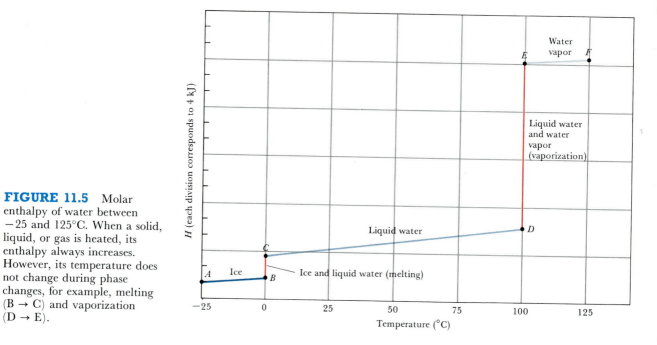

**FIGURE 11.5** Molar enthalpy of water between −25 and 125°C. When a solid, liquid, or gas is heated, its enthalpy always increases. However, its temperature does not change during phase changes, for example, melting (B → C) and vaporization (D → E).

we can construct a graph of enthalpy versus temperature for a mole of water, as shown in Figure 11.5. The vertical line segments correspond to the phase changes. The segment *BC*, at 0°C, corresponds to the molar enthalpy of fusion, and segment *DE*, at 100°C, corresponds to the molar enthalpy of vaporization. Temperature remains constant during phase changes, because the added energy is used to overcome attractive forces between molecules.

---

**SAMPLE EXERCISE 11.2**

Calculate the enthalpy change associated with converting 1.00 mol of ice at −25°C to water vapor at 125°C. Use the data given in Table 11.2.

**Solution:**  The heat required to bring the ice from −25°C to 0°C is given by

$$\Delta H = nC_p \Delta T$$

$$= (1.00 \text{ mol})\left(37.6 \frac{J}{\text{mol-°C}}\right)(25°C)$$

$$= 940 \text{ J} = 0.94 \text{ kJ}$$

The heat required to melt 1.00 mol of ice at 0°C is given by the molar enthalpy of fusion, 6.01 kJ. Heating liquid water from 0°C to 100°C then requires

$$\Delta H = nC_p \Delta T$$

$$= (1.00 \text{ mol})\left(75.2 \frac{J}{\text{mol-°C}}\right)(100°C)$$

$$= 7530 \text{ J} = 7.52 \text{ kJ}$$

The heat required to vaporize the water at 100°C is given by the molar enthalpy of vaporization, 40.67 kJ. Finally, to heat the water vapor from 100°C to 125°C requires

$$\Delta H = nC_p \Delta T$$

$$= (1.00 \text{ mol})\left(33.1 \frac{J}{\text{mol-°C}}\right)(25°C)$$

$$= 828 \text{ J} = 0.83 \text{ kJ}$$

Thus the total enthalpy change is 0.94 kJ + 6.01 kJ + 7.52 kJ + 40.67 kJ + 0.83 kJ = 55.97 kJ.

**PRACTICE EXERCISE**

If 20 kJ of heat is added to a 40.0-g sample of ice that is at 0°C, what is the resulting temperature of the water?

*Answer:*  39.6°C

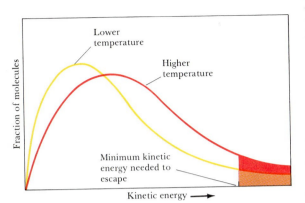

**FIGURE 11.6** Distribution of kinetic energies of surface molecules of a hypothetical liquid at two different temperatures. The same minimum kinetic energy is needed to escape at both temperatures. This energy depends on the magnitude of the attractive forces between molecules. The fraction of molecules having sufficient kinetic energy to escape the liquid (the colored area) is larger at the higher temperature.

## 11.3 PROPERTIES OF LIQUIDS

We have seen that molecules may escape from the surface of a liquid into the gas phase by vaporization, or evaporation. The weaker the attractive forces between molecules in the liquid phase, the more readily vaporization occurs. When two liquids are compared, the one that evaporates more readily is described as being the more **volatile**. For example, alcohol (ethanol) is more volatile than water. The rate of vaporization for any liquid increases with increasing temperature. Figure 11.6 provides a graphic explanation for this behavior. In that figure the distribution of kinetic energies of the particles at the surface of a liquid is compared at two temperatures with the energy required to escape from the surface to the gas phase. Notice that the distribution curves are like those shown earlier for gases (Figure 10.12); average kinetic energy increases with increasing temperature. Therefore, as the temperature of a liquid increases, so does the number of particles having sufficient kinetic energy to escape from the surface. Loss of particles with high kinetic energy causes the average kinetic energy of the particles in the liquid to decrease. Since average kinetic energy is proportional to temperature, the temperature of the remaining liquid decreases. Anyone getting out of a swimming pool, particularly on a windy day, has experienced the cooling that accompanies evaporation.

**Enthalpy of Vaporization**

The heat required to vaporize a mole of any liquid substance is known as the molar enthalpy of vaporization. Table 11.3 lists the molar enthalpies of vaporization, $\Delta H_v$, of some common substances at their boiling points. Although it is not obvious from Table 11.3, it requires more energy to

**TABLE 11.3** Enthalpies of vaporization of some common liquids at their boiling points

| Substance | Formula | $\Delta H_v$ (kJ/mol) | Boiling point (°C) |
|-----------|---------|----------------------|---------------------|
| Benzene | $C_6H_6$ | 30.8 | 80.2 |
| Ethanol | $C_2H_5OH$ | 39.2 | 78.3 |
| Ether | $C_2H_5OC_2H_5$ | 26.0 | 34.6 |
| Mercury | Hg | 59.3 | 356.9 |
| Methane | $CH_4$ | 10.4 | −164 |
| Water | $H_2O$ | 40.7 | 100 |

vaporize a gram of liquid water than to vaporize an equal mass of any other known liquid substance. Comparing water and mercury, we have:

$$H_2O: \quad \left(40.7\,\frac{kJ}{mol}\right)\left(\frac{1\ mol}{18.0\ g}\right) = 2.26\ kJ/g$$

$$Hg: \quad \left(59.3\,\frac{kJ}{mol}\right)\left(\frac{1\ mol}{200.5\ g}\right) = 0.296\ kJ/g$$

The high enthalpy of vaporization of water plays an important part in regulating body temperature. Metabolic activity within the body produces heat. An important mechanism used by the body to maintain temperature at its normal value, around 98°F, is perspiration. We are cooled by the heat absorbed when water evaporates from the skin. The average adult ingests about 2 L of water each day and loses nearly half of this by evaporation. This process accounts for the removal of about 20 percent of the heat produced in the body. When the surrounding air is humid and warm, the rate of evaporation is slowed. Consequently, we feel more "sweaty," and so more uncomfortable, on warm, humid days.

## Vapor Pressure

Liquids vary greatly in their vapor pressures at any given temperature. Gasoline or water evaporates fairly quickly when exposed to moving air, but engine oil exposed to the air at room temperature seems not to be volatile at all. The volatility of a liquid is determined by the magnitude of the intermolecular forces that constrain the molecules to remain together in the liquid. (We will consider the origins of intermolecular forces in Section 11.4.) Volatility increases with increasing temperature, as the average kinetic energy of the molecules increases in relation to the intermolecular forces. Figure 11.7 depicts the variation in vapor pressure with temperature for four common substances that differ greatly in volatility. Note that in all cases the vapor pressure increases quite nonlinearly with increasing temperature.

**FIGURE 11.7**   Vapor pressure of four common liquids shown as a function of temperature. The temperature at which the vapor pressure is 760 mm Hg is the normal boiling point of the liquid.

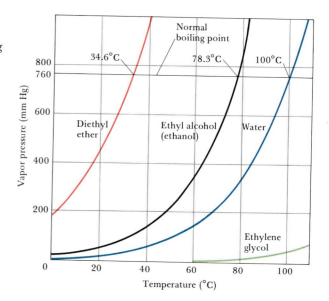

We can represent the dependence of the vapor pressure, $P$, on temperature as a log function of the form

$$\log P = \frac{-\Delta H_v}{2.303RT} + C \qquad [11.1]$$

In this equation $T$ is absolute temperature, $R$ is the gas constant (Table 10.2), $\Delta H_v$ is the enthalpy of vaporization, and $C$ is a constant. This equation, called the Clausius-Clapeyron equation, tells us that a graph of $\log P$ versus $1/T$ should be linear. The slope of the line can be used to calculate the enthalpy of vaporization.

As an example of the application of the Clausius-Clapeyron equation, the vapor pressure data for ethyl alcohol shown in Figure 11.7 are graphed in Figure 11.8 as $\log P$ versus $1/T$ (temperature is in kelvin). Note that a linear relationship results. Because a straight line is easy to extrapolate, we could extend it to obtain values for the vapor pressure of ethyl alcohol at temperatures above and below the temperature range for which we have data.

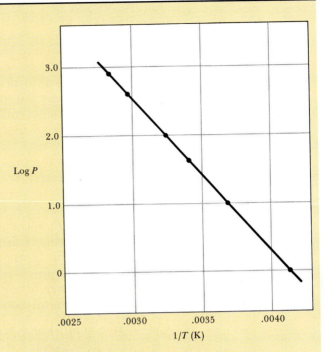

**FIGURE 11.8**  Application of the Clausius-Clapeyron equation, Equation 11.1, to the vapor pressure versus temperature data for ethyl alcohol. The slope of the line equals $-\Delta H_v/2.3R$.

**Boiling Points**

A liquid is said to boil when vapor bubbles form in the interior of the liquid. This condition occurs when the vapor pressure equals the external pressure acting on the liquid's surface. Consequently, the boiling point of a liquid depends on pressure. From Figure 11.7 we see that the boiling point of water at 1 atm pressure (760 mm Hg) is 100°C. The boiling point of a liquid at 1 atm pressure is called its **normal boiling point**. At 650 mm Hg water boils at 96°C.

**SAMPLE EXERCISE 11.3**

Using Figure 11.7, estimate the boiling point of ethanol at 400 mg Hg.

**Solution**  From Figure 11.7 we see that the boiling point must be about 64°.

**PRACTICE EXERCISE** _____

If we wanted to establish a boiling point of 40°C for diethyl ether, what vapor pressure would we need to maintain in the container?
*Answer:*  about 960 mm Hg

The time required to cook food depends on the temperature. As long as water is present, the maximum temperature of the cooking food is the boiling point of water. Pressure cookers are sealed and allow steam to escape only when it exceeds a predetermined pressure; the pressure above the water can therefore increase above atmospheric pressure. The higher pressure causes water to boil at a higher temperature, thereby allowing the food to get hotter and so cook more rapidly. The effect of pressure on boiling point also explains why it takes longer to cook food at higher elevations than at sea level. At higher altitudes the atmospheric pressure is lower, and water boils at a lower temperature.

## Critical Temperature and Pressure

We have seen that the transition from the gaseous to the liquid state is promoted by the intermolecular attractive forces that cause condensation. However, condensation is opposed by the kinetic energies of the molecules, which keep them moving independently of each other.

As temperature rises, the kinetic energies of molecules increase in relation to intermolecular attractions, and a gas becomes more difficult to liquefy. Consequently, the pressure required to condense a gas to a liquid increases. A temperature exists at which no amount of pressure, however great, causes the gas to pass from the gaseous state to the liquid state. The highest temperature at which a substance can exist as a liquid is called its **critical temperature**. The **critical pressure** is the pressure required to bring about liquefaction at this critical temperature. The critical temperatures and pressures of substances are of considerable practical importance to engineers and others working with gases. The relative values of these quantities among different substances also provide some indication of the relative magnitudes of intermolecular forces. Table 11.4 shows the critical temperatures and pressures for the noble gas elements and a few substances composed of nonpolar diatomic molecules. The critical temperature and pressure increase in the same way as the boiling point.

## CHEMISTRY AT WORK

### Supercritical Fluid Extraction

At ordinary pressures, a substance above its critical temperature behaves as an ordinary gas. However, as pressure increases up to several hundred atmospheres, its character changes. Like a gas it still expands to fill the confines of its container, but its density comes to approximate that of a liquid. (For example, the critical temperature of water is 374.4°C and its critical pressure is 217.7 atm. At this temperature and pressure, the density of water is 0.4 g/mL.) It is perhaps more appropriate to speak of a substance at critical temperature and pressure as a *supercritical fluid* rather than as a gas.

Like liquids, supercritical fluids can behave as solvents, dissolving a wide range of substances. This ability forms the basis of a system for separating the components of mixtures, a process known as supercritical fluid extraction. The solvent power of a supercritical fluid increases as its density increases. Conversely, lowering its density (either by decreasing pressure or increasing temperature) causes the

supercritical fluid and the dissolved material to separate. With skillful manipulation of temperature and pressure, it is possible to separate the components of very complicated mixtures.

The process of supercritical fluid extraction is now under extensive study in the chemical, food, drug, and energy industries. A process to remove caffeine from green coffee beans by extraction with supercritical carbon dioxide has been in commercial operation for several years. (The critical temperature of $CO_2$ is 31.1°C and its critical pressure is 73.0 atm.) At the proper temperature and pressure, the $CO_2$ removes caffeine from the beans but leaves the flavor and aroma components, producing decaffeinated coffee. Other applications of supercritical $CO_2$ extraction include removal of nicotine from tobaccos and removal of oil from potato chips, producing a lower-calorie product that is less greasy but has the same flavor and texture.

**TABLE 11.4** Molecular masses, normal boiling points, enthalpies of vaporization, and critical temperatures and pressures of the noble gas elements and a few substances composed of nonpolar diatomic molecules

| Substance | Molecular mass (amu) | Normal boiling point (K) | $\Delta H_v$ (kJ/mol)[a] | Critical temperature (K) | Critical pressure (atm) |
|---|---|---|---|---|---|
| He | 4 | 4.2 | 0.081 | 5.2 | 2.26 |
| Ne | 20 | 27 | 1.76 | 44.4 | 25.9 |
| Ar | 40 | 87 | 6.52 | 151 | 48 |
| Kr | 84 | 121 | 9.03 | 210 | 54 |
| Xe | 131 | 164 | 12.64 | 290 | 58 |
| Rn | 222 | 211 | 16.78 | 377 | 62 |
| $H_2$ | 2 | 20 | 0.903 | 33.2 | 12.8 |
| $N_2$ | 28 | 77 | 5.58 | 126 | 33.5 |
| $O_2$ | 32 | 90 | 6.82 | 154 | 49.7 |
| $Cl_2$ | 71 | 239 | 20.40 | 417 | 76.1 |

[a] Measured at the temperature of the normal boiling point.

## Viscosity

Some liquids literally flow like molasses; others flow quite easily. The resistance of liquids to flow is called their **viscosity**. The larger the viscosity, the more slowly the liquid flows. Liquids such as molasses or motor oils are relatively viscous. Water and organic liquids such as carbon tetrachloride are not. Viscosity can be measured by timing how long it takes a certain amount of the liquid to flow through a thin tube, under gravitational force. More viscous fluids take longer. In another method, the rate of fall of steel spheres through the liquid is measured. The spheres fall more slowly through the more viscous liquids.

Viscosity is related to the ease with which individual molecules of the liquid can move with respect to one another. It thus depends on the attractive forces between molecules, and on whether structural features exist that cause the molecules to become entangled. Viscosity decreases with increasing temperature, because at higher temperature the greater average kinetic energy of the molecules more easily overrides the attractive forces between molecules. The viscosities of some common liquids are listed in Table 11.5.

**TABLE 11.5** Viscosities and surface tensions of some common liquids at 20°C, and of water at several temperatures

| Substance | Formula | Viscosity (N-s/m²)[a] | Surface tension, $\gamma$ (J/m²) |
|---|---|---|---|
| Benzene | $C_6H_6$ | $0.65 \times 10^{-3}$ | $2.89 \times 10^{-2}$ |
| Ethanol | $C_2H_5OH$ | $1.20 \times 10^{-3}$ | $2.23 \times 10^{-2}$ |
| Ether | $C_2H_5OC_2H_5$ | $0.23 \times 10^{-3}$ | $1.70 \times 10^{-2}$ |
| Glycerin | $C_3H_8O_3$ | $1490 \times 10^{-3}$ | $6.34 \times 10^{-2}$ |
| Mercury | Hg | $1.55 \times 10^{-3}$ | $46 \times 10^{-2}$ |
| Water at: | $H_2O$ | | |
| 20°C | | $1.00 \times 10^{-3}$ | $7.29 \times 10^{-2}$ |
| 40°C | | $0.652 \times 10^{-3}$ | $6.99 \times 10^{-2}$ |
| 60°C | | $0.466 \times 10^{-3}$ | $6.70 \times 10^{-2}$ |
| 80°C | | $0.356 \times 10^{-3}$ | $6.40 \times 10^{-2}$ |

[a] This is the SI unit: 1 N-s/m² = 1 kg/m-s.

## Surface Tension

Liquids have a tendency to assume a minimum surface area. This minimum is achieved when the liquid has a spherical shape. For example, when water is placed on a waxy surface, it "beads up," forming distorted spheres. The state of minimum energy is that state of minimum surface area; energy must therefore be supplied to increase the surface area. The energy required to increase the surface area of a liquid by a unit amount is known as its **surface tension**. The origin of surface tension is an imbalance of forces at the surface of the liquid, as shown in Figure 11.9. A net inward pull on the surface contracts the surface and makes the liquid behave almost as if it had a skin. This effect permits a carefully placed needle to float on the surface of water and some insects to "walk" on water even though their densities are greater than that of water (Figure 11.10).

The forces between like molecules that affect a substance's vapor pressure, boiling point, heat of vaporization, viscosity, and surface tension are called **cohesive forces**. The forces between unlike substances, such as water and glass, are called **adhesive forces**. In a glass tube, the adhesive forces between water and glass are sufficiently strong relative to the cohesive forces for water to form a concave-upward surface (Figure 11.11). The curved upper surface of a liquid column is a **meniscus**. For mercury, cohesive forces are greater than the adhesive forces with glass; mercury does not adhere to glass, and a concave-downward meniscus is observed (Figure 11.11).

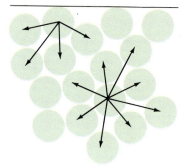

**FIGURE 11.9** Molecular-level view of the unbalanced intermolecular forces acting on a molecule at the surface of a liquid. Because the surface molecules experience fewer attractive intermolecular interactions than those in the bulk liquid, there is a tendency to minimize the amount of surface. We call this tendency surface tension.

**FIGURE 11.10** Surface tension permits an insect such as the water strider to "walk" on water. (© 1977 Michael P. Godomski, National Audubon Society, Photo Researchers, Inc.)

**FIGURE 11.11** Comparison of the shape of the water meniscus in a glass tube and the shape of the mercury meniscus. (Richard Megna/Fundamental Photographs)

When a liquid adheres to or wets the walls of a tube—as water adheres to glass, for example—the surface area of the liquid tends to increase. Surface tension, acting to reduce this area by trying to take the shape of a sphere, pulls liquid up the tube. This phenomenon is known as **capillary rise**, or **capillary action**. Wetting of the walls of a tube tends to increase the surface area of the liquid. Surface tension tends to reduce this area, and consequently the liquid is drawn up the tube. The surface tensions of several common liquids are given in Table 11.5. The surface tension of $7.29 \times 10^{-2}$ J/m$^2$ for water at 20°C indicates that an energy of $7.29 \times 10^{-2}$ J must be supplied to increase the surface area of a given amount of water by 1 m$^2$. Surface tension generally decreases with increasing temperature, as illustrated by the data for water in Table 11.5.

We have made repeated reference to the influence of the strengths of intermolecular forces of attraction on the physical properties of liquids. Indeed, the physical properties of matter ultimately depend on intermolecular attractive forces. For example, enthalpies of vaporization, boiling points, surface tensions, and viscosities of liquids generally increase in magnitude as intermolecular attractive forces increase. Such forces are also directly related to properties of solids, such as enthalpy of fusion, melting point, and hardness. Understanding the origins and relative strengths of these forces, then, is important. Rather than continuing to refer to them in a vague way, then, let's examine these forces before moving on to consider the properties of solids.

We have already seen that covalent bonds hold atoms together *within* molecules (Chapter 8). The fact that a molecular substance such as $H_2O$ exists in the solid and liquid phases indicates that there are also attractive forces *between* molecules. These attractive forces are what we have referred to as **intermolecular forces**. The magnitudes of intermolecular forces vary over a wide range. However, they are generally much smaller than those of the covalent bonds that bind atoms together in the same molecule. For example, only 16 kJ/mol is required to overcome the intermolecular attractions between HCl molecules in liquid HCl in order to vaporize it. By contrast, the energy required to dissociate HCl into H and Cl atoms is 431 kJ/mol. Let's now consider some of the most important types of intermolecular forces.

An **ion-dipole** force exists between an ion and a neutral polar molecule that possesses a permanent dipole moment. Recall that polar molecules are those to which a positive and a negative end can be assigned owing to molecular shape and to charge separation caused by unequal sharing of electrons (Section 9.2). A simple example is HCl ($\mu = 1.03$ D). The end of the dipole possessing an opposite charge from that of the ion is attracted to the ion, as illustrated in Figure 11.12. The energy of the interaction between an ion and a dipole depends on the charge on the ion, $Q$, and the magnitude of the dipole moment of the dipole, $\mu$: $E \propto Q\mu/d^2$, where $d$ is the distance from the center of the ion to the midpoint of the dipole. Ion-dipole forces are especially important in solutions of ionic

## 11.4 INTER-MOLECULAR ATTRACTIVE FORCES

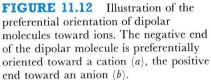

**Ion-Dipole Forces**

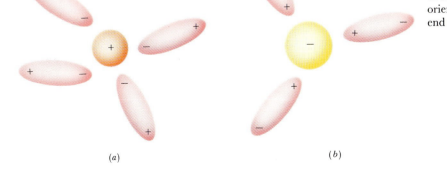

(a)  (b)

**FIGURE 11.12**  Illustration of the preferential orientation of dipolar molecules toward ions. The negative end of the dipolar molecule is preferentially oriented toward a cation (*a*), the positive end toward an anion (*b*).

substances in polar liquids, as, for example, in a solution of NaCl in water. We will have more to say about such solutions later (Section 12.2).

## Dipole-Dipole Forces

**Dipole-dipole** forces exist between polar molecules. The sign and magnitude of the interaction vary with the relative orientations of the two dipoles, as illustrated in Figure 11.13. In crystalline solids, the dipolar molecules tend to orient themselves to maximize the attractive interactions between ends of unlike charge [Figure 11.13(a) and (b)]. However, other factors, such as the shapes of the molecules and location of the dipole within the molecule, also play a role in determining the orientations of the molecules.

In liquids, the dipolar molecules are free to move with respect to one another; they will sometimes be in an orientation that is attractive [Figure 11.13(a) and (b)], sometimes in an orientation that is repulsive [Figure 11.13(c) and (d)]. The net effect, averaged over time, is an attractive energy. Dipole-dipole forces in a liquid are significant only between polar molecules that are very close together. Nevertheless, we find clear evidence that *for molecules of approximately equal molecular mass and size, intermolecular attractions increase with increasing polarity*. Thus the boiling points of the organic liquids listed in Table 11.6 increase with increasing magnitude of the dipole moment.

**FIGURE 11.13** Variation in the dipole-dipole interaction with orientation. In (a) and (b) the dipoles are aligned so as to produce an attractive interaction. In (c) and (d) the interactions are repulsive.

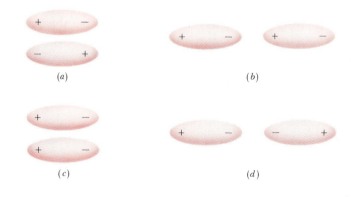

**TABLE 11.6** Molecular masses, dipole moments, and boiling points of several simple organic substances

| Substance | Molecular mass (amu) | Dipole moment, $\mu$ (D) | Boiling point (K) |
|---|---|---|---|
| Propane, $CH_3CH_2CH_3$ | 44 | 0.0 | 231 |
| Dimethyl ether, $CH_3OCH_3$ | 46 | 1.3 | 249 |
| Methyl chloride, $CH_3Cl$ | 50 | 2.0 | 249 |
| Acetaldehyde, $CH_3CHO$ | 44 | 2.7 | 293 |
| Acetonitrile, $CH_3CN$ | 41 | 3.9 | 355 |

## London Dispersion Forces

Table 11.4 contains the masses, normal boiling points, and enthalpies of vaporization of several substances composed either of atoms or simple, nonpolar molecules. The fact that these substances can be liquefied tells us that there must be attractive interactions between the molecules. Yet none of the various types of attractive forces we have discussed can apply here. The solution to this longstanding puzzle was provided in 1930 by Fritz London, who applied the then new theory of quantum mechanics to the problem. London's contribution was essentially to distinguish what we see on the average from what we would see if we could freeze the charge distribution in a collection of molecules at any particular instant. In a collection of helium atoms, the average distribution of electronic charge about each nucleus is spherically symmetrical. But the electrons are in constant motion. Because each electron constantly experiences repulsive interactions with other electrons on the same atom, *and electrons on other adjacent atoms*, the motion of any one electron is at least partly determined by the motions of all its near neighbors. Suppose that at some instant the electrons in a given helium atom are slightly displaced, so that the atom possesses an *instantaneous dipole moment*. This instantaneous dipole moment would induce a similar dipole moment on an adjacent atom because of the somewhat synchronized motions of the electrons, as illustrated in Figure 11.14. The result is an attractive interaction between the two atoms, called the **London dispersion force**. London's analysis showed that dispersion forces between two molecules are significant only when the molecules are very close together.

The ease with which the charge distribution in a molecule can be distorted by an external force is called its **polarizability**. The London dispersion forces between molecules are stronger for molecules of higher polarizability. In general, the larger the molecule, the farther its electrons are from the nuclei, and, consequently, the greater its polarizability. Therefore, the magnitude of the London dispersion forces increases with increase in molecular size. Because molecular size and mass generally parallel each other, *dispersion forces increase in magnitude with increasing molecular mass*. The truth of this generalization is evident in Table 11.4. The boiling points and enthalpies of vaporization of the noble gas elements increase steadily with increase in atomic mass. A similar trend is evident among the diatomic molecules. However, we must expect that the shapes of the molecules involved will also play a role. Note, for example, that $O_2$ has a slightly lower mass than Ar but a slightly higher boiling point and enthalpy of vaporization. In another example, *n*-pentane and neo-

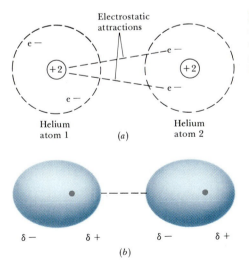

**FIGURE 11.14** Two schematic representations of an instantaneous dipole on two adjacent helium atoms, showing the electrostatic attraction between them.

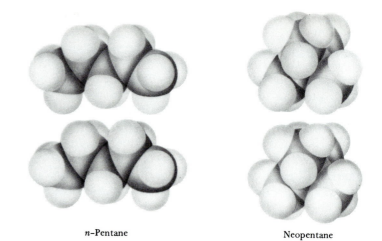

*n*-Pentane          Neopentane

**FIGURE 11.15** Illustration of the effect of molecular shape on intermolecular attraction. The boiling point of *n*-pentane is 36.2°C, and that of neopentane is 9.5°C.

pentane, illustrated in Figure 11.15, have the same molecular formula, $C_5H_{12}$, yet the boiling point of *n*-pentane* is 27°C higher than that of neopentane. The difference can be traced to the different shapes of the two molecules. The overall attraction between molecules is greater in the case of *n*-pentane because the molecules are able to come in contact over the entire length of the molecule. Less contact is possible between molecules of neopentane.

## SAMPLE EXERCISE 11.4

Which of the following substances is most likely to exist as a gas at room temperature and normal atmospheric pressures: $P_4O_{10}$, $Cl_2$, AgCl, or $I_2$?

**Solution:** In essence, the question asks which substance has the weakest intermolecular attractive forces, because the weaker these forces, the more likely the substance is to exist as a gas at any given temperature and pressure. We should therefore select $Cl_2$, because this is a nonpolar molecule and also has the lowest molecular weight. In fact, $Cl_2$ does exist as

a gas at room temperature and normal atmospheric pressure, whereas the others are solids. Of the other substances, AgCl is least likely to be a gas because it exists as $Ag^+$ and $Cl^-$ ions with very strong ionic bonds holding the ions within the solid.

## PRACTICE EXERCISE

Of $Br_2$, Ne, HCl, and $N_2$, which is likely to have **(a)** the largest intermolecular dispersion forces; **(b)** the largest dipole-dipole attractive forces?
*Answers:* **(a)** $Br_2$; **(b)** HCl

Dispersion forces operate between all molecules, whether they are polar or nonpolar. For example, in comparing HCl and HBr we find that HCl has the larger dipole moment ($\mu = 1.03$ D, compared to $\mu = 0.79$ D for HBr). HCl is also smaller, which permits the centers of the dipoles to approach each other more closely. Thus the dipole-dipole forces between HCl molecules are stronger than those between HBr molecules. However, dispersion forces are stronger between the more massive HBr molecules, whose electrons are more polarizable. The fact that the boiling point of HBr (206.2 K) is higher than that of HCl (189.5 K) suggests that the

* The *n* in *n*-pentane is an abbreviation for the word "normal." A normal hydrocarbon is one whose carbon atoms are arranged in a straight chain.

*overall* attractive forces are stronger in the case of HBr. In other cases, the more polar molecule may have the stronger overall attractive forces. It is difficult to make generalizations about the relative strengths of intermolecular attractions unless we restrict ourselves to comparing molecules of either similar size and shape or similar polarity and shape. If molecules are of similar size and shape, dispersion forces are approximately equal, and therefore attractive forces increase with increasing polarity. If molecules are of similar polarity and shape, attractive forces tend to increase with increasing molecular mass because dispersion forces are greater.

Figure 11.16 shows the boiling points of the simple hydrides of the group 4A and 6A elements as a function of molecular mass. In general, the boiling points increase with increasing molecular mass, owing to increased dispersion forces. The notable exception to this trend is $H_2O$. Clearly, the boiling point of this substance is much higher than would be expected on the basis of its molecular mass. We find a similar abnormally high boiling point in $NH_3$ among the boiling points of the group 5A element hydrides, and in HF among boiling points of the group 7A element hydrides. These compounds—$H_2O$, $NH_3$, and HF—also have many other unusual characteristics that distinguish them from other substances of similar molecular mass and polarity. For example, water has a high melting point, high heat capacity, and a high heat of vaporization. Each of these properties indicates that the intermolecular forces between $H_2O$ molecules are abnormally strong.

The origin of these strong intermolecular attractions is the **hydrogen bond**, a special type of intermolecular attraction that exists between the hydrogen atom in a polar bond and an electronegative atom on an adjacent molecule. This intermolecular attractive force is most important in substances in which a hydrogen atom is bonded to nitrogen, oxygen, or

**Hydrogen Bonding**

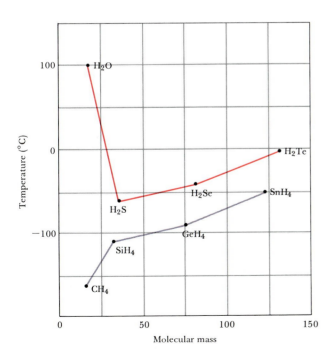

**FIGURE 11.16**  Boiling points of the groups 4A (bottom) and 6A (top) hydrides as a function of molecular mass.

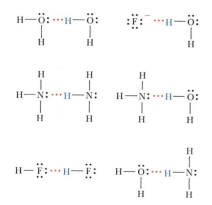

fluorine. The electronegativity of hydrogen is 2.2, much less than the electronegativity of nitrogen (3.0), oxygen (3.4), or fluorine (4.0). As a result, the bond between hydrogen and any of these three elements is quite polar, with hydrogen at the positive end. We can think of each bond as having a bond dipole moment. The bond dipole moments have been estimated for these three bonds as follows:

| Bond: | N—H | O—H | F—H |
|---|---|---|---|
| Dipole moment (D): | 1.0 | 1.6 | 2.3 |

Each of these bond dipoles is capable of interacting with an unshared electron pair on the nitrogen, oxygen, or fluorine atom of an adjacent molecule. It is this electrostatic interaction between the X—H bond dipole of one molecule (where X is an electronegative atom) and the unshared electron pair of an electronegative atom in another molecule that we call hydrogen bonding. Several examples are shown in Figure 11.17. In each case the hydrogen bond is represented by a dotted line.

The energies of hydrogen bonds vary from about 4 or 5 kJ/mol to 25 kJ/mol or so. Thus they are not more than a few percent of the energies of ordinary chemical bonds (see Table 8.3). Nevertheless, hydrogen bonding is generally much stronger than dipole-dipole or dispersion forces and so has important consequences for the properties of many substances, including those in biological systems.

Hydrogen bonding is responsible, for example, for the rather open structure for ice (Figure 11.18), which causes ice to have a *lower* density than liquid water. (By contrast, for most substances the solid phase is *more* dense than the liquid, as illustrated in Figure 11.19.) This unusual property of water has some important consequences. It permits ice to float on water. When ice forms in cold weather, it covers the top of the water, thereby insulating the water below. If ice were more dense than water, ice forming at the top of a lake would fall to the bottom, and the lake could freeze solid. Most aquatic life could not survive under these

FIGURE 11.18 Arrangement of water molecules in ice. Each hydrogen atom on one water molecule is oriented toward a nonbonding pair of electrons on an adjacent water molecule. In the full structure each oxygen atom has a hydrogen bond with two O—H groups.

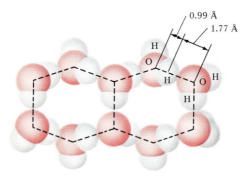

FIGURE 11.19 In common with most substances, the solid phase of cyclohexane is denser than the liquid phase (left). By contrast, the solid phase of water, ice, is less dense than the liquid phase (right). (Donald Clegg and Roxy Wilson)

If the hydrogen bond is indeed the result of an electrostatic interaction between the X—H bond dipole and an unshared electron pair on another atom, Y, then the strength of hydrogen bonding should increase as the X—H bond dipole increases. Thus, for the same Y, we would expect hydrogen-bonding strengths to increase in the series

$$N—H \cdots Y < O—H \cdots Y < F—H \cdots Y$$

This is indeed true. But what property of Y is important? The atom Y must possess an unshared electron pair that attracts the positive end of the X—H dipole. This electron pair must not be too diffuse in space; if the electrons occupy too large a volume, the X—H dipole does not experience a strong, directed attraction. For this reason, we find that hydrogen bonding is not very strong unless Y is one of the smaller atoms: N, O, or F, specifically. Among these three elements, we find that hydrogen bonding is stronger when the electron pair is not attracted too strongly to its own nuclear center. The ionization

energy of the electrons on Y is a good measure of this aspect. For example, the ionization energy of an unshared-pair electron on nitrogen in a covalent molecule is less than the corresponding value for oxygen. Nitrogen is thus a better donor of the electron pair to the X—H bond. For a given X—H, hydrogen-bond strength increases in the order

$$X—H \cdots F < X—H \cdots O < X—H \cdots N$$

When X and Y are the same, the energy of hydrogen bonding increases in the order

$$N—H \cdots N < O—H \cdots O < F—H \cdots F$$

When the Y atom carries a negative charge, the electron pair is able to form especially strong hydrogen bonds. The hydrogen bond in the $F—H \cdots F^-$ ion is among the strongest known; the reaction

$$F^-(g) + HF(g) \longrightarrow FHF^-(g)$$

has a $\Delta H$ value of about $-155$ kJ/mol.

---

circumstances. The expansion of water upon freezing is also what causes water pipes to break in freezing weather.

The low density of ice compared to water can be understood in terms of hydrogen-bonding interactions between water molecules. The interactions in the liquid are random. However, when water freezes, the molecules assume the ordered arrangement shown in Figure 11.18. This structure, which extends in all directions in space, permits the maximum number of hydrogen-bonding interactions between the $H_2O$ molecules. Because the structure has large hexagonal holes, ice is more open and less dense than the liquid. Application of pressure depresses the melting point of ice because the pressure causes the solid to revert more readily to the more dense liquid.

## SAMPLE EXERCISE 11.5

List the substances $BaCl_2$, $H_2$, CO, HF, and Ne in order of increasing boiling points.

**Solution:** The boiling point depends in part on the attractive forces in the liquid. These are stronger for ionic substances than for molecular ones, so $BaCl_2$ has the highest boiling point. The intermolecular forces of the remaining substances depend on molecular weight, polarity, and hydrogen bonding. The other molecular weights are $H_2$ (2), CO (28), HF (20), and Ne (20). The boiling point of $H_2$ should be the lowest because it is nonpolar and has the lowest molecular weight. The molecular weights of CO, HF, and Ne are roughly the same. HF has hydrogen bonding, so it has the highest boiling point of the three. CO, which is slightly polar and has the highest molecular weight, is next. Ne, which is nonpolar, comes last of these three. The predicted boiling points are therefore

$$H_2 < Ne < CO < HF < BaCl_2$$

The actual normal boiling points are $H_2$ (20 K), Ne (27 K), CO (83 K), HF (293 K), and $BaCl_2$ (1813 K).

**PRACTICE EXERCISE**

Which of the following can form hydrogen bonds with $H_2O$: **(a)** $CH_3\ddot{O}CH_3$; **(b)** $CH_4$; **(c)** $NH_4^+$?

*Answer:*   $CH_3OCH_3$ and $NH_4^+$

## 11.5   SOLIDS

Solids are rigid; they cannot be poured like liquids or compressed like gases. Solids such as quartz or diamond possess highly regular crystalline shapes or cleavage planes, as illustrated in Figure 11.20. These facts suggest a regular atomic arrangement within the solid. Indeed, a statement of this regular arrangement is sometimes included in the definition of solids. For our purposes, we shall divide materials that we would ordinarily call solids because of their rigidity into two groups: crystalline solids and amorphous solids. **Crystalline solids** are characterized by the regular three-dimensional arrangement of atoms. **Amorphous solids** lack this regular atomic-level organization. Familiar amorphous solids include rubber and glass, which are composed of large, complicated molecules. In some texts, amorphous solids are referred to as supercooled liquids because they have the molecular disorder of liquids. In fact, glass is capable of flowing, as revealed by a careful examination of the window panes of very old houses. The panes are thicker at the bottom than at the top because the glass has flowed under the continued influence of gravity. In this section our focus is primarily on crystalline solids.

**FIGURE 11.20**   Crystalline solids come in a variety of forms and colors. (*a*) diamond, (*b*) calcite, (*c*) fluorite twin crystals, (*d*) purple fluorite from Montana. [(*a*) Diamond Information Center, (*c*) Runk, Schoenberger/Grant Heilman Photography, (*b*, *d*) Thomas R. Taylor/Photo Researchers]

(*a*)   (*b*)

(*c*)   (*d*)

**TABLE 11.7** Crystal classifications

| Crystal classification | Form of unit particles | Forces between particles | Properties | Examples |
|---|---|---|---|---|
| Atomic | Atoms | London dispersion forces | Soft, very low melting point, poor thermal and electrical conductors | Rare gases—Ar, Kr |
| Molecular | Polar or nonpolar molecules | London dispersion, dipole-dipole forces, hydrogen bonds | Fairly soft, low to moderately highly melting point, poor thermal and electrical conductors | Methane, $CH_4$; sugar, $C_{12}H_{22}O_{11}$; dry ice, $CO_2$ |
| Ionic | Positive and negative ions | Electrostatic attraction | Hard and brittle, high melting point, poor thermal and electrical conductors | Typical salts— for example, NaCl, $Ca(NO_3)_2$ |
| Covalent (network) | Atoms that are connected in covalent-bond network | Covalent bond | Very hard, very high melting point, poor thermal and electrical conductors | Diamond, C; quartz, $SiO_2$. |
| Metallic | Atoms | Metallic bond | Soft to very hard, low to very high melting point, excellent thermal and electrical conductors, malleable and ductile | All metallic elements—for example, Cu, Fe, Al, W |

**Bonding in Solids**

The particles of a crystalline solid, whether they be atoms, molecules, or ions, assume positions that maximize attractive forces between them. Table 11.7 classifies solids according to the types of particles and the types of forces that hold the particles in their positions. It is important to study this table carefully because it contains a great deal of information about solids, relating atomic-level arrangements and forces to the bulk properties of the crystal.

Solid argon and solid methane are examples of **atomic solids** and **molecular solids**, respectively. Intermolecular forces consist of London dispersion forces, dipole-dipole forces, and hydrogen bonds. Because these forces are weak, such solids normally have relatively low melting points and exhibit trends in melting points that are similar to those discussed for the boiling points of molecular substances. Most substances that are gases or liquids at room temperature form molecular solids at low temperature.

The properties of molecular solids depend not only on the magnitudes of the forces that operate between molecules but also on the abilities of the molecular units to pack efficiently in three dimensions. For example, benzene—$C_6H_6$, a highly symmetric molecule—has a higher melting point than substituted benzene compounds such as chlorobenzene or toluene. The lower symmetry of the substituted molecules, shown in Figure 11.21, prevents them from packing as efficiently as benzene. As a result, the intermolecular forces that depend on close contact do not operate as effectively, and the melting point is lower. By contrast, the boiling points of these compounds are higher than that of benzene, indicating that the attractive forces between these molecules in the liquid state are larger than those between liquid benzene molecules.

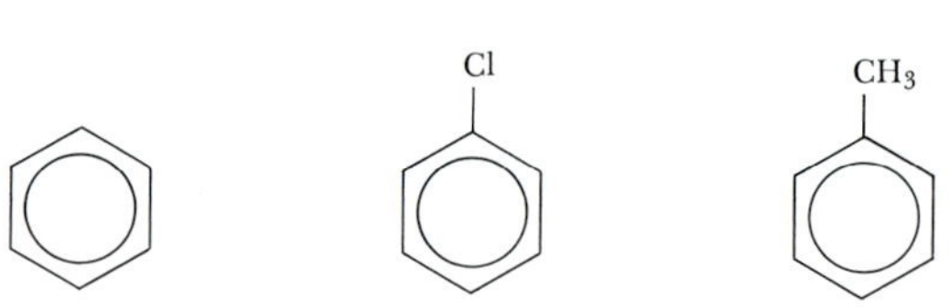

|  | Benzene | Chlorobenzene | Toluene |
|---|---|---|---|
| Melting point (°C) | 5 | −45 | −95 |
| Boiling point (°C) | 80 | 132 | 111 |

**FIGURE 11.21** Comparative melting and boiling points for benzene, chlorobenzene, and toluene.

**Ionic solids** have higher melting points than do atomic and molecular solids because ionic bonds are stronger than intermolecular forces. Ionic solids are also harder, and they fracture when struck rather than simply deforming.

In **covalent**, or **network solids**, the units that make up the three-dimensional solid are joined by covalent bonds. Such materials are consequently much stronger than molecular solids. Diamond, whose structure and bonding was discussed in Section 9.7, is an example of this type of solid (see Figure 9.36).

Bonding in **metals** differs from the bonding in other solids. Each atom in a metallic solid typically has 8 to 12 atoms adjacent to it because of the close packing arrangement. The bonding is too strong to be due to London-dispersion forces, and yet there are not enough valence electrons for ordinary covalent bonds between the atoms. We can visualize the valence electrons as delocalized, somewhat as outlined for the $\pi$ electrons in graphite (Section 9.39). This model allows us to picture metals as composed of metal atoms held together by electrons that are distributed throughout all the spaces between the atoms. The electrons are free to move through the orbitals that extend over the entire metal. However, they maintain a uniform average distribution. Metals vary greatly in the strength of metallic bonding, as evidenced by such physical properties as melting and boiling points. For example, in platinum, with a melting point of 1770°C and a boiling point of 3824°C, the metallic bonding is very strong; in cesium—melting point 29°C and boiling point 678°C—the metallic bonding is comparatively weak. The properties and structures of metals will be examined more closely in Chapter 24.

## Crystal Lattices and Unit Cells

The order characteristic of crystalline solids allows us to convey a picture of an entire crystal by looking at only a small part of it. We can think of the solid as being built up by stacking together identical building blocks, much as a brick wall is formed by stacking individual, identical bricks. The repeating unit of a solid—the crystalline "brick"—is known as the **unit cell**. A simple two-dimensional example appears in the sheet of wallpaper shown in Figure 11.22. There are several ways of choosing the repeat pattern, or unit cell, of the design, but the choice is usually taken to be the smallest one that shows clearly the symmetry characteristic of the entire pattern.

A crystalline solid can be represented by a three-dimensional array of points, each of which represents an identical environment within the

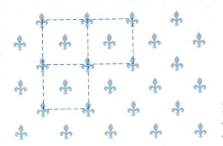

**FIGURE 11.22** Wallpaper design showing a characteristic repeat pattern. Each dashed blue square denotes the unit cell of the repeat pattern.

crystal. Such an array of points is called a **crystal lattice**. We can imagine forming the entire crystal structure by arranging the contents of the unit cell repeatedly in a network of points.

Figure 11.23 shows a crystal lattice and its associated unit cell. In general, unit cells are parallelepipeds (that is, six-sided figures whose faces are parallelograms). Each unit cell can be described in terms of the lengths of the edges of the cell ($a$, $b$, and $c$) and by the angles between these edges ($\alpha$, $\beta$, and $\gamma$), as shown in Figure 11.23. The lattices of millions of compounds can be described in terms of seven basic types of unit cells. The simplest of these is the cubic unit cell, in which all the sides are equal in length and all the angles are 90°.

There are three kinds of cubic unit cells, as illustrated in Figure 11.24. When lattice points are at the corners only, the unit cell is described as

**FIGURE 11.23** Simple crystal lattice and its associated unit cell. Each view of the unit cell shows one of the characteristic angles between unit cell axes. Angle alpha is the angle between the $b$ and $c$ axes; beta is the angle between the $a$ and $c$ axes; gamma is the angle between the $a$ and $b$ axes.

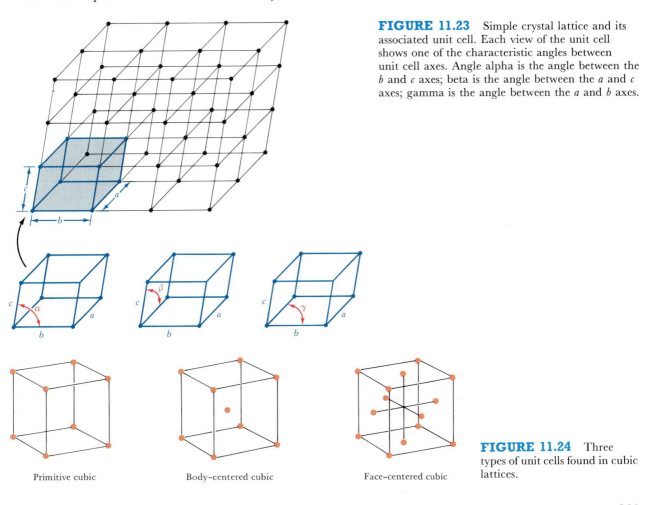

Primitive cubic        Body-centered cubic        Face-centered cubic

**FIGURE 11.24** Three types of unit cells found in cubic lattices.

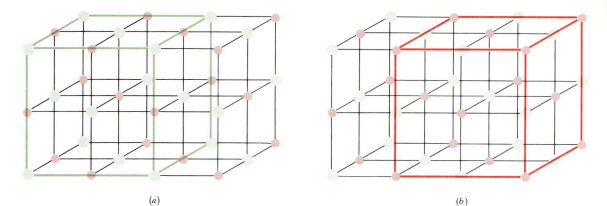

(a)                                                                (b)

**FIGURE 11.25**  Portion of the crystal structure of NaCl illustrating two ways of positioning its crystal lattice. Red spheres represent Na$^+$ ions, and green spheres represent Cl$^-$ ions. Colored lines define the face-centered unit cell of each lattice. In (a) the lattice points are centered on Cl$^-$ ions. In (b) the lattice points are centered on Na$^+$ ions. However, the lattice points need not be centered on ions at all. The only restriction on where lattice points are placed is that the points must be in positions that have identical environments throughout the solid. Regardless of where the lattice points are placed in the NaCl structure, the result is a face-centered-cubic lattice.

**primitive cubic**. When lattice points also occur at the center of the unit cell, the cell is known as **body-centered cubic**. A third type of cubic cell has lattice points at each corner as well as at the center of each face, an arrangement known as **face-centered cubic**.

Figure 11.25 shows a portion of the crystal structure of sodium chloride. Two ways of locating the lattice points so they are in identical environments are shown. In Figure 11.25(a) the points are centered on Cl$^-$. In Figure 11.25(b) they are centered on Na$^+$. In either case the structure is seen to possess a lattice with a face-centered-cubic unit cell. The cubic character of the unit cell is reflected in the shapes of well-formed crystals of NaCl, Figure 11.26.

Figure 11.25 shows "exploded" views of portions of the NaCl structure in which the ions have been moved apart so that the symmetry of the structure can be seen more clearly. In this representation, no attention is paid to the relative sizes of the Na$^+$ and Cl$^-$ ions. In contrast, Figure 11.27 provides a representation that shows the relative sizes of the ions and how they fill the unit cell. Notice that the particles at the corners, edges, and faces do not lie wholly within the unit cell. Instead, these particles are shared by other unit cells. A particle at a corner is shared

**FIGURE 11.26**  Crystals of NaCl, showing well-defined crystal planes based on the underlying cubic structure. (Dr. E. R. Degginger)

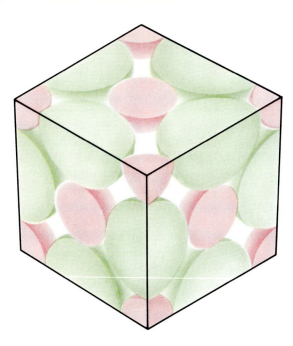

**FIGURE 11.27** Space-filling representation of the unit cell of sodium chloride. The $Cl^-$ ions are the green spheres, the $Na^+$ ions are the red ones. Notice that only a portion of most ions lies within the boundaries of the single unit cell.

by eight unit cells, one at the edge is shared by four, and one at the center of a face is shared by two. The total cation-to-anion ratio of the unit cell must be the same as that for the entire crystal. Thus, if we sum all the $Na^+$ and $Cl^-$ ions within the unit cell of NaCl, there must be one $Na^+$ for each $Cl^-$. Similarly, the unit cell for $CaCl_2$ would have one $Ca^{2+}$ for each two $Cl^-$, and so forth for other crystals. In Sample Exercises 11.6 and 11.7, the contents of the NaCl unit cell are determined and used to calculate the density of the solid.

---

**SAMPLE EXERCISE 11.6**

Determine the net number of $Na^+$ and $Cl^-$ ions in the NaCl unit cell (Figure 11.27).

**Solution:** There is $\frac{1}{8}$ of a $Na^+$ on each corner (each $Na^+$ on a corner is shared by eight cubes which intersect at that point). There is $\frac{1}{2}$ of a $Na^+$ on each face, $\frac{1}{4}$ of a $Cl^-$ on each edge, and a whole $Cl^-$ in the center of the cube. We therefore have the following:

$Na^+$:  $(\frac{1}{8} Na^+ \text{ per corner})(8 \text{ corners}) = 1 Na^+$
$(\frac{1}{2} Na^+ \text{ per face})(6 \text{ faces}) \quad = 3 Na^+$

$Cl^-$:  $(\frac{1}{4} Cl^- \text{ per edge})(12 \text{ edges}) \quad = 3 Cl^-$
$(1 Cl^- \text{ per center})(1 \text{ center}) \quad = 1 Cl^-$

Thus the unit cell contains $4Na^+$ and $4Cl^-$. This result agrees with the compound's stoichiometry: one $Na^+$ for each $Cl^-$.

**PRACTICE EXERCISE**

The element iron crystallizes in a form called $\alpha$-iron, which has a body-centered-cubic unit cell. How many iron atoms are in the unit cell?
*Answer:* two

---

**SAMPLE EXERCISE 11.7**

If the unit cell of NaCl is 5.64 Å on an edge, calculate the density of NaCl.

**Solution:** The volume of the unit cell is $(5.64 \text{ Å})^3$. Because each unit cell contains $4Na^+$ and $4Cl^-$ (Sample Exercise 11.6), its mass is

$$4(23.0 \text{ amu}) + 4(35.5 \text{ amu}) = 234.0 \text{ amu}$$

The density is mass/volume:

$$\text{Density} = \frac{234.0 \text{ amu}}{(5.64 \text{ Å})^3} \left( \frac{1 \text{ g}}{6.02 \times 10^{23} \text{ amu}} \right)$$
$$\times \left( \frac{1 \text{ Å}}{10^{-8} \text{ cm}} \right)^3$$
$$= 2.17 \text{ g/cm}^3$$

This value agrees with that found by simple density measurements: 2.165 g/cm³. Thus the size and contents of the unit cell are consistent with the macroscopic density of the substance.

**PRACTICE EXERCISE**
The body-centered-cubic unit cell of α-iron is 2.8664 Å on each side. Calculate the density of α-iron.
*Answer:* 7.878 g/cm³

## Close Packing

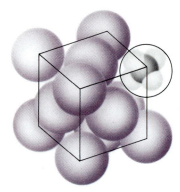

**FIGURE 11.28** Unit cell of solid methane. Each large sphere represents a CH₄ molecule, as shown in the upper right.

The structures adopted by crystalline solids are those that bring particles into closest contact, to maximize the attractive forces between them. In many cases the particles that make up the solids are spherical or approximately so. Such is the case for atoms in metallic and atomic solids. Many molecules can also be approximated as spheres, as seen for CH₄ in Figure 11.28. It is therefore instructive to consider how equal-sized spheres can pack most efficiently (that is, with the minimum amount of empty space).

The most efficient packing arrangements for spheres are called closest-packed structures. These structures are quite common, accounting, for example, for about two-thirds of the structures of the metallic elements. The most efficient arrangement of a layer of equal-sized spheres can be seen in Figure 11.29(a). Each sphere is surrounded by six others in the layer. The most efficient arrangement of the spheres in a second layer is in the depressions of the first yellow layer shown in Figure 11.29(a). This is shown in Figure 11.29(b). The spheres of the third layer sit in depressions in the second layer. However, there are two types of depressions, and they lead to two different structures. If the spheres of the third layer are placed immediately above those of the first layer, as are the blue spheres in Figure 11.29(c), the structure known as **hexagonal close packing** results. If we consider more than three layers of spheres, the arrangement of the layers in this structure can be represented as

**FIGURE 11.29** Closest packing of equal-sized spheres. (a) One layer. Note that each sphere is surrounded by six others in the plane. (b) Two superimposed layers. Each atom of the second layer rest in the triangular pocket formed by three spheres below. (c) Three superimposed layers in hexagonal close-packed arrangement. (d) Three superimposed layers in cubic close-packed arrangement.

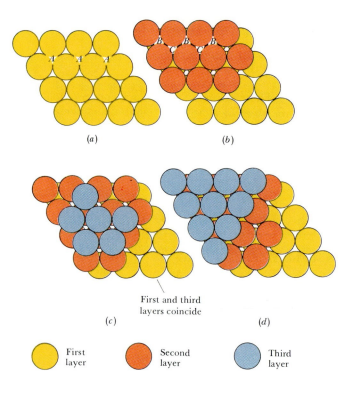

First and third layers coincide

(c)      (d)

First layer    Second layer    Third layer

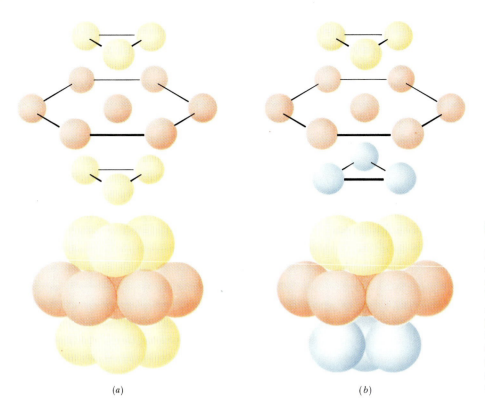

**FIGURE 11.30**

(a) Hexagonal close packing. In this arrangement the atoms of the third layer lie directly over those in the first layer. Thus the order of layers is *ABABAB*. (b) Cubic close packing. In this arrangement the atoms of the third layer are not directly over those in the first, but are offset by a bit. As a result the layers are arranged vertically as *ABCABC*.

(a)      (b)

*ABABAB.* . . . The second type of close-packed structure results if the third-layer spheres are placed in slightly different positions as in Figure 11.29(*d*). The resultant structure is known as the **cubic close packing**. In this case the stacking sequence can be represented as *ABCABC.* . . . Although it is not obvious from Figure 11.29, the cubic-close-packed arrangement of spheres has a face-centered cubic unit cell. In Figure 11.30, the close-packed structures are viewed from a perspective that permits the unit cells to be seen more clearly. In both of the close-packed structures, each sphere has 12 equidistant nearest neighbors: six in one plane, three above that plane, and three below. We say that these spheres have a **coordination number** of 12.

When unequal-sized spheres are packed in a lattice, the large particles sometimes assume one of the close-packed arrangements with small particles occupying the holes between the large spheres. For example, in $Li_2O$ the oxide ions assume a cubic close-packed structure, and the $Li^+$ ions occupy small cavities that exist between oxide ions. We shall consider this particular view of crystals further in Chapter 25 when we discuss the structures of metal oxides.

Until the early part of this century, scientists had no means of visualizing the internal arrangements of atoms, ions, and molecules in solids. They could look at the shapes of different crystals and note the angles made between their faces, but the internal character of the solid was simply not knowable. That situation changed dramatically about twenty years after the 1895 discovery of X rays by the German physicist Wilhelm Roentgen.

**X-Ray Study of Crystal Structures**

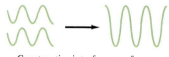

Constructive interference of waves

(a)

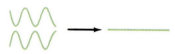

Destructive interference of waves

(b)

**FIGURE 11.31**
Illustration of interference. In (a) the waves are in phase, and the interference is constructive; in (b) the waves are out of phase, and the interference is destructive.

William Bragg and his son Lawrence, working together in England, reasoned that if X rays were short-wavelength, high-energy electromagnetic radiation (Section 6.1), it should be possible to observe diffraction of a beam of X rays by a crystalline solid. **Diffraction** results from the scattering of light waves by a regular arrangement of points. It arises when the wavelength of the light rays is comparable to the distances that separate the points. The scattered waves interfere with each other when they come together. If the waves are in phase—that is, if their peaks and troughs coincide—their amplitudes add, producing a more intense wave. This situation, shown in Figure 11.31(a), is termed constructive interference. If the two waves are out of phase, as shown in Figure 11.31(b), destructive interference occurs, and the resultant wave has no intensity.

The Braggs' genius was in seeing how the diffraction phenomenon could be applied to determining the arrangements of atoms in a solid. When a crystal is irradiated with X rays, each individual atom within the solid scatters the X rays. That is, each atom absorbs the radiation and reradiates it in all directions. The waves from the various scattering centers (the atoms) will be in phase when they come off at certain angles from the direction of the entering X ray beam, and not in phase at other angles, as illustrated in Figure 11.32. Thus a photographic plate placed in back of the crystal shows a pattern of spots corresponding to the angles at which constructive interference occurs, as illustrated in Figure 11.33. It is then possible to work backward from the pattern of spots and their intensities to deduce what arrangement of atoms gives rise to that particular pattern.

A modern X-ray diffraction instrument (Figure 11.34) is capable of measuring in a couple of days the intensities of thousands of reflections from a tiny crystal. From these extensive data, very complex structures can be determined.

The development of X-ray diffraction as a means of "seeing" the

**FIGURE 11.32**   (a) Constructive interference of waves scattered from adjacent atoms in a lattice. For constructive interference to occur, the distances $d_1$ and $d_2$ traveled by the waves must differ by a whole number of wavelengths. (b) Destructive interference of waves from adjacent atoms. The distances traveled by the waves do not differ by a whole number of wavelengths.

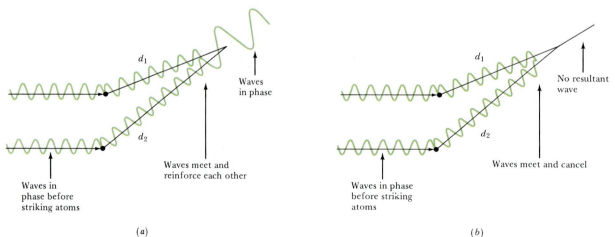

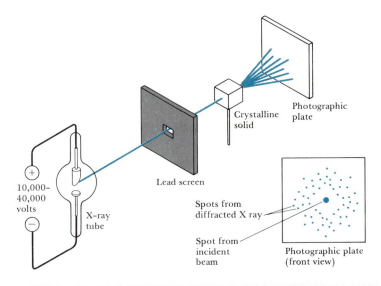

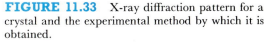

**FIGURE 11.33** X-ray diffraction pattern for a crystal and the experimental method by which it is obtained.

10,000–40,000 volts

X-ray tube

Lead screen

Crystalline solid

Photographic plate

Spots from diffracted X ray

Spot from incident beam

Photographic plate (front view)

*(a)*

**FIGURE 11.34** *(a)* A modern X-ray diffraction spectrometer. The detector head, mounted in the circular arc [close-up, *(b)*], moves under computer control to find the angles at which X rays are constructively scattered from the crystal. The intensity of the scattering at this angle is measured. The data from many points, typically several thousand, are analyzed using a computer associated with the diffractometer to produce a picture of the arrangement of atoms within the crystal. (Courtesy of Nicolet)

*(b)*

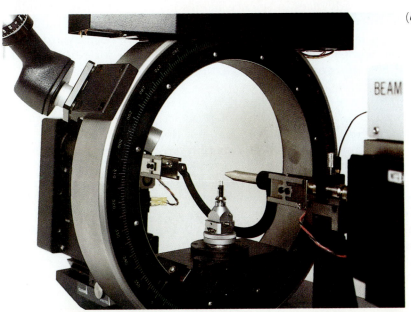

The Bragg equation relates the angles at which X rays are scattered to the spacings between layers of atoms in the solid and the wavelength of the scattered ray. Let's consider a solid with planes of atoms separated by a distance $d$, as shown in Figure 11.35. We are looking at the planes of atoms edge on.

The incoming rays are in phase at $AB$. Wave $ACA'$ is scattered, or reflected by an atom in the first layer of the solid. Wave $BEB'$ is reflected by an atom in the second layer. If these two waves are to be in phase at $A'B'$, the extra distance covered by $BEB'$ must be a whole-number multiple of the wavelength, $\lambda$. In Figure 11.35, the extra distance, $DEF$, is $2\lambda$. Now notice that the triangle $CDE$ is a right triangle. Using trigonometry, it can be shown that the distance $DE$ is $d \sin \theta$, where $d$ is the distance between the planes

and $\theta$ is the angle between the incoming wave and the plane. Because the distance $DEF$ is twice $DE$, we have

$$DEF = 2\lambda = 2d \sin \theta$$

It can be shown that the general equation for constructive interference is

$$n\lambda = 2d \sin \theta \qquad \text{where } n = 1, 2, 3, 4, \ldots \quad [11.2]$$

This relationship, known as the **Bragg equation**, allows determination of the spacing between planes from the known wavelength of the light and from experimentally determined values of $\theta$ at which constructive interference occurs.

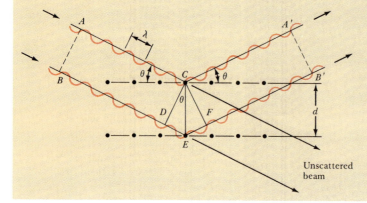

**FIGURE 11.35** Scattering of X rays by atoms in parallel planes. The incoming X rays of wavelength $\lambda$ are diffracted by atoms, which are represented as dots. The atoms are arranged in planes separated by a distance $d$. The incoming X rays make an angle $\theta$ with the planes and are scattered at an angle $2\theta$ from the unscattered beam.

arrangements of atoms and molecules within solids has been one of the most important advances in science in this century. From such studies researchers learn about which atoms are bonded to which, and about bond distances and bond angles. Several Nobel prizes have been awarded for outstanding work in determining crystal structure or for research that depended on results obtained from X-ray diffraction studies.

**Crystal Defects**

Although we have generally pictured crystalline solids as being composed of perfectly ordered arrays of particles, this image is merely a useful abstraction, like that of an ideal gas. Real crystals contain imperfections whose number and type can play an important role in determining the properties of the solid. For example, a crystal lattice that has many sites where particles are missing (vacancies) can be more readily deformed than can a perfect crystal lattice of the same substance. It is not hard to appreciate that structural imperfections can occur readily. In forming a solid the lattice is built up very rapidly, and misplacements can readily occur. Defects can also arise through thermal motion of the

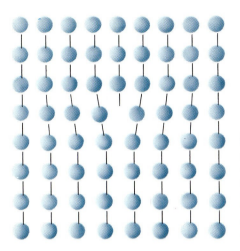

**FIGURE 11.36** Schematic illustration of an edge dislocation. The circles represent repeating units of the solid structure. An edge is created where the plane of units in the center stops. The adjacent planes shift to close up over the gap that is created. In the process, strains are created in the lattice.

units that make up the lattice, or through strains that might be externally imposed on the solid.

An edge dislocation is a common defect in many solids. In Figure 11.36 we see schematically that one of the planes within the solid has discontinued, and the adjacent planes have closed up over the edge of the discontinued plane.

Dislocations and other crystal defects are important because they represent sites where the atoms or molecules within the solid are in strained, high-energy environments. For example, oxidation of graphite would proceed more readily at an edge dislocation than elsewhere on the surface of the carbon particle.

**Amorphous Solids**

Not all solids have a regular arrangement of particles. For example, the material obtained when quartz, $SiO_2$, is melted and then rapidly cooled is amorphous. Quartz has a three-dimensional structure like that of diamond (Section 9.7). When quartz is melted (at approximately 1600°C) it becomes a viscous, tacky liquid. Although the silicon-oxygen network remains largely intact, many Si—O bonds are broken, and the orderliness of the quartz is lost. If the melt is rapidly cooled, the atoms are unable to return to this orderly arrangement, and an **amorphous solid** known as quartz glass or silica glass results. The lack of molecular-level regularity in this glass is shown in Figure 11.37. where its structure is compared with that of quartz.

Even a solid that lacks any long-range order may have small regions, called crystallites, where particles are arranged in an orderly fashion. The extent of such ordering is called the *degree of crystallinity* of the solid. Synthetic *polymers*—large molecules composed of many molecular parts fused together—often have such ordered regions, as shown in Figure 11.38.

The simplest synthetic polymer is polyethylene, a waxy-feeling substance used in packaging films, wire insulation, and molded articles. Polyethylene is formed by causing ethylene, $C_2H_4$, molecules to bond together to form chains. Typically 700 to 2000 $C_2H_4$ molecules combine to form a chain containing 1400 to 4000 carbon atoms:

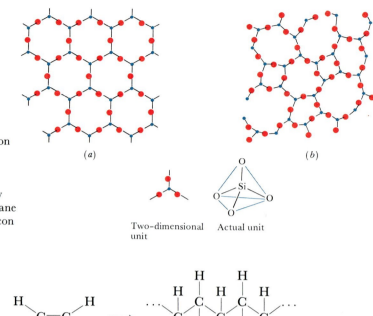

**FIGURE 11.37** Schematic comparisons of (a) crystalline $SiO_2$ (quartz) and (b) amorphous $SiO_2$ (quartz glass). The blue dots represent silicon atoms; the red dots represent oxygen atoms. The structure is actually three-dimensional and not planar as drawn. The unit shown as the basic building block (silicon and three oxygen) actually has four oxygens, the fourth coming out of the plane of the paper and capable of bonding to other silicon atoms.

Two-dimensional unit        Actual unit

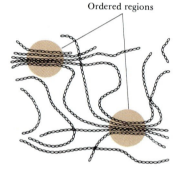

**FIGURE 11.38** Solid composed of large, flexible molecules. This solid lacks the long-range order characteristic of crystalline solids, but nonetheless contains small regions where molecules are arranged in an orderly fashion.

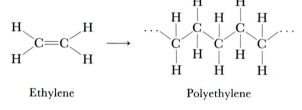

Ethylene          Polyethylene

The properties of a plastic* such as polyethylene are determined by at least three factors: (1) the length of the polymer chain, (2) the degree of crystallinity, and (3) the extent of bonding between chains. As the length of a polymer chain increases, intermolecular attractive forces increase, thus making the polymer mechanically stronger and harder. Mechanical strength and hardness also increase as the degree of crystallinity increases. The regular arrangement of chains in crystallites permits closer approach of the molecules, thereby increasing intermolecular attractions.

The effect of the degree of crystallinity on the properties of polyethylene can be seen in Table 11.8. The degree of crystallinity depends on the conditions of the polymerization.

To soften a plastic or make it more pliable, a substance with low molecular weight known as a plasticizer may be added in the course of fabrication. The plasticizer molecules occupy positions between polymer strands, thereby interfering with intermolecular forces between chains and lowering the degree of crystallinity. Plasticizers account for some of the "new car" odor inside a new automobile. When a new car is left standing in the hot sun the plasticizer is slowly volatilized from the plastic and appears as a fine film on the inside of the car window. Eventually, loss of plasticizer leaves the plastic brittle and subject to cracking.

* Although the term "plastic" has come to mean a certain type of synthetic material, the term is most precisely used to describe any material that changes shape when a force is exerted and maintains that distortion upon removal of the force. Materials such as rubber that distort but return to their original shape when the force is removed are called elastomers.

**TABLE 11.8**   Properties of polyethylene as a function of crystallinity

| | Degree of crystallinity | | | | |
|---|---|---|---|---|---|
| | **55%** | **62%** | **70%** | **77%** | **85%** |
| Melting point (°C) | 109 | 116 | 125 | 130 | 133 |
| Density (g/cm$^3$) | 0.92 | 0.93 | 0.94 | 0.95 | 0.96 |
| Stiffness$^a$ | 25 | 47 | 75 | 120 | 165 |
| Yield stress$^a$ | 1700 | 2500 | 3300 | 4200 | 5100 |

$^a$ These test results show that the mechanical strength of the polymer increases with increased crystallinity. The physical units for the stiffness test are psi $\times 10^{-3}$ (psi = pounds per square inch); those for the yield stress test are psi. Discussion of the exact meaning and significance of these tests is beyond the scope of this text.

## 11.6   PHASE DIAGRAMS

The physical state of a substance depends not only on inherent intermolecular attractive forces, but also on temperature and pressure. Now that we have examined each state, let us conclude our discussions by considering the temperatures and pressures at which the various phases of a substance can exist. Such information can be summarized in a **phase diagram**. The general form for such a diagram for a single substance that exhibits three phases is shown in Figure 11.39. This diagram contains three important curves, each of which represents the conditions of temperature and pressure at which the various phases can coexist at equilibrium.

1.   The line from $A$ to $B$ is the vapor-pressure curve of the liquid. It represents the equilibrium between the liquid and gas phases at various temperatures. This curve ends at $B$, the **critical point**. The temperature at this point is the **critical temperature** (Section 11.3). At this temperature the liquid and gas phases become indistinguishable. The pressure at the critical temperature is the **critical pressure**. Beyond the critical point the substance is described as a **supercritical fluid**. For an actual substance the point

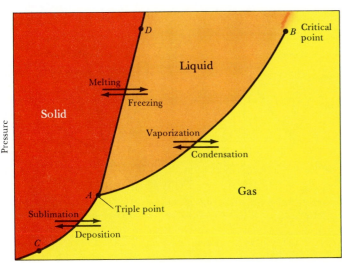

**FIGURE 11.39**   General shape for a phase diagram of a system exhibiting three phases: gas, liquid, and solid.

on curve *AB* where the equilibrium vapor pressure is 1 atm would be, of course, the normal boiling point of the substance.

2. The line *AC* represents the variation in the vapor pressure of the solid as a function of temperature.

3. The line from *A* through *D* represents the change in melting point of the solid with increasing pressure. This line normally slopes slightly to the right as pressure increases. Most solids expand upon melting, and increasing pressure therefore favors formation of the more dense solid phase. Thus higher temperatures are required to melt the solid at higher pressures. The melting point at 1 atm pressure is the normal melting point. (Remember that this is also the normal freezing point.)

Point *A*, where the three curves intersect, is known as the **triple point**. All three phases are at equilibrium at this temperature and pressure. Any other point on the three curves represents an equilibrium between two phases. Any point on the diagram that does not fall on a line corresponds to conditions under which only one phase is present. Notice that the gas phase is the stable phase at low pressures and high temperatures. The conditions under which the solid phase is stable extend to low temperatures and high pressures. The stability range for liquids lies between the other two regions.

Figure 11.40 shows the phase diagrams of $H_2O$ and $CO_2$. Notice that the solid-liquid equilibrium (melting point) line of $CO_2$ is normal; its melting point increases with increasing pressure. On the other hand, the melting point of $H_2O$ *decreases* with increasing pressure. Water is among the very few substances whose liquid form is more compact than the solid (Section 11.4).

The triple point of $H_2O$ (0.0098°C and 4.58 mm Hg) is much lower than that of $CO_2$ (−56.4°C and 5.11 atm). For $CO_2$ to exist as a liquid, the pressure must exceed 5.11 atm. Consequently, solid $CO_2$ does not melt but rather sublimes when heated at 1 atm. Thus $CO_2$ does not have a normal melting point; instead, it has a normal sublimation point, −78.5°C. Because $CO_2$ sublimes rather than melts as it absorbs energy at ordinary pressures, solid $CO_2$ (commonly called "dry ice") is a convenient coolant.

**FIGURE 11.40** Phase diagram of (*a*) $H_2O$ and (*b*) $CO_2$. The axes are not drawn to scale in either case. In (*a*), for water, note the triple point A (0.0098°C, 4.58 mm Hg), the normal melting (or freezing) point B (0°C, 1 atm), the normal boiling point C (100°C, 1 atm), and the critical point D (374.4°C, 217.7 atm). In (*b*), for carbon dioxide, note the triple point X (−56.4°C, 5.11 atm), the normal sublimation point Y (−78.5°C, 1 atm), and the critical point Z (31.1°C, 73.0 atm).

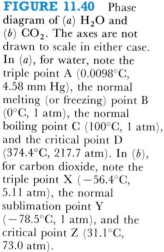

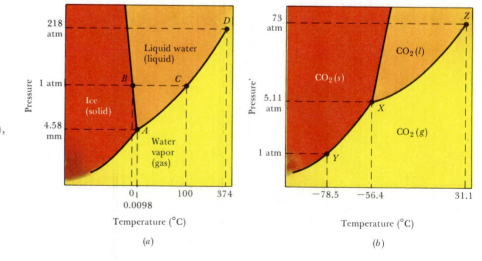

## SAMPLE EXERCISE 11.8

Referring to Figure 11.41, describe any changes in the phases present when $H_2O$ is **(a)** kept at 0°C while the pressure is increased from that at point 1 to that at point 5 (vertical line); **(b)** kept at 1.00 atm while the temperature is increased from that at point 6 to that at point 9 (horizontal line).

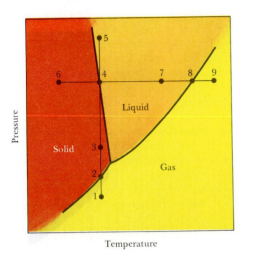

**FIGURE 11.41**   Phase diagram of $H_2O$.

**Solution:**   **(a)** At 1 the $H_2O$ exists totally as a vapor. At 2 a solid-vapor equilibrium exists. Above that pressure, at point 3, all the $H_2O$ is converted to a solid. At 4 some of the solid melts and an equilibrium between solid and liquid is achieved. At still higher pressures, all the $H_2O$ melts, so that only the liquid phase is present at point 5.

**(b)** At point 6 the $H_2O$ exists entirely as a solid. When the temperature reaches 4, the solid begins to melt and an equilibrium condition occurs between the solid and the liquid phases. At a yet higher temperature, 7, the solid has been converted entirely to a liquid. When point 8 is encountered, vapor forms and a liquid-vapor equilibrium is achieved. Upon further heating, to point 9, the $H_2O$ is converted entirely to the vapor phase.

### PRACTICE EXERCISE

Using Figure 11.40(*b*), describe what happens when the following changes are made to a $CO_2$ sample initially at 1 atm and $-60$°C. **(a)** Pressure increases at constant temperature to 60 atm. **(b)** Following (a) temperature increases to $-20$°C at constant 60 atm pressure.
***Answers:***   **(a)**  $CO_2(g) \longrightarrow CO_2(s)$;   **(b)**  $CO_2(s) \longrightarrow CO_2(l)$

# *FOR REVIEW*

## SUMMARY

In this chapter we have concerned ourselves with the properties of liquids and solids. We can understand the physical properties of matter in these states in terms of the kinetic-molecular theory, which we used earlier to explain the properties of gases. In a liquid the intermolecular forces keep molecules in close proximity, though they retain freedom to move with respect to one another. The free volume in a liquid is thus small, and liquids are not very compressible. The particles that make up a solid are even more restrained than in a liquid; they occupy specific locations in a three-dimensional arrangement. Thus solids retain both their shape and their volume. Any substance may exist in more than one state of matter, or phase. The equilibria between phases are dynamic; that is, there is continuous transfer of particles from one phase to the other. **Equilibrium** in such a dynamic system occurs when the rates of transfer between the phases are equal. The change of matter from one state of matter to another is termed a **phase change**. Conversions of a solid to a liquid (**melting**), solid to a gas (**sublimation**), or liquid to a gas (**vaporization**) are all endothermic processes; that is, the enthalpies of melting, sublimation, or vaporization are all positive. The reverse processes, conversion of a liquid to a solid (**freezing**), gas to a solid (**deposition**), or a gas to a liquid (**condensation**) are all exothermic; thus the enthalpy changes for these processes are all negative.

The **vapor pressures** of liquids increase nonlinearly with temperature. Boiling occurs when the vapor pressure equals the externally applied pressure. The **normal boiling point** is the temperature at which the vapor pressure of the liquid equals 1 atm.

Physical properties of liquids, such as their vapor pressure, normal boiling point, **viscosity**, **sur-**

face tension, critical temperature, and critical pressure are related to the intermolecular forces between molecules.

The intermolecular forces that keep the particles of a liquid or solid together include ion-dipole, dipole-dipole, and London dispersion forces. The relative importance of each of these contributions to the intermolecular attractions depends on the character of each substance. In general, the intermolecular attractive forces depend on the polarizability, polarity, and shape of the molecule. The magnitudes of London dispersion forces between nonpolar molecules of similar shape and complexity tend to increase with increasing molecular mass. Hydrogen bonding is an important source of intermolecular attractions in compounds containing O—H, N—H, and F—H bonds, and in mixtures containing these. The unusual properties of water, such as its high boiling point and high heat of vaporization, are due to the extensive O—H···O hydrogen bonding in both the liquid and solid forms.

Solids are characterized on a macroscopic level by their rigidity and on the atomic level by the relatively fixed nature of the particles. Solids whose particles are arranged in a regularly repeating pattern are said to be crystalline, and those whose particles show no such order are said to be amorphous.

Solids may be classified according to the type of bonding between the units of the solid as atomic, molecular, ionic, covalent network, or metallic. The magnitude of the forces operating between the units of the lattice ranges from very small, as with the rare gases, to very large, as in diamond (a covalent network structure), NaCl (ionic), or W (metallic).

The essential structural features of any crystalline solid (crystal) can be represented by its unit cell, the smallest part of the crystal that can, by simple displacement, reproduce the three-dimensional structure. The three-dimensional patterns of crystals can also be represented by their crystal lattices. The points in a crystal lattice represent positions in the structure where there are identical environments. (Although these positions might be centered on atoms, ions, or molecules, this is not necessary.)

In many solid structures the particles have a close-packing arrangement, in which spherical particles are arranged so as to leave the minimal amount of free volume. Two closely related forms of close packing, cubic and hexagonal, are possible. In both, each close-packed unit is in contact with 12 equivalent neighbors.

The structures of crystalline solids can be determined by observations of X-ray diffraction patterns. The Bragg equation, $n\lambda = 2d \sin \theta$, expresses the angles at which constructive interference of the diffracted rays occurs, in terms of the distance $d$ between planes in the crystal.

Crystalline solids have imperfections, or defects, that cause them to have less than perfect order. Conversely, amorphous solids often have some small regions where the particles are arranged in an orderly fashion. The extent to which these ordered regions occur is expressed by the percent crystallinity of the solid.

The equilibria between the solid, liquid, and gas phases of a substance as a function of temperature and pressure are displayed on a phase diagram. Equilibrium between any two phases on such a diagram is indicated by a line. The point on such a diagram at which all three phases coexist in equilibrium is called the triple point.

---

## LEARNING GOALS

Having read and studied this chapter, you should be able to:

1. Distinguish between gases, liquids, and solids on a molecular level.

2. Employ the kinetic-molecular theory and the concept of intermolecular attractions to explain the properties of each phase, such as surface tension, viscosity, vapor pressure, and boiling and melting points.

3. Explain the nature of the equilibria that may exist between phases. Account for the enthalpy changes that accompany phase changes.

4. Calculate the heat absorbed or evolved when a given quantity of a substance changes from one condition to another, given the needed heat capacities and enthalpy changes associated with phase changes.

5. Describe the manner in which the vapor pressure of a substance changes with temperature and the relationship between the pressure on the surface of a liquid and the boiling point of that liquid.

6. Explain the meaning of the terms *critical temperature* and *critical pressure*, and account for the variation in critical temperatures of different substances in terms of intermolecular forces.

7. Describe the various types of intermolecular attractive forces and indicate how each arises.

8. Predict, for any particular substance of known structure, which types of intermolecular forces may be operative and which particular type is of major importance.

9. Describe the nature of the hydrogen bond, and identify those molecular systems in which hydrogen bonding is likely to be important.

10. Distinguish between crystalline and amorphous solids.

11. Predict the type of solid (atomic, molecular, ionic, covalent network, or metallic) formed by a substance, and predict its general properties.

12. Determine the net contents in a cubic unit cell, given a drawing or verbal description of the cell. Use this information, together with the atomic weights of the atoms in the cell and the cell dimensions, to calculate the density of the substances.

13. Explain the origin of the diffraction patterns obtained when X rays impinge on a crystal.

14. Draw a phase diagram of a substance, given appropriate data.

15. Use a phase diagram to predict what phases are present at any given temperature and pressure.

## KEY TERMS

Among the more important terms and expressions used for the first time in this chapter are the following:

An **amorphous solid** (Section 11.5) is a solid whose molecular arrangement lacks a regular and long-range pattern.

The **boiling point** (Section 11.3) of a liquid is the temperature at which its vapor pressure equals the external pressure. The **normal boiling point** is the temperature at which the liquid boils when the external pressure is 1 atm (that is, the temperature at which the vapor pressure of the liquid is 1 atm).

The **Bragg equation** (Section 11.5) relates the angles at which X rays are scattered from a crystal to the spacing between the layers of particles.

**Capillary action** (Section 11.3) is the term used to describe the process by which a liquid rises in a tube because of a combination of adhesion with the walls of the tube and cohesion between liquid particles.

**Close packing** (Section 11.5) refers to the most efficient packing of spheres in a three-dimensional array. There are two closely similar forms, **cubic close packing** and **hexagonal close packing**.

**Critical pressure** (Section 11.3) is the pressure at which a gas at its critical temperature is converted to the liquid state.

**Critical temperature** (Section 11.3) is the highest temperature at which it is possible to convert the gaseous form of a substance to a liquid. The critical temperature increases with an increase in the magnitude of intermolecular forces.

The **crystal lattice** (Section 11.5) of a solid is an imaginary network of points on which the repeating unit of the structure (the contents of the unit cell) may be imagined to be laid down so that the structure of the crystal is obtained. Each point represents an identical environment in the crystal.

A **crystalline solid** (Section 11.5) (or simply a **crystal**) is a solid whose internal arrangement of atoms, molecules, or ions shows a regular repetition in any direction through the solid.

A **dynamic equilibrium** (Section 11.2) is a state of balance in which opposing processes occur at the same rate.

**Hydrogen bonds** (Section 11.4) are intermolecular attractions between molecules containing hydrogen bonded to an electronegative element. The most important examples involve oxygen, nitrogen, or fluorine.

**Intermolecular forces** (Section 11.4) are the short-range attractive forces operating between the particles that make up the units of a liquid or solid substance. These same forces also cause gases to liquefy or solidify at low temperatures.

**London dispersion forces** (Section 11.4) are intermolecular forces resulting from attractions between induced dipoles.

The **melting point** (Section 11.2) of a solid (or the **freezing point** of a liquid) is the temperature at which solid and liquid phases coexist in equilibrium. The **normal melting point** is the melting point at 1 atm pressure.

A **meniscus** (Section 11.3) is the curved upper surface of a liquid column.

A **phase change** (Section 11.2) represents the conversion of a substance from one state of matter to another. The phase changes we consider are **melting** and **freezing** (solid ↔ liquid), **sublimation** and **deposition** (solid ↔ gas), and **vaporization** and **condensation** (liquid ↔ gas).

A **phase diagram** (Section 11.6) is a graphic representation of the equilibria between the solid, liquid, and gaseous phases of a substance as a function of temperature and pressure.

The **polarizability** (Section 11.4) of a molecule

describes the ease with which its electron cloud is distorted by an outside influence, thereby inducing a dipole moment.

**Surface tension** (Section 11.3) is the intermolecular, cohesive attraction that causes a liquid surface to become as small as possible.

The **triple point** (Section 11.6) of a substance is the temperature at which solid, liquid, and gas phases coexist in equilibrium.

A **unit cell** (Section 11.5) is the smallest portion of a crystal that reproduces the structure of the entire crystal when repeated in different directions in space. It is the repeating unit or "building block" of the crystal lattice.

**Vapor pressure** (Section 11.2) is the pressure exerted by a vapor in equilibrium with its liquid or solid phase.

**Viscosity** (Section 11.3) is a measure of the resistance of fluids to flow.

**X-ray diffraction** (Section 11.5) refers to the scattering of X rays by the units of a regular crystalline solid. The scattering patterns obtained can be used to deduce the arrangements of particles in the crystal.

# EXERCISES

## Kinetic-Molecular Theory: Phase Changes

**11.1** Account for the differences in compressibilities of solids, liquids, and gases.

**11.2** Compare the degree of order within a solid with that within a liquid. What accounts for the difference?

**11.3** Describe the ways in which liquids are more similar to solids than to gases.

**11.4** Suppose that a drop of liquid bromine is added to a 1-L vessel containing air and another drop is added to a 1-L vessel containing water (in which the bromine dissolves). Compare the relative rates at which bromine would diffuse through the 1-L volume in each case. Explain in terms of the kinetic-molecular theory.

**11.5** Why does lowering the temperature cause a substance to change in succession from the gaseous to the liquid to the solid state?

**11.6** Compare the relative amounts of free space (that is, the space not actually occupied by the molecules themselves) in solids, liquids, and gases.

**11.7** Both when a gas condenses to a liquid and when a liquid freezes, heat is liberated. What is the origin of the heat evolution in each case?

**11.8** Ethyl chloride, $C_2H_5Cl$, has a normal boiling point of 12°C. When liquid $C_2H_5Cl$ under pressure is sprayed on a surface at atmospheric pressure, the surface is cooled considerably. Explain.

**11.9** The molar enthalpy of sublimation of a solid is always larger than the molar heat of vaporization of the corresponding liquid. Explain.

**11.10** The enthalpy of fusion for any substance is generally smaller than the enthalpy of vaporization. Why is this true?

**11.11** The enthalpy of melting of ice at 0°C is 6.01 kJ/mol, and the enthalpy of vaporization of water at 0°C is 44.94 kJ/mol. Calculate the enthalpy of sublimation of ice at 0°C.

**11.12** (a) How much heat is produced when 60.6 g of steam at 122°C is converted to ice at −20.2°C? (b) How much heat is required to convert 80.0 g of ice at −22.0°C to steam at 150°C?

**11.13** Ethyl alcohol melts at −114°C and boils at 78°C. The enthalpy of fusion at −114°C is 105 J/g, and the enthalpy of vaporization at 78°C is 870 J/g. If the specific heat of solid ethyl alcohol is taken as 0.97 J/g-°C and that for the liquid as 2.3 J/g-°C, how much heat is required to convert 16.0 g of ethyl alcohol at −130°C to the vapor phase at 78°C?

**11.14** Calculate the heat required to convert 10.0 g of propanol, $C_3H_7OH$, from a solid at −140°C into a vapor at 110°C. The normal melting point and boiling point of $C_3H_7OH$ are −127°C and 97°C, respectively. The heat of fusion is 5.18 kJ/mol, and the heat of vaporization is 41.7 kJ/mol. The heat capacities of the solid, liquid, and gas states are 142, 170, and 108 J/mol-°C, respectively.

## Properties of Liquids

**11.15** Explain why the boiling point of a liquid varies substantially with pressure, whereas the melting point of a solid depends little on pressure.

**11.16** Explain how each of the following affects the vapor pressure of a liquid: (a) surface area; (b) temperature; (c) intermolecular attractive forces; (d) volume of the liquid.

**11.17** Explain the following observations. (a) Raindrops that collect on a waxed auto hood take on a nearly spherical shape. (b) A tin can filled with steam at 100° C and then sealed collapses when it is cooled. (c) It takes longer to boil eggs at high altitudes than at lower ones. (d) When heated above 279°C, carbon disulfide, $CS_2$, cannot be liquefied regardless of how great the pressure exerted on the gas.

**11.18** Account for each of the following. (a) The critical temperature and pressure for Xe are higher than for Ar. (b) In contact with a narrow capillary tube made of polyethylene, water forms a concave-downward meniscus. **(c)** At 60°C and 1 atm, ethanol has a higher vapor pressure than water. **(d)** The viscosity of water decreases with increasing temperature (Table 11.4).

## Intermolecular Forces

**11.19** The molar enthalpies of vaporization of some simple hydrocarbons follow: methane, $CH_4$, 8.87 kJ; ethane, $C_2H_6$, 15.6 kJ; propane, $C_3H_8$, 19.0 kJ; butane, $C_4H_{10}$, 24.3 kJ. What do these data suggest regarding variations in intermolecular forces in the series? Explain.

**11.20** The molar enthalpy of vaporization of CO at its boiling point is 6.02 kJ; that of $Cl_2$ is 20.4 kJ. Why is the value higher for $Cl_2$?

**11.21** For each of the following pairs, choose the one with the lower boiling point: **(a)** NaCl and $CH_3Cl$; **(b)** $Cl_2$ and $Br_2$; **(c)** $SO_2$ and $SiO_2$; **(d)** HF and HCl. Give your reasoning in each case.

**11.22** For each of the following pairs, choose the one with the lower boiling point: **(a)** $NH_3$ and $PH_3$; **(b)** $C_6H_6$ and C (graphite); **(c)** $O_2$ and $I_2$; **(d)** $H_2S$ and $H_2Te$. Give your reasoning in each case.

**11.23** What is the nature of the major attractive intermolecular force in each of the following: **(a)** $Xe(l)$; **(b)** $NH_3(l)$; **(c)** $PCl_3(l)$; **(d)** an aqueous solution of $Fe(NO)_3$, between the solvent and $Fe^{3+}$ ions?

**11.24** What is the nature of the major attractive intermolecular force in each of the following: **(a)** $CH_3CN(l)$ **(b)** $CH_3OH(l)$; **(c)** $Cl_2(l)$; **(d)** an aqueous solution of glucose, $C_6H_{12}O_6$, between the solvent and glucose molecules?

**11.25** Using the thermodynamic data listed in Appendix D, calculate $\Delta H$ for the following processes at 25°C:

$$Br_2(l) \longrightarrow Br_2(g)$$
$$Br_2(g) \longrightarrow 2Br(g)$$

Discuss the relative magnitude of these enthalpy changes in terms of the forces involved in each case.

**11.26** Using the thermodynamic data listed in Appendix D, calculate the enthalpy change in the process $POCl_3(l) \longrightarrow POCl_3(g)$. Describe the intermolecular interactions that contribute to this enthalpy change. Indicate some of the ways in which intermolecular forces differ from *intra*molecular forces, which hold atoms together within molecules.

**11.27** How would you expect the heats of vaporization to vary among the members of the following sets of compounds: **(a)** $PH_3$, $AsH_3$, $SbH_3$; **(b)** $CH_4$, $CF_4$, $CCl_4$; **(c)** $CH_4$, $NH_3$, $H_2O$?

**11.28** How would you expect the boiling points to vary among the members of the following sets of compounds: **(a)** $Cl_2$, $Br_2$, $I_2$; **(b)** HF, HCl, HBr; **(c)** $SiH_4$, $SiCl_4$, $SiBr_4$?

**11.29** Account for the difference in critical temperatures in each of the following pairs of compounds: **(a)** $O_2$, −118°C; $O_3$, −5°C; **(b)** $SiH_4$, −3°C; $SiCl_4$, 233°C; **(c)** $NH_3$, 132°C; $PH_3$, 51°C.

**11.30** The critical temperatures for the hydrogen halides vary as follows: HF, 188°C; HCl, 51°C; HBr, 90°C; HI, 151°C. Explain these variations in terms of intermolecular attactive forces.

**11.31** Indicate which of the following substances **(a)** is most likely to exist as a crystalline solid at room temperature, and **(b)** would be the least readily liquefied gas: HF, $PCl_3$, $F_2$, $FeCl_2$, $SO_2$.

**11.32** The boiling points of the fluorides of the second-row elements of the periodic table are LiF, 1717°C; $BeF_2$, 1175°C; $BF_3$, −101.0°C; $CF_4$, −128°C; $NF_3$, −120°C; $OF_2$, −145°C; $F_2$, −188°C. Account for the variations in terms of the nature and strengths of intermolecular forces.

**11.33** Cite three properties of water that can be attributed to the existence of hydrogen bonding in either the solid or liquid phase.

**11.34** Predict the order of increasing viscosity among the following liquids, all at a common temperature: **(a)** propanol, $CH_3CH_2CH_2OH$; **(b)** propane, $CH_3CH_2CH_3$; **(c)** propane-1,3-diol, $HOCH_2CH_2CH_2OH$. Explain your reasoning.

## Solids

**11.35** Indicate the type of crystal (atomic, molecular, metallic, covalent, or ionic) each of the following would form upon solidification: **(a)** HBr; **(b)** Ar; **(c)** Mn; **(d)** $Co(NO_3)_2$; **(e)** C.

**11.36** Compare the following group of substances with respect to hardness, electrical conductivity, and melting point. Relate these properties to the type of crystal formed by each: CaO; $SO_2$; Na; Si.

**11.37** For each of the following pairs of substances, predict which will have the higher melting point and indicate why: **(a)** KBr, $Br_2$; **(b)** $SiO_2$, $CO_2$; **(c)** Se, CO; **(d)** NaF, $MgF_2$.

**11.38** For each of the following compounds, predict which will have the higher melting point and indicate why: **(a)** $C_6Cl_6$, $C_6H_6$; **(b)** HF, HCl; **(c)** $KO_2$, $SiO_2$; **(d)** Ar, Xe.

**11.39** What is a unit cell? What properties does it have?

**11.40** The contents of the unit cell in a crystalline lattice must have the same stoichiometry as that of the solid as a whole. Why is this so?

**11.41** In the face-centered-cubic unit cells that many metals have, the atom in the center of each face is in contact with the corner atoms, as shown in Figure 11.42 (next page). Copper crystallizes in a face-centered-cubic lattice; the unit cell edge is 3.61 Å in length. Calculate the atomic radius of copper and density of the metal.

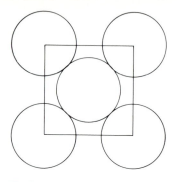

**FIGURE 11.42**

**11.42** Nickel crystallizes in a face-centered-cubic lattice. The atomic radius of nickel is 1.24 Å. What is the length of the edge of the unit cell in nickel, and what is the density of the metal? (Refer to Figure 11.42.)

**11.43** **(a)** Potassium fluoride has the NaCl type of crystal structure (Figure 11.25). The density of KF at 25°C is 2.468 g/cm$^3$. Calculate the dimensions of the KF unit cell. **(b)** Silver chloride, which also has the NaCl type of crystal structure, has a density of 5.57 g/cm$^3$ at 25°C. What is the length of the edge of the AgCl unit cell?

**11.44** An element crystallizes in a body-centered-cubic lattice. The edge of the unit cell is 2.86 Å, and the density of the crystal is 7.92 g/cm$^3$. Calculate the atomic weight of the element.

**11.45** **(a)** The unit cell of aluminum metal is cubic with an edge length of 4.05 Å. Determine the type of unit cell (primitive, face-centered, or body-centered) if the metal has a density of 2.70 g/cm$^3$. **(b)** Platinum has a density of 21.5 g/cm$^3$. How many atoms are in the unit cell if it is cubic with an edge length of 3.914 Å. Is the cell cubic, body-centered cubic, or face-centered cubic?

**11.46** KCl has the same structure as NaCl. The length of the unit cell is 628 pm. The density of KCl is 1.984 g/cm$^3$, and its formula mass is 74.55. Using this information, calculate Avogadro's number.

**11.47** **(a)** In a three-dimensional, close-packed array of equal-sized spheres, what is the coordination number of each sphere? **(b)** What is the coordination number of each sphere in a primitive-cubic structure? **(c)** What is the coordination number of each sphere in a body-centered cubic lattice?

**11.48** Describe the distinction between cubic close packing and hexagonal close packing. Do the two forms of close packing differ in the fraction of free space in the lattice? Explain.

**11.49** A crystal of pure NaCl is transparent to visible light. Irradiating an NaCl crystal with light from a helium-neon laser, wavelength 632 nm, does not result in the appearance of a diffraction pattern. Why is this so? Why does the same crystal diffract X rays of wavelength 1.395 Å?

**11.50** Describe what is meant by "constructive interference" and "destructive interference," and explain how these terms apply to the diffraction of X rays by a crystal.

**11.51** In the diffraction of a crystal using X rays with wavelength of 1.54 Å, a first-order ($n = 1$) reflection was obtained at an angle of 12.5°. What is the distance between the planes of atoms that cause this diffraction?

**11.52** The first-order ($n = 1$) diffraction of X rays from crystal planes separated by 2.81 Å occurs at 11.8°. **(a)** What is the wavelength of the X rays? **(b)** Calculate the angle at which the second-order ($n = 2$) diffraction will appear.

## Phase Diagrams

**11.53** Refer to Figure 11.40(a), and describe all the phase changes that would occur in each of the following cases. **(a)** Water vapor originally at $1.0 \times 10^{-3}$ atm and $-0.10°C$ is slowly compressed at constant temperature until the final pressure is 10 atm. **(b)** Water originally at $-10°C$ and 0.30 atm is heated at constant pressure until the temperature is 80.0°C.

**11.54** Refer to Figure 11.40(b), and describe the phase changes (and the temperature at which they occur) when $CO_2$ is heated from $-80°C$ to $-20°C$ at **(a)** a constant pressure of 3 atm; **(b)** a constant pressure of 6 atm.

**11.55** The normal melting and boiling points of sulfur dioxide, $SO_2$, are $-72.7°C$ and $-10.0°C$, respectively. Its triple point is at $-75.5°C$ and $1.65 \times 10^{-3}$ atm, and its critical point is at 157°C and 78 atm. Sketch the phase diagram for $SO_2$, showing the four points given above and indicating the areas in which each phase is stable.

**11.56** The normal melting and boiling points of xenon are $-112°C$ and $-108°C$, respectively. Its triple point is at $-121°C$, at a pressure of 282 mm Hg. Sketch the phase diagram for xenon, showing the three points given above and indicating the areas in which each phase is stable.

## Additional Exercises

**11.57** For many years drinking water has been cooled in hot climates by the evaporation of water from the surfaces of canvas bags or porous clay pots. How many grams of water can be cooled from 35°C to 22°C by the evaporation of 10 g of water?

**11.58** Differentiate between **(a)** sublimation and vaporization; **(b)** an ionic solid and a covalent network solid; **(c)** the enthalpy of vaporization and the normal boiling point; **(d)** hydrogen bonding and dipole-dipole forces.

**11.59** In dichloromethane, $CH_2Cl_2$ ($\mu = 1.60$ D), the dispersion-force contribution to the intermolecular attractive forces is about five times larger than the dipole-dipole contribution. Would you expect the relative importance of the two kinds of intermolecular attractive forces to differ **(a)** in dibromomethane ($\mu = 1.43$ D); **(b)** in difluoromethane ($\mu = 1.93$ D)? Explain.

**11.60** Account for the observed variation in viscosities of the series of straight-chain hydrocarbons, as listed below:

| Compound | Formula | Viscosity (N-s/m$^2$) |
|----------|---------|----------------------|
| Pentane | $C_5H_{12}$ | $0.225 \times 10^{-3}$ |
| Hexane | $C_6H_{14}$ | $0.313 \times 10^{-3}$ |
| Heptane | $C_7H_{16}$ | $0.397 \times 10^{-3}$ |
| Octane | $C_8H_{18}$ | $0.546 \times 10^{-3}$ |

**11.61** From the following list of substances, indicate which is expected to have the highest boiling point and which the lowest: $CO_2$, Ar, $CF_4$, RbCl, $SiF_4$. Explain your choices.

**11.62** The enthalpy of vaporization of water at 0°C is 44.94 kJ/mol, whereas at 100°C the enthalpy of vaporization of water is 40.67 kJ/mol. Account for the fact that water has a lower enthalpy of vaporization at the higher temperature.

**11.63** Graph the surface tension of water (Table 11.4), as a function of temperature. What is the equation for the relationship between temperature and surface tension for this substance?

**11.64** Test the applicability of the Clausius-Clapeyron equation (Equation 11.1), using the following vapor pressure versus temperature data for mercury:

| Temperature (°C) | Vapor pressure of Hg (mm Hg) |
|---|---|
| 50.0 | 0.01267 |
| 60.0 | 0.02524 |
| 70.0 | 0.04825 |
| 80.0 | 0.0880 |
| 90.0 | 0.1582 |

If the Clausius-Clapeyron equation is obeyed, use the slope of the line to calculate $\Delta H_v$ for mercury in this temperature range.

**[11.65]** We might guess that the Clausius-Clapeyron equation (Equation 11.1), would be applicable also to the vapor pressure data for a solid. Use this equation to estimate the heat of sublimation of ice from the following data:

| Temperature (°C) | Vapor pressure (mm Hg) |
|---|---|
| −20.0 | 0.640 |
| −16.0 | 1.132 |
| −12.0 | 1.632 |
| −8.0 | 2.326 |
| −4.0 | 3.280 |
| 0.0 | 4.579 |

**11.66** The critical temperatures of the boron trihalides vary as follows: $BF_3$, −12°C; $BCl_3$, 179°C; $BBr_3$, 300°C. Explain this variation in terms of the nature of the intermolecular attractive forces in these systems.

**11.67** A clean, very dry silica or glass surface has many Si—OH groups on it. Describe the nature of the adhesive forces that might operate between such a surface and $CH_3OH$.

**11.68** Hydrogen peroxide, $H_2O_2$, melts at −0.4°C and boils at 151°C. In comparison with water, do these data suggest that hydrogen bonding might be important? Draw a diagram that shows the nature of the hydrogen-bonding interactions that could occur between molecules in this substance.

**11.69** From which of the following materials would you expect to obtain well-defined X-ray diffraction patterns: **(a)** a sugar crystal; **(b)** KBr; **(c)** liquid water; **(d)** pure iron; **(e)** ice; **(f)** a section of a rubber stopper?

**11.70** Chromium crystallizes in a body-centered-cubic unit cell whose edge length is 288.4 pm. If the atoms touch along the body diagonal of the unit cell, calculate the atomic radius of a Cr atom.

**[11.71]** The critical temperature and pressure of $CClF_3$ (Freon 13) are 29°C and 39 atm, respectively. For $CClH_3$ (methyl chloride) the corresponding values are 143°C and 66 atm. What do these values tell us about the relative intermolecular attractive forces between molecules in the two substances? Are the results surprising? If so, why?

**[11.72]** It is widely recognized that impurities in solids tend to concentrate at the sites of dislocations. On the basis of Figure 11.36, explain why.

**[11.73]** Assume that 3.000 g of $H_2O$ is introduced into an evacuated flask whose volume is 1.000 L. If the temperature of the water and flask is 30.0°C, what mass of water will evaporate? The vapor pressure of water at 30.0°C is 31.82 mm Hg.

**11.74** The triple point of benzene is 5°C, 21 mm Hg. The density of solid benzene is 1.005 g/cm$^3$, whereas that of the liquid is 0.894 g/cm$^3$. The normal boiling point of benzene is 80°C; its critical point is 289°C, 48 atm. Sketch the phase diagram.

**11.75** Amorphous silica is found to have a density of about 2.2 g/cm$^3$, whereas the density of crystalline quartz is 2.65 g/cm$^3$. Account for this difference in densities.

# Solutions

Very few of the materials that we encounter in everyday life are pure substances; most are mixtures. Many of these mixtures are homogeneous; that is, their components are uniformly intermingled on a molecular level. We have seen that homogeneous mixtures are called solutions (Sections 1.2 and 3.8). Examples of solutions abound in the world around us. The air we breathe is a homogeneous mixture of several gaseous substances. The familiar metal brass is a solution of zinc in copper. The oceans are a solution of many dissolved substances in water. The fluids that run through our bodies are solutions, carrying a great variety of essential nutrients, salts, and so forth.

Solutions may be gaseous, liquid, or solid. Examples of each kind are given in Table 12.1. Recall that the **solvent** is the component whose phase does not change when the solution forms (Section 3.8). If all components remain in the same phase, the one in greatest amount is called the solvent. Other components are called **solutes**. Liquid solutions are the most common, and it is on this type of solution that we focus our attention in this chapter.

In Chapter 11 we considered the various types of intermolecular forces that exist between molecular and ionic particles. In this chapter we will see that these forces are also involved in the interactions between solutes and solvents. We will examine the solution process, the factors that determine the amount of solute that can dissolve in a given quantity of solvent, and some properties of the solutions that result. We will be particularly concerned with aqueous solutions of ionic substances, because of their central importance in chemistry and in our daily lives. Near the end of the chapter we will consider a type of mixture, known as a colloid, that is on the borderline between heterogeneous mixtures and solutions. Before we examine these topics, however, it is useful to discuss ways of describing the concentrations of solutions, that is, the amount of solute dissolved in a given quantity of solvent or solution.

# CONTENTS

**TABLE 12.1**    Examples of solutions

| State of solution | State of solvent | State of solute | Example |
|---|---|---|---|
| Gas | Gas | Gas | Air |
| Liquid | Liquid | Gas | Oxygen in water |
| Liquid | Liquid | Liquid | Alcohol in water |
| Liquid | Liquid | Solid | Salt in water |
| Solid | Solid | Gas | Hydrogen in platinum |
| Solid | Solid | Liquid | Mercury in silver |
| Solid | Solid | Solid | Silver in gold (certain alloys) |

Pt. Reyes National Seashore, California. The world ocean is a complex solution containing about 3.5 percent by weight dissolved salts. (Frans Lanting/Photo Researchers)

## 12.1 WAYS OF EXPRESSING CONCENTRATION

The concentration of a solution can be expressed either qualitatively or quantitatively. The terms "dilute" and "concentrated" are used to describe a solution qualitatively. A solution with a relatively small concentration of solute is said to be **dilute**; one with a large concentration is said to be **concentrated**.

Several quantitative expressions of concentration are employed in chemistry. One of the simplest is the **weight percentage**. The weight percentage of a component of a solution is given by

$$\text{Wt \% of component} = \frac{\text{mass of component in soln}}{\text{total mass of soln}} \times 100 \qquad [12.1]$$

For example, if a solution of hydrochloric acid contains 36 percent HCl by weight, it has 36 g of HCl for each 100 g of solution.

---

**SAMPLE EXERCISE 12.1**

A solution is made containing 6.9 g of $NaHCO_3$ per 100 g of water. What is the weight percentage of solute in this solution?

**Solution:**

$$\text{Wt \% of solute} = \frac{\text{mass solute}}{\text{mass soln}} \times 100$$

$$= \frac{6.9 \text{ g}}{6.9 \text{ g} + 100 \text{ g}} \times 100 = 6.5\%$$

Notice that the mass of solution is the sum of the mass of solvent and the mass of solute. The weight percentage of solvent in this solution is $(100 - 6.5)\% = 93.5\%$.

**PRACTICE EXERCISE**

A commercial bleaching solution contains 3.62 weight percent sodium hypochlorite, NaOCl. What is the mass of NaOCl in a bottle containing 2500 g of bleaching solution?
***Answer:*** 90.5 g of NaOCl

---

Several concentration expressions are based on the number of moles of one or more components of the solution. Three are commonly used in chemistry: mole fraction, molarity, and molality. The **mole fraction** of a component of a solution is given by

$$\text{Mole fraction of component} = \frac{\text{moles component}}{\text{total moles all components}} \qquad [12.2]$$

The symbol $X$ is commonly used for mole fraction, with a subscript to indicate the component on which attention is being focused. For example, the mole fraction of HCl in a hydrochloric acid solution can be represented as $X_{HCl}$. The mole fractions of all components of a solution will total 1.

---

**SAMPLE EXERCISE 12.2**

Calculate the mole fraction of HCl in a solution of hydrochloric acid containing 36 percent HCl by weight.

**Solution:** Assume that there is 100 g of solution. (You can verify for yourself that assuming any other quantity will not change the result, though it can make the arithmetic more difficult.) The solution therefore contains 36 g of HCl and 64 g of $H_2O$.

$$\text{Moles HCl} = (36 \text{ g HCl})\left(\frac{1 \text{ mol HCl}}{36.5 \text{ g HCl}}\right)$$

$$= 0.99 \text{ mol HCl}$$

$$\text{Moles } H_2O = (64 \text{ g } H_2O)\left(\frac{1 \text{ mol } H_2O}{18 \text{ g } H_2O}\right)$$

$$= 3.6 \text{ mol } H_2O$$

$$X_{\text{HCl}} = \frac{\text{moles HCl}}{\text{moles } H_2O + \text{moles HCl}}$$

$$= \frac{0.99}{3.6 + 0.99} = \frac{0.99}{4.6} = 0.22$$

**PRACTICE EXERCISE**
Calculate the mole fraction of NaOCl in a commercial bleach solution containing 3.62 weight percent NaOCl.
*Answer:* $9.00 \times 10^{-3}$

**Molarity** ($M$), which we discussed in Section 3.8, is defined as the number of moles of solute in a liter of solution:

$$\text{Molarity} = \frac{\text{moles solute}}{\text{liters soln}} \qquad [12.3]$$

### SAMPLE EXERCISE 12.3

**(a)** Calculate the molarity of an ascorbic acid ($C_6H_8O_6$, vitamin C) solution prepared by dissolving 1.80 g in enough water to make 125 mL of solution. **(b)** How many milliliters of this solution contain 0.0100 mol of ascorbic acid?

**Solution:** **(a)** We use the molecular weight of $C_6H_8O_6$ (176 amu) to figure the number of moles of the substance:

$$(1.80 \text{ g } C_6H_8O_6)\left(\frac{1 \text{ mol } C_6H_8O_6}{176 \text{ g } C_6H_8O_6}\right)$$

$$= 0.0102 \text{ mol } C_6H_8O_6$$

Using the number of moles of $C_6H_8O_6$ and the volume in liters (125 mL = 0.125 L), we have

$$M = \frac{\text{mol } C_6H_8O_6}{\text{L soln}} = \frac{0.0102 \text{ mol } C_6H_8O_6}{0.125 \text{ L}}$$

$$= 0.0818 \text{ } M$$

**(b)** Rearranging Equation 12.3, solving for liters, we have

$$\text{Liters soln} = \text{moles solute} \times \frac{1}{\text{molarity}}$$

$$= (0.0100 \text{ mol } C_6H_8O_6)$$

$$\times \left(\frac{1 \text{ L soln}}{0.0818 \text{ mol } C_6H_8O_6}\right) = 0.122 \text{ L}$$

$$\text{mL soln} = (0.122 \text{ L})\left(\frac{10^3 \text{ mL}}{1 \text{ L}}\right) = 122 \text{ mL}$$

**PRACTICE EXERCISE**
Calculate the molarity of a commercial bleach solution that contains 9.65 g of CaCl(OCl) per liter.
*Answer:* 0.0760 $M$

The **molality** ($m$) of a solution is defined as the number of moles of solute in a kilogram of solvent:

$$\text{Molality} = \frac{\text{moles solute}}{\text{mass of solvent in kilograms}} \qquad [12.4]$$

Notice the difference between molarity and molality. These two ways of expressing concentration are similar enough to be easily confused. Molality is defined in terms of the mass of solvent, but molarity is defined in terms of the volume of solution. A 1.50 molal (written 1.50 $m$) solution contains 1.50 mol of solute for every kilogram of solvent. (When water is the solvent, the molality and molarity of a dilute solution are numerically about the same, because 1 kg of solvent is nearly the same as 1 kg of solution, and 1 kg of the solution has a volume of about 1 L.)

The molality of a given solution does not vary with temperature, because masses do not vary with temperature. Molarity, however, changes with temperature because of the expansion or contraction of the solution.

---

## SAMPLE EXERCISE 12.4

What is the molality of a solution made by dissolving 5.0 g of toluene ($C_7H_8$) in 225 g of benzene ($C_6H_6$)?

**Solution:** **(a)** We determine the number of moles of solute, $C_7H_8$, by using its molecular weight (92 amu):

$$(5.0 \text{ g } C_7H_8)\left(\frac{1 \text{ mol } C_7H_8}{92 \text{ g } C_7H_8}\right) = 0.054 \text{ mol } C_7H_8$$

Using the number of moles of $C_7H_8$ and the number

of kilograms of solvent (225 g = 0.225 kg), we have

$$m = \frac{\text{mol } C_7H_8}{\text{kg } C_6H_6} = \frac{0.0543 \text{ mol } C_7H_8}{0.225 \text{ kg } C_6H_6} = 0.241 \text{ } m$$

### PRACTICE EXERCISE
Determine the molality of a solution that contains 36.5 g of naphthalene, $C_{10}H_8$, in 420 g of toluene, $C_7H_8$.
**Answer:** 0.678 $m$

---

For most purposes in ordinary chemical laboratory work, molarity is the most useful expression of concentration. However, the other ways we have just considered find use in special situations, and you should be familiar with them. Sample Exercise 12.5 shows by way of example how the various expressions of concentrations are related.

---

## SAMPLE EXERCISE 12.5

Given that the density of a solution of 5.0 g of toluene and 225 g of benzene (Sample Exercise 12.4) is 0.876 g/mL, calculate the concentration of the solution in **(a)** molarity; **(b)** mole fraction of solute; **(c)** weight percentage of solute.

**Solution:** **(a)** The total mass of the solution is equal to the mass of the solvent plus the mass of the solute:

$$\text{Mass soln} = 5.0 \text{ g} + 225 \text{ g} = 230 \text{ g}$$

The density of the solution is used to convert the mass of the solution to its volume:

$$\text{Milliliters soln} = (230 \text{ g})\left(\frac{1 \text{ mL}}{0.876 \text{ g}}\right) = 263 \text{ mL}$$

Density must be known in order to interconvert molarity and molality, because one is based on mass and the other is based on volume. The number of moles of solute must be known to calculate either molarity or molality:

$$\text{Moles } C_7H_8 = (5.0 \text{ g } C_7H_8)\left(\frac{1 \text{ mol } C_7H_8}{92.0 \text{ g } C_7H_8}\right)$$
$$= 0.054 \text{ mol}$$

Molarity is moles of solute per liter of solution:

$$\text{Molarity} = \frac{\text{moles } C_7H_8}{\text{liter soln}}$$
$$= \left(\frac{0.054 \text{ mol } C_7H_8}{263 \text{ mL soln}}\right)\left(\frac{1000 \text{ mL soln}}{1 \text{ L soln}}\right)$$
$$= 0.21 \text{ } M$$

Compare this value with the molality of the same solution calculated in Sample Exercise 12.4.
**(b)** The mole fraction of solute is expressed as

$$X_{C_7H_8} = \frac{\text{moles } C_7H_8}{\text{moles } C_7H_8 + \text{moles } C_6H_6}$$

We have already calculated the number of moles of $C_7H_8$.

$$\text{Moles } C_6H_6 = (225 \text{ g } C_6H_6)\left(\frac{1 \text{ mol } C_6H_6}{78.0 \text{ g } C_6H_6}\right)$$

$$= 2.88 \text{ mol}$$

$$X_{C_7H_8} = \frac{0.054 \text{ mol}}{0.054 \text{ mol} + 2.88 \text{ mol}} = \frac{0.054}{2.93}$$

$$= 0.018$$

(c) The weight percentage of solute is calculated as follows:

$$\text{Wt } \% \text{ } C_7H_8 = \frac{5.0 \text{ g } C_7H_8}{5.0 \text{ g } C_7H_8 + 225 \text{ g } C_6H_6} \times 100$$

$$= \frac{5.0 \text{ g}}{230 \text{ g}} \times 100 = 2.2\%$$

**PRACTICE EXERCISE**

A solution containing equal masses of glycerol, $C_3H_8O_3$, and water has a density of 1.10 g/mL. Calculate the (**a**) molarity; (**b**) mole fraction of glycerol; (**c**) molality of the solution.
***Answers:*** (**a**) 5.97 $M$; (**b**) mole fraction $C_3H_8O_3 = 0.163$; (**c**) 10.9 $m$

The concentration units that we have just discussed appear throughout the rest of this chapter and elsewhere in this text. However, one further concentration expression, called **normality**, may be encountered in other places. Normality (abbreviated $\mathcal{N}$) is defined as the number of **equivalents** of solute per liter of solution.

$$\text{Normality} = \frac{\text{equivalents solute}}{\text{liters soln}} \qquad [12.5]$$

An equivalent is defined according to the type of reaction being examined. For acid-base reactions, an equivalent of an acid is the quantity that supplies 1 mol of $H^+$; an equivalent of a base is the quantity reacting with 1 mol of $H^+$. In an oxidation-reduction reaction, an equivalent is the quantity of substance that gains or loses 1 mol of electrons. The masses of 1 equivalent of several substances are given in Table 12.2. An equivalent is always defined in such a way that 1 equivalent of reagent A will react with 1 equivalent of reagent B. For example, in an oxidation-reduction reaction, 31.6 g (1 equivalent) of $KMnO_4$ will react with 67.0 g (1 equivalent) of $Na_2C_2O_4$ (refer to Table 12.2). Similarly, in an acid-base reaction, 49.0 g of $H_2SO_4$ (1 equivalent) is stoichiometrically equivalent to 26.0 g of $Al(OH)_3$ (1 equivalent).

If $KMnO_4$ is reduced to $Mn^{2+}$, thereby gaining five electrons, we have

$$1 \text{ mol } KMnO_4 = 5 \text{ equivalents of } KMnO_4$$

**TABLE 12.2** Equivalent-mass relationships

| Reactant | Product | Reaction type | Mass of 1 mol of reactant (g) | Mass of 1 equivalent of reactant (g) |
|----------|---------|---------------|-------------------------------|--------------------------------------|
| $KMnO_4$ | $Mn^{2+}$ | Reduction (5 $e^-$) | 158.0 | 158.0/5 = 31.6 |
| $KMnO_4$ | $MnO_2$ | Reduction (3 $e^-$) | 158.0 | 158.0/3 = 52.7 |
| $Na_2C_2O_4$ | $CO_2$ | Oxidation (2 $e^-$) | 134.0 | 134.0/2 = 67.0 |
| $H_2SO_4$ | $SO_4^{2-}$ | Acid (2 $H^+$) | 98.0 | 98.0/2 = 49.0 |
| $Al(OH)_3$ | $Al^{3+}$ | Base (3 $OH^-$) | 78.0 | 78.0/3 = 26.0 |

Therefore, if 1 mol of $KMnO_4$ is dissolved in sufficient water to form 1 L of solution, the concentration of the solution can be expressed as either 1 $M$ or 5 $N$. Normality is always a whole-number multiple of molarity. In oxidation-reduction reactions, the whole number is the number of electrons gained or lost by one formula unit of the substance. In acid-base reactions the whole number is the number of $H^+$ or $OH^-$ available in a formula unit of the substance. If $H_2SO_4$ is used as an acid, forming $SO_4{}^{2-}$ as a product, it will lose *two* $H^+$ ions. Thus its normality will be *twice* its molarity. For example, a 0.255 $M$ solution of $H_2SO_4$ will be $2 \times 0.255 = 0.510\ N$.

---

**SAMPLE EXERCISE 12.6**

How many milliliters of 0.200 $N$ $KMnO_4$ solution are required to oxidize 25.0 mL of 0.120 $N$ $FeSO_4$ solution?

**Solution:**   At the equivalence point, equal numbers of equivalents of $KMnO_4$ and $FeSO_4$ must react. Rearranging Equation 12.5, solving for equivalents, we have

Equivalents solute = normality × liters soln

Thus the number of equivalents of $FeSO_4$ is

Equivalents $FeSO_4$

$$= \left(0.120\ \frac{\text{equivalents}}{L}\right)(0.0250\ L)$$
$$= 3.00 \times 10^{-3}\ \text{equivalent}$$

This must also be the number of equivalents of $KMnO_4$ reacted. Thus rearranging Equation 12.5, solving for liters, we have

$$\text{Liters soln} = \text{equivalents solute} \times \frac{1}{N}$$

The volume of $KMnO_4$ used is therefore

$$\text{Liters soln} = (3.00 \times 10^{-3}\ \text{equivalent})$$
$$\times \left(\frac{1\ L}{0.200\ \text{equivalent}}\right)$$
$$= 0.0150\ L = 15.0\ mL$$

This exercise illustrates one advantage of using normality over molarity in stoichiometric calculations: We do not need to refer to a completely balanced chemical equation.

**PRACTICE EXERCISE**

**(a)** How many mL of 0.100 $N$ $Ba(OH)_2$ solution is required to neutralize 25.00 mL of 0.100 $N$ HCl solution? **(b)** What is the normality of 0.050 $M$ hydrobromic acid, HBr, solution? **(c)** What is the normality of 0.010 $M$ $Ba(IO_3)_2$ solution? (Iodate, $IO_3{}^-$, is reduced to $I^-$, a reduction of six electrons per iodine atom.)
***Answers:*** **(a)** 25.00 mL; **(b)** 0.050 $N$ **(c)** 0.120 $N$

---

## 12.2 THE SOLUTION PROCESS

A solution is formed when one substance disperses uniformly throughout another. With the exception of gas mixtures, all solutions involve substances in a condensed phase. We learned in Chapter 11 that substances in the liquid and solid state experience intermolecular attractive forces that hold the individual particles together. Intermolecular forces also operate between a solute particle and the solvent that surrounds it.

Any of the various kinds of intermolecular forces that we discussed in Chapter 11 can operate between solute and solvent particles in a solution. As a general rule, we expect solutions to form when the attractive forces between solute and solvent are comparable in magnitude with those that exist between the solute particles themselves or between the solvent particles themselves. For example, the ionic substance NaCl dissolves readily in water. When NaCl is added to water, the water molecules orient themselves on the surface of the NaCl crystals as shown in Figure 12.1. The positive end of the water dipole is oriented toward the $Cl^-$ ions, while

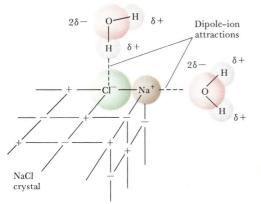

**FIGURE 12.1** Interactions between $H_2O$ molecules and the $Na^+$ and $Cl^-$ ions of a NaCl crystal.

the negative end of the water dipole is oriented toward the $Na^+$ ions. The ion-dipole attractions between $Na^+$ and $Cl^-$ ions and water molecules are sufficiently strong to pull these ions from their positions in the crystal. Notice that the corner $Na^+$ ion is held in the crystal by only three adjacent $Cl^-$ ions. In contrast, an $Na^+$ ion on the edge of the crystal has four nearby $Cl^-$ ions, and a $Na^+$ ion in the interior of the crystal has six surrounding $Cl^-$ ions. The corner $Na^+$ ion is therefore particularly vulnerable to removal from the crystal. Once this $Na^+$ ion has been removed, adjacent $Cl^-$ ions are similarly exposed and are therefore removed more easily than before.

Now separated from the crystal, the $Na^+$ and $Cl^-$ ions are surrounded by water molecules, as shown in Figure 12.2. Such interactions between solute and solvent molecules are known as **solvation**. When the solvent is water the interactions are known as **hydration**.

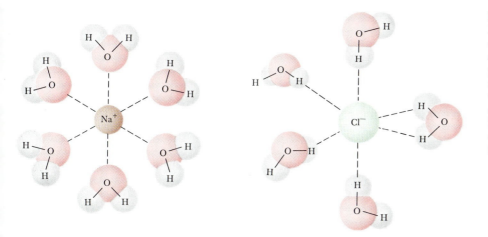

**FIGURE 12.2** Hydrated $Na^+$ and $Cl^-$ ions. The negative ends of the water dipole point toward the positive ion. The positive ends of the water dipole point toward the negative ion. We do not know whether one or both positive hydrogens are oriented toward the negative ion.

Sodium chloride dissolves in water because the water molecules have a sufficient attraction for the $Na^+$ and $Cl^-$ ions to overcome the attraction of these two ions for one another in the crystal. To form an aqueous solution of NaCl, water molecules must also separate from one another to make room for the solute particles. This example suggests that there are three attractive interactions involved in solution formation:

**The Energies of Solution Formation**

Frequently, hydrated ions remain in crystalline salts that are obtained by evaporation of water from aqueous solutions. Common examples include $FeCl_3 \cdot 6H_2O$ [iron(III) chloride hexahydrate] and $CuSO_4 \cdot 5H_2O$ [copper(II) sulfate pentahydrate]. The $FeCl_3 \cdot 6H_2O$ consists of $Fe(H_2O)_6^{3+}$ and $Cl^-$ ions; the $CuSO_4 \cdot 5H_2O$ consists of $Cu(H_2O)_4^{2+}$ and $SO_4(H_2O)^{2-}$ ions. Water molecules can also occur in positions in the crystal lattice that are not specifically associated with either a cation or anion. $BaCl_2 \cdot 2H_2O$ (barium chloride dihydrate) is an example. Compounds such as $FeCl_3 \cdot 6H_2O$, $CuSO_4 \cdot 5H_2O$, and $BaCl_2 \cdot 2H_2O$, which contain a salt and water combined in definite proportions, are known as *hydrates;* the water associated with them is called *water of hydration.* Figure 12.3 shows an example of a hydrate and the corresponding anhydrous (water-free) substances.

**FIGURE 12.3** Samples of hydrated copper sulfate, $CuSO_4 \cdot 5H_2O$ (left) and anhydrous $CuSO_4$. (Dr. E. R. Degginger)

**1.** Solute-solute interactions
**2.** Solvent-solvent interactions
**3.** Solute-solvent interactions

The net enthalpy change in forming a solution is the sum of the enthalpy changes that correspond to each of these three interactions. To appreciate this more clearly, let's think of the solution process as the sum of three separate processes, each corresponding to one of the interactions listed above. The net enthalpy change is then the sum of three enthalpy terms:

$$\Delta H_{soln} = \Delta H_1 + \Delta H_2 + \Delta H_3$$

Figure 12.4 depicts the enthalpy change for each process. Separation of the solute particles from one another means overcoming the attractive interactions between them. This is therefore an *endothermic* process, with enthalpy change $\Delta H_1$. The enthalpy of the system increases. The separation of solvent particles from one another is similarly an endothermic process, with enthalpy change $\Delta H_2$. The enthalpy of the system increases still further. Now, however, having paid the price of readying solvent and solute particles, we have an exothermic process, with enthalpy change $\Delta H_3$, as solute and solvent particles interact to form the solution.

As shown in Figure 12.4, $\Delta H_{soln}$ can be either exothermic or endothermic. For example, when sodium hydroxide, NaOH, is added to water, the resultant solution gets quite warm; $\Delta H_{soln}$ is $-44.48$ kJ/mol. By contrast, the dissolution of ammonium nitrate, $NH_4NO_3$, is endothermic, $\Delta H_{soln} = 26.4$ kJ/mol. Ammonium nitrate has been used to make instant ice packs, which are used to treat athletic injuries (Figure 12.5).

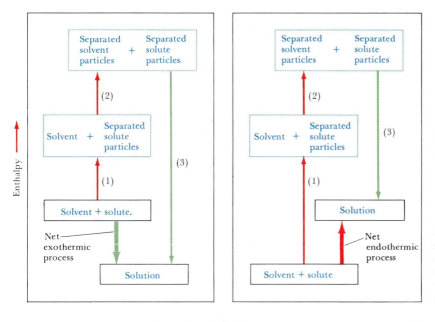

**FIGURE 12.4** Analysis of the enthalpy changes accompanying the solution process: process (1) represents the enthalpy required to separate solute particles; process (2) represents the enthalpy required to separate solvent particles; process (3) represents the enthalpy released when solute and solvent particles interact with each other. The figure on the left shows a net exothermic heat of solution, whereas the one on the right shows a net endothermic heat of solution.

**FIGURE 12.5** Photo of instant ice pack, ammonium nitrate, used to treat athletic injuries. To activate the ice pack, the container is kneaded, breaking the seal separating solid $NH_4NO_3$ from water. (© Richard Megna/ Fundamental Photographs)

A solution will not form if $\Delta H_{soln}$ is greatly endothermic. The solvent-solute interaction must be strong enough to make $\Delta H_3$ comparable in quantity to $\Delta H_1 + \Delta H_2$. Thus we can understand why ionic solutes like NaCl do not dissolve in nonpolar liquids like gasoline. The nonpolar hydrocarbon molecules of the gasoline would experience only weak attractive interactions with the ions, and these interactions would not go very far toward compensating for the energies required to separate the ions from one another.

By similar reasoning we can understand why polar liquids like water do not form solutions with nonpolar liquids like carbon tetrachloride, CCl₄. The water molecules experience strong hydrogen-bonding interactions with one another (Section 11.4); these attractive forces would need to be overcome to disperse the water molecules throughout the nonpolar liquid. The energy that it costs to separate the $H_2O$ molecules is not recovered in the form of attractive interactions between $H_2O$ and $CCl_4$ molecules.

When two nonpolar substances like $CCl_4$ and hexane, $C_6H_{14}$, are mixed, they readily dissolve in one another in all proportions. The attractive forces between molecules in both of these substances are London dispersion forces. The two substances have similar boiling points: $CCl_4$ boils at 77°C and $C_6H_{14}$ boils at 69°C. Thus it is reasonable to suppose that the magnitudes of the attractive forces between molecules are comparable in the two substances. When the two are mixed, there is little or no energy change. Yet the dissolving process occurs spontaneously; that is, it occurs to an appreciable extent without any extra input of energy from outside the system. Two distinct factors are involved in processes that occur spontaneously. The most obvious is energy, the other is disorder.

If you let go of a book you have in hand, it falls to the floor. The book falls because of gravity. At its initial height it has a potential energy higher than when it is on the floor. Unless it is restrained, the book falls and so loses energy. This fact leads us to the first basic principle identifying spontaneous processes and the direction they take: *Processes in which the energy content of the system decreases tend to occur spontaneously.* Spontaneous processes tend to be exothermic. Change occurs in the direction that leads to a lower energy content.

However, we can think of processes that do not result in a lower energy—or that may even be endothermic—and yet still occur spontaneously. The mixing of $CCl_4$ and $C_6H_{14}$ provides a simple example. All such processes are characterized by an increase in the disorder, or randomness, of the system. Suppose that we could suddenly remove a barrier that separated 500 mL of $CCl_4$ from 500 mL of $C_6H_{14}$, as in Figure 12.6(*a*). Before the barrier is removed each liquid occupies a volume of 500 mL. We know that we can find all the $CCl_4$ molecules in the 500 mL to the left, all the $C_6H_{14}$ molecules in the 500 mL to the right of the barrier. When equilibrium has been established after removal of the barrier, the two liquids together occupy a volume of about 1000 mL.* Formation of a homogeneous solution has resulted in an increased disorder, or randomness, in that the molecules of each substance are now distributed in a volume twice as large as that which they occupied before mixing. This example illustrates our second basic principle: *Processes in which the disorder of the system increases tend to occur spontaneously.*

When molecules of different types are brought together, an increase in disorder occurs spontaneously unless the molecules are restrained by sufficiently strong intermolecular forces or by physical barriers. Thus, because of the strong bonds holding the sodium and chloride ions together, sodium chloride does not spontaneously dissolve in gasoline. Conversely, gases

## Solubility and Disorder

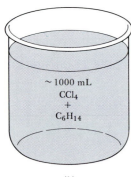

**FIGURE 12.6** Formation of a homogeneous solution between $CCl_4$ and $C_6H_{14}$ upon removal of a barrier separating the two liquids. The solution in (*b*) is more disordered, or random in character, than the separate liquids before solution formation (*a*).

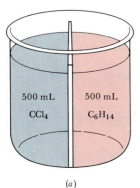

500 mL
CCl₄

500 mL
C₆H₁₄

(*a*)

~ 1000 mL
CCl₄
+
C₆H₁₄

(*b*)

* The slight change in total volume that may occur upon mixing is unimportant for our example.

spontaneously expand unless restrained by their containers; in this case intermolecular forces are too weak to restrain the molecules. We shall discuss spontaneous processes again in Chapter 19. At that time we shall consider the balance between the tendencies toward lower energy and toward increased disorder in greater detail. For the moment we need to be aware that formation of a solution is always favored by the increase in disorder that accompanies mixing. Consequently, a solution will form unless solute-solute or solvent-solvent interactions are too strong relative to the solute-solvent interactions.

In all our discussions of solutions we must be careful to distinguish the *physical* process of solution formation from *chemical* processes that lead to a solution. For example, zinc metal is dissolved on contact with hydrochloric acid solution. The following chemical reaction occurs:

$$Zn(s) + 2HCl(aq) \longrightarrow H_2(g) + ZnCl_2(aq) \qquad [12.6]$$

In this instance the chemical form of the substance being dissolved is changed. If the solution is evaporated to dryness, $Zn(s)$ is not recovered as such; instead, $ZnCl_2(s)$ is recovered. In contrast, when $NaCl(s)$ is dissolved in water, it can be recovered by evaporation of its solution to dryness. Our focus throughout this chapter is on solutions from which the solute can be recovered unchanged from the solution.

As the solution process involving a solid solute proceeds, the concentration of solute particles in solution increases, so the chances of their colliding with the surface of the solid increases (Figure 12.7). Such a collision may result in the solute particle becoming attached to the solid. This process, which is the opposite of the solution process, is called **crystallization**. Thus two opposing processes occur in a solution in contact with undissolved solute. This situation is represented in Equation 12.7 by use of a double arrow:

$$\text{Solute + solvent} \underset{\text{crystallize}}{\overset{\text{dissolve}}{\rightleftharpoons}} \text{solution} \qquad [12.7]$$

When the rates of these opposing processes become equal, no further net increase in the amount of solute in solution occurs. This balance is an example of **dynamic equilibrium**, similar to that discussed in Section 11.2, where the processes of evaporation and condensation were considered. A solution that is in equilibrium with undissolved solute is said to be **saturated**. Additional solute will not dissolve if added to a saturated solution. The amount of solute needed to form a saturated solution in a given quantity of solvent is known as the **solubility** of that solute. For example, the solubility of NaCl in water at 0°C is 35.7 g per 100 mL of water. This is the maximum amount of NaCl that can be dissolved in water to give a stable equilibrium solution at that temperature. If less solute is added than the equilibrium amount, the solution is said to be **unsaturated**. In some cases it is possible to prepare solutions that contain more solute than the equilibrium amount. Such solutions, which are said to be **supersaturated**, are unstable, and under the proper conditions solute will crystallize from them to give saturated solutions. An experimental demonstration of supersaturation is shown in Figure 12.8.

**Solubility**

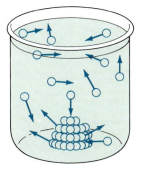

**FIGURE 12.7** Movement of solute particles in solvent containing excess solute. Both dissolution and crystallization occur.

(a)  (b)  (c)  (d)

**FIGURE 12.8**  Sodium acetate readily forms supersaturated solutions in water. (a) The amount of $NaC_2H_3O_2$ shown completely dissolves in hot water. (b) The resulting solution becomes supersaturated when it cools to room temperature. (c) Crystallization of $NaC_2H_3O_2$ from the supersaturated solution begins when a tiny seed crystal is added. (d) Soon all of the excess solution has crystallized from solution. (Donald Clegg and Roxy Wilson)

## 12.3  FACTORS AFFECTING SOLUBILITY

As the discussion in the preceding section indicates, the extent to which one substance dissolves in another depends on the nature of both the solute and the solvent. It also depends on temperature and, at least for gases, on pressure.

### Solubilities and Molecular Structure

Our discussion of the factors that lead to formation of solutions enables us to understand many observations regarding solubilities. As a simple example, consider the data in Table 12.3 for the solubilities of various simple gases in water. Note that solubility increases with increasing molecular mass. The attractive forces between the gas and solvent molecules are mainly of the London-dispersion-force type, which increase with increasing size and mass of the gas molecules. When chemical reaction occurs between the gas and solvent, much higher gas solubilities result. We will encounter instances of this in later chapters, but as a simple example, the solubility of $Cl_2$ in water under the same conditions given in Table 12.3 is $0.102\ M$. This is much higher than would be predicted from the trends in the table, based just on molecular mass. We can infer from this that the dissolving of $Cl_2$ in water is accompanied by some kind of chemical process. The use of chlorine as a bactericide in municipal water supplies and swimming pools is based on its chemical reaction with water (Section 14.6, Equation 14.18).

**TABLE 12.3**  Solubilities of several gases in water at 20°C, with 1 atm gas pressure

| Gas | Solubility ($M$) |
|-----|------------------|
| $N_2$ | $6.9 \times 10^{-4}$ |
| CO | $1.04 \times 10^{-3}$ |
| $O_2$ | $1.38 \times 10^{-3}$ |
| Ar | $1.50 \times 10^{-3}$ |
| Kr | $2.79 \times 10^{-3}$ |

Polar liquids tend to dissolve readily in polar solvents. For example, acetone, a polar molecule whose structure is shown at left, mixes in all proportions with water. Pairs of liquids that mix in all proportions are said to be **miscible**, and liquids that do not mix are termed **immiscible**. Water and hexane, $C_6H_{14}$, for example, are immiscible. Diethyl ketone (see structure at left), which is similar to acetone but has a higher molecular mass, dissolves in water to the extent of about 47 g per liter of water at 20°C, but it is not completely miscible.

Hydrogen-bonding interactions between solute and solvent may lead to high solubility. For example, water is completely miscible with ethanol, $CH_3CH_2OH$. The $CH_3CH_2OH$ molecules are able to form hydrogen bonds with water molecules as well as with each other (see Figure 12.9).

$$\overset{\displaystyle O}{\underset{\displaystyle \|}{}}$$
$$CH_3CCH_3$$

Acetone

$$\overset{\displaystyle O}{\underset{\displaystyle \|}{}}$$
$$CH_3CH_2CCH_2CH_3$$

Diethyl ketone

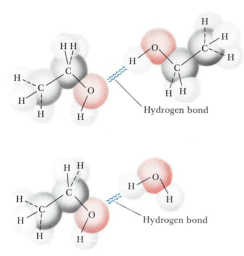

**FIGURE 12.9** Hydrogen-bonding interactions between ethanol molecules and between water and ethanol molecules.

Because of this hydrogen-bonding ability, the solute-solute, solvent-solvent, and solute-solvent forces are not appreciably different within a mixture of $CH_3CH_2OH$ and $H_2O$. There is no significant change in the environment of the molecules as they are mixed. The increase in disorder accompanying mixing therefore plays a significant role in formation of the solution.

The number of carbon atoms in an alcohol affects its solubility in water, as shown in Table 12.4. As the length of the carbon chain increases, the OH group becomes an ever smaller part of the molecule and the molecule becomes more like a hydrocarbon. The solubility of the alcohol decreases correspondingly. If the number of OH groups along the carbon chain increases, more solute-water hydrogen bonding is possible and solubility generally increases. Glucose, $C_6H_{12}O_6$, has five OH groups on a six-carbon framework, which makes the molecule very soluble in water (83 g dissolve in 100 mL of water at 17.5°C). The glucose molecule is shown in Figure 12.10.

Examination of pairs of substances such as those listed in the preceding paragraphs has led to an important generalization: *Substances with similar intermolecular attractive forces tend to be soluble in one another*. This generalization is often simply stated as *"likes dissolve likes."* Nonpolar substances are soluble in nonpolar solvents; ionic and polar solutes are soluble in polar

**TABLE 12.4** Solubilities of some alcohols in water

| Alcohol | Solubility in $H_2O$ (mol/100 g $H_2O$ at 20°C) |
|---------|---------------------------------------|
| $CH_3OH$ (methanol) | $\infty$[a] |
| $CH_3CH_2OH$ (ethanol) | $\infty$ |
| $CH_3CH_2CH_2OH$ (propanol) | $\infty$ |
| $CH_3CH_2CH_2CH_2OH$ (butanol) | 0.11 |
| $CH_3CH_2CH_2CH_2CH_2OH$ (pentanol) | 0.030 |
| $CH_3CH_2CH_2CH_2CH_2CH_2OH$ (hexanol) | 0.0058 |
| $CH_3CH_2CH_2CH_2CH_2CH_2CH_2OH$ (heptanol) | 0.0008 |

[a] The infinity symbol indicates that there is no real limit to the solubility of this alcohol in water.

**FIGURE 12.10** Structure of glucose. Colored spheres indicate sites capable of hydrogen bonding with water.

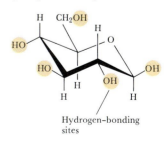

Hydrogen–bonding sites

solvents. Network solids like diamond and quartz are not soluble in either polar or nonpolar solvents because of the strong bonding forces within the solid.

## SAMPLE EXERCISE 12.7

Predict whether each of the following substances is more likely to dissolve in carbon tetrachloride, $CCl_4$, which is a nonpolar solvent, or in water: $C_7H_{16}$, $NaHCO_3$, HCl, $I_2$.

**Solution:** Both $C_7H_{16}$ and $I_2$ are nonpolar. We would therefore predict that they would be more soluble in $CCl_4$ than in $H_2O$. $NaHCO_3$ is ionic and HCl is polar covalent. Water would be a better solvent than $CCl_4$ for these two substances.

## PRACTICE EXERCISE

Which of the following is most likely to be miscible with water?

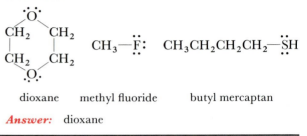

dioxane        methyl fluoride        butyl mercaptan

*Answer:*   dioxane

---

**Effect of Pressure on Solubility**

The solubility of a gas in any solvent is increased as the pressure of the gas over the solvent increases. By contrast, the solubilities of solids and liquids are not appreciably affected by pressure. We can understand the effect of pressure on the solubility of a gas by considering the equilibrium that operates, illustrated in Figure 12.11. Suppose that we have a gaseous substance distributed between the gas and solution phases. When equilibrium is established, the rate at which gas molecules enter the solution equals the rate at which solute molecules escape from the solution to enter the gas phase. The small arrows in Figure 12.11(a) represent the rate of these opposing processes. This is another example of dynamic equilibrium, which we have encountered before (Section 11.2). Now suppose that we exert added pressure on the piston and compress the gas above the solution, as shown in Figure 12.11(b). If we reduce the volume to half its original value, the pressure of the gas would increase to about twice its original value. But this would mean that the rate at which gas molecules strike the surface to enter the solution phase would increase. Thus the solubility of the gas in the solution would increase until there

**FIGURE 12.11**   Effect of pressure on the solubility of a gas. When the pressure is increased, as in (b), the rate at which gas molecules enter the solution increases. The concentration of solute molecules at equilibrium increases in proportion to the pressure.

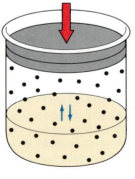
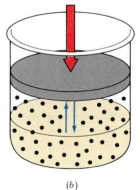

(a)                    (b)

Vitamins B and C are water soluble. Vitamins A, D, E, and K are soluble in nonpolar solvents and in the fatty tissue of the body (which is nonpolar). Because of their water solubility, vitamins B and C are not stored to any appreciable extent in the body, and so foods containing these vitamins should be included in the daily diet. In contrast, the fat-soluble vitamins are stored in sufficient quantities to keep vitamin-deficiency diseases from appearing even after a person has subsisted for a long period on a vitamin-deficient diet. With the ready availability of vitamin supplements, cases of hypervitaminosis, an illness caused by an excessive amount of vitamins, are now being seen by physicians in this country. Because the body can store only the fat-soluble vitamins,

true hypervitaminosis has been observed solely for these vitamins.

The different solubility patterns of the water-soluble vitamins and the fat-soluble ones can be rationalized in terms of the structures of the molecules. The chemical structures of vitamin A (retinol) and of vitamin C (ascorbic acid) are shown below. Note that the vitamin A molecule is an alcohol with a very long carbon chain. It is nearly nonpolar, and because the OH group is such a small part of the molecule, the molecule resembles the long-chain alcohols listed in Table 12.4. In contrast, the vitamin C molecule is smaller and has more OH groups that can form hydrogen bonds with water. It is somewhat like glucose, discussed above.

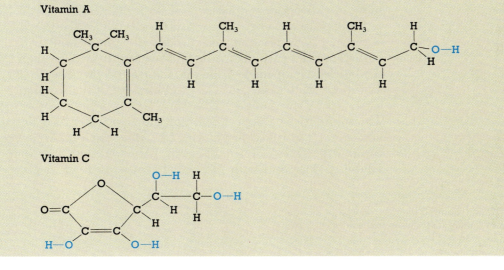

Vitamin A

Vitamin C

is again an equilibrium; that is, solubility increases until the rates at which gas molecules enter the solution equal the rate at which solute molecules escape from the solvent, as indicated by the arrows in Figure 12.11(b). This means that the solubility of the gas should increase in direct proportion to the pressure. The relationship between pressure and solubility is expressed in terms of a simple equation known as **Henry's law**:

$$C_g = kP_g \qquad\qquad [12.8]$$

where $C_g$ is the solubility of the gas in the solution phase, $P_g$ is the pressure of the gas over the solution, and $k$ is a proportionality constant, known as the Henry's law constant. As an example, the solubility of pure nitrogen gas in water at 25°C and 0.78 atm pressure is $5.3 \times 10^{-4}$ $M$. If

Deep-sea divers rely on compressed air for their oxygen supply. According to Henry's law, the solubilities of gases increase with pressure. If a diver is suddenly exposed to atmospheric pressure, where the solubility of gases is less, bubbles form in the bloodstream and in other fluids of the body. These bubbles affect nerve impulses and give rise to the disease known as "the bends", or decompression sickness. Nitrogen is the main problem because it has the highest partial pressure in air and because it can be removed only through the respiratory system. Oxygen is consumed in metabolism. Substitution of helium for nitrogen minimizes this effect,

because helium has a much lower solubility in biological fluids than does $N_2$. Cousteau's divers on *Conshelf III* used a mixture of 98 percent helium and 2 percent oxygen. At the high pressures (10 atm) experienced by the divers, this percentage of oxygen gives an oxygen partial pressure of about 0.2 atm, which is the partial pressure in normal air at 1 atm. If the oxygen partial pressures become too great, the urge to breathe is reduced, $CO_2$ is not removed from the body, and $CO_2$ poisoning occurs. At excessive concentrations in the body, carbon dioxide acts as a neurotoxin, interfering with nerve conduction and transmission.

the partial pressure of $N_2$ is doubled, Henry's law predicts that the solubility in water is also doubled to $1.06 \times 10^{-3}\ M$.

Bottlers use the effect of pressure on solubility in producing carbonated beverages such as champagne, beer, and many soft drinks. These are bottled under a carbon dioxide pressure slightly greater than 1 atm. When the bottles are opened to the air, the partial pressure of $CO_2$ above the solution is decreased, and $CO_2$ bubbles out of the solution.

**Effect of Temperature on Solubility**

The solubilities of several common gases in water as a function of temperature are graphed in Figure 12.12. These solubilities correspond to a constant pressure of 1 atm of gas over the solution. Note that, in general, solubility decreases with increasing temperature. If a glass of cold tap

**FIGURE 12.12**   Solubilities of several gases in water as a function of temperature. Note that solubilities are in units of millimoles per liter, for a constant pressure of 1 atm in the gas phase.

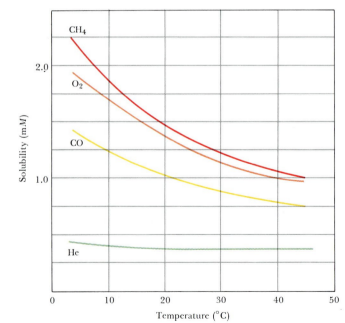

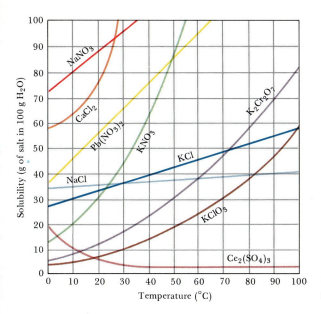

**FIGURE 12.13** Solubilities of several common ionic solids shown as a function of temperature.

water is warmed, bubbles of air are seen on the side of the glass. Similarly, a carbonated beverage like soda pop goes "flat" as it is allowed to warm; as the temperature of the solution increases, $CO_2$ escapes from the solution. The decreased solubility of $O_2$ in water as temperature increases is one of the effects of the thermal pollution of lakes and streams. The effect is particularly serious in deep lakes, because warm water is less dense than cold water. It therefore tends to remain on top of the cold water, at the surface. This situation impedes the dissolving of oxygen into the deeper layers, thus stifling the respiration of all aquatic life needing oxygen. Fish may suffocate and die in these circumstances.

Figure 12.13 shows the effect of temperature on the solubilities of several ionic substances in water. There are few exceptions to the general rule that the solubility of ionic compounds in water increases with increasing temperature.

## 12.4 ELECTROLYTE SOLUTIONS

Solutes can be classified according to whether or not they exist in aqueous solutions as ions. Those that do are called **electrolytes**. A familiar example is sodium chloride, whose solid consists of $Na^+$ and $Cl^-$ ions. This substance dissolves in water to give hydrated $Na^+$ and $Cl^-$ ions (Section 12.2). A solution labeled 1.0 $M$ NaCl actually contains 1.0 mol of $Na^+$ and 1.0 mol of $Cl^-$ per liter of solution. Because all NaCl present in solution is in the form of ions, NaCl is said to be completely ionized and is called a **strong electrolyte**. Not surprisingly, ionic compounds form electrolyte solutions.

Molecular compounds that contain highly polar covalent bonds may also dissolve in water to form ions. For example, hydrogen chloride gas, HCl, dissolves readily in water to form an aqueous solution that we know as hydrochloric acid. The HCl in that solution exists entirely as $H^+$ and $Cl^-$ ions. (We will take a closer look at this ionization process in Section 13.4.) Other compounds ionize incompletely on dissolving in water. An example is $HgCl_2$, which participates in the following equilibrium:

$$HgCl_2(aq) \rightleftharpoons HgCl^+(aq) + Cl^-(aq) \qquad [12.9]$$

Only a small fraction of the $HgCl_2$ that is dissolved is present in solution as ions. Compounds such as this one, which ionize only partially in solution, are called **weak electrolytes**. Finally, those substances that undergo no ionization in solution are called **nonelectrolytes**. Examples include nonpolar gases such as $O_2$ and most organic compounds, such as glucose, $C_6H_{12}O_6$, and ethanol, $C_2H_5OH$. These substances maintain their molecular structures rather than ionize in solution.

We can experimentally determine whether a substance forms an electrolyte solution by testing the ability of the solution to conduct an electrical current. Pure water itself is a poor conductor. However, solute ions carry

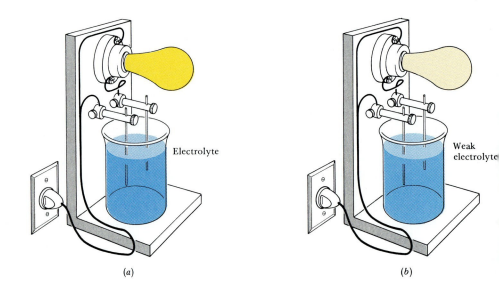

(a)  Electrolyte

(b)  Weak electrolyte

**FIGURE 12.14** A device for distinguishing strong electrolyte, weak electrolyte, and nonelectrolyte solutions. In (a) the bulb glows brightly because the ions present in the strong electrolyte solution provide a large number of electrical current carriers. In (b) the bulb glows weakly because the solution of a weak electrolyte has comparatively few ions to serve as current carriers. In (c) the bulb does not glow at all because the nonelectrolyte solution has no charged species to serve as current carriers.

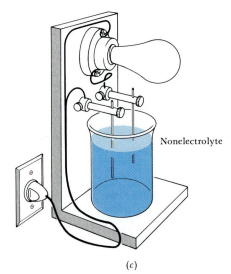

Nonelectrolyte

(c)

electrical charge through the solution when a voltage is placed across the solution. A device such as that shown in Figure 12.14 can be used to test whether or not ions are present, provided that the solutions are not too dilute. If ions are present, the solution completes the electrical circuit, and the light bulb glows. If we use solutions of the same molarity for comparison, it is even possible to roughly distinguish solutions of strong electrolytes from those of weak electrolytes. The bulb glows more brightly for a solution of strong electrolytes, because a higher concentration of ions is present in the solution to carry the current.

It is very important to distinguish a substance's solubility from whether it is a strong or weak electrolyte. For example, silver chloride, AgCl, is only very slightly soluble in water. Therefore, if we shake some solid AgCl with water for a long time until equilibrium is established, most of the AgCl will remain as an insoluble solid. All of the AgCl that does dissolve, however, is present in solution as $Ag^+$ and $Cl^-$ ions. We say, therefore, that AgCl is a strong electrolyte.

In Section 3.2 we introduced acids, bases, and salts. You should review that section at this time. We can distinguish acids and bases according to the degree to which they ionize in solution. Acids and bases that are completely ionized in solution are referred to as **strong acids** and **strong bases**. Those that are partly ionized are referred to as **weak acids** and **weak bases**. The terms weak acid and weak base have no relation to how reactive the acid or base is. Hydrofluoric acid, HF, is a weak electrolyte and is therefore called a weak acid. A 0.1 $M$ solution of HF is 8 percent ionized. However, HF is a very reactive acid that vigorously attacks many substances, including glass.

The following generalizations are useful in recognizing which substances are strong electrolytes and which weak:

1. Most **salts** (that is, ionic compounds) are strong electrolytes. A few of the heavy metals form weak electrolytes, as in the example of $HgCl_2$ cited in Equation 12.9.
2. Most **acids** are weak electrolytes. The common strong acids are HCl, HBr, HI, $HNO_3$, $H_2SO_4$, and $HClO_4$.
3. The common strong **bases** are the hydroxides of Li, Na, K, Rb, and Cs (the alkali metals, group 1A) and the hydroxides of Ca, Sr, and Ba (the heavy alkaline earths, group 2A). $NH_3$ is a weak electrolyte.

## SAMPLE EXERCISE 12.8

Classify each of the following substances as nonelectrolyte, weak electrolyte, or strong electrolyte: $CaCl_2$, $HNO_3$, $CH_3OH$ (methanol), $HC_2H_3O_2$ (acetic acid), KOH, $H_2O_2$.

**Solution:** Only one of the substances, $CaCl_2$, is a salt. It is a strong electrolyte. Two of the substances, $HNO_3$ and $HC_2H_3O_2$, are acids. $HNO_3$ is a common strong acid (strong electrolyte). Because $HC_2H_3O_2$ is not a common strong acid, our best guess would be that it is a weak acid (weak electrolyte). This is correct. There is one base, KOH. It is

one of the common strong bases (a strong electrolyte) because it is a hydroxide of an alkali metal. The remaining compounds, $CH_3OH$ and $H_2O_2$, are neither acids, bases, nor salts. They are nonelectrolytes.

**PRACTICE EXERCISE**
Predict which one of the following 0.1 $M$ solutions in water will cause the light bulb in the apparatus of Figure 12.14 to glow most brightly: $C_2H_5OH$, $HC_2H_3O_2$ (acetic acid), $NaC_2H_3O_2$.
*Answer:* $NaC_2H_3O_2$

## SAMPLE EXERCISE 12.9

What is the concentration of all species in a 0.1 $M$ solution of $Ba(NO_3)_2$?

**Solution:** $Ba(NO_3)_2$ is a salt, so we expect it to be a strong electrolyte. It ionizes into $Ba^{2+}$ and the polyatomic nitrate ion, $NO_3^-$. A 0.1 $M$ solution of $Ba(NO_3)_2$ is 0.1 $M$ in $Ba^{2+}$ and 0.2 $M$ in $NO_3^-$ [since

1 mol of $Ba(NO_3)_2$ supplies 1 mol of $Ba^{2+}$ and 2 mol of $NO_3^-$].

### PRACTICE EXERCISE

Which of the following 0.1 $M$ solutions has the highest total concentration of ions: $H_3PO_4$, LiCl, $Na_3PO_4$, HCl?

***Answer:*** $Na_3PO_4$

---

## 12.5 THE VAPOR PRESSURES OF SOLUTIONS

Some physical properties of solutions differ in important ways from those of the pure solvent. For example, pure water freezes at 0°C, but aqueous solutions freeze at lower temperatures. Ethylene glycol, added to the water in the radiators of cars as an antifreeze, also raises the boiling point of the solution above that for pure water. This effect reduces the loss of coolant in hot weather and permits operation of the engine at a higher temperature.

The influence of a nonvolatile solute on the freezing and boiling points of a solution is related to its effect on the vapor pressure of the solvent. Let's consider the simple experiment illustrated in Figure 12.15. Two beakers are placed side by side in a sealed enclosure called a bell jar. One beaker contains pure water, the other an equal volume of an aqueous solution of sugar. Gradually, the volume of the sugar solution increases, while the volume of the pure water decreases. With time, all the water transfers to the sugar solution, as shown in Figure 12.15(*b*). How do we explain this result?

Let's refer back to Section 11.2, particularly to Figure 11.2. Recall that the vapor pressure over a liquid is the result of a dynamic equilibrium; the rate at which gas-phase molecules return to the surface of the liquid is equal to the rate at which molecules leave the liquid surface for the gas phase. A nonvolatile solute added to the liquid phase reduces the capacity of the solvent molecules to move from the liquid phase to the

**FIGURE 12.15** Experiment showing that a solution possesses a lower vapor pressure than does the pure solvent.

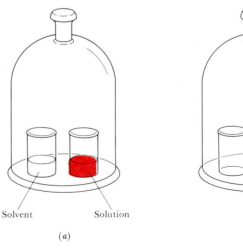

Solvent     Solution

(*a*)

Solution

(*b*)

vapor phase. The result is that solvent molecules leave the liquid phase for the gas phase at a lower rate from the solution than from the pure solvent. At the same time, however, there is no change in the rate at which solvent molecules in the gas phase return to the liquid phase. The shift in equilibrium due to the solute reduces the vapor pressure over the solution. The vapor pressure over the pure solvent is thus higher.

Return to the experiment shown in Figure 12.15. You should now see why the solvent transfers from the pure solvent beaker to the solution beaker. The vapor pressure necessary to achieve equilibrium with the pure solvent is higher than that required with the solution. Thus as the pure solvent seeks to reach equilibrium by forming vapor, the solution seeks to reach equilibrium by removing molecules from the vapor state. A net movement of solvent molecules in the gas phase from the pure solvent to the solution results. The process continues until no free solvent remains.

The *extent* to which a nonvolatile solute lowers the vapor pressure is proportional to its concentration. Doubling the concentration of solute doubles its effect. In fact, the reduction in vapor pressure is roughly proportional to the total concentration of solute particles, whether they are neutral or charged. For example, 1.0 mol of a nonelectrolyte, like glucose, produces essentially the same reduction in vapor pressure in a given quantity of water as does 0.5 mol of NaCl. Both solutions have 1.0 mol of particles because 0.5 mol of dissolved NaCl gives rise to 0.5 mol of $Na^+$ and 0.5 mol of $Cl^-$ ions.

Quantitatively, the vapor pressure of solutions containing nonvolatile solutes is given by **Raoult's law**, Equation 12.10, where $P_A$ is the vapor pressure of the solution, $X_A$ is the mole fraction of the solvent, and $P_A^\circ$ is the vapor pressure of the pure solvent:

**Raoult's Law**

$$P_A = X_A P_A^\circ \qquad [12.10]$$

For example, the vapor pressure of water is 17.5 mm Hg at 20°C. Imagine holding the temperature constant while adding glucose, $C_6H_{12}O_6$, to the water so that the resulting solution has $X_{H_2O} = 0.80$ and $X_{C_6H_{12}O_6} = 0.20$. According to Equation 12.10, the vapor pressure of water over the solution will be 14 mm Hg:

$$P_{H_2O} = (0.80)(17.5 \text{ mm Hg}) = 14 \text{ mm Hg}$$

The reduction in vapor pressure, $\Delta P_A$, is $P_A^\circ - P_A$. Using Equation 12.10, we can write this as

$$\Delta P_A^\circ = P_A^\circ - X_A P_A^\circ$$
$$= P_A^\circ (1 - X_A) \qquad [12.11]$$

Since $X_A + X_B = 1$, where $X_B$ is the mole fraction of solute particles, we can write

$$X_B = 1 - X_A$$
$$\Delta P_A = X_B P_A^\circ \qquad [12.12]$$

In our example, $X_{C_6H_{12}O_6} = 0.200$. The vapor pressure will be lowered by 3.5 mm Hg:

$$\Delta P_A = (0.20)(17.5 \text{ mm Hg}) = 3.5 \text{ mm Hg}$$

## SAMPLE EXERCISE 12.10

Calculate the reduction in vapor pressure caused by the addition of 100 g of sucrose, $C_{12}H_{22}O_{11}$, to 1000 g of water if the vapor pressure of the pure water at 25°C is 23.8 mm Hg.

**Solution:**

$$X_{C_{12}H_{22}O_{11}} = \frac{\text{moles } C_{12}H_{22}O_{11}}{\text{moles } H_2O + \text{moles } C_{12}H_{22}O_{11}}$$

$$\text{Moles } C_{12}H_{22}O_{11} = (100 \text{ g})\left(\frac{1 \text{ mol } C_{12}H_{22}O_{11}}{342 \text{ g } C_{12}H_{22}O_{11}}\right)$$

$$= 0.292 \text{ mol}$$

$$\text{Moles } H_2O = (1000 \text{ g})\left(\frac{1 \text{ mol } H_2O}{18.0 \text{ g } H_2O}\right)$$

$$= 55.5 \text{ mol}$$

$$X_{C_{12}H_{22}O_{11}} = \frac{0.292}{55.5 + 0.292} = \frac{0.292}{55.8}$$

$$\Delta P_A = \left(\frac{0.292}{55.8}\right)(23.8 \text{ mm Hg}) = 0.125 \text{ mm Hg}$$

## PRACTICE EXERCISE

The vapor pressure over pure water at 120°C is 1480 mm Hg. What mole fraction of ethylene glycol must be present in the water to reduce the vapor pressure of water to 1 atm, 760 mm Hg. Assume ideal solution behavior.

*Answer:* 0.486

It is important to keep in mind that Raoult's law applies to ideal solutions. Ideal behavior is approached when the solvent and solute are very much alike both in molecular size and in the strength and type of interactions between their molecules. In reality, solutions depart from ideal behavior to extents that depend on the particular solute and solvent. The solvent vapor pressure over a solution is greater than predicted by Raoult's law when the intermolecular forces between solvent and solute are weaker than between solvent and solvent and between solute and solute. Conversely, when the interactions between solute and solvent are exceptionally strong, as might be the case when hydrogen bonding or ion-solvent forces exist, the solvent vapor pressure is lower than Raoult's law predicts. In applying Raoult's law we will ignore departures from ideal behavior.

## 12.6 COLLIGATIVE PROPERTIES

Many properties of a solution depend not only on the concentration of solute but also on its particular nature. For example, the density of a solution depends on the identity of both solute and solvent as well as on the concentration of solute. However, for certain kinds of solutes, a number of physical properties of the solution depend on the number of solute particles present in solution but not on the identity of the solute. Such properties are called colligative properties. In addition to vapor-pressure lowering, these properties include boiling-point elevation, freezing-point depression, and osmotic pressure.

We saw in Section 11.3 that the vapor pressure of a liquid increases with increasing temperature; it boils when its vapor pressure is the same as the external pressure over the liquid. Because nonvolatile solutes lower the vapor pressure of the solution, a higher temperature is required to cause the solution to boil. Figure 12.16 shows the variation in vapor pressure of a pure liquid with temperature (the black line), as compared with a solution (the colored line).* Because the vapor pressure of the solution is lower than that of the solvent at all temperatures, in accordance with Raoult's law, a higher temperature is required to attain a vapor pressure of 1 atm. The increase in boiling point, $\Delta T_b$ (relative to the boiling point of the pure solvent), is directly proportional to the number of solute particles per mole of solvent molecules. We know that molality expresses the number of moles of solute per 1000 g of solvent, which represents a fixed number of moles of solvent. Thus $\Delta T_b$ is proportional to molality, as shown in

$$\Delta T_b = K_b m \qquad [12.13]$$

The magnitude of $K_b$, which is called the **molal boiling-point-elevation constant**, depends on the solvent. Some typical values for several common solvents are given in Table 12.5.

For water, $K_b$ is $0.52°C/m$; therefore, a $1\ m$ aqueous solution of sucrose or any other aqueous solution that is $1\ m$ in nonvolatile solute particles will boil at a temperature $0.52°C$ higher than pure water. It is important to notice that the boiling-point elevation is proportional to the number of solute particles present in a given quantity of solution. When NaCl dissolves in water, 2 mol of solute particles—1 mol of $Na^+$ and 1 mol of $Cl^-$—are formed for each mole of NaCl that dissolves. Thus a $1\ m$ solution of NaCl in water causes a boiling-point elevation twice as large as a $1\ m$ solution of a nonelectrolyte, like sucrose.

* Both lines refer to the usual laboratory situation, in which we have a constant total gas pressure of 1 atm over the system.

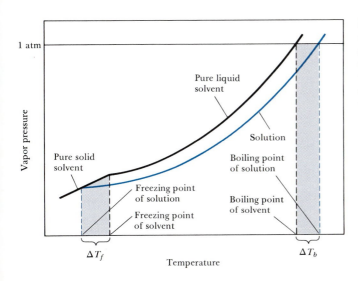

**FIGURE 12.16** Vapor-pressure versus temperature curves of a pure solvent and a solution of a nonvolatile solute, under a constant total pressure of 1 atm. The vapor pressure of the solid solvent is unaffected by the presence of solute if the solid freezes out without containing a significant concentration of solute, as is usually the case.

**TABLE 12.5** Molal boiling-point-elevation and freezing-point-depression constants

| Solvent | Normal boiling point (°C) | $K_b$ (°C/$m$) | Normal freezing point (°C) | $K_f$ (°C/$m$) |
|---------|---------------------------|----------------|----------------------------|----------------|
| Water, $H_2O$ | 100.0 | 0.52 | 0.0 | 1.86 |
| Benzene, $C_6H_6$ | 80.1 | 2.53 | 5.5 | 5.12 |
| Carbon tetrachloride, $CCl_4$ | 76.8 | 5.02 | −22.3 | 29.8 |
| Ethanol, $C_2H_5OH$ | 78.4 | 1.22 | −114.6 | 1.99 |
| Chloroform, $HCCl_3$ | 61.2 | 3.63 | −63.5 | 4.68 |

## Freezing-Point Depression

The freezing point corresponds to the temperature at which the vapor pressures of the solid and liquid phases are the same. The freezing point of a solution is lowered because the solute is not normally soluble in the solid phase of the solvent. For example, when aqueous solutions freeze, the solid that separates out is almost always pure ice. Thus the vapor pressure of the solid is unaffected by the presence of solute. However, if the solute is nonvolatile, the vapor pressure of the solution is reduced in proportion to the mole fraction of solute (Equation 12.12). This means that the temperature at which the solution and solid phases will have the same vapor pressure is reduced, as illustrated in Figure 12.16. Note that the point that represents the equilibrium between solid and liquid lies to the left of the corresponding point for pure solvent, at a lower temperature. Like the boiling-point elevation, the decrease in freezing point, $\Delta T_f$, is directly proportional to the molality of the solute:

$$\Delta T_f = K_f m \qquad [12.14]$$

The values of $K_f$, the **molal freezing-point-depression constant**, for several common solvents are given in Table 12.5. For water $K_f$ is 1.86°C/$m$; therefore, a 0.5 $m$ aqueous solution of NaCl or any aqueous solution that is 1 $m$ in nonvolatile solute particles will freeze 1.86°C lower than pure water. The freezing-point lowering caused by solutes explains the use of antifreeze (Sample Exercise 12.11) and the use of calcium chloride, $CaCl_2$, to melt ice on roads during winter.

## SAMPLE EXERCISE 12.11

Calculate the freezing point and the boiling point of a solution of 100 g of ethylene glycol ($C_2H_6O_2$) antifreeze in 900 g of $H_2O$.

**Solution:**

$$\text{Molality} = \frac{\text{moles } C_2H_6O_2}{\text{kilograms } H_2O}$$

$$= \left(\frac{100 \text{ g } C_2H_6O_2}{900 \text{ g } H_2O}\right)\left(\frac{1 \text{ mol } C_2H_6O_2}{62.0 \text{ g } C_2H_6O_2}\right)$$

$$\times \left(\frac{1000 \text{ g } H_2O}{1 \text{ kg } H_2O}\right)$$

$$= 1.79 \ m$$

$$\Delta T_f = K_f m = \left(1.86 \frac{°C}{m}\right)(1.79 \ m) = 3.33°C$$

Freezing point = (normal f.p. of solvent) − $\Delta T_f$
$$= 0.00°C - 3.33°C = -3.33°C$$

$$\Delta T_b = K_b m = \left(0.52 \frac{°C}{m}\right)(1.79 \ m) = 0.93°C$$

Boiling point = (normal b.p. of solvent) + $\Delta T_b$

= 100.00°C + 0.93°C

= 100.93°C

**PRACTICE EXERCISE** _____
Calculate the freezing point of a solution containing

600 g of $CHCl_3$ and 42 g of eucalyptol, $C_{10}H_{18}O$, a fragrant substance found in the leaves of eucalyptus trees.

*Answer:* Freezing point −65.6°C

## SAMPLE EXERCISE 12.12

List the following solutions in order of their expected freezing points: 0.050 $m$ $CaCl_2$; 0.15 $m$ NaCl; 0.10 $m$ HCl; 0.050 $m$ $HC_2H_3O_2$; 0.10 $m$ $C_{12}H_{22}O_{11}$.

**Solution:** First notice that $CaCl_2$, NaCl, and HCl are strong electrolytes, $HC_2H_3O_2$ is a weak electrolyte, and $C_{12}H_{22}O_{11}$ is a nonelectrolyte. The molality of each solution in total particles is as follows:

| | |
|---|---|
| 0.050 $m$ $CaCl_2$ | (0.15 $m$ in particles) |
| 0.15 $m$ NaCl | (0.30 $m$ in particles) |
| 0.10 $m$ HCl | (0.20 $m$ in particles) |
| 0.050 $m$ $HC_2H_3O_2$ | (between 0.050 and 0.10 $m$ in particles) |

0.10 $m$ $C_{12}H_{22}O_{11}$   (0.10 $m$ in particles)

The freezing points are expected to run from 0.15 $m$ NaCl (lowest freezing point) to 0.10 $m$ HCl to 0.050 $m$ $CaCl_2$ to 0.10 $m$ $C_{12}H_{22}O_{11}$ to 0.050 $HC_2H_3O_2$ (highest freezing point).

**PRACTICE EXERCISE** _____
Which of the following solutes will produce the largest total molality of solute particles upon addition to water: 1 mol of $Co(NO_3)_2$, 2 mol of KCl, or 3 mol of $C_2H_5OH$?

*Answer:* 2 mol of KCl

Because NaCl is a strong electrolyte, the expected freezing-point depression of a 0.100 $m$ aqueous solution of NaCl is (0.200 $m$) × (0.186°C/$m$) = 0.372°C. However, the measured value, 0.348°C, is slightly less. Although strong electrolytes, like NaCl, give nearly the effect expected from the number of ions present in solution, the actual effect is always slightly less than that of a corresponding number of nonelectrolyte particles. The residual electrostatic attraction between ions, even when separated by solvent molecules, makes solutions of electrolytes behave as though their ion concentrations were less than they actually are.

## Osmosis

Certain materials—including many membranes in biological systems and man-made substances such as cellophane—are semipermeable. That is, when in contact with a solution, they permit the passage of some molecules but not others. Often they permit the passage of small solvent molecules such as water but block the passage of larger solute molecules or ionic solutes. This semipermeable character is due to a network of tiny pores within the membrane. Consider a situation in which only solvent molecules are able to pass through a membrane. If such a membrane is placed between two solutions of different concentration, solvent molecules move in both directions through the membrane. However, the concentration of *solvent* is higher in the solution containing less solute than in the more concentrated one. Thus the rate of passage of solvent from the less concentrated to the more concentrated solution side is greater than the

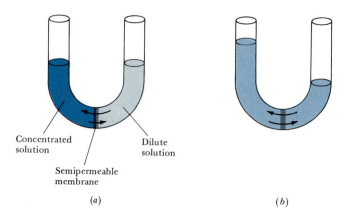

**FIGURE 12.17** Osmosis: (*a*) net movement of solvent from the solution with low solute concentration into the solution with high solute concentration; (*b*) osmosis stops when the column of solution on the left becomes high enough to exert sufficient pressure to stop the osmosis.

Concentrated solution

Dilute solution

Semipermeable membrane

(*a*)

(*b*)

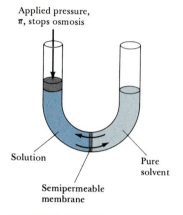

**FIGURE 12.18** Applied pressure on the left arm of the apparatus stops net movement of solvent from the right side of the semipermeable membrane. This applied pressure is known as the osmotic pressure of the solution.

rate in the opposite direction. Thus there is a net movement of solvent molecules from the less concentrated solution into the more concentrated one. This process is called **osmosis**. The important point to remember is that the net movement of solvent is always toward the more concentrated solution.

Figure 12.17(*a*) shows two solutions separated by a semipermeable membrane. Solvent moves through the membrane from right to left, as if the solutions were driven to attain equal concentrations. As a result, the liquid levels in the two arms become uneven. Eventually, the pressure difference resulting from the uneven heights of the liquid in the two arms becomes so large that the net flow of solvent ceases, as shown in Figure 12.17(*b*). Alternatively, we may apply pressure to the left arm of the apparatus, as shown in Figure 12.18, to halt the net flow of solvent. The pressure required to prevent osmosis is known as the osmotic pressure, $\pi$, of the solution. The osmotic pressure is related to concentration as follows:

$$\pi = MRT \qquad [12.15]$$

where $M$ is molarity, $R$ is the ideal gas constant, and $T$ is the temperature on the Kelvin scale.

If two solutions of identical osmotic pressure are separated by a semipermeable membrane, no osmosis will occur. The two solutions are said to be **isotonic**. If one solution is of lower osmotic pressure, it is described as being **hypotonic** with respect to the more concentrated solution. The more concentrated solution is said to be **hypertonic** with respect to the dilute solution.

Osmosis plays a very important role in living systems. For example, the membranes of red blood cells are semipermeable. Placement of red blood cells in a solution that is hypertonic relative to the intracellular solution (the solution within the cells) causes water to move out of the cell, as shown in Figure 12.19. This causes the cell to shrivel, a process known as **crenation**. Placement of the cell in a solution that is hypotonic relative to the intracellular fluid causes water to move into the cell. This causes rupturing of the cell, a process known as **hemolysis**. Persons needing replacement of body fluids or nutrients who cannot be fed orally

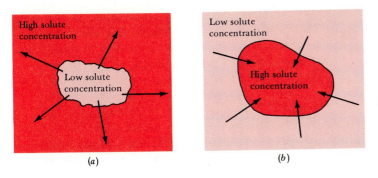

**FIGURE 12.19** Osmosis through the semipermeable membrane of a red blood cell: (*a*) crenation caused by movement of water from the cell; (*b*) hemolysis caused by movement of water into the cell.

are administered solutions by intravenous (or IV) infusion, which feeds nutrients directly into the veins. To prevent crenation or hemolysis of red blood cells, the IV solutions must be isotonic with the intracellular fluids of the cells.

### SAMPLE EXERCISE 12.13

The average osmotic pressure of blood is 7.7 atm at 25°C. What concentration of glucose, $C_6H_{12}O_6$, will be isotonic with blood?

**Solution:**

$$\pi = MRT$$

$$M = \frac{\pi}{RT} = \frac{7.7 \text{ atm}}{\left(0.082 \dfrac{\text{L-atm}}{\text{K-mol}}\right)(298 \text{ K})}$$

$$= 0.31 \ M$$

In clinical situations, the concentrations of solutions are generally expressed in terms of weight percent-ages. The weight percentage of a 0.31 *M* solution of glucose is 5.3 percent. The concentration of NaCl that is isotonic with blood is 0.16 *M* since NaCl ionizes to form two particles, $Na^+$ and $Cl^-$ (a 0.155 *M* solution of NaCl is 0.310 *M* in particles). A 0.16 *M* solution of NaCl is 0.9 percent in NaCl. Such a solution is known as a physiological saline solution.

### PRACTICE EXERCISE

What is the osmotic pressure at 20°C of a 0.0020 *M* sucrose ($C_{12}H_{22}O_{11}$) solution?
*Answer:* 0.048 atm, or 37 mm Hg

There are many interesting examples of osmosis. A cucumber placed in concentrated brine loses water via osmosis and shrivels into a pickle. A carrot that has become limp because of water loss to the atmosphere can be placed in water. Water moves into the carrot through osmosis, making it firm once again. People eating a lot of salty food experience water retention in tissue cells and intercellular spaces because of osmosis. The resultant swelling or puffiness is called edema. Movement of water from soil into plant roots and subsequently into the upper portions of the plant is due at least in part to osmosis. The preservation of meat by salting and of fruit by adding sugar protects against bacterial action. Through the process of osmosis, a bacterium on salted meat or candied fruit loses water, shrivels, and dies.

In osmosis, water moves from an area of high water concentration (low solute concentration) into an area of low water concentration (high solute concentration). Such movement of a substance from an area where its concentration is high to an area where it is low is spontaneous. Biological cells transport not only water but also other select materials through their membrane walls. This permits entry of nutrients and allows for disposal of waste materials. In some cases, substances must be moved from an area of low concentration to one of high concentration. This movement is called *active transport*. It is not spontaneous and so requires expenditures of energy by the cell.

**Determination of Molecular Weight**

The colligative properties of solutions provide a useful means of experimentally determining molecular weights. Any of the four colligative properties could be used to determine molecular weight. The procedures are illustrated in Sample Exercise 12.14.

## SAMPLE EXERCISE 12.14

A solution of an unknown nonvolatile nonelectrolyte was prepared by dissolving 0.250 g in 40.0 g of $CCl_4$. The normal boiling point of the resultant solution was increased by 0.357°C. Calculate the molecular weight of the solute.

**Solution:** Using Equation 12.13, we have

$$\text{Molality} = \frac{\Delta T_b}{K_b} = \frac{0.357°C}{5.02°C/m} = 0.0711 \ m$$

Thus the solution contains 0.0711 mol of solute per kilogram of solvent. The solution was prepared from 0.250 g of solute and 40.0 g of solvent. The number of grams of solute in a kilogram of solvent is therefore

$$\frac{\text{Grams solute}}{\text{Kilograms } CCl_4} = \left(\frac{0.250 \text{ g solute}}{40.0 \text{ g } CCl_4}\right)\left(\frac{1000 \text{ g } CCl_4}{1 \text{ kg } CCl_4}\right)$$

$$= \frac{6.25 \text{ g solute}}{1 \text{ kg } CCl_4}$$

Notice that a kilogram of solvent contains 6.25 g, which from the $\Delta T_b$ measurement must be 0.0711 $m$. Therefore,

$$0.0711 \text{ mol} = 6.25 \text{ g}$$

$$1 \text{ mol} = \frac{6.25 \text{ g}}{0.0711} = 87.9 \text{ g}$$

Therefore, $\mathcal{M} = 87.9$ amu

### PRACTICE EXERCISE

Camphor, $C_{10}H_{16}O$, melts at 179.8°C; it has a particularly large freezing-point depression constant, $K_f = 40°C/m$. When 0.186 g of an organic substance of unknown molecular weight is dissolved in 22.01 g of liquid camphor, the freezing point of the mixture is found to be 176.7°C. What is the approximate molecular weight of the solute?
*Answer:* 108 g/mol

## 12.7 COLLOIDS

When finely divided clay particles are dispersed through water, they do not remain suspended but eventually settle out of the water because of the gravitational pull. The dispersed clay particles are much larger than molecules and consist of many thousands or even millions of atoms. In contrast, the dispersed particles of a solution are of molecular size. Between these extremes is the situation in which dispersed particles are larger than molecules but not so large that the components of the mixture separate under the influence of gravity. These intermediate types of dispersions or suspensions are called **colloidal dispersions**, or simply **colloids**. Thus colloids are on the dividing line between solutions and heterogeneous mixtures. Like solutions, colloids can be gases, liquids, or solids. Examples of each are listed in Table 12.6.

The size of the dispersed particle is the property used to classify a mixture as a colloid. Colloid particles range in diameter from approximately 10 to 2000 Å. Solute particles are smaller. The colloid particle may con-

**TABLE 12.6**  Types of colloids

| Phase of colloid | Dispersing (solventlike) substance | Dispersed (solutelike) substance | Colloid type | Example |
|---|---|---|---|---|
| Gas | Gas | Gas | — | None (all are solutions) |
| Gas | Gas | Liquid | Aerosol | Fog |
| Gas | Gas | Solid | Aerosol | Smoke |
| Liquid | Liquid | Gas | Foam | Whipped cream |
| Liquid | Liquid | Liquid | Emulsion | Milk |
| Liquid | Liquid | Solid | Sol | Paint |
| Solid | Solid | Gas | Solid foam | Marshmallow |
| Solid | Solid | Liquid | Solid emulsion | Butter |
| Solid | Solid | Solid | Solid sol | Ruby glass |

sist of many atoms, ions, or molecules or may even be a single giant molecule. For example, the hemoglobin molecule, which carries oxygen in blood, has molecular dimensions of 65 Å × 55 Å × 50 Å and a molecular weight of 64,500 amu.

Even though colloid particles are so small that the dispersion appears uniform even under a microscope, they are large enough to scatter light very effectively. Consequently, most colloids appear cloudy or opaque unless they are very dilute. Furthermore, because they scatter light, a light beam can be seen as it passes through a colloidal suspension, as shown in Figure 12.20. This scattering of light by colloidal particles, known as the **Tyndall effect**, makes it possible to see the light beam coming from the projection housing in a smoke-filled theater or the light beam from an automobile on a dusty dirt road.

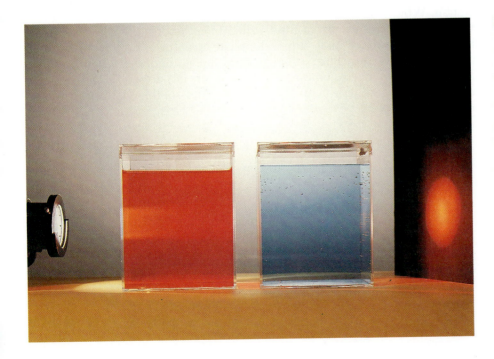

**FIGURE 12.20**
Illustration of the Tyndall effect. The vessel on the left contains a colloidal suspension, that on the right a solution. Note that the path of the beam through the colloidal suspension is clearly seen, because the light is scattered by the colloidal particles. Light is not scattered by the individual solute molecules in the solution. (© Richard Megna/Fundamental Photographs)

## Hydrophilic and Hydrophobic Colloids

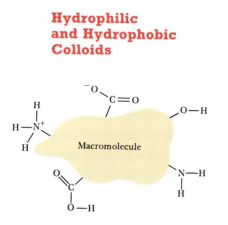

**FIGURE 12.21** Examples of hydrophilic groups at the surface of a giant molecule (macromolecule) that help keep the macromolecule suspended in water.

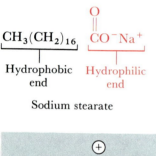

Sodium stearate

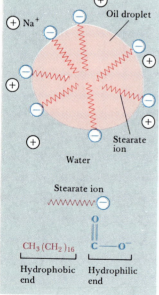

**FIGURE 12.23** Stabilization of an emulsion of oil in water by stearate ions.

The most important colloids are those in which the dispersing medium is water. Such colloids are frequently referred to as **hydrophilic** (water loving) or **hydrophobic** (water hating). Hydrophilic colloids are most like the solutions that we have previously examined. In the human body, the extremely large molecules that make up such important substances as enzymes and antibodies are kept in suspension by interaction with surrounding water molecules. The molecules fold so that polar, or charged, groups can interact with water molecules at the periphery of the molecules. These hydrophilic groups generally contain oxygen or nitrogen. Some examples are shown in Figure 12.21.

Hydrophobic colloids can be prepared in water only if they are stabilized in some way. Otherwise, their natural lack of affinity for water causes them to separate from the water. Hydrophobic colloids can be stabilized by adsorption of ions on their surface, as shown in Figure 12.22. These adsorbed ions can interact with water, thereby stabilizing the colloid. At the same time the mutual repulsion between colloid particles with adsorbed ions of the same charge keeps the particles from colliding and so getting larger.

Hydrophobic colloids can also be stabilized by the presence of other hydrophilic groups on their surfaces. For example, small droplets of oil are hydrophobic. They do not remain suspended in water; instead, they separate, forming an oil slick on the surface of the water. Addition of sodium stearate, whose structure is shown at left, or any similar substance having one end that is hydrophilic (polar, or charged) and one that is hydrophobic (nonpolar), will stabilize a suspension of oil in water, as shown in Figure 12.23. The hydrophobic ends of the stearate ions interact with the oil droplet, and the hydrophilic ends point out toward the water with which they interact.

These concepts have an interesting application in our own digestive system. When fats in our diet reach the small intestine, a hormone causes the gallbladder to excrete a fluid called bile. Among the components of bile are compounds that have chemical structures similar to sodium stearate; that is, they have a hydrophilic (polar) end and a hydrophobic (nonpolar) end. These compounds emulsify the fats present in the intestine and thus permit digestion and absorption of fat-soluble vitamins through the intestinal wall.

**FIGURE 12.22** Schematic representation of the stabilization of a hydrophobic colloid by adsorbed ions.

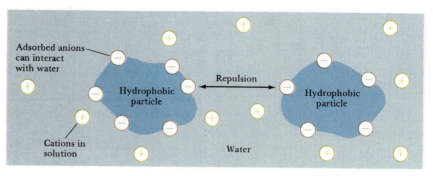

In the genetic disease known as sickle-cell anemia, hemoglobin molecules are abnormal: They have lower solubility, especially in the unoxygenated form. Consequently, as much as 85 percent of the hemoglobin in the red blood cells crystallizes from solution. This distorts the cells into a sickle shape, as shown in Figure 12.24. These clog the capillaries, thus causing gradual deterioration of the vital organs. The disease is hereditary, and if both parents carry the defective genes, it is likely that the child will possess only abnormal hemoglobin. Such children seldom survive more than a few years after birth. The reason for the insolubility of hemoglobin in sickle-cell anemia can be traced to a change in one part of the molecule.

Normal hemoglobin molecules have an amino acid in their makeup that has a side chain protruding from the main body of the molecule:

$$-CH_2-CH_2-\overset{\displaystyle O}{\overset{\|}{C}}-OH$$

**Normal**

Notice that this side chain terminates in a polar group, which contributes to the solubility of the hemoglobin molecule in water. In the hemoglobin molecules of persons suffering from sickle-cell anemia, the side chain is of a different type:

$$-\underset{\displaystyle \underset{|}{CH_3}}{CH}-CH_3$$

**Abnormal**

This abnormal group of atoms is nonpolar (hydrophobic), and its presence leads to the aggregation of this defective form of hemoglobin into particles too large to remain suspended in biological fluids.

**FIGURE 12.24**
Electron micrograph showing a normal red blood cell (left) and sickled red blood cells (right). (Bill Longcore/Photo Researchers)

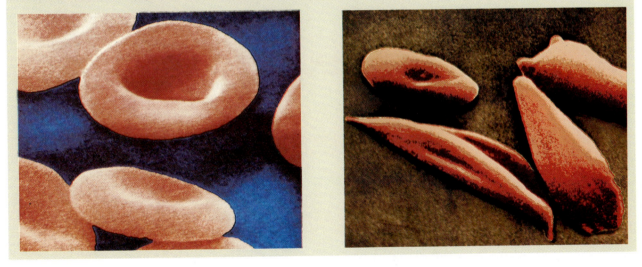

**Removal of Colloidal Particles**

Often colloidal particles must be removed from a dispersing medium, as in the removal of smoke from stacks or butterfat from milk. Because colloidal particles are so small, they cannot be extracted by simple filtration. The colloidal particles must be enlarged, a process called **coagulation**. The resultant, larger particles can then be separated by filtration or merely by allowing them to settle out of the dispersing medium.

Heating or adding an electrolyte to the mixture may bring about coagulation. Heating the colloidal dispersion increases the particle motion and so the number of collisions. The particles increase in size as they stick together following collision. The addition of electrolytes causes neutralization of the surface charges of the particles, thereby removing the electrostatic repulsions that inhibit their coming together. The effect of electrolytes is seen in the depositing of suspended clay in a river as it mixes with salt water. This results in the formation of river deltas wherever rivers empty into oceans or other salty bodies of water.

Semipermeable membranes can also be used to separate ions from colloidal particles; the ions can pass through the membrane, the colloid particles cannot. This type of separation is known as **dialysis**. This process is used in the purification of blood in artificial kidney machines. Our kidneys are responsible for removing the waste products of metabolism from blood. In the kidney machine, blood is circulated through a dialyzing tube immersed in a washing solution. That solution is isotonic in ions that must be retained by the blood but is lacking the waste products. Wastes therefore dialyze out of the blood.

# FOR REVIEW

## SUMMARY

Solutions are homogeneous mixtures of atoms, ions, or molecules. The relative amounts of **solute** and **solvent** in a solution can be described qualitatively (**dilute** or **concentrated** solutions) or quantitatively (in terms of **weight percentage**, **molarity**, **molality**, **normality**, and **mole fraction**).

The extent to which a solute will dissolve in a particular solvent depends on the relative magnitudes of solute-solute, solute-solvent, and solvent-solvent attractive forces as well as on the changes in disorder accompanying mixing. The rule "likes dissolve likes" was found to be useful in rationalizing solubilities. It is possible to change the solubility of a solute by changing its temperature. For most ionic solutes in water, solubility increases with temperature. For gases, solubility generally decreases with increasing temperature and increases with increasing pressure.

Substances that exist in solution as ions are called

**electrolytes**. Those substances that are completely ionized in solution are called **strong electrolytes**.

The presence of a solute in a solvent lowers the vapor pressure and the freezing point and increases the boiling point of the solvent. These changes are termed **colligative properties**. The magnitude of the change depends on the total concentration of solute particles in solution and not on their characteristics. **Osmotic pressure** is the pressure that must be applied to a solution to prevent transfer of solvent molecules from a pure solvent through a semipermeable membrane. It is a colligative property because it is proportional to the total concentration of solute particles. Colligative properties can be used to determine the molecular weights of nonvolatile nonelectrolytes.

True solutions can be differentiated from **colloids** on the basis of particle size. Colloids play an important role in many chemical and biological systems.

## LEARNING GOALS

Having read and studied this chapter, you should be able to:

1. Define molarity, molality, mole fraction, normality, and weight percentage, and calculate concentrations in any of these concentration units.

2. Convert concentration in one concentration unit into any other (given the density of the solution where necessary).

3. Give definitions of the qualitative terms used to describe solutions: dilute, concentrated, saturated, unsaturated, and supersaturated.

4. Describe the solution process, including the bonds made and broken when a substance dissolves.

5. Describe the energy changes that occur in the solution process in terms of the attractive forces that operate in the solvent and solute and the attractive forces between the solvent and solute.

6. Describe the role of disorder in the solution process.

7. Rationalize the solubilities of substances in various solvents in terms of their molecular structures and intermolecular forces.

8. Discuss the effects of pressure and temperature on solubilities of gases.

9. Predict which substances are electrolytes and which are nonelectrolytes; predict which electrolytes are strong electrolytes and which are weak.

10. Describe the effect of solute concentration on solvent vapor pressure; state Raoult's law.

11. Explain the origin of the colligative properties—boiling-point elevation, freezing-point lowering, and osmotic pressure—of solutions.

12. Determine the molecular weight of a solute from the magnitude of the effect of a known concentration of solute on one of the colligative properties of a solvent.

13. Explain the difference between the magnitude of changes in colligative properties caused by electrolytes compared to those caused by nonelectrolytes.

14. Describe how a colloid differs from a true solution.

## KEY TERMS

Among the more important terms and expressions used for the first time in this chapter are the following.

**Colligative properties** (Section 12.6) are those properties of a solvent (vapor-pressure lowering, freezing-point lowering, boiling-point elevation, osmotic pressure) that depend on the total concentration of solute particles present.

**Colloids** (Section 12.7) are particles that are larger than normal molecules but that are nevertheless small enough to remain suspended in a dispersing medium indefinitely.

**Dialysis** (Section 12.7) is the separation of small solute particles from colloid particles by means of a semipermeable membrane.

An **electrolyte** (Section 12.4) is a solute that gives a solution containing ions; such a solution will conduct an electrical current. **Strong electrolytes** (strong acids, strong bases, and most common salts) are completely ionized in solution. **Weak electrolytes** are partially ionized in solution.

**Henry's law** (Section 12.3) states that the solubility of a gas in a liquid, $C_g$, is proportional to the pressure of gas over the solution: $C_g = kP_g$.

**Molality** (Section 12.1) is the concentration of a solution expressed as moles of solute per kilogram of solvent; abbreviated $m$.

**Mole fraction** (Section 12.1) is the ratio of the number of moles of one component of a solution to the total number of moles of all substances present in the solution; abbreviated $X$, with a subscript identifying the component.

**Normality** (Section 12.1) is the concentration of a solution expressed as equivalents of solute per liter of solution; abbreviated $N$. An **equivalent** is defined according to the type of reaction being considered; 1 equivalent of a given reactant will react with 1 equivalent of a second reactant.

**Osmosis** (Section 12.6) is the net movement of solvent through a semipermeable membrane toward the solution with greater solute concentration. The **osmotic pressure** of a solution is the pressure that must be applied to a solution to stop osmosis from pure solvent into the solution.

**Raoult's law** (Section 12.5) states that the partial pressure of a solvent over a solution, $P_A$, is given by the vapor pressure of the pure solvent, $P_A^\circ$, times the mole fraction of solvent in the solution, $X_A$: $P_A = X_A P_A^\circ$.

A **saturated solution** (Section 12.2) is a solution in which undissolved solute and dissolved solute are in equilibrium. The **solubility** of a substance is the amount of solute that dissolves to form a saturated solution. Solutions containing less solute than this are said to be **unsaturated**, and those containing more are said to be **supersaturated**.

**Solvation** (Section 12.2) is the clustering of solvent molecules around a solute particle. When the solvent is water, this clustering is known as **hydration**.

The **Tyndall effect** (Section 12.7) refers to the scattering of a beam of visible light by the particles in a colloidal dispersion.

The **weight percentage** (Section 12.1) of a solution is the number of grams of solute it contains in each 100 g of solution.

# EXERCISES

## Concentrations of Solutions

**12.1** Calculate the weight percentage of solute in the following solutions: (**a**) 2.00 g of NaCl in 40.0 g of $H_2O$; (**b**) 10.0 g of benzene, $C_6H_6$, in 26.0 g of carbon tetrachloride, $CCl_4$.

**12.2** Calculate the weight percentage of solute in the following solutions: (**a**) 120 g of KCl in 2.6 kg of $H_2O$; (**b**) 2.55 g of $Ba(NO_3)_2$ in 80.0 g of $H_2O$.

**12.3** Calculate the mole fraction of methyl alcohol, $CH_3OH$, in the following solutions: (**a**) 6.00 g of $CH_3OH$ in 480 g of $H_2O$; (**b**) 4.13 g of $CH_3OH$ in 48.6 g of $CCl_4$.

**12.4** Calculate the mole fraction of ethylene glycol, $C_2H_6O_2$, in the following solutions: (**a**) 120 g of $C_2H_6O_2$ dissolved in 120 g of water; (**b**) 120 g of $C_2H_6O_2$ dissolved in 1.20 kg of acetone, $C_3H_6O$.

**12.5** Calculate the molarity of each of the following solutions: (**a**) 2.00 g of KBr in 1.60 L of solution; (**b**) 3.00 g of NaOH in 0.650 L of solution; (**c**) 2.58 g of $Ca(NO_3)_2 \cdot 4H_2O$ in 0.400 L of solution; (**d**) 50.0 mL of 0.200 $M$ HCl diluted to 800 mL.

**12.6** Calculate the molarity of each of the following solutions: (**a**) 6.2 g $CaCl_2$ in 600 mL of solution; (**b**) 25.0 g of $Co(NO_3)_2 \cdot 6H_2O$ in 3.4 L of solution; (**c**) 1.65 g of $NH_4NO_3$ in 800 mL of dioxane $(C_4H_8O_2)$ solution; (**d**) 30.0 mL of 0.260 $M$ $CaCl_2$ solution diluted to 75.0 mL.

**12.7** Calculate the molality of each of the following solutions: (**a**) 12.0 g of benzene, $C_6H_6$, dissolved in 38.0 g of carbon tetrachloride, $CCl_4$; (**b**) 2.88 g of NaI dissolved in 0.200 L of water; (**c**) 5.85 g of $KNO_3$ dissolved in 0.386 L of water.

**12.8** Calculate the molality of each of the following solutions: (**a**) 210 g of $NH_4NO_3$ dissolved in 2.6 kg of $H_2O$; (**b**) 2.1 g of elemental sulfur, $S_8$, dissolved in 86.2 g of naphthalene, $C_{10}H_8$; (**c**) 1.50 mol of NaI dissolved in 40.0 mol of $H_2O$.

**12.9** A solution containing 66.0 g of acetone, $C_3H_6O$, and 46.0 g of $H_2O$ has a density of 0.926 g/mL. Calculate (**a**) the weight percentage; (**b**) the mole fraction; (**c**) the molality; (**d**) the molarity of $H_2O$ in this solution.

**12.10** A sulfuric acid solution containing 571.6 g of $H_2SO_4$ per liter of solution has a density of 1.329 g/cm³. Calculate (**a**) the weight percentage; (**b**) the mole fraction; (**c**) the molality; (**d**) the molarity of $H_2SO_4$ in this solution.

**12.11** Calculate the molarity, molality, and mole fraction of (**a**) commercial concentrated hydrochloric acid, which is 38 percent HCl by weight, density 1.19 g/cm³; (**b**) commercial concentrated nitric acid, 69 percent $HNO_3$ by weight, density 1.42 g/cm³.

**12.12** Calculate the molarity, molality, and mole fraction of (**a**) commercial concentrated acetic acid (called "glacial" acetic acid, because it freezes at 11°C), 95 percent $HC_2H_3O_2$ by weight, density 1.06 g/cm³; (**b**) concen-

trated aqueous ammonia, 28 percent $NH_3$ by weight, density 0.90 g/cm³.

**12.13** The density of an aqueous solution containing 20.0 percent $Pb(NO_3)_2$ by weight is 1.202 g/cm³. Calculate the molarity and molality of the solution.

**12.14** The density of an aqueous solution containing 20.0 percent potassium iodide by weight is 1.170 g/cm³. Calculate the molarity and molality of the solution.

**12.15** Calculate the number of moles of solute present in each of the following solutions: (**a**) 356 mL of 0.358 $M$ $Ca(NO_3)_2$; (**b**) $4.60 \times 10^2$ L of 0.582 $M$ HBr; (**c**) 132 mL of 0.0288 $M$ $Al(NO_3)_3$.

**12.16** Calculate the number of moles of solute present in each of the following solutions: (**a**) 60.0 g of an aqueous solution that is 1.25 percent KI by weight; (**b**) 250 g of an aqueous solution that is 0.460 percent NaCl by weight; (**c**) 600 mL of 1.25 $M$ $H_2SO_4$.

**12.17** Describe how you would prepare each of the following aqueous solutions, starting with solid KBr: (**a**) 1.40 L of $1.5 \times 10^{-2}$ $M$ KBr; (**b**) 250 g of 0.400 $m$ KBr; (**c**) 1.50 L of a solution that is 12 percent KBr by weight (the density of the solution is 1.10 g/mL); (**d**) a 0.200 $M$ solution of KBr that contains just enough KBr to precipitate 21.0 g of AgBr from a solution containing 0.480 mol of $AgNO_3$.

**12.18** Describe how you would prepare each of the following aqueous solutions: (**a**) 2.5 L of 1.5 $M$ $HNO_3$, starting from 6.0 $M$ $HNO_3$; (**b**) 150 mL of a solution that is 1.0 $m$ KI; (**c**) a 0.50 $M$ solution of HCl that would just neutralize 6.0 g of $Ba(OH)_2$, starting from 6.0 $M$ HCl.

**12.19** In a certain reaction, $Sn^{4+}$ is reduced to $Sn^{2+}$. (**a**) What is the normality of a 0.38 $N$ solution of $Sn^{4+}$? (**b**) What is the molarity of a 0.222 $N$ solution of $H_2SO_4$ reacted with NaOH solution, in which both of the hydrogens of the $H_2SO_4$ react?

**12.20** In each of the following reactions, indicate the normality of a 0.1 $M$ solution of the reagent underlined:

(**a**) $\underline{H_3PO_3}(aq) + 2NaOH(aq) \longrightarrow$
$$Na_2HPO_3(aq) + 2H_2O(aq)$$

(**b**) $8HNO_3(aq) + \underline{KMnO_4}(aq) + 5Fe(NO_3)_2(aq) \longrightarrow$
$$5Fe(NO_3)_3(aq) + Mn(NO_3)_2(aq) +$$
$$KNO_3(aq) + 4H_2O(l)$$

(**c**) $MnS(s) + 2\underline{HCl}(aq) \longrightarrow MnCl_2(aq) + H_2S(g)$

## The Solution Process; Solubility and Structure

**12.21** Indicate the type of solute-solvent intermolecular attractive force (Section 11.4) that should be most important in each of the following solutions: (**a**) KCl in water; (**b**) benzene, $C_6H_6$, in carbon tetrachloride, $CCl_4$; (**c**) HF in water; (**d**) acetonitrile, $CH_3CN$, in acetone, $C_6H_6O$ (see Table 11.6).

**12.22** Offer an explanation, in terms of the intermolecular forces involved, for each of the following observations. (**a**) $CoCl_2$ is soluble to the extent of 540 g per kilogram of gram of ethyl alcohol. (**b**) Water is completely miscible with dioxane, for which the structural formula is

However, water is not soluble in cyclohexane, for which the structural formula is

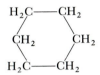

(**c**) Chloroform, $CHCl_3$, is soluble in water to the extent of only 1 g per 100 g of water, but is miscible with ethyl alcohol.

**12.23** Based on the nature of solute-solvent interactions, rank the following substances in order of increasing solubility in hexane $(C_6H_{14})$: $C_2H_6$, $CH_3OH$, KCl.

**12.24** Judging from the nature of the solute-solvent interactions, rank the following substances in order of increasing solubility in water: benzene, erythritol, $HOCH_2$—$CH(OH)$—$CH(OH)$—$CH_2OH$, and butanol, $CH_3CH_2CH_2CH_2OH$.

**12.25** The enthalpies of solution of hydrated salts are generally more positive than those of anhydrous materials. For example, $\Delta H$ of solution for KOH is $-57.3$ kJ/mol and that for KOH $H_2O$ is $-14.6$ kJ/mol. Similarly, $\Delta H_{soln}$ for $NaClO_4$ is $+13.8$ kJ/mol and that for $NaClO_4$ $H_2O$ is $+22.5$ kJ/mol. Explain this effect in terms of the enthalpy contributions to the solution process depicted in Figure 12.4.

**12.26** The enthalpy of solution of KBr in water is about $+19.8$ kJ/mol. The process, then, is endothermic. Nevertheless, the solubility of KBr in water is relatively high. Why does the solution process, although endothermic, proceed?

**12.27** Explain, in terms of the enthalpy changes in the various steps involved (Figure 12.4), why KBr is not soluble in $CCl_4$.

**12.28** Which of the following pairs of liquids would you expect to be miscible and which nonmiscible: (**a**) $H_2O$ and $HOCH_2CH(OH)CH_2OH$; (**b**) $C(CH_2OH)_4$ and $C_7H_{16}$; (**c**) $Hg(l)$ and $NaI(l)$; (**d**) $CH_2Cl_2$ and $CH_3CH_2OCH_2CH_3$? Explain in each case.

## Effect of Temperature and Pressure on Solubility

**12.29** The Henry's law constant for helium gas in water at $30°C$ is $3.7 \times 10^{-4}$ $M$/atm; that for $N_2$ at $30°C$ is $6.0 \times$ $10^{-4}$ $M$/atm. If the two gases are each present at 2.5 atm pressure, calculate the solubility of each gas.

**12.30** The partial pressure of $O_2$ in air at sea level is 0.21 atm. Using the data in Table 12.3, together with Henry's law, calculate the molar concentration of $O_2$ in the surface water of a lake saturated with air at $20°C$.

**12.31** In terms of the kinetic-molecular theory, explain why the solubility of gases in liquids generally decreases with increasing temperature.

**12.32** Explain how the kinetic-molecular theory accounts for Henry's law. What factor is most important in causing variations in the Henry's law constant among different gases in the same solvent?

## Electrolyte Solutions

**12.33** Classify each of the following solutions as a nonelectrolyte, weak electrolyte, or strong electrolyte in water: (**a**) HBrO; (**b**) $Co(NO_3)_2$; (**c**) sucrose, $C_{12}H_{22}O_{11}$; (**d**) $NaC_2H_3O_2$.

**12.34** Classify each of the following substances as a nonelectrolyte, weak electrolyte, or strong electrolyte in water: (**a**) HF; (**b**) methanol, $CH_3OH$; (**c**) HCl; (**d**) $NH_3$.

**12.35** An aqueous solution of an unknown substance tests basic with litmus paper. The solution is weakly conducting, compared with an NaCl solution of the same concentration. Which of the following substances could the unknown be: KCl, NaOH, $NH_3$, $H_2SO_3$, $CH_3OH$?

**12.36** An aqueous solution of an unknown substance tests acidic. The solution is observed to be roughly as conducting as an HCl solution of the same concentration. Which of the following substances could the unknown be: $NaC_2H_3O_2$, HF, $HClO_4$, $H_3BO_3$, $KClO_3$, KOH, $H_2SO_4$?

**12.37** For each of the following solutions, indicate whether the substance exists entirely in molecular form, entirely in ionic form, or as a mixture of molecular and ionic forms in the solution indicated: (**a**) KBr in water; (**b**) $CCl_4$ in $C_6H_{14}$; (**c**) acetonitrile, $CH_3CN$, in water; (**d**) $HClO_2$ in water.

**12.38** For each of the following solutions, indicate whether the substance exists entirely in molecular form, entirely in ionic form, or in a mixture of molecular and ionic forms in the solution indicated: (**a**) HI in water; (**b**) $NH_3$ in water; (**c**) glucose, $C_6H_{12}O_6$, in alcohol; (**d**) $PbCl_2$ in water.

**12.39** Indicate the total concentration of all solute species present in each of the following solutions: (**a**) 0.14 $M$ NaOH; (**b**) 0.25 $M$ $CaBr_2$; (**c**) 0.25 $M$ $CH_3OH$; (**d**) a mixture of 50.0 mL of 0.20 $M$ $KClO_3$ and 25.0 mL of 0.20 $M$ $Na_2SO_4$.

**12.40** Indicate the total concentration of each ion present in the solution formed by mixing (**a**) 20.0 mL of 0.100 $M$ HCl and 10.0 mL of 0.220 $M$ HCl; (**b**) 15.0 mL of 0.300 $M$ $Na_2SO_4$ and 10.0 mL of 0.100 $M$ NaCl; (**c**) 3.50 g of KCl in 60.0 mL of 0.500 $M$ $CaCl_2$ solution (assume no volume change).

## Vapor Pressures of Solutions

**12.41** Explain how the kinetic-molecular theory accounts for Raoult's law in an ideal solution.

**12.42** Explain how the kinetic-molecular theory accounts for Henry's law. (Note that Henry's law can be thought of as the relationship between the vapor pressure of a volatile solute over a solution and the concentration of the solute in the solution.)

**12.43** Consider the experiment illustrated in Figure 12.15. Suppose that one beaker contains 0.10 $M$ $CaCl_2$. What concentration of KCl solution should the second beaker contain if there is to be no net transfer of solvent?

**12.44** Consider the experiment illustrated in Figure 12.15. Suppose that one beaker initially contains 10.0 mL of 0.100 $M$ NaCl, the other 10.0 mL of 0.200 $M$ NaCl. What volume of solution does each beaker have when equilibrium is attained?

**12.45** Calculate the vapor pressure of water above a solution prepared by adding (**a**) 10.00 g of lactose, $C_{12}H_{22}O_{11}$, to 82.0 g of water at 338 K; (**b**) 5.00 g of sodium sulfate, $Na_2SO_4$, to 115 g of water at 338 K; (**c**) 32.5 g of glycerin, $C_3H_8O_3$, to 120 g of water at 338 K. (The vapor pressure of water is given in Appendix C.)

**12.46** Calculate (**a**) the amount of KBr that must be added to 120 g of water to reduce the vapor pressure by 1.50 mm Hg at 25°C; (**b**) the amount of ethylene glycol, $C_2H_6O_2$, that must be added to 1.0 kg of ethanol to reduce its vapor pressure by 10.0 mm Hg from the value of 100 mm Hg due to pure solvent at 34°C.

## Colligative Properties

**12.47** Using Table 12.5, calculate the freezing and boiling points of each of the following solutions: (**a**) 0.2 $m$ glucose in ethanol; (**b**) 2.5 g of $CCl_4$ in 72 g of benzene; (**c**) 1.8 g of $KNO_3$ in 43.6 g of water; (**d**) 2.00 g of $Li_2Cr_2O_7$ in 60.0 g of water.

**12.48** Using Table 12.5, calculate the freezing and boiling points of each of the following solutions: (**a**) 1.80 g of naphthalene, $C_{10}H_8$, in 110 g of $CCl_4$; (**b**) 1.44 g of LiI in 46.0 mL of water; (**c**) 2.15 g of $Ca(NO_3)_2$ in 80.0 g of $H_2O$; (**d**) 26.0 g of glycerol, $C_3H_8O_3$, in 92.0 g of ethanol.

**12.49** List the following aqueous solutions in order of increasing boiling point: 0.030 $m$ glycerin, 0.020 $m$ KBr, 0.030 $m$ benzoic acid, $HC_7H_5O_2$.

**12.50** Rank the following aqueous solutions from highest boiling point to lowest: 0.080 $m$ glucose, 0.060 $m$ LiBr, 0.030 $m$ $Zn(NO_3)_2$.

**12.51** A solution is prepared from 18.0 g of an unknown compound and 110.0 g of acetone, $C_3H_6O$, at 313 K. At this temperature the vapor pressure of pure acetone is 0.526 atm, and that of the solution is 0.506 atm. Calculate the molecular weight of the unknown compound, assuming that the compound is in molecular form in acetone.

**12.52** A dilute sugar solution prepared by mixing 16.0 g of an unknown sugar with sufficient water to form 0.200 L of solution is found to have an osmotic pressure of 2.86 atm at 25°C. What is the molecular weight of the sugar?

**12.53** A sample of seawater taken from the Arctic Ocean freezes at −1.98°C; a sample taken from the middle of the Atlantic Ocean freezes at −2.08°C. What is the total molality of ionic solutes present in each of these solutions?

**12.54** Biphenyl, $C_{12}H_{10}$, an organic substance that melts at 71.0°C, has a molal freezing-point-depression constant of 8.0°C/$m$. When 1.05 g of an unknown organic substance is mixed with 18.6 g of biphenyl, the melting point of the mixture is lowered by 3.3°C. What is the approximate molecular weight of the unknown substance?

## Colloids

**12.55** Explain how observation of the passage of a beam of light through a liquid can be used to distinguish between a solution and a colloidal dispersion. Account for the different behavior in the two cases.

**12.56** Explain how each of the following factors operates in determining the stability or instability of a colloidal dispersion: (**a**) particulate mass; (**b**) hydrophobic character; (**c**) charges on colloidal particles.

**12.57** Indicate whether each of the following is a hydrophilic or a hydrophobic colloid: (**a**) butterfat in homogenized milk; (**b**) hemoglobin in blood; (**c**) vegetable oil in a salad dressing.

**12.58** Glucose, which has the structure shown in Figure 12.10, is not soluble in benzene. Explain how it might be possible to stabilize a colloidal dispersion of glucose in benzene using sodium stearate. Draw a sketch to show how the stabilization would occur.

**12.59** It is possible to precipitate gold from an aqueous solution in such a manner that the gold particles are extremely small. Although the density of gold is 19.3 g/cm³, such a sol is stable indefinitely. In terms of the kinetic-molecular theory, how is it possible for the gold particles to remain dispersed?

**12.60** Why does milk curdle upon addition of acid?

## Additional Exercises

**12.61** Molarity as a concentration unit has the disadvantage that it changes with a solution's temperature. Explain why this is so. Is the same true of molality? Explain.

**12.62** Calculate the molarity of each of the following solutions (**a**) 2.60 g of LiI in 200 mL of ethanol solution; (**b**) one formed by diluting 25.0 mL of 1.60 $M$ $H_2SO_4$ to a total volume of 80.0 mL.

**12.63** Acetonitrile, $CH_3CN$, is a polar organic solvent that dissolves a wide range of solutes, including many salts. The density of a 1.80 $m$ acetonitrile solution of LiBr is 0.826 g/cm³. Calculate the concentration of the solution in (**a**) molality; (**b**) mole fraction of LiBr; (**c**) weight percentage of $CH_3CN$.

**12.64** Copper(II) sulfate, $CuSO_4$, is commonly used to reduce the growth of algae in lakes, ponds, and water reservoirs. An aqueous solution of $CuSO_4$ that is 18.0 percent by weight has a density of $1.208 \text{ g/cm}^3$. Copper(II) sulfate is commonly available as the pentahydrate, $CuSO_4 \cdot 5H_2O$. What mass of this substance must be used to make up 8.00 L of a solution that is 18.0 percent by weight? What is the molarity of this solution?

**12.65** A solution is made by dissolving 2.16 g of benzoic acid, $HC_7H_5O_2$, in 180 mL of $CCl_4$, density $1.59 \text{ g/cm}^3$. Another solution is prepared by dissolving the same quantity of benzoic acid in 180 mL of $C_2H_5OH$, density $0.782 \text{ g/cm}^3$. **(a)** Calculate the mole fractions and molalities in each case. **(b)** Assuming that the density of the solution is the same as that of pure solvent, calculate the molarity in each case. **(c)** Compare the molarities and molalities of these solutions, and comment on their relative magnitudes.

**12.66** In terms of the concepts developed in this chapter, indicate why each of the following commercial products is formulated as it is. **(a)** A gas-line antifreeze additive to the gas tank consists almost entirely of methyl alcohol. **(b)** The solvent in a water-repellent spray for treating shoes is methylene chloride, $CH_2Cl_2$. **(c)** A wax remover for cross-country skis consists of benzene and similar aromatic hydrocarbons (Section 9.4).

**12.67** Indicate whether each of the following processes proceeds with an increase or decrease in randomness, or disorder, of the system: **(a)** evaporation of water from a teakettle; **(b)** the alignment of liquid crystal molecules in a watch display to show the number 7; **(c)** precipitation of AgCl upon mixing solutions of NaCl and $AgNO_3$; **(d)** mixing of concentrated HCl and water.

**12.68** Butylated hydroxytoluene (BHT) has the following molecular structure:

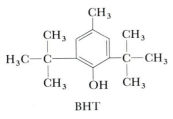

BHT

It is widely used as a preservative in a variety of foods, including dried cereals, cooking oils, and canned goods. The average person in the United States consumes about 2 mg of BHT daily. In terms of its structure, would you expect it to be readily excreted from the body or found stored in body fat? (Incidentally, BHT is not known to have any harmful properties; it is, in fact, a known antiviral agent.)

**12.69** Recently concern has grown regarding a possible health hazard arising from the presence of the radioactive gas radon (Rn) in well water obtained from aquifers that lie in rock deposits in Maine and elsewhere. A sample consisting of various gases contains $3.5 \times 10^{-6}$ mole fraction of radon. This gas at total pressure of 36 atm is shaken with water at 30°C. Assume that the solubility of radon in water with 1 atm pressure of the gas over the solution at 30°C is $7.27 \times 10^{-3}$ M. Calculate the molar concentration of radon in the water.

**12.70** Lysozyme, an enzyme that cleaves bacterial cell walls, has a molecular weight of 13,930 g/mol. What is the osmotic pressure exerted by a solution of 0.150 g of this enzyme in 210 mL of solution at 25°C?

**12.71** A "canned heat" product used to warm chafing dishes consists of a homogeneous mixture of ethyl alcohol, $C_2H_5OH$, and paraffin that has an average formula $C_{24}H_{50}$. What mass of $C_2H_5OH$ should be added to 620 kg of the paraffin in formulating the mixture if the vapor pressure of ethyl alcohol at 35°C over the mixture is to be 8 mm Hg? The vapor pressure of pure ethyl alcohol at 35°C is 100 mm Hg.

**12.72** A constant humidity in a closed container can be achieved by placing in the chamber an aqueous solution that has the desired vapor pressure of water. Provide a recipe for preparing 1 kg of a solution of ethylene glycol, $C_2H_6O_2$, in water that has a vapor pressure of 13.0 mm Hg at 24°C. (Assume that ethylene glycol is nonvolatile; see Appendix C for water vapor pressure.)

**[12.73]** Calculate the vapor pressure over a 2.20 *m* aqueous solution of glucose at 50°C. (See Appendix C; note that you will need to convert from units of molality to units of mole fraction.)

**[12.74]** A lithium salt used in lubricating grease has the formula $LiC_nH_{2n-1}O_2$. The salt is soluble in water to the extent of 0.036 g per 100 g of water at 25°C. The osmotic pressure of this solution is found to be 57.1 mm Hg. Assuming that molality and molarity in such a dilute solution are the same and that the lithium salt is completely dissociated in the solution, determine an appropriate value of *n* in the formula for the salt.

**[12.75]** Pheromones are compounds secreted by the females of many insect species to attract males. One of these compounds contains 80.78 percent C, 13.56 percent H, and 5.66 percent O. A solution of 1.00 g of this substance in 8.50 g of benzene freezes at 3.37°C. What are the molecular weight and molecular formula of the compound?

**[12.76]** Adrenaline, the hormone that triggers release of extra glucose molecules in times of stress or emergency, contains 59.0 percent C, 26.2 percent O, 7.1 percent H, and 7.6 percent N. A solution of 0.64 g of adrenaline in 36.0 g of $CCl_4$ causes an elevation of 0.49°C in the boiling point. What are the molecular weight and molecular formula of adrenaline?

**[12.77]** Suppose that the bell jar in Figure 12.15 were sealed and evacuated after the beakers were put into place. How would this affect the rate at which the experiment went to completion? How would it affect the final result? Explain.

**[12.78]** In the bell jar experiment shown in Figure 12.15, 20.0 mL of a 0.050 *M* aqueous solution of a nonvolatile nonelectrolyte is placed in the left beaker and 20.0 mL of a 0.030 *M* aqueous solution of NaCl is placed in the right beaker. What are the volumes in the two beakers when equilibrium is attained?

## CHAPTER 13

# Reactions in Aqueous Solution

Chemistry is about chemical change. For this reason, chemical reactions are the essence of chemistry. We have already encountered many examples of chemical reactions, and we have seen how they are represented by chemical equations.

Chemical reactions occur under a variety of conditions. For example, gases can react with each other, as when methane burns in air (Equation 3.5). Solids can react with gases, as when a freshly exposed surface of sodium tarnishes in air. It is even possible for solids to react with other solids, as when powdered aluminum reacts with powdered iron in the thermite reaction (Figure 13.1). However, a great fraction of all the most important reactions occur in solution. Particles in solution are free to move throughout the solution volume. Solute molecules or ions may thus collide, and these collisions may lead to reaction.

In this chapter we focus on solution reactions in which water is the solvent. Water, which is so plentiful in our natural environment, is the medium in which many reactions occur in nature. For example, spectac-

**FIGURE 13.1** The thermite reaction, in which powdered aluminum is used to reduce $Fe_2O_3$ to metallic iron. The aluminum-iron oxide mixture is in the upper clay vessel. When initiated by a burning Mg ribbon (a), the reaction begins and generates considerable heat, (b) and (c). Because so much heat is evolved in the reaction, it is self-sustaining once initiated (d). The iron formed in the reaction is white hot as it flows out the bottom of the container and into the lower vessel. (Donald Clegg and Roxy Wilson)

(a)  (b)  (c)  (d)

Mammoth Hot Springs in Yellowstone National Park. The water in the hot springs is rich in dissolved minerals. As the concentrated solution of minerals reaches the surface and cools, much of the mineral content is deposited. Thus the series of terraces and ponds in the springs continues to grow. (Dr. E. R. Degginger)

**FIGURE 13.2** Mammoth Cave in Kentucky. The great vaults in these and similar caves were formed by the action of water containing dissolved $CO_2$ on limestone formations. (Russ Kinne/Photo Researchers)

ular limestone caves, such as Mammoth Cave in Kentucky (Figure 13.2), were formed by the dissolving action of underground water containing dissolved carbon dioxide:

$$CaCO_3(s) + H_2O(l) + CO_2(aq) \longrightarrow Ca(HCO_3)_2(aq) \qquad [13.1]$$

We turn now to some important types of reactions in aqueous solutions, and we will examine some of the principles necessary to understand them.

## 13.1 NET IONIC EQUATIONS

Water is a good solvent for many molecular and ionic compounds. When these substances dissolve in water, they may dissociate or ionize; they are then called electrolytes (Section 12.4). We have seen that acids, bases, and salts are electrolytes. Substances that do not form ions in aqueous solution are called nonelectrolytes.

Reactions in aqueous solution frequently involve ions as reactants or products. For example, in neutralization or precipitation reactions, ions come together to form a nonelectrolyte or an insoluble solid (Section 3.2). In writing chemical equations for reactions in solution, it is often useful to indicate explicitly whether substances are present predominantly as ions or molecules. Consider the neutralization reaction between HCl and NaOH. We have previously written the equation for this reaction in the following form:

$$HCl(aq) + NaOH(aq) \longrightarrow H_2O(l) + NaCl(aq) \qquad [13.2]$$

Equations written in this fashion, showing the complete chemical formulas of reactants and products, are often called **molecular equations**. That term is a bit of a misnomer in this case, because all of the substances in the reaction other than $H_2O$ are present predominantly as ions. HCl, NaOH, and NaCl are strong electrolytes, and we can write the chemical equation to indicate that they are completely ionized in solution:

$$H^+(aq) + Cl^-(aq) + Na^+(aq) + OH^-(aq) \longrightarrow H_2O(l) + Na^+(aq) + Cl^-(aq) \qquad [13.3]$$

An equation written in this form–with all soluble strong electrolytes shown as they exist in aqueous solution, which is as ions—is known as a complete **ionic equation**. Notice that $Na^+$ and $Cl^-$ appear on both sides of the equation. Ions that appear in identical forms and with the same coefficients on both sides of an ionic equation are known as **spectator ions**. When the spectator ions are omitted from both sides of the equation (they cancel out like algebraic quantities), the resultant equation is called the **net ionic equation**. In our present example, omitting $Na^+$ and $Cl^-$ leaves us

$$H^+(aq) + OH^-(aq) \longrightarrow H_2O(l) \qquad [13.4]$$

This net ionic equation expresses the essential feature of the neutralization reaction between *any* strong acid and *any* strong base: $H^+$ and $OH^-$ ions combine to form $H_2O$. Notice that charge is conserved in the net ionic equation. That is, the sum of the charges is the same on both sides of the equation. In this example the sum is zero, but it need not be so.

Net ionic equations are widely used because they can illustrate the similarities between a large number of reactions involving electrolytes. You can write a net ionic equation for any reaction occurring in aqueous solution by following these conventions:

1. Only soluble strong electrolytes are written in ionic form.
2. The chemical formulas for soluble weak electrolytes and nonelectrolytes, as well as all insoluble substances, are written in "molecular" form.
3. Spectator ions are omitted.

---

## SAMPLE EXERCISE 13.1

Write the net ionic equation for neutralization of one of the acidic hydrogens of phosphoric acid, $H_3PO_4$, by sodium hydroxide.

**Solution:** The molecular equation for this reaction is

$$H_3PO_4(aq) + NaOH(aq) \longrightarrow H_2O(l) + NaH_2PO_4(aq)$$

$H_3PO_4$ is not one of the strong acids listed in Section 12.4, where the rules are given for predicting which substances are strong electrolytes. We would therefore predict that it is a weak electrolyte, a prediction that is indeed correct. Both NaOH and $NaH_2PO_4$ are strong electrolytes; $NaH_2PO_4$ ionizes to form $Na^+(aq)$ and $H_2PO_4^-(aq)$ ions. Thus the ionic equation for the reaction is

$$H_3PO_4(aq) + Na^+(aq) + OH^-(aq) \longrightarrow H_2O(l) + Na^+(aq) + H_2PO_4^-(aq)$$

Elimination of the spectator ion, $Na^+$, gives the net ionic equation,

$$H_3PO_4(aq) + OH^-(aq) \longrightarrow H_2O(l) + H_2PO_4^-(aq)$$

In sum, the weak acid $H_3PO_4$ reacts with the basic hydroxide ion to form water and the aqueous dihydrogen phosphate ion.

## PRACTICE EXERCISE

Write the net ionic equation for reaction of the strong acid $HNO_3$ with the weak base ammonia, $NH_3$.
***Answer:*** $NH_3(aq) + H^+(aq) \longrightarrow NH_4^+(aq)$

## SAMPLE EXERCISE 13.2

When aqueous solutions of barium nitrate, $Ba(NO_3)_2$, and sodium chromate, $Na_2CrO_4$, are mixed as shown in Figure 13.3, a precipitate of barium chromate, $BaCrO_4$, forms while sodium nitrate, $NaNO_3$, remains dissolved in the solution. Write the net ionic equation for the reaction.

**Solution:** The balanced molecular equation is

$$Ba(NO_3)_2(aq) + Na_2CrO_4(aq) \longrightarrow$$
$$BaCrO_4(s) + 2NaNO_3(aq)$$

All reactants and products involved in the reaction are ionic compounds ("salts") and are thus strong electrolytes. However, in writing the ionic equation, the formula for $BaCrO_4$ is not written in ionic form because it is a solid:

$$Ba^{2+}(aq) + 2NO_3^-(aq) +$$
$$2Na^+(aq) + CrO_4^{2-}(aq) \longrightarrow$$
$$BaCrO_4(s) + 2Na^+(aq) + 2NO_3^-(aq)$$

Elimination of the spectator ions, $Na^+$ and $NO_3^-$, gives the net ionic equation,

$$Ba^{2+}(aq) + CrO_4^{2-}(aq) \longrightarrow BaCrO_4(s)$$

This equation summarizes the general chemical fact that soluble barium salts will react with soluble chromate salts to form a precipitate of $BaCrO_4$. Thus the same net ionic equation would describe the reaction between $BaCl_2(aq)$ and $K_2CrO_4(aq)$.

### PRACTICE EXERCISE

When aqueous silver nitrate is added to a solution of calcium bromide, a precipitate of silver bromide forms. Write **(a)** the complete ionic equation and **(b)** the net ionic equation.
*Answers:*
**(a)** $2Ag^+(aq) + 2NO_3^-(aq) + Ca^{2+}(aq) + 2Br^-(aq) \longrightarrow$
$$2AgBr(s) + Ca^{2+}(aq) + 2NO_3^-(aq)$$
**(b)** $Ag^+(aq) + Br^-(aq) \longrightarrow AgBr(s)$

## 13.2 SOLUBILITY RULES

Recall from an earlier discussion (Section 3.1) that one of the most common and important metathesis, or double displacement, reactions is the precipitation reaction. We can do an experiment such as that illustrated in Figure 13.3 to determine whether a precipitate forms when two solutions are mixed. If a precipitate does form and we establish its identity, we can write a net ionic equation to describe the results. However, to predict in advance whether precipitation will occur, we must have knowl-

**FIGURE 13.3** Addition of an aqueous solution of sodium chromate, $Na_2CrO_4$, to an aqueous solution of barium nitrate, $Ba(NO_3)_2$, leads to formation of a yellow precipitate of barium chromate, $BaCrO_4$. (Donald Clegg and Roxy Wilson)

edge of solubilities, particularly of ionic compounds. There are no simple rules based on first principles to guide us. For example, although NaCl is quite soluble in water, only $1.3 \times 10^{-5}$ mol of AgCl dissolves in a liter of water at 25°C. For most practical purposes, AgCl is insoluble. In our discussions, any substance whose solubility is less than 0.01 mol/L will be referred to as insoluble.

Experimental observations have led to a set of general solubility rules for the common ions which appear in Table 13.1. The rules are organized according to anions. If you look at the rules closely, however, you will see a generalization that applies to cations: All common salts of the alkali metal ions (group 1A) and of the ammonium ion, $NH_4^+$, are water-soluble.

**TABLE 13.1**    Solubility rules for common ionic compounds in water[a]

| | **Mainly water soluble** |
|---|---|
| $NO_3^-$ | All nitrates are soluble. |
| $C_2H_3O_2^-$ | All acetates are soluble. |
| $ClO_3^-$ | All chlorates are soluble. |
| $Cl^-$ | All chlorides are soluble except AgCl, $Hg_2Cl_2$, and $PbCl_2$. |
| $Br^-$ | All bromides are soluble except AgBr, $Hg_2Br_2$, $PbBr_2$, and $HgBr_2$. |
| $I^-$ | All iodides are soluble except AgI, $Hg_2I_2$, $PbI_2$, and $HgI_2$. |
| $SO_4^{2-}$ | All sulfates are soluble except $CaSO_4$, $SrSO_4$, $BaSO_4$, $PbSO_4$, $Hg_2SO_4$, and $Ag_2SO_4$. |
| | **Mainly water insoluble** |
| $S^{2-}$ | All sulfides are insoluble except those of the 1A and 2A elements and $(NH_4)_2S$. |
| $CO_3^{2-}$ | All carbonates are insoluble except those of the 1A elements and $(NH_4)_2CO_3$. |
| $SO_3^{2-}$ | All sulfites are insoluble except those of the 1A elements and $(NH_4)_2SO_3$. |
| $PO_4^{3-}$ | All phosphates are insoluble except those of the 1A elements and $(NH_4)_3PO_4$. |
| $OH^-$ | All hydroxides are insoluble except those of the 1A elements, $Ba(OH)_2$, $Sr(OH)_2$, and $Ca(OH)_2$. |

[a] The following cations are considered: those of the 1A and 2A families, $NH_4^+$, $Ag^+$, $Al^{3+}$, $Cd^{2+}$, $Co^{2+}$, $Cr^{3+}$, $Cu^{2+}$, $Fe^{2+}$, $Fe^{3+}$, $Hg_2^{2+}$, $Hg^{2+}$, $Mn^{2+}$, $Ni^{2+}$, $Pb^{2+}$, $Sn^{2+}$, and $Zn^{2+}$.

## SAMPLE EXERCISE 13.3

Write balanced molecular, ionic, and net ionic equations for the precipitation reactions (if any) that occur when solutions of the following compounds are mixed: **(a)** $BaCl_2$ and $Na_2SO_4$; **(b)** KCl and $Na_2SO_4$.

**Solution:**    **(a)** Both $BaCl_2$ and $Na_2SO_4$ are soluble and ionize to give $Ba^{2+}$, $Cl^-$, $Na^+$, and $SO_4^{2-}$ ions. The possible precipitation products are $BaSO_4$ and NaCl. The $BaSO_4$ is insoluble according to the rule given in Table 13.1 for $SO_4^{2-}$ compounds. Therefore, the molecular equation is

$$BaCl_2(aq) + Na_2SO_4(aq) \longrightarrow$$
$$BaSO_4(s) + 2NaCl(aq)$$

The ionic equation is

$$Ba^{2+}(aq) + 2Cl^-(aq) + 2Na^+(aq) + SO_4^{2-}(aq) \longrightarrow$$
$$BaSO_4(s) + 2Na^+(aq) + 2Cl^-(aq)$$

Both $Na^+(aq)$ and $Cl^-(aq)$ are spectator ions. The net ionic equation is

$$Ba^{2+}(aq) + SO_4^{2-}(aq) \longrightarrow BaSO_4(s)$$

**(b)** Both reactants are soluble and ionize in solution. There are no possible insoluble salts resulting from reaction. Both NaCl and $K_2SO_4$ are soluble. Therefore, there is no reaction; the solutes merely mix in the solution.

## PRACTICE EXERCISE

Write net ionic equations for the reactions that occur when solutions of the following compounds are mixed: **(a)** NaOH and $Co(NO_3)_2$; **(b)** $Ca(C_2H_3O_2)_2$ and $H_2SO_4$ (acetic acid, $HC_2H_3O_2$, is a weak acid).

*Answers:*
**(a)** $2OH^-(aq) + Co^{2+}(aq) \longrightarrow Co(OH)_2(s)$
**(b)** $Ca^{2+}(aq) + 2C_2H_3O_3^-(aq) + 2H^+(aq) +$
$\qquad SO_4^{2-}(aq) \longrightarrow 2HC_2H_3O_2(aq) + CaSO_4(s)$

## 13.3 FACTORS DRIVING CHEMICAL REACTIONS

For a chemical reaction to take place, a net chemical change in the system must occur. Mixing aqueous solutions of LiCl and $Ba(NO_3)_2$ produces no precipitation because both possible products—$LiNO_3$ and $BaCl_2$—are soluble in water. Neither is there any other type of chemical reaction on mixing LiCl and $Ba(NO_3)_2$ solutions. Chemical reaction occurs only when some driving force is present that leads to chemical change.

A summary of reaction types provides a basis for predicting chemical change and helps us characterize reactions that occur in solution. In general, chemical reactions proceed essentially to completion when some form of driving force is present. This driving force may be formation of a gas, a precipitate, or a nonelectrolyte, but other types of chemical change also possess inherent driving forces. We observe a chemical change in aqueous medium whenever one or more of the following occur:

1. *A precipitate is formed.* We have already seen several examples of this type.

2. *A nonelectrolyte is formed.* The neutralization reaction is an example of this type of reaction; $H_2O(l)$ is an example of a nonelectrolyte.

3. *A weak electrolyte is formed.* As an example, consider the reaction of sodium acetate, $NaC_2H_3O_2$, with hydrochloric acid. The net ionic equation describes the formation of acetic acid, a weak electrolyte:

$$H^+(aq) + C_2H_3O_2^-(aq) \longrightarrow HC_2H_3O_2(aq) \qquad [13.5]$$

We will encounter many more examples of this type of reaction in Chapters 17 and 18, when we discuss aqueous equilibria involving acids, bases, and other substances.

4. *Oxidation and reduction occur.* You will recall that we encountered the idea of oxidation numbers in Chapter 8. Clearly, when substances undergo changes in their oxidation numbers, chemical change has occurred. In a later section of this chapter we will look into the nature of oxidation-reduction reactions and the methods for writing balanced net ionic equations for them.

5. *One or more covalent bonds are formed or broken.* Many reactions involve the rupture and formation of covalent bonds in gaseous or solid substances. Reactions in which covalent bonds are formed and broken also occur among substances in water. For example, ethanol is oxidized by potassium dichromate to form acetaldehyde, as illustrated in Figure 13.4:

$$2H^+(aq) + 3C_2H_5OH(aq) + Cr_2O_7^{2-}(aq) \longrightarrow 3C_2H_4O(aq) + Cr_2O_3(s) + 4H_2O(l) \quad [13.6]$$

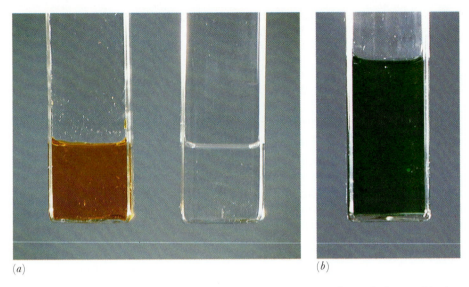

(a)                                                                          (b)

FIGURE 13.4   (a) When ethanol, $C_2H_5OH$, is added to an orange aqueous solution of potassium dichromate, $K_2Cr_2O_7$, the reaction shown in Equation 13.6 occurs, forming acetaldehyde and the green $Cr^{3+}(aq)$ ion (b). (Donald Clegg and Roxy Wilson)

In this reaction not only have new covalent bonds formed, but oxidation numbers have changed and an insoluble substance as well as a nonelectrolyte ($H_2O$) have formed. This example demonstrates that any particular reaction might properly be fitted into more than one of the categories we have just defined.

6. *A gas is formed.* Formation of a product that can escape from the solution as a gas is another important driving force. The net ionic equations for three common examples of this reaction type follow:

**a.** A carbonate reacting with an acid to form $CO_2$ and $H_2O$:

$$CO_3{}^{2-}(aq) + 2H^+(aq) \longrightarrow CO_2(g) + H_2O(l) \qquad [13.7]$$

**b.** A sulfite reacting with an acid to form $SO_2$ and $H_2O$:

$$SO_3{}^{2-}(aq) + 2H^+(aq) \longrightarrow SO_2(g) + H_2O(l) \qquad [13.8]$$

**c.** A sulfide reacting with an acid to form $H_2S$:

$$MnS(s) + 2H^+(aq) \longrightarrow Mn^{2+}(aq) + H_2S(g) \qquad [13.9]$$

---

## SAMPLE EXERCISE 13.4

Write balanced complete ionic and net ionic equations for any reactions that occur when the following compounds are mixed: **(a)** $FeCO_3(s)$ and $HCl(aq)$; **(b)** $NiS(s)$ and $HCl(aq)$.

**Solution:**   **(a)** This reaction is similar to Equation 13.7, but the reactant carbonate is a solid, insoluble in water.

Complete ionic:

$$FeCO_3(s) + 2H^+(aq) + 2Cl^-(aq) \longrightarrow$$
$$Fe^{2+}(aq) + 2Cl^-(aq) + H_2O(l) + CO_2(g)$$

Net ionic:
$$FeCO_3(s) + 2H^+(aq) \longrightarrow$$
$$Fe^{2+}(aq) + H_2O(l) + CO_2(g)$$

**(b)** This reaction is of the type of Equation 13.9. Most sulfides will react with acids even though they are insoluble. Similarly, solid carbonates and hydroxides will generally react with acids, as exemplified in part (a).

Complete ionic:

$$NiS(s) + 2H^+(aq) + 2Cl^-(aq) \longrightarrow$$
$$Ni^{2+}(aq) + 2Cl^-(aq) + H_2S(g)$$

Net ionic:

$$NiS(s) + 2H^+(aq) \longrightarrow Ni^{2+}(aq) + H_2S(g)$$

Notice that in writing the net ionic equation, in both (a) and (b), both solids and gases are included. The only species that goes through the reaction unchanged is the $Cl^-$ ion.

**PRACTICE EXERCISE**

Write the net ionic equation for reaction of solid barium sulfite with aqueous sulfuric acid.

***Answer:*** $BaSO_3(s) + 2H^+(aq) + SO_4^{2-}(aq) \longrightarrow$
$$BaSO_4(s) + SO_2(g) + H_2O(l)$$

## Chemical Synthesis

Formation of new substances is an important aspect of chemistry. The design and execution of a plan for preparing a new drug or for preparing in the laboratory a complex substance found in nature often require great ingenuity and lengthy research. By contrast, the preparation of simple salts from aqueous solution reactions can often be accomplished quite easily by taking advantage of one or another of the driving forces that impel reactions.

Suppose that we need a sample of solid lead sulfate, $PbSO_4$. How might it be prepared? We note from the solubility rules (Table 13.1) that $PbSO_4$ is insoluble in water. Thus we can prepare the solid material by mixing a soluble lead salt with a soluble sulfate salt:

$$Na_2SO_4(aq) + Pb(NO_3)_2(aq) \longrightarrow PbSO_4(s) + 2NaNO_3(aq) \qquad [13.10]$$

In choosing the reagents for precipitation reactions, we normally employ the chloride salt or nitrate salt of the desired cation, because these are generally the least expensive soluble salts of the cations. The common sources of anions are the acids, sodium salts, or potassium salts. To make the most cost-effective use of chemicals, the amounts of solution and their concentrations should be chosen so that stoichiometrically equivalent quantities of the two soluble salts are employed: for example, one mole of $Na_2SO_4$ for each mole of $Pb(NO_3)_2$. If, however, there were an excess of one or the other substance, solid $PbSO_4$ would still be formed in a quantity determined by the amount of limiting reagent (Section 3.7). After formation the solid would be filtered to remove the solution, washed with pure water to remove traces of the solution, and then dried. The net ionic equation for the reaction is

$$Pb^{2+}(aq) + SO_4^{2-}(aq) \longrightarrow PbSO_4(s) \qquad [13.11]$$

We can see from this equation that formation of the desired product is independent of the particular soluble lead salt or sulfate salt we choose, as long as the accompanying ions do not themselves react.

In the preparation of $PbSO_4$ just outlined, we took advantage of the formation of a precipitate as the driving force for reaction. This driving force can be combined with another force to obtain $PbSO_4$ by a different route. We react the metal oxide PbO, which is soluble in acidic solution, with $H_2SO_4$ (see Figure 13.5):

$$PbO(s) + H_2SO_4(aq) \longrightarrow PbSO_4(s) + H_2O(l) \qquad [13.12]$$

In this case there remains no soluble salt in solution; as a result, if we

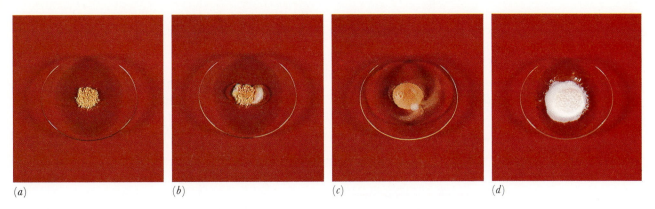

**FIGURE 13.5**  Action of sulfuric acid on lead(II) oxide. Notice that the colored oxide is converted to a colorless solid, $PbSO_4$. (Donald Clegg and Roxy Wilson)

(a)        (b)        (c)        (d)

employ just a tiny excess of $H_2SO_4$ to ensure that all the PbO reacts, the products are solid $PbSO_4$ and a solution containing only the slight excess of $H_2SO_4$. The net ionic equation for this route is

$$PbO(s) + 2H^+(aq) + SO_4{}^{2-}(aq) \longrightarrow PbSO_4(s) + H_2O(l) \quad [13.13]$$

Now suppose we wish to prepare a salt that is water-soluble. In this event we need to find a reaction in which the soluble compound is left as the only solute. We could then use the solution directly in a subsequent reaction or evaporate off the water to recover the desired material as a solid. For example, suppose that we wish to prepare copper(II) bromide, which is very water-soluble. One approach would be to mix solutions of $CuSO_4$ and $BaBr_2$. The driving force for reaction is the formation of insoluble $BaSO_4$:

$$CuSO_4(aq) + BaBr_2(aq) \longrightarrow BaSO_4(s) + CuBr_2(aq) \quad [13.14]$$

(Note that the substance in which we are most interested, $CuBr_2$, would not appear in the net ionic equation for this reaction, because the constituent ions remains in solution. Thus we find at times the complete ionic equation, or the molecular equation, as written above, to be the most useful description of the reaction.) The $BaSO_4$ can be removed from the reaction mixture by filtration, and the $CuBr_2$ is obtained by evaporating the remaining solution to dryness.

Alternatively, we might prepare $CuBr_2$ by taking advantage of the formation of a gas as a driving force; for example,

$$CuCO_3(s) + 2HBr(aq) \longrightarrow CuBr_2(aq) + H_2O(l) + CO_2(g) \quad [13.15]$$

## SAMPLE EXERCISE 13.5

Devise a synthesis for $SrCO_3$. Use this material to prepare $Sr(ClO_3)_2$.

**Solution:**  $SrCO_3$ is an insoluble salt (Table 13.1). We can form it by reacting a soluble carbonate such as $Na_2CO_3$ with a soluble strontium salt such as $SrCl_2$:

$$Na_2CO_3(aq) + SrCl_2(aq) \longrightarrow$$
$$SrCO_3(s) + 2NaCl_2(aq)$$

After filtering and washing the precipitate, it could be reacted with a stoichiometrically equivalent amount of HCl solution, then with a solution of $Pb(ClO_3)_2$:

$$SrCO_3(s) + 2HCl(aq) \longrightarrow$$
$$SrCl_2(aq) + H_2O(l) + CO_2(g)$$

$$SrCl_2(aq) + Pb(ClO_3)_2(aq) \longrightarrow$$
$$Sr(ClO_3)_2(aq) + PbCl_2(s)$$

In the first of these reactions we take advantage of the formation of gas as driving force; in the second, the driving force is the formation of insoluble $PbCl_2$.

**PRACTICE EXERCISE**

Write a balanced molecular equation or equations for synthesis of $Zn(ClO_4)_2(aq)$, beginning with ZnO.
***Answer:*** $ZnO(s) + 2HClO_4(aq) \longrightarrow$
$$Zn(ClO_4)_2(aq) + H_2O(l)$$

## 13.4 ACIDS AND BASES IN WATER

One of the most important classifications of substances is in terms of acids and bases. From the earliest days of chemistry it was recognized that certain substances called acids possess a characteristic sour taste (for example, citric acid in lemon juice) and can dissolve active metals such as zinc or tin. Acids also cause certain vegetable dyes to turn characteristic colors. For example, litmus turns red on contact with acids. Bases likewise possess a set of characteristic properties that can be used to identify them. They have a characteristic bitter taste, as does milk of magnesia, which is a suspension of the base $Mg(OH)_2$. Bases dissolved in water impart a slippery feel (soap is a good example) and cause litmus to turn blue. Bases also react with many dissolved metal salts to form precipitates.

The fact that all acids and all bases show certain characteristic chemical properties suggests that there must be an essential feature common to the members of each class. Antoine Lavoisier (1743–1794) proposed that acids were oxygen-containing substances. In fact, Lavoisier derived the name *oxygen* from the Greek word meaning "acid-former." However, hydrochloric acid, at that time a well-known substance, was eventually shown to contain no oxygen. By 1830 it had become evident that hydrogen was the one element common to all acids.

If hydrogen is the one element present in all acids, how does it give rise to the characteristic properties of acids in aqueous solution? In the 1880s Svante Arrhenius (1859–1927) suggested the existence of ions to explain why acids, bases, and salts in water impart electrical conductivity to the solutions. He was led from this basic hypothesis to further postulate that acids are substances that form $H^+$ ions in water solutions and that bases produce $OH^-$ ions.

We now need to take a closer look at acids and bases to understand more fully their role in reactions occurring in aqueous solution. We will see that the properties of acids and bases that we observe very much depend on water being the solvent.

### The Proton in Water

If you have had much chemical laboratory experience to date, you have probably encountered hydrochloric acid. Hydrochloric acid is an aqueous solution of hydrogen chloride gas. The reagent hydrochloric acid sold commercially is 37 to 38 percent HCl by weight and has a density of $1.19 \text{ g/cm}^3$. From these data we can calculate that this concentrated hydrochloric acid solution is 12 $M$.

What is it about water that causes a molecule such as HCl to come apart, or dissociate, into $H^+$ and $Cl^-$ ions? The first and most obvious point to make is that water is a polar liquid (Section 11.4). Consequently, the oxygen atom, which bears a partial negative charge, is attracted to the partially positive hydrogen end of the polar HCl molecule. Second, the water molecule has unshared electron pairs on the oxygen atom, which are capable of forming a covalent bond to the hydrogen ion in the following manner:

$$Cl-H\cdots\ddot{\underset{\displaystyle H}{O}}-H \longrightarrow Cl^- + \left[H-\overset{\displaystyle\cdot\cdot}{\underset{\displaystyle H}{O}}-H\right]^+ \qquad [13.19]$$

In the reaction between HCl and $H_2O$ shown in Equation 13.19, the proton (an $H^+$ ion) transfers from HCl to the water molecule, forming the hydronium ion.

The situation in water is actually much more complex than Equation 13.19 suggests. We've learned (Section 11.3) that hydrogen bonds exist throughout liquid water. The existence of this hydrogen-bond network is responsible for many of the special properties of water (for example, its high polarity and high melting and boiling points). Much research has been devoted to learning how $H^+$ ions fit into the complex structure of liquid water. Experimental studies show that, in part, the $H^+$ ions must exist as hydronium ions. In fact, it is possible to isolate $H_3O^+Cl^-$, $H_3O^+ClO_4^-$, and other salts in which there is clearly an $H_3O^+$ ion in the solid lattice. But just as water molecules are strongly hydrogen bonded to one another, the $H_3O^+$ ion in solution is hydrogen bonded to other water molecules. Thus ions such as the two shown in Figure 13.6 may exist and have in fact been found.

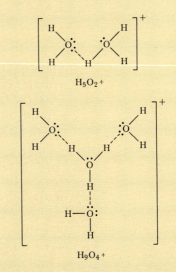

**FIGURE 13.6** Two possible structural forms for the proton in water, in addition to $H_3O^+$. There is good experimental evidence for the existence of both these species.

Why is HCl so soluble in water? Its solubility in many other solvents, such as benzene ($C_6H_6$), is very low. Its high solubility in water is due to its chemical reaction with water, which produces aquated hydrogen and chloride ions:

$$HCl(g) \xrightarrow{\ H_2O\ } H^+(aq) + Cl^-(aq)$$

The "(aq)" that follows both $H^+$ and $Cl^-$ indicates that the ions are separated from one another and aquated, or hydrated, by water molecules, as illustrated in Figure 12.2. The hydrogen ion fits into the water structure differently from any other ion. Keep in mind that the hydrogen ion is simply the proton, with no surrounding valence electron. Because of the unique hydrogen-bonding properties of water and the existence of lone pairs of electrons on the oxygen, the proton fits into the water environment by forming the **hydronium ion**, $H_3O^+$:

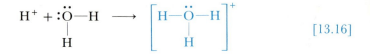

$$\text{H}^+ + :\overset{\displaystyle ..}{\underset{\displaystyle |}{\text{O}}}\text{—H} \longrightarrow \left[ \text{H—}\overset{\displaystyle ..}{\underset{\displaystyle |}{\text{O}}}\text{—H} \right]^+ \qquad [13.16]$$

The hydronium ion is strongly hydrogen-bonded to adjacent water molecules. Both $\text{H}^+(aq)$ and $\text{H}_3\text{O}^+(aq)$ are used to represent the aquated hydrogen ion surrounded by solvent water. Thus we may write the reaction of HCl with water to form hydrochloric acid in either of the following ways:

$$\text{HCl}(aq) + \text{H}_2\text{O}(l) \longrightarrow \text{H}_3\text{O}^+(aq) + \text{Cl}^-(aq) \qquad [13.17]$$

$$\text{HCl}(aq) \longrightarrow \text{H}^+(aq) + \text{Cl}^-(aq) \qquad [13.18]$$

The aquated proton will ordinarily be represented throughout the remainder of this book as $\text{H}^+(aq)$. However, regardless of whether $\text{H}^+(aq)$ or $\text{H}_3\text{O}^+(aq)$ is used to represent the aquated proton, you should keep in mind that *acidic solutions are formed by a chemical reaction in which an acid transfers a proton ($H^+$) to water.*

**Reactions of Nonmetal Oxides with Water**

Hydrogen-containing compounds like HCl are not the only substances that dissolve in water to form acidic solutions. You may recall from Section 5.3 that nonmetal oxides such as $\text{SO}_2$ or $\text{CO}_2$ also dissolve in water to form acidic solutions. These oxides form acids through chemical reaction with water. For example, carbon dioxide dissolves in water and reacts with it to some extent to form carbonic acid, $\text{H}_2\text{CO}_3$:

$$\text{CO}_2(g) + \text{H}_2\text{O}(l) \longrightarrow \text{H}_2\text{CO}_3(aq) \qquad [13.20]$$

$\text{H}_2\text{CO}_3$ is a weak acid. This means that at any instant only a small fraction of the $\text{H}_2\text{CO}_3$ molecules have transferred a proton to water to form $\text{H}^+(aq)$:

$$\text{H}_2\text{CO}_3(aq) \rightleftharpoons \text{H}^+(aq) + \text{HCO}_3{}^-(aq) \qquad [13.21]$$

Notice that we have employed a double arrow in Equation 13.21. This signifies that both the forward reaction *and* the reverse reaction are significant. On the one hand, $\text{H}_2\text{CO}_3$ transfers a proton to water to form $\text{H}^+(aq)$ and $\text{HCO}_3{}^-(aq)$; on the other hand, these two ions react to form $\text{H}_2\text{CO}_3$. When the rate of the forward reaction in Equation 13.21 equals the rate of the reverse reaction, the system is in a state of **equilibrium**. At equilibrium nothing seems to be happening in the solution. In fact, both the forward and reverse reactions are occurring, but their rates are balanced, so there is no overall change in concentrations with time. In the particular case of Equation 13.21, the reaction does not proceed very far to the right before equilibrium is attained; only a few percent of the $\text{H}_2\text{CO}_3$ molecules are converted to $\text{H}^+(aq)$ and $\text{HCO}_3{}^-(aq)$. We will deal with the quantitative aspects of aqueous solution equilibria in Chapters 17 and 18.

## SAMPLE EXERCISE 13.6

When $SO_2(g)$ dissolves in water it forms an acidic solution. **(a)** Write a chemical equation for the reaction of $SO_2$ with water. **(b)** Indicate which species is present in the aqueous solution of $SO_2$ to impart acidic properties to the solution.

**Solution:** **(a)** The reaction of $SO_2(g)$ with water results in the formation of sulfurous acid:

$$SO_2(g) + H_2O(l) \longrightarrow H_2SO_3(aq)$$

Sulfurous acid is a weak acid (Section 12.4). This means that it does not ionize to a large extent in water.

**(b)** The species present in solution that imparts acidic properties to the solution is $H^+(aq)$, as it is for all acids in aqueous solvent. $H^+(aq)$ is formed as a result of the dissociation of $H_2SO_3(aq)$:

$$H_2SO_3(aq) \rightleftharpoons H^+(aq) + HSO_3^-(aq)$$

Because $H_2SO_3(aq)$ is not a strong acid, the dissociation process proceeds to only a slight extent.

### PRACTICE EXERCISE

**(a)** Write the reaction for formation of an aqueous solution of KOH by reaction of $K_2O(s)$ with water. **(b)** What species imparts basic properties to the aqueous solution of KOH in water?
*Answers:*
**(a)** $K_2O(s) + H_2O(l) \longrightarrow 2K^+(aq) + 2OH^-(aq)$
**(b)** $OH^-(aq)$

---

The nature of the reaction between an acid and water as we have just described it was first appreciated by the Danish chemist Johannes Brønsted (1879–1947) and the English chemist Thomas M. Lowry (1874–1936). Brønsted and Lowry recognized that acid-base behavior could be related to the ability of substances to transfer protons. In 1923 Brønsted and Lowry independently proposed that *acids be defined as substances that are capable of donating a proton, and bases as substances capable of accepting a proton.* In these terms, when HCl dissolves in water, it acts as an acid, donating a proton to the solvent. The solvent, $H_2O$, at the same time acts as a base, accepting a proton.

Because the emphasis in the Brønsted-Lowry theory is on proton transfer, the theory is applicable to reactions occurring in other than aqueous solution. For example, the reaction between HCl and $NH_3$ to form $NH_4Cl$ entails a proton transfer from the acid HCl to the base $NH_3$:

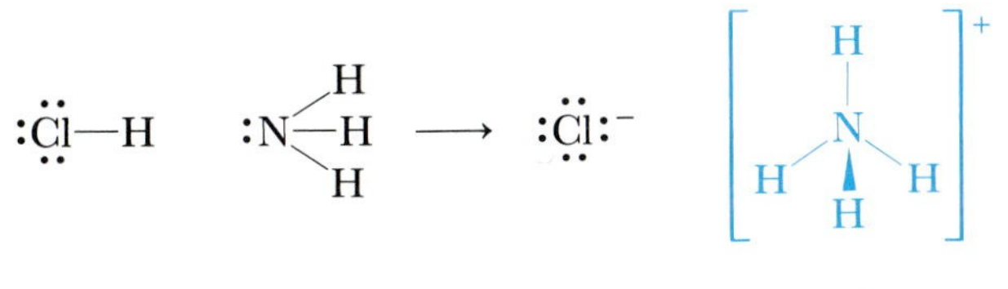

[13.22]

This reaction can occur in solvents other than water, if the reactants are soluble in the medium. It also can occur in the gas phase. The hazy film that forms on the windows of general chemistry laboratories and on glassware in the lab is largely $NH_4Cl$, formed by the gas-phase reaction of HCl and $NH_3$.

In earlier discussions we have applied the term "base" to substances that produce an excess of $OH^-$ ions in aqueous solution, in accordance with Arrhenius's theory of acids and bases. Thus aqueous solutions of ammonia are basic because $NH_3$ reacts with $H_2O$ to form $NH_4^+$ and $OH^-$:

## 13.5 THE BRØNSTED-LOWRY THEORY OF ACIDS AND BASES

$$H_2O(l) + NH_3(aq) \rightleftharpoons NH_4^+(aq) + OH^-(aq) \qquad [13.23]$$

In terms of the Brønsted-Lowry theory we say that $NH_3$ acts as a base in this reaction because it accepts a proton from the solvent water. A great many ions and neutral molecules are capable of similarly acting as proton acceptors, and we will see several examples in the following discussion.

Notice that we used a double arrow in Equation 13.23. This signifies that the reverse reaction is significant. Just as $NH_3$ reacts with water to form $NH_4^+$ and $OH^-$ ions, these two ions react to form $NH_3$ and $H_2O$. For example, when aqueous solutions of $NH_4Cl$ and $NaOH$ are mixed, the $NH_4^+$ and $OH^-$ ions react:

$$NH_4^+(aq) + OH^-(aq) \longrightarrow NH_3(aq) + H_2O(l) \qquad [13.24]$$

This reaction, of course, is just the reverse of the reaction in Equation 13.23. In Equation 13.24, $NH_4^+$ acts as the proton donor and $OH^-$ as the proton acceptor. Thus we see that as the reaction in Equation 13.23 proceeds in the forward direction (to the right), $H_2O$ is the acid and $NH_3$ is the base. As it proceeds in the reverse direction, $NH_4^+$ is the acid and $OH^-$ is the base.

To summarize, then, a **Brønsted acid** is defined as a substance capable of donating a proton; a **Brønsted base** is defined as a substance capable of accepting a proton. Equation 13.23 illustrates a general characteristic of any acid-base reaction as viewed from the perspective of the Brønsted-Lowry theory: Every Brønsted acid has associated with it a **conjugate base**, formed from the acid by loss of a proton; every Brønsted base has associated with it a **conjugate acid**, formed from the base by addition of a proton.* Thus $NH_3$ is the conjugate base of the acid $NH_4^+$, and $H_2O$ is the conjugate acid of the base $OH^-$. An acid and a base such as $H_2O$ and $OH^-$, which differ only in the presence or absence of a proton, are called a **conjugate acid-base pair**.

Notice that an acid can be a charged species, such as $NH_4^+$, or a neutral species, such as $H_2O$. Similarly, bases can be neutral or charged. Note also that some substances can act as an acid in one situation and a base in another. Thus $H_2O$ is a base in its reaction with $HCl$ (Equation 13.17), and an acid in its reaction with $NH_3$ (Equation 13.23).

---

### SAMPLE EXERCISE 13.7

**(a)** Write the formula for the conjugate base of each of the following: HI; $HNO_2$; $PH_4^+$; $HSO_4^-$.
**(b)** Write the formula for the conjugate acid of each of the following: $HSO_4^-$; $H_2O$; $NH_2^-$; $F^-$.

**Solution:** **(a)** The conjugate base of each of the species listed is just the species with one proton removed: $I^-$; $NO_2^-$; $PH_3$; $SO_4^{2-}$.
  **(b)** The conjugate acid of each of the species listed

is just the species with one proton added: $H_2SO_4$; $H_3O^+$; $NH_3$; HF.
  Notice that a species such as $HSO_4^-$ is capable of acting as either an acid or a base.

**PRACTICE EXERCISE** _____
Write the formula for the conjugate acid of each of the following: $S^{2-}$; $ClO_3^-$; $HPO_4^{2-}$; CO.
***Answers:*** $HS^-$; $HClO_3$; $H_2PO_4^-$; $HCO^+$

---

* The word *conjugate* means "joined together as a pair," or "coupled."

A conjugate acid-base pair is always linked by an equilibrium such as the one in Equation 13.23 relating $NH_3$ and $NH_4^+$. As we have already noted, an equilibrium may lie far to the right, toward products, or far to the left, toward reactants. This fact leads to an interesting and important generalization: The more readily a substance gives up a proton, the less readily its conjugate base will accept a proton. Similarly, the more readily a base accepts a proton, the less readily will the conjugate acid transfer a proton. In other words, *the stronger an acid, the weaker its conjugate base; the weaker an acid, the stronger its conjugate base*. For example, if the acid on the left in an equilibrium equation is a strong acid, it will readily transfer a proton to an acceptor. In that case, the conjugate base formed cannot be a strong proton acceptor, so there is little tendency for the reaction to occur in the reverse direction. The result is that the equilibrium will lie far to the right. HCl is a good proton donor toward $H_2O$ because its conjugate base, $Cl^-$, has less attraction for protons than does water. The proton is therefore transferred to $H_2O$ to form $H^+(aq)$.

Figure 13.7 displays some common acids and their conjugate bases. $H^+(aq)$ is the strongest proton donor that can exist at equilibrium in aqueous solution. Thus acids listed above $H^+(aq)$ in Figure 13.7 completely transfer protons to water to form $H^+(aq)$. Likewise, $OH^-(aq)$ is the strongest base that can exist at equilibrium in aqueous solution. Any stronger proton acceptor will completely react with water, removing a proton to form $OH^-$ ions.

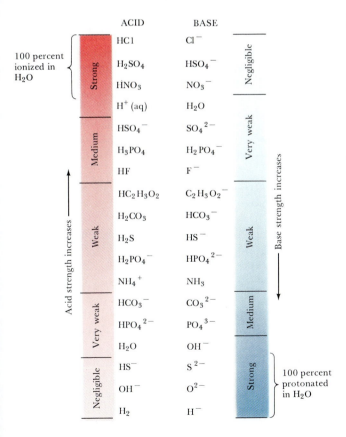

**FIGURE 13.7** Relative strengths of some common conjugate acid-base pairs, which are listed opposite one another in the two columns. The conjugate bases of strong acids have negligible basicity; the conjugate acids of strong bases have negligible acidity. The strongest acid that can exist in water is the aquated proton, $H^+(aq)$. Any substance that gives up a proton more readily than the solvent simply loses that proton to the solvent, forming $H^+(aq)$. Such strong acids, for example, $HNO_3$, are essentially completely ionized in water. Similarly, any base that is a stronger proton acceptor than $OH^-$ removes protons from the water to become essentially completely protonated.

## SAMPLE EXERCISE 13.8

Hydrocyanic acid, HCN, is ionized in water by the reaction

$$HCN(aq) \rightleftharpoons H^+(aq) + CN^-(aq)$$

to a lesser extent than is an HF solution of the same concentration. What is the conjugate base of HCN? Is it a stronger or weaker base than $F^-$?

**Solution:** The conjugate base of HCN is $CN^-$, the ion that remains after a proton has been lost to the solvent. HCN dissociates to a lesser extent than does HF, which means that the tendency of the reverse reaction to occur is greater. In the reverse reaction

the conjugate base, $CN^-$, accepts a proton from the solvent. In other words, $CN^-$ is a stronger conjugate base than is $F^-$.

### PRACTICE EXERCISE

The dimethylammonium ion, $(CH_3)_2NH_2^+$, is a weak acid, ionized to a slight degree in water. **(a)** What is the conjugate base of the dimethylammonium ion? **(b)** How does the strength of this base compare with that of $Cl^-$?
**Answers:** **(a)** dimethylamine, $(CH_3)_2NH$; **(b)** dimethylamine is a stronger base than $Cl^-$

## 13.6 OXIDATION-REDUCTION REACTIONS

In this chapter we have seen that acids and bases can be defined on the basis of whether they are proton donors or acceptors. Acid-base reactions, in the Brønsted-Lowry theory, are viewed as proton transfer processes. A second important class of processes—called oxidation-reduction, or "redox," reactions—involves the transfer of electrons. Oxidation-reduction reactions occur under a variety of reaction conditions, but those occurring in aqueous solution are most numerous and important. An **oxidation-reduction reaction** is one in which one or more substances undergo a change in oxidation number. (Oxidation numbers are discussed in Section 8.10; this might be a good time to review that material.) A simple but dramatic example of such a reaction is the thermite reaction, in which powdered Al reacts with powdered $Fe_2O_3$, as shown in Figure 13.1. The equation for this reaction is

$$2Al(s) + Fe_2O_3(s) \longrightarrow 2Fe(l) + Al_2O_3(s)$$

$$\begin{array}{ccc} & & \\ 0 & +3 \;\; -2 & 0 \;\; +3 \;\; -2 \end{array} \qquad [13.25]$$

In this reaction the oxidation state of Al changes from 0 to $+3$; that of Fe changes from $+3$ to 0, as shown in color above the reactants and products in Equation 13.25.

In some oxidation-reduction reactions a clear transfer of electrons occurs. For example, metallic aluminum, the reactant in Equation 13.25, has all of its valence electrons. In the product, $Al_2O_3$, $Al^{3+}$ ions are distributed between oxide ions ($O^{2-}$) in the solid lattice. Thus in the course of reaction each aluminum atom loses three electrons. At the same time, each iron gains three electrons in going from an $Fe^{3+}$ ion in the reactant $Fe_2O_3$ to iron metal. The overall effect is a transfer of three electrons per mole from Al to $Fe^{3+}$.

In other reactions there are changes in oxidation states, but we can't say that any substance literally gains or loses one or more electrons. As a simple example, consider the oxidation of methyl alcohol, $CH_3OH$, to formaldehyde, HCHO, by hydrogen peroxide, in aqueous solution:

$$CH_3OH(aq) + H_2O_2(aq) \longrightarrow HCHO(aq) + 2H_2O(l) \qquad [13.26]$$

The oxidation number of the atoms in methyl alcohol and formaldehyde are as follows:

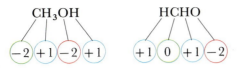

We see that the oxidation number of carbon has increased from $-2$ to $0$. At the same time the oxidation number of the oxygen atoms in hydrogen peroxide has decreased from $-1$ to $-2$. The overall charge on carbon is more positive in formaldehyde than in methyl alcohol, but there is not a literal loss of two electrons from carbon in forming the product.

The substance in an oxidation-reduction reaction that undergoes an increase in oxidation number is said to be oxidized. In the course of being oxidized, a different substance is reduced, so the first substance is called a **reducing agent**, or **reductant**. Aluminum in Equation 13.25 is a reducing agent. A substance that undergoes a reduction in oxidation number is said to be reduced. In the course of being reduced, some other substance is oxidized, so the first substance is called an **oxidizing agent**, or **oxidant**. In Equation 13.26 hydrogen peroxide is an oxidizing agent.

## SAMPLE EXERCISE 13.9

Phosphine, $PH_3$, a highly toxic gas, burns readily in air to form phosphorus(V) oxide and water vapor. The balanced equation for this reaction is

$$2PH_3(g) + 4O_2(g) \longrightarrow P_2O_5(s) + 3H_2O(g)$$

Identify the elements oxidized and reduced, and indicate which substances are the oxidizing and reducing agents.

**Solution:** We first assign oxidation numbers to all atoms in the reaction

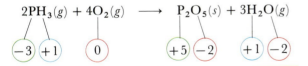

We see that phosphorus undergoes an increase in oxidation number from $-3$ to $+5$ and that oxygen undergoes a reduction from $0$ to $-2$. Thus phosphorus is oxidized and $PH_3$ is the reducing agent. Oxygen, $O_2$, is reduced and is therefore the oxidizing agent.

### PRACTICE EXERCISE
Identify the oxidizing and reducing agents in the following oxidation-reduction equation:

$$2H_2O(l) + Al(s) + MnO_4^-(aq) \longrightarrow$$
$$Al(OH)_4^-(aq) + MnO_2(s)$$

***Answer:*** Al is the reducing agent; $MnO_4^-(aq)$ is the oxidizing agent

Oxidation-reduction reactions are among the most common and important of all chemical processes. They include such diverse chemical processes as the combustion of fuels in automobile engines, the metabolism of foods in our bodies, and the reactions of active metals with acidic solutions. We will be concerned in this section with balancing oxidation-reduction equations. To some extent, balancing equations is no more than a form of bookkeeping. We know that we must obey the law of

## 13.7 BALANCING OXIDATION-REDUCTION EQUATIONS

conservation of mass; the amount of each element must be the same on both sides of the equation. Also, in oxidation-reduction equations we have the additional requirement of balancing the gains and losses of electrons. If a substance loses a certain number of electrons during a chemical process, and is thereby oxidized, another substance must gain that same number of electrons, and be thereby reduced. These gains and losses of electrons occur simultaneously during the reaction.

**Oxidation Number Method**

One simple method for balancing oxidation-reduction equations, called the **oxidation number method**, is to match any increase in oxidation number with a corresponding decrease in oxidation number. As an example, consider the process by which powdered aluminum is used to reduce pyrolusite, a manganese ore, to manganese metal. This process is a variation of the thermite reaction, illustrated in Figure 13.1. The unbalanced equation for the process is

$$Al(l) + MnO_2(s) \longrightarrow Al_2O_3(s) + Mn(l) \qquad [13.27]$$

We proceed by first assigning oxidation numbers to all the elements in the reaction:

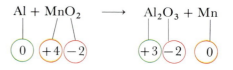

We then note the changes in oxidation number in going from reactants to products:

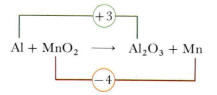

We must now choose coefficients to meet the requirement that the increase in oxidation number by aluminum equal the decrease in oxidation number by manganese. To balance these changes we need 4 Al for every 3 Mn atoms: $4 \times (+3) = 3 \times (-4)$. This gives us

$$4Al + 3MnO_2 \longrightarrow Al_2O_3 + Mn$$

We can now complete the balancing process for each metal by inspection. We need a 2 before $Al_2O_3$ and a 3 before Mn:

$$4Al + 3MnO_2 \longrightarrow 2Al_2O_3 + 3Mn$$

Notice that balancing the metal atoms automatically balances the oxygen atoms.

The procedure for balancing an oxidation-reduction equation by the oxidation number method is summarized as follows:

1. Write the overall unbalanced equation.
2. Assign oxidation numbers and determine which elements undergo changes in oxidation number in the reaction.
3. Assign coefficients to make the total gain in oxidation numbers for the substances being oxidized equal to the total decrease in oxidation numbers for the substances being reduced.
4. Balance the remaining elements by inspection.

**SAMPLE EXERCISE 13.10**

Barium chlorite, $Ba(ClO_2)_2$, can be prepared by reacting hydrogen peroxide, $H_2O_2$, with chlorine dioxide, $ClO_2$, in an aqueous solution of barium hydroxide, $Ba(OH)_2$. The unbalanced complete ionic equation is

$$Ba^{2+}(aq) + 2OH^-(aq) + H_2O_2(aq) + ClO_2(aq) \longrightarrow$$
$$Ba(ClO_2)_2(s) + H_2O(l) + O_2(g)$$

Complete and balance this equation by the oxidation number method.

**Solution:** We already have the first step in the outline of procedure, the unbalanced chemical equation. The next step is to assign oxidation numbers to all elements:

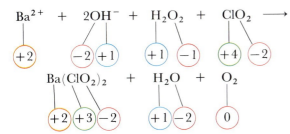

This equation illustrates a situation that is sometimes a source of difficulty for students. A single element, in this case oxygen, appears in more than one place and undergoes an oxidation-state change in one case but not in the other. It is best to focus simply on that portion of the element that undergoes the change in oxidation number; the rest of the element will come into balance later in the process of balancing the equation. Here we see that chlorine undergoes a decrease in oxidation number from $+4$ to $+3$ and that the oxygen atom of the hydrogen peroxide undergoes an increase in oxidation number from $-1$ to $0$:

$$Ba^{2+} + 2OH^- + H_2O_2 + ClO_2 \longrightarrow Ba(ClO_2)_2 + H_2O + O_2$$

To balance the changes in oxidation number we need to insert a 2 before the $ClO_2$:

$$Ba^{2+}(aq) + 2OH^-(aq) + H_2O_2(aq) + 2ClO_2(aq) \longrightarrow$$
$$Ba(ClO_2)_2(s) + H_2O(l) + O_2(g)$$

We note that now the numbers of oxygen, hydrogen, and barium atoms on both sides are in balance. The overall equation is therefore balanced.

**PRACTICE EXERCISE**

Sodium cyanide can be made by the high-temperature reaction of sodium carbonate with carbon under a nitrogen atmosphere. The unbalanced equation for the process is

$$Na_2CO_3(s) + C(s) + N_2(g) \xrightarrow{\Delta} NaCN(s) + CO(g)$$

Complete and balance the equation for this process. (Hint: It helps to notice that because all the Na and N end up in NaCN, there *must* be one $Na_2CO_3$ for each $N_2$.)

***Answer:*** $Na_2CO_3(s) + 4C(s) + N_2(g) \longrightarrow$
$$2NaCN(s) + 3CO(g)$$

Although oxidation and reduction must take place simultaneously, it is often convenient to consider them as separate processes. For example, the oxidation of $Sn^{2+}$ by $Fe^{3+}$,

$$Sn^{2+}(aq) + 2Fe^{3+}(aq) \longrightarrow Sn^{4+}(aq) + 2Fe^{2+}(aq) \qquad [13.28]$$

can be considered to consist of two processes: (1) the oxidation of $Sn^{2+}$ (Equation 13.29) and (2) the reduction of $Fe^{3+}$ (Equation 13.30).

Oxidation: $\quad Sn^{2+}(aq) \longrightarrow Sn^{4+}(aq) + 2e^{-} \qquad [13.29]$

Reduction: $\quad 2Fe^{3+}(aq) + 2e^{-} \longrightarrow 2Fe^{2+}(aq) \qquad [13.30]$

Equations that show either oxidation or reduction alone are called **half-reactions**. As shown in Equations 13.29 and 13.30, the number of electrons lost in an oxidation half-reaction must equal the number of electrons gained in the reduction half-reaction. When this condition is met and each half-reaction is balanced, the two half-reactions can be added to give the overall balanced total oxidation-reduction equation.

The use of half-reactions provides a general method for balancing oxidation-reduction equations. As an example, let's consider the reaction that occurs between permanganate ion, $MnO_4^{-}$, and oxalate ion, $C_2O_4^{2-}$, in acidic water solutions. When $MnO_4^{-}$ is added to an acidified solution of $C_2O_4^{2-}$, the deep purple color of the $MnO_4^{-}$ ion fades, as illustrated in Figure 13.8. Bubbles of $CO_2$ form, and the solution takes on the pale pink color of $Mn^{2+}$. We can therefore write the rough, unbalanced equation as follows:

$$MnO_4^{-}(aq) + C_2O_4^{2-}(aq) \longrightarrow Mn^{2+}(aq) + CO_2(g) \qquad [13.31]$$

Experiments would also show that $H^{+}$ is consumed and $H_2O$ produced in the reaction. We shall see that these facts can be deduced in the course of balancing the equation.

To complete and balance Equation 13.31 by the method of half-reactions, we begin with the unbalanced reaction and write two incomplete half-reactions, one involving the oxidant and the other involving the reductant.

$$MnO_4^{-}(aq) \longrightarrow Mn^{2+}(aq)$$
$$C_2O_4^{2-}(aq) \longrightarrow CO_2(g)$$

**FIGURE 13.8** (a) Reaction between aqueous permanganate solution (deep purple) and an acidic solution of oxalic acid, which contains oxalate ions, $C_2O_4^{2-}$. (b) As reaction proceeds, $MnO_4^{-}$ is reduced to the $Mn^{2+}(aq)$ ion, which has a pink color too pale to be seen. The reaction is very rapid, and no $MnO_4^{-}$ persists in solution until all the $C_2O_4^{2-}$ ion has been reacted. (c) When reaction is complete, the purple color of unreacted $MnO_4^{-}$ is visible in solution. (Donald Clegg and Roxy Wilson)

(a) (b) (c)

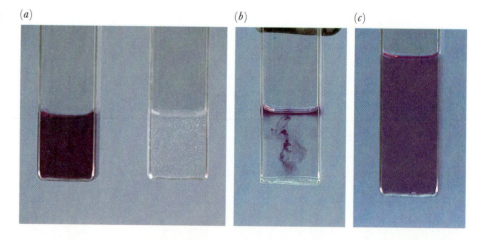

Next the half-reactions are completed and balanced separately. This means that the number of atoms of each element appearing in the half-reaction must be the same on both sides of the equation. First the atoms undergoing oxidation or reduction are balanced by adding coefficients on one side or the other as necessary. Then the remaining elements are balanced in the same way. If the reaction occurs in acidic water solution, $H^+$ and $H_2O$ can be added to either reactants or products to balance hydrogen and oxygen. Similarly, in basic solution the equation can be completed using $OH^-$ and $H_2O$. These species are in large supply in the respective solutions, and their formation as products or their use as reactants can easily go undetected experimentally. In the permanganate half-reaction, we already have one manganese atom on each side of the equation. However, we have four oxygens on the left and none on the right side; four $H_2O$ molecules are needed among the products to balance the four oxygen atoms in $MnO_4^-$:

$$MnO_4^-(aq) \longrightarrow Mn^{2+}(aq) + 4H_2O(l)$$

The eight hydrogen atoms that this introduces among the products can then be balanced by adding $8H^+$ to the reactants:

$$8H^+(aq) + MnO_4^-(aq) \longrightarrow Mn^{2+}(aq) + 4H_2O(l)$$

At this stage there are equal numbers of each type of atom on both sides of the equation, but the charge still needs to be balanced. The total charge of the reactants is $+8 - 1 = +7$, while that of the products is $+2 + 4(0) = +2$. To balance the charge, five electrons are added to the reactant side.*

$$5e^- + 8H^+(aq) + MnO_4^-(aq) \longrightarrow Mn^{2+}(aq) + 4H_2O(l)$$

Proceeding similarly with the oxalate half-reaction, we have

$$C_2O_4^{2-}(aq) \longrightarrow 2CO_2(g)$$

Charge is balanced by adding two electrons among the products:

$$C_2O_4^{2-}(aq) \longrightarrow 2CO_2(g) + 2e^-$$

Next we multiply each equation by an appropriate factor so that the number of electrons gained in one half-reaction equals the number of electrons lost in the other. The half-reactions are then added to give the overall balanced equation. In our example, the $MnO_4^-$ half-reaction must be multiplied by 2 and the $C_2O_4^{2-}$ half-reaction must be multiplied by 5:

---

* Although the oxidation numbers of the elements need not be used in balancing a half-reaction by this method, oxidation numbers can be used as a check. In this example $MnO_4^-$ contains manganese in a $+7$ oxidation state. Because manganese changes from a $+7$ to a $+2$ oxidation state, it must gain five electrons, just as we have already concluded.

$$10e^- + 16H^+(aq) + 2MnO_4^-(aq) \longrightarrow 2Mn^{2+}(aq) + 8H_2O(l)$$

$$5C_2O_4^{2-}(aq) \longrightarrow 10CO_2(g) + 10e^-$$

$$\overline{16H^+(aq) + 2MnO_4^-(aq) + 5C_2O_4^{2-}(aq) \longrightarrow 2Mn^{2+}(aq) + 8H_2O(l) + 10CO_2(g)}$$

The balanced equation is the sum of the balanced half-reactions.

The steps to balance an oxidation-reduction equation by the method of half-reactions when the reaction occurs in acid solution are summarized as follows:

1. Write the overall unbalanced equation for the reaction.
2. Divide the overall reaction into two unbalanced half-reactions, one for oxidation, the other for reduction.
3. Balance the atoms in each half-reaction that undergo oxidation or reduction, and then balance elements other than H and O.
4. Balance the O atoms by adding $H_2O$, and then balance the H atoms by adding $H^+$.
5. Balance charge on each side of the half-reaction equation by adding $e^-$ to the side with the greater positive charge.
6. Multiply the two half-reactions by coefficients such that the overall electron loss equals the overall electron gain.
7. Add the two half-reactions; simplify where possible by eliminating terms appearing on both sides of the equation.

The equations for reactions that occur in basic solution can be balanced initially as if they occurred in acidic solution. The $H^+$ ions can then be "neutralized" by adding an equal number of $OH^-$ ions to both sides of the equation. This procedure is shown in Sample Exercise 13.11.

## SAMPLE EXERCISE 13.11

Complete and balance the following equations:

(a)  $Cr_2O_7^{2-}(aq) + Cl^-(aq) \longrightarrow$
$Cr^{3+}(aq) + Cl_2(g)$    (acidic solution)

(b)  $CN^-(aq) + MnO_4^-(aq) \longrightarrow$
$CNO^-(aq) + MnO_2(s)$    (basic solution)

**Solution:** (a) The incomplete and unbalanced half-reactions are

$$Cr_2O_7^{2-}(aq) \longrightarrow Cr^{3+}(aq)$$
$$Cl^-(aq) \longrightarrow Cl_2(g)$$

The half-reactions are first balanced with respect to the elements. In the first half-reaction the presence of $Cr_2O_7^{2-}$ among the reactants requires two $Cr^{3+}$ among the products. The 7 oxygen atoms in $Cr_2O_7^{2-}$ are balanced by adding $7H_2O$ to the products. The 14 hydrogen atoms in $7H_2O$ are then balanced by adding $14H^+$ among the reactants:

$$14H^+(aq) + Cr_2O_7^{2-}(aq) \longrightarrow 2Cr^{3+}(aq) + 7H_2O(l)$$

Charge is balanced by adding electrons to the left side of the equation so that the total charge is the same on both sides:

$$6e^- + 14H^+(aq) + Cr_2O_7^{2-}(aq) \longrightarrow 2Cr^{3+}(aq) + 7H_2O(l)$$

In the second half-reaction, two $Cl^-$ are required to balance one $Cl_2$:

$$2Cl^-(aq) \longrightarrow Cl_2(g)$$

We add two electrons to the right side to attain charge balance:

$$2Cl^-(aq) \longrightarrow Cl_2(g) + 2e^-$$

This second half-reaction must be multiplied by 3 to equalize electron loss and gain in the two half-reactions. The two half-reactions are then added to give the balanced equation:

$$14H^+(aq) + Cr_2O_7^{2-}(aq) + 6Cl^-(aq) \longrightarrow 2Cr^{3+}(aq) + 7H_2O(l) + 3Cl_2(g)$$

**(b)** The incomplete and unbalanced half-reactions are

$$CN^-(aq) \longrightarrow CNO^-(aq)$$
$$MnO_4^-(aq) \longrightarrow MnO_2(s)$$

The equations may be balanced initially as if they took place in acidic solution. The resultant balanced half-reactions are

$$CN^-(aq) + H_2O(l) \longrightarrow$$
$$CNO^-(aq) + 2H^+(aq) + 2e^-$$

$$3e^- + 4H^+(aq) + MnO_4^-(aq) \longrightarrow$$
$$MnO_2(s) + 2H_2O(l)$$

Because $H^+$ cannot exist in any appreciable concentration in basic solution, it is removed from the equations by the addition of an appropriate amount of $OH^-(aq)$. In the $CN^-$ half-reaction, $2OH^-(aq)$ is added to both sides of the equation to "neutralize" the $2H^+(aq)$. The $2OH^-(aq)$ and $2H^+(aq)$ form $2H_2O(l)$:

$$2OH^-(aq) + H_2O(l) + CN^-(aq) \longrightarrow$$
$$CNO^-(aq) + 2H_2O(l) + 2e^-$$

The half-reaction can be simplified because $H_2O$ occurs on both sides of the equation. The simplified equation is

$$2OH^-(aq) + CN^-(aq) \longrightarrow$$
$$CNO^-(aq) + H_2O(l) + 2e^-$$

For the $MnO_4^-$ half-reaction, $4OH^-(aq)$ is added to both sides of the equation:

$$3e^- + 4H_2O(l) + MnO_4^-(aq) \longrightarrow$$
$$MnO_2(s) + 2H_2O(l) + 4OH^-(aq)$$

Simplifying gives

$$3e^- + 2H_2O(l) + MnO_4^-(aq) \longrightarrow$$
$$MnO_2(s) + 4OH^-(aq)$$

The top equation is multiplied by 3 and the bottom one by 2 to equalize electron loss and gain in the two half-reactions. The half-reactions are then added:

$$6OH^-(aq) + 3CN^-(aq) \longrightarrow$$
$$3CNO^-(aq) + 3H_2O(l) + 6e^-$$

$$6e^- + 4H_2O(l) + 2MnO_4^-(aq) \longrightarrow$$
$$2MnO_2(s) + 8OH^-(aq)$$

$$\overline{\begin{aligned}6OH^-(aq) + 3CN^-(aq) + \\ 4H_2O(l) + 2MnO_4^-(aq) \longrightarrow \\ 3CNO^-(aq) + 3H_2O(l) + 2MnO_2(s) + 8OH^-(aq)\end{aligned}}$$

The overall equation can be simplified, because $H_2O$ and $OH^-$ occur on both sides. The simplified equation is

$$3CN^-(aq) + H_2O(l) + 2MnO_4^-(aq) \longrightarrow$$
$$3CNO^-(aq) + 2MnO_2(s) + 2OH^-(aq)$$

**PRACTICE EXERCISE** _____
Balance the following oxidation-reduction reactions by the method of half-reactions. Reaction **(a)** occurs in acidic solution, reaction **(b)** in basic solution.

**(a)** $Mn^{2+}(aq) + NaBiO_3(s) \longrightarrow$
$$Bi^{3+}(aq) + MnO_4^-(aq)$$

**(b)** $NO_2^-(aq) + Al(s) \longrightarrow$
$$NH_3(aq) + Al(OH)_4^-(aq)$$

*Answers:*

**(a)** $2Mn^{2+}(aq) + 5NaBiO_3(s) + 14H^+(aq) \longrightarrow$
$$2MnO_4^-(aq) + 5Bi^{3+}(aq) + 5Na^+(aq) + 7H_2O(l)$$
**(b)** $NO_2^-(aq) + 2Al(s) + 5H_2O(l) + OH^-(aq) \longrightarrow$
$$NH_3(aq) + 2Al(OH)_4^-(aq)$$

## 13.8 OXIDIZING AND REDUCING AGENTS

A wide variety of substances undergo characteristic oxidation or reduction reactions in aqueous solution. Table 13.2 lists several substances commonly used as oxidizing or reducing agents in aqueous medium and presents their characteristic half-reactions. The reactions shown are the most common but not necessarily the only ones possible in each case. For example, nitric acid may also be reduced to $N_2$, $NH_3$, or other species, depending on reaction conditions. Let us now consider in a little more depth some of the properties of two common oxidizing agents and two common reducing agents.

### Hydrogen Peroxide

Hydrogen peroxide, $H_2O_2$, is a commonly used oxidizing agent because it is quite versatile and relatively inexpensive. Pure hydrogen peroxide is a colorless, rather viscous liquid. The pure material is too reactive for

**TABLE 13.2** Common oxidizing and reducing agents

| Substance | Characteristic half-reaction | Solution condition |
|---|---|---|
| | *Oxidizing agents* | |
| $KMnO_4$ | $MnO_4^-(aq) + 8H^+(aq) + 5e^- \longrightarrow Mn^{2+}(aq) + 4H_2O(l)$ | Acidic |
| | $MnO_4^-(aq) + 2H_2O(l) + 3e^- \longrightarrow MnO_2(s) + 4OH^-(aq)$ | Basic |
| $K_2Cr_2O_7$ | $Cr_2O_7^{2-}(aq) + 14H^+(aq) + 6e^- \longrightarrow 2Cr^{3+}(aq) + 7H_2O(l)$ | Acidic |
| $HNO_3$ | $NO_3^-(aq) + 4H^+(aq) + 3e^- \longrightarrow NO(g) + 2H_2O(l)$ | Acidic |
| $Cl_2$ | $Cl_2(aq) + 2e^- \longrightarrow 2Cl^-(aq)$ | Acidic |
| $H_2O_2$ | $H_2O_2(aq) + 2H^+(aq) + 2e^- \longrightarrow 2H_2O(l)$ | Acidic |
| | $H_2O_2(aq) + 2e^- \longrightarrow 2OH^-(aq)$ | Basic |
| | *Reducing agents* | |
| $H_2O_2$ | $H_2O_2(aq) \longrightarrow O_2(g) + 2H^+(aq) + 2e^-$ | Acidic |
| | $H_2O_2(aq) + 2OH^-(aq) \longrightarrow O_2(g) + 2H_2O(l) + 2e^-$ | Basic |
| $H_2SO_3$ | $H_2SO_3(aq) + H_2O(l) \longrightarrow SO_4^{2-}(aq) + 4H^+(aq) + 2e^-$ | Acidic |
| | $SO_3^{2-}(aq) + 2OH^-(aq) \longrightarrow SO_4^{2-}(aq) + H_2O(l) + 2e^-$ | Basic |
| $NaS_2O_4^a$ | $S_2O_4^{2-}(aq) + 2H_2O(l) \longrightarrow 2SO_3^{2-}(aq) + 4H^+(aq) + 2e^-$ | Acidic |
| $NaI$ | $2I^-(aq) \longrightarrow I_2(s) + 2e^-$ | Acidic |
| $SnCl_2$ | $Sn^{2+}(aq) \longrightarrow Sn^{4+}(aq) + 2e^-$ | Acidic |

$^a$ $Na_2S_2O_4$, named sodium dithionite, is a widely used reducing agent.

easy shipment or use in that form. The article of commerce is a 30 percent aqueous solution. The label from a reagent bottle is shown in Figure 13.9. The warning on the label indicates that the 30 percent solution is also strongly oxidizing.

Note from Table 13.2 that $H_2O_2$ is capable of acting as either an oxidizing agent or a reducing agent. Thus $H_2O_2$ is subject to decomposition by oxidation (yielding $H_2O$) and reduction (yielding $O_2$):

$$2H_2O_2(aq) \longrightarrow O_2(g) + 2H_2O(l) \qquad [13.32]$$

The label shown in Figure 13.9 warns against contact of the reagent with metals, dust, or organic materials, any of which can initiate rapid decomposition, with generation of large quantities of $O_2$. Metal surfaces, $MnO_2$, and transition-metal ions, such as $Cu^{2+}$, are especially effective in promoting the decomposition.

A few examples of oxidation reactions using $H_2O_2$ follow:

$$H_2O_2(aq) + H_2SO_3(aq) \longrightarrow SO_4^{2-}(aq) + H_2O(l) + 2H^+(aq) \qquad [13.33]$$

$$H_2O_2(aq) + 2H^+(aq) + 2I^-(aq) \longrightarrow I_2(s) + 2H_2O(l) \qquad [13.34]$$

$$3H_2O_2(aq) + 2Cr(OH)_4^-(aq) + 2OH^-(aq) \longrightarrow 2CrO_4^{2-}(aq) + 8H_2O(l) \qquad [13.35]$$

**Nitric Acid**

Nitric acid, $HNO_3$, can be obtained as a pure liquid that boils at 83°C. However, it is unstable as a pure liquid. The material of commerce is a 70 percent by weight aqueous solution. The most common product of reactions in which nitric acid acts as an oxidizing agent is NO.

Figure 13.10 shows the dissolution of iron metal by 6 $M$ nitric acid.

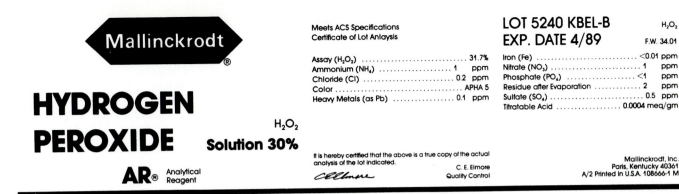

**Mallinckrodt**®

# HYDROGEN PEROXIDE

Solution 30%   $H_2O_2$

**AR**® Analytical Reagent

Meets ACS Specifications
Certificate of Lot Anlaysis

Assay ($H_2O_2$) ............................ 31.7%
Ammonium ($NH_4$) .................... 1    ppm
Chloride (Cl) ...................... 0.2  ppm
Color .................................. APHA 5
Heavy Metals (as Pb) ................ 0.1  ppm

LOT 5240 KBEL-B          $H_2O_2$
EXP. DATE 4/89          F.W. 34.01

Iron (Fe) ............................ <0.01 ppm
Nitrate ($NO_3$) ........................ 1    ppm
Phosphate ($PO_4$) ................... <1    ppm
Residue after Evaporation ............ 2    ppm
Sulfate ($SO_4$) ........................ 0.5  ppm
Titratable Acid ................. 0.0004 meq/gm

It is hereby certified that the above is a true copy of the actual
analysis of the lot indicated.
                                    C. E. Elmore
                                    Quality Control

Mallinckrodt, Inc.
Paris, Kentucky 40361
A/2 Printed in U.S.A. 108666-1 M

Oxidizer

Storage Code
White

**5240**

**LabGuard** ™

Health Hazard: Severe
Flammability: None
Reactivity: Severe
Contact Hazard: Extreme

**Protective Equipment:**
Protective Eyewear
Hand Protection
Safety Clothing
Laboratory Hood

1 GAL. (3.8 Liter)

**DANGER! Strong oxidizer. Contact with other material may cause fire. Cause severe burns. Effects may be delayed. Harmful if swallowed or inhaled. Loosen closure cautiously before opening.**

Avoid contact with skin and eyes. Wash thoroughly after handling. Keep from contact with clothing and other combustible materials as drying on these materials may cause fire. Remove and wash contaminated clothing promptly. Avoid contamination from any source including metals, dust, and organic materials. Such contamination may cause rapid decomposition, generation of large quantities of oxygen gas and high pressure. Never return unused hydrogen peroxide to container. Dilute with plenty of water and discard. Store in original vented container away from combustible materials. Keep container tightly closed. When empty, rinse thoroughly with clean water before discarding. KEEP OUT OF REACH OF CHILDREN.

**NOTE:** Hydrogen Peroxide 30% causes plastic to become brittle. Handle carefully. Store preferably under refrigeration. USE OR DISCARD CONTENTS BEFORE EXPIRATION DATE.

CAS 7722-84-1
Proper Shipping Name-
HYDROGEN PEROXIDE SOLUTION, UN2014

**FIRST AID: In case of contact,** immediately flush skin or eyes with plenty of water for at least 15 minutes. **CALL A PHYSICIAN. If swallowed,** give water or milk to drink. Get medical attention immediately. Never give anything by mouth to an unconscious person. **If inhaled,** remove to fresh air. Get medical attention for any breathing difficulty.

EPA: HWDC - Reactive

*See LabGuard Guide and Material Safety
Data Sheet for additional information.

**FIGURE 13.9**   Label from a bottle of commercial reagent grade 30% hydrogen peroxide (sold in plastic bottles to reduce decomposition, which occurs in glass bottles). (Courtesy of Mallinckrodt, Inc.)

**FIGURE 13.10**   When dilute nitric acid is added to iron wool, a rapid reaction ensues, forming NO gas and iron(III) oxide. The NO, which is colorless, is oxidized to red-brown $NO_2$ when it reaches the top of the flask. (We used a volumetric flask here to provide a long, thin neck to prevent excessive amounts of air from entering the flask.) (Donald Clegg and Roxy Wilson)

The reaction produces a colorless gas within the flask, but as the reaction product, NO, comes into contact with air outside the flask, the distinctive red-brown coloration of $NO_2$ is observed:

$$2NO(g) + O_2(g) \longrightarrow 2NO_2(g) \qquad [13.36]$$

Nitric acid is capable of oxidizing most metals. The oxidizing agent in these reactions is normally the nitrate ion and not $H^+(aq)$. With strongly reducing metals such as Zn, reduction proceeds all the way to the $NH_4^+$ ion:

$$4Zn(s) + 10H^+(aq) + NO_3^-(aq) \longrightarrow 4Zn^{2+}(aq) + NH_4^+(aq) + 3H_2O(l) \qquad [13.37]$$

Nitric acid is not by itself capable of oxidizing gold and other inert metals. However, a mixture consisting of three volumes of concentrated HCl and one volume of concentrated $HNO_3$—called aqua regia (kingly water)—is a strong oxidizing agent, able to oxidize the so-called noble metals such as gold or platinum. The action of aqua regia is based in part on the formation of a complex ion between the metal ion and chloride ions:

$$Au(s) + 3NO_3^-(aq) + 4Cl^-(aq) + 6H^+(aq) \longrightarrow AuCl_4^-(aq) + 3H_2O(l) + 3NO_2(g) \qquad [13.38]$$

**Sulfurous Acid and Sulfite Ion**

Sulfur dioxide gas, when dissolved in water, forms sulfurous acid, $H_2SO_3$, a strong reducing agent. Sulfurous acid is also readily formed by adding sodium sulfite to an acidic solution. This acid is only weakly ionized in water. It is easily oxidized to sulfuric acid, $H_2SO_4$; in the process the oxidation number of sulfur increases from +4 to +6 (Table 13.2). Some examples of sulfite reductions follow:

$$H_2SO_3(aq) + 2Fe^{3+}(aq) + H_2O(l) \longrightarrow SO_4^{2-}(aq) + 2Fe^{2+}(aq) + 4H^+(aq) \qquad [13.39]$$

$$H_2SO_3(aq) + I_2(aq) + H_2O(l) \longrightarrow SO_4^{2-}(aq) + 2I^-(aq) + 4H^+(aq) \qquad [13.40]$$

$$3H_2SO_3(aq) + Cr_2O_7^{2-}(aq) + 2H^+(aq) \longrightarrow 3SO_4^{2-}(aq) + 2Cr^{3+}(aq) + 4H_2O(l) \qquad [13.41]$$

When sulfur-containing oil or coal is burned in electric power plants, sulfur dioxide is emitted. This $SO_2$ dissolves in raindrops to form a sulfurous acid solution. In time the sulfurous acid is oxidized to sulfuric acid, mainly by oxygen in the atmosphere. Both sulfurous and sulfuric acids are major components in acid rain (Section 14.4).

Sulfurous acid is widely used in the paper industry in the production of pulp and paper from trees. The acid is used in the form of the bisulfite ion, which results from removal of one proton from $H_2SO_3$. The $HSO_3^-$ ion reacts with lignin, an undesirable component of the pulverized wood raw material, to form a water-soluble component that can be separated.

**Tin(II) ion**

Tin(II) ion is a good example of a low-valent metal ion that acts as a strong reducing agent because of the tendency of the metal ion to attain a higher oxidation state. When oxidized to tin(IV) the oxidation number of the tin increases from +2 to +4. Some examples of reductions that can be effected with $Sn^{2+}$ follow:

$$Sn^{2+}(aq) + 2HgCl_2(aq) \longrightarrow Sn^{4+}(aq) + Hg_2Cl_2(s) + 2Cl^-(aq) \qquad [13.42]$$

$$Sn^{2+}(aq) + 2Fe^{3+}(aq) \longrightarrow Sn^{4+}(aq) + 2Fe^{2+}(aq) \qquad [13.43]$$

Solutions of tin do not keep well because atmospheric oxygen oxidizes the tin(II):

$$2Sn^{2+}(aq) + O_2(aq) + 4H^+(aq) \longrightarrow 2Sn^{4+}(aq) + 2H_2O(l) \qquad [13.44]$$

The reagent should therefore be prepared fresh before use or stored under $N_2$. A sensitivity to atmospheric oxygen is fairly general for solutions of reducing agents, because $O_2$ is a moderately strong oxidizing agent.

 *FOR REVIEW*

## SUMMARY

For many reasons, water is the solvent of preeminent importance in our lives. In this chapter we have examined the variety of chemical reactions observed in aqueous solutions. A particular reaction may involve ions, nonelectrolytes, gases, immiscible liquids, or solids as well as dissolved substances. The chemical equations we write for such reactions may be **molecular equations** or may explicitly show that strong electrolytes are present as ions. Ions that undergo no change in the course of a reaction are termed **spectator ions**. They appear in the same form on both sides of the equation. The **net ionic equation** shows only those substances that undergo change in the reaction; that is, spectator ions are dropped from both sides of the equation. The net ionic equation expresses the essential process that occurs in the reaction.

Reactions proceed far toward the formation of products when a driving force operates. The driving force may be formation of a nonelectrolyte, as in the neutralization reaction. It may also be formation of a weak electrolyte, gas, or precipitate, an oxidation-reduction process, or formation or rupture of covalent bonds. In many reactions more than one of these factors may be operative. The preparation of desired substances through a chemical reaction is called chemical synthesis.

Acids and bases play important roles in aqueous solution chemistry. The characteristic acid species in water is the proton, $H^+$, and the characteristic base species is the hydroxide ion, $OH^-$. The proton is strongly bound to the solvent. The **hydronium ion**, $H_3O^+$, is often used to represent the predominant form of the proton in water. In general, we represent the proton in water simply as $H^+(aq)$, understanding that this symbol refers to the proton bound to the solvent. A reaction involving an acid or base proceeds toward formation of products until it achieves a state of **equilibrium**, in which no apparent further change occurs. Equilibrium may be attained in some cases when reaction has proceeded only a slight degree toward completion.

The **Brønsted-Lowry theory** of acids and bases emphasizes the transfer of the proton from an acid to a base. A **Brønsted acid** is defined as any substance capable of donating a proton to another substance. A **Brønsted base** is defined as a substance capable of accepting a proton from a Brønsted acid. A **conjugate acid** is a Brønsted acid formed from a Brønsted base by addition of a proton. Similarly, a **conjugate base** is a Brønsted base formed by removal of a proton from a Brønsted acid. The base and the conjugate acid formed from it by addition of a proton (for example, $NH_3$ and $NH_4^+$) are termed a **conjugate acid-base pair**. The acid-base strengths of conjugate acid-base pairs are related: The stronger an acid, the weaker its conjugate base; the weaker an acid, the stronger its conjugate base.

**Oxidation-reduction reactions** are those reactions that involve a change in the oxidation state of one or more elements. Every oxidation-reduction reaction has a substance that is oxidized, that is, that undergoes an increase in oxidation state. The substance undergoing oxidation is referred to as a **reducing agent** or **reductant**, because it causes the reduction of some other substance. Similarly, a substance that undergoes a reduction in oxidation state is referred to as an **oxidizing agent**, or **oxidant**, because it causes the oxidation of some other substance.

In one approach to balancing oxidation-reduction reactions, called the **oxidation number method**, changes in the oxidation numbers of all species are identified and balanced within the overall reaction. The method of **half-reactions** divides the overall process into two half-reactions, one for oxidation, the other for reduction. Each half-reaction is balanced separately, and the two are brought together with proper coefficients to produce a balance of electrons gained and lost.

A wide variety of substances serve as useful oxidizing and reducing agents in aqueous solution. Hydrogen peroxide and nitric acid are examples of oxidizing agents; sulfurous acid and tin(II) ion are examples of reducing agents.

## LEARNING GOALS

Having read and studied this chapter, you should be able to:

1. Identify spectator ions in a complete ionic equation.

2. Convert a molecular equation to a complete ionic equation and to a net ionic equation for a reaction occurring in aqueous solution.

3. Identify the driving forces for completion of reactions occurring in aqueous medium.

4. Describe the general approaches for the synthesis of simple salts by taking advantage of one or more driving forces for reactions in aqueous solution.

5. List the general properties that characterize acidic solutions and basic solutions.

6. Describe the molecular structural forms in which the proton exists in aqueous medium.

7. Define Brønsted acid, Brønsted base, and conjugate acid-base pair.

8. Identify the conjugate base associated with a given Brønsted acid and the conjugate acid associated with a given Brønsted base.

9. Identify the oxidant and reductant in an oxidation-reduction equation.

10. Balance simple oxidation-reduction equations by the oxidation number method.

11. Balance oxidation-reduction equations by the method of half-reactions.

12. Describe the important oxidation-reduction properties of hydrogen peroxide, nitric acid, sulfurous acid, and the tin(II) ion.

## KEY TERMS

Among the more important terms and expressions used for the first time in this chapter are the following:

A **Brønsted acid** (Section 13.5) is any substance capable of acting as a source of protons.

A **Brønsted base** (Section 13.5) is any substance capable of acting as a proton acceptor.

A **conjugate acid** (Section 13.5) is a substance formed by addition of a proton to a Brønsted base.

A **conjugate base** (Section 13.5) is a substance formed by loss of a proton from a Brønsted acid.

**Equilibrium** (Section 13.4) is attained in a chemical reaction when the forward and reverse processes occur at equal rates.

A **half-reaction** (Section 13.7) is an equation for either oxidation or reduction that explicitly shows the electrons involved [for example, $2H^+(aq) + 2e^- \longrightarrow H_2(g)$].

The **hydronium ion**, $H_3O^+$ (Section 13.4), represents the predominant form of the proton in aqueous solution.

A **molecular equation** (Section 13.1) is an equation describing a chemical reaction in which the formula for each substance is written without regard for whether it is an electrolyte or a nonelectrolyte.

A **net ionic equation** (Section 13.1) is an equation for a reaction involving ionic substances in solution and in which only those ions undergoing change in the reaction appear. Other ions—**spectator ions**, which go through the reaction unchanged and which would appear on both sides of the equation—are omitted from the net ionic equation.

An **oxidant**, or **oxidizing agent** (Section 13.6), is a substance that is reduced and thereby causes the oxidation of some other substance in an oxidation-reduction reaction.

An **oxidation-reduction reaction** (Section 13.6) is one in which one or more atoms undergo a change in oxidation state.

A **reductant**, or **reducing agent** (Section 13.6), is a substance that is oxidized and thereby causes the reduction of some other substance in an oxidation-reduction reaction.

## EXERCISES

### Net Ionic Equations; Solubility Rules

**13.1** Write balanced net ionic equations for each of the following reactions. **(a)** Sulfur dioxide gas is bubbled through a solution of potassium hydroxide, forming a solution of potassium sulfite. **(b)** Addition of lithium metal to water produces hydrogen gas and a solution of lithium hydroxide. **(c)** Addition of metallic zinc to a solution of copper(II) sulfate produces copper metal and a solution of zinc sulfate.

**13.2** Write balanced net ionic equations for each of the following reactions: **(a)** neutralization of the weak acid $H_2SO_3$ by aqueous sodium hydroxide; **(b)** precipitation of lead(II) chromate upon mixing solutions of lead nitrate and sodium chromate; **(c)** addition of hydrobromic acid solution to a beaker containing solid calcium carbonate.

**13.3** Using solubility rules or reasonable extensions of them, predict whether each of the following compounds is soluble in water: **(a)** $PbCl_2$; **(b)** $CsBr$; **(c)** $Mn(C_2H_3O_2)_2 \cdot 4H_2O$; **(d)** $SrSeO_4$.

**13.4** Using solubility rules or reasonable extensions of them, predict whether each of the following compounds is soluble in water: **(a)** $NiCl_2$; **(b)** $Hg_2I_2$; **(c)** $Co(OH)_2$; **(d)** $Ca(BrO_3)_2 \cdot H_2O$.

**13.5** Separate samples of a solution of an unknown salt are treated with dilute solutions of $HBr$, $H_2SO_4$, and $NaOH$. A precipitate forms with $H_2SO_4$. Which of the following cations could the solution contain: $K^+$; $Pb^{2+}$; $Ba^{2+}$?

**13.6** Separate samples of a solution of an unknown salt are treated with dilute $AgNO_3$, $Pb(NO_3)_2$, and $Ba(ClO_3)_2$. Precipitates form in all three cases. Which of the following anions could be the anion of the unknown salt: $Br^-$; $SO_3^{2-}$; $ClO_3^-$?

**13.7** Write balanced net ionic equations for the reactions, if any, that occur between **(a)** $ZnS(s)$ and $HCl(aq)$; **(b)** $Na_2CO_3(aq)$ and $BaCl_2(aq)$; **(c)** $Na_3PO_4(aq)$ and $HBr(aq)$; **(d)** $Ba(OH)_2(aq)$ and $HCl(aq)$; **(e)** $Sr(C_2H_3O_2)_2(aq)$ and $NiSO_4(aq)$; **(f)** $ZnSO_3(aq)$ and $HCl(aq)$; **(g)** $Pb(NO_3)_2(aq)$ and $H_2S(aq)$; **(h)** $Fe(OH)_3(s)$ and $HClO_4(aq)$.

**13.8** Write balanced net ionic equations for the reactions, if any, that occur when each of the following pairs is mixed: **(a)** $H_2SO_4(aq)$ and $BaCl_2(aq)$; **(b)** $NaCl(aq)$ and $(NH_4)_2SO_4(aq)$; **(c)** $AgNO_3(aq)$ and $Na_2CrO_4(aq)$; **(d)** $KOH(aq)$ and $HNO_3(aq)$; **(e)** $Ca(OH)_2(aq)$ and $HC_2H_3O_2(aq)$; **(f)** $K_2SO_3(s)$ and $H_2SO_4(aq)$; **(g)** $Pb(ClO_3)_2(aq)$ and $MgSO_4(aq)$.

## Factors Driving Chemical Reactions

**13.9** Explain what is meant by the term "driving force" in accounting for the occurrence of a chemical reaction. Give at least two examples of driving force.

**13.10** Give a specific example of a chemical reaction that occurs because of each of the following driving forces: **(a)** formation of a nonelectrolyte; **(b)** formation of a precipitate; **(c)** formation of a gaseous product.

**13.11** Which of the following substances would you expect to be soluble in dilute hydrochloric acid: **(a)** $SrSO_3$; **(b)** $SrI_2$; **(c)** $AgCl$? In each case, explain your answer.

**13.12** Suppose that 0.01 mol of solid were treated with 100 mL of aqueous reagent as indicated below. In which cases would you expect to find a solid after mixing: **(a)** $Hg_2CO_3$ with 0.1 $M$ $H_2SO_4$; **(b)** $LiCl$ with 0.15 $M$ $NaOH$; **(c)** $CoSO_4$ with 0.25 $M$ $NaOH$; **(d)** $MnSeO_4$ with 0.3 $M$ $Na_2CO_3$?

**13.13** For each reaction in Exercise 13.7 that occurs, indicate the nature of the driving force (that is, formation of a gas, formation of a precipitate, and so on).

**13.14** For each reaction in Exercise 13.8 that occurs, indicate the nature of the driving force (that is, formation of a gas, formation of a precipitate, and so on).

**13.15** Suggest a method for synthesis of each of the following substances, and write a balanced molecular equation or equations for the process: **(a)** $Cu(NO_3)_2$; **(b)** $MnS$; **(c)** $Fe(OH)_2$.

**13.16** Suggest a method for synthesis of each of the following substances, and write a balanced molecular equation or equations for the process: **(a)** $ZnBr_2$; **(b)** $HgI_2$; **(c)** $CoCO_3$.

## Acids and Bases in Water

**13.17** Although $HCl$ and $H_2SO_4$ have very different properties as pure substances, their aqueous solutions possess many common properties. Explain this in terms of Arrhenius's theory of acids and bases.

**13.18** Lavoisier noted that oxides such as $CO_2$ or $SO_2$ form acidic solutions and postulated that all acids must contain oxygen. What examples can you think of to counter Lavoisier's hypothesis? What is the essential feature common to all acids in water?

**13.19** Hydrogen bromide is very soluble in water. Its solutions are good conductors of electricity. Describe the nature of the interaction between $HBr$ and water to account for these observations.

**13.20** Compare the dissolution of $HCl$ in water with the dissolution of $NaOH$ in water. In what respects do the processes differ? In what respects are they similar?

**13.21** When dry $HCl$ is bubbled through the organic liquid toluene, $C_7H_8$, it exhibits only a slight solubility. The solution that results does not react readily with zinc, and it does not conduct electricity. Compare these characteristics to the behavior of a solution of $HCl$ and water. Account for the difference.

**13.22** What characteristics of the water molecule and of liquid water make water an especially good solvent for strong acids and bases?

**13.23** Give the conjugate acid of each of the following bases: **(a)** $NH_3$; **(b)** $Br^-$; **(c)** $NH_2^-$; **(d)** $H_2PO_4^-$; **(e)** $OH^-$.

**13.24** Give the conjugate base of each of the following proton sources: **(a)** $H_3PO_4$; **(b)** $HBr$; **(c)** $H_2C_2O_4$; **(d)** $HS^-$; **(e)** $NH_4^+$; **(f)** $PH_3$.

**13.25** Identify the acid and base in each of the following reactions:
**(a)** $NH_2^-(aq) + H_2O(l) \rightleftharpoons NH_3(aq) + OH^-(aq)$
**(b)** $H_2C_2O_4(aq) + H_2O(l) \rightleftharpoons$
$HC_2O_4^-(aq) + H_3O^+(aq)$
**(c)** $H^+(aq) + HPO_4^{2-}(aq) \rightleftharpoons H_2PO_4^-(aq)$

**13.26** Identify the acid and base in each of the following equibrium systems:
**(a)** $HC_2O_4^-(aq) + CO_3^{2-}(aq) \rightleftharpoons$
$C_2O_4^{2-}(aq) + HCO_3^-(aq)$
**(b)** $PH_4^+(aq) + H_2O(aq) \rightleftharpoons PH_3(aq) + H_3O^+(aq)$
**(c)** $NH_4^+(aq) + CN^-(aq) \rightleftharpoons NH_3(aq) + HCN(aq)$

**13.27** Using Figure 13.7, predict whether the following

reactions proceed to the right to any appreciable extent:

(a) $HSO_4^-(aq) + NH_3(aq) \rightleftharpoons$
$$SO_4^{2-}(aq) + NH_4^+(aq)$$

(b) $H_2PO_4^-(aq) + HCO_3^-(aq) \rightleftharpoons$
$$H_3PO_4(aq) + CO_3^{2-}(aq)$$

(c) $NO_3^-(aq) + H_2O(l) \rightleftharpoons HNO_3(aq) + OH^-(aq)$

**13.28** Using Figure 13.7, predict whether the position of equilibrium in each of the following reactions lies appreciably to the right (that is, in the direction of products):

(a) $HNO_3(aq) + HS^-(aq) \rightleftharpoons$
$$NO_3^-(aq) + H_2S(aq)$$

(b) $HCN(aq) + NO_3^-(aq) \rightleftharpoons$
$$CN^-(aq) + HNO_3(aq)$$

(c) $NH_3(aq) + HS^-(aq) \rightleftharpoons NH_4^+(aq) + S^{2-}(aq)$

## Oxidation-Reduction Equations

**13.29** In each of the following balanced oxidation-reduction equations, identify those elements that undergo changes in oxidation number and indicate the magnitude of the change in each case:

(a) $3H_2S(aq) + 2HNO_3(aq) \longrightarrow$
$$3S(s) + 2NO(g) + 4H_2O(l)$$

(b) $5H_2SO_3(aq) + 2MnO_4^-(aq) \longrightarrow$
$$5SO_4^{2-}(aq) + 2Mn^{2+}(aq) + 4H^+(aq) + 3H_2O(l)$$

(c) $2CrO_2^-(aq) + 3ClO^-(aq) + 2OH^-(aq) \longrightarrow$
$$2CrO_4^{2-}(aq) + 3Cl^-(aq) + H_2O(l)$$

(d) $2Cu^{2+}(aq) + 2H_2O(l) \longrightarrow$
$$2Cu(s) + O_2(g) + 4H^+(aq)$$

**13.30** In each of the following balanced oxidation-reduction equations, identify those elements that undergo changes in oxidation number and indicate the magnitude of the change in each case:

(a) $2KOH(aq) + Cl_2(aq) \longrightarrow$
$$KCl(aq) + KClO(aq) + H_2O(l)$$

(b) $BaSO_4(s) + 4C(s) \longrightarrow BaS(s) + 4CO(g)$

(c) $5PbO_2(s) + 2Mn^{2+}(aq)$
$+ 5SO_4^{2-}(aq) + 4H^+(aq) \longrightarrow$
$$5PbSO_4(s) + 2MnO_4^-(aq) + 2H_2O(l)$$

(d) $CH_4(g) + 2O_2(g) \longrightarrow CO_2(g) + 2H_2O(g)$

**13.31** Hydrazine ($N_2H_4$) and dinitrogen tetraoxide ($N_2O_4$) form a self-igniting mixture that has been used as a rocket propellant. The reaction products are $N_2$ and $H_2O$. **(a)** Write a balanced chemical equation for this reaction. **(b)** Which substance serves as reducing agent and which as oxidizing agent?

**13.32** Solid lead(II) sulfide reacts at elevated temperatures with oxygen in the air to form lead(II) oxide and sulfur dioxide. **(a)** Write a balanced chemical equation for this reaction. **(b)** Which substances are reductants and which are oxidants?

**13.33** Complete and balance each of the following oxidation-reduction half-reactions:

(a) $Ni(s) \longrightarrow Ni^{2+}(aq)$ (acidic solution)
(b) $ClO^-(aq) \longrightarrow ClO_3^-(aq)$ (basic solution)
(c) $H_2S(aq) \longrightarrow S(s)$ (acidic solution)
(d) $SO_4^{2-}(aq) \longrightarrow SO_2(g)$ (acidic solution)

**13.34** Complete and balance each of the following oxidation-reduction half-reactions:

(a) $BrO_3^-(aq) \longrightarrow Br^-(aq)$ (acidic solution)
(b) $PbO_2(s) + Cl^-(aq) \longrightarrow PbCl_2(s)$ (acidic solution)
(c) $MnO_4^-(aq) \longrightarrow MnO_2(s)$ (basic solution)
(d) $Cr(OH)_4^-(aq) \longrightarrow CrO_4^{2-}(aq)$ (basic solution)

**13.35** Complete and balance each of the following equations:

(a) $Cr_2O_7^{2-}(aq) + I^-(aq) \longrightarrow$
$$Cr^{3+}(aq) + IO_3^-(aq)$$ (acidic solution)

(b) $MnO_4^-(aq) + CH_3OH(aq) \longrightarrow$
$$Mn^{2+}(aq) + HCO_2H(aq)$$ (acidic solution)

(c) $As(s) + ClO_3^-(aq) \longrightarrow$
$$H_3AsO_3(aq) + HClO(aq)$$ (acidic solution)

(d) $As_2O_3(s) + NO_3^-(aq) \longrightarrow$
$$H_3AsO_4(aq) + N_2O_3(aq)$$ (acidic solution)

(e) $MnO_4^-(aq) + Br^-(aq) \longrightarrow$
$$MnO_2(s) + BrO_3^-(aq)$$ (basic solution)

(f) $H_2O_2(aq) + Cl_2O_7(aq) \longrightarrow$
$$ClO_2^-(aq) + O_2(g)$$ (basic solution)

**13.36** Complete and balance each of the following equations:

(a) $Pb(OH)_4^{2-}(aq) + ClO^-(aq) \longrightarrow$
$$PbO_2(s) + Cl^-(aq)$$ (basic solution)

(b) $TlOH(s) + NH_2OH(aq) \longrightarrow$
$$Tl_2O_3(s) + N_2(g)$$ (basic solution)

(c) $Cr_2O_7^{2-}(aq) + CH_3OH(aq) \longrightarrow$
$$HCO_2H(aq) + Cr^{3+}(aq)$$ (acidic solution)

(d) $MnO_4^-(aq) + Cl^-(aq) \longrightarrow$
$$Mn^{2+}(aq) + Cl_2(aq)$$ (acidic solution)

(e) $H_2O_2(aq) + ClO_2(aq) \longrightarrow$
$$ClO_2^-(aq) + O_2(g)$$ (basic solution)

(f) $NO_2^-(aq) + Cr_2O_7^{2-}(aq) \longrightarrow$
$$Cr^{3+}(aq) + NO_3^-(aq)$$ (acidic solution)

## Additional Exercises

**13.37** Write balanced net ionic equations for each of the following reactions. **(a)** Manganate ion, $MnO_4^{2-}$, reacts with hydrogen ions in acidic solution to form permanganate ion and solid manganese(IV) oxide. **(b)** Residual gold metal is recovered from mine tailings by reacting it with aqueous, basic sodium cyanide solution, aerated to provide $O_2$. The soluble compound $Na[Au(CN)_2]$ is formed. **(c)** Ozone is measured quantitatively by reacting it with an aqueous, basic solution of potassium iodide to form iodine and $O_2$.

**13.38** An ionic strontium compound is insoluble in water. Upon addition of dilute HCl, no change is evident. Which of the following compounds might the unknown be: $SrCO_3$, $SrBr_2$, $SrSO_4$, $Sr(C_2H_3O_2)_2$?

**13.39** Suppose you have a solution that might contain any or all of the following cations: $Ni^{2+}$, $Ag^+$, $Sr^{2+}$, and $Mn^{2+}$. Addition of HCl solution causes a precipitate to form. After filtering off the precipitate, $H_2SO_4$ solution is added to the resultant solution and another precipitate forms. This is filtered off, and a solution of $Na_2CrO_4$ is added to the resulting solution. No precipitate is observed. Which ions are present in each of the precipitates? Which of the four ions listed above ions must be absent from the original solution?

**13.40** Account for each of the following observations, and write a balanced net ionic chemical equation to represent what occurs. **(a)** A solid forms upon addition of sodium hydroxide solution to an aqueous solution of manganese chloride. **(b)** Limestone formations, composed mainly of $CaCO_3$, dissolve when in contact with acidic groundwater. **(c)** Chloric acid, $HClO_3$, was first prepared by Gay-Lussac, who reacted a barium chlorate solution with dilute $H_2SO_4$.

**13.41** Antacids are often used to relieve pain and promote healing in the treatment of mild ulcers. Write balanced net ionic equations for the reactions that occur between stomach acid, $HCl(aq)$, and each of the following substances used in various antacids: **(a)** $Al(OH)_3(s)$; **(b)** $Mg(OH)_2(s)$; **(c)** $MgCO_3(s)$; **(d)** $NaAl(CO_3)(OH)_2(s)$; **(e)** $CaCO_3(s)$.

**13.42** Why does an aqueous solution of HCl conduct an electrical current? What species actually move through the solution to carry the current? Why does a benzene solution of HCl *not* carry an electrical current?

**13.43** Gaseous hydrogen iodide is very soluble in water. A solution containing 52.4 percent HI by weight has a density of $1.60 \text{ g/cm}^3$. What is the ratio of moles of $H_2O$ to HI in this solution? Suggest an explanation for the high solubility of HI in water. Suggest one or more experiments that would provide a test of your hypothesis.

**13.44** The electrical conductivities of aqueous acid solutions are higher than those of other aqueous solutions of electrolytes at comparable concentrations. Using Lewis structures, indicate a mechanism by which the proton could appear to move through the solution by a jumping process.

**13.45** Identify the Brønsted acid and Brønsted base among the reactants in each of the following reactions:
**(a)** $HNO_2(aq) + OH^-(aq) \rightleftharpoons NO_2^-(aq) + H_2O(l)$

**(b)** $ClO_2^-(aq) + H^+(aq) \rightleftharpoons HClO_2(aq)$
**(c)** $ClO_2^-(aq) + H_2O(aq) \rightleftharpoons HClO_2(aq) + OH^-(aq)$

**(d)** $HCO_2H(aq) + CN^-(aq) \rightleftharpoons HCN(aq) + HCO_2^-(aq)$

**13.46** Complete the net ionic equation for each of the following acid-base reactions:
**(a)** $HClO_3(aq) + Mg(OH)_2(s) \longrightarrow$
**(b)** $HBr(aq) + La_2(CO_3)_3 \cdot 8H_2O(s) \longrightarrow$
**(c)** $Al(OH)_3(s) + H_2SO_4(aq) \longrightarrow$
**(d)** $H_2C_2O_4(aq) + KOH(aq) \longrightarrow$

**13.47** Tartaric acid, $C_4H_6O_6$, has two acidic hydrogens. The acid is often present in wines and precipitates from solution as the wine ages. A solution containing an unknown concentration of the acid is titrated with NaOH. It requires 22.62 mL of $0.200 \ M$ NaOH solution to titrate 40.00 mL of the tartaric acid solution. Write a balanced net ionic equation for the neutralization reaction, and calculate the molarity of the tartaric acid solution.

**13.48** Explain in detail what the symbol $H^+(aq)$ signifies.

**13.49** Hydrogen peroxide, a chemical widely used as a bleach, can act as either an oxidizing agent or a reducing agent. Complete and balance the following equations, and indicate how hydrogen peroxide is functioning in each case:
**(a)** $MnO_4^-(aq) + H_2O_2(aq) \longrightarrow Mn^{2+}(aq) + O_2(g)$ (acidic solution)
**(b)** $I^-(aq) + H_2O_2(aq) \longrightarrow I_2(s)$ (acidic solution)

**13.50** Complete and balance each of the following equations:
**(a)** $Si(s) + HCl(g) \longrightarrow SiHCl_3(g) + H_2(g)$
**(b)** $NH_3(g) + F_2(g) \longrightarrow NF_3(g) + NH_4F(s)$
**(c)** $B_2O_3(s) + BrF_3(l) \longrightarrow BF_3(g) + Br_2(l) + O_2(g)$
**(d)** $Fe^{2+}(aq) + HBrO_3(aq) \longrightarrow Fe_2^+(aq) + Br_2(l)$ (acidic solution)
**(e)** $Cu(s) + NO_3^-(aq) \longrightarrow Cu^{2+}(aq) + NO(g)$ (acidic solution)
**(f)** $IO_3^-(aq) + I^-(aq) \longrightarrow I_3^-(aq)$ (basic solution)

**13.51** Write balanced chemical equations for reaction of **(a)** tin(II) chloride with dilute nitric acid; **(b)** dilute hydrogen peroxide solution with sulfurous acid solution; **(c)** dilute hydrogen peroxide solution with tin(II) chloride.

## CHAPTER 14

# Chemistry of the Environment

In previous chapters we have dealt for the most part with the principles that govern the chemical and physical behavior of matter. We are now in a position to apply these principles to an understanding of the world in which we live. In this chapter we consider some aspects of the chemistry of our environment, focusing on the earth's atmosphere and on the aqueous environment, which we call the **hydrosphere**.

Both the atmosphere and hydrosphere of our planet make life as we know it possible. Management of this environment so as to maintain and enhance the quality of life is one of the most important concerns of our time. It has become evident that major reforms and much stricter standards are required if we are to preserve the quality of life in the world. As a voting citizen you will be called on to help decide bond issues and referenda that may have an impact on your health as well as on your economic security. The better you understand the chemical principles that underlie environmental issues, the better your chances of forming a sound judgment. Our intent in this chapter is to provide an introduction to the nature of the earth's atmosphere and hydrosphere, and to indicate some of the ways in which pollution occurs.

## 14.1 EARTH'S ATMOSPHERE

Because most of us have never been very far from the earth's surface, we tend to take for granted the many ways in which the atmosphere determines the environment in which we live. In this section we will examine some of the important physical characteristics of our planet's atmosphere in light of what we know of the properties of gases.

The temperature of the atmosphere varies in a rather complex manner as a function of altitude, as shown in Figure 14.1. Just above the surface, in the **troposphere**, the temperature normally decreases with increasing altitude and reaches a minimum value at an elevation of about 12 km. In this region, called the **tropopause**, the temperature is about 215 K ($-60°C$). Above this elevation the temperature increases to about 275 K in the region of 50 km and then begins again to decrease. The altitude at which the temperature reaches a maximum is called the **stratopause**. The region between the tropopause and stratopause is known as the **stratosphere**. Above the stratopause the temperature drops to an even lower value than at the tropopause. The region of this second temperature minimum is called the **mesopause**. The region between the stratopause and mesopause is known as the **mesosphere**. Above the mesopause, the temperature rises rapidly in the region called the **thermosphere**. Note that the regions of temperature extremes are denoted by the suffix *-pause*. The regions between these are denoted by the suffix *-sphere*. The bound-

Looking Glass Falls in Pisgah National Forest, North Carolina. In this one scene, we see the three major components of our environment: the lithosphere (rocks), the hydrosphere (water), and the biosphere (plants and animals). (J. H. Robinson/Photo Researchers)

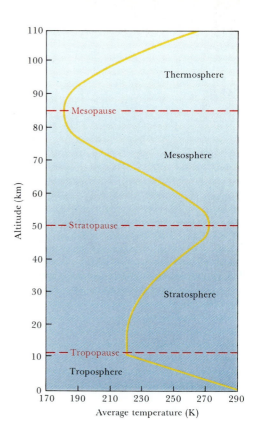

**FIGURE 14.1** Temperature variations in the atmosphere at altitudes below 110 km.

aries between different regions are important because mixing of the atmosphere across the boundaries is relatively slow. For example, pollutant gases generated in the troposphere find their way into the stratosphere only very slowly.

The **troposphere** is the region of the atmosphere in which nearly all of us live out our entire lives. Howling winds and soft breezes, rain, sunny skies—all that we normally think of as weather occurs in this region. Even when we fly in a modern jet aircraft between distant cities, we are still in the troposphere, though we may be near the tropopause.

In contrast to the temperature changes that occur in the atmosphere, the pressure of the atmosphere decreases in a quite regular way with increasing elevation, as shown in Figure 14.2. We see that atmospheric pressure drops off much more rapidly at lower elevations than at higher.

**FIGURE 14.2** Variations in atmospheric pressure with altitude. At 50 km altitude the pressure has declined to about 1 mm Hg. At still higher altitudes the pressure continues to decline, although this cannot be shown on the scale of the figure. For example, at 100 km the pressure is only $2.3 \times 10^{-3}$ mm Hg.

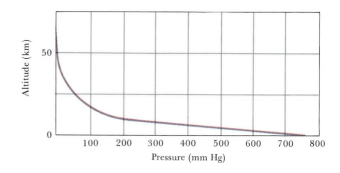

**FIGURE 14.3** The aurora borealis, or northern lights. This luminous display in the northern sky is produced by the collisions of high-speed electrons and protons from the sun with air molecules. The charged particles are channeled toward the polar regions by the earth's magnetic field. (Science Source/Photo Researchers)

The explanation for this characteristic of the atmosphere lies in its compressibility. As a result of the atmosphere's compressibility, the pressure decreases from an average value of 760 mm Hg at sea level to $2.3 \times 10^{-3}$ mm Hg at 100 km, to only $1.0 \times 10^{-6}$ mm Hg at 200 km.

The atmosphere is an extremely complex system. Its temperature and pressure change over a wide range with altitude, as we have already seen. The atmosphere is subjected to bombardment by radiation and energetic particles from the sun and by cosmic radiation from outer space. This barrage of energy has profound chemical effects, especially on the outer reaches of the atmosphere (Figure 14.3). In addition, because of the earth's gravitational field, lighter atoms and molecules tend to rise to the top. As a result of all these factors, the composition of the atmosphere is not constant. However, it is useful to know the composition of the atmosphere in the region near the earth's surface. Table 14.1 shows the com-

## Composition of the Atmosphere

**TABLE 14.1**  Composition of dry air near sea level

| Component[a] | Content (mole fraction) | Molecular weight |
|---|---|---|
| Nitrogen | 0.78084 | 28.013 |
| Oxygen | 0.20948 | 31.998 |
| Argon | 0.00934 | 39.948 |
| Carbon dioxide | 0.000330 | 44.0099 |
| Neon | 0.00001818 | 20.183 |
| Helium | 0.00000524 | 4.003 |
| Methane | 0.000002 | 16.043 |
| Krypton | 0.00000114 | 83.80 |
| Hydrogen | 0.0000005 | 2.0159 |
| Nitrous oxide | 0.0000005 | 44.0128 |
| Xenon | 0.000000087 | 131.30 |

[a] Ozone, sulfur dioxide, nitrogen dioxide, ammonia, and carbon monoxide are present as trace gases in variable amounts.

position of dry air near sea level. We note that although traces of a great many substances are present, only a few dominate. The two diatomic molecules $N_2$ and $O_2$ make up about 99 percent of the entire atmosphere. Essentially all the remainder, with the exception of carbon dioxide, is made up of the monatomic rare gases.

Note that the contribution of each component of the atmosphere listed in Table 14.1 is given in terms of its mole fraction. This is simply the total number of moles of a particular component in a given sample of air, divided by the total number of moles of all the components in that sample (Section 12.1). The partial pressure of a given component in the atmosphere is given by the total atmospheric pressure times the mole fraction of that component. Using $P_t$ to represent the total pressure, and $X_1$, $X_2$, and so on to represent the mole fractions of the components of the gas mixture, we have

$$P_t = X_1 P_t + X_2 P_t + X_3 P_t + \cdots$$
$$= P_t(X_1 + X_2 + X_3 + \cdots) \qquad [14.1]$$

This statement follows from Dalton's law of partial pressures (Section 10.5) and Avogadro's hypothesis (Section 10.3). Of course, the mole fractions must add up to 1.

---

**SAMPLE EXERCISE 14.1**

What is the partial pressure of $CO_2$ in dry air when the total dry air pressure ($P_t$) is 735 mm Hg?

**Solution:** Referring to Table 14.1, we see that the mole fraction of $CO_2$ is $3.30 \times 10^{-4}$.

$$P_{CO_2} = X_{CO_2} P_t$$
$$P_{CO_2} = (735 \text{ mm Hg } P_t)$$

$$\times \left( \frac{3.30 \times 10^{-4} \text{ mm Hg CO}_2}{1 \text{ mm Hg } P_t} \right)$$

$$= 0.243 \text{ mm Hg CO}_2$$

**PRACTICE EXERCISE**

What is the partial pressure of argon in dry air when total pressure is 750 mm Hg?

*Answer:* 7.00 mm Hg

---

In speaking of trace constituents of the atmosphere it is common to use **parts per million**, ppm, as the unit of concentration. In dealing with gases, a part per million refers to 1 part by volume in 1 million volume units of the whole. By virtue of the properties of gases, volume fraction and mole fraction are the same. Thus 1 part per million of a trace constituent amounts to 1 mol of that constituent in 1 million moles total of the gas; that is, it is mole fraction times $10^6$. Note that Table 14.1 lists the mole fraction of $CO_2$ in the atmosphere as 0.000330; thus its concentration in parts per million is $0.000330 \times 10^6 = 330$ ppm.

When applied to substances in solution, parts per million refers to grams of substance per million grams of solution. That is, it can be expressed in any of the following ways:

$$\text{ppm} = \left( \frac{\text{g solute}}{\text{g soln}} \right) \times 10^6 = \left( \frac{\text{mg solute}}{\text{kg soln}} \right) \simeq \left( \frac{\text{mg solute}}{\text{L soln}} \right) \qquad [14.2]$$

The last expression is approximately true for water as solvent, because the density of water is about 1 kg/L. Thus a solution containing 3 ppm of $Cu^{2+}$ contains 3 mg of $Cu^{2+}$ per liter of solution.

Before we begin a consideration of the chemical processes that occur in various regions of the atmosphere, let's remind ourselves about some of the most important chemical properties of the two major components of the atmosphere, $N_2$ and $O_2$. We saw in Chapter 9 that the $N_2$ molecule possesses a triple bond between the nitrogen atoms. This very strong bond is responsible for the exceptional lack of reactivity of $N_2$, which undergoes reaction only under extreme conditions. The O—O bond energy in $O_2$ is much lower than that for $N_2$ (Table 9.6), and $O_2$ is therefore much more reactive than $N_2$. Oxygen reacts with many substances to form oxides, as described in Section 8.11. The oxides of nonmetals—for example, $SO_2$—for the most part form acidic solutions on dissolving in water. The oxides of active metals, and other metals in low oxidation states—for example, CaO—form basic solutions on dissolving in water. Many metal oxides—for example, $Fe_2O_3$ and $Al_2O_3$—are insoluble in water but may dissolve in strongly acidic or strongly basic solutions.

## 14.2 THE OUTER REGIONS

Although the portion of the atmosphere extending above the mesopause contains only a small fraction of the atmospheric mass, it plays an important role in determining the conditions of life at the earth's surface. This upper layer forms the outer bastion of defense against the hail of radiation and high-energy particles that continually bombard the planet. In absorbing these assaults, the molecules and atoms of the atmosphere undergo chemical change.

### Photodissociation

The sun emits radiant energy over a wide range of wavelengths. The shorter-wavelength, higher-energy radiations in the ultraviolet range of the spectrum are sufficiently energetic to cause chemical changes. We have already seen, in Section 6.2, that electromagnetic radiation can be pictured as a stream of photons. The energy of each photon is given by the relationship $E = h\nu$, where $h$ is Planck's constant and $\nu$ is the frequency of the radiation. For a chemical change to occur when radiation falls on the earth's atmosphere, two conditions must be met. First, there must be photons with energy at least as large as that required to accomplish whatever chemical process is being considered. Second, molecules must absorb these photons. This requirement means that the energy of the photons is converted into some other form of energy within the molecule.

The rupture of a chemical bond resulting from absorption of a photon by a molecule is called **photodissociation**. One of the most important processes occurring in the upper atmosphere (that is, above about 120 km elevation) is the photodissociation of the oxygen molecule, as shown in Equation 14.3:

$$O_2(g) + h\nu \longrightarrow 2O(g) \qquad [14.3]$$

The minimum energy required to cause this change is determined by the dissociation energy of $O_2$, 495 kJ/mol. In Sample Exercise 14.2 we calculate the longest-wavelength photon having sufficient energy to dissociate the $O_2$ molecule.

## SAMPLE EXERCISE 14.2

What is the wavelength of a photon corresponding to a molar bond-dissociation energy of 495 kJ/mol?

**Solution:** We must first calculate the energy required per molecule and then determine the wavelength of a photon with that energy:

$$\left(495 \times 10^3 \ \frac{J}{mol}\right)\left(\frac{1 \ mol}{6.022 \times 10^{23} \ molecules}\right)$$

$$= 8.22 \times 10^{-19} \ \frac{J}{molecule}$$

The energy of the photon is given by $E = h\nu$. Rearranging, we have

$$\nu = \frac{E}{h} = \left(\frac{8.22 \times 10^{-19} \ J}{6.625 \times 10^{-34} \ J\text{-}s}\right)$$

$$= 1.24 \times 10^{15}/s$$

Recall from Section 6.1 that the product of frequency and wavelength of radiation equals the velocity of light:

$$\nu\lambda = c = 3.00 \times 10^8 \ m/s$$

Therefore, rearranging this equation, we have

$$\lambda = \frac{c}{\nu} = \left(\frac{3.00 \times 10^8 \ m/s}{1.24 \times 10^{15} \ /s}\right)\left(\frac{10^9 \ nm}{1 \ m}\right)$$

$$= 242 \ nm$$

### PRACTICE EXERCISE

The bond energy in ClO is 268 kJ/mol. What is the longest-wavelength photon that has sufficient energy to cause a Cl—O bond rupture?
*Answer:* 446 nm

The calculations in Sample Exercise 14.2 tell us that any photon of wavelength *shorter* than 242 nm will have sufficient energy to dissociate the $O_2$ molecule. (Remember that shorter wavelength means higher energy!)

The second condition that must be met before dissociation actually occurs is that the photon must be absorbed by $O_2$. Fortunately for us, $O_2$ absorbs much of the high-energy, short-wavelength radiation from the solar spectrum before it reaches the lower atmosphere. As it does so, atomic oxygen (O) is formed. At higher elevations the dissociation of $O_2$ is very extensive. At 400 km only 1 percent of the oxygen is in the form of $O_2$; the other 99 percent is in the form of atomic oxygen. At 130 km, $O_2$ and O are just about equally abundant. Below this elevation $O_2$ is more abundant than O.

Because the bond-dissociation energy of $N_2$ is very high (Table 9.7), only photons of very short wavelength possess sufficient energy to cause dissociation of this molecule. Furthermore, $N_2$ does not readily absorb photons, even when they do possess sufficient energy. The overall result is that very little atomic nitrogen is formed in the upper atmosphere by dissociation of $N_2$.

### Photoionization

In 1901, Guglielmo Marconi carried out a sensational experiment. He received in St. John's, Newfoundland, a radio signal transmitted from Land's End, England, some 2900 km away. Because radio waves were thought to travel in straight lines, it had been assumed that radio communication over large distances on earth would be impossible. Marconi's successful experiment suggested that the earth's atmosphere in some way substantially affects radio-wave propagation, which led to intensive study of the upper atmosphere. In about 1924 the existence of electrons in the upper atmosphere was established by experimental studies.

For each electron present in the upper atmosphere, there is a corre-

**TABLE 14.2** Ionization processes, ionization energies, and maximum wavelength of a photon capable of causing ionization

| Process | Ionization energy (kJ/mol) | $\lambda_{max}$ (nm) |
|---|---|---|
| $N_2 + h\nu \longrightarrow N_2^+ + e^-$ | 1495 | 80.1 |
| $O_2 + h\nu \longrightarrow O_2^+ + e^-$ | 1205 | 99.3 |
| $O + h\nu \longrightarrow O^+ + e^-$ | 1313 | 91.2 |
| $NO + h\nu \longrightarrow NO^+ + e^-$ | 890 | 134.5 |

sponding positively charged ion. The electrons in the upper atmosphere result mainly from the **photoionization** of molecules, caused by solar radiation. For photoionization to occur, a photon must be absorbed by the molecule, and this photon must have enough energy to remove the highest-energy electron. Some of the more important ionization processes occurring in the upper atmosphere—that is, above about 90 km—appear in Table 14.2, together with the ionization energies and $\lambda_{max}$, the maximum wavelength of a photon capable of causing ionization. Photons with energies sufficient to cause ionization have wavelengths in the short, or high-energy, region of the ultraviolet. These wavelengths are completely filtered out of the radiation reaching earth as a result of their absorption by the upper atmosphere.

## 14.3 OZONE IN THE UPPER ATMOSPHERE

At an elevation of about 90 km, most of the short-wavelength solar radiation capable of causing ionization has been absorbed. As a result, the concentration of ions and electrons drops off very rapidly at about this elevation. Radiation capable of causing dissociation of the $O_2$ molecule remains sufficiently intense, however, so that photodissociation of $O_2$ (Equation 14.3) remains important down to 30 km. The chemical processes that occur in the region below about 90 km following photodissociation of $O_2$ are very different from processes that occur at higher elevations. In the mesosphere and stratosphere the concentration of $O_2$ is much greater than that of atomic oxygen. Thus the O atoms that do form in the mesosphere and stratosphere undergo frequent collisions with $O_2$ molecules. These collisions lead to formation of ozone, $O_3$:

$$O(g) + O_2(g) \longrightarrow O_3^*(g) \qquad [14.4]$$

The asterisk over the $O_3$ denotes that the ozone molecule contains an excess of energy. Reaction of O with $O_2$ to form $O_3$ results in release of 105 kJ/mol. This energy must be gotten rid of by the $O_3$ molecule in a very short time, or else it will simply fly apart again into $O_2$ and O. This decomposition is shown in Equation 14.5 as the reverse of the process by which $O_3$ is formed. The double arrows, $\rightleftharpoons$, indicate the reaction is reversible; that is, that it may occur in either direction. The energy-rich $O_3$ molecule can get rid of the excess energy by colliding with another atom or molecule and transferring some of the excess energy to it. Let us represent the atom or molecule with which $O_3$ is colliding as M. (Nearly always M is $O_2$ or $N_2$, because these are the most abundant molecules.) The transfer of energy can then be represented as in the second reaction, Equation 14.6:

$$O(g) + O_2(g) \rightleftharpoons O_3^* \tag{14.5}$$

$$\frac{O_3^*(g) + M(g) \longrightarrow O_3(g) + M^*(g)}{O(g) + O_2(g) + M(g) \longrightarrow O_3(g) + M^*(g)} \tag{14.6}$$

$$\tag{14.7}$$

The rate at which $O_3$ forms depends on the relative rates of the stabilizing collisions between $O_3^*$ and M (Equation 14.6) and the dissociation of $O_3^*$ back to $O_2$ and O (the reverse process in Equation 14.5). Frequent collisions favor formation of $O_3$ via Equation 14.6. Because the concentration of molecules is greater at lower altitudes, the frequency of stabilizing collisions is greater. However, at low altitudes most of the radiation energetic enough to dissociate $O_2$ has been absorbed. The overall result of these opposing factors is a maximum in the rate of $O_3$ production at about 50 km altitude.

The ozone molecule, once formed, does not last long; ozone itself is capable of absorbing solar radiation, with the result that it is decomposed into $O_2$ and O. Because the energy required for this process is only 105 kJ/mol, photons of wavelength shorter than 1140 nm are sufficiently energetic to dissociate $O_3$. The strongest and most important absorptions are of photons with wavelengths from about 200 to 310 nm. Radiation in this wavelength range is not strongly absorbed by any species other than ozone. If it were not for the layer of ozone in the stratosphere, therefore, these short-wavelength, high-energy photons would penetrate to the earth's surface. Plant and animal life as we know it could not survive in the presence of this high-energy radiation. The "ozone shield" is thus essential for our continued well being. It should be noted, however, that the ozone molecules that form this essential shield against radiation represent only a tiny fraction of the oxygen atoms present in the stratosphere. This is so because the ozone molecules are continually destroyed even as they are formed.

The photodecomposition of ozone reverses the reaction leading to its formation. We thus have a cyclic process of ozone formation and decomposition, summarized as follows:

$$O_2(g) + h\nu \longrightarrow O(g) + O(g)$$

$$O(g) + O_2(g) + M(g) \longrightarrow O_3(g) + M^*(g) \qquad \text{(heat released)}$$

$$O_3(g) + h\nu \longrightarrow O_2(g) + O(g)$$

$$O(g) + O(g) + M(g) \longrightarrow O_2(g) + M^*(g) \qquad \text{(heat released)}$$

The first and third processes are photochemical in nature; they represent conversion of a solar photon into chemical energy. The second and fourth processes are exothermic chemical reactions. The net result of all four processes is a cycle in which solar radiant energy is converted into thermal energy. The ozone cycle in the stratosphere is responsible for the temperature rise that reaches its maximum at the stratopause, as illustrated in Figure 14.1.

The scheme described above for the life and death of ozone molecules

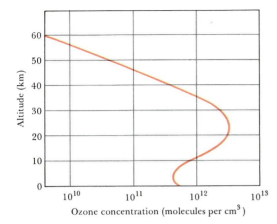

**FIGURE 14.4** Variations in ozone concentration in the atmosphere as a function of altitude.

accounts for some but not all of the known facts about the ozone layer. Many chemical reactions involving substances other than just oxygen occur. In addition, the effects of turbulence and winds in mixing up the stratosphere must be considered. A very complicated picture results.

The overall result of ozone formation and removal reactions, coupled with atmospheric turbulence and other factors, is to produce an ozone profile in the upper atmosphere, as shown in Figure 14.4.

In the preceding sections, we have described the photodissociation of oxygen and ozone. These two processes—and, to a lesser extent, other photodissociations and photoionizations—result in essentially complete absorption of all solar radiation of less than about 300 nm wavelength by the time it reaches the altitude of the tropopause. Because the major constituents of the atmosphere do not interact with radiation of wavelength longer than 300 nm, the photochemical reactions that occur in the troposphere are entirely those of minor atmospheric constituents. Many of the minor constituents occur to only a slight extent in the natural environment but exhibit much higher concentrations in certain areas as a result of human activities. Table 14.3 is a summary of information on

## 14.4 CHEMISTRY OF THE TROPOSPHERE

**TABLE 14.3** Sources and typical concentrations of some minor atmospheric constituents

| Minor constituent | Sources | Typical concentrations |
|---|---|---|
| Carbon dioxide ($CO_2$) | Decomposition of organic matter; release from the oceans; fossil-fuel combustion | 330 ppm throughout troposphere |
| Carbon monoxide ($CO$) | Decomposition of organic matter; industrial processes; fuel combustion | 0.05 ppm in nonpolluted air; 1 to 50 ppm in urban traffic areas |
| Methane ($CH_4$) | Decomposition of organic matter; natural-gas seepage | 1 to 2 ppm throughout troposphere |
| Nitric oxide ($NO$) | Electrical discharges; internal-combustion engines; combustion of organic matter | 0.01 ppm in nonpolluted air; 0.2 ppm in smog atmospheres |
| Ozone ($O_3$) | Electrical discharges; diffusion from stratosphere; photochemical smog | 0 to 0.01 ppm in nonpolluted air; 0.5 ppm in photochemical smog |
| Sulfur dioxide ($SO_2$) | Volcanic gases; forest fires; bacterial action; fossil-fuel combustion; industrial processes (roasting of ores, and so on) | 0 to 0.01 ppm in nonpolluted air; 0.1 to 2 ppm in polluted urban environment |

The oxides of nitrogen are important in the ozone cycle. Nitric oxide, NO, and its close relative nitrogen dioxide, $NO_2$, are present in the stratosphere in low concentrations. Ozone reacts with NO to form $NO_2$ and $O_2$; then $NO_2$ reacts with atomic oxygen to regenerate NO and form $O_2$. The NO is then ready again to react with $O_3$. The overall reaction involving NO is simply

$$O_3(g) + NO(g) \longrightarrow NO_2(g) + O_2(g)$$
$$\underline{NO_2(g) + O(g) \longrightarrow NO(g) + O_2(g)}$$
$$O_3(g) + O(g) \longrightarrow 2O_2(g) \qquad \text{[14.8]}$$

We see from this sequence of reactions that NO serves the function of increasing the rate of decomposition of $O_3$. There is no net change in the chemical state of the NO. We have here a very simple example of a **catalyst**, a substance that has the effect of increasing the rate of a chemical reaction, without itself undergoing a net chemical change.

A few years ago scientists recognized that the **chlorofluoromethanes**, principally $CF_2Cl_2$ and $CFCl_3$, may deplete the ozone layer. These substances have been widely used as propellent gases in spray cans and as refrigerant gases. They are quite inert chemically; there seems to be no relatively rapid chemical process that removes chlorofluoromethanes from the lower atmosphere. The lifetimes of these molecules in the atmosphere are controlled by the rate at which they diffuse into the stratosphere and become subject to the action of ultraviolet light. The action of high-energy light with

**FIGURE 14.5** Map of the total ozone present in the southern hemisphere, taken in October 1986 from an orbiting satellite. The data were gathered using a special instrument called the Total Ozone Mapping Spectrometer (TOMS). The ozone "hole" is the oval feature generally covering the Antarctic, portrayed in gray and violet colors. The ozone level in the most depleted portion of the hole is only about 50 percent of its average value during most of the year, when the hole is not present. The hole is surrounded by a ring of higher-than-average total ozone, shown as yellow, green, and brown. (NASA/Goddard Space Flight Center).

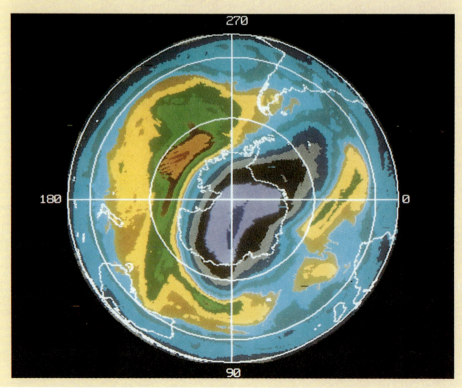

wavelengths in the range 190 to 225 nm results in **photolysis**, or light-induced rupture, of a carbon-chlorine bond:

$$CF_xCl_{4-x}(g) + h\nu \longrightarrow CF_xCl_{3-x}(g) + Cl(g) \qquad [14.9]$$

Photochemical breakdown of the $CF_xCl_{3-x}$ fragment may also occur. Calculations suggest that the rate of chlorine-atom formation will be maximized at an altitude of about 30 km. The atomic chlorine produced by photolysis is capable of rapid reaction with ozone to form chlorine oxide and molecular oxygen. Chlorine oxide is capable of reaction with atomic oxygen to re-form atomic chlorine:

$$\begin{aligned} Cl(g) + O_3(g) &\longrightarrow ClO(g) + O_2(g) \qquad [14.10] \\ ClO(g) + O(g) &\longrightarrow Cl(g) + O_2(g) \qquad [14.11] \\ \hline O_3(g) + O(g) &\longrightarrow 2O_2(g) \qquad [14.8] \end{aligned}$$

This pair of reactions is analogous to those involving nitric oxide to produce the net reaction shown in Equation 14.8. In both cases the original species is regenerated. The overall result is reaction of ozone with atomic oxygen to form molecular oxygen. Many uncertainties are involved in any quantitative estimate of how much ozone destruction might be due to chlorofluoromethanes. Because rates of diffusion of molecules into the stratosphere from the earth's surface are likely to be very slow, several decades may pass before the full impact of chlorofluoromethanes is felt. Additional research on this problem is in progress in many laboratories. In the meantime, use of the chlorofluoromethanes as aerosol propellent gases has been substantially curtailed.

During the past decade researchers have discovered an annual thinning of the ozone layer over the South Pole. It occurs during the austral (southern hemisphere) spring. Ozone levels in October (Figure 14.5) drop to almost 50 percent of the levels in August. Such a thinning over populated parts of the earth would cause severely adverse effects. It is not known at this time whether the effect originates from a seasonal change in global winds, from chlorofluoromethanes in the atmosphere, or from some other cause, such as solar activity.

several minor atmospheric constituents. We shall discuss the most important characteristics of a few of these minor constituents and their chemical role as air pollutants.

**Sulfur Compounds**

Sulfur-containing compounds are present to some extent in the natural, unpolluted atmosphere. They originate in the bacterial decay of organic matter, in volcanic gases, and from other sources listed in Table 14.3. Some scientists think that a certain amount of sulfur dioxide may also originate in the oceans. The concentration of sulfur-containing compounds in the atmosphere resulting from natural sources is very small in comparison with the concentrations built up in urban and industrial environments as a result of human activities. Sulfur compounds, chiefly sulfur dioxide, $SO_2$, are among the most unpleasant and harmful of the common pollutant gases. Table 14.4 shows the concentrations of several

**TABLE 14.4** Concentrations of atmospheric pollutants likely to be exceeded about 50 percent of the time in a typical urban atmosphere

| Pollutant | Concentration (ppm) |
|---|---|
| Carbon monoxide | 10 |
| Hydrocarbons | 3 |
| Sulfur dioxide | 0.08 |
| Nitrogen oxides | 0.05 |
| Total oxidants (ozone and others) | 0.02 |

pollutant gases in a *typical* urban environment (not one that is particularly affected by smog). According to these data, the level of sulfur dioxide would be 0.08 ppm or higher about half the time. This concentration is considerably lower than that of other pollutants, notably carbon monoxide. Nevertheless, sulfur dioxide is regarded as the most serious health hazard among the pollutants shown, especially for persons with respiratory difficulties. Studies of the medical case histories of large population segments in urban environments have shown clearly that those living in the most heavily polluted parts of cities have higher levels of respiratory· disease and shorter life expectancies.

One industrial process that may produce very high local levels of $SO_2$ is the roasting, or smelting, of ores. By this process a metal sulfide is oxidized, driving off $SO_2$, as in the following example:

$$2ZnS(s) + 3O_2(g) \longrightarrow 2ZnO(s) + 2SO_2(g) \qquad [14.12]$$

Smelting operations account for about 8 percent of the total $SO_2$ released in the United States. About 80 percent of the $SO_2$ comes from the combustion of coal and oil. The extent to which $SO_2$ emissions are a problem in the burning of fossil fuels depends on the level of sulfur concentration in the coal or oil. Oil that is burned in the power plants of electrical generating stations is the nonvolatile residue that remains after the low boiling fractions have been distilled off. Some oil, such as that from the Middle East, is relatively low in sulfur, whereas Venezuelan oil is relatively high. Because of concern about $SO_2$ pollution, low-sulfur oil is in greater demand and is consequently more expensive.

Coals vary considerably in their sulfur content. Much of the coal lying in beds east of the Mississippi is relatively high in sulfur content, up to 6 percent by weight. Much of the coal lying in the western states has a lower sulfur content. (This coal, however, also has a lower heat content per unit weight of coal, so that the difference in sulfur content on the basis of a unit amount of heat produced is not as large as is often assumed.)

Altogether, more than 30 million tons of $SO_2$ are released into the atmosphere in the United States each year. This material does a great deal of damage to both property and human health. Not all the damage, however, is caused by $SO_2$ itself; it is likely, in fact, that $SO_3$, formed by oxidation of $SO_2$, is the major culprit. Sulfur dioxide may be oxidized to $SO_3$ by any of several pathways, depending on the particular nature of the atmosphere. Once $SO_3$ is formed it dissolves in water droplets, forming sulfuric acid, $H_2SO_4$:

$$SO_3(g) + H_2O(l) \longrightarrow H_2SO_4(aq) \qquad [14.13]$$

The phenomenon of **acid rain** has been known for a long time in the Scandinavian countries and other parts of northern Europe. The dominant contributor to the acidity in the rain is sulfuric acid. The acidity has caused fish populations in many freshwater lakes to decline, and it has measurably affected other parts of the network of interdependent living things within the lakes and surrounding forests (see Figure 14.6). Acid rain is now affecting many northern lakes in the United States.

**FIGURE 14.6** Severe forest damage in the Erz Gebirge region of Czechoslovakia due to air pollutants from heavy industry in the area. (Tom McHugh/Photo Researchers)

Acid rain is strongly corrosive when it comes in contact with metals, paints, and similar substances (see Figure 14.7). Billions of dollars each year are lost as a result of corrosion resulting from $SO_2$ pollution. Obviously we all want this noxious gas removed from the environment. But removal of sulfur from coal or oil is difficult and, therefore, expensive. Rather than attempt to remove sulfur from fuel before it is burned, the sulfur dioxide formed when the fuel is combusted may be removed in many possible ways. For example, powdered limestone, $CaCO_3$, may

**FIGURE 14.7** The erosion of this stone railway marker post, erected in the nineteenth century, is the result mainly of the reaction of acidic rainfall with the limestone. (Anne LaBastille/Photo Researchers)

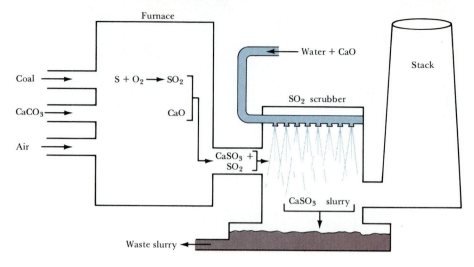

**FIGURE 14.8** Common method for removing $SO_2$ from combusted fuel. Powdered limestone decomposes into CaO, which reacts with $SO_2$ to form $CaSO_3$. The $CaSO_3$ and any unreacted $SO_2$ enter a purification chamber, where a shower of CaO and water converts the remaining $SO_2$ into $CaSO_3$ and precipitates the $CaSO_3$ into a watery residue called slurry.

be blown into the combustion chamber. The carbonate (limestone) is decomposed into lime (CaO) and carbon dioxide:

$$CaCO_3(s) \longrightarrow CaO(s) + CO_2(g) \qquad [14.14]$$

The lime then reacts with $SO_2$ to form calcium sulfite:

$$CaO(s) + SO_2(g) \longrightarrow CaSO_3(s) \qquad [14.15]$$

Only about half the $SO_2$ is removed by contact with the dry solid. The furnace gas must then be "scrubbed" with an aqueous suspension of lime to remove the $CaSO_3$ and any unreacted $SO_2$. This process, which is illustrated in Figure 14.8, is difficult to engineer, reduces the heat effectiveness of the fuel, and leaves an enormous solid waste disposal problem. An electric power plant that would serve the needs of a population of about 150,000 people would produce about 160,000 tons per year of solid waste if it were equipped with the purification system just described. This is three times the normal fly-ash waste from a plant of this size. Various schemes may be employed to recover elemental sulfur or some other industrially useful chemical from the $SO_2$, but as yet no process has been found sufficiently attractive from an economic point of view to warrant large-scale development. Pollution by sulfur dioxide remains a major problem and will probably continue to for some time.

### Nitrogen Oxides; Photochemical Smog

The atmospheric chemistry of the nitrogen oxides is interesting because these substances are components of smog, a phenomenon with which city dwellers are all too familiar. The term **smog** refers to a particularly unpleasant condition of pollution in certain urban environments, that occurs when weather conditions produce a relatively stagnant air mass. The smog made famous by Los Angeles, but now common in many other urban areas as well, is more accurately described as a **photochemical smog**, because photochemical processes play an essential role in its formation (Figure 14.9).

Nitric oxide, NO, is formed in small quantities in the cylinders of

**FIGURE 14.9** A typical urban photochemical smog produced by action of sunlight on automobile exhaust gases. (Susan McCartney/Photo Researchers)

internal-combustion engines via the direct combination of nitrogen with oxygen. Prior to installation of control measures, typical emission levels of $NO_x$ were 4 g/mi. (The $x$ is either 1 or 2; both NO and $NO_2$ are formed, though NO predominates.) Present auto-emission standards call for $NO_x$ emission levels of less than 1 g/mi.

Nitric oxide may be oxidized to $NO_2$, in air or in the automobile engine. The dissociation of $NO_2$ into NO and O requires 304 kJ/mol. This requirement corresponds to a photon wavelength of 393 nm. In sunlight, $NO_2$ undergoes dissociation to NO and O:

$$NO_2(g) + h\nu \longrightarrow NO(g) + O(g) \qquad [14.16]$$

The atomic oxygen formed undergoes several possible reactions, one of which forms ozone, as described earlier:

$$O(g) + O_2(g) + M(g) \longrightarrow O_3(g) + M^*(g) \qquad [14.7]$$

Ozone is capable of rapidly oxidizing NO to $NO_2$:

$$O_3(g) + NO(g) \longrightarrow NO_2(g) + O_2(g) \qquad [14.17]$$

In addition to nitrogen oxides and carbon monoxide, an automobile engine also emits as pollutants unburned and partially burned hydrocarbons, compounds made up entirely of carbon and hydrogen. A typical engine without effective emission controls emits about 10 to 15 g of such organic compounds per mile. Current standards require that hydrocarbon emissions be less than 0.4 g/mi.

Table 14.5 shows typical concentrations of trace constituents in photochemical smog. The most important organic compounds in this list for smog formation are olefins and aldehydes. An **olefin** is an organic compound, a type of hydrocarbon, containing a double bond between carbon atoms. Ethylene, $C_2H_4$ (Section 9.4), is the simplest member of the series. **Aldehydes** are compounds containing a carbon-oxygen double bond

**TABLE 14.5** Typical concentrations of trace pollutants in a photochemical smog (levels are subject to wide variation from one situation to another)

| Constituent | Concentration (ppm in air) |
|---|---|
| $NO_x$ | 0.2 |
| $NH_3$ | 0.02 |
| CO | 40 |
| $O_3$ | 0.5 |
| $CH_4$ | 2 |
| $C_2H_4$ | 0.5 |
| Higher olefins[a] | 0.25 |
| $C_2H_2$ (acetylene) | 0.25 |
| Aldehydes | 0.6 |
| $SO_2$ | 0.2 |

[a] Higher olefins are compounds that contain a hydrocarbon chain attached to one of the carbons of the C=C bond.

on a carbon atom at the end of a hydrocarbon chain. Formaldehyde, acetaldehyde, and propionaldehyde are examples:

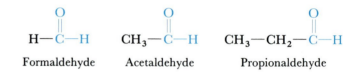

Formaldehyde          Acetaldehyde          Propionaldehyde

Both $NO_2$ and $O_3$ are capable of reaction with olefins to eventually yield aldehydes.

Many of the compounds formed by reaction of atomic oxygen and ozone with organic compounds are **free radicals**, molecular fragments that contain an unpaired electron. They are very reactive and lead to a complex chemistry in the polluted atmosphere. One group of molecules formed in this way is the peroxyacylnitrates (PAN), especially unpleasant substances that cause eye irritation and breathing difficulties:

In this diagram, R represents an organic group such as $CH_3$, $C_6H_5$, and so on.

Reduction or elimination of smog requires that the essential ingredients for its formation be removed from automobile exhaust. Catalytic mufflers are designed to reduce drastically the levels of two of the major ingredients of smog: $NO_x$ and hydrocarbons. However, emission-control systems are not notably successful in poorly maintained autos.

**Carbon Monoxide**

In terms of total mass, carbon monoxide is the most abundant of all the pollutant gases. The level of CO present in fresh, nonpolluted air is small, probably on the order of 0.05 to 0.1 ppm. The estimated total amount of CO in the earth's atmosphere is about $5.2 \times 10^{14}$ g. In the United

States alone, however, about $1 \times 10^{14}$ g of CO is produced each year. The CO is formed mostly in the incomplete combustion of fossil fuels. The major sources of CO in the United States are automobile and power-plant emissions.

Carbon monoxide is a relatively unreactive molecule, and it might be supposed that it would not be a health hazard. It does have the unusual ability, however, of binding very strongly to **hemoglobin**, the iron-containing protein that is responsible for oxygen transport in the blood. Hemoglobin consists of four protein chains loosely held together in a cluster. Each chain has within its folds a heme molecule. The structure of heme is shown in Figure 14.10. Note that iron is situated in the center of a plane of four nitrogen atoms. A hemoglobin molecule in the lungs picks up an oxygen molecule, which reacts with the iron atom to form a species called **oxyhemoglobin**. The equilibrium between hemoglobin and oxyhemoglobin is shown in Figure 14.11. As the blood circulates, the oxygen molecule is released in tissues as needed for cell metabolism, that is, for the chemical processes occurring in the cell.

Carbon monoxide also happens to bind very strongly to the iron in hemoglobin. The complex is called **carboxyhemoglobin** and is represented as COHb. The affinity of human hemoglobin for CO is about 210 times greater than for $O_2$. As a result, a relatively small quantity of CO can inactivate a substantial fraction of the hemoglobin in the blood for oxygen transport. For example, a person breathing air that contains only 0.1 percent of CO takes in enough CO after a few hours of breathing to convert up to 60 percent of the hemoglobin into COHb, thus reducing the blood's normal oxygen-carrying capacity by 60 percent.

Under normal conditions, a nonsmoker breathing unpolluted air has about 0.3 to 0.5 percent carboxyhemoglobin, COHb, in the bloodstream. This amount arises mainly from the production of small quantities of CO in the course of normal body chemistry and from the small amount of CO present in clean air. Exposure to higher concentrations of CO causes the COHb level to increase. This doesn't happen instantly, but requires several hours. Similarly, when the CO level is suddenly decreased, it requires several hours for the COHb concentration to level off at a lower value. Table 14.6 shows the percentages of COHb in blood that are typical of various groups of people.

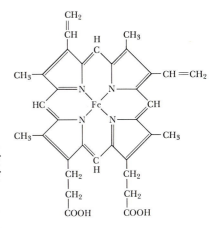

**FIGURE 14.10** Structure of the heme molecule.

**FIGURE 14.11** Equilibrium between hemoglobin and oxyhemoglobin. The tan regions represent the protein chain. The heme molecule is attached to the protein.

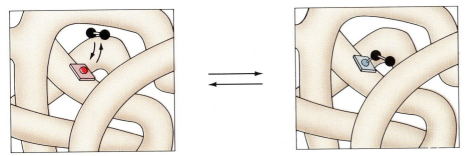

**TABLE 14.6** Carboxyhemoglobin (COHb) percentages in the blood of persons under various conditions

|  | COHb (%) |
| --- | --- |
| Continuous exposure, 10 ppm CO | 2.0 |
| Continuous exposure, 30 ppm CO | 5.0 |
| Nonsmokers, Chicago (1970) | 2.0 |
| Smokers, Chicago (1970) | 5.8 |
| Nonsmokers, Milwaukee (1969–1971) | 1.1 |
| Smokers, Milwaukee (1969–1971) | 5.0 |
| Nonsmokers, clean air | 0.3–0.5 |

## Water Vapor, Carbon Dioxide, and Climate

We have seen how the atmosphere makes life as we know it possible on earth by screening out harmful short-wavelength radiation. In addition, the atmosphere is essential in maintaining a reasonably uniform and moderate temperature on the surface of the planet. The two atmospheric components of major importance in maintenance of the earth's surface temperature are carbon dioxide and water.

The earth is in overall thermal balance with its surroundings. This means that the planet radiates energy into space at a rate equal to the rate at which it absorbs energy from the sun. The sun has a temperature of about 6000 K. As seen from outer space the earth is relatively cold, with a temperature of about 254 K. The distribution of wavelengths in the radiation emitted from an object is determined by its temperature. The radiation emitted by relatively cold objects is in the low-energy, or long-wavelength, region of the spectrum. The maximum in the wavelength of radiation from the earth is in the far-infrared region, around 12,000 nm (Figure 6.3). The troposphere, transparent to visible light, is not at all transparent to infrared radiation. Figure 14.12 shows the dis-

---

### CHEMISTRY AT WORK        Smoking and CO

The CO concentration in city traffic often reaches 50 ppm and may go as high as 140 ppm in traffic jams. The most serious source of carbon monoxide poisoning, however, comes from cigarette smoking. The inhaled smoke from cigarettes contains about 400 ppm of CO. The effect of smoking on COHb percentage is evident from the data in Table 14.6. A study of a group of San Francisco dockworkers presented further proof of the dramatic relationship between smoking and COHb percentage. Nonsmokers in the group averaged 1.3 percent COHb; light smokers (less than half a pack per day) averaged 3.0 percent; moderate smokers averaged 4.7 percent; and heavy smokers (two packs or more per day) averaged 6.2 percent COHb.

Widespread evidence indicates that chronic exposure to CO impairs performance on standardized tests. Thus it is most definitely not a good idea to smoke heavily before and during a test. In addition, motor performance is also impaired by high COHb percentages. For example, evidence shows that drivers responsible for traffic accidents have, on the average, higher than normal percentages of COHb in their blood. As has been mentioned, a chronically high level of COHb means that a certain fraction of the hemoglobin in the blood is not available for oxygen transport. This in turn means that the heart must work that much harder to ensure an adequate supply of oxygen. It is not surprising, therefore, that many medical researchers believe that chronic exposure to CO is a contributing factor in heart disease and in heart attacks.

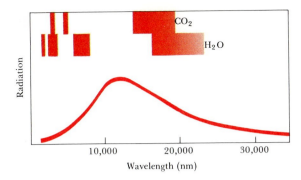

Radiation

10,000    20,000    30,000
Wavelength (nm)

**FIGURE 14.12**
Long-wavelength radiation from earth compared with the absorption of infrared radiation by carbon dioxide and water.

tribution of radiation from the earth's surface and, on the same scale, the wavelengths absorbed by atmospheric water vapor and carbon dioxide. Clearly, these atmospheric gases absorb much of the outgoing radiation from the earth's surface. It is indeed fortunate for us that they do so; they serve to maintain a livably uniform temperature at the surface by holding in, as it were, the infrared radiation from the surface, which we feel as heat. The influence of water and carbon dioxide on earth's climate is often called the "greenhouse effect."

The partial pressure of water vapor in the atmosphere varies greatly from place to place and time to time, but, in general, it is highest near the surface and drops off very sharply with increased elevation. Carbon dioxide, by contrast, is uniformly distributed throughout the atmosphere, at a concentration of about 330 ppm. Because water vapor absorbs infrared radiation so strongly, it plays the major role in maintaining the atmospheric temperature at night, when the surface is emitting radiation into space and not receiving energy from the sun. In very dry desert climates, where the water-vapor concentration is unusually low, it may be extremely hot during the day but very cold at night. In the absence of an extensive layer of water vapor to absorb and then radiate back to the earth part of the infrared radiation, the surface loses this radiation into space and cools off very rapidly.

Carbon dioxide plays a secondary, but very important, role in maintaining the surface temperature. The worldwide combustion of fossil fuels, principally coal and oil, on a prodigious scale in the modern era has materially increased the carbon dioxide level of the atmosphere. From measurements carried out over several decades it is clear that the $CO_2$ concentration in the atmosphere is steadily increasing. From a knowledge of the infrared-absorbing characteristics of $CO_2$ and water, and using a theoretical model for the atmosphere, it has been estimated that if $CO_2$ were to double from its present level, the average surface temperature of the planet would increase 3°C. On the basis of present and expected future rates of fossil-fuel use, the atmospheric $CO_2$ level is expected to just about double during the century ending 2050. If the calculated effect of a doubling of $CO_2$ on surface temperature is correct, the earth's temperature will rise about 3°C during that time. Such a small change may seem insignificant, but it is not. Major changes in global climate could result from a temperature change of this or even smaller magnitude. Because so many factors go into determining climate, it is not possible to predict with certainty what changes will occur. It is clear, however, that humanity has acquired the potential, by changing the $CO_2$ concentration in the atmosphere, for substantially altering the climate of the planet.

## 14.5 THE WORLD OCEAN

Let's now change our focus from the atmosphere to the hydrosphere. However, before we begin to examine our water environment, we should review some of water's most important characteristics as a solvent and reactant molecule. We have learned that water possesses many exceptional properties because of extensive hydrogen bonding (Section 11.4). Hydrogen bonding is responsible for the high melting and boiling points of water and for its high heat capacity. The highly polar character of water is responsible for its exceptional ability to dissolve a wide range of ionic substances (Chapter 13). Water reacts with many substances such as oxides (Section 8.11) to form acidic or basic solutions. It is also the medium in which important types of reactions such as neutralization and precipitation occur (Section 3.2). We will see that all of these key characteristics of water come into play as we consider natural waters, beginning with the world ocean.

### Seawater

All of the vast layer of salty water that covers so much of the earth is connected and is more or less constant in composition. For this reason, oceanographers (scientists whose major interest is the sea) speak in terms of a world ocean, rather than of the separate oceans we learn about in geography books. The world ocean is indeed huge. Its volume is 1.35 billion cubic kilometers. It covers about 72 percent of the earth's surface. Almost all the water on earth, 97.2 percent, is in the world ocean. About 2.1 percent is in the form of ice caps and glaciers. All of the fresh water—in lakes, rivers, and ground water—amount to only 0.6 percent. The remaining 0.1 percent is in the form of brine wells and brackish (salty) waters.

Seawater is often referred to as saline water. The **salinity** of seawater is defined as the mass in grams of dry salts present in 1 kg of seawater. In the world ocean, the salinity varies from 33 to 37, with an average of about 35. To put it another way, seawater contains about 3.5 percent dissolved salts. The list of elements present in seawater is very long. However, most are present only in very low concentrations. Table 14.7 lists the 11 ionic species present in seawater at concentrations greater than 0.001 g/kg, or 1 part per million (ppm) by weight. (To convert from g/kg to ppm, multiply by 1000; see Equation 14.2.) In a lower range of concentration, the elements nitrogen, lithium, rubidium, phosphorus, iodine, iron, zinc, and molybdenum are present in amounts ranging from 1 to 0.01 ppm. At least 50 other elements have been identified at still lower concentrations.

The sea is a vast storehouse of chemicals. Each cubic mile of seawater contains $1.5 \times 10^{11}$ kg of dissolved solids. The sea is so vast that if a substance is present in seawater to the extent of only 1 part per billion by mass, there are still $5 \times 10^9$ kg of it in the world ocean. Nevertheless, the ocean is not used very much as a source of raw materials, because the costs of extracting the desired substances from the water are too high.

**TABLE 14.7**  Ionic constituents of seawater present in concentrations greater than 0.001 g/kg (1 ppm) by weight

| Ionic constituent | g/kg seawater | Concentration ($M$) |
|---|---|---|
| Chloride, $Cl^-$ | 19.35 | 0.55 |
| Sodium, $Na^+$ | 10.76 | 0.47 |
| Sulfate, $SO_4^{2-}$ | 2.71 | 0.028 |
| Magnesium, $Mg^{2+}$ | 1.29 | 0.054 |
| Calcium, $Ca^{2+}$ | 0.412 | 0.010 |
| Potassium, $K^+$ | 0.40 | 0.010 |
| Carbon dioxide[a] | 0.106 | $2.3 \times 10^{-3}$ |
| Bromide, $Br^-$ | 0.067 | $8.3 \times 10^{-4}$ |
| Boric acid, $H_3BO_3$ | 0.027 | $4.3 \times 10^{-4}$ |
| Strontium, $Sr^{2+}$ | 0.0079 | $9.1 \times 10^{-5}$ |
| Fluoride, $F^-$ | 0.001 | $7 \times 10^{-5}$ |

[a] $CO_2$ is present in seawater as $HCO_3^-$ and $CO_3^{2-}$

Only three substances are recovered from seawater in commercially important amounts: sodium chloride, bromine, and magnesium.

**Desalination**

Because of its high salt content, seawater is unfit for human consumption and indeed for most of the uses to which we put water. In the United States, the salt content of municipal water supplies is restricted by health codes to no more than about 500 ppm. This amount is much lower than the 3.5 percent dissolved salts present in seawater and the 0.5 percent or so present in brackish water found underground in some regions. The removal of salts from seawater or brackish water to the extent that the water becomes usable is called **desalination**.

Water can be separated from dissolved salts by *distillation* (described in Section 1.2) since water is a volatile substance and the salts are nonvolatile. The principle of distillation is simple enough, but there are many problems associated with carrying out the process on a large scale. For example, as water is distilled from a vessel containing seawater, the salts become more and more concentrated and eventually precipitate out.

Seawater can also be desalinated using **reverse osmosis**. Recall from Section 12.6 that osmosis is the net movement of solvent molecules, but not solute molecules, through a semipermeable membrane. In osmosis, solvent passes from the more dilute solution into the more concentrated one. However, if a sufficient external pressure is applied, osmosis can be stopped and, at still higher pressures, reversed. When this occurs, solvent passes from the more concentrated into the more dilute solution. In a modern reverse-osmosis facility, tiny hollow fibers are used as the semipermeable membrane. Water is introduced under pressure into the fibers, and desalinated water is recovered on the outside, as illustrated in Figure 14.13.

The island of Malta, located in the Mediterranean Sea, is formed from limestone, and there is little underground fresh water. Part of the island's water supply is obtained from a reverse-osmosis desalination plant (Figure 14.14) that produces 5.3 million gallons per day of desalinated water. The salt content is reduced from 36,000 ppm of dissolved salts to less than 500 ppm, within the acceptable limits for drinkable water.

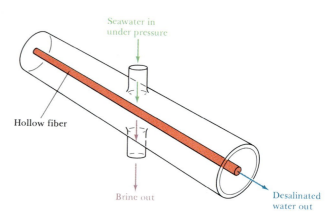

Seawater in
under pressure

Hollow fiber

Brine out

Desalinated
water out

**FIGURE 14.13**   Schematic view of a hollow-fiber
reverse-osmosis unit. Water is introduced under pressure
around the hollow fibers. Desalinated water is recovered
from the inside of the fiber. In practice each unit contains
more than three million fibers, each of which is about the
diameter of a human hair. The fibers are bundled together
in units such as those shown in Figure 14.14(*b*).

**FIGURE 14.14**
(*a*) Seawater reverse osmosis
desalination plant at Ghar
Lapsi on the island of Malta,
in the Mediterranean. (*b*) A
permeator room at Ghar Lapsi.
Each of the cylinders shown
contains several million tiny
hollow fibers. When seawater
is introduced under pressure
into the cylinder, water passes
through the fiber wall to the
desalinated, or largely salt-free,
side. (Courtesy Polymetrics,
Inc., and DuPont)

(*a*)

(*b*)

We've seen that the total amount of fresh water on earth is not a large fraction of the total water present. Fresh water results from evaporation from the oceans and the land and transpiration through the leaves of plants. The water vapor that accumulates in the atmosphere is transported via global atmospheric circulation to other latitudes, where it falls as rain or snow. The water that falls on land runs off in rivers or collects in lakes or underground caverns. Eventually, it is evaporated or carried via streams and rivers back to the oceans.

Fresh water, of course, is not simply $H_2O$. It contains dissolved gases (principally $O_2$, $N_2$, and $CO_2$), a variety of cations (mainly $Na^+$, $K^+$, $Mg^{2+}$, $Ca^{2+}$, and $Fe^{2+}$) and a variety of anions (mainly $Cl^-$, $SO_4^{2-}$, and $HCO_3^-$). Suspended solids such as tiny clay particles are also likely to be present.

The amount of dissolved oxygen present is an important indicator of the quality of the water. Water fully saturated with air at 1 atm and 20°C contains about 9 ppm of $O_2$. Oxygen is necessary for fish and much other aquatic life. Cold-water fish require about 5 ppm of dissolved oxygen for survival. Aerobic bacteria consume dissolved oxygen in order to oxidize organic materials and so meet their energy requirements. The organic material that the bacteria are able to oxidize is said to be **biodegradable**. This oxidation occurs by a complex set of chemical reactions, and the organic material disappears gradually. The carbon, hydrogen, oxygen, nitrogen, sulfur, and phosphorus in the biodegradable material end up mainly as $CO_2$, $H_2O$, $NO_3^-$, $SO_4^{2-}$, and phosphates. These oxidation reactions sometimes reduce the amount of dissolved oxygen to the point where the aerobic bacteria can no longer survive. Anaerobic bacteria then take over the decomposition process, forming $CH_4$, $NH_3$, $H_2S$, $PH_3$, and other products that contribute to the offensive odors of some polluted waters.

The water needed for domestic uses, for agriculture, or for industrial processes is taken from naturally occurring lakes, rivers, and underground sources or from reservoirs. Much of the water that finds its way into municipal water systems is "used" water; it has already passed through one or more sewage systems or industrial plants. Consequently, this water must be treated before it is distributed to our faucets. Municipal water treatment usually involves five steps: coarse filtration, sedimentation, sand filtration, aeration, and sterilization. Figure 14.15 shows a typical treatment process.

After coarse filtration through a screen, the water is allowed to stand in large settling tanks, in which finely divided sand and other minute particles can settle out. To aid removal of very small particles, the water may first be made slightly basic by adding $CaO$. Then $Al_2(SO_4)_3$ is added. The aluminum sulfate reacts with $OH^-$ ions to form a spongy, gelatinous precipitate of $Al(OH)_3$. This precipitate settles slowly, carrying suspended particles down with it, thereby removing nearly all finely divided matter and most bacteria. The water is then filtered through a sand bed. Following filtration the water may be sprayed into the air to hasten the oxidation of dissolved organic substances.

**Treatment of Municipal Water Supplies**

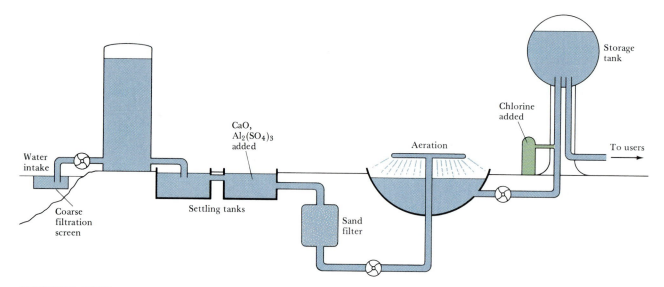

**FIGURE 14.15**  Common steps in treating water for a public water system.

The final stage of the operation normally involves treating the water with a chemical agent to ensure the destruction of bacteria. Ozone, $O_3$, is most effective, but it must be generated at the place where it is used. Chlorine, $Cl_2$, is therefore more convenient to use. Chlorine can be shipped in tanks as the liquefied gas and dispensed from the tanks through a metering device directly into the water supply. The amount used depends on the presence of other substances with which the chlorine might react and on the concentrations of bacteria and viruses to be removed. The sterilizing action of chlorine is probably due not to $Cl_2$ itself but to hypochlorous acid, which forms when chlorine reacts with water:

$$Cl_2(aq) + H_2O(l) \longrightarrow HClO(aq) + H^+(aq) + Cl^-(aq) \quad [14.18]$$

**Water Softening**

The water treatment described thus far should remove all the substances potentially harmful to health. Sometimes additional treatment is used to reduce the concentrations of $Ca^{2+}$ and $Mg^{2+}$, which are responsible for **water hardness**. These ions react with soaps to form an insoluble material. Although they do not form precipitates with detergents, they adversely affect the performance of such cleaning agents. In addition, mineral deposits may form when water containing these ions is heated. When water containing $Ca^{2+}$ and bicarbonate ions is heated, some carbon dioxide is driven off. As a result, the solution becomes less acidic, and insoluble calcium carbonate forms:

$$Ca^{2+}(aq) + 2HCO_3{}^-(aq) \xrightarrow{\text{heat}} CaCO_3(s) + CO_2(g) + H_2O(l) \quad [14.19]$$

The solid $CaCO_3$ coats the surfaces of hot-water systems and the insides of teakettles, thereby reducing heating efficiency. Deposits of scale can be especially serious in boilers in which water is heated under pressure in pipes running through a furnace. Formation of scale reduces the efficiency of heat transfer and may result in the pipes melting.

Not all municipal water supplies require water softening. In those that do, the water is generally taken from underground sources in which the water has had considerable contact with limestone $(CaCO_3)$ and other minerals containing $Ca^{2+}$, $Mg^{2+}$, and $Fe^{2+}$. The **lime-soda process** is used for large-scale municipal water-softening operations. The water is treated with "lime," CaO [or "slaked lime," $Ca(OH)_2$], and "soda ash," $Na_2CO_3$. These chemicals cause precipitation of calcium as $CaCO_3$ and of magnesium as $Mg(OH)_2$:

$$Ca^{2+}(aq) + CO_3{}^{2-}(aq) \longrightarrow CaCO_3(s) \qquad [14.20]$$

$$Mg^{2+}(aq) + 2OH^-(aq) \longrightarrow Mg(OH)_2(s) \qquad [14.21]$$

The role of $Na_2CO_3$ is to provide a source of $CO_3{}^{2-}$, if needed. If the water already contains a high concentration of bicarbonate ion, calcium can be removed as $CaCO_3$ simply by addition of $Ca(OH)_2$:

$$Ca^{2+}(aq) + 2HCO_3{}^-(aq) + [Ca^{2+}(aq) + 2OH^-(aq)] \longrightarrow 2CaCO_3(s) + 2H_2O(l) \qquad [14.22]$$

Lime is used only to the extent that bicarbonate is present: 1 mol of $Ca(OH)_2$ for each 2 mol of $HCO_3{}^-$. When bicarbonate is not present, addition of $Na_2CO_3$ causes removal of $Ca^{2+}$ as $CaCO_3$. The carbonate ion also serves to cause precipitation of $Mg(OH)_2$:

$$Mg^{2+}(aq) + 2CO_3{}^{2-}(aq) + 2H_2O(l) \longrightarrow 2HCO_3{}^-(aq) + Mg(OH)_2(s) \qquad [14.23]$$

There are two problems with the lime-soda process. First, formation of $CaCO_3$ and $Mg(OH)_2$ may take a long time, and they may not settle out very well. Second, the solution that results after removal of the precipitates is too strongly basic. This is so because the chemicals added, $Ca(OH)_2$ and $Na_2CO_3$, are both bases. (We will discuss the reason that $Na_2CO_3$ is a base in Chapter 17.) Alum, $Al_2(SO_4)_3$, is added to remove the precipitates. Because the solution is basic, the $Al^{3+}$ ion forms $Al(OH)_3(s)$, a gelatinous precipitate that carries finely divided solids with it out of solution. To prevent any further precipitation of $Mg(OH)_2$ or $CaCO_3$ by the ions remaining, the solution is made less basic by bubbling $CO_2$ through the water.

## Sewage Treatment

Municipal sewage treatment is generally divided into primary, secondary, and tertiary treatments. About 10 percent of sewage handled by public sewers receives no treatment, about 30 percent receives only primary treatment, and about 60 percent is subject to primary and secondary treatment. Tertiary treatment is presently rare, but it is expected to become more common as communities upgrade their treatment facilities to meet federal water-pollution standards.

Primary treatment consists first of screening the incoming sewage to filter out debris and larger suspended solids. Then the sewage is passed

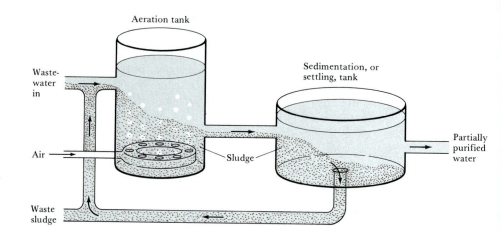

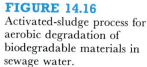

**FIGURE 14.16**
Activated-sludge process for aerobic degradation of biodegradable materials in sewage water.

into settling or sedimentation tanks where suspended solids, called sludge, settle out. If the water receives no secondary treatment it is then often treated with chlorine before being dumped back into the natural water system. Primary treatment removes about 60 percent of the suspended solids and 35 percent of dissolved biodegradable materials.

Secondary treatment is based on aerobic decomposition of organic material. The most common type of secondary treatment is known as the activated-sludge method. In this method the waste from primary treatment is passed into an aeration tank, where air is blown through it (see Figure 14.16). This aeration results in rapid growth of aerobic bacteria that feed on the organic wastes in the water. The bacteria form a mass called activated sludge. This sludge settles out in sedimentation tanks, and the liquid effluent is discharged, often after chlorination. Most of the activated sludge is returned to the aeration tank, where it aids in decomposing the organic wastes in the incoming water. After secondary treatment, about 90 percent of suspended solids and 90 percent of dissolved biodegradable material has been removed.

## FOR REVIEW

### SUMMARY

In this chapter we've examined the physical and chemical properties of the earth's atmosphere. The complex temperature variations in the atmosphere give rise to several regions, each with characteristic properties. The lowest of these regions, the **troposphere**, extends from the surface to about 11 km. Above the troposphere, in order of increasing altitude, are the **stratosphere**, **mesosphere**, and **thermosphere**. In the upper reaches of the atmosphere,

only the simplest chemical species can survive the bombardment of highly energetic particles and radiation from the sun. The average molecular weight of the atmosphere at high elevations is lower than at the earth's surface, because the lightest atoms and molecules diffuse upward and because of **photodissociation**. Absorption of radiation may also lead to **photoionization**.

Ozone is produced in the mesosphere and stratosphere as a result of reaction of atomic oxygen with $O_2$. Ozone is itself decomposed by absorption of a photon or by reaction with an active species such as NO. Human activities could result in the addition to the stratosphere of atomic chlorine, which is capable of reacting with ozone in a catalytic cycle to convert ozone to $O_2$. A marked reduction in the ozone level in the upper atmosphere would have serious adverse consequences, because the ozone layer filters out certain wavelengths of ultraviolet light that are not taken out by any other atmospheric component.

In the troposphere, the chemistry of trace atmospheric components is of major importance. Many of these minor components are pollutants; sulfur dioxide is one of the more noxious and prevalent. It is oxidized in air to form sulfur trioxide, which upon dissolving in water forms sulfuric acid. One method of preventing $SO_2$ from escaping from industrial operations involves reacting the $SO_2$ with $CaO$ to form calcium sulfite, $CaSO_3$.

**Photochemical smog** is a complex mixture of components in which both nitrogen oxides and ozone play important roles. The smog components are generated mainly in automobile engines, and smog control consists largely of controlling emissions from automobiles.

Carbon monoxide is found in high concentrations in the exhaust of automobile engines and in cigarette smoke. This compound is a health hazard because of its ability to form a strong bond with hemoglobin and thus reduce the capacity of blood for oxygen transfer from the lungs.

Carbon dioxide and water vapor are the only major components of the atmosphere that strongly absorb infrared radiation. The level of carbon dioxide in the atmosphere is thus important in determining worldwide climate. As a result of the extensive combustion of fossil fuels (coal, oil, and natural gas) the carbon dioxide level of the atmosphere is steadily increasing.

Seawater contains about 3.5 percent by weight of dissolved salts. Because most of the world's water is in the oceans, humankind may eventually look to the seas for fresh water. **Desalination** is the removal of dissolved salts from seawater, brine, or brackish water to make it fit for human consumption. Among the means by which desalination may be accomplished are distillation and reverse osmosis.

Fresh water that is available from rivers, lakes, and underground sources may require treatment. The several steps which may be used in water treatment are coarse filtration, sedimentation, sand filtration, aeration, sterilization, and softening.

Wastewater treatment is applied to sewage waters or to water that has been used in an industrial operation. Municipal wastewaters are given a primary treatment to remove debris and larger suspended solids. Secondary treatment consists of aeration of sewage sludge to promote the growth of microorganisms that feed on the organic compounds present in sewage. Eventually, clear water is separated from the mass of microorganisms. The water has a lower content of dissolved biodegradable substances than before treatment. The many substances that remain in waters after secondary treatment can be removed only by extensive additional processing, referred to as tertiary treatment.

# LEARNING GOALS

Having read and studied this chapter, you should be able to:

1. Sketch the manner in which the atmospheric temperature varies with altitude, and list the names of the various regions of the atmosphere and the boundaries between them.

2. Sketch the manner in which atmospheric pressure decreases with elevation, and explain in general terms the reason for the decrease.

3. Describe the composition of the atmosphere with respect to the four most abundant components.

4. Explain what is meant by the term *photodissociation*, and calculate the maximum wavelength of a photon that is energetically capable of producing photodissociation, given the dissociation energy of the bond to be broken in the process.

5. Explain what is meant by *photoionization*.

6. Explain the presence of ozone in the mesosphere and stratosphere in terms of appropriate chemical reactions.

7. List the names and chemical formulas of the more important pollutant substances present in the troposphere and in urban atmospheres.

8. List the major sources of sulfur dioxide as an atmospheric pollutant.

9. List the more important reactions of nitrogen oxides and ozone that occur in smog formation.

10. Explain why carbon monoxide constitutes a health hazard.

**11.** Explain why the concentration of carbon dioxide in the troposphere has an effect on the average temperature at the earth's surface.

**12.** List the more abundant ionic species present in seawater.

**13.** Explain the desalination of water by distillation and by reverse osmosis.

**14.** List and explain the various stages of treatment that may be applied to a freshwater supply.

**15.** Describe the chemical principles involved in the lime-soda process for reducing water hardness.

**16.** List and explain the stages in treatment of wastewater.

---

## KEY TERMS

Among the more important terms and expressions used for the first time in this chapter are the following:

**Acid rain** (Section 14.4) refers to rainwater that has become excessively acidic because of absorption of pollutant oxides, notably $SO_3$, produced by human activities.

**Carboxyhemoglobin** (Section 14.4) is a complex formed between carbon monoxide and hemoglobin, in which CO is bound to the iron atom.

A **catalyst** (Section 14.3) is a substance that affects the rate of a chemical reaction but does not itself undergo a net, overall chemical change.

**Chlorofluoromethanes** (Section 14.3) are compounds of the general formula $CF_xCl_{4-x}$, $x = 1, 2$, or 3, used as propellant gases in aerosol spray cans and in refrigeration units.

**Desalination** (Section 14.5) is the removal of salts from seawater, brine, or brackish water to make it fit for human consumption.

**Hemoglobin** (Section 14.4) is an iron-containing protein responsible for oxygen transport in the blood.

The **lime-soda process** (Section 14.6) is a method for removal of $Mg^{2+}$ and $Ca^{2+}$ ions from water to reduce water hardness. The substances added to the water are "lime," CaO [or "slaked lime," $Ca(OH)_2$], and "soda ash," $Na_2CO_3$, in amounts determined by the concentrations of the offending ions.

**Photochemical smog** (Section 14.4) is a complex mixture of undesirable substances produced by the action of sunlight on an urban atmosphere polluted with automobile emissions. The major starting ingredients are nitrogen oxides and organic substances, notably olefins and aldehydes.

**Photodissociation** (Section 14.2) refers to the breaking of a molecule into two or more neutral fragments as a result of absorption of light.

**Photoionization** (Section 14.2) refers to the removal of an electron from an atom or molecule by absorption of light.

**Reverse osmosis** (Section 14.5) is the process by which water molecules move under high pressure through a semipermeable membrane from the more concentrated to the less concentrated solution.

**Salinity** (Section 14.5) is a measure of the salt content of seawater, brine, or brackish water. It is equal to the mass in grams of dissolved salts present in 1 kg of seawater.

The **troposphere** (Section 14.1) is the region of earth's atmosphere extending from the surface to about 11 km altitude. The regions of the atmosphere extending above the troposphere are, in order of increasing altitude, the **stratosphere**, **mesosphere**, and **thermosphere**.

**Water hardness** (Section 14.6) refers to the presence in a water supply of $Ca^{2+}$ and $Mg^{2+}$ ions. These ions form insoluble precipitates with soaps and are responsible for scale formation when the water is heated.

---

## EXERCISES

### Earth's Atmosphere; The Outer Regions

**14.1** Name the regions of the atmosphere, indicate the altitude interval for each region, and describe the variation in temperature in that region.

**14.2** Name the boundaries between the regions of the atmosphere, and indicate the temperatures in the boundary regions.

**14.3** From the data in Table 14.1, calculate the partial pressures in mm Hg of argon and neon when the total pressure is 750 mm Hg.

**14.4** In a particular sample of atmosphere the partial pressure of $O_2$ is found to be 136 mm Hg. Assuming that the sample has the overall composition indicated in Table 14.1, what is the total pressure in the sample?

**14.5**  The dissociation energy of a carbon-bromine bond is typically about 210 kJ/mol. What is the maximum wavelength of photon that can cause C—Br bond dissociation?

**14.6**  In $CF_3Cl$ the C—Cl bond dissociation energy is 339 kJ/mol. In $CCl_4$ the C—Cl bond dissociation energy is 293 kJ/mol. What is the range of wavelengths of photons that can cause C—Cl bond rupture in one molecule but not in the other?

**14.7**  What conditions need to be met for radiant energy to cause photodissociation of NO: $NO(g) \longrightarrow N(g) + O(g)$?

**14.8**  In terms of the energy requirements, explain why photodissociation of oxygen is more important than photoionization of oxygen at altitudes below about 90 km.

## Chemistry of the Stratosphere

**14.9**  Explain why oxygen atoms exist in the atomic state for much longer average times at 120 km elevation than at 50 km elevation.

**14.10**  Explain the means by which ozone, $O_3$, is formed in the stratosphere. What is the biological significance at the earth's surface of the ozone layer in the stratosphere?

**14.11**  Using the thermodynamic data in Appendix D, calculate the overall enthalpy change in each step in the catalytic cycle that converts $O_3$ to $O_2$ (see Equation 14.8).

$$NO(g) + O_3(g) \longrightarrow NO_2(g) + O_2(g)$$
$$NO_2(g) + O(g) \longrightarrow O_2(g) + NO(g)$$

**14.12**  The standard enthalpies of formation of ClO and $ClO_2$ are 101 and 102 kJ/mol, respectively. Using these data and the thermodynamic data in Appendix D, calculate the overall enthalpy change for each step in the following catalytic cycle:

$$ClO(g) + O_3(g) \longrightarrow ClO_2(g) + O_2(g)$$
$$ClO_2(g) + O(g) \longrightarrow ClO(g) + O_2(g)$$

On the basis of your results, indicate whether the ClO—$ClO_2$ pair is at least a possible catalyst for decomposition of ozone in the atmosphere.

## Chemistry of the Troposphere

**14.13**  In a particular urban environment, the ozone concentration is 0.26 ppm. Assuming a temperature of 16°C and an atmospheric pressure of 750 mm Hg at the time, calculate the partial pressure of ozone and the number of $O_3$ molecules per cubic meter.

**14.14**  In a particular urban environment the NO concen-

tration is 0.75 ppm. If the atmospheric pressure at the time is 730 mm Hg and the temperature is 20°C, calculate the partial pressure of NO and the number of NO molecules per cubic meter.

**14.15**  Compare typical concentrations of CO, $SO_2$, and NO in nonpolluted air (Table 14.3) and urban air (Table 14.4), and indicate in each case at least one possible source of the higher values in Table 14.4.

**14.16**  For each of the following gases, make a list of known or possible naturally occurring sources: (**a**) $CH_4$; (**b**) $SO_2$; (**c**) NO; (**d**) CO.

**14.17**  A recent study carried out in Canada found that at a particular location far from industrial activity the sulfate in rainfall amounted to 210 mg/$m^2$ per year. Calculate the total number of moles of sulfate falling in a square mile per year.

**14.18**  Assuming an overall efficiency of about 30 percent, how much calcium carbonate would be required to remove the $SO_2$ formed in burning a ton of coal containing 2.7 percent sulfur by weight?

## The World Ocean

**14.19**  What is the molarity of $Na^+$ in a solution of NaCl whose salinity is 5 if the solution has a density of 1.0 g/mL?

**14.20**  Phosphorus is present in seawater to the extent of 0.07 ppm by weight (that is, 0.07 g of P per $10^6$ g of $H_2O$). If the phosphorus is present as phosphate, $PO_4^{3-}$, calculate the corresponding molar concentration of phosphate.

**14.21**  Assuming a 10 percent efficiency of recovery, how many liters of seawater must be processed to obtain $10^8$ kg of bromine in a commercial production process, assuming the bromide ion concentration listed in Table 14.7?

**14.22**  A first stage in recovery of magnesium from seawater is precipitation of $Mg(OH)_2$ by use of CaO:

$$Mg^{2+}(aq) + CaO(s) + H_2O(l) \longrightarrow \\ Mg(OH)_2(s) + Ca^{2+}(aq)$$

What mass of CaO is needed to precipitate $4.0 \times 10^7$ g of $Mg(OH)_2$?

## Fresh Water

**14.23**  Explain why the concentration of dissolved oxygen in fresh water is an important indicator of the quality of the water.

**14.24**  What forms of decomposition occur when dissolved oxygen levels in water drop below the levels at which aerobic bacteria can survive? Give examples of the decomposition products that form.

**14.25** The following organic anion is found in most detergents:

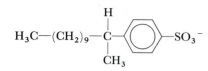

Assume that this anion undergoes aerobic decomposition in the following manner:

$$2C_{18}H_{29}O_3S^-(aq) + 51O_2(aq) \longrightarrow$$
$$36CO_2(aq) + 28H_2O(l) + 2H^+(aq) + 2SO_4^{2-}(aq)$$

What is the total mass of $O_2$ required to biodegrade 1.0 g of this substance dissolved in 100 L water?

**14.26** The average daily mass of $O_2$ taken up by sewage discharged in the United States is 59 g per person. How many liters of water at 9 ppm $O_2$ are totally depleted of oxygen in 1 day by a population of 50,000 people?

**14.27** Write a balanced chemical equation to describe what occurs when hard water is heated.

**14.28** Write a balanced chemical equation to describe how magnesium ion is removed in water treatment by the addition of slaked lime, $Ca(OH)_2$.

**14.29** In a particular water supply, the concentration of $Ca^{2+}$ is $2.2 \times 10^{-3}$ $M$, and the concentration of bicarbonate ion, $HCO_3^-$, is $1.3 \times 10^{-3}$ M. What weights of $Ca(OH)_2$ and $Na_2CO_3$ are needed to reduce the level of $Ca^{2+}$ to one-fourth its original level if $1.0 \times 10^7$ L of water must be treated?

**14.30** How many moles of $Ca(OH)_2$ and of $Na_2CO_3$ should be added to soften $10^3$ L of water in which $[Ca^{2+}] = 5.0 \times 10^{-4}$ $M$ and $[HCO_3^-] = 7.0 \times 10^{-4}$ $M$?

## Additional Exercises

**14.31** Describe the major factors that lead to a maximum in the ozone concentration at about 22 km elevation.

**14.32** The temperature profile in the atmosphere shows a maximum at about 50 km (Figure 14.1). Explain the origin of this maximum.

**14.33** It is said that atomic chlorine serves as a catalyst for the decomposition of ozone according to the reaction $O_3 + O \longrightarrow 2O_2$. What is a catalyst? How does Cl serve as a catalyst in this reaction? What overall chemical change occurs for Cl?

**14.34** Experiments have been performed in which metals such as sodium or barium have been released into the atmosphere at altitudes of about 120 km. Assuming that the metals are present in the atomic form, what reactions would you expect to occur with the ionic species present? Explain.

**14.35** Except for two substances, the components of the earth's atmosphere are transparent to long-wavelength, infrared radiation. What are these two substances? In what way does the absorption of infrared radiation affect the earth's climate? Explain how increased levels of infrared-absorbing substances in the atmosphere could lead to a higher average surface temperature.

**14.36** Suppose that on another planet the atmosphere consisted of 20 percent Ar, 35 percent $CH_4$, and 45 percent $O_2$. What would be the average molecular weight at the surface? What would be the average molecular weight at 200 km, assuming that all the $O_2$ is photodissociated?

**14.37** Explain why the stratosphere, which extends from 11 km to about 50 km, contains a smaller total atmospheric mass than the troposphere, which extends from the surface to 11 km.

**14.38** Describe the bonding in $N_2$ and $O_2$. Using data from earlier chapters, discuss the reasons for the relative reactivities of these two molecules in the earth's upper atmosphere.

**14.39** Beginning with the intact chlorofluoromethane $CF_2Cl_2$, write equations showing how a catalytic effect for destruction of ozone may be established in the stratosphere.

**14.40** We have noted that the affinity of carbon monoxide for hemoglobin is about 210 times that of $O_2$. Assume that a person is inhaling air that contains 86 ppm of CO. If all the hemoglobin leaving the lungs carries either oxygen or CO, calculate the fraction in the form of carboxyhemoglobin.

**14.41** (a) What ions are commonly responsible for the hardness of water? (b) What makes these ions objectionable?

**14.42** Explain what is meant by the following terms: (a) biodegradable; (b) aerobic decay; (c) anaerobic decomposition.

**14.43** Suppose that seawater to be processed in a reverse osmosis plant has the concentrations of dissolved salts listed in Table 14.7. If the concentration of salts on the pure desalinated water side is taken as zero, what osmotic pressure must be applied before water begins to flow from the seawater through the semipermeable membrane to the desalinated water side? (Refer to Section 12.6.)

**14.44** Complete and balance a chemical equation corresponding to each of the following verbal descriptions. (a) The nitric oxide molecule undergoes photodissociation in the upper atmosphere. (b) The nitric oxide molecule undergoes photoionization in the upper atmosphere. (c) Nitric oxide undergoes oxidation by ozone in the stratosphere. (d) Sulfur dioxide is formed when lead sulfide, PbS, is roasted. (e) Hypochlorous acid is formed when chlorine is added to water. (f) Hypochlorous acid reacts with aqueous ammonia to form chloramine, $NH_2Cl$. (g) A sample of water containing $Ca^{2+}$ and bicarbonate ion forms a precipitate when heated.

**14.45** The hollow fibers used in the reverse-osmosis desalination plant shown in Figure 14.14 are bundled together in modules, each containing about 3 million fibers of very small diameter. Why is this approach superior to employing a few fibers of much larger diameter?

**14.46** Distinguish between primary and secondary sewage treatment. Do these modes of treatment remove the phosphorus that might be present in detergents? Explain.

**[14.47]** It has recently been pointed out that there may be increased amounts of NO in the troposphere as compared with the past because of massive use of nitrogen-containing compounds in fertilizers. Assuming that NO can eventually diffuse into the stratosphere, what role might it play in affecting the conditions of life on earth? Using the index to this text, look up the chemistry of nitrogen oxides. What chemical pathways might NO in the troposphere follow?

**14.48** Each of the following equations represents a reaction that occurs over platinum catalysts in automobile mufflers. Complete and balance each equation.

(a) $NH_3(g) + NO(g) + O_2(g) \longrightarrow N_2(g) + H_2O(g)$

(b) $NH_3(g) + NO_2(g) + O_2(g) \longrightarrow N_2(g) + H_2O(g)$

(c) $NH_3(g) + NO(g) + O_2(g) \longrightarrow N_2O(g) + H_2O(g)$

## CHAPTER 15

# Chemical Kinetics

Chemistry is by its very nature concerned with change. Substances with well-defined properties are converted by chemical reactions into other materials with different properties. Chemists want to know which new substances are formed from a given set of starting reactants. However, it is equally important to know how rapidly chemical reactions occur and to understand the factors that control their speeds. For example, what factors are important in determining how rapidly foods spoil? What determines the rate at which steel rusts? How does one design a rapidly setting material for dental fillings? What biochemical reaction rates determine the contraction and relaxation of smooth muscle in the arteries of the heart? What factors control the rate at which fuel burns in an auto engine?

The area of chemistry concerned with the speeds, or rates, at which chemical reactions occur is called **kinetics**. In this chapter we learn how to express and determine the rates at which reactions occur. We shall learn how reaction rates are affected by variables such as concentration, temperature, and the presence of catalysts. We shall also consider what we can infer from chemical kinetics about the detailed pathways by which chemical reactions occur.

## CONTENTS

## 15.1 REACTION RATE

The speed of any event is measured by the change occurring in a given interval of time. For example, the speed of an automobile expresses its change in position in a certain time; the associated units are usually miles per hour (mi/h). Similarly, the speed or rate of a reaction is expressed as the change in the concentration of a reactant or product in a certain time; the units with which we express this speed are usually molarity per second ($M$/s). As an example, consider the reaction that occurs when butyl chloride, $C_4H_9Cl$, is placed in water. The resulting reaction produces butyl alcohol, $C_4H_9OH$, and hydrochloric acid:

$$C_4H_9Cl(l) + H_2O(l) \longrightarrow C_4H_9OH(aq) + HCl(aq) \qquad [15.1]$$

Suppose that we begin with a 0.1000 $M$ solution of $C_4H_9Cl$ in water and then measure the $C_4H_9Cl$ concentration at different times after the solution is mixed. We could in this way collect the data shown in the first two columns of Table 15.1. The average rate of the reaction over any time interval is a positive quantity given by the change in the concentration of $C_4H_9Cl$ divided by the time in which that change occurs:

$$\text{Average rate} = -\frac{\text{change in concentration of } C_4H_9Cl}{\text{corresponding time interval}} = -\frac{\Delta[C_4H_9Cl]}{\Delta t}$$

$$[15.2]$$

The ability of one runner to race at a faster rate than another is due in part to variations in the rates of many chemical reactions occurring within each runner. (Auscape/Photo Researchers)

**TABLE 15.1**  Rate data for reaction of $C_4H_9Cl$ with water

| Time (s) | [$C_4H_9Cl$] ($M$) | Average rate ($M$/s) |
|---|---|---|
| 0 | 0.1000 | |
| 50 | 0.0905 | $1.90 \times 10^{-4}$ |
| 100 | 0.0820 | $1.70 \times 10^{-4}$ |
| 150 | 0.0741 | $1.58 \times 10^{-4}$ |
| 200 | 0.0671 | $1.40 \times 10^{-4}$ |
| 300 | 0.0549 | $1.22 \times 10^{-4}$ |
| 400 | 0.0448 | $1.01 \times 10^{-4}$ |
| 500 | 0.0368 | $0.80 \times 10^{-4}$ |
| 800 | 0.0200 | $0.56 \times 10^{-4}$ |
| 10,000 | 0 | |

The brackets around $C_4H_9Cl$ in this equation indicate the concentration of that substance. The Greek capital letter delta, $\Delta$, is read "change in"; $\Delta[C_4H_9Cl]$ is the change in the concentration of $C_4H_9Cl$:

$$\Delta[C_4H_9Cl] = [C_4H_9Cl]_{final} - [C_4H_9Cl]_{initial} \qquad [15.3]$$

Similarly, $\Delta t$, the corresponding time interval, is the amount of time between the beginning and the end of the interval. The negative sign in Equation 15.2 indicates that the concentration of $C_4H_9Cl$ is decreasing with time.

Notice in Table 15.1 that during the first interval of 50 s the concentration of $C_4H_9Cl$ decreases from 0.1000 $M$ to 0.0905 $M$. Thus the average rate over this 50-s interval is

$$\text{Average rate} = -\frac{(0.0905 - 0.1000)\ M}{(50 - 0)\ s} = 1.90 \times 10^{-4}\ M/s$$

Average rates over other intervals calculated similarly are shown in the third column of Table 15.1. Notice that the average rate steadily decreases as the reaction proceeds. At some point the reaction stops; that is, there is no longer any change in concentration with time.

We can also see this decrease in rate if we display the data graphically, as in Figure 15.1. The dots represent the experimental data from the first two columns of Table 15.1. Using this curve we can determine the instantaneous rate; this is the rate at a particular time as opposed to the average rate over an interval of time. The instantaneous rate is obtained from the straight-line tangent that touches the curve at the point of interest. We have drawn two such tangents on Figure 15.1, one at $t = 0$ and the other at $t = 600$ s. The slopes of these tangents give the instantaneous rates at these times.* For example, at 600 s we have

* You may wish to briefly review the idea of graphical determination of slopes by referring to Appendix A. If you are familiar with calculus, you may recognize that the average rate approaches the instantaneous rate as the time interval approaches zero. This limit, in the notation of calculus, is represented as $-d[C_4H_9Cl]/dt$.

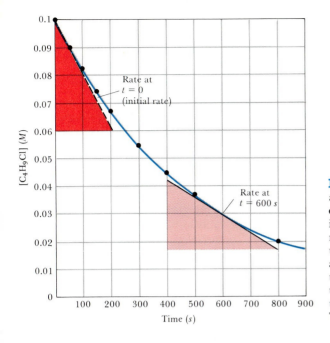

**FIGURE 15.1** Concentration of butyl chloride, $C_4H_9Cl$, as a function of time. The dots represent the experimental data from the first two columns of Table 15.1; the blue line is the smooth curve drawn to connect the data points. The reaction rate at any time is given by the slope of the tangent to the curve at that time. The slope of the tangent is defined as the vertical change divided by the horizontal change of the tangent, that is, $\Delta[C_4H_9Cl]/\Delta t$. Tangents have been drawn that touch the curve at $t = 0$ and $t = 600$ s. The calculation of the slope at $t = 600$ s is performed in the body of the text. The slope at $t = 0$ is calculated in Sample Exercise 15.1.

$$\text{Instantaneous rate} = -\frac{(0.017 - 0.042)\ M}{(800 - 400)\ s} = 6.2 \times 10^{-5}\ M/s$$

We will usually refer to the instantaneous rate merely as the rate.

Whenever we discuss either the rate or average rate of a reaction we need to define which substance we are using as a reference. In the reaction

$$2HI(g) \longrightarrow H_2(g) + I_2(g)$$

we can measure the rate of disappearance of **HI** or the appearance of either $H_2$ or $I_2$. Because 2 mol of **HI** disappears for each mole of $H_2$ or $I_2$ that forms, the rate of disappearance of **HI** is twice the rate of appearance of $H_2$ or $I_2$:

$$-\frac{\Delta[HI]}{\Delta t} = 2\frac{\Delta[H_2]}{\Delta t} = 2\frac{\Delta[I_2]}{\Delta t} \qquad [15.4]$$

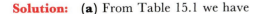

## SAMPLE EXERCISE 15.1

**(a)** Using the data in Table 15.1, calculate the average rate of disappearance of $C_4H_9Cl$ over the time interval from 50 to 150 s. **(b)** Using Figure 15.1, estimate the instantaneous rate of disappearance of $C_4H_9Cl$ at $t = 0$ (the initial rate). **(c)** How is the rate of disappearance of $C_4H_9Cl$ related to the rate of appearance of $C_4H_9OH$?

**Solution:** **(a)** From Table 15.1 we have

$$\text{Average rate} = -\frac{\Delta[C_4H_9Cl]}{\Delta t}$$
$$= -\frac{(0.0741 - 0.0905)\ M}{(150 - 50)\ s}$$
$$= 1.64 \times 10^{-4}\ M/s$$

**(b)** The initial rate is given by the slope of the dashed line in Figure 15.1. The slope of a straight line is given by the change in the vertical axis divided by the corresponding change in the horizontal axis. The straight line falls from $[C_4H_9Cl] = 0.100$ to $0.060\ M$ in the time change from 0 to 200 s. Thus the instantaneous rate is

$$\text{Rate} = -\frac{(0.060 - 0.100)\ M}{(200 - 0)\ s}$$

$$= 2.0 \times 10^{-4}\ M/s$$

(c) Because 1 mol of $C_4H_9OH$ forms for each mole of $C_4H_9Cl$ that disappears,

$$-\frac{\Delta[C_4H_9Cl]}{\Delta t} = \frac{\Delta[C_4H_9OH]}{\Delta t}$$

**PRACTICE EXERCISE**
Using Figure 15.1, estimate the instantaneous rate of disappearance of $C_4H_9Cl$ at $t = 300$ s.
*Answer:* $1.10 \times 10^{-4}\ M/s$

## 15.2 DEPENDENCE OF REACTION RATE ON CONCENTRATIONS

The decreasing rate of reaction with passing time that is evident in Figure 15.1 is quite typical of reactions. Reaction rates diminish as the concentrations of reactants diminish. Conversely, rates generally increase when reactant concentrations are increased.

One way of studying the effect of concentration on reaction rate is to determine the way in which the rate at the beginning of a reaction depends on the starting concentrations. To illustrate this approach, consider the following reaction:

$$NH_4^+(aq) + NO_2^-(aq) \longrightarrow N_2(g) + 2H_2O(l) \qquad [15.5]$$

We might study the rate of this reaction by measuring the concentration of $NH_4^+$ or $NO_2^-$ as a function of time or by measuring the volume of $N_2$ collected. Because of the $1:1$ stoichiometry of the reaction, all of these rates will be equal. If we determine the initial reaction rate (the instantaneous rate at $t = 0$) for various starting concentrations of $NH_4^+$ and $NO_2^-$ we could collect the data shown in Table 15.2. These data indicate that changing either $[NH_4^+]$ or $[NO_2^-]$ changes the reaction rate. Notice that if we double $[NO_2^-]$ while holding $[NH_4^+]$ constant, the rate doubles (compare experiments 1 and 2). If $[NO_2^-]$ is increased by a factor of 4 (compare experiments 1 and 3) the rate changes by a factor of 4, and so forth. These results indicate that the rate is directly proportional to $[NO_2^-]$. When $[NH_4^+]$ is similarly varied while $[NO_2^-]$ is held constant, the rate is affected in the same manner. We conclude that the rate

**TABLE 15.2** Rate data for the reaction of ammonium and nitrite ions in water at 25°C

| Experiment number | Initial $NO_2^-$ concentration ($M$) | Initial $NH_4^+$ concentration ($M$) | Observed initial rate ($M/s$) |
|---|---|---|---|
| 1 | 0.0100 | 0.200 | $5.4 \times 10^{-7}$ |
| 2 | 0.0200 | 0.200 | $10.8 \times 10^{-7}$ |
| 3 | 0.0400 | 0.200 | $21.5 \times 10^{-7}$ |
| 4 | 0.0600 | 0.200 | $32.3 \times 10^{-7}$ |
| 5 | 0.200 | 0.0202 | $10.8 \times 10^{-7}$ |
| 6 | 0.200 | 0.0404 | $21.6 \times 10^{-7}$ |
| 7 | 0.200 | 0.0606 | $32.4 \times 10^{-7}$ |
| 8 | 0.200 | 0.0808 | $43.3 \times 10^{-7}$ |

is also directly proportional to the concentration of $NH_4^+$. We can express the overall concentration dependence in the following way:

$$\text{Rate} = k[NH_4^+][NO_2^-] \qquad [15.6]$$

The proportionality constant, $k$, in Equation 15.6 is called the **rate constant**. We can evaluate the magnitude of $k$ using the data in Table 15.2. Using the results of experiment 1, and substituting into Equation 15.6, we have

$$5.4 \times 10^{-7} \ M/s = k(0.0100 \ M)(0.200 \ M)$$

Solving for $k$ gives

$$k = \frac{5.4 \times 10^{-7} \ M/s}{(0.0100 \ M)(0.200 \ M)} = 2.7 \times 10^{-4}/M\text{-s}$$

You may wish to satisfy yourself that this same value of $k$ is obtained using any of the other experimental results given in Table 15.2. You might also note that given $k = 2.7 \times 10^{-4}/M$-s and using Equation 15.6 we can calculate the rate for any concentration of $NH_4^+$ and $NO_2^-$. Suppose that $[NH_4^+] = 0.100 \ M$ and $[NO_2^-] = 0.100 \ M$; then

$$\text{Rate} = (2.7 \times 10^{-4}/M\text{-s})(0.100 \ M)(0.100 \ M) = 2.7 \times 10^{-6} \ M/s$$

An equation like 15.6 that relates the rate of a reaction to concentration is called a **rate law**. *The rate law for any chemical reaction must be determined experimentally; it cannot be predicted by merely looking at the chemical equation.* The following are some additional examples of rate laws:

$$2N_2O_5(g) \longrightarrow 4NO_2(g) + O_2(g) \quad \text{Rate} = k[N_2O_5] \qquad [15.7]$$

$$CHCl_3(g) + Cl_2(g) \longrightarrow CCl_4(g) + HCl(g) \quad \text{Rate} = k[CHCl_3][Cl_2]^{1/2} \quad [15.8]$$

$$H_2(g) + I_2(g) \longrightarrow 2HI(g) \quad \text{Rate} = k[H_2][I_2] \qquad [15.9]$$

The rate laws for a great many reactions have the general form

$$\text{Rate} = k[\text{reactant 1}]^m[\text{reactant 2}]^n \ldots \qquad [15.10]$$

As noted above, the proportionality constant, $k$, in a rate law is called the rate constant. For a given set of reactant concentrations, the reaction rate increases as $k$ increases. The exponents $m$ and $n$ in Equation 15.10 are called the **reaction orders** and their sum is the **overall reaction order**. For the reaction of $NH_4^+$ with $NO_2^-$, the rate law (Equation 15.6) contains the concentration of $NH_4^+$ raised to the first power. Thus the reaction is said to be first order in $NH_4^+$. Similarly, it is also first order in $NO_2^-$. The overall reaction order is two.

## SAMPLE EXERCISE 15.2

What are the overall reaction orders for the reactions described in Equations 15.7, 15.8, and 15.9?

**Solution:** The overall reaction order is the sum of the powers to which all the concentrations of reactants are raised in the rate law, when the rate law is in the form shown in Equation 15.10. The three rate laws given in Equations 15.7, 15.8, and 15.9 are in this form. The overall reaction orders are one for 15.7, three halves for 15.8, and two for 15.9.

**PRACTICE EXERCISE**
**(a)** What is the reaction order of the reactant $H_2$ in Equation 15.9? **(b)** What is the reaction order of $Cl_2$ in Equation 15.8?
*Answers:* **(a)** one; **(b)** one half

In a great many rate laws the reaction orders are zero, one, or two. However, reaction orders can be fractional or even negative. If a reaction is zero order in a particular reactant, changing its concentration will have no influence on rate as long as some of that reactant is present. If the reaction is first order in a reactant, changes in the concentration of that substance will produce proportional changes in the rate; doubling the concentration will double the rate, and so forth. When the rate law is second order in a particular reactant, doubling its concentration changes the rate by a factor of $2^2 = 4$; tripling its concentration causes the rate to increase by a factor of $3^2 = 9$.

In working with rate laws, be careful not to confuse the rate constant for a reaction with the reaction rate. The rate of a reaction depends on the concentrations of reactants; the rate constant does not. As we shall see later in this chapter, the rate constant and consequently the reaction rate are affected by temperature and the presence of a catalyst.

## SAMPLE EXERCISE 15.3

The initial rate of a reaction $A + B \longrightarrow C$ was measured for several different starting concentrations of A and B, with the results given below:

| Experiment number | [A] (*M*) | [B] (*M*) | Initial rate (*M*/s) |
|---|---|---|---|
| 1 | 0.100 | 0.100 | $4.0 \times 10^{-5}$ |
| 2 | 0.100 | 0.200 | $4.0 \times 10^{-5}$ |
| 3 | 0.200 | 0.100 | $16.0 \times 10^{-5}$ |

Using these data, determine **(a)** the rate law for the reaction; **(b)** the magnitude of the rate constant; **(c)** the rate of the reaction when $[A] = 0.050\ M$ and $[B] = 0.100\ M$.

**Solutions:** **(a)** We may assume that the rate law has the form: rate $= k[A]^m[B]^n$. Our task is to deduce the values of $m$ and $n$. Experiments 1 and 2 indicate that the concentration of B has no influence on the reaction rate. The reaction is therefore zero order in B. Experiments 1 and 3 indicate that doubling A increases the rate fourfold. This result indicates that rate is proportional to $[A]^2$; the reaction is second order in A. The rate law is

$$\text{Rate} = k[A]^2[B]^0 = k[A]^2$$

This same conclusion could be reached in a more formal way by taking the ratio of the rates from two experiments:

$$\frac{\text{Rate 1}}{\text{Rate 2}} = \frac{4.0 \times 10^{-5}\ M/s}{4.0 \times 10^{-5}\ M/s} = 1$$

Using the rate law, then, we have

$$1 = \frac{\text{rate 1}}{\text{rate 2}}$$

$$= \frac{k[0.100\ M]^m[0.100\ M]^n}{k[0.100\ M]^m[0.200\ M]^n} = \frac{[0.100]^n}{[0.200]^n} = \left(\frac{1}{2}\right)^n$$

But $(\frac{1}{2})^n$ can equal 1 only if $n = 0$.

We can deduce the value of $m$ in a similar fashion:

$$\frac{\text{Rate 1}}{\text{Rate 3}} = \frac{4.0 \times 10^{-5} \ M/s}{16.0 \times 10^{-5} \ M/s} = \frac{1}{4}$$

Using the rate law gives us

$$\frac{1}{4} = \frac{\text{rate 1}}{\text{rate 3}}$$

$$= \frac{k[0.100 \ M]^m[0.100 \ M]^n}{k[0.200 \ M]^m[0.100 \ M]^n} = \frac{[0.100]^m}{[0.200]^m} = \left(\frac{1}{2}\right)^m$$

The fact that $\left(\frac{1}{2}\right)^m = \frac{1}{4}$ indicates that $m = 2$.

**(b)** Using the rate law and the data from experiment 1, we have

$$k = \frac{\text{rate}}{[A]^2} = \frac{4.0 \times 10^{-5} \ M/s}{(0.100 \ M)^2}$$

$$= 4.0 \times 10^{-3}/M\text{-s}$$

**(c)** Using the rate law from part (a) and the rate constant from part (b), we have

$$\text{Rate} = k[A]^2$$

$$= (4.0 \times 10^{-3}/M\text{-s})(0.050 \ M)^2$$

$$= 1.0 \times 10^{-5} \ M/s$$

Because [B] is not part of the rate law, its concentration is immaterial to the rate, provided that there is at least some B present to react with A.

**PRACTICE EXERCISE**

A particular reaction was found to depend on the concentration of the hydrogen ion, $[H^+]$. The initial rates varied as a function of $[H^+]$ as follows:

| $[H^+]$ ($M$) | 0.0500 | 0.100 | 0.200 |
|---|---|---|---|
| Initial rate ($M/s$) | $6.4 \times 10^{-7}$ | $3.2 \times 10^{-7}$ | $1.6 \times 10^{-7}$ |

**(a)** What is the order of the reaction in $[H^+]$?
**(b)** Predict the initial reaction rate when $[H^+] = 0.400 \ M$.

*Answers:* **(a)** $-1$ (the rate is *inversely* proportional to $[H^+]$); **(b)** $0.8 \times 10^{-7} \ M/s$

---

## 15.3 RELATION BETWEEN REACTANT CONCENTRATION AND TIME

The rate law (Equation 15.10) tells us how the rate of a reaction changes as we change reactant concentrations. Rate laws can be converted into equations that tell us what the concentrations of the reactants or products are at any time during the course of a reaction. The mathematics required involve calculus. We don't expect you to be able to perform the calculus operations; however, you should be able to use the resulting equations. We will apply this conversion to two of the simplest rate laws—those that are first order overall and those that are second order overall.

### First-Order Reactions

If the rate of a reaction of the sort A $\longrightarrow$ products is first order in A, we can write the following rate law:

$$\text{Rate} = -\frac{\Delta[A]}{\Delta t} = k[A] \qquad \text{[15.11]}$$

Using calculus, this equation can be transformed into an equation that relates the concentration of A at the start of the reaction, $[A]_0$, to its concentration at any other time $t$, $[A]$.*

$$\log [A] - \log [A]_0 = \log \frac{[A]}{[A]_0} = -\frac{kt}{2.30} \qquad \text{[15.12]}$$

* The factor 2.30 in the term on the right in Equation 15.12 arises because we have used base 10 log rather than natural log. Recall that $2.30 \log x = \ln x$ (Appendix A-2).

Rearranging this equation slightly gives

$$\log [A] = \left(-\frac{k}{2.30}\right)t + \log [A]_0 \qquad [15.13]$$

One important fact about Equation 15.13 is that it is in the form of an equation for a straight line. Straight-line equations are of the type

$$y = ax + b \qquad [15.14]$$

where $a$ is the slope and $b$ is the intercept of the line (see Appendix A.4). Equation 15.13 has this form with $y = \log [A]$, $a = -k/2.30$, $x = t$, and $b = \log [A]_0$. Thus a graph of $\log [A]$ versus time gives a straight line with a slope of $-k/2.30$ and an intercept of $\log [A]_0$.

The conversion of methyl isonitrile to acetonitrile provides a simple example of a first-order reaction:

$$H_3C—N≡C: \longrightarrow H_3C—C≡N: \qquad [15.15]$$

Methyl isonitrile            Acetonitrile

Figure 15.2(a) shows how the pressure of the isonitrile varies with time as it rearranges in the gas phase at 198.9°C. Figure 15.2(b) shows the same data in straight-line form, as the logarithm of pressure as a function of time. Pressure is a legitimate unit of concentration for a gas because the number of moles per unit volume is directly proportional to pressure. The slope of the linear plot is $-2.22 \times 10^{-5}$/s. (You should verify this for yourself, remembering that your result may vary slightly from ours because of the inaccuracies associated with reading the graph.) We have that

$$\frac{-k}{2.30} = \text{slope}$$

$$k = -2.30(-2.22 \times 10^{-5}/s) = 5.11 \times 10^{-5}/s$$

**FIGURE 15.2**

(a) Variation in the pressure of methyl isonitrile, $CH_3NC$, with time at 198.9°C during the reaction $CH_3NC \longrightarrow CH_3CN$. (b) The data from (a) plotted in a linear form as log of $CH_3NC$ pressure as a function of time.

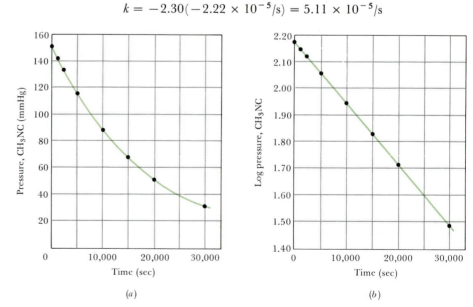

(a)                    (b)

For a first-order reaction, Equations 15.12 and 15.13 can be used to determine (1) the concentration of a reactant remaining at any time after the reaction has started, (2) the time required for a given fraction of sample to react, or (3) the time required for a reactant concentration to reach a certain level.

## SAMPLE EXERCISE 15.4

The first-order rate constant for hydrolysis of a certain insecticide in water at 12°C is 1.45/yr. A quantity of this insecticide is washed into a lake in June, leading to an overall concentration of $5.0 \times 10^{-7}$ $g/cm^3$ of water. Assume that the effective temperature of the lake is 12°C. **(a)** What is the concentration of the insecticide in June of the following year? **(b)** How long will it take for the concentration of the insecticide to drop to $3.0 \times 10^{-7}$ $g/cm^3$?

**Solution:** **(a)** Substituting $k = 1.45/yr$, $t = 1$ yr, and $[\text{insecticide}]_0 = 5.0 \times 10^{-7}$ $g/cm^3$ into Equation 15.13 gives

$$\log [\text{insecticide}] = -\frac{1.45/yr}{2.30} (1.00 \text{ yr})$$
$$+ \log(5.0 \times 10^{-7})$$

We use the log function on a calculator to take the log of the second term on the right:

$$\log [\text{insecticide}] = -0.630 - 6.30 = -6.93$$

To obtain [insecticide] we employ the antilog function on the calculator. The antilog function is usually labeled as $10^x$. Enter $-6.93$ followed by the $10^x$ function.

$$[\text{insecticide}] = 10^{-6.93} = 1.2 \times 10^{-7} \text{ g/cm}^3$$

(The concentration units for $[A]_0$ and $[A]$ must be the same.)

**(b)** Again substituting into Equation 15.13, with $[\text{insecticide}] = 3.0 \times 10^{-7}$ $g/cm^3$, gives

$$\log (3.0 \times 10^{-7}) = -\frac{1.45/yr}{2.30} t + \log (5.0 \times 10^{-7})$$

Solving for $t$ gives

$$t = -\frac{2.30}{1.45/yr} [\log (3.0 \times 10^{-7}) - \log (5.0 \times 10^{-7})]$$

$$= -\frac{2.30}{1.45/yr} (-6.52 + 6.30) = 0.35 \text{ yr}$$

## PRACTICE EXERCISE

The decomposition of $N_2O_5$ according to the equation

$$2N_2O_5(g) \longrightarrow 4NO_2(g) + O_2(g)$$

has a simple rate law:

$$\text{Rate} = -\frac{\Delta[N_2O_5]}{\Delta t} = k[N_2O_5]$$

Assume an initial pressure of 240 mm Hg for $N_2O_5$. What is the pressure of $N_2O_5$ after 300 s if $k$ has the value $8.5 \times 10^{-3}/s$?
*Answer:* 18.7 mm Hg

The **half-life** of a reaction, $t_{1/2}$, is the time required for the concentration of the reactant to decrease to halfway between its initial and final values. In the cases we'll be considering, the final concentration is zero. To obtain an expression for $t_{1/2}$ for a first-order reaction, we begin with Equation 15.12. The half-life corresponds to the time when $[A] = \frac{1}{2}[A]_0$. Inserting these quantities into the equation, we have

**Half-Life**

$$\log \frac{\frac{1}{2}[A]_0}{[A]_0} = \left(\frac{-k}{2.30}\right) t_{1/2}$$

$$\log \frac{1}{2} = \left(\frac{-k}{2.30}\right) t_{1/2}$$

$$t_{1/2} = \frac{-(2.30) \log \frac{1}{2}}{k} = \frac{0.693}{k} \qquad [15.16]$$

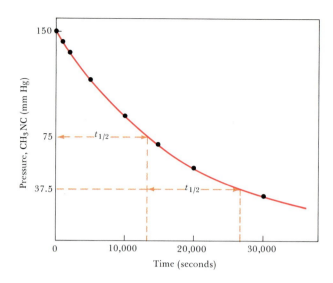

**FIGURE 15.3** Pressure of methyl isonitrile as a function of time. Two successive half-lives of the rearrangement reaction, Equation 15.15, are shown.

Notice that $t_{1/2}$ is independent of the initial concentration of reactant. This result tells us that if we measure reactant concentration at *any* time in the course of a first-order reaction, the concentration of reactant will be half of that measured value at a time $0.693/k$ later. The concept of half-life is widely used in describing radioactive decay. This application is discussed in detail in Section 21.4.

The data for the first-order rearrangement of methyl isonitrile at 198.9°C are graphed in Figure 15.3. The first half-life is shown at 13,320 s. At a time 13,320 s later, the isonitrile concentration has decreased to $\frac{1}{2}$ of $\frac{1}{2}$, or $\frac{1}{4}$ the original concentration. *It is a characteristic of a first-order reaction that the concentration of the reactant decreases by factors of $\frac{1}{2}$ in a series of regularly spaced time intervals.*

**SAMPLE EXERCISE 15.5**
From Figure 15.1 estimate the half-life of the reaction of $C_4H_9Cl$ with water.

**Solution:** From the figure we see that the initial value of $[C_4H_9Cl]$ is 0.100 *M*. The half-life for this first-order reaction is the time required for $[C_4H_9Cl]$ to decrease to 0.0500 *M*. This point occurs at approximately 340 s. At the end of the second half-life, which should occur at 680 s, the concentration should have decreased by yet another factor of 2, to 0.025 *M*. Inspection of the graph shows that this is indeed the case.

**PRACTICE EXERCISE**
Calculate $t_{1/2}$ for the reaction described in Practice Exercise 15.4.
*Answer:* 81.5 s

**Second-Order Reactions**

For a reaction that is second order in just one reactant, A, the rate is given by

$$\text{Rate} = k[A]^2$$

Relying on calculus, this rate law can be used to derive the following equation:

$$\frac{1}{[A]} = \frac{1}{[A]_0} + kt \qquad\qquad [15.17]$$

In this case, a linear form of the data is obtained by plotting $1/[A]$ versus $t$. The resultant line has a slope of $k$ and an intercept of $1/[A]_0$. One way to distinguish between first- and second-order rate laws is to graph both $\log [A]$ and $1/[A]$ against $t$. If the $\log [A]$ plot is linear, the reaction is first order; if the $1/[A]$ plot is linear, the reaction is second order.

Using Equation 15.17, we can show that the half-life of a second-order reaction is given by the expression $t_{1/2} = 1/k[A]_0$. It is *not* independent of the initial concentration of reactant as is $t_{1/2}$ for a first-order reaction. Thus a constant half-life is indicative of a first-order reaction, but not a second-order one.

## SAMPLE EXERCISE 15.6

The following data were obtained for the gas-phase decomposition of nitrogen dioxide at 300°C, $2NO_2(g) \longrightarrow 2NO(g) + O_2(g)$:

| Time (s) | $[NO_2]$ ($M$) |
|---|---|
| 0 | 0.0100 |
| 50 | 0.0079 |
| 100 | 0.0065 |
| 200 | 0.0048 |
| 300 | 0.0038 |

Is the reaction first or second order in $NO_2$?

**Solution:** To test whether the reaction is first or second order, we can construct plots of $\log [NO_2]$ and $1/[NO_2]$ against time. In doing so, we will find it useful to prepare the following table from the data given:

| Time (s) | $[NO_2]$ | $\log [NO_2]$ | $1/[NO_2]$ |
|---|---|---|---|
| 0 | 0.0100 | −2.00 | 100 |
| 50 | 0.0079 | −2.10 | 127 |
| 100 | 0.0065 | −2.19 | 154 |
| 200 | 0.0048 | −2.32 | 208 |
| 300 | 0.0038 | −2.42 | 263 |

As Figure 15.4 shows, only the plot of $1/[NO_2]$ versus time is linear. Thus the reaction obeys a second-order rate law: rate = $k[NO_2]^2$. From the slope of this straight-line graph we have that $k = 0.543/M$-s.

A reaction may also be second order by having a first-order dependence of the rate on each of two reagents, that is, rate = $k[A][B]$. It is possible to derive an expression for the variation in concentrations of A and B with time. However, we will not consider this and other more complicated rate laws in this text.

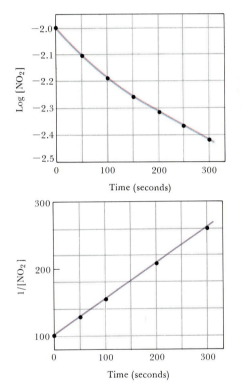

**FIGURE 15.4** Kinetic data from Sample Exercise 15.6 for the reaction $2NO_2(g) \longrightarrow 2NO(g) + O_2(g)$ at 300°C. The plot of $\log [NO_2]$ versus time is not linear; consequently, the reaction is not first order in $NO_2$. The plot of $1/[NO_2]$ versus time is linear; the reaction is second order in $NO_2$.

## PRACTICE EXERCISE
What is the half-life measured from time $t = 0$ for the decomposition of $NO_2$, as represented by the tabular data above?
*Answer:* 184 s

## 15.4 THE TEMPERATURE DEPENDENCE OF REACTION RATES

The rates of most chemical reactions increase as the temperature rises. We see examples of this generalization in many biological processes around us. The rate at which grass grows and the metabolic activity of the common housefly are both greater in warm weather than in the cold of winter. As another example, food cooks more rapidly in boiling water than in merely hot water. However, it is dangerous to place too much emphasis on the apparent rates at which biological systems operate as a function of temperature, because they are very complex and are adapted to operate optimally in a narrow temperature range. To obtain a clear understanding of how temperature affects reaction rates, we must examine simple reaction systems. As an example, let us consider the reaction about which we have already learned quite a bit, the first-order rearrangement of methyl isonitrile (Equation 15.15). Figure 15.5 shows the experimentally determined rate constant for this reaction as a function of temperature. The rate of the reaction increases rapidly with temperature. Furthermore, the increase is nonlinear.

As we seek an explanation for this behavior, perhaps the first question to ask is, why do *any* reactions go slowly? What keeps reactions from simply occurring immediately? If the methyl isonitrile molecules are going eventually to rearrange into acetonitrile, why don't they all do it at once?

### Activation Energy

We know from the kinetic-molecular theory of gases that with increasing temperature the average energy of the gas molecules increases. The fact that the rate of the methyl isonitrile reaction increases with increasing temperature suggests that perhaps the rearrangement is related to the kinetic energies of the molecules. Svante Arrhenius suggested in 1888 that before reaction can occur, a certain minimum amount of energy must be available to "propel" the molecules from one chemical state into another. The situation is rather like that shown in Figure 15.6. The boulder will be in a lower (or more stable) potential-energy state in valley

**FIGURE 15.5** Variation in the first-order rate constant for rearrangement of methyl isonitrile as a function of temperature. (The four points indicated are used in connection with Sample Exercise 15.7.)

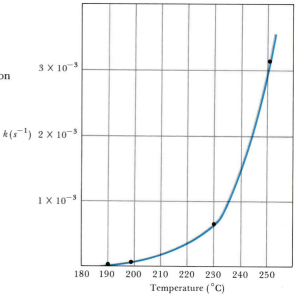

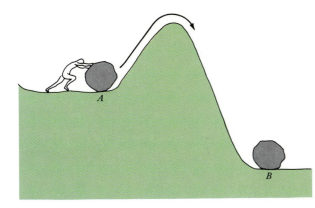

**FIGURE 15.6** Illustration of the potential-energy profile for a boulder. The boulder must be moved over the energy barrier before it can come to rest in the lower-energy location, *B*.

*B* than in valley *A*. Before it can come to rest in *B*, however, it must acquire the energy to overcome the barrier blocking its passage from the one state into the other. In the same way, molecules may require a certain minimum energy to overcome the forces that tend to keep them as they are, if they are to form the new chemical bonds that will result in a different arrangement. In our methyl isonitrile example, we might imagine that for rearrangement to occur the N≡C portion of the molecule must turn over:

$$H_3C-N\equiv C: \longrightarrow \left[H_3C\cdots\overset{\overset{\cdot\cdot}{C}}{\underset{\underset{\cdot\cdot}{N}}{|||}}\right] \longrightarrow H_3C-C\equiv N: \qquad [15.18]$$

Even though the bonding may be more stable in the product acetonitrile than in the starting compound, energy is required to force the molecule through the relatively unstable intermediate state to the final result. The energy of the molecule as it proceeds along this reaction pathway is shown in Figure 15.7. Arrhenius called the energy barrier between the starting molecule and the highest energy along the reaction pathway the **activation energy**, $E_a$. The particular arrangement of atoms that has the maximum energy is often called the **activated complex**.

The conversion of $H_3C-N\equiv C$ to $H_3C-C\equiv N$ is exothermic; Figure 15.7 therefore shows the product as having a lower energy than the reactant. Notice that the reverse reaction is then endothermic; for that

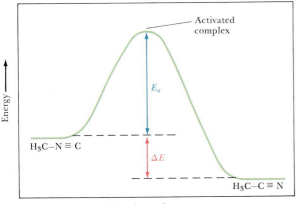

**FIGURE 15.7** Energy profile for the rearrangement of methyl isonitrile. The molecule must surmount the activation-energy barrier before it can form the product, acetonitrile.

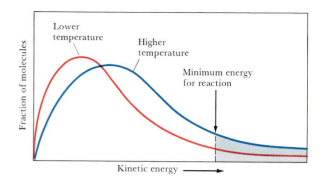

**FIGURE 15.8** Distribution of kinetic energies in a sample of gas molecules at two different temperatures. At the higher temperature a larger number of molecules have higher energies. Thus a larger fraction in any one instant will have more than the minimum energy required for reaction.

reaction the activation barrier is equal to the sum of $\Delta E$ and $E_a$ for the forward reaction.

Energy is transferred between molecules through collisions. Thus, within a certain period of time, any particular isonitrile molecule might acquire enough energy to overcome the energy barrier and be converted into acetonitrile. At any given temperature only a small fraction of collisions will occur with sufficient energy to overcome the barrier to reaction. However, as shown in Figure 10.11, the distribution of molecular speeds is more spread out toward higher values when the gas is at a higher temperature. The distribution of kinetic energies of molecules changes in a similar way, as shown in Figure 15.8. This graph shows that at the higher temperature a larger fraction of molecules possess the minimum energy needed for reaction.

In addition to the requirement that the reactant species collide with sufficient energy to begin to rearrange bonds, an orientational requirement exists. The relative orientations of the molecules during their collisions may determine whether the energy gets to the right place for reaction or whether atoms are suitably oriented to form new bonds. Only a fraction of the collisions possessing enough energy for reaction actually produce products.

In a mixture of $H_2$ and $I_2$ at ordinary temperatures and pressures, each molecule undergoes about $10^{10}$ collisions per second. If every collision between $H_2$ and $I_2$ resulted in formation of HI, the reaction would be over in much less than a second. Instead, at room temperature the reaction proceeds very slowly. Obviously, every collision does not lead to reaction. In fact, only about 1 in every $10^{13}$ collisions is effective. Only a small fraction of the collisions occur with suitable orientation and with sufficient energy to carry the molecule over the energy barrier to products. As the temperature increases, the number of collisions increases, as does the fraction that are sufficiently energetic for reaction. With each 10°C rise in temperature, the rate of this reaction triples.

**The Arrhenius Equation**

Arrhenius noted that the increase in rate with increasing temperature for most reactions is nonlinear, as in the example shown in Figure 15.5. He found that most reaction-rate data obeyed the equation

$$k = Ae^{-E_a/RT}$$

[15.19]

where $k$ is the rate constant. This equation is called the **Arrhenius equation**. The term $E_a$ is the activation energy, which we have already defined; $R$ is the gas constant (8.314 J/K-mol); and $T$ is absolute temperature. The term $A$ is constant, or nearly so, as temperature is varied. Called the **frequency factor**, $A$ is related to the frequency of collisions and the probability that the collisions are favorably oriented for reaction. Notice that as the magnitude of $E_a$ increases, $k$ becomes smaller. Thus reaction rates decrease as the energy barrier increases. Taking the log of both sides of Equation 15.19 and converting to base 10, we have

$$\log k = \log A - \frac{E_a}{2.30RT} \qquad [15.20]$$

Equation 15.20 has the form of a straight line, in which one variable is $\log k$ and the other is $1/T$. The slope of the line is given by $-E_a/2.30R$; the intercept, at $1/T = 0$, is $\log k = \log A$. Equation 15.20 can be used to determine $E_a$ from a graph of $\log k$ versus $1/T$.

It is sometimes convenient to manipulate Equation 15.20 further to give the relationship between the rate constants at two different temperatures, $T_1$ and $T_2$. At $T_1$ we have

$$\log k_1 = \log A - \frac{E_a}{2.30RT_1}$$

At $T_2$,

$$\log k_2 = \log A - \frac{E_a}{2.30RT_2}$$

Subtracting $\log k_2$ from $\log k_1$ gives

$$\log k_1 - \log k_2 = \left( \log A - \frac{E_a}{2.30RT_1} \right) - \left( \log A - \frac{E_a}{2.30RT_2} \right)$$

Simplifying this equation and rearranging it gives

$$\log \frac{k_1}{k_2} = \frac{E_a}{2.30R} \left( \frac{1}{T_2} - \frac{1}{T_1} \right) \qquad [15.21]$$

Equation 15.21 provides a convenient means of calculating the rate constant at some temperature, $T_1$, when we know the activation energy and the rate constant, $k_2$, at some other temperature, $T_2$.

## SAMPLE EXERCISE 15.7

The following table shows the rate constants for rearrangement of methyl isonitrile at various temperatures (these are the data that are graphed in Figure 15.5).

| Temperature (°C) | $k$ (s$^{-1}$) |
| --- | --- |
| 189.7 | $2.52 \times 10^{-5}$ |
| 198.9 | $5.25 \times 10^{-5}$ |
| 230.3 | $6.30 \times 10^{-4}$ |
| 251.2 | $3.16 \times 10^{-3}$ |

**(a)** From these data calculate the activation energy for the reaction. **(b)** What is the magnitude of the rate constant at 430.0 K?

**Solution:** **(a)** We must first convert temperatures to the absolute temperature scale, K. We then take the inverse of these temperatures and obtain the corresponding log values for $k$. This gives us the following table:

| $T$ (K) | $1/T$ (K) | $\log k$ |
|---------|-----------|----------|
| 462.7 | $2.161 \times 10^{-3}$ | $-4.60$ |
| 471.9 | $2.119 \times 10^{-3}$ | $-4.29$ |
| 503.3 | $1.987 \times 10^{-3}$ | $-3.20$ |
| 524.2 | $1.908 \times 10^{-3}$ | $-2.50$ |

A graph of these data results in a straight line, as shown in Figure 15.9. The data points in the graph

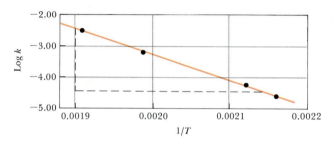

**FIGURE 15.9** Log of the rate constant for rearrangement of methyl isonitrile as a function of $1/T$. The linear relationship is predicted from the Arrhenius equation.

lie very close to the best straight line through all four points. The slope of the line is obtained by choosing two well-separated points, as shown, and using the coordinates of each:

$$\text{Slope} = \frac{-2.45 - (-4.45)}{0.00190 - 0.00214} = -8330$$

The numerator in this equation has no units, because logs have no units. The denominator has the units of $1/T$, that is, $1/\text{K}$. Thus the overall units for the slope are K. The slope is equal to $-E_a/2.30R$. We want the value for the molar gas constant $R$ in J/K-mol (Table 10.2), which is 8.314. Thus we obtain

$$\text{Slope} = \frac{-E_a}{2.30R}$$

$$E_a = -(\text{slope})(2.30R)$$

$$= -(-8330 \text{ K})(2.30)\left(8.31 \frac{\text{J}}{\text{K mol}}\right)\left(\frac{1 \text{ kJ}}{1000 \text{ J}}\right)$$

$$= 159 \text{ kJ/mol}$$

**(b)** To determine the rate constant, $k_1$, at 430 K, we apply Equation 15.21 with $E_a = 159$ kJ/mol, $k_2 = 2.52 \times 10^{-5}$/s, $T_2 = 462.7$ K, and $T_1 = 430.0$ K:

$$\log \frac{k_1}{2.52 \times 10^{-5}/\text{s}} = \frac{159 \text{ kJ/mol}}{(2.30)(8.31 \text{ J/K-mol})}$$

$$\times \left(\frac{1}{462.7 \text{ K}} - \frac{1}{430.0 \text{ K}}\right)\left(\frac{10^3 \text{ J}}{1 \text{ kJ}}\right)$$

$$= -1.367$$

$$\frac{k_1}{2.52 \times 10^{-5}/\text{s}} = 10^{-1.367} = 4.30 \times 10^{-2}$$

$$k_1 = (2.52 \times 10^{-5}/\text{s})(4.30 \times 10^{-2})$$

$$= 1.08 \times 10^{-6}/\text{s}$$

**PRACTICE EXERCISE**
Estimate the rate constant for rearrangement of methyl isonitrile at 280°C.
*Answer:* $2.14 \times 10^{-2}$/s

## 15.5  REACTION MECHANISM

A balanced equation for a chemical reaction indicates the substances that are present at the start of the reaction and those produced at the end. However, it provides no information about how the reaction occurs. The process by which a reaction occurs is called the **reaction mechanism**. At the most sophisticated level, a reaction mechanism will describe in great detail the order in which bonds are broken and formed and the changes in relative positions of the atoms in the course of the reaction. But we will begin with more rudimentary descriptions of how reactions occur.

### Elementary Reactions

We have seen (Section 15.4) that reactions take place as a result of collisions among reacting molecules. For example, the collision between methyl isonitrile, $CH_3NC$, and some other molecule can provide the energy to allow the $CH_3NC$ to rearrange:

$$H_3C-N\equiv C: \longrightarrow \left[ H_3C \cdots \overset{\overset{\displaystyle ..}{C}}{\underset{\underset{\displaystyle ..}{N}}{\parallel\parallel}} \right] \longrightarrow H_3C-C\equiv N:$$

Similarly, the reaction of $O_3$ and NO to form $O_2$ and $NO_2$ appears to occur as a result of a single collision involving suitably oriented and sufficiently energetic NO and $O_3$ molecules:

$$NO(g) + O_3(g) \longrightarrow NO_2(g) + O_2(g) \qquad [15.22]$$

Both of these processes occur in a single event or step and are called **elementary reactions** (or elementary steps).

The net change represented by a balanced chemical equation often occurs by a sequence of elementary reactions. For example, consider the reaction of $NO_2$ and CO:

$$NO_2(g) + CO(g) \longrightarrow NO(g) + CO_2(g) \qquad [15.23]$$

Below 225°C this reaction appears to proceed in two elementary steps. First, two $NO_2$ molecules collide and an oxygen atom is transferred from one to the other. The resultant $NO_3$ then transfers an oxygen atom to CO during a collision between these molecules:

$$NO_2(g) + NO_2(g) \longrightarrow NO_3(g) + NO(g) \qquad [15.24]$$
$$NO_3(g) + CO(g) \longrightarrow NO_2(g) + CO_2(g) \qquad [15.25]$$

*The elementary reactions in a multistep mechanism must always add to give the chemical equation of the overall process.* In the present example, the sum of the elementary reactions is

$$2NO_2(g) + NO_3(g) + CO(g) \longrightarrow NO_2(g) + NO_3(g) + NO(g) + CO_2(g)$$

Simplifying this equation by eliminating substances that appear on both sides of the arrow gives the net equation for the process, Equation 15.23. Because $NO_3$ is neither a reactant nor a product in the overall reaction—it is formed in one elementary reaction and consumed in the next—it is called an **intermediate**. Multistep mechanisms involve one or more intermediates.

The number of molecules that participate as reactants in an elementary reaction defines the **molecularity** of the reaction. If a single molecule is involved, the reaction is said to be **unimolecular**. The rearrangement of methyl isonitrile (Equation 15.18) is a unimolecular process. Elementary reactions involving two reactant molecules are said to be **bimolecular**. The reaction between NO and $O_3$ (Equation 15.22) is bimolecular. Elementary reactions involving three molecules are said to be **termolecular**. Termolecular reactions are less probable than unimolecular or bimolecular reactions and are rarely encountered. The chance that four or more molecules will collide simultaneously with any regularity is even more remote; consequently, such collisions are never proposed as part of a reaction mechanism.

## SAMPLE EXERCISE 15.8

It has been proposed that the conversion of ozone into $O_2$ proceeds in two steps:

$$O_3(g) \rightleftharpoons O_2(g) + O(g)$$
$$O_3(g) + O(g) \longrightarrow 2O_2(g)$$

(a) Write the equation for the overall reaction. (b) Identify the intermediate, if any. (c) Describe the molecularity of each step in the mechanism.

**Solution:** (a) Adding the two elementary reactions gives

$$2O_3(g) + O(g) \longrightarrow 3O_2(g) + O(g)$$

Because $O(g)$ appears in equal amounts on both sides of the equation, it can be eliminated to give the net equation for the chemical process:

$$2O_3(g) \longrightarrow 3O_2(g)$$

(b) The intermediate is $O(g)$. It is neither an original reactant nor a final product, but is formed in one step and consumed in another.

(c) The first elementary reaction involves a single reactant and is consequently unimolecular. The second step, which involves two reactant molecules, is bimolecular.

## PRACTICE EXERCISE

For the reaction of $Mo(CO)_6$,

$$Mo(CO)_6 + P(CH_3)_3 \longrightarrow$$
$$Mo(CO)_5P(CH_3)_3 + CO$$

the proposed mechanism is

$$Mo(CO)_6 \longrightarrow Mo(CO)_5 + CO$$
$$Mo(CO)_5 + P(CH_3)_3 \longrightarrow Mo(CO)_5P(CH_3)_3$$

(a) Is the proposed mechanism consistent with the equation for the overall reaction? (b) Identify the intermediate or intermediates.

*Answers:* (a) yes, the two equations add to yield the equation for the reaction (b) $Mo(CO)_5$

## Rate Laws of Elementary Reactions

In discussing rate laws in Section 15.2, we stressed that they must be determined experimentally; they cannot, in general, be predicted from the coefficients of balanced chemical reactions. It is not the chemical equation for the overall process that determines the rate law; rather, it is the elementary reactions and their relative speeds.

The rate law of any elementary reaction is based directly on its molecularity. For example, consider the general unimolecular process

$$A \longrightarrow products \qquad [15.26]$$

As the number of A molecules increases, the number that decompose in a given interval of time will increase. Thus the rate of a unimolecular process will be first order:

$$Rate = k[A] \qquad [15.27]$$

In the case of bimolecular reactions, the rate law will be second order, as in the following examples:

$$A + B \longrightarrow products \qquad Rate = k[A][B] \qquad [15.28]$$
$$A + A \longrightarrow products \qquad Rate = k[A]^2 \qquad [15.29]$$

The second-order rate law follows from the fact that the rate of collision between A and B molecules is proportional to the concentrations of A and B.

In general, the order for each reactant in an elementary reaction is equal to its coefficient in the chemical equation for that step. Of course, we cannot tell by merely looking at a balanced chemical equation whether that reaction occurs in a single step or in a number of steps. That brings us back to the need for experimental studies.

---

**SAMPLE EXERCISE 15.9**

If the following reaction occurs in a single elementary step, predict the rate law:

$$O_3(g) + NO(g) \longrightarrow NO_2(g) + O_2(g)$$

**Solution:**   The rate law is first order each in $O_3$ and NO, corresponding to the coefficients (understood to be one) in the chemical equation:

$$Rate = k[O_3][NO]$$

**PRACTICE EXERCISE** ———————

Write the rate law for the reaction $2NO(g) + Br_2(g) \longrightarrow 2NOBr(g)$, if it proceeds in a single step.

***Answer:***   $-\Delta[NO]/\Delta t = k[NO]^2[Br_2]$

---

When chemical reactions occur by a mechanism involving a number of sequential elementary steps, often one step is much slower than others. The overall rate of a reaction cannot exceed the rate of the slowest elementary step of its mechanism. The slow step limits the overall reaction rate and is consequently called the **rate-determining step**.

Think of a toll road along which there are several plazas at which a toll must be paid, as illustrated in Figure 15.10. Assume that traffic is slowest at one particular plaza (perhaps its gates are malfunctioning). The total rate of flow of traffic from one end of the toll road to the other is limited by the flow through that slowest plaza. Traffic backs up before the slowest plaza. The rate at which traffic flows through the plazas that come after the slowest one does not affect the overall rate of traffic flow.

In the same way, the slowest step in a multistep reaction determines the overall rate. If the slow step is not the first one, the products of the faster preceding steps will produce intermediate products that accumulate before being consumed in the slow step. Furthermore, the overall rate is not affected by the rates of any faster steps that follow the slowest one.

Consider the situation where the first step of a mechanism is the rate-determining one, as occurs in the reaction of $NO_2$ and CO, mentioned earlier in this section:

$$NO_2(g) + NO_2(g) \longrightarrow NO_3(g) + NO(g) \qquad \text{(slow)}$$
$$NO_3(g) + CO(g) \longrightarrow NO_2(g) + CO_2(g) \qquad \text{(fast)}$$

**Rate Laws of Multistep Mechanisms**

**FIGURE 15.10**   The flow of traffic on a toll road is limited by the flow through the slowest toll plaza. For example, if only 5 cars per minute are able to pass through plaza C, only 5 cars per minute can exit from the toll road further on, regardless of how many cars per minute plaza D is able to handle, provided it is faster than plaza C.

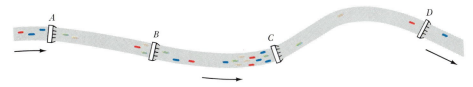

The rate law found experimentally corresponds to that for the first step in the mechanism, which is the slow, rate-determining step:

$$\text{Rate} = k[NO_2]^2 \qquad [15.30]$$

In general, steps that occur after the rate-determining one do not affect the rate law for the overall process.

It is not so easy to derive the rate law for a mechanism in which an intermediate is a reactant in the rate-determining step. This situation arises in multistep mechanisms when the first step is *not* rate determining. Let's consider one example, the gas-phase reaction of chlorine, $Cl_2$, with chloroform, $CHCl_3$:

$$Cl_2(g) + CHCl_3(g) \longrightarrow HCl(g) + CCl_4(g) \qquad [15.31]$$

Experiments indicate that the rate law for this reaction is as follows:

$$\text{Rate} = k[CHCl_3][Cl_2]^{1/2} \qquad [15.32]$$

Clearly, the reaction cannot occur in a single elementary step; in that case the rate law would be first order in both $CHCl_3$ and $Cl_2$, corresponding to the coefficients in the balanced chemical equation for the process. To explain the rate law and other experimental features of this reaction, the following sequence of elementary reactions is proposed as the mechanism:*

$$Cl_2(g) \underset{k_{-1}}{\overset{k_1}{\rightleftharpoons}} 2Cl(g) \qquad \text{(fast)} \qquad [15.33]$$

$$Cl(g) + CHCl_3(g) \overset{k_2}{\longrightarrow} HCl(g) + CCl_3(g) \qquad \text{(slow)} \qquad [15.34]$$

$$Cl(g) + CCl_3(g) \overset{k_3}{\longrightarrow} CCl_4(g) \qquad \text{(fast)} \qquad [15.35]$$

Because the second step of the mechanism is the slow, rate-determining step, the rate of the overall reaction should be governed by the rate law for that step:

$$\text{Rate} = k_2[CHCl_3][Cl] \qquad [15.36]$$

However, Cl is an intermediate generated by dissociation of $Cl_2$ molecules. Intermediates are normally encountered in low, unknown concentrations. Experimental rate laws are generally expressed in terms of substances present in measurable concentrations, not in terms of intermediates. Thus the experimental rate law, Equation 15.32, is expressed in terms of the reactant, $Cl_2$, from which the intermediate Cl atoms form. To see how the concentration of Cl depends on the concentration of $Cl_2$, we assume that the first step begins to reverse itself soon after the start of the reaction, before step two has a chance to form much product. That

---

* The subscript 1 on $k$ identifies this rate constant as that for the first elementary reaction of the mechanism. Similarly, $k_{-1}$ is the rate constant for the reverse of the first reaction, and $k_2$ is that for the second reaction, and so forth.

The basic procedure in establishing the mechanism of a chemical reaction is first to determine the rate law experimentally. One or more elementary reactions are then postulated that account for that rate law and for other experimental observations. That a mechanism gives a rate law that agrees with the observed one does not prove that that mechanism is correct. Often several mechanisms can be envisioned that give rise to the same rate law. To distinguish between such mechanisms, chemists might search for proposed intermediates or carry out other types of studies. As an example, consider the following reaction:

$$O^+(g) + NO(g) \longrightarrow NO^+(g) + O(g) \quad [15.37]$$

This reaction occurs in the upper atmosphere. The rate law for this reaction is: rate $= k[O^+][NO]$, and

the reaction is believed to occur in a single elementary step. However, we may ask whether the reaction involves the breaking of the NO bond and the formation of a new bond between N and $O^+$, or whether it involves merely the transfer of an electron from NO to $O^+$. These two possibilities are shown in Figure 15.11. The question of which of these mechanisms is operative is answered by performing the reaction using $O^+$ that has been highly enriched in the rare isotope $^{18}O$. Because this isotope is present to the extent of only 0.2 percent in nature, the NO contains only 0.2 percent $N^{18}O$. By measuring the location of $^{18}O$ in the reaction products, the two mechanisms can be distinguished (see Figure 15.11). The experiment indicates that no $^{18}O$ is incorporated into $NO^+$; thus the experimental results are consistent with the electron-transfer pathway but not the atom-transfer pathway.

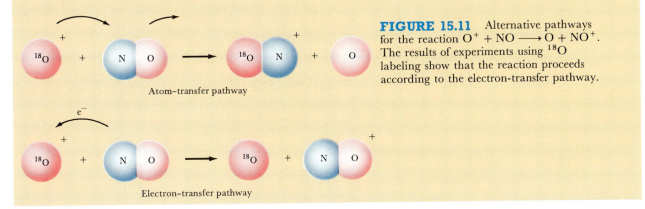

Atom–transfer pathway

Electron–transfer pathway

**FIGURE 15.11** Alternative pathways for the reaction $O^+ + NO \longrightarrow O + NO^+$. The results of experiments using $^{18}O$ labeling show that the reaction proceeds according to the electron-transfer pathway.

is, both the forward and reverse reactions in Equation 15.33 are rapid relative to reaction 15.34. If we further assume that the forward and reverse reactions of the first step are approximately equal in rate (that is, they are in equilibrium), then we can write

$$\text{Rate}_1 = k_1[Cl_2] = \text{rate}_{-1} = k_{-1}[Cl]^2 \quad [15.38]$$

Solving for Cl, we have

$$[Cl]^2 = \frac{k_1}{k_{-1}}[Cl_2]$$

Thus,

$$[Cl] = \left(\frac{k_1}{k_{-1}}[Cl_2]\right)^{1/2} \quad [15.39]$$

Substituting this relationship into the rate law for the rate-determining step (Equation 15.36), we have

$$\text{Rate} = k_2[\text{CHCl}_3]\left(\frac{k_1}{k_{-1}}[\text{Cl}_2]\right)^{1/2} \qquad [15.40]$$

Combining constants and defining $k = k_2(k_1/k_{-1})^{1/2}$, we have the experimentally observed rate law:

$$\text{Rate} = k[\text{CHCl}_3][\text{Cl}_2]^{1/2}$$

## 15.6 CATALYSIS

A **catalyst** is a substance that acts to change the speed of a chemical reaction without itself undergoing a permanent chemical change in the process. Catalysts are very common; most reactions occurring in the human body, the atmosphere, the oceans, or in industrial chemical processes are affected by catalysts.

If you have been exposed to chemical laboratory work, you probably have carried out the reaction in which oxygen is produced by heating potassium chlorate, $\text{KClO}_3$:

$$2\text{KClO}_3(s) \xrightarrow{\Delta} 2\text{KCl}(s) + 3\text{O}_2(g) \qquad [15.41]$$

In the absence of a catalyst, $\text{KClO}_3$ does not readily decompose in this manner, even on strong heating. However, mixing black manganese dioxide, $\text{MnO}_2$, with the $\text{KClO}_3$ before heating causes the reaction to occur much more readily. The $\text{MnO}_2$ can be recovered largely unchanged from this reaction, so the overall chemical process is clearly still the same. Thus $\text{MnO}_2$ acts as a catalyst for decomposition of $\text{KClO}_3$. As another example, we know that a cube of sugar, when dissolved in water at 37°C, does not undergo oxidation at a significant rate. The sugar could be recovered essentially unchanged from the solution after several days. Yet sugar ingested into the human body at about 37°C is rapidly oxidized and soon ends up mostly as carbon dioxide and water:

$$\text{C}_{12}\text{H}_{22}\text{O}_{11}(aq) + 12\text{O}_2(aq) \longrightarrow 12\text{CO}_2(aq) + 11\text{H}_2\text{O}(l) \qquad [15.42]$$

The oxidation of sugar in the biochemical system is greatly speeded up by the presence of one or more catalysts. These biochemical catalysts are **enzymes**, protein molecules that act to catalyze specific biochemical reactions. (Enzymes are discussed at length in Chapter 28.)

Much industrial chemical research is devoted to the search for new and more effective catalysts for reactions of commercial importance. Extensive research efforts also are devoted to finding means of inhibiting or removing certain catalysts that promote undesirable reactions, such as those involved in corrosion of metals, aging, and tooth decay.

### Homogeneous Catalysis

A catalyst that is present in the same phase as the components of a chemical reaction is a **homogeneous catalyst**. For example, a homogeneous catalyst for a reaction occurring in solution would itself be dissolved in the solution.

In Chapter 14 we considered a simple example of homogeneous catalysis: the action of NO in promoting the decomposition of ozone, $O_3$. The NO acts as a catalyst by reaction with $O_3$ to form $NO_2$ and $O_2$. The $NO_2$ thus formed then reacts with atomic oxygen present in the stratosphere to re-form NO and yield $O_2$ as the other product. The sequence of reactions and the overall result are as follows:

$$NO(g) + O_3(g) \longrightarrow NO_2(g) + O_2(g)$$
$$NO_2(g) + O(g) \longrightarrow NO(g) + O_2(g)$$
$$\overline{O_3(g) + O(g) \longrightarrow 2O_2(g)}$$

In this example, NO acts as a catalyst for $O_3$ decomposition because it speeds up the rate of the overall reaction without itself undergoing any net, or overall, chemical change; it is used in one step and re-formed in the next.

As another example, hydrogen peroxide, $H_2O_2$, when dissolved in water undergoes slow decomposition, forming oxygen and water:

$$2H_2O_2(aq) \longrightarrow 2H_2O(l) + O_2(g) \qquad [15.43]$$

In the absence of a catalyst, this reaction occurs at an extremely slow rate. Many different substances are capable of catalyzing the reaction; among these is bromine, $Br_2$. The bromine reacts with hydrogen peroxide in acidic solution, forming bromide ion and liberating oxygen:

$$Br_2(aq) + H_2O_2(aq) \longrightarrow 2Br^-(aq) + 2H^+(aq) + O_2(g) \qquad [15.44]$$

If this were the complete reaction, bromine would not be a catalyst, because it undergoes chemical change in the reaction. It happens, however, that hydrogen peroxide reacts with bromide ion in acidic solution to form bromine:

$$2Br^-(aq) + H_2O_2(aq) + 2H^+(aq) \longrightarrow Br_2(l) + 2H_2O(l) \qquad [15.45]$$

The overall sum of reaction Equations 15.44 and 15.45 is just Equation 15.43. (Add these two reactions together yourself to make sure you see that reaction Equation 15.43 results.) We see that bromine (or bromide ion) is indeed a catalyst in the reaction, because it speeds the overall reaction without itself undergoing any net, or overall, change. The effect of adding bromide ion on the decomposition of an aqueous hydrogen peroxide solution is shown in Figure 15.12.

On the basis of the Arrhenius expression for a chemical reaction, Equation 15.19, the rate constant $k$ is determined by the activation energy $E_a$ and the frequency factor $A$. A catalyst may affect the rate of reaction by altering the value for either $E_a$ or $A$. The most dramatic catalytic effects come from lowering $E_a$. As a general rule, *a catalyst lowers the overall activation energy for chemical reaction.* The lowering of $E_a$ by a catalyst is shown schematically in Figure 15.13.

A catalyst usually lowers the overall activation energy for reaction by providing a completely different pathway for reaction. The two examples

**FIGURE 15.12** Effect of added NaBr on the decomposition of $H_2O_2(aq)$ solution (*a*). Very soon after adding colorless NaBr solution to an $H_2O_2$ solution, the solution turns brown, because $Br_2$ is formed (Equation 15.45). (*b*) After a time, when the $Br_2$ concentration has built up, decomposition of $H_2O_2$ (Equation 15.44) occurs at a rapid pace with evolution of $O_2$. The bromide ion that formed is recycled (Equation 15.45). Bromine is thus a true catalyst in the reaction, undergoing no net change. (Donald Clegg and Roxy Wilson)

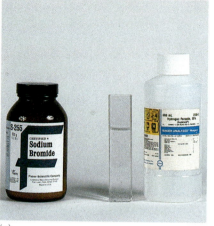

(*a*)

(*b*)

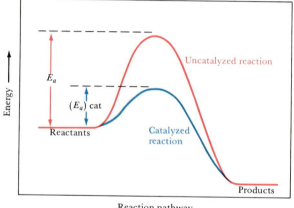

**FIGURE 15.13** Energy profiles for a catalyzed and uncatalyzed reactions. The catalyst functions in this example to lower the activation energy for reaction. Notice that the energies of reactants and products are unchanged by the catalyst.

given above involve a reversible, cyclic reaction of the catalyst with the reactants. For example, in the decomposition of hydrogen peroxide, two successive reactions of $H_2O_2$, with bromine and then with bromide, take place. Because these two reactions together serve as a catalytic pathway for hydrogen peroxide decomposition, *both* of these reactions must have significantly lower activation energies than the uncatalyzed decomposition, as shown schematically in Figure 15.14.

### Heterogeneous Catalysis

A **heterogeneous catalyst** exists in a phase different from the reactant molecules. For example, a reaction between molecules in the gas phase might be catalyzed by a finely divided metal oxide. In the absence of a catalyst, the reaction would occur slowly in the gas phase. However, when the catalyst is present, the reaction occurs more rapidly on the surface of the solid catalyst.

Many industrially important reactions occurring in the gas phase are catalyzed by solid surfaces. For example, hydrocarbon molecules are rearranged to form gasoline with the aid of what are called "cracking" catalysts (Section 27.2). Reactions occurring in solution may also be catalyzed by solids. Heterogeneous catalysts are often composed of finely

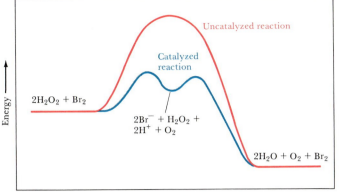

FIGURE 15.14 Energy profiles for the uncatalyzed decomposition of hydrogen peroxide and for the reaction as catalyzed by $Br_2$. The catalyzed reaction involves two successive steps, each of which has a lower activation energy than the uncatalyzed reaction.

divided metal or metal oxides. Because the catalyzed reaction occurs on the surface, special methods are often used to prepare catalysts so that they have very large surface areas. Some examples of heterogeneous catalysts used in industrially important processes are shown in Figure 15.15.

The initial step in heterogeneous catalysis is usually **adsorption** of reactants. The term *ad*sorption should be distinguished from *ab*sorption. Adsorption refers to the binding of molecules to a surface, whereas absorption refers to the uptake of molecules into the interior of another substance. Adsorption occurs because the atoms or ions at the surface of a solid are extremely reactive. Unlike their counterparts in the interior of the substance, they have unfulfilled valence requirements. The unused bonding capability of surface atoms or ions may be used to bond molecules

FIGURE 15.15 Samples of industrially important heterogeneous catalysts. Each catalyst is specially tailored for use in a particular type of reaction. For example, the blue material in the front beaker consists of a cobalt-molybdenum catalyst supported on alumina, $Al_2O_3$. The catalyst is used for removal of sulfur and nitrogen from crude oil. (Photo courtesy of the Harshaw/Filtrol Partnership)

from the gas or solution phase to the surface of the solid. In practice, not all the atoms or ions of the surface are reactive; various impurities may be adsorbed at the surface, and these may occupy many potential reaction sites and block further reaction. The places where reacting molecules may become adsorbed are called **active sites**. The number of active sites per unit amount of catalyst depends on the nature of the catalyst, on its method of preparation, and on its treatment before use.

As an example of heterogeneous catalysis, consider the hydrogenation of ethylene to form ethane:

$$\underset{\text{Ethylene}}{\underset{H}{\overset{H}{\phantom{.}}}C=C\underset{H}{\overset{H}{\phantom{.}}} + H_2} \longrightarrow \underset{\text{Ethane}}{\underset{H}{\overset{H}{\phantom{.}}}H-C-C-H\underset{H}{\overset{H}{\phantom{.}}}} \qquad [15.46]$$

In the absence of a catalyst, this reaction occurs slowly. However, in the presence of a very finely divided metal such as nickel, palladium, or platinum, the reaction occurs rather easily at room temperature, under a few hundred atmospheres of hydrogen pressure. The mechanism by which reaction occurs is shown diagrammatically in Figure 15.16. Both ethylene and hydrogen are adsorbed at the metal surface [Figure 15.16(a)]. The adsorption of hydrogen results in breaking of the H—H bond and formation of two M—H bonds, where M represents the metal surface [Figure 15.16(b)]. The hydrogen atoms are relatively free to move about the surface. When they encounter an adsorbed ethylene, the hydrogen may become bound to the carbon [Figure 15.16(c)]. The carbon thus acquires four $\sigma$ bonds about it, which reduces its tendency to remain adsorbed at the metal. When the other carbon also acquires a hydrogen, the ethane molecule is released from the surface [Figure 15.16(d)]. The active site is ready to adsorb another ethylene molecule and thus begin the cycle again.

**FIGURE 15.16**

Mechanism for reaction of ethylene with hydrogen on a catalytic surface. (a) The hydrogen and ethylene are adsorbed at the metal surface. (b) The H—H bond is broken to give adsorbed hydrogen atoms. (c) These migrate to the adsorbed ethylene and bond to the carbon atoms. (d) As C—H bonds are formed, the adsorption of the molecule to the metal surface is decreased, and ethane is released.

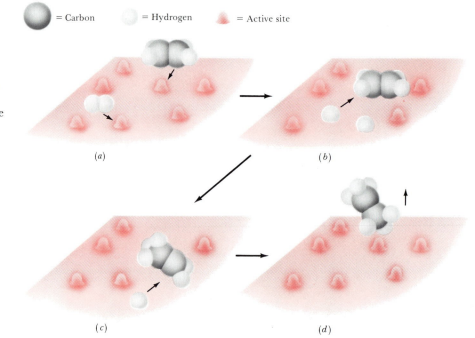

Heterogeneous catalysis plays a major role in the fight against urban air pollution. Two components of automobile exhausts that are involved in the formation of photochemical smog are nitrogen oxides and unburned hydrocarbons of various types (Section 14.4). In addition, automobile exhausts may contain considerable quantities of carbon monoxide. Even with the most careful attention to engine design and fuel characteristics, it is not possible under normal driving conditions to reduce the contents of these pollutants to an acceptable level in the exhaust gases coming from the engine. It is therefore necessary somehow to remove them from the exhaust gases before they are vented to the air. This removal is accomplished in the *catalytic converter*.

The catalytic converter, illustrated in Figure 15.17, must perform two distinct functions: (1) oxidation of CO and unburned hydrocarbons to carbon dioxide and water, and (2) reduction of nitrogen oxides to nitrogen gas:

$$CO, \text{hydrocarbons } (C_xH_y) \xrightarrow{O_2} CO_2 + H_2O$$

$$NO, NO_2 \longrightarrow N_2$$

These two functions require two distinctly different catalysts, so the development of a successful catalyst system is a very difficult challenge. The catalysts must be effective over a wide range of operating temperatures; they must continue to be active in spite of the poisoning action of various gasoline additives emitted along with the exhaust; they must be physically rugged enough to withstand gas turbulence and the mechanical shocks of driving under various conditions for thousands of miles.

Catalysts that promote the combustion of CO and

hydrocarbons are, in general, the transition-metal oxides and noble metals such as platinum. As an example, a mixture of two different metal oxides—CuO and $Cr_2O_3$, for example—might be used. These materials are supported on a structure (Figure 15.18) that allows the best possible contact between the flowing exhaust gas and the catalyst surface. Either bead or honeycomb structures made from alumina, $Al_2O_3$, and impregnated with the catalyst may be employed. Such catalysts operate by first adsorbing oxygen gas, also present in the exhaust gas. This adsorption weakens the O—O bond in $O_2$, so that oxygen atoms are in effect available for reaction with adsorbed CO to form $CO_2$. Hydrocarbon oxidation probably proceeds somewhat similarly, with the hydrocarbons first being adsorbed by rupture of a C—H bond.

The most effective catalysts for reduction of NO to yield $N_2$ and $O_2$ are transition-metal oxides and noble metals, the same kinds of materials that catalyze the oxidation of CO and hydrocarbons. The catalysts that are most effective in one reaction, however, are usually much less effective in the other. It is therefore necessary to have two different catalytic components.

The activity of catalytic converters decreases with use, because of loss of active catalyst, cracking and fractures due to repeated heating and cooling, and poisoning of the catalysts. One of the most active poisons is the lead that comes from the tetramethyl lead, $Pb(CH_3)_4$, or tetraethyl lead, $Pb(C_2H_5)_4$, added to gasoline. Because of the severe catalyst poisoning that results from use of leaded fuels, cars built since 1975 have been engineered for use with unleaded gas.

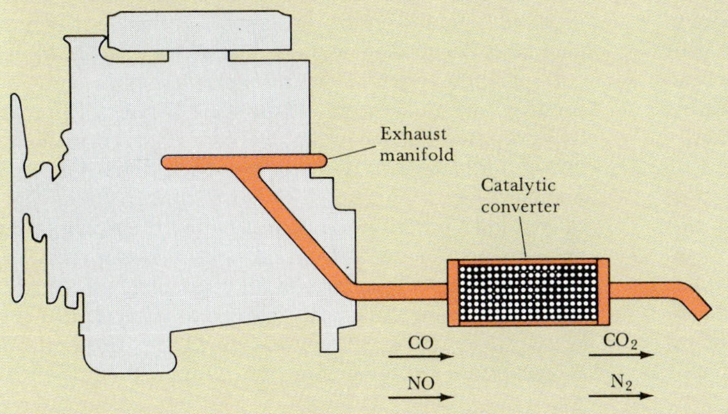

**FIGURE 15.17** Illustration of the arrangement and functions of a catalytic converter.

Exhaust manifold

Catalytic converter

CO

NO

CO$_2$

N$_2$

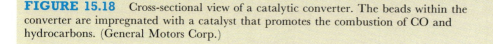

**FIGURE 15.18** Cross-sectional view of a catalytic converter. The beads within the converter are impregnated with a catalyst that promotes the combustion of CO and hydrocarbons. (General Motors Corp.)

# FOR REVIEW

## SUMMARY

In this chapter we've discussed ways of expressing reaction rates and the factors that influence these rates: namely, concentration, temperature, and catalysts. Rates are expressed as changes of concentration per unit of time; typically, for reactions in solution, the units are $M/s$. The quantitative relationship between rate and concentration is expressed by the **rate law**, which often has the following form: rate = $k$[reactant 1]$^m$[reactant 2]$^n \cdots$. The constant $k$ in the rate law is called the **rate constant**; the exponents $m$, $n$, and so forth, are called **reaction orders**. The rate law depends not on the overall reaction, but on the mechanism by which it occurs. Thus rate laws cannot ordinarily be determined from the coefficients in the balanced chemical equation, but must be determined by experiment.

In a first-order reaction, the reaction rate is proportional to the concentration of a single reactant raised to the first power: rate = $k$[A]. In such cases $\log [A] = (-k/2.30)t + \log [A]_0$, where [A] is the concentration of reactant A at time $t$, $k$ is the rate constant, and $[A]_0$ is the initial concentration of reactant A. Thus a graph of $\log [A]$ versus time yields a straight line of slope $-k/2.30$. First-order reactions are also characterized by having a constant **half-life**, which is related to the rate constant: $t_{1/2} = 0.693/k$. Reactions more complex than first order yield different expressions for the rate constant and half-life. For example, for second-order reactions, rate = $k$[A]$^2$, the following relationship holds: $1/[A] = 1/[A]_0 + kt$; thus in this case a graph of $1/[A]$ versus time yields a straight line.

Reactions occur as a result of collisions between molecules. Collisions bring atoms together so that new chemical bonds can form. In addition, when molecules collide, part of their kinetic energies may be used to cause old bonds to break. Of course, not all collisions lead to reaction; in order for reaction to occur, molecules must collide with sufficient energy and proper orientation. The minimum energy required for a reaction to occur is called the **activation energy**. When the activation energy is appreciable, only a very small fraction of all collisions between reactants provide the energy required to surmount the activation-energy barrier and form the products of the reaction. By increasing the temperature of a system we can increase the fraction of molecules whose kinetic energies exceed the activation energy and thus increase the reaction rate. From the manner in which the rate constant for a reaction varies with temperature, it is possible to determine the activation energy, $E_a$, for a reaction using the Arrhenius equation: $\log k = \log A - E_a/2.30RT$. A graph of $\log k$ versus $1/T$ yields a straight line whose slope is $-E_a/2.30R$.

From a knowledge of the rate law for a reaction and with other experimental information, a picture of how the reaction proceeds may be formulated. The overall reaction may occur in a series of **elementary steps** or in a single elementary step. Elementary steps (or **elementary reactions**) may be **unimolecular**, **bimolecular**, or **termolecular**, depending on whether one, two, or three molecules, respectively, are involved as reactants. The rate law of an elementary reaction is related directly to its molecularity. Thus unimolecular reactions have first-order rate laws, and bimolecular reactions have second-order rate laws.

The sequence of elementary steps by which a reaction proceeds is known as its **mechanism**. If a reaction proceeds by a multistep mechanism, one of the elementary reactions may be much slower than the others and so be the **rate-determining step**. The rate-determining step governs the rate law for the overall process.

A **catalyst** speeds a reaction without itself undergoing a net chemical change. It does so by providing a different mechanism for the reaction, one having a lower activation-energy barrier. Catalysts may be either **homogeneous**—that is, in the same phase with the reactants—or **heterogeneous**—in a separate phase. Heterogeneous catalysts are particularly important in large-scale industrial chemical processes and in applications such as the catalytic converter of an automobile.

## LEARNING GOALS

Having read and studied this chapter, you should be able to:

1. Express the rate of a given reaction in terms of the variation in concentration of a reactant or product substance with time.

2. Calculate the average rate over a time interval, given the concentrations of a reactant or product at the beginning and end of that interval.

3. Calculate instantaneous rate from a graph of reactant or product concentrations as a function of time.

4. Explain the meaning of the term *rate constant*, and state the units associated with rate constants for first- and second-order reactions.

5. Determine the rate law for a reaction from experimental results that show how concentration affects rate.

6. Calculate rate, rate constants, or reactant concentration, given two of these together with the rate law.

7. Use Equation 15.12 or 15.13 (for first-order reactions) or 15.17 (for second-order reactions) to determine (a) the concentration of a reactant or product at any time after a reaction has started, (b) the time required for a given fraction of sample to react, or (c) the time required for a reactant concentration to reach a certain level.

8. Use Equations 15.13 and 15.17 to determine graphically whether the rate law for a reaction is first or second order.

9. Explain the concept of reaction half-life, and describe the relationship between half-life and rate constant for a first-order reaction.

10. Explain the concept of activation energy and how it relates to the variation of reaction rate with temperature.

11. Determine the activation energy for a reaction from a knowledge of how the rate constant varies with temperature (the Arrhenius equation).

12. Using the terms *elementary steps*, *rate-determining step*, and *intermediate*, explain what is meant by the mechanism of a reaction.

13. Derive the rate law for a reaction that has a rate-determining step, given the elementary steps and their relative speeds; conversely, choose a plausible mechanism for a reaction given the rate law.

14. Describe the effect of a catalyst on the energy requirements for a reaction.

15. Explain the factors that are important in determining the activity of a heterogeneous catalyst.

# KEY TERMS

Among the more important terms and expressions used for the first time in this chapter are the following:

The **activated complex** (Section 15.4) is the particular arrangement of reactant and product molecules at the point of maximum energy in the rate-determining step of a reaction.

**Activation energy** (Section 15.4), $E_a$, is the minimum energy that must be supplied by the reactants in a chemical reaction to overcome the barrier to formation of products.

The **Arrhenius equation** (Section 15.4) relates the rate constant for a reaction to the frequency factor, $A$, the activation energy, $E_a$, and the temperature, $T$: $k = Ae^{-E_a/RT}$. In its log form it is written: $\log k = \log A - E_a/2.30RT$.

A **catalyst** (Section 15.6) is a substance that acts to change the speed of a chemical reaction without itself undergoing a permanent chemical change in the process.

An **elementary reaction** (Section 15.5) is a single-step **unimolecular**, **bimolecular**, or **termolecular** reaction. A multistep mechanism involves two or more such reactions.

An **enzyme** (Section 15.6) is a protein molecule that acts to catalyze specific biochemical reactions.

A **first-order reaction** (Section 15.3) is one in which the reaction rate is proportional to the concentration of a single reactant, raised to the first power.

The **half-life** (Section 15.3) of a reaction is the time required for the concentration of a reactant substance to decrease to halfway between its initial and final values.

A **Heterogeneous catalyst** (Section 15.6) is a catalyst that is in a different phase from that of the reactant substances.

A **homogeneous catalyst** (Section 15.6) is a catalyst that is in the same phase as the reactant substances.

An **intermediate** (Section 15.5) is a substance formed in one elementary step of a multistep mechanism and consumed in another; it is neither a reactant nor an ultimate product of the overall reaction.

A **mechanism** (Section 15.5) for a chemical reaction is a detailed picture, or model, of how the reaction occurs; that is, the order in which bonds are broken and formed, and the changes in relative positions of the atoms as reaction proceeds.

The **rate constant** (Section 15.2) is a constant of proportionality between the reaction rate and the concentrations of reactants that appear in the rate law.

The **rate-determining step** (Section 15.5) in a chemical reaction is the slowest step in a reaction that proceeds via a series of elementary steps from reactants to products.

A **rate law** (Section 15.2) is an equation in which the reaction rate is set equal to a mathematical expression involving the concentrations of reactants (and sometimes of products also).

The **reaction order** (Section 15.2) is the sum of the powers to which all the reactants appearing in the rate expression are raised.

**Reaction rate** (Section 15.1) is defined in terms of the decrease in concentration of a reactant molecule or the increase in concentration of a product molecule with time. It can be expressed as either the average rate over a period of time or the instantaneous rate at a particular time.

# EXERCISES

## Reaction Rates

**15.1** The rate of disappearance of $H^+$ was measured for the following reaction:

$$CH_3OH(aq) + HCl(aq) \longrightarrow CH_3Cl(aq) + H_2O(l)$$

The following data were collected:

| Time (min) | $[H^+]$ ($M$) |
|------------|---------------|
| 0 | 1.85 |
| 79 | 1.67 |
| 158 | 1.52 |
| 316 | 1.30 |
| 632 | 1.00 |

Calculate the average rate of reaction for the time interval between each measurement.

**15.2** The rearrangement of methyl isonitrile, $CH_3NC$, was studied in the gas phase at 215°C, and the following data were obtained:

| Time (s) | $[CH_3NC]$ ($M$) |
|----------|------------------|
| 0 | 0.0165 |
| 2,000 | 0.0110 |
| 5,000 | 0.00591 |
| 8,000 | 0.00314 |
| 12,000 | 0.00137 |
| 15,000 | 0.00074 |

Calculate the average rate of reaction for the time interval between each measurement.

**15.3** Using the data provided in Exercise 15.1, make a graph of $[H^+]$ versus time. Draw tangents to the curve at $t = 100$ and $t = 500$ min. Determine the rates at these times.

**15.4** Using the data provided in Exercise 15.2, make a graph of $[CH_3NC]$ versus time. Draw tangents to the curve at $t = 3500$ and $t = 13,500$ s. Determine the rates at these times.

**15.5** In each of the following reactions, how is the rate of appearance of the product that is underlined related to the rate of disappearance of the reactant that is underlined?
**(a)** $2\underline{H_2O_2(aq)} \longrightarrow 2H_2O(l) + O_2(aq)$
**(b)** $2\underline{NO(g)} + Cl_2(g) \longrightarrow 2\underline{NOCl(g)}$
**(c)** $\underline{BrO_3^-(aq)} + 5Br^-(aq) + 6H^+(aq) \longrightarrow$
$\qquad\qquad\qquad\qquad 3\underline{Br_2(l)} + 3H_2O(l)$

**15.6** In each of the following reactions, how is the rate of disappearance of each reactant related to the rate of appearance of each product?
**(a)** $2NOCl(g) \longrightarrow 2NO(g) + Cl_2(g)$
**(b)** $HI(g) + CH_3I(g) \longrightarrow CH_4(g) + I_2(g)$
**(c)** $Ag^+(aq) + 2NH_3(aq) \longrightarrow Ag(NH_3)_2^+(aq)$
**(d)** $2H_2O_2(aq) \longrightarrow 2H_2O(l) + O_2(g)$

**15.7** **(a)** Consider the combustion of $NH_3$, $4NH_3(g) + 5O_2(g) \longrightarrow 4NO(g) + 6H_2O(g)$. If ammonia is burning at the rate of 6 mol/s, what is the rate of formation of NO? of $H_2O$? **(b)** In the reaction $2O_3(g) \longrightarrow 3O_2(g)$, if the pressure of $O_3$ in a closed reaction vessel is decreasing at the rate of 30 mm Hg/min, what is the rate of change of the total pressure in the vessel?

**15.8** **(a)** Consider the combustion of methane, $CH_4(g) + 2O_2(g) \longrightarrow CO_2(g) + 2H_2O(g)$. If the concentration of $CH_4$ is decreasing at the rate of 0.40 $M$/s, what are the rates of change of the concentrations of $CO_2$ and of $H_2O$? **(b)** If the rate of increase of $NH_3$ pressure in a closed reaction vessel from the reaction $N_2(g) + 3H_2(g) \longrightarrow 2NH_3(g)$ is 100 mm Hg/h, what is the rate of change of the total pressure in the vessel?

## Rate Laws

**15.9** The decomposition of $N_2O_5$ in carbon tetrachloride proceeds as follows: $2N_2O_5 \longrightarrow 4NO_2 + O_2$. The rate law for this reaction is first order in $N_2O_5$. At 45°C, the rate constant is $6.08 \times 10^{-4}$/s. Calculate the rate of reaction when **(a)** $[N_2O_5] = 0.100$ $M$; **(b)** $[N_2O_5] = 0.305$ $M$.

**15.10** Consider the following reaction:

$$2NO(g) + 2H_2(g) \longrightarrow N_2(g) + 2H_2O(g)$$

**(a)** The rate law for this reaction is first order in $H_2$ and second order in NO. Write the rate law. **(b)** If the rate constant for this reaction at 1000 K is $6.0 \times 10^4/M^2$-s, what is the reaction rate when $[NO] = 0.050$ $M$ and

$[H_2] = 0.010$ $M$? **(c)** What is the reaction rate at 1000 K when the concentration of NO is doubled, to 0.10 $M$, while the concentration of $H_2$ is 0.010 $M$?

**15.11** Consider the hypothetical reaction $A + B \longrightarrow$ products. For each of the possible rate laws listed below, indicate the reaction order with respect to A, with respect to B, and the overall reaction order: **(a)** rate = $k[A][B]$; **(b)** rate = $k[A]^2$; **(c)** rate = $k[A][B]^2$.

**15.12** Suppose the following rate laws are being considered for the reaction $2NO(g) + O_2(g) \longrightarrow 2NO_2(g)$:

(i) $\quad -\dfrac{\Delta[NO]}{\Delta t} = k[NO]$

(ii) $\quad -\dfrac{\Delta[NO]}{\Delta t} = k[NO]^2[O_2]$

(iii) $\quad -\dfrac{\Delta[NO]}{\Delta t} = \dfrac{k[NO]}{[NO_2]}$

Which of these rate laws is **(a)** first order in [NO]; **(b)** first order overall; **(c)** third order overall?

**15.13** **(a)** What units would each of the rate constants in Exercise 15.11 have if concentration is expressed as $M$ and rate as $M/s$? **(b)** For each rate law in Exercise 15.11, indicate how the rate changes (i) if the concentration of A is doubled while the concentration of B remains constant; (ii) if the concentration of B is doubled while the concentration of A is held constant.

**15.14** **(a)** What units would each of the rate constants in Exercise 15.12 have if the concentrations of reactants are expressed in atm? **(b)** For each rate law in Exercise 15.12, indicate how the rate of loss of NO changes (i) if the pressure of NO is doubled; (ii) if the pressure of oxygen is halved; (iii) if the pressure of $NO_2$ is doubled.

**15.15** Consider the reaction of peroxydisulfate ion, $S_2O_8^{2-}$, with iodide ion, $I^-$, in aqueous solution:
$$S_2O_8^{2-}(aq) + 3I^-(aq) \longrightarrow 2SO_4^{2-}(aq) + I_3^-(aq)$$

At a particular temperature the rate of this reaction varies with reactant concentrations in the following manner:

| Experiment | $[S_2O_8^{2-}]$ ($M$) | $[I^-]$ ($M$) | $-\dfrac{\Delta[S_2O_8^{2-}]}{\Delta t}$ ($M$/s) |
| --- | --- | --- | --- |
| 1 | 0.038 | 0.060 | $1.4 \times 10^{-5}$ |
| 2 | 0.076 | 0.060 | $2.8 \times 10^{-5}$ |
| 3 | 0.076 | 0.030 | $1.4 \times 10^{-5}$ |

**(a)** Write the rate law for the rate of disappearance of $S_2O_8^{2-}$. **(b)** What is the numerical value of the rate constant for the disappearance of $S_2O_8^{2-}$? **(c)** What is the rate of disappearance of $S_2O_8^{2-}$ when $[S_2O_8^{2-}] = 0.025$ $M$ and $[I^-] = 0.100$ $M$? **(d)** What is the rate of appearance of $SO_4^{2-}$ when $[S_2O_8^{2-}] = 0.025$ $M$ and $[I^-] = 0.050$ $M$?

**15.16** The following data were collected for the gas-phase reaction between nitric oxide and bromine at 273°C:

$$2NO(g) + Br_2(g) \longrightarrow 2NOBr(g)$$

| Experiment | [NO] ($M$) | [Br$_2$] ($M$) | Initial rate of appearance of NOBr ($M$/s) |
|---|---|---|---|
| 1 | 0.10 | 0.10 | 12 |
| 2 | 0.10 | 0.20 | 24 |
| 3 | 0.20 | 0.10 | 48 |
| 4 | 0.30 | 0.10 | 108 |

**(a)** Determine the rate law. **(b)** Calculate the value of the rate constant for the appearance of NOBr. **(c)** How is the rate of appearance of NOBr related to the rate of disappearance of Br$_2$? **(d)** What is the rate of appearance of NOBr when [NO] = 0.15 $M$ and [Br$_2$] = 0.25 $M$? **(e)** What is the rate of disappearance of Br$_2$ when [NO] = 0.075 $M$ and [Br$_2$] = 0.185 $M$?

## Concentration and Time; Half-Lives

**15.17** The decomposition of a substance is found to be first order. If it takes $4.2 \times 10^4$ s for the concentration of that substance to fall to half its original value, what is the magnitude of the rate constant?

**15.18** The decomposition of a substance is found to be first order. If $k = 1.5 \times 10^{-2}$/s, what is the half-life for the reaction?

**15.19** The reaction

$$SO_2Cl_2(g) \longrightarrow SO_2(g) + Cl_2(g)$$

is first order in SO$_2$Cl$_2$. Using the following kinetic data, determine the magnitude of the first-order rate constant:

| Time (s) | Pressure, SO$_2$Cl$_2$ (atm) |
|---|---|
| 0 | 1.000 |
| 2,500 | 0.947 |
| 5,000 | 0.895 |
| 7,500 | 0.848 |
| 10,000 | 0.803 |

**15.20** From the following data for the first-order gas-phase rearrangement of CH$_3$NC at 215°C, calculate the first-order rate constant and half-life for the reaction:

| Time (s) | Pressure CH$_3$NC (mm Hg) |
|---|---|
| 0 | 502 |
| 2,000 | 335 |
| 5,000 | 180 |
| 8,000 | 95.5 |
| 12,000 | 41.7 |
| 15,000 | 22.4 |

**15.21** Sucrose, C$_{12}$H$_{22}$O$_{11}$, which is more commonly known as table sugar, reacts in dilute acid solutions to form two simpler sugars, glucose and fructose. Both of these sugars have the molecular formula C$_6$H$_{12}$O$_6$, though they differ in molecular structure. The reaction is

$$C_{12}H_{22}O_{11}(aq) + H_2O(l) \longrightarrow 2C_6H_{12}O_6(aq)$$

The rate of this reaction was studied at 23°C and in 0.5 $M$ HCl; the following data were obtained:

| Time (min) | [C$_{12}$H$_{22}$O$_{11}$] ($M$) |
|---|---|
| 0 | 0.316 |
| 39 | 0.274 |
| 80 | 0.238 |
| 140 | 0.190 |
| 210 | 0.146 |

Is the reaction first order or second order with respect to the concentration of sucrose? What is the numerical value for the rate constant?

**15.22** The decomposition of N$_2$O$_5$ has been studied in carbon tetrachloride solution at 45°C. The following data are obtained under these conditions:

| Time (min) | [N$_2$O$_5$] ($M$) |
|---|---|
| 0 | 1.40 |
| 6.67 | 1.10 |
| 13.33 | 0.87 |
| 20.00 | 0.68 |
| 26.67 | 0.53 |
| 33.33 | 0.42 |

Is the reaction first order or second order with respect to N$_2$O$_5$? What is the numerical value of the rate constant?

**15.23** The first-order rate constant for the gas-phase decomposition of N$_2$O$_5$ to NO$_2$ and O$_2$ at 70°C is $6.82 \times 10^{-3}$/s. **(a)** If we start with 0.500 mol of N$_2$O$_5$ in a 500-mL container, how many moles of N$_2$O$_5$ will remain after 3.00 min? **(b)** How long will it take for the quantity of N$_2$O$_5$ to drop to 0.100 mol?

**15.24** The first-order rate constant for the decomposition of a certain antibiotic in water at 20°C is 2.06/yr. **(a)** If a $3.0 \times 10^{-3}$ $M$ solution of this antibiotic is stored at 20°C for 1 month, what is the concentration of the antibiotic? After 1 year? **(b)** How long will it take for the concentration of the antibiotic to reach $1.0 \times 10^{-3}$ $M$? **(c)** What is the half-life of the antibiotic at 20°C?

**15.25** A reaction that is second order in A is 50 percent complete after 450 min. If [A]$_0$ = 1.35 $M$, what is the rate constant?

**15.26** The rate of the gas-phase decomposition of NO$_2$ to form NO and O$_2$, $2NO_2(g) \longrightarrow 2NO(g) + O_2(g)$, is second order in NO$_2$. If the concentration of NO$_2$ at 383°C varies with time as shown in the table that follows, what is the numerical value of the rate constant?

| Time (s) | $[NO_2]$ $(M)$ |
|----------|----------------|
| 0.0 | 0.100 |
| 5.0 | 0.017 |
| 10.0 | 0.0090 |
| 15.0 | 0.0062 |
| 20.0 | 0.0047 |

## Effects of Temperature; Activation Energy

**15.27** For the reaction $2N_2O_5(g) \longrightarrow 4NO_2(g) + O_2(g)$, the activation energy, $E_a$ and overall $\Delta E$ are 100 kJ/mol and $-23$ kJ/mol, respectively. **(a)** Sketch the energy profile for this reaction. **(b)** What is the activation energy for the reverse reaction?

**15.28** For the uncatalyzed decomposition of $H_2O_2(aq)$ to form $H_2O(l)$ and $O_2(g)$, the activation energy and overall $\Delta E$ are 75.3 kJ/mol and $-98.1$ kJ/mol, respectively. **(a)** Sketch the energy profile for this reaction. **(b)** What is the activation energy for the reverse reaction?

**15.29** Two reactions have identical values of $E_a$. Does this ensure that they will have the same rate constant if run at the same temperature? Explain.

**15.30** Two similar reactions have the same rate constant at 25°C, but at 35°C one of the reactions has a higher rate constant than the other. Account for these observations.

**15.31** The rate of the reaction

$$CH_3COOC_2H_5(aq) + OH^-(aq) \longrightarrow$$
$$CH_3COO^-(aq) + C_2H_5OH(aq)$$

was measured at several temperatures and the following data collected:

| Temperature (°C) | $k$ (1/$M$-s) |
|------------------|----------------|
| 15 | 0.0521 |
| 25 | 0.101 |
| 35 | 0.184 |
| 45 | 0.332 |

Using these data, construct a graph of log $k$ versus $1/T$. Using your graph, determine the value of $E_a$.

**15.32** The temperature dependence of the rate constant for the reaction

$$CO(g) + NO_2(g) \longrightarrow CO_2(g) + NO(g)$$

is tabulated below. Calculate $E_a$ and $A$.

| Temperature (K) | $k$ (1/$M$-s) |
|-----------------|----------------|
| 600 | 0.028 |
| 650 | 0.22 |
| 700 | 1.3 |
| 750 | 6.0 |
| 800 | 23 |

**15.33** The gas-phase decomposition of HI into $H_2$ and $I_2$ is found to have $E_a = 182$ kJ/mol. The rate constant at 700°C is $1.57 \times 10^{-3}$ $M$-s. What is the value of $k$ at **(a)** 600°C; **(b)** 800°C?

**15.34** The rate constant, $k$, for a reaction is $1.0 \times 10^{-3}$/s at 25°C. Calculate the rate constant for the same reaction at 100°C if **(a)** $E_a = 67.0$ kJ/mol; **(b)** $E_a = 134$ kJ/mol.

## Reaction Mechanisms; Catalysis

**15.35** The following mechanism has been proposed for the reaction of NO with $Br_2$ to form NOBr:

$$NO(g) + Br_2(g) \rightleftharpoons NOBr_2(g)$$
$$NOBr_2(g) + NO(g) \longrightarrow 2NOBr(g)$$

**(a)** Show that the elementary steps of the proposed mechanism add to provide a properly balanced equation for the reaction. **(b)** Write a rate law for each elementary step in the mechanism. **(c)** Identify any intermediates in the mechanism. **(d)** The observed rate law is: Rate = $k[NO]^2[Br_2]$. If the proposed mechanism is correct, what can we conclude about the relative speeds of the first and second steps?

**15.36** The balanced equation for the oxidation of HBr is $4HBr(g) + O_2(g) \longrightarrow 2H_2O(g) + 2Br_2(g)$. The following mechanism has been proposed:

$$HBr + O_2 \longrightarrow HOOBr$$
$$HOOBr + HBr \longrightarrow 2HOBr$$
$$HOBr + HBr \longrightarrow H_2O + Br_2$$

**(a)** Indicate whether the elementary steps of the proposed mechanism add to give the balanced equation for the reaction. (Hint: You may need to multiply all the coefficients of a given elementary step by some integer before adding.) **(b)** Write the rate law for each elementary step. **(c)** Identify any intermediates in the mechanism. **(d)** The reaction is found to be first order with respect to both HBr and $O_2$. Neither HOBr nor HOOBr is detected among the products. What can you conclude about the rate-determining step?

**15.37** Consider the following reaction: $H_2(g) + 2ICl(g) \longrightarrow 2HCl(g) + I_2(g)$. The rate law for this reaction is first order in both $H_2$ and ICl: Rate = $k[H_2][ICl]$. Which of the following mechanisms are consistent with the observed rate law?

**(a)** $2ICl(g) + H_2(g) \longrightarrow 2HCl(g) + I_2(g)$
(termolecular reaction)

**(b)** $H_2(g) + ICl(g) \longrightarrow HI(g) + HCl(g)$ (slow)
$HI(g) + ICl(g) \longrightarrow HCl(g) + I_2(g)$ (fast)

**(c)** $H_2(g) + ICl(g) \longrightarrow HI(g) + HCl(g)$ (fast)
$HI(g) + ICl(g) \longrightarrow HCl(g) + I_2(g)$ (slow)

**(d)** $H_2(g) + ICl(g) \longrightarrow HClI(g) + H(g)$ (slow)
$H(g) + ICl(g) \longrightarrow HCl(g) + I(g)$ (fast)
$HClI(g) \longrightarrow HCl(g) + I(g)$ (fast)
$I(g) + I(g) \longrightarrow I_2(g)$ (fast)

**15.38** The decomposition of hydrogen peroxide is catalyzed by iodide ion. The catalyzed reaction is thought to proceed by a two-step mechanism:

$$H_2O_2(aq) + I^-(aq) \longrightarrow H_2O(l) + IO^-(aq)$$
$$\text{(slow)}$$

$$IO^-(aq) + H_2O_2(aq) \longrightarrow H_2O(l) + O_2(g) + I^-(aq)$$
$$\text{(fast)}$$

**(a)** Assuming that the first step of the mechanism is rate determining, predict the rate law for the overall process. **(b)** Write the chemical equation for the overall process. **(c)** Identify the intermediate, if any, in the mechanism.

**15.39** What distinguishes an intermediate from a catalyst in a chemical reaction?

**15.40** In older texts, catalysts were sometimes defined as substances that speed up a chemical reaction without taking part in the reaction. In what way is this definition misleading? How might the definition be modified to correct it?

**15.41** The activity of a heterogeneous catalyst is highly dependent on its method of preparation and prior treatment. Explain why this is so.

**15.42** Many metallic catalysts, particularly the precious metal ones, are often deposited as very thin films on a substance of high surface area per unit mass, such as alumina ($Al_2O_3$) or silica ($SiO_2$). Why is this an effective way of utilizing the catalyst material?

**15.43** When $D_2$ is reacted with ethylene, $C_2H_4$, in the presence of a finely divided catalyst, ethane with two deuteriums, $CH_2D{-}CH_2D$, is formed. (Deuterium, D, is an isotope of hydrogen of mass 2.) Very little ethane forms in which two deuteriums are bound to one carbon, for example, $CH_3{-}CHD_2$. Explain why this is so in terms of the sequence of steps involved in hydrogenation.

**15.44** Suppose that the hydrogens on ethylene (Equation 15.46) were replaced by bulky organic groups such as $-CH_3$. What effect do you think this would have on the ease with which the compound underwent hydrogenation using a nickel metal catalyst? Explain.

## Additional Exercises

**15.45** **(a)** List three factors that can be varied to change the speed of a particular reaction. **(b)** Explain, on a molecular level, how each exerts its influence.

**15.46** What factors determine whether a collision will lead to chemical reaction?

**15.47** Explain why the rate law for a reaction cannot in general be written using the coefficients of the balanced chemical equation as the reaction orders.

**15.48** A particular reaction is found to have the following rate law: Rate $= k[A]^2[B]$. Which terms in this rate law are made different by each of the following changes? **(a)** The concentration of A is doubled. **(b)** A catalyst is added. **(c)** The concentration of A is increased by a factor of 2, and the concentration of B is decreased by a factor of 4. **(d)** The temperature is increased.

**15.49** The oxidation of NO by $O_3$, a reaction of importance in formation of photochemical smog (Section 14.4), is first order in each of the reactants. The second-order rate constant at 28°C is $1.5 \times 10^7/M$-s. If the concentrations of NO and $O_3$ are each $2 \times 10^{-8}$ $M$, what is the rate of oxidation of NO?

**15.50** Hydrogen sulfide, $H_2S$, is a common and troublesome pollutant in industrial waste waters. One way to remove this substance is to treat the water with chlorine, in which case the following reaction occurs:

$$H_2S(aq) + Cl_2(aq) \longrightarrow S(s) + 2H^+(aq) + 2Cl^-(aq)$$

The rate of this reaction is first order in each reactant. The rate constant for disappearance of $H_2S$ at 28°C is $3.5 \times 10^{-2}/M$-s. If at a given time the concentration of $H_2S$ is $1.6 \times 10^{-4}$ $M$ and that of $Cl_2$ is $0.070$ $M$, what is the rate of formation of $Cl^-$?

**15.51** The reaction of $(CH_3)_3CBr$ with hydroxide ion proceeds with the formation of $(CH_3)_3COH$:

$$(CH_3)_3CBr(aq) + OH^-(aq) \longrightarrow$$
$$(CH_3)_3COH(aq) + Br^-(aq)$$

This reaction was studied at 55°C, and the following data were obtained:

| Experiment | [(CH₃)₃CBr] ($M$) | [OH⁻] ($M$) | Initial rate ($M$/s) |
|---|---|---|---|
| 1 | 0.10 | 0.10 | $1.0 \times 10^{-3}$ |
| 2 | 0.20 | 0.10 | $2.0 \times 10^{-3}$ |
| 3 | 0.10 | 0.20 | $1.0 \times 10^{-3}$ |

Write the rate law for this reaction and compute the rate constant.

**15.52** Consider the following reaction between mercury(II) chloride and oxalate ion:

$$2HgCl_2(aq) + C_2O_4^{2-}(aq) \longrightarrow$$
$$2Cl^-(aq) + 2CO_2(g) + Hg_2Cl_2(s)$$

The initial rate of this reaction was determined for several concentrations of $HgCl_2$ and $C_2O_4^{2-}$ and the following rate data obtained:

| Experiment | [HgCl₂] ($M$) | [C₂O₄²⁻] ($M$) | Rate ($M$/s) |
|---|---|---|---|
| 1 | 0.105 | 0.15 | $1.8 \times 10^{-5}$ |
| 2 | 0.105 | 0.30 | $7.1 \times 10^{-5}$ |
| 3 | 0.052 | 0.30 | $3.5 \times 10^{-5}$ |
| 4 | 0.052 | 0.15 | $8.9 \times 10^{-6}$ |

**(a)** What is the rate law for this reaction? **(b)** What is the numerical value of the rate constant? **(c)** What is the reaction rate when the concentration of $HgCl_2$ is $0.080$ $M$ and that of $C_2O_4^{2-}$ is $0.10$ $M$ if the temperature is the same as that used to obtain the data shown above?

**15.53** Urea, $NH_2CONH_2$, is the end product in protein metabolism in animals. The decomposition of urea in 0.1 $M$ HCl occurs according to the reaction

$$NH_2CONH_2(aq) + H^+(aq) + 2H_2O(l) \longrightarrow$$
$$2NH_4^+(aq) + HCO_3^-(aq)$$

The reaction is first order in urea. When [urea] = 0.200 $M$, the rate at 61.05°C is $8.56 \times 10^{-5}$ $M$/s. **(a)** What is the numerical value for the rate constant, $k$? **(b)** What is the concentration of urea in this solution after $5.00 \times 10^3$ s if the starting concentration is 0.500 $M$? **(c)** What is the half-life for this reaction at 61.05°C?

**15.54** The reaction

$$F(g) + H_2(g) \longrightarrow HF(g) + H(g)$$

has been studied at various temperatures and has been found to have an activation energy of 22 kJ/mol. The overall change in energy in the reaction, $\Delta E$, is $-130$ kJ/mol. Draw a diagram of the energy profile of the system as a function of the reaction coordinate.

**15.55** The first-order rate constant for hydrolysis of a particular organic compound in water varies with temperature as follows:

| Temperature (K) | Rate constant $(s^{-1})$ |
| --- | --- |
| 300 | $1.0 \times 10^{-5}$ |
| 320 | $5.0 \times 10^{-5}$ |
| 340 | $2.0 \times 10^{-4}$ |
| 355 | $5.0 \times 10^{-4}$ |

From these data calculate the activation energy in units of kJ/mol.

**15.56** The activation energy for the reaction

$$2NO_2(g) \longrightarrow 2NO(g) + O_2(g)$$

is 114 kJ/mol. If $k = 0.75/M$-s at 600°C, what is the value of $k$ at 500°C?

**15.57** Outline briefly the advantages you can think of for the use of catalysts as opposed to high temperatures to promote chemical reactions.

**[15.58]** The gas-phase reaction of chlorine with carbon monoxide to form phosgene, $Cl_2(g) + CO(g) \longrightarrow COCl_2(g)$, obeys the following rate law:

$$Rate = \frac{\Delta[COCl_2]}{\Delta t} = k[Cl_2]^{3/2}[CO]$$

A mechanism involving the following series of steps is consistent with the rate law:

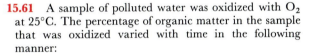

Assuming that this mechanism is correct, which of the steps above is the slow, or rate-determining, step? Explain.

**15.59** The primary focus of this chapter has been on homogeneous reactions, yet many significant processes are heterogeneous. Explain the basis for the following observations involving heterogeneous systems. **(a)** Bulk flour is hard to burn, yet elevators in which flour is stored have been known to explode. **(b)** Solids dissolve more rapidly in water when they are first crushed to give a small particle size.

**[15.60]** Derive the rate law for the reaction

$$OCl^-(aq) + I^-(aq) \longrightarrow OI^-(aq) + Cl^-(aq)$$

if its mechanism proceeds as follows:

$$OCl^-(aq) + H_2O(l) \rightleftharpoons HOCl(aq) + OH^-(aq)$$
$$\text{(fast)}$$

$$I^-(aq) + HOCl(aq) \longrightarrow HOI(aq) + Cl^-(aq)$$
$$\text{(slow)}$$

$$HOI(aq) + OH^-(aq) \longrightarrow H_2O(l) + OI^-(aq)$$
$$\text{(fast)}$$

**15.61** A sample of polluted water was oxidized with $O_2$ at 25°C. The percentage of organic matter in the sample that was oxidized varied with time in the following manner:

| Time (days) | 1 | 2 | 3 | 4 | 5 | 6 | 7 | 10 | 20 |
| --- | --- | --- | --- | --- | --- | --- | --- | --- | --- |
| Organic matter oxidized (%) | 21 | 37 | 50 | 60 | 68 | 75 | 80 | 90 | 99 |

Is the oxidation process first or second order? What is the rate constant for this reaction?

**[15.62]** We have seen in this chapter that the hydrogenation of ethylene to form ethane is heterogeneously catalyzed by nickel metal. In one particular study it was found that the rate of hydrogenation of $C_2H_4$ was first order with respect to $C_2H_4$ pressure at low $C_2H_4$ pressure. However, with all other factors remaining constant, the rate of hydrogenation was found to be zero order with respect to ethylene pressure at high pressures of $C_2H_4$. **(a)** What is the significance of a zero-order dependence, insofar as the rate of reaction is concerned? **(b)** Suggest an explanation for why the reaction rate has a zero-order dependence on ethylene pressure at high ethylene pressures.

**[15.63]** Beginning with Equation 15.17, derive the expression for $t_{1/2}$ for a second-order reaction.

## CHAPTER 16

# Chemical Equilibrium

In Chapter 13 we discussed the idea of a driving force that causes chemical reactions to proceed toward completion. We also noted (Section 13.4) that chemical reactions often come to an apparent halt before complete reaction has occurred. The condition in which the concentrations of all reactants and products cease to change with time is a state of **chemical equilibrium**. Equilibrium occurs when opposing reactions are occurring at equal rates; the rate at which products are formed from reactants equals the rate at which reactants are formed from products.

We have already encountered instances of simple equilibria. For example, in a closed container, the vapor above a liquid achieves an equilibrium with the liquid phase (Section 11.2). The rate at which molecules escape from the liquid to the gas phase equals the rate at which molecules of the gas phase strike the surface and become part of the liquid. As another example, in a saturated solution of sodium chloride, solid sodium chloride is in equilibrium with the ions dispersed in water (Section 12.2). The rate at which ions leave the solid surface equals the rate at which other ions are removed from the liquid to become part of the solid.

In this and the next three chapters we will explore chemical equilibria in some detail. We will learn in this chapter how to express the equilibrium position of a reaction in quantitative terms, and we will study the factors that determine the relative concentrations of reactants and products at equilibrium. To demonstrate the importance of equilibrium conditions for chemical reactions, we begin with a discussion of the Haber process for synthesizing ammonia.

## CONTENTS

## 16.1 THE HABER PROCESS

Of all the chemical reactions that humans have learned to carry out and control for their own purposes, the synthesis of ammonia from hydrogen and atmospheric nitrogen is the most important. Plant growth requires a substantial store of nitrogen in the soil, in a form usable by plants. The quantity of food required to feed the ever-increasing human population far exceeds that which could be produced if we relied solely on naturally available nitrogen in the soil. Ever-larger quantities of fertilizer rich in nitrogen will be needed in the decades ahead.

The only widely available source of nitrogen is the $N_2$ present in the atmosphere. The problem thus becomes one of "fixing" atmospheric $N_2$, that is, converting it to compounds that plants can use. This process is called *nitrogen fixation*.

The $N_2$ molecule is exceptionally unreactive, due in large measure to the strong triple bond between the nitrogen atoms (Section 8.4). For this

Liquid ammonia can be added directly to the soil as a fertilizer. Agricultural use is the largest single application of manufactured $NH_3$. (Farmland Industries, Inc.)

reason, fixation is not easy to achieve. In nature the fixation of $N_2$ is carried out by a special group of nitrogen-fixing bacteria that grow on the roots of certain plants, for example, clover or alfalfa. In our discussion we are interested in the particular fixation reaction called the **Haber process**.

Fritz Haber, a German chemist, investigated the energy relations in the reaction between nitrogen and hydrogen and convinced himself that forming ammonia in a reasonable yield from these two starting substances was possible. The chemical reaction is

$$N_2(g) + 3H_2(g) \rightleftharpoons 2NH_3(g) \qquad [16.1]$$

The double arrow indicates the reversible character of the reaction. The $NH_3$ can form from $N_2$ and $H_2$, but it can also decompose into these elements.

Haber's research was of great interest to the German chemical industry. Germany was preparing for World War I, and nitrogen compounds figured heavily in the manufacture of explosives. Without a synthetic source of these nitrogen compounds Germany would be greatly handicapped. By 1913, Haber had designed a process that worked, and for the first time ammonia was produced on a large scale from atmospheric nitrogen. In the following year, World War I began.

From these unhappy beginnings as a major factor in international warfare the Haber process has become the world's principal source of fixed nitrogen. It is estimated that 14 million tons of ammonia were formed via the Haber process in the United States in 1986.* The ammonia produced in the Haber process can be applied directly to the soil as shown in the chapter-opening photo. It may also be converted into ammonium salts—for example, ammonium sulfate, $(NH_4)_2SO_4$, or ammonium hydrogen phosphate, $(NH_4)_2HPO_4$—that are then used as fertilizers.

In designing the process that bears his name, Haber had to face two separate questions. First, is there a catalyst that will allow the reaction to occur at a reasonable speed under practically attainable conditions? After much long and difficult searching, Haber found a suitable catalyst; we shall return to this aspect of his work later, in Section 16.6. Second, assuming that a catalyst can be found, to what extent will nitrogen be converted into ammonia? Let us now consider the latter question, which relates to chemical equilibrium.

## 16.2
## THE EQUILIBRIUM
## CONSTANT

The Haber process consists of putting together $N_2$ and $H_2$ in a high-pressure tank at a total pressure of several hundred atmospheres, in the presence of a catalyst, and at a temperature of a few hundred degrees Celsius. Under these conditions, the two gases react to form ammonia. But the reaction does not lead to complete consumption of the $N_2$ and $H_2$. Rather, at some point the reaction appears to stop, with all three components of the reaction mixture present at the same time. The manner in which the concentration of $H_2$, $N_2$, and $NH_3$ vary with time in this situa-

---

* The industrial fixation of nitrogen, chiefly by the Haber process, now accounts for more than 30 percent of all the nitrogen fixed on the planet.

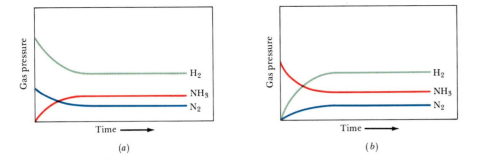

tion is shown in Figure 16.1(a). The condition of the system in which all concentrations have achieved steady values is referred to as chemical equilibrium. The relative amounts of $N_2$, $H_2$, and $NH_3$ present at equilibrium do not depend on the amount of catalyst present. However, they do depend on the relative amounts of $H_2$ and $N_2$ with which the reaction was begun. Furthermore, if only ammonia is placed into the tank under the usual reaction conditions, at equilibrium there is again a mixture of $N_2$, $H_2$, and $NH_3$. The variations in concentrations as a function of time for this situation are shown in Figure 16.1(b). By comparing the two parts of Figure 16.1, we can see that at equilibrium the relative concentrations of $H_2$, $N_2$, and $NH_3$ are the same, regardless of whether the starting mixture was a 3:1 molar ratio of $H_2$ and $N_2$ or pure $NH_3$. Equilibrium is thus a condition of the system that can be approached from either direction. This suggests to us that equilibrium is *not* a static condition. On the contrary, both the forward reaction in Equation 16.1, in which ammonia is formed, and the reverse reaction, in which $H_2$ and $N_2$ are formed from ammonia, are going on at precisely the same rates. Thus the net rate of change in the system is zero.

If we were systematically to change the relative amounts of $N_2$, $H_2$, and $NH_3$ in the starting mixture of gases and then analyze the gas mixtures at equilibrium, it would be possible to determine what sort of "law" governs the equilibrium state. Chemists carried out studies of this kind on other chemical systems in the nineteenth century, long before Haber's work. In 1864, Cato Maximilian Guldberg and Peter Waage proposed their **law of mass action**. This law expresses the relative concentrations of reactants and products at equilibrium in terms of a quantity called the equilibrium constant. Suppose that we have the general reaction

$$j A + k B \rightleftharpoons p R + q S \qquad [16.2]$$

where A, B, R, and S are the chemical species involved, and $j$, $k$, $p$, and $q$ are their coefficients in the balanced chemical equation. According to the law of mass action, the equilibrium condition is expressed by the equation

$$K = \frac{[R]^p[S]^q}{[A]^j[B]^k} \qquad [16.3]$$

where $K$ is a constant, called the **equilibrium constant**, and the square brackets signify the *concentration* of the species within the brackets. The

**FIGURE 16.2** The sealed tube contains a mixture of $NO_2(g)$, $N_2O_4(g)$, and $N_2O_4(l)$. $NO_2$ is red-brown in color, and $N_2O_4$ is colorless. The brown color in both the gas and liquid phases is due to $NO_2$. In both phases $NO_2$ is in equilibrium with $N_2O_4$. (Donald Clegg and Roxy Wilson)

law of mass action applies only to a system that has attained equilibrium. In general, the equilibrium constant is given by the concentrations of all reaction products multiplied together, each raised to the power of its coefficient in the balanced equation, divided by the concentrations of all reactants multiplied together, each raised to the power of its coefficient in the balanced equation. (Remember that the convention is to write the concentration terms for the *products* in the *numerator* and those for the *reactants* in the *denominator*.)

The equilibrium constant is a true constant. Its value at any given temperature does not depend on the initial concentrations of reactants and products. It also does not matter whether there are other substances present, as long as they do not consume a reactant or product through chemical reaction. The value of the equilibrium constant does, however, vary with temperature.

As an illustration of the law of mass action, consider the gas-phase equilibrium between dinitrogen tetroxide and nitrogen dioxide:

$$N_2O_4(g) \rightleftharpoons 2NO_2(g) \qquad [16.4]$$

Figure 16.2 shows a sealed tube containing a mixture of $NO_2$ and $N_2O_4$. Because $NO_2$ is a dark brown gas and $N_2O_4$ is colorless, the amount of $NO_2$ in the mixture can be learned by measuring the intensity of the brown color of the gas mixture.

Following the rule given above, the equilibrium-constant expression for reaction Equation 16.4 is

$$K = \frac{[NO_2]^2}{[N_2O_4]} \qquad [16.5]$$

Suppose that to determine the numerical value for $K$ and to verify that it is indeed constant as the concentrations of $NO_2$ and $N_2O_4$ change, three samples of $NO_2$ were placed in sealed glass vessels. In addition, a sample of $N_2O_4$ was placed in a fourth vessel. The vessels were allowed to remain at 100°C until no further change in the color of the gas was noted. The mixture of gases was then analyzed to determine the concentrations of both $NO_2$ and $N_2O_4$. The results are given in Table 16.1.

To evaluate the equilibrium constant, $K$, the equilibrium concentrations are inserted into the equilibrium-constant expression, Equation 16.5. When the concentration unit is molarity, as in the present case, we label the equilibrium constant as $K_c$.

**TABLE 16.1** Initial and equilibrium concentrations (molarities) of $NO_2$ and $N_2O_4$ in the gas phase at 100°C

| Experiment | Initial $N_2O_4$ concentration ($M$) | Initial $NO_2$ concentration ($M$) | Equilibrium $N_2O_4$ concentration ($M$) | Equilibrium $NO_2$ concentration ($M$) | $K_c$ |
|---|---|---|---|---|---|
| 1 | 0.0 | 0.0200 | 0.00140 | 0.0172 | 0.211 |
| 2 | 0.0 | 0.0300 | 0.00280 | 0.0243 | 0.211 |
| 3 | 0.0 | 0.0400 | 0.00452 | 0.0310 | 0.213 |
| 4 | 0.0200 | 0.0 | 0.00452 | 0.0310 | 0.213 |

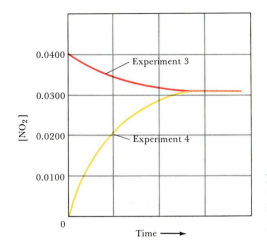

**FIGURE 16.3**   The same equilibrium mixture is produced starting with either 0.0400 $M$ NO$_2$ (experiment 3) or 0.0200 $M$ N$_2$O$_4$ (experiment 4).

For example, using the first set of data,

$$[NO_2] = 0.0172 \ M \qquad [N_2O_4] = 0.00140 \ M$$

$$K_c = \frac{[NO_2]^2}{[N_2O_4]} = \frac{(0.0172)^2}{0.00140} = 0.211$$

Proceeding in the same way, the values of $K_c$ for the other samples were calculated, as listed in Table 16.1. Note that the value for $K_c$ is essentially constant, even though the initial concentrations vary. Furthermore, the results of experiment 4 show that equilibrium can be attained beginning with N$_2$O$_4$ as well as with NO$_2$. That is, equilibrium can be approached from either direction. Figure 16.3 shows how both experiments 3 and 4 result in the same equilibrium mixture even though one begins with 0.0400 $M$ NO$_2$ and the other with 0.0200 $M$ N$_2$O$_4$.

## SAMPLE EXERCISE 16.1

Write the equilibrium-constant expression for each of the following reactions:
(a)   N$_2(g)$ + 3H$_2(g)$ $\rightleftharpoons$ 2NH$_3(g)$
(b)   2NH$_3(g)$ $\rightleftharpoons$ N$_2(g)$ + 3H$_2(g)$

**Solution:**   (a) As indicated by Equation 16.3, the equilibrium-constant expression has the form of a quotient. The numerator of this quotient is obtained by multiplying the equilibrium concentrations of products, each raised to a power equal to its coefficient in the balanced equation. The denominator is obtained similarly using the equilibrium concentrations of reactants:

$$K = \frac{[NH_3]^2}{[N_2][H_2]^3}$$

(b) This reaction is just the reverse of that given

in part (a). Placing products over reactants, we have

$$K = \frac{[N_2][H_2]^3}{[NH_3]^2}$$

Notice that this expression is just the reciprocal of that given in part (a). It is a general rule that the equilibrium-constant expression for a reaction written in one direction is the reciprocal of the one for the reverse reaction. This also means that the numerical value of the equilibrium constant for a reaction written in one direction is the reciprocal of the value of the equilibrium constant for the reaction written in the reverse direction.

### PRACTICE EXERCISE
Write the equilibrium-constant expression for the reaction H$_2(g)$ + I$_2(g)$ $\rightleftharpoons$ 2HI$(g)$.
*Answer:*   $K = [HI]^2/[H_2][I_2]$

## 16.3
## DETERMINATION OF THE VALUE OF THE EQUILIBRIUM CONSTANT

One of the first tasks confronting Haber and his co-workers when they approached the problem of ammonia synthesis was the determination of the numerical value of the equilibrium constant for the synthesis of $NH_3$ at various temperatures. If the value of $K$ for this reaction is very small, then the amount of $NH_3$ formed would be small relative to the initial amounts of $N_2$ and $H_2$ used. This situation can be described by saying that the equilibrium for

$$N_2(g) + 3H_2(g) \rightleftharpoons 2NH_3(g)$$

lies to the left, that is, toward the reactant side. Clearly, if the equilibrium were too far to the left, it would not be possible to develop a satisfactory synthesis for ammonia.

### SAMPLE EXERCISE 16.2

The reaction of $N_2$ with $O_2$ to form NO might be considered as a means of "fixing" nitrogen:

$$N_2(g) + O_2(g) \rightleftharpoons 2NO(g)$$

The value for the equilibrium constant for this reaction at 25°C is $K_c = 1 \times 10^{-30}$. Describe the feasibility of this reaction for nitrogen fixation.

**Solution:** Because $K_c$ is so small, very little NO will form at 25°C. The equilibrium is said to lie to the left, favoring the reactants. Consequently, this reaction is

an extremely poor choice for nitrogen fixation, at least at 25°C.

### PRACTICE EXERCISE

**(a)** Using the value for $K_c$ given in Sample Exercise 16.2, determine the magnitude of the equilibrium constant for decomposition of NO: $2NO(g) \rightleftharpoons N_2(g) + O_2(g)$. **(b)** Is it feasible to try to convert undesirable NO in automobile exhausts into $N_2$ and $O_2$ using this reaction?
**Answers:** **(a)** $K = 10^{30}$; **(b)** yes ($K$ is large; it is necessary only to find a way to make the reaction proceed at an acceptable rate).

The value of $K$ can be calculated if we know the equilibrium concentrations of all reactants and products in the reaction. These concentrations might be obtained by a direct experimental measurement. This was the method described in constructing Table 16.1.

### SAMPLE EXERCISE 16.3

In one of their experiments, Haber and co-workers introduced a mixture of hydrogen and nitrogen into a reaction vessel and allowed the system to attain chemical equilibrium at 472°C. The equilibrium mixture of gases was analyzed and found to contain 0.1207 $M$ $H_2$, 0.0402 $M$ $N_2$, and 0.00272 $M$ $NH_3$. From these data, calculate the equilibrium constant, $K_c$, for

$$N_2(g) + 3H_2(g) \rightleftharpoons 2NH_3(g)$$

**Solution:**

$$K_c = \frac{[NH_3]^2}{[N_2][H_2]^3} = \frac{(0.00272)^2}{(0.0402)(0.1207)^3} = 0.105$$

### PRACTICE EXERCISE

Nitryl chloride, $NO_2Cl$, is in equilibrium in a closed container with $NO_2$ and $Cl_2$:

$$2NO_2Cl(g) \rightleftharpoons 2NO_2(g) + Cl_2(g)$$

At equilibrium the concentrations of the substances in the equilibrium are

$$[NO_2Cl] = 0.00106\ M \qquad [NO_2] = 0.0108\ M$$
$$[Cl_2] = 0.00538\ M$$

From these data, calculate the equilibrium constant.

**Answer:** 0.558

We often don't know the equilibrium concentrations of all chemical species in an equilibrium. However, the equilibrium constant can be determined from a knowledge of the initial concentrations of all species and some information about the equilibrium condition, such as the concentration of at least one species, as illustrated in Sample Exercise 16.4.

## SAMPLE EXERCISE 16.4

A mixture of $5.00 \times 10^{-3}$ mol of $H_2$ and $1.00 \times 10^{-2}$ mol of $I_2$ is placed in a 5.00-L container at 448°C and allowed to come to equilibrium. Analysis of the equilibrium mixture shows that the concentration of HI is $1.87 \times 10^{-3}$ $M$. Calculate $K_c$ at 448°C for the reaction

$$H_2(g) + I_2(g) \rightleftharpoons 2HI(g)$$

**Solution:** We are given the balanced chemical equation for the equilibrium of interest. From this we first set up the equilibrium-constant expression:

$$K_c = \frac{[HI]^2}{[H_2][I_2]}$$

We next build a table listing the initial and equilibrium concentrations of all the species in the equilibrium and the changes in each case in going from initial to final concentrations. The initial concentrations of $H_2$ and $I_2$ must be calculated:

$$[H_2]_i = \frac{5.00 \times 10^{-3} \text{ mol}}{5.00 \text{ L}} = 1.00 \times 10^{-3} \text{ } M$$

$$[I_2]_i = \frac{1.00 \times 10^{-2} \text{ mol}}{5.00 \text{ L}} = 2.00 \times 10^{-3} \text{ } M$$

The first entries to our table are:

| | $H_2(g)$ + | $I_2(g)$ $\rightleftharpoons$ | $2HI(g)$ |
|---|---|---|---|
| **Initial** | $1.00 \times 10^{-3}$ $M$ | $2.00 \times 10^{-3}$ $M$ | $0$ $M$ |
| **Change** | | | |
| **Equilibrium** | | | $1.87 \times 10^{-3}$ $M$ |

The equilibrium concentrations of $H_2$ and $I_2$ can be calculated from these initial concentrations and the equilibrium concentration of HI. During the course of the reaction the concentration of HI changes from 0 to $1.87 \times 10^{-3}$ $M$. The balanced equation indicates that 2 mol of HI form from each mole of $H_2$. Thus the amount of $H_2$ consumed is

$$\left(1.87 \times 10^{-3} \frac{\text{mol HI}}{\text{L}}\right)\left(\frac{1 \text{ mol } H_2}{2 \text{ mol HI}}\right)$$
$$= 0.935 \times 10^{-3} \text{ mol } H_2/\text{L}$$

The equilibrium concentration of $H_2$ is the initial concentration minus that consumed:

$$[H_2] = 1.00 \times 10^{-3} \text{ } M - 0.935 \times 10^{-3} \text{ } M$$
$$= 0.065 \times 10^{-3} \text{ } M$$

The same line of argument gives the equilibrium concentration of $I_2$:

$$[I_2] = 2.00 \times 10^{-3} \text{ } M - 0.935 \times 10^{-3} \text{ } M$$
$$= 1.065 \times 10^{-3} \text{ } M$$

The filled-in table now looks like this:

| | $H_2(g)$ + | $I_2(g)$ $\rightleftharpoons$ | $2HI(g)$ |
|---|---|---|---|
| **Initial** | $1.00 \times 10^{-3}$ $M$ | $2.00 \times 10^{-3}$ $M$ | $0$ $M$ |
| **Change** | $-0.935 \times 10^{-3}$ $M$ | $-0.935 \times 10^{-3}$ $M$ | $+1.87 \times 10^{-3}$ $M$ |
| **Equilibrium** | $0.065 \times 10^{-3}$ $M$ | $1.065 \times 10^{-3}$ $M$ | $1.87 \times 10^{-3}$ $M$ |

Once we have the equilibrium concentrations of each reactant and product we can calculate the equilibrium constant:

$$K_c = \frac{[HI]^2}{[H_2][I_2]}$$

$$= \frac{(1.87 \times 10^{-3})^2}{(0.065 \times 10^{-3})(1.065 \times 10^{-3})}$$

$$= 50$$

**PRACTICE EXERCISE**

Sulfur trioxide decomposes at high temperature in a sealed container: $2SO_3(g) \rightleftharpoons 2SO_2(g) + O_2(g)$. Initially the vessel is charged at 1000 K with $SO_3(g)$ at a concentration of $6.09 \times 10^{-3}$ M. At equilibrium the $SO_3$ concentration is $2.44 \times 10^{-3}$ M. Calculate the value for $K_c$ at 1000 K.

*Answer:* $4.07 \times 10^{-3}$

## Concentration Units and Equilibrium Constants

As we have seen, the square brackets around a chemical symbol, as in $[NH_3]$, represent the concentration of that substance. Molarity is the common concentration unit for reactions occurring in solution. For gas-phase reactions, the concentration units used are either molarity or atmospheres of pressure. When the concentration is expressed in molarity, we denote the equilibrium constant as $K_c$. When the units are atmospheres, we write $K_p$. Since the numerical values of $K_c$ and $K_p$ will generally be different, we must take care to indicate which we are using by means of these subscripts.

The ideal-gas equation permits us to convert between atmospheres and molarity and therefore to convert between $K_p$ and $K_c$:

$$PV = nRT$$

$$P = \left(\frac{n}{V}\right)RT = MRT \qquad [16.6]$$

where $n/V$ (the number of moles per liter) is concentration in molarity, $M$. As a result of the relationship between pressure and molarity expressed in Equation 16.6, a general expression relating $K_p$ and $K_c$ can be written

$$K_p = K_c(RT)^{\Delta n} \qquad [16.7]$$

The quantity $\Delta n$ in this equation is the change in the number of moles of gas upon going from reactants to products; $\Delta n$ is equal to the number of moles of gaseous products minus the number of moles of gaseous reactants. For example, in the reaction

$$H_2(g) + I_2(g) \rightleftharpoons 2HI(g)$$

there is 2 mol of HI (the coefficient in the balanced equation); there is also 2 mol of gaseous reactants ($1H_2 + 1I_2$). Therefore, $\Delta n = 2 - 2 = 0$, and $K_p = K_c$ for this reaction.

**SAMPLE EXERCISE 16.5**

Using the value of $K_c$ obtained in Sample Exercise 16.3, calculate $K_p$ for

$$N_2(g) + 3H_2(g) \rightleftharpoons 2NH_3(g)$$

at 472°C.

**Solution:** There are 2 mol of gaseous products ($2NH_3$) and 4 mol of gaseous reactants ($1N_2 + 3H_2$). Therefore, $\Delta n = 2 - 4 = -2$. (Remember that $\Delta$ functions are always based on products minus reactants.) The temperature, $T$, is $273 + 472 = 745$ K.

The value for the ideal-gas constant, $R$, is 0.0821 L-atm/K-mol. The value of $K_c$ from Sample Exercise 16.3 is 0.105. We therefore have

$$K_p = \frac{P_{NH_3}^2}{P_{N_2}P_{H_2}^3} = K_c(RT)^{\Delta n}$$
$$= (0.105)(0.0821 \times 745)^{-2}$$
$$= 2.81 \times 10^{-5}$$

**PRACTICE EXERCISE**

For the equilibrium $2SO_3(g) \rightleftharpoons 2SO_2(g) + O_2(g)$ at temperature 1000 K, $K_c$ has the value $4.07 \times 10^{-3}$. Calculate the value for $K_p$.

*Answer:* 0.334

Concentration units can be carried through the calculation of the equilibrium constant to give units for $K$. For example, for the reaction $N_2O_4(g) \rightleftharpoons 2NO_2(g)$ we have $K = [NO_2]^2/[N_2O_4]$. When concentration is in molarity, the units of the equilibrium constant are $M^2/M = M$; when concentration is in atmospheres, the units are $atm^2/atm = atm$. Attaching units to the equilibrium constant has the advantage of clearly indicating the units in which concentration is expressed. Nevertheless, the more common practice is to write equilibrium constants as dimensionless quantities. We have adopted this practice in this text.

## 16.4 HETEROGENEOUS EQUILIBRIA

Many equilibria of importance, such as the hydrogen-nitrogen-ammonia system, involve substances all in the same phase. Such equilibria are said to be **homogeneous**. On the other hand, the substances in equilibrium may be in different phases, giving rise to **heterogeneous equilibria**. As an example, consider the decomposition of calcium carbonate:

$$CaCO_3(s) \rightleftharpoons CaO(s) + CO_2(g) \qquad [16.8]$$

This system involves a gas in equilibrium with two solids. If we write the equilibrium-constant expression for this process in the usual way, we obtain

$$K = \frac{[CaO][CO_2]}{[CaCO_3]} \qquad [16.9]$$

This example presents us with a problem we have not encountered previously: How do we express the concentration of a solid substance? The concentration of a pure substance, liquid or solid, equals its density divided by its molecular weight, $\mathscr{M}$:

$$\frac{\text{Density}}{\mathscr{M}} = \frac{\text{g/cm}^3}{\text{g/mol}} = \frac{\text{mol}}{\text{cm}^3}$$

The density of a pure liquid or solid is a constant at any given temperature and in fact changes very little with temperature. Thus the effective concentration of a pure liquid or solid is a constant. Because the calcium carbonate and calcium oxide of our example are present as pure solids, their concentrations are constant. The number of moles per liter of both solids is not changed, whether we have a large amount of solid present or just a little. The equilibrium-constant expression for Equation 16.8 then simplifies to

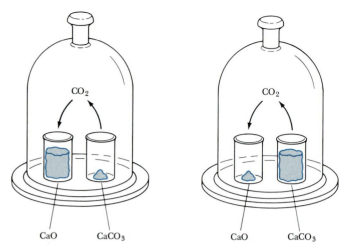

**FIGURE 16.4** The decomposition of $CaCO_3$ is an example of a heterogeneous equilibrium. At the same temperature the equilibrium pressure of $CO_2$ is the same in the two bell jars, even though the relative amounts of pure $CaCO_3$ and $CaO$ differ greatly.

CaO      $CaCO_3$          CaO      $CaCO_3$

$$K = \frac{[CO_2](\text{constant } 1)}{(\text{constant } 2)}$$

where constant 1 is the concentration of $CaO$ and constant 2 is the concentration of $CaCO_3$. Moving the constants to the left-hand side of the equation, we have

$$K' = K\frac{(\text{constant } 2)}{(\text{constant } 1)} = [CO_2] \qquad\qquad [16.10]$$

As a practical matter, the overall effect is the same as if we set the concentrations of both solids equal to one in the equilibrium-constant expression.

Equation 16.10 tells us that the equilibrium position doesn't depend on how much $CaCO_3$ or $CaO$ is present, as long as there is some of each in the system. As shown in Figure 16.4, we would have the same pressure of $CO_2$ in the system when we have an excess of $CaO$ as when we have an excess of $CaCO_3$. On the other hand, if one of the three ingredients is missing, we cannot have an equilibrium.

Thus we see that any pure solid or liquid that might be in an equilibrium has the same effect on the equilibrium no matter how much solid or liquid is present. The "concentrations" of pure solids and liquids are incorporated into the equilibrium constant. What this means when we write equilibrium-constant expressions is that the concentrations of pure solids and liquids are absent from the expression.

## SAMPLE EXERCISE 16.6

Each of the mixtures listed below was placed into a closed container and allowed to stand. Which of these mixtures is capable of attaining the equilibrium expressed by Equation 16.8: **(a)** pure $CaCO_3$; **(b)** $CaO$ and a pressure of $CO_2$ greater than the value of $K_c$;

**(c)** some $CaCO_3$ and a pressure of $CO_2$ greater than the value of $K_c$; **(d)** $CaCO_3$ and $CaO$?

**Solution:** Equilibrium can be reached in all cases except (c). In **(a)**, $CaCO_3$ simply decomposes until the equilibrium pressure of $CO_2$ is attained. In (b), $CO_2$ combines with the CaO present until its pressure decreases to the equilibrium value. In (c), equilibrium can't be attained, because there is no way in which the $CO_2$ pressure can decrease so as to attain its equilibrium value. In (d), the situation is essentially the same as in (a); $CaCO_3$ decomposes until equilibrium is attained. The presence of CaO initially makes no difference.

**PRACTICE EXERCISE** _____

Which of the following substances—$H_2(g)$, $H_2O(g)$, $O_2(g)$—when added to $Fe_3O_4(s)$ in a closed container at high temperature, permits attainment of equilibrium in the reaction $3Fe(s) + H_2O(g) \rightleftharpoons Fe_3O_4(s) + 2H_2(g)$?

*Answer:* Only $H_2(g)$

## SAMPLE EXERCISE 16.7

Write the equilibrium-constant expression for each of the following reactions:
(a) $CO_2(g) + H_2(g) \rightleftharpoons CO(g) + H_2O(l)$
(b) $SnO_2(s) + 2CO(g) \rightleftharpoons Sn(s) + 2CO_2(g)$

**Solution:** (a) The equilibrium-constant expression is

$$K = \frac{[CO]}{[CO_2][H_2]}$$

(Because $H_2O$ is a pure liquid its concentration does not appear in the equilibrium-constant expression.)

(b) The equilibrium-constant expression is

$$K = \frac{[CO_2]^2}{[CO]^2}$$

(Because $SnO_2$ and Sn are both pure solids, they do not appear in the equilibrium-constant expression.)

**PRACTICE EXERCISE** _____

Write the equilibrium-constant expression for the reaction $3Fe(s) + 4H_2O(g) \rightleftharpoons Fe_3O_4(s) + 4H_2(g)$.

*Answer:* $K = [H_2]^4/[H_2O]^4$

---

We have seen that the magnitude of $K$ indicates the extent to which a reaction will proceed. If $K$ is very large, the reaction will tend to proceed far to the right; if $K$ is very small, very little reaction will occur and the equilibrium mixture will contain mainly reactants. The equilibrium constant also allows us (1) to predict the direction in which a reaction mixture will proceed to achieve equilibrium, and (2) to calculate the concentrations of reactants and products once equilibrium has been reached.

## 16.5 APPLICATIONS OF EQUILIBRIUM CONSTANTS

### Prediction of the Direction of Reaction

Suppose that we place a mixture of 2.00 mol of $H_2$, 1.00 mol of $N_2$, and 2.00 mol of $NH_3$ in a 1-L container at 472 K. Will there be a reaction between $N_2$ and $H_2$ to form more $NH_3$? If we insert the starting concentrations of $N_2$, $H_2$, and $NH_3$ into the equilibrium-constant expression, we have

$$\frac{[NH_3]^2}{[N_2][H_2]^3} = \frac{(2.00)^2}{(1.00)(2.00)^3} = 0.500$$

We have seen (Sample Exercise 16.3) that at this temperature $K_c = 0.105$. Therefore, the quotient $[NH_3]^2/[N_2][H_2]^3$ will need to change from 0.500 to 0.105 to move the system toward equilibrium. This change can happen only if $[NH_3]$ decreases and $[N_2]$ and $[H_2]$ increase. Thus the reaction proceeds toward equilibrium with the formation of $N_2$ and $H_2$ from the $NH_3$; the reaction proceeds from right to left.

When we substitute reactant and product concentrations into the equilibrium-constant expression as we did above, the result is known as the

| Relationship | Direction |
|---|---|
| $Q > K$ | $\longleftarrow$ |
| $Q = K$ | Equilibrium |
| $Q < K$ | $\longrightarrow$ |

**reaction quotient** and is represented by the letter $Q$. The reaction quotient will equal the equilibrium constant, $K$, only if the concentrations are ones for the system at equilibrium: $Q = K$ only at equilibrium. We have seen that when the reaction quotient is larger than $K$, substances on the right side of the chemical equation will react to form substances on the left; the reaction moves from right to left in approaching equilibrium: if $Q > K$, the reaction moves right to left. Conversely, if $Q < K$, the reaction will move toward equilibrium with the formation of more products (from left to right). These relationships are summarized in Table 16.2.

## SAMPLE EXERCISE 16.8

At 448°C the equilibrium constant, $K_c$, for the reaction

$$H_2(g) + I_2(g) \rightleftharpoons 2HI(g)$$

is 50.5. Predict the direction in which the reaction will proceed to reach equilibrium at 448°C if we start with $2.0 \times 10^{-2}$ mol of HI, $1.0 \times 10^{-2}$ mol of $H_2$, and $3.0 \times 10^{-2}$ mol of $I_2$ in a 2.0-L container.

**Solution:** The starting concentrations are

$[HI] = 2.0 \times 10^{-2}$ mol/2.0 L $= 1.0 \times 10^{-2}$ $M$
$[H_2] = 1.0 \times 10^{-2}$ mol/2.0 L $= 5.0 \times 10^{-3}$ $M$
$[I_2] = 3.0 \times 10^{-2}$ mol/2.0 L $= 1.5 \times 10^{-2}$ $M$

The reaction quotient is

$$Q = \frac{[HI]^2}{[H_2][I_2]} = \frac{(1.0 \times 10^{-2})^2}{(5.0 \times 10^{-3})(1.5 \times 10^{-2})} = 1.3$$

Because $Q < K_c$, [HI] will need to increase and [$H_2$] and [$I_2$] decrease to reach equilibrium; the reaction will proceed from left to right.

## PRACTICE EXERCISE

At 1000 K the value of $K_c$ for the reaction $2SO_3(g) \rightleftharpoons 2SO_2(g) + O_2(g)$ is $4.12 \times 10^{-3}$. Calculate the value for $Q$ and predict the direction in which the reaction will proceed toward equilibrium if the initial concentrations of reactants are [$SO_3$] $= 2 \times 10^{-3}$ $M$; [$SO_2$] $= 5 \times 10^{-3}$ $M$; [$O_2$] $= 3 \times 10^{-2}$ $M$.

*Answer:* $Q = 0.2$; the reaction will proceed from right to left, forming $SO_3$

## Calculation of Equilibrium Concentrations

In many problem-solving situations involving equilibrium constants, we must often be content with incomplete information. For example, in determining the value of an equilibrium constant we may know only the initial concentrations of all species and some information about the equilibrium condition, such as the concentration of a single reactant or product. In many equilibrium calculations we know the value for $K$ and must use it to calculate some other property of the equilibrium system, such as total pressure or the concentration of a given reactant or product. The following sample exercises illustrate some of the problems encountered and the techniques employed in solving them.

## SAMPLE EXERCISE 16.9

What is the partial pressure of $NH_3$ that is in equilibrium with $N_2$ and $H_2$ at 500°C if the equilibrium partial pressure of $H_2$ is 0.733 atm and that of $N_2$ is 0.527 atm [$K_p = 1.45 \times 10^{-5}$ at 500°C for $N_2(g) + 3H_2(g) \rightleftharpoons 2NH_3(g)$]?

**Solution:**

$$N_2(g) \quad + \quad 3H_2(g) \quad \rightleftharpoons \quad 2NH_3(g)$$

Equil.     0.527 atm     0.733 atm          $x$

$$K_p = \frac{P^2_{NH_3}}{P_{N_2}P^3_{H_2}} = 1.45 \times 10^{-5}$$

$$K_p = \frac{x^2}{(0.527)(0.733)^3}$$

Rearrange to solve for $x$:

$$x^2 = K_p P_{N_2}P^3_{H_2} = (1.45 \times 10^{-5})(0.527)(0.733)^3$$
$$= 3.01 \times 10^{-6}$$
$$x = \sqrt{3.01 \times 10^{-6}} = 1.73 \times 10^{-3} \text{ atm} = P_{NH_3}$$

**PRACTICE EXERCISE**

At 500 K, the equilibrium constant $K_p$ for the reaction

$$PCl_5(g) \rightleftharpoons PCl_3(g) + Cl_2(g)$$

has the value 0.497. If the pressure of $PCl_5$ is 0.62 atm and that of $PCl_3$ is 0.15 atm in a particular equilibrium mixture, what is the equilibrium pressure of $Cl_2$?

*Answer:*  2.05 atm

## SAMPLE EXERCISE 16.10

A 1-L container is filled with 0.50 mol of HI at 448°C. The value of the equilibrium constant, $K_c$, for the reaction

$$H_2(g) + I_2(g) \rightleftharpoons 2HI(g)$$

at this temperature is 50.5. What are the concentrations of $H_2$, $I_2$, and HI in the vessel at equilibrium?

**Solution:**  In this case we are not given any of the equilibrium concentrations, only the starting concentrations: $[H_2] = [I_2] = 0$ and $[HI] = 0.50\ M$. The problem is related to the one we worked in Sample Exercise 16.4: We must use the balanced chemical equation to write an expression for equilibrium concentrations in terms of initial concentrations. It is useful to set up a table as in Sample Exercise 16.4. We can define $2x$ as the amount of HI that reacts forming $H_2$ and $I_2$. For each $2x$ HI that decomposes, $x$ $H_2$ and $x$ $I_2$ form

$$H_2(g) + I_2(g) \rightleftharpoons 2HI(g)$$

|  | | | |
|---|---|---|---|
| **Initial** | $0\ M$ | $0\ M$ | $0.50\ M$ |
| **Change** | $x\ M$ | $x\ M$ | $-2x\ M$ |
| **Equilibrium** | $x\ M$ | $x\ M$ | $(0.50 - 2x)\ M$ |

We can substitute the equilibrium concentrations into the equilibrium expression and solve for the single unknown, $x$:

$$K_c = \frac{[HI]^2}{[H_2][I_2]} = \frac{(0.50 - 2x)^2}{x^2} = 50.5$$

This equation is second order in $x$ (it contains $x^2$ as the highest power of $x$); such equations can always be solved by use of the quadratic formula (Appendix A.3). However, a quicker solution can be obtained in this particular case by taking the square root of both sides of the equation:

$$\frac{0.50 - 2x}{x} = \sqrt{50.5} = 7.11$$

Solving for $x$ yields

$$0.50 - 2x = 7.11x$$
$$0.50 = 2x + 7.11x = 9.11x$$
$$x = \frac{0.50}{9.11} = 0.055\ M$$

Thus the equilibrium concentrations are as follows:

$$[H_2] = x = 0.055\ M$$
$$[I_2] = x = 0.055\ M$$
$$[HI] = 0.50\ M - 0.11\ M = 0.39\ M$$

**PRACTICE EXERCISE**

For the equilibrium $PCl_5(g) \rightleftharpoons PCl_3(g) + Cl_2(g)$, the equilibrium constant $K_p$ has the value 0.497 at 500 K. A gas cylinder at 500 K is charged with $PCl_5(g)$ at an initial pressure of 2.20 atm. What are the equilibrium pressures of $PCl_5$, $PCl_3$, and $Cl_2$ at this temperature? (You will need to use the quadratic formula.)

*Answer:*  $P_{PCl_5} = 1.38$ atm; $P_{PCl_3} = P_{Cl_2} = 0.825$ atm

In working equilibrium problems you will find it helpful to proceed in a systematic way in approaching a solution. The following steps correspond to those used in the foregoing sample exercises:

1. Write the balanced chemical equation for the equilibrium reaction.
2. Make up a table of all reactants and products, and list whatever you know about initial concentrations and equilibrium concentrations.
3. Write the expression for the equilibrium constant, using the law of mass action.
4. Using the balanced equation and the given initial and equilibrium concentrations, either deduce the equilibrium concentrations of all reactants and products (if $K$ is to be calculated) or express these concentrations in terms of a single unknown.
5. If your goal is to determine the value of the equilibrium constant, you should have at this point all the concentration information you need to substitute into the expression for $K$ and solve for the numerical value.
6. If your goal is to solve for an equilibrium concentration, that concentration must be set up as a variable that relates to the equilibrium constant expression, as illustrated in Sample Exercises 16.9 and 16.10.

## 16.6 FACTORS AFFECTING EQUILIBRIUM: LE CHÂTELIER'S PRINCIPLE

In developing his process for making ammonia from $N_2$ and $H_2$, Haber sought to know the factors that might be varied to increase the yield of $NH_3$. Using the values of the equilibrium constant at various temperatures, he calculated the equilibrium amounts of $NH_3$ formed under a variety of conditions. The results of some of his calculations are shown in Table 16.3. Notice that the percent of $NH_3$ present at equilibrium decreases with increasing temperature and increases with increasing pressure. We can understand these effects in terms of a principle first put forward by Henri-Louis Le Châtelier (1850–1936), a French industrial chemist. **Le Châtelier's principle** can be stated as follows: *If a system at equilibrium is disturbed by a change in temperature, pressure, or the concentration of one of the components, the system will tend to shift its equilibrium position so as to counteract the effect of the disturbance.*

In this section we will use Le Châtelier's principle to make qualitative predictions about the response of a system at equilibrium to various changes in external conditions. We will consider three ways that a chemical equilibrium can be shifted: (1) adding or removing a reactant or product, (2) changing the pressure, and (3) changing the temperature.

**TABLE 16.3** Effect of temperature and total pressure on the percentage of ammonia present at equilibrium, beginning with a 3:1 molar $H_2/N_2$ mixture

| Temperature (°C) | Total pressure (atm) | | | |
| --- | --- | --- | --- | --- |
| | 200 | 300 | 400 | 500 |
| 400 | 38.7 | 47.8 | 54.9 | 60.6 |
| 450 | 27.4 | 35.9 | 42.9 | 48.8 |
| 500 | 18.9 | 26.0 | 32.2 | 37.8 |
| 600 | 8.8 | 12.9 | 16.9 | 20.8 |

### Change in Reactant or Product Concentrations

A system at equilibrium is in a dynamic state; the forward and reverse processes are occurring at equal rates, and the system is in a state of balance. An alteration in the conditions of the system may cause the state of balance to be disturbed. If this occurs, the equilibrium shifts until a new state of balance is attained. Le Châtelier's principle states that the shift will be in the direction that minimizes or reduces the effect of the change. Therefore, *if a chemical system is at equilibrium, and we add a substance (either a reactant or a product), the reaction will shift so as to reestablish equilibrium by consuming part of that added substance. Conversely, removal of a substance will result in the reaction moving in the direction that forms more of that substance.*

For example, addition of hydrogen to an equilibrium mixture of $H_2$, $N_2$, and $NH_3$ would cause the system to shift in such a way as to reduce the hydrogen pressure toward its original value. This can occur only if the equilibrium is shifted in the direction of forming more $NH_3$. At the same time, the quantity of $N_2$ would also be reduced slightly. This situation is illustrated in Figure 16.5. Addition of more $N_2$ to an equilibrium system would similarly cause a shift in the direction of forming more ammonia. On the other hand, Le Châtelier's principle tells us that if we add $NH_3$ to the system at equilibrium, the shift will be in such a direction as to reduce the $NH_3$ concentration toward its original value; that is, some of the added ammonia will decompose to form $N_2$ and $H_2$.

We can reach the same conclusions by considering the effect that adding or removing a substance has on the reaction quotient (Section 16.5). For example, removal of $NH_3$ from an equilibrium mixture gives

$$\frac{[NH_3]^2}{[N_2][H_2]^3} = Q < K$$

Because $Q < K$, the reaction shifts from left to right, forming more $NH_3$ and decreasing $[N_2]$ and $[H_2]$, to restore a new equilibrium that is still governed by $K$.

If the products of a reaction can be removed continuously, the reacting system can be continuously shifted to form more products. The yield of $NH_3$ in the Haber process can be increased dramatically by liquefying the $NH_3$; the liquid $NH_3$ is removed and the $N_2$ and $H_2$ are recycled to form more $NH_3$. The general way this is accomplished is shown in Figure

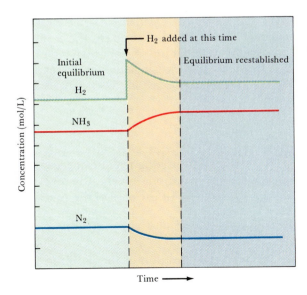

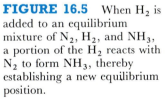

**FIGURE 16.5** When $H_2$ is added to an equilibrium mixture of $N_2$, $H_2$, and $NH_3$, a portion of the $H_2$ reacts with $N_2$ to form $NH_3$, thereby establishing a new equilibrium position.

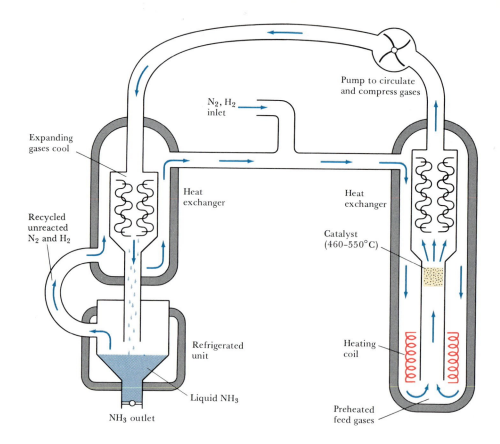

**FIGURE 16.6** Schematic diagram summarizing the industrial production of ammonia. Incoming $N_2$ and $H_2$ gases are heated to approximately 500°C and passed over a catalyst. The resultant gas mixture is allowed to expand and cool, causing $NH_3$ to liquefy. Unreacted $N_2$ and $H_2$ gases are recycled.

Pump to circulate and compress gases

$N_2$, $H_2$ inlet

Expanding gases cool

Heat exchanger

Heat exchanger

Recycled unreacted $N_2$ and $H_2$

Catalyst (460–550°C)

Refrigerated unit

Heating coil

Liquid $NH_3$

$NH_3$ outlet

Preheated feed gases

16.6. If a reaction is operated so that equilibrium cannot be achieved because of the escape of products, or if the equilibrium constant is very large, the reaction will proceed essentially to completion. In such instances the chemical equation for the reaction is usually given with a single arrow: reactants $\longrightarrow$ products.

### Effect of Pressure and Volume Changes

*If a system is at equilibrium and the total pressure is increased by application of an external pressure, the system responds by a shift in equilibrium in the direction that reduces the pressure.* If the system is gaseous in whole or part, the equilibrium shifts in the direction that reduces the total number of moles of gas. Conversely, decreasing the pressure by increasing the volume causes a shift in the direction that produces more gas molecules. For the reaction

$$N_2(g) + 3H_2(g) \rightleftharpoons 2NH_3(g)$$

there is 2 mol of gas on the right side of the chemical equation ($2NH_3$) and 4 mol of gas on the left ($1N_2 + 3H_2$). Consequently, an increase in pressure (decrease in volume) leads to the formation of $NH_3$; the reaction shifts toward the side with fewer gas molecules. In the case of the reaction

$$H_2(g) + I_2(g) \rightleftharpoons 2HI(g)$$

changing the pressure will not influence the position of the equilibrium.

In this case the number of gaseous product molecules is the same as the gaseous reactant molecules.

It is important to keep in mind that pressure-volume changes do not change the value of $K$, as long as the temperature remains constant. Rather, they change the concentrations of the gaseous substances. In Sample Exercise 16.3, we calculated $K_c$ for an equilibrium mixture at 472°C that contained $[H_2] = 0.1207\ M$, $[N_2] = 0.0402\ M$, and $[NH_3] = 0.00272\ M$. The value of $K_c$ is 0.105. Consider what happens if we suddenly reduce the volume of the system by half. If there were no shift in equilibrium, this volume change would cause the concentrations of all substances to double, giving $[H_2] = 0.2414\ M$, $[N_2] = 0.0804\ M$, and $[NH_3] = 0.00544\ M$. The reaction quotient would then no longer equal the equilibrium constant:

$$Q = \frac{[NH_3]^2}{[N_2][H_2]^3} = \frac{(0.00544)^2}{(0.0804)(0.2414)^3} = 2.62 \times 10^{-2}$$

Because $Q < K_c$, the system is no longer at equilibrium. Equilibrium will be reestablished by increasing $[NH_3]$ and decreasing $[N_2]$ and $[H_2]$ until $Q = K_c = 0.105$. Therefore, the equilibrium shifts to the right as Le Châtelier's principle predicts.

It is possible to change the total pressure of the system without changing its volume. For example, pressure increases if additional amounts of any of the reacting components are added to the system. We have already seen how to deal with a change in concentration of a reactant or product. The total pressure within the reaction vessel might also be increased by addition of a gas that is not involved in the equilibrium. For example, argon might be added to the ammonia equilibrium system. This addition would not alter the partial pressures of any of the reacting components and therefore would not cause a shift in equilibrium.

## Effect of Temperature Changes

Changes in concentrations or total pressure can cause shifts in equilibrium without changing the equilibrium constant. In contrast, almost every equilibrium constant changes in value with change in temperature. Consider the decomposition of carbon dioxide into carbon monoxide and oxygen:

$$2CO_2(g) \rightleftharpoons 2CO(g) + O_2(g) \qquad [16.11]$$

From a knowledge of the heats of formation of $CO_2(g)$ and $CO(g)$, we can conclude that the forward reaction is highly endothermic; using the values of $\Delta H_f^\circ$ from Appendix D, $\Delta H^\circ$ for the overall reaction is calculated to be 566 kJ. At room temperature and thereabouts, $CO_2$ has no observable tendency to dissociate into CO and $O_2$. However, at high temperatures the equilibrium shifts to the right, as shown in Table 16.4. Clearly, the equilibrium constant for reaction Equation 16.11 is very dependent on temperature and increases with increasing temperature.

The equilibrium constants for exothermic reactions—that is, those in which heat is evolved—decrease with increase in temperature. By con-

**TABLE 16.4** Percent dissociation of $CO_2$ into CO and $O_2$ as a function of temperature

| Temperature (K) | Percent dissociation |
| --- | --- |
| 1500 | 0.048 |
| 2000 | 2.05 |
| 2500 | 17.6 |
| 3000 | 54.8 |

trast, equilibrium constants for endothermic reactions increase with increase in temperature.

We can deduce the rules for the temperature dependence of the equilibrium constant by applying Le Châtelier's principle. When heat is added to a system by increasing the temperature, the equilibrium should shift in such a direction as to undo partially the effect of the added heat. It shifts, therefore, in the direction in which heat is absorbed. If a reaction is exothermic in the forward direction, it must be endothermic in the reverse direction. Thus, when heat is added to an equilibrium system that is exothermic in the forward direction, the equilibrium shifts in the reverse direction, in the direction of reactants. In summary, the rule is that *when heat is added at constant pressure to an equilibrium system, the equilibrium shifts in the direction that absorbs heat.* Conversely, if heat is removed from an equilibrium system, the equilibrium shifts in the direction that evolves heat.

## SAMPLE EXERCISE 16.11

Consider the following reaction:

$$N_2O_4(g) \rightleftharpoons 2NO_2(g) \qquad \Delta H° = 58.0 \text{ kJ}$$

In what direction will the equilibrium shift when each of the following changes is made to a system at equilibrium: **(a)** add $N_2O_4$; **(b)** remove $NO_2$; **(c)** increase pressure; **(d)** increase volume; **(e)** decrease temperature?

**Solution:** Le Châtelier's principle can be applied to determine the effects of each of these changes.

**(a)** The system will adjust so as to decrease the concentration of the added $N_2O_4$; the reaction consequently shifts toward the formation of more products (the right side of the equation).

**(b)** The system will adjust to this change by forming more $NO_2$; the equilibrium shifts toward the product side of the equation, to the right.

**(c)** The system will establish a new equilibrium that has a smaller volume (fewer gas molecules); consequently, the reaction shifts to the left.

**(d)** The system will shift in the direction that oc-

cupies a larger volume (more gas molecules); it moves to the right.

**(e)** The system will adjust to a new equilibrium position by shifting in the direction that produces heat. The reaction is endothermic in the forward direction (left to right). It therefore shifts to the left, with the formation of more $N_2O_4$ in order to produce heat. Note that only this last change affects the numerical value of the equilibrium constant, $K$.

## PRACTICE EXERCISE

For the reaction

$$PCl_5(g) \rightleftharpoons PCl_3(g) + Cl_2(g) \qquad \Delta H° = 87.9 \text{ kJ}$$

in what direction will equilibrium shift when **(a)** $Cl_2(g)$ is added; **(b)** temperature is increased; **(c)** the volume of the reaction system is decreased; **(d)** $PCl_5(g)$ is added?

*Answer:* **(a)** left: additional $PCl_5(g)$ is formed; **(b)** right: increased dissociation of $PCl_5(g)$ occurs; **(c)** left: more $PCl_5(g)$ is formed; **(d)** right: additional $PCl_3(g)$ and $Cl_2(g)$ is formed.

## SAMPLE EXERCISE 16.12

Using the standard heat of formation data in Appendix D, determine the enthalpy change for the reaction:

$$N_2(g) + 3H_2(g) \rightleftharpoons 2NH_3(g)$$

From this determine how the equilibrium constant for the reaction should change with temperature.

**Solution:** Recall that the standard enthalpy change for a reaction is given by the standard molar enthalpies of formation of the products, each multi-

plied by its coefficient in the balanced chemical equation, less the same quantities for the reactants. $\Delta H_f°$ for $NH_3(g)$ at 25°C is $-46.19$ kJ/mol. The $\Delta H_f°$ values for $H_2(g)$ and $N_2(g)$ are zero by definition, because the enthalpies of formation of the elements in their normal states at 25°C are defined as zero (Section 5.5). Because 2 mol of $NH_3$ is formed, the total enthalpy change is

$$2 \text{ mol}(-46.19 \text{ kJ/mol}) - 0 = -92.38 \text{ kJ}$$

The reaction in the forward direction is exothermic. An increase in temperature causes the reaction to shift in the *reverse* direction—in our example, in the direction of less $NH_3$ and more $N_2$ and $H_2$. This is what occurs, as reflected in the values for $K_p$ presented in Table 16.5. Notice that $K_p$ changes very markedly with change in temperature and that it is larger at lower temperatures. This is a matter of great practical importance. To form ammonia at a reasonable rate requires higher temperatures. Yet at higher temperatures, the equilibrium constant is smaller, so the percentage conversion to ammonia is smaller. To compensate for this, higher pressures are needed, because high pressure favors ammonia formation.

### PRACTICE EXERCISE

Using the thermodynamic data in Appendix D, determine the enthalpy change for the reaction

$$2POCl_3(g) \rightleftharpoons 2PCl_3(g) + O_2(g)$$

From this determine how the equilibrium constant for the reaction should change with temperature.
***Answer:*** $\Delta H° = 620$ kJ; the equilibrium constant for the reaction will increase with increasing temperature.

It is very important to be clear on the distinction between a system at equilibrium and the rate at which the system approaches equilibrium in the first place. Imagine that we have a system initially containing only reactant molecules and no products. As the system begins to change, the only reaction occurring is formation of products. As products accumulate, however, the reverse reaction also begins to occur. In many systems, this reverse reaction is so slow that it is completely unimportant. The system then just keeps on changing until essentially all the reactants are converted to products. Reactions of this sort are said to proceed to completion. Even though the reverse reaction—conversion of products to reactants—is possible in principle, it is not observed.

In many other reactions, on the other hand, the reverse reaction occurs more rapidly. After a time, reactant molecules are being formed just as rapidly as they are themselves reacting to form products. When the rates of the two opposing processes are equal, the overall rate of change in the system is zero, and we have chemical equilibrium. A true chemical equilibrium always involves a balancing of equal and opposite rate processes. This is true regardless of the particular mechanism or pathway by which the reaction proceeds. Figure 16.7 shows an energy profile for the single-step, bimolecular reaction between reactants A and B and products C and D:

## 16.7 THE RELATIONSHIP BETWEEN CHEMICAL EQUILIBRIUM AND CHEMICAL KINETICS

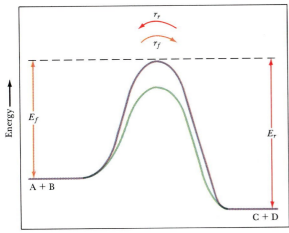

Progress of reaction

**FIGURE 16.7** Schematic illustration of chemical equilibrium in the reaction $A + B \rightleftharpoons C + D$. When equilibrium is attained, the rate of the forward reaction, $r_f$, equals the rate of the reverse reaction, $r_r$. The green line refers to the energy profile for a catalyzed reaction in which the activation energy is lowered. The rates of forward and reverse reactions in the catalyzed reaction are increased to the same degree.

$$A + B \;\rightleftharpoons\; C + D \qquad\qquad [16.12]$$

At equilibrium, the rate at which reactants cross the energy barrier to form products equals the rate at which products cross the energy barrier in the reverse direction to form reactants. Because we have chosen the situation where the reaction occurs in a single, bimolecular step in each direction, the rate law for each direction is second order. Thus

$$r_f = \text{rate of forward reaction} = k_f[\text{A}][\text{B}]$$

$$r_r = \text{rate of reverse reaction} = k_r[\text{C}][\text{D}]$$

(Remember that we use a lowercase $k$ to represent rate constants and a capital $K$ to represent equilibrium constants.) At equilibrium these two rate processes must be equal:

$$k_f[\text{A}][\text{B}] = k_r[\text{C}][\text{D}]$$

If we rearrange this equation we obtain

$$\frac{k_f}{k_r} = \frac{[\text{C}][\text{D}]}{[\text{A}][\text{B}]} = K \qquad\qquad [16.13]$$

Thus the equilibrium constant is just the ratio of rate constants for the forward and reverse reactions.* Recall that the rate of a reaction decreases as the height of the energy barrier increases (Section 15.4). The barrier for the forward reaction ($E_f$) in Figure 16.7 is lower than for the reverse reaction ($E_r$). Thus $k_f$ must be larger than $k_r$, so that $K$ should be a large number. This is in keeping with the fact that the energies of the products are lower than the energies of the reactants.

**The Effects of Catalysts**

Suppose that we add a catalyst to the reaction system described in Equation 16.12 so that the energy barrier to reaction is lowered, as shown by the dashed line in Figure 16.7. The rates of *both* the forward and reverse reactions are increased by the presence of a catalyst. In fact, the two rate constants are affected to precisely the same degree. In other words, it is impossible for a catalyst to lower the barrier for just the forward reaction and not the reverse reaction. Because the forward and reverse reaction rate constants are affected to exactly the same degree, their ratio is unchanged. We therefore have the rule that *a catalyst may change the rate of approach to equilibrium*, but it does not change the value of the equilibrium constant.

The rate at which a reaction approaches equilibrium is an important practical consideration. As an example, let us again consider the synthesis of ammonia from $N_2$ and $H_2$. In designing a process for ammonia synthesis, Haber had to deal with a rather serious problem. He wished to

---

* The relationship between rate constants and the equilibrium constant is somewhat more complicated if the reaction occurs by a multistep mechanism. However, even in that case the equilibrium constant can be related to a ratio of rate constants. Regardless of the mechanism, the equilibrium-constant expression can be written directly from the balanced equation for the overall reaction. By contrast, rate laws must be determined by experiment.

The formation of NO from $N_2$ and $O_2$ provides another interesting example of the practical importance of changes in equilibrium constant and reaction rate with temperature. The balanced equation and the standard enthalpy change for the reaction are

$$\tfrac{1}{2}N_2(g) + \tfrac{1}{2}O_2(g) \rightleftharpoons NO(g) \qquad \Delta H^\circ = 90.4 \text{ kJ} \qquad [16.14]$$

The reaction is endothermic, that is, heat is absorbed when NO is formed from the elements. By applying Le Châtelier's principle, we deduce that an increase in temperature will shift the equilibrium in the direction of more NO. The equilibrium constant, $K_c$, for formation of 1 mol of NO from the elements at 300 K is only about $10^{-15}$. On the other hand, at a much higher temperature, about 2400 K, the equilibrium constant is much larger, about 0.05. The manner in which $K_c$ for reaction Equation 16.14 varies with temperature is shown in Figure 16.8.

This graph helps to explain why NO is a pollution problem. In the cylinder of a modern high-compression auto engine, the temperatures during the fuel-burning part of the cycle may be on the order of 2400 K. Also, there is a fairly large excess of air in the cylinder. These conditions provide an opportunity for the formation of some NO. After the combustion, however, the gases are quickly cooled. As the temperature drops, the equilibrium of Equation 16.14 shifts strongly to the left, that is, in the direction of $N_2$ and $O_2$. But the lower temperatures also mean that the rate of the reaction is decreased. The NO formed at high temperatures is essentially "frozen" in that form as the gas cools.

The gases exhausting from the cylinder are still quite hot, perhaps 1200 K. At this temperature, as shown in Figure 16.8, the equilibrium constant for formation of NO is much smaller. However, the rate of conversion of NO to $N_2$ and $O_2$ is too slow to permit much loss of NO before the gases are cooled still further. Getting the NO out of the exhaust gases, described in Chapter 14, depends on finding a catalyst that will work at the temperatures of the exhaust gases and that will cause conversion of NO into something harmless. If a catalyst could be found which would convert the NO back into $N_2$ and $O_2$, the equilibrium would be sufficiently favorable. It has not proved possible to find a catalyst capable of withstanding the grueling conditions found in automobile exhaust systems that can catalyze the conversion of NO into $N_2$ and $O_2$. Instead, the catalysts used are designed to catalyze the reaction of NO with $H_2$ or CO.

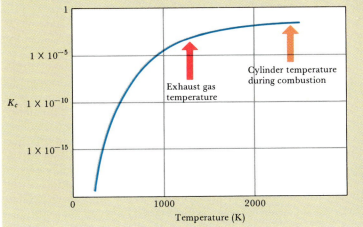

**FIGURE 16.8** Variation of the equilibrium constant for the reaction $\tfrac{1}{2}N_2(g) + \tfrac{1}{2}O_2(g) \rightleftharpoons NO(g)$ as a function of temperature. It is necessary to use a log scale for $K_c$ because the values of $K_c$ vary over such a huge range.

synthesize ammonia at the lowest temperature possible, consistent with a reasonable reaction rate. But in the absence of a catalyst, hydrogen and nitrogen do not react with one another at a significant rate either at room temperature or even at much higher temperatures. On the other hand, Haber had to cope with a rapid decrease in equilibrium constant with in-

**TABLE 16.5** Variation in $K_p$ for the equilibrium $N_2 + 3H_2 \rightleftharpoons 2NH_3$ as a function of temperature

| Tempera-ture (°C) | $K_p$ |
|---|---|
| 300 | $4.34 \times 10^{-3}$ |
| 400 | $1.64 \times 10^{-4}$ |
| 450 | $4.51 \times 10^{-5}$ |
| 500 | $1.45 \times 10^{-5}$ |
| 550 | $5.38 \times 10^{-6}$ |
| 600 | $2.25 \times 10^{-6}$ |

creasing temperature, as shown in Table 16.5. At temperatures sufficiently high to give a satisfactory reaction rate, the amount of ammonia formed was too small. The solution to this dilemma was to develop a catalyst that would produce a reasonably rapid approach to equilibrium, at a sufficiently low temperature so that the equilibrium constant was still reasonably large. The development of a catalyst thus became the focus of Haber's research efforts.

After trying different substances to see which would be most effective, Haber finally settled on iron mixed with metal oxides. Variants of the original catalyst formulations are still used. These catalysts make it possible to obtain a reasonably rapid approach to equilibrium at temperatures around 400 to 500°C, and with gas pressures of 200 to 600 atm. The high pressures are needed to obtain a satisfactory degree of conversion at equilibrium. You can see from Table 16.3 that if an improved catalyst could be found—one that would lead to sufficiently rapid reaction at temperatures lower than 400 to 500°C—it would be possible to obtain the same degree of equilibrium conversion at much lower pressures. This would result in a great savings in the cost of equipment for ammonia synthesis. In view of the growing need for nitrogen as fertilizer, the fixation of nitrogen is a process of ever-increasing importance, worthy of additional research effort.

# *FOR REVIEW*

## SUMMARY

If the reactants and products of a reaction are kept in contact, a chemical reaction can achieve a state of dynamic balance in which the forward and reverse reactions are occurring at equal rates. This condition is known as **chemical equilibrium**. A system at equilibrium does not change with time. For such a system we may define an **equilibrium constant**, **K**. The equilibrium constant is equal to the product of concentrations of all reaction products, each raised to the power of its coefficient in the balanced equation, divided by the product of concentrations of reactants, each raised to the power of its coefficient in the balanced chemical equation. The equilibrium constant changes with temperature but is not affected by changes in relative concentrations of any reacting substances or by pressure or the presence of catalysts. In **heterogeneous equilibria**, the concentrations of pure solids or liquids are absent from the equilibrium-constant expression.

If concentrations are expressed in molarity, we label the equilibrium constant $K_c$; if the concentration units are atmospheres, we use $K_p$. $K_c$ and $K_p$ are related by the following equation: $K_p = K_c(RT)^{\Delta n}$. A

large value for $K_c$ or $K_p$ indicates that the equilibrium mixture contains more products than reactants. A small value for the equilibrium constant means the equilibrium lies toward the reactant side.

The **reaction quotient**, **Q**, is found by substituting reactant and product concentrations into the equilibrium-constant expression. If the system is at equilibrium, $Q = K$. However, if $Q \neq K$, nonequilibrium conditions apply; if $Q < K$, the reaction will move toward equilibrium by forming more products (the reaction moves from left to right); if $Q > K$, the reaction will proceed from right to left. Knowledge of the value of $K_c$ or $K_p$ permits calculation of the equilibrium concentrations of reactants and products.

**Le Châtelier's principle** indicates that if we disturb a system that is at equilibrium, the equilibrium will shift to minimize the disturbing influence. The effects of adding (or removing) reactants or products and of changing pressure, volume, or temperature can be predicted using this principle. Catalysts affect the speed at which equilibrium is reached but do not affect $K$ (do not affect the position of equilibrium).

# LEARNING GOALS

Having read and studied this chapter, you should be able to:

1. Write the equilibrium-constant expression for a balanced chemical equation, whether heterogeneous or homogeneous.
2. Numerically evaluate $K_c$ (or $K_p$) from a knowledge of the equilibrium concentrations (or pressures) of reactants or products, or from the initial concentrations and the equilibrium concentration of at least one substance.
3. Interconvert $K_c$ and $K_p$.
4. Calculate the reaction quotient, $Q$, and by comparison with the value of $K_c$ or $K_p$ determine whether a reaction is at equilibrium. If it is not at equilibrium, you should be able to predict in which direction it will shift to reach equilibrium.

5. Use the equilibrium constant to calculate equilibrium concentrations. (You will also need to know either the equilibrium concentrations of all but one substance or the initial concentrations together with the equilibrium concentration of one substance.)

6. Explain how the relative equilibrium quantities of reactants and products are shifted by changes in temperature, pressure, or the concentrations of substances in the equilibrium reaction.

7. Explain how the change in equilibrium constant with change in temperature is related to the enthalpy change in the reaction.

8. Describe the effect of a catalyst on a system as it approaches equilibrium.

# KEY TERMS

Among the more important terms and expressions used for the first time in this chapter are the following:

**Chemical equilibrium** (Section 16.2) is a state of a chemical system in which the rate of formation of products equals the rate of formation of reactants from products.

The **Haber process** (Section 16.1) refers to the catalyst system and conditions of temperature and pressure developed by Fritz Haber and co-workers for the formation of $NH_3$ from $H_2$ and $N_2$.

**Heterogeneous equilibrium** (Section 16.4) refers to the equilibrium state established between substances in two or more different phases, for example, between a gas and solid or between a solid and liquid.

**Homogeneous equilibrium** (Section 16.4) is the state of equilibrium established between reactant and product substances all in the same phase, for example,

all gases, or all dissolved in solution.

The **law of mass action** (Section 16.2) provides the rules according to which the equilibrium constant is expressed in terms of the concentrations of reactants and products, in accordance with the balanced chemical equation for the reaction.

**Le Châtelier's principle** (Section 16.6) tells us that when we bring to bear some disturbing influence on a system at chemical equilibrium, the relative concentrations of reactants and products will shift so as to undo partially the effects of the disturbance.

The **reaction quotient**, $Q$ (Section 16.5), is the value that is obtained when concentrations of reactants and products are inserted into the equilibrium-constant expression. If the concentrations are equilibrium concentrations, $Q = K$; otherwise, $Q \neq K$.

# EXERCISES

## Equilibrium-Constant Expressions

**16.1** Write the equilibrium-constant expression for each of the following reactions. In each case indicate whether the reaction is homogeneous or heterogeneous.
   (a) $2SO_3(g) \rightleftharpoons 2SO_2(g) + O_2(g)$
   (b) $NH_4NO_2(s) \rightleftharpoons N_2(g) + 2H_2O(g)$
   (c) $3Fe(s) + 4H_2O(g) \rightleftharpoons Fe_3O_4(s) + 4H_2(g)$
   (d) $SiHCl_3(g) + 3H_2O(g) \rightleftharpoons$
   $\qquad\qquad\qquad SiH(OH)_3(s) + 3HCl(g)$
   (e) $4NH_3(g) + 3O_2(g) \rightleftharpoons 2N_2(g) + 6H_2O(g)$

**16.2** Write the equilibrium-constant expression for each of the following reactions. In each case indicate whether

the reaction is homogeneous or heterogeneous.
   (a) $2H_2(g) + O_2(g) \rightleftharpoons 2H_2O(g)$
   (b) $Fe_3O_4(s) + H_2(g) \rightleftharpoons 3FeO(s) + H_2O(g)$
   (c) $4HCl(g) + O_2(g) \rightleftharpoons 2H_2O(l) + 2Cl_2(g)$
   (d) $PbO(s) + CO_2(g) \rightleftharpoons PbCO_3(s)$
   (e) $3O_2(g) \rightleftharpoons 2O_3(g)$

**16.3** What observations regarding chemical systems can you cite to support the idea that chemical equilibrium is a dynamic rather than a static state of the system?

**16.4** How do the data of Table 16.1 support the hypothesis that the equilibrium constant for Equation 16.4 is a constant at $100°C$?

## Calculation of $K_c$ and $K_p$

**16.5** A sample of chlorine gas is placed in a vessel and heated to 1400 K. The chlorine dissociates: $Cl_2(g) \rightleftharpoons 2Cl(g)$. When equilibrium is reached at this temperature it is found that $P_{Cl_2} = 1.00$ atm and $P_{Cl} = 2.97 \times 10^{-2}$ atm. What is the $K_p$ value at 1400 K?

**16.6** Gaseous hydrogen iodide is introduced into a container at 425°C, where it partly decomposes to hydrogen and iodine: $2HI(g) \rightleftharpoons H_2(g) + I_2(g)$. At equilibrium the resultant mixture is analyzed, and it is found that $[H_2] = 4.79 \times 10^{-4}$ $M$, $[I_2] = 4.79 \times 10^{-4}$ $M$, and $[HI] = 3.53 \times 10^{-3}$ $M$. What is the value of $K_c$ at this temperature?

**16.7** The equilibrium constant for the reaction

$$SO_2(g) + \tfrac{1}{2}O_2(g) \rightleftharpoons SO_3(g)$$

is $K_c = 20.4$ at 700°C. **(a)** What is the value of $K_c$ for

$$SO_3(g) \rightleftharpoons SO_2(g) + \tfrac{1}{2}O_2(g)$$

**(b)** What is the value of $K_c$ for

$$2SO_2(g) + O_2(g) \rightleftharpoons 2SO_3(g)$$

**(c)** What is the value of $K_p$ for

$$2SO_2(g) + O_2(g) \rightleftharpoons 2SO_3(g)$$

**16.8** The equilibrium constant for the reaction

$$2NO(g) + O_2(g) \rightleftharpoons 2NO_2(g)$$

is $K_p = 1.48 \times 10^4$ at 184°C. **(a)** What is the value of $K_p$ for

$$2NO_2(g) \rightleftharpoons 2NO(g) + O_2(g)$$

**(b)** What is the value of $K_p$ for

$$NO(g) + \tfrac{1}{2}O_2(g) \rightleftharpoons NO_2(g)$$

**(c)** What is the value of $K_c$ for

$$2NO(g) + O_2(g) \rightleftharpoons 2NO_2(g)$$

**16.9** At temperatures near 800°C, steam passed over hot coke (a form of carbon obtained from coal) reacts to form CO and $H_2$:

$$C(s) + H_2O(g) \rightleftharpoons CO(g) + H_2(g)$$

The mixture of gases that results is an important industrial fuel called water gas. When equilibrium is achieved at 800°C, $[H_2] = 4.0 \times 10^{-2}$ $M$, $[CO] = 4.0 \times 10^{-2}$ $M$, and $[H_2O] = 1.0 \times 10^{-2}$ $M$. Calculate $K_c$ and $K_p$ at this temperature.

**16.10** **(a)** At 700 K, $K_c = 0.11$ for the following reaction:

$$CO_2(g) + H_2(g) \rightleftharpoons CO(g) + H_2O(g)$$

What is the value of $K_p$? **(b)** At 1000 K, $K_c = 278$ for the reaction

$$2SO_2(g) + O_2(g) \rightleftharpoons 2SO_3(g)$$

What is the value of $K_p$? **(c)** For the equilibrium

$$C(s) + CO_2(g) \rightleftharpoons 2CO(g)$$

$K_p = 167.5$ at 1000°C. What is the value of $K_c$?

**16.11** A mixture of 0.100 mol of NO, 0.050 mol of $H_2$, and 0.100 mol of $H_2O$ is placed in a 1.00-L vessel. The following equilibrium is established:

$$2NO(g) + 2H_2(g) \rightleftharpoons N_2(g) + 2H_2O(g)$$

At equilibrium, $[NO] = 0.062$ $M$. **(a)** Calculate the equilibrium concentrations of $H_2$, $N_2$, and $H_2O$. **(b)** Calculate $K_c$.

**16.12** A mixture of 1.000 mol of $SO_2$ and 1.000 mol of $O_2$ is placed in a 1.000-L vessel and kept at 1000 K until equilibrium is reached. At equilibrium the vessel is found to contain 0.828 mol of $SO_3$. Assume the following reaction:

$$2SO_2(g) + O_2(g) \rightleftharpoons 2SO_3(g)$$

**(a)** Calculate the equilibrium concentrations of $SO_2$ and $O_2$. **(b)** Calculate $K_c$.

**16.13** A sample of nitrosyl bromide, NOBr, decomposes according to the following equation:

$$2NOBr(g) \rightleftharpoons 2NO(g) + Br_2(g)$$

An equilibrium mixture in a 5.00-L vessel at 100°C contains 3.22 g of NOBr, 3.08 g of NO, and 4.19 g of $Br_2$. **(a)** Calculate $K_c$. **(b)** Calculate $K_p$. **(c)** What is the total pressure exerted by the mixture of gases?

**16.14** A mixture of 1.374 g of $H_2$ and 70.31 g of $Br_2$ is heated together in a 2.00-L vessel at 700 K. These substances react as follows:

$$H_2(g) + Br_2(g) \rightleftharpoons 2HBr(g)$$

At equilibrium, the vessel is found to contain 0.566 g of $H_2$. What is the numerical value for $K_c$ at 700 K?

## Reaction Quotient

**16.15** As shown in Table 16.5, $K_p$ for the equilibrium

$$N_2(g) + 3H_2(g) \rightleftharpoons 2NH_3(g)$$

is $4.51 \times 10^{-5}$ at 450°C. Each of the mixtures listed below may or may not be at equilibrium at 450°C. Indicate in each case whether the mixture is at equilibrium; if it is not at equilibrium, indicate the direction (toward product or toward reactants) in which the mixture must shift to achieve equilibrium: **(a)** 100 atm $NH_3$, 30 atm $N_2$, 500 atm $H_2$; **(b)** 30 atm $NH_3$, 600 atm $H_2$, no $N_2$; **(c)** 26 atm $NH_3$, 42 atm $H_2$, 202 atm $N_2$; **(d)** 100 atm $NH_3$, 60 atm $H_2$, 5 atm $N_2$.

**16.16** At 122°C, the equilibrium constant for decomposition of nitryl chloride, $2NO_2Cl(g) \rightleftharpoons 2NO_2(g) + Cl_2(g)$, has the value $K_c = 0.558$. Are the following mixtures at 122°C at equilibrium? If they are not at equilibrium, indicate the direction the reaction must proceed to reach equilibrium. **(a)** $[NO_2Cl] = 0.0120$ $M$, $[NO_2] = 0.0344$ $M$, $[Cl_2] = 0.00452$ $M$; **(b)** $[NO_2Cl] = 0.130$ $M$, $[NO_2] = 0.0280$ $M$, $[Cl_2] = 0.0260$ $M$; **(c)** $[NO_2Cl] = 0.00127$ $M$, $[NO_2] = 0.00162$ $M$, $[Cl_2] = 0.343$ $M$.

## Equilibrium Concentrations

**16.17** At 1495°C, $K_c$ for the equilibrium

$$H_2(g) + Br_2(g) \rightleftharpoons 2HBr(g)$$

is $3.5 \times 10^4$. If equilibrium concentrations at this temperature are 0.050 $M$ $H_2$ and 0.010 $M$ $Br_2$, what is the concentration of HBr?

**16.18** The equilibrium constant, $K_c$, is $4.1 \times 10^{-4}$ at 2000°C for the reaction

$$N_2(g) + O_2(g) \rightleftharpoons 2NO(g)$$

If 1.4 g of $N_2$ and 0.015 g of NO are in equilibrium with $O_2$ in a 0.500-L vessel at this temperature, how many grams of $O_2$ are present?

**16.19** For the equilibrium

$$C(s) + CO_2(g) \rightleftharpoons 2CO(g)$$

$K_p = 167.5$ at 1000°C. What is the partial pressure of $CO_2$ that is in equilibrium with CO whose partial pressure is 0.500 atm?

**16.20** For the equilibrium

$$2NOBr(g) \rightleftharpoons 2NO(g) + Br_2(g)$$

$K_p = 0.416$ at 373 K. If the pressures of NOBr$(g)$ and NO$(g)$ are equal, what is the equilibrium pressure of $Br_2(g)$?

**16.21** The equilibrium constant for the reaction

$$I_2(g) + Br_2(g) \rightleftharpoons 2IBr(g)$$

has a value of 280 at 150°C. Suppose that a quantity of IBr is placed in a closed reaction vessel and the system allowed to come to equilibrium. When equilibrium is attained, the pressure of IBr is 0.20 atm. What are the pressures of $I_2(g)$ and $Br_2(g)$ at this point?

**16.22** The following reaction can occur when $N_2$ and $O_2$ come in contact at high temperatures:

$$N_2(g) + O_2(g) \rightleftharpoons 2NO(g)$$

At 2400 K, the equilibrium constant, $K_c$, for this reaction is $2.5 \times 10^{-3}$. What are the equilibrium concentrations of $N_2$ and $O_2$ if equilibrium at 2400 K is reached beginning with pure NO and ending with an NO concentration of $4.30 \times 10^{-3}$ $M$?

**16.23** At 21.8°C, the equilibrium constant, $K_c$, is $1.2 \times 10^{-4}$ for the following reaction:

$$NH_4HS(s) \rightleftharpoons NH_3(g) + H_2S(g)$$

Calculate the equilibrium concentrations of $NH_3$ and $H_2S$ if a sample of solid $NH_4HS$ is placed in a closed vessel and allowed to decompose until equilibrium is reached at 21.8°C.

**16.24** At 80°C, the equilibrium constant $K_p$ is 1.57 for the following reaction:

$$PH_3BCl_3(s) \rightleftharpoons PH_3(g) + BCl_3(g)$$

**(a)** Calculate the equilibrium pressures of $PH_3(g)$ and $BCl_3(g)$ if a sample of $PH_3BCl_3$ is placed in a closed vessel at 80°C and allowed to decompose until equilibrium is attained. **(b)** What is the minimum amount of $PH_3BCl_3$ that must be placed in a 0.500-L vessel at 80°C if equilibrium is to be attained?

## Le Châtelier's Principle

**16.25** Consider the following equilibrium system:

$$C(s) + CO_2(g) \rightleftharpoons 2CO(g) \qquad \Delta H° = 119.8 \text{ kJ}$$

If the reaction is at equilibrium, what would be the effect of **(a)** adding $CO_2(g)$; **(b)** adding $C(s)$; **(c)** adding heat; **(d)** increasing the pressure on the system by decreasing the volume; **(e)** adding a catalyst; **(f)** removing CO$(g)$?

**16.26** In the reaction

$$6CO_2(g) + 6H_2O(l) \rightleftharpoons C_6H_{12}O_6(s) + 6O_2(g)$$
$$\Delta H° = 2816 \text{ kJ}$$

how is the equilibrium yield of $C_6H_{12}O_6$ affected by **(a)** increasing $P_{CO_2}$; **(b)** increasing temperature; **(c)** removing $CO_2$; **(d)** increasing the total pressure; **(e)** removing part of the $C_6H_{12}O_6$; **(f)** adding a catalyst?

**16.27** How does each of the following changes affect the numerical value of the equilibrium constant for an exothermic reaction: **(a)** removal of a reactant or product; **(b)** increase in the total pressure; **(c)** decrease in the temperature; **(d)** addition of a catalyst?

**16.28** Indicate whether each of the following statements is true or false. If the statement is false, correct it so that it is true. **(a)** For a gas-phase reaction in which there is no change in the number of moles of gas in going from reactants to products ($\Delta n = 0$), increasing the total pressure does not markedly affect the equilibrium. **(b)** The value of the equilibrium constant for a reaction is affected only by change in temperature or by the addition of a catalyst. **(c)** The position of equilibrium for an exothermic reaction is shifted to the left by an increase in temperature. **(d)** Increasing the concentration of a reactant in an equilibrium always results in a decrease in the equilibrium concentrations of all other reactants.

## Chemical Equilibrium and Kinetics

**16.29** The hypothetical reaction $A + B \rightleftharpoons C$ occurs in the forward direction in a single step. The energy profile for the reaction is shown in Figure 16.9. **(a)** Which is

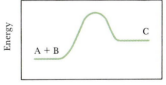

**FIGURE 16.9**  Progress of reaction

faster at equilibrium the forward or the reverse reaction? **(b)** In general, how would a catalyst affect the energy profile shown? **(c)** How would a catalyst affect the ratio of rate constants for the forward and reverse reactions? **(d)** How would you expect the equilibrium constant for this reaction to change with increasing temperature?

**16.30** For the reaction

$$2SO_2(g) + O_2(g) \longrightarrow 2SO_3(g)$$

the standard enthalpy change is $\Delta H° = -196.6$ kJ. The activation energy for the uncatalyzed reaction is about

160 kJ/mol. Sketch the energy profile for this reaction, as in Figure 16.7. Sulfur dioxide is produced in the cylinder of an auto engine by oxidation of the small quantity of sulfur present in gas. The catalytic mufflers installed in cars since 1975 result in conversion of a large portion of this $SO_2$ to $SO_3$ (Section 15.6). Using a dotted line, sketch on your figure an energy profile which might apply for $SO_2$ oxidation in a catalytic muffler.

**16.31** In the reaction

$$NO(g) + O_3(g) \rightleftharpoons NO_2(g) + O_2(g)$$

the rate law for the forward reaction is

$$r_f = k_f[NO][O_3]$$

and for the reverse reaction is

$$r_r = k_r[NO_2][O_2]$$

Using these expressions, write the equilibrium condition in terms of opposing rates, and show how $K_c$ relates to $k_f$ and $k_r$.

**16.32** Consider the following reaction, which occurs in a single step and is reversible:

$$CO(g) + Cl_2(g) \rightleftharpoons COCl(g) + Cl(g)$$

Kinetic studies indicate that the rate constant for the forward reaction, $k_f$, is $1.38 \times 10^{-28}/M\text{-s}$; for the reverse reaction, $k_r = 9.3 \times 10^{10}/M\text{-s}$ (both rate constants at 25°C). What is the value for the equilibrium constant for this reaction at 25°C?

## Additional Exercises

**16.33** Throughout this chapter we have used gas-phase reactions to illustrate the concept of equilibrium. However, solution reactions could also be used. Write the equilibrium-constant expressions for each of the following aqueous reactions:
(a) $Ag^+(aq) + 2NH_3(aq) \rightleftharpoons Ag(NH_3)_2^+(aq)$
(b) $Ag_2CrO_4(s) \rightleftharpoons 2Ag^+(aq) + CrO_4^{2-}(aq)$
(c) $HNO_2(aq) \rightleftharpoons H^+(aq) + NO_2^-(aq)$
(d) $Zn(s) + Cu^{2+}(aq) \rightleftharpoons Zn^{2+}(aq) + Cu(s)$
(e) $NH_3(aq) + H_2O(l) \rightleftharpoons NH_4^+(aq) + OH^-(aq)$

**16.34** For the reaction $N_2O_4(g) \rightleftharpoons 2NO_2(g)$, an equilibrium mixture is found to contain $4.27 \times 10^{-2}$ mol/L of $N_2O_4$ and $1.41 \times 10^{-2}$ mol/L of $NO_2$ at 25°C. What is the value of $K_c$ for this temperature?

**16.35** A mixture of 3.0 mol of $SO_2$, 4.0 mol of $NO_2$, 1.0 mol of $SO_3$, and 4.0 mol of NO is placed in a 2.0-L vessel. The following reaction takes place:

$$SO_2(g) + NO_2(g) \rightleftharpoons SO_3(g) + NO(g)$$

When equilibrium is reached at 700°C, the vessel is found to contain 1.0 mol of $SO_2$. (a) Calculate the equilibrium concentrations of $SO_2$, $NO_2$, $SO_3$, and NO. (b) Calculate the value of $K_c$ for this reaction at 700°C.

**16.36** For the equilibrium

$$PH_3BCl_3(s) \rightleftharpoons PH_3(g) + BCl_3(g)$$

$K_p = 0.052$ at 60°C. (a) Calculate $K_c$. (b) Some solid $PH_3BCl_3$ is added to a closed 0.500-L vessel at 60°C; the vessel is then charged with 0.0216 mol of $BCl_3(g)$. What is the equilibrium concentration of $PH_3$?

**16.37** A mixture of $CH_4$ and $H_2O$ is passed over a nickel catalyst at 1000 K. The emerging gas is collected in a 5.00-L flask and found to contain 8.62 g of CO, 2.60 g of $H_2$, 43.0 g of $CH_4$, and 48.4 g of $H_2O$. Assuming that equilibrium has been reached, calculate $K_c$ for the reaction

$$CH_4(g) + H_2O(g) \rightleftharpoons CO(g) + 3H_2(g)$$

**16.38** A 1.00-g sample of $PCl_5$ is introduced into a 250-mL flask, which is sealed and then heated to 250°C. The $PCl_5$ dissociates as follows:

$$PCl_5(g) \rightleftharpoons PCl_3(g) + Cl_2(g)$$

If 0.250 g of $Cl_2$ is present at equilibrium, what is the numerical value of $K_c$ at 250°C?

**16.39** At 1558 K the equilibrium constant, $K_c$, for the reaction

$$Br_2(g) \rightleftharpoons 2Br(g)$$

is $1.04 \times 10^{-3}$. If a 0.200-L vessel contains $4.53 \times 10^{-2}$ mol of $Br_2$ at equilibrium, how many moles of Br are present?

**[16.40]** Write the equilibrium-constant expression for the equilibrium

$$C(s) + CO_2(g) \rightleftharpoons 2CO(g)$$

The table below shows the relative mole percentages of $CO_2(g)$ and $CO(g)$ at a total pressure of 1 atm for several temperatures. Calculate the value of $K_c$ at each temperature. Is the reaction exothermic or endothermic? Explain.

| Temperature (°C) | CO₂ (%) | CO (%) |
| --- | --- | --- |
| 850 | 6.23 | 93.77 |
| 950 | 1.32 | 98.68 |
| 1050 | 0.37 | 99.63 |
| 1200 | 0.06 | 99.94 |

**[16.41]** A 0.831-g sample of $SO_3$ is placed in a 1.00-L container and heated to 1100 K. The $SO_3$ undergoes decomposition to $SO_2$ and $O_2$:

$$2SO_3(g) \rightleftharpoons 2SO_2(g) + O_2(g)$$

At equilibrium, the total pressure in the container is 1.300 atm. Find the values of $K_p$ and $K_c$ for this reaction at 1100 K.

**[16.42]** $PCl_5$ is placed in a 2.00-L flask at 250°C. The following reaction occurs:

$$PCl_5(g) \rightleftharpoons PCl_3(g) + Cl_2(g)$$

At equilibrium, the total pressure of the mixture is 2.00 atm. The partial pressure of $PCl_5$ at equilibrium is 0.37 atm. Calculate $K_p$ at 250°C.

**16.43** A mixture of 1.000 mol of $N_2$ and 3.000 mol of $H_2$ is placed in a 1.00-L vessel at 600°C. At equilibrium it is found that the mixture contains 0.371 mol of $NH_3$. Calculate $K_c$ at 600°C for the reaction

$$N_2(g) + 3H_2(g) \rightleftharpoons 2NH_3(g)$$

**16.44** Nitric oxide, NO, rapidly oxidizes to nitrogen

dioxide, $NO_2$, even at room temperature. At 1000 K, 0.0400 mol of NO and 0.0600 mol of $O_2$ are placed in a 2.00-L vessel. At equilibrium the concentration of $NO_2$ is $2.2 \times 10^{-3}$ M. (a) Calculate the equilibrium concentrations of NO and $O_2$. (b) Calculate the equilibrium constant, $K_c$, for the reaction

$$2NO(g) + O_2(g) \rightleftharpoons 2NO_2(g)$$

**16.45** Calculate $K_p$ for the reaction

$$2SO_2(g) + O_2(g) \rightleftharpoons 2SO_3(g)$$

if at a particular temperature and a total pressure of 112.0 atm the equilibrium mixture consists of 56.6 mole percent $SO_2$, 10.6 mole percent $O_2$, and 32.8 mole percent $SO_3$.

**16.46** Nitric oxide, NO, reacts readily with chlorine gas as follows:

$$2NO(g) + Cl_2(g) \rightleftharpoons 2NOCl(g)$$

At 700 K, the equilibrium constant, $K_p$, for this reaction is 0.26. Predict the behavior of each of the following mixtures at this temperature: (a) $P_{NO} = 0.15$ atm, $P_{Cl_2} = 0.31$ atm, and $P_{NOCl} = 0.11$ atm; (b) $P_{NO} = 0.12$ atm, $P_{Cl_2} = 0.10$ atm, and $P_{NOCl} = 0.050$ atm; (c) $P_{NO} = 0.15$ atm, $P_{Cl_2} = 0.20$ atm, and $P_{NOCl} = 5.10 \times 10^{-3}$ atm.

**16.47** Consider the reaction

$$2CO(g) + O_2(g) \rightleftharpoons 2CO_2(g) \quad \Delta H° = -514.2 \text{ kJ}$$

In which direction will the equilibrium move if (a) $CO_2$ is added; (b) $CO_2$ is removed; (c) the volume is increased; (d) the pressure is increased; (e) the temperature is increased?

**16.48** For the reaction shown in Exercise 16.47, what effect does increasing temperature have on the magnitude of the equilibrium constant?

**16.49** NiO is to be reduced to nickel metal in an industrial process by use of the reaction

$$NiO(s) + CO(g) \rightleftharpoons Ni(s) + CO_2(g)$$

At 1600 K the equilibrium constant for the reaction is 600. If a CO pressure of 150 mm Hg is to be employed in the furnace, and total pressure never exceeds 760 mm Hg, will reduction occur?

**16.50** Consider the reaction

$$CO(g) + 2H_2(g) \rightleftharpoons CH_3OH(l)$$

Using the thermochemical data in Appendix D, determine whether the equilibrium constant for this reaction increases or decreases with increasing temperature. Assuming equal pressures of CO and $H_2$, how would the extent of conversion of the gas mixture to methanol ($CH_3OH$) vary with total pressure?

**[16.51]** Suppose that there is a region in outer space where initially the hydrogen molecule concentration is $10^2$ molecules/$cm^3$, the $N_2$ concentration is 1 molecule/$cm^3$, and the temperature is 100 K. At this temperature, $K_p$ for the reaction

$$N_2(g) + 3H_2(g) \rightleftharpoons 2NH_3(g)$$

is approximately $6 \times 10^{37}$. Assuming that equilibrium is

attained, is a significant fraction of the $N_2$ converted to $NH_3$?

**16.52** Are any of the following statements false? For those that are, discuss the sense in which they are incorrect. (a) If a catalyst increases the rate of a forward reaction by a factor of 1000 over the uncatalyzed rate, it increases the rate of the reverse reaction by a factor of 1000 also. (b) A catalyst can promote the formation of product in some reactions by inhibiting the reverse reaction in an equilibrium. (c) Although heterogeneous catalysts must affect the rates of both forward and reverse reactions to an equal extent, homogeneous catalysts can be made to affect the rate of just the forward or just the reverse step.

**16.53** At 1200 K, the approximate temperature of automobile exhaust gases (Figure 16.8), the equilibrium constant for the reaction

$$2CO_2(g) \rightleftharpoons 2CO(g) + O_2(g)$$

is about $1 \times 10^{-13}$ atm. Assuming that the exhaust gas (total pressure 1 atm) contains 0.2 percent CO by volume, 12 percent $CO_2$, and 3 percent $O_2$, is the system at equilibrium with respect to the above reaction? Based on your conclusion, would the CO concentration in the exhaust be lowered or increased by a catalyst that speeded up the reaction above?

**16.54** For the single-step reaction

$$NO(g) + O_3(g) \xrightleftharpoons[k_r]{k_f} NO_2(g) + O_2(g)$$

$K_p = 1.32 \times 10^{10}$ at 1000 K. If $k_f = 6.26 \times 10^8$/M-s at this temperature, calculate $k_r$.

**16.55** Suppose that you worked at the U.S. Patent Office and a patent application came across your desk in which it was claimed that a newly developed catalyst was much superior to the Haber catalyst for ammonia synthesis, because the catalyst led to much greater equilibrium conversion of $N_2$ and $H_2$ into $NH_3$ than the Haber catalyst under the same conditions. What would be your response?

**[16.56]** At 1558 K the equilibrium constant, $K_c$, for the reaction

$$Br_2(g) \rightleftharpoons 2Br(g)$$

is $1.04 \times 10^{-3}$. (a) Calculate the equilibrium concentration of Br atoms if the initial concentration of $Br_2$ is 1.00 M. (b) Calculate the fraction of the initial concentration of $Br_2$ that is dissociated into atoms.

**16.57** At 400 K, the equilibrium constant, $K_p$, for the following reaction is $6.0 \times 10^{-9}$:

$$NH_4Cl(s) \rightleftharpoons NH_3(g) + HCl(g)$$

What are the equilibrium vapor pressures of $NH_3$ and HCl that are produced by the decomposition of solid $NH_4Cl$ at 400 K?

**[16.58]** An equilibrium mixture of $H_2$, $I_2$, and HI at 458°C contains $2.24 \times 10^{-2}$ M $H_2$, $2.24 \times 10^{-2}$ M $I_2$, and 0.155 M HI in a 5.00-L vessel. What are the equilibrium concentrations when equilibrium is reestablished following the addition of 0.100 mol of HI?

## CHAPTER 17

# Aqueous Equilibria: Acids and Bases

Water is the most common and most important solvent on this planet. All living matter contains water as the major constituent of cells. Because water is a highly polar liquid, it has an exceptional ability to dissolve a wide range of ionic materials; in addition, many substances react with water to form ions in solution (Section 12.2). In this chapter we will apply the principles of chemical equilibrium to aqueous solutions of acids and bases. We will see how to describe aqueous equilibria quantitatively. We will also examine the relationship between acid or base strength and chemical structure, and we will discuss the theory of acids and bases put forward by G. N. Lewis. The Lewis theory of acids and bases is more general than the Brønsted-Lowry theory (Section 13.5); it enables us to extend acid-base concepts to a wider range of substances and solvents.

## 17.1 THE DISSOCIATION OF WATER AND THE pH SCALE

Recall that in the Brønsted-Lowry theory of acids and bases (Section 13.5) an acid is defined as a substance that is capable of donating a proton, and a base as a substance capable of accepting a proton. We have seen that water can act as a base toward acids, such as HCl, by accepting the proton to form $H^+(aq)$. Water is also capable of donating a proton to a base, such as $NH_3$, forming $OH^-(aq)$ in the process (Equation 13.23). An important feature of water is that it is also capable of acting as a proton donor and proton acceptor toward itself. The process by which this occurs is called **autoionization**:

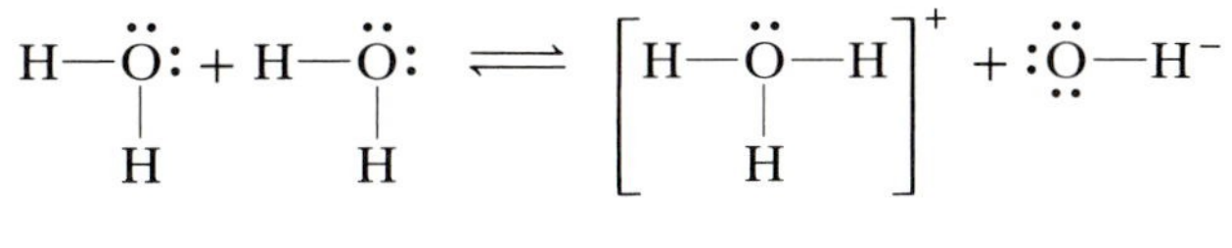

$$H-\overset{\cdot\cdot}{O}: + H-\overset{\cdot\cdot}{O}: \rightleftharpoons \left[ H-\overset{\cdot\cdot}{O}-H \right]^+ + :\overset{\cdot\cdot}{\underset{\cdot\cdot}{O}}-H^- \qquad [17.1]$$

This reaction amounts to a spontaneous ionization of solvent. The reaction occurs to only a small extent and gives rise to a very small electrical conductivity for pure water. At room temperature only about one out of every $10^8$ molecules is in the ionic form at any one instant. No one molecule remains in the ionic condition for long; the equilibria are extremely rapid. On the average, a proton transfers from one molecule to another in water at a rate of about 1000 times per second.

By expressing the hydrated proton as $H^+(aq)$ rather than $H_3O^+(aq)$, we can rewrite Equation 17.1 as

Acids and bases are important components of many common household products, foods, and medications. All of the items shown here, except for plaster of Paris, are notably acidic or basic. Plaster of Paris, $CaSO_4 \cdot \frac{1}{2}H_2O$, is a salt, the product of reaction of an acid with a base. (Richard Megna/Fundamental Photographs)

$$H_2O(l) \rightleftharpoons H^+(aq) + OH^-(aq) \qquad \text{[17.2]}$$

The equilibrium expression for this autoionization reaction can be written as

$$K = \frac{[H^+][OH^-]}{[H_2O]}$$

The concentration of water in aqueous solutions is typically very large, about 55 $M$, and remains essentially constant for dilute solutions. It is therefore customary to exclude the concentration of water from equilibrium-constant expressions for aqueous solutions, just as we exclude the concentrations of pure solids and liquids from the equilibrium-constant expressions for heterogeneous reactions (Section 16.4). Thus we can write the equilibrium-constant expression for the autoionization of water as

$$K[H_2O] = K_w = [H^+][OH^-]$$

The product of two constants, $K$ and $[H_2O]$, defines a new constant, $K_w$. This important equilibrium constant is called the **ion-product constant** for water. $K_w$ has the value of $1.0 \times 10^{-14}$ at 25°C. This is an important equilibrium constant. You should memorize this expression:

$$K_w = [H^+][OH^-] = 1.0 \times 10^{-14} \qquad \text{[17.3]}$$

Equation 17.3 is valid for aqueous solutions as well as for pure water. A solution for which $[H^+] = [OH^-]$ is said to be *neutral*. In most solutions $H^+$ and $OH^-$ concentrations are not equal. As the concentration of one of these ions increases, the concentration of the other must decrease so that the ion product equals $1.0 \times 10^{-14}$. In acidic solutions, $[H^+]$ exceeds $[OH^-]$. In basic solutions, the reverse is true: $[OH^-]$ exceeds $[H^+]$.

## SAMPLE EXERCISE 17.1

Calculate the values of $[H^+]$ and $[OH^-]$ in a neutral solution at 25°C.

**Solution:** By definition, in a neutral solution, $[H^+]$ equals $[OH^-]$. Let us call the concentration of each of these species in neutral solution $x$. Using Equation 17.3, we have

$$[H^+][OH^-] = (x)(x) = 1.0 \times 10^{-14}$$
$$x^2 = 1.0 \times 10^{-14}$$
$$x = 1.0 \times 10^{-7} = [H^+] = [OH^-]$$

In an acid solution, $[H^+]$ is greater than $1.0 \times 10^{-7}$ $M$; in a basic solution it is less than $1.0 \times 10^{-7}$ $M$.

### PRACTICE EXERCISE
Indicate whether each of the following solutions is acidic or basic: **(a)** $[H^+] = 2 \times 10^{-5}$ $M$; **(b)** $[OH^-] = 3 \times 10^{-9}$ $M$; **(c)** $[OH^-] = 1 \times 10^{-7}$ $M$.

***Answers:*** **(a)** acidic; **(b)** acidic; **(c)** neutral

## SAMPLE EXERCISE 17.2

Calculate the concentration of $H^+(aq)$ in **(a)** a solution in which $[OH^-]$ is 0.010 $M$; **(b)** a solution in which $[OH^-]$ is $2.0 \times 10^{-9}$ $M$.

**Solution:** In this problem and all that follow, we assume, unless stated otherwise, that temperature is 25°C.

**(a)** Using Equation 17.3, we have

$$[H^+][OH^-] = 1.0 \times 10^{-14}$$

$$[H^+] = \frac{1.0 \times 10^{-14}}{[OH^-]} = \frac{1.0 \times 10^{-14}}{0.010}$$

$$= 1.0 \times 10^{-12} \, M$$

This solution is basic because $[H^+] < [OH^-]$.

**(b)** In this instance

$$[H^+] = \frac{1.0 \times 10^{-14}}{[OH^-]} = \frac{1.0 \times 10^{-14}}{2.0 \times 10^{-9}}$$

$$= 5.0 \times 10^{-6} \, M$$

This solution is acidic because $[H^+] > [OH^-]$.

**PRACTICE EXERCISE**

Calculate the concentration of $OH^-(aq)$ in a solution in which **(a)** $[H^+] = 2 \times 10^{-6} \, M$; **(b)** $[H^+] = [OH^-]$; **(c)** $[H^+] = 10^2[OH^-]$.

**Answers:** **(a)** $5 \times 10^{-9} \, M$; **(b)** $1 \times 10^{-7} \, M$; **(c)** $1 \times 10^{-8} \, M$

**pH**

In almost every area of pure and applied chemistry the acid-base properties of water are important. As examples, the fate of pollutant chemicals in a water body, the rapidity with which a metal object immersed in water corrodes, and the suitability of an aquatic environment for fish and plant life are all critically dependent on the acidity or basicity of the water. The concentration of $H^+(aq)$ in such solutions is often expressed in terms of **pH**. The pH is defined as the negative log in base 10 of the hydrogen-ion concentration.*

$$pH = -\log [H^+] = \log \left(\frac{1}{[H^+]}\right) \qquad [17.4]$$

Notice that a change in $[H^+]$ by a factor of 10 results in a unit change in pH. (If you need a review of exponential notation and of the use of logs, see Appendix A.) As an example of the use of Equation 17.4, let us calculate the pH of a neutral solution, that is, one in which $[H^+] = [OH^-] = 1.0 \times 10^{-7}$ (Sample Exercise 17.1). The pH is given by

$$pH = -\log [H^+] = -\log (1.0 \times 10^{-7}) = 7.00$$

Thus the pH of a neutral solution is 7.00.

Because pH is simply another means of expressing $[H^+]$, acidic and basic solutions can be distinguished on the basis of their pH values:

pH < 7 in acidic solutions
pH > 7 in basic solutions
pH = 7 in neutral solutions

You should keep in mind that the pH is a measure only of the equilibrium concentration of dissociated hydrogen ion present as $H^+(aq)$. The pH values characteristic of several familiar solutions are shown in Figure 17.1

With a log table (Appendix B) or log function on a calculator, converting from concentration of $H^+(aq)$ to pH, and vice versa, is a simple matter, as outlined in Sample Exercises 17.3 and 17.4.

---

* Usually you will see pH defined as $-\log [H^+]$, occasionally as $-\log [H_3O^+]$. As discussed in Section 13.4, the same species is involved in all cases.

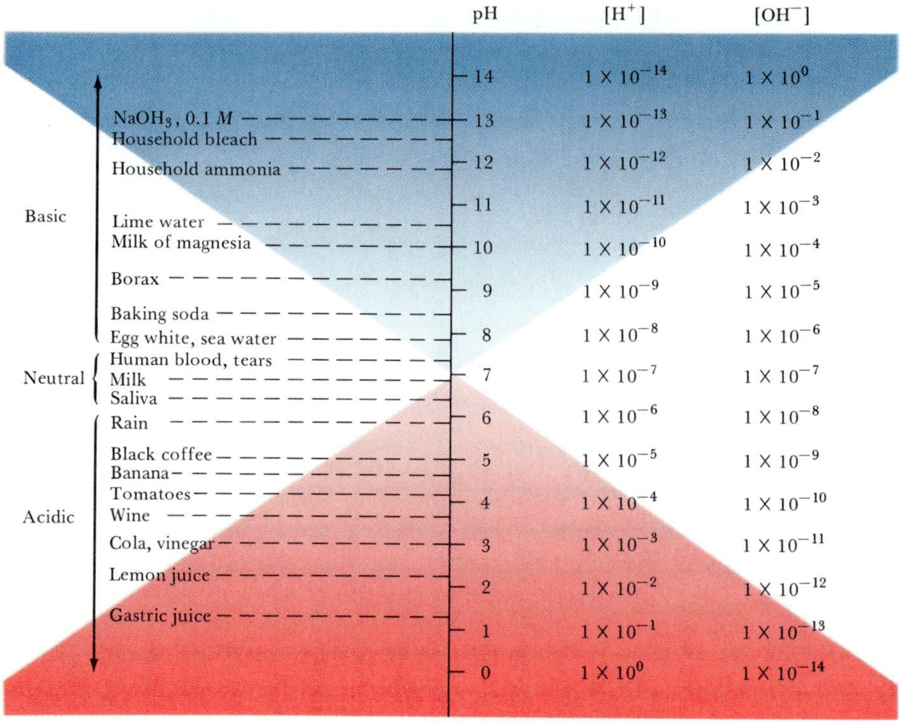

**FIGURE 17.1** Values of pH for some more common solutions. The pH scale in this figure is shown to extend from 0 to 14, because nearly all solutions commonly encountered have pH values in that range. In principle, however, the pH values for strongly acidic solutions can be less than 0, and for strongly basic solutions can be greater than 14.

---

## SAMPLE EXERCISE 17.3

Calculate the pH values for the two solutions described in Sample Exercise 17.2.

**Solution:** In the first instance we found $[H^+]$ to be $1.0 \times 10^{-12}$. The pH of this solution is given by

$$pH = -\log (1.0 \times 10^{-12}) = -(-12.00) = 12.00$$

The pH of the second solution is given by

$$pH = -\log (5.0 \times 10^{-6}) = 5.30$$

This example illustrates an important point about significant figures: *The number of decimal places in the log quantity equals the number of significant figures in the original number.*

### PRACTICE EXERCISE
**(a)** In a sample of lemon juice, $[H^+]$ is $3.8 \times 10^{-4}\ M$. What is the pH? **(b)** A commonly available window-cleaning solution has a $[H^+]$ of $5.3 \times 10^{-9}\ M$. What is the pH?

***Answers:*** **(a)** 3.42; **(b)** 8.28

---

## SAMPLE EXERCISE 17.4

A sample of freshly pressed apple juice has a pH of 3.76. Calculate $[H^+]$.

**Solution:** From the equation defining pH (Equation 17.4), we have $-\log [H^+] = 3.76$. Thus $\log [H^+] = -3.76$. To find $[H^+]$ we need to find the antilog of $-3.76$. That is, we want the number whose log is $-3.76$. Some calculators have an antilog function (usually labeled INV log or $\log^{-1}$), which makes the calculation quite simple. Other calculators rely on $10^x$ or $y^x$ functions to find antilogs: antilog $(-3.76) = 10^{-3.76}$. If you are relying on a log table,

the simplest way to take the antilog is to write the log as a sum of an integer and a positive decimal fraction: $-3.76 = -4.00 + 0.24$. The antilog of $-4$ is $1 \times 10^{-4}$. From a log table we find that the antilog of 0.24 is approximately 1.7. Thus $[H^+] = 1.7 \times 10^{-4}\ M$.

### PRACTICE EXERCISE
A solution formed by dissolving an antacid tablet has a pH of 9.18. Calculate $[H^+]$.

***Answer:*** $[H^+] = 6.6 \times 10^{-10}\ M$

The negative log is a convenient way of expressing the magnitudes of numbers that are generally very small. We use the convention that the negative log of a quantity is labeled p(quantity). For example, one can express the concentration of $OH^-$ as pOH:

$$pOH = -\log[OH^-]$$

By taking the log of both sides of Equation 17.3 and multiplying through by $-1$, we can obtain

$$pH + pOH = -\log K_w = 14.00 \qquad [17.5]$$

This expression is often convenient to use. We will see later (Section 17.5) that the pX notation is also useful in dealing with equilibrium constants.

## CHEMISTRY AT WORK

**The pH meter**

The pH meter is a widely used, simple instrument for rapid and accurate determination of pH. The first pH meter of the kind we encounter today was developed in 1934 by Arnold O. Beckman. The pH meter is so common that if you go on to further study in chemistry or in an applied science you are almost certain to encounter one.

A complete understanding of how a pH meter works requires a knowledge of electrochemistry, a subject we take up in Chapter 20. However, we can say at this point that a pH meter consists of a pair of electrodes that are placed in the solution to be measured and a sensitive meter for measuring small voltages, on the order of millivolts. A typical pH meter is shown in Figure 17.2. When the electrodes are placed in the solution, they form an electrochemical cell (something like a battery) that has a voltage. The voltage of the cell is dependent on $(H^+)$; thus, by measuring the voltage we obtain a measure of $[H^+]$. Electrodes that can be used with pH meters come in all shapes and sizes, depending on their intended use, but fundamentally they are nearly all

the same. One of the electrodes is a reference electrode. The one that is actually sensitive to $H^+ (aq)$ is almost always a so-called glass electrode. The wire in the inner compartment of the electrode is in contact with a solution of known and fixed $H^+(aq)$ concentration. The wall of the compartment is formed of a special thin glass that is permeable to $H^+(aq)$. As a result, the voltage that this electrode, together with a reference electrode, generates when placed in a solution depends on $[H^+]$ in the solution.

To extend the range of possible pH measurements, much research has gone into the development of electrodes that can be used with very small quantities of solution. It is now possible to insert electrodes into single living cells in order to monitor pH of the cell medium. The pH meter is also widely used outside the laboratory. Pocket-sized models are available for use in environmental studies, in monitoring of industrial effluents, and in agricultural work.

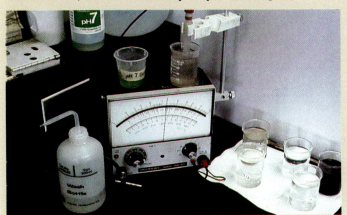

**FIGURE 17.2** A pH meter of the type normally used for student work. (Runk, Schoenberger/Grant Heilman Photography)

## Indicators

Various means are available for quantitatively estimating pH. The simplest is the use of an indicator. An **indicator** is a colored substance, usually derived from plant material, that can exist in either an acid or base form. The two forms are differently colored. By adding a small amount of an indicator to a solution and noting its color, it is possible to determine whether it is in the acid or base form. If one knows the pH at which the indicator turns from one form to the other, one can then determine from the observed color whether the solution has a higher or lower pH than this value. For example, litmus, one of the most common indicators, changes color in the vicinity of pH 7. However, the color change is not very sharp. Red litmus indicates a pH of about 5 or lower, and blue litmus indicates a pH of about 8.2 or higher. Many other indicators change color at various pH values between 1 and 14. Some of the more commonly used are listed in Table 17.1. We see from this table that methyl orange, for example, changes color over the pH interval from 3.1 to 4.4. Below pH 3.1 it is in the acid form, which is red. In the interval from pH 3.1 to 4.4 it is gradually converted to its basic form, which has a yellow color. By pH 4.4 the conversion is complete, and the solution is yellow. Paper tape that is impregnated with various indicators and that comes complete with a comparator color scale is widely used for approximate determinations of pH.

**TABLE 17.1**   Common acid-base indicators

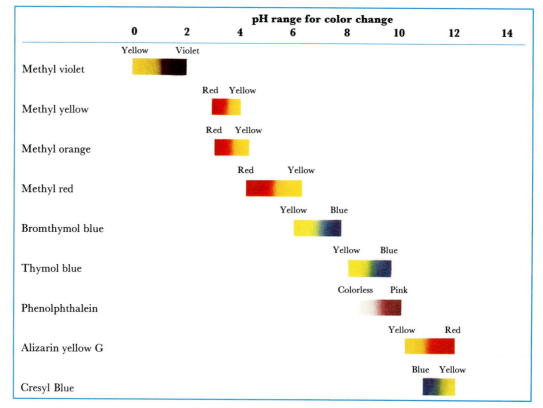

The pH of an aqueous solution depends on the ability of the solute to give protons to, or remove them from, water. In this section and the following ones we will examine the extent to which these proton-transfer reactions occur.

In terms of the Brønsted-Lowry concept, a strong aqueous acid is any substance that reacts completely with water to form $H^+(aq)$; a weak acid is a substance that only partly reacts in this fashion. The number of strong acids is not very large; the six most important ones are listed in Table 17.2. We recommend that you commit them to memory if you have not already done so. In the case of $H_2SO_4$, only the first proton is completely ionized in aqueous solution.

We can consider aqueous solutions of all strong acids to consist entirely of ions, with no significant concentration of neutral solute molecules remaining; such acids are said to be completely ionized or dissociated. For example, a 0.10 $M$ aqueous solution of $HNO_3$, nitric acid, contains $[H^+] = 0.10\ M$, and $[NO_3^-] = 0.10\ M$; the concentration of $HNO_3$ is virtually zero.

Strong bases are strong proton acceptors. The most common strong bases are NaOH and KOH, which are ionic compounds containing $OH^-$ ions in the solid. They are strong electrolytes, dissolving in water as would any other ionic substance (Section 12.4). For example, a 0.10 $M$ aqueous solution of NaOH contains 0.10 $M$ $Na^+(aq)$ and 0.10 $M$ $OH^-(aq)$ with no undissociated NaOH:

$$NaOH(s) \xrightarrow{H_2O} Na^+(aq) + OH^-(aq) \qquad [17.6]$$

All of the hydroxides of the alkali metals (family 1A) are strong electrolytes, but the compounds of Li, Rb, and Cs are too expensive to be encountered commonly in the laboratory. The hydroxides of all of the alkaline earths (family 2A) except Be are also strong electrolytes. However, they have limited solubilities and are consequently used only when high solubility is not critical. $Mg(OH)_2$ has an especially low solubility ($9 \times 10^{-3}$ g/L of water at 25°C). The least expensive and most common of the alkaline earth hydroxides is $Ca(OH)_2$, with a solubility of 0.97 g/L at 25°C.

Basic solutions are also created when substances react with water to form $OH^-(aq)$. The most common of these is the oxide ion. Ionic metal oxides, especially $Na_2O$ and CaO, are often used in industry when a strong base is needed. Each mole of $O^{2-}$ reacts with water to form 2 mol

**TABLE 17.2**  Common strong acids and bases

| Acids | Bases |
|---|---|
| HCl (hydrochloric acid) | Hydroxides and oxides of 1A metals |
| HBr (hydrobromic acid) | Hydroxides and oxides of 2A metals (except Be) |
| HI (hydroiodic acid) | |
| $HNO_3$ (nitric acid) | |
| $HClO_4$ (perchloric acid) | |
| $H_2SO_4$ (sulfuric acid) | |

of $OH^-$, leaving virtually no $O^{2-}$ remaining in the solution:

$$O^{2-}(aq) + H_2O(l) \longrightarrow 2OH^-(aq) \qquad [17.7]$$

Similarly, ionic hydrides and nitrides, such as $NaH(s)$ and $Mg_3N_2(s)$, react to form basic solutions:

$$NaH(s) + H_2O(l) \longrightarrow Na^+(aq) + H_2(g) + OH^-(aq) \qquad [17.8]$$

$$Mg_3N_2(s) + 6H_2O(l) \longrightarrow 3Mg^{2+}(aq) + 2NH_3(aq) + 6OH^-(aq) \qquad [17.9]$$

In these examples the anions, $H^-$ and $N^{3-}$, are stronger bases than $OH^-(aq)$. They therefore remove a proton from $H_2O$, which is the conjugate acid of $OH^-$.

---

## SAMPLE EXERCISE 17.5

What is the pH of a solution of **(a)** 0.010 $M$ HCl; **(b)** 0.010 $M$ $Ca(OH)_2$?

$$[H^+] = \frac{1.00 \times 10^{-14}}{[OH^-]} = \frac{1.00 \times 10^{-14}}{0.020}$$
$$= 5.0 \times 10^{-13}$$
$$pH = -\log{(5.0 \times 10^{-13})} = 12.30$$

**Solution:** **(a)** HCl is a strong acid. Consequently, $[H^+] = 0.010$ $M$ and $pH = -\log{(0.010)} = 2.00$.
   **(b)** $Ca(OH)_2$ is a strong base; each $Ca(OH)_2$ forms $2OH^-$ ions. Consequently, $[OH^-] = 0.020$ $M$ and

**PRACTICE EXERCISE**

What is the pH of a solution formed by dissolving 0.010 mol of $Na_2O(s)$ in 1.00 L of water?
*Answer:* 12.30

---

## 17.3 WEAK ACIDS

Most substances that are acidic in water are actually weak acids. The extent to which an acid ionizes in an aqueous medium can be expressed by the equilibrium constant for the ionization reaction. In general, we can represent any acid by the symbol HX, where $X^-$ is the formula for the conjugate base that remains when the proton ionizes. The ionization equilibrium is then given by Equation 17.10:

$$HX(aq) \rightleftharpoons H^+(aq) + X^-(aq) \qquad [17.10]$$

The corresponding equilibrium-constant expression is

$$K_a = \frac{[H^+][X^-]}{[HX]} \qquad [17.11]$$

The equilibrium constant is often given the symbol $K_a$ and is called the **acid-dissociation constant**.

   Table 17.3 shows the names, structures, and values of $K_a$ for several weak acids. A more complete table is given in Appendix E. Note that many weak acids are compounds composed largely of carbon and hydrogen. Generally speaking, hydrogen atoms bound to carbon are not ionized in an aqueous medium. The ionizable hydrogens are in most instances bound to oxygen. The smaller the value for $K_a$, the weaker the acid. For example, phenol is the weakest acid listed in Table 17.3.

**TABLE 17.3** Some weak acids in water at 25°C[a]

| Acid | Molecular formula | Structural formula[a] | Conjugate base | $K_a$ |
|------|-------------------|----------------------|----------------|-------|
| Hydrofluoric | HF | H—F | $F^-$ | $6.8 \times 10^{-4}$ |
| Nitrous | $HNO_2$ | H—O—N—O | $NO_2^-$ | $4.5 \times 10^{-4}$ |
| Benzoic | $HC_7H_5O_2$ | $H-O-\overset{\overset{\displaystyle O}{\|\|}}{C}-\bigcirc$ | $C_7H_5O_2^-$ | $6.5 \times 10^{-5}$ |
| Acetic | $HC_2H_3O_2$ | $H-O-\overset{\overset{\displaystyle O}{\|\|}}{C}-\overset{\overset{\displaystyle H}{\|}}{\underset{\underset{\displaystyle H}{\|}}{C}}-H$ | $C_2H_3O_2^-$ | $1.8 \times 10^{-5}$ |
| Hypochlorous acid | HClO | H—O—Cl | $ClO^-$ | $3.0 \times 10^{-8}$ |
| Hydrocyanic | HCN | H—C≡N | $C≡N^-$ | $4.9 \times 10^{-10}$ |
| Phenol | $HOC_6H_5$ | $H-O-\bigcirc$ | $C_6H_5O^-$ | $1.3 \times 10^{-10}$ |

[a] The proton that ionizes is shown in color.

---

## SAMPLE EXERCISE 17.6

A student prepared a 0.10 $M$ solution of formic acid, $HCHO_2$, and measured its pH using a pH meter of the type illustrated in Figure 17.2. The pH at 25°C was found to be 2.38. **(a)** Calculate $K_a$ for formic acid at this temperature. **(b)** What percentage of the acid is dissociated in this 0.10 $M$ solution?

**Solution:** **(a)** The first step in solving any equilibrium problem is to write the equation for the equilibrium reaction. The ionization equilibrium for formic acid can be written as follows:

$$HCHO_2(aq) \rightleftharpoons H^+(aq) + CHO_2^-(aq)$$

The equilibrium-constant expression for this equilibrium is

$$K_a = \frac{[H^+][CHO_2^-]}{[HCHO_2]}$$

From the measured pH we can calculate $[H^+]$:

$$pH = -\log[H^+] = 2.38$$
$$\log[H^+] = -2.38$$
$$[H^+] = 4.2 \times 10^{-3}\ M$$

(If you are using a calculator you would have used the INV log, or $\log^{-1}$ function, as appropriate, to obtain the antilog of $-2.38$.)

The Lewis structure for formic acid is as follows:

$$H-\overset{\overset{\displaystyle :O:}{\|\|}}{C}-\overset{\displaystyle ..}{O}-H$$

The proton that ionizes is the one bound to oxygen, shown in color.

We can do a little accounting to determine the concentrations of the species involved in the equilibrium. For each $H^+$ produced in solution, one $CHO_2^-$ anion also forms, and there is a loss of one $HCHO_2$ molecule. Let's write under the equilibrium expression the initial and equilibrium concentrations of the species involved:

| | $HCHO_2$ | $\rightleftharpoons$ $H^+$ | + $CHO_2^-$ |
|------|----------|------|------|
| Initial | 0.10 $M$ | 0 $M$ | 0 $M$ |
| Equilibrium | $(0.10 - 4.2 \times 10^{-3})\ M$ | $4.2 \times 10^{-3}\ M$ | $4.2 \times 10^{-3}\ M$ |

Notice that the amount of $HCHO_2$ that dissociates is very small in comparison with the initial concentration of the acid. To the number of significant figures we are using, the subtraction yields just 0.10 $M$.

We can now insert the equilibrium concentrations into the expression for $K_a$:

$$K_a = \frac{(4.2 \times 10^{-3})(4.2 \times 10^{-3})}{0.10}$$

$$= 1.8 \times 10^{-4}$$

**(b)** The percentage of acid that dissociates is given by the concentration of $H^+$ or $CHO_2^-$ at equilibrium, divided by the initial acid concentration, times 100:

$$\text{Percent dissociation} = \frac{[H^+] \times 100}{[HCHO_2]}$$

$$= \frac{(4.2 \times 10^{-3}) \times 100}{0.10}$$

$$= 4.2 \text{ percent}$$

**PRACTICE EXERCISE**
Niacin, one of the B vitamins, has the following molecular structure:

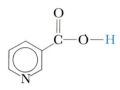

A 0.020 $M$ solution of niacin has a pH of 3.26. What is the acid dissociation constant, $K_a$, for the ionizable proton?

***Answer:*** $1.5 \times 10^{-5}$

## pH in Weak Acid Solutions

From the value for $K_a$ we can calculate the concentration of $H^+(aq)$ in a solution of a weak acid. For example, consider acetic acid, $HC_2H_3O_2$, the substance that gives the characteristic odor and acidic properties to vinegar. Let us calculate the concentration of $H^+(aq)$ in a 0.10 $M$ solution of acetic acid.

Our first step is to write the ionization equilibrium for acetic acid:

$$HC_2H_3O_2(aq) \rightleftharpoons H^+(aq) + C_2H_3O_2^-(aq) \qquad [17.12]$$

Note from the Lewis structure for acetic acid, shown in Table 17.3, that the hydrogen that ionizes is attached to the oxygen atom. We write this hydrogen separate from the others in the formula to emphasize that this one hydrogen is readily ionized.

The second step is to write the equilibrium-constant expression and the value for the equilibrium constant, if that is known. From Table 17.3 we have $K_a = 1.8 \times 10^{-5}$. Thus we can write the following:

$$K_a = \frac{[H^+][C_2H_3O_2^-]}{[HC_2H_3O_2]} = 1.8 \times 10^{-5} \qquad [17.13]$$

As the third step, we need to express the concentrations that make up the equilibrium-constant expression. This can be done with a little accounting, as described in Sample Exercise 17.6:

|  | $HC_2H_3O_2(aq) \rightleftharpoons$ | $H^+(aq) +$ | $C_2H_3O_2^-(aq)$ |
|---|---|---|---|
| **Initial** | 0.10 $M$ | 0 $M$ | 0 $M$ |
| **Equilibrium** | $(0.10 - x)$ $M$ | $x$ $M$ | $x$ $M$ |

Because we seek to find the equilibrium value for $[H^+]$, let us call this quantity $x$. The concentration of acetic acid before any of it dissociates is 0.10 $M$. The equation for the equilibrium tells us that for each molecule of $HC_2H_3O_2$ that dissociates, one $H^+(aq)$ and one $C_2H_3O_2^-(aq)$ are formed. Thus, if $x$ moles per liter of $H^+(aq)$ form at equilibrium, $x$

moles per liter of $C_2H_3O_2^-(aq)$ must also form, and $x$ moles per liter of $HC_2H_3O_2$ must be dissociated. This gives rise to the equilibrium concentrations shown above.

As the fourth step of the problem, we need to substitute the equilibrium concentrations into the equilibrium-constant expression. The substitutions give the following equation:

$$K_a = \frac{[H^+][C_2H_3O_2^-]}{[HC_2H_3O_2]} = \frac{(x)(x)}{0.10 - x} = 1.8 \times 10^{-5} \qquad [17.14]$$

Because this equation has only one unknown it can be solved using algebra. However, the solution is a little tedious, because it requires use of the quadratic formula (Appendix A.3). By taking account of what is actually occurring in the solution, we can make things simpler for ourselves. Because the value of $K_a$ is small, we might guess that $x$ will be quite small. (In other words, perhaps only a small fraction of the $HC_2H_3O_2$ is actually ionized.) Indeed, if we solve the problem using the quadratic formula we find that $x = 1.3 \times 10^{-3}$ $M$. Now you know that if a small number is subtracted from a much larger one, the result is approximately equal to the larger number. In our example we have

$$0.10 - x = 0.10 - 0.0013 \simeq 0.10$$

We can therefore make the approximation of ignoring $x$ relative to $0.10$ in the denominator of Equation 17.14. This leads us to the following simplified expression:

$$K_a = \frac{(x)(x)}{0.10} = 1.8 \times 10^{-5}$$

Solving for $x$, we have

$$x^2 = (0.10)(1.8 \times 10^{-5}) = 1.8 \times 10^{-6}$$
$$x = \sqrt{1.8 \times 10^{-6}} = 1.3 \times 10^{-3} \ M = [H^+]$$

From the value calculated for $x$ we see that our simplifying approximation is quite reasonable. This type of approximation can be used whenever conditions in solution are such that only a small fraction of acid ionizes. As a general rule, if the quantity $x$, which is subtracted from the initial concentration of the acid, is more than about 5 percent of the initial value, it is better to use the quadratic formula. In cases of doubt, assume that the approximation is valid and solve for $x$ in the simplified equation. Compare this approximate value of $x$ with the initial concentration of acid. For example, if the initial concentration of acid were $0.050$ $M$, and $x$ turned out in a given case to be $0.0016$ $M$, then

$$\left(\frac{0.0016 \ M}{0.050 \ M}\right)(100) = 3.2\%$$

If the result is more than about 5 percent, the problem should be reworked using the quadratic formula.

## SAMPLE EXERCISE 17.7

Calculate the pH of a 0.20 $M$ solution of HCN. (Refer to Table 17.3 or Appendix E for the value of $K_a$.)

**Solution:** Proceeding as in the example worked out above, we write

$$HCN(aq) \rightleftharpoons H^+(aq) + CN^-(aq)$$

$$K_a = \frac{[H^+][CN^-]}{[HCN]} = 4.9 \times 10^{-10}$$

Let $x = [H^+]$ at equilibrium. Then we have the following concentrations:

| | HCN(aq) $\rightleftharpoons$ | H$^+$(aq) + | CN$^-$(aq) |
|---|---|---|---|
| **Initial** | 0.20 $M$ | 0 $M$ | 0 $M$ |
| **Equilibrium** | (0.20 − x) $M$ | x $M$ | x $M$ |

Substituting into the equilibrium-constant expression yields

$$K_a = \frac{(x)(x)}{0.20 - x} = 4.9 \times 10^{-10}$$

We next make the simplifying approximation that $x$, the amount of acid that dissociates, is small in comparison with the initial concentration of acid; that is,

$$0.20 - x \simeq 0.20$$

Thus

$$\frac{x^2}{0.20} = 4.9 \times 10^{-10}$$

Solving for $x$, we have

$$x^2 = (0.20)(4.9 \times 10^{-10})$$
$$= 0.98 \times 10^{-10}$$
$$x = \sqrt{0.98 \times 10^{-10}}$$
$$= 9.9 \times 10^{-6} = [H^+]$$

Notice that $9.9 \times 10^{-6}$ is much smaller than 5 percent of 0.20, the initial HCN concentration. Our simplifying approximation is therefore appropriate.

$$pH = -\log [H^+] = -\log (9.9 \times 10^{-6})$$
$$= 5.00$$

**PRACTICE EXERCISE**

The $K_a$ for niacin (Practice Exercise 17.6) is $1.5 \times 10^{-5}$. What is the pH of a 0.010 $M$ solution of niacin?
*Answer:* 3.41

The result obtained in Sample Exercise 17.7 is typical of the behavior of weak acids; the concentration of $H^+(aq)$ is only a small fraction of the concentration of the acid in solution. Of course, those properties of the acid solution that relate directly to the concentration of $H^+(aq)$, such as electrical conductivity or rate of reaction with an active metal, are much less in evidence for a solution of a weak acid than for a solution of a strong acid. Figure 17.3 illustrates an experiment often carried out in the

**FIGURE 17.3**
Demonstration of the relative rates of two acid solutions of the same concentration with an active metal. (a) The flask on the left contains 1 $M$ acetic acid; the one on the right contains 1 $M$ HCl. Each balloon contains the same amount of magnesium metal. (b) When the Mg metal is dumped into the acid, $H_2$ is formed. The rate of $H_2$ formation is clearly higher for the 1 $M$ HCl solution on the right. (Donald Clegg and Roxy Wilson)

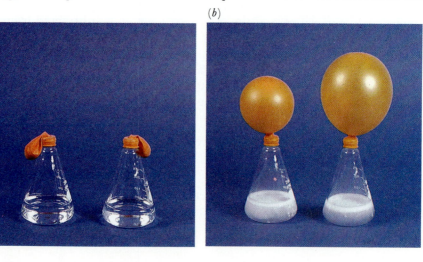

(a)    (b)

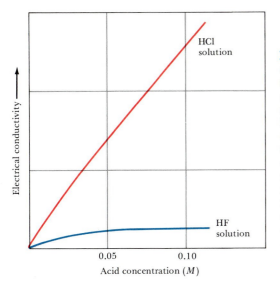

FIGURE 17.4 Electrical conductivity versus concentration for solutions of HCl, a strong acid, and HF, a weak acid. The conductivity of the HCl solution is not completely linear with concentration because of attractive forces between the ions at higher concentrations. The conductivity for the HF solution is quite nonlinear with concentration and much lower than for HCl because only a fraction of the HF molecules ionize. It is nonlinear with concentration because the fraction of molecules ionizing changes with concentration.

chemistry laboratory to demonstrate the difference in concentration of $H^+(aq)$ in weak and strong acid solutions of the same concentration. The rate of reaction with the metal is much faster for the solution of a strong acid. Reactions in which the rate depends on $[H^+]$ are common.

Figure 17.4 illustrates an experiment in which the electrical conductivity of an HCl solution is compared with the conductivity of an HF solution. The conductivity of the solution of the strong acid increases approximately in proportion to the concentration. This is what one would expect; because all the acid molecules ionize, the concentration of ions in solution is directly proportional to the concentration of acid. The conductivity of the solution of the weak acid is very much less than that for the strong acid and does not vary linearly with the acid concentration. The nonlinearity of the graph arises from the fact that the percentage of acid ionized varies with the acid concentration. This is illustrated in Sample Exercise 17.8.

## SAMPLE EXERCISE 17.8

Calculate the percentage of HF molecules ionized in a 0.10 $M$ HF solution; in a 0.010 $M$ HF solution.

**Solution:** The equilibrium reaction and equilibrium concentrations can be written as follows:

| | $HF(aq)$ $\rightleftharpoons$ | $H^+(aq)$ + | $F^-(aq)$ |
|---|---|---|---|
| **Initial** | 0.10 $M$ | 0 $M$ | 0 $M$ |
| **Equilibrium** | $(0.10 - x)$ $M$ | $x$ $M$ | $x$ $M$ |

The equilibrium-constant expression is as follows:

$$K_a = \frac{[H^+][F^-]}{[HF]} = \frac{(x)(x)}{0.10 - x} = 6.8 \times 10^{-4}$$

We might be tempted to try solving this equation using the same approximation used in earlier examples, that is, by neglecting the concentration of acid that ionizes in comparison with the initial concentration (by neglecting $x$ in comparison with 0.10). However, $K_a$ is large enough in this case to make that a poor approximation. We must therefore rearrange our equation and write it in standard quadratic form:

$$x^2 = (0.10 - x)(6.8 \times 10^{-4})$$
$$= 6.8 \times 10^{-5} - (6.8 \times 10^{-4})x$$
$$x^2 + (6.8 \times 10^{-4})x - 6.8 \times 10^{-5} = 0$$

Solution of this equation by use of the standard quadratic formula,

$$x = \frac{-b \pm (b^2 - 4ac)^{1/2}}{2a}$$

gives

$$x = \frac{-6.8 \times 10^{-4} \pm [(6.8 \times 10^{-4})^2 + 4(6.8 \times 10^{-5})]^{1/2}}{2}$$

$$= \frac{-6.8 \times 10^{-4} \pm 1.65 \times 10^{-2}}{2}$$

Of the two solutions, only the one that gives a positive value for $x$ is physically reasonable. Thus

$$x = 7.9 \times 10^{-3}$$

(You might also see what answer you would get by making the simplifying approximation of neglecting $x$ with respect to 0.10.)

From our result we can calculate the percent of molecules ionized:

$$\text{Percent ionized} = \left(\frac{\text{concentration ionized}}{\text{original concentration}}\right)(100)$$

$$= \left(\frac{7.9 \times 10^{-3}\ M}{0.10\ M}\right)(100) = 7.9\%$$

Proceeding similarly for the 0.010 $M$ solution, we have

$$\frac{x^2}{0.010 - x} = 6.8 \times 10^{-4}$$

Solving the resultant quadratic expression, we obtain

$$x = [H^+] = [F^-] = 2.3 \times 10^{-3}\ M$$

The percentage of molecules ionized is

$$\left(\frac{0.0023}{0.010}\right)(100) = 23\%$$

Notice that in diluting the solution by a factor of 10, the percentage of molecules ionized increases by a factor of 3. We could have arrived at this conclusion qualitatively by applying Le Châtelier's principle (Section 16.6) to the equilibrium. There are more "particles" or reaction components on the right side of the equation than on the left. Dilution causes the reaction to shift in the direction of the larger number of particles, because this counters the effect of the decreasing concentration of particles.

### PRACTICE EXERCISE

Calculate the percentage of niacin molecules ionized in (a) the solution of Practice Exercise 17.7; (b) a $1.0 \times 10^{-3}\ M$ solution of niacin.

*Answers:* (a) 3.9%; (b) 11.5%

## Polyprotic Acids

Many substances are capable of furnishing more than one proton to water (see, for example, the Lewis structure for citric acid, in the margin). Substances of this type are called **polyprotic acids**. As an example, sulfurous acid, $H_2SO_3$, may react with water in two successive steps:

$$H_2SO_3(aq) \rightleftharpoons H^+(aq) + HSO_3^-(aq) \qquad K_a = 1.7 \times 10^{-2} \qquad [17.15]$$

$$HSO_3^-(aq) \rightleftharpoons H^+(aq) + SO_3^{2-}(aq) \qquad K_a = 6.4 \times 10^{-8} \qquad [17.16]$$

The values for $K_a$ in each case show that the reactions are incomplete. The smaller value for $K_a$ in the second reaction shows that loss of the second proton occurs much less readily than the first. This trend is intuitively reasonable; on the basis of electrostatic attractions we would expect the positively charged proton to be lost more readily from the neutral $H_2SO_3$ molecule than from the negatively charged $HSO_3^-$ ion.

The successive acid-dissociation constants of polyprotic acids are sometimes labeled $K_{a1}$, $K_{a2}$, and so forth. This notation is often simplified to $K_1$, $K_2$, and so forth. For example, the equilibrium constant for the loss of a proton from $HSO_3^-$ (Equation 17.16) can be labeled as $K_{a2}$ or $K_2$, because this proton is the second one removed from the neutral acid,

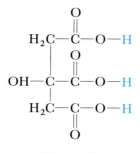

Citric acid

**TABLE 17.4** Acid-dissociation constants of some common polyprotic acids

| Name | Formula | $K_{a1}$ | $K_{a2}$ | $K_{a3}$ |
|------|---------|----------|----------|----------|
| Ascorbic | $H_8C_6O_6$ | $8.0 \times 10^{-5}$ | $1.6 \times 10^{-12}$ | |
| Carbonic | $H_2CO_3$ | $4.3 \times 10^{-7}$ | $5.6 \times 10^{-11}$ | |
| Citric | $H_8C_6O_7$ | $7.4 \times 10^{-4}$ | $1.7 \times 10^{-5}$ | $4.0 \times 10^{-7}$ |
| Oxalic | $H_2C_2O_4$ | $5.9 \times 10^{-2}$ | $6.4 \times 10^{-5}$ | |
| Phosphoric | $H_3PO_4$ | $7.5 \times 10^{-3}$ | $6.2 \times 10^{-8}$ | $4.2 \times 10^{-13}$ |
| Sulfurous | $H_2SO_3$ | $1.7 \times 10^{-2}$ | $6.4 \times 10^{-8}$ | |
| Sulfuric | $H_2SO_4$ | Large | $1.2 \times 10^{-2}$ | |
| Tartaric | $H_6C_4O_6$ | $1.0 \times 10^{-3}$ | $4.6 \times 10^{-5}$ | |

Tartaric acid

$H_2SO_3$. The acid-dissociation constants for a few common polyprotic acids are given in Table 17.4; a more complete list is provided in Appendix E. Notice that the $K_a$ values for successive losses of protons from these acids usually differ by at least a factor of $10^3$. The Lewis structures for tartaric and ascorbic acid are shown in the margin.

Because $K_{a1}$ is so much larger than subsequent dissociation constants for these polyprotic acids, almost all the $H^+(aq)$ in the solution comes from the first ionization reaction. As long as successive $K_a$ values differ by a factor of $10^3$ or more, it is possible to obtain a satisfactory estimate of the pH of polyprotic acid solutions by considering only $K_{a1}$.

Ascorbic acid
(vitamin C)

## SAMPLE EXERCISE 17.9

The solubility of $CO_2$ in pure water at 25°C and 0.1 atm pressure is 0.0037 $M$. The common practice is to assume that all of the dissolved $CO_2$ is in the form of $H_2CO_3$, which is produced by reaction between the $CO_2$ and $H_2O$:

$$CO_2(aq) + H_2O(l) \rightleftharpoons H_2CO_3(aq)$$

What is the pH of a 0.0037 $M$ solution of $H_2CO_3$?

**Solution:** $H_2CO_3$ is a polyprotic acid; the two acid dissociation constants, $K_{a1}$ and $K_{a2}$ (Table 17.4), differ by more than a factor of $10^3$. Consequently, the pH can be determined by considering only $K_{a1}$, thereby treating the acid as if it were a monoprotic acid. Proceeding as in Sample Exercises 17.7 and 17.8, we can write the equilibrium reaction and equilibrium concentrations as follows:

| | $H_2CO_3(aq) \rightleftharpoons$ | $H^+(aq) +$ | $HCO_3^-(aq)$ |
|---|---|---|---|
| **Initial** | 0.0037 $M$ | 0 $M$ | 0 $M$ |
| **Equilibrium** | $(0.0037 - x)$ $M$ | $x$ $M$ | $x$ $M$ |

The equilibrium-constant expression is as follows:

$$K_{a1} = \frac{[H^+][HCO_3^-]}{[H_2CO_3]} = \frac{(x)(x)}{0.0037 - x} = 4.3 \times 10^{-7}$$

Because $K_{a1}$ is small, we make the simplifying approximation that $x$ is small so that $0.0037 - x \simeq 0.0037$. Thus

$$\frac{(x)(x)}{0.0037} = 4.3 \times 10^{-7}$$

Solving for $x$, we have

$$x^2 = (0.0037)(4.3 \times 10^{-7}) = 1.6 \times 10^{-9}$$

$$x = \sqrt{1.6 \times 10^{-9}} = 4.0 \times 10^{-5} \, M = [H^+]$$
$$= [HCO_3^-]$$

The small value of $x$ indicates that our simplifying assumption was justified. The pH is therefore

$$pH = -\log [H^+] = -\log (4.0 \times 10^{-5}) = 4.40$$

If we had been asked to solve for $[CO_3^{2-}]$, we would need to use $K_{a2}$. Let's illustrate that calculation. Using the values of $[HCO_3^-]$ and $[H^+]$ calculated above, and setting $[CO_3^{2-}] = y$, we have the following initial and equilibrium concentration values:

$$HCO_3^-(aq) \rightleftharpoons H^+(aq) + CO_3^{2-}(aq)$$

| | $HCO_3^-(aq)$ | $H^+(aq)$ | $CO_3^{2-}(aq)$ |
|---|---|---|---|
| **Initial** | $4.0 \times 10^{-5}\ M$ | $4.0 \times 10^{-5}\ M$ | $0\ M$ |
| **Equilibrium** | $(4.0 \times 10^{-5} - y)\ M$ | $(4.0 \times 10^{-5} + y)\ M$ | $y\ M$ |

Assuming that $y$ is small compared to $4.0 \times 10^{-5}$, we have

$$K_{a2} = \frac{[H^+][CO_3^{2-}]}{[HCO_3^-]} = \frac{(4.0 \times 10^{-5})(y)}{4.0 \times 10^{-5}}$$
$$= 5.6 \times 10^{-11}$$
$$y = 5.6 \times 10^{-11}\ M = [CO_3^{2-}]$$

The value calculated for $y$ is indeed very small in comparison with $4.0 \times 10^{-5}$, showing that our assumption was justified. It also shows that the ionization of the $HCO_3^-$ is negligible in comparison with that of $H_2CO_3$ as far as production of $H^+$ is concerned. However, it is the *only* source of $CO_3^{2-}$, which has a very low concentration in the solution.

Our calculations thus tell us that in a solution of carbon dioxide in water most of the $CO_2$ is in the form of $CO_2$ or $H_2CO_3$, a small fraction ionizes to form $H^+$ and $HCO_3^-$, and an even smaller fraction ionizes to give $CO_3^{2-}$.

**PRACTICE EXERCISE**

Calculate the pH and concentration of oxalate ion, $[C_2O_4^{2-}]$, in a 0.020 $M$ solution of oxalic acid, $H_2C_2O_4$ (see Table 17.4).

*Answer:* pH = 1.80; $[C_2O_4^{2-}] = 6.4 \times 10^{-5}\ M$

## 17.4 WEAK BASES

Many substances behave as weak bases in water. Such substances react with water, removing protons from $H_2O$, thereby forming the conjugate acid of the base and $OH^-$ ions:

$$\text{Weak base} + H_2O \rightleftharpoons \text{conjugate acid} + OH^- \qquad [17.17]$$

The most commonly encountered weak base is ammonia:

$$NH_3(aq) + H_2O(l) \rightleftharpoons NH_4^+(aq) + OH^-(aq) \qquad [17.18]$$

The equilibrium-constant expression for this reaction can be written as

$$K = \frac{[NH_4^+][OH^-]}{[NH_3][H_2O]} \qquad [17.19]$$

Because the concentration of water is essentially constant even when moderate concentrations of other substances are present, the $[H_2O]$ term is incorporated into the equilibrium constant, giving

$$K[H_2O] = K_b = \frac{[NH_4^+][OH^-]}{[NH_3]} \qquad [17.20]$$

The constant $K_b$ is called the **base-dissociation constant**, by analogy with the acid-dissociation constant, $K_a$, for weak acids. However, as we can see from Equations 17.17 and 17.18, the term "dissociation" in the case of weak bases has a slightly different meaning. It refers to the dissociation of *water* as a result of reaction with the base. Table 17.5 lists

**TABLE 17.5**  Weak bases and their aqueous solution equilibria

| Base | Lewis structure | Conjugate acid | Equilibrium reaction | $K_b$ |
|------|-----------------|----------------|----------------------|-------|
| Ammonia $(NH_3)$ | H—N̈—H<br>│<br>H | $NH_4^+$ | $NH_3 + H_2O \rightleftharpoons NH_4^+ + OH^-$ | $1.8 \times 10^{-5}$ |
| Pyridine $(C_5H_5N)$ | (ring)N: | $C_5H_5NH^+$ | $C_5H_5N + H_2O \rightleftharpoons C_5H_5NH^+ + OH^-$ | $1.7 \times 10^{-9}$ |
| Hydroxylamine $(H_2NOH)$ | H—N̈—OH<br>│<br>H | $H_3NOH^+$ | $H_2NOH + H_2O \rightleftharpoons H_3NOH^+ + OH^-$ | $1.1 \times 10^{-8}$ |
| Methylamine $(NH_2CH_3)$ | H—N̈—CH₃<br>│<br>H | $NH_3CH_3^+$ | $NH_2CH_3 + H_2O \rightleftharpoons NH_3CH_3^+ + OH^-$ | $4.4 \times 10^{-4}$ |
| Nicotine $(C_{10}H_{14}N_2)$ | (structure) | $HC_{10}H_{14}N_2^+$ | $C_{10}H_{14}N_2 + H_2O \rightleftharpoons C_{10}H_{14}N_2H^+ + OH^-$ | $7 \times 10^{-7}$<br>$1.4 \times 10^{-11}$ |
| Hydrosulfide ion $(HS^-)$ | $[H—\ddot{S}:]^-$ | $H_2S$ | $HS^- + H_2O \rightleftharpoons H_2S + OH^-$ | $1.8 \times 10^{-7}$ |
| Carbonate ion $(CO_3^{2-})$ | (structure) | $HCO_3^-$ | $CO_3^{2-} + H_2O \rightleftharpoons HCO_3^- + OH^-$ | $1.8 \times 10^{-4}$ |
| Hypochlorite $(ClO^-)$ | $[:\ddot{Cl}—\ddot{O}:]^-$ | $HClO$ | $ClO^- + H_2O \rightleftharpoons HClO + OH^-$ | $3.3 \times 10^{-7}$ |

the names, formulas, Lewis structures, equilibrium reactions, and values of $K_b$ for several weak bases in water. Appendix E includes a more extensive list. Notice that these bases contain one or more unshared pairs of electrons. An unshared pair is necessary to form the bond with $H^+$.

## SAMPLE EXERCISE 17.10

Calculate the concentration of $OH^-$ in a 0.15 $M$ solution of $NH_3$.

**Solution:**  We use essentially the same procedure here as used in solving problems involving the dissociation of acids. The first step is to write the equilibrium expression and the corresponding equilibrium-constant expression:

$$NH_3(aq) + H_2O(l) \rightleftharpoons NH_4^+(aq) + OH^-(aq)$$

$$K_b = \frac{[NH_4^+][OH^-]}{[NH_3]} = 1.8 \times 10^{-5}$$

We then tabulate the equilibrium concentrations involved in the equilibrium:

| | $NH_3(aq) + H_2O(l) \rightleftharpoons$ | $NH_4^+(aq)$ + | $OH^-(aq)$ |
|---|---|---|---|
| **Initial** | 0.15 $M$ | 0 $M$ | 0 $M$ |
| **Equilibrium** | $(0.15 - x)$ $M$ | $x$ $M$ | $x$ $M$ |

(Notice that we ignore the concentration of $H_2O$, because this is not involved in the equilibrium-constant expression.) Inserting these quantities into the equilibrium-constant expression gives the following:

$$K_b = \frac{[NH_4^+][OH^-]}{[NH_3]} = \frac{(x)(x)}{0.15 - x} = 1.8 \times 10^{-5}$$

Because $K_b$ is small we can neglect the small amount of $NH_3$ that reacts with water, as compared with the total $NH_3$ concentration; that is, we can neglect $x$ in comparison with 0.15 $M$. Then we have

$$\frac{x^2}{0.15} = 1.8 \times 10^{-5}$$

$$x^2 = (0.15)(1.8 \times 10^{-5}) = 0.27 \times 10^{-5}$$
$$x = \sqrt{2.7 \times 10^{-6}} = 1.6 \times 10^{-3}\ M = [OH^-]$$

Notice that the value obtained for $x$ is only about 1 percent of the $NH_3$ concentration, 0.15 $M$. Therefore, our neglect of $x$ in comparison with 0.15 is justified.

## PRACTICE EXERCISE

Which of the following compounds should produce the highest pH as a 0.05 $M$ solution: pyridine, methylamine and nitrous acid?

***Answer:*** methylamine

---

**Amines**

The weak nitrogen bases listed in Table 17.5 belong to a family known as **amines**. These compounds can be thought of as being formed by replacing one or more of the N—H bonds in $NH_3$ with N—C bonds. (In hydroxylamine, $H_2NOH$, one of the N—H bonds of $NH_3$ has been replaced by an N—OH bond.) Like ammonia, such amines are able to extract a proton from the water molecule by forming an N—H bond. The following equation illustrates this behavior:

$$H-\overset{\overset{\displaystyle ..}{}}{\underset{\underset{\displaystyle H}{|}}{N}}-CH_3(aq) + H_2O(l) \rightleftharpoons \left[ H-\overset{\overset{\displaystyle H}{|}}{\underset{\underset{\displaystyle H}{|}}{N}}-CH_3 \right]^+ (aq) + OH^-(aq) \qquad [17.21]$$

**Anions of Weak Acids**

A second common class of weak bases is composed of the anions of weak acids. Consider, for example, an aqueous solution of sodium acetate, $NaC_2H_3O_2$. This salt dissolves in water to give $Na^+$ and $C_2H_3O_2^-$ ions. The $Na^+$ ion is always a spectator ion in acid-base reactions. However, the $C_2H_3O_2^-$ ion is the conjugate base of a weak acid, acetic acid. Consequently, the $C_2H_3O_2^-$ ion is basic and reacts to a slight extent with water ($K_b = 5.6 \times 10^{-10}$):

$$C_2H_3O_2^-(aq) + H_2O(l) \rightleftharpoons HC_2H_3O_2(aq) + OH^-(aq) \qquad [17.22]$$

---

## SAMPLE EXERCISE 17.11

Calculate the pH of a 0.010 $M$ solution of sodium hypochlorite, $NaClO$.

**Solution:** $NaClO$ is an ionic compound consisting of $Na^+$ and $ClO^-$ ions. As such it is a strong electrolyte. The hypochlorite ion, $ClO^-$, supplied by this salt is a weak base. The base dissociation constant for $ClO^-$ is given in Table 17.5: $K_b = 3.3 \times 10^{-7}$. We can write the reaction between $ClO^-$ and water, and the equilibrium concentrations present in this solution, as follows on page 566.

Many amines with low molecular weights have unpleasant, often "fishy" odors. Amines and $NH_3$ are produced by the anaerobic (absence of $O_2$) decomposition of dead animal or plant matter. One such amine, $H_2N(CH_2)_5NH_2$, is known as cadaverine.

Many drugs, including quinine, codeine, caffeine, and amphetamine (Benzedrine), are amines. Like other amines, these substances are weak bases; the amine nitrogen is readily protonated by treatment with an acid. The resulting products are called acid salts. If we use A as the abbreviation for an amine, the acid salt formed by reaction with hydrochloric acid would be written as $AH^+Cl^-$. (It is sometimes written as $A \cdot HCl$ and referred to as a hydrochloride.) For example, amphetamine hydrochloride is the acid salt formed by treating amphetamine with HCl:

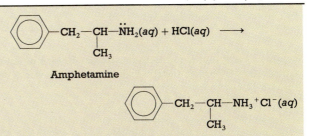

Amphetamine

Amphetamine hydrochloride

Such acid salts are less volatile, more stable, and generally more water soluble than the corresponding neutral amines. Many drugs that are amines are sold and administered as acid salts. Some examples of over-the-counter medications that contain amine hydrochlorides as active ingredients are shown in Figure 17.5.

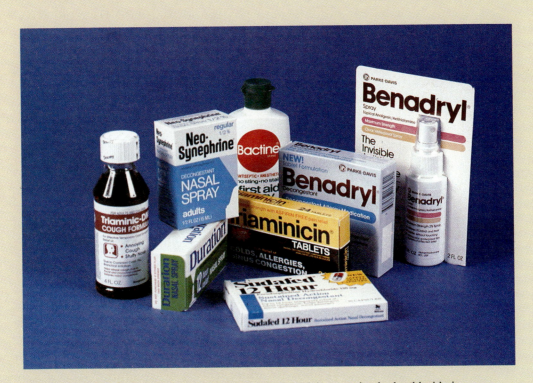

**FIGURE 17.5** Over-the-counter medications in which an amine hydrochloride is a major active ingredient. (Donald Clegg and Roxy Wilson)

| | $ClO^-(aq) + H_2O(l) \rightleftharpoons$ | $HClO(aq)$ + | $OH^-(aq)$ |
|---|---|---|---|
| **Initial** | 0.010 $M$ | 0 $M$ | 0 $M$ |
| **Equilibrium** | $(0.010 - x)$ $M$ | $x$ $M$ | $x$ $M$ |

Because $K_b$ is small, we anticipate that $x$ will be small, so that $(0.010 - x) \simeq 0.010$. Using this approximation, we have

$$K_b = \frac{[HClO][OH^-]}{[ClO^-]} = \frac{(x)(x)}{0.010} = 3.3 \times 10^{-7}$$

Solving for $x$ yields

$$x^2 = (0.010)(3.3 \times 10^{-7}) = 3.3 \times 10^{-9}$$
$$x = [OH^-] = \sqrt{3.3 \times 10^{-9}} = 5.7 \times 10^{-5} \, M$$

$[H^+]$ can be obtained using the ion-product constant for water:

$$[H^+] = \frac{1.0 \times 10^{-14}}{[OH^-]} = \frac{1.0 \times 10^{-14}}{5.7 \times 10^{-5}}$$
$$= 1.8 \times 10^{-10} \, M$$
$$pH = -\log [H^+] = -\log (1.8 \times 10^{-10}) = 9.75$$

Thus we see that this solution of NaClO is slightly basic.

**PRACTICE EXERCISE**

$K_b$ for $BrO^-$ is $5.0 \times 10^{-6}$. Calculate the pH of a 0.050 $M$ solution of NaBrO.
*Answer:* 10.70

---

# 17.5 RELATION BETWEEN $K_a$ AND $K_b$

We've seen in a qualitative way that the stronger acids have the weaker conjugate bases (Section 13.5). The fact that this qualitative relationship exists suggests that we might be able to find a quantitative relationship. Let's explore this matter by considering the $NH_4^+$ and $NH_3$ conjugate acid-base pair. Each of these species reacts with water as follows:

$$NH_4^+(aq) \rightleftharpoons NH_3(aq) + H^+(aq) \qquad [17.23]$$
$$NH_3(aq) + H_2O(l) \rightleftharpoons NH_4^+(aq) + OH^-(aq) \qquad [17.24]$$

Each of these equilibria is expressed by a characteristic dissociation constant:

$$K_a = \frac{[NH_3][H^+]}{[NH_4^+]} \qquad K_b = \frac{[NH_4^+][OH^-]}{[NH_3]}$$

Now notice something very interesting and important. When Equations 17.23 and 17.24 are added together, the $NH_4^+$ and $NH_3$ species cancel, and we are left with just the autoionization of water:

$$NH_4^+(aq) \rightleftharpoons NH_3(aq) + H^+(aq)$$
$$\underline{NH_3(aq) + H_2O(l) \rightleftharpoons NH_4^+(aq) + OH^-(aq)}$$
$$H_2O(l) \rightleftharpoons H^+(aq) + OH^-(aq)$$

To determine what we should do about the equilibrium constants for the added reactions, we make use of a rule that can be derived from the general principles governing chemical equilibria: *When two reactions are added to give a third reaction, the equilibrium constant for the third reaction is given by the product of the equilibrium constants for the two added reactions.* Thus in general,

$$\text{If reaction 1} + \text{reaction 2} = \text{reaction 3}$$

$$\text{then } K_1 \times K_2 = K_3$$

Applying this to our present example, if we multiply $K_a$ and $K_b$, we obtain the following result:

$$K_a \times K_b = \left(\frac{[\cancel{NH_3}][H^+]}{[\cancel{NH_4}^+]}\right)\left(\frac{[\cancel{NH_4}^+][OH^-]}{[\cancel{NH_3}]}\right)$$

$$= [H^+][OH^-] = K_w$$

Thus the result of multiplying $K_a$ times $K_b$ is just the ion-product constant, $K_w$ (Equation 17.3). This is, of course, just what we would expect, because addition of Equations 17.23 and 17.24 gave us just the autoionization equilibrium for water, for which the equilibrium constant is $K_w$.

The relationship we have just found is so important that it should be called to special attention: *The product of the acid-dissociation constant for an acid and the base-dissociation constant for its conjugate base is the ion-product constant for water:*

$$K_a \times K_b = K_w \qquad [17.25]$$

As the strength of an acid increases (larger $K_a$), the strength of its conjugate base must decrease (smaller $K_b$), so that the product $K_a \times K_b$ remains equal to $1.0 \times 10^{-14}$. This point is illustrated by the data shown in Table 17.6.

Because of Equation 17.25 we can calculate $K_a$ for any weak acid if we know $K_b$ for its conjugate base. Similarly, we can calculate $K_b$ for a weak base if we know $K_a$ for its conjugate acid. As a practical consequence, ionization constants are often listed for only one member of a conjugate acid-base pair. For example, Appendix E does not contain $K_b$ values for the anions of weak acids because these can be readily calculated from the tabulated $K_a$ values for their conjugate acids.

If you have the occasion to look up the values for acid- or base-dissociation constants in a chemistry handbook, you may find them expressed as $pK_a$ or $pK_b$, that is, as $-\log K_a$ or $-\log K_b$ (Section 17.1). Equation 17.25 can be put in terms of $pK_a$ and $pK_b$ by taking the negative log of both sides:

$$pK_a + pK_b = pK_w = 14.00 \qquad [17.26]$$

**TABLE 17.6**  Some conjugate acid-base pairs

| Acid | $K_a$ | Base | $K_b$ |
|------|-------|------|-------|
| $HNO_3$ | (Strong acid) | $NO_3^-$ | (Negligible basicity) |
| $HF$ | $6.8 \times 10^{-4}$ | $F^-$ | $1.5 \times 10^{-11}$ |
| $HC_2H_3O_2$ | $1.8 \times 10^{-5}$ | $C_2H_3O_2^-$ | $5.6 \times 10^{-10}$ |
| $H_2CO_3$ | $4.3 \times 10^{-7}$ | $HCO_3^-$ | $2.3 \times 10^{-8}$ |
| $NH_4^+$ | $5.6 \times 10^{-10}$ | $NH_3$ | $1.8 \times 10^{-5}$ |
| $HCO_3^-$ | $5.6 \times 10^{-11}$ | $CO_3^{2-}$ | $1.8 \times 10^{-4}$ |
| $OH^-$ | (Negligible acidity) | $O^{2-}$ | (Strong base) |

This form is particularly useful when the tabulated p$K$ value is that for the conjugate acid or base of the substance of interest. Often the dissociation constants for bases are tabulated as p$K_a$ values for the corresponding conjugate acids. An an example, morphine, a nitrogen-containing base, is listed as the protonated cation, with p$K_a = 7.87$. This means that for the reaction

$$C_{17}H_{19}O_3NH^+(aq) \rightleftharpoons C_{17}H_{19}O_3N(aq) + H^+(aq)$$

the equilibrium constant, $K_a$, has the value $K_a = $ antilog $(-7.87) = 10^{-7.87} = 1.3 \times 10^{-8}$. The reaction

$$C_{17}H_{19}O_3N(aq) + H_2O(l) \rightleftharpoons C_{17}H_{19}O_3NH^+(aq) + OH^-(aq)$$

is described by equilibrium constant $K_b$. Using Equation 17.26 and p$K_a = 7.87$, we have

$$pK_b = 14.00 - pK_a = 14.00 - 7.87 = 6.13$$

Thus $K_b = $ antilog $(-6.13) = 10^{-6.13} = 7.4 \times 10^{-7}$.

---

**SAMPLE EXERCISE 17.12**

Calculate **(a)** the base-dissociation constant, $K_b$, for the fluoride ion, $F^-$; **(b)** the acid-dissociation constant, $K_a$, for the ammonium ion, $NH_4^+$.

**Solution:** **(a)** $K_b$ for $F^-$ is not included in Table 17.5 or in Appendix E. However, $K_a$ for its conjugate acid, HF, is given in Table 17.3 and Appendix E as $K_a = 6.8 \times 10^{-4}$. We can therefore use Equation 17.25 to calculate $K_b$:

$$K_b = \frac{K_w}{K_a} = \frac{1.0 \times 10^{-14}}{6.8 \times 10^{-4}} = 1.5 \times 10^{-11}$$

**(b)** $K_b$ for $NH_3$ is listed in Table 17.5 and in Appendix E as $K_b = 1.8 \times 10^{-5}$. Using Equation 17.25, we can calculate $K_a$ for the conjugate acid, $NH_4^+$:

$$K_a = \frac{K_w}{K_b} = \frac{1.0 \times 10^{-14}}{1.8 \times 10^{-5}} = 5.6 \times 10^{-10}$$

**PRACTICE EXERCISE**
**(a)** Which of the following anions has the largest base-dissociation constant: $NO_2^-$, $PO_4^{3-}$; $N_3^-$?
**(b)** The base quinoline, which has the structure

is listed in handbooks as having a p$K_a$ of 4.90. What is the base-dissociation constant for quinoline?
**Answers:** **(a)** $PO_4^{3-}$; **(b)** $7.9 \times 10^{-10}$

---

## 17.6 ACID-BASE PROPERTIES OF SALT SOLUTIONS

Even before you began this chapter you were undoubtably aware of many substances that are acidic, such as $HNO_3$, HCl, and $H_2SO_4$, and others that are basic, such as NaOH and $NH_3$. However, our recent discussions have indicated that ions can also exhibit acidic or basic properties. For example, we calculated $K_a$ for $NH_4^+$ and $K_b$ for $F^-$ in Sample Exercise 17.12. Such behavior implies that salt solutions can be acidic or basic. Before proceeding with further discussions of acids and bases, let's summarize some features of salts that should bring their acid and base properties into sharper focus.

We can assume that when salts dissolve in water they are completely ionized; nearly all salts are strong electrolytes. Consequently the acid-base properties of salt solutions are due to the behavior of the cations and anions. Many ions are able to react with water to generate $H^+(aq)$ or $OH^-(aq)$. This type of reaction is often called **hydrolysis**.

The anions of weak acids, HX, are basic, and consequently they react with water to produce $OH^-$ ions:

$$X^-(aq) + H_2O(l) \rightleftharpoons HX(aq) + OH^-(aq) \qquad [17.27]$$

In contrast, the anions of strong acids, such as the $NO_3^-$ ion, exhibit no significant basicity; these ions do not hydrolyze and consequently do not influence pH.

Anions of polyprotic acids, such as $HCO_3^-$, that still have ionizable protons are capable of acting as either proton donors or proton acceptors (that is, either acids or bases). Their behavior toward water will be determined by the relative magnitudes of $K_a$ and $K_b$ for the ion, as shown in Sample Exercise 17.13.

---

**SAMPLE EXERCISE 17.13**

Predict whether the salt $Na_2HPO_4$ will form an acidic or basic solution on dissolving in water.

**Solution:** The two possible reactions that $HPO_4^{2-}$ may undergo on addition to water are

$$HPO_4^{2-}(aq) \rightleftharpoons H^+(aq) + PO_4^{3-}(aq) \qquad [17.28]$$

$$HPO_4^{2-}(aq) + H_2O \rightleftharpoons$$
$$H_2PO_4^-(aq) + OH^-(aq) \qquad [17.29]$$

Depending on which of these has the larger equilibrium constant, the ion will cause the solution to be acidic or basic. The value of $K_a$ for reaction Equation 17.28, as shown in Table 17.4, is $4.2 \times 10^{-13}$. We must calculate the value of $K_b$ for reaction Equation 17.29 from the value of $K_a$ for the conjugate acid formed, $H_2PO_4^-$. We make use of the relationship shown in Equation 17.25:

$$K_a \times K_b = K_w$$

We want to know $K_b$ for the base $HPO_4^{2-}$, knowing the value of $K_a$ for the conjugate acid $H_2PO_4^-$:

$$K_b(HPO_4^{2-}) \times K_a(H_2PO_4^-) = K_w = 1.0 \times 10^{-14}$$

Because $K_a$ for $H_2PO_4^-$ is $6.2 \times 10^{-8}$ (Table 17.4), we calculate $K_b$ for $HPO_4^{2-}$ to be $1.6 \times 10^{-7}$. This is considerably larger than $K_a$ for $HPO_4^{2-}$; thus the reaction shown in Equation 17.29 predominates over that in Equation 17.28 and the solution is basic.

**PRACTICE EXERCISE**
Predict whether the potassium salt of citric acid, $K_2H_6C_6O_7$, will form an acidic or basic solution in water (see Table 17.4 for data).

*Answer:* acidic

---

All cations except those of the alkali metals and the heavier alkaline earths ($Ca^{2+}$, $Sr^{2+}$, and $Ba^{2+}$) act as weak acids in water solution. Because the alkali metal and alkaline earth cations do not hydrolyze in water, the presence of any of these ions in solution does not influence pH. It may surprise you that metal ions, such as $Al^{3+}$, and the transition metal ions form weakly acidic solutions. We can take this observation for now as a point of fact. The reasons for this behavior are discussed in Section 17.8.

Among the cations that produce an acidic solution is, of course, $NH_4^+$, which is the conjugate acid of the base $NH_3$. The $NH_4^+$ ion dissociates in water as follows:

$$NH_4^+(aq) \rightleftharpoons H^+(aq) + NH_3(aq) \qquad [17.30]$$

The pH of a solution of a salt can be qualitatively predicted by considering the cation and anion of which the salt is composed. A convenient way to do this is to consider the relative strengths of the acids and bases from which the salt is derived:*

1.  *Salt derived from a strong base and a strong acid:* Examples are NaCl and $Ca(NO_3)_2$, which are derived from NaOH and HCl and from $Ca(OH)_2$ and $HNO_3$, respectively. Neither cation nor anion hydrolyzes. The solution has a pH of 7.

2.  *Salt derived from a strong base and a weak acid:* In this case the anion is a relatively strong conjugate base. Examples are NaClO and $Ba(C_2H_3O_2)_2$. The anion hydrolyzes to produce $OH^-(aq)$ ions. The solution has a pH above 7.

3.  *Salt derived from a weak base and a strong acid:* In this case the cation is a relatively strong conjugate acid. Examples are $NH_4Cl$ and $Al(NO_3)_3$. The cation hydrolyzes to produce $H^+(aq)$ ions. The solution has a pH below 7.

4.  *Salt derived from a weak base and a weak acid:* Examples are $NH_4C_2H_3O_2$, $NH_4CN$, and $FeCO_3$. Both cation and anion hydrolyze. The pH of the solution depends upon the extent to which each ion hydrolyzes. The pH of a solution of $NH_4CN$ is greater than 7 because $CN^-$ ($K_b = 2.0 \times 10^{-5}$) is more basic than $NH_4^+$ ($K_a = 5.6 \times 10^{-10}$) is acidic. Consequently, $CN^-$ hydrolyzes to a greater extent than $NH_4^+$ does.

## SAMPLE EXERCISE 17.14

List the following solutions in the order of increasing pH: (i) 0.1 $M$ $Co(ClO_4)_3$; (ii) 0.1 $M$ RbCN; (iii) 0.1 $M$ $Sr(NO_3)_2$; (iv) 0.1 $M$ $KC_2H_3O_2$.

**Solution:**   The most acidic solution will be (i), which has a metal ion that undergoes hydrolysis and an anion derived from a strong acid. Solution (iii) should have pH of about 7, because it is derived from an alkaline earth cation and the anion of a strong acid. Solutions (ii) and (iv) are both derived from an alkali metal ion, which does not undergo hydrolysis, and an anion of a weak acid. Anion hydrolysis should lead to a basic solution in both cases, but

solution (ii) will be more strongly basic because $CN^-$ is a stronger base than is $C_2H_3O_2^-$. We see that the order of pH is 0.1 $M$ $Co(ClO_4)_3 < 0.1$ $M$ $Sr(NO_3)_2 < 0.1$ $M$ $KC_2H_3O_2 < 0.1$ $M$ RbCN.

### PRACTICE EXERCISE
In each case below, indicate which salt will form the more acidic (or less basic) 0.010 $M$ solution: (a) $NaNO_3$, $Fe(NO_3)_3$; (b) KBr, KBrO; (c) $CH_3NH_3Cl$, $BaCl_2$; (d) $NH_4NO_2$, $NH_4NO_3$.

*Answers:* (a) $Fe(NO_3)_3$; (b) KBr; (c) $CH_3NH_3Cl$; (d) $NH_4NO_3$

## 17.7 ACID-BASE CHARACTER AND CHEMICAL STRUCTURE

From our discussion to this point, we have seen that when any substance is dissolved in water one of three things can happen. The $H^+(aq)$ concentration might increase, in which case the substance behaves as an acid; it might decrease (with corresponding increase in $OH^-$), in which case the substance acts as a base; or, there might be no change in $[H^+]$, an indication that the substance possesses neither acid nor base character. It would be very helpful to have further guidelines as to how acid or base characteristics relate to chemical structure, so that we might be better able to predict how a compound will behave on dissolving in water.

---

* These rules apply to what can be called normal salts. These salts are ones that contain no ionizable protons on the anion. The pH of an acid salt (such as $NaHCO_3$ and $NaH_2PO_4$) is affected not only by the hydrolysis of the anion but also by its acid dissociation as well, as shown in Sample Exercise 17.13.

However, we must expect that any simple rules we might formulate won't always work. Many different factors contribute to ionization in a polar solvent such as water. The best we can hope for are a few rules that are *almost always* obeyed.

When a substance HX transfers a proton to the solvent, an ionic rupture of the H—X bond occurs. Such a reaction will occur most readily when the H—X bond is already polarized in the following sense:

For example, compare $NH_4^+$ and $CH_4$. These two species have the same electronic structure; that is, they are isoelectronic. Both consist of a central atom with an octet of electrons bonding four hydrogens. The difference is in the nuclear charge of the central atom. Because the nuclear charge of N is one greater than for C, the electron pairs shared with the hydrogens are more closely attracted to N in $NH_4^+$ than to C in $CH_4$. That is, the N—H bonds are more polarized than the C—H bonds. Correspondingly, ammonium ion is an acid in water and methane is not:

$$NH_4^+(aq) \rightleftharpoons H^+(aq) + NH_3(aq) \qquad K_a = 5.6 \times 10^{-10} \qquad [17.31]$$
$$CH_4(aq) \rightleftharpoons H^+(aq) + CH_3^-(aq) \qquad \text{no reaction} \qquad [17.32]$$

One other factor of major importance in determining whether a substance acts as an acid is the strength of the H—X bond. Very strong bonds are less easily ionized than weaker ones. This factor is of importance in the case of the hydrogen halides. The H—F bond is the most polar of any H—X bond. One might therefore expect that HF would be a very strong acid, if the first rule were all that mattered. However, the energy required to dissociate HF into H and F atoms is much higher than for the other hydrogen halides, as shown in Table 8.3. As a result, HF is a weak acid, whereas all the other hydrogen halides are strong acids in water.

The factors we have just considered can be used to relate the acid-base properties of the hydride of an element to its position in the periodic table. In any horizontal row of the table, the most basic hydrides are on the left, the most acidic hydrides on the right. For example, in the second row of the table, NaH is a basic hydride. On addition to water it reacts to form $OH^-(aq)$, as described by Equation 17.8. On the right-hand side of the row, the acidity increases in the order $PH_3 < H_2S < HCl$. This general trend is related to the increasing electronegativity of the element as we move from left to right in any horizontal row. In general, *metal hydrides are either basic or show no pronounced acid-base properties in water, whereas nonmetal hydrides range from acidic to showing no pronounced acid-base properties.*

In any vertical row of nonmetallic elements there is a tendency toward increasing acidity with increasing atomic number. For example, among the group 6A elements the acid dissociation constants vary in the order $H_2O < H_2S < H_2Se < H_2Te$. This order arises primarily because the bond strengths steadily decrease in this series as the central atom grows

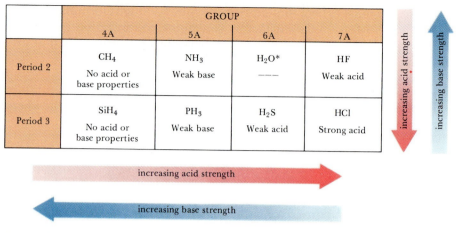

| | GROUP | | | |
|---|---|---|---|---|
| | 4A | 5A | 6A | 7A |
| Period 2 | CH$_4$ <br> No acid or base properties | NH$_3$ <br> Weak base | H$_2$O* <br> ---- | HF <br> Weak acid |
| Period 3 | SiH$_4$ <br> No acid or base properties | PH$_3$ <br> Weak base | H$_2$S <br> Weak acid | HCl <br> Strong acid |

increasing acid strength →

← increasing base strength

**FIGURE 17.6** Acid-base properties of the nonmetal hydrides of periods 2 and 3.

*Water is the solvent in which the acid–base properties of the other substances are measured. Thus, the other compounds are acidic or basic *relative to* water.

larger and the overlaps of atomic orbitals grow smaller. Figure 17.6 illustrates the application of these general correlations to the nonmetal hydrides of periods 2 and 3.

## Oxyacids

Many of the acids commonly encountered involve one or more O—H bonds. For example, H$_2$SO$_4$ contains two such bonds:

$$H-\overset{..}{\underset{..}{O}}-\overset{\overset{\overset{..}{O}}{|}}{\underset{\underset{..}{O:}}{S}}-\overset{..}{\underset{..}{O}}-H$$

Substances in which OH groups and possibly additional oxygen atoms are bound to a central atom are called **oxyacids**. Let's consider, then, an OH group bound to some other atom Y, which might in turn have other groups attached to it:

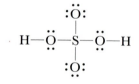

At one extreme, Y might be a metal, such as Na, K, or Mg. The pair of electrons shared between Y and O is then completely transferred to oxygen, and an ionic compound involving OH$^-$ is formed. Because of the charge that surrounds it, the oxygen of the OH$^-$ ion does not strongly attract to itself the electron pair it shares with hydrogen. That is, the O—H bond in OH$^-$ is not strongly polarized. Therefore, the hydrogen of OH$^-$ has no tendency to transfer to the solvent as H$^+$(aq). Such compounds therefore behave as bases.

When Y is an element of intermediate electronegativity, around 2.0, the bond to O is more covalent in character, and the substance does not readily lose OH$^-$. Elements with electronegativities in this range include B, C, P, As, and I (Figure 8.10). Examples of acids of such elements

**TABLE 17.7** Acid-dissociation constants for orthoboric acid, hypoiodous acid, and methanol

| Acid | $K_a$ |
|---|---|
| $\begin{array}{c} O-H \\ \vert \\ H-O-B-O-H \end{array}$<br>Orthoboric acid | $6.5 \times 10^{-10}$ |
| $I-O-H$<br>Hypoiodous acid | $2.3 \times 10^{-11}$ |
| $CH_3-O-H$<br>Methanol | Not measurable |

include orthoboric acid, hypoiodous acid, and methanol, the structures of which are shown in Table 17.7.

Such substances might behave as acids in water, depending on the ease with which the proton is lost from oxygen. As a general rule, the more strongly the group Y attracts the electron pair it shares with the oxygen, the more polar the OH bond will be and the more acidic the substance. In the three examples just given, the central atom does not strongly attract the electron pair it shares with oxygen. As a result, the acid-dissociation constants are small, as shown in Table 17.7.

As the electronegativity of Y increases, or as groups with greater electron-attracting ability are placed on Y, the acidic properties of the substance increase. Two simple rules relate the acid strengths of oxyacids to the electronegativity of Y and to the number of groups attached to Y:

1. *For acids that have the same structure but differ in the electronegativity of the central atom, Y, acid strength increases with increasing electronegativity.* Examples are shown in Table 17.8.

**TABLE 17.8** Acid-dissociation constants ($K_a$) of oxyacids in comparison with electronegativity values (EN) of atom Y

| H—O—Y | $K_a$ | EN of Y | $\begin{array}{c} O \\ \parallel \\ H-O-Y-O-H \end{array}$ | $K_{a1}$ | EN of Y |
|---|---|---|---|---|---|
| HClO | $3 \times 10^{-8}$ | 3.2 | $H_2SO_3$ | $1.7 \times 10^{-2}$ | 2.6 |
| HBrO | $2 \times 10^{-9}$ | 3.0 | $H_2SeO_3$ | $3.5 \times 10^{-3}$ | 2.6 |
| HIO | $2 \times 10^{-11}$ | 2.7 | $H_2CO_3$ | $4.3 \times 10^{-7}$ | 2.5 |
| HOCH$_3$ | ~0 | 2.5[a] | | | |

[a] This value is the electronegativity for carbon.

2. *In a series of acids that have the same central atom, Y, but differ in the number of attached groups, the acid strength increases with increasing oxidation number of the central atom.* For example, in the series of oxyacids of chlorine extending from hypochlorous to perchloric acid, acid strength steadily increases:

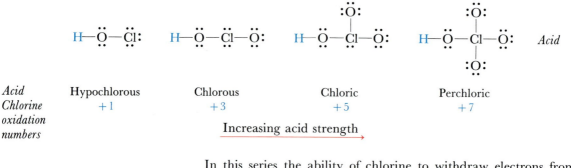

| Acid | Hypochlorous | Chlorous | Chloric | Perchloric | *Acid* |
| --- | --- | --- | --- | --- | --- |
| *Chlorine oxidation numbers* | +1 | +3 | +5 | +7 | |

Increasing acid strength

In this series the ability of chlorine to withdraw electrons from the OH group—and thus make the O—H bond even more polar—increases as electron-withdrawing oxygen atoms are added to the chlorine. It is interesting to note that this correlation can be stated in another, but equivalent, way: In a series of oxyacids, the acidity increases with the number of non-protonated oxygens bound to the central atom.

---

**SAMPLE EXERCISE 17.15**

Arrange the compounds in each of the following series in the order of increasing acid strength: **(a)** $AsH_3$, HI, NaH, $H_2O$; **(b)** $H_2SeO_3$, $H_2SeO_4$, $H_2O$.

**Solution:** **(a)** We have seen that the elements from the left side of the periodic table are the most basic hydrides, because the hydrogen in these compounds carries a negative charge. Thus NaH should be the most basic hydride in the list. Because arsenic is a less electronegative element than oxygen, we might expect that $AsH_3$ would be a weak base toward water. That is also what we would predict by an extension of the trends shown in Figure 17.6. Further, we expect that the hydrides of the halogens, as the most electronegative element in each period, will be acidic relative to water. Finally, we know that HI is one of the strong acids in water. Thus the order of increasing acidity is NaH < $AsH_3$ < $H_2O$ < HI.

**(b)** The rule for oxyacids is that the acidity in-

creases with increasing oxidation state on the central atom. $H_2SeO_4$, as the acid representing the highest oxidation state of Se, should be a comparatively strong acid, much like $H_2SO_4$. Because it is the product of the reaction of a nonmetal oxide with water ($SeO_2 + H_2O \longrightarrow H_2SeO_3$), we expect that $H_2SeO_3$ will be acidic in water. That is, $H_2SeO_3$ is a stronger acid than $H_2O$. However, it should be a weaker acid than $H_2SeO_4$, because Se is in a lower oxidation state in $H_2SeO_3$. Thus the order of increasing acidity is $H_2O$ < $H_2SeO_3$ < $H_2SeO_4$.

**PRACTICE EXERCISE**
In each of the following pairs, choose the compound that leads to the more acidic (or less basic) solution: **(a)** HBr, HF; **(b)** $PH_3$, $H_2S$; **(c)** $HNO_2$, $HNO_3$; **(d)** $H_2SO_3$, $H_4SiO_4$.

*Answers:* **(a)** HBr; **(b)** $H_2S$; **(c)** $HNO_3$; **(d)** $H_2SO_3$

---

## 17.8 THE LEWIS THEORY OF ACIDS AND BASES

For a substance to be a proton acceptor (that is, a base in the Brønsted-Lowry sense), that substance must possess an unshared pair of electrons for binding the proton. For example, we have seen that $NH_3$ acts as a proton acceptor. Using Lewis structures we can write the reaction between $H^+$ and $NH_3$ as follows:

$$H^+ + \overset{H}{\underset{H}{:N-H}} \longrightarrow \left[ \overset{H}{\underset{H}{H-N-H}} \right]^+ \qquad [17.33]$$

G. N. Lewis was the first to notice this aspect of acid-base reactions. He proposed a definition of acid and base that emphasizes the shared electron pair: A **Lewis acid** is defined as an electron-pair acceptor and a **Lewis base** as an electron-pair donor.

Every base that we have discussed thus far—whether it be $OH^-$, $H_2O$, an amine, or an anion—is an electron-pair donor. Everything that is a base in the Brønsted-Lowry sense (a proton acceptor) is also a base in the Lewis sense (an electron-pair donor). However, in the Lewis theory, a base can donate its electron pair to something other than $H^+$. The Lewis definition therefore greatly increases the number of species that can be considered as acids; $H^+$ is a Lewis acid, but not the only one. For example, consider the reaction between $NH_3$ and $BF_3$. This reaction occurs because $BF_3$ has a vacant orbital in its valence shell (Section 8.7). It therefore acts as an electron-pair acceptor (a Lewis acid) toward $NH_3$, which donates the electron pair:

$$
\underset{\text{Base}}{\begin{array}{c} H \\ | \\ H-N: \\ | \\ H \end{array}} + \underset{\text{Acid}}{\begin{array}{c} F \\ | \\ B-F \\ | \\ F \end{array}} \longrightarrow \begin{array}{cc} H & F \\ | & | \\ H-N-B-F \\ | & | \\ H & F \end{array} \qquad [17.34]
$$

Our emphasis throughout this chapter has been on water as the solvent and on the proton as the source of acidic properties. In such cases we find the Brønsted-Lowry definition of acids and bases to be the most useful one to use. In fact, when we speak of a substance as being acidic or basic, we are usually thinking of aqueous solutions and using these terms in the Arrhenius or Brønsted-Lowry sense (Sections 13.4 and 13.5). The advantage of the Lewis theory is that it allows us to treat a wider variety of reactions, including ones that do not involve proton transfer, as acid-base reactions. To avoid confusion, a substance like $BF_3$ is rarely called an acid unless it is clear from the context that we are using the term in the sense of the Lewis definition. Instead, substances that function as electron-pair acceptors are referred to explicitly as "Lewis acids."

Lewis acids include molecules that like $BF_3$ have an incomplete octet of electrons. In addition, many simple cations can function as Lewis acids. For example, $Fe^{3+}$ interacts strongly with cyanide ions to form the ferri-cyanide ion, $Fe(CN)_6^{3-}$:

$$ Fe^{3+} + 6 : C \equiv N : ^- \longrightarrow [Fe(C \equiv N :)_6]^{3-} $$

The $Fe^{3+}$ ion has vacant orbitals to accept the electron pairs donated by the $CN^-$ ions; we will learn more in Chapter 26 about just which orbitals are used by the $Fe^{3+}$. That the metal ion is highly charged also contributes to the interaction with $CN^-$ ions.

Some compounds with multiple bonds can behave as Lewis acids. For example, the reaction of carbon dioxide with water to form carbonic acid, $H_2CO_3$, can be pictured as an attack by a water molecule on $CO_2$, in which the water acts as an electron-pair donor, and the $CO_2$ as an electron-pair acceptor, as shown in the right margin:

The electron pair of one of the carbon-oxygen double bonds is moved onto the oxygen to leave a vacant orbital on the carbon, which can act as electron-pair acceptor. We have shown the shift of these electrons with arrows. After forming the initial acid-base product, a proton moves from one oxygen to another, thereby forming carbonic acid:

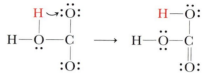

A similar kind of Lewis acid-base reaction takes place when any oxide of a nonmetal dissolves in water to form an acidic solution.

## Hydrolysis of Metal Ions

The Lewis theory is also helpful in explaining why solutions of many metal ions show acidic properties (Section 17.6). For example, a solution of $Cr(NO_3)_3$ is quite acidic. A solution of $ZnCl_2$ is also acidic, though to a lesser extent. To understand why this is so, we must examine the interaction between a metal ion and water molecules.

Because metal ions are positively charged, they attract the unshared electron pairs of water molecules. It is primarily this interaction, referred to as **hydration**, that causes salts to dissolve in water, as explained in Section 12.2. The strength of attraction increases with the charge of the ion and is strongest for the smallest ions. The ratio of ionic charge to ionic radius provides a good measure of the extent of hydration. This ratio is listed in Table 17.9 for a selection of metal ions. The process of hydration is a Lewis acid-base interaction, in which the metal ion acts as a Lewis acid and the water molecules as Lewis bases. When the water molecule interacts with the positively charged metal ion, electron density is drawn from the oxygen, as illustrated in Figure 17.7. This flow of electron density causes the O—H bond to become more polarized; as a result, water molecules bound to the metal ion, M, are more acidic than those in the bulk solvent. The hydrated metal ion thus acts as a source of protons:

$$M(H_2O)_n{}^{z+} \rightleftharpoons M(H_2O)_{(n-1)}(OH)^{(z-1)+} + H^+(aq) \qquad [17.35]$$

In this equation $z$ is the charge on the metal ion, and $n$ is the number of hydrating water molecules. For the 3+ metal ions listed in Table 17.9, $n$ is 6; for the other ions it is probably closer to 4, although the exact number is difficult to determine. The hydrolysis reaction shown in Equation 17.35 represents the behavior of an acid in just the same way as Equation 17.10, which applies to HX. The "acid" in Equation 17.35 is not just a single molecule, but a collection of molecules. For example it might be $Fe(H_2O)_6{}^{3+}$, which we usually represent merely as $Fe^{3+}(aq)$. The acid hydrolysis constants (that is, the equilibrium constants for Equation 17.35) are listed in Table 17.9 for several ions whose charge/ionic radius ratios are also given. Notice that there is a general

**FIGURE 17.7** Interaction of a water molecule with a cation of 1+ charge or 3+ charge. The interaction is much stronger with the smaller ion of higher charge.

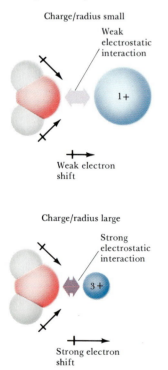

Charge/radius small

Weak electrostatic interaction

1+

Weak electron shift

Charge/radius large

Strong electrostatic interaction

3+

Strong electron shift

**TABLE 17.9**  Ionic charge/ionic radius ratio and acid hydrolysis constant for metal ions of various charges

| Metal ion | Ionic charge Ionic radius | Acid hydrolysis constant, $K_h$ |
|-----------|---------------------------|----------------------------------|
| $Na^+$ | 1.0 | Negligible |
| $Li^+$ | 1.5 | $2 \times 10^{-14}$ |
| $Ca^{2+}$ | 2.1 | $2 \times 10^{-13}$ |
| $Mg^{2+}$ | 3.1 | $4 \times 10^{-12}$ |
| $Zn^{2+}$ | 2.7 | $1 \times 10^{-9}$ |
| $Cu^{2+}$ | 2.8 | $1 \times 10^{-8}$ |
| $Al^{3+}$ | 6.7 | $1 \times 10^{-5}$ |
| $Cr^{3+}$ | 4.8 | $1 \times 10^{-4}$ |
| $Fe^{3+}$ | 4.7 | $2 \times 10^{-3}$ |

trend toward larger acid hydrolysis constants as the ionic charge/ionic radius ratio increases. The tendency to transfer a proton to water is greatest for the smallest and most highly charged ions, as demonstrated in Figure 17.8. Note that as the charge on the metal ion increases, the solution becomes increasingly more acidic.

**FIGURE 17.8**  pH values of 1.0 $M$ solutions of a series of nitrate salts. From left to right, $NaNO_3$, $Ca(NO_3)_2$, $Zn(NO_3)_2$, and $Al(NO_3)_3$. (Donald Clegg and Roxy Wilson).

| Salt | $NaNO_3$ | $Ca(NO_3)_2$ | $Zn(NO_3)_2$ | $Al(NO_3)_3$ |
|------|----------|--------------|--------------|--------------|
| Indicator | Bromthymol Blue | Bromthymol Blue | Methyl Red | Methyl Orange |
| Estimated pH | 7.0 | 6.9 | 5.5 | 3.5 |

# FOR REVIEW

## SUMMARY

In this chapter we have considered several aspects of solutions of acids and bases in water. These have included the quantitative aspects of acid-base equilibria and the relationship between molecular structure and acid-base properties. We have also looked briefly at Lewis acid-base theory, which encompasses acid-base reactions beyond those involving proton transfer.

Water spontaneously ionizes to a slight degree (**autoionization**), forming $H^+(aq)$ and $OH^-(aq)$. The extent of ionization is expressed by the **ion-product constant** for water:

$$K_w = [H^+][OH^-] = 1.0 \times 10^{-14}$$

This relationship describes not only pure water, but aqueous solutions as well. Because the concentration of water is effectively constant in dilute solutions, $[H_2O]$ is omitted from this equilibrium-constant expression as well as from others associated with reactions in aqueous solution.

The concentration of $H^+(aq)$ is often expressed on the **pH scale**: $pH = -\log [H^+]$. Solutions of pH less than 7 are acidic; those with pH greater than 7 are basic.

Through most of this chapter we have relied on the **Brønsted-Lowry theory** of acids and bases. According to this theory, an acid is a proton $(H^+)$ donor, a base is a proton acceptor. Reaction of an acid with water results in the formation of $H^+(aq)$ and the **conjugate base** of the acid. **Strong acids** have conjugate bases that are weaker proton acceptors than $H_2O$. Such acids are strong electrolytes, ionizing completely in solution. The common strong acids are HCl, HBr, HI, $HNO_3$, $HClO_4$, and $H_2SO_4$. **Weak acids** are substances for which the reaction with water is incomplete, and an equilibrium is established. The extent to which the reaction proceeds is expressed by the **acid-dissociation constant**, $K_a$. **Polyprotic acids** are acids such as $H_2SO_3$ that have more than one ionizable proton. These acids have more than one acid-dissociation constant: $K_{a1}$,

$K_{a2}$, and so forth, which decrease in magnitude in the order $K_{a1} > K_{a2} > K_{a3}$.

Aside from the ionic hydroxides such as NaOH, bases produce an increase of $OH^-$ by reaction with water. **Strong bases** have **conjugate acids** that are no stronger as proton donors than $H_2O$. The common strong bases are the hydroxides and oxides of the alkali metals and alkaline earths. **Weak bases** include $NH_3$, amines, and the anions of weak acids. The extent to which a weak base reacts with water to generate $OH^-$ and the conjugate acid of the base is measured by the **base-dissociation constant**, $K_b$.

The stronger an acid, the weaker its conjugate base; the weaker an acid, the stronger its conjugate base. This qualitative observation is expressed quantitatively by the expression $K_a \times K_b = K_w$ (where $K_a$ and $K_b$ are dissociation constants for conjugate acid-base pairs).

The acid-base properties of salts can be ascribed to the behavior of their respective cations and anions. The reaction of ions with water with a resultant change in pH is called **hydrolysis**. The cations of strong bases (the alkali metal ions and alkaline earth metal ions) and the anions of strong acids do not undergo hydrolysis.

The tendency of a substance to show acidic or basic characteristics in water can be correlated reasonably well with chemical structure. Acid character requires the presence of a highly polar H—X bond, promoting loss of hydrogen as $H^+$ on reaction with water. Basic character, on the other hand, requires the presence of an available pair of electrons. By considering the effects of changes in structure, it is possible to predict how a given structural change is likely to alter the acidity or basicity.

The **Lewis theory** of acids and bases emphasizes the shared electron pair rather than the proton. An acid is defined as an electron-pair acceptor, a base as an electron-pair donor. The Lewis theory is more general than the Brønsted-Lowry model, because it applies to cases in which the proton is the acid and to others as well.

# LEARNING GOALS

Having read and studied this chapter, you should be able to:

1. Explain what is meant by the autoionization of water, and write the ion-product-constant expression.

2. Explain what is meant by pH; calculate pH from a knowledge of $[H^+]$ or $[OH^-]$, and perform the reverse operation.

3. Calculate $[OH^-]$ from pOH and $K$ from $pK$, and be able to perform the reverse operations.

4. Identify the common strong acids and bases.

5. Write the acid-dissociation-constant expression for any weak acid in water.

6. Calculate $[H^+]$ for a weak acid solution in water, knowing acid concentration and $K_a$; calculate $K_a$ for a weak acid, knowing the acid concentration and $[H^+]$.

7. Write the base-dissociation-constant expression for a weak base in water.

8. Calculate $[H^+]$ for any weak base solution in water, knowing the base concentration and $K_b$; calculate $K_b$ for a weak base, knowing the base concentration and $[H^+]$.

9. Calculate the percent dissociation for an acid or base, knowing its concentration in solution and the acid- or base-dissociation constant.

10. Explain the relationship between an acid and its conjugate base, and between a base and its conjugate acid; calculate $K_b$ from a knowledge of $K_a$, and vice versa.

11. Predict whether a particular salt solution will be acidic, basic, or neutral.

12. Explain how acid strength relates in a general way to the nature of the H—X bond.

13. Predict the relative acid strengths of oxyacids and oxyanions.

14. Define an acid or base in terms of the Lewis acid-base theory.

15. Predict the relative acidities of solutions of metal salts from a knowledge of metal-ion charges and ionic radii.

# KEY TERMS

Among the more important terms and expressions used for the first time in this chapter are the following:

An **acid-base indicator** (Section 17.1) is a substance whose color changes in passing from an acidic to a basic form, or vice versa.

The **acid-dissociation constant**, $K_a$ (Section 17.3) is an equilibrium constant that expresses the extent to which an acid transfers a proton to solvent water.

**Autoionization** of water (Section 17.1) is the process whereby water spontaneously forms low concentration of $H^+(aq)$ and $OH^-(aq)$ ions by proton transfer from one water molecule to another.

The **base-dissociation constant**, $K_b$ (Section 17.4) is an equilibrium constant that expresses the extent to which a base reacts with solvent water, accepting a proton and forming $OH^-(aq)$.

**Hydrolysis** (Section 17.6) is a process in which a cation or anion reacts with water so as to change the pH.

The **ion-product constant** (Section 17.1) for water, $K_w$, is the product of the aquated hydrogen ion and hydroxide ion concentrations: $[H^+][OH^-] = K_w = 1.0 \times 10^{-14}$.

A **Lewis acid** (Section 17.8) is defined as an electron-pair acceptor.

A **Lewis base** (Section 17.8) is defined as an electron-pair donor.

An **oxyacid** (Section 17.7) is a compound in which one or more OH groups, and possibly additional oxygen atoms, are bonded to a central atom.

The term **pH** (Section 17.1) is defined as the negative log in base 10 of the aquated hydrogen-ion concentration: $pH = -\log[H^+]$.

A **polyprotic acid** (Section 17.3) is a substance capable of dissociating more than one proton in water; $H_2SO_4$ is an example.

# EXERCISES

## Autoionization of Water: pH

**17.1** What experimental evidence can you cite to support the idea that water autoionizes?

**17.2** What are the characteristic acid and base species that result from the autoionization of water?

**17.3** Indicate whether each of the following solutions is acidic, basic, or neutral: **(a)** $5.0 \times 10^{-5}\ M\ H^+$; **(b)** $3.5 \times 10^{-6}\ M\ OH^-$; **(c)** $1 \times 10^{-7}\ M\ H^+$; **(d)** $2.2 \times 10^{-11}\ M\ H^+$.

**17.4** Indicate whether each of the following solutions is acidic, basic, or neutral: **(a)** $8.0 \times 10^{-6}\ M\ OH^-$; **(b)** $1 \times 10^{-7}\ M\ OH^-$; **(c)** $6.6 \times 10^{-10}\ M\ OH^-$; **(d)** $0.0020\ M\ H^+$.

**17.5** Calculate the pH corresponding to each of the following concentrations of $H^+(aq)$ or $OH^-(aq)$: **(a)** $2.0 \times 10^{-4}\ M\ H^+$; **(b)** $0.0034\ M\ H^+$; **(c)** $3.3 \times 10^{-3}\ M\ OH^-$; **(d)** $7.4 \times 10^{-10}\ M\ OH^-$.

**17.6** Calculate the pH corresponding to each of the following concentrations of $H^+(aq)$ or $OH^-(aq)$: **(a)** $5.3 \times 10^{-12}\ M\ H^+$; **(b)** $0.0210\ M\ OH^-$; **(c)** $9.2 \times 10^{-3}\ M\ H^+$; **(d)** $5.0 \times 10^{-8}\ M\ OH^-$.

**17.7** Calculate $[H^+]$ for solutions with each of the following pH values: **(a)** 3.4; **(b)** 8.8; **(c)** 0.0; **(d)** 11.9.

**17.8** Calculate $[OH^-]$ for solutions with each of the following pH values: **(a)** 2.2; **(b)** 11.44; **(c)** 6.95; **(d)** 13.0.

**17.9** **(a)** The $[OH^-]$ in a given solution is $5.2 \times 10^{-4}\ M$; calculate pOH and pH. **(b)** What are the pH and pOH when $[H^+]/[OH^-] = 1.0 \times 10^3$?

**17.10** **(a)** The $[H^+]$ in a given solution is $2.0 \times 10^{-5}\ M$; calculate pOH and pH. **(b)** What are the pH and pOH when $[H^+]/[OH^-] = 1.6 \times 10^{-3}$?

**17.11** By what factor does $[H^+]$ change for a pH change of **(a)** 1.00 unit; **(b)** 3.00 units; **(c)** 0.15 unit?

**17.12** What is the ratio of $[H^+]$ at the acidic end of the change range for methyl orange to that at the basic end of the change range (Table 17.1)?

## Strong Acids and Bases; Weak Acids

**17.13** Calculate the pH of each of the following solutions: **(a)** $3.0 \times 10^{-4}\ M\ HBr$; **(b)** $0.0035\ M\ Ca(OH)_2$; **(c)** 3.00 g of NaOH in 300 mL of solution; **(d)** 0.424 g of $HClO_4$ in 600 mL of solution.

**17.14** Calculate the pH of each of the following solutions: **(a)** $4.2 \times 10^{-3}\ M\ KOH$; **(b)** $2.3 \times 10^{-4}\ M\ Ba(OH)_2$; **(c)** 2.00 mL of 6.0 $M\ HNO_3$ solution added to 398 mL of water; **(d)** 10.0 ml of $1 \times 10^{-3}\ M\ HCl$ solution plus 10.0 mL of $1.2 \times 10^{-1}\ M\ HNO_3$ solution.

## Weak Acids

**17.15** Using the data in Table 17.3, indicate which substance in each of the following pairs is the stronger acid in aqueous solution: **(a)** $HOC_6H_5$ or $HNO_2$; **(b)** HCN or HClO; **(c)** $HC_2H_3O_2$ or HClO.

**17.16** Using the data in Appendix E, indicate which substance in each of the following pairs is the stronger acid in aqueous solution: **(a)** formic acid or benzoic acid; **(b)** hydrofluoric acid or hydrocyanic acid; **(c)** nitrous acid or acetic acid.

**17.17** Write the acid-dissociation equation and acid-dissociation-constant expression for **(a)** chloric acid, $HClO_3$; **(b)** acetic acid, $HC_2H_3O_2$.

**17.18** Write the acid-dissociation equation and acid-dissociation-constant expression for **(a)** hydrogen sulfate ion, $HSO_4^-$; **(b)** hypoiodous acid, HIO.

**17.19** Lactic acid, $HC_3H_5O_3$, has one dissociable hydrogen. A 0.10 $M$ solution of lactic acid has a pH of 2.44. Calculate $K_a$.

**17.20** A 0.050 $M$ solution of $KHCrO_4$ has a pH of 3.80. Calculate $K_a$ for $HCrO_4^-$.

**17.21** Calculate the concentration of $H^+(aq)$ in each of the following solutions ($K_a$ values are given in Appendix E): **(a)** 0.025 $M$ hypochlorous acid, HClO; **(b)** 0.120 $M$ hydrazoic acid, $HN_3$; **(c)** 0.0068 $M$ phenol, $HOC_6H_5$.

**17.22** Calculate the pH of each of the following solutions ($K_a$ values are given in Appendix E): **(a)** 0.010 $M$ propionic acid; **(b)** 0.035 $M$ hypoiodous acid; **(c)** 0.040 $M$ benzoic acid.

**17.23** Calculate the percent ionization of hydrazoic acid in solutions of each of the following concentrations ($K_a$ given in Appendix E): **(a)** 0.400 $M$; **(b)** 0.100 $M$; **(c)** 0.0400 $M$.

**17.24** Calculate the percent ionization of $HCrO_4^-$ in solutions of each of the following concentrations ($K_a$ given in Appendix E): **(a)** 0.250 $M$; **(b)** 0.0800 $M$; **(c)** 0.0200 $M$.

**17.25** Calculate the percent ionization of chloroacetic acid in 0.0200 $M$ solution ($K_a$ given in Appendix E).

**17.26** Calculate the percent ionization of cyanic acid in 0.0150 $M$ solution ($K_a$ given in Appendix E).

**[17.27]** Show that for a weak acid the percent ionization should vary as the inverse square root of the acid concentration.

**[17.28]** For solutions of a weak acid a graph of pH versus the log of the initial acid concentration should be a straight line. What is the magnitude of the slope of that line?

**17.29** A 0.200 $M$ solution of a weak acid HX is 9.4 percent ionized. Using this information, calculate $[H^+]$, $[X^-]$, [HX], and $K_a$ for HX.

**17.30** A 0.10 $M$ solution of bromoacetic acid, $CH_2BrCOOH$, is 13.2 percent ionized. Using this information, calculate $[CH_2BrCOO^-]$, $[H^+]$, $[CH_2BrCOOH]$, and $K_a$ for bromoacetic acid.

**[17.31]** Citric acid is present in citrus fruits. As indicated in Table 17.4, it is a triprotic acid. Calculate the pH of a 0.050 $M$ solution of citric acid. Explain any approximations or assumptions that you make in your calculations.

**[17.32]** Tartaric acid is found in many plants (grapes, for example). It is partly responsible for the dry texture of certain wines. Calculate the pH of a 0.025 $M$ solution of tartaric acid, for which the acid dissociation constants are listed in Table 17.4. Explain any approximations or assumptions that you make in your calculation.

## Weak Bases; $K_a$-$K_b$ Relationship

**17.33** Write the balanced net ionic equation for the reaction of each of the following bases with water. Also write the base-dissociation-constant expression for each substance: **(a)** propyl amine, $C_3H_7NH_2$; **(b)** cyanide ion, $CN^-$; **(c)** formate ion, $CHO_2^-$.

**17.34** Write the balanced net ionic equation for the reaction of each of the following bases with water. Also write the base-dissociation-constant expression for each substance: **(a)** hydrazine, $H_2NNH_2$; **(b)** methyl amine, $CH_3NH_2$; **(c)** benzoate ion, $C_6H_5CO_2^-$.

**17.35** Calculate $[OH^-]$ and pH for each of the following solutions ($K_b$ values in Appendix E): **(a)** 0.050 $M$ pyridine; **(b)** 0.020 $M$ hydroxylamine; **(c)** $3.0 \times 10^{-3}$ $M$ $NH_3$.

**17.36** Calculate $[OH^-]$ and pH for each of the following solutions ($K_a$ or $K_b$ values in Appendix E): **(a)** 0.020 $M$ NaOCl; **(b)** 0.100 $M$ methylamine; **(c)** $2.0 \times 10^{-3}$ $M$ dimethylamine ($K_b = 5.4 \times 10^{-4}$).

**17.37** Using the values of $K_a$ from Appendix E, calculate the base-dissociation constant for each of the following species: **(a)** nitrite ion, $NO_2^-$; **(b)** azide ion, $N_3^-$; **(c)** hydrogen phosphate ion, $HPO_4^{2-}$; **(d)** formate ion, $CHO_2^-$.

**17.38** Using the values of $K_b$ from Appendix E, calculate $K_a$ for each of the following species: **(a)** dimethylammonium ion, $(CH_3)_2NH_2^+$; **(b)** hydrazinium ion, $H_3NNH_2^+$; **(c)** pyridinium ion, $C_5H_5NH^+$; **(d)** hydroxylammonium ion, $HONH_3^+$.

**17.39** The $pK_a$ values for $H_2SeO_3$ at 25°C are 2.64 and 8.27. Write the base-hydrolysis-equilibrium equations for the $SeO_3^{2-}(aq)$ and $HSeO_3^-(aq)$ ions, and calculate the equilibrium-constant values for each equilibrium.

**17.40** *Ortho*-phthalic acid, $C_6H_4(CO_2H)_2$, has the following structure:

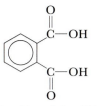

The $K_a$ values for this acid are $1.3 \times 10^{-3}$ and $3.9 \times 10^{-8}$. Calculate the base-dissociation constant for the phthalate, $C_6H_4(CO_2)_2^{2-}$, and for the hydrogen phthalate, $C_6H_4(CO_2H)(CO_2)^-$, ions.

**17.41** Sodium carbonate, $Na_2CO_3$, sodium ascorbate, $NaC_6H_7O_6$, and sodium bicarbonate, $NaHCO_3$, are all used in antacid tablets. For equal molar amounts, which of these substances will produce the most basic solution on dissolving in water?

**17.42** Arrange the following substances in the order of increasing acidity of their 0.100 $M$ solutions in water: $(CH_3)_2NH_2Cl$, $HN_3$, HCN, $H_3BO_3$.

## Salt Solutions

**17.43** Indicate whether each of the following substances would form an acidic, basic, or neutral solution in water: **(a)** $KC_2H_3O_2$; **(b)** $NaHCO_3$; **(c)** $CH_3NH_3Br$; **(d)** $KNO_2$.

**17.44** Indicate whether each of the following substances would form an acidic, basic, or neutral solution in water: **(a)** $NaHC_2O_4$; **(b)** CsI; **(c)** $Al(NO_3)_3$; **(d)** $NH_4CN$.

**17.45** Arrange the following substances in the order of increasing base strength: $N_3^-$; $NO_3^-$; $H_3O^+$; $HPO_4^-$.

**17.46** Arrange the following substances in order of increasing pH of a 0.010 $M$ solution of each: sodium acetate, $NaC_2H_3O_2$; chloroacetic acid, $HC_2H_2O_2Cl$; pyridinium nitrate, $C_5H_5NHNO_3$; potassium nitrate, $KNO_3$.

**17.47** Of these salts—KCNO, $CH_3NH_3Br$, $Ba(C_2H_3O_2)_2$, $Zn(NO_3)_2$, $Na_2HPO_4$—identify those that **(a)** form an acidic solution; **(b)** form a basic solution. **(c)** Identify the most basic substance in the list.

**17.48** Of these salts—$Na_2SO_3$, $Fe(ClO_4)_3$, CaO, $NH_4IO_3$—identify those that **(a)** form an acidic solution; **(b)** form a basic solution. **(c)** Identify the most acidic substance in the list.

## Acid-Base Character and Chemical Structure

**17.49** List the following hydrides in the order of increasing acidity in aqueous solution: $PH_3$; HBr; $MgH_2$; $H_2Se$.

**17.50** List the following oxides in the order of increasing acidity in aqueous solution: $SO_3$; $B_2O_3$; CaO; $CO_2$.

**17.51** Indicate whether each of the following statements is true or false. For those that are false, correct the statement so that it is true. **(a)** In general, the acidity of hydrides increases from left to right in a given row of the periodic table. **(b)** In a series of acids that have the same central atom, acid strength increases with the number of hydrogen atoms bonded to the central atom. **(c)** $H_2Te$ is a stronger acid than $H_2S$ because Te has a higher electronegativity than S.

**17.52** Indicate whether each of the following statements is true or false. For those that are false, correct the statement so that it is true. **(a)** Acid strength in a series $H_nX$ increases with increasing size of X. **(b)** For acids of the same structure but differing electronegativity of the central atom, acid strength decreases with increasing electronegativity of the central atom. **(c)** HF is the strongest acid of all because fluorine is the most electronegative element.

**17.53** The Lewis structure for acetic acid is shown in Table 17.3. Replacement of hydrogen atoms on the carbon by chlorine atoms causes an increase in acidity, as follows:

| Acid | Formula | $K_a(25°C)$ |
|---|---|---|
| Acetic | $CH_3COOH$ | $1.8 \times 10^{-5}$ |
| Chloroacetic | $CH_2ClCOOH$ | $1.4 \times 10^{-3}$ |
| Dichloroacetic | $CHCl_2COOH$ | $3.3 \times 10^{-2}$ |
| Trichloroacetic | $CCl_3COOH$ | $2 \times 10^{-1}$ |

Using Lewis structures as the basis of your discussion, explain the observed trend in acidities in the series. Calculate the pH of a 0.10 $M$ solution of each acid.

**17.54** The $pK_a$ values for the hypohalous acids vary as follows: HClO, 7.54; HBrO, 8.62; HIO, 10.64. Draw the general Lewis structure for the hypohalous acid, HXO. Account for the variation in acid strength in the series. Calculate the pH of a 0.010 $M$ solution of each of the acids.

## Lewis Acids and Bases

**17.55** Prepare a table in which you compare the definitions of acids and bases according to the Lewis, Brønsted, and Arrhenius theories. Which is the most general; that is, which includes the others within its scope? Explain.

**17.56** For each of the following descriptive statements, provide an interpretation in terms of the Brønsted-Lowry theory, the Lewis theory, or both, as appropriate. **(a)** HBr dissolves in water to form an acidic solution. **(b)** Sodium hydride, NaH, reacts with water to form a basic solution. **(c)** Pyridine reacts with sulfur dioxide in a dry organic solvent to form pyridine-$SO_2$:

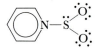

**(d)** $SO_2$ dissolves in water to form an acidic solution.

**17.57** Identify the Lewis acid and Lewis base in each of the following reactions:
**(a)** $Fe(ClO_4)_3(s) + H_2O(l) \longrightarrow$
$$Fe(H_2O)_6^{3+}(aq) + 3ClO_4^-(aq)$$

**(b)** $CN^-(aq) + H_2O(l) \rightleftharpoons HCN(aq) + OH^-(aq)$
**(c)** $(CH_3)_3N(g) + BF_3(g) \rightleftharpoons (CH_3)_3NBF_3(s)$
**(d)** $HIO(lq) + NH_2^-(lq) \rightleftharpoons NH_3(l) + IO^-(lq)$
       (*lq* denotes liquid ammonia as solvent)

**17.58** Identify the Lewis acid and Lewis base in each of the following reactions:
**(a)** $HNO_2(aq) + OH^-(aq) \rightleftharpoons NO_2^-(aq) + H_2O(l)$
**(b)** $FeBr_3(s) + Br^-(aq) \rightleftharpoons FeBr_4^-(aq)$
**(c)** $Zn^{2+}(aq) + 4NH_3(aq) \rightleftharpoons Zn(NH_3)_4^{2+}(aq)$
**(d)** $SO_2(g) + H_2O(l) \rightleftharpoons H_2SO_3(aq)$

**17.59** Which member of each of the following pairs would you expect to produce the more acidic solution: **(a)** LiI or $CdI_2$; **(b)** $Fe(NO_3)_3$ or $Ca(NO_3)_2$; **(c)** $CoCl_2$ or $CoCl_3$?

**17.60** Which member of each of the following pairs would you expect to be the stronger Lewis acid: **(a)** $BH_4^-$ or $BH_3$; **(b)** $S_8$ or $SO_3$; **(c)** NaH or HBr; **(d)** $Mo^{4+}$ or $Zn^{2+}$?

## Additional Exercises

**17.61** Calculate $[H^+]$ for each of the following solutions: **(a)** urine, pH 6.1; **(b)** lemon juice, pH 2.1; **(c)** gastric juice, pH 1.4; **(d)** household ammonia, pH 11.9. Indicate in each case whether the solution is acidic, basic, or neutral.

**17.62** Arrange the following 0.10 $M$ solutions in the order of decreasing pH: $KNO_2$, $KClO_4$, $HClO_4$, KCN, $NH_4Br$.

**17.63** What is the pH of each of the following solutions: **(a)** 0.030 $M$ NaOH; **(b)** $2.8 \times 10^{-2}$ $M$ $HClO_4$; **(c)** 0.0400 $M$ HIO; **(d)** 0.150 $M$ KCNO; **(e)** 0.020 $M$ $CH_3NH_2$; **(f)** 0.0400 $M$ $CH_3NH_3Cl$?

**17.64** If $K_w = 5.47 \times 10^{-14}$ at 50°C, what is the value of $[H^+]$, and what is the pH, for a neutral solution at this temperature?

**17.65** A 0.100 $M$ solution of $NH_3$ at 50°C has a pH of 10.40. Using the data in Exercise 17.64, calculate $K_b$ for ammonia at this temperature.

**17.66** Calculate the number of moles of each of the following substances that must be present in 200 mL of solution to form a solution with pH 3.25: **(a)** HCl; **(b)** $HC_7H_5O_2$; **(c)** HF.

**17.67** Calculate the percent ionization of each of the following substances in the solutions indicated: **(a)** 0.050 $M$ $HNO_2$; **(b)** 0.050 $M$ $CH_3NH_2$.

**17.68** Liquid ammonia undergoes autoionization analogous to that of water:
$$2NH_3(l) \rightleftharpoons NH_4^+(lq) + NH_2^-(lq)$$
(Here we use *lq* as we would *aq*; it means that the ions are present in the solvent, which in this case is liquid ammonia.) At −50°C, the ion product for this reaction, $[NH_4^+][NH_2^-]$, is $1.0 \times 10^{-33}$. What is the concentration of $NH_4^+$ in a liquid ammonia solution at −50°C **(a)** if $2.0 \times 10^{-5}$ mol of $NH_4Cl$ is dissolved to form 500 mL of solution; **(b)** if $1.2 \times 10^{-5}$ mol of $KNH_2$ is dissolved to form 500 mL of solution?

**17.69** The dye bromthymol blue is a weak acid whose ionization can be represented as
$$HBb(aq) \rightleftharpoons H^+(aq) + Bb^-(aq)$$

Which way will this equilibrium shift when NaOH is added? The acid form of the dye is yellow, whereas its conjugate base is blue. What color is the NaOH solution containing this dye?

**17.70** Predict the effect that each of the following added substances would have on the pH of an aqueous solution of $HNO_2$: **(a)** $KNO_2$; **(b)** $HClO_4$; **(c)** NaCN; **(d)** KOH.

**17.71** Hemoglobin plays a part in a series of equilibria involving protonation-deprotonation and oxygenation-deoxygenation. The overall reaction is approximately as follows:

$$HbH^+(aq) + O_2(aq) \rightleftharpoons HbO_2(aq) + H^+(aq)$$

(where Hb stands for hemoglobin and $HbO_2$ for oxyhemoglobin). **(a)** $[O_2]$ is higher in the lungs and lower in the tissues. What effect does high $[O_2]$ have on the position of this equilibrium? **(b)** The normal pH of blood is 7.4. What is $[H^+]$ in normal blood? Is the blood acidic, basic, or neutral? **(c)** If the blood pH is lowered by the presence of large amounts of acidic metabolism products, a condition known as acidosis results. What effect does lowering blood pH have on the ability of hemoglobin to transport $O_2$?

**17.72** Although we think of $NH_3$ as a base, it is capable of donating a proton when a sufficiently strong base is present. **(a)** Could such a reaction occur in aqueous solution? Explain. **(b)** Calculate the pH of a solution obtained by adding 0.30 g of $NaNH_2$ to sufficient water to form 0.400 L of solution.

**[17.73]** Pure sulfuric acid, $H_2SO_4$, is a colorless liquid that melts at 10°C and boils at 338°C. This pure substance undergoes autoionization to a much larger extent than does water. **(a)** Write the equilibrium expression for the autoionization of sulfuric acid, and identify the acid, base, conjugate acid, and conjugate base. **(b)** Write the Lewis structures for the conjugate acid and conjugate base formed in the autoionization. **(c)** Write expressions for the reactions you would expect to occur when a small amount of each of the following substances is added to pure sulfuric acid: $H_2O$; $HClO_4$ (a stronger acid than $H_2SO_4$); $K_2SO_4$.

**17.74** Saccharin, a sugar substitute, is a weak acid with $pK_a = 11.68$ at 25°C. It ionizes in aqueous solution as follows:

$$HNC_7H_4SO_3(aq) \rightleftharpoons H^+(aq) + NC_7H_4SO_3^-(aq)$$

What is the pH of a 0.10 M solution of this substance?

**[17.75]** What are the concentrations of $H^+$, $HSO_4^-$, and $SO_4^{2-}$ in a 0.025 M solution of $H_2SO_4$?

**[17.76]** What are the concentrations of $H^+$, $H_2PO_4^-$, $HPO_4^{2-}$, and $PO_4^{3-}$ in a 0.10 M solution of $H_3PO_4$?

**17.77** Ephedrine, a central nervous system stimulant, is used in nasal sprays as a decongestant. This compound is a weak organic base:

$$C_{10}H_{15}ON(aq) + H_2O(l) \rightleftharpoons$$
$$C_{10}H_{15}ONH^+(aq) + OH^-(aq)$$

$K_b$ has the value $1.4 \times 10^{-4}$. What pH would you expect for a 0.035 M solution of ephedrine, assuming that no other substances are present? What is the value of $pK_a$ for the conjugate acid, ephedrine hydrochloride?

**17.78** Morphine, $C_{17}H_{19}NO_3$, is a weak base containing a nitrogen atom, with $pK_b = 6.1$. **(a)** What is the pH of a 0.050 M solution of morphine? **(b)** What is the value of $pK_a$ for the conjugate acid, morphine hydrochloride? What is the pH of a 0.050 M solution of this substance?

**17.79** Codeine, $C_{18}H_{21}NO_3$, is a weak organic base. A $5.0 \times 10^{-3}$ M solution of codeine has a pH of 9.95; calculate the value of $K_b$ for this substance.

**[17.80]** Many moderately large organic molecules containing basic nitrogen atoms are not very soluble in water as the neutral molecule, but they are frequently much more soluble as the acid salt. Assuming that the pH in the stomach is 2.5, indicate whether each of the following compounds would be present in the stomach as the neutral base or in the protonated form: nicotine, $K_b = 7 \times 10^{-7}$; caffeine, $K_b = 4 \times 10^{-14}$; strychnine, $K_b = 1 \times 10^{-6}$; quinine, $K_b = 1.1 \times 10^{-6}$.

**[17.81]** Amino acids contain an amino group, $-NH_2$, located on the carbon atom that also contains a carboxylic acid group, $-COOH$. Glycine, the simplest amino acid, could exist in water in either form I or II below:

$$\underset{\text{I}}{H_2N-CH_2-\overset{\overset{\displaystyle O}{\|}}{C}-OH} \qquad \underset{\text{II}}{{}^+H_3N-CH_2-\overset{\overset{\displaystyle O}{\|}}{C}-O^-}$$

$K_a$ for the carboxylic acid group of glycine is $4.3 \times 10^{-3}$, and $K_b$ for the amino group is $6.0 \times 10^{-5}$. **(a)** What is the pH of a 0.10 M aqueous solution of glycine? **(b)** In what forms, other than I and II, can glycine exist in solution, depending on pH? **(c)** What form of glycine would you expect to be present in a solution with pH 10; with pH 2?

**[17.82]** A 1.00 m solution of HF freezes at −1.90°C. Based on the extent of freezing-point lowering (Section 12.6), calculate the fraction of HF dissociated at this temperature. What is the value of $K_a$ for HF at −1.90°C?

**[17.83]** In an aqueous solution containing only a weak diprotic acid $H_2X$, the concentration of $X^{2-}$ is numerically equal to $K_{a2}$. Show why this is so.

**[17.84]** What is the pH of a $1.0 \times 10^{-9}$ M solution of HBr?

**17.85** Fibers of $Al_2O_3$ and $ZrO_2$ have many useful properties as high-temperature materials. They are impervious to attack by hot, concentrated base, but are successfully attacked by hot, concentrated $H_2SO_4$ or aqueous HF. ($Al^{3+}$ and $Zr^{4+}$ both form fluoride complexes: $AlF_6^{3-}$ and $ZrF_6^{2-}$, respectively.) Account for the observed chemical properties of $Al_2O_3$ and $ZrO_2$ in terms of acid-base concepts. Write balanced chemical equations for the reactions with $H_2SO_4$ and HF.

## CHAPTER 18

# Applications of Aqueous Equilibria

Water is the solvent of preeminent importance on our planet. In a sense, it is the solvent of life. It is difficult to imagine how living matter in all its complexity could exist with any liquid other than water as solvent. Water occupies its position of importance not only because of its abundance, but also because of its exceptional ability to dissolve a wide variety of substances. Aqueous solutions encountered in nature, such as biological fluids and seawater, contain many solutes. Consequently, many equilibria take place simultaneously in these solutions. To understand these equilibria better, we will deal in this chapter with more applications of acid-base equilibria. We will broaden our discussion to include two further types of aqueous equilibria, those involving slightly soluble salts and those forming metal complex ions in solution.

## 18.1 THE COMMON-ION EFFECT

In Chapter 17 we were concerned with determining the equilibrium concentrations of ions in solutions containing a weak acid or a weak base. Let's now consider solutions that contain not only a weak acid, such as acetic acid, $HC_2H_3O_2$, but also a soluble salt of that acid, such as $NaC_2H_3O_2$. When $NaC_2H_3O_2$ is added to a solution of $HC_2H_3O_2$, the pH of the solution increases ([$H^+$] is reduced). This result isn't surprising because $C_2H_3O_2^-$ is a weak base; like any other base it should increase the pH. However, it is instructive to view this effect from the perspective of Le Châtelier's principle. Like most salts, $NaC_2H_3O_2$ is a strong electrolyte. Consequently, it completely ionizes in aqueous solution to form $Na^+$ ions and $C_2H_3O_2^-$ ions. $HC_2H_3O_2$ is a weak electrolyte that dissociates as follows:

$$HC_2H_3O_2(aq) \rightleftharpoons H^+(aq) + C_2H_3O_2^-(aq) \qquad [18.1]$$

The addition of $C_2H_3O_2^-$, from $NaC_2H_3O_2$, causes this equilibrium, Equation 18.1, to shift to the left, thereby decreasing the equilibrium concentration of $H^+(aq)$. The $C_2H_3O_2^-$ is said to decrease or repress the dissociation of $HC_2H_3O_2$. In general, the dissociation of a weak electrolyte is decreased by adding to the solution a strong electrolyte that has an ion in common with the weak electrolyte. This shift in equilibrium position, which occurs when we add an ion that is a component of an equilibrium reaction, is called the **common-ion effect**. Sample Exercises 18.1 and 18.2 illustrate how equilibrium concentrations may be calculated when a solution contains a mixture of a weak electrolyte and a strong electrolyte that share a common ion.

A mollusk photographed on the reefs of Palau, a Pacific island in Micronesia. The mollusk shell is composed mainly of calcium carbonate. (Stan Wayman/Photo Researchers)

## SAMPLE EXERCISE 18.1

Suppose that we add 8.20 g, or 0.100 mol, of sodium acetate, $NaC_2H_3O_2$, to 1 L of a 0.100 $M$ solution of acetic acid, $HC_2H_3O_2$. What is the pH of the resultant solution?

**Solution:** Our first step in a problem of this kind should always be to identify the important equilibrium. Having done that (Equation 18.1), we need to find a value or expression for the concentration of each species in that equilibrium. Because $NaC_2H_3O_2$ is a strong electrolyte, the only equilibrium that we need to consider in our present problem is that for the dissociation of acetic acid. The equilibrium reaction, the initial concentrations of the species involved, and their equilibrium concentrations are summarized as follows:

| | $HC_2H_3O_2(aq)$ | $\rightleftharpoons$ $H^+(aq)$ | $+$ $C_2H_3O_2^-(aq)$ |
|---|---|---|---|
| **Initial** | 0.100 $M$ | 0 $M$ | 0.100 $M$ |
| **Equilibrium** | $(0.100 - x)$ $M$ | $x$ $M$ | $(0.100 + x)$ $M$ |

Notice that the equilibrium concentration of $C_2H_3O_2^-$ is the initial amount coming from the added $NaC_2H_3O_2$ (0.100 $M$), plus the amount ($x$) formed by dissociation of acetic acid. The $Na^+$ ion from $NaC_2H_3O_2$ is merely a spectator ion; it has no influence on pH (Section 17.6).

The equilibrium-constant expression is

$$K_a = 1.8 \times 10^{-5} = \frac{[H^+][C_2H_3O_2^-]}{[HC_2H_3O_2]}$$

(The value of the equilibrium constant is taken from Appendix E.) Addition of the acetate salt does not change the value of the equilibrium constant. Substituting the equilibrium concentrations in this equilibrium-constant equation, we obtain

$$\frac{x(0.100 + x)}{0.100 - x} = 1.8 \times 10^{-5}$$

We can simplify this equation by ignoring $x$ relative to 0.100. This simplification gives us

$$\frac{x(0.100)}{0.100} = 1.8 \times 10^{-5}$$

$$x = 1.8 \times 10^{-5} \, M = [H^+]$$

The resulting value of $x$ is indeed small relative to 0.100, justifying the approximation made in simplifying the problem. The pH, then, is

$$pH = -\log(1.8 \times 10^{-5}) = 4.74$$

Earlier (Section 17.3) we calculated that in a 0.10 $M$ solution of $HC_2H_3O_2$, $[H^+]$ is $1.3 \times 10^{-3}$ $M$, corresponding to a pH value of 2.89. Thus the addition of $NaC_2H_3O_2$ has substantially increased the solution pH, as expected.

### PRACTICE EXERCISE

Calculate the pH of a solution containing 0.060 $M$ formic acid, $HCHO_2$ ($K_a = 1.8 \times 10^{-4}$), and 0.030 $M$ potassium formate, $KCHO_2$.

*Answer:* 3.44

## SAMPLE EXERCISE 18.2

Calculate the fluoride concentration and pH of a solution containing 0.10 mol of HCl and 0.20 mol of HF in a liter of solution.

**Solution:** The equilibrium of interest here is the dissociation of the weak acid HF, shown below. If we let $x$ equal the number of moles per liter of HF that dissociate, then the equilibrium $F^-$ concentration is equal to $x$, because the dissociation of HF is the only source of $F^-$. The equilibrium concentration of $H^+$ is the sum of $x$ (the amount coming from HF dissociation) and 0.10 $M$ (the amount supplied by the strong acid HCl). Thus we have

| | $HF(aq)$ | $\rightleftharpoons$ $H^+(aq)$ | $+$ $F^-(aq)$ |
|---|---|---|---|
| **Initial** | 0.20 $M$ | 0.10 $M$ | 0 |
| **Equilibrium** | $(0.20 - x)$ $M$ | $(0.10 + x)$ $M$ | $x$ $M$ |

The equilibrium constant, from Appendix E, is $6.8 \times 10^{-4}$.

$$K_a = 6.8 \times 10^{-4} = \frac{(0.10 + x)x}{0.20 - x}$$

If we assume that $x$ is small relative to 0.10 or 0.20 $M$, this expression simplifies to give

$$\frac{(0.10)x}{0.20} = 6.8 \times 10^{-4}$$

$$x = \frac{0.20}{0.10}(6.8 \times 10^{-4})$$

$$= 1.4 \times 10^{-3} \, M = [F^-]$$

This $F^-$ concentration is substantially smaller than it would be in a 0.200 $M$ solution of HF with no added HCl. The common ion, $H^+$, has repressed the dissociation of HF. The concentration of $H^+(aq)$ is

$$[H^+] = (0.10 + x) \, M = 0.10 \, M$$

Thus pH = 1.00. Notice that $[H^+]$ is for practical purposes due entirely to the HCl; the HF makes a negligible contribution by comparison.

**PRACTICE EXERCISE**

Calculate the formate ion concentration and pH of a solution that is 0.050 $M$ in formic acid, $HCHO_2$ ($K_a = 1.8 \times 10^{-4}$), and 0.100 $M$ in $HNO_3$.
***Answer:*** $[CHO_2^-] = 9.0 \times 10^{-5}$; pH = 1.00

Sample Exercises 18.1 and 18.2 both involve weak acids. However, we might just as easily have used a weak-base equilibrium to illustrate the common-ion effect. For example, a mixture of aqueous $NH_3$ and $NH_4Cl$ can be considered from the perspective of dissociation of the weak base, $NH_3$, or the weak acid, $NH_4^+$. The weak-base dissociation equation is

$$NH_3(aq) + H_2O(l) \;\rightleftharpoons\; NH_4^+(aq) + OH^-(aq) \qquad [18.2]$$

This equilibrium expression contains the weak base and its conjugate acid. The acid-dissociation constant for $NH_4^+$ also contains both these quantities:

$$NH_4^+(aq) \;\rightleftharpoons\; NH_3(aq) + H^+(aq) \qquad [18.3]$$

You will see as we move through the chapter that we can calculate the pH of a solution involving this and other conjugate acid-base pairs by working with either the equation for weak-base dissociation or the equation for dissociation of the conjugate acid.

The common ion that affects a weak-acid or weak-base equilibrium may be present because it is added as a salt, as in the examples above. However, it might also arise as a result of an acid-base reaction. Indeed, this is a common and important situation. Consider a solution formed by mixing a weak acid, such as $HC_2H_3O_2$, and a strong base, like NaOH. The net ionic equation for the reaction that occurs is

**Common Ions Generated by Acid-Base Reactions**

$$HC_2H_3O_2(aq) + OH^-(aq) \;\longrightarrow\; C_2H_3O_2^-(aq) + H_2O(l) \qquad [18.4]$$

This reaction proceeds virtually to completion. In general, *reactions between (1) strong acids and strong bases, (2) strong acids and weak bases, and (3) strong bases and weak acids proceed essentially to completion.*

If the number of moles of $HC_2H_3O_2$ exceeds the number of moles of NaOH, the reaction between them produces $C_2H_3O_2^-$ ions, leaving the excess $HC_2H_3O_2$ unreacted. For example, consider a 1-L solution that initially contains 0.20 mol of $HC_2H_3O_2$ and 0.10 mol of NaOH. The reaction (Equation 18.4) consumes 0.10 mol of $HC_2H_3O_2$, producing 0.10 mol of $C_2H_3O_2^-$ and leaving 0.10 mol of $HC_2H_3O_2$ unreacted:

$$HC_2H_3O_2(aq) + OH^-(aq) \;\rightleftharpoons\; C_2H_3O_2^-(aq) + H_2O(l)$$

| | | | | |
|---|---|---|---|---|
| **Before rxn:** | 0.20 mol | 0.10 mol | 0 | — |
| **Change:** | −0.10 mol | −0.10 mol | +0.10 mol | — |
| **After rxn:** | 0.10 mol | 0 | 0.10 mol | — |

The resultant solution, which is 0.10 $M$ in $C_2H_3O_2^-$ and 0.10 $M$ in $HC_2H_3O_2$, is identical to a solution produced by mixing 0.10 mol of $NaC_2H_3O_2$ and 0.10 mol of $HC_2H_3O_2$ to form a liter of solution. As shown in Sample Exercise 18.1, such a solution has a pH of 4.74.

This example suggests a general approach to determining the pH of any acid-base mixture:

1. If the solution mixture consists only of a conjugate acid-base pair, then you need consider only the proton-transfer equilibrium between these two species, as in Sample Exercise 18.1.

2. If the solution mixture contains a strong acid or strong base component, you must first consider the complete reaction of this strong acid or strong base with a weak base or weak acid present in the solution. If a conjugate acid-base pair is present after reaction, as in the example outlined above, then consider the proton-transfer equilibrium involving that acid-base pair.

---

## SAMPLE EXERCISE 18.3

Calculate the pH of a solution produced by mixing 0.60 L of 0.10 $M$ $NH_4Cl$ with 0.40 L of 0.10 $M$ NaOH.

**Solution:** This solution mixture contains the strong base $OH^-$ in addition to the weak acid $NH_4^+$. We first must consider the reaction between these two species. The molar quantities of these ions present in solution before any reaction occurs are

$$\text{Moles } NH_4^+ = \text{moles } Cl^- = M_{NH_4Cl} \times V_{NH_4Cl}$$

$$= \left(0.10 \frac{\text{mol}}{\text{L}}\right)(0.60 \text{ L}) = 0.060 \text{ mol}$$

$$\text{Moles } Na^+ = \text{moles } OH^- = M_{NaOH} \times V_{NaOH}$$

$$= \left(0.10 \frac{\text{mol}}{\text{L}}\right)(0.40 \text{ L}) = 0.040 \text{ mol}$$

The weakly acidic $NH_4^+$ and the basic $OH^-$ react as follows:

$$NH_4^+(aq) + OH^-(aq) \rightleftharpoons NH_3(aq) + H_2O(l)$$

This reaction proceeds essentially to completion. Thus the 0.040 mol of $OH^-$ consumes 0.040 mol of $NH_4^+$, producing 0.040 mol of $NH_3$. The remaining $NH_4^+$ is the original amount minus the amount consumed.

These considerations are summarized in the following table:

$$NH_4^+(aq) + OH^-(aq) \rightleftharpoons NH_3(aq) + H_2O(l)$$

| | | | | |
|---|---|---|---|---|
| **Before rxn:** | 0.060 mol | 0.040 mol | 0 | — |
| **Change:** | −0.040 mol | −0.040 mol | +0.040 mol | — |
| **After rxn:** | 0.020 mol | 0 | 0.040 mol | — |

The total volume of the solution is the sum of the two original solution volumes: 0.060 L + 0.040 L = 1.00 L. Thus the concentrations of $NH_3$ and $NH_4^+$ after reaction are

$$[NH_3] = \frac{0.040 \text{ mol}}{1.00 \text{ L}} = 0.040 \ M$$

$$[NH_4^+] = \frac{0.020 \text{ mol}}{1.00 \text{ L}} = 0.020 \ M$$

The pH of a solution containing both $NH_4^+$ and $NH_3$ can be calculated by setting up the equilibrium-constant expression for Equation 18.5:

$$NH_4^+(aq) \rightleftharpoons H^+(aq) + NH_3(aq) \qquad [18.5]$$

$$K_a = \frac{[H^+][NH_3]}{[NH_4^+]}$$

The initial and equilibrium values of the pertinent species are as follows:

| | $NH_4^+(aq)$ $\rightleftharpoons$ | $H^+(aq)$ + | $NH_3(aq)$ |
|---|---|---|---|
| **Initial** | 0.020 $M$ | 0 | 0.040 $M$ |
| **Equilibrium** | $(0.020 - x) \ M$ | $x \ M$ | $(0.040 + x) \ M$ |

$K_a$ for $NH_4^+$ is related to $K_b$ for $NH_3$ through the relationship $K_a \times K_b = K_w$. From Appendix E we have $K_b = 1.8 \times 10^{-5}$.

$$K_a = \frac{K_w}{K_b} = \frac{1.0 \times 10^{-14}}{1.8 \times 10^{-5}} = 5.6 \times 10^{-10}$$

We can assume that $x$ is small relative to 0.020 and 0.040. Then

$$K_a = 5.6 \times 10^{-10} = \frac{(x)(0.040)}{0.020}$$

$$x = [H^+] = 2.7 \times 10^{-10}$$

$$pH = -\log(2.7 \times 10^{-10}) = 9.56$$

### PRACTICE EXERCISE
Calculate the pH of a solution formed by mixing 0.50 L of 0.015 $M$ NaOH solution with 0.50 L of 0.030 $M$ benzoic acid solution.
*Answer:* 4.19

## 18.2 BUFFER SOLUTIONS

Many aqueous solutions resist a change in pH upon addition of small amounts of acid or base. Such solutions, called **buffer solutions**, are said to be buffered. Human blood, for example, is a complex aqueous medium with a pH buffered at about 7.4. Any significant variation of the pH from this value results in a severe pathological response and, eventually, death. As another example, the chemical behavior of seawater is determined in very important respects by its pH, buffered at about 8.1 to 8.3 near the surface. Addition of a small amount of an acid or base to either blood or seawater does not result in a large change in pH. Buffer solutions find many important applications in the laboratory and in medicine (Figure 18.1).

Buffers ordinarily require two species, an acidic one to react with added $OH^-$ and a basic one to react with added $H^+$. It is, of course, necessary that these acidic and basic species not consume each other through a neutralization reaction. These requirements are fulfilled by an acid-base conjugate pair such as $HC_2H_3O_2$–$C_2H_3O_2^-$ or $NH_4^+$–$NH_3$. The $HC_2H_3O_2$–$C_2H_3O_2^-$ buffer mixture can be prepared by adding sodium

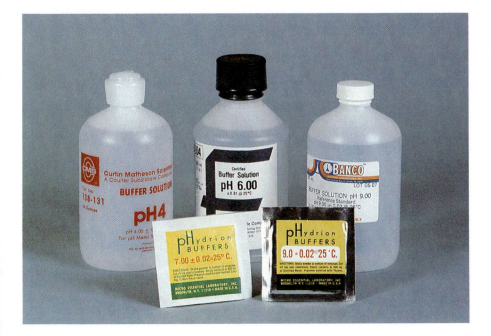

### FIGURE 18.1
Prepackaged buffer solutions and ingredients for forming buffer solutions of predetermined pH. (Donald Clegg and Roxy Wilson)

acetate, $NaC_2H_3O_2$, to a solution of acetic acid, $HC_2H_3O_2$. The $NH_4{}^+$–$NH_3$ buffer mixture can be prepared by adding ammonium chloride, $NH_4Cl$, to a solution of ammonia, $NH_3$. In general, a buffer mixture consists of an aqueous solution of an acid-base conjugate pair prepared by mixing a weak acid or weak base with a salt of that acid or base.

To understand how a buffer works, let's consider a solution of $HC_2H_3O_2$ and $NaC_2H_3O_2$. Ionization of $HC_2H_3O_2$ is governed by the following equilibrium reaction:

$$HC_2H_3O_2(aq) \rightleftharpoons H^+(aq) + C_2H_3O_2{}^-(aq) \qquad [18.6]$$

The $C_2H_3O_2{}^-$ in this equilibrium comes from both $HC_2H_3O_2$ and $NaC_2H_3O_2$. This mixture can either react with surplus $H^+$ ions or release them, according to the circumstances. For example, if a small quantity of acid is added to the solution, the equilibrium shifts to the left; acetate ion reacts with the added $H^+$. The solution thereby limits pH change due to added acid. On the other hand, if a small quantity of a base is added, it reacts with $H^+$. This reaction causes the equilibrium of Equation 18.6 to shift to the right; $HC_2H_3O_2$ dissociates to form more $H^+$. The solution thereby also resists change in pH due to added base.

For a better understanding of how buffers work, let's consider a buffer mixture consisting of a weak acid HX and a corresponding salt MX, where M could be $Na^+$, $K^+$, and so forth. The acid-dissociation equilibrium is

$$HX(aq) \rightleftharpoons H^+(aq) + X^-(aq) \qquad [18.7]$$

and the corresponding acid-dissociation-constant expression is

$$K_a = \frac{[H^+][X^-]}{[HX]} \qquad [18.8]$$

Solving this expression for $[H^+]$, we have

$$[H^+] = K_a \frac{[HX]}{[X^-]} \qquad [18.9]$$

We see from this expression that $[H^+]$, and thus the pH, is determined by two factors: the value of $K_a$ for the weak-acid component of the buffer, and the ratio of the concentrations of the conjugate acid-base pair, $[HX]/[X^-]$. When an acid is added to the solution, some of the $X^-$ is converted to HX, and the ratio changes. However, if the amounts of HX and $X^-$ present are large compared with the amount of acid added, the ratio doesn't change by much, and thus the change in pH is small. Similarly, if a strong base were added to the buffer solution, it would react with HX to form $X^-$. Once again, if the amount of added base were small in comparison with the concentrations of HX and $X^-$, the change in the $[HX]/[X^-]$ ratio, and thus the change in pH, would be small. The solution is most effective in buffering against a change in pH in *either* direction when the concentrations of HX and $X^-$ are about the same. Notice that under these conditions $[H^+]$ is approximately equal to $K_a$. For this reason one usually tries to select a buffer whose acid form has a $pK_a$ close to the desired pH.

Two important characteristics of a buffer are buffering capacity and pH. Buffering capacity is the amount of acid or base the buffer can neutralize before the pH begins to change to an appreciable degree. This capacity depends on the amount of acid and base from which the buffer is made. The pH of the buffer depends on $K_a$ for the acid and on the relative concentrations of the acid and base that comprise the buffer. For example, we can see from Equation 18.9 that $[H^+]$ for a 1-L solution that is 1 $M$ in $HC_2H_3O_2$ and 1 $M$ in $NaC_2H_3O_2$ will be the same as for a 1-L solution that is 0.1 $M$ in $HC_2H_3O_2$ and 0.1 $M$ in $NaC_2H_3O_2$. However, the first solution has a greater buffering capacity because it contains more $HC_2H_3O_2$ and $C_2H_3O_2^-$.

Let us now take the negative log of both sides of Equation 18.9:

$$-\log [H^+] = -\log K_a - \log \frac{[HX]}{[X^-]}$$

Because $-\log [H^+] = pH$ and $-\log K_a = pK_a$, we have

$$pH = pK_a - \log \frac{[HX]}{[X^-]} = pK_a + \log \frac{[X^-]}{[HX]} \qquad [18.10]$$

In general,

$$pH = pK_a + \log \frac{[base]}{[acid]} \qquad [18.11]$$

This relationship, which is called the **Henderson-Hasselbalch equation**, is very useful in dealing with buffers. As we indicated earlier, buffer solutions are most effective when the concentrations of the two components of the conjugate acid-base pair are about equal. This means that pH is close to the $pK_a$ of the acid component.

---

### SAMPLE EXERCISE 18.4

What is the pH of a buffer mixture composed of equal concentrations of $NH_4Cl$ and $NH_3$?

**Solution:** The dominant equilibrium is that involving the acid $NH_4^+$ and its conjugate base, $NH_3$:

$$NH_4^+(aq) \rightleftharpoons NH_3(aq) + H^+(aq)$$

$K_a$ for $NH_4^+$ is related to $K_b$ for $NH_3$ through the

relationship $K_a \times K_b = K_w$. From Appendix E we have $K_b = 1.8 \times 10^{-5}$.

$$K_a = \frac{K_w}{K_b} = \frac{1.0 \times 10^{-14}}{1.8 \times 10^{-5}} = 5.6 \times 10^{-10}$$

The initial and equilibrium concentrations are as follows:

| | $NH_4^+(aq)$ | $\rightleftharpoons$ $NH_3(aq)$ | $+$ $H^+(aq)$ |
|---|---|---|---|
| **Initial** | $y$ $M$ | $y$ $M$ | 0 |
| **Equilibrium** | $(y-x)$ $M$ | $(y+x)$ $M$ | $x$ |

Note that we have used $y$ for the concentrations of $NH_4^+$ and $NH_3$ because we were not given specific

concentrations for these two species. Substituting into the expression for $K_a$, we have

$$K_a = \frac{[NH_3][H^+]}{[NH_4^+]} = \frac{(y+x)(x)}{y-x} \approx \frac{(y)(x)}{y}$$

Because $NH_4^+$ is a weak acid, $x$ should be much smaller than $y$, so the approximation made in arriving at the last expression is justified. The $y$'s cancel, leaving $x = [H^+] = K_a = 5.6 \times 10^{-10}$.

$$pH = pK_a = -\log(5.6 \times 10^{-10}) = 9.25$$

Notice that this is just the result we would have reached from direct application of Equation 18.11. Because $[NH_4^+] = [NH_3]$, $\log \dfrac{[NH_3]}{[NH_4^+]} = \log 1 = 0$, and $pH = pK_a$.

**PRACTICE EXERCISE** ——————
Calculate the pH of a solution formed from 0.10 $M$ formic acid and 0.20 $M$ potassium formate.
*Answer:* 4.04

## SAMPLE EXERCISE 18.5

**(a)** How is the pH of an aqueous solution of $NH_3$ affected by the addition of $NH_4Cl$? **(b)** What must be the concentration of $NH_4Cl$ in a 0.10 $M$ solution of $NH_3$ if the pH is to be 9.00?

**Solution:** Let's approach this problem from the viewpoint of Equation 18.2, to show that we can employ this equilibrium just as well as Equation 18.3. **(a)** $NH_3$ is a weak base that reacts with water as follows:

$$NH_3(aq) + H_2O(l) \rightleftharpoons NH_4^+(aq) + OH^-(aq)$$

$NH_4Cl$ is a strong electrolyte, supplying both $NH_4^+$ and $Cl^-$ ions. The presence of $NH_4^+$, which is a weak acid, shifts the equilibrium to the left, lowering $[OH^-]$ and thereby lowering pH. We have here simply another example of the common-ion effect; aqueous solutions of $NH_3$ and $NH_4Cl$ have the $NH_4^+$ ion in common.

**(b)** The equilibrium constant for the reaction of $NH_3$ with water is

$$K_b = \frac{[NH_4^+][OH^-]}{[NH_3]}$$

Let's rearrange the expression for $K_b$ to solve for $[NH_4^+]$:

$$[NH_4^+] = K_b \frac{[NH_3]}{[OH^-]}$$

The value of $K_b$ (Appendix E) is $1.8 \times 10^{-5}$. If pH is to be 9.00, then pOH is 5.00, and $[OH^-]$ is $1.0 \times 10^{-5}$.

We can assume that $[NH_3]$ is 0.10 $M$. The amount that reacts with water to form $NH_4^+$ is very small, especially in the presence of added $NH_4^+$. Thus,

$$[NH_4^+] = \frac{1.8 \times 10^{-5} \times 0.10}{1.0 \times 10^{-5}} = 0.18 \ M$$

Since the amount of $NH_4^+$ formed from $NH_3$ is very small, this concentration represents the concentration of $NH_4Cl$ that we must have to achieve a pH of 9.00.

**PRACTICE EXERCISE** ——————
Calculate the concentration of sodium formate that must be present in a 0.10 $M$ solution of formic acid to produce a pH of 3.80.
*Answer:* 0.11 $M$

**Addition of Acids or Bases to Buffers**

Let us now consider in a more quantitative way the response of a buffer solution to addition of an acid or base.

## SAMPLE EXERCISE 18.6

A liter of solution containing 0.100 mol of $HC_2H_3O_2$ and 0.100 mol of $NaC_2H_3O_2$ is being prepared to provide a buffer of pH 4.74. Calculate the pH of this solution **(a)** after 0.020 mol of NaOH is added (neglect any volume changes); **(b)** after 0.020 mol of HCl is added (again, neglect any volume changes).

**Solution:** **(a)** The $OH^-$ provided by the strong base NaOH reacts nearly completely with the weak acid $HC_2H_3O_2$:

$$HC_2H_3O_2(aq) + OH^-(aq) \longrightarrow$$
$$H_2O(l) + C_2H_3O_2^-(aq)$$

The 0.020 mol of added $OH^-$ therefore reacts with 0.020 mol of $HC_2H_3O_2$, producing 0.020 mol of $C_2H_3O_2^-$. Thus the final solution contains 0.080 mol of $HC_2H_3O_2$ (the original 0.100 mol minus that which reacts, 0.020 mol), and 0.120 mol of $C_2H_3O_2^-$ (the original 0.100 mol plus that formed in the reaction, 0.020 mol). In summary,

$$HC_2H_3O_2(aq) + OH^-(aq) \longrightarrow H_2O(l) + C_2H_3O_2^-(aq)$$

| | $HC_2H_3O_2$ | $OH^-$ | | $C_2H_3O_2^-$ |
|---|---|---|---|---|
| **Before rxn:** | 0.100 M | 0.020 M | | 0.100 M |
| **Change:** | −0.020 | −0.020 | | +0.020 |
| **After rxn:** | 0.080 M | 0.0 M | | 0.120 M |

The acid-base equilibrium that concerns us in determining the pH is that of the weak acid $HC_2H_3O_2$ and its conjugate base $C_2H_3O_2^-$. Proceeding in the usual way to set up the problem, we obtain:

$$HC_2H_3O_2(aq) \rightleftharpoons C_3H_3O_2^-(aq) + H^+(aq)$$

| | $HC_2H_3O_2(aq)$ | $C_3H_3O_2^-(aq)$ | $H^+(aq)$ |
|---|---|---|---|
| **Initial** | 0.080 M | 0.120 M | 0 |
| **Equilibrium** | (0.080 − x) M | (0.120 + x) M | x M |

$$K_a = \frac{[C_2H_3O_2^-][H^+]}{[HC_2H_3O_2]}$$
$$= \frac{(0.120 + x)(x)}{0.080 - x} \approx \frac{(0.120)(x)}{0.080}$$

$$x = [H^+] = \frac{(0.080)K_a}{(0.120)} = 0.67 \times 1.8 \times 10^{-5}$$

$$= 1.2 \times 10^{-5}$$
$$pH = -\log(1.2 \times 10^{-5}) = 4.92$$

We could have obtained this answer directly by applying Equation 18.11, but until you are sure you have a thorough understanding of the equilibria of importance in these problems, it is best to set the problem up in full.

**(b)** The $H^+$ provided by the HCl reacts completely with the weak base $C_2H_3O_2^-$ to form $HC_2H_3O_2$:

$$C_2H_3O_2^-(aq) + H^+(aq) \longrightarrow HC_2H_3O_2(aq)$$

The 0.020 mol of $H^+$ therefore reacts with 0.020 mol of $C_2H_3O_2^-$ to form 0.020 mol of $HC_2H_3O_2$. The final solution then contains 0.120 mol of $HC_2H_3O_2$ and 0.080 mol of $C_2H_3O_2^-$. In summary,

$$C_2H_3O_2^-(aq) + H^+(aq) \rightleftharpoons HC_2H_3O_2(aq)$$

| | $C_2H_3O_2^-$ | $H^+$ | $HC_2H_3O_2$ |
|---|---|---|---|
| **Before rxn:** | 0.100 M | 0.020 M | 0.100 M |
| **Change:** | −0.020 | −0.020 | +0.020 |
| **After rxn:** | 0.080 M | 0.0 M | 0.120 M |

We set up the equilibrium problem as before:

$$HC_2H_3O_2(aq) \rightleftharpoons C_2H_3O_2^-(aq) + H^+(aq)$$

Equilibrium: $(0.120 - x)$ M $\quad (0.080 + x)$ M $\quad x$ M

Substituting into the acid-dissociation-equilibrium expression as before, and simplifying, we obtain

$$x = [H^+] = \frac{(0.120)K_a}{0.080} = 1.5 \times 1.8 \times 10^{-5}$$
$$= 2.7 \times 10^{-5}$$
$$pH = -\log(2.7 \times 10^{-5}) = 4.57$$

The pH decreases, as expected for addition of a strong acid to a solution. However, the magnitude of the decrease is small, because the solution is a buffer with a capacity to absorb the added acid.

**PRACTICE EXERCISE**
Calculate the pH of the buffer solution described in Sample Exercise 18.6 following addition of 0.040 mol of HCl, with no volume change.

*Answer:* 4.38

To appreciate more fully the buffer action of the solution of acetic acid and sodium acetate that we've just considered, let's compare its behavior with the action of a solution that is not a buffer. We saw in Sample Exercise 18.1 that the pH of a solution that is 0.100 M in acetic acid and 0.100 M in sodium acetate is 4.74. A solution of this same pH is obtained by addition of $1.8 \times 10^{-5}$ mol of HCl to a liter of water. Because HCl is a strong acid, $1.8 \times 10^{-5}$ M HCl has a $H^+(aq)$ concentration of $1.8 \times 10^{-5}$ M and thus a pH of 4.74. Now suppose that 2.0 mL of 10 M HCl solution—that is, 0.020 mol of HCl—is added to a liter of each of these two solutions of pH 4.74.

Blood is an important example of a buffered solution. Human blood is slightly basic with a pH of about 7.39 to 7.45. In a healthy person the pH never departs more than perhaps 0.2 pH unit from the average value. Whenever pH falls below about 7.4, the condition is called *acidosis*; when pH rises above 7.4, the condition is called *alkalosis*. Death may result if the pH falls below 6.8 or rises above 7.8. Acidosis is the more common tendency, because ordinary metabolism produces several acids.

The body uses three primary methods to control blood pH: (1) The blood contains several buffers, including $H_2CO_3$–$HCO_3^-$ and $H_2PO_4^-$–$HPO_4^{2-}$ pairs, and hemoglobin-containing conjugate acid-base pairs. (2) The kidneys serve to absorb or release $H^+(aq)$. The pH of urine is normally about 5.0 to 7.0. Acidosis is accompanied by increased loss of body fluids as the kidneys work to reduce $H^+(aq)$. (3) The concentration of $H^+(aq)$ is also altered by the rate at which $CO_2$ is removed from the lungs. The pertinent equilibria are

$$H^+(aq) + HCO_3^-(aq) \rightleftharpoons H_2CO_3(aq) \rightleftharpoons$$
$$H_2O(l) + CO_2(g)$$

Removal of $CO_2$ shifts these equilibria to the right, thereby reducing $H^+(aq)$.

Acidosis or alkalosis disrupts the mechanism by which hemoglobin transports oxygen in blood. Hemoglobin (Hb) is involved in a series of equilibria whose overall result is approximately

$$HbH^+(aq) + O_2(aq) \rightleftharpoons HbO_2(aq) + H^+(aq)$$

In acidosis, this equilibrium is shifted to the left, and the ability of hemoglobin to form oxyhemoglobin, $HbO_2$, is decreased. The lesser amount of $O_2$ thereby available to cells in the body causes fatigue and headaches; if great enough, it also triggers air hunger (the feeling of being "out of breath" that causes deep breathing).

Temporary acidosis occurs during strenuous exercise, when energy demands exceed the oxygen available for complete oxidation of glucose to $CO_2$. In this case the glucose is converted to an acidic metabolism product, lactic acid, $CH_3CHOHCOOH$. Acidosis also occurs when glucose is unavailable to the cells. This situation can arise, for example, during starvation or as a result of diabetes. In the case of diabetes, glucose is unable to enter the cells because of inadequate insulin, the substance responsible for passage of glucose from the bloodstream to the interior of cells. When glucose is unavailable, the body relies for energy on stored fats, which produce acidic metabolism products.

Figure 18.2 illustrates the effect of the added HCl. As we've seen in Sample Exercise 18.6(b), the added HCl reacts with $C_2H_3O_2^-$ to form $HC_2H_3O_2$. From the calculations carried out there we found that the pH of the buffer solution decreased by 0.17 pH unit. Note, however, that addition of the 2.0 mL of HCl to a liter of HCl solution that has a pH of 4.74 causes the pH to drop to 1.7. In this case, the added HCl provides a much larger total amount of acid than is present in the very dilute HCl solution.

We might have chosen to add 0.02 mol of hydroxide to the solutions illustrated in Figure 18.2. This addition would have caused the pH of the dilute HCl solution to go from 4.74 to 12.3 (you should be able to explain why this is so), whereas the pH of the buffer solution would have increased by 0.18 pH unit. Thus a buffer solution responds to approximately the same degree, but in opposite direction, to acid or base addition.

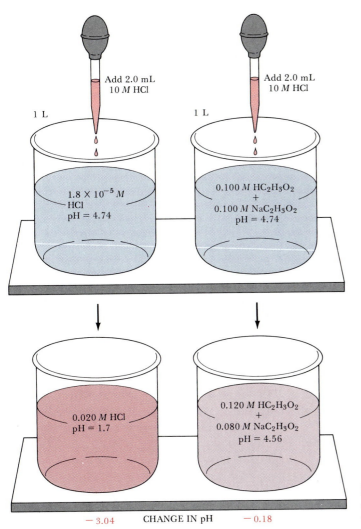

Add 2.0 mL
10 M HCl

Add 2.0 mL
10 M HCl

1 L

1 L

$1.8 \times 10^{-5} M$
HCl
pH = 4.74

$0.100 M$ HC$_2$H$_3$O$_2$
+
$0.100 M$ NaC$_2$H$_3$O$_2$
pH = 4.74

0.020 M HCl
pH = 1.7

0.120 M HC$_2$H$_3$O$_2$
+
0.080 M NaC$_2$H$_3$O$_2$
pH = 4.56

$-3.04$     CHANGE IN pH     $-0.18$

**FIGURE 18.2**   Comparison of the effect of added acid on a buffer solution of pH 4.74 compared with an HCl solution of pH 4.74.

Many acid-base reactions are used in chemical analyses. For example, the carbonate content in a sample can be determined by a procedure that involves titration with a strong acid, such as HCl. Titrations were described briefly in Chapter 3 (Section 3.8). In that earlier discussion we noted that acid-base indicators can be used to signal the equivalence point (that is, the point at which stoichiometrically equivalent quantities of acid and base have been brought together). But given the variety of indicators, changing colors at different pH values, which indicator is best for a particular titration? This question can be answered by examining a graph of pH changes during a titration. A graph of pH as a function of the volume of added titrant is called a **titration curve**.

The titration curve produced when a strong base is added to a strong acid has the general shape shown in Figure 18.3. This curve depicts the pH change that occurs as 0.100 $M$ NaOH is added to 50.0 mL of 0.100 $M$

**Strong Acid-Strong Base**

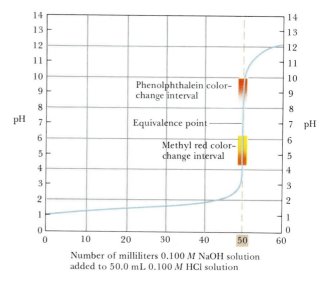

**FIGURE 18.3** pH curve for titration of a solution of a strong acid with a solution of a strong base, in this case HCl and NaOH.

Number of milliliters 0.100 $M$ NaOH solution added to 50.0 mL 0.100 $M$ HCl solution

HCl. The pH can be calculated at various stages in the course of the titration. The pH starts out low; pH = 1.00 for 0.100 $M$ HCl. As NaOH is added, the pH increases slowly at first and then rapidly in the vicinity of the equivalence point. *The pH at the equivalence point in any acid-base titration is the pH of the resultant salt solution.* In this example, NaCl is formed, which does not hydrolyze (Section 17.6); the equivalence point therefore occurs at pH 7. Because the pH change is very large near the equivalence point, the indicator for the titration need not change color precisely at 7.0. Most strong acid-strong base titrations are carried out using phenolphthalein as an indicator because its color change is dramatic (Figure 18.4). From Table 17.1 we see that this indicator changes color in the

**FIGURE 18.4** Change in appearance of a solution containing phenolphthalein indicator (*a*) as base is added in the pH range 8.3 to 10.0. In (*b*) the indicator has changed to the base form, which has a pink color. (Fundamental Photographs)

(*a*)

(*b*)

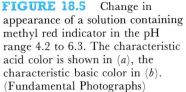

**FIGURE 18.5**  Change in appearance of a solution containing methyl red indicator in the pH range 4.2 to 6.3. The characteristic acid color is shown in (a), the characteristic basic color in (b). (Fundamental Photographs)

pH range 8.3 to 10. Thus a slight excess of NaOH must be present to cause the observed color change. However, it requires such a tiny excess of base to make the color change occur that no serious error is introduced. Similarly, methyl red, which changes color in the slightly acid range (Figure 18.5), could also be used. The pH intervals of color change for these two indicators are shown in Table 17.1.

Titration of a solution of a strong base with a solution of a strong acid would yield an entirely analogous curve of pH versus added acid. In this case, however, the pH would be high at the outset of the titration and low at its completion.

## SAMPLE EXERCISE 18.7

Calculate the pH when the following quantities of 0.100 $M$ NaOH solution have been added to 50.00 mL of 0.100 $M$ HCl solution: **(a)** 49.00 mL; **(b)** 49.90 mL; **(c)** 50.10 mL; **(d)** 51.00 mL.

**Solution:**  **(a)** The number of moles of OH$^-$ in 49.00 mL of 0.100 $M$ NaOH is

$$0.04900 \text{ L soln}\left(\frac{0.100 \text{ mol OH}^-}{1 \text{ L soln}}\right)$$
$$= 4.90 \times 10^{-3} \text{ mol OH}^-$$

The number of moles of H$^+$ in the original solution is $5.00 \times 10^{-3}$. Thus there remains

$$(5.00 \times 10^{-3}) - (4.90 \times 10^{-3})$$
$$= 1.0 \times 10^{-4} \text{ mol H}^+(aq)$$

in 0.0990 L of solution. The concentration of H$^+(aq)$ is thus

$$\frac{1.0 \times 10^{-4} \text{ mol}}{0.0990 \text{ L soln}} = 1.0 \times 10^{-3} \ M$$

The corresponding pH is 3.0.

To answer parts (b), (c), and (d) of the exercise we proceed in the same way. In each case, we calculate the amount of added OH$^-$, compare that with the amount of H$^+$ originally present, and determine what amount of H$^+$ or OH$^-$ exists in excess over the other. The concentration is then calculated from a knowledge of the total volume of the solution. By proceeding in this way we can construct a table of pH versus volume of added NaOH solution, as shown in Table 18.1. By graphing the pH as a function of the volume of added NaOH, we obtain the titration curve shown in Figure 18.3.

## PRACTICE EXERCISE
Calculate the pH when the following quantities of 0.100 $M$ HCl have been added to 25.00 mL of 0.100 $M$ NaOH solution: **(a)** 24.90 mL; **(b)** 25.00 mL; **(c)** 25.10 mL.

***Answers:*** **(a)** 10.30; **(b)** 7.00; **(c)** 3.70

**TABLE 18.1**   Titration of 50.00 mL of 0.100 $M$ HCl solution with 0.100 $M$ NaOH solution

| Volume of HCl | Volume of NaOH | Total volume | Moles H$^+$ | Moles OH$^-$ | Molarity of excess ion | pH |
|---|---|---|---|---|---|---|
| 50.00 | 0.00 | 50.00 | $5.00 \times 10^{-3}$ | 0.00 | 0.100 | 1.00 |
| 50.00 | 49.00 | 99.00 | $5.00 \times 10^{-3}$ | $4.90 \times 10^{-3}$ | $1.00 \times 10^{-3}$(H$^+$) | 3.00 |
| 50.00 | 49.90 | 99.90 | $5.00 \times 10^{-3}$ | $4.99 \times 10^{-3}$ | $1.00 \times 10^{-4}$(H$^+$) | 4.00 |
| 50.00 | 50.10 | 100.10 | $5.00 \times 10^{-3}$ | $5.01 \times 10^{-3}$ | $1.00 \times 10^{-4}$(OH$^-$) | 10.00 |
| 50.00 | 51.00 | 101.00 | $5.00 \times 10^{-3}$ | $5.10 \times 10^{-3}$ | $1.00 \times 10^{-3}$(OH$^-$) | 11.00 |
| 50.00 | 60.00 | 110.00 | $5.00 \times 10^{-3}$ | $6.00 \times 10^{-3}$ | $9.09 \times 10^{-3}$(OH$^-$) | 11.96 |

## Titrations Involving a Weak Acid or Weak Base

Titration of a weak acid by a strong base results in pH curves that look similar to those for strong acid-strong base titrations. Figure 18.6 shows the pH curve produced when 0.100 $M$ NaOH is added to 50.0 mL of 0.100 $M$ acetic acid. There are three noteworthy differences between this curve and that for a strong acid-strong base titration: (1) The weak-acid solution has a higher initial pH. In Section 17.3 we calculated that the pH of a 0.100 $M$ solution of HC$_2$H$_3$O$_2$ is 2.89; by contrast, the pH of a 0.100 $M$ solution of HCl is 1.00. (2) The pH rises more rapidly in the early part of the titration, but more slowly near the equivalence point. The weaker the acid, the less marked the pH change near the equivalence point. This aspect is illustrated in Figure 18.7, which shows a family of

**FIGURE 18.6**   The green line shows the variation in pH as 0.10 $M$ NaOH solution is added in the titration of 0.10 $M$ acetic acid solution. The blue line segment shows the graph of pH versus added base for the titration of 0.10 $M$ HCl.

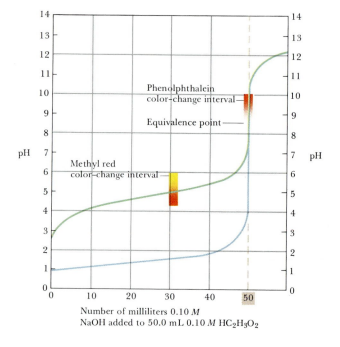

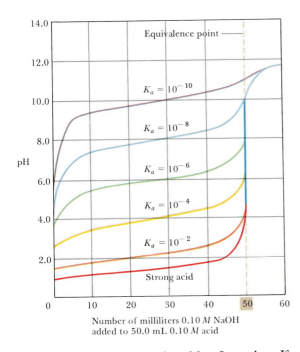

**FIGURE 18.7** Influence of acid strength on the shape of the curve for titration with NaOH. Each curve represents titration of 50.0 mL of 0.10 $M$ acid with 0.10 $M$ NaOH.

titration curves for weak acids of varying $K_a$. (3) The pH at the equivalence point is *not* 7. Remember that the pH at the equivalence point in any acid-base titration is the pH of the resultant salt solution. In the titration illustrated in Figure 18.6, the solution contains 0.050 $M$ $NaC_2H_3O_2$ at the equivalence point (remember that the total volume of solution is doubled by addition of the NaOH solution to $HC_2H_3O_2$). We saw in Chapter 17 how to calculate the pH of a solution of a salt containing a weak base as anion (see, for example, Sample Exercise 17.11). The pH of a 0.050 $M$ $NaC_2H_3O_2$ solution is 8.71.

The following sample exercise shows how to calculate the pH of the solution at any intermediate point along the titration curve. You will see that this problem has two parts. The first part involves simple stoichiometry. We determine how much base has reacted with acid (or vice versa) in the neutralization reaction, and how much acid (or base) remains. The concentrations of the salt formed in the neutralization reaction and the unreacted acid or base are then calculated. The second part of the problem is to calculate the pH of a solution that is a mixture of an acid or base and its conjugate. This is the same problem we have just dealt with in connection with buffer solutions.

## SAMPLE EXERCISE 18.8

Calculate the pH in the titration of acetic acid by sodium hydroxide after 30.0 mL of 0.100 $M$ NaOH solution has been added to 50.0 mL of 0.100 $M$ acetic acid solution.

**Solution:** The total number of moles of $HC_2H_3O_2$ originally in solution is

$$\left(\frac{0.100 \text{ mol } HC_2H_3O_2}{1 \text{ L soln}}\right)(0.0500 \text{ L soln})$$

$$= 5.00 \times 10^{-3} \text{ mol } HC_2H_3O_2$$

Similarly, 30.0 mL of 0.100 $M$ NaOH solution contains $3.00 \times 10^{-3}$ mol of $OH^-$. During the titration, this $OH^-$ reacts with the acetic acid, forming $3.00 \times 10^{-3}$ mol of $C_2H_3O_2^-$ and leaving $2.00 \times 10^{-3}$ mol of $HC_2H_3O_2$, in 80 mL of solution:

$$OH^-(aq) \; + \; HC_2H_3O_2(aq) \; \rightleftharpoons \; C_2H_3O_2^-(aq) \; + \; H_2O(l)$$

| | | | | |
|---|---|---|---|---|
| **Before rxn:** | $3.00 \times 10^{-3}$ mol | $5.00 \times 10^{-3}$ mol | 0 | — |
| **Change:** | $-3.00 \times 10^{-3}$ mol | $-3.00 \times 10^{-3}$ mol | $+3.00 \times 10^{-3}$ mol | — |
| **After rxn:** | 0 | $2.00 \times 10^{-3}$ mol | $3.00 \times 10^{-3}$ mol | — |

The resulting molarities are thus

$$[HC_2H_3O_2] = \frac{2.00 \times 10^{-3} \text{ mol } HC_2H_3O_2}{0.0800 \text{ L}}$$
$$= 0.0250 \; M$$

$$[C_2H_3O_2^-] = \frac{3.00 \times 10^{-3} \text{ mol } C_2H_3O_2^-}{0.0800 \text{ L}}$$
$$= 0.0375 \; M$$

For the weak-acid dissociation equilibrium of $HC_2H_3O_2(aq)$ we have

$$K_a = \frac{[H^+][C_2H_3O_2^-]}{[HC_2H_3O_2]} = 1.8 \times 10^{-5}$$

$$[H^+] = K_a \frac{[HC_2H_3O_2]}{[C_2H_3O_2^-]} = (1.8 \times 10^{-5}) \left( \frac{0.0250}{0.0375} \right)$$
$$= 1.2 \times 10^{-5}$$
$$pH = -\log (1.2 \times 10^{-5}) = 4.92$$

By proceeding in similar fashion, other points on the titration curve shown in Figure 18.6 could be calculated. Incidentally, you might note that at the halfway point in the titration, when $[C_2H_3O_2^-]$ equals $[HC_2H_3O_2]$, the pH equals $pK_a$, which in this case is 4.74.

## PRACTICE EXERCISE

(a) Calculate the pH in the titration of ammonia by hydrochloric acid after 15.00 mL of 0.100 $M$ HCl has been added to 25.00 mL of 0.100 $M$ $NH_3(aq)$ solution. (b) What is the pH at the equivalence point in this titration?

*Answers:* (a) 9.08; (b) 5.28

It is important that the pH at the equivalence point in the titration of the weak acid is considerably higher than it is in the titration of the strong acid. Acid-base titrations in an analytical laboratory are usually carried out using a pH meter (Section 17.1) to indicate progress of the titration. It is essential for the analyst to know what the pH should be at the equivalence point, so the titration can be stopped. For more occasional titrations, it is common to use an indicator to signal the equivalence point. We saw earlier that in titrating 0.10 $M$ HCl with 0.10 $M$ NaOH, either phenolphthalein or methyl red could be used as indicator. Although the pH of color change does not correspond precisely to the equivalence point for either indicator, both were close enough that no significant error would be introduced by using either of them (Figure 18.3). In a titration of acetic acid with NaOH, phenolphthalein is an ideal indicator, because it changes color just at the pH of the equivalence point. However, methyl red is not a good choice. The pH at the midrange of its color change is 5.2. At this pH we are still far short of the equivalence point, as shown in Figure 18.6.

If the acid to be titrated were weaker than acetic acid, the titration curve would look similar to that shown in Figure 18.6. However, the pH at the beginning of the reaction would be higher, and the pH at the equivalence point would also be higher. This means that the pH change at the equivalence point would not be as sharp, and the end point would be more difficult to detect (see Figure 18.7). With very weak acids, for which $K_a$ is less than about $1 \times 10^{-8}$, an acid-base titration is not really a good quantitative procedure when using indicators.

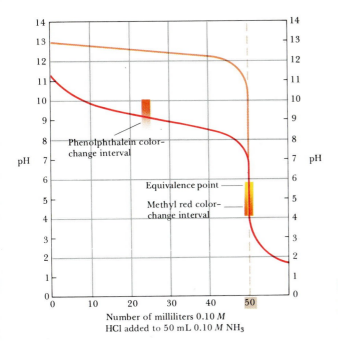

Phenolphthalein color-
change interval

pH

Equivalence point

Methyl red color-
change interval

Number of milliliters 0.10 $M$
HCl added to 50 mL 0.10 $M$ NH₃

**FIGURE 18.8**   The red line shows pH versus volume
of added HCl in the titration of 0.10 $M$ ammonia with
0.10 $M$ HCl. The orange line segment shows the graph of
pH versus added acid for the titration of 0.10 $M$ NaOH.

Titration of a weak base (for example, 0.10 $M$ NH₃) with a strong
acid (such as 0.10 $M$ HCl) solution leads to the titration curve shown in
Figure 18.8. In this particular example, the equivalence point occurs
at pH 5.3. In this case methyl red would be an ideal indicator, but
phenolphthalein would be a poor choice.

In the case of weak acids containing more than one ionizable proton,
reaction with OH⁻ occurs in a series of steps. An important example
of the kind of equilibria that occur is carbonic acid, $H_2CO_3$. We saw
in Sample Exercise 17.9 that in an aqueous solution of $CO_2$ the relative concentrations of the species involved are $[H_2CO_3] \gg [HCO_3^-] \gg$
$[CO_3^{2-}]$. Neutralization of $H_2CO_3$ proceeds in two stages:

Titration
of Polyprotic Acids

$$H_2CO_3(aq) + OH^-(aq) \rightleftharpoons H_2O(l) + HCO_3^-(aq) \qquad [18.12]$$
$$HCO_3^-(aq) + OH^-(aq) \rightleftharpoons H_2O(l) + CO_3^{2-}(aq) \qquad [18.13]$$

Figure 18.9 shows how the relative abundances of $H_2CO_3$, $HCO_3^-$, and
$CO_3^{2-}$ vary as a function of the solution pH. As OH⁻ is added to $H_2CO_3$
(as pH increases), $[H_2CO_3]$ decreases while $[HCO_3^-]$ increases. By pH
8.5 $H_2CO_3$ has been nearly completely converted to $HCO_3^-$. At this
stage Figure 18.9 shows $[H_2CO_3] = 0$, and $[CO_3^{2-}] = 0$; the concentrations of these species are too small to register on the figure. As pH increases beyond 8.5, $[HCO_3^-]$ decreases while $[CO_3^{2-}]$ increases. Thus
it is only after the first reaction goes nearly to completion that the second
reaction starts.

   When the neutralization steps of a polyprotic acid or polybasic base
are sufficiently separated, the substance exhibits a titration curve with
multiple equivalence points. Figure 18.10 shows the titration curve for
the $H_2CO_3$–$HCO_3^-$–$CO_3^{2-}$ system. Notice that there are two distinct
equivalence points along the titration curve.

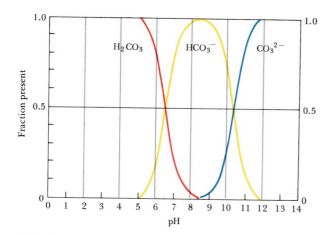

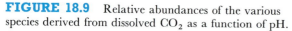

**FIGURE 18.9**   Relative abundances of the various species derived from dissolved $CO_2$ as a function of pH.

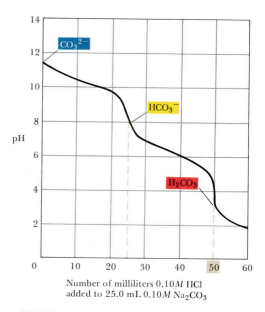

Number of milliliters $0.10M$ HCl
added to 25.0 mL $0.10M$ $Na_2CO_3$

**FIGURE 18.10**   Titration curve for the reaction of 25.0 mL of 0.10 $M$ $Na_2CO_3$ with 0.10 $M$ HCl.

## 18.4 SOLUBILITY EQUILIBRIA

The equilibria that we have considered thus far in this chapter have involved acids and bases. Furthermore, they have been homogeneous; that is, all the species involved in an equilibrium have been in the same phase. In this section we will consider the equilibria involved in another important type of solution reaction: the dissolution or precipitation of slightly soluble salts. Such reactions are heterogeneous.

For an equilibrium to exist between a solid substance and its solution, the solution must be saturated and in contact with undissolved solid. As an example, consider a saturated solution of $BaSO_4$ that is in contact with solid $BaSO_4$. The chemical equation for the relevant equilibrium can be written as

$$BaSO_4(s) \rightleftharpoons Ba^{2+}(aq) + SO_4^{2-}(aq) \qquad [18.14]$$

The solid is an ionic compound (as are others that we will discuss in this section). Such compounds are almost invariably strong electrolytes; to the extent that they dissolve, these compounds are present in solutions as ions. We observed in Section 16.4 that for heterogeneous reactions the concentration of the solid is constant. It is thus incorporated into the equilibrium constant. Thus the equilibrium-constant expression for the dissolution of $BaSO_4$ can be written as

$$K = [Ba^{2+}][SO_4^{2-}]$$

When concentration is expressed in molarity, the equilibrium constant is called the **solubility-product constant** and designated as $K_{sp}$:

$$K_{sp} = [Ba^{2+}][SO_4^{2-}] \qquad [18.15]$$

The rules for writing the solubility-product expression are the same as those for the writing of any equilibrium-constant expression: *The solubility product is equal to the product of the concentrations of the ions involved in the equilibrium, each raised to the power of its coefficient in the equilibrium equation.*

## SAMPLE EXERCISE 18.9

Write the expression for the solubility-product constant for $Ca_3(PO_4)_2$.

**Solution:** We first write the equation for the solubility equilibrium:

$$Ca_3(PO_4)_2(s) \rightleftharpoons 3Ca^{2+}(aq) + 2PO_4^{3-}(aq)$$

Following the rule stated above, the power to which the $Ca^{2+}$ concentration is raised is 3 and the power to which the $PO_4^{3-}$ concentration is raised is 2. The resulting expression for $K_{sp}$ is

$$K_{sp} = [Ca^{2+}]^3[PO_4^{3-}]^2$$

**PRACTICE EXERCISE**

The $K_{sp}$ for $Cu(N_3)_2$ is $6.3 \times 10^{-10}$. What is the solubility of $Cu(N_3)_2$ in water in grams per liter?
*Answer:* 0.080 g/L

It is important to distinguish carefully between solubility and solubility product. The solubility is the quantity of substance that dissolves in a given quantity of water. It is often expressed as grams of solute per 100 g of water, or in terms of molarity. The solubility of a substance can be changed by adding some other substance to the solution. For example, the addition of $Na_2SO_4$ lowers the solubility of $BaSO_4$; the $SO_4^{2-}$ supplied by the $Na_2SO_4$ shifts the solubility equilibrium (Equation 18.14) to the left. This behavior is another example of the common-ion effect. However, the presence of $Na_2SO_4$ does not affect $K_{sp}$ for $BaSO_4$.* The value of $K_{sp}$ at a particular temperature is the same, regardless of whether other substances are present in solution. Nevertheless, $K_{sp}$ and solubility are related, and one can be calculated from the other. Sample Exercise 18.10 shows the calculation of $K_{sp}$ from solubility data; Sample Exercise 18.11 shows the calculation of molar solubility from $K_{sp}$.

## SAMPLE EXERCISE 18.10

Solid barium sulfate is shaken in contact with pure water at 25°C for several days. Each day a sample is withdrawn and analyzed for its barium concentration. After several days, the value of $[Ba^{2+}]$ is constant, indicating that equilibrium has been reached. The concentration of $Ba^{2+}$ is $1.04 \times 10^{-5}$ $M$. What is $K_{sp}$ for $BaSO_4$?

**Solution:** The analysis provides values for both $[Ba^{2+}]$ and $[SO_4^{2-}]$. Because all of the $[Ba^{2+}]$ and $SO_4^{2-}$ ions come from $BaSO_4$, there must be one $Ba^{2+}$ for each $[SO_4^{2-}]$. Consequently, $[Ba^{2+}] =$ $[SO_4^{2-}] = 1.04 \times 10^{-5}$ $M$. Thus we have

$$\begin{aligned} K_{sp} &= [Ba^{2+}][SO_4^{2-}] \\ &= (1.04 \times 10^{-5})(1.04 \times 10^{-5}) \\ &= 1.08 \times 10^{-10} \end{aligned}$$

**PRACTICE EXERCISE**

When $MgF_2$ is shaken with water at 27°C until the solution is saturated, the concentration of $Mg^{2+}$ ion is found to be $1.2 \times 10^{-3}$ $M$. What is $K_{sp}$ for $MgF_2$ at this temperature?
*Answer:* $6.9 \times 10^{-9}$

---

* This is strictly true only for very dilute solutions. The values of equilibrium constants are somewhat altered when the total concentration of ionic substances in water is increased. However, we shall ignore these effects, which are taken into consideration only for very accurate work.

## SAMPLE EXERCISE 18.11

The $K_{sp}$ for $CaF_2$ is $3.9 \times 10^{-11}$. What is the solubility of $CaF_2$ in water in grams per liter?

**Solution:** The solubility equilibrium involved is

$$CaF_2(s) \rightleftharpoons Ca^{2+}(aq) + 2F^-(aq)$$

For each mole of $CaF_2$ that dissolves, 1 mol of $Ca^{2+}$ and 2 mol of $F^-$ enter the solution. Thus if we let $x$ equal the solubility of $CaF_2$ in moles per liter, the molar concentrations of $Ca^{2+}$ and $F^-$ will be $[Ca^{2+}] = x$ and $[F^-] = 2x$.

The $K_{sp}$ expression for this reaction is

$$K_{sp} = [Ca^{2+}][F^-]^2 = 3.9 \times 10^{-11}$$

Substituting $[Ca^{2+}] = x$ and $[F^-] = 2x$ and solving for $x$, we have

$$x(2x)^2 = 3.9 \times 10^{-11}$$
$$4x^3 = 3.9 \times 10^{-11}$$
$$x = 2.1 \times 10^{-4} \ M$$

Thus the molar solubility of $CaF_2$ is $2.1 \times 10^{-4}$ mol/L. The mass of $CaF_2$ that dissolves in a liter of solution is

$$\left(\frac{2.1 \times 10^{-4} \text{ mol } CaF_2}{1 \text{ L soln}}\right)\left(\frac{78.1 \text{ g } CaF_2}{1 \text{ mol } CaF_2}\right)$$
$$= 1.6 \times 10^{-2} \text{ g } CaF_2/\text{L soln}$$

## PRACTICE EXERCISE

The value of $K_{sp}$ for $Cu(N_3)_2$ is $6.3 \times 10^{-10}$. What is the concentration of azide ion in a solution that is $0.050 \ M$ in $CuSO_4$ and saturated with $Cu(N_3)_2$?
*Answer:* $1.1 \times 10^{-4} \ M$

---

Appendix E contains $K_{sp}$ values for a variety of salts at 25°C. From these equilibrium constants we can calculate solubilities under a variety of conditions If a single solute is dissolved in water, the relative concentrations of the ions are determined by the formula of the solute. In the case of $CaF_2$, the concentration of $F^-$ must be twice the concentration of $Ca^{2+}$.

$$CaF_2(s) \rightleftharpoons Ca^{2+}(aq) + 2F^-(aq) \qquad [18.16]$$

However, it is possible to alter the relative concentrations of $Ca^{2+}$ and $F^-$ by adding a soluble salt containing one of these ions. For example, we might add either $Ca(NO_3)_2$ or NaF. Either will lower the solubility of the $CaF_2$. If $Ca(NO_3)_2$ is added, the equilibrium concentration of $Ca^{2+}$ is the sum of the $Ca^{2+}$ concentrations from the two sources. This is another example of the common-ion effect, introduced in Section 18.1.

## SAMPLE EXERCISE 18.12

What is the molar solubility of $CaF_2$ in a solution containing $0.010 \ M$ NaF?

**Solution:** As in Sample Exercise 18.11, we have

$$K_{sp} = [Ca^{2+}][F^-]^2 = 3.9 \times 10^{-11}$$

The value of $K_{sp}$ is unchanged by the fact that the solution initially contains $0.010 \ M$ NaF. In Sample Exercise 18.11, the relative concentrations of $Ca^{2+}$ and $F^-$ were determined entirely by the solubility of $CaF_2$. In the present exercise we must take account of a second source of $F^-$. Let's again let the molar solubility of $CaF_2$ be $x$. Then the concentration of $Ca^{2+}$ is $x$ and the concentration of $F^-$ derived from $CaF_2$ is $2x$. But there is in addition a $0.010 \ M$ con-

tribution to the $F^-$ concentration from the dissolved NaF. We thus have

$$[Ca^{2+}] = x \qquad [F^-] = 0.010 + 2x$$
$$K_{sp} = [Ca^{2+}][F^-]^2 = (x)(0.010 + 2x)^2$$
$$= 3.9 \times 10^{-11}$$

This would be a messy problem to solve exactly, but fortunately it is possible to greatly simplify matters. Even without the common-ion effect of $0.010 \ M \ F^-$, the solubility of $CaF_2$ in water is not very great, as illustrated by the small value of $x$ obtained in Sample Exercise 18.11. We know from application of Le Châtelier's principle that the solubility of $CaF_2$ will be even smaller in the presence of $0.010 \ M$ NaF. We can therefore safely assume that the 0.010 $M$ $F^-$ concentration from the NaF is much greater than the small additional contribution resulting from the solubility of $CaF_2$. That is, we can neglect $2x$ in comparison with 0.010 (that is, $0.010\ M + 2x \simeq 0.010\ M$). We then have

$$3.9 \times 10^{-11} = x(0.010)^2$$
$$x = 3.9 \times 10^{-7}\ M = [Ca^{2+}]$$

This value for $x$ represents the solubility of $CaF_2$ in a solution that is 0.010 $M$ in NaF. Note that it is much smaller than $2.1 \times 10^{-4}\ M$, the solubility of $CaF_2$ in pure water.

**PRACTICE EXERCISE**
The value of $K_{sp}$ for $Li_3PO_4$ is $3.2 \times 10^{-9}$. What is the molar solubility of $Li_3PO_4$ in a solution that contains 0.050 $M$ $LiNO_3$?
*Answer:* $2.6 \times 10^{-5}\ M$

Equilibrium can be achieved starting with the substances on either side of the chemical equation. The equilibrium between $BaSO_4(s)$, $Ba^{2+}(aq)$, and $SO_4^{2-}(aq)$ (Equation 18.14) can be achieved starting with solid $BaSO_4$. It can also be achieved starting with solutions of salts containing $Ba^{2+}$ and $SO_4^{2-}$, say $BaCl_2$ and $Na_2SO_4$. When these two solutions are mixed, a precipitate of $BaSO_4$ will form if the product of ion concentrations, $Q = [Ba^{2+}][SO_4^{2-}]$, is greater than $K_{sp}$.*

The possible relationships between $Q$ and $K_{sp}$ are summarized as follows:

If $Q > K_{sp}$ precipitation occurs until $Q = K_{sp}$.
If $Q = K_{sp}$ equilibrium exists (saturated solution).
If $Q < K_{sp}$ solid dissolves until $Q = K_{sp}$.

## 18.5 CRITERIA FOR PRECIPITATION OR DISSOLUTION

## SAMPLE EXERCISE 18.13

Will a precipitate form when 0.100 L of $3.0 \times 10^{-3}\ M$ $Pb(NO_3)_2$ is added to 0.400 L of $5.0 \times 10^{-3}\ M$ $Na_2SO_4$?

**Solution:** The possible reaction products are $PbSO_4$ and $NaNO_3$. Sodium salts are quite soluble; however, $PbSO_4$ has a $K_{sp}$ of $1.6 \times 10^{-8}$ (Appendix E). To determine whether the $PbSO_4$ precipitates, we must calculate $Q = [Pb^{2+}][SO_4^{2-}]$ and compare it with $K_{sp}$.

When the two solutions are mixed, the total volume becomes $0.100\ L + 0.400\ L = 0.500\ L$. The number of moles of $Pb^{2+}$ in 0.100 L of $3.0 \times 10^{-3}\ M$ $Pb(NO_3)_2$ is

$$(0.100\ L)\left(3.0 \times 10^{-3}\ \frac{mol}{L}\right) = 3.0 \times 10^{-4}\ mol$$

The concentration of $Pb^{2+}$ in the 0.500-L mixture is therefore

$$[Pb^{2+}] = \frac{3.0 \times 10^{-4}\ mol}{0.500\ L} = 6.0 \times 10^{-4}\ M$$

The number of moles of $SO_4^{2-}$ is

$$(0.400\ L)\left(5.0 \times 10^{-3}\ \frac{mol}{L}\right) = 2.0 \times 10^{-3}\ mol$$

Therefore, $[SO_4^{2-}]$ in the 0.500-L mixture is

$$[SO_4^{2-}] = \frac{2.0 \times 10^{-3}\ mol}{0.500\ L} = 4.0 \times 10^{-3}\ M$$

We then have

$$Q = [Pb^{2+}][SO_4^{2-}] = (6.0 \times 10^{-4})(4.0 \times 10^{-3})$$
$$= 2.4 \times 10^{-6}$$

Because $Q > K_{sp}$, precipitation of $PbSO_4$ will occur.

* The use of the reaction quotient, $Q$, to determine the direction a reaction must proceed to reach equilibrium was discussed earlier, in Section 16.5. In the present case the equilibrium expression contains no denominator and therefore is really not a quotient. Thus $Q$ is often referred to simply as the ion product.

## SAMPLE EXERCISE 18.14

What concentration of $OH^-$ must be exceeded in a 0.010 $M$ solution of $Ni(NO_3)_2$ in order to precipitate $Ni(OH)_2$? (Assume that the added $OH^-$ does not change the concentration of $Ni^{2+}$.)

**Solution:**   From Appendix E we have $K_{sp} = 1.6 \times 10^{-14}$ for $Ni(OH)_2$. Any $OH^-$ in excess of that in a saturated solution of $Ni(OH)_2$ will cause some $Ni(OH)_2$ to precipitate. For a saturated solution we have

$$K_{sp} = [Ni^{2+}][OH^-]^2 = 1.6 \times 10^{-14}$$

Thus if $Q = [Ni^{2+}][OH^-]^2 > 1.6 \times 10^{-14}$ precipitation will occur. Letting $[OH^-] = x$ and using $[Ni^{2+}] = 0.010\ M$, we have

$$(0.010)x^2 > 1.6 \times 10^{-14}$$
$$x^2 > \frac{1.6 \times 10^{-14}}{0.010} = 1.6 \times 10^{-12}$$
$$x > \sqrt{1.6 \times 10^{-12}} = 1.3 \times 10^{-6}\ M$$

This concentration of $OH^-$ corresponds to a solution pH of 8.11  (pH = 14.00 − pOH = 14.00 − 5.89 = 8.11). Thus $Ni(OH)_2$ will precipitate when the solution pH is 8.11 or higher.

**Solubility and pH**

The solubility of any substance whose anion is basic will be affected to some extent by the pH of the solution. For example, consider $Mg(OH)_2$, for which the solubility equilibrium is

$$Mg(OH)_2(s) \ \rightleftharpoons\ Mg^{2+}(aq) + 2OH^-(aq) \qquad [18.17]$$

The value of $K_{sp}$ for $Mg(OH)_2$ is $1.8 \times 10^{-11}$. Suppose that solid $Mg(OH)_2$ is equilibrated with a solution buffered at a pH of 9.0. Then pOH is 5.0, that is, $[OH^-] = 1.0 \times 10^{-5}$. Inserting this value for $[OH^-]$ into the solubility-product expression, we have

$$K_{sp} = [Mg^{2+}][OH^-]^2 = 1.8 \times 10^{-11}$$
$$[Mg^{2+}][1.0 \times 10^{-5}]^2 = 1.8 \times 10^{-11}$$
$$[Mg^{2+}] = 0.18\ M$$

Thus $Mg(OH)_2$ is quite soluble in a buffered, slightly basic medium. If the solution were made more acidic, the solubility of $Mg(OH)_2$ would increase, because the $OH^-$ concentration decreases with increasing acidity. The $Mg^{2+}$ concentration would thus increase to maintain the equilibrium condition.

The solubility of almost any salt is affected if the solution is made sufficiently acidic or basic. The effects are very noticeable, however, only when one or both ions involved is moderately acidic or basic. The metal hydroxides we've just discussed are good examples of compounds with a strong base, the hydroxide ion. As an additional example, the fluoride ion of $CaF_2$ is a weak base; it is the conjugate base of the weak acid HF. As a result, $CaF_2$ is more soluble in acidic solutions than in neutral or basic ones, because of the reaction of $F^-$ with $H^+$ to form HF. The solution process can be considered as two consecutive reactions:

$$CaF_2(s) \rightleftharpoons Ca^{2+}(aq) + 2F^-(aq) \qquad [18.18]$$

$$F^-(aq) + H^+(aq) \rightleftharpoons HF(aq) \qquad [18.19]$$

The equation for the overall process is

$$CaF_2(s) + 2H^+(aq) \rightleftharpoons Ca^{2+}(aq) + 2HF(aq) \qquad [18.20]$$

Qualitatively, we can understand what occurs in terms of Le Châtelier's principle: The solubility equilibrium is driven to the right because the free $F^-$ concentration is reduced by reaction with $H^+$. The reduction of $[F^-]$ causes $Q$ to be reduced so that it becomes smaller than $K_{sp}$. Thus more $CaF_2$ dissolves.

These examples illustrate a general rule: *The solubility of slightly soluble salts containing basic anions increases as $[H^+]$ increases (as pH is lowered)*. Salts with anions of negligible basicity (the anions of strong acids) are largely unaffected by pH.

---

**SAMPLE EXERCISE 18.15**

Which of the following substances will be more soluble in acidic solution than in basic solution: **(a)** $Ni(OH)_2(s)$; **(b)** $CaCO_3(s)$; **(c)** $BaSO_4(s)$; **(d)** $AgCl(s)$?

**Solution:** **(a)** $Ni(OH)_2(s)$ will be more soluble in acidic solution, because of reaction of $H^+$ with the $OH^-$ ion, forming water:

$$Ni(OH)_2(s) \rightleftharpoons Ni^{2+}(aq) + 2OH^-(aq)$$
$$2OH^-(aq) + 2H^+(aq) \rightleftharpoons 2H_2O(l)$$

Overall: $\overline{Ni(OH)_2(s) + 2H^+(aq) \rightleftharpoons}$
$$Ni^{2+}(aq) + 2H_2O(l)$$

**(b)** Similarly, $CaCO_3(s)$ reacts with acid, liberating gaseous $CO_2$:

$$CaCO_3(s) \rightleftharpoons Ca^{2+}(aq) + CO_3{}^{2-}(aq)$$
$$CO_3{}^{2-}(aq) + 2H^+(aq) \rightleftharpoons$$

$$H_2CO_3(aq) \longrightarrow CO_2(g) + H_2O(l)$$
Overall: $\overline{CaCO_3(s) + 2H^+(aq) \longrightarrow}$
$$Ca^{2+}(aq) + CO_2(g) + H_2O(l)$$

**(c)** The solubility of $BaSO_4$ is largely unaffected by changes in solution pH, because $SO_4{}^{2-}$ is a rather weak base and thus has little tendency to combine with a proton. However, $BaSO_4$ is slightly more soluble in strongly acidic solutions.

**(d)** The solubility of AgCl is unaffected by changes in pH, because $Cl^-$ is the anion of a strong acid.

**PRACTICE EXERCISE**
Write the net ionic equation for the reaction of the following copper(II) compounds with acid: **(a)** $CuCrO_4$; **(b)** $Cu(N_3)_2$.

*Answers:*
**(a)** $CuCrO_4(s) + H^+(aq) \rightleftharpoons Cu^{2+}(aq) + HCrO_4{}^-(aq)$
**(b)** $Cu(N_3)_2(s) + 2H^+(aq) \rightleftharpoons Cu^{2+}(aq) + 2HN_3(aq)$

---

Ions can be separated from each other on the basis of their solubilities. One widely used procedure for separating metal ions is based on the relative solubilities of their sulfides. Metals that form insoluble metal salts will not precipitate unless $[S^{2-}]$ is sufficiently high for $Q$ to exceed $K_{sp}$. The sulfide concentration can be adjusted by regulating the pH of the

**Precipitations of Metal Sulfides**

Tooth enamel consists mainly of a mineral called hydroxyapatite, $Ca_{10}(PO_4)_6(OH)_2$. It is the hardest substance in the body. Tooth cavities are caused by the dissolving action of acids on tooth enamel:

$$Ca_{10}(PO_4)_6(OH)_2(s) + 8H^+(aq) \longrightarrow$$
$$10Ca^{2+}(aq) + 6HPO_4^{2-}(aq) + 2H_2O(l)$$

The resultant $Ca^{2+}$ and $HPO_4^{2-}$ ions diffuse out of the tooth enamel and are washed away by saliva. The acids that attack the hydroxyapatite are formed by the action of specific bacteria on sugars and other carbohydrates present in the plaque adhering to the teeth.

Fluoride ion, present in drinking water, toothpaste, or other sources can react with hydroxyapatite to form fluoroapatite, $Ca_{10}(PO_4)_6F_2$. This mineral, in which $F^-$ has replaced $OH^-$, is much more resistant to attack by acids, because the fluoride ion is a much weaker Brønsted base than the hydroxide ion is.

Because of the success of fluoride ion in preventing cavities, fluoride ion is added to the public water supply in many places to give a concentration of 1 mg/L (that is, 1 ppm). The compound added may be NaF or $Na_2SiF_6$. The latter compound reacts with water to release fluoride ion by the following reaction:

$$SiF_6^{2-}(aq) + 2H_2O(l) \longrightarrow$$
$$6F^-(aq) + 4H^+(aq) + SiO_2(s)$$

About 80 percent of all toothpastes now sold in the United States contain fluoride compounds, usually at the level of 0.1 percent fluoride by weight. The most common compounds in toothpastes are stannous fluoride, $SnF_2$, sodium monofluorophosphate, $NaPO_3F$, and sodium fluoride, NaF.

---

solution. For example, CuS can be precipitated from a mixture of $Cu^{2+}$ and $Zn^{2+}$ by bubbling $H_2S$ gas into a properly acidified solution of these ions. CuS ($K_{sp} = 6.3 \times 10^{-36}$) is less soluble than ZnS ($K_{sp} = 1.1 \times 10^{-21}$). Consequently, $[S^{2-}]$ can be regulated to cause CuS to precipitate while not exceeding $K_{sp}$ for ZnS.

When $H_2S$ is bubbled through an aqueous solution at 25°C and 1 atm, a saturated solution of $H_2S$ forms that is approximately 0.1 $M$ in $H_2S$. $H_2S$ is a weak diprotic acid:

$$H_2S(aq) \rightleftharpoons H^+(aq) + HS^-(aq) \qquad K_{a1} = 5.7 \times 10^{-8} \qquad [18.21]$$

$$HS^-(aq) \rightleftharpoons H^+(aq) + S^{2-}(aq) \qquad K_{a2} = 1.3 \times 10^{-13} \qquad [18.22]$$

These equations can be combined to give an equation for the overall dissociation of $H_2S$ into $S^{2-}$:

$$H_2S(aq) \rightleftharpoons 2H^+(aq) + S^{2-}(aq) \qquad [18.23]$$

The equilibrium-constant expression for this overall dissociation process is

$$K = \frac{[H^+]^2[S^{2-}]}{[H_2S]} = K_{a1} \times K_{a2} = 7.4 \times 10^{-21} \qquad [18.24]$$

Substituting the solubility of $H_2S$, 0.1 $M$, into this expression gives

$$\frac{[H^+]^2[S^{2-}]}{(0.1)} = 7.4 \times 10^{-21}$$

$$[H^+]^2[S^{2-}] = (0.1)(7.4 \times 10^{-21})$$
$$= 7 \times 10^{-22} \qquad \text{[18.25]}$$

Equation 18.25 can be used to calculate the concentration of $S^{2-}$ in saturated solutions of $H_2S$ at various pH values.

Now consider how we can calculate the pH that will allow us to prevent precipitation of ZnS while precipitating CuS from a solution that is 0.10 $M$ in $Zn^{2+}$, 0.10 $M$ in $Cu^{2+}$, and saturated with $H_2S$. The maximum concentration of $S^{2-}$ that can be present in the solution before precipitation of ZnS occurs can be calculated from the solubility-product expression for this substance:

$$[Zn^{2+}][S^{2-}] = K_{sp} = 1.1 \times 10^{-21}$$

$$[S^{2-}] = \frac{1.1 \times 10^{-21}}{[Zn^{2+}]} = \frac{1.1 \times 10^{-21}}{0.10}$$
$$= 1.1 \times 10^{-20} \; M$$

The concentration of hydrogen ions necessary to give $[S^{2-}] = 1.1 \times 10^{-20} \; M$ can be calculated using Equation 18.25:

$$[H^+]^2[S^{2-}] = 7 \times 10^{-22}$$

$$[H^+]^2 = \frac{7 \times 10^{-22}}{[S^{2-}]} = \frac{7 \times 10^{-22}}{1.1 \times 10^{-20}} = 6 \times 10^{-2}$$

$$[H^+] = \sqrt{6 \times 10^{-2}} = 0.24 \; M$$

$$\text{pH} = -\log{(2.4 \times 10^{-1})} = 0.6$$

Thus ZnS will not precipitate if the pH is 0.6 or lower. However, at pH 0.6, where $[S^{2-}] = 1.1 \times 10^{-20}$, the ion product for a 0.10 $M$ $Cu^{2+}$ solution would be

$$Q = [Cu^{2+}][S^{2-}] = (0.10)(1.1 \times 10^{-20}) = 1.1 \times 10^{-21}$$

Because $Q$ exceeds the $K_{sp}$ of CuS (that is, $6.3 \times 10^{-36}$), CuS will precipitate under these conditions. Indeed, since $[Cu^{2+}] = 0.10 \; M$, $[S^{2-}]$ must be less than $6.3 \times 10^{-35} \; M$ to prevent the precipitation of CuS from this solution. Even under strongly acidic conditions $[S^{2-}]$ will be high enough to cause precipitation of CuS. Thus a fairly broad range of sulfide concentration will allow for effective separation of $Cu^{2+}$ (as CuS) from $Zn^{2+}$.

A characteristic property of metal ions is that they are able to act as Lewis acids, or electron-pair acceptors, toward water molecules, which act as Lewis bases, or electron-pair donors (Section 17.8). Lewis bases other than water can also interact with metal ions, particularly with transition metal ions. Such interactions can have a dramatic effect on the solubility

**Effect of Complex Formation on Solubility**

**FIGURE 18.11** (a) A precipitate of AgCl dissolves (b) upon addition of concentrated $NH_3(aq)$. (Fundamental Photographs)

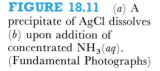

(a)                                    (b)

of a metal salt. For example, AgCl, whose $K_{sp} = 1.82 \times 10^{-10}$, will dissolve in the presence of aqueous ammonia because of the interaction between $Ag^+$ and the Lewis base $NH_3$, as shown in Figure 18.11. This process can be viewed as the sum of two reactions, the solubility equilibrium of AgCl and the Lewis acid-base interaction between $Ag^+$ and $NH_3$:

$$AgCl(s) \rightleftharpoons Ag^+(aq) + Cl^-(aq) \qquad [18.26]$$

$$Ag^+(aq) + 2NH_3(aq) \rightleftharpoons Ag(NH_3)_2{}^+(aq) \qquad [18.27]$$

Overall: $\quad AgCl(s) + 2NH_3(aq) \rightleftharpoons Ag(NH_3)_2{}^+(aq) + Cl^-(aq) \qquad [18.28]$

The presence of $NH_3$ drives the top reaction, the solubility equilibrium of AgCl, to the right as $Ag^+(aq)$ is removed to form $Ag(NH_3)_2{}^+$.

For a Lewis base such as $NH_3$ to increase the solubility of a metal salt, it must be able to interact more strongly with the metal ion than water does. The $NH_3$ must displace solvating $H_2O$ molecules (Sections 12.2 and 17.8) in order to form $Ag(NH_3)_2{}^+$:

$$Ag^+(aq) + 2NH_3(aq) \rightleftharpoons Ag(NH_3)_2{}^+(aq) \qquad [18.29]$$

An assembly of a metal ion and the Lewis bases bonded to it, such as $Ag(NH_3)_2{}^+$, is called a **complex ion**. The stability of a complex ion in aqueous solution can be judged by the magnitude of the equilibrium constant for formation of the complex ion from the hydrated metal ion. For example, the equilibrium constant for formation of $Ag(NH_3)_2{}^+$ (Equation 18.29) is $1.7 \times 10^7$:

$$K_f = \frac{[\text{Ag(NH}_3)_2{}^+]}{[\text{Ag}^+][\text{NH}_3]^2} = 1.7 \times 10^7 \qquad [18.30]$$

Such equilibrium constants are called **formation constants**, $K_f$. The formation constants for several complex ions are shown in Table 18.2.

**TABLE 18.2** Formation constants for some metal complex ions in water at 25°C

| Complex ion | $K_f$ | Equilibrium equation |
|---|---|---|
| $\text{Ag(NH}_3)_2{}^+$ | $1.7 \times 10^7$ | $\text{Ag}^+(aq) + 2\text{NH}_3(aq) \rightleftharpoons \text{Ag(NH}_3)_2{}^+(aq)$ |
| $\text{Ag(CN)}_2{}^-$ | $1 \times 10^{21}$ | $\text{Ag}^+(aq) + 2\text{CN}^-(aq) \rightleftharpoons \text{Ag(CN)}_2{}^-(aq)$ |
| $\text{Ag(S}_2\text{O}_3)_2{}^{3-}$ | $2.9 \times 10^{13}$ | $\text{Ag}^+(aq) + 2\text{S}_2\text{O}_3{}^{2-}(aq) \rightleftharpoons \text{Ag(S}_2\text{O}_3)_2{}^{3-}(aq)$ |
| $\text{CdBr}_4{}^{2-}$ | $5 \times 10^3$ | $\text{Cd}^{2+}(aq) + 4\text{Br}^-(aq) \rightleftharpoons \text{CdBr}_4{}^{2-}(aq)$ |
| $\text{Cr(OH)}_4{}^-$ | $8 \times 10^{29}$ | $\text{Cr}^{3+}(aq) + 4\text{OH}^-(aq) \rightleftharpoons \text{Cr(OH)}_4{}^-(aq)$ |
| $\text{Co(SCN)}_4{}^{2-}$ | $1 \times 10^3$ | $\text{Co}^{2+}(aq) + 4\text{SCN}^-(aq) \rightleftharpoons \text{Co(SCN)}_4{}^{2-}(aq)$ |
| $\text{Cu(NH}_3)_4{}^{2+}$ | $5 \times 10^{12}$ | $\text{Cu}^{2+}(aq) + 4\text{NH}_3(aq) \rightleftharpoons \text{Cu(NH}_3)_4{}^{2+}(aq)$ |
| $\text{Cu(CN)}_4{}^{2-}$ | $1 \times 10^{25}$ | $\text{Cu}^{2+}(aq) + 4\text{CN}^-(aq) \rightleftharpoons \text{Cu(CN)}_4{}^{2-}(aq)$ |
| $\text{Ni(NH}_3)_6{}^{2+}$ | $5.5 \times 10^8$ | $\text{Ni}^{2+}(aq) + 6\text{NH}_3(aq) \rightleftharpoons \text{Ni(NH}_3)_6{}^{2+}(aq)$ |
| $\text{Fe(CN)}_6{}^{4-}$ | $1 \times 10^{35}$ | $\text{Fe}^{2+}(aq) + 6\text{CN}^-(aq) \rightleftharpoons \text{Fe(CN)}_6{}^{4-}(aq)$ |
| $\text{Fe(CN)}_6{}^{3-}$ | $1 \times 10^{42}$ | $\text{Fe}^{3+}(aq) + 6\text{CN}^-(aq) \rightleftharpoons \text{Fe(CN)}_6{}^{3-}(aq)$ |

## SAMPLE EXERCISE 18.16

Calculate the concentration of $\text{Ag}^+$ present in solution at equilibrium when concentrated ammonia is added to a 0.010 $M$ solution of $\text{AgNO}_3$ to give an equilibrium concentration of $[\text{NH}_3] = 0.20\ M$. Neglect the small volume change that occurs on addition of $\text{NH}_3$.

**Solution:** Because $K_f$ is quite large, we begin with the assumption that essentially all of the $\text{Ag}^+$ is converted to $\text{Ag(NH}_3)_2{}^+$, in accordance with Equation 18.29. Thus $[\text{Ag}^+]$ will be small at equilibrium. If $[\text{Ag}^+]$ was 0.010 $M$ initially, then $[\text{Ag(NH}_3)_2{}^+]$ will be 0.010 $M$ following addition of the $\text{NH}_3$. Let the concentration of $\text{Ag}^+$ at equilibrium be $x$. Then

$$\text{Ag}^+(aq) + 2\text{NH}_3(aq) \rightleftharpoons \text{Ag(NH}_3)_2{}^+(aq)$$
$$x\ M \qquad 0.20\ M \qquad (0.010 - x)\ M$$

Because the concentration of $\text{Ag}^+$ is very small, we can ignore $x$ in comparison with 0.010. Thus $0.010 - x \approx 0.010\ M$. Substituting these values into the equilibrium-constant expression, Equation 18.30, we obtain

$$\frac{[\text{Ag(NH}_3)_2{}^+]}{[\text{Ag}^+][\text{NH}_3]^2} = \frac{0.010}{x(0.20)^2} = 1.7 \times 10^7$$

Solving for $x$, we obtain $x = 1.4 \times 10^{-8}\ M = [\text{Ag}^+]$. It is evident that formation of the $\text{Ag(NH}_3)_2{}^+$ complex drastically reduces the concentration of free $\text{Ag}^+$ ion in solution.

## PRACTICE EXERCISE

Calculate the concentration of free $\text{Cr}^{3+}$ ion when 0.010 mol of $\text{Cr(NO}_3)_3$ is dissolved in a liter of solution buffered at pH 10.0.
***Answer:*** $[\text{Cr}^{3+}] = 1.2 \times 10^{-16}\ M$

The general rule is that metal salts will dissolve in the presence of a suitable Lewis base, such as $\text{NH}_3$, $\text{CN}^-$, or $\text{OH}^-$, if the metal forms a sufficiently stable complex with the base. The ability of metal ions to form complexes is an extremely important aspect of their chemistry. In Chapter 26 we will take a much closer look at complex ions. In that chapter and others we shall see applications of complex ions to areas such as biochemistry, metallurgy, and photography.

## Amphoterism

Many metal hydroxides and oxides that are relatively insoluble in neutral water dissolve in a strongly acidic *and* in a strongly basic medium. Such substances are said to be **amphoteric** (from the Greek word *amphoteros*, which means "both"; see Section 8.11). Examples include the hydroxides and oxides of $Al^{3+}$, $Cr^{3+}$, $Zn^{2+}$, and $Sn^{2+}$.

The dissolution of these species in acidic solutions should be anticipated based on the earlier discussions in this section. We have seen that acids promote the dissolving of compounds with basic anions. What makes amphoteric oxides and hydroxides special is that they also dissolve in strongly basic solutions. This behavior results from the formation of complex anions containing several (typically four) hydroxides bound to the metal ion:

$$Al(OH)_3(s) + OH^-(aq) \rightleftharpoons Al(OH)_4^-(aq) \qquad [18.31]$$

Amphoterism is often interpreted in terms of the behavior of the water molecules that surround the metal ion and that are bonded to it by Lewis acid-base interactions (Section 17.8). For example, $Al^{3+}(aq)$ is more accurately represented as $Al(H_2O)_6^{3+}(aq)$; six water molecules are bonded to the $Al^{3+}$ in aqueous solution. As discussed in Section 17.8, this hydrated ion is a weak acid. As a strong base is added, the $Al(H_2O)_6^{3+}$ loses protons in a stepwise fashion, eventually forming the neutral and water-insoluble $Al(H_2O)_3(OH)_3$. This substance then dissolves upon removal of an additional proton to form the anion $Al(H_2O)_2(OH)_4^-$. The reactions that occur are as follows:

$$Al(H_2O)_6^{3+}(aq) + OH^-(aq) \rightleftharpoons Al(H_2O)_5(OH)^{2+}(aq) + H_2O(l)$$

$$Al(H_2O)_5(OH)^{2+}(aq) + OH^-(aq) \rightleftharpoons Al(H_2O)_4(OH)_2^+(aq) + H_2O(l)$$

$$Al(H_2O)_4(OH)_2^+(aq) + OH^-(aq) \rightleftharpoons Al(H_2O)_3(OH)_3(s) + H_2O(l)$$

$$Al(H_2O)_3(OH)_3(s) + OH^-(aq) \rightleftharpoons Al(H_2O)_2(OH)_4^-(aq) + H_2O(l)$$

Further proton removals are possible, but each successive reaction occurs less readily than the one before. As the charge on the ion becomes more negative, it becomes increasingly difficult to remove a positively charged proton. Addition of an acid reverses these reactions. The proton adds in a stepwise fashion to convert the $OH^-$ groups to $H_2O$, eventually reforming $Al(H_2O)_6^{3+}$. The common practice is to simplify the equations for these reactions by excluding the bound $H_2O$ molecules. Thus we usually write $Al^{3+}$ instead of $Al(H_2O)_6^{3+}$, $Al(OH)_3$ instead of $Al(H_2O)_3(OH)_3$, $Al(OH)_4^-$ instead of $Al(H_2O)_2(OH)_4^-$, and so forth.

The extent to which an insoluble metal hydroxide reacts with either acid or base varies with the particular metal ion involved. Many metal hydroxides—for example, $Ca(OH)_2$, $Fe(OH)_2$, and $Fe(OH)_3$—are capable of dissolving in acidic solution but do not react with excess base. These hydroxides are not amphoteric. In Sample Exercise 18.17 we work through calculations useful in considering such equilibria from a quantitative point of view.

## SAMPLE EXERCISE 18.17

The solubility product for zinc hydroxide, $Zn(OH)_2$, is $1.2 \times 10^{-17}$. The formation constant for the hydroxo complex, $Zn(OH)_4{}^{2-}$, is $4.6 \times 10^{17}$. What concentration of $OH^-$ is required to dissolve 0.010 mol of $Zn(OH)_2$ in a liter of solution?

**Solution:** Let us first write the equilibria with which we will be concerned:

$$Zn(OH)_2(s) \rightleftharpoons Zn^{2+}(aq) + 2OH^-(aq)$$
$$Zn^{2+}(aq) + 4OH^-(aq) \rightleftharpoons Zn(OH)_4{}^{2-}(aq)$$

We know that if 0.010 mol of $Zn(OH)_2$ dissolves in the liter of solution, we will have a total of 0.010 mol of $Zn(II)$ present *in some form*. Because the $Zn(OH)_2$ is considerably more soluble in the $OH^-$ solution than it would be in pure water, we can assume that essentially all of the $Zn^{2+}$ is present as $Zn(OH)_4{}^{2-}$. Thus $[Zn(OH)_4{}^{2-}] = 0.010$ $M$. Let us now set up the equilibrium-constant expression for complex formation:

$$K_f = \frac{[Zn(OH)_4{}^{2-}]}{[Zn^{2+}][OH^-]^4} = 4.6 \times 10^{17} \quad [18.32]$$

As we've already indicated, we can assume that $[Zn(OH)_4{}^{2-}] = 0.010$ $M$. However, it would appear that we still have two unknowns in the expression, $[Zn^{2+}]$ and $[OH^-]$. We can eliminate one of these by making use of the solubility-product expression. At the point at which all the $Zn(OH)_2$ just

dissolves, the solubility-product expression must hold:

$$[Zn^{2+}][OH^-]^2 = 1.2 \times 10^{-17}$$

Let's solve this expression for $[Zn^{2+}]$ and substitute into Equation 18.32:

$$[Zn^{2+}] = \frac{1.2 \times 10^{-17}}{[OH^-]^2}$$

$$4.6 \times 10^{17} = \frac{[Zn(OH)_4{}^{2-}]}{\left(\dfrac{1.2 \times 10^{-17}}{[OH^-]^2}\right)[OH^-]^4}$$

Substituting in $[Zn(OH)_4{}^{2-}] = 0.010$ and solving for $[OH^-]$, we obtain

$$[OH^-]^2 = \frac{0.010}{(1.2 \times 10^{-17})(4.6 \times 10^{17})}$$

$$[OH^-] = 0.042 \ M$$

That is, at a hydroxide ion concentration of 0.042 $M$ or larger, 0.010 mol of $Zn(OH)_2$ will completely dissolve in a liter of solution, because of formation of the $Zn(OH)_4{}^{2-}$ complex.

### PRACTICE EXERCISE

The solubility-product constant for chromium(III) hydroxide, $Cr(OH)_3$, is $6.3 \times 10^{-31}$. The formation constant for the hydroxo complex, $Cr(OH)_4{}^-$, is $8.0 \times 10^{29}$. What concentration of $OH^-$ is required to dissolve 0.010 mol of $Cr(OH)_3$ in a liter of solution?
*Answer:* $2.0 \times 10^{-2}$ $M$

---

The purification of aluminum ore in the manufacture of aluminum metal provides an interesting application of the property of amphoterism. As we have seen, $Al(OH)_3$ is amphoteric, whereas $Fe(OH)_3$ is not. Aluminum occurs in large quantities as the ore **bauxite**, which is essentially $Al_2O_3$ with additional water molecules. The ore is contaminated with $Fe_2O_3$ as an impurity. When bauxite is added to a strongly basic solution, the $Al_2O_3$ dissolves, because the aluminum forms complex ions, such as $Al(OH)_4{}^-$. The $Fe_2O_3$ impurity, however, is not amphoteric and remains as a solid. The solution is filtered, getting rid of the iron impurity. Aluminum hydroxide is then precipitated by addition of acid. The purified hydroxide receives further treatment and eventually yields aluminum metal.

When writing reactions of amphoteric substances, we often show them as the oxides rather than the hydrated hydroxides. Although the various ways of showing the reactions are confusing at first, the only real differences are in the number of water molecules shown. Sample Exercise 18.18 shows how we can view the reaction of a metal oxide with base.

## SAMPLE EXERCISE 18.18

Write the balanced net-ionic equation for the reaction between $Al_2O_3(s)$ and aqueous solutions of $NaOH$ that causes the $Al_2O_3(s)$ to dissolve. ($H_2O$ is also present as a reactant.)

**Solution:** $NaOH$ is a strong electrolyte that provides the necessary $OH^-$ ions. The reaction product can be taken to be $Al(OH)_4^-$, an assembly of $Al^{3+}$ and four $OH^-$ ions. The unbalanced equation is

$$Al_2O_3(s) + OH^-(aq) \rightleftharpoons Al(OH)_4^-(aq)$$

Two Al are needed among the products:

$$Al_2O_3(s) + OH^-(aq) \rightleftharpoons 2Al(OH)_4^-(aq)$$

To balance the charge on both sides of the equation requires $2OH^-$. Sufficient $H_2O$ is then added to balance the O and H counts. The balanced equation is

$$Al_2O_3(s) + 2OH^-(aq) + 3H_2O(l) \rightleftharpoons 2Al(OH)_4^-(aq)$$

## PRACTICE EXERCISE

Write balanced net-ionic equations for the dissolution of **(a)** $Zn(OH)_2(s)$ and **(b)** $ZnO(s)$ in excess base.
***Answers:***
**(a)** $Zn(OH)_2(s) + 2OH^-(aq) \rightleftharpoons Zn(OH)_4^{2-}(aq)$
**(b)** $ZnO(s) + H_2O(l) + 2OH^-(aq) \rightleftharpoons Zn(OH)_4^{2-}(aq)$

## 18.6 QUALITATIVE ANALYSES FOR METALLIC ELEMENTS

In this chapter we have seen several examples of equilibria involving metal ions in aqueous solution. In this final section we look very briefly at how solubility equilibria and complex formation can be used to detect the presence of particular metal ions in solution. Before the development of modern analytical instrumentation, it was necessary to analyze mixtures of metals in a sample by so-called "wet" chemical methods. For example, a metallic sample that might contain several metallic elements was dissolved in a concentrated acid solution. This solution was then tested in a systematic way for the presence of various metallic ions.

**Qualitative analysis** determines only the presence or absence of a particular metal ion. It should be distinguished from **quantitative analysis**, which determines how much of a given substance is present. Wet methods of qualitative analysis are no longer so important as a means of analysis. However, they are frequently used in general chemistry laboratory programs as a means of illustrating equilibria, teaching the properties of common metal ions in solution, and developing laboratory skills. Typically, such analyses proceed in three stages. (1) The ions are separated into broad groups on the basis of solubility properties. (2) The individual ions within each group are then separated by selectively dissolving members in the group. (3) The ions are then identified by means of specific tests.

A scheme in common use divides the common cations into five groups, as shown in Figure 18.12. The order of addition of reagents is important. The most selective separations—that is, those that involve the smallest number of ions—are carried out first. The reactions that are used must proceed so far toward completion that any concentration of cations remaining in the solution is too small to interfere with subsequent tests. Let's take a closer look at each of these five groups of cations, examining briefly the logic used in this qualitative analysis scheme.

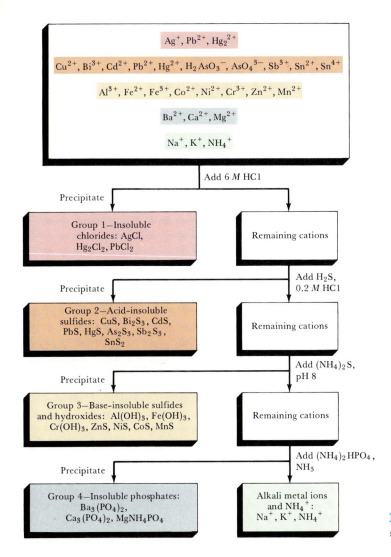

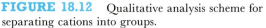

**FIGURE 18.12** Qualitative analysis scheme for separating cations into groups.

1.  *Insoluble chlorides:* Of the common metal ions only $Ag^+$, $Hg_2^{2+}$, and $Pb^{2+}$ form insoluble chlorides. Thus when dilute HCl is added to a mixture of cations, only AgCl, $Hg_2Cl_2$, and $PbCl_2$ will precipitate, leaving the other cations in solution. The absence of a precipitate indicates that the starting solution contains no $Ag^+$, $Hg_2^{2+}$, or $Pb^{2+}$.

2.  *Acid-insoluble sulfides:* After any insoluble chlorides have been removed, the remaining solution, now acidic, is treated with $H_2S$. As we saw in Section 18.5, the dissociation of $H_2S$ is repressed in acidic solutions so that the concentration of free $S^{2-}$ is very low. Consequently, only the most insoluble metal sulfides, CuS, $Bi_2S_3$, CdS, PbS, HgS, $As_2S_3$, $Sb_2S_3$, and $SnS_2$, can precipitate. (Note the very small values of $K_{sp}$ for some of these sulfides in Appendix E.) Those metal ions whose sulfides are somewhat more soluble—for example, ZnS or NiS—remain in solution.

3.  *Base-insoluble sulfides:* After the solution is filtered to remove any acid-insoluble sulfides, the remaining solution is made slightly basic, and $(NH_4)_2S$

is added. In basic solutions the concentration of $S^{2-}$ is higher than in acidic solutions. Thus the ion products for many of the more soluble sulfides are caused to exceed their $K_{sp}$ values and precipitation occurs. The metal ions precipitated at this stage are $Al^{3+}$, $Cr^{3+}$, $Fe^{3+}$, $Zn^{2+}$, $Ni^{2+}$, $Co^{2+}$, and $Mn^{2+}$. (Actually, the $Al^{3+}$, $Fe^{3+}$, and $Cr^{3+}$ ions do not form insoluble sulfides; instead they are precipitated as insoluble hydroxides at the same time.)

4. *Insoluble phosphates:* At this point the solution contains only metal ions from periodic table groups 1A and 2A. Addition of $(NH_4)_2HPO_4$ to a basic solution causes precipitation of the group 2A elements $Mg^{2+}$, $Ca^{2+}$, $Sn^{2+}$, and $Ba^{2+}$, because these metals form insoluble phosphates.

5. *The alkali metal ions and $NH_4^+$:* The ions that remain after removal of the insoluble phosphates form a small group. We can test for each ion individually. For example, the flame test is useful to show the presence of $K^+$, because the flame turns a characteristic violet color if $K^+$ is present.

Additional separation and testing is necessary to determine which ions are present within each of the groups. As an example, consider the ions of the insoluble chloride group. The precipitate containing the metal chlorides is boiled in water. It happens that $PbCl_2$ is relatively soluble in hot water, whereas $AgCl$ and $Hg_2Cl_2$ are not. The hot solution is filtered, and a solution of $Na_2CrO_4$ added to the filtrate. If $Pb^{2+}$ is present, a yellow precipitate of $PbCrO_4$ forms. The test for $Ag^+$ consists of treating the metal chloride precipitate with dilute ammonia. Only $Ag^+$ forms an ammonia complex. If $AgCl$ is present in the precipitate, it will dissolve in the ammonia solution:

$$AgCl(s) + 2NH_3(aq) \rightleftharpoons Ag(NH_3)_2^+(aq) + Cl^-(aq) \qquad [18.33]$$

After treatment with ammonia, the solution is filtered and the filtrate made acidic by adding nitric acid. The nitric acid removes ammonia from solution by forming $NH_4^+$, thus releasing $Ag^+$, which should re-form the $AgCl$ precipitate:

$$Ag(NH_3)_2^+(aq) + Cl^-(aq) + 2H^+(aq) \rightleftharpoons AgCl(s) + 2NH_4^+(aq) \qquad [18.34]$$

The analyses for individual ions in the acid-insoluble and base-insoluble sulfides are a bit more complex, but the same general principles are involved. The detailed procedures for carrying out such analyses are given in laboratory manuals.

# *FOR REVIEW*

## SUMMARY

In this chapter we've considered several types of important equilibria occurring in aqueous solution. Our primary emphasis has been on acid-base equilibria in solutions containing two or more solutes and on solubility equilibria. We observed that the dissociation of a weak acid or weak base is repressed by the presence of a strong electrolyte that provides an ion common to the equilibrium. This phenomenon is an example of the **common-ion effect**.

A particularly important type of acid-base mixture is a weak conjugate acid-base pair. Such mixtures function as **buffers**. Addition of small amounts of additional acid or base to a buffered solution causes only small changes in pH, because the buffer reacts with the added acid or base. (Recall that strong acid-strong base, strong acid-weak base, and weak acid-strong base reactions proceed essentially to completion.) Buffer solutions are usually prepared from a weak acid and a salt of that acid or from a weak base and a salt of that base. Two important characteristics of a buffer solution are its buffering capacity and its pH.

The plot of the pH of an acid (or base) as a function of the volume of added base (or acid) is called a **titration curve**. Titration curves aid in selecting a proper pH indicator for an acid-base titration. The titration curve of a strong acid-strong base titration exhibits a large change in pH in the immediate vicinity of the equivalence point; at the equivalence point for this titration, pH = 7. For strong acid-weak base or weak acid-strong base titrations, the pH change in the vicinity of the equivalence point is not as large. Furthermore, the pH at the equivalence point is not 7 in either of these cases. Rather, it is the pH of the salt solution that results from the neutralization reaction.

The equilibrium between a solid salt and its ions in solution provides an example of heterogeneous equilibrium. The **solubility-product constant**, $K_{sp}$, is an equilibrium constant that expresses quantitatively the extent to which the salt dissolves. Addition to the solution of an ion common to a solubility equilibrium causes the solubility of the salt to decrease. This phenomenon is another example of the common-ion effect.

Comparison of the ion product, $Q$, with the value of $K_{sp}$ can be used to judge whether a precipitate will form when solutions are mixed or whether a slightly soluble salt will dissolve under various conditions. Solubility is affected by the common-ion effect, by pH, and by the presence of certain Lewis bases that react with metal ions to form stable **complex ions**. Solubility is affected by pH when one or more of the ions in the solubility equilibrium is an acid or base. For example, the solubility of MnS is increased on addition of acid, because $S^{2-}$ is basic. **Amphoteric metal hydroxides** are those slightly soluble metal hydroxides that dissolve on addition of either acid or base. The reactions that give rise to the amphoterism are acid-base reactions involving the $OH^-$ or $H_2O$ groups bound to the metal ion. Complex-ion formation in aqueous solution involves the displacement by Lewis bases (such as $NH_3$ and $CN^-$) of water molecules attached to the metal ion. The extent to which such complex formation occurs is expressed quantitatively by the **formation constant** for the complex ion.

The fact that the ions of different metallic elements vary a great deal in the solubilities of their salts, in their acid-base behavior, and in their tendencies to form complexes can be used to separate and detect the presence of metal ions in mixtures. **Qualitative analysis** determines the presence or absence of a metal ion in a mixture of metal ions in solution. The analysis usually proceeds by separating the ions into groups on the basis of precipitation reactions and then analyzing each group for individual metal ions.

# LEARNING GOALS

Having read and studied this chapter, you should be able to:

1. Predict qualitatively and calculate quantitatively the effect of an added common ion on the pH of an aqueous solution of a weak acid or weak base.

2. Calculate the concentrations of each species present in a solution formed by mixing an acid and a base.

3. Describe how a buffer solution of a particular pH is made and how it operates to control pH.

4. Calculate the change in pH of a simple buffer solution of known composition caused by adding a small amount of strong acid or strong base.

5. Describe the form of the titration curves for titration of a strong acid by a strong base, a weak acid by a strong base, or a weak base by a strong acid.

6. Calculate the pH at any point, including the equivalence point, in an acid-base titration.

7. Set up the expression for the solubility-product constant for a salt.

8. Calculate $K_{sp}$ from solubility data and solubility from the value for $K_{sp}$.

9. Calculate the effect of an added common ion on the solubility of a slightly soluble salt.

10. Predict whether a precipitate will form when two solutions are mixed, given appropriate $K_{sp}$ values.

11. Explain the effect of pH on a solubility equilibrium involving a basic or acidic ion.

12. Formulate the equilibrium between a metal ion and a Lewis base to form a complex ion of a metal.

13. Describe how complex formation can affect the solubility of a slightly soluble salt.

14. Calculate the concentration of metal ion in equilibrium with a Lewis base with which it forms a soluble complex ion, from a knowledge of initial concentrations and $K_f$.

15. Explain the origin of amphoteric behavior, and write equations describing the dissolution of an amphoteric metal hydroxide in either an acidic or a basic medium.

16. Explain the general principles that apply to the groupings of metal ions in the qualitative analysis of an aqueous mixture.

## KEY TERMS

Among the more important terms and expressions used for the first time in this chapter are the following:

**Amphoterism** (Section 18.5) is a term used to describe the ability of certain slightly soluble metal hydroxides to dissolve in either an acidic or a basic medium. The solubility results from formation of a complex ion via an acid-base reaction.

A **buffer solution** (Section 18.2) is one that undergoes a limited change in pH upon addition of a small amount of acid or base.

The **common-ion effect** (Section 18.1) refers to the effect of an ion common to an equilibrium in shifting the equilibrium. For example, added $Na_2SO_4$ decreases the solubility of the slightly soluble salt $BaSO_4$, or added $NaC_2H_3O_2$ decreases the percent ionization of $HC_2H_3O_2$.

A **complex ion** (Section 18.5) consists of a metal ion and a well-defined group of ions or neutral molecules bound to the ion via a Lewis acid-base interaction.

The **formation constant** (Section 18.5) for a metal ion complex is the equilibrium constant for formation of the complex from the metal ion and base species present in solution. It is a measure of the tendency of the complex to form.

**Qualitative analysis** (Section 18.6) determines the presence or absence of a particular substance in a mixture.

The **solubility-product constant** (Section 18.4) is an equilibrium constant related to the equilibrium between a solid salt and its ions in solution. It provides a quantitative measure of the solubility of a slightly soluble salt.

## EXERCISES

### Common-ion Effect; Buffers

**18.1** Describe the effect on pH (increase, decrease, or no change) that results from each of the following additions: (a) sodium formate, $NaCHO_2$, to a solution of formic acid, $HCHO_2$; (b) ammonium perchlorate, $NH_4ClO_4$, to a solution of ammonia, $NH_3$; (c) potassium bromide to a solution of potassium nitrite, $KNO_2$; (d) hydrochloric acid, HCl, to a solution of sodium acetate, $NaC_2H_3O_2$.

**18.2** Describe the effect on pH (increase, decrease, or no

change) that results from each of the following additions: (a) ammonia to a solution of HCl; (b) ammonium chloride to a solution of HCl; (c) sodium cyanide, NaCN, to a solution of HBr; (d) pyridinium nitrate, $C_5H_5NHNO_3$, to a solution of pyridine, $C_5H_5N$.

**18.3** What is the pH of a solution that is (a) 0.050 $M$ in sodium formate $NaCHO_2$, and 0.100 $M$ in formic acid, $HCHO_2$; (b) 0.060 $M$ in sodium benzoate, $NaC_7H_5O_2$, and 0.090 $M$ in benzoic acid, $HC_7H_5O_2$; (c) 0.10 $M$ in sodium propionate, $NaC_3H_5O_2$, and 0.20 $M$ in propionic acid, $HC_3H_5O_2$?

**18.4** Calculate the pH of each of the following solutions: **(a)** 0.25 $M$ hydroazoic acid, $HN_3$, and 0.125 $M$ sodium azide, $NaN_3$; **(b)** 0.20 $M$ benzoic acid, $HC_7H_5O_2$, and 0.20 $M$ sodium benzoate, $NaC_7H_5O_2$; **(c)** 0.15 $M$ pyridine, $C_5H_5N$, and 0.20 $M$ pyridinium chloride, $C_5H_5NH^+Cl^-$.

**18.5** A certain organic compound that is used as an indicator for acid-base reactions exists in aqueous solution as equal concentrations of the acid form, HB, and the base form, $B^-$, at a pH of 7.80. What is p$K_a$ for the acid form of this indicator, HB?

**18.6** Dimethylglyoxime, $C_4H_8N_2O_2$, is a weak base with $K_b = 4.0 \times 10^{-4}$. At what pH does a solution of dimethylglyoxime and its conjugate acid contain equal concentrations of the two species?

**18.7** Assume that $5.0 \times 10^{-3}$ mol of NaOH is added to each of the following solutions. Calculate the change in pH in each case: **(a)** 1 L of 0.050 $M$ acetic acid, $HC_2H_3O_2$; **(b)** 1 L of 0.050 $M$ hydrochloric acid, HCl; **(c)** 1 L of 0.050 $M$ sodium acetate, $NaC_2H_3O_2$.

**18.8** Assume that 0.0030 mol of $HClO_4$ is added to 1 L of each of the following solutions. Calculate the pH change in each case: **(a)** 0.0050 $M$ NaOH solution; **(b)** 0.0050 $M$ sodium acetate solution; **(c)** 0.0050 $M$ acetic acid solution.

**18.9** **(a)** Write the net ionic equation for the reaction that occurs when a solution of perchloric acid, $HClO_4$, is mixed with a solution of sodium benzoate, $NaC_7H_5O_2$. **(b)** Calculate the equilibrium constant for this reaction. **(c)** Calculate the concentrations of $Na^+$, $ClO_4^-$, $H^+$, $C_7H_5O_2^-$, and $HC_7H_5O_2$ when 0.50 L of 0.20 $M$ $HClO_4$ is mixed with 0.50 L of 0.20 $M$ $NaC_7H_5O_2$.

**18.10** **(a)** Write the net ionic equation for the reaction that occurs when a solution of potassium hydroxide, KOH, is mixed with a solution of dimethylammonium bromide, $(CH_3)_2NH_2Br$. **(b)** Calculate the equilibrium constant for this reaction. **(c)** Calculate the concentrations of $K^+$, $Br^-$, $H^+$, and $(CH_3)_2NH_2^+$ ions and of $(CH_3)_2NH$ when 0.40 L of 0.15 $M$ KOH is mixed with 0.40 L of 0.15 $M$ $(CH_3)_2NH_2Br$.

**18.11** Suppose that you have two solutions of pH 9.0, one a buffer solution, the other a solution of KOH. Explain how you would test a small sample from either of these solutions to determine which is the buffer solution.

**18.12** Explain what is meant in speaking of the buffering *capacity* of a buffer solution.

**18.13** What is the pH of the solution after 0.015 mol of NaBrO has been added to 0.500 L of a 0.100 $M$ solution of hypobromous acid, HOBr?

**18.14** How many moles of sodium hypobromite, NaBrO, should be added to 1.00 L of 0.200 $M$ hypobromous acid, HBrO, to form a buffer solution of pH 8.80? Assume that no volume change occurs when the NaBrO is added.

**18.15** Calculate the pH of the solution formed by mixing **(a)** 100 mL of 0.10 $M$ benzoic acid, $HC_7H_5O_2$, and 50 mL of 0.10 $M$ sodium benzoate, $NaC_7H_5O_2$; **(b)** 100 mL of 0.10 $M$ $HNO_3$ and 100 mL of 0.10 $M$ $HC_2H_3O_2$.

**18.16** Calculate the pH of a solution formed by mixing **(a)** 50 mL of 0.10 $M$ $NaC_2H_3O_2$ and 50 mL of 0.10 $M$ NaOH; **(b)** 25.0 mL of 0.100 $M$ NaOH and 75.0 mL of 0.100 $M$ $HC_2H_3O_2$.

**18.17** One liter of a buffer solution contains 0.15 mol of acetic acid and 0.10 mol of sodium acetate. **(a)** What is the pH of this buffer? **(b)** What is the pH after addition of 0.010 mol of $HNO_3$? **(c)** What is the pH after addition of 0.010 mol of NaOH?

**18.18** One liter of a buffer solution contains 0.020 mol of $NaH_2AsO_4$ and 0.020 mol of $Na_2HAsO_4$. **(a)** What is the pH of this buffer? (See Appendix E for $K_a$ values.) **(b)** What is the pH after addition of 0.0050 mol of HCl? **(c)** What is the pH after addition of 0.0050 mol of KOH?

**18.19** **(a)** What is the ratio of $HCO_3^-$ to $H_2CO_3$ in blood of pH 7.4? **(b)** What is the ratio of $HCO_3^-$ to $H_2CO_3$ in an exhausted marathon runner whose blood pH is 7.1?

**18.20** A phosphate buffer, consisting of $H_2PO_4^-$ and $HPO_4^{2-}$, helps control the pH of blood. Many carbonated soft drinks also use this buffer system. What is the pH of a soft drink in which the major buffer ingredients are 6.5 g of $NaH_2PO_4$ and 8.0 g of $Na_2HPO_4$ per 355 mL of solution?

## Titration Curves

**18.21** How does titration of a strong acid with a strong base differ from titration of a weak acid with a strong base with respect to the following points: **(a)** quantity of base required to reach the equivalence point; **(b)** pH at the beginning of the titration; **(c)** pH at the equivalence point; **(d)** pH after addition of a slight excess of base; **(e)** choice of indicator for determining the equivalence point?

**18.22** Assume that 30.0 mL of a 0.10 $M$ solution of a weak base B that accepts one proton is titrated with a 0.10 $M$ solution of the monoprotic strong acid HX. **(a)** How many moles of HX have been added at the equivalence point? **(b)** What is the predominant form of B at the equivalence point? **(c)** What factor determines the pH at the equivalence point? **(d)** Which indicator, phenolphthalein or methyl red, is likely to be the better choice for this titration?

**18.23** How many milliliters of 0.048 $M$ NaOH are required to reach the equivalence point in titrating each of the following solutions: (**a**) 30.0 mL of 0.038 $M$ HBr; (**b**) 28.0 mL of 0.018 $M$ $H_2SO_4$; (**c**) 32.0 mL of 0.034 $M$ $HC_2H_3O_2$?

**18.24** Four 20.0-mL samples of different HBr solutions were titrated with 0.100 $M$ NaOH solution. The volumes of base required to reach the equivalence point in each case were (**a**) 27.5 mL; (**b**) 21.8 mL; (**c**) 48.9 mL; (**d**) 25.5 mL. Calculate the concentrations of the four HBr solutions.

**18.25** A 20.00-mL sample of 0.200 $M$ HBr solution is titrated with 0.200 $M$ NaOH solution. Calculate the pH of the solution after the following volumes of base have been added: (**a**) 15.00 mL; (**b**) 19.9 mL; (**c**) 20.0 mL; (**d**) 20.1 mL; (**e**) 35.0 mL.

**18.26** A 30.00-mL sample of 0.200 $M$ KOH is titrated with 0.150 $M$ $HClO_4$ solution. Calculate the pH after the following volumes of acid have been added: (**a**) 30.00 mL; (**b**) 39.5 mL; (**c**) 39.9 mL; (**d**) 40.00 mL; (**e**) 40.10 mL.

**18.27** Calculate the pH at the equivalence point for titrating 0.200 $M$ solutions of each of the following bases with 0.200 $M$ HBr: (**a**) sodium hydroxide, NaOH; (**b**) hydroxylamine, $NH_2OH$; (**c**) aniline, $C_6H_5NH_2$.

**18.28** Calculate the pH at the equivalence point in titrating 0.100 $M$ solutions of each of the following with 0.080 $M$ NaOH: (**a**) hydrobromic acid, HBr; (**b**) lactic acid, $HC_3H_5O_3$; (**c**) sodium hydrogen chromate, $NaHCrO_4$.

## Solubility Equilibria

**18.29** Write the expression for the solubility-product constant for the solubility equilibrium of each of the following compounds (**a**) CdS; (**b**) $MgC_2O_4$; (**c**) $CeF_3$; (**d**) $Fe_3(AsO_4)_2$.

**18.30** Write the expression for the solubility-product constant for the solubility equilibrium of each of the following compounds: (**a**) $Cu_2S$; (**b**) $Co(OH)_3$; (**c**) $La_2(MoO_4)_3$; (**d**) $Ce(OH)_4$.

**18.31** If the molar solubility of $CaF_2$ at 35°C is $1.24 \times 10^{-3}$ mol/L, what is its $K_{sp}$ at this temperature?

**18.32** The $K_{sp}$ for $SrCO_3$ is $1.1 \times 10^{-10}$. What is the molar solubility (that is, moles per liter) for this substance in water?

**18.33** For each salt listed below we have given the solubility at 25°C. From these data calculate the value for $K_{sp}$ in each case: (**a**) $AgIO_3$, 0.0283 g/L; (**b**) $Cd(CN)_2$, 0.22 g/L; (**c**) $Ag_2CO_3$, $3.5 \times 10^{-2}$ g/L; (**d**) $CrF_3$, 0.14 g/L.

**18.34** Using the $K_{sp}$ values listed in Appendix E, calculate the solubility in grams per liter of solution for each of the following substances: (**a**) $MnCO_3$; (**b**) $Mg(OH)_2$; (**c**) $CeF_3$.

**18.35** Calculate molar solubility of AgI in (**a**) pure water; (**b**) $3.0 \times 10^{-3}$ $M$ NaI solution; (**c**) $3.0 \times 10^{-3}$ $M$ $AgNO_3$ solution.

**18.36** Calculate the molar solubility of $Ag_2SO_4$ in (**a**) pure water; (**b**) 0.10 $M$ $Na_2SO_4$; (**c**) 0.10 $M$ $AgNO_3$.

**18.37** Will $Mn(OH)_2$ precipitate from solution if the pH of a 0.050 $M$ solution of $MnCl_2$ is adjusted to 8.0?

**18.38** The $K_{sp}$ for $AgIO_3$ is $3.0 \times 10^{-8}$. Should a precipitate of $AgIO_3$ form when 100 mL of 0.010 $M$ $AgNO_3$ solution is mixed with 10 mL of 0.010 $M$ $NaIO_3$ solution?

**18.39** The solubility product for $CrF_3$ is $6.6 \times 10^{-11}$; that for $CeF_3$ is $8.0 \times 10^{-16}$. A 0.010 $M$ solution of NaF is shaken with a mixture of solid $CrF_3$ and $CeF_3$. At equilibrium what is the ratio of the concentration of $Cr^{3+}(aq)$ to that of $Ce^{3+}(aq)$?

**18.40** What is the ratio of $[Ca^{2+}]$ to $[Fe^{2+}]$ in a lake in which the water is in equilibrium with deposits of both $CaCO_3$ and $FeCO_3$?

## Precipitation; Dissolution

**18.41** Which of the following salts will be substantially more soluble in acid solution than in pure water: (**a**) $AgCO_3$, (**b**) $CeF_3$; (**c**) $Ag_2SO_4$; (**d**) $PbI_2$; (**e**) $Cd(OH)_2$?

**18.42** For each of the following slightly soluble salts, write the net ionic equation for reaction, if any, with acid: (**a**) $Pb(N_3)_2$; (**b**) $LaF_3$; (**c**) $BiI_3$; (**d**) AgCN; (**e**) $SrSO_3$.

**18.43** Calculate the molar solubility of $Cd(OH)_2$ (**a**) at pH 7.0; (**b**) at pH 9.0.

**18.44** Calculate the molar solubility of $Mg(OH)_2$ (**a**) at pH 9.5; (**b**) at pH 11.8.

**18.45** Calculate the solubility of ZnS (in moles per liter) in a 0.10 $M$ solution of $H_2S$ if the pH of the solution is 2.40.

**18.46** Calculate the solubility of CdS (in moles per liter) in a 0.10 $M$ solution of $H_2S$ if the pH of the solution is 1.0.

**18.47** The solubility-product constant for barium permanganate, $Ba(MnO_4)_2$, is $2.5 \times 10^{-10}$. Suppose that solid $Ba(MnO_4)_2$ is in equilibrium with a solution of $KMnO_4$. What concentration of $KMnO_4$ is required to establish a concentration of $2.0 \times 10^{-8}$ $M$ for the $Ba^{2+}$ ion in solution?

**18.48** (**a**) The $K_{sp}$ for cerium iodate, $Ce(IO_3)_3$, is $3.2 \times 10^{-10}$. Calculate the molar solubility of $Ce(IO_3)_3$ in pure water. **(b)** What concentration of $NaIO_3$ in solution would be necessary to reduce the $Ce^{3+}$ concentration in a saturated solution of $Ce(IO_3)_3$ by a factor of 10 below that calculated in part (a)?

**18.49** Complete and balance each of the following reactions. In each case indicate which species in the reaction can be considered a Lewis base and which a Lewis acid:
**(a)** $Zn(OH)_2(s) + H^+(aq) \longrightarrow$
**(b)** $Cd(CN)_2(s) + CN^-(aq) \longrightarrow$
**(c)** $CrF_3(s) + OH^-(aq) \longrightarrow$

**18.50** Complete and balance each of the following reactions. In each case indicate which species in the reaction can be considered a Lewis base and which a Lewis acid:
**(a)** $Cu^{2+}(aq) + CN^-(aq) \longrightarrow$
**(b)** $AgCl(s) + S_2O_3{}^{2-}(aq) \longrightarrow$
**(c)** $Cu(OH)_2(s) + NH_3(aq) \longrightarrow$

**18.51** From the value for $K_f$ listed in Table 18.2 calculate the concentration of $Cu^{2+}$ in 1 L of a solution that contains a total of $1 \times 10^{-3}$ mol of copper(II) ion and that is 0.10 $M$ in $NH_3$.

**18.52** To what final concentration of $NH_3$ must a solution be adjusted to just dissolve 0.020 mol of $NiC_2O_4$ in a liter of solution? (Hint: You can neglect the hydrolysis of $C_2O_4{}^{2-}$, because the solution will be quite basic.)

**18.53** Using the value of $K_{sp}$ for AgCl and $K_f$ for $Ag(CN)_2{}^-$, calculate the equilibrium constant for the following reaction:

$$AgCl(s) + 2CN^-(aq) \longrightarrow Ag(CN)_2{}^-(aq) + Cl^-(aq)$$

**18.54** Using the value of $K_{sp}$ for $Ag_2S$, $K_{a1}$ and $K_{a2}$ for $H_2S$, and $K_f = 1.1 \times 10^5$ for $AgCl_2{}^-$, calculate the equilibrium constant for the following reaction:

$$Ag_2S(s) + 4Cl^-(aq) + 2H^+(aq) \longrightarrow 2AgCl_2{}^-(aq) + H_2S(aq)$$

## Qualitative Analysis

**18.55** A solution containing an unknown number of metal ions is treated with dilute HCl; no precipitate forms. The pH is adjusted to about 1, and $H_2S$ is bubbled through. Again no precipitate forms. The pH of the solution is then adjusted to about 8. Again $H_2S$ is bubbled through. This time a precipitate forms. The filtrate from this solution is treated with $(NH_4)_2HPO_4$. No precipitate forms. Which metal ions discussed in Section 18.6 are possibly present? Which are definitely absent within the limits of these tests?

**18.56** An unknown solid is entirely soluble in water. On addition of dilute HCl, a precipitate forms. After filtering off the precipitate, the pH is adjusted to about 1 and $H_2S$ is bubbled in; a precipitate again forms. After filtering

off this precipitate, the pH is adjusted to 8 and $H_2S$ is again added; no precipitate forms. No precipitate forms upon addition of $(NH_4)_2HPO_4$. The remaining solution shows a yellow color in a flame test. Based on these observations, which of the following compounds might be present, which are definitely present, and which are definitely absent: CdS, $Pb(NO_3)_2$, HgO, $ZnSO_4$, $Cd(NO_3)_2$, and $Na_2SO_4$?

**18.57** In the course of various qualitative analysis procedures the following mixtures are encountered: **(a)** $Zn^{2+}$ and $Cd^{2+}$; **(b)** $Cr(OH)_3$ and $Fe(OH)_3$; **(c)** $Mg^{2+}$ and $K^+$; **(d)** $Ag^+$ and $Mn^{2+}$. Suggest how each mixture might be separated.

**18.58** Suggest how the cations in each of the following solution mixtures can be separated: **(a)** $Na^+$ and $Cd^{2+}$; **(b)** $Cu^{2+}$ and $Mg^{2+}$; **(c)** $Pb^{2+}$ and $Al^{3+}$; **(d)** $Ag^+$ and $Hg^{2+}$.

**18.59** **(a)** Precipitation of the group 4 cations requires a basic medium. Why is this so? **(b)** What is the most significant difference between the sulfides precipitated in group 2 and those precipitated in group 3? **(c)** Suggest a procedure that would serve to redissolve all of the group 3 cations following their precipitation and separation.

**18.60** A student who is in a great hurry to finish his laboratory work decides that his qualitative analysis unknown contains a metal ion from the insoluble phosphate group, group 4. He therefore tests his sample directly with $(NH_4)_2HPO_4$, skipping earlier tests for the metal ions in groups 1–3. He observes a precipitate and concludes that a metal ion from group 4 is indeed present. Why is this possibly an erroneous conclusion?

## Additional Exercises

**18.61** Which of the following solutions has the greatest buffering capacity, and which has the least: **(a)** 0.10 $M$ $HC_2H_3O_2$ and 0.10 $M$ $NaC_2H_3O_2$, pH 4.74; **(b)** $1.8 \times 10^{-5}$ $M$ HBr, pH 4.74; **(c)** 0.01 $M$ $HC_2H_3O_2$ and 0.01 $M$ $NaC_2H_3O_2$, pH 4.74? Explain.

**18.62** Write balanced net-ionic equations for the reactions between **(a)** sodium hydroxide and benzoic acid; **(b)** sodium hydroxide and hydrobromic acid; **(c)** sodium formate and ammonium chloride; **(d)** nitrous acid and potassium hydroxide.

**18.63** A hypothetical weak acid, HA, was combined with NaOH in the following proportions: 0.20 mol of HA, 0.080 mol of NaOH. The mixture was diluted to total volume of 1 L, and the pH measured. **(a)** If pH = 4.80, what is the $pK_a$ of the acid? **(b)** How many additional moles of NaOH would need to be added to the solution to increase the pH to 5.00?

**18.64** A solution containing 0.050 mol of a weak acid, HX, and 0.030 mol of a salt, KX, is diluted to a total volume of 0.500 L. The pH of the solution is 4.86. What is the acid-dissociation constant $K_a$?

**18.65** A 0.30 $M$ solution of dimethylamine, $(CH_3)_2NH$, containing an unknown concentration of dimethylammonium chloride, $(CH_3)_2NH_2Cl$, has a pH of 10.40. What is the concentration of dimethylammonium ion in the solution?

**18.66** The acid-base indicator bromcresol green is a weak acid. The yellow acid and blue base forms of the indicator are present in equal concentrations in a solution when the pH is 4.68. What is p$K_a$ for bromcresol green?

**18.67** When asked as an examination question to give an example of an alkaline buffer solution, a student listed a solution of $NaHCO_3$. If you were the instructor, would you give no credit, partial credit, or full credit for this answer? Explain.

**18.68** You have 1.0 $M$ solutions of $H_3PO_4$ and NaOH. Describe how you would prepare a buffer solution of pH 7.20 with the highest possible buffering capacity from these reagents. Describe the composition of the buffer solution.

**18.69** Indicate whether each of the following statements is true or false, and explain your reasoning. **(a)** In a solution of equimolar weak base and strong base, the pOH is determined by the strong base alone. **(b)** Optimal buffer action is achieved when the conjugate acid-base pair is present in equal concentrations. **(c)** The capacity of a buffer solution to resist changes in pH is related to the ratio of acid and conjugate base (or base and conjugate acid), and independent of the total concentration of these reagents.

**18.70** What is the pH at the equivalence point in the following titrations: **(a)** 0.100 $M$ HCl with 0.100 $M$ methylamine; **(b)** 0.100 $M$ NaOH with 0.100 $M$ ascorbic acid; **(c)** 0.100 $M$ HBr with 0.100 $M$ KOH?

**18.71** If 50.00 mL of 0.100 $M$ $Na_2SO_3$ is titrated with 0.100 $M$ HCl, calculate **(a)** the pH at the start of the titration; **(b)** the volume of HCl required to reach the first equivalence point. What is the predominant species present? **(c)** the volume of HCl required to reach the second equivalence point, and the pH at the second equivalence point.

**18.72** A biochemist needs 750 mL of an acetic acid–sodium acetate buffer with pH 4.50. Solid sodium acetate, $NaC_2H_3O_2$, and glacial acetic acid, $HC_2H_3O_2$, are available. Glacial acetic acid is 99 percent $HC_2H_3O_2$ by weight and has a density of 1.05 g/mL. If the buffer is to be 0.30 $M$ in $HC_2H_3O_2$, how many grams of $NaC_2H_3O_2$ and how many milliliters of glacial acetic acid must be used?

**[18.73]** Equivalent quantities of 0.10 $M$ solutions of an acid HA and a base B are mixed. The pH of the resulting solution is 8.8. **(a)** Write the equilibrium equation and equilibrium-constant expression for the reaction between HA and B. **(b)** If $K_a$ for HA is $5.0 \times 10^{-6}$, what is the value of the equilibrium constant for the reaction between HA and B?

**[18.74]** What should be the pH of a buffer solution that will result in a $Mg^{2+}$ concentration of $3.0 \times 10^{-2}$ $M$ in equilibrium with solid magnesium oxalate?

**[18.75]** Calculate the solubility of $CuCO_3$ in a strongly buffered solution of pH 7.5.

**18.76** What concentration of $Pb^{2+}$ remains in a solution after $PbCl_2$ has been precipitated from a solution that is 0.1 $M$ in $Cl^-$ at 25°C? Will PbS form if the filtrate from the $PbCl_2$ precipitate is saturated with $H_2S$ at pH 1?

**18.77** Write the net ionic equation for the reaction that occurs when **(a)** hydrogen sulfide gas is bubbled through a solution of mercury(II) chloride; **(b)** solid magnesium sulfite is added to a solution of hydrochloric acid; **(c)** aqueous ammonia is added to a solution of nickel(II) nitrate; **(d)** sodium carbonate solution is added to a solution of manganese(II) sulfate; **(e)** aluminum fluoride dissolves in a solution containing excess sodium fluoride (log $K_f$ for the $AlF_6^{3-}$ ion is 19.8).

**18.78** Germanic acid, $H_2GeO_3$, has p$K_{a1}$ and p$K_{a2}$ values of 9.0 and 12.4, respectively. **(a)** Draw the Lewis structure for this acid. **(b)** What is the pH of a 0.010 $M$ solution of the acid? **(c)** Would titration of this acid with NaOH solution, using phenolphthalein as indicator (Table 17.1), be a good quantitative procedure? Explain.

**18.79** What is the concentration of $HS^-$ in a saturated (0.1 $M$) solution of $H_2S$ if the pH is **(a)** 3.0; **(b)** 5.5?

**[18.80]** The value of $K_{sp}$ for $Mg_3(AsO_4)_2$ is $2.1 \times 10^{-20}$. The $AsO_4^{3-}$ ion is, of course, derived from the weak acid $H_3AsO_4$, p$K_{a1} = 2.22$, p$K_{a2} = 6.98$, p$K_{a3} = 11.50$. When asked to calculate the molar solubility of $Mg_3(AsO_4)_2$ in water, a student used the $K_{sp}$ expression and assumed that $[Mg^{2+}] = 1.5[AsO_4^{3-}]$. Why was this a mistake?

**[18.81]** Cadmium sulfide, CdS, is used commercially as a yellow paint pigment known as cadmium yellow. It is prepared by saturating a slightly acidic solution of $Cd^{2+}$ with $H_2S$ gas. The presence of $Fe^{2+}$, a common impurity, can result in the precipitate being contaminated with black FeS, thereby ruining the color of the pigment. **(a)** If a solution containing 0.10 $M$ $Cd^{2+}$ and $1 \times 10^{-4}$ $M$ $Fe^{2+}$ is saturated with $H_2S$ ($[H_2S] = 0.10$ $M$), what pH is needed to keep the $Fe^{2+}$ from precipitating? **(b)** If a $HC_2H_3O_2$–$C_2H_3O_2^-$ buffer is used to control pH, what ratio of $HC_2H_3O_2$ to $C_2H_3O_2^-$ is required?

**18.82** The pH of a solution that is 0.020 $M$ in $Hg^{2+}$, 0.020 $M$ in $Ni^{2+}$, and 0.020 $M$ in $Pb^{2+}$ is adjusted to 2.0 and then saturated with $H_2S$ gas. Which metal ions, if any, will precipitate?

**[18.83]** The solubility of manganese oxalate dihydrate, $MnC_2O_4 \cdot 2H_2O$, is 0.035 g per 100 mL of solution. The acid-dissociation constants of oxalic acid are listed in Appendix E. Calculate the pH of a saturated solution of $MnC_2O_4 \cdot 2H_2O$.

[18.84] Calculate the molar solubility of $Fe(OH)_3$ in water. (This problem is not as simple as it might first appear; consider carefully the concentration of $OH^-$.)

[18.85] What pH range will permit PbS to precipitate while leaving $Mn^{2+}$ in solution if the solution is 0.010 $M$ in $Pb^{2+}$ and 0.010 $M$ in $Mn^{2+}$ prior to saturation with $H_2S$?

[18.86] Using the $K_f$ value for $Ag(NH_3)_2{}^+$ that is listed in Table 18.2, calculate the concentration of $Ag^+$ present in solution at equilibrium when concentrated $NH_3$ is added to a 0.010 $M$ solution of $AgNO_3$ until the ammonia concentration is 0.20 $M$. (Neglect the small concentration change in $Ag^+$ that accompanies the change in the volume of the solution upon addition of $NH_3$).

CHAPTER *19*

# Chemical Thermodynamics

Two of the major questions concerning any chemical reaction are: How far toward completion does the reaction proceed? How rapidly does the reaction approach equilibrium? We get the answer to the first question from a knowledge of the equilibrium constant. We learn about the second from a study of the reaction rate.

If we want to carry out a reaction that has a favorable equilibrium constant—that is, a reaction that proceeds with conversion of a sizable fraction of reactants into products—we need worry only about getting the reaction rate into a convenient range. This may not be easy, but hard work and ingenious research might eventually lead to a catalyst that can speed up a slow reaction or to some means of controlling a reaction that is too rapid. Haber's discovery of a suitable catalyst for ammonia synthesis (Chapter 16) provides an excellent example of successful research of this type. However, if Haber had attempted to fix nitrogen by reacting $N_2$ and $O_2$ at 400 to 500°C, instead of reacting $N_2$ and $H_2$, he would never have developed a successful process. The equilibrium constant for the reaction of $N_2$ with $O_2$ to form NO is so small in this temperature range that the amount of NO present at equilibrium would have been too small for practical purposes. Even if one had a catalyst that would enable rapid reaction of $N_2$ with $O_2$ to form NO, the reaction would still have no practical consequences. We see that a need exists for some way to predict in advance whether a reaction can proceed to any significant extent before coming to equilibrium.

When we first introduced chemical equilibrium, in Chapter 16, we defined it from a kinetic point of view: Equilibrium occurs when opposing reactions occur at equal rates. However, equilibrium also has a basis in thermodynamics, the area of science dealing with energy relationships. In this chapter we shall see that we may predict the position of an equilibrium using certain principles and concepts from thermodynamics.

## CONTENTS

## 19.1 SPONTANEOUS PROCESSES

Thermodynamics is based on several fundamental laws that summarize our experience with energy changes. The first law of thermodynamics, which we discussed at length in Chapter 4, states that energy is conserved. By this we mean that energy is neither created nor destroyed in any process, such as the falling of a brick, the melting of an ice cube, or a chemical reaction. Energy flows from one part of nature to another, or is converted from one form to another, but the total remains constant. We saw in Chapter 4 that we could express the first law in the form $\Delta E = q + w$, where $\Delta E$ is the change in energy of a system, $q$ is the heat absorbed by the system from its surroundings, and $w$ is the work done on the system by its surroundings.

---

The uncontrolled burning of gas shown here is an example of a highly exothermic, spontaneous reaction. (SYGMA)

Once we specify a particular process or change, the first law helps us to balance the books, so to speak, on the heat released, work done, and so forth. However, it says nothing about whether the process or change we specify can in fact occur. That question is encompassed in the second law of thermodynamics.

The **second law of thermodynamics** expresses the notion that there is an inherent direction in which any system not at equilibrium moves. For example, if you hold a brick in your hand and let it go, the brick falls to the floor. Water placed in a freezer compartment converts to ice. A shiny nail left outdoors turns to rust. Every one of these processes occurs without outside intervention; such processes are said to be **spontaneous**.

For every spontaneous process, we can imagine a reverse process occurring. For example, we can imagine a brick moving from the floor into your hand, ice cubes melting at $-10°C$, or a rusty iron nail being transformed into a shiny one. It is inconceivable that any of these occurrences is spontaneous. If we saw a film in which these things happened, we would conclude that the film was being run backward. Our years of observing nature at work have impressed us with a simple rule: *Processes that are spontaneous in one direction are not spontaneous in the reverse direction.*

Consider a reaction about which we had much to say in Chapter 16:

$$N_2(g) + 3H_2(g) \rightleftharpoons 2NH_3(g) \qquad\qquad [19.1]$$

When we mix $N_2$ and $H_2$ at any temperature, say 472 K, the reaction proceeds in the forward direction; this process is spontaneous. However, if we place a mixture of 1.00 mol of $N_2$, 3.00 mol of $H_2$, and 1.00 mol of $NH_3$ in a 1-L container at 472 K, it is not immediately obvious whether or not the formation of more $NH_3$ should be spontaneous. Nevertheless, if we have the equilibrium constant, which at this temperature happens to be $K_c = 0.105$, we can predict the direction in which the reaction proceeds to reach equilibrium. In this case

$$Q = \frac{[NH_3]^2}{[N_2][H_2]^3} = \frac{(1.00)^2}{(1.00)(3.00)^3} = 0.0370$$

Because the reaction quotient, $Q$, is smaller than $K_c$, the system moves spontaneously toward equilibrium by formation of $NH_3$ (Section 16.4). The opposite process, conversion of $NH_3$ into $N_2$ and $H_2$, is not spontaneous for this particular reaction mixture at 472 K.

It is important to realize that just because a process is spontaneous does not mean that it will occur at an observable rate. A spontaneous reaction may be very fast, as in the case of an acid-base neutralization, or very slow, as in the case of rusting of iron. Thermodynamics can tell us the *direction* and *extent* of a reaction, but it can say nothing about its *speed*. Reaction rates are the subject of chemical kinetics.

What factors make a process spontaneous? We considered that question briefly in Section 12.2. In that earlier discussion we noted that two basic principles are involved: Processes in which the energy content of the system decreases (exothermic processes) tend to occur spontaneously.

Furthermore, spontaneity characterizes processes in which the randomness or disorder of the system increases. In the next section we consider these factors, especially the matter of disorder, in greater detail.

The spontaneous motion of a brick released from your hand is toward the ground. As the brick falls, it loses potential energy. This potential energy is first converted into kinetic energy, the energy of motion of the brick. When the brick hits the floor, its kinetic energy is converted into heat. The overall result of the brick's fall is thus a conversion of the potential energy of the brick into heat in its surroundings. Our experience with other simple mechanical systems is similar: objects fall, clocks run down, stretched rubber bands contract. All of these phenomena can be summarized by saying that such systems seek a resting place of minimum energy.

It is clear that the tendency for a system to achieve the lowest possible energy is one of the driving forces that determines the behaviour of molecular systems. For example, just as a brick possesses potential energy because of its position relative to the floor, so also a chemical substance possesses potential energy relative to other substances because of the arrangements of nuclei and electrons. When these arrangements change, energy may be released. For example, the combustion of propane (bottled gas), which is clearly a spontaneous process, is strongly exothermic:

$$C_3H_8(g) + 5O_2(g) \longrightarrow 3CO_2(g) + 4H_2O(l) \qquad \Delta H° = -2202 \text{ kJ} \qquad [19.2]$$

The rearrangements in space of nuclei and electrons in going from propane and oxygen to carbon dioxide and water lead to a lower chemical potential energy, so heat is evolved. Reactions that are exothermic are generally spontaneous. However, it is clear that the tendency toward minimum enthalpy cannot be the only factor that determines spontaneity in molecular processes. It is instructive to consider some spontaneous processes that are not exothermic.

**Spontaneity and
Entropy Change**

A bit of thinking brings to mind several processes that are spontaneous even though they are not exothermic. For example, consider an ideal gas confined at 1 atm pressure to a 1-L flask, as shown in Figure 19.1. The flask is connected via a closed stopcock to another 1-L flask, which is evacuated. Now suppose the stopcock is opened. Is there any doubt about what would happen? We intuitively recognize that the gas would expand into the second flask until the pressure is equally distributed in both flasks, at 0.5 atm. In the course of expanding from the 1-L flask into the larger volume, the ideal gas neither absorbs nor emits heat. Nevertheless, the process is spontaneous. The reverse process, in which the gas that is evenly distributed between the two flasks suddenly moves entirely into one of the flasks, leaving the other vacant, is inconceivable. Yet this is also a process that would involve no emission or absorption of heat. Evidently some factor other than heat emitted or absorbed is important in making the process of gas expansion spontaneous.

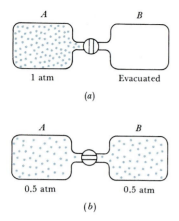

**FIGURE 19.1** Expansion of an ideal gas into an evacuated space. In (*a*), flask *A* holds an ideal gas at 1 atm pressure and flask *B* is evacuated. In (*b*), the stopcock connecting the flasks has been opened. The ideal gas expands to occupy both flasks *A* and *B* at a pressure of 0.5 atm.

As another example, consider the melting of ice cubes at room temperature. The process

$$H_2O(s) \longrightarrow H_2O(l) \qquad [19.3]$$

at 27°C is highly spontaneous, as we all know. Yet this is an endothermic change. The melting of ice above 0°C thus represents an example of a spontaneous, endothermic process.

A similar type of process, discussed in Chapter 12, is the endothermic dissolving of many salts in water. If we add solid potassium chloride, KCl, to a glass of water at room temperature and stir, we can feel the solution growing colder as the salt dissolves. The process is endothermic and yet spontaneous.

The three processes just described have something in common that accounts for their being spontaneous. In each instance, the products of the process are in a more random or disordered state than the reactants. Let's consider each case in turn.

When we have a gas confined to a 1-L volume, as in Figure 19.1(*a*), we can specify the location of each and every gas molecule as being in that liter of space. After the gas has expanded, we can't be sure which gas molecules are at any one instant in the original volume and which are on the other side. We must therefore say that the location of each and every gas molecule is specified as being in the entire 2-L space. In other words, the gas molecules, because they can be anywhere within a 2-L space, are more randomized than when they are confined to a 1-L space.

The molecules of water that make up an ice crystal are held rigidly in place in the ice crystal lattice (Figure 19.2). When the ice melts, the water molecules are free to move about with respect to one another and to turn over. Thus in liquid water the individual water molecules are more randomly distributed than in the solid. The highly ordered solid structure is replaced by the highly disordered liquid structure.

A similar situation applies when KCl dissolves in water, although here we must be a little careful not to take too much for granted. In solid

**FIGURE 19.2** Structure of ice.

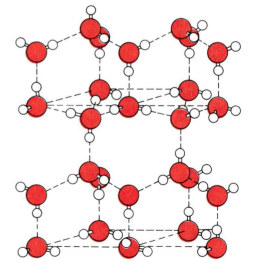

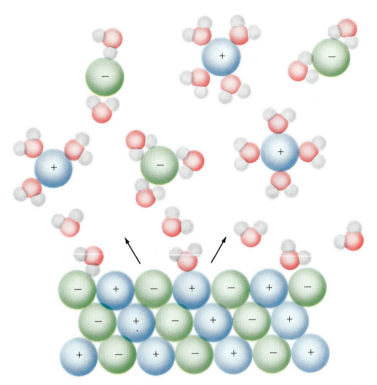

**FIGURE 19.3** Changes in degree of order in the ions and solvent molecules on dissolving an ionic solid in water. The ions themselves become more randomized, but the water molecules that hydrate the ions become less randomized.

KCl, the $K^+$ and $Cl^-$ are in a highly ordered, crystalline state. When the solid dissolves, the ions are free to move about in the water. They are obviously in a much more random and disordered state than before. At the same time, though, water molecules are held around the ions, as water of hydration (Section 12.2), shown in Figure 19.3. These water molecules are in a *more* ordered state than before, because they are confined to the immediate environment of the ions. Thus the dissolving of a salt involves both ordering and disordering processes. It happens that the disordering processes are dominant, so the overall effect is an increase in disorder upon dissolving a salt in water.

As these examples illustrate, spontaneity is associated with an increase in randomness or disorder of a system. The randomness is expressed by a thermodynamic quantity called **entropy**, given the symbol $S$. The more random a system, the larger its entropy. Like enthalpy, entropy is a state function (Section 4.2). The change in entropy of a system, $\Delta S = S_{final} - S_{initial}$, depends only on the initial and final states of the system and not on the particular pathway by which the system changes. Like other thermodynamic quantities, $\Delta S$ has two parts, a number giving the magnitude of the change and a sign giving its direction. A positive $\Delta S$ indicates an increase in randomness or disorder. A negative $\Delta S$ indicates a decrease in randomness or disorder.

### SAMPLE EXERCISE 19.1

By considering the relative extents of randomness or disorder in the reactants and products, predict whether $\Delta S$ is positive or negative for each of the following processes:

**(a)** $H_2O(l) \longrightarrow H_2O(g)$

**(b)** $Ag^+(aq) + Cl^-(aq) \longrightarrow AgCl(s)$

**(c)** $4Fe(s) + 3O_2(g) \longrightarrow 2Fe_2O_3(s)$

**Solution:** **(a)** The evaporation of a liquid is accompanied by a large increase in volume. One mole of water (18 g) occupies about 18 mL as a liquid and 22.4 L as a gas at STP. Because the molecules are distributed throughout a much larger volume in the gaseous state than in the liquid state, an increase in disorder accompanies vaporization. Thus $\Delta S$ is positive.

**(b)** In this process the ions that are free to move about in the larger volume of the solution form a solid in which the ions are confined to highly regular positions. Thus there is a decrease in disorder, and $\Delta S$ is negative.

**(c)** The particles of a solid are much more highly ordered and confined to specific locations than are the molecules of a gas. Because a gas is converted into part of a solid product, disorder decreases and $\Delta S$ is negative.

### PRACTICE EXERCISE

Indicate whether each of the following reactions produces an increase or decrease in the entropy of the system:

**(a)** $CO_2(s) \longrightarrow CO_2(g)$
**(b)** $CaO(s) + CO_2(g) \longrightarrow CaCO_3(s).$

*Answers:* **(a)** increase; **(b)** decrease

---

## The Second Law of Thermodynamics

Our introduction of the concept of entropy allows us to reexamine the second law of thermodynamics and its implications. In the discussion of spontaneity in Section 19.1, we noted that the second law concerns the direction in which processes move; it is associated with the idea that processes that are spontaneous in one direction are not spontaneous in the opposite direction. This idea applies not only to chemical changes but to other processes as well.

We all know that heat flows spontaneously from a hot object to a cold one. We also know that to cause heat to flow in the reverse direction—from a cold object to a hotter one or from a system at some temperature to surroundings at a higher temperature—requires an input of energy. For example, it requires electrical energy to maintain a refrigerator at a lower temperature than the surrounding kitchen.

There are many ways to state the second law. In chemical contexts it's usually expressed in terms of entropy. To develop such a statement, let's think in terms of an **isolated system**, one that doesn't exchange energy or matter with its surroundings. When a process occurs spontaneously in an isolated system, the system always ends up in a more random state. For example, when a gas expands under the conditions shown in Figure 19.1, there is no exchange of heat, work, or matter with the surroundings; this is an isolated system. The spontaneous expansion corresponds to an increase in entropy.

In the real world we rarely deal with isolated systems. We are usually concerned with systems that exchange energy with their surroundings in the form of heat or work. When such a system changes spontaneously, it may undergo either an increase or decrease in its entropy. However, the second law tells us that *in any spontaneous process there is always an increase in the entropy of the universe.* As an example, consider the oxidation of iron to $Fe_2O_3(s)$:

$$4Fe(s) + 3O_2(g) \longrightarrow 2Fe_2O_3(s) \qquad [19.4]$$

As discussed in Sample Exercise 19.1, this chemical process results in a decrease in the degree of randomness; that is, $\Delta S$ for the process is negative. But when the process occurs, some change also occurs in the surroundings. For example, the reaction is exothermic; heat is therefore

The consequences of the statement of the second law that is given above are quite profound. We humans, for example, are very complex, highly organized, and well-ordered systems. We have a very low entropy content as compared with the same amount of carbon dioxide, water, and several other simple chemicals into which our bodies might be decomposed. But all of the thousands of chemical reactions necessary to produce one adult human have caused a very large increase in entropy of the rest of the universe. Thus the overall entropy change necessary to form and maintain a human, or for that matter any other living system, is positive.

In a similar way, the human activities that produce such an impressive ordering of the world around us—formation of copper metal from a widely dispersed copper ore; production from sand of silicon used in transistors; production of the paper on which this book is printed from trees—have along the way used a great deal of energy that has been converted, in a sense, to disorder—coal and oil burned to form $CO_2$ and $H_2O$; a sulfide ore roasted to form $SO_2$ that pollutes the atmosphere; various waste products scattered in the environment. Modern human society is, in effect, using up its limited storehouse of energy-rich materials in its headlong rush to exploit technology. We must eventually learn to live within the bounds of the energy supply that reaches earth daily from the sun, because we will soon have exhausted the supply of readily available energy of other sorts.

evolved and absorbed by the surroundings. In fact, the change that occurs in the surroundings causes an increase in the entropy of the surroundings that is larger than the decrease in entropy that occurs in the system itself. For any spontaneous process the sum of the entropy change in the system and that in the surroundings (which is the entropy change in the universe caused by that process) must be positive:

$$\Delta S_{universe} = \Delta S_{system} + \Delta S_{surroundings} > 0 \qquad [19.5]$$

No process that produces order (lower entropy) in a system can proceed without producing an even larger disorder (higher entropy) in its surroundings. That is, the disorder introduced into the surroundings will always exceed the order achieved in the system. Thus while energy is conserved (the first law), entropy continues to increase (the second law).

## 19.3 A MOLECULAR INTERPRETATION OF ENTROPY

It is useful to develop a qualitative sense of how entropy changes in a system depend on changes in structure, physical state, and so forth. In Section 19.2 some examples of spontaneous, endothermic processes were considered. We saw, for example, that the increase in volume that occurs when a gas expands results in an increase in the randomness of the system (positive $\Delta S$). In a similar fashion, the distribution of a liquid or solid solute in a solution is accompanied by an increase in entropy. For example, $\Delta S$ is positive when KCl dissolves in water. However, the dissolving of a gas, such as $CO_2$ in $H_2O$, causes the gas molecules to move in a much smaller volume; consequently, for this process the entropy of the system decreases (negative $\Delta S$). Similarly, a decrease in the number of gaseous particles as the result of a reaction causes a decrease in entropy (negative $\Delta S$). For example, $\Delta S$ for the following reaction is negative:

$$2NO(g) + O_2(g) \longrightarrow 2NO_2(g) \qquad [19.6]$$

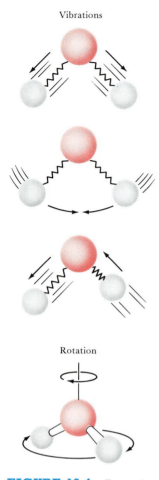

**Vibrations**

**Rotation**

**FIGURE 19.4** Examples of vibrational and rotational motion, illustrated for the water molecule. Vibrational motions involve periodic displacements of the atoms with respect to one another, a phenomenon similar to the vibrations of the arms of a tuning fork. Rotational motions involve the spinning of a molecule about an axis.

Entropy changes can also be associated with molecular motions within a substance. A molecule consisting of more than one atom can engage in several types of motion. The entire molecule can move in one direction or another as in the movements of gas molecules. We call such movement **translational motion**. The atoms within a molecule may also undergo **vibrational motion** in which they move periodically toward and away from each other, much as a tuning fork vibrates about its equilibrium shape. Figure 19.4 shows the vibrational motions possible for the water molecule. In addition, molecules may possess **rotational motion**, as though they were spinning like a top. The rotational motion of the water molecule is also illustrated in Figure 19.4. These forms of motion are ways that the molecule has of storing energy. As the temperature of a system increases, the amounts of energy stored in these forms of motion increase.

To see what this has to do with entropy, let's imagine that we begin with a pure substance that forms a perfect crystalline lattice at the lowest temperature possible, absolute zero. At this stage, none of the kinds of motion that we have been talking about are present. The individual atoms and molecules are as well defined in position and in terms of energy as they can ever be. The **third law of thermodynamics** states that *the entropy of a pure crystalline substance at absolute zero is zero:* $S$ (0 K) = 0. As the temperature is raised, the units of the solid lattice begin to acquire energy. In a crystalline solid, the molecules or atoms that occupy the lattice sites are constrained to remain more or less in place. Nevertheless, they may store energy in the form of vibrational motion about their lattice positions. Instead of all the molecules necessarily being in the lowest possible energy state, the number of possible energies that the lattice atoms or molecules may have expands. This increase in possible energy states is not unlike the expansion of the gas illustrated in Figure 19.1. The entropy of the gas increases on expansion because the volume through which the gas molecules move is larger. The entropy of the lattice increases with temperature because the number of possible energy states in which the molecules or atoms are distributed is larger.

It is instructive to follow what happens to the entropy of our substance as we continue to heat it. Let's suppose that at some temperature a phase change occurs, converting the substance from one solid form to another. This means that the arrangement of the lattice units changes in some way, possibly so that the lattice is less regular.* This type of phase change occurs sharply at one temperature, just as do other types of phase changes, (for example the change from a solid to a liquid). When the phase change occurs, entropy changes, because the two lattice arrangements do not have precisely the same degrees of randomness.

Figure 19.5 shows the variation in entropy with temperature for our sample. Note that the change in $S$ with temperature is gradual up to the solid-state phase change and that there is then a sharp increase in $S$ at that

---

* As an example of a solid-state phase change, gray tin converts at 13°C to another solid form called white tin. White tin is stable above the transition temperature, gray tin below it. White tin has a higher entropy than that of gray tin.

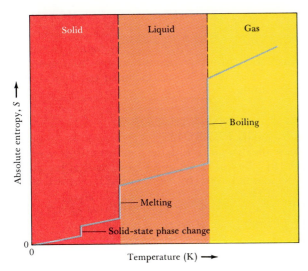

FIGURE 19.5 Entropy changes that occur as the temperature of a substance rises from absolute zero.

temperature. At temperatures above the phase change, the entropy increases with increasing temperature, up to the melting point of the solid.

When the solid melts, the units of the lattice are no longer confined to specific locations relative to other units, but are free to move about the entire volume of the substance. This added freedom of motion for the individual molecules adds greatly to the entropy content of the substance. At the temperature of melting, we therefore see a large increase in entropy content. After all the solid has melted, the temperature again increases, and with it the entropy.

---

**SAMPLE EXERCISE 19.2**

Figure 19.5 shows that the entropy of a liquid increases as its temperature increases. What factors are responsible for the increase in entropy?

**Solution:** The average kinetic energy of the molecules in a liquid increases with temperature. As temperature increases, more molecules at any given instant possess higher energies. This "expansion" in the energies that the molecules possess is measured by the increase in entropy. The increase in $S$ with increasing temperature results from increased energy of motion of all kinds within the liquid.

**PRACTICE EXERCISE**
Based on Figure 19.5, what general relationship exists among the magnitudes of the entropies of the gas, liquid, and solid phases of a substance?
*Answer:* $S_{gas} > S_{liquid} > S_{solid}$

---

At the boiling point of the liquid another big increase in entropy occurs. The increase in this case results largely from the increased volume in which the molecules may be found. This is intuitively in line with our earlier ideas about entropy, because an increase in volume means an increase in randomness.

As the gas is heated, the entropy increases steadily, because more and more energy is being stored in the gas molecules. The distribution of molecular speeds is spread out toward higher values, as illustrated in Figure 10.11. Again, the idea of an expansion in the range of energies in which molecules may be found helps us remember that increased average energy means increased entropy.

## SAMPLE EXERCISE 19.3

Choose the substance with greater entropy in each pair, and explain your choice: **(a)** 1 mol of NaCl($s$) and 1 mol of HCl($g$) at 25°C; **(b)** 2 mol of HCl($g$) and 1 mol of HCl($g$) at 25°C; **(c)** 1 mol of HCl($g$) and 1 mol of Ar($g$) at 25°C; **(d)** 1 mol of N$_2$($s$) at 24 K and 1 mol of N$_2$($g$) at 298 K.

**Solution:** **(a)** Gaseous HCl has the higher entropy per mole, because it has acquired a high degree of randomness as a result of being in the gaseous state. **(b)** The sample containing 2 mol of HCl has twice the entropy of the sample containing 1 mol. **(c)** The HCl sample has the higher entropy, because the HCl molecule is capable of storing energy in more ways than is Ar. It may rotate, or the H—Cl distance may change periodically in a vibrational motion.

**(d)** The gaseous N$_2$ sample has the higher entropy, because the entropy increases resulting from melting and then boiling of N$_2$ are included in its total entropy content.

### PRACTICE EXERCISE

Choose the substance with the greater entropy in each case: **(a)** 1 mol of H$_2$($g$) at STP or 1 mol of H$_2$($g$) at 100°C and 0.5 atm; **(b)** 1 mol of H$_2$O($s$) at 0°C and 1 mol of H$_2$O($l$) at 25°C; **(c)** 1 mol of H$_2$($g$) at STP and 1 mol of SO$_2$($g$) at STP; **(d)** 1 mol of N$_2$O$_4$($g$) at STP and 2 mol of NO$_2$($g$) at STP.
**Answers:** **(a)** 1 mol of H$_2$($g$) at 100°C; **(b)** 1 mol of H$_2$O($l$) at 25°C; **(c)** 1 mol of SO$_2$($g$) at STP; **(d)** 2 mol of NO$_2$($g$) at STP

In general, entropy *increases* are expected to accompany processes in which

Pure liquids or solutions are formed from solids.
Gases are formed from either solids or liquids.
The number of molecules of gases increases during a chemical reaction.
The temperature of a substance is increased.

## SAMPLE EXERCISE 19.4

Predict whether the entropy change of the system in each of the following reactions is positive or negative.

**(a)** CaCO$_3$($s$) $\longrightarrow$ CaO($s$) + CO$_2$($g$)
**(b)** N$_2$($g$) + 3H$_2$($g$) $\longrightarrow$ 2NH$_3$($g$)
**(c)** N$_2$($g$) + O$_2$($g$) $\longrightarrow$ 2NO($g$)

**Solution:** **(a)** The entropy change here is positive, because a solid is converted into a solid and a gas. Gaseous substances generally possess more entropy than solids, so whenever the products contain more moles of gas than the reactants, the entropy change is probably positive.
 **(b)** The entropy change in formation of ammonia from nitrogen and hydrogen is negative, because there are fewer moles of gas in the product than in the reactants.

**(c)** This represents a case in which the entropy change will be small, because the same number of moles of gas is involved in the reactants and in the product. The sign of $\Delta S$ is impossible to predict based on our discussions thus far, but we can predict that $\Delta S$ will be small.

### PRACTICE EXERCISE

Predict whether $\Delta S$ is positive or negative in each of the following processes:

**(a)** HCl($g$) + NH$_3$($g$) $\longrightarrow$ NH$_4$Cl($s$)
**(b)** 2SO$_2$($g$) + O$_2$($g$) $\longrightarrow$ 2SO$_3$($g$)
**(c)** cooling of nitrogen gas from 20°C to −50°C.

**Answers:** **(a)** negative; **(b)** negative; **(c)** negative

## 19.4 CALCULATION OF ENTROPY CHANGES

The enthalpy change in a chemical reaction is often easily measured in a calorimeter, as described in Section 4.7. No comparable, easy means exists for measuring the change in entropy. Nevertheless, various types of measurements have made it possible to determine the absolute entropy, $S$, at any temperature for a great number of substances. These entropies are based on the reference point of zero entropy for perfect crystalline solids. The entropy values for a variety of substances at 298 K and 1 atm are given in Appendix D. These standard entropies, $S°$, are expressed in units of joules per Kelvin per mole; J/K-mol.

The entropy change in a chemical reaction is given by the sum of the entropies of the products less the sum of entropies of reactants. Thus, in the overall reaction

$$aA + bB + \cdots \rightleftharpoons pP + qQ + \cdots \qquad [19.7]$$

the standard entropy change, $\Delta S°$, is given by

$$\Delta S° = [pS°(P) + qS°(Q) + \cdots] - [aS°(A) + bS°(B) + \cdots] \qquad [19.8]$$

In other words, we sum the standard entropies of all the products, multiplying each by the coefficient of the product in the balanced equation, and then subtract the sum of entropies of the reactants, multiplied by their coefficients.

---

### SAMPLE EXERCISE 19.5

Calculate $\Delta S°$ for the synthesis of ammonia from $N_2(g)$ and $H_2(g)$:

$$N_2(g) + 3H_2(g) \longrightarrow 2NH_3(g)$$

**Solution:** Using Equation 19.8, we have

$$\Delta S° = 2S°(NH_3) - [S°(N_2) + 3S°(H_2)]$$

Substituting the appropriate $S°$ values from Appendix D yields $\Delta S° =$

$$(2 \text{ mol})\left(192.5 \frac{J}{K\text{-mol}}\right) - \left[(1 \text{ mol})\left(191.5 \frac{J}{K\text{-mol}}\right)\right.$$
$$\left. + (3 \text{ mol})\left(130.58 \frac{J}{K\text{-mol}}\right)\right]$$

$$= -198.2 \text{ J/K}$$

The value for $\Delta S°$ is negative, as we predicted in Sample Exercise 19.4(b).

**PRACTICE EXERCISE**

Using the standard entropies in Appendix D, calculate the standard entropy change, $\Delta S°$, for the following reaction at 298 K: $Al_2O_3(s) + 3H_2(g) \longrightarrow 2Al(s) + 3H_2O(g)$.

*Answer:* 179.9 J/K

---

We still haven't attempted to use thermodynamics to predict whether a given reaction will be spontaneous. We have seen that spontaneity involves two thermodynamic concepts, entropy and enthalpy. Before we can make the predictions we would like, we must introduce a third function that interrelates entropy and enthalpy. That function is called **free energy**, or Gibbs free energy after the American mathematician J. Willard Gibbs (1839–1903), who first proposed it (Figure 19.6). The free energy, $G$, is related to enthalpy and entropy by the expression

## 19.5 THE FREE-ENERGY FUNCTION

$$G = H - TS \qquad [19.9]$$

where $T$ is the absolute temperature. Free energy, like the enthalpy and entropy functions to which it is related, is a state function.

For a process occurring at constant temperature and pressure, the change in free energy is given by the expression

$$\Delta G = \Delta H - T\,\Delta S \qquad\qquad [19.10]$$

A process that is driven spontaneously toward equilibrium both by decreasing energy (negative $\Delta H$) and increasing randomness (positive $\Delta S$) will have a negative $\Delta G$. Indeed, there is a simple relationship between the sign of $\Delta G$ for a reaction and the spontaneity of that reaction operated at constant temperature and pressure:

1. If $\Delta G$ is negative, the reaction is spontaneous in the forward direction.
2. If $\Delta G$ is zero, the reaction is at equilibrium; there is no driving force tending to make the reaction go in either direction.
3. If $\Delta G$ is positive, the reaction in the forward direction is nonspontaneous; work must be supplied from the surroundings to make it occur. However, the reverse reaction will be spontaneous.

An analogy is often drawn between the free-energy change in a spontaneous reaction and the potential energy change in a boulder rolling down a hill. Potential energy in a gravitational field "drives" the boulder until it reaches a state of minimum potential energy in the valley [Figure 19.7(a)]. Similarly, the free energy of a chemical system decreases (negative $\Delta G$) until it reaches a minimum value [Figure 19.7(b)]. When this minimum is reached, a state of equilibrium exists. In any spontaneous process at constant temperature and pressure, the free energy always decreases. As shown in Figure 19.7(b), the equilibrium condition can be approached by a spontaneous change from either direction, from the product side or the reactant side.

As an example of these ideas, let's return to the synthesis of ammonia from nitrogen and hydrogen. Imagine that we have a certain number of moles of nitrogen and three times that number of moles of hydrogen in a reaction vessel that permits us to maintain a constant temperature and pressure. We know from our earlier discussions that the formation of ammonia will not be complete; an equilibrium will be reached in which the reaction vessel contains some mixture of $N_2$, $H_2$, and $NH_3$. The free energy of the system decreases (negative $\Delta G$) until this equilibrium is

**FIGURE 19.6** Josiah Willard Gibbs (1839–1903) was the first person to be awarded a Ph.D. in science from an American university (Yale, 1863). He went on to become one of the foremost mathematical scientists of his day. From 1871 until his death he held the chair of mathematical physics at Yale University. He made significant contributions to thermodynamics and is credited with laying much of the theoretical foundation that led to the development of chemical thermodynamics. (Culver Pictures)

**FIGURE 19.7** Analogy between the potential-energy change of a boulder rolling down a hill (a) and the free-energy change in a spontaneous reaction (b). The equilibrium position in (a) is given by the minimum potential energy available to the system. The equilibrium position in (b) is given by the minimum free energy available to the system.

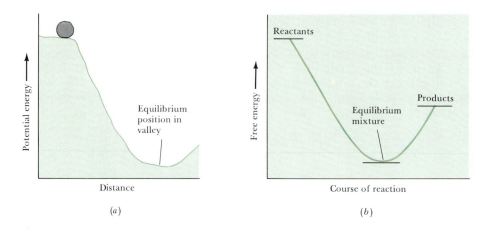

attained. Once the equilibrium has been reached, there is no further spontaneous formation of $NH_3$. The equilibrium condition is the minimum free energy available to the system at this temperature and pressure. To form more $NH_3$ from $N_2$ and $H_2$ once equilibrium has been reached requires an increase in free energy (positive $\Delta G$).

It is not necessary that we reach equilibrium beginning only with $N_2$ and $H_2$. We could reach the same equilibrium by beginning with an appropriate amount of $NH_3$. Ammonia held at a constant temperature and pressure will decompose to form $N_2$ and $H_2$ until an equilibrium is attained. This process is also spontaneous, involving a decrease in free energy as the system approaches equilibrium. The equilibrium can be approached from either the reactant side or product side, as indicated in a general fashion in Figure 19.7(b).

We have noted that free energy is a state function. This means that it is possible to tabulate standard free energies of formation for substances, just as it is possible to tabulate standard enthalpies of formation. It is important to remember that standard values for these functions imply a particular set of conditions, or standard states (Section 4.6). The standard state for gaseous substances is 1 atm pressure. For solid substances the standard state is the pure solid; for liquids, the pure liquid. For substances in solution, the standard state is normally a concentration of 1 $M$; in accurate work it may be necessary to make certain corrections, but we need not worry about these. The temperature usually chosen for purposes of tabulating data is 25°C. Just as for the standard heats of formation, the free energies of elements in their standard states are arbitrarily set to zero. This arbitrary choice of reference point has no effect on the quantity in which we are really interested, namely the *difference* in free energy between reactants and products. The rules about standard states are summarized in Table 19.1. A table of standard free energies of formation appears in Appendix D.

The standard free energies of formation for substances are useful in calculating the standard free-energy change in a chemical process. For the general reaction

$$a\text{A} + b\text{B} + \cdots \longrightarrow p\text{P} + q\text{Q} + \cdots \qquad [19.11]$$

the standard free-energy change is

$$\Delta G° = [p\,\Delta G_f°(\text{P}) + q\,\Delta G_f°(\text{Q}) + \cdots] - [a\,\Delta G_f°(\text{A}) + b\,\Delta G_f°(\text{B}) + \cdots] \qquad [19.12]$$

In this expression $\Delta G_f°(\text{P})$ represents the standard free energy of formation of product P, and all the other $\Delta G°$ have similar meanings. Stated verbally, the standard free-energy change for a reaction equals the sum of the standard free-energy values per mole of each product, each multiplied by the corresponding coefficient in the balanced equation, less the corresponding sum for the reactants.

What use can be made of this standard free-energy change for a chemical reaction? The quantity $\Delta G°$ tells us whether a mixture of reactants and products, each present under standard conditions, would spontaneously react in the forward direction to produce more products ($\Delta G°$

**Calculation of $\Delta G°$**

**TABLE 19.1** Conventions used in establishing standard free-energy values

| State of matter | Standard state |
|---|---|
| Solid | Pure solid |
| Liquid | Pure liquid |
| Gas | 1 atm pressure[a] |
| Solution | Usually 1 $M$[b] |
| Elements | Standard free energy of formation of the element in its normal state is defined as zero |

[a] Neglecting nonideal gas behavior.
[b] Neglecting nonideality of solutions.

negative) or in the reverse direction to form more reactants ($\Delta G°$ positive). Because standard free-energy values are readily available for a large number of substances, the standard free-energy change is easy to calculate for many reaction systems of interest.

Determine the standard free-energy change for the following reaction at 298 K:

$$N_2(g) + 3H_2(g) \longrightarrow 2NH_3(g)$$

**Solution:** Using Appendix D, we find that the standard free energies for the three substances of interest are as follows: $N_2(g)$, $\Delta G_f° = 0.0$; $H_2(g)$, $\Delta G_f° = 0.0$; $NH_3(g)$, $\Delta G_f° = -16.66$ kJ/mol.

The standard free-energy change for the reaction of interest is

$$\Delta G° = 2\,\Delta G_f°(NH_3) - [3\,\Delta G_f°(H_2) + \Delta G_f°(N_2)]$$

Inserting numerical quantities we obtain

$$\Delta G° = -33.32 \text{ kJ}$$

The fact that $\Delta G°$ is negative tells us that a mixture of $H_2$, $N_2$, and $NH_3$ at 25°C, each present at a pressure of 1 atm, would react spontaneously to form more ammonia. (Remember, however, that this says nothing about the rate at which the reaction occurs.)

**PRACTICE EXERCISE**

Using the standard free energies of formation tabulated in Appendix D, calculate $\Delta G°$ for the following reaction at 298 K: $2CH_3OH(l) + 3O_2(g) \longrightarrow 2CO_2(g) + 4H_2O(g)$.
*Answer:* $-1370.8$ kJ

## Free Energy and Temperature

It is worthwhile to examine Equation 19.10 closely to see how the free-energy function depends on both the enthalpy and the entropy changes for a given process. If it were not for entropy effects, all exothermic reactions—those in which $\Delta H$ is negative—would be spontaneous. The entropy contribution, represented by the quantity $-T\Delta S$, may increase or decrease the tendency of the reaction to proceed spontaneously. When $\Delta S$ is positive, meaning that the final state is more random or disordered than the initial state, the term $-T\Delta S$ makes $\Delta G$ less positive (or more negative), increasing the tendency of the reaction to occur spontaneously. When $\Delta S$ is negative, however, the term $-T\Delta S$ decreases the tendency of the reaction to occur spontaneously.

When $\Delta H$ and $-T\Delta S$ are of opposite sign, the relative importance of the two terms determines whether $\Delta G$ is negative or positive. In these instances, temperature is an important consideration. Both $\Delta H$ and $\Delta S$ are in principle capable of changing with temperature. However, the only quantity in the equation

$$\Delta G = \Delta H - T\Delta S$$

that changes markedly with temperature is $-T\Delta S$. Thus at high temperatures the $-T\Delta S$ term becomes relatively more important in determining the sign and magnitude of $\Delta G$.

Various possible situations for the relative signs of $\Delta H$ and $\Delta S$ are shown in Table 19.2, with examples of each. By applying the concepts we have developed for predicting entropy changes, we may often predict how $\Delta G$ will change with change in temperature.

**TABLE 19.2**  Effect of temperature on reaction spontaneity

| $\Delta H$ | $\Delta S$ | $\Delta G$ | Reaction characteristics | Example |
|---|---|---|---|---|
| − | + | Always negative | Reaction is spontaneous at all temperatures; reverse reaction is always nonspontaneous | $2O_3(g) \longrightarrow 3O_2(g)$ |
| + | − | Always positive | Reaction is nonspontaneous at all temperatures; reverse reaction occurs | $3O_2(g) \longrightarrow 2O_3(g)$ |
| − | − | Negative at low temperatures; positive at high temperatures | Reaction is spontaneous at low temperatures but becomes nonspontaneous at high temperatures | $CaO(s) + CO_2(g) \longrightarrow CaCO_3(s)$ |
| + | + | Positive at low temperatures; negative at high temperatures | Reaction is nonspontaneous at low temperatures but becomes spontaneous as temperature is raised | $CaCO_3(s) \longrightarrow CaO(s) + CO_2(g)$ |

## SAMPLE EXERCISE 19.7

**(a)** Predict the direction in which $\Delta G°$ for the equilibrium

$$N_2(g) + 3H_2(g) \rightleftharpoons 2NH_3(g)$$

will change with increase in temperature. **(b)** Calculate $\Delta G°$ at 500°C, assuming that $\Delta H°$ and $\Delta S°$ do not change with temperature.

**Solution:**  **(a)** In Sample Exercise 19.5 we saw that the change in $\Delta S°$ for the equilibrium of interest is negative. This means that the term $-T\Delta S°$ is positive and grows larger with increasing temperature. The standard free-energy change, $\Delta G°$, is the sum of the negative quantity $\Delta H°$ and the positive quantity, $-T\Delta S°$. Because only the latter grows larger with increasing temperature, $\Delta G°$ grows less negative. Thus the equilibrium constant becomes smaller with increasing temperature, indicating less conversion to products.

**(b)** The $\Delta H_f°$ and $S°$ values necessary to calculate $\Delta H°$ and $\Delta S°$ for this reaction can be taken from Appendix D. We have previously performed these calculations in Sample Exercise 16.12 (Section 16.6) and in Sample Exercise 19.5, giving $\Delta H° = -92.38$ kJ and $\Delta S° = -198.2$ J/K. To calculate $\Delta G°$ given $\Delta H°$ and $\Delta S°$, we use the following relationship:

$$\Delta G° = \Delta H° - T\Delta S°$$

(Recall that the superscript $°$ indicates that the process is operated under standard-state conditions.) Assuming that $\Delta H°$ and $\Delta S°$ do not change with temperature, and using $T = 500 + 273 = 773$ K, we have

$$\Delta G° = -92.38 \text{ kJ} - (773 \text{ K})\left(-198.2 \frac{J}{K}\right)\left(\frac{1 \text{ kJ}}{10^3 \text{ J}}\right)$$
$$= -92.38 \text{ kJ} + 153.21 \text{ kJ} = 60.83 \text{ kJ}$$

Notice that we changed $T\Delta S°$ from units of joules to kilojoules so it could be added to $\Delta H°$, which is in units of kilojoules.

In Sample Exercise 19.6 we calculated $\Delta G°$ for this reaction at 298 K: $\Delta G°_{298} = -33.32$ kJ. Thus we see that increasing the temperature from 298 K to 773 K changes $\Delta G°$ from $-33.32$ kJ to $+60.83$ kJ. Of course, the result at 773 K is not as accurate as that at 298 K, because $\Delta H°$ and $\Delta S°$ do change slightly with temperature. Nevertheless, the result should be a reasonable approximation. The positive increase in $\Delta G°$ with increasing temperature is in agreement with the qualitative prediction made in part (a) of this exercise. Our result indicates that a mixture of $N_2(g)$, $H_2(g)$, and $NH_3(g)$, each at 1 atm pressure (standard-state conditions), will react spontaneously at 298 K to form more $NH_3(g)$; however, this same reaction is not spontaneous at 773 K. In fact, at 773 K the reaction will proceed in the opposite direction, to form more $N_2(g)$ and $H_2(g)$.

### PRACTICE EXERCISE
**(a)** Using the standard enthalpies of formation and standard entropies in Appendix D, calculate $\Delta H°$ and $\Delta S°$ at 298 K for the following reaction: $2SO_2(g) + O_2(g) \longrightarrow 2SO_3(g)$ **(b)** Using the values for $\Delta H°$ and $\Delta S°$ from part (a), estimate $\Delta G°$ at 400 K.
*Answers:*  **(a)**  $\Delta H° = -196.6$ kJ,  $\Delta S° = -189.6$ J/K;
**(b)**  $\Delta G° = -120.8$ kJ

## 19.6 FREE ENERGY AND THE EQUILIBRIUM CONSTANT

Although it is very valuable to have a ready means of determining $\Delta G°$ for a reaction from tabulated values, we usually want to know about the direction of spontaneous change for systems that are not at standard conditions. For any chemical process the general relationship between the free-energy change under standard conditions, $\Delta G°$, and the free-energy change under any other conditions, $\Delta G$, is given by the following expression.

$$\Delta G = \Delta G° + 2.303 RT \log Q \qquad [19.13]$$

$R$ in this expression is the ideal-gas-equation constant, 8.314 J/K-mol; $T$ is the absolute temperature; and $Q$ is the reaction quotient (Section 16.5) that corresponds to the chemical reaction and particular reaction mixture of interest.

### SAMPLE EXERCISE 19.8

Calculate $\Delta G$ at 298 K for the following reaction if the reaction mixture consists of 1.0 atm $N_2$, 3.0 atm $H_2$, and 1.0 atm $NH_3$:

$$N_2(g) + 3H_2(g) \longrightarrow 2NH_3(g)$$

**Solution:**  For the balanced equation and set of concentrations given, the reaction quotient $Q$ is

$$Q = \frac{P_{NH_3}^2}{P_{N_2} P_{H_2}^3} = \frac{(1.0)^2}{(1.0)(3.0)^3} = 3.7 \times 10^{-2}$$

$\Delta G_{298}°$ was calculated in Sample Exercise 19.6: $\Delta G_{298}° = -33.32$ kJ. Thus we have

$$\Delta G = \Delta G° + 2.303 \, RT \log Q$$

$$= (-33.32 \text{ kJ}) + 2.303 \left(8.314 \, \frac{J}{K}\right) (298 \text{ K})$$

$$\times \left(\frac{1 \text{ kJ}}{10^3 \text{ J}}\right) \log (3.7 \times 10^{-2})$$

$$= -33.32 \text{ kJ} + (-8.17 \text{ kJ})$$

$$= -41.49 \text{ kJ}$$

The free-energy change becomes more negative, changing from $-33.32$ kJ to $-41.49$ kJ, as the pressures of $N_2$, $H_2$, and $NH_3$ are changed from 1.0 atm each (standard-state conditions, $\Delta G°$) to 1.0 atm, 3.0 atm, and 1.0 atm, respectively. The larger negative value for $\Delta G$ when the pressure of $H_2$ is increased from 1.0 atm to 3.0 atm indicates a larger "driving force" to produce $NH_3$. This result bears out the prediction of Le Châtelier's principle, which indicates that increasing $P_{H_2}$ should shift the reaction more to the product side, thereby forming more $NH_3$.

### PRACTICE EXERCISE

Calculate $\Delta G$ at 298 K for the reaction of nitrogen and hydrogen to form ammonia if the reaction mixture consists of 3.0 atm $NH_3$, 1.0 atm $N_2$, and 0.50 atm $H_2$.

*Answer:* $-22.72$ kJ

When a system is at equilibrium, $\Delta G$ must be zero, and the reaction quotient $Q$ must by definition equal $K$ (Section 16.5). Thus for a system at equilibrium (when $\Delta G = 0$ and $Q = K$), Equation 19.13 transforms as follows:

$$\Delta G = \Delta G° + 2.303 RT \log Q \qquad [19.13]$$

$$0 = \Delta G° + 2.303 RT \log K$$

$$\Delta G° = -2.303 RT \log K \qquad [19.14]$$

From Equation 19.14 we can readily see that if $\Delta G°$ is negative, $\log K$ must be positive. A positive value for $\log K$ means that $K > 1$. Thus the more negative $\Delta G°$ is, the larger the equilibrium constant, $K$. Conversely, if $\Delta G°$ is positive, log $K$ is negative, which means that $K < 1$. To summarize:

$\Delta G°$ negative:    $K > 1$
$\Delta G°$ zero:    $K = 1$
$\Delta G°$ positive:    $K < 1$

It is possible from a knowledge of $\Delta G°$ for a reaction to calculate the value for the equilibrium constant, using Equation 19.14. Some care is necessary, however, in the matter of units. When dealing with gases, the concentrations of reactants should be expressed in units of atmospheres. The concentrations of solids are 1 if they are pure solids. The concentrations of pure liquids are 1 if they are pure liquids; if a mixture of liquids is involved, the concentration of each is expressed as mole fraction. For substances in solution, concentrations in moles per liter are appropriate.

---

**SAMPLE EXERCISE 19.9**

From standard free energies of formation, calculate the equilibrium constant for the reaction

$$N_2(g) + 3H_2(g) \rightleftharpoons 2NH_3(g)$$

at 25°C.

**Solution:** The equilibrium constant for this reaction is written as

$$K_p = \frac{P_{NH_3}^2}{P_{N_2} P_{H_2}^3}$$

where the gas concentrations are expressed in atmospheres pressure. The standard free-energy change for the reaction was determined in Sample Exercise 19.6 to be $-33.32$ kJ. Inserting this into Equation 19.14, we obtain

$$-33,320 \text{ J} = -2.303(8.314 \text{ J/K-mol})(298 \text{ K}) \log K_p$$

$$\log K_p = 5.84$$

Taking the antilog, we have

$$K_p = 6.9 \times 10^5$$

This is a large equilibrium constant. Compare its magnitude with the equilibrium constants at higher temperature, as listed in Table 16.5. If a catalyst could be found that would permit reasonably rapid reaction of $N_2$ with $H_2$ at room temperature, high pressures would not be required to force the equilibrium toward $NH_3$.

**PRACTICE EXERCISE**
Using data given in Appendix D, calculate first the standard free-energy change, $\Delta G°$, and then the equilibrium constant at 298 K for the following reaction: $2H_2(g) + O_2(g) \longrightarrow 2H_2O(g)$.

*Answer:* $-457.2$ kJ; $1.4 \times 10^{80}$

---

One of the important ways we use chemical reactions is to produce energy for performing work. Naturally, we want to accomplish the work as efficiently and economically as possible. Any process that occurs spontaneously can be utilized to perform work, at least in principle. For example, the burning of gasoline in the cylinders of a car produces the work accomplished in moving the car. How much work is extracted from a particular process depends on how it is carried out. For example, we might burn gasoline in an open container and extract no useful work at all. In an automobile engine, less than 20 percent of the released energy is used to accomplish work. If the gasoline were burned under more favorable conditions, we could extract more work. However, there is a theoretical limit to the amount of work we can obtain from a spontaneous

**19.7**
**FREE ENERGY AND WORK**

process. In practice, we always obtain less than this maximum possible amount. Nevertheless, it is useful to know the maximum for a process, so we can measure our success in extracting work from it.

Thermodynamics tells us that *the change in free energy for a process, $\Delta G$, equals the maximum useful work that can be done by the system on its surroundings in a spontaneous process occurring at constant temperature and pressure:*

$$w_{max} = \Delta G \qquad [19.15]$$

This relationship explains why $\Delta G$ is called the *free* energy. It is the portion of the energy change of a spontaneous reaction that is free to do useful work. The remainder of the energy enters the environment as heat.

For processes that are not spontaneous ($\Delta G > 0$) the free-energy change is a measure of the *minimum* amount of work that must be done to cause the process to occur. In actual cases, we always need to do more than this theoretical minimum amount, because of the inefficiencies in the way the changes occur.

## CHEMISTRY AT WORK　　Driving Nonspontaneous Reactions

Free-energy considerations are very important in thinking about many nonspontaneous reactions that we might wish to carry out for our own purposes or that occur in nature. For example, we might wish to extract a metal from an ore. If we look at a reaction such as

$$Cu_2S(s) \longrightarrow 2Cu(s) + S(s) \qquad \Delta G° = +86.2 \text{ kJ} \qquad [19.16]$$

we find that it is highly nonspontaneous. Clearly, then, we cannot hope to obtain copper metal from $Cu_2S$ merely by trying to catalyze the reaction shown in Equation 19.16. Instead, we must "do work" on the reaction in some way, in order to force it to occur as we wish. We might do this by coupling the reaction we've written with another reaction, so that we arrive at an overall reaction that *is* spontaneous. Consider, for example, the reaction

$$S(s) + O_2(g) \longrightarrow SO_2(g) \qquad \Delta G° = -300.1 \text{ kJ} \qquad [19.17]$$

This is a spontaneous reaction. Adding reaction Equations 19.16 and 19.17, we obtain

| | | |
|---|---|---|
| $Cu_2S(s) \longrightarrow 2Cu(s) + S(s)$ | $\Delta G° =$ | $+86.2$ kJ |
| $S(s) + O_2(g) \longrightarrow SO_2(g)$ | $\Delta G° =$ | $-300.1$ kJ |
| $Cu_2S(s) + O_2(g) \longrightarrow 2Cu(s) + SO_2(g)$ | $\Delta G° =$ | $-213.9$ kJ |

The free-energy change for the overall reaction is the sum of the free-energy changes of the two reactions. Because the negative free-energy change for the second reaction is larger than the positive free-energy change for the first, the overall reaction has a large and negative standard free-energy change.

The coupling of two or more reactions together to cause a nonspontaneous chemical process to occur is very important in biochemical systems. Many of the reactions that are essential to the maintenance of life do not occur spontaneously within the human body. These necessary reactions are made to occur, however, by coupling them with reactions that are spontaneous and release energy. The energy releases that accompany the metabolism of foodstuffs provide the primary source of necessary free energy. For example, the compound glucose, $C_6H_{12}O_6$, is oxidized in the body, and a substantial amount of energy is released.

$$C_6H_{12}O_6(s) + 6O_2(g) \longrightarrow 6CO_2(g) + 6H_2O(l)$$
$$\Delta G° = -2880 \text{ kJ} \qquad [19.18]$$

This energy "does work" in the body. However, some means is necessary to couple, or connect, the energy released by glucose oxidation to the reactions that require energy. One means of accomplishing this is shown graphically in Figure 19.8.

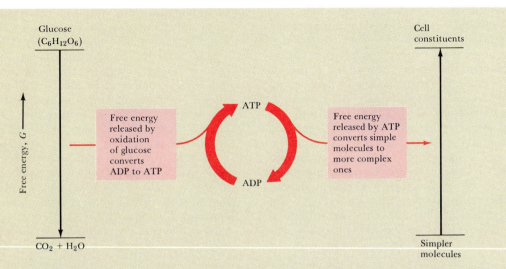

**FIGURE 19.8** Schematic representation of a part of the free-energy changes that occur in cell metabolism. The oxidation of glucose to $CO_2$ and $H_2O$ produces free energy. This released free energy is used to convert ADP into the more energetic ATP. The ATP is then used, as needed, as an energy source to convert simple molecules into more complex cell constituents. When ATP releases its free energy, it is converted to ADP.

Adenosine triphosphate (ATP) is a high-energy molecule. When ATP is converted to a lower-energy molecule, adenosine diphosphate (ADP), the energy to drive other chemical reactions becomes available. The energy released in glucose oxidation is used in part to reconvert ADP back to ATP. Thus the ATP-ADP interconversions act as a means of storing energy and releasing it to drive needed reactions. The coupling of reactions so that the free energy released in one may be used in another way requires particular enzymes as catalysts. In Chapter 28, which deals with biochemistry, we will examine the energy relationships in living systems in more detail.

# *FOR REVIEW*

## SUMMARY

In this chapter we have examined the concept of equilibrium from a thermodynamic point of view. The enthalpy change of a system, $\Delta H$, is a measure of the potential energy change in a process. Exothermic processes ($\Delta H < 0$) tend to occur spontaneously. The spontaneous character of a reaction is also determined by the change in randomness or disorder of the system, measured by the **entropy**, $S$. Processes that produce an increase in randomness or disorder of the system ($\Delta S > 0$) tend to occur spontaneously.

Entropy changes in a system are associated with an increase in the number of ways the particles of the system can be distributed among possible energy states or spatial arrangements. For example, an increase in volume, in translational energy, or in number of particles all lead to an increase in entropy. In any process the total entropy change is the sum of the entropy change of the system and the surroundings. The **second law of thermodynamics** tells us that in any spontaneous process the entropy of the universe increases. That is,

$$\Delta S_{system} + \Delta S_{surroundings} > 0$$

The standard entropy change in a system, $\Delta S^\circ$, can be calculated from tabulated standard entropy values, $S^\circ$. Entropies are determined with the aid of the **third law of thermodynamics**, which states that the entropy of a pure crystalline solid at 0 K is zero.

The **free energy**, $G$, is a thermodynamic state function that combines the two state functions enthalpy and entropy: $G = H - TS$. For processes that occur at constant temperature and pressure, $\Delta G = \Delta H - T\Delta S$. The free-energy change for a process occurring at constant temperature and pressure relates directly to reaction spontaneity. $\Delta G$ is negative for all spontaneous processes, whereas a positive value for $\Delta G$ indicates a nonspontaneous process (one that is spontaneous in the reverse direction). $\Delta G$ is zero at equilibrium. The free energy also has the important property that it is a measure of the maximum useful work that can be performed by a system in a spontaneous process. In practice this amount of useful work is never realized, because processes in the real world have inherent inefficiencies. If a process is nonspontaneous, the value of $\Delta G$ is a measure of the minimum work that must be done on the system to cause the process to occur. In practice we always need to do more than this minimum.

The standard free-energy change, $\Delta G°$, for any process can be calculated from tabulated standard free energies of formation, $\Delta G_f°$; it can also be calculated from standard enthalpy and entropy changes: $\Delta G° = \Delta H° - T\Delta S°$. Temperature changes will change the value of $\Delta G$ and can also change its sign.

The free-energy change under nonstandard conditions is related to the standard free-energy change: $\Delta G = \Delta G° + 2.303RT \log Q$. At equilibrium ($\Delta G = 0$, $Q = K$), $\Delta G° = -2.303RT \log K$. Thus the standard free-energy change is related to the equilibrium constant. Consequently, we can understand the position of a chemical equilibrium in terms of the $\Delta H°$ and $T\Delta S°$ functions of which $\Delta G°$ is composed.

## LEARNING GOALS

Having read and studied this chapter, you should be able to:

1. Define the term *spontaneity* and identify spontaneous processes.

2. Describe how entropy is related to randomness or disorder.

3. State the second law of thermodynamics.

4. Predict whether the entropy change in a given process is positive, negative, or near zero.

5. State the third law of thermodynamics.

6. Describe how and why the entropy of a substance changes with increasing temperature or when a phase change occurs, starting with the substance as a pure solid at $0\ K$.

7. Calculate $\Delta S°$ for any reaction from tabulated standard entropy values, $S°$.

8. Define free energy in terms of enthalpy and entropy.

9. Explain the relationship between the sign of the free-energy change, $\Delta G$, and whether a process is spontaneous in the forward direction.

10. Calculate the standard free-energy change at constant temperature and pressure, $\Delta G°$, for any process from tabulated values for the standard free energies of reactants and products.

11. List the usual conventions regarding standard states in setting the values for standard free energies.

12. Predict how $\Delta G$ will change with temperature, given the signs for $\Delta H$ and $\Delta S$.

13. Estimate $\Delta G°$ at any temperature given $\Delta S_{298}°$ and $\Delta H_{298}°$.

14. Calculate the free-energy change under nonstandard conditions, $\Delta G$, given $\Delta G°$, temperature, and the data needed to calculate the reaction quotient.

15. Calculate $\Delta G°$ from $K$ and perform the reverse operation.

16. Describe the relationship between $\Delta G$ and the maximum useful work that can be derived from a spontaneous process, or the minimum work required to accomplish a nonspontaneous process.

## KEY TERMS

Among the more important terms and expressions used for the first time in this chapter are the following:

**Entropy** (Section 19.2) is a thermodynamic function associated with the number of different, equivalent energy states or spatial arrangements in which a system may be found. It is a thermodynamic state function, which means that once we specify the conditions for a system—that is, the temperature, pressure, and so on—the entropy is defined.

**Free energy** (Section 19.5) is a thermodynamic state function that combines enthalpy and entropy, in the form $G = H - TS$. For a change occurring at constant temperature and pressure, the change in free energy is $\Delta G = \Delta H - T\Delta S$.

An **isolated system** (Section 19.2) is one that does not exchange energy or matter with its surroundings.

The **second law of thermodynamics** (Section 19.1) is a statement of our experience that there is a direction to the way events occur in nature: When a process occurs spontaneously in one direction, it is nonspontaneous in the reverse direction. It is possible to state the second law in many different forms, but they all relate back to the same idea about spontaneity. One of the most common statements found in chemical contexts is that in any spontaneous process the entropy of the universe increases.

A **spontaneous** process (Section 19.1) is one that is capable of proceeding in a given direction, as written or described, without needing to be driven by an outside source of energy. A process may be spontaneous even though it is very slow. $\Delta G$ is negative for all spontaneous processes.

The **third law of thermodynamics** (Section 19.3) states that the entropy of a pure, crystalline solid at absolute zero temperature is zero: $S$ (0 K) = 0.

**Vibrational and rotational energies** (Section 19.3) refer to the storing of energies in molecules in the form of vibrational motions between the atoms of the molecules or rotational motions of the molecule as a whole.

# EXERCISES

## Spontaneity and Entropy

**19.1** Which of the following processes are spontaneous and which are nonspontaneous: (a) spreading of the fragrance of perfume through a room; (b) separation of $N_2$ and $O_2$ molecules in air from each other; (c) mending a broken clock; (d) the reaction of sodium metal with chlorine gas to form sodium chlorine; (e) the dissolution of $HCl(g)$ in water to form concentrated hydrochloric acid?

**19.2** Which of the following processes are spontaneous and which are nonspontaneous: (a) the melting of ice cubes at $-5°C$ and 1 atm pressure; (b) dissolution of sugar in a hot cup of coffee; (c) the reaction of nitrogen atoms to form $N_2$ molecules at 25°C and 1 atm; (d) alignment of iron filings in a magnetic field; (e) formation of $CH_4$ and $O_2$ molecules from $CO_2$ and $H_2O$ at room temperature and 1 atm pressure?

**19.3** A nineteenth-century chemist, Marcellin Berthelot, suggested that all chemical processes that proceed spontaneously are exothermic. Is this correct? If you think not, offer some counterexamples.

**19.4** In a living cell, large molecules are assembled from smaller ones. Is this process consistent with the second law of thermodynamics?

**19.5** How does the entropy of the system change when the following processes occur: (a) a solid is melted; (b) a liquid is vaporized; (c) a solid is dissolved in water; (d) a gas is liquefied?

**19.6** Why is the increase in entropy of the system greater for the vaporization of a substance than for its melting?

**19.7** For each of the following pairs, choose the substance with the higher entropy (per mole) at a given temperature: (a) $N_2(g)$ at 1 atm or $N_2(g)$ at 0.1 atm; (b) $Br_2(l)$ or $Br_2(g)$; (c) $O(g)$ or $O_2(g)$; (d) $NH_4Cl(s)$ or $NH_4Cl(aq)$.

**19.8** For each of the following pairs, indicate which substance you would expect to possess the larger standard entropy: (a) 1 mol of $H_2(g)$, 298 K, 1 atm pressure, or 1 mol of $H_2(g)$, 298 K, 10 atm pressure; (b) 1 mol of $H_2O(s)$ at 5°C or 1 mol of $H_2O(l)$ at 5°C; (c) 1 mol of $Br_2(l)$ at 1 atm, 58.8°C, or 1 mol of $Br_2(g)$ at 1 atm, 58.8°C; (d) 1 mol of $KNO_3(s)$, or 1 mol of $KNO_3(aq)$ at 30°C.

**19.9** Using Appendix D, compare the standard entropies at 298 K for the substances in each of the following pairs: (a) $Hg(l)$ and $Hg(g)$; (b) $F(g)$ and $F_2(g)$; (c) $FeO(s)$ and $Fe_2O_3(s)$; (d) $H_2O(l)$ and $H_2O_2(l)$. Explain the origin of the difference in $S°$ values in each case.

**19.10** Using Appendix D, compare the standard entropies at 298 K for the substances in each of the following pairs; (a) $NaCl(s)$ and $NaCl(aq)$; (b) $Cu(s)$ and $Cu(g)$; (c) $P_2(g)$ and $P_4(g)$; (d) $KCl(s)$ and $KClO_3(s)$. Explain the origin of the difference in $S°$ values in each case.

**19.11** Predict the sign of $\Delta S$ for the system in each of the following processes: (a) freezing of 1 mol of $H_2O(l)$; (b) evaporation of 1 mol of $Br_2(l)$; (c) precipitation of $BaSO_4$ upon mixing $Ba(NO_3)_2(aq)$ and $H_2SO_4(aq)$; (d) oxidation of magnesium metal: $2Mg(s) + O_2(g) \longrightarrow 2MgO(s)$.

**19.12** Predict whether the entropy change in the system is positive or negative for each of the following processes:
(a) $2C(s) + O_2(g) \longrightarrow 2CO(g)$
(b) $2K(s) + Br_2(l) \longrightarrow 2KBr(s)$
(c) $2MnO_2(s) \longrightarrow 2MnO(s) + O_2(g)$
(d) $O(g) + O_2(g) \longrightarrow O_3(g)$

**19.13** Using tabulated $S°$ values from Appendix D, calculate $\Delta S°$ for each of the following reactions:
(a) $2HBr(g) + F_2(g) \longrightarrow 2HF(g) + Br_2(g)$
(b) $2NO(g) + O_2(g) \longrightarrow 2NO_2(g)$
(c) $2CH_3OH(g) + 3O_2(g) \longrightarrow 2CO_2(g) + 4H_2O(g)$
(d) $4FeO(s) + O_2(g) \longrightarrow 2Fe_2O_3(s)$
In each case, account for the sign of $\Delta S°$.

**19.14** Using standard entropies tabulated in Appendix D, calculate $\Delta S°$ for each of the following processes:
(a) $I_2(s) \longrightarrow I_2(g)$
(b) $NaCl(s) \longrightarrow NaCl(aq)$
(c) $NH_4NO_3(s) \longrightarrow N_2O(g) + 2H_2O(g)$
(d) $2C_2H_2(g) + 5O_2(g) \longrightarrow 4CO_2(g) + 2H_2O(l)$

## Free Energy

**19.15** (a) Express the free-energy change in a process in terms of the changes that occur in the enthalpy and entropy of the system. (b) What is the relationship between $\Delta G$ for a process and the speed at which it occurs?

**19.16** (a) What is the significance of $\Delta G = 0$ for any process in a system? (b) What is the meaning of the *standard* free-energy change in a process, $\Delta G°$, as contrasted with simply the free-energy change, $\Delta G$?

**19.17** Calculate $\Delta G°_{298}$ for

$$H_2O_2(g) \longrightarrow H_2O(g) + \tfrac{1}{2}O_2(g)$$

given that $\Delta H°_{298} = -106$ KJ and $\Delta S°_{298} = +58$ J/K for this process. Would you expect $H_2O_2(g)$ to be very stable at 298 K? Explain briefly.

**19.18** Using the data in Appendix D, calculate $\Delta H°$, $\Delta S°$, and $\Delta G°$ for each of the following reactions. In each case show that $\Delta G° = \Delta H° - T\Delta S°$.
(a) $BaO(s) + CO_2(g) \longrightarrow BaCO_3(s)$
(b) $2KClO_3(s) \longrightarrow 2KCl(s) + 3O_2(g)$
(c) $2CH_3OH(l) + 3O_2(g) \longrightarrow 2CO_2(g) + 4H_2O(g)$
(d) $NOCl(g) + Cl(g) \longrightarrow NO(g) + Cl_2(g)$

**19.19** Using the data from Appendix D, calculate the standard free-energy change for each of the following processes. In each case indicate whether the reaction is spontaneous under standard conditions.
(a) $MnO_2(s) + 2CO(g) \longrightarrow Mn(s) + 2CO_2(g)$
(b) $H_2(g) + Br_2(g) \longrightarrow 2HBr(g)$
(c) $6Cl_2(g) + 2Fe_2O_3(g) \longrightarrow 4FeCl_3(s) + 3O_2(g)$
(d) $CaO(s) + H_2O(l) \longrightarrow Ca(OH)_2(s)$

**19.20** Using the standard free energies of formation tabulated in Appendix D, calculate $\Delta G°$ at 298 K for each of the following reactions. In each case indicate whether the reaction is spontaneous under standard conditions.
(a) $N_2(g) + O_2(g) \longrightarrow 2NO(g)$
(b) $2SO_2(g) + O_2(g) \longrightarrow 2SO_3(g)$
(c) $2H_2S(g) + SO_2(g) \longrightarrow 3S(s) + 2H_2O(g)$
(d) $2C_6H_6(l) + 15O_2(g) \longrightarrow 12CO_2(g) + 6H_2O(l)$

**19.21** Classify each of the following reactions as belonging to one of the four possible types summarized in Table 19.2:
(a) $N_2(g) + 3F_2(g) \longrightarrow 2NF_3(g)$
$\quad\quad\quad \Delta H° = -249$ kJ; $\Delta S° = -278$ J/K
(b) $N_2(g) + 3Cl_2(g) \longrightarrow 2NCl_3(g)$
$\quad\quad\quad \Delta H° = +460$ kJ; $\Delta S° = -275$ J/K
(c) $N_2F_4(g) \longrightarrow 2NF_2(g)$
$\quad\quad\quad \Delta H° = 85$ kJ: $\Delta S° = 198$ J/K
(d) $2H_2O(l) \longrightarrow 2H_2(g) + O_2(g)$
$\quad\quad\quad \Delta H° = 572$ kJ; $\Delta S° = 329$ J/K

**19.22** From the values given for $\Delta H°$ and $\Delta S°$, calculate $\Delta G°$ for each reaction at 298 K. If the reaction is not spontaneous under standard conditions at 298 K, at what temperature (if any) would the reaction become spontaneous?
(a) $2PbS(s) + 3O_2(g) \longrightarrow 2PbO(s) + 2SO_2(g)$
$\quad\quad\quad \Delta H° = -844$ kJ: $\Delta S° = -0.165$ kJ/K

(b) $2POCl_3(g) \longrightarrow 2PCl_3(g) + O_2(g)$
$\quad\quad\quad \Delta H° = 572$ kJ; $\Delta S° = 179$ J/K

**19.23** (a) Using the data of Appendix D, predict how $\Delta G°$ for the following process will change with increasing temperature:

$$P_4(g) \longrightarrow 2P_2(g)$$

(b) Calculate $\Delta G°$ at 900 K, assuming that $\Delta H°$ and $\Delta S°$ do not change with temperature.

**19.24** (a) Using the data of Appendix D, predict how $\Delta G°$ for the following process will change with increasing temperature:

$$SO_3(g) + H_2(g) \longrightarrow SO_2(g) + H_2O(g)$$

(b) Calculate $\Delta G°$ at 600 K, assuming that $\Delta H°$ and $\Delta S°$ do not change with variation in temperature.

## Free Energy, Equilibrium, and Work

**19.25** Explain qualitatively how $\Delta G$ changes for each of the following reactions as the partial pressure of $N_2$ is increased:
(a) $N_2H_4(g) \longrightarrow N_2(g) + 2H_2(g)$
(b) $N_2(g) + 3F_2(g) \longrightarrow 2NF_3(g)$
(c) $NH_4NO_2(s) \longrightarrow N_2(g) + 2H_2O(g)$

**19.26** Indicate whether $\Delta G$ increases, decreases, or does not change when the partial pressure of $H_2$ is increased in each of the following reactions:
(a) $N_2(g) + 3H_2(g) \longrightarrow 2NH_3(g)$
(b) $2HBr(g) \longrightarrow H_2(g) + Br_2(g)$

**19.27** Consider the reaction $2NO_2(g) \longrightarrow N_2O_4(g)$. (a) Using data from Appendix D, calculate $\Delta G°$ at 298 K for this reaction. (b) Calculate $\Delta G$ at 298 K if the partial pressures of $NO_2$ and $N_2O_4$ are 2.00 atm and 0.10 atm, respectively.

**19.28** Consider the reaction $2CO(g) + O_2(g) \longrightarrow 2CO_2(g)$. (a) Using data from Appendix D, calculate $\Delta G°$ for this reaction at 298 K. (b) Calculate $\Delta G$ at 298 K if the reaction mixture consists of 6.0-atm CO, 300-atm $O_2$, and 0.10-atm $CO_2$.

**19.29** Calculate $\Delta G°$ and $K_p$ at 298 K for each of the following reactions:
(a) $2H_2(g) + 2I_2(g) \longrightarrow 2HI(g)$
(b) $2H_2O(g) \longrightarrow 2H_2(g) + O_2(g)$

**19.30** Write the equilibrium-constant expression and calculate the magnitude of the equilibrium constant at 298 K for each of the following reactions, using data from Appendix D:
(a) $2HBr(g) + Cl_2(g) \rightleftharpoons 2HCl(g) + Br_2(g)$
(b) $CaSO_4(s) + CO_2(g) \rightleftharpoons CaCO_3(s) + SO_3(g)$

**[19.31]** Consider the decomposition of calcium carbonate:

$$CaCO_3(s) \longrightarrow CaO(s) + CO_2(g)$$

Given that $\Delta H° = 177.8$ kJ and $\Delta S° = 160.5$ J/K, calculate the equilibrium pressure of $CO_2$ at **(a)** 298 K and **(b)** 600°C.

**[19.32]** Consider the following reaction:

$$PbCO_3(s) \longrightarrow PbO(s) + CO_2(g)$$

Using data in Appendix D, calculate the equilibrium pressure of $CO_2$ in the system at **(a)** 25°C and **(b)** 500°C.

**19.33** **(a)** Given $K_a$ for benzoic acid at 298 K (Appendix E), calculate $\Delta G°$ for the dissociation of this substance in aqueous solution. **(b)** What is the value of $\Delta G$ at equilibrium? **(c)** What is the value of $\Delta G$ when $[H^+] = 3.0 \times 10^{-3}$ $M$, $[C_7H_5O_2^-] = 2.0 \times 10^{-5}$ $M$, and $[HC_7H_5O_2] = 0.10$ $M$?

**19.34** **(a)** Given $K_b$ for ammonia at 298 K (Appendix E), calculate $\Delta G°$ for the following reaction:

$$NH_3(aq) + H_2O(l) \longrightarrow NH_4^+(aq) + OH^-(aq)$$

**(b)** what is the value of $\Delta G$ at equilibrium? **(c)** What is the value of $\Delta G$ when $[NH_3] = 0.10$ $M$, $[NH_4^+] = 0.10$ $M$, and $[OH^-] = 0.050$ $M$?

**19.35** Natural gas consists primarily of methane, $CH_4$. **(a)** How much heat is produced in burning a mole of $CH_4$ under standard conditions if reactants and products are brought to 298 K and $H_2O(l)$ is formed? **(b)** What is the maximum amount of useful work that can be accomplished under standard conditions by this system?

**19.36** Acetylene gas, $C_2H_2(g)$, is used in welding. **(a)** How much heat is produced in burning a mole of $C_2H_2$ under standard conditions if both reactants and products are brought to 298 K and $H_2O(l)$ is formed? **(b)** What is the maximum amount of useful work that can be accomplished under standard conditions by this system?

## Additional Exercises

**19.37** Indicate whether each of the following statements is true or false. If it is false, correct it. **(a)** The feasibility of manufacturing $NH_3$ from $N_2$ and $H_2$ depends entirely on the value of $\Delta H$ for the process $N_2(g) + 3H_2(g) \longrightarrow 2NH_3(g)$. **(b)** The reaction of $H_2(g)$ with $Cl_2(g)$ to form $HCl(g)$ is an example of a spontaneous process. **(c)** A spontaneous process is one that occurs rapidly. **(d)** A process that is nonspontaneous in one direction is spontaneous in the opposite direction. **(e)** Spontaneous processes are those that are exothermic and that lead to a higher degree of order in the system.

**19.38** For each of the following processes, indicate whether the sign of $\Delta S$ and $\Delta H$ is expected to be positive, negative, or about zero. **(a)** A solid sublimes. **(b)** The temperature of a solid is lowered by 25°C. **(c)** Ethyl alcohol evaporates from a beaker. **(d)** A diatomic molecule dissociates into atoms. **(e)** A piece of charcoal is combusted to form $CO_2(g)$ and $H_2O(g)$.

**19.39** When applied, the insecticide DDT is sprayed over large areas. It later moves in the environment through soil into plants and water supplies. In lakes it concentrates in the fatty tissue of fish. If we think of DDT as the "system," describe the entropy change associated with each of the processes above. In terms of cleaning up DDT in the environment, what entropy change to the DDT system is required? How can such an entropy change be brought about? Is the process feasible?

**19.40** The reaction $2Mg(s) + O_2(g) \longrightarrow 2MgO(s)$ has associated with it a negative value for $\Delta S°$, and it is a highly spontaneous process. The second law of thermodynamics states that in any spontaneous process there is always an increase in the entropy of the universe. Is there an inconsistency between the second law and what we know about the oxidation of magnesium? Explain.

**19.41** The melting and boiling points of HCl are $-115$°C and $-84$°C, respectively. Using this information and the value of $S°$ listed in Apppendix D, draw a rough sketch of $S°$ for HCl from absolute zero to 298 K.

**19.42** **(a)** Calculate $\Delta H°$, $\Delta S°$, and $\Delta G°$ for each of the following processes at 25°C, using the data of Appendix D:
  (i)   $4Cr(s) + 3O_2(g) \longrightarrow 2Cr_2O_3(s)$
  (ii)  $2F_2(g) + 2CaO(s) \longrightarrow 2CaF_2(g) + O_2(g)$
  (iii) $C_2H_2(g) + 4Cl_2(g) \longrightarrow 2CCl_4(g) + H_2(g)$
**(b)** For each of these processes, predict the manner in which the free-energy change varies with an increase in temperature.

**19.43** On the basis of the $\Delta H°$ and $\Delta S°$ values given for each reaction below, indicate **(a)** which are spontaneous under standard conditions at 25°C and **(b)** which can be expected to be spontaneous at high temperatures.
  (i)   $KClO_3(s) \longrightarrow KCl(s) + \frac{3}{2}O_2(g)$
        $\Delta H° = -44.7$ kJ; $\Delta S° = +59.1$ J/K
  (ii)  $2Al(s) + 3Cl_2(g) \longrightarrow 2AlCl_3(s)$
        $\Delta H° = -332$ kJ; $\Delta S° = -93.4$ J/K
  (iii) $NOCl(g) \longrightarrow NO(g) + \frac{1}{2}Cl_2(g)$
        $\Delta H° = +37.6$ kJ; $\Delta S° = +58.5$ J/K

**[19.44]** **(a)** Below are given the normal boiling points (Section 10.3) for several substances, and their enthalpies of vaporization at those temperatures. From these data calculate $\Delta S$ of vaporization for each substance. (Hint: The gaseous and liquid states are in equilibrium; what does this imply about $\Delta G$?)

| Substance | Normal boiling point (°C) | Enthalpy of vaporization (kJ/mol) |
|---|---|---|
| Acetone, $(CH_3)_2O$ | 56.2 | 30.3 |
| Benzene, $C_6H_6$ | 80.1 | 30.7 |
| Ammonia, $NH_3$ | $-33.4$ | 23.4 |
| Water, $H_2O$ | 100.0 | 40.6 |

**(b)** What significance can be attached to the variations in $\Delta S$ among these compounds? Explain.

**19.45** Using the data in Appendix D and given the pressures listed, calculate $\Delta G$ for each of the following reactions:

**(a)** $N_2(g) + 3H_2(g) \longrightarrow 2NH_3(g)$
$P_{N_2} = 12.0$ atm, $P_{H_2} = 2.0$ atm, $P_{NH_3} = 4.0$ atm

**(b)** $N_2H_4(g) \longrightarrow N_2(g) + 2H_2(g)$
$P_{N_2H_4} = 6.0$ atm, $P_{N_2} = 1 \times 10^{-3}$ atm,
$P_{H_2} = 2 \times 10^{-4}$ atm

**19.46** For the following equilibria, calculate $\Delta G°$ at the temperature indicated:

**(a)** $H_2(g) + I_2(g) \longrightarrow 2HI(g)$
$K_p = 50.2$ at $445°C$

**(b)** $2SO_2(g) + O_2(g) \longrightarrow 2SO_3(g)$
$K_p = 9.1 \times 10^2$ at $800°C$

**19.47** Evaporating $NH_3$ and HCl from aqueous ammonia ("ammonium hydroxide") and hydrochloric acid, respectively, produce the white haze often seen on laboratory windows and glassware:

$$NH_3(g) + HCl(g) \longrightarrow NH_4Cl(s)$$

Calculate the equilibrium constant, $K_p$, for this reaction at $25°C$.

**19.48** **(a)** In each of the following reactions, predict the sign of $\Delta H°$ and $\Delta S°$, and discuss briefly how these factors determine the magnitude of $K$. **(b)** Based on your general chemical knowledge, predict which of these reactions will have $K > 1$. **(c)** In each case, indicate whether $K$ should increase or decrease with increasing temperature.
(i) $2Mg(s) + O_2(g) \rightleftharpoons 2MgO(s)$
(ii) $2KI(l) \rightleftharpoons 2K(g) + I_2(g)$
(iii) $Na_2(g) \rightleftharpoons 2Na(g)$
(iv) $V_2O_5(s) \rightleftharpoons 2V(s) + \frac{5}{2}O_2(g)$

**[19.49]** The relationship between the temperature of a reaction, its standard enthalpy change, and the equilibrium constant at that temperature can be expressed in terms of the following linear equation:

$$\log K = \frac{-\Delta H°}{2.30RT} + \text{constant}$$

**(a)** Explain how this equation can be used to determine $\Delta H°$ experimentally from the equilibrium constants at several different temperatures. **(b)** Derive the equation above using relationships given in this chapter. What is the constant equal to?

**[19.50]** The potassium-ion concentration in blood plasma is about $5.0 \times 10^{-3}$ $M$ while the concentration in muscle-cell fluid is much greater, $0.15$ $M$. The plasma and intracellular fluid are separated by the cell membrane, which we assume is permeable only to $K^+$. **(a)** What is $\Delta G$ for the transfer of 1 mol of $K^+$ from blood plasma to the cellular fluid at body temperature, $37°C$? **(b)** What is the minimum amount of work that must be used to transfer this $K^+$?

**[19.51]** The standard free-energy change at absolute temperature $T$ (K) for the equilibrium

$$CaCO_3(s) \rightleftharpoons CaO(s) + CO_2(g)$$

is given by the expression

$$\Delta G°_T = 177.1 \text{ kJ/mol} - (158 \times T) \text{ J/mol-K}$$

**(a)** Account for the negative sign before the temperature-dependent term of this expression. **(b)** At what temperature does the equilibrium pressure of $CO_2(g)$ in the system equal $0.100$ atm?

**19.52** The oxidation of glucose, $C_6H_{12}O_6$, in body tissue produces $CO_2$ and $H_2O$. In contrast, anaerobic decomposition, which occurs during fermentation, produces ethyl alcohol, $C_2H_5OH$. **(a)** Compare the equilibrium constants for the following reactions:

$$C_6H_{12}O_6(s) + 6O_2(g) \rightleftharpoons 6CO_2(g) + 6H_2O(l)$$
$$C_6H_{12}O_6(s) \rightleftharpoons 2C_2H_5OH(l) + 2CO_2(g)$$

The standard free energy of formation of $C_6H_{12}O_6(s)$ is $-912$ kJ/mol. **(b)** Compare the maximum amounts of work that can be obtained from these two processes under standard conditions.

**[19.53]** Consider the apparatus shown in Figure 19.9, which is much like that shown in Figure 19.1. $N_2(g)$ is initially confined in flask $A$, of 1 L volume, at 2.0 atm pressure. Chamber $B$, of volume 1 L, is evacuated. When the stopcock is opened, $N_2$ expands to both sides. **(a)** Assuming that $N_2$ behaves as an ideal gas, what is $\Delta H$ for this process? **(b)** Describe the process by which the gas can be returned to its original condition. What change in $\Delta H$ occurs in the system in this process? **(c)** What are the changes in $\Delta G$ and $\Delta H$ of the system for the overall processes of expansion followed by compression to the original condition state? **(d)** In this overall process is work done by the system on the surroundings, or by the surroundings on the system? Explain. **(e)** Describe how this example illustrates the second law of thermodynamics.

**FIGURE 19.9**

$A$          $B$

**[19.54]** Cells use the hydrolysis of adenosine triphosphate, ATP, as a source of energy (Figure 19.8). The conversion of ATP to ADP has a standard free-energy change of $-30.5$ kJ/mol. If all the free energy from the metabolism of glucose,

$$C_6H_{12}O_6(s) + 6O_2(g) \longrightarrow 6CO_2(g) + 6H_2O(l)$$

goes into the conversion of ADP to ATP, how many moles of ATP can be produced for each mole of glucose? The standard free energy of formation of $C_6H_{12}O_5(s)$ is $-912$ kJ/mol.

**[19.55]** The reaction

$$SO_2(g) + 2H_2S(g) \rightleftharpoons 3S(s) + 2H_2O(g)$$

is the basis of a suggested method for removal of $SO_2$ from power-plant stack gases. The standard free energy for each substance is given in Appendix D. **(a)** What is the equilibrium constant at 298 K for the reaction as written? **(b)** Is this reaction at least in principle a feasible method of removing $SO_2$? **(c)** Assuming that the $H_2O$ vapor pressure is 25 mm Hg and adjusting the conditions so that $P_{SO_2}$ equals $P_{H_2S}$, calculate the equilibrium $SO_2$ pressure in this system. **(d)** The reaction as written is exothermic. Would you expect the process to be more or less effective at higher temperatures?

# CHAPTER 20

# Electrochemistry

Many important chemical processes use electricity; other processes produce it. Because of the importance of electricity in modern society, it is useful for us to examine the subject area of electrochemistry, which deals with the relationships that exist between electricity and chemical reactions. As we shall see, our discussions of electrochemistry will provide insight into such diverse topics as the construction and operation of batteries, spontaneity of chemical reactions, electroplating, and the corrosion of metals. Because electricity involves the flow of electrons, electrochemistry focuses on oxidation-reduction, or redox, reactions (Section 13.6), in which electrons are transferred from one substance to another.

## CONTENTS

## 20.1 VOLTAIC CELLS

In principle, the energy released in any spontaneous redox reaction can be directly harnessed to perform electrical work. This task is accomplished through a **voltaic** (or **galvanic**) **cell**, which is merely a device in which electron transfer is forced to take place through an external pathway rather than directly between reactants.

One such spontaneous reaction occurs when a piece of zinc is placed in contact with a solution containing $Cu^{2+}$. As the reaction proceeds, the blue color that is characteristic of $Cu^{2+}(aq)$ ions fades, and copper metal begins to deposit on the zinc. At the same time, the zinc begins to dissolve. These transformations are shown in Figure 20.1 and are summarized by Equation 20.1:

$$Zn(s) + Cu^{2+}(aq) \longrightarrow Zn^{2+}(aq) + Cu(s) \qquad [20.1]$$

Figure 20.2 shows a voltaic cell that uses the oxidation-reduction reaction between Zn and $Cu^{2+}$ expressed in Equation 20.1. Although the experimental design shown in Figure 20.2 is more complex than that in Figure 20.1, it is important to recognize that the chemical reaction is the same in both cases. The major difference between the two arrangements is that in Figure 20.2 the zinc metal and $Cu^{2+}(aq)$ are no longer in direct contact. Consequently, reduction of the $Cu^{2+}$ can occur only by a flow of electrons through the wire that connects Zn and Cu (the external circuit).

The two solid metals that are connected by the external circuit are called electrodes. By definition, the electrode at which oxidation occurs is called the **anode**; the electrode at which reduction occurs is called the

---

Steel pipes used in dredging operations show extensive corrosion after lying on a North Carolina beach for a time. Corrosion of iron is an electrochemical process of great economic importance. The annual cost of metallic corrosion in the U.S. economy is estimated to be $70 billion. (Jack Deraid/Photo Researchers)

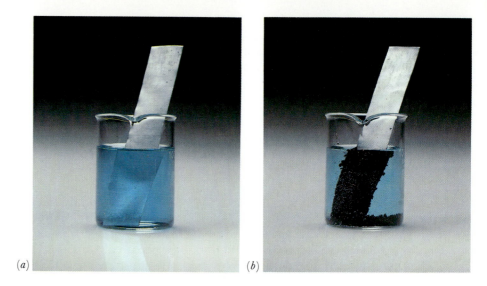

**FIGURE 20.1** (*a*) A strip of zinc is placed in a solution of copper(II) sulfate. (*b*) Electrons are transferred from the zinc to the $Cu^{2+}$ ion, forming $Zn^{2+}$ ions and copper. Thus as the reaction proceeds, the zinc dissolves, the blue color due to $Cu^{2+}(aq)$ fades, and copper deposits. (Fundamental Photographs)

(*a*)　　　　　　(*b*)

**FIGURE 20.2** An electrochemical cell based on the reaction shown in Equation 20.1. The compartment on the left contains 1 *M* $CuSO_4$ and a copper electrode. The one on the right contains 1 *M* $ZnSO_4$ and a zinc electrode. The solutions are connected electrically through the bridge, which has a fritted glass disc that permits contact of the solutions in the two compartments. The metal electrodes are connected through the digital voltmeter, which reads the potential of the cell, in this case 1.074 V. (Donald Clegg and Roxy Wilson)

**cathode**.* In our present example, Zn is the anode and Cu is the cathode:

Oxidation; anode: $\qquad\qquad\qquad\qquad Zn(s) \longrightarrow Zn^{2+}(aq) + 2e^-$

Reduction; cathode: $\qquad Cu^{2+}(aq) + 2e^- \longrightarrow Cu(s)$

We may regard the voltaic cell as two half-cells, one corresponding to the oxidation process and one corresponding to the reduction process. Electrons become available as zinc metal is oxidized at the anode. They flow through the external circuit to the cathode, where they are consumed as $Cu^{2+}(aq)$ is reduced.

---

* To help remember these definitions, it is useful to note that anode and oxidation both begin with a vowel and cathode and reduction both begin with a consonant.

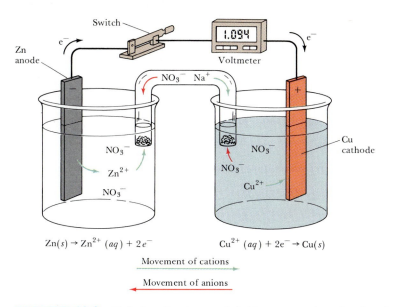

$$Zn(s) \rightarrow Zn^{2+}(aq) + 2e^-$$ $$Cu^{2+}(aq) + 2e^- \rightarrow Cu(s)$$

Movement of cations →

← Movement of anions

**FIGURE 20.3**  Voltaic cell using a salt bridge to complete the electrical circuit.

We must be careful about the signs we attach to the electrodes in a voltaic cell. We have seen that electrons are released at the anode, as the zinc is oxidized. Electrons are thus flowing out of the anode and into the external circuit, as illustrated in Figure 20.3. Because the electrons are negatively charged, we assign a negative sign to the anode. Conversely, electrons flow into the cathode, where they are consumed in the reduction of copper. A positive sign is thus assigned to the cathode, because it appears to attract the negative electrons.

As the cell pictured in Figure 20.2 operates, oxidation of Zn introduces additional $Zn^{2+}$ ions into the anode compartment. Unless a means is provided to neutralize this positive charge, no further oxidation can take place. Similarly, the reduction of $Cu^{2+}$ at the cathode leaves an excess of negative charge in solution in that compartment. Electrical neutrality is maintained by a migration of ions through the fritted glass disc separating the two compartments, or through a "salt bridge," as illustrated in Figure 20.3. A salt bridge consists of a U-shaped tube that contains an electrolyte solution, such as $NaNO_3(aq)$, whose ions will not react with other ions in the cell or with the electrode materials. The ends of the U-tube may be loosely plugged with glass wool or the electrolyte may be incorporated into a gel so that the electrolyte solution does not run out when the U-tube is inverted. As oxidation and reduction proceed at the electrodes, ions from the salt bridge migrate to neutralize charge in the anode and cathode compartments. Anions migrate toward the anode and cations toward the cathode. In fact, no measurable electron flow will occur through the external circuit unless a means is provided for ions to migrate through the solution from one electrode compartment to another, thereby completing the circuit.

## SAMPLE EXERCISE 20.1

The following oxidation-reduction reaction is spontaneous in the direction indicated:

$$Cr_2O_7^{2-}(aq) + 14H^+(aq) + 6I^-(aq) \longrightarrow$$
$$2Cr^{3+}(aq) + 3I_2(s) + 7H_2O(l)$$

A solution containing $K_2Cr_2O_7$ and $H_2SO_4$ is poured into one beaker, and a solution of KI is poured into another. A salt bridge is used to join the beakers. A metallic conductor that will not react with either solution (such as platinum foil) is suspended in each solution, and the two conductors are connected with wires through a voltmeter or some other device to detect an electric current. The resultant voltaic cell generates an electric current. Indicate the reaction occurring at the anode, the reaction at the cathode, the direction of electron and ion migrations, and the signs of the electrodes.

**Solution:**  The half-reactions for the given chemical equation are

$$Cr_2O_7^{2-}(aq) + 14H^+(aq) + 6e^- \longrightarrow$$
$$2Cr^{3+}(aq) + 7H_2O(l)$$
$$6I^-(aq) \longrightarrow 3I_2(s) + 6e^-$$

The first half-reaction is the reduction process, which by definition occurs at the cathode. The second half-reaction is an oxidation, which occurs at the anode.

The $I^-$ ions are the source of electrons and the $Cr_2O_7^{2-}$ ions are the receptors. Consequently, the electrons flow through the external circuit from the electrode immersed in the KI solution (the anode) to the electrode immersed in the $K_2Cr_2O_7$–$H_2SO_4$ solution (the cathode). The electrodes themselves do not react in any way; they merely provide a means of transferring electrons from or to the solutions. The cations move through the solutions toward the cathode, while the anions move toward the anode. The anode is the negative electrode and the cathode the positive one.

### PRACTICE EXERCISE

The two half-reactions in a voltaic cell are

$$Zn(s) \longrightarrow Zn^{2+}(aq) + 2e^-$$
$$ClO_3^-(aq) + 6H^+(aq) + 6e^- \longrightarrow$$
$$Cl^-(aq) + 3H_2O(l)$$

**(a)** Indicate which reaction occurs at the anode and which at the cathode. **(b)** Which electrode is consumed in the cell reaction? **(c)** Which electrode is positive?

*Answers:*  **(a)** The first reaction occurs at the anode, the second reaction at the cathode. **(b)** The anode (Zn) is consumed in the cell reaction. **(c)** The cathode is positive.

## 20.2  CELL EMF

A voltaic cell may be thought to possess a "driving force" or "electrical pressure" that pushes electrons through the external circuit. This driving force is called the **electromotive force** (abbreviated **emf**), or cell potential. The emf, or cell potential, is measured in units of volts. One volt (V) is the emf required to impart 1 J of energy to a charge of 1 C (coulomb):

$$1 \text{ V} = 1 \frac{J}{C} \qquad [20.2]$$

The accurate determination of cell emf requires the use of special apparatus. The measurement must be made in such a manner that only a negligible current flows. If a significant current is allowed to flow, the apparent voltage of the cell is lowered as a result of the internal resistance of the cell and because of changes in concentrations around the electrodes.

The voltaic cell pictured in Figure 20.2 generates an emf of 1.10 V when operated under standard-state conditions. You will recall from Section 19.5 (see Table 19.1) that standard-state conditions include 1 $M$ concentrations for reactants and products that are in solution and 1 atm pressure for those that are gases. In the present example, the cell would be operated with $[Cu^{2+}]$ and $[Zn^{2+}]$ both at 1 $M$.* The emf generated by a cell is given the symbol $E$. Under standard-state conditions emf is called the **standard emf**, $E°$ (sometimes called *standard cell potential*):

$$Zn(s) + Cu^{2+}(aq) \longrightarrow Zn^{2+}(aq) + Cu(s) \qquad E° = 1.10 \text{ V} \qquad [20.3]$$

The emf of any voltaic cell depends on the nature of the chemical reactions taking place in the cell, the concentrations of reactants and products, and the temperature of the cell, which we shall take to be 25°C unless otherwise noted.

Just as we can think of an overall cell reaction as the sum of two half-reactions, we can think of the emf of a cell as the sum of two half-cell potentials: that due to the loss of electrons at the anode (the oxidation potential, $E_{ox}$) and that due to electron gain at the cathode (the reduction potential, $E_{red}$):

$$E_{cell} = E_{ox} + E_{red} \qquad [20.4]$$

Direct measurement of an isolated oxidation or reduction potential is impossible. However, if one half-reaction is arbitrarily assigned a standard half-cell potential, standard potentials of other half-reactions can be determined relative to that reference. The half-reaction that corresponds to reduction of $H^+$ to form $H_2$ has been chosen as the reference and assigned a standard reduction potential of exactly 0 volts:

$$2H^+(1 \text{ } M) + 2e^- \longrightarrow H_2(1 \text{ atm}) \qquad E°_{red} = 0 \text{ V} \qquad [20.5]$$

Figure 20.4 illustrates a voltaic cell constructed to use the following redox reaction between Zn and $H^+$:

$$Zn(s) + 2H^+(aq) \longrightarrow Zn^{2+}(aq) + H_2(g) \qquad [20.6]$$

Oxidation of zinc occurs in the anode compartment, and reduction of $H^+$ occurs in the cathode compartment. The standard hydrogen electrode, which is designed to operate under standard-state conditions ($[H^+] =$ 1 $M$ and $P_{H_2} = 1$ atm), consists of a platinum wire and a piece of platinum foil covered with finely divided platinum that serves as an inert surface for the cathode reaction. This electrode is encased in a glass tube so that hydrogen gas can bubble over the platinum. The voltaic cell generates a standard emf ($E°_{cell}$) of 0.76 V. By using the defined standard reduction potential of $H^+$ ($E°_{red} = 0$), we may calculate the standard oxidation potential of Zn:

$$E°_{cell} = E°_{ox} + E°_{red}$$
$$0.76 \text{ V} = E°_{ox} + 0$$

**Standard Electrode Potentials**

---

* Actually, solutions of 1 $M$ represent the standard state only if they behave ideally. Salts dissolved in an aqueous medium at concentrations on the order of 1 $M$ do not in general behave ideally; corrections need to be made to allow for nonideal behavior. However, in this introduction to electrochemistry we will not concern ourselves with these corrections.

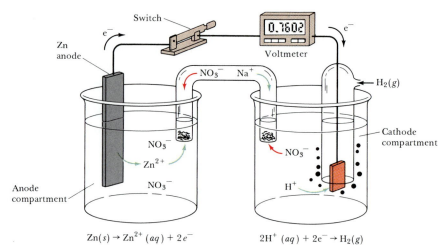

Switch

0.7802

Voltmeter

Zn
anode

$NO_3^-$   $Na^+$

$H_2(g)$

Cathode
compartment

$NO_3^-$

$Zn^{2+}$

$NO_3^-$

$NO_3^-$

$H^+$

Anode
compartment

**FIGURE 20.4**   Voltaic cell using
a standard hydrogen electrode.

$$Zn(s) \rightarrow Zn^{2+}(aq) + 2e^-$$

$$2H^+(aq) + 2e^- \rightarrow H_2(g)$$

Thus a standard oxidation potential of 0.76 V can be assigned to Zn:

$$Zn(s) \quad \longrightarrow \quad Zn^{2+}(aq) + 2e^- \qquad E^\circ_{ox} = 0.76 \text{ V} \qquad [20.7]$$

The standard potentials for other half-reactions can be established from other cell emfs in a similar fashion.

*The half-cell potential for any oxidation is equal in magnitude but opposite in sign to that of the reverse reduction.* For example,

**TABLE 20.1**   Standard electrode potentials in water at 25°C

| Standard potential (V) | Reduction half-reaction |
|---|---|
| 2.87 | $F_2(g) + 2e^- \longrightarrow 2F^-(aq)$ |
| 1.51 | $MnO_4^-(aq) + 8H^+(aq) + 5e^- \longrightarrow Mn^{2+}(aq) + 4H_2O(l)$ |
| 1.36 | $Cl_2(g) + 2e^- \longrightarrow 2Cl^-(aq)$ |
| 1.33 | $Cr_2O_7^{2-}(aq) + 14H^+(aq) + 6e^- \longrightarrow 2Cr^{3+}(aq) + 7H_2O(l)$ |
| 1.23 | $O_2(g) + 4H^+(aq) + 4e^- \longrightarrow 2H_2O(l)$ |
| 1.06 | $Br_2(l) + 2e^- \longrightarrow 2Br^-(aq)$ |
| 0.96 | $NO_3^-(aq) + 4H^+(aq) + 3e^- \longrightarrow NO(g) + 2H_2O(l)$ |
| 0.80 | $Ag^+(aq) + e^- \longrightarrow Ag(s)$ |
| 0.77 | $Fe^{3+}(aq) + e^- \longrightarrow Fe^{2+}(aq)$ |
| 0.68 | $O_2(g) + 2H^+(aq) + 2e^- \longrightarrow H_2O_2(aq)$ |
| 0.59 | $MnO_4^-(aq) + 2H_2O(l) + 3e^- \longrightarrow MnO_2(s) + 4OH^-(aq)$ |
| 0.54 | $I_2(s) + 2e^- \longrightarrow 2I^-(aq)$ |
| 0.40 | $O_2(g) + 2H_2O(l) + 4e^- \longrightarrow 4OH^-(aq)$ |
| 0.34 | $Cu^{2+}(aq) + 2e^- \longrightarrow Cu(s)$ |
| 0 | $2H^+(aq) + 2e^- \longrightarrow H_2(g)$ |
| −0.28 | $Ni^{2+}(aq) + 2e^- \longrightarrow Ni(s)$ |
| −0.44 | $Fe^{2+}(aq) + 2e^- \longrightarrow Fe(s)$ |
| −0.76 | $Zn^{2+}(aq) + 2e^- \longrightarrow Zn(s)$ |
| −0.83 | $2H_2O(l) + 2e^- \longrightarrow H_2(g) + 2OH^-(aq)$ |
| −1.66 | $Al^{3+}(aq) + 3e^- \longrightarrow Al(s)$ |
| −2.71 | $Na^+(aq) + e^- \longrightarrow Na(s)$ |
| −3.05 | $Li^+(aq) + e^- \longrightarrow Li(s)$ |

$$Zn^{2+}(aq) + 2e^- \longrightarrow Zn(s) \qquad E^\circ_{red} = -0.76 \text{ V} \qquad [20.8]$$

By convention, half-cell potentials are tabulated as standard reduction potentials, also called **standard electrode potentials**. Table 20.1 contains a selection of electrode potentials; a more complete list is found in Appendix F. These standard electrode potentials may be combined to calculate the standard emfs of a large variety of voltaic cells.

## SAMPLE EXERCISE 20.2

Given that $E^\circ_{cell}$ is 1.10 V for the Zn–Cu$^{2+}$ cell shown in Figure 20.3, and that $E^\circ_{ox}$ is 0.76 V for the oxidation of zinc (Equation 20.7), calculate $E^\circ_{red}$ for the reduction of Cu$^{2+}$ to Cu:

$$Cu^{2+}(aq) + 2e^- \longrightarrow Cu(s)$$

**Solution:**

$$E^\circ_{cell} = E^\circ_{ox} + E^\circ_{red}$$
$$1.10 \text{ V} = 0.76 \text{ V} + E^\circ_{red}$$
$$E^\circ_{red} = 1.10 \text{ V} - 0.76 \text{ V} = 0.34 \text{ V}$$

**PRACTICE EXERCISE**

A voltaic cell employs the following half-reactions:

$$In^+(aq) \longrightarrow In^{3+}(aq) + 2e^-$$
$$Br_2(l) + 2e^-(aq) \longrightarrow 2Br^-(aq)$$

The standard cell potential for this cell is 1.46 V. Using the data in Table 20.1, calculate $E^\circ_{ox}$ for the first half-cell reaction.

*Answer:* +0.40 V

## SAMPLE EXERCISE 20.3

Using standard electrode potentials tabulated in Table 20.1, calculate the standard emf for the cell described in Sample Exercise 20.1:

$$Cr_2O_7^{2-}(aq) + 14H^+(aq) + 6I^-(aq) \longrightarrow$$
$$2Cr^{3+}(aq) + 3I_2(s) + 7H_2O(l)$$

**Solution:** The standard emf of the cell, $E^\circ_{cell}$, is the sum of the standard oxidation and reduction potentials for the appropriate half-reactions:

$$E^\circ_{cell} = E^\circ_{ox} + E^\circ_{red}$$

The half-reactions and their potentials are as follows:

$$Cr_2O_7^{2-}(aq) + 14H^+(aq) + 6e^- \longrightarrow$$
$$2Cr^{3+}(aq) + 7H_2O(l) \qquad E^\circ_{red} = 1.33 \text{ V}$$

$$\underline{6I^-(aq) \longrightarrow 3I_2(s) + 6e^- \qquad E^\circ_{ox} = -0.54 \text{ V}}$$
$$Cr_2O_7^{2-}(aq) + 14H^+(aq) + 6I^-(aq) \longrightarrow$$
$$2Cr^{3+}(aq) + 7H_2O(l) + 3I_2(s) \qquad E^\circ_{cell} = 0.79 \text{ V}$$

Notice that $E^\circ_{ox}$ for I$^-$ has the opposite sign from

the reduction potential of I$_2$ listed in Table 20.1. Also notice that even though the iodide half-reaction must be multiplied by 3 in order to obtain the balanced equation for the reaction, the half-cell potential is *not* multiplied by 3. The standard potential is an intensive property; it does not depend on the quantities of reactants and products, but only on their concentrations. Thus it does not matter whether there are 6 mol of I$^-$ or 1 mol as long as the concentration of iodide is 1 $M$.

Just as the sum of the half-reactions gives the chemical equation for the overall cell reaction, the sum of the corresponding oxidation and reduction potentials gives the cell emf. In this case $E^\circ_{cell} = 0.79$ V; a positive emf is characteristic of all voltaic cells.

**PRACTICE EXERCISE**

Using Table 20.1, calculate the standard emf for a cell that employs the following overall cell reaction:

$$2Al(s) + 3I_2(s) \longrightarrow 2Al^{3+}(aq) + 6I^-(aq)$$

*Answer:* 2.20 V

In Section 13.8 we considered briefly the characteristic reactions and properties of some common oxidizing and reducing agents. The half-cell potential provides us with a tool for quantitatively expressing the ease with which a species is oxidized or reduced. *The more positive the $E^\circ$ value for a half-reaction, the greater the tendency for that reaction to occur as written.* A

**Oxidizing and Reducing Agents**

negative reduction potential indicates that the species is more difficult to reduce than $H^+(aq)$, whereas a negative oxidation potential indicates that the species is more difficult to oxidize than $H_2$. Examination of the half-reactions in Table 20.1 shows that $F_2$ is the most easily reduced species and consequently the strongest oxidizing agent listed:

$$F_2(g) + 2e^- \longrightarrow 2F^-(aq) \qquad E°_{red} = 2.87 \text{ V} \qquad [20.9]$$

Lithium ion, $Li^+$, is the most difficult to reduce and therefore the poorest oxidizing agent:

$$Li^+(aq) + e^- \longrightarrow Li(s) \qquad E°_{red} = -3.05 \text{ V} \qquad [20.10]$$

The most frequently encountered oxidizing agents are the halogens, oxygen, and oxyanions such as $MnO_4^-$, $Cr_2O_7^{2-}$, and $NO_3^-$, whose central atoms have high positive oxidation states. Metal ions in high positive oxidation states—such as $Ce^{4+}$, which is readily reduced to $Ce^{3+}$—are also employed as oxidizing agents.

Among the substances listed in Table 20.1, lithium is the most easily oxidized and consequently the strongest reducing agent:

$$Li(s) \longrightarrow Li^+(aq) + e^- \qquad E°_{ox} = 3.05 \text{ V} \qquad [20.11]$$

Fluoride ion, $F^-$, is the most difficult to oxidize and therefore the poorest reducing agent:

$$2F^-(aq) \longrightarrow F_2(g) + 2e^- \qquad E°_{ox} = -2.87 \text{ V} \qquad [20.12]$$

$H_2$ and a variety of metals having positive oxidation potentials (such as Zn and Fe) are employed as reducing agents. Some metal ions in low oxidation states—such as $Sn^{2+}$, which is oxidized to $Sn^{4+}$—also function as reducing agents. Solutions of reducing agents are difficult to store for extended periods because of the ubiquitous presence of $O_2$, a good oxidizing agent. For example, developer solutions used in photography are mild reducing agents; they have only a limited shelf life because they are readily oxidized by $O_2$ from the air.

## SAMPLE EXERCISE 20.4

Using Table 20.1, determine which of the following species is the strongest oxidizing agent: $MnO_4^-$ (in acid solution), $I_2(s)$, $Zn^{2+}(aq)$.

**Solution:**  The strongest oxidizing agent will be the species that is most readily reduced. Therefore, we should compare reduction potentials. From Table 20.1 we have

$$MnO_4^-(aq) + 8H^+(aq) + 5e^- \longrightarrow$$
$$Mn^{2+}(aq) + 4H_2O(l) \qquad E°_{red} = 1.51 \text{ V}$$

$$I_2(s) + 2e^- \longrightarrow 2I^-(aq) \qquad E°_{red} = 0.54 \text{ V}$$

$$Zn^{2+}(aq) + 2e^- \longrightarrow Zn(s) \qquad E°_{red} = -0.76 \text{ V}$$

Because the reduction of $MnO_4^-$ has the highest positive potential, $MnO_4^-$ is the strongest oxidizing agent of the three.

## PRACTICE EXERCISE

Using Table 20.1, determine which of the following species is the strongest reducing agent: $F^-(aq)$, $Zn(s)$, $I^-(aq)$.

***Answer:***  $Zn(s)$

We have observed that voltaic cells use redox reactions that proceed spontaneously. Conversely, any reaction that can occur in a voltaic cell to produce a positive emf must be spontaneous. Consequently, it is possible to decide whether a redox reaction will be spontaneous by using half-cell potentials to calculate the emf associated with it: *A positive emf indicates a spontaneous process, and a negative emf indicates a nonspontaneous one.*

### SAMPLE EXERCISE 20.5

Using the standard electrode potentials listed in Table 20.1, determine whether the following reactions are spontaneous under standard conditions:

**(a)** $Cu(s) + 2H^+(aq) \longrightarrow Cu^{2+}(aq) + H_2(g)$
**(b)** $Cl_2(g) + 2I^-(aq) \longrightarrow 2Cl^-(aq) + I_2(s)$

**Solution:** **(a)** We use Table 20.1 to obtain the necessary half-reaction potentials. Because the overall reaction converts Cu to $Cu^{2+}$, we use the standard oxidation potential for Cu. Because the overall reaction converts $H^+$ to $H_2$, we use the reduction potential for $H^+$. Adding these, we obtain the standard emf for the overall reaction:

$$Cu(s) \longrightarrow Cu^{2+}(aq) + 2e^- \qquad E^{\circ}_{ox} = -0.34 \text{ V}$$
$$2e^- + 2H^+(aq) \longrightarrow H_2(g) \qquad E^{\circ}_{red} = 0 \text{ V}$$
$$\overline{Cu(s) + 2H^+(aq) \longrightarrow Cu^{2+}(aq) + H_2(g)}$$
$$E^{\circ} = -0.34 \text{ V}$$

Because the standard emf is negative, the reaction is not spontaneous in the direction written. Copper does not react with acids in this fashion. However, the reverse reaction is spontaneous: $Cu^{2+}$ can be reduced by $H_2$.

**(b)** We write equations to obtain the standard emf for the overall reaction:

$$2e^- + Cl_2(g) \longrightarrow 2Cl^-(aq) \qquad E^{\circ}_{red} = 1.36 \text{ V}$$
$$2I^-(aq) \longrightarrow I_2(s) + 2e^- \qquad E^{\circ}_{ox} = -0.54 \text{ V}$$
$$\overline{Cl_2(g) + 2I^-(aq) \longrightarrow 2Cl^-(aq) + I_2(s)}$$
$$E^{\circ} = 0.82 \text{ V}$$

This reaction is spontaneous and could be used to build a voltaic cell. It is often used as a qualitative test for the presence of $I^-$ in aqueous solution. The solution is treated with a solution of $Cl_2$. If $I^-$ is present, $I_2$ forms. If $CCl_4$ is added, the $I_2$ dissolves in the $CCl_4$, imparting a characteristic purple color to the solution (see Figure 23.2).

**PRACTICE EXERCISE**
Using the standard electrode potentials listed in Appendix F, determine which of the following reactions are spontaneous under standard conditions:

**(a)** $I_2(s) + 5Cu^{2+}(aq) + 6H_2O(l) \longrightarrow$
$$2IO_3^-(aq) + 5Cu(s) + 12H^+(aq)$$
**(b)** $Hg^{2+}(aq) + 2I^-(aq) \longrightarrow Hg(l) + I_2(s)$
**(c)** $H_2SO_3(aq) + 2Mn(s) + 4H^+(aq) \longrightarrow$
$$S(s) + 2Mn^{2+}(aq) + 3H_2O(l)$$

***Answer:*** Reactions **(b)** and **(c)** are spontaneous.

We've seen that the free-energy change, $\Delta G$, accompanying a chemical process is a measure of its spontaneity (Chapter 19). Because the cell emf indicates whether a redox reaction is spontaneous, we might expect some relationship to exist between the cell emf and the free-energy change. Indeed, this is the case; the emf, $E$, and the free-energy change, $\Delta G$, are related by Equation 20.13:

$$\Delta G = -n\mathfrak{F}E \qquad [20.13]$$

**EMF and Free-Energy Change**

In this equation, $n$ is the number of moles of electrons transferred in the reaction and $\mathfrak{F}$ is Faraday's constant, named after Michael Faraday (Figure 20.5). Faraday's constant is the electrical charge on 1 mol of electrons. This quantity of charge is called a **faraday**:

$$1 \mathfrak{F} = 96{,}500 \frac{C}{\text{mol } e^-} = 96{,}500 \frac{J}{\text{V-mol } e^-} \qquad [20.14]$$

**FIGURE 20.5** Michael Faraday (1791–1867) was born in England, one of ten children of a poor blacksmith. At the age of 14 he was apprenticed to a bookbinder who, with uncommon leniency, gave the young man time to read and even to attend lectures. In 1812 he became an assistant in Humphry Davy's laboratory in the Royal Institution. He eventually succeeded Davy as the most famous and influential scientist in England. During his scientific career he made an amazing number of important discoveries in chemistry and physics. He developed methods for liquefying gases, discovered benzene, and formulated the quantitative relationships between electrical current and the extent of chemical reaction in electrochemical cells that either produce or use electricity. In addition, he worked out the design of the first electric generator and laid the theoretical foundations for the development of our modern theory of electricity. (Culver Pictures)

For the situation in which reactants and products are in their standard states, Equation 20.13 is modified to relate $\Delta G^\circ$ and $E^\circ$:

$$\Delta G^\circ = -n\mathfrak{F}E^\circ \qquad [20.15]$$

---

### SAMPLE EXERCISE 20.6

Use the standard electrode potentials given in Table 20.1 to calculate the standard free-energy change, $\Delta G^\circ$, for the following reaction:

$$2Br^-(aq) + F_2(g) \longrightarrow Br_2(l) + 2F^-(aq)$$

**Solution:**

$$
\begin{array}{ll}
2Br^-(aq) \longrightarrow Br_2(l) + 2e^- & E^\circ_{ox} = -1.06 \text{ V} \\
F_2(g) + 2e^- \longrightarrow 2F^-(aq) & E^\circ_{red} = \phantom{-}2.87 \text{ V} \\
\hline
2Br^-(aq) + F_2(g) \longrightarrow Br_2(l) + 2F^-(aq) & \\
\phantom{xxxxxxxxxxxxxxxxxxxxxx} E^\circ = \phantom{-}1.81 \text{ V}
\end{array}
$$

Notice that two electrons are transferred in the reaction, so $n = 2$.

$$\Delta G^\circ = -n\mathfrak{F}E^\circ$$

$$= -(2 \text{ mol } e^-)\left(96{,}500 \; \frac{\text{J}}{\text{V-mol } e^-}\right)(1.81 \text{ V})$$

$$= -3.49 \times 10^5 \text{ J} = -349 \text{ kJ}$$

Notice also that whereas a negative sign for $\Delta G^\circ$ indicates spontaneity under standard conditions, a positive sign for $E^\circ$ indicates the same.

### PRACTICE EXERCISE

Use the standard electrode potentials given in Appendix F to calculate the free-energy change for the following reaction:

$$I_2(s) + 5Cu^{2+}(aq) + 6H_2O(l) \longrightarrow$$
$$2IO_3^-(aq) + 5Cu(s) + 12H^+(aq)$$

***Answer:*** +828 kJ

---

In Chapter 19 we discussed the significance of $\Delta G$ for a chemical system. Among other things, we saw that the standard free-energy change is related to the equilibrium constant, $K$, through Equation 20.16:

$$\Delta G^\circ = -2.30RT \log K \qquad [20.16]$$

This relationship indicates that $E^\circ$ should also be related to the equilibrium constant. Substituting the relationship $\Delta G^\circ = -n\mathfrak{F}E^\circ$ (Equation 20.15) into Equation 20.16 gives:

$$-n\mathfrak{F}E^\circ = -2.30RT \log K$$

$$E^\circ = \frac{+2.30RT}{n\mathfrak{F}} \log K \qquad [20.17]$$

When $T = 298$ K, this equation can be simplified by substituting the numerical values for $R$ and $\mathfrak{F}$:

$$E^\circ = \frac{+2.30(8.314 \text{ J/K-mol})(298 \text{ K})}{n(96,500 \text{ J/V-mol})} \log K$$

$$= \frac{0.0591 \text{ V}}{n} \log K \qquad [20.18]$$

Thus the standard emf generated by a cell increases as the equilibrium constant for the cell reaction increases.

## SAMPLE EXERCISE 20.7

Using standard electrode potentials listed in Appendix F, calculate the equilibrium constant at 25°C for the reaction

$$O_2(g) + 4H^+(aq) + 4Fe^{2+}(aq) \longrightarrow$$
$$4Fe^{3+}(aq) + 2H_2O(l)$$

**Solution:**

$$O_2(g) + 4H^+(aq) + 4e^- \longrightarrow 2H_2O(l)$$
$$E^\circ_{red} = 1.23 \text{ V}$$

$$4Fe^{2+}(aq) \longrightarrow 4Fe^{3+}(aq) + 4e^-$$
$$E^\circ_{ox} = -0.77 \text{ V}$$

$$\overline{\phantom{xxxxxxxxxxxxxxxxxxxxxxxxxxxx}}$$

$$O_2(g) + 4H^+(aq) + 4Fe^{2+}(aq) \longrightarrow$$
$$4Fe^{3+}(aq) + 2H_2O(l) \quad E^\circ = 0.46 \text{ V}$$

From the half-reactions given above we see that $n = 4$. Using Equation 20.18, we have

$$\log K = \frac{nE^\circ}{0.0591 \text{ V}}$$
$$= \frac{4(0.46 \text{ V})}{0.0591 \text{ V}}$$
$$= 31.1$$
$$K = 1 \times 10^{31}$$

Thus $Fe^{2+}$ ions are stable in acidic solutions only in the absence of $O_2$ (unless a suitable reducing agent is present.)

## PRACTICE EXERCISE

Using standard electrode potentials (Appendix F), calculate the equilibrium constant at 25°C for the reaction

$$2IO_3^-(aq) + 5Cu(s) + 12H^+(aq) \longrightarrow$$
$$I_2(s) + 5Cu^{2+}(aq) + 6H_2O(l)$$

***Answer:*** $K = 1 \times 10^{145}$

In practice, voltaic cells are unlikely to be operated under standard-state conditions. However, the emf generated under nonstandard conditions can be calculated from $E°$, temperature, and the concentrations of reactants and products in the cell. The equation that permits this calculation can be derived from the relationship between $\Delta G$ and $\Delta G°$, given in Section 19.6:

$$\Delta G = \Delta G° + 2.30RT \log Q \qquad \text{[20.19]}$$

Because $\Delta G = -n\mathcal{F}E$ (Equation 20.13), we can write

$$-n\mathcal{F}E = -n\mathcal{F}E° + 2.30RT \log Q$$

Solving this equation for $E$ gives

$$E = E° - \frac{2.30RT}{n\mathcal{F}} \log Q \qquad \text{[20.20]}$$

This relationship is known as the **Nernst equation** after Walther Hermann Nernst (1864–1941), a German chemist who established a good part of the theoretical foundations of electrochemistry. At 298 K, the quantity $2.30RT/\mathcal{F}$ is equal to 0.0591 V-mol, so the Nernst equation can be written in the following simplified form:

$$E = E° - \frac{0.0591}{n} \log Q \qquad \text{[20.21]}$$

As an example of how Equation 20.21 might be used, consider the following reaction:

$$\text{Zn}(s) + \text{Cu}^{2+}(aq) \longrightarrow \text{Zn}^{2+}(aq) + \text{Cu}(s) \qquad E° = 1.10 \text{ V}$$

In this case $n = 2$, and the Nernst equation gives

$$E = 1.10 \text{ V} - \frac{0.0591 \text{ V}}{2} \log \frac{[\text{Zn}^{2+}]}{[\text{Cu}^{2+}]} \qquad \text{[20.22]}$$

Recall that $Q$ includes expressions for the species in solution but not for solids. Experimentally it is found that the emf generated by a cell is independent of the size or shape of the solid electrodes used. From Equation 20.22 it is evident that the emf of a cell based on this chemical reaction increases as $[\text{Cu}^{2+}]$ increases and as $[\text{Zn}^{2+}]$ decreases. For example, when $[\text{Cu}^{2+}]$ is 5.0 $M$ and $[\text{Zn}^{2+}]$ is 0.050 $M$, we have

$$E = 1.10 \text{ V} - \frac{0.0591 \text{ V}}{2} \log \left( \frac{0.050}{5.0} \right)$$

$$= 1.10 \text{ V} - \frac{0.0591 \text{ V}}{2} (-2.00) = 1.16 \text{ V}$$

We could have anticipated this result by applying Le Châtelier's principle

(Section 16.6). If the concentrations of reactants increase relative to the concentrations of products, the cell reaction becomes more highly spontaneous and the emf increases. Conversely, if the concentrations of products increase relative to reactants, emf decreases. As a cell operates, reactants are consumed and products form. Resultant decreases in reactant concentrations and increases in product concentrations cause the emf to decrease.

## SAMPLE EXERCISE 20.8

Calculate the emf generated by the cell described in Sample Exercise 20.1 when $[Cr_2O_7^{2-}] = 2.0\ M$, $[H^+] = 1.0\ M$, $[I^-] = 1.0\ M$, and $[Cr^{3+}] = 1.0 \times 10^{-5}\ M$:

$$Cr_2O_7^{2-}(aq) + 14H^+(aq) + 6I^-(aq) \longrightarrow$$
$$2Cr^{3+}(aq) + 3I_2(s) + 7H_2O(l)$$

**Solution:** The standard emf for this reaction was calculated in Sample Exercise 20.3: $E° = 0.79$ V. As you will see if you refer back to that exercise, $n$ is 6. The reaction quotient, $Q$, is

$$Q = \frac{[Cr^{3+}]^2}{[Cr_2O_7^{2-}][H^+]^{14}[I^-]^6} = \frac{(1.0 \times 10^{-5})^2}{(2.0)(1.0)^{14}(1.0)^6}$$
$$= 5.0 \times 10^{-11}$$

Using Equation 20.21, we have

$$E = 0.79\ \text{V} - \frac{0.0591\ \text{V}}{6}\log(5.0 \times 10^{-11})$$

$$= 0.79\ \text{V} - \frac{0.0591\ \text{V}}{6}(-10.30)$$

$$= 0.79\ \text{V} + 0.10\ \text{V}$$

$$= 0.89\ \text{V}$$

This result is qualitatively what we expect: Because the concentration of $Cr_2O_7^{2-}$ (a reactant) is above 1 $M$ and the concentration of $Cr^{3+}$ (a product) is below 1 $M$, the emf is greater than $E°$.

## PRACTICE EXERCISE

Calculate the emf generated by the cell described in the practice exercise accompanying Sample Exercise 20.3 when $[Al^{3+}] = 4.0 \times 10^{-3}\ M$ and $[I^-] = 0.010\ M$.
*Answer:* $E = 2.36$ V

## SAMPLE EXERCISE 20.9

If the measured voltage in a $Zn-H^+$ cell (such as that shown in Figure 20.4) is 0.45 V at 25°C when $[Zn^{2+}]$ is 1 $M$ and $P_{H_2}$ is 1 atm, what is the concentration of $H^+$?

**Solution:** The cell reaction is

$$Zn(s) + 2H^+(aq) \longrightarrow Zn^{2+}(aq) + H_2(g)$$

The standard emf is $E° = 0.76$ V. Applying Equation 20.21 with $n = 2$ gives

$$0.45 = 0.76 - \frac{0.0591}{2}\log\frac{[Zn^{2+}]P_{H_2}}{[H^+]^2}$$

$$= 0.76 - \frac{0.0591}{2}\log\frac{1}{[H^+]^2}$$

Since

$$\log\frac{1}{x^2} = -\log x^2 = -2\log x$$

we can write

$$0.45 = 0.76 - \frac{0.0591}{2}(-2\log[H^+])$$

Solving for $\log[H^+]$ gives us

$$\log[H^+] = \frac{0.45 - 0.76}{0.0591} = -5.25$$
$$[H^+] = 10^{-5.25} = 5.6 \times 10^{-6}\ M$$

This example shows how a voltaic cell whose cell reaction involves $H^+$ can be used to measure $[H^+]$ or pH. A pH meter (Section 17.1) is merely a specially designed voltaic cell with a voltmeter calibrated to read pH directly.

## PRACTICE EXERCISE

What is the pH of the solution in the cathode compartment of the cell pictured in Figure 20.4 when $P_{H_2} = 1$ atm, $[Zn^{2+}]$ in the anode compartment is 0.10 $M$, and cell emf is 0.542 V?
*Answer:* pH = 4.19

## 20.4 SOME COMMERCIAL VOLTAIC CELLS

Voltaic cells that receive wide use are convenient energy sources whose primary virtue is portability. Although any spontaneous redox reaction can serve as the basis of a voltaic cell, making a commercial cell that utilizes a particular redox reaction can require considerable ingenuity. The salt-bridge cells that we have been discussing provide us with considerable insight into the operation of voltaic cells. However, they are generally unsuitable for commercial use, because they have high internal resistances. This means that the flow of current within the cell, due to movement of the ions within the cell compartments and the electrolyte bridge, is restricted. Because the flow is restricted, there is resistance. As a result, if we attempt to draw a large current, voltage drops sharply. Furthermore, the cells that we have pictured so far lack the compactness and ruggedness required for portability.

Voltaic cells cannot yet compete with other common energy sources on the basis of cost alone. The cost of electricity from a common flashlight battery is on the order of $80 per kilowatt-hour. By comparison, electrical energy from power plants normally costs the consumer less than 10¢ per kilowatt-hour.

In this section we shall consider some common batteries. A battery consists of one or more voltaic cells. When the cells are connected in series (that is, with the positive terminal of one attached to the negative terminal of another), the battery produces an emf that is the sum of the emfs of the individual cells.

### Lead Storage Battery

One of the most common batteries is the lead storage battery used in automobiles. A 12-V lead storage battery consists of six cells, each producing 2 V. The anode of each cell is composed of lead; the cathode is composed of lead dioxide, $PbO_2$, packed on a metal grid. Both electrodes are immersed in sulfuric acid. The electrode reactions that occur during discharge are as follows:

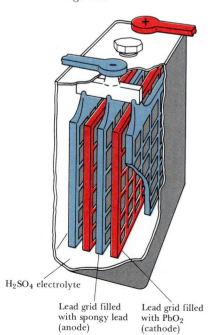

**FIGURE 20.6** Lead storage cell.

$H_2SO_4$ electrolyte

Lead grid filled with spongy lead (anode)

Lead grid filled with $PbO_2$ (cathode)

Anode: $Pb(s) + SO_4^{2-}(aq) \longrightarrow PbSO_4(s) + 2e^- \qquad E° = +0.356\ V$

Cathode: $PbO_2(s) + SO_4^{2-}(aq) + 4H^+(aq) + 2e^- \longrightarrow$
$$PbSO_4(s) + 2H_2O(l) \qquad E° = +1.685\ V$$

$$\overline{Pb(s) + PbO_2(s) + 4H^+(aq) + 2SO_4^{2-}(aq) \longrightarrow}$$
$$2PbSO_4(s) + 2H_2O(l) \qquad E° = +2.041\ V$$

[20.23]

The reactants Pb and $PbO_2$, between which electron transfer occurs, serve as the electrodes. Because the reactants are solids, there is no need to separate the cell into anode and cathode compartments; the Pb and $PbO_2$ cannot come into direct physical contact unless one electrode plate touches another. To keep the electrodes from touching, wood or glass-fiber spacers are placed between them. To increase the current output, each cell contains a number of anode and cathode plates, as shown in Figure 20.6.

The cell emf of a lead storage battery varies during use, because the concentration of $H_2SO_4$ varies with the extent of cell discharge. As Equation 20.23 indicates, $H_2SO_4$ is used up during the discharge of a lead storage battery.

One advantage of the lead storage battery is that it can be recharged. During recharging, an external source of energy is used to reverse the direction of the spontaneous redox reaction, Equation 20.23. Thus the overall process during charging is as follows:

$$2PbSO_4(s) + 2H_2O(l) \longrightarrow Pb(s) + PbO_2(s) + 4H^+(aq) + 2SO_4^{2-}(aq) \quad [20.24]$$

The energy necessary for recharging the battery is provided in an automobile by a generator that is driven by the engine. The recharging is possible because $PbSO_4$ formed during discharge adheres to the electrodes. Thus as the external source forces electrons from one electrode to another, the $PbSO_4$ is converted to Pb at one electrode and to $PbO_2$ at the other; these, of course, are the materials of a fully charged cell. If the battery is charged too rapidly, water may be decomposed to form $H_2$ and $O_2$. Besides the explosive potential of $H_2$–$O_2$ mixtures, this secondary reaction can shorten the lifetime of the battery. The evolution of these gases can dislodge Pb, $PbO_2$, or $PbSO_4$ from the plates. Solids accumulate as a sludge at the bottom of the battery. In time they may form a short circuit that renders the cell useless.

The problem of short-circuiting can be substantially reduced by adding calcium (about 0.07 percent by weight) to the lead in forming the electrodes. The presence of the calcium reduces the extent to which water is decomposed during the charging cycle. In newer batteries, electrolysis of water is sufficiently low so that the battery is "sealed"; that is, no provision is made for addition of water and escape of gases, as was necessary in the older designs.

## Dry Cell

The common dry cell is widely used in flashlights, portable radios, and the like. In fact, the dry cell is often called a flashlight battery. It is also known as the Leclanché cell after its inventor, who patented it in 1866. In the acid version, the anode consists of a zinc can that is in contact with a paste of $MnO_2$, $NH_4Cl$, and carbon. An inert cathode, consisting of a graphite rod, is immersed in the center of the paste, as shown in Figure 20.7. The cell has an exterior layer of cardboard or metal to seal the cell against the atmosphere. The electrode reactions are complex, and the cathode reaction appears to vary with the rate of discharge. The reactions at the electrodes are generally represented as shown below:

Anode: $\qquad Zn(s) \longrightarrow Zn^{2+}(aq) + 2e^- \qquad [20.25]$

Cathode: $\quad 2NH_4^+(aq) + 2MnO_2(s) + 2e^- \longrightarrow$
$$Mn_2O_3(s) + 2NH_3(aq) + H_2O(l) \qquad [20.26]$$

Only a fraction of the cathode material, that near the electrode, is electrochemically active because of the limited mobility of the chemicals in the cell.

In the alkaline version of the dry cell, $NH_4Cl$ is replaced by KOH. The anode reaction still involves oxidation of Zn, but the zinc is present as a powder, mixed with the electrolyte in a gel formulation. As in the common dry cell, the cathode reaction involves reduction of $MnO_2$. Figure 20.8 shows a cutaway view of a miniature alkaline cell, of the type

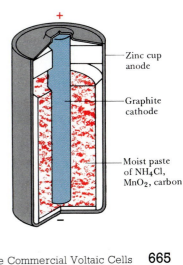

**FIGURE 20.7**   Cutaway view of a dry cell.

+

Zinc cup anode

Graphite cathode

Moist paste of $NH_4Cl$, $MnO_2$, carbon

−

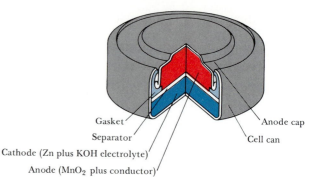

Gasket
Separator
Cathode (Zn plus KOH electrolyte)
Anode (MnO$_2$ plus conductor)
Anode cap
Cell can

**FIGURE 20.8** Cutaway view of a miniature alkaline Zn–MnO$_2$ dry cell.

used in camera exposure controls, calculators, and some watches. Although more costly than the common dry cell, alkaline cells provide improved performance. They maintain usable voltage over a longer fraction of consumption of anode and cathode materials, and they provide up to 50 percent more total energy than a common dry cell of the same size.

## Nickel-Cadmium Batteries

Because dry cells are not rechargeable, they have to be replaced frequently. Thus a rechargeable cell, the nickel-cadmium battery, has become increasingly popular, especially for use in battery-operated tools and calculators. Cadmium metal acts as the anode. NiO$_2$(s) serves as cathode; it is reduced to Ni(OH)$_2$(s). The electrode reactions occurring within this cell during discharge are as follows:

Anode: $$Cd(s) + 2OH^-(aq) \longrightarrow Cd(OH)_2(s) + 2e^-$$

[20.27]

Cathode: $$NiO_2(s) + 2H_2O(l) + 2e^- \longrightarrow Ni(OH)_2(s) + 2OH^-(aq)$$

[20.28]

As in the lead storage cell, the reaction products adhere to the electrodes. This permits the reactions to be readily reversed during charging. Because no gases are produced during either charging or discharging, the battery can be sealed.

## Fuel Cells

Many substances are used as fuels. The thermal energy released by combustion is often converted to electrical energy. The heat may convert water to steam, which drives a turbine that in turn drives the generator. Typically a maximum of only 40 percent of the energy from combustion is converted to electricity; the remainder is lost as heat. The *direct* production of electricity from fuels via a voltaic cell could, in principle, yield a higher rate of conversion of the chemical energy of the reaction. Voltaic cells that perform this conversion using conventional fuels, such as H$_2$ and CH$_4$, are called **fuel cells**.

A great deal of research has gone into attempts to develop practical fuel cells. One of the problems encountered is the high operating temperatures of most cells, which not only siphons off energy but also accelerates corrosion of cell parts. A low-temperature cell has been developed that uses H$_2$, but the present cost of the cell makes it too expensive for large-scale use. However, it has been used in special situations, such as

All of the batteries we have discussed until now have contained an aqueous solution or water-based paste as the electrolyte medium connecting the two half-cells. In recent years much research effort has been devoted to developing batteries in which the electrolyte, or ion carrier, within the battery is a solid. A solid that is capable of conducting a current through the motion of ions within it is called a **solid electrolyte** or **fast-ion conductor**.

The existence of solid ionic conductors has been known for a long time. Figure 20.9 shows schematically an experiment performed in 1910 by the German scientist C. Tubandt. It was known that when a pair of electrodes is placed in contact with solid silver iodide, AgI, the solid readily conducts a current. However, it was not known whether the current was carried by electrons moving into and through the solid or by the movement of ions. Tubandt placed Ag electrodes of known mass on each end of a block of AgI. He allowed current to flow for a period of time and measured the total number of faradays of electricity passed. At the end of the experiment he weighed the two Ag electrodes. One had gained weight, the other had lost an equal amount of weight. The weight changes were consistent with the reduction of $Ag^+$ at the cathode, adding to the weight of the cathode, and with the oxidation of Ag at the anode, corresponding to the weight loss at this electrode. For these electrode processes to occur, $Ag^+$ ions produced at the anode must travel through the solid to the cathode to be reduced there. Thus, AgI is an example of a solid electrolyte.

Figure 20.10 shows a schematic diagram of a battery incorporating a solid electrolyte. The anode reaction is the oxidation of lithium:

$$Li(s) \longrightarrow Li^+(s) + e^-$$

The cathode reaction is the reduction of $TiS_2$:

$$TiS_2(s) + e^- \longrightarrow TiS_2^-(s)$$

The cell potential for these reactions is greater than 2 V. Electrons released at the anode flow through the external circuit to the cathode. The cathode electrode, $TiS_2$, is a solid electrolyte of a special type,

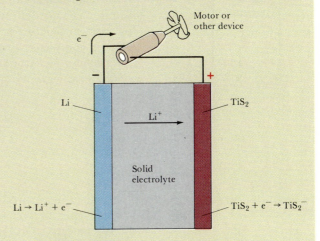

**FIGURE 20.10** A schematic drawing of a solid electrolyte battery. Lithium metal is the anode, $TiS_2$ is the cathode. Lithium ions are conducted within the battery by movement through a polymer solid electrolyte and through the cathode material, where $LiTiS_2$ is formed.

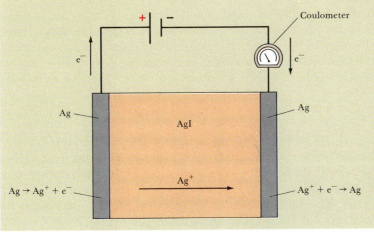

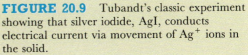

**FIGURE 20.9** Tubandt's classic experiment showing that silver iodide, AgI, conducts electrical current via movement of $Ag^+$ ions in the solid.

called an **ion-insertion compound**. It is a good electronic conductor as well as an ionic conductor. When $TiS_2$ is reduced, lithium ions move into the electrode from the solid electrolyte to form $LiTiS_2$:*

$$Li^+(s) + TiS_2(s) + e^- \longrightarrow LiTiS_2(s)$$

The solid electrolyte separating the anode and cathode must be capable of conducting $Li^+$ ions but not

* This description is a bit oversimplified chemically, but the principle is correct.

conducting electrons. (Conducting electrons would create an internal short circuit of the external path.) The most promising materials in use today are solid polymers (molecules of high molecular weight) that permit the flow of ions through the material.

A battery such as the one outlined in Figure 20.10 could be recharged in the same way as a nickel-cadmium or lead-acid cell. The reactions for recharging the half-cells are the reverse of the above reactions. The big advantage expected for the solid electrolyte batteries over other cells is that they will have from two to five times as much energy per unit weight and volume.

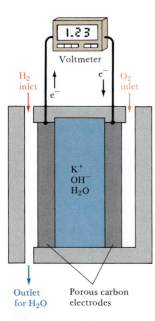

FIGURE 20.11 Cross section of a $H_2$–$O_2$ fuel cell.

in space vehicles. For example, a $H_2$–$O_2$ fuel cell was used as the primary source of electrical energy on the *Apollo* moon flights. The weight of the fuel cell sufficient for 11 days in space was approximately 500 lb. This may be compared to the several tons that would have been required for an engine-generator set.

The electrode reactions in the $H_2$–$O_2$ fuel cell are as follows:

| | |
|---|---|
| Anode: | $2H_2(g) + 4OH^-(aq) \longrightarrow 4H_2O(l) + 4e^-$ |
| Cathode: | $4e^- + O_2(g) + 2H_2O(l) \longrightarrow 4OH^-(aq)$ |

$$2H_2(g) + O_2(g) \longrightarrow 2H_2O(l) \qquad [20.29]$$

The cell is illustrated in Figure 20.11. The electrodes are composed of hollow tubes of porous, compressed carbon impregnated with catalyst; the electrolyte is KOH. Because the reactants are supplied continuously, a fuel cell does not "go dead."

## 20.5 ELECTROLYSIS AND ELECTROLYTIC CELLS

We have seen that spontaneous oxidation-reduction reactions are used as the basis for voltaic cells, electrochemical devices that generate electricity. Conversely, it is possible to use electrical energy to cause nonspontaneous oxidation-reduction reactions to occur. For example, electricity can be used to decompose molten sodium chloride into its component elements:

$$2NaCl(l) \longrightarrow 2Na(l) + Cl_2(g)$$

Such processes, which are driven by an outside source of electrical energy, are called **electrolysis** reactions and take place in **electrolytic cells**.

An electrolytic cell consists of two electrodes in a molten salt or aqueous solution. The cell is driven by a battery or some other source of direct electrical current. The battery acts as an electron pump, pushing electrons into one electrode and pulling them from the other. Withdrawing

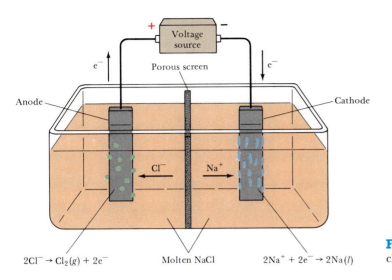

Anode

Cathode

$2Cl^- \rightarrow Cl_2(g) + 2e^-$

Molten NaCl

$2Na^+ + 2e^- \rightarrow 2Na(l)$

**FIGURE 20.12** Electrolysis of molten sodium chloride.

electrons from an electrode gives it a positive charge, and adding electrons to an electrode makes it negative. In the electrolysis of molten NaCl, shown in Figure 20.12, $Na^+$ ions pick up electrons at the negative electrode and are thereby reduced. As the $Na^+$ ions in the vicinity of this electrode are depleted, additional $Na^+$ ions migrate in. In a related fashion, there is a net movement of $Cl^-$ ions to the positive electrode, where they give up electrons and are thereby oxidized. Just as in voltaic cells, the electrode at which reduction occurs is called the cathode, and the electrode at which oxidation occurs is called the anode.

Anode:     $2Cl^-(l) \longrightarrow Cl_2(g) + 2e^-$

Cathode:   $\dfrac{2Na^+(l) + 2e^- \longrightarrow 2Na(l)}{2Na^+(l) + 2Cl^-(l) \longrightarrow 2Na(l) + Cl_2(g)}$     [20.30]

Notice that the sign convention for the electrodes in an electrolytic cell is just the opposite of that for a voltaic cell. The cathode in the electrolytic cell is negative, because electrons are being forced onto it by the external voltage source. The anode is positive because electrons are being withdrawn by the external source.

Electrolyses of molten salts and molten salt solutions for the production of active metals such as sodium and aluminum are important industrial processes. We shall have more to say about them in Chapter 24, when processes for obtaining metals are discussed.

Sodium cannot be prepared by electrolysis of aqueous solutions of NaCl, because water is more easily reduced than $Na^+(aq)$:

**Electrolysis
of Aqueous Solutions**

$$2H_2O(l) + 2e^- \longrightarrow H_2(g) + 2OH^-(aq) \qquad E^\circ_{red} = -0.83 \text{ V}$$

$$Na^+(aq) + e^- \longrightarrow Na(s) \qquad E^\circ_{red} = -2.71 \text{ V}$$

Consequently, $H_2$ is produced at the cathode.

The possible anode reactions are the oxidation of $Cl^-$ and of $H_2O$:

$$2Cl^-(aq) \longrightarrow Cl_2(g) + 2e^- \qquad E^\circ_{ox} = -1.36 \text{ V}$$

$$2H_2O(l) \longrightarrow 4H^+(aq) + O_2(g) + 4e^- \qquad E^\circ_{ox} = -1.23 \text{ V}$$

These standard oxidation potentials are not greatly different, but they do suggest that $H_2O$ should be oxidized more readily than $Cl^-$. However, the voltage required for a reaction is sometimes much greater than that indicated by the electrode potentials. The additional voltage required to cause electrolysis is called the **overvoltage**. The overvoltage is believed to be caused by slow reaction rates at the electrodes. Overvoltages for the deposition of metals are low, but those required for the liberation of hydrogen gas or oxygen gas are usually high. In the present case the overvoltage for $O_2$ formation is sufficiently high to permit oxidation of $Cl^-$ rather than $H_2O$. Consequently, electrolysis of aqueous solutions of NaCl, known as brines, produces $H_2$ and $Cl_2$ unless the concentration of $Cl^-$ is quite low:

Anode: $\qquad\qquad\qquad 2Cl^-(aq) \longrightarrow Cl_2(g) + 2e^-$

Cathode: $\qquad\qquad 2H_2O(l) + 2e^- \longrightarrow H_2(g) + 2OH^-(aq)$

$$\overline{2Cl^-(aq) + 2H_2O(l) \longrightarrow Cl_2(g) + H_2(g) + 2OH^-(aq)}$$

$$[20.31]$$

The $Na^+$ ion is merely a spectator ion (Section 13.1) in the electrolysis. This process is used commercially because all of the products ($H_2$, $Cl_2$, and NaOH) are commercially important chemicals.

Electrode potentials can be used to determine the minimum emf required for an electrolysis. In the case of the formation of $H_2$ and $Cl_2$ from a brine solution under standard conditions, a minimum emf of 2.19 V is required.

$$E^\circ = E^\circ_{ox}(Cl^-) + E^\circ_{red}(H_2O)$$
$$= -1.36 \text{ V} + (-0.83 \text{ V}) = -2.19 \text{ V}$$

The emf calculated above is negative, reminding us that the process is not spontaneous but must be driven by an outside source of energy. Higher voltages than those calculated are invariably needed. One reason is the internal resistance of the cell; another is the overvoltage phenomenon discussed above.

## SAMPLE EXERCISE 20.10

Explain why the electrolysis of an aqueous solution of $CuCl_2$ produces $Cu(s)$ and $Cl_2(g)$. What is the minimum emf required for this process under standard conditions?

**Solution:** At the cathode we can envision reduction of either $Cu^{2+}$ or $H_2O$:

$$Cu^{2+}(aq) + 2e^- \longrightarrow Cu(s) \qquad E^\circ_{red} = 0.34 \text{ V}$$

$$2H_2O(l) + 2e^- \longrightarrow H_2(g) + 2OH^-(aq)$$
$$E^\circ_{red} = -0.83 \text{ V}$$

The electrode potentials indicate that reduction of $Cu^{2+}$ occurs more readily. Reduction of $H_2O$ is made even more difficult because of the overvoltage for $H_2$ formation.

At the anode, we can envision the oxidation of either $Cl^-$ or $H_2O$. As in the case of NaCl solutions,

$Cl_2$ is usually produced because of the overvoltage for $O_2$ formation.

The minimum emf required for this electrolysis under standard conditions is 1.02 V:

$$E° = E°_{red}(Cu^{2+}) + E°_{ox}(Cl^-)$$
$$= 0.34\ V + (-1.36\ V) = -1.02\ V$$

**PRACTICE EXERCISE** ——————
**(a)** What are the expected products of electrolysis of a 1.0 $M$ aqueous HBr solution? **(b)** What is the minimum emf required to produce these products?

***Answers:*** **(a)** at the cathode, $H_2(g)$; at the anode, $Br_2(l)$; **(b)** $-1.06$ V

In our discussion of the electrolysis of molten NaCl and of NaCl solutions, we considered the electrodes to be inert. Consequently, the electrodes did not undergo reaction but merely served as the surface at which oxidation and reduction of solvent or solute occurred. However, often the electrodes themselves participate in the electrolysis process.

When aqueous solutions are electrolyzed using metal electrodes, the electrode will be oxidized if its oxidation potential is more positive than that for water. For example, nickel is oxidized more readily than water:

$$Ni(s) \longrightarrow Ni^{2+}(aq) + 2e^- \qquad E°_{ox} = +0.28\ V$$
$$2H_2O(l) \longrightarrow 4H^+(aq) + O_2(g) + 4e^- \qquad E°_{ox} = -1.23\ V$$

If nickel is made the anode in an electrolytic cell, nickel metal is oxidized as the anode reaction. If $Ni^{2+}(aq)$ is present in the solution, it is reduced at the cathode in preference to reduction of water. An electrolytic cell of this kind is illustrated in Figure 20.13. As current flows, nickel dissolves from the anode and deposits on the cathode:

| | | | |
|---|---|---|---|
| Anode: | $Ni(s) \longrightarrow Ni^{2+}(aq) + 2e^-$ | | [20.32] |
| Cathode: | $Ni^{2+}(aq) + 2e^- \longrightarrow Ni(s)$ | | [20.33] |

Electrolytic processes involving active metal electrodes—that is, in which the metal electrodes participate in the cell reaction—have several important applications. We will see in Chapter 24 that electrolysis affords a means of purifying crude metals such as copper, zinc, cobalt, and nickel. An additional important application is in **electroplating**, in which one

**Electrolysis with Active Electrodes**

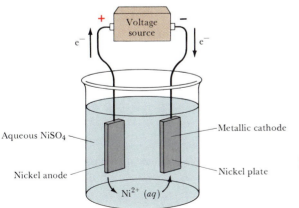

**FIGURE 20.13** Electrolytic cell with an active metal electrode. Nickel dissolves from the anode to form $Ni^{2+}(aq)$. At the cathode $Ni^{2+}(aq)$ is reduced and forms a nickel "plate."

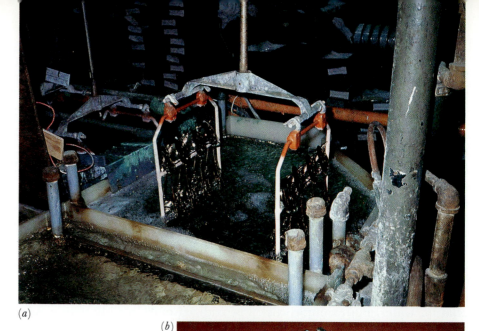

(a)

(b)

**FIGURE 20.14** Electroplating of silverware. Part (a) shows the silverware being withdrawn from the electroplating bath. In part (b) we see the polished final product. (Courtesy Oneida Silversmiths)

metal is "plated," or deposited, on another. Electroplating is used to protect objects against corrosion and to improve their appearance. For example, fine dinnerware is silver-plated electrochemically by making the dinnerware utensils the cathode in an electrolytic plating bath (Figure 20.14).

## 20.6 QUANTITATIVE ASPECTS OF ELECTROLYSIS

The quantity of chemical reaction occurring in an electrolytic cell is directly proportional to the quantity of electricity passed into the cell. For example, 1 mol of electrons will plate out 1 mol of Na metal, and 2 mol of electrons will plate out 2 mol of Na metal:

$$Na^+ + e^- \longrightarrow Na$$

Similarly, it requires 2 mol of electrons to produce 1 mol of copper from $Cu^{2+}$ and 3 mol of electrons to produce 1 mol of aluminum from $Al^{3+}$:

$$Cu^{2+} + 2e^- \longrightarrow Cu$$
$$Al^{3+} + 3e^- \longrightarrow Al$$

The quantity of charge passing through an electrical circuit, such as that in an electrolytic cell, is generally measured in **coulombs**. As noted in Section 20.3, there are 96,500 C in a faraday:

$$1 \, \mathfrak{F} = 96{,}500 \text{ C} = \text{charge of 1 mol of electrons} \qquad [20.34]$$

In terms of other, perhaps more familiar electrical units, a coulomb is the quantity of electrical charge passing a point in a circuit in 1 s when the current is 1 ampere (A).* Therefore, the number of coulombs passing through a cell can be obtained by multiplying the amperage and the elapsed time in seconds:

$$\text{Coulombs} = \text{amperes} \times \text{seconds} \qquad [20.35]$$

These ideas are applied in Sample Exercise 20.11. Although the exercise involves electrolytic cells, the same relationships can be applied to voltaic cells.

---

### SAMPLE EXERCISE 20.11

Calculate the amount of aluminum produced in 1.00 h by the electrolysis of molten $AlCl_3$ if the current is 10.0 A.

**Solution:**  Using Equation 20.35, we can write

$$\text{Coulombs} = (10.0 \text{ A})(1.00 \text{ h})\left(\frac{3600 \text{ s}}{1 \text{ h}}\right)\left(\frac{1 \text{ C}}{1 \text{ A-s}}\right)$$

$$= 3.60 \times 10^4 \text{ C}$$

The half-reaction for the reduction of $Al^{3+}$ is

$$Al^{3+} + 3e^- \longrightarrow Al$$

The amount of aluminum produced depends on the number of available electrons: 1 mol $Al \rightleftharpoons 3 \, \mathfrak{F}$. We can therefore write

$$\text{Grams Al} = (3.60 \times 10^4 \text{ C})\left(\frac{1 \, \mathfrak{F}}{96{,}500 \text{ C}}\right)$$

$$\times \left(\frac{1 \text{ mol Al}}{3 \, \mathfrak{F}}\right)\left(\frac{27.0 \text{ g Al}}{1 \text{ mol Al}}\right)$$

$$= 3.36 \text{ g}$$

### PRACTICE EXERCISE

The half-reaction for formation of magnesium metal upon electrolysis of molten $MgCl_2$ is $Mg^{2+} + 2e^- \longrightarrow Mg$. Calculate the amount of magnesium formed upon passage of a current of 60.0 A for a period of $4.00 \times 10^3$ s.
***Answer:***   30.2 g of Mg

---

This is a good point at which to consider the relationship between electrochemical processes and work. We have already seen that a positive value for $E$ is associated with a negative value for the free-energy change, and thus with a spontaneous process. We also know that for any spontaneous process $\Delta G$ is a measure of the maximum useful work, $w_{max}$, that can be extracted from the process: $\Delta G = w_{max}$ (Section 19.7). Since $\Delta G = -n\mathfrak{F}E$, the maximum useful electrical work obtainable from a voltaic cell should be simply

**Electrical Work**

$$w_{max} = -n\mathfrak{F}E \qquad [20.36]$$

---

* Conversely, current is the rate of flow of electricity. An ampere is the current associated with the flow of 1 C past a point each second.

A word about sign conventions is in order here. Remember that work done *by* the system *on* its surroundings is indicated by a negative sign for $w$ (see Section 4.2). Thus a negative value for $w_{max}$ corresponds to a spontaneous process, for which $E$ is positive.

Notice that the maximum work obtainable is proportional to the cell potential $E$. We can think of $E$ as a kind of pressure, a measure of the driving force for the process. Recall from Section 20.2 that the units of $E$ are J/C; $w_{max}$ is also proportional to the number of coulombs that flow, measured by $n\mathcal{F}$. When we put these quantities together we can see how the units cancel to leave us with units of energy:

$$w_{max} = -n \times \mathcal{F} \times E$$
$$\text{(J)} = \text{(mol)} \times \left(\frac{\text{C}}{\text{mol}}\right) \times \left(\frac{\text{J}}{\text{C}}\right)$$

In an electrolytic cell we employ an external source of energy to cause a nonspontaneous electrochemical process to occur. In this case, $\Delta G$ for the cell process is positive and the cell potential is negative. The *minimum* amount of work done on the system to cause the nonspontaneous cell reaction to occur, $w_{min}$, is given by

$$w_{min} = -n\mathcal{F}E \qquad [20.37]$$

where $E$ is the calculated cell potential. In practice, we always need to expend more than this minimum amount of work, because of inefficiencies in the process. In addition, a higher potential $E'$ may be required to cause the cell reaction to occur, because of overvoltage. In this case the minimum work required is given by

$$w_{min} = -n\mathcal{F}E' \qquad [20.38]$$

Electrical work is usually expressed in energy units of watts times time. The **watt** is a unit of electrical power, that is, the rate of energy expenditure:

$$1 \text{ watt (W)} = \frac{1 \text{ J}}{\text{s}} \qquad [20.39]$$

Thus a watt-second is a joule. The unit employed by electric utilities is the kilowatt-hour, which works out to be $3.6 \times 10^6$ J:

$$1 \text{ kWh} = (1000 \text{ W})(1 \text{ h})\left(\frac{3600 \text{ s}}{1 \text{ h}}\right)\left(\frac{1 \text{ J/s}}{1 \text{ W}}\right) = 3.6 \times 10^6 \text{ J} \qquad [20.40]$$

Using these considerations, we can calculate the maximum work obtainable from voltaic cells and the minimum work required to bring about desired electrolysis reactions.

## SAMPLE EXERCISE 20.12

Calculate the minimum number of kilowatt-hours of electricity required to produce 1000 kg of aluminum by electrolysis of $Al^{3+}$ if the required emf is 4.5 V.

**Solution:** We need to employ Equation 20.38 to calculate $w_{min}$ for an applied potential of 4.5 V. First we need to calculate $n\mathfrak{F}$, the number of coulombs required.

$$\text{Coulombs} = (1000 \text{ kg Al})\left(\frac{1000 \text{ g Al}}{1 \text{ kg Al}}\right)\left(\frac{1 \text{ mol Al}}{27.0 \text{ g Al}}\right)$$
$$\times \left(\frac{3 \mathfrak{F}}{1 \text{ mol Al}}\right)\left(\frac{96,500 \text{ C}}{1 \mathfrak{F}}\right)$$
$$= 1.07 \times 10^{10} \text{ C}$$

We can now employ Equation 20.38 to calculate $w_{min}$. In doing so we must apply the unit conversion factor of Equation 20.40:

$$\text{Kilowatt-hours} = (1.07 \times 10^{10} \text{ C})(4.5 \text{ V})$$
$$\times \left(\frac{1 \text{ J}}{1 \text{ C-V}}\right)\left(\frac{1 \text{ kWh}}{3.6 \times 10^6 \text{ J}}\right)$$
$$= 1.34 \times 10^4 \text{ kWh}$$

(The negative sign on the right in Equation 20.38 is canceled by the negative sign for $E'$, because it is a voltage applied to make the cell reaction occur.)

This quantity of energy does not include the energy used to mine, transport, and process the aluminum ore, and to keep the electrolysis bath molten during electrolysis. A typical electrolytic cell used to reduce aluminum is only 40 percent efficient, 60 percent of the electrical energy being dissipated as heat. It therefore requires on the order of 33 kWh of electricity to produce 1 kg of aluminum. The aluminum industry consumes about 2 percent of the electrical energy generated in the United States. Because this is used mainly for reduction of aluminum, recycling this metal saves large quantities of energy.

### PRACTICE EXERCISE

Calculate the minimum number of kilowatt-hours of electricity required to produce 1.00 kg of Mg from electrolysis of molten $MgCl_2$, if the applied emf is 5.0 V.
*Answer:* 11.0 kWh

## SAMPLE EXERCISE 20.13

A 12-V lead storage battery contains 410 g of lead in its anode plates and a stoichiometrically equivalent amount of $PbO_2$ in the cathodes. **(a)** What is the maximum number of coulombs of electrical charge it can deliver without being recharged? **(b)** For how many hours could the battery deliver a steady current of 1.0 A assuming that the current does not fall during discharge? **(c)** What is the maximum electrical work that the battery can accomplish in kilowatt-hours?

**Solution:** **(a)** The lead anode undergoes a two-electron oxidation:

$$\text{Pb} \longrightarrow \text{Pb}^{2+} + 2e^-$$

Consequently, $2 \mathfrak{F} \backsimeq 1$ mol Pb. Using this relationship, we have

$$\text{Coulombs} = (410 \text{ g Pb})\left(\frac{1 \text{ mol Pb}}{207 \text{ g Pb}}\right)$$
$$\times \left(\frac{2 \mathfrak{F}}{1 \text{ mol Pb}}\right)\left(\frac{96,500 \text{ C}}{1 \mathfrak{F}}\right)$$
$$= 3.8 \times 10^5 \text{ C}$$

Although the emf of a cell is independent of the masses of the solid reactants involved in the cell, the total electrical charge the cell can deliver does depend on these quantities. The size and surface area further affects the *rate* at which electrical charge can be delivered.

**(b)** We calculate the number of hours of operation at a current level of 1.0 A by recalling that a coulomb corresponds to a current of 1 A flowing for 1 s:

$$\text{Hours} = (3.8 \times 10^5 \text{ C})\left(\frac{1 \text{ A-s}}{1 \text{ C}}\right)\left(\frac{1 \text{ h}}{3600 \text{ s}}\right)\left(\frac{1}{1.0 \text{ A}}\right)$$
$$= 1.1 \times 10^2 \text{ h}$$

This battery might be described as a 110 amp-hour battery.

**(c)** The maximum work is given by the product $-n\mathfrak{F}E$, Equation 20.36:

$$\text{Kilowatt-hours} = -(3.8 \times 10^5 \text{ C})(12 \text{ V})$$
$$\times \left(\frac{1 \text{ J}}{1 \text{ C-V}}\right)\left(\frac{1 \text{ kWh}}{3.6 \times 10^6 \text{ J}}\right)$$
$$= -1.3 \text{ kWh}$$

(The negative sign indicates merely that the system does work on its surroundings.)

### PRACTICE EXERCISE

A "deep discharge" lead-acid battery for marine use is advertised as having an 80 ampere-hour capacity. What amount of Pb would be oxidized if this battery were discharged so as to consume 80 percent of its capacity?
*Answer:* 247 g of Pb

## 20.7 CORROSION

Before we close our discussion of electrochemistry, let's apply some of what we have learned to a very important problem, the **corrosion** of metals. Corrosion reactions are redox reactions in which a metal is attacked by some substance in its environment and converted to an unwanted compound.

All metals except gold and platinum are thermodynamically capable of undergoing oxidation in air at room temperature. When the oxidation process is not inhibited in some way, it can be very destructive. However, oxidation can result in formation of an insulating, protective oxide layer that prevents further reaction of the underlying metal. On the basis of the standard oxidation potential for aluminum ($E_{ox}^{\circ} = 1.66$ V), we would expect aluminum to be very readily oxidized. The many aluminum soft drink and beer cans that litter the environment are ample evidence, however, that aluminum undergoes only very slow chemical corrosion. The exceptional stability of this active metal in air is due to formation of a thin, protective coat of oxide—a hydrated form of $Al_2O_3$—on the surface of the metal. The oxide coat is impermeable to the passage of $O_2$ or $H_2O$, and so protects the underlying metal from further corrosion. Magnesium, which also has a high oxidation potential, is similarly protected. Some metal alloys, such as stainless steel and nichrome (Section 24.7), also form protective, impervious oxide coats.

### Corrosion of Iron

One of the most familiar corrosion processes is the rusting of iron (Figure 20.15). From an economic standpoint this is a significant process. It is estimated that up to 20 percent of the iron produced annually in this country is used to replace iron objects that have been discarded because of rust damage.

The rusting of iron is known to require oxygen; iron does not rust in water unless $O_2$ is present. Rusting also requires water; iron does not rust in oil, even if it contains $O_2$, unless $H_2O$ is also present. Other factors—such as the pH of the solution, the presence of salts, contact with metals more difficult to oxidize than iron, and stress on the iron—can accelerate rusting.

The corrosion of iron is generally believed to be electrochemical in nature. A region on the surface of the iron serves as an anode at which the iron undergoes oxidation:

$$Fe(s) \longrightarrow Fe^{2+}(aq) + 2e^- \qquad E_{ox}^{\circ} = 0.44 \text{ V} \qquad [20.41]$$

The electrons so produced migrate through the metal to another portion of the surface that serves as the cathode. Here oxygen can be reduced:

$$O_2(g) + 4H^+(aq) + 4e^- \longrightarrow 2H_2O(l) \qquad E_{red}^{\circ} = 1.23 \text{ V} \qquad [20.42]$$

Notice that $H^+$ takes part in the reduction of $O_2$. As the concentration of $H^+$ is lowered (that is, as pH is increased), the reduction of $O_2$ becomes less favorable. It is observed that iron in contact with a solution whose pH is above 9 does not corrode. In the course of the corrosion, the $Fe^{2+}$

**FIGURE 20.15** A "tin" can (actually a steel can which originally had a thin tin plate), found on the beach of Sanibel Island, Florida. In contact with an aqueous salt solution such as seawater, iron and most forms of steel corrode very rapidly, forming hydrated iron(III) oxide, or rust. (Donald Clegg and Roxy Wilson)

formed at the anode is further oxidized to $Fe^{3+}$. The $Fe^{3+}$ forms the hydrated iron(III) oxide known as rust:*

$$4Fe^{2+}(aq) + O_2(g) + 4H_2O(l) + 2xH_2O(l) \longrightarrow 2Fe_2O_3 \cdot xH_2O(s) + 8H^+(aq) \qquad [20.43]$$

Because the cathode is generally the area having the largest supply of $O_2$, the rust often deposits there. If you look closely at a shovel after it has stood outside in the moist air with wet dirt adhered to its blade, you may notice that pitting has occurred under the dirt but that rust has formed elsewhere, where $O_2$ is more readily available. The corrosion process is summarized by way of illustration in Figure 20.16.

The enhanced corrosion caused by the presence of salts is usually quite evident on autos in areas where heavy salting of roads occurs during winter. The effect of salts is readily explained by the voltaic mechanism: The ions of a salt provide the electrolyte necessary for completion of the electrical circuit.

The presence of anodic and cathodic sites on the iron requires two different chemical environments on the surface. These can occur through the presence of impurities or lattice defects (perhaps introduced by strain on the metal). At the sites of such impurities or defects, the atomic-level environment around the iron atom may permit the metal to be either more or less oxidized than at normal lattice sites. Thus these sites may serve as either anodes or cathodes. Ultrapure iron, prepared in such a way as to minimize lattice defects, is far less susceptible to corrosion than is ordinary iron.

Iron is often covered with a coat of paint or another metal such as tin, zinc, or chromium to protect its surface against corrosion. Sheet steel used in beverage and food cans can be coated by dipping the sheets in molten tin or by depositing a thin coat (1 to 20 $\mu$m) of tin electrochemically. The tin protects the iron only as long as the protective layer remains

---

* Frequently, metal compounds that are obtained from aqueous solution have water associated with them. For example, copper(II) sulfate crystallizes from water with 5 mol of water per mole of $CuSO_4$. We represent this formula as $CuSO_4 \cdot 5H_2O$. Such compounds are called hydrates (Section 12.2). Rust is a hydrate of iron(III) oxide with a variable amount of water of hydration. We represent the variable water content by writing the formula as $Fe_2O_3 \cdot xH_2O$.

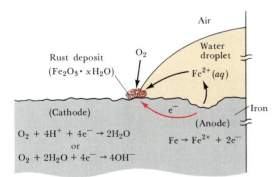

**FIGURE 20.16** Corrosion of iron in contact with water.

intact. Once it is broken and the iron exposed to air and water, tin actually promotes the corrosion of the iron. It does so by serving as the cathode in the electrochemical corrosion. As shown by the following half-cell potentials, iron is more readily oxidized than tin:

$$Fe(s) \longrightarrow Fe^{2+}(aq) + 2e^- \qquad E^\circ_{ox} = 0.44 \text{ V} \qquad [20.44]$$

$$Sn(s) \longrightarrow Sn^{2+}(aq) + 2e^- \qquad E^\circ_{ox} = 0.14 \text{ V} \qquad [20.45]$$

The iron therefore serves as the anode and is oxidized as shown in Figure 20.17.

"Galvanized iron" is produced by coating iron with a thin layer of zinc. The zinc protects the iron against corrosion even after the surface coat is broken. In this case the iron serves as the cathode in the electrochemical corrosion, because zinc is oxidized more easily than iron:

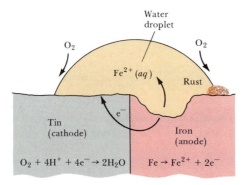

**FIGURE 20.17**  Corrosion of iron in contact with tin.

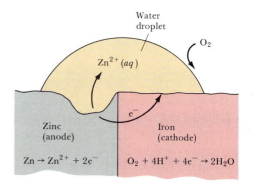

**FIGURE 20.18**  Cathodic protection of iron in contact with zinc.

**FIGURE 20.19**  Cathodic protection of an iron water pipe. The magnesium anode is surrounded by a mixture of gypsum, sodium sulfate, and clay to promote conductivity of ions. The pipe, in effect, is the cathode of a voltaic cell.

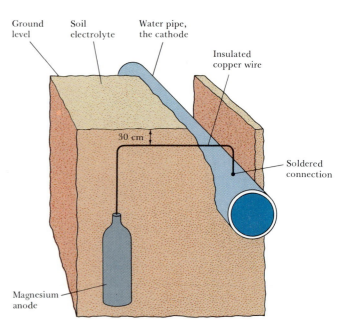

$$Zn(s) \longrightarrow Zn^{2+}(aq) + 2e^- \qquad E^\circ_{ox} = 0.76 \text{ V} \qquad [20.46]$$

The zinc therefore serves as the anode and is corroded instead of the iron, as shown in Figure 20.18. Protection of a metal by making it the cathode in an electrochemical cell is known as **cathodic protection**. The metal that oxidizes while protecting the cathode is called the sacrificial anode. Underground pipelines are often protected against corrosion by making the pipeline the cathode of a voltaic cell. Pieces of an active metal such as magnesium are buried along the pipeline and connected to it by wire, as shown in Figure 20.19. In moist soil, where corrosion can occur, the active metal serves as the anode and the pipe experiences cathodic protection.

## SAMPLE EXERCISE 20.14

Predict the nature of the corrosion that would take place if an iron gutter were nailed to a house using aluminum nails.

**Solution:** A voltaic cell can be formed at the point of contact of the two metals. The metal that is more easily oxidized will serve as the anode, and the other metal serves as the cathode. By comparing standard oxidation potentials of Al and Fe we see that Al will be the anode:

$$Al(s) \longrightarrow Al^{3+}(aq) + 3e^- \qquad E^\circ_{ox} = 1.66 \text{ V}$$
$$Fe(s) \longrightarrow Fe^{2+}(aq) + 2e^- \qquad E^\circ_{ox} = 0.44 \text{ V}$$

The gutter will thus be protected against corrosion in the vicinity of the nail, because the iron serves as the cathode. However, the nail would quickly corrode, leaving the gutter on the ground.

What do you think would happen if aluminum siding were nailed to a house using iron nails?

## PRACTICE EXERCISE

From examination of Table 20.1, indicate which of the following metals could provide cathodic protection to iron: Al, Cu, Ni, Zn.
*Answer:* Al, Zn

# FOR REVIEW

## SUMMARY

Spontaneous redox reactions can be used to generate electricity in **voltaic cells**. Conversely, electricity can be used to bring about nonspontaneous reactions in **electrolytic cells**. In either type of cell, the electrode at which oxidation occurs is called the **anode**, and the electrode at which reduction occurs is called the **cathode**.

A voltaic cell may be thought to possess a "driving force" that moves the electrons through the external circuit, from anode to cathode. This driving force is called the **electromotive force (emf)** and is measured in volts. The emf of a cell can be regarded as being composed of two parts: that due to oxidation at the anode and that due to reduction at the cathode $(E_{cell} = E_{ox} + E_{red})$.

Oxidation potentials $(E_{ox})$ and reduction potentials $(E_{red})$ can be assigned to half-reactions by defining the standard hydrogen electrode as a reference:

$$2H^+(1 \ M) + 2e^- \longrightarrow H_2(1 \text{ atm}) \qquad F^\circ = 0$$

Standard reduction potentials are referred to merely as **standard electrode potentials** and are tabulated for a great variety of reduction half-reactions. The oxidation potential for an oxidation half-reaction will be of the same magnitude as, but opposite in sign to, the electrode potential for the reverse reduction process. The more positive the potential associated with a half-reaction, the greater the tendency for that reaction to occur as written. Electrode potentials can be used to determine the maximum voltages generated by voltaic cells or the minimum voltages required in electrolytic cells. They can also be used to predict whether certain redox reactions are spontaneous (positive $E$). The emf is related to free-energy changes: $\Delta G = -n\mathfrak{F}E$, where $\mathfrak{F}$ is **Faraday's constant**, 96,500 J/V-mol = 96,500 C/mol.

The emf of a cell varies in magnitude with temperature and with the concentrations of reactants and products. The Nernst equation relates emf under nonstandard conditions to the standard emf:

$$E = E^\circ - \frac{2.30RT}{n\mathfrak{F}} \log Q$$

At equilibrium

$$E = 0 \quad \text{and} \quad E^\circ = \frac{2.30RT}{n\mathfrak{F}} \log K$$

Thus standard emfs are related to equilibrium constants. The maximum electrical work that can be obtained from a voltaic cell is the product of the total charge it delivers, $n\mathfrak{F}$, and its emf, $E$: $w_{max} = -n\mathfrak{F}E$. To illustrate the principles involved in voltaic cells, simple cells utilizing salt bridges were discussed. Commercial cells need to be more rugged. Four common batteries were discussed: the lead storage battery, the nickel-cadmium battery, the common dry cell, and the alkaline dry cell. The first two are rechargeable, the latter two are not. Fuel cells, still largely experimental, are voltaic cells that utilize redox reactions involving conventional fuels, such as $H_2$ and $CH_4$.

In an electrolytic cell an external source of electricity is used to "pump" electrons from the anode to the cathode. The current-carrying medium within the cell may be either a molten salt or an aqueous electrolyte solution. The products of electrolysis can generally be predicted by comparing electrode potentials associated with possible oxidation and reduction processes. However, because of the **overvoltage** phenomenon, some reactions (such as those that generate $H_2$ or $O_2$) occur less readily than electrode potentials would suggest.

The extent of chemical reaction in either an electrolytic or voltaic cell can be related to the quantity of electricity that passes through the external circuit. The amount of electrical charge possessed by a mole of electrons is known as a faraday (abbreviated $\mathfrak{F}$), 96,500 C; 1 coulomb equals 1 ampere-second: 1 C = 1 A-s.

Our knowledge of electrochemistry allows us to design batteries and to bring about desirable redox reactions, such as those used in electroplating and in the reduction and refining of metals. Electrochemical principles also help us to understand and combat corrosion. Corrosion of a metal such as iron is electrochemical in origin. A metal can be protected against corrosion by putting it in contact with another metal that more readily undergoes oxidation. This process is known as **cathodic protection**.

---

## LEARNING GOALS

Having read and studied this chapter, you should be able to:

1. Diagram simple voltaic and electrolytic cells, labeling anode, cathode, the directions of ion and electron movements, and the signs of the electrodes.

2. Calculate the emf generated by a voltaic cell or the minimum emf required to cause an electrolytic cell reaction to proceed, having been given appropriate electrode potentials.

3. Use electrode potentials to predict whether a reaction will be spontaneous.

4. Interconvert $E^\circ$, $\Delta G^\circ$, and $K$ for redox reactions.

5. Use the Nernst equation to calculate an emf under nonstandard conditions.

6. Use the Nernst equation to calculate the concentration of an ion, given $E$, $E^\circ$, and the concentrations of the remaining ions.

7. Describe the lead storage battery, the dry cell, and the nickel-cadmium cell.

8. Describe the principles underlying the operation of a fuel cell, and diagram an $H_2$–$O_2$ fuel cell.

9. Calculate the third quantity, having been given any two of the following: time, current, and the amount of substance produced or consumed in an electrolysis reaction.

10. Calculate the maximum electrical work performed by a voltaic cell (or the minimum electrical work to cause a reaction in an electrolytic

cell), having been given emf and electrical charge or information from which they can be determined.

**11.** Describe the phenomenon of corrosion in terms of electrochemical principles, and explain the principle that underlies cathodic protection.

## KEY TERMS

Among the more important terms and expressions used for the first time in this chapter are the following:

An **anode** (Section 20.1) is an electrode at which oxidation occurs.

A **cathode** (Section 20.1) is an electrode at which reduction occurs.

**Cathodic protection** (Section 20.7) is a means of protecting a metal against corrosion by making it the cathode in a voltaic cell. This can be done by attaching a more active metal.

**Corrosion** (Section 20.7) is the process by which a metal is oxidized by substances in its environment.

An **electrolytic cell** (Section 20.5) is a device in which a nonspontaneous reaction is caused to occur by passage of current under a sufficient external electrical potential.

The **electromotive force**, **emf** (Section 20.2), is a measure of the driving force or "electrical pressure" for completion of a chemical reaction via passage of electrons through an external circuit. It is measured in volts, which is equivalent to joules per coulomb.

A **faraday** (Section 20.3) is the total charge of a mole of electrons, 96,500 C.

A **fuel cell** (Section 20.4) is a voltaic cell that utilizes the oxidation of a conventional fuel such as $H_2$ or $CH_4$ in the cell reaction.

The **Nernst equation** (Section 20.3) relates the cell emf, $E$, to the standard cell emf, $E°$, and the reaction quotient, $Q$:

$$E = E° - \frac{2.30RT}{n\mathfrak{F}} \log Q$$

A **standard electrode potential**, $E°$ (Section 20.2), is the reduction potential of a half-reaction with all solution species at 1 $M$ concentration and all gaseous species at 1 atm, measured relative to the hydrogen electrode, for which $E°$ is exactly 0 V.

A **voltaic cell** (Section 20.1) is a device in which a spontaneous chemical reaction is made to occur via passage of electrons through an external circuit.

## EXERCISES

### Voltaic Cells; EMF; Spontaneity

**20.1** What is the role of the salt bridge in the voltaic cell diagrammed in Figure 20.3?

**20.2** Using the cell depicted in Figure 20.4 as an example, explain the distinction between an inert and an active electrode in a voltaic cell.

**20.3** The two half-cell reactions in a voltaic cell are

$$NO_3^-(aq) + 4H^+(aq) + 3e^- \longrightarrow NO(g) + 2H_2O(l)$$
$$Sn^{4+}(aq) + 2e^- \longrightarrow Sn^{2+}(aq)$$

**(a)** Using Appendix F, determine which reaction (or its reverse) occurs at the cathode and which at the anode. **(b)** What is the standard cell potential?

**20.4** Given the following half-cell reactions and associated standard half-cell potentials:

$$AuBr_4^-(aq) + 3e^- \longrightarrow Au(s) + 4Br^-(aq)$$
$$E° = -0.858 \text{ V}$$
$$Eu^{3+}(aq) + e^- \longrightarrow Eu^{2+}(aq)$$
$$E° = -0.43 \text{ V}$$
$$IO^-(aq) + H_2O(l) + 2e^- \longrightarrow I^-(aq) + 2OH^-(aq)$$
$$E° = 0.49 \text{ V}$$
$$Sn^{2+}(aq) + 2e^- \longrightarrow Sn(s)$$
$$E° = -0.14 \text{ V}$$

**(a)** Write the cell reaction for the combination of these half-cell reactions that leads to the largest cell emf, and calculate the value. **(b)** Write the cell reaction for the combination of half-cell reactions that leads to the smallest cell emf, and calculate that value.

**20.5** A 1 $M$ solution of $Cu(NO_3)_2$ is placed in a beaker with a strip of Cu metal. A 1 $M$ solution of $SnSO_4$ is placed in a second beaker with a strip of Sn metal. The two beakers are connected by a salt bridge, and the two metal electrodes are linked by wires to a voltmeter. **(a)** Which electrode serves as anode, and which as cathode? **(b)** Which electrode gains mass and which loses mass as the cell reaction proceeds? **(c)** Indicate the sign of each electrode. **(d)** What is the emf generated by the cell under standard conditions? **(e)** Write the chemical equation for the overall cell reaction.

**20.6** A voltaic cell consists of a strip of lead metal in a solution of $Pb(NO_3)_2$ in one beaker, and in the other beaker a platinum electrode immersed in an NaCl solution, with $Cl_2$ gas bubbled around the electrode. The two beakers are connected with a salt bridge. **(a)** Which electrode serves as anode, and which as cathode? **(b)** Which electrode, if either, gains or loses mass as the cell reaction proceeds? **(c)** Indicate the sign of each electrode. **(d)** What is the emf generated by the cell under standard conditions? **(e)** Write the overall cell reaction.

**20.7** A voltaic cell that uses the reaction

$$Tl^{3+}(aq) + 2Cr^{2+}(aq) \longrightarrow Tl^+(aq) + 2Cr^{3+}(aq)$$

has a measured emf of 1.19 V under standard conditions. **(a)** What is $E°$ for the half-cell reaction: $Tl^{3+}(aq) + 2e^- \longrightarrow Tl^+(aq)$? **(b)** What is meant by the expression "under standard conditions"? **(c)** Sketch the voltaic cell; label the anode and cathode, and indicate the direction of electron flow.

**20.8** A voltaic cell that uses the reaction

$$PdCl_4{}^{2-}(aq) + Cd(s) \longrightarrow Pd(s) + 4Cl^-(aq) + Cd^{2+}(aq)$$

has a measured emf under standard conditions of 1.03 V. **(a)** Write the two half-cell reactions. **(b)** From the known value of $E°$ for the half-cell involving Cd (Appendix F), determine the half-cell potential for the reaction involving Pd. **(c)** Sketch the cell, label the anode and cathode, and indicate the direction of electron flow.

**20.9** **(a)** Sketch a voltaic cell based on the following reaction:

$$Ni(s) + PtCl_4{}^{2-}(aq) \longrightarrow Ni^{2+}(aq) + Pt(s) + 4Cl^-(aq)$$

**(b)** Label the anode and cathode, and identify the positive and the negative terminals. **(c)** Indicate the composition of the electrolytic solution in each compartment and the composition of the electrodes. **(d)** Show the directions of ion and electron movements. **(e)** Calculate the emf generated by the cell under standard conditions.

**20.10** **(a)** Sketch a voltaic cell based on the following reaction:

$$Sn(s) + ClO^-(aq) + H_2O(l) \longrightarrow$$
$$Sn^{2+}(aq) + Cl^-(aq) + 2OH^-(aq)$$

**(b)** In each case label the anode and cathode, and identify the positive and the negative terminal. **(c)** Indicate the composition of the electrolytic solution in each compartment and the composition of the electrodes. **(d)** Show the directions of ion and electron movements. **(e)** Calculate the emf generated by the cell under standard conditions.

**20.11** **(a)** Arrange the following species in order of increasing strength as oxidizing agents: $Cr_2O_7{}^{2-}$, $H_2O_2$, $Cu^{2+}$, $Cl_2$, $O_2$. **(b)** Arrange the following species in order of increasing strength as reducing agents: Zn, $I^-$, $Sn^{2+}$, $H_2O_2$, Al.

**20.12** Based on the data in Appendix F, **(a)** which of the following is the strongest oxidizing agent, and which the weakest: $Ce^{4+}$, $Br_2$, $H_2O_2(aq$ acid), Zn? **(b)** Which of the following is the strongest reducing agent, and which is the weakest: $F^-$, Zn, $N_2H_5{}^+$, $I_2$, NO?

**20.13** Using Table 20.1, suggest one or more agents capable of reducing $Eu^{3+}(aq)$ to $Eu^{+2}(aq)$ $(E° = -0.43$ V).

**20.14** Based on Table 20.1, suggest one or more agents capable of oxidizing $RuO_4{}^{2-}(aq)$ to $RuO_4{}^-(aq)$. $(E°$ for the reaction $RuO_4{}^{2-}(aq) \longrightarrow RuO_4{}^-(aq) + e^-$ is $-0.59$ V.)

**20.15** Which of the following processes are spontaneous under standard conditions?
**(a)** $H_2S(aq) + Cl_2(aq) \longrightarrow$
$$S(s) + 2Cl^-(aq) + 2H^+(aq)$$
**(b)** $5H_2O_2(l) + Cl_2(aq) \longrightarrow$
$$4H_2O(l) + 2H^+(aq) + 2ClO_3{}^-(aq)$$
**(c)** $2Cl^-(aq) + Cu^{2+}(aq) \longrightarrow Cu(s) + Cl_2(g)$

**20.16** **(a)** What is the electrochemical criterion that a reaction be spontaneous in the direction written? **(b)** Using data from Appendix F, write two complete spontaneous redox reactions in which $H_2O_2$ is oxidized in acidic solution.

## Relationships between $E°$, $\Delta G°$, and $K$

**20.17** Indicate whether each of the following statements is true or false. Correct those that are false. **(a)** For a given set of concentrations, the cell emf is independent of the number of moles of reacting substances present. **(b)** The free-energy change for a cell reaction is also independent of the number of moles of substance reacting. **(c)** $E°$ and $\Delta G°$ have the same units.

**20.18** Indicate whether each of the following statements is true or false. Correct those that are false. **(a)** The standard cell potential $E°$ is directly proportional to the equilibrium constant for the cell reaction. **(b)** The cell potential is independent of the concentrations of solution species that enter into the cell reaction. **(c)** The cell potential is independent of the size of the zinc electrode in the cell in Figure 20.3.

**20.19** From standard potentials calculate $\Delta G°$ for each of the reactions listed in Exercise 20.15.

**20.20** Given the following half-cell potentials:

$$Fe^{2+}(aq) \longrightarrow Fe^{3+}(aq) + e^- \longrightarrow E° = -0.771 \text{ V}$$

$$S_2O_6^{2-}(aq) + 4H^+(aq) + 2e^- \longrightarrow 2H_2SO_3(aq)$$
$$E° = 0.60 \text{ V}$$

$$N_2O(aq) + 2H^+(aq) + 2e^- \longrightarrow N_2(g) + H_2O(l)$$
$$E° = -1.77 \text{ V}$$

$$VO_2^+(aq) + H^+(aq) + e^- \longrightarrow VO^{2+}(aq) + H_2O(l)$$
$$E° = 1.00 \text{ V}$$

**(a)** Write balanced chemical equations for oxidation of $Fe^{2+}(aq)$ by the other three reagents for which the half-cell reaction is given. **(b)** Calculate $\Delta G°$ for each reaction at 298 K.

**20.21** Using the standard half-cell potentials listed in Appendix F, calculate the equilibrium constant for each of the following reactions at 298 K:
**(a)** $Zn(s) + Sn^{2+}(aq) \longrightarrow Zn^{2+}(aq) + Sn(s)$
**(b)** $Co(s) + 2H^+(aq) \longrightarrow Co^{2+}(aq) + H_2(g)$
**(c)** $10Br^-(aq) + 2MnO_4^-(aq) + 16H^+(aq) \longrightarrow$
$2Mn^{2+}(aq) + 8H_2O(l) + 5Br_2(l)$

**20.22** Using the standard half-cell potentials listed in Appendix F, calculate the equilibrium constant for each of the following reactions at 298 K:
**(a)** $2VO_2^+(aq) + 4H^+(aq) + Ni(s) \longrightarrow$
$2VO^{2+}(aq) + H_2O(l) + Ni^{2+}(aq)$
**(b)** $3Ce^{4+}(aq) + Bi(s) + H_2O(l) \longrightarrow$
$3Ce^{3+}(aq) + BiO^+(aq) + 2H^+(aq)$
**(c)** $N_2H_5^+(aq) + 4Fe(CN)_6^{3-}(aq) \longrightarrow$
$N_2(g) + 5H^+(aq) + 4Fe(CN)_6^{4-}(aq)$

**20.23** Assuming that $n = 2$, what magnitude of equilibrium constant is associated with a standard cell emf of **(a)** 1.00 V; **(b)** $-0.20$ V (each at 298 K)?

**20.24** The equilibrium constant for the reaction

$$2Cu(OH)_2(s) + Zn(s) + 2OH^-(aq) \longrightarrow$$
$$Cu_2O(s) + 3H_2O(l) + ZnO_2^{2-}(aq)$$

at 298 K is $1.7 \times 10^{38}$. What is the value of the standard emf for this reaction?

## Nernst Equation

**20.25** In a cell that utilizes the reaction

$$Sn(s) + Br_2(aq) \longrightarrow Sn^{2+}(aq) + 2Br^-(aq)$$

what is the effect on cell emf of each of the following changes? **(a)** NaBr is dissolved in the cathode compartment. **(b)** $SnSO_4$ is dissolved in the anode compartment. **(c)** The electrode in the cathode compartment is changed from platinum to graphite. **(d)** The area of the anode is doubled.

**20.26** What is the effect on the emf of the cell shown in Figure 20.4 of each of the following changes? **(a)** The $H_2$ gas is diluted with an equal volume of Ar. **(b)** The area of the anode is doubled. **(c)** Sulfuric acid is added to the cathode compartment. **(d)** Sodium nitrate is added to the anode compartment.

**20.27** Calculate the emf of the cell described in Exercise 20.25 under the following conditions: $[Br^-] = 0.10 \text{ } M$; $[Sn^{2+}] = 0.050 \text{ } M$.

**20.28** Calculate the emf of a cell that utilizes the reaction

$$Co(s) + I_2(s) \longrightarrow Co^{2+}(aq) + 2I^-(aq)$$

at 298 K when $[Co^{2+}] = 0.010 \text{ } M$ and $[I^-] = 0.0080 \text{ } M$.

**20.29** The cell in Figure 20.4 could be used to provide a measure of the pH in the cathode compartment. Calculate the pH of the cathode compartment solution if the cell emf is measured to be 0.720 V when $[Zn^{2+}] = 0.1 \text{ } M$, $P_{H_2} = 1$ atm.

**20.30** A voltaic cell is constructed that is based on the following reaction:

$$Sn^{2+}(aq) + Pb(s) \longrightarrow Pb^{2+}(aq) + Sn(s)$$

If the concentration of $Sn^{2+}$ in the cathode compartment is $1.00 \text{ } M$ and the cell generates an emf of 0.22 V, what is the concentration of $Pb^{2+}$ in the anode compartment? If the anode compartment contains $[SO_4^{2-}] = 1.00 \text{ } M$ in equilibrium with $PbSO_4(s)$, what is the $K_{sp}$ of the $PbSO_4$?

## Commercial Voltaic Cells

**20.31** If 120 g of zinc is employed in the casing of a particular Leclanché dry cell, and if all of this is consumed in the cell reaction, how many grams of $MnO_2$ undergo reaction?

**20.32** During a period of discharge of a lead-acid battery, 600 g of Pb from the anode is converted into $PbSO_4(s)$. What mass of $PbO_2(s)$ is reduced at the cathode during this same period?

**20.33** **(a)** Write the reactions for discharge and charge of the nickel-cadmium rechargeable cell. **(b)** Given the following half-cell potentials, calculate the standard emf of the cell:

$$Cd(OH)_2(s) + 2e^- \longrightarrow Cd(s) + 2OH^-(aq)^-$$
$$E° = -0.76 \text{ V}$$

$$NiO_2(s) + 2H_2O(l) + 2e^- \longrightarrow Ni(OH)_2(s) + 2OH^-(aq)$$
$$E° = +0.49 \text{ V}$$

**20.34** Mercuric oxide dry-cell batteries are often used where a high energy density is required; for example, in hearing aids, watches, cameras, and electronic systems. The two half-cell reactions that occur in the battery are as follows:

$$HgO(s) + H_2O(l) + 2e^- \longrightarrow Hg(l) + 2OH^-(aq)$$
$$Zn(s) + 2OH^-(aq) \longrightarrow ZnO(s) + H_2O(l) + 2e^-$$

**(a)** Write the overall cell reaction. **(b)** $E°$ for the reduction half-reaction written above is 0.098 V. The measured potential for a fresh mercury dry cell is 1.35 V. Assuming that both half-cells are effectively operating under standard conditions, what is the standard potential for the oxidation half-cell reaction? Why is this potential different from that for oxidation of Zn in aqueous acidic medium?

**20.35** Suppose that an alkaline dry cell were manufactured using cadmium metal rather than zinc. What effect would this have on the cell emf?

**20.36** The most promising experimental batteries that might someday find their way into commercial use employ lithium or sodium as the anodic metal. What advantages might be realized by using these metals rather than zinc, cadmium, lead, or nickel?

## Electrolysis

**20.37** Why are different products obtained when molten $MgCl_2$ and aqueous $MgCl_2$ are electrolyzed with inert electrodes? Predict the products in each case.

**20.38** What are the expected half-reactions at both anode and cathode upon electrolysis of **(a)** molten $CdCl_2$, using inert electrodes; **(b)** aqueous $CdCl_2$ solution, using inert electrodes; **(c)** aqueous $CdCl_2$ solution, using a cadmium metal anode and iron cathode?

**20.39** Sketch a cell for electrolysis of a $CoCl_2$ solution using inert electrodes. Indicate the directions in which ions and electrons move. Give the electrode reactions, and label the anode and cathode, indicating which is positive and which is negative.

**20.40** Sketch a cell for the electrolysis of aqueous HCl using copper electrodes. Give the electrode reactions, labeling the anode and cathode. Calculate the minimum applied voltage required to cause electrolysis to occur, assuming standard conditions.

**20.41** A $Zn^{2+}(aq)$ solution is electrolyzed using a current of 0.600 A. What mass of Zn is plated out after 300 min?

**20.42** Chromium is reduced from $CrO_4^{2-}(aq)$ to $Cr(OH)_4^-(aq)$ in an electrolytic cell. What mass of $Cr(OH)_4^-$ is formed by passage of 4.50 A for a period of 8000 s?

**20.43** **(a)** In the electrolysis of aqueous NaCl, how many liters of $Cl_2(g)$ (measured at STP) are generated by a current of 5.50 A for a period of 100 min? **(b)** How many moles of $NaOH(aq)$ have formed in the solution in this period?

**20.44** If 0.500 L of a 0.600 $M$ $SnSO_4$ solution is electrolyzed for a period of 30.0 min using a current of 4.60 A, and if inert electrodes are used, what is the final concentration of each ion remaining in the solution? (Assume that the volume of the solution does not change.)

## Electrical Work

**20.45** Indicate whether each of the following statements is true or false. Correct those that are false. **(a)** The theoretical maximum work obtainable from a voltaic cell is exactly equal to the theoretical minimum work required to drive the cell reaction in the opposite direction under the same conditions. **(b)** For a given cell reaction the maximum work obtainable is proportional to the total mass of reactants that pass into products. **(c)** The minimum work required to cause an electrolytic cell reaction to proceed is often less than the theoretical value, because of overvoltage.

**20.46** Provide a simple, direct argument to support the statement that the maximum work obtainable from a given voltaic cell is a state function; that is, it depends only on the initial and final states of the cell. Is this true also of the *actual* work obtainable from the cell? Give examples to illustrate your answer.

**20.47** What is the maximum electrical work, in joules, that a cell employing the cell reaction

$$Sn(s) + Br_2(aq) \longrightarrow Sn^{2+}(aq) + 2Br^-(aq)$$

can accomplish under standard conditions if 0.650 mol of Sn is consumed?

**20.48** What is the maximum electrical work, in joules, that a cell employing the cell reaction

$$Co(s) + I_2(s) \longrightarrow Co^{2+}(aq) + 2I^-(aq)$$

can accomplish under standard conditions if 82.0 g of $I_2$ is consumed?

**20.49** **(a)** Calculate the mass of magnesium produced in a large facility by electrolysis of molten $MgCl_2$, by a current of 90,000 A flowing for 16 h. Assume that the cell is 50 percent efficient. **(b)** What is the total energy requirement for this electrolysis if the applied emf is 4.20 V?

**20.50** **(a)** Calculate the mass of Li formed by electrolysis of molten LiCl by a current of $6.6 \times 10^4$ A flowing for a period of 12 h. Assume the cell is 85 percent efficient. **(b)** What is the energy requirement for this electrolysis per mole of Li formed if the applied emf is 5.50 V?

## Corrosion

**20.51**  An iron object is plated with a coating of cobalt to protect against corrosion. Does the cobalt protect iron by cathodic protection? Explain.

**20.52**  When an iron object is plated with tin, does the tin act as a sacrificial anode in protecting against corrosion? Explain.

**20.53**  Amines are compounds related to ammonia. One of their characteristics is their ability to function as Brønsted bases (Sections 13.5 and 17.4). Suggest how they protect against corrosion when added to antifreeze as corrosion inhibitors.

**20.54**  The following quotation is taken from a recent article dealing with corrosion of electronic materials: "Sulfur dioxide, its acidic oxidation products, and moisture are well established as the principal causes of outdoor corrosion of many metals." Using Ni as an example, explain why the factors cited affect the rate of corrosion. Write chemical equations to illustrate your points. Note that $NiO(s)$ is soluble in acidic solution.

## Additional Exercises

**20.55**  The following oxidation-reduction reaction is spontaneous in the direction indicated:

$$5Fe^{2+}(aq) + MnO_4^-(aq) + 8H^+(aq) \longrightarrow$$
$$5Fe^{3+}(aq) + Mn^{2+}(aq) + 4H_2O(l)$$

A solution containing $KMnO_4$ and $H_2SO_4$ is poured into one beaker while a solution of $FeSO_4$ is poured into another. A salt bridge is used to join the beakers. A platinum foil is placed in each solution, and the two solutions are connected by a wire that passes through a voltmeter. **(a)** Indicate the reactions occurring at the anode and at the cathode, the direction of electron movement through the external circuit, the direction of ion migrations through the solutions, and the signs of the electrodes. **(b)** Calculate the emf of the cell under standard conditions.

**20.56**  A common shorthand way of representing a voltaic cell is to list the reactants and products from left to right in the following form:

anode | anode solution || cathode solution | cathode

A double vertical line represents a salt bridge or porous barrier. A single vertical line represents a change of phase, such as from solid to solution. **(a)** Write the half-reactions and overall cell reaction represented by $Fe \,|\, Fe^{2+} \,||\, Ag^+ \,|\, Ag$; sketch the cell. **(b)** Write the half-reactions and overall cell reaction represented by $Zn \,|\, Zn^{2+} \,||\, H^+ \,|\, H_2$; sketch the cell. **(c)** Using the notation just described, represent a cell based on the following reaction.

$$ClO_3^-(aq) + Cu(s) + 6H^+(aq) \longrightarrow$$
$$Cl^-(aq) + Cu^{2+}(aq) + 3H_2O(l)$$

Sketch the cell.

**20.57**  The zinc-silver oxide cell used in hearing aids and electrical watches is based on the following half-reactions:

$$Zn^{2+}(aq) + 2e^- \longrightarrow Zn(s)$$
$$E° = -0.763 \text{ V}$$
$$Ag_2O(s) + H_2O(l) + 2e^- \longrightarrow 2Ag(s) + 2OH^-(aq)$$
$$E° = \phantom{-}0.344 \text{ V}$$

**(a)** What substance is oxidized and what is reduced in the cell during discharge? **(b)** What is the positive electrode and what is the negative electrode? **(c)** What emf does this cell generate under standard conditions?

**20.58**  A voltaic cell consists of Co and a 0.1 $M$ solution of $Co^{2+}(aq)$ in one compartment, and Cu and a 0.1 $M$ solution of $Cu^{2+}(aq)$ in the other. **(a)** Sketch the cell and label the anode and cathode. **(b)** What is the emf of the cell? **(c)** What would be the effect on the emf of reducing $[Co^{2+}]$ in the cobalt compartment?

**20.59**  Predict whether the following reactions will be spontaneous in acidic solution under standard conditions: **(a)** oxidation of Sn to $Sn^{2+}$ by $I_2$ (to form $I^-$); **(b)** reduction of $Ni^{2+}$ to Ni by $I^-$ (to form $I_2$); **(c)** reduction of $Ce^{4+}$ to $Ce^{3+}$ by $Br^-$ (to form $Br_2$); **(d)** reduction of $Ag^+$ to Ag by $H_2O_2$.

**20.60**  Cytochrome, a complicated molecule that we shall represent as $CyFe^{2+}$, reacts with the air we breathe to supply energy required to synthesize adenosine triphosphate, ATP. The body uses ATP as an energy source to drive other reactions (Section 19.7). At pH 7 the following electrode potentials pertain to this oxidation of $CyFe^{2+}$:

$$O_2(g) + 4H^+(aq) + 4e^- \longrightarrow 2H_2O(l) \qquad E = 0.82 \text{ V}$$
$$CyFe^{3+}(aq) + e^- \longrightarrow CyFe^{2+}(aq) \qquad E = 0.22 \text{ V}$$

**(a)** What is $\Delta G$ for the oxidation of $CyFe^{2+}$ by air? **(b)** If the synthesis of 1 mol of ATP from adenosine diphosphate, ADP, requires a $\Delta G$ of 37.7 kJ, how many moles of ATP are synthesized per mole of $O_2$?

**20.61**  In a concentration cell the same reactants are present in both the anode and the cathode compartments, but at different concentrations. Calculate the emf of a cell containing 0.040 $M$ $Cr^{3+}$ in one compartment and 1.0 $M$ $Cr^{3+}$ in the other if Cr electrodes are used in both. Which is the anode compartment?

**[20.62]**  **(a)** Write the cell reaction for the cell shown in Figure 20.20 on page 686. **(b)** Write the full Nernst equation expression for this cell, and rearrange it so that the pH of the solution is given as a function of the other cell variables. **(c)** If the cell emf is 0.660 V, the pressure of $H_2$ is 1.0 atm, and $[Cl^-]$ is $1.0 \times 10^{-3}$ $M$, what is the pH of the solution?

**20.63**  From the following half-cell reactions and their potentials, construct a cell with the largest possible standard emf; calculate the equilibrium constant for the cell reaction.

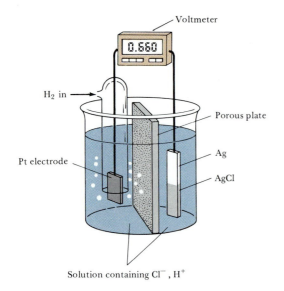

Voltmeter

0.660

H₂ in →

Porous plate

Pt electrode

Ag

AgCl

Solution containing Cl⁻, H⁺

**FIGURE 20.20**

$$PbO_2(s) + H_2O(l) + 2e^- \longrightarrow$$
$$PbO(s) + 2OH^-(aq) \quad\quad E° = +0.28 \text{ V}$$

$$IO_3^-(aq) + 2H_2O(l) + 4e^- \longrightarrow$$
$$IO^-(aq) + 4OH^-(aq) \quad\quad E° = +0.56 \text{ V}$$

$$PO_4^{3-}(aq) + 2H_2O(l) + 2e^- \longrightarrow$$
$$HPO_3^{2-} + 3OH^-(aq) \quad\quad E° = -1.05 \text{ V}$$

**[20.64]** The standard potential for the reduction of AgSCN is 0.0895 V:

$$AgSCN(s) + e^- \longrightarrow Ag(s) + SCN^-(aq)$$

Using this value together with another electrode potential, calculate $K_{sp}$ for AgSCN.

**20.65** **(a)** How many coulombs are required to plate a layer of chromium metal 0.23 mm thick on an auto bumper with a total area of 0.32 m² from a solution containing $CrO_4^{2-}$? The density of chromium metal is 7.20 g/cm³. **(b)** What current flow is required for this electroplate if the bumper is to be plated in 6.0 s?

**20.66** The element indium is to be obtained by electrolysis of a molten halide of the element. Passage of a current of 3.20 A for a period of 40.0 min results in formation of 4.57 g of In. What is the oxidation state of indium in the halide melt?

**20.67** Explain why electrolysis of an aqueous solution of $Na_2SO_4$ containing litmus develops a blue color at the cathode and a red color at the anode.

**20.68** Peroxyborate bleaches, such as found in Borateem, have replaced older "chlorine" bleaches in many bleaching agents. Sodium peroxyborate, $NaBO_3$, can be prepared by electrolytic oxidation of borax ($Na_2B_4O_7$)

solutions:

$$Na_2B_4O_7(aq) + 10NaOH(aq) \longrightarrow$$
$$4NaBO_3(aq) + 5H_2O(l) + 8Na^+(aq) + 8e^-$$

How many grams of $NaBO_3$ can be prepared in 24.0 h if the current is 20.0 A?

**20.69** **(a)** Calculate the minimum applied voltage required to cause the following electrolysis reaction to occur, assuming that the anode is platinum and the cathode is nickel:

$$Ni^{2+}(aq) + 2Br^-(aq) \longrightarrow Ni(s) + Br_2(l)$$

**(b)** In practice, a large voltage than this calculated minimum is required to produce the electrode reactions. Why is this so?

**20.70** How long could an alkaline dry cell be used to power a hearing aid if a current of $3.0 \times 10^{-3}$ A is required? Assume the cell life is limited by the amount of Zn that can react (1.68 g).

**20.71** **(a)** What is the maximum amount of work that a 6-V golf-cart lead storage battery can accomplish if it is rated at 300 A-h? **(b)** List some of the reasons why this amount of work is never realized.

**20.72** The type of lead storage cell used in automobiles does not tolerate "deep discharge" (in which the cell reaction is allowed to proceed to near completion) very well. Typically, the battery fails after 20 to 30 such cycles. **(a)** Why does deep discharge lead to battery failure? **(b)** How is deep discharge avoided during normal auto use?

**20.73** Write the reaction that might be expected to occur if a block of Mg is attached to a steel ship hull. Explain how the Mg acts to protect the ship against corrosion.

**20.74** If you were going to apply a small potential to a steel ship resting in the water as a means of inhibiting corrosion, would you apply a negative or a positive charge? Explain.

**20.75** Describe the role of atmospheric oxygen in the corrosion of iron.

**20.76** Considering the following standard half-cell potentials:

$$Ti(s) \longrightarrow Ti^{2+}(aq) + 2e^-$$
$$E° = +1.63 \text{ V}$$

$$Ti(s) + 2H_2O(l) \longrightarrow TiO_2(s) + 4H^+ + 4e^-$$
$$E° = +0.86 \text{ V}$$

Why is titanium metal quite corrosion-resistant?

**20.77** A family owns an antique set of silverware that has a fine, dark coating of $Ag_2S$ in the crevices of the pattern that adds to its beauty. The set is placed in a galvanized container together with soap and water in order to be cleaned. The $Ag_2S$ disappears as the set sits in the container, leaving the silverware with the appearance of a new rather than an antique set. Explain the electrochemical processes that have occurred.

**[20.78]** Several years ago, a unique proposal was made to raise the *Titanic*. The plan involved placing pontoons within the ship using a surface-controlled submarine-type vessel. The pontoons would contain cathodes and would be filled with hydrogen gas formed by the electrolysis of water. It has been estimated that it would require about $7 \times 10^8$ mol of $H_2$ to provide the bouyancy to lift the ship (*Journal of Chemical Education*, vol. 50, p. 61, 1973). **(a)** How many coulombs of electrical charge would be required? **(b)** What is the minimum voltage required to generate $H_2$ and $O_2$ if the pressure on the gases at the depth of the wreckage (2 mi) is 300 atm? **(c)** What is the minimum electrical energy required to raise the *Titanic* by electrolysis? **(d)** What is the minimum cost of the electrical energy required to generate the necessary $H_2$ if the electricity costs 23¢ per kilowatt-hour?

**[20.79]** Edison's invention of the light bulb and its public demonstration in December 1879 generated considerable demand for the distribution of electricity to homes. One problem was how to measure the amount of electricity consumed by each household. Edison invented a coulometer (described in the *Journal of Chemical Education*, vol. 49, p. 627, 1972) that could be used with alternating current. Zinc plated out at the cathode of the coulometer. Every month the cathode was removed and weighed to determine the quantity of electricity used. If the cathode increased in mass by 1.62 g and the coulometer drew 0.35 percent of the current entering the home, how many coulombs of electricity were used in that month?

## CHAPTER 21

# Nuclear Chemistry

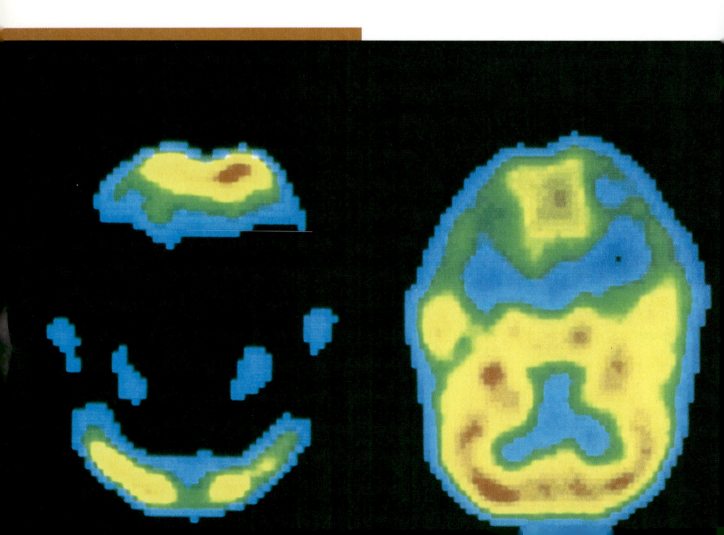

As we have progressed through this book, our focus has been on chemical reactions, specifically reactions in which electrons play a dominant role. In this chapter, we shall consider nuclear reactions, changes in matter originating in the nucleus of the atom. Some experts predict that we shall have to depend more and more on nuclear energy to replace our dwindling supplies of fossil fuels and to meet our rising energy demands. Thus our consideration of nuclear chemistry continues a minor theme of energy generation, started in Chapter 20. Even before we begin, you should have some awareness of the controversy surrounding nuclear energy—how do you feel about having a nuclear power plant in your town? Because the topic of nuclear energy evokes such emotional reactions, it is difficult to sift fact from opinion and begin to weigh pros and cons rationally. Therefore educated people of our time ought to have some understanding of nuclear reactions and the uses of radioactive substances.

However, before we get too deeply involved in our discussions, let's review and extend slightly some ideas introduced in Section 2.3. First, we should recall that there are two subatomic particles that reside in the nucleus, the **proton** and the **neutron**. We shall refer to these particles as **nucleons**. Recall also that all atoms of a given element have the same number of protons; this number is known as the element's **atomic number**. However, the atoms of a given element can have different numbers of neutrons and therefore different **mass numbers**; the mass number is the total number of nucleons in the nucleus. Atoms with the same atomic number but different mass numbers are known as **isotopes**. The different isotopes of an element are distinguished by citing their mass numbers. For example, the three naturally occurring isotopes of uranium are identified as uranium-233, uranium-235, and uranium-238, where the numbers given are the mass numbers. These isotopes are also labeled, using chemical symbols, as $^{233}_{92}U$, $^{235}_{92}U$, and $^{238}_{92}U$. The superscript is the mass number, the subscript is the atomic number. Different isotopes have different natural abundances. For example, 99.3 percent of naturally occurring uranium is uranium-238, 0.7 percent is uranium-235, and only a trace is uranium-233. One reason that we must now distinguish between different isotopes is that the nuclear properties of an atom depend on the number of both protons and neutrons in its nucleus. In contrast, we have found that an atom's chemical properties are unaffected by the number of neutrons in the nucleus. Now let's begin to discuss the reactions that a nucleus can undergo.

Output from a PET (positron emission transaxial tomography) scanner showing two human brains. The brain on the right is normal, that on the left is of a person suffering from dementia. The images are created by measuring positron emissions from compounds administered to a patient. Detailed comparisons of the images from healthy and ill patients provide important clues to the nature of mental illness and can be used to diagnose other illnesses, such as Alzheimer's disease and brain cancers. The synthesis of compounds that have incorporated short-lived radionuclides that emit positrons is a difficult art, essential to the use of PET scanners. (NIH Science Source/Photo Researchers)

## 21.1 NUCLEAR REACTIONS: AN OVERVIEW

A nucleus can undergo a reaction that changes its identity. Some nuclei are unstable and spontaneously emit particles and electromagnetic radiation. Such spontaneous emission from the nucleus of the atom is known as **radioactivity**. The discovery of this phenomenon by Henri Becquerel in 1896 was described in Section 2.2. Those isotopes that are radioactive are known as **radioisotopes**. An example is uranium-238, which spontaneously emits **alpha rays**; these rays consist of streams of helium-4 nuclei known as **alpha particles**. When a uranium-238 nucleus loses an alpha particle, the remaining fragment has an atomic number of 90 and a mass number of 234. It is therefore a thorium-234 nucleus. We can represent this reaction by the following nuclear equation:

$$^{238}_{92}\text{U} \longrightarrow {}^{234}_{90}\text{Th} + {}^{4}_{2}\text{He} \qquad [21.1]$$

When a nucleus spontaneously decomposes in this way, it is said to have decayed, or undergone **radioactive decay**.

Notice in Equation 21.1 that the sum of the mass numbers is the same on both sides of the equation ($238 = 234 + 4$). Likewise, the sum of the atomic numbers of nuclear charges on both sides of the equation is equal ($92 = 90 + 2$). Mass numbers and atomic numbers are similarly balanced in all nuclear equations. In writing nuclear equations, we are not concerned with the chemical form of the atom in which the nucleus resides. The radioactive properties of the nucleus are essentially independent of the state of chemical combination of the atom. It makes no difference whether we are dealing with the atom in the form of an element or of one of its compounds.

Another way a nucleus can change identity is to be struck by a neutron or by another nucleus. Nuclear reactions that are induced in this way are known as **nuclear transmutations**. Such a transmutation occurs when the chlorine-35 nucleus is struck by a neutron ($^{1}_{0}\text{n}$); this collision produces a sulfur-35 nucleus and a proton ($^{1}_{1}\text{p}$ or $^{1}_{1}\text{H}$). The nuclear equation for this reaction is shown in Equation 21.2:

$$^{35}_{17}\text{Cl} + {}^{1}_{0}\text{n} \longrightarrow {}^{35}_{16}\text{S} + {}^{1}_{1}\text{H} \qquad [21.2]$$

Notice again that the sum of mass numbers and of atomic numbers is the same on both sides of the equation. By bombarding nuclei with various particles, it is possible to prepare nuclides* not found in nature. The sulfur-35 produced in Equation 21.2 is an example.

**SAMPLE EXERCISE 21.1**

Write a balanced nuclear equation for the nuclear transmutation in which an aluminum-27 nucleus is struck by a helium-4 nucleus, producing a phosphorus-30 nucleus and a neutron.

**Solution:** By referring to a periodic table or a list of elements, we find that aluminum has an atomic number of 13; its chemical symbol is therefore $^{27}_{13}\text{Al}$. The atomic number of phosphorus is 15; its chemical

---

* Recall that the term nuclide (Section 2.3) applies to a nucleus with a specified number of protons and neutrons.

symbol is therefore $^{30}_{15}\text{P}$. The balanced equation is

$$^{27}_{13}\text{Al} + ^{4}_{2}\text{He} \longrightarrow ^{30}_{15}\text{P} + ^{1}_{0}\text{n}$$

**PRACTICE EXERCISE** _____
Carbon-11, a radioactive nuclide used in medical imaging, is formed by reaction of a proton with nitrogen-14, yielding carbon-11 and helium-4. Write the nuclear reaction.

*Answer:* $^{14}_{7}\text{N} + ^{1}_{1}\text{H} \longrightarrow ^{11}_{6}\text{C} + ^{4}_{2}\text{He}$

---

## 21.2 RADIOACTIVITY

As we were discussing radioactivity in the preceding section, two general questions might have occurred to you: Which nuclei are radioactive, and what types of radiation do they emit? These are important questions, and we shall now examine them. Let us first consider the types of radiation involved in the phenomenon of radioactivity.

### Types of Radioactive Decay

Emission of radiation is one of the ways by which an unstable nucleus is transformed into a stable one with less energy. The emitted radiation is the carrier of the excess energy. In Section 2.2, we discussed the three most common types of radiation emitted by radioactive substances: alpha ($\alpha$), beta ($\beta$), and gamma ($\gamma$) rays.

As we noted in Section 21.1, alpha rays consist of streams of helium-4 nuclei known as **alpha particles**. Equation 21.3 gives another example of this type of radioactive decay:

$$^{222}_{86}\text{Rn} \longrightarrow ^{218}_{84}\text{Po} + ^{4}_{2}\text{He} \qquad [21.3]$$

Beta rays consist of streams of electrons. Because the **beta particles** are electrons, they are represented as $^{0}_{-1}\text{e}$. The superscript zero indicates the exceedingly small mass of the electron in comparison to the mass of a nucleon. The subscript $-1$ represents the negative charge of the particle, which is opposite that of the proton. Iodine-131 is an example of an isotope that undergoes decay by beta emission. This reaction is summarized by Equation 21.4:

$$^{131}_{53}\text{I} \longrightarrow ^{131}_{54}\text{Xe} + ^{0}_{-1}\text{e} \qquad [21.4]$$

Emission of a beta particle has the effect of converting a neutron within the nucleus into a proton, thereby increasing the atomic number of the nucleus by one:

$$^{1}_{0}\text{n} \longrightarrow ^{1}_{1}\text{p} + ^{0}_{-1}\text{e} \qquad [21.5]$$

However, just because an electron is ejected from the nucleus, we need not think that the nucleus is composed of these particles, any more than we consider a match to be composed of sparks simply because it gives them off when struck. The electron comes into being only when the nucleus is disrupted.

**Gamma rays** consist of electromagnetic radiation of very short wavelength (that is, high-energy photons). The position of gamma rays in the electromagnetic spectrum is shown in Figure 6.3. Gamma rays can be represented as $^{0}_{0}\gamma$. Such radiation changes neither the atomic number nor the mass number of a nucleus. It almost always accompanies other radio-

**TABLE 21.1** Particles common to radioactive decay and nuclear transformations

| Particle | Symbol |
|---|---|
| Neutron | $^1_0n$ |
| Proton | $^1_1p$ or $^1_1H$ |
| Electron | $^0_{-1}e$ |
| Alpha particle | $^4_2\alpha$ or $^4_2He$ |
| Beta particle | $^0_{-1}\beta$ or $^0_{-1}e$ |
| Positron | $^0_1e$ |

active emission, because it represents the energy lost when the remaining nucleons reorganize into more stable arrangements. Generally, we shall not show the gamma rays when writing nuclear equations.

Two other types of radioactive decay that occur are positron emission and electron capture. A **positron** is a particle that has the same mass as an electron but an opposite charge.* The positron is represented as $^0_1e$. Carbon-11 is an example of an isotope that decays by positron emission:

$$^{11}_6C \longrightarrow {}^{11}_5B + {}^0_1e \qquad [21.6]$$

Emission of a positron can be thought of as converting a proton into a neutron, as shown in Equation 21.7. The atomic number of the nucleus is thereby decreased by one:

$$^1_1p \longrightarrow {}^1_0n + {}^0_1e \qquad [21.7]$$

**Electron capture** is the capture by the nucleus of an inner-shell electron from the electron cloud surrounding the nucleus. Rubidium-81 undergoes decay in this fashion, as shown in Equation 21.8:

$$^{81}_{37}Rb + {}^0_{-1}e \text{ (orbital electron)} \longrightarrow {}^{81}_{36}Kr \qquad [21.8]$$

Electron capture has the effect of converting a proton within the nucleus into a neutron as shown in Equation 21.9

$$^1_1p + {}^0_{-1}e \longrightarrow {}^1_0n \qquad [21.9]$$

Table 21.1 summarizes the symbols used to represent the various elementary particles in radioactive decay and nuclear transformations.

---

**SAMPLE EXERCISE 21.2**

Write balanced nuclear equations for the following reactions: **(a)** thorium-230 undergoes alpha decay; **(b)** thorium-231 undergoes decay to form protactinium-231.

**Solution:** **(a)** The information given in the problem can be summarized as

$$^{230}_{90}Th \longrightarrow {}^4_2He + X$$

The remaining product, X, must be deduced. Because mass numbers must have the same sum on both sides of the equation, we deduce that X has a mass number of 226. Similarly, the atomic number of X must be 88. Element number 88 is radium (refer to the periodic table or list of elements). The equation is therefore as follows:

$$^{230}_{90}Th \longrightarrow {}^4_2He + {}^{226}_{88}Ra$$

**(b)** In this case we must determine what type of particle is emitted in the course of the radioactive decay. We can write the following equation:

$$^{231}_{90}Th \longrightarrow {}^{231}_{91}Pa + X$$

The atomic numbers are obtained from a list of elements such as that given on the inside front cover. In order for the mass numbers to balance, X must have a mass number of 0. Its atomic number must be −1. The particle with these characteristics is the beta particle (electron). We therefore write the following:

$$^{231}_{90}Th \longrightarrow {}^{231}_{91}Pa + {}^0_{-1}e$$

**PRACTICE EXERCISE**

Write a balanced nuclear equation for the reaction in which oxygen-15 undergoes positron emission.
*Answer:* $^{15}_8O \longrightarrow {}^{15}_7N + {}^0_1e$

---

* The positron has a very short life, because it is annihilated when it collides with an electron, producing gamma rays: $^0_1e + {}^0_{-1}e \longrightarrow 2^0_0\gamma$.

There is no single rule that will allow us to predict whether a particular nucleus is radioactive and how it might decay. However, we can list some empirical observations that are helpful in making predictions:

1. All nuclei with 84 or more protons are unstable. For example, all isotopes of uranium, atomic number 92, are radioactive.
2. Nuclei with a total of 2, 8, 20, 50, 82, or 126 protons or neutrons are generally more stable than nuclei found near them in the periodic table. For example, there are three stable nuclei with an atomic number of 18, two with 19, five with 20, and one with 21; there are three stable nuclei with 18 neutrons, none with 19, four with 20, and none with 21. Thus there are more stable nuclei with 20 protons or 20 neutrons than with 18, 19, or 21. The numbers 2, 8, 20, 50, 82, and 126 are called **magic numbers**. Just as enhanced chemical stability is associated with the presence of 2, 10, 18, 36, 54, or 86 electrons, the noble gas configurations, enhanced nuclear stability is associated with the magic number of nucleons.
3. Nuclei with even numbers of both protons and neutrons are generally more stable than those with odd numbers of nucleons, as shown in Table 21.2.
4. The stability of a nucleus can be correlated to a certain degree with its neutron-to-proton ratio. All nuclei with two or more protons contain neutrons. Neutrons apparently help to hold protons together within the nucleus. As shown in Figure 21.1, the number of neutrons necessary to create a stable nucleus increases rapidly as the number of protons increases; the neutron-to-proton ratios of stable nuclei increase with increasing atomic number. The area within which all stable nuclei are found is known as the **belt of stability**.

**TABLE 21.2**  The number of stable isotopes with even and odd numbers of protons and neutrons

| Number of stable isotopes | Protons | Neutrons |
|---|---|---|
| 157 | Even | Even |
| 52 | Even | Odd |
| 50 | Odd | Even |
| 5 | Odd | Odd |

## SAMPLE EXERCISE 21.3

Would you expect the following nuclei to be radioactive: $^{4}_{2}\text{He}$, $^{39}_{20}\text{Ca}$, $^{210}_{85}\text{At}$?

**Solution:**  Helium-4 has a magic number of protons and of neutrons (two each). We would therefore expect $^{4}_{2}\text{He}$ to be stable.

Calcium-39 has an even number of protons (20) and an odd number of neutrons (19); 20 is one of the magic numbers. Nevertheless, we should suspect that this nuclide is radioactive, because the neutron-to-

proton ratio is less than 1. This ratio would place it below the belt of stability.

Astatine-210 is radioactive. Recall that there are no stable nuclei beyond atomic number 83.

## PRACTICE EXERCISE

Which of the following nuclides of group 6A elements are likely to be unstable: $^{14}_{8}\text{O}$, $^{32}_{16}\text{S}$, $^{78}_{34}\text{Se}$, $^{84}_{34}\text{Se}$, $^{115}_{52}\text{Te}$, $^{208}_{84}\text{Po}$?

***Answer:***  $^{14}_{8}\text{O}$, $^{84}_{34}\text{Se}$, $^{115}_{52}\text{Te}$, $^{208}_{84}\text{Po}$

The type of radioactive decay that a particular radioisotope will undergo depends to a large extent on its neutron-to-proton ratio compared to those of nearby nuclei that are within the belt of stability. Consider a nucleus whose high neutron-to-proton ratio places it above the belt of stability. This nucleus can lower its ratio and move toward the belt of stability by emitting a beta particle. Beta emission decreases the number of neutrons and increases the number of protons in a nucleus, as shown in Equation 21.5.

Nuclei that have low neutron-to-proton ratios and thus lie below the belt of stability either emit positrons or undergo electron capture. Both modes of decay decrease the number of protons and increase the number

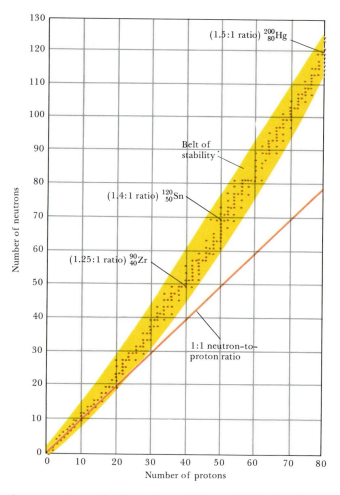

**FIGURE 21.1** Plot of the number of neutrons versus the number of protons in stable nuclei. As the atomic number increases, the neutron-to-proton ratio of the stable nuclei increases. The stable nuclei are located in the yellow area of the graph known as the belt of stability. The majority of radioactive nuclei occur outside this belt.

of neutrons in the nucleus, as shown in Equations 21.7 and 21.9. Positron emission is more common than electron capture among the lighter nuclei; however, electron capture becomes increasingly common as nuclear charge increases.

Alpha emission is found primarily among nuclei with an atomic number greater than 83. These nuclei would lie beyond the upper right edge of Figure 21.1, outside the belt of stability. Emission of an alpha particle moves the nucleus diagonally toward the belt of stability by decreasing both the number of protons and the number of neutrons by two. The result of each type of radioactive decay relative to a stable nucleus is shown in Figure 21.2.

## SAMPLE EXERCISE 21.4

By referring to Figure 21.1, predict the mode of radioactive decay of the following nuclei: **(a)** $^{20}_{11}$Na; **(b)** $^{97}_{40}$Zr; **(c)** $^{235}_{92}$U.

**Solution:** To answer this question we use the guidelines given above in the text.

**(a)** This nucleus has a neutron-to-proton ratio below 1. It therefore lies below the belt of stability. It can gain stability either by positron emission or electron capture. Because the atomic number is small we might predict that the nucleus undergoes positron emission. If we refer to a standard reference like the *Handbook of Chemistry and Physics*, we find that this prediction is correct. The nuclear reaction is

$$^{20}_{11}Na \longrightarrow \ ^{0}_{1}e + \ ^{20}_{10}Ne$$

**(b)** In referring to Figure 21.1, we find that this nucleus has a neutron-to-proton ratio that is too high. We would therefore predict that it undergoes beta decay. Again the prediction is correct. The nuclear reaction is

$$^{97}_{40}Zr \longrightarrow \ ^{0}_{-1}e + \ ^{97}_{41}Nb$$

**(c)** This nucleus lies outside the belt of stability to the upper right. We might therefore predict that it would undergo alpha emission. Again the prediction is correct. The nuclear equation is

$$^{235}_{92}U \longrightarrow \ ^{4}_{2}He + \ ^{231}_{90}Th$$

**PRACTICE EXERCISE**

Predict the mode of decay of the following unstable nuclei: **(a)** $^{15}O$; **(b)** $^{139}Xe$; **(c)** $^{212}Po$.

***Answers:*** **(a)** positron emission; **(b)** beta emission; **(c)** alpha emission

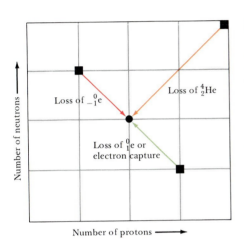

**FIGURE 21.2** Result of alpha emission ($^{4}_{2}He$), beta emission ($^{0}_{-1}e$), positron emission ($^{0}_{1}e$), and electron capture on the number of protons and neutrons in a nucleus. The squares represent unstable nuclei, and the circle represents a stable one. Moving from left to right or from bottom to top, each tick mark represents an additional proton or neutron, respectively. Moving in the reverse direction indicates the loss of a proton or neutron.

At this point we should note that our guidelines don't always work. For example, thorium-233, $^{233}_{90}Th$, which we might expect to undergo alpha decay, undergoes beta decay instead. Furthermore, a few radioactive nuclei actually lie within the belt of stability. For example, both $^{146}_{60}Nd$ and $^{148}_{60}Nd$ are stable and lie in the belt of stability; however, $^{147}_{60}Nd$, which lies between them, is radioactive.

Some nuclei, like uranium-238, cannot gain stability by a single emission. Consequently, a series of successive emissions occurs. As shown in Figure 21.3, uranium-238 decays to thorium-234, which is radioactive and decays to protactinium-234. This nucleus is also unstable and subsequently decays. Such successive reactions continue until a stable nucleus, lead-206, is formed. A series of nuclear reactions that begins with an unstable nucleus and terminates with a stable one is known as a **radioactive series** or a **nuclear disintegration series**. Three such series occur in nature. In addition to the series that begins with uranium-238 and terminates with lead-206, there is one that begins with uranium-235 and ends with lead-207. The third series begins with thorium-232 and ends with lead-208.

**Radioactive Series**

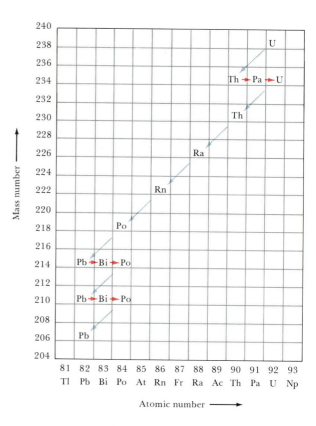

**FIGURE 21.3** Nuclear disintegration series for uranium-238. The $^{238}_{92}U$ nucleus decays to $^{234}_{90}Th$. Subsequent decay processes eventually form the stable $^{206}_{82}Pb$ nucleus. Each of the blue arrows corresponds to the loss of an alpha particle. Each red arrow corresponds to the loss of a beta particle.

## 21.3 PREPARATION OF NEW NUCLEI

In 1919, Rutherford performed the first artificial conversion of one nucleus into another. He succeeded in converting nitrogen-14 into oxygen-17, using the high-velocity alpha ($\alpha$) particles emitted by radium. The reaction is

$$^{14}_{7}N + ^{4}_{2}He \longrightarrow ^{17}_{8}O + ^{1}_{1}H \qquad [21.10]$$

(The symbols $^{1}_{1}H$ and $^{1}_{1}p$ are equivalent; both represent the proton.) This reaction demonstrated that nuclear reactions can be induced by striking nuclei with particles such as alpha particles. Such reactions have permitted synthesis of hundreds of radioisotopes in the laboratory. As noted in Section 21.1, these conversions of one nucleus into another are called nuclear transmutations. It is common to represent such conversions by listing, in order, the target nucleus, the bombarding particle, the ejected particle, and the product nucleus. Written in this fashion, Equation 21.10 is $^{14}_{7}N(\alpha, p)^{17}_{8}O$. The alpha particle, proton, and neutron are abbreviated as $\alpha$, p, and n, respectively.

**SAMPLE EXERCISE 21.5**

Write the balanced nuclear equation for the process summarized as $^{27}_{13}Al(n, \alpha)^{24}_{11}Na$.

**Solution:** The n is the abbreviation for a neutron, and $\alpha$ represents an alpha particle. The neutron is the bombarding particle, and the alpha particle is a product. Therefore, the nuclear equation is

$$^{27}_{13}Al + ^{1}_{0}n \longrightarrow ^{4}_{2}He + ^{24}_{11}Na$$

**PRACTICE EXERCISE**

Write the nuclear reaction
$$^{16}_{8}O + ^{1}_{1}H \longrightarrow ^{13}_{7}N + ^{4}_{2}He$$
in a shorthand notation.
*Answer:* $^{16}_{8}O(p, \alpha)^{13}_{7}N$

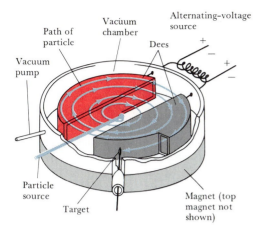

Path of particle
Vacuum chamber
Alternating-voltage source
Dees
Vacuum pump
Particle source
Target
Magnet (top magnet not shown)

**FIGURE 21.4**   Schematic drawing of a cyclotron. Charged particles are accelerated around the ring by the application of an alternating voltage to the dees.

Charged particles, such as alpha particles, must be moving very fast in order to overcome the electrostatic repulsion between them and the target nucleus. The higher the nuclear charge on either the projectile or the target, the faster the projectile must be moving to bring about a nuclear reaction. Many methods have been devised to accelerate charged particles using strong magnetic and electrostatic fields. These **particle accelerators** bear such names as the **cyclotron** and **synchrotron**. The cyclotron is illustrated in Figure 21.4. The hollow D-shaped electrodes are called "dees." The projectile particles are introduced into a vacuum chamber within the cyclotron. The particles are then accelerated by making the dees alternately positively and negatively charged. Magnets placed above and below the dees keep the particles moving in a spiral path until they are finally deflected out of the cyclotron and emerge to strike a target substance. Particle accelerators have been used mainly to synthesize heavy elements and to investigate the fundamental structure of matter. Figure 21.5 shows an aerial view of FermiLab, the National

**FIGURE 21.5**   An aerial view of the Fermi National Accelerator Laboratory at Batavia, Illinois. Particles are accelerated to very high energies by circulating them through magnets in the ring, which has a circumference of 6.3 km. (Fermilab Photo Dept.)

Accelerator Laboratory near Chicago. This facility will one day be supplanted by the even larger and higher-energy Superconducting Supercollider.

Most synthetic isotopes used in quantity in medicine and scientific research are made using neutrons as projectiles. Because neutrons are neutral, they are not repelled by the nucleus; consequently, they do not need to be accelerated, as do the charged particles, in order to cause nuclear reactions (indeed, they cannot be so accelerated). The necessary neutrons are produced by the reactions that occur in nuclear reactors (Section 21.7). Cobalt-60, used in radiation therapy for cancer, is produced by neutron capture. Iron-58 is placed in a nuclear reactor, where it is bombarded by neutrons. The following sequence of reactions takes place:

$$\ce{^{58}_{26}Fe} + \ce{^{1}_{0}n} \longrightarrow \ce{^{59}_{26}Fe} \qquad [21.11]$$

$$\ce{^{59}_{26}Fe} \longrightarrow \ce{^{59}_{27}Co} + \ce{^{0}_{-1}e} \qquad [21.12]$$

$$\ce{^{59}_{27}Co} + \ce{^{1}_{0}n} \longrightarrow \ce{^{60}_{27}Co} \qquad [21.13]$$

**Transuranium Elements**

Artificial transmutations have been used to produce the elements from atomic number 93 to 109. These are known as the **transuranium elements**, because they occur immediately following uranium in the periodic table. Elements 93 (neptunium) and 94 (plutonium) were first discovered in 1940. They were produced by bombarding uranium-238 with neutrons, as shown in Equations 21.14 and 21.15:

$$\ce{^{238}_{92}U} + \ce{^{1}_{0}n} \longrightarrow \ce{^{239}_{92}U} \longrightarrow \ce{^{239}_{93}Np} + \ce{^{0}_{-1}e} \qquad [21.14]$$

$$\ce{^{239}_{93}Np} \longrightarrow \ce{^{239}_{94}Pu} + \ce{^{0}_{-1}e} \qquad [21.15]$$

Elements with larger atomic numbers are normally formed in small quantities in particle accelerators. For example, curium-242 is formed when a plutonium-239 target is struck with accelerated alpha particles:

$$\ce{^{239}_{94}Pu} + \ce{^{4}_{2}He} \longrightarrow \ce{^{242}_{96}Cm} + \ce{^{1}_{0}n} \qquad [21.16]$$

Controversies have arisen over competing claims for the discovery and the right to name new elements. To provide a noncontroversial means for naming new elements, the International Union of Pure and Applied Chemistry has recommended that the names and corresponding three-letter symbols be based on the following roots:

| | | | |
|---|---|---|---|
| 0 = nil | | 5 = pent |
| 1 = un | | 6 = hex |
| 2 = bi | | 7 = sept |
| 3 = tri | | 8 = oct |
| 4 = quad | | 9 = enn |

The roots corresponding to the atomic number of the element are joined together and *ium* is added. For example, element 107 would have the name un-nil-sept-ium = unnilseptium. The symbol for the element would be Uns.

Again our discussions may have raised some questions in your mind. For example, why is it that some radioisotopes, like uranium-238, are found in nature, whereas others are not and must be synthesized? The key to this question is the fact that different nuclei undergo decay at different speeds. Uranium-238 undergoes decay very slowly, whereas many other nuclei, such as sulfur-35, decay rapidly. To understand the phenomenon of radoactivity, it is important to consider the rates of radioactive decay.

Radioactive decay is a first-order process. As shown in Section 15.3, first-order processes have characteristic half-lives. The **half-life** is the time required for half of any given quantity of a substance to react. The rates of decay of nuclei are commonly discussed in terms of their half-lives.

Each isotope has its own characteristic half-life. For example, the half-life of strontium-90 is 29 yr. If we started with 10.0 g of strontium-90, only 5.0 g of that isotope would remain after 29 yr. The other half of the strontium-90 would have been converted to yttrium-90, as shown in Equation 21.17:

$$\ce{^{90}_{38}Sr} \longrightarrow \ce{^{90}_{39}Y} + \ce{^{0}_{-1}e} \qquad [21.17]$$

After another 29-yr period, half the remaining 5.0 g of strontium-90 would likewise decay. The loss of strontium-90 as a function of time is shown in Figure 21.6.

Half-lives as short as millionths of a second and as long as billions of years have been observed. The half-lives of some radioisotopes are listed in Table 21.3. One important feature of half-lives for nuclear decay is that they are unaffected by external conditions such as temperature, pressure, or state of chemical combination. Therefore, unlike chemical toxins, radioactive atoms cannot be rendered harmless by chemical reac-

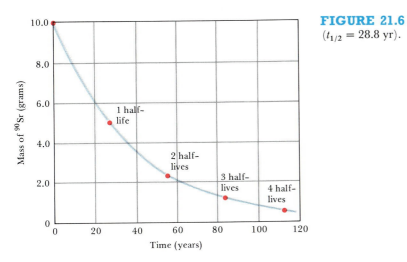

**FIGURE 21.6**  Decay of a 10.0-g sample of $\ce{^{90}_{38}Sr}$ ($t_{1/2} = 28.8$ yr).

**TABLE 21.3** Some radioactive isotopes and their half-lives and type of decay

|  | Isotope | Half-life (yr) | Type of decay |
|---|---|---|---|
| Natural radioisotopes | $^{238}_{92}U$ | $4.5 \times 10^9$ | Alpha |
|  | $^{235}_{92}U$ | $7.1 \times 10^8$ | Alpha |
|  | $^{232}_{90}Th$ | $1.4 \times 10^{10}$ | Alpha |
|  | $^{40}_{19}K$ | $1.3 \times 10^9$ | Beta |
|  | $^{14}_{6}C$ | 5730 | Beta |
| Synthetic radioisotopes | $^{239}_{94}Pu$ | 24,000 | Alpha |
|  | $^{137}_{55}Cs$ | 30 | Beta |
|  | $^{90}_{38}Sr$ | 28.8 | Beta |
|  | $^{131}_{53}I$ | 0.022 | Beta |

tion or by any other practical treatment. As far as we know now we can do nothing but allow these nuclei to lose radioactivity at their own characteristic rates. In the meantime, of course, we must take precautions to isolate the radioisotopes as much as possible because of the damage radiation can cause (see Section 21.9).

## SAMPLE EXERCISE 21.6

The half-life of cobalt-60 is 5.3 yr. How much of a 1.000-mg sample of cobalt-60 is left after a 15.9-yr period?

**Solution:** A period of 15.9 yr is three half-lives for cobalt-60. At the end of one half-life, 0.500 mg of cobalt-60 remains; 0.250 mg remains at the end of two half-lives, and 0.125 mg at the end of three half-lives.

## PRACTICE EXERCISE

Carbon-11, used in medical imaging, has a half-life of 20.4 minutes. The carbon-11 nuclides are formed, then incorporated into a desired compound. The resulting sample is injected into a patient and the medical image obtained. The entire process lasts five half-lives. What percentage of the original carbon-11 activity remains at this time?
*Answer:* 3.1 percent

### Dating

Because the half-life of any particular nuclide is constant, it can serve as a molecular clock and can be used to determine the ages of different objects. For example, carbon-14 has been used to determine the age of organic materials (see Figure 21.7). The procedure is based on the formation of carbon-14 by neutron capture in the upper atmosphere:

$$^{14}_{7}N + ^{1}_{0}n \longrightarrow ^{14}_{6}C + ^{1}_{1}H \qquad [21.18]$$

This reaction provides a small but reasonably constant source of carbon-14. The carbon-14 is radioactive, undergoing beta decay with a half-life of 5730 yr:

$$^{14}_{6}C \longrightarrow ^{14}_{7}N + ^{0}_{-1}e \qquad [21.19]$$

In using radiocarbon dating, it is generally assumed that the ratio of carbon-14 to carbon-12 in the atmosphere has been constant for at least 50,000 yr. The carbon-14 is incorporated into carbon dioxide, which is in turn incorporated, through photosynthesis, into more complex carbon-containing molecules within plants. The plants are eaten by animals, and the carbon-14 thereby becomes incorporated within them as well. Because a living plant or animal has a constant intake of carbon compounds, it is able to maintain a ratio of carbon-14 to carbon-12 that is identical with that of the atmosphere. However, once the organism dies, it no longer ingests carbon compounds to replenish the carbon-14 that is lost through radioactive decay. The ratio of carbon-14 to carbon-12 therefore decreases. If the ratio diminishes to half that of the atmosphere, we can conclude that the object is one half-life, or 5730 yr, old. This method cannot be used to date objects older than about 20,000 to 50,000 yr. After this length of time, the radioactivity is too low to be measured accurately.

The radiocarbon-dating technique has been checked by comparing the ages of trees determined by counting their rings and by radiocarbon analysis. As the tree grows, it adds a ring each year. In the old growth, the carbon-14 decays while the concentration of carbon-12 remains constant. The two dating methods agree to within about 10 percent. Most of the wood used in these tests was from California bristle-cone pines, which reach ages up to 2000 yr. By using trees that died at a known time thousands of years ago, it is possible to make comparisons back to about 5000 B.C.

Other isotopes can be similarly used to date other types of objects. For example, it takes $4.5 \times 10^9$ yr for half of a sample of uranium-238 to decay to a stable product, lead-206. The age of rocks containing uranium can be determined by measuring the ratio of lead-206 to uranium-238. If the lead-206 had somehow become incorporated into the rock by normal chemical processes instead of by radioactive decay, the rock would also contain large amounts of the more abundant isotope lead-208. In the absence of large amounts of this "geonormal" isotope of lead, it is assumed that all of the lead-206 was at one time uranium-238.

The oldest rocks found on the earth are approximately $3 \times 10^9$ yr old. This age indicates that the crust of the earth has been solid for at least this length of time. It is estimated that it required $1-1.5 \times 10^9$ yr for the earth to cool and its surface to become solid. This places the age of the earth at $4.0-4.5 \times 10^9$ yr.

**FIGURE 21.7** Egyptian mummy of a boy. The object dates from the second century A.D., as determined by carbon-14 dating. (The Granger Collection)

So far our discussion has been mainly qualitative. We now consider the topic of half-lives from a more quantitative point of view. This permits us to answer questions of the following types: How do we determine the half-life of uranium-238? We certainly don't sit around for $4.5 \times 10^9$ yr waiting for half of it to decay! Similarly, how do we quantitatively determine the age of an object that can be dated by radiometric means?

The rate of radioactive decay of any radioisotope is first order. It can therefore be described by Equation 21.20:

$$\text{Rate} = kN \qquad [21.20]$$

where $N$ is the number of nuclei of a particular radioisotope, and $k$ is the first-order rate constant. This equation can be transformed into Equation 21.21. (Note that the equations that follow are the same as those we encountered in Section 15.3, in dealing with first-order chemical processes.)

**Calculations Based on Half-Life**

$$\frac{-kt}{2.30} = \log \frac{N_t}{N_0} \qquad [21.21]$$

In this equation, $t$ is the time interval of decay, $k$ is the rate constant, $N_0$ is the initial number of nuclei (at zero time), and $N_t$ is the number remaining after the time interval. The relationship between the rate constant, $k$, and half-life, $t_{1/2}$, is given by Equation 21.22:

$$k = \frac{0.693}{t_{1/2}} \qquad [21.22]$$

## SAMPLE EXERCISE 21.7

A rock contains 0.257 mg of lead-206 for every milligram of uranium-238. How old is the rock if the half-life for decay of uranium-238 to lead-206 is $4.5 \times 10^9$ yr?

**Solution:**    The amount of uranium-238 in the rock when it was first formed is assumed to be 1.000 mg present at the time of the analysis plus the quantity that decayed to lead-206. We obtain the latter quantity by multiplying the mass of lead-206 present by the ratio of the atomic mass of uranium to the atomic mass of lead, into which it has decayed. The total original $^{238}_{92}U$ was thus

$$\text{Original } ^{238}_{92}U = 1.000 \text{ mg} + \frac{238}{206}(0.257 \text{ mg})$$

$$= 1.297 \text{ mg}$$

Using Equation 21.22, we can calculate the rate constant for the process from its half-life:

$$k = \frac{0.693}{4.5 \times 10^9 \text{ yr}} = 1.5 \times 10^{-10}/\text{yr}$$

Rearranging Equation 21.21 to solve for time, $t$, and substituting known quantities gives

$$t = \frac{-2.30}{k} \log \frac{N_t}{N_0}$$

$$= \frac{-2.30}{1.5 \times 10^{-10}/\text{yr}} \log \frac{1.000}{1.297} = 1.7 \times 10^9 \text{ yr}$$

### PRACTICE EXERCISE
A wooden object from an archeological site is reduced to carbon. The radioactivity of the sample due to $^{14}C$ is measured to be 12.4 disintegrations per second. The radioactivity of a carbon sample of equal mass from fresh wood is 19.5 disintegrations per second. The half-life of $^{14}C$ is 5730 yr. What is the age of the archeological sample?
*Answer:*   3740 yr

## SAMPLE EXERCISE 21.8

If we start with 1.000 g of strontium-90, 0.953 g will remain after 2.00 yr. **(a)** What is the half-life of strontium-90? **(b)** How much strontium-90 would remain after 5.00 yr?

**Solution:**   **(a)** Equation 21.21 can be solved for the rate constant, $k$, and then Equation 21.22 used to calculate half-life, $t_{1/2}$:

$$k = \frac{-2.30}{t} \log \frac{N_t}{N_0} = \frac{-2.30}{2.00 \text{ yr}} \log \frac{0.953 \text{ g}}{1.000 \text{ g}}$$

$$= \frac{-2.30}{2.00 \text{ yr}}(-0.0209) = 0.0240/\text{yr}$$

$$t_{1/2} = \frac{0.693}{k} = \frac{0.693}{0.0240/\text{yr}} = 28.9 \text{ yr}$$

**(b)** Again using Equation 21.21, with $k = 0.0240/\text{yr}$, we have

$$\log \frac{N_t}{N_0} = \frac{-kt}{2.30}$$

$$= \frac{-(0.0240/\text{yr})(5.00 \text{ yr})}{2.30} = -0.0522$$

Calculation of $N_t/N_0$ from $\log(N_t/N_0) = -0.0522$ is readily accomplished using the INV log or $10^x$ function of a calculator:

$$\frac{N_t}{N_0} = 10^{-0.0522} = 0.887$$

Because $N_0 = 1.000$ g, we have

$$N_t = 0.887 N_0 = 0.887(1.000 \text{ g}) = 0.887 \text{ g}$$

A sample to be used for medical imaging is labeled with $^{18}$F, which has a half-life of 110 minutes. What percentage of the original activity in the sample remains after 300 minutes?

*Answer:* 15.1 percent

## 21.5 Detection of Radioactivity

A variety of methods have been devised to detect emissions from radioactive substances. Becquerel discovered radioactivity because of the effect of radiation on photographic plates. Photographic plates and film have long been used to detect radioactivity. The radiation affects photographic film in the same way as ordinary light does. With care, film can be used to give a quantitative measure of activity. The greater the extent of exposure to radiation, the darker the area of the developed negative. People who work with radioactive substances carry film badges to record the extent of their exposure to radiation.

Radioactivity can also be detected and measured using a device known as a **Geiger counter**. The operation of the Geiger counter is based on the ionization of matter caused by radiation (Section 21.9). The ions and electrons produced by the ionizing radiation permit conduction of an electrical current. The basic design of a Geiger counter is shown in Figure 21.8. It consists of a metal tube filled with gas. The cylinder has a "window" made of material that can be penetrated by alpha, beta, or gamma rays. In the center of the tube is a wire. The wire is connected to one terminal of a source of direct current, and the metal cylinder is attached to the other terminal. Current flows between the wire and metal cylinder whenever ions are produced by entering radiation. The current pulse created when radiation enters the tube is amplified; each pulse is counted as a measure of the amount of radiation.

Certain substances that are electronically excited by radiation can also be used as means for detecting and measuring radiation. Some substances excited by radiation give off light (fluoresce) as electrons return to their lower-energy states. For example, dials of luminous watches are painted with a mixture of ZnS and a tiny quantity of $RaSO_4$. The zinc sulfide fluoresces when struck by the radioactive emissions from the radium. An instrument known as a **scintillation counter** can be used to detect and measure fluorescence and thereby the radiation that causes it.

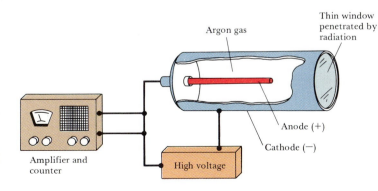

Argon gas

Thin window penetrated by radiation

Anode (+)

Cathode (−)

Amplifier and counter

High voltage

**FIGURE 21.8** Schematic representation of a Geiger counter.

## Radiotracers

Because radioisotopes can be detected so readily, they can be used to follow an element through its chemical reactions. For example, the incorporation of carbon atoms from $CO_2$ into glucose in photosynthesis has been studied using $CO_2$ containing carbon-14:

$$6CO_2 + 6H_2O \xrightarrow[\text{chlorophyll}]{\text{sunlight}} C_6H_{12}O_6 + 6O_2 \qquad [21.23]$$

The $CO_2$ is said to be labeled with the carbon-14. Detection devices such as scintillation counters follow the carbon-14 as it moves from the $CO_2$ through the various intermediate compounds to glucose.

Such use of radioisotopes is possible because all isotopes of an element have essentially identical chemical properties. When a small quantity of a radioisotope is mixed with the naturally occurring stable isotopes of the same element, all of the isotopes go through the same reactions together. The element's path is revealed by the radioactivity of the radioisotope. Because the radioisotope can be used to trace the path of the element, it is called a **radiotracer**.

## 21.6 MASS-ENERGY CONVERSIONS

So far we have said little about the energies associated with nuclear reactions. These energies can be considered with the aid of Einstein's famous equation relating mass and energy:

$$E = mc^2 \qquad [21.24]$$

In this equation $E$ stands for energy, $m$ for mass, and $c$ for the speed of light, $3.00 \times 10^8$ m/s. This equation states that the mass and energy of an object are proportional. The greater an object's mass, the greater its energy. Because the proportionality constant in the equation, $c^2$, is such a large number, even small changes in mass are accompanied by large changes in energy.

The mass changes in chemical reactions are too small to detect easily. For example, the mass change associated with the combustion of a mole of $CH_4$ is $9.9 \times 10^{-9}$ g. For this reason, it is possible to speak of the conservation of mass in chemical reactions.

The mass changes and the associated energy changes in nuclear reactions are much greater than in chemical reactions. The energy released through the nuclear fission of only about a pound of uranium (Section 21.7) is equivalent to that released by combustion of 1500 tons of coal.

### Nuclear Binding Energies

Scientists discovered in the 1930s that the masses of nuclei are always less than the masses of the individual nucleons of which they are composed. For example, the helium-4 nucleus has a mass of 4.00150 amu. The mass of a proton is 1.00728 amu, and that of a neutron is 1.00867 amu. Consequently, two protons and two neutrons have a total mass of 4.03190 amu:

$$
\begin{aligned}
\text{Mass of two protons} &= 2(1.00728 \text{ amu}) = 2.01456 \text{ amu} \\
\text{Mass of two neutrons} &= 2(1.00867 \text{ amu}) = \underline{2.01734 \text{ amu}} \\
\text{Total mass} &= \overline{4.03190 \text{ amu}}
\end{aligned}
$$

Radiotracers have found wide use as a diagnostic tool in medicine. For example, iodine-131 has been used to test the activity of the thyroid gland. This gland is the only important user of iodine in the body. The patient drinks a solution of NaI containing iodine-131. Only a very small amount is used so that the patient does not receive a harmful dose of radioactivity. A Geiger tube placed close to the thyroid, in the neck region, determines the ability of the thyroid to take up the iodine. A normal thyroid will absorb about 12 percent of the iodine within a few hours.

In 1977, Rosalyn S. Yalow received the Nobel Prize in medicine for her pioneering work in developing the technique of radioimmunoassay (RIA). This technique is an extraordinarily sensitive method involving radiotracers for detecting the presence of minute amounts of drugs, hormones, peptides, antibiotics, and a host of other substances. RIA methods can be used, for example, to provide an early indication of pregnancy and to detect substances that signal the early stages of a disease. The tests are carried out on a small sample of tissue, blood, or other fluid taken from the subject.

A new, very promising tool for clinical diagnosis of many diseases is called positron emission transaxial tomography, or PET. In this method, compounds that contain radionuclides that decay by positron emission are injected into a patient. These compounds are chosen to enable researchers to monitor blood flow, oxygen and glucose metabolic rates, and other biological functions. Some of the most interesting work involves study of the brain, which depends on glucose for most of its energy. Changes in how this sugar is metabolized or used by the brain may signal a disease such as cancer, epilepsy, Parkinson's disease, or schizophrenia.

The compound to be detected in the patient must be labeled with a radionuclide that is a positron emitter. The most widely used nuclides are carbon-11 (half-life 20.4 minutes), fluorine-18 (half-life 110 minutes), oxygen-15 (half-life 2 minutes), and nitrogen 13 (half-life 10 minutes). As an example, glucose can be labeled with $^{11}C$. The half-lives of these positron emitters are short so the chemist must quickly incorporate the radionuclide into the sugar (or other appropriate) molecule and inject the compound immediately. The patient is placed in an elaborate instrument [Figure 21.9(a)] that measures the positron emission and constructs a computer-based image of the organ in which the emitting compound is localized. The nature of this image [Figure 21.9(b)] provides clues as to the presence of disease or other abnormality and helps medical researchers understand how a particular disease affects the functioning of the brain.

**FIGURE 21.9** (a) An instrument for determining positron emission transaxial tomography (PET). The patient is injected with a solution of radiolabeled compound that quickly moves to the brain. Radioactive nuclei within the compound emit positrons. The PET instrument measures the positron emissions and develops a three-dimensional image of the brain. The level of radioactivity employed in such clinical work is not harmful to the patient. (b) PET images of human brains. The red line in the figure at lower right shows the two-dimensional slice through the brain that is shown in the images. The brain of a normal person and those of patients suffering from schizophrenia and manic depression show different image patterns, suggesting different patterns of metabolism. (Brookhaven National Laboratory and New York University Medical Center)

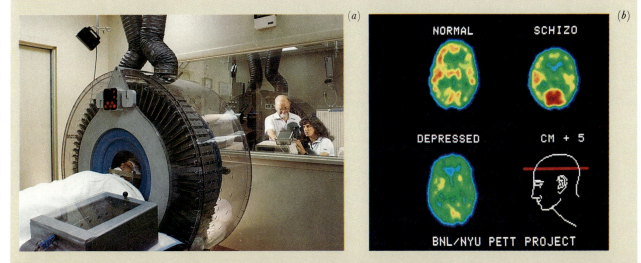

The individual nucleons weigh 0.03040 amu more than the helium-4 nucleus:

$$\text{Mass of two protons and two neutrons} = 4.03190 \text{ amu}$$
$$\text{Mass of } {}^4_2\text{He nucleus} = 4.00150 \text{ amu}$$
$$\text{Mass difference} = \overline{0.03040 \text{ amu}}$$

The mass difference between a nucleus and its constituent nucleons is called the **mass defect**. The origin of the mass defect is readily understood if we consider that energy must be added to a nucleus in order to break it into separated protons and neutrons:

$$\text{Energy} + {}^4_2\text{He} \longrightarrow 2{}^1_1\text{p} + 2{}^1_0\text{n} \qquad [21.25]$$

According to Einstein's mass-energy equivalence relationship (Equation 21.24), the addition of energy to a system must be accompanied by a proportional increase of mass. The mass change, $\Delta m$, is defined as the total mass of the products minus the total mass of the reactants. The mass change for the conversion of helium-4 into separated nucleons is $\Delta m = 0.03040$ amu, as shown in the calculations above. The associated energy change, therefore, is readily calculated:

$$\Delta E = c^2 \, \Delta m$$
$$= (3.00 \times 10^8 \text{ m/s})^2 (0.0304 \text{ amu})$$
$$\times \left( \frac{1.00 \text{ g}}{6.02 \times 10^{23} \text{ amu}} \right) \left( \frac{1 \text{ kg}}{1000 \text{ g}} \right)$$
$$= 4.54 \times 10^{-12} \, \frac{\text{kg-m}^2}{\text{s}^2} = 4.54 \times 10^{-12} \text{ J}$$

Decomposition of a mole of helium-4 in this fashion would require a tremendous quantity of energy:

$$\left( 6.02 \times 10^{23} \, \frac{\text{nuclei}}{\text{mol}} \right) \left( 4.54 \times 10^{-12} \, \frac{\text{J}}{\text{nucleus}} \right) = 2.73 \times 10^{12} \text{ J/mol}$$

The energy change calculated from the mass defect of a nucleus is called the **binding energy** of the nucleus. It is the energy required to decompose the nucleus into separated protons and neutrons. Thus the larger the binding energy, the more stable the nucleus is toward such decomposition. The mass defects and binding energies of three nuclei (helium-4, iron-56, and uranium-238) are compared in Table 21.4. The

**TABLE 21.4**  Mass differences and binding energies for three nuclei

| Nucleus | Mass of nucleus (amu) | Mass of individual nucleons (amu) | Mass difference (amu) | Binding energy (J) | Binding energy per nucleon (J) |
|---|---|---|---|---|---|
| ${}^4_2\text{He}$ | 4.00150 | 4.03190 | 0.0304 | $4.54 \times 10^{-12}$ | $1.14 \times 10^{-12}$ |
| ${}^{56}_{26}\text{Fe}$ | 55.92066 | 56.44938 | 0.52872 | $7.90 \times 10^{-11}$ | $1.41 \times 10^{-12}$ |
| ${}^{238}_{92}\text{U}$ | 238.0003 | 239.9356 | 1.9353 | $2.89 \times 10^{-10}$ | $1.22 \times 10^{-12}$ |

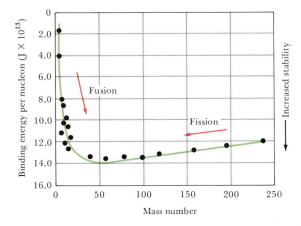

Binding energy per nucleon ($J \times 10^{13}$)

Mass number

Fusion

Fission

Increased stability

**FIGURE 21.10** The average binding energy per nucleon increases to a maximum at a mass number of 50 to 60 and decreases slowly thereafter. As a result of these trends, fusion of light nuclei and fission of heavy nuclei are exothermic processes.

binding energies per nucleon (that is, the binding energy of each nucleus divided by the total number of nucleons in that nucleus) are also compared in the table. Similar calculations for other nuclei indicate that the binding energy per nucleon increases in magnitude to about $1.4 \times 10^{-12}$ J for nuclei whose mass numbers are in the vicinity of iron-56. It then decreases slowly to about $1.2 \times 10^{-12}$ J for very heavy nuclei. This trend is shown in Figure 21.10. These results indicate that heavy nuclei will gain stability, and therefore give off energy, if they are fragmented. This process, known as **fission**, occurs in the atomic bomb and in nuclear power plants. Figure 21.10 also indicates that even greater amounts of energy should be available if very light nuclei are fused together. Such **fusion reactions** take place in the hydrogen bomb and are the essential energy-producing reactions in the sun. We shall look more closely at fission and fusion in Sections 21.7 and 21.8.

In calculating energy changes for nuclear reactions thus far, we have used nuclear masses. However, the masses of nuclides are normally expressed in tables as atomic masses (that is, the masses of nuclei plus electrons). It is a simple matter to calculate nuclear masses from atomic masses. For example, the mass of the $^{56}_{26}Fe$ atom is given in a table of nuclides as 55.93492 amu. This includes the mass of 26 electrons. When the mass of 26 electrons (26 electrons $\times 5.486 \times 10^{-4}$ amu/electron = 0.01426 amu) is subtracted from the atomic mass, we obtain the mass of the nucleus listed in Table 21.4, 55.92066 amu. In calculating the mass change in a nuclear reaction, it is usually acceptable to employ the masses of the atoms containing the nuclides of interest, because the number of electrons in the reactants and products is usually the same. Thus the difference in atomic masses is usually the same as the difference in nuclear masses. If you have any doubt about whether you can use atomic masses like this in a particular situation, it is never wrong to subtract the mass of electrons from the atomic masses and use the resultant nuclear masses.

## SAMPLE EXERCISE 21.9

How much energy is lost or gained when a mole of cobalt-60 undergoes beta decay: $^{60}_{27}Co \longrightarrow ^{0}_{-1}e + ^{60}_{28}Ni$? The mass of the $^{60}_{27}Co$ atom is 59.9338 amu, that of a $^{60}_{28}Ni$ atom is 59.9308 amu, and that of an electron is 0.000549 amu.

**Solution:** The products of the nuclear reaction are $^{60}_{28}Ni^+$ (the 27 electrons of cobalt are carried along in the reaction) and an electron (the beta particle). The mass of these products is just the mass of the neutral $^{60}_{28}Ni$ atom. The mass change in the reaction, therefore, is given by

$$\Delta m = \text{mass of } {}^{60}_{28}\text{Ni atom} - \text{mass of } {}^{60}_{27}\text{Co atom}$$

$$= 59.9308 \text{ amu} - 59.9338 \text{ amu}$$

$$= -0.0030 \text{ amu}$$

Thus for a mol of cobalt-60, $\Delta m = -0.0030$ g. The energy produced by the reaction can be calculated from this mass:

$$\Delta E = c^2 \Delta m$$

$$= (3.00 \times 10^8 \text{ m/s})^2 (-0.0030 \text{ g}) \left( \frac{1 \text{ kg}}{1000 \text{ g}} \right)$$

$$= -2.7 \times 10^{11} \frac{\text{kg-m}^2}{\text{s}^2} = -2.7 \times 10^{11} \text{ J}$$

By comparison, it takes only $9 \times 10^5$ J to break all the chemical bonds in a mole of water.

**PRACTICE EXERCISE**

Positron emission from $^{11}$C,

$$^{11}_{6}\text{C} \longrightarrow {}^{11}_{5}\text{B} + {}^{0}_{1}\text{e}$$

occurs with release of $2.87 \times 10^{11}$ J per mole of $^{11}$C. What is the mass change per mole of $^{11}$C in this nuclear reaction?

*Answer:* $3.19 \times 10^{-3}$ g

## 21.7  NUCLEAR FISSION

Our discussion of the energy changes in nuclear reactions (Section 21.6) revealed an important observation: Both the splitting of heavy nuclei (fission) and the union of light nuclei (fusion) are exothermic. Commercial nuclear power plants and the most common forms of nuclear weaponry depend for their operation on the process of nuclear fission. The first nuclear fission to be discovered was that of uranium-235. This nucleus, as well as those of uranium-233 and plutonium-239, undergoes fission when struck by a slow-moving neutrons.* The fission process is illustrated in Figure 21.11. A heavy nucleus can split in many different ways, just as can a piece of glass. Two different ways that the uranium-235 nucleus splits are shown in Equations 21.26 and 21.27:

$$^{1}_{0}\text{n} + {}^{235}_{92}\text{U} \diagup \begin{array}{l} {}^{137}_{52}\text{Te} + {}^{97}_{40}\text{Zr} + 2{}^{1}_{0}\text{n} \qquad [21.26] \\ \\ {}^{142}_{56}\text{Ba} + {}^{91}_{36}\text{Kr} + 3{}^{1}_{0}\text{n} \qquad [21.27] \end{array}$$

Over 200 different isotopes of 35 different elements have been found among the fission products of uranium-235. In general, these are radioactive.

On the average, 2.4 neutrons are produced by every fission of uranium-235. If one fission produces two neutrons, these two neutrons can cause two fissions. The four neutrons thereby released produce four fissions, and so forth, as shown in Figure 21.12. The number of fissions and

* There are other heavy nuclei that can be induced to undergo fission. However, these three are the only ones of practical importance.

**FIGURE 21.11** Schematic representation of the fission of uranium-235 showing one of its many fission patterns. In this process, $3.5 \times 10^{-11}$ J of energy is produced.

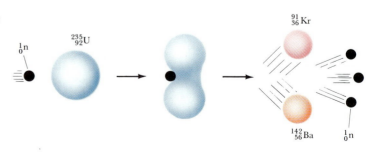

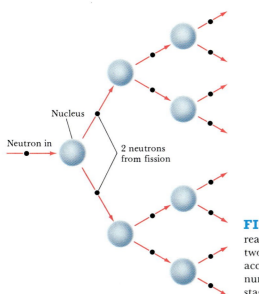

**FIGURE 21.12** Chain fission reaction in which each fission produces two neutrons. The process leads to an accelerating rate of fission, with the number of fissions doubling at each stage.

their associated energies quickly escalate, and, if the process is unchecked, the result is a violent explosion. Reactions that multiply in this fashion are called **branching chain reactions**.

In order for a chain fission reaction to occur, the sample of fissionable material must have a minimum size. Otherwise, neutrons escape from the sample before they have an opportunity to strike a nucleus and cause fission. The chain stops if enough neutrons are lost. The reaction is then said to be **subcritical**. If the mass is large enough to maintain the chain reaction with a constant rate of fission, the reaction is said to be **critical**. This situation results if only one neutron from each fission is subsequently effective in producing another fission. If the mass is larger still, few of the neutrons produced are able to escape. The chain reaction then multiplies the number of fissions, and the reaction is said to be **supercritical**. The effect of mass size on whether a reaction is subcritical or supercritical is illustrated in Figure 21.13.

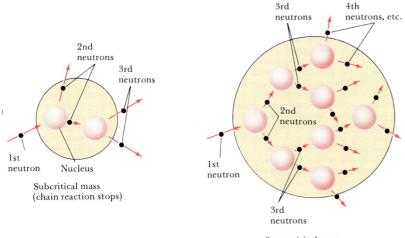

**FIGURE 21.13** The chain reaction in a subcritical mass soon stops because neutrons are lost from the mass without causing fission. As the size of the mass increases, fewer neutrons are able to escape. In a supercritical mass, the chain reaction is able to expand.

The fission of uranium-235 was first achieved in the late 1930s by Enrico Fermi and his colleagues in Rome and shortly thereafter by Otto Hahn and his co-workers in Berlin. Both groups were trying to produce transuranium elements. In 1938, Hahn identified barium among his reaction products. He was puzzled by this observation and questioned the identification because the presence of barium was so unexpected. He sent a detailed letter describing his experiments to Lise Meitner, a former co-worker. Meitner had been forced to leave Germany because of the anti-Semitism of the Third Reich, and she had settled in Sweden. She surmised that Hahn's experiment indicated that a new nuclear process was occurring in which the uranium-235 split. She called this process "nuclear fission." Meitner passed word of this discovery to her nephew, Otto Frisch, a physicist working at Niels Bohr's institute in Copenhagen. He repeated the experiment, verifying Hahn's observations and finding that tremendous energies were involved. In January 1939, Meitner and Frisch published a short article describing this new reaction. In March 1939, Leo Szilard and Walter Zinn at Columbia University discovered that more neutrons are produced than were used in each fission. As we have seen, this allows a branching chain reaction. News of these discoveries and an awareness of their potential use in explosive devices spread rapidly within the scientific community. Several scientists finally persuaded Albert Einstein, the most famous physicist of the time, to write a letter to President Roosevelt outlining the implications of these discoveries. Einstein's letter, written in August 1939, outlined the possible military applications of nuclear fission and emphasized the danger that weapons based on fission would pose if they were to be developed by the Nazis. Roosevelt judged it imperative that the United States investigate the possibility of such weapons. Late in 1941 the decision was made to build a bomb based on the fission reaction. An enormous research project, known as the "Manhattan Project," began. This undertaking led to the development of the atomic bomb and the dawning of the nuclear age.

One of the ways that an atomic bomb is triggered to produce a supercritical mass is shown in Figure 21.14. As shown, two subcritical masses are brought together by use of chemical explosives to form a supercritical mass. The basic design of such a bomb is quite simple. The fissionable materials are potentially available to any nation with a nuclear reactor. This simplicity has already resulted in the proliferation of atomic weapons. It is possible that such weapons could come into the hands of terrorist groups or groups intent on using the weapons to extort money from corporations or nations. Such events would add a frightening aspect to the nuclear age in which we live.

**FIGURE 21.14** One design used in atomic bombs. A conventional explosive is used to bring two subcritical masses together to form a supercritical mass.

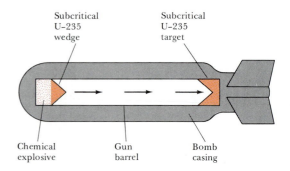

Nuclear fission produces the energy generated by nuclear power plants. The "fuel" of the nuclear reactor is a fissionable substance, such as uranium-235. Typically, uranium is enriched to about 3 percent uranium-235 and then used in the form of $UO_2$ pellets. These enriched uranium pellets are encased in zirconium or stainless-steel tubes. Rods composed of materials such as cadmium or boron control the fission process by absorbing neutrons. These **control rods** regulate the flux of neutrons to keep the reaction chain self-sustaining but yet prevent the reactor core from overheating.*

The reactor is started up by a neutron-emitting source; it is stopped by inserting the control rods more deeply into the reactor core, the site of the fission (Figure 21.15). The reactor core also contains a **moderator**, which acts to slow down neutrons, so that they can be captured more readily by the fuel. A **cooling liquid** circulates through the reactor core to carry off the heat generated by the nuclear fission. The cooling liquid can also serve as the neutron moderator.

The design of the power plant is basically the same as that of a power plant that burns fossil fuel (except that the burner is replaced by the reactor core). In both instances, steam is used to drive a turbine that is connected to an electrical generator. The steam must be condensed, so additional cooling water, generally obtained from a large source such as a river or lake, is needed. The water is returned to its source at a higher temperature than when it was removed. Power plants are therefore a significant source of thermal pollution. The nuclear power plant design shown in Figure 21.16 is currently the most popular. The primary coolant, which passes through the core, is in a closed system. Other coolants never pass through the reactor core at all. This lessens the chance that radioactive products could escape the core. Additionally, the reactor is

* The reactor core cannot reach supercritical levels and explode with the violence of an atomic bomb because the concentration of uranium-235 is too low. However, if the core overheats, sufficient damage might be done to release materials into the environment.

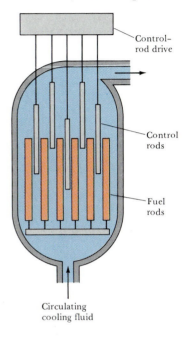

**FIGURE 21.15**  Reactor core showing fuel elements, control rods, and cooling fluid.

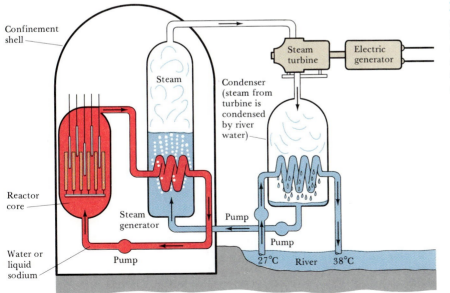

**FIGURE 21.16**  Basic design of a nuclear power plant. Heat produced by the reactor core is carried by a cooling fluid such as water or liquid sodium to a steam generator. The steam so produced is used to drive an electrical generator.

surrounded by a concrete shell to shield personnel and nearby residents from radiation.

Fission products accumulate as the reactor operates. These products lessen the efficiency of the reactor by capturing neutrons. The reactor must be stopped periodically, so that the nuclear fuel can be reprocessed. When the fuel rods are removed from the reactor they are initially very radioactive. It was originally intended that they be stored for several months in pools at the reactor site to allow decay of short-lived radioactive nuclei. They were then to be transported in shielded containers to reprocessing plants, where the fuel would be separated from the fission products. However, reprocessing plants have been plagued with operational difficulties, and there is intense opposition to the transport of nuclear wastes on the nation's highways. Even if the transportation difficulties could be overcome, the high radioactivity of the spent fuel makes reprocessing a hazardous operation. At present the spent fuel rods are simply being kept in storage at reactor sites.

In reprocessing, the uranium is separated from nuclear waste products and repackaged into new fuel rods. The problem then becomes one of disposal of the leftover fission products. Storage poses a major problem because the fission products are extremely radioactive. It is estimated that 20 half-lives are required for their radioactivity to reach levels acceptable for biological exposure. Based on the 28.8-yr half-life of strontium-90, one of the longer-lived and most dangerous of the products, the wastes must be stored for 600 yr. If plutonium-239 is not removed, storage must be for longer periods; plutonium-239 has a half-life of 24,000 yr. It is advantageous, however, to remove plutonium-239 because it can be used as a fissionable fuel.

A considerable amount of research has been and will continue to be devoted to disposal of radioactive wastes. At present the most attractive possibilities appear to be formation of a glass, ceramic, or synthetic rock from the wastes, as a means of immobilizing them. The solid materials could then be buried deep underground. Because the radioactivity will persist for a long time, there must be assurances that the solids will not crack from the heat generated by nuclear decay, possibly allowing radioactivity to find its way into underground water supplies.

**Breeder Reactors**

Uranium-235 is rare, so its supply can readily be exhausted. Methods of generating other fissionable materials are being actively investigated. Fissionable plutonium-239 and uranium-233 can be produced in nuclear reactors from nuclides that are far more abundant than uranium-235. The reactions involved are as follows:

$$^{238}_{92}U + ^{1}_{0}n \longrightarrow ^{239}_{92}U \tag{21.28}$$

$$^{239}_{92}U \longrightarrow ^{239}_{93}Np + ^{0}_{-1}e \quad (t_{1/2} = 24 \text{ min}) \tag{21.29}$$

$$^{239}_{93}Np \longrightarrow ^{239}_{94}Pu + ^{0}_{-1}e \quad (t_{1/2} = 2.3 \text{ days}) \tag{21.30}$$

$$^{232}_{90}Th + ^{1}_{0}n \longrightarrow ^{233}_{90}Th \tag{21.31}$$

$$^{233}_{90}Th \longrightarrow ^{233}_{91}Pa + ^{0}_{-1}e \quad (t_{1/2} = 22 \text{ min}) \tag{21.32}$$

$$^{233}_{91}Pa \longrightarrow ^{233}_{92}U + ^{0}_{-1}e \quad (t_{1/2} = 27 \text{ days}) \tag{21.33}$$

It is theoretically possible to build a reactor that both produces energy and converts uranium-238 or thorium-232 into fissionable fuel. We can envision a fission of a uranium-235 nucleus producing two neutrons, one causing another fission, the second initiating the change of uranium-238 into plutonium-239. Plutonium-239 is produced in ordinary reactors because of the presence of uranium-238. However, the hope is to be able to produce more fuel than the reactor uses. Reactors able to do this are called **breeder reactors**.

The design of breeder reactors poses many difficult technical problems. In addition, the breeder reactor program has become the subject of intense political debate. Because both breeder reactors and nuclear fuel reprocessing plants produce plutonium-239, any nation that acquired a breeder reactor or reprocessing technology would have in hand the raw materials for atomic weaponry. Fear of nuclear proliferation or the possibility of thefts of nuclear materials by terrorist groups has resulted in uncertainty as to whether the United States should proceed with breeder reactor development and fuel reprocessing. Breeder reactors are, however, in operation in other countries.

## 21.8 NUCLEAR FUSION

As shown in Section 21.6, energy is produced when light nuclei are fused to form heavier ones. Reactions of this type are responsible for the energy produced by the sun. Spectroscopic studies indicate that the sun is composed of 73 percent H, 26 percent He, and only 1 percent of all other elements, by mass. Among the several fusion processes that are believed to occur are the following:

$$^1_1\text{H} + ^1_1\text{H} \longrightarrow ^2_1\text{H} + ^0_1\text{e} \qquad [21.34]$$

$$^1_1\text{H} + ^2_1\text{H} \longrightarrow ^3_2\text{He} \qquad [21.35]$$

$$^3_2\text{He} + ^3_2\text{He} \longrightarrow ^4_2\text{He} + 2^1_1\text{H} \qquad [21.36]$$

$$^3_2\text{He} + ^1_1\text{H} \longrightarrow ^4_2\text{He} + ^0_1\text{e} \qquad [21.37]$$

Theories have been proposed for the generation of the other elements via fusion processes.

Fusion is appealing as an energy source because of the availability of light isotopes and because fusion products are generally not radioactive. It is therefore potentially a cleaner process than is fission. The problem is that high energies are needed to overcome the repulsion between nuclei. The required energies are achieved by high temperatures. Fusion reactions are therefore also known as **thermonuclear reactions**. The lowest temperature required for any fusion is that needed to fuse $^2_1\text{H}$ and $^3_1\text{H}$, shown in Equation 21.38, which requires a temperature of 40,000,000 K:

$$^2_1\text{H} + ^3_1\text{H} \longrightarrow ^4_2\text{He} + ^1_0\text{n} \qquad [21.38]$$

Such high temperatures have been achieved by using an atomic bomb to initiate the fusion process. This is done in the thermonuclear, or hydrogen, bomb. Clearly, this approach is unacceptable for controlled power generation.

Numerous problems must be overcome before fusion becomes a practical energy source. Besides the high temperatures necessary to initiate the reaction, there is the problem of confining the reaction. No known structural material is able to withstand the enormous temperatures necessary for fusion. Research has centered on the use of strong magnetic fields to contain the reaction. Use of powerful lasers to generate the temperatures required for fusion has also been the subject of much recent research. Although there is reason for some degree of optimism, it is impossible to tell if and when the tremendous technical difficulties involving nuclear fusion will be overcome. It is therefore not yet clear whether fusion will ever be a practical source of energy for humankind.

## 21.9 BIOLOGICAL EFFECTS OF RADIATION

The increased pace of synthesis and use of radioisotopes has led to increased concern over the effects of radiation on matter, particularly in biological systems. We therefore close this chapter by examining the health hazards associated with radioisotopes.

Alpha, beta, and gamma rays (as well as X rays) possess energies far in excess of ordinary bond energies and ionization energies. Consequently, these forms of radiation are able to fragment and ionize molecules, generating unstable, highly reactive particles as they pass through matter. For example, gamma rays are able to ionize water molecules, forming unstable $H_2O^+$ ions. An $H_2O^+$ ion can react with another water molecule to form an $H_3O^+$ ion and a neutral OH molecule:

$$H_2O^+ + H_2O \longrightarrow H_3O^+ + OH \qquad [21.39]$$

The unstable and highly reactive OH molecule is an example of a **free radical**, a substance with one or more unpaired electrons. In a biological system such particles can attack a host of other compounds to produce new free radicals, which in turn attack yet other compounds. Thus the formation of a single free radical can instigate a large number of chemical reactions that are ultimately able to disrupt the normal operations of cells.

The resultant radiation damage to living systems can be classified as either **somatic** or **genetic**. Somatic damage is that which affects the organism during its own lifetime. Genetic damage is that which has a genetic effect; it harms offspring through damage to genes and chromosomes, the body's reproductive material. It is more difficult to study genetic effects than somatic ones because it may take several generation for genetic damage to become apparent. Somatic damage includes "burns," molecular disruptions similar to those produced by high temperatures. It also includes cancer. Cancer is brought about by damage to the growth-regulation mechanism of cells, which causes them to reproduce in an uncontrolled manner. In general, the tissues that show the greatest damage from radiation are those that reproduce at a rapid rate, such as bone marrow, blood-forming tissues, and lymph nodes. Leukemia is probably the major cancer problem associated with radiation.

The clinical symptoms of acute (short-term) exposure to radiation include a decrease in the number of white blood cells, fatigue, nausea, and diarrhea. Sufficient exposure can result in death from blood disorderrs, gastrointestinal failure, and damage to the central nervous system. In light of these effects, it is important to determine whether there are any safe levels of exposure to radiation. What are the maximum levels of radiation that we should permit from various human activities? Unfortunately, we are hampered in our attempts to set realistic standards by our lack of understanding of the effects of chronic (long-term) exposure to radiation. Most scientists presently believe that the effects of radiation are proportional to exposure, even down to low exposures. This means that *any* amount of radiation causes some finite risk of injury.

**Radiation Doses**

The SI unit of nuclear radioactivity is called the **becquerel**, named after Henri Becquerel, the discoverer of radioactivity (Section 2.2). A becquerel is defined as one nuclear disintegration per second. The older and more widely used unit of activity is the **curie** (Ci), which is the number of nuclear disintegrations per second from 1 g of radium, $3.7 \times 10^{10}$ disintegrations/s. For example, a 5.0-millicurie sample of cobalt-60 undergoes $(5.0 \times 10^{-3})(3.7 \times 10^{10}) = 1.8 \times 10^8$ disintegrations per second.

The damage produced by radiation outside the body depends not only on the nuclear activity, but also on the energy and penetrating power of the radiation. Gamma rays are particularly dangerous because they penetrate human tissue very effectively, just as do X rays. Consequently, their damage is not limited to the skin. In contrast, most alpha rays are stopped by skin, and beta rays are able to penetrate only about 1 cm beyond the surface of the skin. Hence, neither is as dangerous as gamma rays unless the radiation source somehow enters the body. Within the body, alpha rays are particularly dangerous because they leave a very dense trail of damaged molecules as they move through matter.

Two units, the rad and rem, are commonly used to measure radiation doses. (A third unit, the roentgen, is essentially the same as the rad.) A **rad** (radiation *a*bsorbed *d*ose) is the amount of radiation that deposits $1 \times 10^{-2}$ J of energy per kilogram of tissue. A rad of alpha rays can produce more damage than a rad of beta rays. Consequently, the rad is often multiplied by a factor that measures the relative biological damage caused by the radiation. This factor is known as the relative biological effectiveness of the radiation, abbreviated **RBE**. The RBE is approximately 1 for beta and gamma rays and 10 for alpha rays. The exact value for the RBE varies with dose rate, total dose, and type of tissue affected. The product of the number of rads and the RBE of the radiation gives the effective dosage in **rems** (*r*oentgen *e*quivalent for *m*an):

$$\text{Number of rems} = (\text{number of rads})(\text{RBE}) \qquad [21.40]$$

The effects of some short-time exposures to radiation appear in Table 21.5. For most persons the effects of long-term exposure to low levels of radiation are more important. Each of us receives an average exposure of 0.1 to 0.2 rem each year as background radiation from natural sources, such as naturally occurring radioisotopes and cosmic rays coming in from outer space. This amount may vary widely from one person to the next, depending on living situation. For example, persons who live in stone or brick houses normally receive more exposure than those who live in wooden houses. Those who live at higher elevations receive more cosmic ray exposure, because the earth's atmosphere absorbs cosmic rays to some extent.

**TABLE 21.5**  Effects of short-time exposures to radiation

| Dose (rem) | Effect |
|---|---|
| 0–25 | No detectable clinical effects |
| 25–50 | Slight, temporary decrease in white blood cell counts |
| 100–200 | Nausea; marked decrease in white blood cells |
| 500 | Death of half the exposed population within 30 days after exposure |

In addition to natural sources, we all receive some exposure to radiation resulting from human activity. Table 21.6 shows radiation levels and the percent contributions of the more important sources to the average exposure of the U.S. population. It must be emphasized that these are merely averages; the relative contributions of the various sources will vary widely from one person to another. For example, a chest X ray involves a radiation dose of 20 to 40 millirem. Some people have never been X-rayed; others have had many exposures. Uranium miners clearly receive larger exposures than the general populace.

Radiation exposure limits are presently set by the Environmental Protection Agency at 500 millirem each year for the general population and 5 rem for occupational exposure, exclusive of background radiation. As more scientists have become convinced that even low radiation dosage poses some risk, there has been increasing pressure to set tougher standards for radiation exposure. As with many other aspects of human activity, the question at issue is one of risk versus benefit. Before we can make intelligent decisions we must have a deeper understanding of the risks than we do at present.

**TABLE 21.6**  Contributions to the average exposure of the U.S. population to ionizing radiation

| Source | Percent contribution | Radiation level (millirem) |
|---|---|---|
| Natural background | 52 | 100 |
| X rays, other medical | 43 | 80 |
| Uranium mining, other technology-related sources | 2 | 4 |
| Fallout from nuclear weapons testing | 3 | 6 |
| Nuclear power plants | 0.14 | 0.3 |
| Consumer products— watch dials, color TV, etc. | 0.02 | 0.04 |

# FOR REVIEW

## SUMMARY

Certain nuclei are **radioactive**. Most of these nuclei gain stability by emitting **alpha particles** ($_2^4$He), **beta particles** ($_{-1}^{0}$e), and/or **gamma radiation** ($_0^0\gamma$). Some nuclei undergo decay by **positron** ($_1^0$e) emission or by **electron capture**. The **neutron-to-proton** ratio is one factor determining nuclear stability. The presence of **magic numbers** of **nucleons** and an even number of protons and neutrons are also important. **Nuclear transmutations** can be induced by bombarding nuclei with charged particles using particle accelerators or with neutrons in a nuclear reactor.

**Radioisotopes** have characteristic rates of decay. These rates are generally expressed in terms of **half-lives**. The constant half-lives of nuclides permit their use in dating objects. The ease of detection of radioisotopes also permits their use as **tracers**, to follow elements through their reactions. Three methods of detection were discussed: use of **photographic film**, **scintillation counters**, and **Geiger counters**.

The energy produced in nuclear reactions is accompanied by measurable losses of mass in accordance with Einstein's relationship, $\Delta E = c^2 \Delta m$. The difference in mass between nuclei and the nucleons of which they are composed is known as the **mass defect**. The mass defect of a nuclide allows calculation of its nuclear **binding energy**, the energy required to separate the nucleus into individual nucleons. Examination of binding energies per nucleon reveals that energy is produced when heavy nuclei split (**fission**) and when light nuclei fuse (**fusion**).

Uranium-235, uranium-233, and plutonium-239 undergo fission when they capture a neutron. The resulting nuclear reaction is a **chain reaction**. If the reaction maintains a constant rate it is said to be **critical**. If it slows it is said to be **subcritical**. In the atomic bomb, subcritical masses are brought together to form a mass that is **supercritical**. In nuclear reactors the fission is controlled to generate a constant power. The reactor core consists of fission-able fuel, control rods, moderator, and cooling fluid. The nuclear power plant resembles a conventional power plant except that the core replaces the fuel burner. In **breeder reactors** more nuclear fuel is produced than is used to generate energy. There is concern regarding the safety with which nuclear power plants can be operated. Beyond this, the reprocessing of spent fuel rods and disposal of highly radioactive nuclear wastes are unsolved problems.

Nuclear fusion requires high temperatures because nuclei must have large kinetic energies to overcome their mutual repulsions. It has not yet been possible to generate a controlled fusion process.

The SI unit of radioactivity is the **becquerel**, defined as one nuclear disintegration per second. Radioactivity is more often measured in **curies**. One curie corresponds to $3.7 \times 10^{10}$ disintegrations per second. The amount of energy deposited in biological tissue by radiation is measured in terms of the **rad**; one rad corresponds to $1 \times 10^{-2}$ J per kilogram of tissue. The **rem** is a more useful measure of the biological damage created by the deposited energy. We receive radiation from both naturally occurring and human sources in roughly equal amounts for the general population. The effects of long-term exposure to low levels of radiation are not completely understood, but the most commonly accepted hypothesis is that the extent of biological damage varies in direct proportion to the level of exposure.

## LEARNING GOALS

Having read and studied this chapter, you should be able to:

1. Write the nuclear symbols for protons, neutrons, electrons, alpha particles, and positrons.

2. Determine the effect of different types of decay on the proton-neutron ratio, and predict the type of decay that a nucleus will undergo based on its position relative to the belt of stability.

3. Complete and balance nuclear equations, having been given all but one of the particles involved.

4. Use the half-life of a substance to predict the amount of radioisotope present after a given period of time.

5. Calculate half-life, age of an object, or the remaining amount of radioisotope, having been given any two of these pieces of information.

6. Explain how radioisotopes can be used in dating objects and as radiotracers.

7. Explain how radioactivity is detected, including a description of the basic design of a Geiger counter.

8. Use Einstein's relation, $\Delta E = c^2 \Delta m$, to calculate the energy change or the mass change of a reaction, having been given one of these quantities.

9. Calculate the binding energies of nuclei, having been given their masses and the masses of protons and neutrons.

10. Explain what fission and fusion are, and state what types of nuclei produce energy when undergoing these processes.

11. Describe the design of a nuclear power plant, including an explanation of the role of fuel elements, control rods, moderator, and cooling fluid.

12. Explain the role played by the mode of radioactivity of a radioisotope in determining its ability to damage biological systems.

**13.** Define the units used to describe the level of radioactivity (becquerel, curie) and to measure the effects of radiation on biological systems (rem and rad).

**14.** Describe various sources of radiation to which the general population is exposed, and indicate the relative contributions of each.

## KEY TERMS

Among the more important terms and expressions used for the first time in this chapter are the following:

**Alpha particles** (Section 21.1) are identical to helium-4 nuclei, consisting of two protons and two neutrons, symbol $^4_2\text{He}$.

The **becquerel** (Section 21.9) is the SI unit of radioactivity. It corresponds to one nuclear disintegration per second.

**Beta particles** (Section 21.2) are energetic electrons emitted from the nucleus, symbol $^0_{-1}\text{e}$.

The **binding energy** (Section 21.6) of a nucleus is the energy required to decompose that nucleus into nucleons; it is usually calculated from the **mass defect** of the nucleus.

A **breeder reactor** (Section 21.7) is a nuclear-fission reactor that produces more fissionable fuel than it consumes.

A **chain reaction** (Section 21.7) is a series of reactions in which one reaction initiates the next.

A **critical mass** (Section 21.7) is the amount of fissionable material necessary to maintain a chain reaction. Smaller masses are said to be **subcritical**; larger ones are **supercritical**.

A **curie** (Section 21.9) is a measure of radioactivity: 1 curie $= 3.7 \times 10^{10}$ nuclear disintegrations per second.

**Electron capture** (Section 21.2) is a mode of radioactive decay in which an inner-shell orbital electron is captured by the nucleus.

**Fission** (Section 21.7) is the splitting of a large nucleus into two intermediate-sized ones.

A **free radical** (Section 21.9) is a substance with one or more unpaired electrons.

**Fusion** (Section 21.8) is the joining of two light nuclei to form a more massive one.

**Gamma rays** (Section 21.2) are energetic electromagnetic radiation emanating from the nucleus of a radioactive atom.

**Half-life** (Section 21.4) is the time required for half of a sample of a particular radioisotope to decay.

The **mass defect** (Section 21.6) of a nucleus is the difference between the mass of the nucleus and the total masses of the individual nucleons that it contains.

A **nuclear transmutation** (Section 21.1) is a conversion of one kind of nucleus to another.

A **positron** (Section 21.2) is a particle with the same mass as an electron but with a positive charge, symbol $^0_1\text{e}$.

A **rad** (Section 21.9) is a measure of the energy absorbed from radiation by tissue or other biological material; 1 rad = transfer of $1 \times 10^{-2}$ J of energy per kilogram of material.

A **radioisotope** (Section 21.1) is an isotope that is radioactive; that is, it is undergoing nuclear changes with emission of nuclear radiation.

**RBE** (Section 21.9) (relative biological effectiveness) is an adjustment factor used to convert rads to rems; it accounts for differences in biological effects of different particles having the same energy.

A **rem** (Section 21.9) is a measure of the biological damage caused by radiation; rems = rads × RBE.

## EXERCISES

### Nuclear Reactions

**21.1** Indicate the number of protons and neutrons in each of the following nuclei: **(a)** oxygen-17; **(b)** $^{99}_{42}\text{Mo}$; **(c)** cesium-136; **(d)** $^{115}\text{Ag}$.

**21.2** Indicate the number of protons, neutrons, and nucleons in each of the following: **(a)** cerium-137; **(b)** $^{234}\text{Pu}$; **(c)** cadmium-113; **(d)** $^{37}_{17}\text{Cl}$.

**21.3** Write balanced nuclear equations for the following transformations. **(a)** $^{181}\text{Hf}$ undergoes beta decay. **(b)** Radium-226 decays to a radon isotope. **(c)** Lead-205 undergoes positron emission. **(d)** Tungsten-179 undergoes orbital electron capture.

**21.4** Write balanced nuclear equations for each of the following nuclear transformations. **(a)** Zirconium-93 undergoes beta decay. **(b)** Neptunium-233 undergoes alpha decay **(c)** Francium-218 is formed by decay of an actinium nuclide. **(d)** Einsteinium-246 undergoes orbital electron capture.

**21.5** Complete and balance the following nuclear equations by supplying the missing particle:

**(a)** $^{32}_{16}\text{S} + ^1_0\text{n} \longrightarrow ^1_1\text{H} + ?$

**(b)** $^7_4\text{Be} + ^0_{-1}\text{e}$ (orbital electron) $\longrightarrow$ ?

**(c)** ? $\longrightarrow ^{187}_{76}\text{Os} + ^0_{-1}\text{e}$

**(d)** $^{98}_{42}\text{Mo} + ^2_1\text{H} \longrightarrow ^1_0\text{n} + ?$

**(e)** $^{235}_{92}\text{U} + ^1_0\text{n} \longrightarrow ^{135}_{54}\text{Xe} + ? + 2^1_0\text{n}$

**21.6** Complete and balance the following nuclear equations by supplying the missing particle:

(a) $^{252}_{98}\text{Cf} + ^{10}_{5}\text{B} \longrightarrow 3^{1}_{0}\text{n} + ?$

(b) $^{2}_{1}\text{H} + ^{3}_{2}\text{He} \longrightarrow ^{4}_{2}\text{He} + ?$

(c) $^{1}_{1}\text{H} + ^{11}_{5}\text{B} \longrightarrow 3?$

(d) $^{122}_{53}\text{I} \longrightarrow ^{122}_{54}\text{Xe} + ?$

(e) $^{59}_{26}\text{Fe} \longrightarrow ^{0}_{-1}\text{e} + ?$

**21.7** The naturally occurring radioactive decay series that begins with $^{235}_{92}\text{U}$ stops with formation of the stable $^{207}_{82}\text{Pb}$. The decays proceed through a series of alpha particle and beta particle emissions. How many of each type of emission are involved in this series?

**21.8** A radioactive decay series that begins with $^{237}_{93}\text{Np}$ ends with formation of the stable nuclide $^{209}_{83}\text{Bi}$. How many alpha particle emissions and how many beta particle emissions are involved in the sequence of radioactive decays?

**21.9** Write balanced equations for the following nuclear reactions: **(a)** $^{238}_{92}\text{U}(n, \gamma)^{239}_{92}\text{U}$; **(b)** $^{14}_{7}\text{N}(p, \alpha)^{11}_{6}\text{C}$; **(c)** $^{18}_{8}\text{O}(n, \beta)^{19}_{9}\text{F}$.

**21.10** Write balanced equations for each of the following nuclear reactions: **(a)** $^{14}_{7}\text{N}(p, \alpha)^{11}_{6}\text{C}$; **(b)** $^{14}_{7}\text{N}(\alpha, p)^{17}_{8}\text{O}$; **(c)** $^{59}_{26}\text{Fe}(\alpha, \beta)^{63}_{29}\text{Cu}$.

## Nuclear Stability

**21.11** Which of the following nuclides would you expect to be radioactive: **(a)** $^{17}_{8}\text{O}$; **(b)** $^{176}_{74}\text{W}$; **(c)** $^{108}_{50}\text{Sn}$; **(d)** $^{92}_{40}\text{Zr}$; **(e)** $^{238}_{94}\text{Pu}$? Justify your choices.

**21.12** In each of the following pairs, which nuclide would you expect to be the more abundant in nature: **(a)** $^{19}_{9}\text{F}$ or $^{18}_{9}\text{F}$; **(b)** $^{80}_{34}\text{Se}$ or $^{81}_{34}\text{Se}$; **(c)** $^{56}_{26}\text{Fe}$ or $^{57}_{26}\text{Fe}$; **(d)** $^{118}_{50}\text{Sn}$ or $^{118}_{51}\text{Sb}$? Justify your choices.

**21.13** Which of the following nuclides is most likely to be a positron emitter: **(a)** $^{53}_{24}\text{Cr}$; **(b)** $^{51}_{25}\text{Mn}$; **(c)** $^{59}_{26}\text{Fe}$? Explain.

**21.14** All of the following nuclides are radioactive and undergo either beta or positron emission: **(a)** $^{66}_{32}\text{Ge}$; **(b)** $^{105}_{45}\text{Rh}$; **(c)** $^{137}_{53}\text{I}$; **(d)** $^{133}_{58}\text{Ce}$. Indicate which nuclei undergo which type of emission.

**21.15** It has been suggested that strontium-90 (derived from nuclear testing) deposited in the hot desert will undergo radioactive decay more rapidly because it will be exposed to much higher average temperatures. Is this a reasonable suggestion?

**21.16** Sulfur-35 is radioactive and undergoes beta decay. What differences would you expect in the chemical behavior of atoms containing sulfur-35 as compared with those containing sulfur-32, which is nonradioactive? Explain.

**21.17** Indicate whether each of the following nuclides lies within the belt of stability in Figure 21.1: **(a)** $^{108}_{49}\text{In}$; **(b)** $^{102}_{47}\text{Ag}$; **(c)** $^{17}_{7}\text{N}$; **(d)** $^{210}_{86}\text{Rn}$. For any that does not, describe a nuclear decay process that would alter the neutron-to-proton ratio in the direction of increased stability.

**21.18** Indicate whether each of the following nuclides lies within the belt of stability in Figure 21.1: **(a)** $^{70}_{33}\text{As}$; **(b)** $^{34}_{15}\text{P}$; **(c)** $^{74}_{32}\text{Ge}$; **(d)** $^{248}_{98}\text{Cf}$. For any that does not, describe a nuclear decay process that would alter the neutron-to-proton ratio in the direction of increased stability.

## Half-life; Dating

**21.19** Germanium-66 decays by positron emission, with a half-life of 2.5 h. Write the equation for the nuclear reaction. How much $^{66}\text{Ge}$ remains from a 25.0-mg sample after 10.0 h?

**21.20** The half-life of tritium (hydrogen-3) is 12.3 yr. If 48.0 mg of tritium is released from a nuclear power plant during the course of an accident, what mass of this nuclide will remain after 12.3 yr? after 49.2 yr?

**21.21** A sample of the synthetic nuclide curium-243 was prepared. After 1 yr the radioactivity of the sample had declined from 3012 disintegrations per second to 2921 disintegrations per second. What is the half-life of the decay process?

**22.22** A sample of a radioactive nuclide prepared in a cyclotron exhibits 6332 disintegrations per second immediately upon its formation. After 100.0 minutes the number of disintegrations per second is 378. What is the half-life of the radionuclide?

**21.23** The half-life for decay of $^{139}\text{Ba}$ is 85 min. What mass of $^{139}\text{Ba}$ remains after 16 h from a sample that originally contained 24.0 $\mu\text{g}$ of the nuclide?

**21.24** The half-life of $^{239}\text{Pu}$ is 24,000 yr. What fraction of the $^{239}\text{Pu}$ present in nuclear wastes generated today will be present in the year 3000?

**21.25** An experiment was designed to determine whether an aquatic plant absorbed iodide ion from water. Iodine-131 ($t_{1/2} = 8.1$ days) was added as a tracer in the form of iodide ion to a tank containing the plants. The initial activity of a 1.00-$\mu\text{L}$ sample of the water was 89 counts per minute. After 32 days the level of activity in a 1.00-$\mu\text{L}$ sample was 5.7 counts per minute. Did the plants absorb iodide from the water?

**21.26** A sample of strontium-89 has an initial activity of 4600 counts per minute on a device that measures the level of radioactivity. After exactly 30 days the activity has declined to 3130 counts per minute. What is the half-life for decay of strontium-89?

**21.27** The half-life for the process $^{238}\text{U} \longrightarrow ^{206}\text{Pb}$ is $4.5 \times 10^{9}$ yr. A mineral sample contains 50.0 mg of $^{238}\text{U}$ and 14.0 mg of $^{206}\text{Pb}$. What is the age of the mineral?

**21.28** Potassium-40 decays to argon-40 with a half-life of $1.27 \times 10^{9}$ yr. What is the age of a rock in which the weight ratio of $^{40}\text{Ar}$ to $^{40}\text{K}$ is 3.6?

**21.29** A wooden artifact from a Chinese temple has a $^{14}\text{C}$ activity of 25.8 counts per minute as compared with an activity of 31.7 counts per minute for a standard of zero age. From the half-life for $^{14}\text{C}$ decay, $5.73 \times 10^{3}$ yr, determine the age of the artifact.

**21.30** An ancient wooden object is found to have an activity of 9.6 disintegrations per minute per gram of carbon. By contrast, the carbon in a living tree undergoes 18.4 disintegrations per minute per gram of carbon. Based on the activity of carbon-14 (with a half-life of $5.73 \times 10^3$ yr) in the object, calculate its age.

## Mass-Energy Relationships

**21.31** Calculate the binding energy per nucleon for the following nuclei: (a) $^{12}_{6}C$ (atomic mass, 12.00000 amu); (b) $^{61}_{28}Ni$ (atomic mass, 60.93106 amu); (c) $^{206}_{82}Pb$ (atomic mass, 205.97447 amu).

**21.32** Calculate the total binding energy and the binding energy per nucleon for each of the following nuclei: (a) $^{64}_{30}Zn$ (atomic mass, 63.92914); (b) $^{37}_{17}Cl$ (atomic mass, 36.96590 amu); (c) $^{4}_{2}He$ (atomic mass, 4.00260 amu).

**21.33** How much energy must be supplied to break a single $^{6}_{3}Li$ nucleus into separated protons and neutrons if the nucleus has a mass of 6.01347 amu? What does this energy correspond to for 1 mol of $^{6}Li$ nuclei?

**21.34** How much energy is absorbed or released when a single $^{7}_{4}Be$ nucleus of nuclear mass 7.0147 amu is separated into protons and neutrons? What is the amount of energy corresponding to this process for 1 mol of $^{7}_{4}Be$ nuclei?

**21.35** The solar radiation falling on earth amounts to $1.07 \times 10^{16}$ kJ/min. What is the mass equivalence of the solar energy falling on earth in a 24-h period? If the energy released in the reaction

$$^{235}_{92}U + {}^{1}_{0}n \longrightarrow {}^{141}_{56}Ba + {}^{92}_{36}Kr + 3{}^{1}_{0}n$$

($^{235}U$ atomic mass, 235.0439 amu; $^{141}Ba$ atomic mass, 140.9140 amu; $^{92}Kr$ atomic mass, 91.9218 amu)

is taken as typical of that occurring in a nuclear reactor, what mass of uranium-235 is required to equal 0.10 percent of the solar energy that falls on earth in 1 day?

**21.36** Based on the following atomic mass values—$^{1}_{1}H$, 1.00782; $^{2}_{1}H$, 2.01410 amu; $^{3}_{1}H$, 3.01605 amu; $^{3}_{2}He$, 3.01603 amu; $^{4}_{2}He$, 4.00260 amu—and the mass of the neutron given in the text, calculate the energy released in each of the following nuclear reactions, all of which are possibilities for a controlled fusion process:
(a) $^{2}_{1}H + {}^{3}_{1}H \longrightarrow {}^{4}_{2}He + {}^{1}_{0}n$
(b) $^{2}_{1}H + {}^{2}_{1}H \longrightarrow {}^{3}_{2}H + {}^{1}_{0}n$
(c) $^{2}_{1}H + {}^{3}_{2}He \longrightarrow {}^{4}_{2}He + {}^{1}_{1}H$

**21.37** Which of the following nuclei is likely to have the largest mass defect per nucleon: (a) $^{59}_{27}Co$; (b) $^{11}_{5}B$; (c) $^{118}_{50}Sn$; $^{243}_{96}Cm$? Explain your answer.

**21.38** Based on Figure 21.10, explain why we should expect energy to be released in the course of the fission of heavy nuclei that occurs in a nuclear reactor.

## Effects and Uses of Radioisotopes

**21.39** Complete and balance the nuclear equations for the following fission reactions:

(a) $^{235}_{92}U + {}^{1}_{0}n \longrightarrow {}^{160}_{62}Sm + {}^{72}_{30}Zn + \underline{\hspace{1cm}} {}^{1}_{0}n$
(b) $^{239}_{94}Pu + {}^{1}_{0}n \longrightarrow {}^{144}_{58}Ce + \underline{\hspace{1cm}} + 2{}^{1}_{0}n$

**21.40** Complete and balance the nuclear equation for the following fission or fusion reactions:
(a) $^{2}_{1}H + {}^{2}_{1}H \longrightarrow {}^{3}_{2}He + \underline{\hspace{1cm}}$
(b) $^{233}_{92}U + {}^{1}_{0}n \longrightarrow {}^{133}_{51}Sb + {}^{98}_{41}Nb + \underline{\hspace{1cm}} {}^{1}_{0}n$

**21.41** The stresses placed on materials in nuclear reactor cores are very severe, especially in breeder reactors. In terms of what we have learned in this chapter, what sorts of damage would you expect materials in the reactor core to undergo?

**21.42** Fusion energy is generally advanced as potentially a cleaner form of nuclear power than fission energy. It is anticipated, however, that the materials used to construct the inner components of a fusion reactor will become intensively radioactive and will need frequent replacement. What is the origin of this problem?

**21.43** Explain how one might use radioactive $^{59}Fe$ (a beta emitter with $t_{1/2} = 46$ days) to determine the extent to which rabbits are able to convert a particular iron compound in their diet into blood hemoglobin, which contains an iron atom.

**21.44** Chlorine-36 is a convenient radiotracer. It is a weak beta emitter, with $t_{1/2} = 3 \times 10^5$ yr. Describe how you would use this radiotracer to carry out each of the following experiments. (a) Determine whether trichloroacetic acid, $CCl_3COOH$, undergoes any ionization of its chlorines as chloride ion in aqueous solution. (b) Demonstrate that the equilibrium between dissolved $BaCl_2$ and solid $BaCl_2$ in a saturated solution is a dynamic process. (c) Determine the effects of soil pH on the uptake of chloride ion from the soil by soybeans.

**21.45** A portion of the sun's energy comes from the reaction

$$4{}^{1}_{1}H \longrightarrow {}^{4}_{2}He + 2{}^{0}_{1}e$$

This reaction requires a temperature of about $10^6$ to $10^7$ K. Why is such a high temperature required?

**21.46** Rutherford was able to carry out the first nuclear transmutation reactions by bombarding nitrogen-14 nuclei with alpha particles. However, in the famous experiment on scattering of alpha particles by gold foil (Section 2.2), a nuclear transmutation reaction did not occur. What is the difference in the two experiments? What would one need to do to carry out a successful nuclear transmutation reaction involving gold nuclei and alpha Particles?

## Additional Exercises

**21.47** Distinguish between the terms in each of the following pairs: (a) electron and positron; (b) binding energy and mass defect; (c) curie and rem; (d) gamma rays and beta rays.

**21.48** Figure 21.3 shows the stepwise decay of uranium-238 to form the stable lead-206 nucleus. Write balanced nuclear equations for each step in this sequence.

**21.49** Harmful chemicals are often destroyed by chemical treatment. For example, an acid can be neutralized by a base. Why can't chemical treatment be applied to destroy the fission products produced in a nuclear reactor?

**21.50** A sample of cobalt-60 was purchased in 1968 for use as a source of beta rays in some biomedical experiments. The half-life for decay of $^{60}$Co is 5.25 yr. What fraction of the activity present in the original sample will remain in 1992?

**21.51** We have seen that one possible mode of nuclear transformation is orbital electron capture. Which electron in a many-electron atom is most likely to be captured in such a process? How would you expect the likelihood of orbital electron capture to change with atomic number?

**21.52** Describe or define the following: (**a**) rem; (**b**) curie; (**c**) moderator; (**d**) breeder reactor; (**e**) critical mass.

**21.53** Compare the characteristics of alpha, beta, and gamma rays as they pertain to the health hazards associated with radioactivity.

**21.54** Plutonium-239 emits alpha particles that possess energies of about $5 \times 10^8$ kJ/mol. The half-life for the decay is about 24,000 yr. It has been said that the main danger from plutonium is inhalation of plutonium-containing dust. Why is inhalation, rather than mere exposure to plutonium in the environment, the major concern?

**21.55** During the past 30 yr nuclear scientists have synthesized approximately 1600 nuclei not known in nature. Many more might be discovered by using heavy-ion bombardment, which is possible only if high-energy instruments are used to accelerate the ions. Complete and balance the following reactions, which involve heavy-ion bombardments:

(**a**) $^{6}_{3}\text{Li} + ^{63}_{28}\text{Ni} \longrightarrow$ ?

(**b**) $^{48}_{20}\text{Ca} + ^{248}_{96}\text{Cm} \longrightarrow$ ?

(**c**) $^{88}_{38}\text{Sr} + ^{84}_{36}\text{Kr} \longrightarrow ^{116}_{46}\text{Pd} +$ ?

(**d**) $^{48}_{20}\text{Ca} + ^{238}_{92}\text{U} \longrightarrow ^{70}_{20}\text{Ca} + 4^{1}_{0}\text{n} + 2$?

**21.56** The 13 known nuclides of zinc range from $^{60}_{30}\text{Zn}$ to $^{72}_{30}\text{Zn}$. The naturally occurring nuclides have mass numbers 64, 66, 67, 68, and 70. What mode or modes of decay would you expect for the least massive radioactive nuclides of zinc? What mode for the most massive nuclides?

**21.57** The sun radiates energy into space at the rate of $3.9 \times 10^{26}$ J/s. Calculate the rate of mass loss from the sun.

**[21.58]** The synthetic radioisotope technetium-99, which decays by beta emission, is the most widely used isotope in nuclear medicine. The following data were collected on a sample of $^{99}\text{Tc}$:

| Disintegrations per minute | Time (h) |
| --- | --- |
| 180 | 0 |
| 130 | 2.5 |
| 104 | 5.0 |
| 77 | 7.5 |
| 59 | 10.0 |
| 46 | 12.5 |
| 24 | 17.5 |

Make a graph of these data similar to Figure 21.5 and determine the half-life. (You may wish to make a graph of the log of the disintegration rate versus time; a little rearranging of Equation 21.21 will produce an equation for a linear relation between log $N_t$ and $t$; from the slope you can obtain $k$.)

**[21.59]** According to current regulations, the maximum permissible dose of strontium-90 in the body of an adult is 1 microcurie ($1 \times 10^{-6}$ Ci). Using the relationship

$$\text{Rate} = kN$$

calculate the number of atoms of strontium-90 to which this corresponds. To what mass of strontium-90 does this correspond ($t_{1/2}$ for strontium-90 is 27.6 yr)?

**[21.60]** The half-life for decay of $^{230}_{90}\text{Th}$ is $8.0 \times 10^4$ yr. Because one cannot take the time to collect data for a graph such as that in Figure 21.5, how might one determine the half-life for such a long-lived isotope?

**[21.61]** Suppose you had a detection device that could count every decay from a radioactive sample of plutonium-239 ($t_{1/2}$ is 24,000 yr). How many counts per second would you obtain from a sample that contained 0.500 g of plutonium-239? (Hint: Look at Equations 21.20 and 21.22.)

**[21.62]** Tests on human subjects in Boston in 1965 and 1966, following the era of atomic bomb testing, revealed average quantities of about 2 picocuries of plutonium radioactivity in the average person. How many disintegrations per second does this level of activity imply? If each alpha particle deposits $8 \times 10^{-13}$ J of energy and if the average person weighs 75 kg, calculate the number of rads of radiation dose in 1 yr from such a level of plutonium, and also calculate the number of rems.

**[21.63]** A 26.00-g sample of water containing tritium, $^{3}_{1}\text{H}$, emits $1.50 \times 10^3$ beta particles per second. Tritium is a weak beta emitter, with a half-life of 12.26 yr. What fraction of all the hydrogen in the water sample is tritium? (Hint: Use Equations 21.20 and 21.22.)

**[21.64]** When a positron is annihilated by combination with an electron, two photons of equal energy result. What is the wavelength of these photons? Are they gamma-ray photons?

## CHAPTER 22

# Chemistry of Hydrogen, Oxygen, Nitrogen, and Carbon

For the most part, the previous chapters of this book have involved chemical principles, such as rules for bonding, the laws of thermodynamics, the factors influencing reaction rates, and so forth. In the course of explaining these principles, we have described the chemical and physical properties of many substances. However, we have done little to systematically examine the chemical elements and the compounds they form. This aspect of chemistry, often referred to as **descriptive chemistry**, is the subject of the next several chapters.

In this chapter we examine four important nonmetals: hydrogen, oxygen, nitrogen, and carbon. These nonmetals form many commercially important compounds and are the primary elements in biological systems. In Chapter 23 we will examine the remaining nonmetals on a group-by-group basis.

In studying descriptive chemistry, it is important to look for trends and general types of behavior, rather than trying to memorize all the facts presented. The periodic table is, of course, an invaluable tool in this task. Before we begin our examination of particular nonmetals, it is useful to review briefly some general periodic trends.

## CONTENTS

## 22.1 PERIODIC TRENDS

Figure 22.1 summarizes the way in which several important properties of elements vary in relation to the periodic chart. One of the most useful features shown is the division of elements into the broad categories of metals and nonmetals. We have on previous occasions discussed the general properties of metals and nonmetals (see Section 5.3).

The trend in electronegativity is also significant. As shown in Figure 22.1, electronegativity decreases as we move down a given group and increases from left to right across the table. As a result, nonmetals have higher electronegativities than do metals. Consequently, compounds formed between strongly metallic and strongly nonmetallic elements tend to be ionic (for example, metal fluorides and metal oxides). These substances are solids at room temperature. In contrast, compounds formed between nonmetals are molecular substances. Molecular substances with low molecular weight tend to be gases, liquids, or volatile solids at room temperature.

Among the nonmetals, the chemistry of the first member of each family often differs in several important ways from that of subsequent members. The differences are due in part to the smaller size and greater electronegativity of the first member. In addition, the first member is restricted to forming a maximum of four bonds because it has only the $2s$ and the three $2p$ orbitals available for bonding. Subsequent members of the family

Lightning storm over Verde Valley, Arizona. Atmospheric nitrogen, $N_2$, combines with oxygen, $O_2$, during lightning flashes to form nitric oxide, NO. (Deeks-Barber/Photo Researchers)

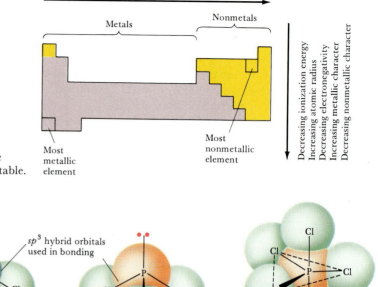

Increasing ionization energy
Decreasing atomic radius
Increasing nonmetallic character and electronegativity
Decreasing metallic character

Metals

Nonmetals

Decreasing ionization energy
Increasing atomic radius
Decreasing electronegativity
Increasing metallic character
Decreasing nonmetallic character

Most
metallic
element

Most
nonmetallic
element

**FIGURE 22.1** Trends in key properties of the elements as a function of position in the periodic table.

**FIGURE 22.2** Comparison of $NCl_3$, $PCl_3$, and $PCl_5$. Nitrogen is unable to form $NCl_5$ because it does not have $d$ orbitals available for bonding.

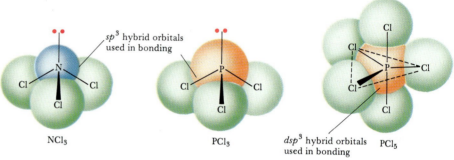

$sp^3$ hybrid orbitals used in bonding

$NCl_3$

$PCl_3$

$dsp^3$ hybrid orbitals used in bonding

$PCl_5$

**FIGURE 22.3** Comparison of $\pi$-bond formation by sideways overlap of $p$ orbitals between two carbon atoms and between two silicon atoms. The distance between nuclei increases as we move from carbon to silicon because of the larger size of the silicon atom. The $p$ orbitals do not overlap as effectively between two silicon atoms because of this greater separation.

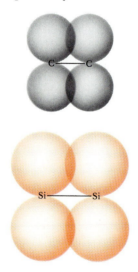

are able to use $d$ orbitals in bonding in addition to $s$ and $p$ orbitals; they can therefore form more than four bonds. As an example, consider the chlorides of nitrogen and phosphorus, the first two members of group 5A. Nitrogen forms a maximum of three bonds with chlorine, $NCl_3$. Although phosphorus can form the trichloride compound, $PCl_3$, it is also able to form five bonds with chlorine, $PCl_5$. These compounds are shown in Figure 22.2. Because hydrogen has only the $1s$ orbital available for bonding, it is restricted to forming one bond.

Another difference between the first member of any family and subsequent members of the same family is the greater ability of the former to form $\pi$ bonds. We can understand this, in part, in terms of atomic size. As atoms increase in size, the sideways overlap of $p$ orbitals, which form the strongest type of $\pi$ bond, becomes less effective. This is shown in Figure 22.3. As an illustration of this effect, consider two differences in the chemistry of carbon and silicon, the first two members of group 4A. Carbon has two crystalline allotropes, diamond and graphite. In diamond there are $\sigma$ bonds between carbon atoms but no $\pi$ bonds. In graphite, $\pi$ bonds result from sideways overlap of $p$ orbitals (Section 9.7). Silicon occurs only in the diamondlike crystal form. Silicon does not exhibit a graphitelike structure because of the low stability of $\pi$ bonds between silicon atoms.

We see the same type of difference in the dioxides of these elements (Figure 22.4). In $CO_2$, carbon forms double bonds to oxygen, thereby

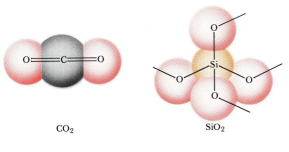

**FIGURE 22.4** Comparison of the structures of $CO_2$ and $SiO_2$; $CO_2$ has double bonds, whereas $SiO_2$ has only single bonds.

CO₂

SiO₂

achieving its valence by $\pi$ bonding. In contrast, $SiO_2$ contains no double bonds. Instead, four oxygens are bonded to each silicon, forming an extended structure.*

## SAMPLE EXERCISE 22.1

Sulfur forms a fluoride compound, $SF_6$, that has six sulfur-fluorine bonds. In contrast, oxygen is able to form at most two bonds to fluorine, $OF_2$. **(a)** Rationalize this difference. **(b)** Predict the molecular geometry of each compound and describe the hybridization employed by S and O.

**Solution:** **(a)** The electron configuration of oxygen is $[He]2s^2 2p^4$. The next-highest-energy orbitals available for bonding are in the third shell. Thus oxygen uses only the $2s$ and $2p$ orbitals in bonding. These orbitals can accommodate only eight electrons. In $OF_2$, oxygen already has eight electrons, six of its own and one from each fluorine atom:

$$F—\overset{..}{\underset{..}{O}}—F$$

Consequently, it is unable to accommodate additional fluorine atoms.

The electron configuration of sulfur is $[Ar]3s^2 3p^4$. In this case the $3d$ orbitals are available for bonding. Thus sulfur is able to expand its octet to accommodate more than eight electrons:

$$\begin{array}{c} F \\ | \ \ F \\ F—S—F \\ F \ \ | \\ F \end{array}$$

**(b)** According to the VSEPR theory (Section 9.1), the four electron pairs around oxygen are disposed at the corners of a tetrahedron. The four electron pairs are accommodated in $sp^3$ hybrid orbitals. The geometric arrangement of atoms associated with two bonded pairs and two nonbonded pairs is nonlinear.

In the case of $SF_6$, the six electron pairs around sulfur are disposed at the corners of an octahedron. The associated hybridization is $d^2 sp^3$. Because all six electrons are bonded pairs, the geometric arrangement of atoms is also octahedral.

### PRACTICE EXERCISE
Consider the following list of elements: Li, K, N, P, Ne. From this list, select the element that **(a)** is most electronegative; **(b)** has the greatest metallic character; **(c)** can bond to more than four surrounding atoms in a molecule; **(d)** forms $\pi$ bonds most readily.
*Answers:* **(a)** N; **(b)** K; **(c)** P; **(d)** N

Throughout this chapter and subsequent ones, you will find a large number of chemical reactions presented. To remember all of them is an overwhelming task. However, you will probably need to remember some, especially those of a more general nature. In earlier discussions we have encountered several general categories of reactions: combustion reactions (Section 3.2), metathesis reactions (Sections 3.2 and 13.1), single dis-

## 22.2 CHEMICAL REACTIONS

---

* The formula $SiO_2$ is consistent with this structure, because each oxygen is shared by two silicon atoms (not shown in Figure 22.4). For bookkeeping purposes, we may therefore count a half of each of the four oxygens that are bound to a given silicon atom as belonging to that silicon. We shall consider silicon-oxygen compounds in some detail in Chapter 23.

placement reactions (Section 5.5), Brønsted acid-base (proton transfer) reactions (Section 13.5), Lewis acid-base reactions (Section 17.8), and redox reactions (Section 13.6). Because $O_2$ and $H_2O$ are abundant and widespread in our environment, it is particularly important to consider the possible reactions of these substances with other compounds. About a third of the reactions discussed in this chapter involve either $O_2$ (oxidation or combustion reactions) or $H_2O$ (especially proton-transfer reactions).

In oxidation reactions with $O_2$, hydrogen-containing compounds produce $H_2O$. Carbon-containing ones produce $CO_2$ (unless $O_2$ is in short supply, in which case CO or even C can form). Nitrogen-containing compounds tend to form $N_2$, although NO can form in special cases. The following reactions are illustrative of these generalizations:

$$2CH_3OH(l) + 3O_2(g) \longrightarrow 2CO_2(g) + 4H_2O(l) \qquad [22.1]$$

$$4CH_3NH_2(g) + 9O_2(g) \longrightarrow 4CO_2(g) + 10H_2O(l) + 2N_2(g) \qquad [22.2]$$

The formation of $H_2O$, $CO_2$, and $N_2$ reflects the high thermodynamic stabilities of these compounds, which are indicated by the large bond energies for the O—H, C=O, and N≡N bonds that they contain (463, 799, and 941 kJ/mol, respectively). The formation of the stable $H_2O$ and $CO_2$ molecules also occurs when $H_2$ and C are used to reduce metal oxides:

$$NiO(s) + H_2(g) \longrightarrow Ni(s) + H_2O(l) \qquad [22.3]$$

$$2CuO(s) + C(s) \longrightarrow 2Cu(s) + CO_2(g) \qquad [22.4]$$

In dealing with proton-transfer reactions, you should remember that the weaker a Brønsted acid, the stronger its conjugate base. For example, $H_2$, $OH^-$, $NH_3$, $CH_4$, and $C_2H_2$, which we encounter in this chapter, are exceedingly weak proton donors. In fact, they have *no* tendency to act as acids in water. Thus species formed from them by removing one or more protons (such as $H^-$, $O^{2-}$, $NH_2^-$, $N^{3-}$, $CH_3^-$, $C^{4-}$, and $C_2^{2-}$) are extremely strong bases. All react readily with water, removing protons from the $H_2O$ to form $OH^-$. The following reactions are illustrative:

$$CH_3^-(aq) + H_2O(l) \longrightarrow CH_4(g) + OH^-(aq) \qquad [22.5]$$

$$N^{3-}(aq) + 3H_2O(l) \longrightarrow NH_3(aq) + 3OH^-(aq) \qquad [22.6]$$

Of course, substances that are stronger proton donors than $H_2O$, such as HCl, $H_2SO_4$, $HC_2H_3O_2$, and other acids, also react readily with basic anions.

---

**SAMPLE EXERCISE 22.2**

Predict the products formed in each of the following reactions and write a balanced chemical equation:

(a) $CH_3NHNH_2(g) + O_2(g) \longrightarrow$
(b) $Mg_3P_2(s) + H_2O(l) \longrightarrow$
(c) $NaCN(s) + HCl(aq) \longrightarrow$

**Solution:** (a) This combustion reaction should produce $CO_2$, $H_2O$, and $N_2$:

$$2CH_3NHNH_2(g) + 5O_2(g) \longrightarrow$$
$$2CO_2(g) + 6H_2O(l) + 2N_2(g)$$

**(b)** The $Mg_3P_2$ consists of $Mg^{2+}$ and $P^{3-}$ ions. The $P^{3-}$ ion, like $N^{3-}$, has a strong affinity for protons and thus reacts with $H_2O$ to form $OH^-$ and $PH_3$ ($PH^{2-}$, $PH_2^-$, and $PH_3$ are all exceedingly weak proton donors):

$$Mg_3P_2(s) + 6H_2O(l) \longrightarrow$$
$$2PH_3(g) + 3Mg(OH)_2(aq)$$

The $Mg(OH)_2$ is of low solubility in water and may precipitate.

**(c)** The NaCN consists of $Na^+$ and $CN^-$ ions. The $CN^-$ ion is basic (HCN is a weak acid). Thus $CN^-$ reacts with protons to form its conjugate acid:

$$NaCN(s) + HCl(aq) \longrightarrow HCN(aq) + NaCl(aq)$$

or

$$NaCN(s) + H^+(aq) \longrightarrow HCN(aq) + Na^+(aq)$$

The HCN has limited solubility in water and readily escapes as a gas.

**PRACTICE EXERCISE** _____

Write a balanced chemical equation for the reaction of solid sodium hydride with water.

***Answer:*** $NaH(s) + H_2O(l) \longrightarrow NaOH(aq) + H_2(g)$

---

## 22.3 HYDROGEN

Formation of elemental hydrogen was first recorded in the sixteenth century by the alchemist Paracelsus (1493–1541), who observed the formation of an "air" (gas) produced by the action of acids on iron. However, it was the English chemist Henry Cavendish (1731–1810) who first isolated pure hydrogen and distinguished it from other gases. Because the element produces water when burned in air, the French chemist Lavoisier gave it the name "hydrogen," which means "water producer" (Greek: *hydor*, water; *gennao*, to produce).

Hydrogen is the most abundant element in the universe. It is the nuclear fuel consumed by our sun and other stars to produce energy (Section 21.8). Although about 70 percent of the universe is composed of hydrogen, it constitutes only 0.87 percent of the earth's mass. Most of the hydrogen on our planet is found associated with oxygen. Water, which is 11 percent hydrogen by mass, is the most abundant hydrogen compound. Hydrogen is also an important part of petroleum, cellulose, starch, fats, alcohols, acids, and a wide variety of other materials.

**Isotopes of Hydrogen**

The most common isotope of hydrogen, $^1_1H$, has a nucleus consisting of a single proton. This isotope, sometimes referred to as **protium**,* comprises 99.9844 percent of naturally occurring hydrogen.

Two other isotopes are known: $^2_1H$, whose nucleus contains a proton and a neutron, and $^3_1H$, whose nucleus contains a proton and two neutrons. The $^2_1H$ isotope is called **deuterium**. Deuterium comprises 0.0156 percent of naturally occurring hydrogen. It is not radioactive. In writing the chemical formulas of compounds containing deuterium, that isotope is often given the symbol D, as in $D_2O$. $D_2O$, which is called deuterium oxide or **heavy water**, can be obtained by electrolysis of ordinary water. The heavier $D_2O$ undergoes electrolysis at a slower rate than does the lighter $H_2O$ and is thus concentrated during the electrolysis. Electrolysis of 2400 L of water will produce about 83 mL of 99 percent $D_2O$. $D_2O$, which is presently available in ton quantities, is used as a moderator and coolant in certain nuclear reactors. Some of the physical properties of

---

* Giving unique names to isotopes is limited to hydrogen. Because of the proportionally large differences in their masses, the isotopes of H show appreciably more differences in their chemical and physical properties than do isotopes of heavier elements.

**TABLE 22.1**  Comparison of properties of $H_2O$ and $D_2O$

| Property | $H_2O$ | $D_2O$ |
|---|---|---|
| Melting point (°C) | 0.00 | 3.81 |
| Boiling point (°C) | 100.00 | 101.42 |
| Density at 25°C (g/mL) | 0.997 | 1.104 |
| Heat of fusion (kJ/mol) at m.p. | 6.008 | 6.276 |
| Heat of vaporization (kJ/mol) at b.p. | 40.67 | 41.61 |
| Ion product, ($K_w$) at 25°C | $1.01 \times 10^{-14}$ | $1.95 \times 10^{-15}$ |

$H_2O$ and $D_2O$ are compared in Table 22.1. Notice the small but discernible difference in properties.

The third isotope, $^3_1H$, is known as **tritium** and is often given the symbol T. It is radioactive, with a half-life of 12.3 yr:

$$^3_1H \longrightarrow \ ^3_2He + \ ^0_{-1}e \qquad t_{1/2} = 12.3 \text{ yr} \qquad [22.7]$$

It is formed continuously in the upper atmosphere in nuclear reactions induced by cosmic rays; however, because of its short half-life, only trace quantities exist naturally. The isotope can be synthesized in nuclear reactors. The preferred method of production is neutron bombardment of lithium-6:

$$^6_3Li + \ ^1_0n \longrightarrow \ ^3_1H + \ ^4_2He \qquad [22.8]$$

Each isotope of hydrogen contains a single electron and consequently undergoes identical chemical reactions. However, the heavier deuterium and tritium generally undergo reactions at a somewhat slower rate than protium.

Deuterium and tritium have proved valuable in studying the reactions of compounds containing hydrogen. A compound can be "labeled" by replacing one or more ordinary hydrogen atoms at specific locations within a molecule with deuterium or tritium. By comparing the location of the heavy hydrogen label in the reactants with that in the products, the mechanism of the reaction can often be inferred. As an example of the chemical insight that can be gained by using deuterium, consider what happens when methyl alcohol, $CH_3OH$, is placed in $D_2O$. The H atom of the O—H bond exchanges rapidly with the D atoms in $D_2O$, forming $CH_3OD$. The H atoms of the $CH_3$ group do not exchange. This experiment demonstrates the kinetic stability of C—H bonds and reveals the speed at which the O—H bond in the molecule breaks and reforms.

**Properties of Hydrogen**

Hydrogen is the only element that is not a member of any family in the periodic table. Because of its $1s^1$ electron configuration, it is generally placed above lithium in the periodic table. However, it is definitely not an alkali metal. It forms a positive ion much less readily than any alkali metal; the ionization energy of the hydrogen atom is 1310 kJ/mol, whereas that of lithium is 517 kJ/mol. Furthermore, the hydrogen atom has no electrons below its valence shell; the $H^+$ ion is just a bare proton. The simple $H^+$ ion is not known to exist in any compound. Its small size gives it a strong attraction for electrons; either it strips electrons from surrounding matter (forming hydrogen atoms that combine to form $H_2$), or it shares electron pairs, forming covalent bonds with other atoms. While we may represent the aquated hydrogen ion as $H^+(aq)$, that proton is bonded to one or more water molecules and is thus often represented as $H_3O^+(aq)$ (Section 13.4).

Hydrogen is also sometimes placed above the halogens in the periodic table, because the hydrogen atom can also pick up one electron to form the **hydride ion**, $H^-$. However, the electron affinity of hydrogen, $\Delta H_{EA} = -71$ kJ/mol, is not as large as that of any halogen; the electron affinity of fluorine is $-332$ kJ/mol, and that of iodine is $-295$ kJ/mol. In general, hydrogen shows no closer resemblance to the halogens than it does to the alkali metals.

In its elemental form, hydrogen exists at room temperature as a colorless, odorless, tasteless gas composed of diatomic molecules, $H_2$. We can call $H_2$ dihydrogen, but it is more commonly referred to as molecular hydrogen or merely hydrogen. Because $H_2$ is nonpolar and has only two electrons, attractive forces between molecules are extremely weak. Consequently, the melting point ($-259°C$) and boiling point ($-253°C$) of $H_2$ are very low.

The H—H bond-dissociation energy (436 kJ/mol) is high for a single bond (see Table 8.3). By comparison, the Cl—Cl bond-dissociation energy is only 242 kJ/mol. As a result of its strong bond, most reactions of $H_2$ are slow at room temperature. However, the molecule is readily activated by heating, irradiation, or catalysis. The activation process generally produces hydrogen atoms, which are very reactive. The activation of $H_2$ by finely divided nickel, palladium, and platinum was considered briefly in our earlier discussions of heterogeneous catalysis (Section 15.6). Once $H_2$ is activated, it reacts rapidly and exothermally with a wide variety of substances.

Hydrogen forms strong covalent bonds with many elements, including oxygen; the H—O bond-dissociation energy is 464 kJ/mol. The strong H—O bond makes hydrogen an effective reducing agent for many metal oxides. For example, when $H_2$ is passed over heated CuO, copper is produced:

$$CuO(s) + H_2(g) \longrightarrow Cu(s) + H_2O(g) \qquad [22.9]$$

When $H_2$ is ignited in air, a vigorous reaction occurs, forming $H_2O$:

$$2H_2(g) + O_2(g) \longrightarrow 2H_2O(l) \qquad \Delta H = -571.7 \text{ kJ} \qquad [22.10]$$

Air containing as little as 4 percent $H_2$ (by volume) is potentially explosive. The disastrous burning of the hydrogen-filled airship the *Hindenberg* in 1937 (Figure 22.5) dramatically demonstrated the high flammability of $H_2$.

When a small quantity of $H_2$ is needed in the laboratory, it is usually obtained by the reaction between an active metal such as zinc and a dilute acid, such as HCl or $H_2SO_4$:

**Preparation of Hydrogen**

$$Zn(s) + 2H^+(aq) \longrightarrow Zn^{2+}(aq) + H_2(g) \qquad [22.11]$$

**FIGURE 22.5** Burning of the airship *Hindenburg* while landing at Lakehurst, New Jersey, on May 6, 1937. This picture was taken only 22 seconds after the first explosion occurred. (UPI)

Because $H_2$ has an extremely low solubility in water, it can be collected by displacement of water, as shown in Figure 22.6.

When commercial quantities of $H_2$ are needed, the raw materials are usually hydrocarbons (from either natural gas or petroleum) or water. Hydrocarbons are substances that consist of carbon and hydrogen, like $CH_4$ and $C_8H_{18}$. Much hydrogen is presently obtained in the course of refining petroleum. In the refining process, large hydrocarbons are catalytically broken into smaller molecules with the accompanying production of $H_2$ as a by-product.

Hydrogen is also produced by the reaction of methane, $CH_4$, the principal component of natural gas, with steam at 1100°C:

**FIGURE 22.6** Apparatus commonly used in the laboratory for preparation of hydrogen.

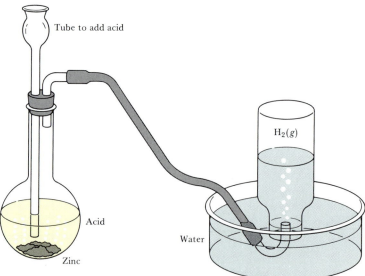

$$CH_4(g) + H_2O(g) \longrightarrow CO(g) + 3H_2(g) \qquad [22.12]$$

$$CO(g) + H_2O(g) \longrightarrow CO_2(g) + H_2(g) \qquad [22.13]$$

When heated to about $1000°C$, carbon also reacts with steam to produce a mixture of $H_2$ and $CO$ gases:

$$C(s) + H_2O(g) \longrightarrow H_2(g) + CO(g) \qquad [22.14]$$

Both $CO$ and $H_2$ burn in air to produce heat, and this mixture, known as **water gas**, is used as an industrial fuel.

Simple electrolysis of water consumes too much energy and is consequently too costly a process to be used commercially to produce $H_2$. However, hydrogen is produced as a by-product in the electrolysis of brine (NaCl) solutions in the course of $Cl_2$ and NaOH manufacture:

$$2NaCl(aq) + 2H_2O(l) \xrightarrow{\text{electrolysis}} H_2(g) + Cl_2(g) + 2NaOH(aq) \qquad [22.15]$$

**Uses of Hydrogen**

Hydrogen is a commercially important substance; about $2 \times 10^8$ kg (200,000 tons) is produced annually in the United States. Over two-thirds of the annual production is consumed in the synthesis of ammonia by the Haber process (Section 16.1). Hydrogen is also used to manufacture methanol, $CH_3OH$. As shown in Equation 22.16, the synthesis involves the catalytic combination of carbon monoxide and hydrogen under high pressures and temperatures:

$$CO(g) + 2H_2(g) \xrightarrow[\substack{200-300 \text{ atm} \\ \text{catalyst}}]{300-400°C} CH_3OH(l) \qquad [22.16]$$

Hydrogenation of vegetable oils in the manufacture of margarine and vegetable shortening is another important use. In this process $H_2$ is added to carbon-carbon double bonds in the oil. A simple example of a hydrogenation occurs in the conversion of ethylene to ethane:

[22.17]

Ethylene                  Ethane

Because organic compounds with double bonds have the ability to add additional hydrogen atoms, they are said to be unsaturated. The term "polyunsaturated," which often appears in food advertisements, refers to molecules that have several (poly) double bonds between carbon atoms. The following molecule is a polyunsaturated hydrocarbon:

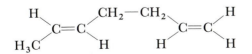

The polyunsaturated molecules that occur in vegetable oils are much more complex (Section 28.5).

## Binary Hydrogen Compounds

Hydrogen reacts with other elements to form compounds of three general types: (1) ionic hydrides, (2) metallic hydrides, and (3) molecular hydrides.

The **ionic hydrides** are formed by the alkali metals and by the heavier alkaline earths (Ca, Sr, and Ba). These active metals are much less electronegative than hydrogen. Consequently, hydrogen acquires electrons from them to form hydride ions, $H^-$, as shown in Equations 22.18 and 22.19:

$$2Li(s) + H_2(g) \longrightarrow 2LiH(s) \qquad [22.18]$$

$$Ca(s) + H_2(g) \longrightarrow CaH_2(s) \qquad [22.19]$$

The resultant ionic hydrides are solids with high melting points (LiH melts at 680°C).

The hydride ion is very basic and reacts readily with compounds having even weakly acidic protons to form $H_2$. For example, $H^-$ reacts readily with $H_2O$:

$$H^-(aq) + H_2O(l) \longrightarrow H_2(g) + OH^-(aq) \qquad [22.20]$$

Thus ionic hydrides can be used as convenient (although expensive) sources of $H_2$. Calcium hydride, $CaH_2$, is sold in commercial quantities and used for inflation of life rafts, weather balloons, and the like, where a simple, compact means of $H_2$ generation is desired. $CaH_2$ is also used to remove $H_2O$ from organic liquids. The reaction of $CaH_2$ with $H_2O$ is shown in Figure 22.7.

The reaction between $H^-$ and $H_2O$ (Equation 22.20) is not only an acid-base reaction, but a redox reaction as well. The $H^-$ ion can be viewed not only as a good base, but also as a good reducing agent. In fact, hydrides are able to reduce $O_2$ to $H_2O$:

$$2NaH(s) + O_2(g) \longrightarrow Na_2O(s) + H_2O(l) \qquad [22.21]$$

Thus hydrides are normally stored in an environment that is free of both moisture and air.

**Metallic hydrides** are formed when hydrogen reacts with transition metals. These compounds are so named because they retain their metallic conductivity and other metallic properties. In many metallic hydrides the ratio of metal atoms to hydrogen atoms is not a ratio of small whole numbers, nor is it a fixed ratio. The composition can vary within a range, depending on the conditions of synthesis. For example, although $TiH_2$ can be produced, preparations usually yield substances with about 10 percent less hydrogen than this, $TiH_{1.8}$. These nonstoichiometric metallic hydrides are sometimes called **interstitial hydrides**. They may be considered to be solutions of hydrogen atoms in the metal, with the hydrogen atoms occupying the holes or interstices between metal atoms in the solid lattice. However, this description is an oversimplification; there is evidence for chemical interaction between metal and hydrogen.

**FIGURE 22.7** Reaction of $CaH_2$ with water in the presence of phenolphthalein. The red color of the phenolphthalein is due to the presence of $OH^-$ ions. The gas bubbles are $H_2$. (Donald Clegg and Roxy Wilson)

The ready absorption of $H_2$ by palladium metal has been used to separate $H_2$ from other gases and in purifying $H_2$ on an industrial scale. At 300 to 400 K $H_2$ dissociates into atomic hydrogen on the Pd surface. The H atoms dissolve in the Pd, and under $H_2$ pressure they diffuse through, recombining to form $H_2$ on the opposite surface. Because no other molecules exhibit this property, absolutely pure $H_2$ results.

Research is in progress investigating metallic hydrides as storage media for hydrogen. In the case of many metals, hydride formation occurs directly upon contact between the metal and $H_2$. Furthermore, the reaction is often readily reversible so that $H_2$ can be obtained from the hydride by reducing the pressure of the hydrogen gas above the metal. Metals accommodate an extremely high density of hydrogen, because the hydrogen atoms are closely packed into the interstitial sites between metal atoms. Indeed, the number of hydrogen atoms per unit volume is greater in some metallic hydrides than in liquid $H_2$. In some hydrides the metal can accommodate two or three times as many hydrogen atoms as there are metal atoms; that is, the stoichiometry approaches $MH_3$ (where M is the metal).

The absorption of hydrogen by metals also has detrimental effects. The hydrides tend to be more brittle than the metal. Thus absorption of hydrogen can weaken and embrittle steel and other structural metals (Figure 22.8).

**FIGURE 22.8** Cracking in niobium metal due to hydride formation when the metal is stressed under a 1 percent $H_2$ atmosphere. The raised lines running nearly vertically are due to formation of NbH at certain planes within the metal. The cracks begin in these regions because NbH is very brittle. (Photo courtesy of Professor Howard Birnbaum, University of Illinois; electron microscope photo, about 200 × magnification. Reprinted with permission from *Acta Metallurgica*, vol. 25, M. L. Grossbeck and H. K. Birnbaum, "Low Temperature Hydrogen Embrittlement of Niobium II—Microscopic Observations," copyright 1977, Pergamon Press Ltd.)

The **molecular hydrides**, formed by nonmetals and semimetals, are either gases or liquids under standard conditions. The simple molecular hydrides are listed in Figure 22.9 together with their standard free energies of formation, $\Delta G_f^\circ$. In each family, the thermal stability (measured by $\Delta G_f^\circ$) decreases as we move down the family. (Recall that the more

**FIGURE 22.9** Standard free energies of formation (kJ/mol) of molecular hydrides.

| 4A | 5A | 6A | 7A |
|---|---|---|---|
| $CH_4(g)$ −50.8 | $NH_3(g)$ −16.7 | $H_2O(l)$ −237 | $HF(g)$ −271 |
| $SiH_4(g)$ +56.9 | $PH_3(g)$ +18.2 | $H_2S(g)$ −33.0 | $HCl(g)$ −95.3 |
| $GeH_4(g)$ +117 | $AsH_3(g)$ +111 | $H_2Se(g)$ +71 | $HBr(g)$ −53.2 |
| | $SbH_3(g)$ +187 | $H_2Te(g)$ +138 | $HI(g)$ +1.30 |

stable a compound with respect to its elements under standard conditions, the more negative $\Delta G_f^\circ$ is.) We will discuss the molecular hydrides further in the course of examining the other nonmetallic elements.

## 22.4  OXYGEN

By the middle of the seventeenth century it was recognized that air contained a component associated with burning and breathing. Not until 1774, however, did Joseph Priestley discover oxygen (Figure 22.10). Lavoisier subsequently named the element "oxygen," meaning "acid former."

Oxygen plays an important role in the chemistries of most other elements and is found in combination with other elements in a great variety of compounds. Indeed, oxygen is the most abundant element both in the earth's crust and in the human body. It constitutes 89 percent of water by mass and 20.9 percent of air by volume (23 percent by mass). It also comprises 50 percent by mass of the sand, clay, limestone, and igneous rocks that make up the bulk of the earth's crust.

### Properties of Oxygen

Oxygen has two allotropes, $O_2$ and $O_3$. When we speak of elemental or molecular oxygen, it is usually understood that we are speaking of $O_2$, the normal form of the element; $O_3$ is called **ozone**.

Molecular oxygen, $O_2$, exists at room temperature as a colorless, odorless, and tasteless gas. It melts at $-218°C$ and has a normal boiling point of $-183°C$. It is only slightly soluble in water, but its presence in water is essential to marine life.

The electron configuration of the oxygen atom is $[He]2s^2 2p^4$. Thus oxygen can complete its octet of electrons either by picking up two electrons to form the oxide ion, $O^{2-}$, or by sharing two electrons. In its covalent compounds it tends to form two bonds: either two single bonds, as in $H_2O$, or a double bond, as in formaldehyde, $H_2C{=}O$. The $O_2$ molecule itself contains a double bond.

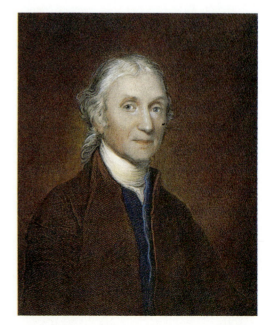

**FIGURE 22.10**  Joseph Priestley (1733–1804). Priestley became interested in chemistry at the age of 39, perhaps through his personal acquaintance with Benjamin Franklin. Because Priestley lived next door to a brewery from which he could obtain carbon dioxide, his studies focused on this gas first and were later extended to other gases. Because he was suspected of sympathizing with the American and French Revolutions, his church, home, and laboratory in Birmingham, England, were burned by a mob in 1791. Priestley had to flee in disguise. He eventually emigrated to the United States in 1794, where he lived his remaining years in relative seclusion in Pennsylvania. Although his discovery of "dephlogisticated air" (oxygen) eventually led to the downfall of the phlogiston theory, Priestley stubbornly continued to support this theory even after strong evidence had brought it into serious question. Priestley was a scientific conservative, but he was very liberal in his religious and political views. (The Granger Collection)

The bond in $O_2$ is very strong (the bond-dissociation energy is 495 kJ/mol). Oxygen also forms strong bonds with many other elements. Consequently, many oxygen-containing compounds are thermodynamically more stable than $O_2$. However, in the absence of a catalyst most reactions of $O_2$ have high activation energies and thus require high temperatures to proceed at a suitable rate. Once a sufficiently exothermic reaction begins, it may accelerate rapidly, producing a reaction of explosive violence.

**Preparation of Oxygen**

Oxygen can be obtained either from air or from certain oxygen-containing compounds. Nearly all commercial oxygen is obtained by fractional distillation of liquefied air. The normal boiling point of $O_2$ is $-183°C$, whereas that of $N_2$, the other principal component of air, is $-196°C$. Thus when liquefied air is warmed, the $N_2$ boils off, leaving liquid $O_2$ contaminated mainly with small amounts of $N_2$ and Ar.

The common laboratory preparation of $O_2$ is the thermal decomposition of potassium chlorate, $KClO_3$, with manganese dioxide, $MnO_2$, added as a catalyst:

$$2KClO_3(s) \longrightarrow 2KCl(s) + 3O_2(g) \qquad [22.22]$$

Like $H_2$, $O_2$ can be collected by displacement of water because of its relatively low solubility.

**Uses of Oxygen**

Oxygen is one of the most widely used industrial chemicals. In 1985, it ranked behind only sulfuric acid, $H_2SO_4$, nitrogen, $N_2$, ammonia, $NH_3$, and CaO. About $1.4 \times 10^{10}$ kg (15 million tons) of $O_2$ is used annually in the United States. It is shipped and stored either as liquid or in steel containers as compressed gas. However, about 70 percent of $O_2$ output is generated at the site where it is needed.

Oxygen is by far the most widely used oxidizing agent. Over half of the $O_2$ produced is used in the steel industry, mainly to remove impurities from steel (Section 24.3). It is also used to bleach pulp and paper. (Oxidation of intensely colored compounds often gives colorless products.) In medicine oxygen eases breathing difficulties. It is also used together with acetylene, $C_2H_2$, in oxyacetylene welding (Figure 22.11). The reaction between $C_2H_2$ and $O_2$ is highly exothermic, producing temperatures in excess of $3000°C$:

**FIGURE 22.11**
Oxyacetylene welding. (Lowell Georgia/Photo Researchers)

$$2C_2H_2(g) + 5O_2(g) \longrightarrow 4CO_2(g) + 2H_2O(g) \qquad \Delta H° = -2510 \text{ kJ} \qquad [22.23]$$

**Ozone**

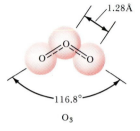

**FIGURE 22.12** Structure of the ozone molecule.

Ozone, $O_3$, is a pale blue gas with a sharp, irritating odor. It is poisonous but no human deaths have been attributed to it. Most people can detect about 0.01 ppm in air. Exposure to 0.1 to 1 ppm produces headaches, burning of the eyes, and irritation to the respiratory passages.

The structure of the $O_3$ molecule is shown in Figure 22.12. The molecule possesses a $\pi$ bond that is delocalized over the three oxygen atoms (Section 9.7). The molecule dissociates readily, forming reactive oxygen atoms:

$$O_3(g) \longrightarrow O_2(g) + O(g) \qquad \Delta H° = 107 \text{ kJ} \qquad [22.24]$$

Not surprisingly, ozone is a stronger oxidizing agent than dioxygen. One measure of this oxidizing power is the high reduction potential of $O_3$, compared to that of $O_2$:

$$O_3(g) + 2H^+(aq) + 2e^- \longrightarrow O_2(g) + H_2O(l) \qquad E° = 2.07 \text{ V} \qquad [22.25]$$

$$O_2(g) + 4H^+(aq) + 4e^- \longrightarrow 2H_2O(l) \qquad E° = 1.23 \text{ V} \qquad [22.26]$$

Ozone forms oxides with many elements under conditions where $O_2$ will not react; indeed, it oxidizes all of the common metals except gold and platinum.

Ozone can be prepared by passing electricity through dry $O_2$:

$$3O_2(g) \xrightarrow{\text{electricity}} 2O_3(g) \qquad \Delta H = 287 \text{ kJ} \qquad [22.27]$$

The pungent odor of ozone gas can sometimes be detected when a spark jumps during lightning storms. The preparation of $O_3$ may be accomplished in an apparatus such as that shown in Figure 22.13. The gas cannot be stored for long except at low temperature because it readily decomposes to $O_2$. The decomposition is catalyzed by certain metals, such as Ag, Pt, and Pd, and by many transition metal oxides.

---

**SAMPLE EXERCISE 22.3**

Using $\Delta G_f°$ for ozone from Appendix D, calculate the equilibrium constant $K_p$, for Equation 22.27 at 298.0 K.

**Solution:** From Appendix D we have $\Delta G_f°(O_3) = 163.4$ kJ/mol. Thus for Equation 22.27. $\Delta G° = (2 \text{ mol } O_3)(163.4 \text{ kJ/mol } O_3) = 326.8$ kJ. From Equation 19.14, we have $\Delta G° = -2.303RT \log K$. Thus

$$\log K = \frac{-\Delta G°}{2.303RT}$$

$$= \frac{-326.8 \times 10^3 \text{ J}}{(2.303)(8.314 \text{ J/K-mol})(298.0 \text{ K})}$$

$$= -57.27$$

$$K = 5.3 \times 10^{-58}$$

In spite of the unfavorable equilibrium constant, ozone can be prepared from $O_2$ as described in the text above. The unfavorable free energy of formation is overcome by energy from the electrical discharge, and $O_3$ is removed before the reverse reaction can occur, so a nonequilibrium mixture results.

**PRACTICE EXERCISE**

Using the data in Appendix D, calculate $\Delta G°$ and the equilibrium constant, $K_p$, for Equation 22.24 at 298.0 K.

**Answer:** $\Delta G° = 66.7$ kJ; $K_p = 2 \times 10^{-12}$

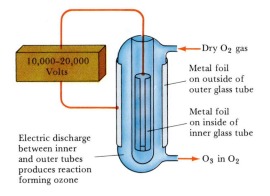

FIGURE 22.13 Apparatus for producing ozone from $O_2$.

Labels in figure:
- Dry $O_2$ gas
- Metal foil on outside of outer glass tube
- Metal foil on inside of inner glass tube
- 10,000–20,000 Volts
- Electric discharge between inner and outer tubes produces reaction forming ozone
- $O_3$ in $O_2$

At the present time, uses of ozone as an industrial chemical are relatively small. It is sometimes used in treatment of domestic water in place of chlorine. Like $Cl_2$, it serves to kill bacteria and oxidize organic compounds. The largest use of ozone, however, is in the preparation of pharmaceuticals, synthetic lubricants, and other commercially useful organic compounds, where $O_3$ is used to sever carbon-carbon double bonds.

Ozone is an important component of the upper atmosphere, where it serves to screen out ultraviolet radiation (Section 14.3) In this way ozone protects the earth from the effects of these high-energy rays. However, in the lower atmosphere ozone is considered an air pollutant. It is a major constituent of smog. Because of its oxidizing power, it causes damage to living systems and structural materials, especially rubber.

The electronegativity of oxygen is second only to that of fluorine. Consequently, oxygen exhibits negative oxidation states in all compounds except those with fluorine, $OF_2$ and $O_2F_2$. The −2 oxidation state is by far the most common. As we have seen, compounds in this oxidation state are called *oxides*.

Nonmetals form covalent oxides (Section 8.11). Most of these oxides are simple molecules with low melting and boiling points. However, $SiO_2$ and $B_2O_3$ have polymeric structures. Most nonmetal oxides combine with water to give oxyacids. For example, sulfur dioxide, $SO_2$, dissolves in water to give sulfurous acid, $H_2SO_3$:

$$SO_2(g) + H_2O(l) \longrightarrow H_2SO_3(aq) \qquad [22.28]$$

Such oxides are called **acidic anhydrides** (anhydride means "without water") or **acidic oxides**. A few nonmetal oxides, especially ones with the nonmetal in a low oxidation state—such as $N_2O$, $NO$, and $CO$—are not acidic anhydrides. These oxides do not react with water.

Most metal oxides are ionic compounds. Those ionic oxides that dissolve in water form hydroxides and are consequently called **basic anhydrides** or **basic oxides**. For example, barium oxide, BaO, dissolves in water to form barium hydroxide, $Ba(OH)_2$:

$$BaO(s) + H_2O(l) \longrightarrow Ba(OH)_2(aq) \qquad [22.29]$$

This reaction is shown in Figure 22.14. Such reactions can be attributed

## Oxides

FIGURE 22.14 Barium oxide, BaO, the white solid at the bottom of the container, reacts with water to produce barium hydroxide, $Ba(OH)_2$. The red color of the solution is caused by phenolthalein and indicates the presence of $OH^-$ ions in solution. (Donald Clegg and Roxy Wilson)

to the high basicity of the $O^{2-}$ ion and consequently its virtually complete hydrolysis in water:

$$O^{2-}(aq) + H_2O(l) \longrightarrow 2OH^-(aq) \qquad [22.30]$$

Even those ionic oxides that are water insoluble tend to dissolve in acids. Iron(III) oxide, for example, dissolves in acids:

$$Fe_2O_3(s) + 6H^+(aq) \longrightarrow 2Fe^{3+}(aq) + 3H_2O(l) \qquad [22.31]$$

This reaction is used to remove rust $(Fe_2O_3 \cdot nH_2O)$ from iron or steel prior to application of a protective coat of zinc or tin.

Oxides that are borderline in acidic and basic character are said to be *amphoteric* (Section 8.11). If a metal forms more than one oxide, the basic character of the oxide decreases as the oxidation state of the metal increases:

| Compound | Oxidation state of Cr | Nature of oxide |
|---|---|---|
| CrO | +2 | Basic |
| $Cr_2O_3$ | +3 | Amphoteric |
| $CrO_3$ | +6 | Acidic |

## Peroxides and Superoxides

Compounds containing O—O bonds and oxygen in an oxidation state of $-1$ are called **peroxides**. Oxygen has an oxidation state of $-\frac{1}{2}$ in $O_2^-$, which is called the **superoxide** ion. The most active metals (Cs, Rb, and K) react with $O_2$ to give superoxides ($CsO_2$, $RbO_2$, and $KO_2$). Their active neighbors in the periodic table (Na, Ca, Sr, and Ba) react with $O_2$, producing peroxides ($Na_2O_2$, $CaO_2$, $SrO_2$, and $BaO_2$). Less active metals and nonmetals produce normal oxides.

When superoxides dissolve in water, $O_2$ is produced:

$$2KO_2(s) + 2H_2O(l) \longrightarrow 2K(aq) + 2OH^-(aq) + O_2(g) + H_2O_2(aq) \qquad [22.32]$$

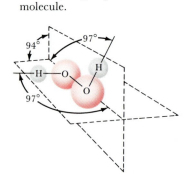

**FIGURE 22.15** Structure of the hydrogen peroxide molecule.

Because of this reaction, potassium superoxide, $KO_2$, is used as an oxygen source in masks worn for rescue work. Moisture in the breath causes the compound to decompose to form $O_2$ and KOH. The KOH so formed serves to remove $CO_2$ from the exhaled breath:

$$2OH^-(aq) + CO_2(g) \longrightarrow H_2O(l) + CO_3^{2-}(aq) \qquad [22.33]$$

Sodium peroxide, $Na_2O_2$, is used commercially as an oxidizing agent. When dissolved in water it produces hydrogen peroxide:

$$Na_2O_2(s) + 2H_2O(l) \longrightarrow 2Na^+(aq) + 2OH^-(aq) + H_2O_2(aq) \qquad [22.34]$$

Hydrogen peroxide, $H_2O_2$, is the most familiar and commercially important peroxide. The structure of $H_2O_2$ is shown in Figure 22.15.

The peroxide linkage (—O—O—) exists in species other than $H_2O_2$ and $O_2^{2-}$ ion. For example, the peroxydisulfate ion, $S_2O_8^{2-}$ or $O_3SOOSO_3^{2-}$, contains this linkage. The $S_2O_8^{2-}$ ion is produced by electrolysis of a 50 percent aqueous solution of sulfuric acid. The reaction of this ion with water produces hydrogen peroxide:

$$2H_2O(l) + S_2O_8^{2-}(aq) \longrightarrow H_2O_2(aq) + 2H^+(aq) + 2SO_4^{2-}(aq) \qquad [22.35]$$

Pure hydrogen peroxide is a clear syrupy liquid, density 1.47 g/cm$^3$ at 0°C. It melts at −0.4°C, and its normal boiling point is 151°C. These properties are characteristic of a highly polar, strongly hydrogen-bonded liquid such as water. Concentrated hydrogen peroxide is a dangerously reactive substance, because the decomposition to form water and oxygen gas is very exothermic:

$$2H_2O_2(l) \longrightarrow 2H_2O(l) + O_2(g) \qquad \Delta H° = -196.0 \text{ kJ} \qquad [22.36]$$

The decomposition can occur with explosive violence if highly concentrated hydrogen peroxide comes in contact with substances that can catalyze the reaction. Hydrogen peroxide is marketed as a chemical reagent in aqueous solutions of up to about 30 percent by weight. A solution containing about 3 percent by weight $H_2O_2$ is commonly used as a mild antiseptic; somewhat more concentrated solutions are employed to bleach fabrics such as cotton, wool, or silk.

The peroxide ion is a by-product of metabolism that results from the reduction of molecular oxygen, $O_2$. The body disposes of this reactive species with enzymes with such names as peroxidase and catalase. The fizzing that occurs when a dilute $H_2O_2$ solution is applied to an open wound is due to decomposition of the $H_2O_2$ into $O_2$ and $H_2O$, a reaction catalyzed by the aforementioned enzymes.

Hydrogen peroxide is capable of acting as either an oxidizing or a reducing agent. Equations 22.37 and 22.38 show the half-reactions for reaction in acid solution.

$$2H^+(aq) + H_2O_2(aq) + 2e^- \longrightarrow 2H_2O(l) \qquad E° = 1.77 \text{ V} \qquad [22.37]$$
$$H_2O_2(aq) \longrightarrow O_2(g) + 2H^+(aq) + 2e^- \qquad E° = -0.67 \text{ V} \qquad [22.38]$$

**The Oxygen Cycle**

Oxygen accounts for about one-fourth of the atoms in living matter. The number of oxygen atoms is fixed; as $O_2$ is removed from air through respiration and other processes, it is replenished. The major nonliving sources of oxygen other than $O_2$ are $CO_2$ and $H_2O$. A simplified picture of the movement of oxygen in our environment is shown in Figure 22.16. This figure points out both how $O_2$ is removed from the atmosphere and how it is replenished. Oxygen, $O_2$, is reformed mainly from $CO_2$ through the process of photosynthesis. Energy is produced when $O_2$ is converted to $CO_2$; energy must therefore be supplied to reform $O_2$ from $CO_2$. This energy is provided by the sun. Thus life on earth depends on chemical recycling made possible by solar energy.

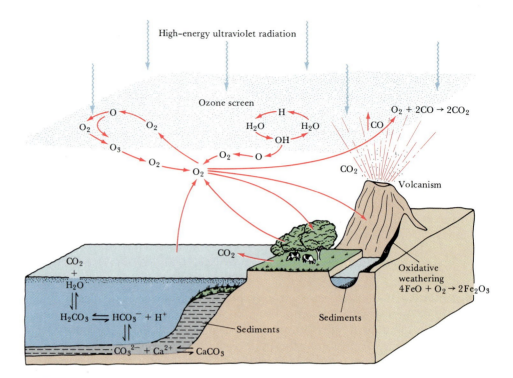

**FIGURE 22.16** Simplified view of the oxygen cycle, showing some of the primary reactions involving oxygen in nature. The atmosphere, which contains $O_2$, is one of the primary sources of the element. Some $O_2$ is produced by radiation-induced dissociation of $H_2O$ in the upper atmosphere. Some $O_2$ is produced by green plants from $H_2O$ and $CO_2$ in the course of photosynthesis. Atmospheric $CO_2$, in turn, results from combustion reactions, animal respiration, and the dissociation of bicarbonate in water. The $O_2$ is used to produce ozone in the upper atmosphere, in oxidative weathering of rocks, in animal respiration, and in combustion reactions.

## 22.5  NITROGEN

Nitrogen was discovered in 1772 by the Scottish botanist Daniel Rutherford. He found that when a mouse was enclosed in a sealed jar, the animal quickly consumed the life-sustaining component of air (oxygen) and died. When the "fixed air" ($CO_2$) in the container was removed, a "noxious air" remained that would not sustain combustion or life. That gas is known to us now as nitrogen.

Nitrogen constitutes 78 percent by volume of the earth's atmosphere, where it occurs as $N_2$ molecules. Although nitrogen is a key element in living creatures, compounds of nitrogen are not abundant in the earth's crust. The major natural deposits of nitrogen compounds are those of $KNO_3$ (saltpeter) in India and $NaNO_3$ (Chile saltpeter) in Chile and other desert regions of South America.

**Properties of Nitrogen**

Nitrogen is a colorless, odorless, and tasteless gas composed of $N_2$ molecules. Its melting point is $-210°C$ and its normal boiling point is $-196°C$.

The $N_2$ molecule is very unreactive because of the strong triple bond between nitrogen atoms (the $N\equiv N$ bond-dissociation energy is 941 kJ/mol, nearly twice that for the bond in $O_2$; see Table 8.3). When substances burn in air they normally react with $O_2$ but not with $N_2$. However, when magnesium burns in air, reaction with $N_2$ also occurs to form magnesium nitride, $Mg_3N_2$. A similar reaction occurs with lithium:

$$3Mg(s) + N_2(g) \longrightarrow Mg_3N_2(s) \qquad [22.39]$$

$$6Li(s) + N_2(g) \longrightarrow 2Li_3N(s) \qquad [22.40]$$

The nitride ion is a strong Brønsted base. It reacts with water to form ammonia, $NH_3$:

$$Mg_3N_2(s) + 6H_2O(l) \longrightarrow 2NH_3(aq) + 3Mg(OH)_2(s) \qquad [22.41]$$

The electron configuration of the nitrogen atom is $[He]2s^2 2p^3$. The element exhibits all formal oxidation states from $+5$ to $-3$, as shown in Table 22.2. Because nitrogen is the fourth most electronegative element after fluorine, oxygen, and chlorine, it exhibits positive oxidation states only in combination with those three elements.

Figure 22.17 summarizes the standard electrode potentials for interconversion of several common nitrogen species. The potentials in the diagram are large and positive, which indicates that the nitrogen oxides and oxyanions shown are strong oxidizing agents.

**TABLE 22.2**   Oxidation states of nitrogen

| Oxidation state | Examples |
|---|---|
| $+5$ | $N_2O_5$, $HNO_3$, $NO_3^-$ |
| $+4$ | $NO_2$, $N_2O_4$ |
| $+3$ | $HNO_2$, $NO_2^-$, $NF_3$ |
| $+2$ | $NO$ |
| $+1$ | $N_2O$, $H_2N_2O_2$, $N_2O_2^{2-}$, $HNF_2$ |
| $0$ | $N_2$ |
| $-1$ | $NH_2OH$, $H_2NF$ |
| $-2$ | $N_2H_4$ |
| $-3$ | $NH_3$, $NH_4^+$, $NH_2^-$ |

**FIGURE 22.17**   Standard electrode potentials in acid solution for reduction of some common nitrogen-containing compounds in various oxidation states. For example, reduction of $NO_3^-$ to $NO_2$ in acid solution has a standard electrode potential of 0.79 V (leftmost entry). You should be able to write the complete, balanced half-reaction for this reduction using $H^+$, $H_2O$, and $e^-$, as discussed in Section 13.7.

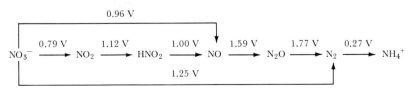

## Preparation and Uses of Nitrogen

Elemental nitrogen is obtained in commercial quantities by fractional distillation of liquid air. About $1.6 \times 10^{10}$ kg (18 million tons) of $N_2$ is produced annually in the United States.

Because of its low reactivity, large quantities of $N_2$ are used as an inert gaseous blanket to exclude $O_2$ during the processing and packaging of foods, the manufacture of chemicals, the fabrication of metals, and the production of electronic devices. Liquid $N_2$ is employed as a coolant to freeze foods rapidly.

The largest use of nitrogen is in the manufacture of nitrogen-containing chemicals. The formation of nitrogen compounds from $N_2$ is known as **nitrogen fixation**. The demand for fixed nitrogen is high because the element is required in maintaining soil fertility. Although we are immersed in an ocean of air that contains abundant $N_2$, our supply of food is limited more by the availability of fixed nitrogen than by that of any other plant nutrient. Thus $N_2$ is used primarily for the manufacture of nitrogen-containing fertilizers. It is also used to manufacture explosives, plastics, and many important chemicals.

The chain of conversion of $N_2$ into a variety of useful, simple nitrogen-containing species is given in Figure 22.18. The processes shown are discussed in more detail in later portions of the section.

## Hydrogen Compounds of Nitrogen

**Ammonia** is one of the most important compounds of nitrogen. It is a colorless, toxic gas that has a characteristic, irritating odor. As we have noted in previous discussions (Section 17.4), the $NH_3$ molecule is basic ($K_b = 1.8 \times 10^{-5}$).

In the laboratory, $NH_3$ is prepared by the action of NaOH on an ammonium salt. The $NH_4^+$ ion, which is the conjugate acid of $NH_3$, loses a proton to $OH^-$. The resultant $NH_3$ is volatile and is driven from the solution by mild heating:

$$NH_4Cl(aq) + NaOH(aq) \longrightarrow NH_3(g) + H_2O(l) + NaCl(aq) \qquad [22.42]$$

Commercial production of $NH_3$ is achieved by the **Haber process** (Section 16.1), in which $N_2$ and $H_2$ are catalytically combined at high pressure and high temperature:

**FIGURE 22.18**  Conversion of $N_2$ into common nitrogen compounds.

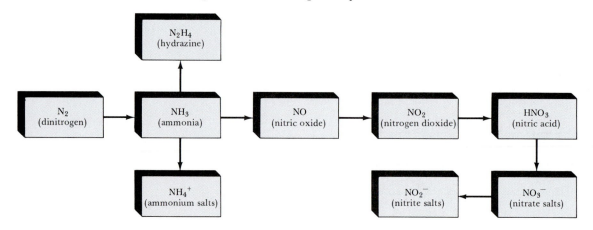

$$N_2(g) + 3H_2(g) \longrightarrow 2NH_3(g) \qquad [22.43]$$

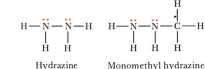

**FIGURE 22.19** Lewis structures of hydrazine, $N_2H_4$, and monomethyl hydrazine, $CH_3NHNH_2$.

About $1.5 \times 10^{10}$ kg (17 million tons) of ammonia is produced annually in the United States. About 75 percent is used for fertilizer.

**Hydrazine**, $N_2H_4$, bears the same relationship to ammonia that hydrogen peroxide does to water. As shown in Figure 22.19, the hydrazine molecule contains an N—N single bond. Hydrazine is quite poisonous. It can be prepared by the reaction of ammonia with hypochlorite ion, $OCl^-$, in aqueous solution:

$$2NH_3(aq) + OCl^-(aq) \longrightarrow N_2H_4(aq) + Cl^-(aq) + H_2O(l) \qquad [22.44]$$

The possible formation of $N_2H_4$ from household ammonia and chlorine bleach, which contains $OCl^-$, is one reason for the oft-cited warning not to mix household cleaning agents.

Pure hydrazine is an oily, colorless liquid with a melting point of $1.5°C$ and a boiling point of $113°C$. The pure substance explodes on heating and is a highly reactive reducing agent. $N_2H_4$ is normally employed in aqueous solution, where it can be handled safely. The substance is weakly basic, and salts of $N_2H_5^+$ can be formed:

$$N_2H_4(aq) + H_2O(l) \rightleftharpoons N_2H_5^+(aq) + OH^-(aq) \qquad K_b = 1.3 \times 10^{-6} \qquad [22.45]$$

The combustion of hydrazine is highly exothermic:

$$N_2H_4(l) + O_2(g) \longrightarrow N_2(g) + 2H_2O(g) \qquad \Delta H° = -534 \text{ kJ} \qquad [22.46]$$

Hydrazine and compounds derived from it, such as monomethyl hydrazine (Figure 22.19), are used as rocket fuels. Monomethyl hydrazine is one of the fuels used in the space shuttles (Figure 22.20).

---

### SAMPLE EXERCISE 22.4

Hydroxylamine reduces copper(II) to the free metal in acid solutions. Write a balanced equation for the reaction, assuming that $N_2$ is the oxidation product.

**Solution:** The unbalanced and incomplete half-reactions are

$$Cu^{2+}(aq) \longrightarrow Cu(s)$$
$$NH_2OH(aq) \longrightarrow N_2(g)$$

Balancing these equations as described in Section 18.1 gives

$$Cu^{2+}(aq) + 2e^- \longrightarrow Cu(s)$$
$$2NH_2OH(aq) \longrightarrow N_2(g) + 2H_2O(l) + 2H^+(aq) + 2e^-$$

Adding these half-reactions gives the balanced equation:

$$Cu^{2+}(aq) + 2NH_2OH(aq) \longrightarrow Cu(s) + N_2(g) + 2H_2O(l) + 2H^+(aq)$$

### PRACTICE EXERCISE

**(a)** In power plants hydrazine is used to prevent corrosion of the metal parts of steam boilers by the $O_2$ dissolved in the water. The hydrazine reacts with $O_2$ in water to give $N_2$ and $H_2O$. Write a balanced chemical equation for this reaction. **(b)** Monomethyl hydrazine, $N_2H_3CH_3(l)$, is used with the oxidizer dinitrogen tetraoxide, $N_2O_4(l)$, to power the steering rockets of the space shuttle orbiter. The reaction of these two substances produces $N_2$, $CO_2$, and $H_2O$. Write a balanced chemical equation for this reaction.
*Answers:*
**(a)** $N_2H_4(aq) + O_2(aq) \longrightarrow N_2(g) + 2H_2O(l)$
**(b)** $5N_2O_4(l) + 4N_2H_3CH_3(l) \longrightarrow 9N_2(g) + 4CO_2(g) + 12H_2O(g)$

**FIGURE 22.20** Launch of the first space shuttle, April 12, 1981. Monomethyl hydrazine is one of the fuels used in the space shuttle. (NASA)

## Oxides and Oxyacids of Nitrogen

Nitrogen forms three common oxides: $N_2O$ (nitrous oxide), NO (nitric oxide), and $NO_2$ (nitrogen dioxide). It also forms two unstable oxides that we will not discuss, $N_2O_3$ (dinitrogen trioxide) and $N_2O_5$ (dinitrogen pentoxide).

**Nitrous oxide**, $N_2O$, is also known as laughing gas because a person becomes somewhat giddy after inhaling only a small amount of it. This colorless gas was the first substance used as a general anesthetic. It is used as the compressed gas propellant in several aerosols and foams, such as in whipped cream. It can be prepared in the laboratory by carefully heating ammonium nitrate to about 200°C:

$$NH_4NO_3(s) \xrightarrow{\Delta} N_2O(g) + 2H_2O(g) \qquad [22.47]$$

**Nitric oxide**, NO, is also a colorless gas, but unlike $N_2O$ it is slightly toxic. It can be prepared in the laboratory by reduction of dilute nitric acid, using copper or iron as a reducing agent, as shown in Figure 22.21.

**FIGURE 22.21** (a) Nitric oxide, NO, can be prepared by reacting copper with 6 $M$ nitric acid. In this photo a jar containing 6 $M$ $HNO_3$ has been inverted over some pieces of copper. Colorless NO, which is only slightly soluble in water, is collected in the jar. The blue color of the solution is due to the presence of $Cu^{2+}$ ions. (b) Colorless NO gas, collected as shown on the left. (c) When the stopper is removed from the jar of NO, the NO reacts with oxygen in the air to form yellow-brown $NO_2$. (Donald Clegg and Roxy Wilson)

(a)  (b)  (c)

$$(s) + 2NO_3^-(aq) + 8H^+(aq) \longrightarrow 3Cu^{2+}(aq) + 2NO(g) + 4H_2O(l) \quad [22.48]$$

It is also produced by direct combination of $N_2$ and $O_2$ at elevated temperatures. As we saw in Section 14.4, this reaction is a significant source of nitrogen oxide air pollutants, which form during combustion reactions in air. However, the direct combination of $N_2$ and $O_2$ is not presently used for commercial production of NO because the yield is low; the equilibrium constant $K_c$ at 2400 K is only 0.05.

The commercial route to NO (and hence to other oxygen-containing compounds of nitrogen) is by means of the catalytic oxidation of $NH_3$ (Figure 22.22):

$$4NH_3(g) + 5O_2(g) \xrightarrow[800°C]{Pt\ catalyst} 4NO(g) + 6H_2O(g) \quad [22.49]$$

In the absence of the platinum catalyst, $NH_3$ is converted to $N_2$ instead of NO:

$$4NH_3(g) + 3O_2(g) \xrightarrow{1000°C} 2N_2(g) + 6H_2O(g) \quad [22.50]$$

The catalytic conversion of $NH_3$ to NO is the first step in a three-step process known as the **Ostwald process**, by which $NH_3$ is converted commercially into nitric acid, $HNO_3$. Nitric oxide reacts readily with $O_2$, forming $NO_2$ when exposed to air (see Figure 22.21):

$$2NO(g) + O_2(g) \longrightarrow 2NO_2(g) \quad [22.51]$$

When dissolved in water, $NO_2$ forms nitric acid:

$$3NO_2(g) + H_2O(l) \longrightarrow 2H^+(aq) + 2NO_3^-(aq) + NO(g) \quad [22.52]$$

**FIGURE 22.22** The oxidation of ammonia during the production of nitric acid is one of the most important chemical processes undertaken on a large industrial scale. The reaction is carried out in the presence of a platinum-rhodium catalyst, which is usually in the form of a woven gauze. Here new catalyst gauzes are being installed in an ammonia oxidation plant. (Johnson-Matthey Metals Limited)

Note that nitrogen is both oxidized and reduced in this reaction. The $NO_2$ is said to have undergone **disproportionation**. The reduction product $NO$ can be converted back into $NO_2$ by exposure to air and thereafter dissolved in water to prepare more $HNO_3$.

**Nitrogen dioxide** is a yellow-brown gas. It is poisonous and has a choking odor. Below room temperature two $NO_2$ molecules combine to form the colorless $N_2O_4$:

$$2NO_2(g) \longrightarrow N_2O_4(g) \qquad \Delta H° = -58 \text{ kJ} \qquad [22.53]$$

The two common oxyacids of nitrogen are nitric acid, $HNO_3$, and nitrous acid, $HNO_2$ (Figure 22.23). **Nitric acid** is a colorless, corrosive liquid. Nitric acid solutions often take on a slightly yellow color (Figure 22.24) as a result of small amounts of $NO_2$ formed by photochemical decomposition:

$$4HNO_3(aq) \longrightarrow 4NO_2(g) + O_2(g) + 2H_2O(l) \qquad [22.54]$$

Nitric acid is a strong acid. It is also a powerful oxidizing agent, as the following electrode potentials indicate:

$$NO_3^-(aq) + 2H^+(aq) + e^- \longrightarrow NO_2(g) + H_2O(l) \qquad E° = 0.79 \text{ V} \qquad [22.55]$$
$$NO_3^-(aq) + 4H^+(aq) + 3e^- \longrightarrow NO(g) + 2H_2O(l) \qquad E° = 0.96 \text{ V} \qquad [22.56]$$

Concentrated nitric acid will attack and oxidize most metals, except Au, Pt, Rh, and Ir.

About $7.1 \times 10^9$ kg (8 million tons) of nitric acid is produced annually in the United States. The largest use of nitric acid is in manufacture of $NH_4NO_3$ for fertilizers, which accounts for about 80 percent of the production. The acid is also used in the production of plastics, drugs, and explosives.

The development of the Haber and Ostwald processes in Germany just prior to World War I permitted Germany to make munitions even though naval blockades prevented access to traditional sources of nitrates. Among the explosives made from nitric acid are nitroglycerin, trinitrotoluene (TNT), and nitrocellulose. The reaction of nitric acid with glycerin to form nitroglycerin is shown in Equation 22.57:

**FIGURE 22.23** Structures of nitric acid and nitrous acid.

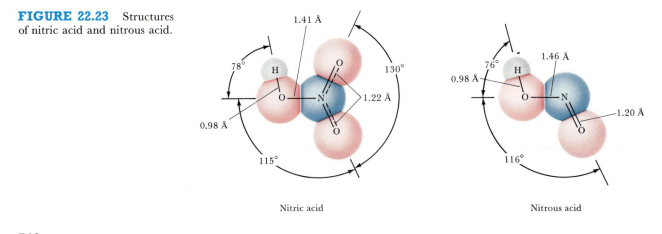

Nitric acid

Nitrous acid

**FIGURE 22.24** Colorless nitric acid solution (left) becomes yellow upon standing in sunlight (right). (Fundamental Photographs)

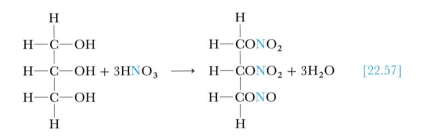

$$
\begin{array}{c}
\hspace{1em}\text{H} \hspace{8em} \text{H} \\
\hspace{1em}| \hspace{8.5em} | \\
\text{H}-\text{C}-\text{OH} \hspace{5em} \text{H}-\text{CONO}_2 \\
\hspace{1em}| \hspace{8.5em} | \\
\text{H}-\text{C}-\text{OH} + 3\text{HNO}_3 \longrightarrow \text{H}-\text{CONO}_2 + 3\text{H}_2\text{O} \hspace{2em} [22.57] \\
\hspace{1em}| \hspace{8.5em} | \\
\text{H}-\text{C}-\text{OH} \hspace{5em} \text{H}-\text{CONO} \\
\hspace{1em}| \hspace{8.5em} | \\
\hspace{1em}\text{H} \hspace{8em} \text{H}
\end{array}
$$

When nitroglycerin explodes, the reaction summarized in Equation 22.58 occurs:

$$
4\text{C}_3\text{H}_5\text{N}_3\text{O}_9(l) \longrightarrow 6\text{N}_2(g) + 12\text{CO}_2(g) + 10\text{H}_2\text{O}(g) + \text{O}_2(g) \hspace{2em} [22.58]
$$

A considerable amount of gaseous products form from the liquid. The sudden formation of these gases, together with their expansion resulting from the heat generated by the reaction, produces the explosion.

**Nitrous acid**, $HNO_2$ (Figure 22.23), is considerably less stable than $HNO_3$ and tends to disproportionate into NO and $HNO_3$. It is normally made by action of a strong acid such as $H_2SO_4$ on a cold solution of a nitrite salt such as $NaNO_2$. Nitrous acid is a weak acid ($K_a = 4.5 \times 10^{-4}$).

**The Nitrogen Cycle in Nature**

Two primary routes for nitrogen fixation exist in nature: Lightning causes the formation of NO from $N_2$ and $O_2$ in air. In addition, the root nodules of certain leguminous plants, such as peas, beans, peanuts, and alfalfa, contain nitrogen-fixing bacteria. Both iron and molybdenum are part of the enzyme system responsible for the nitrogen fixation in these root nodules. It is of interest to understand this process and to develop catalysts that, like enzymes, fix nitrogen at ambient pressure and temperature (and hence potentially at an energy savings over the Haber process).

Nitrogen is found in many compounds that are vital to life, including proteins, enzymes, nucleic acids, vitamins, and hormones. Plants employ

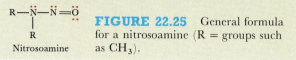
very simple compounds as starting materials from which such complex, biologically necessary compounds are formed. Plants are able to use several forms of nitrogen, especially $NH_3$, $NH_4^+$, and $NO_3^-$. Liquid ammonia, ammonium nitrate, $NH_4NO_3$, and urea, $(NH_2)_2CO$, are among the most commonly applied nitrogen fertilizers. Urea is made by the reaction of ammonia and carbon dioxide:

$$2NH_3(aq) + CO_2(aq) \rightleftharpoons \overset{\displaystyle O}{\overset{\displaystyle \|}{H_2NCNH_2}}(aq) + H_2O(l) \qquad [22.59]$$

Urea

$NH_3$ is slowly released as the urea reacts with water in the soil.

Animals are unable to synthesize the complex nitrogen compounds they require from the simple substances used by plants. Instead, they rely on more complicated precursors present in foods. Those nitrogen compounds not needed by the animal are excreted as nitrogenous waste. Certain microorganisms are able to convert this waste back into $N_2$. Nitrogen is recycled in this fashion. As in the case of the cycling of oxygen, energy from the sun is required. A simplified picture of the nitrogen cycle is shown in Figure 22.26.

Large-scale cultivation of nitrogen-fixing legumes and industrial fixation have increased the quantity of fixed nitrogen in the biosphere. However, only a portion of the added fertilizer ends up being used by the plants for which it was intended. One effect of this intrusion into the nitrogen cycle has been increased water pollution. Much fixed nitrogen ends up as nitrates in the soil. These compounds are highly water soluble. They are therefore readily washed from the soil into water bodies. Once in a lake, nitrates stimulate plant growth, encouraging, for example, rapid growth of algae. When these plants die, their decay consumes $O_2$ in the water, thereby killing fish and other oxygen-dependent organisms. The resultant anaerobic environment leads to the foul odors that we associate with highly polluted bodies of water.

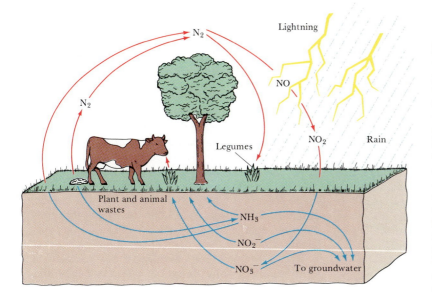

**FIGURE 22.26** Simplified picture of the nitrogen cycle, showing some of the primary reactions involved in the utilization and formation of nitrogen in nature. The main reservoir of nitrogen is the atmosphere, which contains $N_2$. This nitrogen is fixed through the action of lightning and leguminous plants. The compounds of nitrogen reside in the soil as $NH_3$ (and $NH_4^+$), $NO_2^-$, and $NO_3^-$. All are water soluble and can be washed out of the soil by ground water. These nitrogen compounds are utilized by plants in their growth and are incorporated into animals that eat the plants. Animal waste and dead plants and animals are attacked by certain bacteria that free $N_2$, which escapes into the atmosphere, thereby completing the cycle.

## 22.6 CARBON

Carbon is not an abundant element; it constitutes only 0.027 percent of the earth's crust. Although some occurs in elemental form as graphite and diamond, most is found in combined form. Over half occurs in carbonate compounds, such as $CaCO_3$. Carbon is also found in coal, petroleum, and natural gas. The importance of the element stems in large part from its occurrence in all living organisms; life as we know it is based on carbon compounds. About 150 years ago scientists believed that these life-sustaining compounds could be made only within living systems. They were consequently called **organic compounds**. It is now known that organic compounds can be synthesized in the laboratory from simple inorganic (nonorganic) substances. Although the name "organic chemistry" persists, it is now used to describe the portion of chemistry that focuses on hydrocarbons and compounds derived from them by substituting other atoms for some hydrogen atoms. In this section we will take a brief look at carbon and its most common inorganic compounds. Organic chemistry and its application to biological systems are considered in Chapters 27 and 28.

**Elemental Forms of Carbon**

We have seen from earlier discussions (Section 9.7) that carbons exists in two allotropic forms, graphite and diamond. Charcoal, carbon black, and coke are microcrystalline or amorphous forms of graphite.

**Graphite** is mined in several parts of the world, but it can also be produced synthetically from amorphous carbon. The word "graphite" comes from the Greek and means "to write." The so-called lead of pencils contains graphite together with a binder, usually clay, to make it harder. Graphite is a soft, black, slippery solid that has a metallic luster and conducts electricity. Graphite is used commercially as a lubricant and also to construct electrodes. The relative softness and electrical conductivity of graphite can be related to its structure. Crystalline graphite consists of parallel sheets of carbon atoms, each sheet containing hexag-

The properties of graphite are anisotropic (that is, they differ in different directions through the solid). Along the carbon planes graphite possesses great strength because of the number and strength of the carbon-carbon bonds in this direction. In contrast, we have seen that the bonds between planes are relatively weak, making graphite weak in that direction.

Fibers of graphite can be prepared in which the carbon planes are aligned to varying extents parallel to the fiber axis. These fibers are also lightweight (density of about 2 g/cm$^3$) and chemically quite unreactive. The oriented fibers are made by first slowly pyrolyzing organic fibers to about 150 to 300°C. These fibers are then heated to about 2500°C to graphitize them (that is, to convert amorphous carbon to graphite). Stretching the fiber during pyrolysis helps orient the graphite planes parallel to the fiber axis. More amorphous carbon fibers are formed by pyrolysis of organic fibers at 1200 to 1400°C. These amorphous materials [Figure 22.27(a)] commonly called carbon fibers, are the type most commonly used in commercial materials (Figure 22.27).

Composite materials that take advantage of the strength, stability, and low density of carbon fibers are widely used. Composites are combinations of two or more materials; these materials are present as separate phases and are combined to form structures that take advantage of certain desirable properties of each component. In carbon composites, the graphite fibers are often woven into a fabric that is embedded in a matrix that binds them into a solid structure. The fibers transmit loads evenly throughout the matrix. The finished composite is thus stronger than any one of its components.

Epoxy systems are useful matrices because of their excellent adherence, but they are costly and limited to service temperatures below 150°C. More heat-resistent resins are required for many aerospace applications, where carbon composites find wide use. Some parts of high-performance aircraft are now made out of such carbon composites. Figure 22.27(b) shows a photomicrograph of a section through a carbon-epoxy composite.

**FIGURE 22.27** (a) Carbon fibers used to make composite materials. (Courtesy American Cyanamid Co.) (b) A section through a carbon-epoxy composite, showing the round carbon fibers imbedded in an epoxy matrix (magnification 1000 times). (Mark Warmkessel)

(a)                                        (b)

onal arrays of carbon atoms (Figure 9.36). Each atom exhibits $sp^2$ hybridization and takes part in delocalized $\pi$ bonding with other atoms in the sheet. This delocalized $\pi$ system is responsible for the electrical conductivity of the graphite; the interaction of $\pi$ electrons with light is responsible for the black color. The softness and lubricity arise from the weak binding, by London-dispersion forces, that exists between sheets.

**Diamonds** are found in many parts of the world, but most come from South Africa. The conversion of graphite to diamond is endothermic:

$$C(graphite) \longrightarrow C(diamond) \qquad \Delta H^\circ = 1.87 \text{ kJ} \qquad [22.60]$$

Furthermore, diamond is the more compact allotrope ($d = 2.25 \text{ g/cm}^3$ for graphite; $d = 3.51 \text{ g/cm}^3$ for diamond). Thus its formation from graphite is favored by high temperatures and high pressures. Such conditions are used in synthesizing industrial-grade diamonds (Figure 22.28) employed in cutting, grinding, and polishing tools. Diamonds are one of the hardest known substances. Because of this property, diamonds are used as styluses in high-quality record players. The hardness of diamond can be attributed to its structure (Figure 9.35). Each carbon atom has four nearest neighbors to which it is bonded by $\sigma$ bonds. The bond angles are all 109°, typical of $sp^3$ hybridization. The resultant interlocking network of covalent bonds makes the structure very rigid. All valence electrons take part in $\sigma$ bonds so diamond is a nonconductor.

**Carbon black** is formed when hydrocarbons are heated in a very limited supply of oxygen:

$$CH_4(g) + O_2(g) \longrightarrow C(s) + 2H_2O(g) \qquad [22.61]$$

It is used as a pigment in black inks; large amounts are also used in making automobile tires.

**Charcoal** is formed when wood is heated strongly in the absence of air. Charcoal has a very open structure, giving it an enormous surface area per unit mass. Activated charcoal, a pulverized form whose surface is cleaned by heating with steam, is widely used to adsorb molecules. It is used in filters to remove offensive odors from air and colored or bad-tasting impurities from water.

**Coke** is an impure form of carbon formed when coal is heated strongly in the absence of air. It is widely used as a reducing agent in metallurgical operations (Chapter 24).

**FIGURE 22.28** Graphite and synthetic diamond prepared from graphite. Most synthetic diamonds lack the size, color, and clarity of natural diamonds and are therefore not used in jewelry. The conversion of graphite into diamond is usually carried out in the presence of catalysts at about 2000°C and at pressures exceeding 40,000 atm. (General Electric Research and Development Center)

**Oxides of Carbon**

Carbon forms two principal oxides: carbon monoxide, CO, and carbon dioxide, $CO_2$. **Carbon monoxide** is formed when carbon or hydrocarbons are burned in a limited supply of oxygen:

$$2C(s) + O_2(g) \longrightarrow 2CO(g) \qquad [22.62]$$

It is a colorless, odorless, and tasteless gas (m.p. $= -199°C$; b.p. $= -192°C$). It is toxic because of its ability to bind to hemoglobin and thus interfere with oxygen transport (Section 14.4). Low-level poisoning results in headaches and drowsiness; high-level poisoning can cause death.

CO is an unusual carbon compound because it has a lone pair of electrons on carbon: $:C\equiv O:$. One might imagine that CO would be unreactive, like the isoelectronic $N_2$ molecule. Both substances have high bond energies (1072 kJ/mol for $C\equiv O$ and 941 kJ/mol for $N\equiv N$). However, because of the lower nuclear charge on carbon (compared with either N or O), the lone pair on carbon is not held as strongly as that on N or O. Consequently, CO is better able to function as an electron-pair donor (Lewis base) than is $N_2$. CO is able to form a wide variety of covalent compounds, known as metal carbonyls, with transition metals. An example of such a compound is $Ni(CO)_4$, a volatile, very toxic compound that is formed by simply warming metallic nickel in the presence of CO. The formation of such metal carbonyls is the first step in the transition metal catalysis of a variety of reactions of CO.

Carbon monoxide has several commercial uses. Because it burns readily, forming $CO_2$, it is employed as a fuel:

$$2CO(g) + O_2(g) \longrightarrow 2CO_2(g) \qquad \Delta H = -566 \text{ kJ} \qquad [22.63]$$

It is also an important reducing agent, widely used in metallurgical operations to reduce metal oxides. For example, it is the most important reducing agent in the blast furnace reduction of iron(III) oxide:

$$Fe_2O_3(s) + 3CO(g) \longrightarrow 2Fe(s) + 3CO_2(g) \qquad [22.64]$$

This reaction is discussed in greater detail in Chapter 24. Carbon monoxide is also used in the preparation of several organic compounds. In Section 22.3 we saw that it can be combined catalytically with $H_2$ to manufacture methanol, $CH_3OH$ (Equation 22.16).

**Carbon dioxide** is produced when carbon-containing substances are burned in excess oxygen:

$$C(s) + O_2(g) \longrightarrow CO_2(g) \qquad [22.65]$$

$$CH_4(g) + 2O_2(g) \longrightarrow CO_2(g) + 2H_2O(l) \qquad [22.66]$$

$$C_2H_5OH(l) + 3O_2(g) \longrightarrow 2CO_2(g) + 3H_2O(l) \qquad [22.67]$$

It is also produced when many carbonates are heated:

$$CaCO_3(s) \longrightarrow CaO(s) + CO_2(g) \qquad [22.68]$$

Large quantities are also obtained as a by-product of the fermentation of sugar during the production of alcohol:

$$C_6H_{12}O_6(aq) \xrightarrow{\text{yeast}} 2C_2H_5OH(aq) + 2CO_2(g) \qquad [22.69]$$

$$\underset{\text{Glucose}}{\phantom{C_6H_{12}O_6(aq)}} \qquad \underset{\text{Ethanol}}{\phantom{2C_2H_5OH(aq)}}$$

In the laboratory $CO_2$ is normally produced by the action of acids on carbonates, as shown in Figure 22.29:

$$CO_3^{2-}(aq) + 2H^+(aq) \longrightarrow CO_2(g) + H_2O(l) \qquad [22.70]$$

Carbon dioxide is a colorless and odorless gas. It is not toxic, but high concentrations increase respiration rate and can cause suffocation. It is readily liquefied by compression. However, when cooled at atmospheric pressure it condenses as a solid rather than as a liquid. The solid sublimes at atmospheric pressure at $-78°C$. This property makes solid $CO_2$ valuable as a refrigerant that is always free of the liquid form; solid $CO_2$ is thus known as **dry ice**. About half of the $CO_2$ consumed annually is used for refrigeration. The other major use is in the production of carbonated beverages. Large quantities are also used to manufacture **washing soda**, $Na_2CO_3 \cdot 10H_2O$, and **baking soda**, $NaHCO_3$. Baking soda is so named because the following reaction occurs in baking:

$$NaHCO_3(s) + H^+(aq) \longrightarrow Na^+(aq) + CO_2(g) + H_2O(l) \qquad [22.71]$$

The $H^+(aq)$ is provided by vinegar, sour milk, or by hydrolysis of certain salts. The bubbles of $CO_2$ that form are trapped in the dough, causing it to rise. Washing soda is used to precipitate metal ions that interfere with the cleansing action of soap (Section 12.7).

Carbon dioxide is moderately soluble in $H_2O$ at atmospheric pressure. The resultant solutions are moderately acidic as a result of the formation of carbonic acid, $H_2CO_3$:

$$CO_2(aq) + H_2O(l) \rightleftharpoons H_2CO_3(aq) \qquad [22.72]$$

Carbonic acid is a weak diprotic acid. Its acidic character causes carbonated beverages to have a sharp, slightly acidic taste.

Although carbonic acid cannot be isolated as a pure compound, two types of salts can be obtained by neutralization of carbonic acid solutions: hydrogen carbonates (bicarbonates) and carbonates. Partial neutralization produces $HCO_3^-$, and complete neutralization gives $CO_3^{2-}$.

The $HCO_3^-$ ion is a stronger base than acid ($K_a = 5.6 \times 10^{-11}$; $K_b = 2.3 \times 10^{-8}$). Consequently, aqueous solutions of $HCO_3^-$ are weakly alkaline:

$$HCO_3^-(aq) + H_2O(l) \rightleftharpoons H_2CO_3(aq) + OH^-(aq) \qquad [22.73]$$

The carbonate ion is much more strongly basic ($K_b = 1.8 \times 10^{-4}$):

**FIGURE 22.29** Solid $CaCO_3$ reacts with a solution of hydrochloric acid to produce $CO_2$ gas, seen here as the bubbles evident in the test tube. (Paul Silverman/Fundamental Photographs)

**Carbonic Acid and Carbonates**

$$CO_3{}^{2-}(aq) + H_2O(l) \rightleftharpoons HCO_3{}^{-}(aq) + OH^{-}(aq) \qquad [22.74]$$

Minerals containing the carbonate ion are plentiful.* The principal carbonate minerals are calcite ($CaCO_3$), magnesite ($MgCO_3$), dolomite [$MgCa(CO_3)_2$], and siderite ($FeCO_3$). Calcite is the principal mineral in limestone rock, large deposits of which occur in many parts of the world. It is also the main constituent of marble, chalk, pearls, coral reefs, and the shells of marine animals such as clams and oysters. Although $CaCO_3$ has low solubility in pure water, it dissolves readily in acidic solutions, with evolution of $CO_2$:

$$CaCO_3(s) + 2H^{+}(aq) \rightleftharpoons Ca^{2+}(aq) + H_2O(l) + CO_2(g) \qquad [22.75]$$

Because water containing $CO_2$ is slightly acidic (Equation 22.72), $CaCO_3$ dissolves slowly in this medium:

$$CaCO_3(s) + H_2O(l) + CO_2(g) \rightleftharpoons Ca^{2+}(aq) + 2HCO_3{}^{-}(aq) \qquad [22.76]$$

This reaction occurs when surface waters move underground through limestone deposits. It is the principal way that $Ca^{2+}$ enters groundwater, producing "hard water" (Section 14.6). If the dissolving limestone underlies a comparatively thin layer of earth, sinkholes, such as that shown in Figure 22.30, are produced. If the limestone deposit is deep enough underground, the dissolution of the limestone produces a cave; two well-known limestone caves are the Mammoth Cave in Kentucky and the Carlsbad Caverns in New Mexico (Figure 22.31).

One of the most important reactions of $CaCO_3$ is its decomposition into CaO and $CO_2$ at elevated temperatures, given earlier in Equation

* *Minerals* are solid substances that occur in nature. They are usually known by their common names rather than by their chemical names. What we know as *rock* is merely an aggregate of different kinds of minerals.

**FIGURE 22.30** Sinkhole filled with water. (Lawrence E. Naylor/Photo Researchers)

**FIGURE 22.31**  Carlsbad Caverns, New Mexico. (Frances Current/Photo Researchers)

22.68. Over $1.3 \times 10^{10}$ kg (15 million tons) of calcium oxide, known as lime or quicklime, is used in the United States each year. Because calcium oxide reacts with water to form $Ca(OH)_2$, it is an important commercial base. It is also important in making mortar, which is a mixture of sand, water, and CaO used in construction to bind bricks, blocks, and rocks together. CaO reacts with water and $CO_2$ to form $CaCO_3$, which binds the sand in the mortar:

$$CaO(s) + H_2O(l) \rightleftharpoons Ca^{2+}(aq) + 2OH^-(aq) \qquad [22.77]$$

$$Ca^{2+}(aq) + 2OH^-(aq) + CO_2(aq) \longrightarrow CaCO_3(s) + H_2O(l) \qquad [22.78]$$

## Carbides

The binary compounds of carbon with metals, metalloids, and certain nonmetals are called carbides. There are three types: ionic, interstitial, and covalent. The ionic, or saltlike, carbides are formed by the more active metals. The most common ionic carbides contain $C_2^{2-}$ ions: $:C{\equiv}C:^{2-}$. This ion hydrolyzes to form acetylene:

$$CaC_2(s) + 2H_2O(l) \longrightarrow Ca(OH)_2(aq) + C_2H_2(g) \qquad [22.79]$$

Carbides containing $C_2^{2-}$ are thus called **acetylides**. The most important ionic carbide is calcium carbide, $CaC_2$, which is produced industrially by the reduction of CaO with carbon at high temperature:

$$2CaO(s) + 5C(s) \longrightarrow 2CaC_2(s) + CO_2(g) \qquad [22.80]$$

The $CaC_2$ is used to prepare acetylene, which is used in welding.

Interstitial carbides are formed by many transition metals. The carbon atoms occupy open spaces (interstices) between metal atoms, in a manner analogous to the interstitial hydrides (Section 22.3). An example is tungsten carbide, WC, which is very hard and heat resistant and consequently is used in making cutting tools.

Covalent carbides are formed by boron and silicon. Silicon carbide, SiC, is known as carborundum. It is made by heating $SiO_2$ (sand) and carbon to high temperatures:

$$SiO_2(s) + 3C(s) \longrightarrow SiC(s) + 2CO(g) \qquad [22.81]$$

SiC has the same structure as diamond except that silicon atoms alternate with carbon atoms. Like industrial diamond, carborundum is very hard and thus used as an abrasive and in cutting tools.

**Other Inorganic Compounds of Carbon**

Hydrogen cyanide, HCN (Figure 22.32), is an extremely toxic gas that has the odor of bitter almonds. In the laboratory it is made by heating a cyanide salt, such as NaCN, with an acid:

$$CN^-(aq) + H^+(aq) \longrightarrow HCN(g) \qquad [22.82]$$

Aqueous solutions of HCN are known as hydrocyanic acid. Neutralization with a base, such as NaOH, produces cyanide salts, such as NaCN. Cyanides find use in the manufacture of several well-known plastics, including nylon and Orlon. The $CN^-$ ion forms very stable complexes with most transition metals (Sections 24.4 and 26.1). The toxic action of $CN^-$ is caused by its combination with iron(III) in cytochrome oxidase, a key enzyme that promotes respiration.

Carbon disulfide, $CS_2$, is an important industrial solvent for waxes, greases, celluloses, and other nonpolar substances. It is a colorless, volatile liquid (b.p. 46.3°C). The vapor is very poisonous and highly flammable. The compound is formed by direct reaction of carbon and sulfur at high temperature.

**FIGURE 22.32** Structures of hydrogen cyanide and carbon disulfide.

Hydrogen cyanide          Carbon disulfide

## FOR REVIEW

### SUMMARY

Metallic and nonmetallic elements are distinguished by several physical and chemical properties. Nonmetals lack luster, are not malleable or ductile, and are not good conductors of heat or electricity. They possess higher ionization energies and electronegativities than do metallic elements.

Among elements of a given family, size increases with increasing atomic number. Correspondingly, electronegativity and ionization energy decrease. Metallic character parallels electronegativity trends. Among the nonmetallic elements, the first member of each family differs dramatically from the other members; it forms a maximum of four bonds to other atoms (that is, it is confined to an octet of valence-shell electrons). Also, it exhibits a much greater tendency to form $\pi$ bonds than do the heavier elements in that family.

Hydrogen has three isotopes: **protium** ($_1^1H$), **deuterium** ($_1^2H$), and **tritium** ($_1^3H$). Hydrogen is not a member of any periodic family, although it is usually placed above lithium. The hydrogen atom can either lose an electron, forming $H^+$, or gain one, forming $H^-$, the **hydride** ion. Because the H—H bond is relatively strong, $H_2$ is fairly unreactive unless activated by heat, irradiation, or catalysis. Industrially, $H_2O$ or hydrocarbons are used as sources of hydrogen; in the lab $H_2$ is usually obtained by the action of acids on active metals, such as Zn. The binary compounds of hydrogen are of three general types: ionic hydrides (formed by active metals), metallic hydrides (formed by transition metals), and molecular hydrides (formed by nonmetals).

Oxygen exhibits two allotropes, $O_2$ and $O_3$ (ozone). $O_2$ is separated from air; small amounts can be obtained by heating $KClO_3$. Ozone is obtained by passing an electrical discharge through $O_2$. Both $O_2$ and $O_3$ are good oxidizing agents, but $O_3$ is the stronger. Oxygen normally exhibits an oxidation state of $-2$ in compounds (oxides). The soluble oxides of nonmetals generally produce acidic aqueous solutions; they are called **acidic anhydrides**. By contrast, soluble metal oxides produce basic solutions and are called **basic anhydrides**. **Peroxides** contain O—O bonds and oxygen in a $-1$ oxidation state. In superoxides, which contain the $O_2^-$ ion,

oxygen has an oxidation state of $-\frac{1}{2}$.

The primary source of nitrogen is the atmosphere, where it occurs as $N_2$ molecules. Molecular nitrogen is chemically very stable because of the strong N≡N bond. In its compounds nitrogen exhibits oxidation states ranging from $-3$ to $+5$. The most important process for converting $N_2$ into compounds is the **Haber process**, used to prepare ammonia. Another commercially important process is the **Ostwald process**, which is the preparation of $HNO_3$ beginning with the catalytic oxidation of $NH_3$. Nitrogen has three common oxides: $N_2O$ (**nitrous oxide**), NO (**nitric oxide**), and $NO_2$ (**nitrogen dioxide**). It has two common oxyacids, $HNO_2$ (**nitrous acid**) and $HNO_3$ (**nitric acid**). Nitric acid is both a strong acid and a good oxidizing agent. Other important nitrogen compounds include **hydrazine**, $N_2H_4$, and **hydrogen azide**, $HN_3$. Nitrogen compounds are important fertilizers.

Carbon exhibits two allotropes, diamond and graphite. Amorphous forms of graphite include charcoal, carbon black, and coke. Carbon exhibits two common oxides, CO and $CO_2$. Aqueous solutions of $CO_2$ produce **carbonic acid**, $H_2CO_3$, which is the parent acid of carbonate salts. Binary compounds of carbon are called **carbides**. Carbides may be ionic, interstitial, or covalent. Calcium carbide, $CaC_2$, is used to prepare acetylene. Carbon also forms a vast number of **organic compounds**, which are the subject of a later chapter.

## LEARNING GOALS

Having read and studied this chapter, you should be able to:

1. Identify an element as a metal, semimetal, or nonmetal, based on either its position in the periodic table or its properties.

2. Give examples of how the first member in each family of nonmetallic elements differs from the other elements of the same family, and account for these differences.

3. Predict the relative electronegativities and metallic character of any two members of a periodic family or a horizontal row of the periodic table.

4. Cite the most common occurrences of each element discussed in this chapter and how each is obtained in its elemental form.

5. Cite at least two uses for each element discussed in this chapter.

6. Cite the instances where allotropy is observed among the elements discussed in this chapter, and cite the most common molecular form of each element.

7. Describe and name the three isotopes of hydrogen.

8. Distinguish among ionic, metallic, and molecular hydrides.

**9.** Explain what is meant by the terms "basic anhydride" and "acidic anhydride," and give examples of each.

**10.** Distinguish among oxides, peroxides, and superoxides.

**11.** Describe the chemical and physical properties of hydrogen peroxide and its method of preparation.

**12.** Write balanced chemical equations for formation of nitric acid via the Ostwald process starting from $NH_3$.

**13.** Cite at least one example of a nitrogen compound for each of nitrogen's oxidation states (from $-3$ to $+5$), and name these compounds.

**14.** Account for the differences in physical properties of diamond and graphite based on their structures.

**15.** Be able to identify and name the important inorganic compounds of carbon discussed in this chapter, and describe their salient chemical and physical properties.

**16.** Distinguish among ionic, interstitial, and covalent carbides.

---

## KEY TERMS

Among the more important terms and expressions used for the first time in this chapter are the following:

An **acidic anhydride** (Section 22.4) is an oxide that forms an acid when added to water; soluble nonmetal oxides are acidic anhydrides.

A **basic anhydride** (Section 22.4) is an oxide that forms a base when added to water; soluble metal oxides are basic anhydrides.

A **disproportionation reaction** (Section 22.5) is one in which a species undergoes simultaneous oxidation and reduction [as in $N_2O_3(g) \longrightarrow NO(g) + NO_2(g)$].

The **Ostwald process** (Section 22.5) is used to make nitric acid from ammonia. The $NH_3$ is catalytically oxidized by $O_2$ to form NO; NO in air is oxidized to $NO_2$; $HNO_3$ is formed in a disproportionation reaction when $NO_2$ dissolves in water.

---

## EXERCISES

### Periodic Trends and Chemical Reactions

**22.1** Identify each of the following elements as a metal nonmetal, or semimetal: (**a**) scandium; (**b**) gallium; (**c**) argon; (**d**) strontium; (**e**) arsenic; (**f**) iodine.

**22.2** Identify each of the following elements as a metal, nonmetal, or semimetal: (**a**) germanium; (**b**) rubidium; (**c**) xenon; (**d**) selenium; (**e**) antimony; (**f**) zirconium.

**22.3** Which element of group 5A would you expect to have the most metallic character? Describe two properties of the elements in the group that should relate to this degree of metallic character.

**22.4** Which element of group 3A would you expect to have the most nonmetallic character? Describe two properties of the elements that should demonstrate the greater nonmetallic character of this element compared to the others of the group.

**22.5** Consider the following list of elements: Li, K, N, P, Ne, Ar. From this list select the elements that (**a**) is most electronegative; (**b**) has the greatest metallic character; (**c**) most readily forms a positive ion; (**d**) has the smallest atomic radius; (**e**) forms $\pi$ bonds most readily.

**22.6** Consider the following list of elements: O, Ba, I, Be, Cu, Se. From this list select the element that (**a**) is most electronegative; (**b**) exhibits a maximum oxidation state of $+7$; (**c**) loses an electron most readily; (**d**) forms $\pi$ bonds most readily; (**e**) is a transition metal.

**22.7** In each of the following pairs of substances, select the one that has the more polar bonds: (**a**) $NF_3$ and $IF_3$; (**b**) $N_2O_3$ and $B_2O_3$; (**c**) BN and $BF_3$; (**d**) $H_2O$ and $H_2Se$. Explain your choice in each case.

**22.8** In each of the following pairs of substances, select the one that has the more polar bonds: (**a**) $CO_2$ and $SnO_2$; (**b**) $P_2O_3$ and $V_2O_3$; (**c**) $SiO_2$ and $SO_2$; (**d**) $CCl_4$ and $CBr_4$. Explain your choice in each case.

**22.9** Explain the following observations: (**a**) The highest fluoride compound formed by nitrogen is $NF_3$, whereas phosphorus readily forms $PF_5$. (**b**) Although CO is a well-known compound, SiO doesn't exist under ordinary conditions. (**c**) $AsH_3$ is a stronger reducing agent than $NH_3$.

**22.10** Explain the following observations: (**a**) $HNO_3$ is a stronger oxidizing agent than $H_3PO_4$. (**b**) Silicon is able to form a compound with six fluorides, $SF_6$, whereas carbon is able to bond to a maximum of four, $CF_4$. (**c**) There are three compounds formed by carbon and hydrogen that contain two carbon atoms each ($C_2H_2$, $C_2H_4$, and $C_2H_6$), whereas silcon forms only one analogous compound ($Si_2H_6$).

**22.11** Complete and balance the following equations:
(a) $NaNH_2(s) + H_2O(l) \longrightarrow$
(b) $C_3H_7OH(l) + O_2(g) \longrightarrow$
(c) $NiO(s) + C(s) \longrightarrow$
(d) $AlP(s) + H_2O(l) \longrightarrow$
(e) $Na_2S(s) + HCl(aq) \longrightarrow$

**22.12** Complete and balance the following equations:
(a) $NaOCH_3(s) + H_2O(l) \longrightarrow$
(b) $Na_2O(s) + HC_2H_3O_2(aq) \longrightarrow$
(c) $WO_3(s) + H_2(g) \longrightarrow$
(d) $NH_2OH(l) + O_2(g) \longrightarrow$
(e) $Al_4C_3(s) + H_2O(l) \longrightarrow$

## Hydrogen

**22.13** Give the names and chemical symbols for the three isotopes of hydrogen.

**22.14** Which isotope of hydrogen is radioactive? Write the nuclear equation for the radioactive decay of this isotope.

**22.15** Why is hydrogen placed in either group 1A or 7A of the periodic table?

**22.16** Why are the properties of hydrogen different from those of either the group 1A or 7A elements?

**22.17** Give a balanced chemical equation for the preparation of $H_2$ using (a) Mg and an acid; (b) carbon and steam; (c) methane and steam.

**22.18** List (a) three commercial means of producing $H_2$; (b) three industrial uses of $H_2$.

**22.19** Complete and balance the following equations:
(a) $NaH(s) + H_2O(l) \longrightarrow$
(b) $Fe(s) + H_2SO_4(aq) \longrightarrow$
(c) $H_2(g) + Br_2(g) \longrightarrow$
(d) $Na(l) + H_2(g) \longrightarrow$
(e) $PbO(s) + H_2(g) \longrightarrow$

**22.20** Write balanced chemical equations for each of the following reactions (some of these are analogous to reactions shown in the chapter). (a) Aluminum metal reacts with acids to form hydrogen gas. (b) Steam reacts with magnesium metal to give magnesium oxide and hydrogen (magnesium fires cannot be put out with water). (c) Manganese(IV) oxide is reduced to manganese(II) oxide by hydrogen gas. (d) Calcium hydride reacts with water to generate hydrogen gas.

**22.21** Identify the following hydrides as ionic, metallic, or molecular: (a) $BaH_2$; (b) $H_2Te$; (c) $TiH_{1.7}$.

**22.22** Identify the following hydrides as ionic, metallic, or molecular: (a) $B_2H_6$; (b) RbH; (c) $Th_4H_{1.5}$.

**22.23** About $1.1 \times 10^{10}$ kg of ethylene, $C_2H_4$, is produced annually in the United States by the pyrolysis of ethane, $C_2H_6$: $C_2H_6(g) \longrightarrow C_2H_4(g) + H_2(g)$. How many kilograms of $H_2$ are produced as a by-product?

**22.24** Hydrogen gas has a higher fuel value than natural gas on a weight basis but not on a volume basis. Thus hydrogen is not competitive with natural gas as a fuel transported long distance through pipelines. Calculate the heat of combustion of $H_2$ and of $CH_4$ (the principal component of natural gas) (a) per mole of each; (b) per gram of each; (c) per cubic meter of each at STP. Assume $H_2O(l)$ as a product.

## Oxygen

**22.25** List three industrial uses of $O_2$.

**22.26** List two industrial uses of $O_3$.

**22.27** Give the structure of ozone. Explain why the O—O bond length in ozone (1.28 Å) is longer than that in $O_2$ (1.21 Å).

**22.28** How is $O_2$ normally prepared in the laboratory?

**22.29** Complete and balance the following equations:
(a) $CaO(s) + H_2O(l) \longrightarrow$
(b) $Al_2O_3(s) + H^+(aq) \longrightarrow$
(c) $Na_2O_2(s) + H_2O(l) \longrightarrow$
(d) $N_2O_3(g) + H_2O(l) \longrightarrow$
(e) $KO_2(s) + H_2O(l) \longrightarrow$
(f) $NO(g) + O_3(g) \longrightarrow$

**22.30** Write the balanced chemical equations for each of the following reactions. (a) When mercury(II) oxide is heated, it decomposes to form $O_2$ and mercury metal. (b) When copper(II) nitrate is heated strongly, it decomposes to form copper(II) oxide, nitrogen dioxide, and oxygen. (c) Lead(II) sulfide, PbS(s), reacts with ozone to form $PbSO_4(s)$ and $O_2(g)$. (d) When heated in air, ZnS(s) is converted to ZnO. (e) Potassium peroxide reacts with $CO_2(g)$ to give the carbonate ion and $O_2$. (f) Although silver does not react with oxygen at room temperature, it reacts with ozone, forming $Ag_2O$.

**22.31** Predict whether each of the following oxides is acidic, basic, amphoteric, or neutral: (a) CO, (b) $CO_2$; (c) CaO; (d) $Al_2O_3$.

**22.32** Select the more acidic member of each of the following pairs: (a) $Mn_2O_7$ and $MnO_2$; (b) SnO and $SnO_2$; (c) $SO_2$ and $SO_3$; (d) $SiO_2$ and $SO_2$; (e) $Ga_2O_3$ and $In_2O_3$; (f) $SO_2$ and $SeO_2$.

**22.33** Hydrogen peroxide is capable of oxidizing (a) $K_2S$ to S; (b) $SO_2$ to $SO_4^{2-}$; (c) $NO_2^-$ to $NO_3^-$; (d) $As_2O_3$ to $AsO_4^{3-}$; (e) $Fe^{2+}$ to $Fe^{3+}$. Write balanced net ionic equations for each of these oxidations.

**22.34** Hydrogen peroxide reduces (a) $MnO_4^-$ to $Mn^{2+}$; (b) $Cl_2$ to $Cl^-$; (c) $Ce^{4+}$ to $Ce^{3+}$; (d) $O_3$ to $H_2O$. Write balanced net ionic equations for each of these reductions.

## Nitrogen

**22.35** List three industrial uses for $N_2$.

**22.36** What is meant by the term "nitrogen fixation"? Why is the $N_2$ molecule so unreactive?

**22.37** Write the chemical formulas for each of the following compounds, and indicate the oxidation state of nitrogen in each: (a) nitrous acid; (b) hydrazine;

(c) potassium cyanide: (d) sodium nitrate; (e) ammonium chloride; (f) lithium nitride.

**22.38** Write the chemical formulas for each of the following compounds, and indicate the oxidation state of nitrogen in each: (a) sodium nitrite; (b) ammonia; (c) nitrous oxide; (d) sodium cyanide; (e) nitric acid; (f) nitrogen dioxide.

**22.39** Write the Lewis structure for each of the following species, and describe its molecular geometry: (a) $NH_4^+$; (b) $HNO_3$; (c) $N_2O$; (d) $NO_2$.

**22.40** Write the Lewis structure for each of the following species, and describe its molecular geometry: (a) $HNO_2$; (b) $N_3^-$; (c) $N_2H_5^+$; (d) $NO_3^-$.

**22.41** Complete and balance the following equations:
(a) $Mg_3N_2(s) + H_2O(l) \longrightarrow$
(b) $NO(g) + O_2(g) \longrightarrow$
(c) $NH_3(g) + O_2(g) \xrightarrow{\Delta}$ (no catalyst)
(d) $NaNH_2(s) + H_2O(l) \longrightarrow$

**22.42** Complete and balance the following equations:
(a) $N_2O_5(g) + H_2O(l) \longrightarrow$
(b) $Li_3N(s) + H_2O(l) \longrightarrow$
(c) $NH_3(aq) + H^+(aq) \longrightarrow$
(d) $N_2H_4(l) + O_2(g) \longrightarrow$

**22.43** Write balanced net ionic equations for each of the following reactions. (a) Dilute nitric acid reacts with zinc metal with formation of nitrous oxide. (b) Concentrated nitric acid reacts with sulfur with formation of nitrogen dioxide. (c) Concentrated nitric acid oxidizes sulfur dioxide with formation of nitric oxide. (d) Urea reacts with water to form ammonia and carbon dioxide.

**22.44** Write balanced net ionic equations for each of the following reactions. (a) Hydrazine is burned in excess fluorine gas, forming $NF_3$. (b) Hydrazine reduces $CrO_4^{2-}$ to $Cr(OH)_4^-$ (hydrazine is oxidized to $N_2$). (c) Hydroxylamine is oxidized to $N_2$ by $Cu^{2+}$ in aqueous solutions ($Cu^{2+}$ is reduced to copper metal). (d) Aqueous azide ion, $N_3^-$, reacts with $Cl_2$, forming $N_2$.

**22.45** Write complete balanced half-reactions for (a) reduction of nitrate ion to $N_2$ in acidic solution; (b) oxidation of $NH_4^+$ to $N_2$ in acidic solution. What is the standard electrode potential in each case? (See Figure 22.17.)

**22.46** Write complete balanced half-reactions for (a) reduction of nitrate ion to NO in acidic solution; (b) oxidation of $HNO_2$ to $NO_2$ in acidic solution. What is the standard electrode potential in each case? (See Figure 22.17.)

**22.47** It is estimated that 95 percent of the ammonia production in the United States uses $H_2$ obtained from natural gas (Equations 22.12 and 22.13). If 50 percent of the $CH_4$ used to manufacture $H_2$ is burned to maintain proper temperature for ammonia synthesis, how many kilograms of $CH_4$ are consumed in supplying the $1.5 \times 10^{10}$ kg of $NH_3$ produced annually?

**22.48** From the thermodynamic data in Appendix D, calculate $\Delta H°$ and $\Delta G°$ at 25°C for the oxidation of $NH_3$ in the following reactions:

$$4NH_3(g) + 5O_2(g) \longrightarrow 4NO(g) + 6H_2O(g)$$
$$4NH_3(g) + 3O_2(g) \longrightarrow 2N_2(g) + 6H_2O(g)$$

## Carbon

**22.49** Give three industrial uses of carbon dioxide.

**22.50** Give three industrial uses of carbon monoxide.

**22.51** Give the chemical formulas for (a) hydrocyanic acid; (b) carborundum; (c) calcium carbonate; (d) calcium acetylide.

**22.52** Give the chemical formulas for (a) carbonic acid; (b) sodium cyanide; (c) potassium hydrogen carbonate; (d) acetylene.

**22.53** Write the Lewis structures of each of the following species: (a) $CN^-$; (b) CO; (c) $C_2^{2-}$; (d) $CS_2$; (e) $CO_2$; (f) $CO_3^{2-}$.

**22.54** Indicate the geometry and the types of hybrid orbitals used by each carbon atom in the following species: (a) $CH_3C\equiv CH$; (b) NaCN; (c) $CS_2$; (d) $C_2H_6$.

**22.55** Complete and balance the following equations:
(a) $ZnCO_3(s) \xrightarrow{\Delta}$
(b) $BaC_2(s) + H_2O(l) \longrightarrow$
(c) $C_2H_4(g) + O_2(g) \longrightarrow$
(d) $CH_3OH(l) + O_2(g) \longrightarrow$
(e) $NaCN(s) + HCl(aq) \longrightarrow$

**22.56** Complete and balance the following equations:
(a) $CO_2(g) + OH^-(aq) \longrightarrow$
(b) $NaHCO_3(s) + H^+(aq) \longrightarrow$
(c) $CaO(s) + C(s) \xrightarrow{\Delta}$
(d) $C(s) + H_2O(g) \xrightarrow{\Delta}$
(e) $CuO(s) + CO(g) \longrightarrow$

**22.57** Write balanced chemical equations for each of the following reactions. (a) Burning magnesium metal in a carbon dioxide atmosphere reduces the $CO_2$ to carbon. (b) In photosynthesis, solar energy is used to produce glucose, $C_6H_{12}O_6$, and $O_2$ out of carbon dioxide and water. (c) When carbonate salts dissolve in water, they produce basic solutions.

**22.58** Write balanced chemical equations for each of the following reactions. (a) Hydrogen cyanide is formed commercially by passing a mixture of methane, ammonia, and air over a catalyst at 800°C. Water is a by-product of the reaction. (b) Baking soda reacts with acids to produce carbon dioxide gas. (c) When barium carbonate reacts in air with sulfur dioxide, barium sulfate and carbon dioxide form.

**22.59** What volume of dry acetylene, measured at 27°C and 720 mm Hg, is formed when 10.0 g of calcium carbide is placed in water?

**22.60** A certain carbide of magnesium is reacted with water to form $Mg^{2+}(aq)$ and a volatile hydrocarbon. Hydrolysis of 0.3052 g of the carbide produces 0.1443 g of the hydrocarbon. The hydrocarbon consists of 90.0 percent C and 10.0 percent H. The density of the hydrocarbon gas at 25°C and 742 mm Hg is 1.60 g/L. What is the formula for the carbide, and what is the molecular formula for the hydrocarbon?

## Additional Exercises

**22.61** State two chemical and two physical characteristics that distinguish a nonmetallic element from a metallic one.

**22.62** State two important ways in which the first member of each family of nonmetals differs from subsequent members.

**22.63** Starting with $D_2O$, suggest a preparation for (a) $ND_3$; (b) $D_2SO_4$; (c) $NaOD$; (d) $DNO_3$; (e) $C_2D_2$; (f) $DCN$.

**22.64** The annual production of $H_2$, $N_2$, and $O_2$ in the United States is generally reported in cubic feet, measured at STP. If $2.0 \times 10^8$ kg of $H_2$, $1.1 \times 10^{10}$ kg of $N_2$, and $1.4 \times 10^{10}$ kg of $O_2$ are produced in a particular year, how many cubic feet are produced of (a) $H_2$; (b) $N_2$; (c) $O_2$?

**22.65** (a) How many grams of $H_2$ can be stored in 1.00 kg of the alloy FeTi if the hydride $FeTiH_2$ is formed? (b) What volume does this quantity of $H_2$ occupy at STP?

**22.66** Write a balanced chemical equation for the reaction of each of the following compounds with water: (a) $SO_2(g)$; (b) $Cl_2O(g)$; (c) $Na_2O(s)$; (d) $BaC_2(s)$; (e) $RbO_2(s)$; (f) $Mg_3N_2(s)$; (g) $Na_2O_2(s)$; (h) $NaH(s)$.

**22.67** What is the anhydride of each of the following acids: (a) $H_2SO_3$; (b) $HNO_3$; (c) $H_3PO_4$; (d) $HClO_4$?

**22.68** Write a series of chemical equations that describes how $Na_2CO_3$ could be prepared starting with only $H_2O$, NaCl, and $CaCO_3$ as raw materials.

**22.69** In acidic solution the ferrocyanide ion, $Fe(CN)_6^{4-}$, is oxidized by $H_2O_2$ to the ferricyanide ion, $Fe(CN)_6^{3-}$. In basic solution, the ferricyanide ion is reduced by $H_2O_2$ to the ferrocyanide ion. Write balanced net ionic equations for both reactions.

**22.70** Hydrogen peroxide can be prepared in the laboratory by air oxidation of barium metal followed by treatment of the resultant product with dilute sulfuric acid. Write balanced chemical equations for the two reactions that take place.

**22.71** Both dimethylhydrazine, $(CH_3)_2NNH_2$, and monomethylhydrazine, $CH_3HNNH_2$, have been used as rocket fuels. When dinitrogen tetraoxide, $N_2O_4$, is used as the oxidizer, the products are $H_2O$, $CO_2$, and $N_2$. If the thrust of the rocket depends on the volume of the products produced, which of the substituted hydrazines produces a greater thrust per gram total weight of oxidizer plus fuel? [Assume that both fuels generate the same temperature and that $H_2O(g)$ is formed.]

**22.72** Suggest why each of the following compounds is either unstable or does not exist: (a) $NCl_5$; (b) $H_3NO_4$; (c) $(CH_3)_2Si{=}O$; (d) $P_2O$; (e) $H_3$.

**22.73** Each of the following compounds is used as a nitrogen fertilizer: $(NH_2)_2CO$ (urea), $NH_3$, $(NH_4)_2SO_4$, and $NaNO_3$. Calculate the percentage of nitrogen in each compound.

**22.74** Hydrazine has been employed as a reducing agent for metals. Using standard electrode potentials, predict whether the following metals can be reduced to the metallic state by hydrazine under standard conditions in

acidic solution: (a) $Fe^{2+}$; (b) $Sn^{2+}$; (c) $Cu^{2+}$; (d) $Ag^+$; (e) $Cr^{3+}$.

**[22.75]** Thermodynamic data for $HNO_3(aq)$ at 25°C follow: $\Delta H_f^\circ = -207.3$ kJ/mol; $\Delta G_f^\circ = -111.3$ kJ/mol. Calculate $\Delta H^\circ$ and $\Delta G^\circ$ for the reaction

$$\tfrac{1}{2}N_2(g) + \tfrac{1}{2}H_2O(g) + \tfrac{5}{2}O_2(g) \longrightarrow HNO_3(aq)$$

It has been suggested that if a suitable catalyst were present, the atmospheric gases could react to make the oceans a dilute nitric acid solution. In thermodynamic terms, is this a possible process?

**22.76** The Haber process is carried out at high temperatures even though the equilibrium constant for the reaction decreases with increasing temperature. Explain why a high temperature is necessary.

**22.77** If the lunar lander on the Apollo moon missions used 4.0 tons of dimethyl hydrazine, $(CH_3)_2NNH_2$, as fuel, how many tons of $N_2O_4$ oxidizer were required to react with it? (The reaction produces $N_2$, $CO_2$, and $H_2O$.)

**22.78** Rationalize the general trend in Lewis basicities in the following isoelectronic series: $N_2 < CO < CN^-$.

**22.79** Complete and balance the following equations:
(a) $Li_3N(s) + H_2O(l) \longrightarrow$
(b) $NH_3(aq) + H_2O(l) \longrightarrow$
(c) $NO_2(g) + H_2O(l) \longrightarrow$
(d) $2NO_2(g) \longrightarrow$
(e) $NH_3(g) + O_2(g) \xrightarrow{\text{catalyst}}$
(f) $CO(g) + O_2(g) \longrightarrow$
(g) $H_2CO_3(aq) \xrightarrow{\Delta}$
(h) $Ni(s) + CO(g) \longrightarrow$
(i) $CS_2 + O_2(g) \longrightarrow$
(j) $CaO(s) + SO_2(g) \longrightarrow$
(k) $Na(s) + H_2O(l) \longrightarrow$
(l) $CH_4(g) + H_2O(g) \xrightarrow{\Delta}$
(m) $LiH(s) + H_2O(l) \longrightarrow$
(n) $Fe_2O_3(s) + 3H_2(g) \longrightarrow$

**22.80** Complete and balance the following equations:
(a) $MnO_4^-(aq) + H_2O_2(aq) + H^+(aq) \longrightarrow$
(b) $Fe^{2+}(aq) + H_2O_2(aq) \longrightarrow$
(c) $I^-(aq) + H_2O_2(aq) + H^+(aq) \longrightarrow$
(d) $MnO_2(s) + H_2O_2(aq) + H^+(aq) \longrightarrow$
(e) $I^-(aq) + O_3(g) \longrightarrow I_2(s) + O_2(g) + OH^-(aq)$

**[22.81]** Using this chapter and material elsewhere in the text (use the index), compile a list of physical and chemical properties of the elements oxygen and zinc. How do these properties justify the classification of one as a nonmetal and the other as a metal?

**[22.82]** The electronegativity of a given element can be regarded as varying with its oxidation state. How would you expect the electronegativity to change as a function of oxidation state? Although manganese (Mn) is completely different from chlorine in properties as an element, the characteristics of $MnO_4^-$ rather closely resemble those of $ClO_4^-$. Cite at least two instances of this similarity (you may need to look in a handbook; think in terms of acid-base properties, oxidation-reduction behavior, solubility, and so on). Discuss your observations in terms of the oxidation states, electron configurations, and electronegativities of the central atoms.

## CHAPTER 23

# Chemistry of Other Nonmetallic Elements

In Chapter 22 we discussed the chemistry of four important nonmetals: hydrogen, oxygen, nitrogen, and carbon. In this chapter we take a more panoramic view of nonmetals. We will begin with the noble gases and then move, group by group, from right to left through the periodic table. We will, of course, have little to say about those elements discussed in Chapter 22. Of the remaining nonmetals, the most important are the halogens, sulfur, phosphorus, and silicon. Thus we will spend most of our time in this chapter considering these elements.

## 23.1 THE NOBLE GAS ELEMENTS

We have mentioned at several points in the text that the elements of group 8A are chemically unreactive. Indeed, most of our references to these elements have been in relation to their physical properties, as when we discussed intermolecular forces in Section 11.4. According to the Lewis theory of chemical bonding, the relative inertness of these elements is due to the formation of a completed octet of valence-shell electrons. The stability of such an arrangement is reflected in the high ionization energies of the group 8A elements (Section 7.7).

The group 8A elements are all gases at room temperature. All are components of the earth's atmosphere, except for radon, which exists only as a short-lived radioisotope. Only argon is relatively abundant (Table 14.1). Neon, argon, krypton, and xenon are recovered from liquid air by fractional distillation. Argon is used as a blanketing atmosphere in electric light bulbs. The gas conducts heat away from the filament but does not react with it. It is also used as a protective atmosphere to prevent oxidation in welding and certain high-temperature metallurgical processes. Neon is used in electric signs; the gas is caused to radiate by passing an electric discharge through the tube. Krypton, xenon, and radon are not commercially important.

Helium is, in many ways, the most important of the noble gases. Liquid helium is used as a coolant to conduct experiments at very low temperatures. Helium boils at 4.2 K under 1 atm pressure, the lowest boiling point of any substance. Because helium has such a low abundance in the atmosphere and boils at such a low temperature, recovery of the gas from the atmosphere would require an immense expenditure of energy. Helium is found in relatively high concentrations in many natural-gas wells. Some of this helium is separated to meet current demands, and a little is kept for later use. Unfortunately, however, most of the helium escapes.

| 8A |
|:--:|
| 2<br>He |
| 10<br>Ne |
| 18<br>Ar |
| 36<br>Kr |
| 54<br>Xe |
| 86<br>Rn |

A micrograph of sulfur crystals taken with polarized light. (Mike McNamee, Science Photo Library/Photo Researchers)

## Noble Gas Compounds

Because the noble gases are exceedingly stable, it is reasonable to expect that they will undergo reaction only under rather rigorous conditions. Furthermore, we might expect that the heavier noble gases would be most likely to undergo chemical transformation, because the ionization energies are lower for the heavier elements, as illustrated in Figure 7.9. A lower ionization energy would suggest the possibility of losing an electron in formation of an ionic bond. In addition, since the group 8A elements already contain eight electrons in their valence shell (except, of course, for helium, which contains just two), formation of covalent bonds requires an expanded valence shell. We have seen (Section 8.7) that valence-shell expansion occurs most readily with larger atoms.

The first noble gas compound was prepared by Neil Bartlett in 1962. His work caused a sensation, because it undercut the belief that the noble gas elements were truly chemically inert. Bartlett initially worked with xenon in combination with fluorine, the element we would expect to be most reactive. Since then several xenon compounds of fluorine and oxygen have been prepared. Some properties of these substances are listed in Table 23.1. The three fluorides $XeF_2$, $XeF_4$, and $XeF_6$ are made by direct reaction of the elements. By varying the ratio of reactants and altering reaction conditions, one or the other of the three compounds can be obtained. The oxygen-containing compounds are formed when the fluorides are reacted with water, as in Equations 23.1 and 23.2:

$$XeF_6(s) + H_2O(l) \longrightarrow XeOF_4(l) + 2HF(g) \qquad [23.1]$$

$$XeF_6(s) + 3H_2O(l) \longrightarrow XeO_3(aq) + 6HF\ (aq) \qquad [23.2]$$

**TABLE 23.1**  Properties of xenon compounds

| Compound | Oxidation state of Xe | Melting point (°C) | $\Delta H_f^\circ$ (kJ/mol)[a] |
|---|---|---|---|
| $XeF_2$ | +2 | 129 | $-109(g)$ |
| $XeF_4$ | +4 | 117 | $-218(g)$ |
| $XeF_6$ | +6 | 49 | $-298(g)$ |
| $XeOF_4$ | +6 | $-41$ to $-28$ | $+146(l)$ |
| $XeO_3$ | +6 | —[b] | $+402(s)$ |
| $XeO_2F_2$ | +6 | 31 | $+145(s)$ |
| $XeO_4$ | +8 | —[c] | — |

[a] At 25°C, for the compound in the state indicated.
[b] A solid; decomposes at 40°C.
[c] A solid; decomposes at $-40$°C.

### SAMPLE EXERCISE 23.1

Predict the structure of $XeF_4$.

**Solution:**  To predict the structure, we must first write the Lewis structure for the molecule. The total number of valence-shell electrons involved is 36 (8 from xenon and 7 from each of the four fluorines). This leads to the Lewis structure shown in Figure 23.1(a). We see that Xe has 12 electrons in its valence

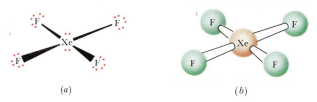

(a)  (b)

**FIGURE 23.1**  Lewis and geometrical structures of $XeF_4$.

shell. We thus expect an octahedral disposition of six electron pairs. Two of these are nonbonded pairs. Because nonbonded pairs have a larger volume requirement than do bonded pairs (Section 9.1), it is reasonable to expect these nonbonded pairs to be opposite one another. The expected structure is square planar, as shown in Figure 23.1(*b*). The experimentally determined structure agrees with this prediction.

**PRACTICE EXERCISE** _____
Describe the electron-pair geometry and the molecular geometry of $XeF_2$.
*Answer:*   trigonal bipyramidal; linear

The enthalpies of formation of the xenon fluorides are negative, which suggests that these compounds should be reasonably stable, and this is indeed found to be the case. They are, however, powerful fluorinating agents and must be handled in containers that do not readily react to form fluorides. Notice that the enthalpies of formation of the oxyfluorides and oxides of xenon are positive (Table 23.1); these compounds are quite unstable.

As we might expect, formation of compounds by the other noble gas elements occurs much less readily than in the case of xenon. Only one binary krypton compound, $KrF_2$, is known with certainty; it decomposes to the elements at $-10°C$.

## 23.2 THE HALOGENS

The elements of group 7A are known as the halogens, after the Greek words *halos* and *gennao*, meaning "salt formers." The halogens have played an important part in the development of chemistry as a science. Chlorine was first prepared by the Swedish chemist Karl Wilhelm Scheele in 1774, but it was not until 1810 that the English chemist Humphry Davy identified it as an element. The discovery of iodine followed in 1811, and bromine was discovered in 1825. Compounds of fluorine were known for a long time, but not until 1886 did the French chemist Henri Moissan succeed in preparing the very reactive free element.

The general valence-electron configuration of the halogens is $ns^2np^5$, where *n* may have values ranging from 2 through 6. When a halogen atom shares all seven of its electrons with a more electronegative atom, it is assigned the $+7$ oxidation state. The halogens also achieve a noble gas configuration by gaining an electron, which results in a $-1$ oxidation state. Thus we expect that the halogens might occur in oxidation states ranging from $+7$ at one extreme to $-1$ at the other. Oxidation states over this entire range are observed for all the halogens except fluorine; this most electronegative of all the elements is observed only in the 0 or $-1$ oxidation states.

| 9 F |
|---|
| 17 Cl |
| 35 Br |
| 53 I |
| 85 At |

**Occurrences of the Halogens**

Table 23.2 summarizes the occurrences of the halogens in nature. Both fluorine and chlorine are fairly abundant, but they are quite differently distributed. This happens because the salts of chlorine are generally quite soluble, whereas some of those of fluorine are not. Bromine is comparatively must less abundant than chlorine or fluorine, and iodine is much rarer still.

Chlorine, bromine, and iodine occur as the halides in seawater and in salt deposits. The concentration of iodine in these sources is generally

**TABLE 23.2**  Occurrences of the halogens

| Element | Occurrences |
|---|---|
| Fluorine | Fluorspar, $CaF_2$; fluorapatite, $Ca_5(PO_4)_3F$; cryolite, $Na_3AlF_6$ |
| | Biologically: teeth, bones |
| Chlorine | Seawater (0.55 $M$, chiefly as NaCl); salt beds, salt lakes (for example, Dead Sea, Great Salt Lake); NaCl deposits |
| | Biologically: gastric juice [as $HCl(aq)$], tissue fluids |
| Bromine | Seawater ($8.3 \times 10^{-4}$ $M$); underground brines; salt beds, salt lakes |
| | Biologically: minor concentrations of bromide along with chloride |
| Iodine | Seawater ($4 \times 10^{-7}$ $M$); seaweeds; $NaIO_3$ in minor amounts in nitrate deposits; oil-well brines |
| | Biologically: in human thyroid gland |

very small. However, iodine is concentrated by certain seaweeds, which are harvested, dried, and burned. Iodine is then extracted from the ashes. The element is also extracted commercially from oil-well brines in California. Fluorine occurs in the minerals fluorspar ($CaF_2$), cryolite ($Na_3AlF_6$), and fluorapatite [$Ca_5(PO_4)_3F$]. Only the first of these is an important commercial source of fluorine for the chemical industry. All isotopes of astatine are radioactive. The longest-lived isotope is astatine-210, which has a half-life of 8.3 h and decays mainly by electron capture. Astatine was first synthesized by bombarding bismuth-209 with high-energy alpha particles, as shown in Equation 23.3:

$$^{209}_{83}\text{Bi} + ^{4}_{2}\text{He} \longrightarrow ^{211}_{85}\text{At} + 2^{1}_{0}\text{n} \qquad [23.3]$$

A cyclotron must be used to synthesize astatine, which makes it very expensive and limits its application and study.

**Properties and Preparation of the Halogens**

Some of the properties of the halogens are summarized in Table 23.3. These properties vary in a regular fashion as a function of atomic number. Within each horizontal row of the periodic table each halogen has a high ionization energy, second only to the noble gas element adjacent to it in the table. Each halogen has the highest electronegativity of all the elements in its horizontal row. Within the halogen family, atomic and ionic radii increase with increasing atomic number. Correspondingly, the ionization energy and electronegativity steadily decrease as we go down the family, from fluorine to iodine.

**TABLE 23.3**  Some properties of the halogen atoms

| Property | F | Cl | Br | I |
|---|---|---|---|---|
| Atomic radius (Å) | 0.72 | 1.00 | 1.15 | 1.40 |
| Ionic radius, $X^-$ (Å) | 1.33 | 1.81 | 1.96 | 2.17 |
| First ionization energy (kJ/mol) | 1680 | 1250 | 1140 | 1010 |
| Electron affinity (kJ/mol) | −332 | −349 | −325 | −295 |
| Electronegativity | 4.0 | 3.2 | 3.0 | 2.7 |
| X—X single-bond energy (kJ/mol) | 155 | 242 | 193 | 151 |
| Reduction potential: $\frac{1}{2}X_2(aq) + e^- \longrightarrow X^-(aq)$ | 2.87 | 1.36 | 1.07 | 0.54 |

**FIGURE 23.2** Aqueous solutions of NaF, NaBr, and NaI (from left to right) to which $Cl_2$ has been added. Each solution is in contact with carbon tetrachloride, $CCl_4$, which forms the lower layer in each container. The halogens are more soluble in $CCl_4$. The $F^-$ ion in the NaF solution (left) does not react with $Cl_2$, and thus both the aqueous and $CCl_4$ layers remain colorless. The $Br^-$ ion (center) is oxidized by $Cl_2$ to form $Br_2$, producing a yellow color in the water layer and an orange color in the $CCl_4$ layer. The $I^-$ ion (right) is oxidized to $I_2$, producing an amber color in the water layer and a violet color in the $CCl_4$ layer. (Donald Clegg and Roxy Wilson)

Under ordinary conditions the halogens exist as diatomic molecules. The molecules are held together in the solid and liquid states by London dispersion forces (Section 11.4). Because $I_2$ is the largest and most polarizable of the halogen molecules, it is not surprising that the intermolecular forces between $I_2$ molecules are the strongest. Thus $I_2$ has the highest melting point and boiling point. At room temperature and 1 atm pressure, $I_2$ is a solid, $Br_2$ is a liquid, and $Cl_2$ and $F_2$ are gases. Chlorine readily liquifies upon compression at room temperature and is normally stored and handled in liquid form in steel containers.

The comparatively low bond energy in $F_2$ (155 kJ/mol) accounts in part for the extreme reactivity of elemental fluorine. Because of its high reactivity, $F_2$ is very difficult to work with. Certain metals, such as copper and nickel, can be used to contain $F_2$ because their surfaces form a protective coating of metal fluoride. Chlorine and the heavier halogens are also reactive, although less so than fluorine. They combine directly with most elements except the rare gases.

Because of the high electronegativities of the halogens compared with other elements, the halogens tend to gain electrons from other substances and thereby serve as oxidizing agents. The oxidizing ability of the halogens, which is indicated by their reduction potentials, decreases going down the group. As a result, we find that a given halogen is able to oxidize the anions of the halogens below it in the family. For example, $Cl_2$ will oxidize $Br^-$ and $I^-$, but not $F^-$, as seen in Figure 23.2.

## SAMPLE EXERCISE 23.2

Write the balanced chemical equation for the reaction, if any, that occurs between (a) $I^-(aq)$ and $Br_2(l)$; (b) $Cl^-(aq)$ and $I_2(s)$.

**Solution:** (a) $Br_2$ is able to oxidize (remove electrons) from the anions of the halogens below it in the periodic table. Thus it will oxidize $I^-$:

$$2I^-(aq) + Br_2(l) \longrightarrow I_2(s) + 2Br^-(aq)$$

(b) $Cl^-$ is the anion of a halogen above iodine in the periodic table. Thus $I_2$ cannot oxidize $Cl^-$; there is no reaction.

### PRACTICE EXERCISE

Write the balanced chemical equation for the reaction that occurs between $Br^-(aq)$ and $Cl_2(aq)$.
*Answer:* $2Br^-(aq) + Cl_2(aq) \longrightarrow Br_2(l) + 2Cl^-(aq)$

Notice from Table 23.2 that the reduction potential of $F_2$ is exceptionally high. Fluorine gas readily oxidizes water according to the reaction

$$F_2(aq) + H_2O(l) \longrightarrow 2HF(aq) + \tfrac{1}{2}O_2(g) \qquad E° = +1.64 \text{ V} \qquad [23.4]$$

Fluorine cannot be prepared by electrolytic oxidation in water, because water itself is oxidized more readily than $F^-$, with formation of $O_2(g)$. In practice the element is formed by electrolytic oxidation of a solution of KF in anhydrous HF. The KF reacts with HF to form a salt, $K^+HF_2^-$, which acts as the current carrier in the liquid. (The $HF_2^-$ ion is stable because of the very strong hydrogen bond between the two fluoride ions, as described in Section 11.4.) The overall cell reaction is

$$2KHF_2(l) \longrightarrow H_2(g) + F_2(g) + 2KF(l) \qquad [23.5]$$

Chlorine is produced mainly by electrolysis of either molten or aqueous sodium chloride, as described in Sections 20.5 and 24.5. Both bromine and iodine are obtained commercially from brines containing the halide ions by oxidation with $Cl_2$.

### Uses of the Halogens

Fluorine has become an important industrial chemical. It is used, for example, to prepare fluorocarbons, very stable carbon-fluorine compounds. An example is $CF_2Cl_2$, known as Freon 12, which is used as a refrigerant and as a propellant for aerosol cans. As we noted in Section 14.3, the effects of these substances on the ozone layer have been under recent investigation. Fluorocarbons are also used as lubricants and in plastics; Teflon (Figure 23.3) is a polymeric fluorocarbon noted for its high thermal stability and lack of chemical reactivity.

Chlorine is by far the most important halogen commercially. About $9.5 \times 10^9$ kg (10.7 million tons) of $Cl_2$ is produced in the United States each year. In addition, hydrogen chloride production is about $2.7 \times 10^9$ kg (3.0 million tons) annually. About half of this inorganic chlorine finds its way eventually into vinyl chloride, $C_2H_3Cl$, used in polyvinyl chloride (PVC) plastics manufacture; ethylene dichloride, $C_2H_2Cl_2$, an organic solvent; and other chlorine-containing organic compounds. Much of the remainder of the chlorine is used as a bleach in the paper and textile industries. When $Cl_2$ dissolves in cold dilute base it forms $Cl^-$ and the hypochlorite ion, $ClO^-$:

**FIGURE 23.3** Structure of Teflon, a fluorocarbon polymer.

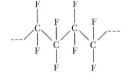

$$2OH^-(aq) + Cl_2(aq) \longrightarrow Cl^-(aq) + ClO^-(aq) + H_2O(l) \qquad [23.6]$$

Sodium hypochlorite, NaClO, is the active ingredient in many liquid bleaches (Figure 23.4). Chlorine is also used in water treatment to oxidize and thereby destroy bacteria (Section 14.6).

Neither bromine nor iodine is as widely used as fluorine and chlorine. One familiar application of bromine is in the production of silver bromide used in photographic film. A common use of iodine is its addition, as KI, to salt to form iodized salt. Table salt contains about 0.02 percent potassium iodide by weight. Iodized salt is able to provide the small amount of iodine necessary in our diets; it is essential for the formation of thyroxin, a hormone secreted by the thyroid gland. Lack of iodine in the diet results in an enlarged thyroid gland, a condition called goiter.

**FIGURE 23.4** Aqueous solutions of sodium hypochlorite, NaClO, are sold as liquid household bleaches. (Diane Schlumo/Fundamental Photographs)

## The Hydrogen Halides

All of the halogens form stable diatomic molecules with hydrogen. These are very important compounds, in part because aqueous solutions of the hydrogen halides other than HF are strongly acidic. Table 23.4 lists some of the more important properties of the hydrogen halides.

Notice that the melting and boiling points of HF are abnormally high as compared with the other hydrogen halides. The cause of this unusual behavior is the strong hydrogen bonding that exists between HF molecules in the liquid state, as described in Section 11.4. Because HF is a small molecule and is capable of strong hydrogen-bonding interactions, it is completely miscible with water. The other hydrogen halides, though not capable of strong hydrogen-bonding interactions, are also very soluble in water. The aqueous solutions of the hydrogen halides are acidic; HF $(aq)$ is a weak acid, but the other hydrogen halide solutions are strong acids.

The hydrogen halides can be formed by direct reaction of the elements. However, the most important means of their preparation is through reaction of a salt of the halide with a strong, nonvolatile acid. Hydrogen fluoride and hydrogen chloride are prepared in this manner by reaction of a cheap, readily available salt with concentrated sulfuric acid:

$$CaF_2(s) + H_2SO_4(l) \longrightarrow 2HF(g) + CaSO_4(s) \qquad [23.7]$$
$$NaCl(s) + H_2SO_4(l) \longrightarrow HCl(g) + NaHSO_4(s) \qquad [23.8]$$

**TABLE 23.4**   Properties of the hydrogen halides

| Property | HF | HCl | HBr | HI |
|---|---|---|---|---|
| Molecular weight | 20.01 | 36.46 | 80.92 | 127.91 |
| Melting point (°C) | −83 | −114 | −87 | −51 |
| Boiling point (°C) | 19.9 | −85 | −67 | −35 |
| Bond-dissociation energy (kJ/mol) | 565 | 431 | 364 | 297 |
| H—X distance (Å) | 0.92 | 1.27 | 1.41 | 1.61 |
| Solubility in H$_2$O (g/100 g H$_2$O, 10°C) | ∞ | 78 | 210 | 234 |

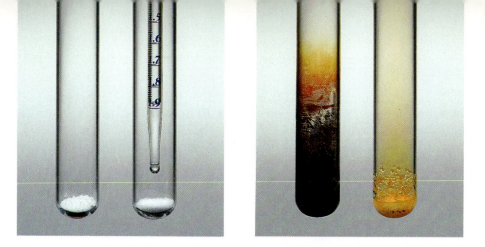

**FIGURE 23.5** (a) Sodium iodide in the left test tube, and sodium bromide on the right. Sulfuric acid is in the pipet. (b) Addition of sulfuric acid to the test tubes oxidizes sodium iodide to form the dark-colored iodine on the left. Sodium bromide is oxidized to the yellow-brown bromine on the right. When more concentrated, bromine has a reddish-brown color. (Richard Megna/Fundamental Photographs)

Because the hydrogen halide is the only volatile component in the mixture, it can be easily removed. The hydrogen halide is usually absorbed in water and marketed as the corresponding acid. In both cases the reaction mixtures must be heated to cause reaction to occur.

Neither hydrogen bromide nor hydrogen iodide can be prepared by analogous reactions of salts with $H_2SO_4$, because HBr and HI undergo oxidation by $H_2SO_4$ at the higher temperatures of the reaction, as seen in Figure 23.5. The overall reactions are described by Equations 23.9 and 23.10:

$$2NaBr(s) + 2H_2SO_4(l) \longrightarrow Br_2(g) + SO_2(g) + Na_2SO_4(s) + 2H_2O(g) \qquad [23.9]$$

$$8NaI(s) + 9H_2SO_4(l) \longrightarrow 8NaHSO_4(s) + H_2S(g) + 4I_2(g) + 4H_2O(g) \qquad [23.10]$$

Notice that in the case of the bromide, part of the $H_2SO_4$ is reduced to $SO_2$, in which sulfur is in the +4 oxidation state. In the reaction with iodide, sulfur is reduced all the way to $H_2S$, in which sulfur is in the −2 oxidation state. This difference in products reflects the greater ease of oxidation of the iodide. The difficulties associated with use of $H_2SO_4$ can be avoided by using a nonvolatile acid that is a poorer oxidizing agent than $H_2SO_4$; concentrated phosphoric acid, $H_3PO_4$, serves well.

The hydrogen halides are also formed when certain molecular (covalent) halides are hydrolyzed, as in the following examples:

$$PBr_3(l) + 3H_2O(l) \longrightarrow H_3PO_3(l) + 3HBr(g) \qquad [23.11]$$

$$SeCl_4(s) + 3H_2O(l) \longrightarrow H_2SeO_3(s) + 4HCl(g) \qquad [23.12]$$

---

## SAMPLE EXERCISE 23.3

Write a balanced chemical equation for the formation of hydrogen bromide gas from the reaction of solid sodium bromide with phosphoric acid.

**Solution:** The chemical formulas of sodium bromide and phosphoric acid are NaBr and $H_3PO_4$, respectively. Let us assume that only one of the dissociable hydrogens of $H_3PO_4$ undergoes reaction. (The actual number depends on the reaction conditions.)

The balanced equation is then

$$NaBr(s) + H_3PO_4(aq) \longrightarrow NaH_2PO_4(s) + HBr(g)$$

### PRACTICE EXERCISE

Write the balanced chemical equation for the preparation of HI from NaI and $H_3PO_4$.
*Answer:*
$$NaI(s) + H_3PO_4(aq) \longrightarrow NaH_2PO_4(s) + HI(g)$$

---

## SAMPLE EXERCISE 23.4

Write the balanced chemical equation between $SiCl_4$ and $H_2O$.

**Solution:** The oxidation state of Si in $SiCl_4$ is $+4$. Because the hydrolysis of nonmetal chlorides is not an oxidation-reduction reaction, the product acid must also have Si in the $+4$ oxidation state. That acid is $H_4SiO_4$, which we can also write as $Si(OH)_4$. Thus the equation is

$$SiCl_4(l) + 4H_2O(l) \longrightarrow Si(OH)_4(s) + 4HCl(g)$$

Because $Si(OH)_4$ readily undergoes loss of water to form $SiO_2$, the reaction is also written in the following way:

$$SiCl_4(l) + 2H_2O(l) \longrightarrow SiO_2(s) + 4HCl(g)$$

### PRACTICE EXERCISE

What is the oxidation state of P in $PCl_5$? What is the chemical formula of the oxyacid of P in this oxidation state? Write the balanced chemical equation for the reaction of solid $PCl_5$ with water.
*Answer:* $+5$; $H_3PO_4$; $PCl_5(s) + 4H_2O(l) \longrightarrow H_3PO_4(aq) + 5HCl(aq)$

Because the hydrogen halides form acidic solutions in water, they exhibit the characteristic properties of acids in their reactions in water. For example, they may react with an active metal to produce hydrogen gas or with a base in a neutralization reaction.

Hydrofluoric acid is unusual among the hydrohalic acids because it reacts readily with silica, $SiO_2$, and with various silicates to form hexafluorosilic acid, as in these examples:

$$SiO_2(s) + 6HF(aq) \longrightarrow H_2SiF_6(aq) + 2H_2O(l) \qquad [23.13]$$

$$CaSiO_3(s) + 8HF(aq) \longrightarrow H_2SiF_6(aq) + CaF_2(aq) + 3H_2O(l) \qquad [23.14]$$

Glass consists mostly of silicate structures (Section 23.5), and these reactions allow HF to etch or frost glass (Figure 23.6). It is also the reason that HF is stored in wax or plastic containers rather than glass.

**FIGURE 23.6** Etched or frosted glass. Designs such as this are produced by first coating the glass with wax. The wax is then removed in the areas to be etched. When treated with hydrofluoric acid, the exposed areas of the glass are attacked, producing the etching effect. The hydrofluoric acid is then washed from the surface and the remaining wax is removed. (Richard Megna/ Fundamental Photographs)

Some nonmetal halides, such as $NF_3$, $CCl_4$, and $SF_6$, are quite unreactive toward water. The lack of reactivity of these halides is not due to thermodynamic factors, but rather to the kinetic features of the reaction. For example, $\Delta G°$ for the hydrolysis of $CCl_4$ is highly negative, $-377$ kJ/mol $CCl_4$, which implies that the reaction should proceed very nearly to completion (Section 19.6). However, no low-energy reaction mechanism is available; the hydrolysis reaction has a very high activation barrier.

Generally, a molecular halide is unreactive when the central atom is unable to expand further the number of electrons in its valence shell (that is, the central atom has its maximum stable coordination number). For example, carbon can accommodate a maximum of eight electrons in its valence shell. Silicon, on the other hand, can accommodate a greater number of electrons by using $d$ orbitals in bonding. The hydrolysis of $SiCl_4$ is believed to occur by attack of the silicon atom by the $H_2O$ molecule, followed by loss of $H^+$ and $Cl^-$ ions, as shown in Figure 23.7. This reaction pathway is not available to $CCl_4$.

**FIGURE 23.7** Proposed mechanism for the first stage of the reaction between $SiCl_4$ and $H_2O$. After the $H_2O$ forms an adduct with $SiCl_4$, that adduct loses $H^+$ and $Cl^-$ ions. The process of $H_2O$ attack is then repeated three additional times, with $Si(OH)_4$ ultimately formed.

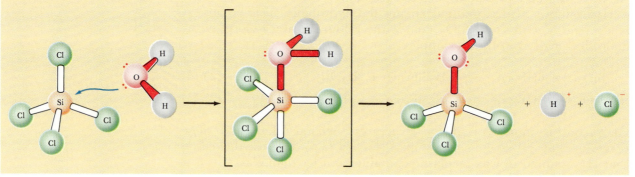

Because the halogens form diatomic molecules in their most stable state at ordinary temperatures and pressures, it is not surprising to discover that diatomic molecules consisting of two different halogen atoms exist. These compounds are the simplest example of **interhalogens**, that is, compounds formed between two different halogen elements. Some properties of the diatomic interhalogens are listed in Table 23.5. The table is incomplete because certain of the interhalogens are not stable;

**TABLE 23.5**    Properties of interhalogen compounds, XX′

| Compound | ClF | BrF | BrCl | IF | ICl | IBr |
|---|---|---|---|---|---|---|
| Molecular weight | 54.6 | 98.9 | 115.4 | 145.9 | 162.4 | 206.8 |
| Melting point (°C) | −155 | −33 | −54 | — | 27 | 42 |
| Boiling point (°C) | −90 | 20 | 5 | — | 98 | 116 |
| Bond distance (Å) | 1.63 | 1.76 | 2.16 | 1.91 | 2.32 | — |
| Dipole moment (D) | 0.9 | 1.3 | 0.6 | — | 0.5 | — |
| Bond-dissociation energy (kJ/mol) | 253 | 237 | 218 | 278 | 208 | 175 |

they undergo decomposition to form the diatomic halogen elements or to form more complex interhalogen compounds.

The higher interhalogen compounds have formulas of the form $XX'_3$, $XX'_5$, or $XX'_7$, where X is chlorine, bromine, or iodine, and X' is fluorine. (The one exception is $ICl_3$, in which X' is chlorine.) Using the VSEPR model (Section 9.1), we can predict the geometrical structures of these compounds. We can also describe the bonding about the central atom in terms of a hybrid orbital description (Section 9.3).

---

### SAMPLE EXERCISE 23.5

Account for the valence-shell electron distribution and geometrical structure in $BrF_3$. What hybrid orbital description is most suitable for the central atom in this molecule?

**Solution:** Bromine has seven valence-shell electrons. When the Br atom is singly bonded to three fluorine atoms, there are an additional three electrons from this source. According to the VSEPR model, these ten electrons are disposed as five electron pairs about the central atom at the vertices of a trigonal bipyramid (Table 9.3). Three of the electron pairs are used in bonding to fluorine; the other two are unshared electron pairs. These unshared pairs require a larger space, so they are placed in the equatorial plane of the trigonal bipyramid:

Because the unshared pairs push the bonding pairs back a little, the molecule should have the shape of a bent T. In terms of a hybridization description, we need to employ one of the bromine valence-shell $d$ orbitals (the $4d$) in addition to the $4s$ and three $4p$ orbitals to provide the five atomic orbitals to contain the five electron pairs in the valence shell. Thus the appropriate hybrid orbital description is $sp^3d$, which results in orbitals directed toward the vertices of a trigonal bipyramid.

### PRACTICE EXERCISE

Predict the molecular geometry and hybridization of orbitals on I in $IF_5$.

*Answer:* square pyramidal; $d^2sp^3$

---

The central atom in these higher interhalogen compounds has valence-shell $d$ orbitals available for bonding. Thus the valence shell can be expanded beyond the octet. The central atom in all cases is relatively large compared to the atoms grouped around it. Because fluorine is small and forms very strong bonds, it is ideally suited as the X' atom. Only when the central atom is very large, as in the case of iodine, can the larger chlorine atom form an interhalogen, as in $ICl_3$. The importance of size is also seen in the fact that iodine is capable of forming $IF_7$, but with bromine a maximum of five fluorines can be fitted around the central atom, in $BrF_5$. Chlorine and fluorine can form the $ClF_5$ molecule, but only with great difficulty.

The interhalogen compounds are exceedingly reactive. They attack glass very readily and must be placed in special metal containers. They are very active fluorinating agents, as in these examples:

$$2CoCl_2(s) + 2ClF_3(g) \longrightarrow 2CoF_3(s) + 3Cl_2(g) \qquad [23.15]$$

$$Se(s) + 3BrF_5(l) \longrightarrow SeF_6(l) + 3BrF_3(l) \qquad [23.16]$$

The *polyhalide ions* are closely related to the interhalogens. Many of these ions are stable as salts of the alkali metal ions—for example, $KI_3$, $CsIBr_2$, $KICl_4$, and $KBrF_4$. Some of them, notably $I_3^-$, are also stable in aqueous solution.

**Oxyacids and Oxyanions**

Table 23.6 summarizes the formulas of the known oxyacids of the halogens and the way they are named.* The oxyacids are rather unstable; they generally decompose (sometimes explosively) when one attempts to isolate them. All of the oxyacids are strong oxidizing agents. The oxyanions, formed on removal of the proton from the oxyacids, are generally more stable than the oxyacids themselves. (Review the nomenclature of the oxyanions, Section 2.6.) Hypochlorite salts are used as bleaches and disinfectants because of the powerful oxidizing capabilities of the hypochlorite ion. Sodium chlorite, which can be isolated as the trihydrate, $NaClO_2 \cdot 3H_2O$, is used as a bleaching agent. Chlorite salts form potentially explosive mixtures with organic materials. Chlorate salts are similarly very reactive. For example, a mixture of potassium chlorate and sulfur may explode when struck. Potassium chlorate is used in making matches and fireworks.

Perchloric acid and its salts are the most stable of the oxyacids and oxyanions. Dilute perchloric acid solutions are quite safe, and most perchlorate salts are stable except when heated with organic materials. When heated, perchlorates can become vigorous, even violent oxidizers. Thus considerable caution should be exercised when handling these substances, and it is crucial to avoid contact between perchlorates and readily oxidized materials such as active metals and combustible organic compounds. The use of ammonium perchlorate as the oxidizer in the solid booster rockets for the space shuttle demonstrates the oxidizing power of perchlorates. The solid propellant contains a mixture of $NH_4ClO_4$ and powdered aluminum, the reducing agent. Each shuttle launch requires about $6 \times 10^5$ kg (700 tons) of $NH_4ClO_4$.

There are two oxyacids that have iodine in the +7 oxidation state. These periodic acids are $HIO_4$ (called metaperiodic acid) and $H_5IO_6$ (called paraperiodic acid). The two forms exist in equilibrium in aqueous solution:

$$H_5IO_6(aq) \;\rightleftharpoons\; H^+(aq) + IO_4^-(aq) + 2H_2O(l) \qquad K = 0.015 \qquad [23.17]$$

**TABLE 23.6**   The oxyacids of the halogens

| Oxidation state of halogen | Formula of acid | | | Type of name |
|---|---|---|---|---|
| | **Cl** | **Br** | **I** | |
| +1 | HClO | HBrO | HIO | *Hypo*hal* ous* acid |
| +3 | $HClO_2$ | — | — | Hal*ous* acid |
| +5 | $HClO_3$ | $HBrO_3$ | $HIO_3$ | Hal*ic* acid |
| +7 | $HClO_4$ | $HBrO_4$ | $HIO_4, H_5IO_6$ | *Per*hal*ic* acid |

* Fluorine forms one oxyacid, HFO. Because the electronegativity of fluorine is greater than that of oxygen, we must consider fluorine to be a $-1$ oxidation state and oxygen to be in the 0 oxidation state.

$HIO_4$ is a strong acid, and $H_5IO_6$ is a weak one; the first two acid-dissociation constants for $H_5IO_6$ are $K_{a1} = 2.8 \times 10^{-2}$ and $K_{a2} = 4.9 \times 10^{-9}$. Crystalline $H_5IO_6$ is obtained when periodic acid solutions are evaporated at low temperatures. Mild heating of $H_5IO_6$ under vacuum produces $HIO_4$. The structure of $H_5IO_6$ is given in Figure 23.8. A molecule forms in which iodine is surrounded by six oxygen atoms because of the large size of the iodine. The smaller halogens do not form acids of this type.

The acid strengths of the oxyacids increase with increasing oxidation state of the central atom; the origin of this trend is discussed in Section 17.7. The stability of the oxyacids and of the corresponding oxyanions toward reduction increases with increasing oxidation state of the central halogen atom. Because the halogens are relatively electronegative elements, we might expect that compounds in which the halogen has an increasingly positive oxidation number would be *less* stable. The origins of this unexpected trend are rather complicated and beyond the scope of our survey.

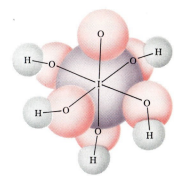

**FIGURE 23.8**
Paraperiodic acid, $H_5IO_6$.

## 23.3 THE GROUP 6A ELEMENTS

The elements of group 6A are oxygen, sulfur, selenium, tellurium, and polonium. We have already discussed oxygen in Section 22.4. In this section we will examine the group as a whole and then look at sulfur, selenium, and tellurium, focusing on sulfur. We will not have much to say about polonium, an element produced by radioactive decay of radium. There are no stable isotopes of this element. As a result, it is found only in trace quantities in radium-containing minerals.

### General Characteristics of the Group 6A Elements

The group 6A elements possess the general outer electron configuration $ns^2np^4$, where $n$ may have values ranging from 2 through 6. These elements thus may attain a noble gas electron configuration by the addition of two electrons, which results in a $-2$ oxidation state. Because the group 6A elements are nonmetals, this is a common oxidation state. Except for oxygen, however, the group 6A elements are also commonly found in positive oxidation states up to $+6$, which corresponds to sharing of all six valence-shell electrons with atoms of a more electronegative element. Sulfur, selenium, and tellurium also differ from oxygen in being able to use $d$ orbitals in bonding. Thus compounds with expanded valence shells such as $SF_6$, $SeF_6$, and $TeF_6$ occur.

Some of the more important properties of the atoms of the group 6A elements are summarized in Table 23.7. The energy of the X—X single bond is estimated from data for the elements, except for oxygen. In this case, because the O—O bond in $O_2$ is not a single bond (Section 9.6), the estimated O—O bond energy in hydrogen peroxide is employed. The reduction potential listed in the last line of the table refers to the reduction of the element in its standard state to form $H_2X(aq)$ in acidic solution. In most of the properties listed in Table 23.7, we again see a regular variation as a function of atomic number. Atomic and ionic radii increase and ionization potentials decrease as expected as we move down the family.

The electron affinities listed apply to the process shown in Equation 23.18:

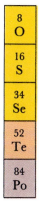

**TABLE 23.7** Some properties of the atoms of the group 6A elements

| Property | O | S | Se | Te |
|---|---|---|---|---|
| Atomic radius (Å) | 0.73 | 1.02 | 1.17 | 1.35 |
| $X^{2-}$ ionic radius (Å) | 1.45 | 1.90 | 2.02 | 2.22 |
| First ionization energy (kJ/mol) | 1312 | 1004 | 946 | 870 |
| Electron affinity (kJ/mol) | −141 | −201 | −195 | −186 |
| Electronegativity | 3.4 | 2.6 | 2.6 | 2.1 |
| X—X single-bond energy (kg/mol) | 142[a] | 266 | 172 | 126 |
| Reduction potential to $H_2X$ in acidic solution (V) | 1.23 | 0.14 | −0.40 | −0.72 |

[a] Based on O—O bond energy in $H_2O_2$.

$$X(g) + e^- \longrightarrow X^-(g) \qquad [23.18]$$

This, of course, does not produce the commonly observed stable ionic form of these elements, $X^{2-}$, but it is the first step in its formation. It is interesting to note that addition of an electron to oxygen is less exothermic than addition of an electron to one of the other group 6A elements. This effect is due to the relatively small size of the oxygen atom. Addition of a single extra electron to the smaller atom results in larger electron-electron repulsions, offsetting the gain in stability that results from the closer approach of the electron to the nucleus.

The decrease in ionization energy is the factor mainly responsible for the decrease in electronegativity in the series. In this connection it is interesting to note that sulfur and selenium are closely similar in many respects, whereas tellurium is substantially less electronegative. Note that the ease of reduction of the free element to form $H_2X$ varies greatly throughout the series. Whereas oxygen is very readily reduced to the −2 oxidation state, the potential for reduction of tellurium is actually quite strongly negative. These observations indicate an increasingly metallic character in the group 6A elements with increasing atomic number. The physical properties of the elements show a corresponding trend. At the top of the family we have oxygen, a diatomic molecule, and sulfur, a nonconducting solid that melts at 114°C. Near the bottom we have tellurium, whose stable form has a bright luster, low electrical conductivity, and a melting point of 452°C.

## Occurrences and Preparation of Sulfur, Selenium, and Tellurium

Large underground deposits serve as the principal source of the elemental sulfur. The Frasch process, illustrated in Figure 23.9, is used to obtain the element from these deposits. The method is based on the low melting point and low density of sulfur. Superheated water is forced into the deposit, where it melts the sulfur. Compressed air then forces the molten sulfur up a pipe that is concentric with the ones that introduce the hot water and compressed air into the deposit.

Sulfur also occurs widely as sulfide and sulfate minerals. Its presence as a minor component of coal and petroleum poses a major problem. As we saw in Section 14.4, combustion of these "unclean" fuels leads to serious sulfur oxide pollution. Also, operations that use sulfide minerals as sources of metals liberate sulfur oxides. Much effort has been directed

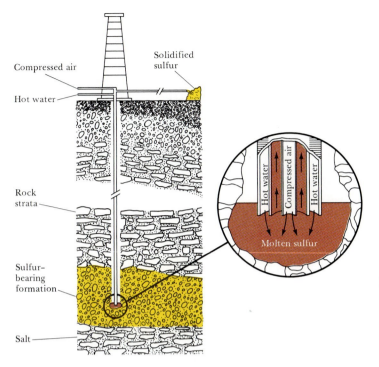

Compressed air
Hot water
Solidified sulfur
Rock strata
Sulfur-bearing formation
Salt

Hot water
Compressed air
Hot water
Molten sulfur

**FIGURE 23.9**   Mining of sulfur by the Frasch process. The process is named after Herman Frasch, who invented the process in the early 1890s. The process is particularly useful for recovering sulfur from deposits located under quicksand or water.

at removing this sulfur, and these efforts have increased the availability of sulfur. The sale of this sulfur helps partially to offset the costs of the desulfurizing processes and equipment. However, sulfur obtained from sulfur deposits by the Frasch process is about 99.5 percent pure and can be used for most commercial processes without purification. It is therefore relatively cheap. Consequently, sulfur from desulfurizing processes must be sold at prices below its cost in order to compete on the market. Nevertheless, about half the sulfur used in the United States each year is produced by means other than the Frasch process.

Selenium and tellurium occur in rare minerals as $Cu_2Se$, $PbSe$, $Ag_2Se$, $Cu_2Te$, $PbTe$, $Ag_2Te$, and $Au_2Te$. They also occur as minor constituents in sulfide ores of copper, iron, nickel, and lead. However, these elements are not very important commercially, so we will not consider the details of how they are obtained.

As we normally encounter it, sulfur is yellow (Figure 23.10), tasteless, and nearly odorless. It is insoluble in water and exists in several allotropic forms. The thermodynamically stable form at room temperature is rhombic sulfur, which consists of puckered $S_8$ rings, as shown in Figure 23.11. When heated above its melting point, at 113°C, sulfur undergoes a variety of changes. The molten sulfur first contains $S_8$ molecules and is fluid because the rings readily slip over each other. Further heating of this straw-colored liquid causes rings to break, the fragments joining to form very long molecules that can become entangled. The sulfur consequently becomes highly viscous. This change is marked by a color change to dark reddish brown (Figure 23.12). Further heating breaks the chains and the viscosity again decreases.

**Properties and Uses of Sulfur, Selenium, and Tellurium**

**FIGURE 23.10** Liquid sulfur is extracted from below ground by the Frasch process and pumped into large ponds where it is allowed to cool. Here the solid sulfur, with its characteristic yellow color, is shown being moved from a cooling pond and placed in a large pile in preparation for shipping. (Farrell Grehan/Photo Researchers)

**FIGURE 23.11** Top view (*a*) and side view (*b*) of the sulfur molecule in rhombic sulfur.

**FIGURE 23.12** When sulfur is heated above its melting point, 113°C, it becomes dark and viscous. Here the liquid is shown falling into cold water, where it again solidifies. (Lawrence Migdale, Science Source/ Photo Researchers)

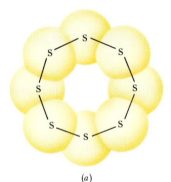

(*a*)

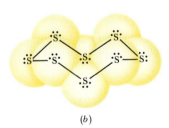

(*b*)

Most of the $1.3 \times 10^{10}$ kg (14 million tons) of sulfur produced in the United States each year is used in the manufacture of sulfuric acid. Sulfur is also used in vulcanizing rubber, a process that toughens rubber by introducing cross-linking between polymer chains.

The most stable allotropes of both selenium and tellurium are crystalline substances containing helical chains of atoms, as illustrated in Figure 23.13. Each atom of the chain, however, is close to atoms in adjacent chains, and it appears that there is some sharing of electron pairs between these atoms, as indicated by the red lines in Figure 23.13. The electrical conductivity of selenium is very small in the dark but increases greatly upon exposure to light. This property of the element is used in photoelectric cells and in light meters. Xerox photocopiers also depend on the photoconductivity of selenium for their operation.

Sulfur is widely distributed in biological systems. It is present in most proteins, as a component of the amino acids cysteine and methionine (Section 28.2). In contrast, selenium is rare in biological systems. Only recently has the human nutritional requirement for the element been established. Selenium is present in trace quantities in most vegetables, especially spinach. The amount of selenium required for adequate nutrition is very small. In quantities much larger than this small nutritional requirement, the element is toxic. Tellurium does not have a known role in human nutrition. Its compounds are poisonous and, if they are volatile, usually have highly offensive odors.

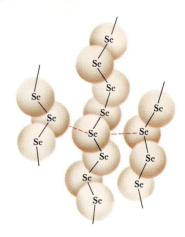

**FIGURE 23.13**  Portion of the structure of crystalline selenium. The dashed lines represent weak bonding interactions between atoms in adjacent chains. Tellurium has the same structure.

## Oxides, Oxyacids, and Oxyanions of Sulfur

Sulfur dioxide was first discovered by Joseph Priestley in 1774, when he heated mercury with concentrated sulfuric acid. The reaction that occurs is shown in Equation 23.19:

$$Hg(l) + 2H_2SO_4(l) \longrightarrow HgSO_4(s) + SO_2(g) + 2H_2O(l) \qquad [23.19]$$

In the laboratory, $SO_2$ is prepared by the action of aqueous acid on a sulfite salt, as shown in Equation 23.20:

$$2H^+(aq) + SO_3^{2-}(aq) \longrightarrow SO_2(g) + H_2O(l) \qquad [23.20]$$

Sulfur dioxide is formed when sulfur is combusted in air; it has a choking odor and is poisonous. The gas is particularly toxic to lower organisms, such as fungi, and is consequently used for sterilizing dried fruit. At 1 atm pressure and room temperature, $SO_2$ dissolves in water to the extent of 45 volumes of gas per volume of water, to produce a solution of about 1.6 $M$ concentration. The solution is acidic, and we describe it as $H_2SO_3(aq)$. Actually, there is evidence that much of the sulfur dioxide exists in solution as hydrated $SO_2$. When the saturated solution is cooled, crystals of the hydrate, $SO_2 \cdot 6H_2O$, can be recovered. It is convenient, however, to assume that all of the $SO_2$ that dissolves is in the form of the weak acid, $H_2SO_3$, which ionizes according to Equations 23.21 and 23.22:

$$H_2SO_3(aq) \rightleftharpoons H^+(aq) + HSO_3^-(aq) \qquad K_{a1} = 1.7 \times 10^{-2}(25°C) \qquad [23.21]$$
$$HSO_3^-(aq) \rightleftharpoons H^+(aq) + SO_3^{2-}(aq) \qquad K_{a2} = 6.4 \times 10^{-8}(25°C) \qquad [23.22]$$

As we have seen, combustion of sulfur in air produces mainly $SO_2$, but small amounts of $SO_3$ are also formed. The reaction produces mainly $SO_2$ because the activation-energy barrier for further oxidation to $SO_3$

**FIGURE 23.14** The reaction between sucrose, $C_{12}H_{22}O_{11}$, and concentrated sulfuric acid. Sucrose is a carbohydrate, containing two H atoms for each O atom. Sulfuric acid, which is an excellent dehydrating agent, removes $H_2O$ from the sucrose to form carbon, the black mass remaining at the end of the reaction. (Kristen Brochman/Fundamental Photographs)

is very high unless the reaction is catalyzed. Sulfur trioxide is of great commercial importance because it is the anhydride of sulfuric acid. In the manufacture of sulfuric acid, $SO_2$ is first obtained by burning sulfur. The $SO_2$ is then oxidized to $SO_3$ using a catalyst such as $V_2O_5$ or platinum. The $SO_3$ is dissolved in $H_2SO_4$ because it does not dissolve quickly in water. The reaction is shown in Equation 23.23. The $H_2S_2O_7$ formed in this reaction, called pyrosulfuric acid, is then added to water to form $H_2SO_4$, as shown in Equation 23.24:

$$SO_3(g) + H_2SO_4(l) \longrightarrow H_2S_2O_7(l) \qquad [23.23]$$

$$H_2S_2O_7(l) + H_2O(l) \longrightarrow 2H_2SO_4(l) \qquad [23.24]$$

Commercial sulfuric acid is generally 98 percent $H_2SO_4$ and boils at 340°C. It is a dense, colorless, oily liquid. Sulfuric acid has many useful properties. Most important, it is a strong acid, a good dehydrating agent,* and a moderately good oxidizing agent. Its dehydrating ability is demonstrated in Figure 23.14.

Year after year, the output of sulfuric acid has been the largest of any chemical produced in the United States. About $3.4 \times 10^{10}$ kg (3.7 million tons) is produced annually in this country. Sulfuric acid is employed in some way in almost all manufacturing. Consequently, its consumption is considered a standard measure of industrial activity. The primary uses for sulfuric acid are shown in Figure 23.15.

* Considerable heat is given off when sulfuric acid is diluted with water. Consequently, dilution must always be done carefully by pouring the acid into water to distribute the heat as uniformly as possible and to avoid spattering of the acid.

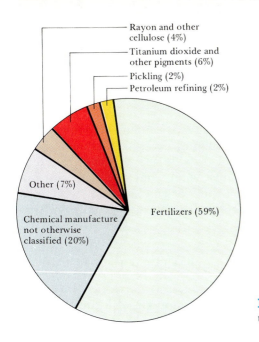

Rayon and other
cellulose (4%)

Titanium dioxide and
other pigments (6%)

Pickling (2%)

Petroleum refining (2%)

Other (7%)

Chemical manufacture
not otherwise
classified (20%)

Fertilizers (59%)

**FIGURE 23.15**    Sulfuric acid use in the United States.

Only the first proton in sulfuric acid is completely ionized in aqueous solution. The second proton ionizes only partially:

$$H_2SO_4(aq) \longrightarrow H^+(aq) + HSO_4^-(aq)$$

$$HSO_4^-(aq) \rightleftharpoons H^+(aq) + SO_4^{2-}(aq) \qquad K_a = 1.1 \times 10^{-2}$$

Consequently, sulfuric acid forms two series of compounds: sulfates and bisulfates (or hydrogen sulfates).

Related to the sulfate ion is the thiosulfate ion, $S_2O_3^{2-}$, formed by boiling an alkaline solution of $SO_3^{2-}$ with elemental sulfur:

$$8SO_3^{2-}(aq) + S_8(s) \longrightarrow 8S_2O_3^{2-}(aq) \qquad [23.25]$$

The term "thio" indicates substitution of sulfur for oxygen. The structures of the sulfate and thiosulfate ions are compared in Figure 23.16. When acidified, the thiosulfate ion decomposes to form sulfur and $H_2SO_3$. The pentahydrated salt of sodium thiosulfate, $Na_2S_2O_3 \cdot 5H_2O$, is known as "hypo." It is used in photography to furnish thiosulfate ion as a complexing agent for silver ion. Photographic film consists of a suspension of microcrystals of AgBr in gelatin. In those portions of the film that are exposed to light the AgBr is "activated." This means that when the film is treated with a mild reducing agent (called the developer) the exposed AgBr is reduced. Thus black, metallic silver is formed in the film in concentrations proportional to the intensity of light exposure. The film is then treated with sodium thiosulfate solution to remove the remaining silver bromide, by forming the soluble silver thiosulfate complex:

$$AgBr(s) + 2S_2O_3^{2-}(aq) \rightleftharpoons Ag(S_2O_3)_2^{3-}(aq) + Br^-(aq) \qquad [23.26]$$

This step in the process is called "fixing." Thiosulfate ion is also used in quantitative analysis as a reducing agent for iodine:

$$2S_2O_3^{2-}(aq) + I_2(s) \longrightarrow 2I^-(aq) + S_4O_6^{2-}(aq) \qquad [23.27]$$

**FIGURE 23.16**

Comparison of the structures of the sulfate, $SO_4^{2-}$, and thiosulfate, $S_2O_3^{2-}$, ions.

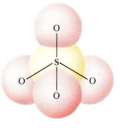

$SO_4^{2-}$

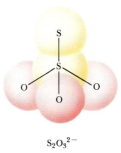

$S_2O_3^{2-}$

## Oxides, Oxyacids, and Oxyanions of Se and Te

Both selenium and tellurium form dioxides upon burning the element in air or oxygen. Selenium dioxide is very soluble in water; it forms an acidic solution from which crystals of selenous acid, $H_2SeO_3$, can be isolated. Neutralization of selenous acid solutions yields salts of hydrogen selenite or selenite ion, for example, $NaHSeO_3$ or $Na_2SeO_3$. Tellurium dioxide is insoluble in water; thus the aqueous acid $H_2TeO_3$ is unknown. However, it is possible to prepare tellurite salts by dissolving $TeO_2$ in an aqueous base.

Both selenium trioxide and tellurium trioxide are also known. These are the anhydrides of selenic and telluric acids, but the acids are not formed by dissolving the trioxides in water. Selenic acid, $H_2SeO_4$, can be formed by oxidation of aqueous $SeO_2$ with 30 percent hydrogen peroxide:

$$H_2SeO_3(aq) + H_2O_2(aq) \longrightarrow HSeO_4^-(aq) + H_2O(l) + H^+(aq) \qquad [23.28]$$

Notice that we represent $H_2SeO_3(aq)$ in the nonionized form; the acid-dissociation constants for $H_2SeO_3$ at 25°C are $K_{a1} = 2.3 \times 10^{-3}$ and $K_{a2} = 5.3 \times 10^{-9}$. The nonionized acid is thus probably most representative of the species present in solution. However, selenic acid, $H_2SeO_4$, is a strong acid in its first dissociation constant; the second dissociation has an equilibrium constant $K_{a2} = 2.2 \times 10^{-2}$ at 25°C.

Telluric acid is formed when $TeO_2(s)$ is oxidized by various aqueous oxidizing agents. We might expect the formula for telluric acid to be $H_2TeO_4$, by analogy with the corresponding acid of sulfur or selenium. However, the acid which is recovered from solution as a white crystalline material is $H_6TeO_6$. [The formula can also be written as $Te(OH)_6$.] This substance is called orthotelluric acid. Notice the close similarity to iodine, which forms paraperiodic acid, $H_5IO_6$. In both cases, the large size of the central atom permits bonding with six surrounding oxygens. Orthotelluric acid is a weak acid, capable of dissociating up to two protons (at 25°C, $K_{a1} = 2.4 \times 10^{-8}$, $K_{a2} = 1.0 \times 10^{-11}$).

## Sulfides, Selenides, and Tellurides

Sulfur forms compounds by direct combination with many elements. When the element is less electronegative than sulfur, sulfides, which contain $S^{2-}$, form. For example, iron(II) sulfide, FeS, forms by direct combination of iron and sulfur. Many metallic elements are found in the form of sulfide ores, for example, PbS (galena) and HgS (cinnabar). A series of related ores containing the $S_2^{2-}$ ion (analogous to the peroxide ion) are known as *pyrites*. Iron pyrite, $FeS_2$, occurs as golden-yellow cubic crystals (Figure 23.17). Because it has been on occasion mistaken for gold by overeager gold miners, it is often called "fool's gold."

One of the most important sulfides is hydrogen sulfide, $H_2S$. This substance is not normally produced by direct union of the elements because it is unstable at elevated temperature and decomposes into the elements. It is normally prepared by action of dilute sulfuric acid on iron(II) sulfide:

$$FeS(s) + 2H^+(aq) \longrightarrow H_2S(aq) + Fe^{2+}(aq) \qquad [23.29]$$

**FIGURE 23.17** Iron pyrite, $FeS_2$, is also known as fool's gold because its color has fooled people into thinking it was gold. Gold is much more dense and much softer than iron pyrite. (Charles R. Belinky/ Photo Researchers)

A common laboratory source of $H_2S$ is the reaction of thioacetamide with water:

$$\underset{\text{CH}_3\overset{\displaystyle S}{\overset{\displaystyle \|}{\text{C}}}\text{NH}_2(aq)}{} + H_2O(l) \longrightarrow \underset{\text{CH}_3\overset{\displaystyle O}{\overset{\displaystyle \|}{\text{C}}}\text{NH}_2(aq)}{} + H_2S(aq) \qquad [23.30]$$

Hydrogen sulfide is often used in the laboratory for qualitative analysis of certain metal ions (Section 18.6).

One of hydrogen sulfide's most readily recognized properties is its odor; $H_2S$ is largely responsible for the offensive odor of rotten eggs. Hydrogen sulfide is actually quite toxic; it has about the same level of toxicity as hydrogen cyanide, the gas that has been used in gas chambers. The volatile hydrides $H_2Se$ and $H_2Te$ are similar to $H_2S$ in many respects. Both compounds possess very offensive, lingering odors and are toxic. In aqueous solutions the acid strength increases in the order $H_2S < H_2Se < H_2Te$.

## 23.4 THE GROUP 5A ELEMENTS

The elements of group 5A are nitrogen, phosphorus, arsenic, antimony, and bismuth. We have already discussed the chemistry of nitrogen (Section 22.5). In our present discussion we will find it convenient to discuss the general characteristics of the group, then to consider phosphorus, and finally to comment briefly on the heavier elements of the group.

### General Characteristics of the Group 5A Elements

The group 5A elements possess the outer electron configuration $ns^2np^3$, where $n$ may have values ranging from 2 to 6. A noble gas configuration results from addition of three electrons to form the $-3$ oxidation state. Ionic compounds containing $X^{3-}$ ions are not common, however, except for salts of the more active metals, for example, $Na_3N$. More commonly the group 5A element acquires an octet of electrons via covalent bonding. The oxidation number may range from $-3$ to $+5$, depending on the nature and number of the atoms to which it is bound.

Because of its lower electronegativity, phosphorus is found more frequently in positive oxidation states than is nitrogen. Furthermore, compounds in which phosphorus has the $+5$ oxidation state are not strong oxidizing agents, as are corresponding compounds of nitrogen. Conversely, compounds in which phosphorus has a $-3$ oxidation state are

**TABLE 23.8**  Properties of the atoms of group 5A elements

| Property | N | P | As | Sb | Bi |
|---|---|---|---|---|---|
| Atomic radius (Å) | 0.75 | 1.10 | 1.22 | 1.43 | — |
| First ionization energy (kJ/mol) | 1402 | 1012 | 947 | 834 | 703 |
| Electron affinity (kJ/mol) | +6.8 | −72 | −77 | −101 | −106 |
| Electronegativity | 3.0 | 2.2 | 2.2 | 2.0 | 2.0 |
| X—X single-bond energy (kJ/mol)ᵃ | 163 | 200 | 150 | 120 | — |
| X≡X triple-bond energy (kJ/mol) | 941 | 480 | 380 | 295 | 192 |

ᵃ Approximate values only.

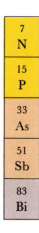

much stronger reducing agents than are corresponding compounds of nitrogen.

Some of the important properties of the atoms of the group 5A elements are listed in Table 23.8. The general pattern that emerges from these data is similar to what we have seen before; size and metallic character increase as atomic number increases within the group.

The variation in properties among the elements of group 5A is more striking than that seen in the case of groups 6A and 7A. Nitrogen at the one extreme exists as a gaseous diatomic molecule; it is clearly nonmetallic in character. At the other extreme, bismuth is a reddish-white, metallic-looking substance that has most of the characteristics of a metal.

The values listed for X—X single-bond energies are not very reliable, because it is difficult to obtain such data from thermochemical experiments. However, there is no doubt about the general trend: a low value for the N—N single bond, an increase at phosphorus, then a gradual decline to arsenic and antimony. From observations of the group 5A elements in the gas phase (for all except $N_2$ this requires high temperatures), it is possible to estimate the X≡X triple-bond energy, as listed in Table 23.8. Here we see a trend that is different than that for the X—X single bond. Nitrogen forms a much stronger bond than do the other elements, and there is a steady decline in the triple-bond energy down through the group. These data help us to appreciate why nitrogen alone of the group 5A elements exists as a diatomic molecule in its stable state at 25°C. All the other elements exist in structural forms with single bonds between the atoms.

### Occurrence, Isolation, and Properties of Phosphorus

Phosphorus occurs mainly in the form of phosphate minerals. The principal source of phosphorus is phosphate rock, which contains phosphate mainly in the form of $Ca_3(PO_4)_2$. Deposits of phosphate rock occur mostly in Florida, the western United States, North Africa, and parts of the USSR. The element is produced commercially by reduction of phosphate with coke* in the presence of $SiO_2$, as shown in Equation 23.31:

$$2Ca_3(PO_4)_2(s) + 6SiO_2(s) + 10C(s) \xrightarrow{1500°C} P_4(g) + 6CaSiO_3(l) + 10CO(g) \quad [23.31]$$

* Coke is a form of carbon made by heating coal in the absence of air.

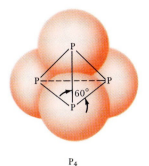

$P_4$

**FIGURE 23.18**
Tetrahedral structure of the $P_4$ molecule of white phosphorus.

**FIGURE 23.19** White and red allotropes of phosphorus, both stored under water. White phosphorus is very reactive and is normally stored under water to protect it from oxygen. The yellowish cast on the white phosphorus in this photo is due to reactions with air. Red phosphorus is much less reactive than white phosphorus, and it is not necessary to store it under water. (Donald Clegg and Roxy Wilson)

The phosphorus produced in this fashion is the allotrope known as white phosphorus. This form distills from the reaction mixture as the reaction proceeds.

White phosphorus consists of $P_4$ tetrahedra, as shown in Figure 23.18. The 60° bond angles in $P_4$ are unusually small for molecules. There must consequently be much strain in the bonding, a fact that is consistent with the high reactivity of white phosphorus. This allotrope bursts spontaneously into flames if exposed to air. It is a white, waxlike solid that melts at 44.2°C and boils at 280°C. When heated in the absence of air to about 400°C, it is converted to a more stable allotrope known as red phosphorus. This form does not ignite on contact with air. It is also considerably less poisonous than the white form. Both allotropes are shown in Figure 23.19.

**Phosphorus Halides**

Phosphorus forms a wide range of compounds with the halogens. A few properties of the more important phosphorus-halogen compounds are listed in Table 23.9.

Phosphorus trifluoride is prepared by reaction with another fluoride, as in Equation 23.32:

$$PCl_3(l) + AsF_3(l) \longrightarrow PF_3(g) + AsCl_3(l) \qquad [23.32]$$

The driving force for this reaction is the greater strength of the phosphorus-fluorine bond as compared with the other bond energies (490 kJ/mol for P—F as compared with 406 kJ/mol for As—F, 326 kJ/mol for P—Cl, and 322 kJ/mol for As—Cl). Phosphorus pentafluoride can be prepared

**TABLE 23.9**  Properties of phosphorus-halogen compounds

| Property | $PF_3$ | $PF_5$ | $PCl_3$ | $PCl_5$ | $PBr_3$ | $PBr_5$ | $PI_3$ |
|---|---|---|---|---|---|---|---|
| Melting point (°C) | −160 | −94 | −112 | 167 | −40 | — | 61 |
| Boiling point (°C) | −95 | −85 | 76 | —[a] | 173 | 106 | 120[b] |
| Dipole moment (D) | 1.0 | 0 | 0.78 | 0 | 0.5 | 0 | 0 |

[a] Sublimes at 163°C at 1 atm pressure.
[b] At 15 mm Hg pressure.

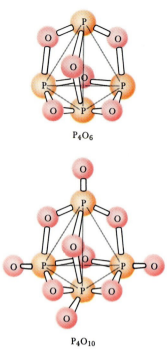

**FIGURE 23.20** Structures of $P_4O_6$ and $P_4O_{10}$.

by direct combination of the elements, by reaction of $PCl_5$ with $AsF_3$, or by heating $P_4O_{10}$ in a sealed tube with $CaF_2$:

$$4P_4O_{10}(l) + 15CaF_2(s) \longrightarrow 6PF_5(l) + 5Ca_3(PO_4)_2(s) \qquad [23.33]$$

Phosphorus trichloride is formed by passing a stream of dry chlorine gas over white or red phosphorus. In an excess of $Cl_2$, the $PCl_3$ reacts further to form $PCl_5$. Phosphorus tribromide is formed by reaction of liquid bromine on red phosphorus. Addition of still more bromine results in formation of $PBr_5$. However, this compound readily dissociates in the vapor state:

$$PBr_5(g) \rightleftharpoons PBr_3(g) + Br_2(g) \qquad [23.34]$$

Phosphorus triiodide can be formed by reacting white phosphorus and iodine in an appropriate solvent.

The phosphorus halides hydrolyze on contact with water. The reactions occur readily, and most of the phosphorus halides fume in air as a result of reaction with water vapor. In the presence of excess water, the products are the corresponding phosphorus oxyacid and hydrogen halide, as in the examples shown in Equations 23.35 and 23.36:

$$PF_3(g) + 3H_2O(l) \longrightarrow H_3PO_3(aq) + 3HF(aq) \qquad [23.35]$$

$$PCl_5(l) + 4H_2O(l) \longrightarrow H_3PO_4(aq) + 5HCl(aq) \qquad [23.36]$$

When only a little water is used, the hydrogen halide can be obtained as a gas; these reactions thus provide convenient laboratory preparations of the hydrogen halides.

**Oxy Compounds of Phosphorus**

Probably the most significant compounds of phosphorus are those in which the element is combined in some way with oxygen. Phosphorus(III) oxide, $P_4O_6$, is obtained by allowing white phosphorus to oxidize in a limited supply of oxygen. When oxidation takes place in the presence of excess oxygen, phosphorus(V) oxide, $P_4O_{10}$, forms. This compound is also readily formed by oxidation of $P_4O_6$. Phosphorus(III) oxide is often called phosphorus trioxide after its empirical formula; similarly, phosphorus(V) oxide is often called phosphorus pentoxide, and written as the empirical formula $P_2O_5$. These two oxides represent the two most common oxidation states for phosphorus, $+3$ and $+5$. The structural relationship between $P_4O_6$ and $P_4O_{10}$ is shown in Figure 23.20. Notice the resemblance these molecules have to the $P_4$ molecule, Figure 23.18.

---

**SAMPLE EXERCISE 23.6**

The reactive chemicals on the tip of a "strike any-where" match are usually $P_4S_3$ and an oxidizing agent such as $KClO_3$. When the match is struck on a rough surface, the heat generated by the friction ignites the $P_4S_3$, and the oxidizing agent brings about rapid combustion. The products of the combustion are $P_4O_{10}$ and $SO_2$. Calculate the standard enthalpy for the combustion of $P_4S_3$ in air,

given the following standard enthalpies of formation: $P_4S_3$ ($-154.4$ kJ/mol); $P_4O_{10}$ ($-2940$ kJ/mol); $SO_2$ ($-296.9$ kJ/mol).

**Solution:** The chemical equation for the combustion is

$$P_4S_3(s) + 8O_2(g) \longrightarrow P_4O_{10}(s) + 3SO_2(g)$$

Recalling that the standard enthalpy of formation of any element in its standard states is zero, we have $\Delta H_f^\circ(O_2) = 0$. Thus we can write

$$\begin{aligned}\Delta H^\circ &= \Delta H_f^\circ(P_4O_{10}) + 4\Delta H_f^\circ(SO_2) - \Delta H_f^\circ(P_4S_3)\\ &= -2940 \text{ kJ} + 4(-296.9) \text{ kJ} - (-154.4 \text{ kJ})\\ &= -3973 \text{ kJ}\end{aligned}$$

The reaction is strongly exothermic, making it evident why $P_4S_3$ is used on match tips.

**PRACTICE EXERCISE**
Write the balanced equation for the reaction of $P_4O_{10}$ with water, and calculate $\Delta H^\circ$ for this reaction using data from Appendix D.
***Answer:*** $P_4O_{10}(s) + 6H_2O(l) \longrightarrow 4H_3PO_4(aq)$; $-498.0$ kJ

Phosphorus(V) oxide is the anhydride of phosphoric acid, $H_3PO_4$, a weak triprotic acid. In fact, $P_4O_{10}$ has a very high affinity for water and is consequently used as a drying agent. Phosphorus(III) oxide is the anhydride of phosphorous acid, $H_3PO_3$, a weak diprotic acid. The structures of $H_3PO_4$ and $H_3PO_3$ are shown in Figure 23.21. The hydrogen atom that is attached directly to phosphorus in $H_3PO_3$ is not acidic.

One characteristic of phosphoric and phosphorous acids is their tendency to undergo condensation reactions when heated. A **condensation reaction** is one in which two or more molecules combine to form a larger molecule by eliminating a small molecule, for example, $H_2O$. The reaction in which two $H_3PO_4$ molecules are joined by the elimination of one $H_2O$ molecule to form $H_4P_2O_7$ is shown in Equation 23.37:

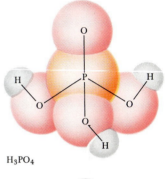

$H_3PO_4$

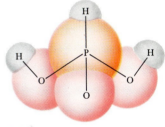

$H_3PO_3$

**FIGURE 23.21** Structures of $H_3PO_4$ and $H_3PO_3$.

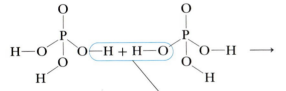

These atoms are eliminated as $H_2O$

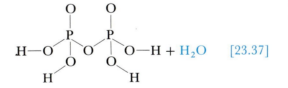 [23.37]

Further condensation produces phosphates having an empirical formula of $HPO_3$:

$$n\text{H}_3\text{PO}_4 \longrightarrow (\text{HPO}_3)_n + n\text{H}_2\text{O} \qquad [23.38]$$

Two phosphates having this empirical formula, one cyclic and the other polymeric, are shown in Figure 23.22. The three acids $H_3PO_4$, $H_4P_2O_7$, and $(HPO_3)_n$ all contain phosphorus in its $+5$ oxidation state, and all are therefore called phosphoric acids. To differentiate them, the prefixes *ortho-*, *pyro-*, and *meta-* are used: $H_3PO_4$ is orthophosphoric acid, $H_4P_2O_7$ is pyrophosphoric acid, and $HPO_3$ is metaphosphoric acid.

Phosphoric acid and its salts find their most important uses in detergents and fertilizers. The phosphates in detergents are often in the form

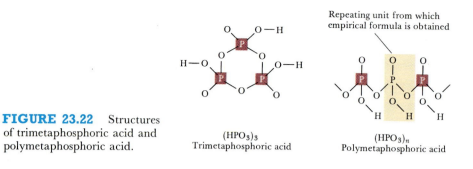

**FIGURE 23.22** Structures of trimetaphosphoric acid and polymetaphosphoric acid.

$(HPO_3)_3$
Trimetaphosphoric acid

$(HPO_3)_n$
Polymetaphosphoric acid

Repeating unit from which empirical formula is obtained

of sodium tripolyphosphate, $Na_5P_3O_{10}$ (see Section 26.2). A typical detergent formulation contains 47 percent phosphate, 16 percent bleaches, perfumes, and abrasives, and 37 percent linear alkylsulfonate (LAS) surfactant (shown as follows):

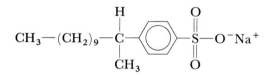

(We have used the notation for the benzene ring described in Section 9.4.) The detergent action of such molecules has been described in Section 12.7. The phosphate ions form bonds with metal ions that contribute to the hardness of water. This keeps the metal ions from interfering with the action of the surfactant. The phosphates also keep the pH above 7 and thus prevent the surfactant molecules from becoming protonated (gaining an $H^+$ ion).

Most phosphate rock mined is converted to fertilizers. The mined phosphate rock contains large amounts of sand and clay. At a treatment plant, sand, clay, and organic materials are removed from the raw ore. The resultant concentrate is shipped to plants that convert it to phosphoric acid or water-soluble phosphate fertilizers. The $Ca_3(PO_4)_2$ in phosphate rock is insoluble ($K_{sp} = 2.0 \times 10^{-29}$). It is converted to a soluble form for use in fertilizers by treating the concentrated phosphate rock with sulfuric or phosphoric acid:

$$Ca_3(PO_4)_2(s) + 4H^+(aq) + 3SO_4^{2-}(aq) \longrightarrow 3CaSO_4(s) + 2H_2PO_4^-(aq) \qquad [23.39]$$

$$Ca_3(PO_4)_2(s) + 4H^+(aq) \longrightarrow 3Ca^{2+}(aq) + 2H_2PO_4^-(aq) \qquad [23.40]$$

The mixture formed when ground phosphate rock is treated with sulfuric acid and then dried and pulverized is known as superphosphate. The $CaSO_4$ formed in this process is of little use in soil except when deficiencies in calcium or sulfur exist. It also dilutes the phosphorus, which is the nutrient of interest. If the phosphate rock is treated with phosphoric acid, the product contains no $CaSO_4$ and has a higher percentage of phosphorus. This product is known as triple superphosphate. Although the solubility of $Ca(H_2PO_4)_2$ allows it to be assimilated by plants, it also allows it to be washed from the soil and into water bodies, thereby contributing to water pollution problems.

Phosphorus compounds are important in biological systems. The element occurs, for example, in phosphate groups in RNA and DNA, the molecules responsible for control of protein biosynthesis and transmission of genetic information. It also occurs in adenosine triphosphate (ATP), which stores energy within biological cells:

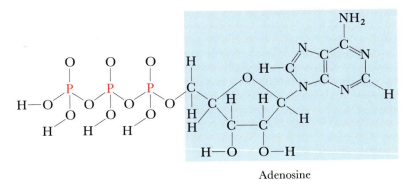

Adenosine

The P—O—P bond of the end phosphate group is broken by hydrolysis with water, forming adenosine diphosphate (ADP). This reaction produces 33 kJ of energy:

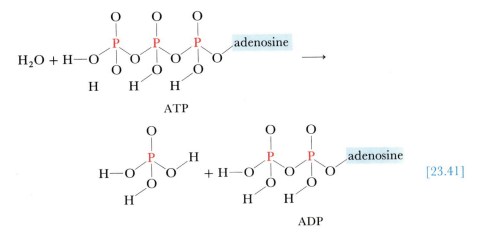

[23.41]

This energy is used to perform the mechanical work in muscle contraction and in many other biochemical reactions (see Section 19.7 and Figure 19.8). More on the biological properties of phosphorus-containing substances appears in Chapter 28.

## Arsenic, Antimony, and Bismuth

Arsenic, antimony, and bismuth occur in nature in the form of sulfide minerals, such as $As_2S_3$, $Sb_2S_3$, and $Bi_2S_3$. In addition, the elements are found as minor components in ores of various metals, such as Cu, Pb, Ag, and Hg. Arsenic and antimony exhibit allotropy similar to that of phosphorus. Both elements can be prepared as soft, yellow, nonmetallic solids by quickly cooling the high-temperature vapor of the element. In this allotropic form, analogous to the white allotrope of phosphorus, the elements are present as $As_4$ or $Sb_4$ tetrahedra. Heating or the action of light converts the substances into the gray, more metallic forms containing sheets of atoms. In its common form bismuth has a reddish-white, rather metallic appearance. The element forms alloys with many metallic elements. Alloys of lead, bismuth, and tin are used in the construction of plugs in sprinkler systems. Water is discharged when the fusible metal plug melts.

Arsenic and antimony resemble phosphorus in much of their chemical behavior. For example, both elements form halides with the formulas $MX_3$ and $MX_5$, with structures and chemical properties similar to the halides of phosphorus. The oxy compounds of these two elements are also rather similar to those of phosphorus, except that the higher oxidation state is not so easily attained. Thus the product of burning arsenic in oxygen is $As_4O_6$, not $As_4O_{10}$. The higher oxide can be obtained by oxidation of $As_4O_6$ with a strong oxidizing agent, such as nitric acid:

$$As_4O_6(s) + 4NO_3{}^-(aq) + 6H_2O(l) + 4H^+(aq) \longrightarrow 4H_3AsO_4(aq) + 4HNO_2(aq) \qquad [23.42]$$

In terms of the criteria discussed in Section 5.3, bismuth is more logically considered a metal rather than a nonmetal. Bismuth usually appears in the $+3$ oxidation state; there is little tendency to attain the higher $+5$ oxidation state that is so common for phosphorus. The common oxide of bismuth is $Bi_2O_3$. This substance is insoluble in water or basic solution but is soluble in acidic solution. It thus is classified as a basic anhydride. As we have seen, the oxides of metals characteristically behave as basic anhydrides.

## 23.5 THE GROUP 4A ELEMENTS

The elements of group 4A are carbon, silicon, germanium, tin, and lead. The general trend from nonmetallic to metallic as we go down a family is strikingly evident in group 4A. Carbon is strictly nonmetallic; silicon is essentially nonmetallic, although it does exhibit some characteristics of a metalloid, particularly in its electrical and physical properties; germanium is a metalloid; tin and lead are both metallic. The chemistry of carbon was discussed in Section 22.5. In the present discussion we will consider a few general characteristics of group 4A and then look more thoroughly at silicon.

### General Characteristics of the Group 4A Elements

Some properties of the group 4A elements are given in Table 23.10. The elements possess the outer electron configuration $ns^2np^4$. The electronegativities of the elements are generally low; carbides that formally contain $C^{4-}$ ions are observed only in the case of a very few compounds of carbon with very active metals. Formation of $+4$ ions by electron loss is not observed for any of these elements; the ionization energies are too high. However, the $+2$ state is found in the chemistry of germanium, tin, and lead; it is the principal oxidation state for lead. The vast majority of the compounds of the group 4A elements are covalently bonded. Carbon forms a maximum of four bonds. The other members of the family are able to form higher coordination numbers because of the availability of $d$ orbitals for bonding.

Carbon differs from the other group 4A elements in its pronounced ability to form multiple bonds both with itself and with other nonmetals,

| 6 |
| C |

| 14 |
| Si |

| 32 |
| Ge |

| 50 |
| Sn |

| 82 |
| Pb |

**TABLE 23.10**   Some properties of the group 4A elements

| Property | C | Si | Ge | Sn | Pb |
|---|---|---|---|---|---|
| Atomic radius (Å) | 0.77 | 1.17 | 1.22 | 1.41 | 1.54 |
| First ionization energy (kJ/mol) | 1090 | 780 | 782 | 704 | 714 |
| Electronegativity | 2.5 | 1.9 | 2.0 | 1.8 | 2.3 |
| X—X single-bond energy (kJ/mol) | 348 | 226 | 188 | 151 | — |

especially N, O, and S. The origin of this behavior was considered earlier, in Section 22.1.

Table 23.10 shows that the strength of a bond between two atoms of a given element decreases as we go down group 4A. Carbon-carbon bonds are quite strong. As a consequence, carbon has a striking ability to form compounds in which carbon atoms are bonded to each other. This property, called **catenation**, permits the formation of extended chains and rings of carbon atoms and accounts for the large number of organic compounds that exist. Catenation is also exhibited by other elements, especially ones in the vicinity of carbon in the periodic table, such as boron, nitrogen, phosphorus, oxygen, sulfur, silicon, and germanium. However, such self-linkage is far less important in the chemistries of these other elements. For example, the Si—Si bond strength (226 kJ/mol) is far smaller than the Si—O bond strength (368 kJ/mol). As a result, the chemistry of silicon is dominated by the formation of Si—O bonds, and Si—Si bonds play a rather minor role.

**Occurrence and Preparation of Silicon**

Silicon is the second most abundant element, after oxygen, in the earth's crust. It occurs in $SiO_2$ and in an enormous variety of silicate minerals. The element is obtained by the reduction of silicon dioxide with carbon at high temperature:

$$SiO_2(l) + 2C(s) \longrightarrow Si(l) + 2CO(g) \qquad [23.43]$$

Elemental silicon has a diamond type of structure (see Figure 9.35);

**FIGURE 23.23**   Elemental silicon. To prepare electronic devices, silicon (powder, left) is melted, drawn into a single crystal, and purified by zone refining. Wafers of silicon, cut from the crystal, are subsequently treated by a series of elegant techniques to produce various electronic devices. (Courtesy of Texas Instruments)

2 in.

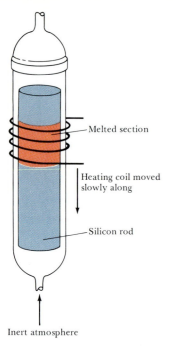

Melted section

Heating coil moved slowly along

Silicon rod

Inert atmosphere

**FIGURE 23.24**
Zone-refining apparatus.

a graphitelike allotrope is not known, presumably because of the weakness of $\pi$ bonds between silicon atoms. Crystalline silicon is a gray, metallic-looking solid that melts at 1410°C (Figure 23.23). The element is a semiconductor (Section 24.6) and is thus used in making transistors and solar cells. To be used as a semiconductor, the element must be extremely pure. One method of purification is to treat the element with $Cl_2$ to form $SiCl_4$. The $SiCl_4$ is a volatile liquid that is purified by fractional distillation and then reduced by $H_2$ to elemental silicon:

$$SiCl_4(g) + 2H_2(g) \longrightarrow Si(s) + 4HCl(g) \qquad [23.44]$$

The element can be further purified by the process of zone refining. In the zone-refining process a heated coil is passed slowly along a silicon rod, as shown in Figure 23.24. A narrow band of the element is thereby melted. As the molten area is swept slowly along the length of the rod the impurities concentrate in the molten region, following it to the end of the rod. The end in which the impurities are collected is cut off and recycled through the purification process starting with formation of $SiCl_4$. The purified top portion of the rod is retained for manufacture of electronic devices.

## Silicates

It is estimated that over 90 percent of the earth's crust consists of **silicates**, if $SiO_2$ is included. The basic structural unit of these silicates consists of a silicon atom surrounded in a tetrahedral fashion by four oxygens, as shown in Figure 23.25. Because the silicon has an oxidation state of +4 and oxygen −2, a simple ion of this composition has a charge of $+4 + 4(-2) = -4$. The simple $SiO_4^{4-}$ ion, which is known as the orthosilicate ion, is found in very few silicate minerals. Usually, silicate tetrahedra share oxygen atoms to build up more complex structures containing Si—O—Si linkages. When two tetrahedra share a single oxygen, the $Si_2O_7^{6-}$ ion shown in Figure 23.26 results.

It is possible to form a chain of $SiO_4$ tetrahedra by sharing oxygens at two corners of each tetrahedron, as shown in Figure 23.27(a). The simplest formula associated with this single-chain silicate anion is $SiO_3^{2-}$. This anion occurs, for example, in the mineral enstatite, $MgSiO_3$, which consists of silicate chains (hence the $SiO_3^{2-}$ portion of the mineral's simplest formula) with $Mg^{2+}$ ions between strands to balance charge.

The other structures shown in Figure 23.27 are also found in silicates. For example, it is possible for each tetrahedron to share three corners, giving rise to sheet structures. Such an arrangement results in a simplest

**FIGURE 23.25**   Structure of the $SiO_4$ tetrahedron of the $SiO_4^{4-}$ ion. This ion is found in several minerals, such as zircon, $ZrSiO_4$.

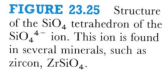

**FIGURE 23.26**
Geometrical structure of the $Si_2O_7^{6-}$ ion, which is formed by the sharing of an oxygen atom by two silicon atoms. This ion occurs in several minerals, such as hardystonite, $Ca_2Zn(Si_2O_7)$.

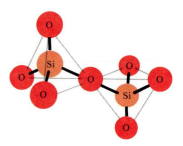

(a) Single–strand chain:

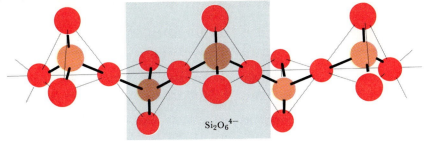

$Si_2O_6^{4-}$

Repeating unit of chain

(b) Double–strand chain:

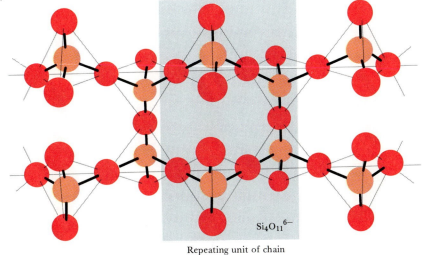

$Si_4O_{11}^{6-}$

Repeating unit of chain

(c) Sheet (or layer) structure:

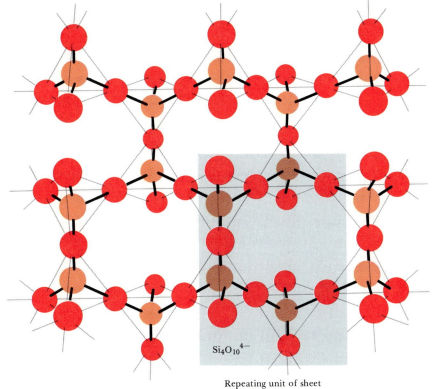

$Si_4O_{10}^{4-}$

Repeating unit of sheet

**FIGURE 23.27** Schematic representations of chain and sheet silicates formed by linking together silicate tetrahedra: (a) single-strand silicate chain, which has an empirical formula of $SiO_3^{2-}$, as in the mineral enstatite, $MgSiO_3$; (b) double-strand silicate chain, which has an empirical formula of $Si_4O_{11}^{6-}$, as in the mineral tremolite, $Ca_2Mg_5(Si_4O_{11})_2(OH)_2$; (c) sheet silicate, which has an empirical formula $Si_2O_5^{2-}$, as in the mineral talc, $Mg_3(Si_2O_5)_2(OH)_2$. Note that in parts (a) and (c) the repeating unit is double the empirical formula.

Asbestos is a general term applied to a group of fibrous silicate minerals. These minerals possess chainlike arrangements of the silicate tetrahedra, or sheet structures in which the sheets are formed into rolls. The result is that the minerals have a fibrous character, as shown in Figure 23.28. Asbestos has been widely used as thermal insulation material, especially in high-temperature applications where the high chemical stability of the silicate structure serves well. However, in recent years there has been a growing awareness that certain forms of asbestos pose a health hazard. Tiny fibers readily penetrate the tissues of the lungs and digestive tract, and they remain there over a long period of time. Eventually, lesions, including cancer, may result.

(a)  (b)

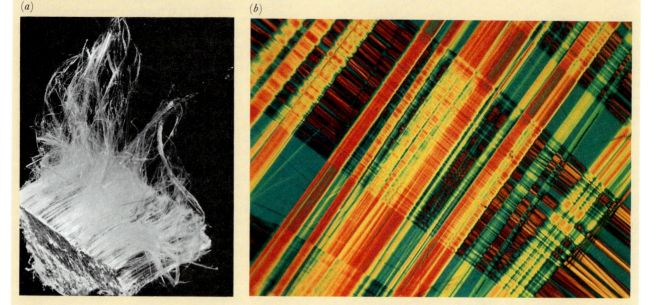

**FIGURE 23.28**  (a) Serpentine asbestos; note the fibrous character of this mineral. (b) Asbestos fibers viewed under a polarizing microscope. (Don Thomson/Photo Researchers)

formula $Si_2O_5{}^{2-}$ as in talc, $Mg_3(Si_2O_5)_2(OH)_2$. When all four oxygens of each $SiO_4$ unit have Si—O—Si linkages, the structure extends in all three dimensions in space, forming quartz, $SiO_2$. Because the structure is locked together in a three-dimensional array much like diamond, quartz is harder than the strand or sheet-type silicates.

**Aluminosilicates**    In many silicate minerals $Si^{4+}$ ions are replaced by $Al^{3+}$ ions within the silicate tetrahedra. This replacement produces **aluminosilicates**. In order to maintain charge balance, an extra cation, such as $K^+$, must accompany each of these substitutions. Muscovite, $KAl_2(AlSi_3O_{10})(OH)_2$,* a mica mineral, is an aluminosilicate.

* Aluminum is found in this mineral in two different environments. The first two $Al^{3+}$ ions are located between the aluminosilicate sheets. The aluminum that is shown in parentheses is located within the sheets. It has replaced a silicon and is therefore located in an $AlO_4$ tetrahedron.

**FIGURE 23.29** Piece of mica, showing its cleavage into thin sheets.

Replacement of a quarter of the silicon atoms in a sheet silicate, $Si_4O_{10}^{4-}$, with aluminum produces the $AlSi_3O_{10}^{5-}$ sheets found in muscovite. In both the silicate-sheet and aluminosilicate-sheet minerals, cations are located between sheets to balance the charge. The electrostatic attraction between these cations and the charges in the sheets is greater for the aluminosilicate than for the corresponding silicate, because there are higher charges in the aluminosilicates. Thus the sheets in the silicate mineral talc slide readily over each other, but the sheets in the aluminosilicate mica do not. Nevertheless, mica does cleave readily into sheets, as illustrated in Figure 23.29.

When aluminum replaces up to half of the silicon atoms in $SiO_2$, the *feldspar* minerals result. Orthoclase, $KAlSi_3O_8$, is a common feldspar mineral. The cations that compensate the extra negative charge accompanying this replacement are usually $Na^+$, $K^+$, or $Ca^{2+}$ ions. The feldspars are the most abundant rock-forming silicates, comprising about 50 percent of the minerals in the earth's crust.

---

**SAMPLE EXERCISE 23.7**

The mineral anorthite is a feldspar mineral formed by replacing half of the silicon atoms in $SiO_2$ with aluminum and maintaining charge balance with $Ca^{2+}$ ions. What is the simplest formula for this mineral?

**Solution:** If we write $SiO_2$ as $Si_2O_4$, the replacement of half of the $Si^{4+}$ with $Al^{3+}$ produces $AlSiO_4^-$. One $Ca^{2+}$ therefore requires two $AlSiO_4^-$ units to maintain charge balance, and the empirical formula of the mineral is $CaAl_2Si_2O_8$.

**PRACTICE EXERCISE** ──────────
The mineral albinite may be considered to form by replacement of one quarter of the Si atoms of $SiO_2$ with Al and maintaining charge balance with $Na^+$ ions. What is the chemical formula of this mineral?
*Answer:* $NaAlSi_3O_8$

---

The **clay minerals** are hydrated aluminosilicates having sheet-type structures. For example kaolinite, $Al_2Si_2O_5(OH)_4$, is a clay mineral. These minerals have small particle size and correspondingly large surface areas. They have the ability to adsorb cations on their surfaces. Often the metal ions displace $H^+$ ions from the OH groups on the surfaces of the clay particle:

$$M^+(aq) + H\!-\!O\!-\!clay \rightleftharpoons H^+(aq) + M\!-\!O\!-\!clay \qquad [23.45]$$

This situation gives rise to pH-dependent equilibria. Notice that the higher the concentration of $H^+(aq)$, the more the equilibrium is shifted to the left. If the soil is basic, the equilibrium lies to the right, and $M^+(aq)$ is not available to plants. Thus the pH of the soil plays an important role in determining the soil's fertility (that is, its ability to supply plants with essential nutrients).

## Glass

Quartz melts at approximately 1600°C, forming a tacky liquid. In the course of melting, many silicon-oxygen bonds are broken. When the liquid is rapidly cooled, silicon-oxygen bonds are reformed before the atoms are able to arrange themselves in a regular fashion. An amorphous solid, known as quartz glass or silica glass, results (see Figure 11.36). Many different substances can be added to $SiO_2$ to cause it to melt at a lower temperature. The common glass used in windows and bottles is known as soda-lime glass. It contains CaO and $Na_2O$ in addition to $SiO_2$ from sand. The CaO and $Na_2O$ are produced by heating two inexpensive chemicals: limestone, $CaCO_3$, and soda ash, $Na_2CO_3$. These carbonates decompose at elevated temperatures, as shown in Equations 23.46 and 23.47.

$$CaCO_3(s) \longrightarrow CaO(s) + CO_2(g) \qquad [23.46]$$

$$Na_2CO_3(s) \longrightarrow Na_2O(s) + CO_2(g) \qquad [23.47]$$

Other substances can be added to soda-lime glass to produce color or to change the properties of the glass in various ways. For example, addition of CoO produces the deep blue color of "cobalt glass." Replacement of $Na_2O$ by $K_2O$ results in a harder glass that has a higher melting point. Replacement of CaO by PbO results in a denser glass with a higher refractive index. Addition of nonmetal oxides, such as $B_2O_3$ or $P_2O_5$, which form network structures related to the silicates, also causes a change in properties of the glass. Addition of $B_2O_3$ results in a glass with a higher melting point and a lower coefficient of thermal expansion. Such glasses, sold commercially under trade names such as Pyrex or Kimex, are used where resistance to thermal shock is important, for example, in laboratory glassware or coffee makers. Table 23.11 lists the formulations and properties of several representative types of glass.

Photochromic glass contains a dispersion of AgCl or AgBr. The glass darkens when exposed to sunlight because the silver halide decomposes in light to form silver and halogen atoms. The finely divided silver is black. The silver and halogen atoms are kept in close proximity by the glass matrix and reform the clear AgCl or AgBr in the dark.

**TABLE 23.11**   Compositions, properties, and uses of various types of glass

| Type of glass | Composition by weight | Properties and uses |
|---|---|---|
| Soda-lime | 12% $Na_2O$, 12% $CaO$, 76% $SiO_2$ | Window glass, bottles |
| Aluminosilicate | 5% $B_2O_3$, 10% $MgO$, 10% $CaO$, 20% $Al_2O_3$, 55% $SiO_2$ | High melting—used in cooking ware |
| Lead alkali | 10% $Na_2O$, 20% $PbO$, 70% $SiO_2$ | High refractive index—used in lenses, decorative glass |
| Borosilicate | 5% $Na_2O$, 3% $CaO$, 16% $B_2O_3$, 76% $SiO_2$ | Low coefficient of thermal expansion—used in laboratory ware, cooking utensils |
| Bioglass | 24% $Na_2O$, 24% $CaO$, 6% $P_2O_5$, 46% $SiO_2$ | Compatible with bone—used as coating on surgical implants |

## 23.6   BORON

At this point there is only one additional element to consider in our survey of nonmetallic elements: boron. Boron is the only element of group 3A that can be considered to be nonmetallic. The element has an extended network structure. The melting point of boron, 2300°C, is intermediate between that for carbon, 3550°C, and that for silicon, 1410°C. The electronic configuration of boron is $[He]2s^2 2p^1$. The element exhibits a valence of 3 in all its common chemical compounds. We have seen (Section 8.7) that the electron configuration about boron in the boron halides provides an exception to the octet rule, in that there are but six electrons in the boron valence shell. The boron halides are thus strong Lewis acids (Section 17.8).

Salts of the borohydride ion, $BH_4^-$, are widely used as reducing agents. This ion can be thought of as an isoelectronic analog of $CH_4$ or $NH_4^+$. The lower charge of the central atom in $BH_4^-$ means that the hydrogens of the $BH_4^-$ are "hydridic," that is, they carry a partial negative charge. Thus it is not surprising that borohydrides are good reducing agents, as illustrated by the reaction in Equation 23.48:

$$3BH_4^-(aq) + 4IO_3^-(aq) \longrightarrow 4I^-(aq) + 3H_2BO_3^-(aq) + 3H_2O(l) \quad \text{[23.48]}$$

The only important oxide of boron is boric oxide, $B_2O_3$. This substance is the anhydride of boric acid, which we may write as $H_3BO_3$ or $B(OH)_3$. Boric acid is so weak an acid that solutions of boric acid are used as an eyewash. Upon warming to 100°C, orthoboric acid loses water by a condensation reaction similar to that described in Section 23.4. The product of the reaction, as shown in Equation 23.49, is metaboric acid, a polymeric substance of formula $HBO_2$:

$$H_3BO_3(s) \longrightarrow HBO_2(s) + H_2O(g) \quad \text{[23.49]}$$

Heating of metaboric acid results in still more loss of water, as in Equations 23.50 and 23.51:

$$4HBO_2(s) \longrightarrow H_2B_4O_7(s) + H_2O(l) \quad \text{[23.50]}$$

$$H_2B_4O_7(s) \longrightarrow 2B_2O_3(s) + H_2O(g) \quad \text{[23.51]}$$

| 5 B |
|---|
| 13 Al |
| 31 Ga |
| 49 In |
| 81 Tl |

The acid $H_2B_4O_7$ is called tetraboric acid. The sodium salt, $Na_2B_4O_7 \cdot 10H_2O$, called borax, occurs in dry lake deposits in California and can also be readily prepared from other borate minerals. Solutions of borax are alkaline; the substance is used in various laundry and cleaning products.

## CHEMISTRY AT WORK

### Optical Fibers

Optical fibers (Figure 23.30) represent a relatively new, high-technology application of glassmaking. An optical fiber is a glass thread that conducts light, just as a copper wire conducts electrons. Optical fibers transmit information, using the light waves that pass through the fiber. Optical fibers are capable of carrying much more information for a given cross section of fiber than can be transmitted via the conventional coaxial cable.

The key to long-distance transmission of signals via optical fibers is glass purity. Impurity ions, such as $Fe^{2+}$, cause absorption of the light waves, with resulting signal loss. Today it is possible to obtain optical fibers that undergo a loss of only about 1 percent of signal over a distance of 1 km. To achieve that level of performance, impurities must be reduced to the level of 1 part per billion. This ultrahigh purity is achieved by distilling high-purity liquid glass to separate it from any remaining impurities.

**FIGURE 23.30**  An array of optical fibers. (John Walsh, Science Photo Library/Photo Researchers)

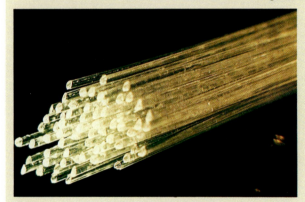

# FOR REVIEW

## SUMMARY

The noble gas elements exhibit a very limited chemical behavior because of the exceptional stability of their electronic configuration. The xenon fluorides and oxides and $KrF_2$ are the best established examples of chemical reactivity among these elements.

The halogens occur as diatomic molecules. These elements possess the highest electronegativities of the elements in each row of the periodic table. All except fluorine exhibit oxidation states varying from $-1$ to $+7$. Fluorine, being the most electronegative element, is restricted to the oxidation states 0 and $-1$. The tendency to form the $-1$ oxidation state from the free element (that is, the oxidizing power of the element) decreases with increasing atomic number in the family. The halogens form compounds with one another, called **interhalogens**. In the higher interhalogens—$XX'_n$—the element X may be Cl, Br, or I, and X' is nearly always F; $n$ may have values 3, 5, or 7.

The group 6A elements range from the very abundant and strongly nonmetallic oxygen to the rare and rather metallic tellurium. Sulfur occurs widely in the form of sulfide ores and in elemental sulfur beds. The element has several allotropic forms; the most stable one consists of $S_8$ rings. The most important compound of the element is sulfuric acid, a strong acid that is a good dehydrating agent and has a high boiling point. Selenium and tellurium are chemically rather similar to sulfur, especially with respect to formation of oxides and oxyanions.

The group 5A elements exhibit a wide range of behavior, from strongly nonmetallic in the case of nitrogen to distinctly metallic in the case of bismuth. Phosphorus occurs in nature in certain phosphate minerals. The element exhibits several allotropes, including one known as white phosphorus, a reactive form consisting of $P_4$ tetrahedra. Phosphorus forms compounds of formula $PX_3$ and $PX_5$ with the halogens. These undergo hydrolysis in water to produce the corresponding oxyacid of phosphorus and HX. Phosphorus forms two oxides, $P_4O_6$ and $P_4O_{10}$. Their corresponding acids, phosphorus acid and phosphoric acid, show a strong tendency to undergo **conden-**

**sation** reactions when heated. Compounds of phosphorus are important components of fertilizers.

The group 4A elements show great diversity in physical and chemical properties. Carbon excels in being able to form multiple bonds and in undergoing **catenation**. Silicon is noteworthy as a semiconductor and for its tendency to form Si—O bonds. Silicon is the second most abundant element, and it occurs in a wide variety of **silicates**. Silicates are composed of $SiO_4$ tetrahedra, which, through sharing oxygen atoms, are able to link together to form chains, sheets, and three-dimensional arrays. In many minerals, $Si^{4+}$ ions are replaced by $Al^{3+}$ ions, thus forming **aluminosilicates**. Silicates are important components of glass.

Boron commonly exhibits an oxidation state of $+3$, as in boric oxide, $B_2O_3$, and boric acid, $H_3BO_3$. The acid readily undergoes condensation reactions.

## LEARNING GOALS

Having read and studied this chapter, you should be able to:

1. Cite the most common occurrences of each nonmetallic element discussed in this chapter (for example, fluorine is found in $CaF_2$).

2. Cite the most common molecular form of each nonmetallic element discussed in the chapter (for example, selenium is found as chains of Se atoms).

3. Explain why xenon forms several compounds with fluorine and oxygen, krypton forms only $KrF_2$, and no chemical reactivity is known for the lighter noble gas elements.

4. Write the formulas of the known fluorides, oxyfluorides, and oxides of xenon, and describe the relative stabilities of the oxides as compared with the fluorides.

5. Describe the electronic and geometrical structures of the known compounds of xenon.

6. Write balanced chemical equations describing at least one means of preparation of each halogen from naturally occurring sources.

7. Describe at least one important use of each halogen element.

8. Write a balanced chemical equation describing the preparation of each of the hydrogen halides.

9. Give examples of diatomic and higher interhalogen compounds, and describe their electronic and geometrical structures.

10. Name the oxyacids or oxyanions of the halogens, given their formula, or vice versa.

11. Describe the variation in acid strength and oxidizing strength of the oxyacids of chlorine.

12. Indicate the formulas of the common oxides of sulfur and the properties of their aqueous solutions.

13. Write balanced chemical equations for formation of sulfuric acid from sulfur, and describe the important properties of the acid.

14. Compare the chemical behaviors of selenium and tellurium with that of sulfur, with respect to common oxidation states and formulas of oxides and oxyacids.

15. Describe the preparation of elemental phosphorus from its ores, using balanced chemical equations.

16. Describe the formulas and structures of the stable halides and oxides of phosphorus.

17. Write balanced chemical equations for the reactions of the halides and oxides of phosphorus with water.

18. Describe a condensation reaction, and give examples using compounds of phosphorus or boron.

19. Describe the structures possible for silicates and their empirical formulas (for example, silicate tetrahedra can combine through bridging oxygens to form a single-string silicate chain whose empirical formula is $SiO_3^{2-}$).

20. Correlate the physical properties of certain silicate minerals, such as asbestos, with their structures.

21. Explain the changes in composition and properties that accompany substitution of $Al^{3+}$ for $Si^{4+}$ in a silicate.

**22.** Describe what is meant by a clay mineral.

**23.** Describe the role of clay minerals in soil fertility and the effects of soil pH.

**24.** Describe the composition of soda-lime glass.

**25.** Describe the manufacture of glass.

## KEY TERMS

Among the more important terms and expressions used for the first time in this chapter are the following:

**Aluminosilicates** (Section 23.5) are compounds that are structurally related to silicates and in which some $Si^{4+}$ ions are replaced by $Al^{3+}$.

**Catenation** (Section 23.5) refers to the linking of like atoms to form chains or rings.

**Clay minerals** (Section 23.5) are hydrated aluminosilicates.

A **condensation reaction** (Section 23.4) is one in which two or more molecules combine to form larger ones by elimination of small molecules, such as $H_2O$.

**Glass** (Section 23.5) is an amorphous solid formed by fusion of $SiO_2$, $CaO$, and $Na_2O$. Other oxides may also be used to form glasses with differing characteristics.

An **interhalogen compound** (Section 23.2) is one formed between two different halogen elements. Examples include IBr and $BrF_3$.

**Silicates** (Section 23.5) are compounds containing silicon and oxygen, structurally based on $SiO_4$ tetrahedra.

## EXERCISES

### The Noble Gases and Halogens

**23.1** Write the chemical formula for each of the following compounds, and indicate the oxidation state of the halogen or noble gas atom in each: (**a**) iodate ion; (**b**) bromic acid; (**c**) bromine trifluoride; (**d**) sodium hypochlorite; (**e**) iodous acid; (**f**) xenon trioxide.

**23.2** Write the chemical formula for each of the following compounds, and indicate the oxidation state of the halogen or noble gas atom in each: (**a**) calcium bromide; (**b**) perchloric acid; (**c**) xenon oxytetrafluoride; (**d**) chlorite ion; (**e**) hypobromous acid; (**f**) iodine pentafluoride

**23.3** Name the following compounds: (**a**) $KClO_3$; (**b**) $Ca(IO_3)_2$; (**c**) $AlCl_3$; (**d**) $HBrO_3$; (**e**) $H_5IO_6$; (**f**) $XeF_4$.

**23.4** Name the following compounds: (**a**) $Fe(ClO_4)_2$; (**b**) $HClO_2$; (**c**) $XeF_2$; (**d**) $IF_5$; (**e**) $XeO_3$; (**f**) HBr (named as an acid).

**23.5** Predict the geometrical structures of the following: (**a**) $I_3^-$; (**b**) $ICl_4^-$; (**c**) $ClO_3^-$; (**d**) $H_5IO_6$; (**e**) $XeF_4$.

**23.6** The interhalogen compound $BrF_3(l)$ reacts with antimony(V) fluoride to form the salt $(BrF_2^+)(SbF_6^-)$. Write the Lewis structure for both the cation and anion in this substance, and describe the likely geometrical structure of each.

**23.7** What are the major factors responsible for the fact that xenon forms stable compounds with fluorine, whereas argon does not?

**23.8** List the halogens in order of increasing X—X halogen bond energies. Suggest a reason for the low F—F bond energy.

**23.9** Explain each of the following observations. (**a**) At room temperature, $I_2$ is a solid, $Br_2$ is a liquid, and $Cl_2$ and $F_2$ are both gases. (**b**) $F_2$ cannot be prepared by electrolytic oxidation of aqueous $F^-$ solutions. (**c**) The boiling point of HF is much higher than those of the other hydrogen halides. (**d**) The halogens decrease in oxidizing power in the order $F_2 > Cl_2 > Br_2 > I_2$.

**23.10** Explain the following observations: (**a**) For a given oxidation state the acid strength of the oxyacid in aqueous solution decreases in the order chlorine > bromine > iodine. (**b**) Hydrofluoric acid cannot be stored in glass bottles. (**c**) HI cannot be prepared by treating NaI with sulfuric acid. (**d**) The interhalogen $ICl_3$ is known but $BrCl_3$ is not. (**e**) $I_2$ is more soluble in aqueous solutions of $I^-$ than in pure water.

**23.11** List one commercial use of each of the halogens.

**23.12** Write a balanced chemical equation for the commercial preparation of each halogen element.

**23.13** Write balanced net ionic equations for the reaction of each of the following substances with water: (**a**) $PBr_5$; (**b**) $IF_5$; (**c**) $SiBr_4$; (**d**) $F_2$; (**e**) $ClO_2$ (chloric acid is a product); (**f**) HI(g).

**23.14** Write a balanced chemical equation that describes a suitable means of preparing each of the following substances: (**a**) HF; (**b**) $I_2$; (**c**) $XeF_4$; (**d**) $Ca(OCl)Cl$; (**e**) $SF_6$; (**f**) NaClO.

**23.15** Write balanced chemical equations for each of the following reactions (some of which are analogous but not identical to reactions shown in this chapter). (**a**) Bromine forms hypobromite ion on addition to aqueous base. (**b**) Chlorine reacts with an aqueous solution of sodium bromide. (**c**) Bromine reacts with an aqueous solution of hydrogen peroxide, liberating $O_2$.

**23.16** Write balanced chemical equations for each of the following reactions (some of which are analogous but not identical to reactions shown in this chapter). (**a**) Hydrogen bromide is produced upon heating calcium bromide with phosphoric acid. **(b)** Hydrogen bromide is formed upon hydrolysis of aluminum bromide. **(c)** Aqueous hydrogen fluoride reacts with solid calcium carbonate, forming water-insoluble calcium fluoride.

**23.17** **(a)** Write the balanced net ionic equation for the reduction of $ClO_3^-$ to $Cl_2$ by $Fe^{2+}$ in acidic aqueous solution. **(b)** Calculate the standard emf for this reaction.

**23.18** Chloride ion is oxidized in aqueous solution to $Cl_2(aq)$ by each of the following reagents: **(a)** $MnO_2(s)$; **(b)** $MnO_4^-(aq)$; **(c)** $Cr_2O_7^{2-}(aq)$. In each case, write a complete, balanced net ionic equation.

**[23.19]** Using the thermochemical data of Table 23.1 and Appendix D, calculate the average Xe—F bond energies in $XeF_2$, $XeF_4$, and $XeF_6$, respectively. What is the significance of the trend in these quantities?

**[23.20]** The solubility of $Cl_2$ in 100 g of water at STP is 310 cm$^3$. Assume that this quantity of $Cl_2$ is dissolved and equilibrated as follows:

$$Cl_2(aq) + H_2O(l) \rightleftharpoons Cl^-(aq) + HClO(aq) + H^+(aq)$$

If the equilibrium constant for this reaction is $4.7 \times 10^{-4}$, calculate the equilibrium concentration of HClO formed.

## Group 6A

**23.21** Write the chemical formula for each of the following compounds, and indicate the oxidation state of the group 6A element in each: **(a)** selenous acid; **(b)** potassium hydrogen sulfite; **(c)** hydrogen telluride; **(d)** carbon disulfide; **(e)** calcium sulfate; **(f)** sodium thiosulfate.

**23.22** Write the chemical formula for each of the following compounds, and indicate the oxidation state of the group 6A element in each: **(a)** selenium trioxide; **(b)** orthotelluric acid; **(c)** zinc selenate; **(d)** sulfur tetrafluoride; **(e)** hydrogen sulfide; **(f)** sulfurous acid.

**23.23** Name each of the following compounds: **(a)** $K_2S_2O_3$; **(b)** $Al_2S_3$; **(c)** $NaHSeO_3$; **(d)** $SeF_6$.

**23.24** Name each of the following compounds: **(a)** $H_2Se$; **(b)** $FeS_2$; **(c)** $NaHSO_4$; **(d)** $Na_2SeO_4$.

**23.25** Write the Lewis structures for each of the following species, and indicate the geometrical structure of each: **(a)** $SeO_4^{2-}$; **(b)** $H_6TeO_6$; **(c)** $TeO_2(g)$; **(d)** $S_2Cl_2$; **(e)** chlorosulfonic acid, $HSO_3Cl$ (chlorine is bonded to sulfur).

**23.26** The $SF_5^-$ ion is formed when $SF_4(g)$ reacts with fluoride salts containing large cations, such as $CsF(s)$. Draw the Lewis structures for $SF_4$ and $SF_5^-$, and predict the molecular structure of each.

**23.27** Write a balanced chemical equation for each of the following reactions. **(a)** Selenium dioxide dissolves in water. **(b)** Solid zinc sulfide reacts with hydrochloric acid. **(c)** Elemental sulfur reacts with sulfite ion to form thiosulfate. **(d)** Elemental selenium is heated with sulfuric acid. **(e)** Sulfur trioxide is dissolved in sulfuric acid. **(f)** Selenous acid is oxidized by hydrogen peroxide.

**23.28** Write balanced chemical equations for each of the following reactions. (You may have to guess at one or more of the reaction products, but you should be able to make a reasonable guess based on your study of this chapter.) **(a)** Selenous acid is reduced by hydrazine in aqueous solution to yield elemental selenium (see Chapter 22 for a discussion of hydrazine). **(b)** Heating orthotelluric acid to temperatures over 200°C yields the acid anhydride. **(c)** Hydrogen selenide can be prepared by reaction of aqueous acid solution on aluminum selenide. **(d)** Sodium thiosulfate is used to remove excess $Cl_2$ from chlorine-bleached fabrics. The thiosulfate ion forms $SO_4^{2-}$ and elemental sulfur while $Cl_2$ is reduced to $Cl^-$.

**23.29** An aqueous solution of $SO_2$ acts as a reducing agent to reduce **(a)** aqueous $KMnO_4$ to $MnSO_4(s)$; **(b)** acidic aqueous $K_2Cr_2O_7$ to aqueous $Cr^{3+}$; **(c)** aqueous $Hg_2(NO_3)_2$ to mercury metal. Write balanced chemical equations for these reactions.

**23.30** In aqueous solution, hydrogen sulfide reduces **(a)** $Fe^{3+}$ to $Fe^{2+}$; **(b)** $Br_2$ to $Br^-$; **(c)** $MnO_4^-$ to $Mn^{2+}$; **(d)** $HNO_3$ to $NO_2$. In all cases, under appropriate conditions, the product is elemental sulfur. Write a balanced net ionic equation for each reaction.

**[23.31]** Suggest an explanation for the fact that telluric acid is a weak acid, whereas sulfuric acid and selenic acids are strong.

**[23.32]** $SF_4$ is very reactive; for example, it reacts readily with water to produce HF and $SO_2$. By contrast, $SF_6$ is stable. It has even been used, with $O_2$, for X-ray examination of the lungs. **(a)** What features of the molecules make attack of sulfur by Lewis bases much more likely for $SF_4$ than for $SF_6$? **(b)** What feature of $SF_4$ makes attack of sulfur by Lewis acids possible? **(c)** Why is $SF_4$ subject to oxidation, whereas $SF_6$ is not?

## Group 5A

**23.33** Write formulas for each of the following compounds, and indicate the oxidation state of the group 5A element in each: **(a)** orthophosphoric acid; **(b)** arsenous acid; **(c)** antimony(III) sulfide; **(d)** calcium dihydrogen phosphate; **(e)** potassium phosphide.

**23.34** Write formulas for each of the following compounds, and indicate the oxidation state of the group 5A element in each: **(a)** phosphorous acid; **(b)** pyrophosphoric acid; **(c)** antimony trichloride; **(d)** magnesium arsenate; **(e)** diphosphorus hexaoxide.

**23.35** Name each of the following compounds: **(a)** $Na_3P$; **(b)** $H_3AsO_4$; **(c)** $P_4O_{10}$; **(d)** $AsF_5$.

**23.36** Name each of the following compounds: **(a)** $K_3As$; **(b)** $PBr_3$; **(c)** $Sb_2O_3$; **(d)** $NaH_2AsO_4$.

**23.37** Phosphorus pentachloride exists in one form in the solid state as an ionic lattice of $PCl_4^+$ and $PCl_6^-$ ions. **(a)** Draw the Lewis structures of these ions and predict their geometries. **(b)** What set of hybrid orbitals is employed by phosphorus in each case? **(c)** Why should

PCl$_5$ exist as an ionic substance in the solid state, whereas it is stable as the neutral molecule in the gas phase?

**23.38** Sodium trimetaphosphate (Na$_3$P$_3$O$_9$) and sodium tetrametaphosphate (Na$_4$P$_4$O$_{12}$) are used as water-softening agents. They contain cyclic P$_3$O$_9^{3-}$ and P$_4$O$_{12}^{4-}$ ions, respectively. Propose reasonable structures for these ions.

**23.39** Account for the following observations (a) H$_3$PO$_3$ is a diprotic acid. (b) Nitric acid is a strong acid, whereas phosphoric acid is weak. (c) Phosphate rock is not effective as a phosphate fertilizer. (d) Phosphorus does not exist at room temperature as diatomic molecules, but nitrogen does.

**23.40** Account for the following observations. (a) Phosphorus forms a pentachloride, but nitrogen does not. (b) H$_3$PO$_2$ is a monoprotic acid. (c) Phosphonium salts, such as PH$_4$Cl, can be formed under anhydrous conditions, but they can't be made in aqueous solution. (d) Whereas PCl$_3$ hydrolyzes readily in water to form H$_3$PO$_3$, SbCl$_3$ hydrolyzes only in part, forming SbOCl.

**23.41** Write balanced chemical equations for each of the following reactions: (a) preparation of white phosphorus from calcium phosphate; (b) hydrolysis of PCl$_3$; (c) preparation of PCl$_3$ from P$_4$; (d) reaction of P$_4$O$_{10}$ with water.

**23.42** Write balanced chemical equations for each of the following reactions: (a) Preparation of PF$_5$ from P$_4$O$_{10}$; (b) reaction of As$_2$O$_3$ with water; (c) dehydration of orthophosphoric acid to form pyrophosphoric acid; (d) reaction of elemental arsenic with dilute nitric acid to form NO and arsenous acid, H$_3$AsO$_3$.

## Group 4A and Boron

**23.43** Write the formulas for each of the following compounds, and indicate the oxidation state of the group 4A element or of boron in each: (a) silicon dioxide; (b) germanium tetrachloride; (c) sodium borohydride; (d) stannous chloride.

**23.44** Write the formulas for each of the following compounds, and indicate the oxidation state of the group 4A element or of boron in each: (a) boric acid; (b) silicon tetrabromide; (c) lead chloride; (d) sodium tetraborate decahydrate (borax).

**23.45** Covalent silicon-hydrogen compounds are called silanes. The silane Si$_2$H$_6$, known as disilane, exists, but no Si$_2$H$_4$ and Si$_2$H$_2$ compounds are known. In contrast, carbon forms C$_2$H$_6$, C$_2$H$_4$, and C$_2$H$_2$. Explain why silicon doesn't form Si$_2$H$_4$ and Si$_2$H$_2$.

**23.46** Suggest some reasons why carbon is more suitable than silicon as the major structural element in living systems.

**23.47** Select the member of group 4A that best fits each of the following descriptions: (a) forms the most acidic oxide; (b) is most commonly found in the +2 oxidation state; (c) is a component of sand.

**23.48** Select the member of group 4A that best fits each of the following descriptions: (a) catenates to the great-

est extent; (b) forms the most basic oxide; (c) is a metalloid that can form +2 ions.

**23.49** Both GeCl$_4$ and SiCl$_4$ fume in moist air because of hydrolysis to GeO$_2$ and SiO$_2$. Write balanced equations for these reactions.

**23.50** Germanium differs markedly from silicon in that the +2 halides are fairly stable. These halides can be prepared by the reduction of the tetrahalide with germanium metal. (a) Write the balanced chemical equation for the formation of GeCl$_2$. (b) Predict the geometrical structure of the gaseous GeCl$_2$ molecule.

**23.51** The mineral orthoclase is a feldspar mineral formed by replacing a quarter of the Si$^{4+}$ ions in SiO$_2$ with Al$^{3+}$ and maintaining charge balance with K$^+$ ions. What is the empirical formula for this mineral?

**23.52** How is the mica mineral KMg$_3$(AlSi$_3$O$_{10}$)(OH)$_2$ related structurally to the mica mineral muscovite mentioned in the text?

**23.53** What empirical formula and unit charge are associated with each of the following structural types: (a) isolated SiO$_4$ tetrahedra; (b) a chain structure of SiO$_4$ tetrahedra joined at corners to adjacent units; (c) a structure consisting of tetrahedra joined at corners to form a six-membered ring of alternating Si and O atoms?

**23.54** Propose a reasonable description of the structure of each of the following minerals: (a) albite, NaAlSi$_3$O$_8$; (b) leucite, KAlSi$_2$O$_6$; (c) zircon, ZrSiO$_4$; (d) sphene, CaTiSiO$_5$.

## Additional Exercises

**23.55** Name each of the following compounds: (a) H$_2$B$_4$O$_7$; (b) SiC; (c) HPO$_3$; (d) XeF$_2$; (e) Na$_2$S; (f) KClO$_3$.

**23.56** Explain each of the following observations. (a) H$_2$S is a better reducing agent than H$_2$O. (b) H$_2$SO$_4$ is a stronger acid than H$_2$SeO$_4$. (c) Astatine is generally not considered in any detail in discussion of halogens. (d) White phosphorus is quite volatile, whereas red phosphorus is not. (e) Xenon hexafluoride is a stable compound, whereas krypton hexafluoride is unknown. (f) Addition of SF$_4$ to water results in an acidic solution. (g) Silicate-sheet minerals are softer than aluminosilicate-sheet ones.

**23.57** Xenon trioxide disproportionates in strongly alkaline solution to form the thermally stable perxenate ion, XeO$_6^{4-}$. Predict the geometry of this ion. Describe the bonding in terms of the hybridization of xenon valence-shell orbitals.

**23.58** What pressure of gas is formed when 0.654 g of XeO$_3$ decomposes completely to the free elements at 48°C in a 0.452-L volume?

**23.59** Write balanced chemical equations to account for the following observations. (There may not be closely similar reactions shown in the chapter; however, you should be able to make reasonable guesses at the likely products.) (a) When burning sodium metal is immersed in a pure HCl atmosphere, it continues to burn. (b) Bubbling $SO_2$ gas through liquid bromine that is covered with a layer of water results in formation of a strongly acidic solution; upon distillation, an aqueous HBr solution is collected. The remaining liquid is still strongly acidic. **(c)** When bromine is added to a basic solution containing potassium hypochlorite, insoluble potassium bromate is formed. **(d)** When bromic acid is reacted with $SO_2$, $Br_2(aq)$ is formed. **(e)** Uranium(VI) fluoride is formed by the action of $ClF_3$ on uranium(IV) chloride.

**23.60** The cyano group behaves in some ways like a halogen. Thus cyanogen gas, $(CN)_2$, has been called a pseudohalogen and the cyanide ion, $CN^-$, a pseudohalide. Cyanogen reacts with an aqueous solution of NaOH in a fashion analogous to $Cl_2$. **(a)** Write a balanced chemical equation for this reaction. **(b)** Write the Lewis structure for $(CN)_2$, and describe its geometrical structure.

**23.61** Calculate the molarity of an aqueous hydrofluoric acid solution that is 50.0 percent HF by weight and has a density of 1.155 $g/cm^3$.

**23.62** What structural feature do the molecules $P_4$, $P_4O_6$, and $P_4O_{10}$ have in common? What is the common structural feature of all of the acids containing phosphorus(V)?

**23.63** Compare the first ionization energies of phosphorus and sulfur, and account for the relative magnitudes in terms of the electronic structures of the atoms.

**23.64** Elemental sulfur is capable of reacting under suitable conditions with Fe, $F_2$, $O_2$, or $H_2$. Write balanced chemical equations to describe the reaction in each case. In which reactions is sulfur acting as a reducing agent and in which as an oxidizing agent?

**23.65** Sodium sulfide is used in the leather industry to remove hair from hides. It can be synthesized by reducing sodium sulfate with carbon. (Carbon monoxide forms together with the sodium sulfide.) Write a balanced chemical equation for this reaction.

**23.66** Sulfur and the group 6A elements below it in the periodic chart exhibit oxidation states ranging from $-2$ to $+6$. What factors control the lowest and highest oxidation states? Explain.

**23.67** Draw the Lewis structure for the following species and predict the relative S—O bond lengths in each: $SO_2$, $SO_3$, $SO_4^{2-}$.

**23.68** One method proposed for removal of $SO_2$ from the flue gases of power plants involves reaction with aqueous $H_2S$. Elemental sulfur is the product. **(a)** Write a balanced chemical equation for the reaction. **(b)** What volume of $H_2S$ at 27°C and 740 mm Hg would be required to remove the $SO_2$ formed by burning 1.0 ton of coal containing 3.5 percent S by weight? **(c)** What mass of elemental sulfur is produced? Assume that all reactions are 100 percent efficient.

**23.69** Although $H_2Se$ is toxic, no deaths have been attributed to it. One reason is its vile odor, which serves as a sensitive warning of its presence. In addition, $H_2Se$ is readily oxidized by $O_2$ in air to nontoxic elemental red selenium before harmful amounts of $H_2Se$ can enter the body. **(a)** Write a balanced chemical equation for this oxidation. **(b)** Calculate $\Delta G°$ and the equilibrium constant (at 298 K) for the reaction.

**23.70** A sulfuric-acid plant produces a considerable amount of heat. This heat is used to generate electricity, which helps reduce operating costs. The synthesis of $H_2SO_4$ consists of three main chemical processes: (1) oxidation of S to $SO_2$; (2) oxidation of $SO_2$ to $SO_3$; (3) the dissolving of $SO_3$ in $H_2SO_4$ and its reaction with water to form $H_2SO_4$. If the third process produces 130 kJ/mol, how much heat is produced in preparing a mole of $H_2SO_4$ from a mole of S? How much heat is produced in preparing a ton of $H_2SO_4$?

**23.71** Ultrapure germanium, like silicon, is used in semiconductors. Germanium of "ordinary" purity is prepared by the high-temperature reduction of $GeO_2$ with carbon. The Ge is converted to $GeCl_4$ by treatment with $Cl_2$ and then purified by distillation; $GeCl_4$ is then hydrolyzed in water to $GeO_2$ and reduced to the elemental form with $H_2$. The element is then zone-refined. Write a balanced chemical equation for each of the chemical transformations in the course of forming ultrapure Ge from $GeO_2$.

**23.72** Describe the fundamental structural unit present in all silicate minerals. How is this unit modified in aluminosilicates?

**23.73** Draw the Lewis structure for the cyclic $Si_3O_9^{6-}$ ion.

**[23.74]** The maximum allowable concentration of $H_2S(g)$ in air is 20 mg per kilogram of air (20 ppm by weight). How many grams of FeS would be required to react with hydrochloric acid to produce this concentration in an average room measuring 2.7 m × 4.3 m × 4.3 m?

**[23.75]** The standard heats of formation of $H_2O(g)$, $H_2S(g)$, $H_2Se(g)$, and $H_2Te(g)$ are $-241.8$, $-20.17$, $+29.7$, and $99.6$ kJ/mol, respectively. The enthalpies necessary to convert the elements in their standard states to one mole of gaseous atoms are 248, 277, 227, and 197 kJ/mol of atoms for O, S, Se, and Te, respectively. The enthalpy for dissociation of $H_2$ is 436 kJ/mol. Calculate the average H—O, H—S, H—Se, and H—Te bond energies and comment on their trend.

**[23.76]** **(a)** Calculate the P-to-P distance in both $P_4O_6$ and $P_4O_{10}$ from the following data: the P—O—P bond angle for $P_4O_6$ is 127.5°, while that for $P_4O_{10}$ is 124.5°. The P—O distance (to bridging oxygens) is 0.165 nm in $P_4O_6$ and 0.160 nm in $P_4O_{10}$. **(b)** Rationalize the relative P-to-P distances in the two compounds.

**[23.77]** The N—X bond distances in the nitrosyl halides, NOX, are 1.52, 1.98, and 2.14 Å for NOF, NOCl, and NOBr, respectively. Compare the distances with the atomic radii for the halogens (Table 23.3). Is the variation in N—X distance what you expect from these covalent radii? If not, account for the deviations.

**[23.78]** Considering the chemical formulas and structures of clay minerals, suggest an explanation for why they can be molded and then become hard and brittle when heated.

**[23.79]** Boron nitride has a graphitelike structure with B—N bond distances of 1.45 Å within sheets and a separation of 3.30 Å between sheets. At high temperatures the BN assumes a diamondlike form that is harder than diamond. Rationalize the similarity between BN and elemental carbon.

CHAPTER *24*

# Metals
# and Metallurgy

In Chapters 22 and 23 we examined the chemistry of nonmetallic elements. In this chapter and the next two we turn our attention to metals. Metals have played a major role in the development of civilization. Early history is often divided into the Stone Age, Bronze Age, and Iron Age, based on the material used in each era for making tools. Modern society relies on many different metals for making the vast variety of tools, machines, instruments, and other items that we encounter every day.

In this chapter we will consider the chemical forms in which the metallic elements occur in nature and the means by which we obtain metals from these sources. We will also examine the bonding in solids and see how metals and mixtures of metals, called alloys, are employed in modern technology. Finally, we will look at some of the properties and uses of an important group of substances called intermetallic compounds. Much of this subject matter can be described as **metallurgy**, the science and technology of extracting metals from their natural sources and preparing them for practical use.

The portion of our environment that is the solid earth beneath our feet is called the **lithosphere**. The lithosphere provides us with most of the materials with which we feed, clothe, shelter, support, and entertain ourselves. Although the bulk of the earth is solid, we have access to only a small region near the surface. The deepest well ever drilled is only about 8 km deep, and the deepest mine extends between 3 and 4 km into the earth. In comparison, the earth has a radius of 6370 km.

Many of the substances most useful to us are not especially abundant in that portion of the lithosphere to which we have ready access. It is interesting to compare the order of abundance of the elements in the lithosphere with the estimated order of their global consumption. The twelve elements listed in Table 24.1 compose 99.5 percent of the lithosphere by mass. The most widely used elements are listed in Table 24.2. The ranking of elements in the two lists shows little correlation. Some elements that are not abundant are widely used (chromium and copper, for example). Consequently, the occurrence and distribution of *concentrated* deposits of some of these elements often play a role in international politics.

Deposits that contain metals in economically exploitable quantities are known as **ores**. Usually, the compounds or elements that we desire must be separated from a large quantity of unwanted material and then chemically processed to render them useful. By processing large quantities of substances we have literally changed the surface of our planet (see Figure 24.1). Experts estimate that about $2.3 \times 10^4$ kg (25 tons) of materials are extracted from the lithosphere and processed annually to support each

## 24.1 OCCURRENCE AND DISTRIBUTION OF METALS

**TABLE 24.1**  The twelve most abundant elements in the lithosphere

| Element | Percent by weight |
|---|---|
| Oxygen | 50 |
| Silicon | 26 |
| Aluminum | 7.5 |
| Iron | 4.7 |
| Calcium | 3.4 |
| Sodium | 2.6 |
| Potassium | 2.4 |
| Magnesium | 1.9 |
| Hydrogen | 0.9 |
| Titanium | 0.6 |
| Chlorine | 0.2 |
| Phosphorus | 0.1 |

Steel is poured from a basic oxygen-argon furnace. (M. E. Warren/Photo Researchers)

**TABLE 24.2**  Estimated annual world consumption of elements[a]

| Element | Annual consumption (kg) |
| --- | --- |
| C | $10^{12}-10^{13}$ |
| Na, Fe | $10^{11}-10^{12}$ |
| N, O, S, K, Ca | $10^{10}-10^{11}$ |
| H, F, Mg, Al, P, Cl, Cr, Mn, Cu, Zn, Ba, Pb | $10^{9}-10^{10}$ |
| B, Ti, Ni, Zr, Sn | $10^{8}-10^{9}$ |
| Ar, Co, As, Mo, Sb, W, U | $10^{7}-10^{8}$ |
| Li, V, Se, Sr, Nb, Ag, Cd, I, rare earths, Au, Hg, Bi | $10^{6}-10^{7}$ |
| He, Be, Te, Ta | $10^{5}-10^{6}$ |

[a] Elements in color also appear in Table 24.1.

**FIGURE 24.1**  Large open-pit mining operation. (Georg Gerster/Photo Researchers)

person in our country. Because the richest sources of many substances are becoming exhausted, it will be necessary in the future to process larger volumes and lower-quality raw materials. Thus the extraction of the compounds and elements we need will cost more in terms of energy and environmental impact.

**Minerals**

Most elements occur in nature in solid inorganic compounds called **minerals**. Table 24.3 lists several common minerals. Some of these are pictured in Figure 24.2. Minerals are usually known by their common names rather than by their chemical names. They are often named after a place, person, or some characteristic, such as color.

Commercially, the most important sources of metals are oxide, sulfide, and carbonate minerals. Silicate minerals (Section 23.5) are very abundant, but they are generally hard to concentrate and reduce. Thus most silicates are not economical sources of metals.

A few metals occur in nature in their elemental form. Examples include silver, gold, platinum, ruthenium, and iridium. These metals, from

**TABLE 24.3**  Some common minerals

| Mineral name | Chemical formula |
| --- | --- |
| Calcite | $CaCO_3$ |
| Chalcopyrite | $CuFeS_2$ |
| Cinnabar | $HgS$ |
| Corundum | $Al_2O_3$ |
| Fluorite | $CaF_2$ |
| Galena | $PbS$ |
| Gypsum | $CaSO_4 \cdot 2H_2O$ |
| Halite | $NaCl$ |
| Hematite | $Fe_2O_3$ |
| Malachite | $Cu_2(CO_3)(OH)_2$ |
| Perovskite | $CaTiO_3$ |
| Pyrite | $FeS_2$ |
| Quartz | $SiO_2$ |
| Talc | $Mg_3(Si_4O_{10})(OH)_2$ |
| Turquoise | $CuAl_6(PO_4)_4(OH)_2 \cdot 4H_2O$ |
| Wulfenite | $PbMoO_4$ |

**FIGURE 24.2**  Four common minerals: (*a*) cinnabar, (*b*) malachite, (*c*) quartz, (*d*) fluorite. Note the variety of colors and shapes. [(a,d) George Whiteley/Photo Researchers; (b) Tom McHugh/Photo Researchers; (c) Thomas R. Taylor/Photo Researchers]

(*a*)

(*b*)

(*c*)

(*d*)

groups 8B and 1B of the periodic table, are known as **noble metals**, because of their lack of reactivity. They have high electrode potentials and are thus difficult to oxidize. For example:

$$\text{Ag}^+(aq) + \text{e}^- \longrightarrow \text{Ag}(s) \qquad E^\circ = 0.799 \text{ V} \qquad [24.1]$$

$$\text{Ir}^{3+}(aq) + 3\text{e}^- \longrightarrow \text{Ir}(s) \qquad E^\circ = 1.15 \text{ V} \qquad [24.2]$$

## 24.2 EXTRACTIVE METALLURGY

**Extractive metallurgy** is the science and technology concerned with all the steps taken in obtaining a metal from its ore. A great many different processes, both physical and chemical, may be used. The first steps are generally physical processes that increase the concentration of the desired mineral and prepare it for subsequent operations. The concentrated ore is then subjected to chemical processes that may further concentrate it and that eventually lead to reduction to the free metal.

### Physical Concentration Processes

An ore is first crushed and ground. Then the desired mineral is sorted from the undesired material, called **gangue** (pronounced "gang"), by taking advantage of some difference in the properties of the mineral and its associated gangue. For example, differences in density could be used. The denser material would settle more rapidly when the finely ground ore is stirred in water or some other fluid medium. Also, certain minerals are magnetic; that is, particles of the mineral are attracted to a magnet. The magnetic material can be separated from the nonmagnetic gangue by moving finely ground ore on a conveyor belt past a series of magnets.

**Flotation** is a very important method for concentrating ores, particularly sulfides. A suspension of ground ore in water is agitated while air is blown through, as shown in Figure 24.3. Certain chemicals are added so that a froth or foam is created. Particles of the desired mineral become attached to the air bubbles and are floated out with the froth, which is

**FIGURE 24.3** Schematic diagram of froth flotation, by means of which a mineral can be separated from a larger amount of gangue.

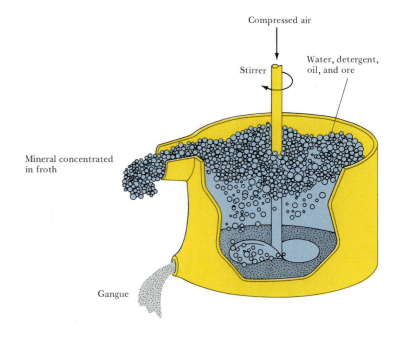

skimmed off the top. At the same time, the gangue settles to the bottom of the tank. Flotation methods work only if the surfaces of the desired mineral particles are hydrophobic (not wetted by water). Chemicals are usually added that selectively adsorb on the surface of the desired mineral particles to reduce wetting by water. The molecules of such substances usually contain a polar end that binds to the mineral surface and a nonpolar end that protrudes into the water. The mode of action of flotation agents is similar to that described in Section 12.7 for soaps and detergents. One of the most successful flotation agents is potassium ethyl xanthate (pronounced "zanthate"):

$$K^+\left[S{=}C\overset{\displaystyle O{-}C_2H_5}{\underset{\displaystyle S^-}{\diagup}}\right]$$

The negative end of this molecule attaches to the surface of a sulfide mineral, such as $Cu_2S$, while the nonpolar organic end of the molecule extends away from the surface to inhibit the approach of water molecules.

Once all the appropriate physical methods have been employed, the ore is ready for further enrichment in the desired metal by chemical means. These chemical processes may conveniently be considered under three headings: pyrometallurgy, hydrometallurgy, and electrometallurgy.

## 24.3 PYROMETALLURGY

In **pyrometallurgy**, heat converts the desired mineral in an ore from one chemical to another, and eventually to the free metal. We can group these chemical reactions according to the type of products formed. **Calcination** is the heating of an ore to bring about its decomposition and the elimination of a volatile product. The volatile product could, for example, be water or $CO_2$. Carbonate ores are frequently calcined to drive off $CO_2$, forming the metal oxide. For example,

$$PbCO_3(s) \longrightarrow PbO(s) + CO_2(g) \qquad [24.3]$$

Most carbonates decompose reasonably rapidly at temperatures in the range of 400 to 500°C, although $CaCO_3$ requires a temperature of about 1000°C. Most hydrated minerals lose water at temperatures on the order of 100 to 300°C.

**Roasting** is a thermal treatment that causes chemical reactions between the ore and the furnace atmosphere. Roasting may produce oxidation or reduction and may be accompanied by calcination. An important roasting process is the oxidation of sulfide ores, in which the metal is converted to the oxide, as in these examples:

$$2ZnS(s) + 3O_2(g) \longrightarrow 2ZnO(s) + 2SO_2(g) \qquad [24.4]$$
$$2MoS_2(s) + 7O_2(g) \longrightarrow 2MoO_3(s) + 4SO_2(g) \qquad [24.5]$$

In such processes, the metal oxide formed is often relatively volatile at the temperature of the roast and sublimes from the furnace.

Depending on the conditions of the roasting process, the metal may form a sulfate rather than an oxide. For example,

$$CoS(s) + 2O_2(g) \longrightarrow CoSO_4(s) \qquad [24.6]$$

Formation of the sulfate can sometimes be used to advantage; because sulfates are generally water soluble, the material containing the desired metal can be leached away from insoluble residue.

The ore of a less active metal, such as mercury, can be roasted to the free metal:

$$HgS(s) + O_2(g) \longrightarrow Hg(g) + SO_2(g) \qquad [24.7]$$

The free metal can also be produced when a reducing atmosphere is present during the roast. Carbon monoxide, CO, provides such an atmosphere:

$$PbO(s) + CO(g) \longrightarrow Pb(l) + CO_2(g) \qquad [24.8]$$

Roasting is not always a feasible method of reduction. Some metals that are difficult to obtain as the free metal by roasting are best converted to the metal halide, which can then be reduced. To obtain the metal halide, the metal oxide or another compound, such as metal carbide, is roasted with an atmosphere of the halogen, usually chlorine. For example,

$$TiC(s) + 4Cl_2(g) \longrightarrow TiCl_4(g) + CCl_4(g) \qquad [24.9]$$

**Smelting** is a melting process in which the materials formed in the course of chemical reactions separate into two or more layers. Smelting often involves a roasting stage in the same furnace. Two of the important types of layers formed in smelters are molten metal and slag. The molten metal many consist almost entirely of a single metal, or it may be a solution of two or more metals.

**Slag** consists mainly of molten silicate minerals, with aluminates, phosphates, fluorides, and other ionic compounds as constituents. A slag is formed when a metal oxide such as CaO racts with molten silica, $SiO_2$. Slags are classified as acidic or basic, according to the acidity or basicity of the oxides added to the silica to form the slag. In this way

## A CLOSER LOOK          Acid-Base Reactions in Molten Oxides

The chemical changes that occur when oxides are melted together can best be understood as acid-base reactions. However, since there is no water or other proton-donating solvent present, we cannot think of these reactions in terms of the Brønsted acid-base theory. Rather, we must view them in terms of the Lewis acid-base theory. Recall that in the Lewis theory, discussed in Section 17.8, a base is an electron-pair donor, and an acid is an electron-pair acceptor. In the molten oxides the electron-pair donor is the oxide ion, $O^{2-}$. A base is a substance that furnishes oxide ions to the medium. In contrast, an acid is a substance that can react with an oxide ion by forming a bond with one of its unshared electron pairs. The oxides of highly charged metal or nonmetal ions are acidic because the central ion is capable of forming a polar covalent bond with the oxide ion. In the oxides of the alkali metal or alkaline earth metals, the oxide ion is readily available to donate an electron pair because the bond to the metal ion is essentially ionic in character; thus these oxides are basic.

of looking at such reactions, the most basic oxides are those of the alkali and alkaline earth metal ions. The most acidic oxides are those of the nonmetals, (for example, $SiO_2$ and $P_2O_5$). Somewhat less acidic are the oxides of the more highly charged metal ions, such as $TiO_2$, $Fe_2O_3$, and $Al_2O_3$. Thus a slag containing mainly CaO is basic in character. A slag containing mainly $SiO_2$, or $SiO_2$ with a nonmetallic oxide such as $P_2O_5$, is acidic in character. The reaction of a basic oxide such as CaO with an acidic oxide such as $SiO_2$ can be thought of as a reaction of an acid with a base to form a "salt" as a reaction product:

$$CaO(l) + SiO_2(l) \longrightarrow CaSiO_3(l) \qquad [24.10]$$

We can also think of such reactions as occurring between ions in the molten mixture:

$$Ca^{2+}(l) + O^{2-}(l) + SiO_2(l) \longrightarrow Ca^{2+}(l) + SiO_3^{2-}(l)$$

We will see several applications of these ideas in the sections on metallurgy that follow.

---

**SAMPLE EXERCISE 24.1**

Predict the product of the reaction between $Na_2O$ and $TiO_2$. Write a balanced chemical solution for the reaction. Which oxide is acting as base? Which as acid?

**Solution:** The salt that is formed in this reaction is likely to be $Na_2TiO_3$, in which $Na^+$ and $TiO_3^{2-}$ ions are present. Another possibility would be $Na_4TiO_4$; indeed, some compounds containing the $TiO_4$ unit are known (for example, $Mg_2TiO_4$ and $Zn_2TiO_4$).

The balanced equation for the reaction is

$$Na_2O(l) + TiO_2(s) \longrightarrow Na_2TiO_3(l)$$

In this reaction $Na_2O$ acts as the base, and $TiO_2$ acts as the acid. The equation could be written in ionic form as

$$2Na^+(l) + O^{2-}(l) + TiO_2(l) \longrightarrow$$
$$2Na^+(l) + TiO_3^{2-}(l)$$

**PRACTICE EXERCISE**
Predict the product of the reaction of $P_2O_5$ and CaO in the molten phase.
*Answer:* $Ca_3(PO_4)_2$

---

**Refining** is the treatment of a crude, relatively impure metal product from a metallurgical process to improve its purity and to better define its composition. Sometimes the goal of the refining process is the metal itself in as pure a form as possible. However, the goal may also be the production of a mixture with well-defined composition, as in the production of steels from crude iron.

The major sources of iron are ores rich in either hematite, $Fe_2O_3$, or magnetite, $Fe_3O_4$. Presently the most actively mined ores contain mainly $Fe_3O_4$ and silica, $SiO_2$. A good-quality ore may contain 30 to 40 percent iron. The ore is enriched by grinding it to a fine particle size so that the $Fe_3O_4$ can be separated from the gangue by a magnetic separation process. The concentrate from this separation step has an iron content of about 60 to 65 percent, and contains 2 to 8 percent silica. The ore is then formed into pellets 6 to 25 mm in diameter, a size suitable as feed for the blast furnace.

**The Pyrometallurgy of Iron**

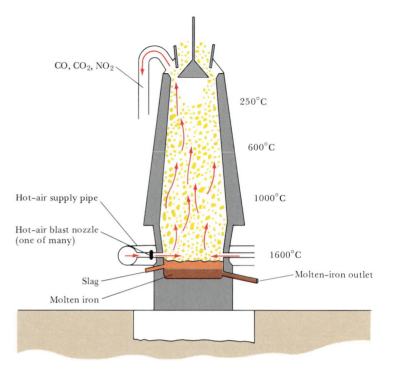

CO, CO₂, NO₂

250°C

600°C

Hot–air supply pipe

1000°C

Hot–air blast nozzle
(one of many)

1600°C

Molten–iron outlet

Slag

Molten iron

**FIGURE 24.4** Blast furnace used for reduction of iron ore. Notice the approximate temperatures in the various regions of the furnace.

Reduction of the iron ore occurs in a blast furnace, illustrated in Figure 24.4. A blast furnace is essentially a huge chemical reactor, capable of continuous operation. The largest furnaces are over 60 m high and 14 m wide. When operating at full capacity they produce up to 10,000 tons of iron per day. The blast furnace is charged at the top with a mixture of iron ore, coke, and limestone. Coke is coal that has been heated in the absence of air to drive off volatile components. It is about 85 to 90 percent carbon. Coke serves as the fuel, producing heat as it is burned in the lower part of the furnace. It is also the source of the reducing gases, CO and $H_2$. Limestone, $CaCO_3$, serves as the source of basic oxide in slag formation. Air, which enters the blast furnace at the bottom after preheating, is also an important raw material; it is required for combustion of the coke. Production of 1 kg of crude iron (called **pig iron**) requires about 2 kg of ore, 1 kg of coke, 0.3 kg of limestone, and 1.5 kg of air.

Coke is burned mainly in the lower part of the furnace, where the temperatures are highest, exceeding 1600°C. At this temperature $CO_2$ reacts with the coke to form CO:

$$C(s) + CO_2(g) \longrightarrow 2CO(g) \qquad [24.11]$$

Thus the overall reaction in the combustion of coke is

$$2C(s) + O_2(g) \longrightarrow 2CO(g) \qquad \Delta H = -110 \text{ kJ} \qquad [24.12]$$

Water vapor present in the air also reacts with carbon:

$$C(s) + H_2O(g) \longrightarrow CO(g) + H_2(g) \qquad \Delta H = +113 \text{ kJ} \qquad [24.13]$$

Note that the reaction with oxygen is exothermic and provides heat for furnace operation, but the reaction with water vapor is endothermic. Addition of water to the air thus provides a means of controlling furnace temperature.

In the upper part of the furnace limestone is calcined (Equation 22.68). Here also the iron oxides are reduced by CO and $H_2$. For example, the important reactions for $Fe_3O_4$ are

$$Fe_3O_4(s) + 4CO(g) \longrightarrow 3Fe(s) + 4CO_2(g) \qquad \Delta H = -19 \text{ kJ} \qquad [24.14]$$

$$Fe_3O_4(s) + 4H_2(g) \longrightarrow 3Fe(s) + 4H_2O(g) \qquad \Delta H = 149 \text{ kJ} \qquad [24.15]$$

Reduction of other elements present in the ore also occurs in the hottest parts of the furnace, where carbon is the major reducing agent. Among the most important of these other elements are manganese, silicon, and phosphorus:

$$MnO(s) + C(s) \longrightarrow Mn(l) + CO(g) \qquad [24.16]$$

$$SiO_2(l) + 2C(s) \longrightarrow Si(l) + 2CO(g) \qquad [24.17]$$

$$P_2O_5(l) + 5C(s) \longrightarrow 2P(l) + 5CO(g) \qquad [24.18]$$

Molten iron collects at the base of the furnace, as shown in Figure 24.4. It is overlaid with a layer of molten slag, formed by the reaction of CaO with the silica present in the ore, as described by Equation 24.10. The layer of slag over the molten iron helps to protect it from reaction with the incoming air. Periodically, the furnace is tapped to drain off slag and molten iron. The iron produced in the furnace may be cast into solid ingots. However, most is used directly in the manufacture of steel. For this purpose it is transported, while still liquid, to the steelmaking shop.

**Refining of Iron; Formation of Steel**

Iron is refined in a vessel called a **converter**, which has a capacity of about 200 tons. In a typical process the converter is charged with about 75 tons of scrap steel and CaO to form a basic slag, then filled with molten iron fresh from the blast furnace. This iron from the blast furnace contains 0.6 to 1.2 percent silicon, about 0.2 percent phosphorus, 0.4 to 2.0 percent manganese, and about 0.03 percent sulfur. In addition, there is considerable dissolved carbon. All of these impurity elements are removed by oxidation. In modern steelmaking the oxiding agent is pure $O_2$, or $O_2$ diluted with argon. The presence of $N_2$ is avoided, because the incorporation of nitrogen in steel causes brittleness.

A cross-sectional view of one converter design appears in Figure 24.5. In this design $O_2$, diluted with argon, is blown directly into the molten metal. The oxygen reacts exothermally with carbon, silicon, and impurity metals, reducing the concentrations of these elements in the iron. Carbon and sulfur are expelled as CO and $SO_2$, respectively. Silicon is oxidized to $SiO_2$ and adds to whatever slag may have been present initially in the melt. Metal oxides react with the $SiO_2$ to form silicates. The presence of a basic slag is important for removal of phosphorus:

$$3CaO(l) + P_2O_5(l) \longrightarrow Ca_3(PO_4)_2(l) \qquad [24.19]$$

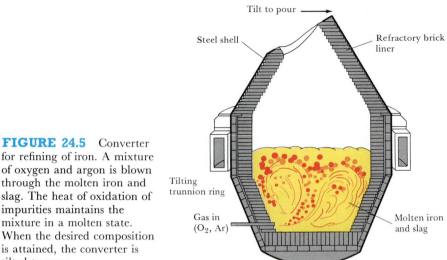

**FIGURE 24.5** Converter for refining of iron. A mixture of oxygen and argon is blown through the molten iron and slag. The heat of oxidation of impurities maintains the mixture in a molten state. When the desired composition is attained, the converter is tilted to pour.

Nearly all of the $O_2$ blown into the converter is consumed in the oxidation reactions. By monitoring the $O_2$ concentration in the gas coming from the converter it is possible to tell when the oxidation is essentially complete. It normally requires only about 20 minutes for oxidation of the impurities present in the iron. At this point a sample of the molten metal is withdrawn and subjected to analysis for all of the elements of importance in determining the quality of the steel produced: carbon, manganese, phosphorus, sulfur, silicon, nickel, chromium, copper, and other metallic elements that might be present. Depending on the results, additional oxygen may be blown in or additional ore might be added. When the desired composition is attained, the contents of the converter are dumped into a large ladle (Figure 24.6). As the ladle is being filled,

**FIGURE 24.6** Molten iron being poured into a converter (also called a basic oxygen furnace) where it is converted to steel. A typical converter is charged with 200 tons of molten pig iron, 100 tons of scrap iron, and 20 tons of limestone (to form slag). Oxygen gas is blown into the molten mixture to oxidize impurities. (Samsung America, Inc.)

alloying elements are added. These added elements produce steels with various kinds of properties. (We will have more to say about alloys in Section 24.7.) The still-molten mixture is then poured into molds for solidification.

**24.4 HYDRO-METALLURGY**

Until now we have considered only high-temperature metallurgical reactions that characterize pyrometallurgy. However, many metals of great economic importance are produced in large quantities by hydrometallurgical techniques. **Hydrometallurgy** is the selective separation of a mineral or group of minerals from the rest of the ore by an aqueous chemical process. The most important hydrometallurgical process is **leaching**, in which the desired mineral is selectively dissolved.

**Hydrometallurgy of Gold**

In the recovery of gold from low-grade ores, the ores are finely ground, then treated with an aerated, aqueous solution of NaCN. Because of formation of the stable, water-soluble cyano complex, gold is oxidized:

$$4Au(s) + 8CN^-(aq) + O_2(aq) + 2H_2O(l) \longrightarrow 4Au(CN)_2^-(aq) + 4OH^-(aq) \qquad [24.20]$$

Following this leaching operation the solution is treated with CaO to cause the precipitation of hydroxides of several metallic species that might be present and to promote settling of finely divided material. The solution is filtered; then deaerated by vacuum pumping to remove $O_2$. Next Zn dust is added to precipitate Au:

$$2Au(CN)_2^-(aq) + Zn(s) \longrightarrow Zn(CN)_4^{2-}(aq) + 2Au(s) \qquad [24.21]$$

A slimy mixture of Zn and Au is recovered by filtration. Zinc is removed from this mixture by heating in air; ZnO forms and is sublimed away. The crude gold that remains is smelted with a charge of borax ($Na_4B_4O_7 \cdot 10H_2O$) and silica ($SiO_2$). The slag that forms helps to remove any remaining metal oxides.

This example illustrates how a mineral, in this case the native metal, is leached from the gangue by taking advantage of the fact that a very stable complex is formed [$K_f$ for $Au(CN)_2^-$ is $2 \times 10^{38}$]. Equilibria in metal complexes play a very important role in hydrometallurgy. The second reaction in the process involves an oxidation-reduction reaction in which the more active and less valuable zinc replaces gold. Hydrometallurgical reactions are characterized by this two-reaction sequence: dissolution, then reprecipitation. The dissolution step might be an oxidation, as in the example we have just considered, but other types of reaction, such as metathesis or complex formation, might also be employed. Similarly, the reprecipitation process might be a reduction, but formation of an insoluble salt is also a common process. We can best illustrate these points by considering the hydrometallurgy of some additional metals.

**Hydrometallurgy of Aluminum**

Among metals, aluminum is second only to iron in commercial use. World production of the metal is about $1.5 \times 10^{10}$ kg (16 million tons) per year. The most useful ore of aluminum is *bauxite*, in which Al is present as hydrated oxides, $Al_2O_3 \cdot xH_2O$. The value of $x$ varies, depending on the particular mineral present. Large bauxite deposits are found in Guiana, Australia, Jamaica, Mexico, and Brazil. Because bauxite deposits in the United States are very limited, the nation imports most of the ore used in production of aluminum.

The major impurities found in bauxite are silica, $SiO_2$, and iron(III) oxide, $Fe_2O_3$. It is essential to separate alumina, $Al_2O_3$, from these impurities before the metal is recovered by electrochemical reduction, as described in Section 24.5. The process used to purify bauxite, called the **Bayer process**, is a hydrometallurgical procedure. The ore is first crushed and ground, then digested in a concentrated aqueous NaOH solution, about 30 percent NaOH by weight, at a temperature in the range 150 to 230°C. Sufficient pressure, between 4 and 30 atm, is maintained to prevent boiling. The trihydrate, $Al_2O_3 \cdot 3H_2O$, dissolves more readily than the monohydrate, $Al_2O_3 \cdot H_2O$. The reaction in either case leads to a complex aluminate anion of the form $Al(H_2O)_2(OH)_4^-$ (see Section 18.5). For example,

$$Al_2O_3 \cdot H_2O(s) + 6H_2O(l) + 2OH^- \longrightarrow 2Al(H_2O)_2(OH)_4^-(aq) \qquad [24.22]$$

Silica dissolves in the strongly basic medium, then forms insoluble aluminosilicate salts with the aluminate ion. These settle out with a red "mud" consisting mostly of iron(III) oxides, which are not soluble in the strong base. Notice that this procedure works because $Al^{3+}$ is amphoteric, but $Fe^{3+}$ is not. After filtration, the solution is diluted to reduce the concentration of hydroxide. In effect, this causes a partial reversal of the equilibrium shown in Equation 24.22, although the product is not the original mineral, but a highly hydrated aluminum hydroxide.

After the aluminum hydroxide precipitate has been filtered, it is calcined in preparation for the electroreduction of the ore to the metal. The solution recovered from the filtration must be reconcentrated so that it can be used again. This is accomplished by heating to evaporate water from the solution. The energy requirements of this evaporative stage are high, and it is the most costly part of the process.

## Hydrometallurgy of Copper

Most copper is obtained from chalcopyrite, $CuFeS_2$, using pyrometallurgical methods. However, the pyrometallurgical procedures do have distinct disadvantages. The necessity for conducting reactions at high temperatures means that the energy costs for the process are high. Also, the roasting and smelting of sulfide ores generate substantial air pollution, because of the large quantities of waste slag and $SO_2$ produced. For each kilogram of Cu produced, about 1.5 kg of iron slag and 2 kg of $SO_2$ are also formed. To absorb all this $SO_2$ in the form of $CaSO_3$ or $CaSO_4$, in the manner described in Section 14.4, is impractical because of the enormous quantities of material involved. Hydrometallurgical procedures offer the possibility of avoiding some of these waste material problems.

One promising hydrometallurgical approach to recovery of copper is the aqueous oxidation of the chalcopyrite ore following concentration by froth flotation methods. When a slurry of the ore in aqueous sulfuric acid is agitated with oxygen, oxidation of the sulfide occurs, with dissolution of copper:

$$2CuFeS_2(s) + 2H^+(aq) + 4O_2(g) \longrightarrow$$
$$2Cu^{2+}(aq) + SO_4{}^{2-}(aq) + Fe_2O_3(s) + 3S(s) + H_2O(l) \qquad [24.23]$$

The resulting solution can be electrolyzed to recover the copper. The solution that remains from the electrolysis is simply aqueous sulfuric acid; it is recycled back to the leaching step with fresh ore.

## 24.5 ELECTROMETALLURGY

The term **electrometallurgy** refers to the use of electrolysis methods to obtain free metal from one of its compounds or to purify a crude form of the metal. The principles of electrolysis were discussed in Section 20.5, which you should review at this time if necessary. Sample Exercise 24.2 is illustrative of what you need to recall.

**SAMPLE EXERCISE 24.2**

Write the electrode reactions for electrolysis of the solution that results from the process described in Equation 24.23.

**Solution:** After filtering to remove solids formed in the reaction, the solution contains just aqueous copper sulfate, with any excess acid that may be present. Upon electrolysis, $Cu^{2+}$ ion is preferentially reduced at the cathode, because $E°$ for reduction of $Cu^{2+}$ is $+0.34$ V. At the anode, water is oxidized. The two half-reactions are:

$$Cu^{2+}(aq) + 2e^- \longrightarrow Cu(s)$$
$$\underline{H_2O(l) \longrightarrow 2H^+(aq) + \tfrac{1}{2}O_2(g) + 2e^-}$$
$$Cu^{2+}(aq) + H_2O(l) \longrightarrow Cu(s) + 2H^+(aq) + \tfrac{1}{2}O_2(g)$$

The overall effect is that copper is replaced by $H^+$, thus forming sulfuric acid.

**PRACTICE EXERCISE**

Based on electrode potentials (Appendix F), which of the following metal ions is most readily reduced to the free metal: $Sn^{2+}$, $Ni^{2+}$, $Cr^{3+}$, $Pb^{2+}$?
*Answer:* $Pb^{2+}$

Electrometallurgical procedures can be broadly differentiated according to whether they involve electrolysis of a molten salt or of an aqueous solution. Electrolytic methods are very important as a means of obtaining the more active metals, such as sodium, magnesium, and aluminum, in the free state. Such active metals could not be reduced in aqueous medium, because the solvent itself is reduced at a lower voltage. The standard potentials for reduction of water under acidic and basic conditions are both more positive than the standard potentials for reductions of such active metals as Na ($E° = -2.71$ V), Mg ($E° = -2.37$ V), or Al ($E° = -1.66$ V):

$$2H^+(aq) + 2e^- \longrightarrow H_2(g) \qquad E° = 0.00 \text{ V} \qquad [24.24]$$
$$H_2O(l) + 2e^- \longrightarrow H_2(g) + 2OH^-(aq) \qquad E° = -0.83 \text{ V} \qquad [24.25]$$

To form such metals by electrochemical reduction, therefore, we must employ a molten salt medium, in which the metal ion of interest is the most readily reduced species.

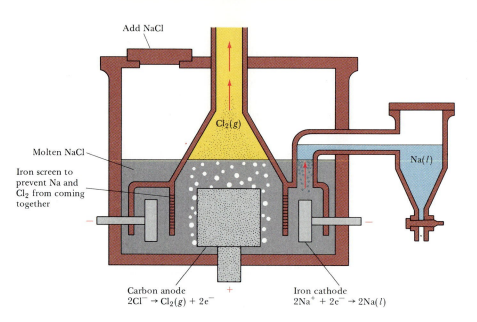

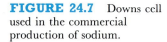

**FIGURE 24.7** Downs cell used in the commercial production of sodium.

Add NaCl

$Cl_2(g)$

Molten NaCl

Iron screen to prevent Na and $Cl_2$ from coming together

Na($l$)

Carbon anode
$2Cl^- \rightarrow Cl_2(g) + 2e^-$

Iron cathode
$2Na^+ + 2e^- \rightarrow 2Na(l)$

## Electrometallurgy of Sodium

In the commercial preparation of sodium, molten NaCl is electrolyzed in a specially designed cell, called the **Downs cell**, illustrated in Figure 24.7. The electrolyte medium through which current flows is molten NaCl. Calcium chloride, $CaCl_2$, is added to lower the melting point of the cell medium from the normal boiling point of NaCl, 804°C, to around 600°C. The Na($l$) and $Cl_2(g)$ produced in the electrolysis are kept from coming in contact and reforming NaCl. In addition, the Na must be kept from contact with oxygen, since the metal would quickly oxidize under the high-temperature conditions of the cell reaction.

## Electrometallurgy of Aluminum

In Section 24.4 we discussed the Bayer process, in which bauxite or another ore of aluminum is concentrated to produce a relatively pure hydrous aluminum hydroxide. When this concentrate is calcined at several hundred degrees Celsius, a partially hydrated aluminum oxide, $Al_2O_3 \cdot xH_2O$, called alumina, is formed. At temperatures in excess of 1000°C, anhydrous aluminum oxide is formed. Anhydrous aluminum oxide melts at over 2000°C. This is too high to permit its use as a molten medium for electrolytic formation of free aluminum. The electrolytic process used com-

**FIGURE 24.8** Typical Hall-process electrolysis cell used to reduce aluminum. Because molten aluminum is more dense than the molten mixture of $Na_3AlF_6$ and $Al_2O_3$, the metal collects at the bottom of the cell.

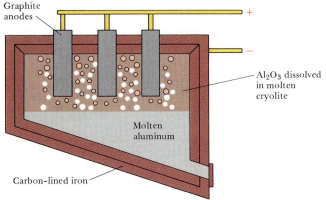

Graphite anodes

$Al_2O_3$ dissolved in molten cryolite

Molten aluminum

Carbon-lined iron

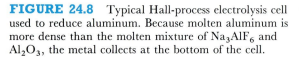

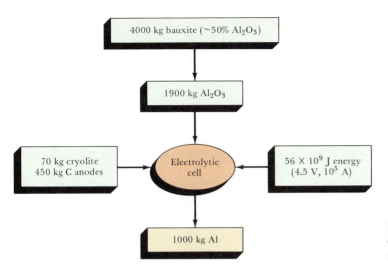

**FIGURE 24.9** The quantities of bauxite, cryolite, graphite, and energy required to produce 1000 kg of aluminum.

mercially to produce aluminum is known as the **Hall process**, named after its inventor, Charles M. Hall (see A Historical Perspective, next page). The purified $Al_2O_3$ is dissolved in molten cryolite, $Na_3AlF_6$, which has a melting point of 1012°C and is an effective conductor of electric current. A schematic diagram of the electrolysis cell is shown in Figure 24.8. Graphite rods are employed as anodes and are consumed in the electrolysis process. The electrode reactions are:

Anode: $\quad C(s) + 2O^{2-}(l) \longrightarrow CO_2(g) + 4e^-$ [24.26]

Cathode: $\quad 3e^- + Al^{3+}(l) \longrightarrow Al(l)$ [24.27]

The amounts of raw materials and energy required to produce 1000 kg of aluminum metal from bauxite by this procedure are summarized in Figure 24.9.

The crude copper produced from $CuFeS_2$ or $Cu_2S$ by pyrometallurgical methods is not sufficiently free of impurities to serve in electrical applications. Further purification is achieved by the electrochemical method, illustrated in Figure 24.10. Crude copper is the anode in a cell in which an acidic solution of copper sulfate is the electrolyte medium. Pure copper sheets are used as cathodes. Application of a suitable voltage across such

**Electrometallurgy of Copper**

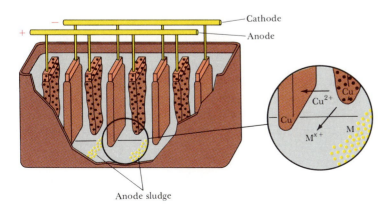

**FIGURE 24.10** Electrolysis cell for refining of copper. Notice that as the anodes dissolve away, the cathodes on which the pure metal is deposited grow in size.

Charles M. Hall (Figure 24.11) began work on the problem of reducing aluminum in about 1885, after he had learned from a professor of the difficulty of reducing ores of very active metals. Prior to the development of his electrolytic process, aluminum was obtained by a chemical reduction using sodium or potassium as reducing agent. The procedure was very costly; as late as 1852 the cost of aluminum was $545 per pound. During the Paris exposition in 1855, aluminum was exhibited as a rare metal in spite of the fact that it is the third most abundant element in the earth's crust. Hall, who was 21 years old when he began his research, utilized handmade and borrowed equipment in his studies, and his laboratory was a woodshed near his home. In about a year's time he was able to solve the problem. The solution consisted of finding an ionic compound that could be melted to form a conducting medium that would dissolve $Al_2O_3$ and that would not interfere in the electrolysis reactions. The relatively rare mineral cryolite, $Na_3AlF_6$, found in Greenland, met these criteria. Ironically, Paul Héroult, who was the same age as Hall, made the same discovery in France at about the same time. As a result of the discovery of Hall and Héroult, large-scale production of aluminum became commercially feasible, and aluminum became a common and familiar metal.

**FIGURE 24.11**  Charles M. Hall (1863–1914) as a young man. (Courtesy of ALCOA)

a cell results in oxidation of copper metal at the anode and reduction of $Cu^{2+}$ at the cathode to form copper metal. These reactions occur because copper is oxidized more readily than water:

$$Cu(s) \longrightarrow Cu^{2+}(aq) + 2e^- \qquad E° = -0.34 \text{ V} \qquad [24.28]$$

$$2H_2O(l) \longrightarrow 4H^+(aq) + O_2(g) + 4e^- \qquad E° = -1.23 \text{ V} \qquad [24.29]$$

Similarly, $Cu^{2+}$ is reduced at the cathode in preference to $H^+$. The crude copper employed as anode material in the electrolytic cells contains many impurities, including lead, zinc, nickel, arsenic, selenium, tellurium, and several precious metals, such as gold, silver, and platinum. Metallic impurities that are more active than copper are readily oxidized at the anode but do not plate out at the cathode, because their reduction potentials are more negative than that for copper. However, the less active metals are not oxidized at the anode. Instead, they collect below the anode as a sludge that is collected and processed to recover the valuable metals. The anode sludges from copper refining cells provide one-fourth of U.S. silver production and about one-eighth of U.S. gold production.

## SAMPLE EXERCISE 24.3

Nickel is one of the chief impurities in the crude copper that is subjected to electrorefining. What happens to this nickel in the course of the electrolytic process?

**Solution:** The standard potential for oxidation of nickel is more positive than that for copper:

$$Ni(s) \longrightarrow Ni^{2+}(aq) + 2e^- \qquad E° = 0.28 \text{ V}$$
$$Cu(s) \longrightarrow Cu^{2+}(aq) + 2e^- \qquad E° = -0.34 \text{ V}$$

Nickel will thus be more readily oxidized than copper at the anode, assuming standard conditions. Of course, we do not have standard conditions in the electrolytic cell. The crude copper anode is nearly all copper, so we can assume that the activity, or effective concentration, of copper in the anode is essentially the same as that for pure metal. However, the nickel is present as a highly dilute impurity in the copper. Thus the activity, or effective concentration, of nickel in the anode is much lower than it would be if we had a pure nickel anode. Nonetheless, the nickel is sufficiently more electropositive than copper to preferentially undergo oxidation at the anode. The reduction of $Ni^{2+}$, which is just the reverse of the oxidation process, occurs less readily than the reduction of $Cu^{2+}$. The $Ni^{2+}$ thus accumulates in the electrolyte solution while the $Cu^{2+}$ is reduced at the cathode. After a time it is necessary to recycle the electrolyte solution to remove the accumulated metal ion impurities, such as $Ni^{2+}$.

### PRACTICE EXERCISE

Zinc is another common impurity in copper. Using electrode potentials, determine whether zinc will accumulate in the anode sludge or in the electrolyte solution during the electrorefining of copper.
*Answer:* electrolyte solution

---

Electrochemical methods such as those described for copper are also employed in the purification of the commercially important metals, zinc, cobalt, and nickel, even though the electrode potentials for reduction of these metals in aqueous solution are negative. We might expect that reduction of hydrogen ions would occur at the cathode in preference to reduction of the metal ion, especially when the solution is acidic. In practice, very little hydrogen is formed at the cathode, because of the phenomenon of *overvoltage*, described in Section 20.5. The overvoltage for formation of $H_2$ at the cathode when the electrode is the purified metal is generally on the order of several tenths of a volt. Thus formation of $H_2$ requires a voltage several tenths of a volt above the calculated value. By contrast, the overvoltage for deposition of the metallic elements is very low. By careful control of the voltage applied to the cell, it is possible to minimize gas formation. It is important that this be done, because the hydrogen produced at the cathode where the pure metal is being deposited could adversely affect the quality of the metal and because any reduction or oxidation of solvent represents wasted energy.

## 24.6 METALLIC BONDING

In our discussions of metallurgy we have so far confined ourselves to discussing the methods employed for obtaining metals in pure form. Metallurgy is also concerned with understanding the properties of metals and alloys, and with the development of useful new materials. As with any branch of science and engineering, our ability to make advances is coupled to our understanding of the fundamental properties of the systems with which we work. At several places in the text we have referred to the differences between metals and nonmetals with regard to both physical and chemical behavior. Let us now consider the distinctive properties of metals and then relate these properties to a model for metallic bonding.

## Properties of Metals

You have no doubt at some time held in your hand a length of copper wire or an iron bolt. Perhaps you have had occasion to observe the surface of a freshly cut piece of sodium metal. These substances, although distinct from one another, share certain similarities that make it easy to classify them as metallic. A fresh metal surface has a characteristic luster. In addition, we know that those metals that we can handle with bare hands have a characteristic cold, metallic feeling, related to their high heat conductivity. Metals also have high electrical conductivities; when a voltage is applied across a length of metal, current flows. The current flow occurs without any displacement of atoms within the metal structure and is due to the flow of electrons within the metal. Current flow in metals is different from current flow in an aqueous ionic solution or molten salt, in which ions rather than electrons move under the influence of the electrical potential. The heat conductivity of a metal usually parallels its electrical conductivity. For example, silver and copper, which possess the highest electrical conductivities, also possess the highest heat conductivities. This observation suggests that the two types of conductivity have the same origin in metals.

Most metals are *malleable*, which means that they can be hammered into thin sheets, and *ductile*, which means that they can be drawn into wires. These properties indicate that the atoms of the metallic lattice are capable of slipping with respect to one another. Ionic solids or crystals of most covalent compounds do not exhibit such behavior. These types of solids are typically brittle and fracture easily along certain planes. Consider, for example, the difference between dropping an ice cube and a block of aluminum metal on a concrete floor.

The metals form solid structures in which the atoms are arranged as close-packed spheres or in a similar packing arrangement. For example, copper possesses a close-packing arrangement called cubic close packing (Section 11.5), in which each copper atom is in contact with 12 other copper atoms. The number of valence-shell electrons available for bond formation is insufficient for a copper atom to form a localized electron-pair bond to each of its neighbors. As another example, magnesium has only two valence electrons, yet each Mg atom in the metal is surrounded by 12 other Mg atoms. If each atom is to share its bonding electrons with all of its neighbors, these electrons must be able to move from one bonding region to another.

## Electron-Sea Model for Metallic Bonding

One very simple model that accounts for some of the most important characteristics of metals is the *electron-sea model*. In this model the metal is pictured as an array of metal cations in a "sea" of electrons, as illustrated in Figure 24.12. The electrons are confined to the metal by electrostatic attractions to the cations, and they are uniformly distributed throughout the structure. However, the electrons are mobile, and no individual electron is confined to any particular metal ion. When a metal wire is connected to the terminals of a battery, electrons migrate through the metal toward the positive terminal, while electrons flow into the metal from the battery at the negative terminal. The high heat conductivity of metals is also accounted for by the mobility of the electrons, which permits ready transfer of kinetic energy throughout the solid. The ability of metals to deform (their malleability and ductility) can be explained by the fact that

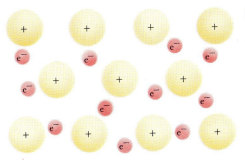

**FIGURE 24.12** Schematic illustration of the electron-sea model for the electronic structure of metals.

the metal ions can move without any specific bonds being broken; the change in the positions of the ions brought about in reshaping the metal is accommodated by a redistribution of electrons.

The electron-sea model accounts in a qualitative way for some properties of metals, such as their conductivity and malleability, but it does not adequately account for others, such as melting points, heats of atomization, and hardness. To understand these properties, we must consider the number of electrons involved in metallic bonding. In the electron-sea model, we expect the strength of bonding to increase as the number of bonding (valence) electrons increases.

If you look back at Figure 5.5 (Section 5.3), you will see that the melting points of the transition metals tend to increase with increasing atomic number up to group 6B, near the center of the transition metals, and then to decrease beyond that point. Curves for heats of fusion, hardness, boiling points, and heats of atomization have similar appearances, although the maximum point in each period is often reached in group 8B. These results imply that the strength of metallic bonding first increases with increasing numbers of electrons, then decreases. This trend can be understood by applying the concepts of molecular orbital theory to metals.

**Molecular-Orbital Model for Metals**

In considering the structures of molecules such as benzene (Section 9.4), we saw that in some cases electrons are delocalized, or distributed, over several atoms. This happens when the atomic orbitals on one atom are able to interact with atomic orbitals on several neighboring atoms. In graphite (Section 9.7) the electrons are delocalized over an entire plane. It is useful to think of the bonding in metals in a similar way. The valence atomic orbitals on one metal atom overlap with those on the several nearest neighbors, which in turn overlap with the atomic orbitals on still other atoms.

We saw in Section 9.5 that overlap of atomic orbitals can lead to formation of molecular orbitals. The number of molecular orbitals is equal to the number of atomic orbitals that overlap. In a metal the number of atomic orbitals that interact or overlap is very large. Thus the number of molecular orbitals is also very large. Figure 24.13 shows schematically what happens as increasing numbers of metals' atoms come together to form molecular orbitals. As overlap of atomic orbitals occurs, bonding and antibonding molecular-orbital combinations are formed. The energies of these molecular orbitals lie at closely spaced intervals in the energy range between the highest- and lowest-energy orbitals. Thus the interaction of all the valence atomic orbitals of each metal atom with the orbitals of

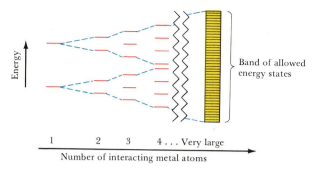

**FIGURE 24.13** Schematic illustration of the interactions of atomic metal orbitals to form the delocalized orbitals of the metal lattice. The two atomic orbitals on each metal atom in this example might represent an *s* orbital and *p* orbital. Whatever the orbitals, a large number of molecular orbitals with closely spaced energy levels arise. The number of electrons available does not completely fill these orbitals.

adjacent metal atoms gives rise to a huge number of very closely spaced molecular orbitals that extend over the entire metal structure. The energy separations between these metal orbitals are so tiny that for all practical purposes we may think of the orbitals as forming a continuous *band* of allowed energy states, as shown in Figure 24.13.

The electrons available for metallic bonding do not completely fill the available molecular orbitals; we can think of the energy band as a partially filled container for electrons. The incomplete filling of the energy band gives rise to characteristic metallic properties. The electrons in orbitals near the top of the occupied levels require very little energy input to be "promoted" to still higher energy orbitals, which are unoccupied. Under the influence of any source of excitation, such as an applied electrical potential or an input of thermal energy, electrons move into previously vacant levels and are thus freed to move through the lattice, giving rise to electrical and thermal conductivity.

Trends in properties such as the melting point can readily be explained in terms of the molecular-orbital model. In transition metals the valence-shell *s*, *p*, and *d* orbitals may overlap. If they do, the resultant energy band can accommodate up to $2 + 6 + 10 = 18$ electrons per atom. The lower-energy half of the band, which contributes to the net bonding in the metal, can hold nine of these electrons. The higher-energy half of the energy band, which is the antibonding portion of the band, can accommodate the other nine electrons. Thus if the *s*, *p*, and *d* orbitals interact, nine electrons per atom gives the maximum bond order and thus the highest bond strength. If only the *s* and *d* orbitals overlap, the maximum bond order would correspond to the presence of six electrons per atom. This analysis agrees well with what we observe. The highest melting points, heats of atomization, hardness, and density in each transition-metal series are usually encountered with metals possessing six to nine valence electrons per atom (groups 6B to 8B). Of course, factors other than the number of electrons (such as atomic radius, nuclear charge, and the particular packing structure of the metal) also play a role in determining the properties of metals.

The molecular-orbital model (or band theory, as it is also called) is not so different in some respects from the electron-sea model. In both models the electrons are free to move about in the solid. However, the molecular-orbital model is more quantitative than the simple electron-sea model. Although we cannot go into the details here, many properties of metals can be accounted for by quantum-mechanical calculations using molecular-orbital theory.

A solid exhibits metallic character because it has a partially filled energy band, as shown in Figure 24.14(a); there are more molecular orbitals in the band than are needed to accommodate all of the bonding electrons of the structure. Thus an excited electron may easily move to the nearby higher orbital. In some solids, however, the electrons completely fill the allowed levels in a band. Such a situation applies, for example, to diamond. When we apply molecular-orbital theory to this solid, we find that the bands of allowed energies are as shown in Figure 24.14(b). The carbon 2s and 2p atomic orbitals combine to form two energy bands. One of these is completely filled with electrons. The other is completely empty. There is a large energy gap between the two bands. Because there is no readily available vacant orbital into which the highest-energy electrons can move under the influence of an applied electrical potential, diamond is not a good electrical conductor. Solids in which the energy bands are either completely filled or completely empty are electrical *insulators*.

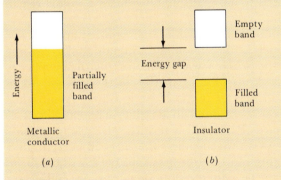

**FIGURE 24.14** Metallic conductors have partly filled energy bands, as shown in (*a*). Insulators have filled or empty energy bands, as in (*b*).

Silicon and germanium have electronic structures like diamond. However, the energy gap between the filled and empty bands becomes smaller as we move from carbon (diamond) to silicon to germanium, as seen in Table 24.4. For silicon and germanium, the energy gap is small enough that at ordinary temperatures, a few electrons have sufficient energy to jump from the filled band (called the valence band) to the empty band (called the conduction band). As a result, there are some empty orbitals in the valence band, permitting electrical conductivity; the electrons in the upper energy band also serve as carriers of electrical

**TABLE 24.4**  Energy gaps in diamond, silicon, and germanium

| Element | Energy gap (kJ/mol) |
|---------|---------------------|
| C | 502 |
| Si | 100 |
| Ge | 67 |

current. Thus silicon and germanium are *semiconductors*, solids with electrical conductivities between those of metals and those of insulators. Other substances, for example GaAs, also behave as semiconductors.

The electrical conductivity of a semiconductor or insulator can be modified by adding small amounts of other substances. This process, called *doping*, causes the solid to have either too few or too many electrons to fill the valence band. Consider what happens to silicon when a small amount of phosphorus or another element from group 5A is added. The phosphorus atoms substitute for silicon atoms at random sites in the structure. Phosphorus, however, possesses five valence-shell electrons per atom, as compared to four for silicon. There is no room for these extra electrons in the valence band. They must therefore occupy the conduction band, as illustrated in Figure 24.15. These higher-energy electrons have access to many vacant orbitals within the energy band they occupy and serve as carriers of electrical current. Silicon doped with phosphorus in this manner is called an *n-type* semiconductor, because this doping introduces extra *negative* charges (electrons) into the system.

If the silicon is doped instead with a group 3A element, such as gallium, the atoms that substitute for silicon have one electron too few to meet their bonding requirements to the four neighboring silicon atoms. The valence band is thus not completely filled, as illustrated in Figure 24.15(c). Under the influence of an applied field, electrons can move from occupied molecular orbitals to the few that are vacant in the valence band. A semiconductor that is formed by doping silicon with a group 3A element is called a *p-type* semiconductor, because this doping creates electron vacancies that can be thought of as *positive* holes in the system. The modern electronics industry is based on integrated circuitry formed from silicon or germanium doped with various elements to create the desired electronic characteristics. (Figure 24.16)

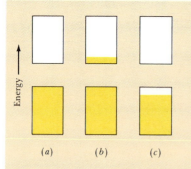

**FIGURE 24.15** Effect of doping on the occupancy of the allowed energy levels in silicon. (*a*) Pure silicon. The valence-shell electrons just fill the lower-energy allowed energy levels. (*b*) Doped with phosphorus. Excess electrons occupy the lowest-energy orbitals in the higher-energy band of allowed energies. These electrons are capable of conducting current. (*c*) Silicon doped with gallium. There are not quite enough electrons to fully occupy the orbitals of the lower-energy allowed band. The presence of vacant orbitals in this band permits current flow.

**FIGURE 24.16** Semiconductors permit tremendous miniaturization of electronic devices. A single computer memory chip may contain over a million transistors. (Courtesy of Hewlett Packard)

## 24.7 ALLOYS

An **alloy** is a material that contains more than one element and has the characteristic properties of metals. The alloying of metals is of great importance because it is one of the primary ways of modifying the properties of the pure metallic elements. For example, nearly all the uses we make of iron involve alloy compositions of one sort or another. As another example, pure gold is too soft to be used in jewelry, whereas alloys of gold and copper are quite hard. Pure gold is termed 24 karat; the common alloy used in jewelry is 14 karat, meaning that it is 58 percent gold ($\frac{14}{24} \times 100$). A gold alloy of this composition has suitable hardness to be used in jewelry. The alloy can be either yellow or white, depending on the elements added. Some further examples of alloys are given in Table 24.5.

Alloys can be classified as solution alloys, heterogeneous alloys, and intermetallic compounds. **Solution alloys** are homogeneous mixtures with the components dispersed randomly and uniformly. Atoms of the solute can take positions normally occupied by a solvent atom, thereby forming a **substitutional alloy**, or they can occupy interstitial positions, thereby forming an **interstitial alloy**. These types are diagrammed in Figure 24.17.

Substitutional alloys are formed when the two metallic components have similar atomic radii and chemical-bonding characteristics. For example, silver and gold form such an alloy over the entire range of possible compositions. When two metals differ in radii by more than about 15 percent, solubility is more limited. For an interstitial alloy to form, the component present in the interstitial positions between the solvent atoms

**TABLE 24.5**   Some common alloys

| Primary element | Name of alloy | Composition by weight | Properties | Uses |
|---|---|---|---|---|
| Bismuth | Wood's metal | 50% Bi, 25% Pb, 12.5% Sn, 12.5% Cd | Low melting point (70°C) | Fuse plugs, automatic sprinklers |
| Copper | Yellow brass | 67% Cu, 33% Zn | Ductile, takes polish | Hardware items |
| Iron | Stainless steel | 80.6% Fe, 0.4% C, 18% Cr, 1% Ni | Resists corrosion | Tableware |
| Lead | Plumber's solder | 67% Pb, 33% Sn | Low melting point (275°C) | Soldering joints |
| Silver | Sterling silver | 92.5% Ag, 7.5% Cu | Bright surface | Tableware |
|  | Dental amalgam | 70% Ag, 18% Sn, 10% Cu, 2% Hg | Easily worked | Dental fillings |
| Tin | Pewter | 85% Sn, 6.8% Cu, 6% Bi, 1.7% Sb |  | Utensils |

must have a relatively much smaller covalent radius. Typically, the interstitial element is a nonmetal that participates in bonding to neighboring atoms. The presence of the extra bonds provided by the interstitial component causes the metal lattice to become harder, stronger, and less ductile. For example, iron containing less than 3 percent carbon is much harder than pure iron and has a much higher tensile strength and other desirable physical properties. *Mild steels* contain less than 0.2 percent carbon; they are malleable and ductile and are used to make cables, nails, and chains. *Medium steels* contain 0.2 to 0.6 percent carbon; they are tougher than the mild steels and are used to make girders and rails. *High-carbon steel*, used in cutlery, tools, and springs, contains 0.6 to 1.5 percent carbon. In all these cases other elements may be added to form *alloy steels*. Vanadium and chromium may be added to impart strength and to increase resistance to fatigue and corrosion. For example, a rail steel used in Sweden on lines bearing heavy ore carriers contains 0.7 percent carbon, 1 percent chromium, and 0.1 percent vanadium.

One of the most important iron alloys is stainless steel, which contains 0.4 percent carbon, 18 percent chromium, and 1 percent nickel. The chromium is obtained by carbon reduction of chromite, $FeCr_2O_4$, in an electric furnace. The product of the reduction is ferrochrome, $FeCr_2$, which is then added in the appropriate amount to the molten iron that comes from the converter to achieve the desired steel composition. Alloys of the type we have been discussing differ from ordinary chemical compounds in that the composition is not fixed. The ratio of elements present may vary over a wide range, imparting a variety of specific physical and chemical properties to the materials.

In **heterogeneous alloys**, the components are not dispersed uniformly. For example, in the form of steel known as pearlite, two distinct phases—essentially pure iron and the compound $Fe_3C$, known as cementite—are present in alternating layers. In general, the properties of heterogeneous alloys depend not only on the composition but also on the manner in which the solid is formed from the molten mixture. Rapid quenching leads to distinctly different properties than does slow cooling.

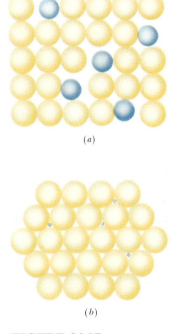

(a)

(b)

**FIGURE 24.17**
(a) Substitutional and (b) interstitial alloys. The yellow spheres are host metal; the blue spheres, the other component of the alloy.

We are literally surrounded with examples of inter-metallic compounds that play important roles in modern society because of their special properties. High-performance aircraft are composed of a wide range of alloy materials, few of which contain a significant amount of iron, the traditional structural material. The intermetallic compound $Ni_3Al$ is a major component of jet aircraft engines, chosen because of its light weight and strength. The coating on the new, very hard razor blades that are so widely advertised is an intermetallic compound, $Cr_3Pt$.

Materials that can be made into permanent magnets have many important uses. Intermetallic compounds of formula $Co_5M$, where M is a rare-earth element (atomic number between 58 and 71), possess unusual magnetic properties. The most useful of these compounds is $Co_5Sm$, from which extraordinarily strong magnets per unit weight of metal can be formed. The lightweight headsets that are used with portable stereo systems employ $Co_5Sm$ permanent magnets (Figure 24.18).

Intermetallic compounds have played an important role in the development of superconducting materials. *Superconductivity* refers to the ability of a substance to conduct current with no apparent resistance whatever. Thus far, materials are known to become superconducting only at low temperatures. One of the most important of these materials is

$Nb_3Sn$, which is superconducting below 18 K and which can be made into filaments to form the coil of a magnet. By passing a very large current through the superconducting material (which can be done because the resistance of the metal is essentially zero), a very high magnetic field can be developed. Such superconducting magnets are important in many areas of basic research and are used in magnetic-resonance imaging (MRI), a medical technique for observing tissues within the human body.

Considerable energy is required to keep every

**FIGURE 24.19** Superconductivity, the loss of all resistance to an electrical current, is demonstrated as a pellet of a superconducting ceramic floats above a magnet. The magnet induces superconducting currents in the pellet, previously cooled to 77 K with liquid nitrogen. The currents create a counter-magnetic force to levitate the pellet. Any resistance in the sample will kill the current, and the pellet will not levitate. The beaker keeps the pellet from slipping off the side. (Courtesy Argonne National Laboratory)

**FIGURE 24.18** Interior of a lightweight audio headset. The assembly can be made small because of the very strong magnetism of the $Co_5Sm$ alloy used. (Fundamental Photographs)

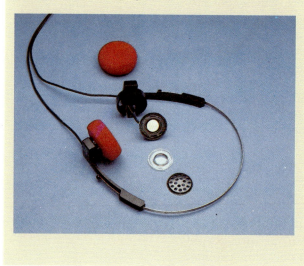

currently known intermetallic semiconductor below its transition temperature (the temperature at which superconductivity sets in). Thus there is great interest in finding materials that are superconducting at higher temperatures. In 1987, a new class of ceramic materials was discovered that becomes superconducting at about 90 K (see Figure 24.19). These materials are mixed oxides containing copper, lanthanum (La) or yttrium (Y), and sometimes other metals. For example, a material with the composition of $YBa_2Cu_3O_x$ (where $x$ is between 6.5 and 7.2) becomes superconducting at 94 K. These higher operating temperatures may provide vast improvements in the cost and convenience of superconductivity. At this writing, however, these new superconducting materials are far from practical use. They are brittle and thus hard to fabricate into wires. Furthermore, they are able to carry only small currents. Nevertheless, their discovery is a significant scientific breakthrough, hailed as one of the most significant discoveries since the invention of the transistor.

## Intermetallic Compounds

**Intermetallic compounds** are homogeneous alloys that have definite properties and composition. For example, copper and aluminum form a compound, $CuAl_2$, known as duralumin. Intermetallic compounds made their appearance very early in the history of technology. For example, when a copper or bronze object is exposed to arsenic or an appropriate arsenic compound, the object becomes coated with $Cu_3As$, which can be polished to a beautiful silvery color. Examples of this technique are found in objects from Anatolia dated from 2100 B.C. (Figure 24.20). As another example, a recipe for forming a dental amalgam from the elements silver, tin, and mercury is found in the Chinese literature of the seventh century A.D. The prescribed mixture results in formation of the two intermetallic compounds $Ag_2Hg_3$ and $Sn_8Hg$.

These examples of early intermetallic compounds illustrate some rather strange ratios of combining elements. Nothing that we have yet discussed in this text would lead us to predict such compositions. Also, any two elements may form several intermetallic compounds. For example, potassium and mercury form $KHg$, $KHg_2$, $KHg_3$, $K_2Hg_9$, and $KHg_{10}$. When we further consider that one cannot readily isolate and purify intermetallic compounds by many of the techniques used frequently in working

**FIGURE 24.20** Bronze figurine in the Boston Museum of Fine Art found at Horoztebe, in Anatolia (from about 2100 B.C.). (Courtesy Boston Museum of Fine Art)

with ordinary ionic and covalent substances, it is perhaps not surprising that there was for a long time considerable doubt whether intermetallic compositions ever amounted to pure compounds. However, by the early part of this century it was clear that many intermetallic compounds did indeed exist, and it was left for future research to understand their compositions. Although much progress has been made since then, that research still goes on today. At the same time, intermetallic compounds have many very important applications in current high technology.

# FOR REVIEW

## SUMMARY

In this chapter our focus has been on the metallic elements: their occurrences, the means by which they can be obtained from their ores, the nature of metallic bonding and its relationship to the properties of metals, and the characteristics of alloys.

The metallic elements occur in nature as **minerals**, inorganic substances found in various kinds of deposits. **Metallurgy** is concerned with how to obtain the desired metals from these natural sources and with the properties and structures of metallic elements. **Extractive metallurgy** is concerned with the processes that must be carried out to obtain the metal from the crude ore. These processes include physical concentration steps, such as froth flotation and magnetic separation. The chemical processes for obtaining the metal from the enriched ore are grouped according to the type of chemical process involved. **Pyrometallurgy** is the use of heat to bring about chemical reactions that convert an ore from one chemical form to another. One such process is **calcination**, in which an ore is heated to drive off a volatile substance. For example, a carbonate ore might be heated to drive off $CO_2$. In **roasting**, the ore is heated under conditions that bring about reaction with the furnace atmosphere. For example, a sulfide ore such as ZnS might be roasted to cause oxidation of the sulfur to $SO_2$. In the course of pyrometallurgical operations two or more layers of mutually insoluble materials often form in the furnace. Processes in which this happens are called **smelting**. One of the layers often formed in this way, called **slag**, is composed of molten silicate minerals, with other ionic materials such as fluorides or phosphates as constituents.

Iron is a most important metal in modern society. The metal is obtained from oxide ores, $Fe_2O_3$ and $Fe_3O_4$, by reduction in a **blast furnace**. The reducing agent is carbon, in the form of coke. Limestone, $CaCO_3$, is added to react with the silica present in the crude ore to form a slag. The raw iron from the blast furnace, called **pig iron**, is usually taken directly to a **converter**, in which it is refined to form various kinds of steel. In the converter, the crude molten iron is reacted with pure oxygen to oxidize impurity elements such as manganese, sulfur, phosphorus, and others.

**Hydrometallurgy** is the use of chemical processes occurring in aqueous solution to effect separation of a mineral from its ore or of one particular element from others. In **leaching**, an ore is treated with an aqueous reagent to selectively dissolve a component. For example, gold metal can be leached from raw ore by treatment with aqueous cyanide solution in air. In the **Bayer process**, aluminum is selectively dissolved from bauxite as the $Al(H_2O)_2(OH)_4^-$ ion by treatment with concentrated NaOH solution. **Electrometallurgy** is the use of electrolytic methods for preparation or purification of a metallic element. Sodium is prepared by electrolysis of molten NaCl solution, in a **Downs cell**. Aluminum is obtained in the **Hall process** by electrolysis of $Al_2O_3$ in molten cryolite, $Na_3AlF_6$. Copper is purified by electrolysis of aqueous copper sulfate solution, using crude copper anodes.

The properties of metals can be accounted for in general by a model in which the electrons are free to move throughout the metal atom structure. The chemical bonds between metal atoms are delocalized over the entire solid. In the molecular-orbital model for metallic bonding, the available electrons do not completely fill the band of closely spaced allowed energy orbitals. As a result, promotion of electrons to higher-energy orbitals requires very little energy, giving rise to high electrical and thermal conductivity, as well as other characteristic metallic properties. Metals are conductors. By contrast, in an insulator all the orbitals of the band of allowed orbitals are completely filled.

**Alloys** are materials composed of more than one element and possessing characteristic metallic properties. Usually, one or more metallic elements are major components. **Solution alloys** are homogeneous alloys, with the components distributed uniformly throughout. In **heterogeneous alloys** the components are not distributed uniformly; instead, two or more distinct phases with characteristic compositions are present. **Intermetallic compounds** are homogeneous alloys that have definite properties and compositions. For example, the alloy duralumin is an intermetallic compound of composition $CuAl_2$. The intermetallic compounds possess a wide variety of interesting properties that have important applications in modern technology.

## LEARNING GOALS

Having read and studied this chapter, you should be able to:

1. Describe what is meant by the term *mineral*, and provide a few examples of common minerals.
2. Define various terms employed in discussions of metallurgy, notably *gangue, calcination, roasting, smelting, refining* and *leaching*.
3. Describe the process of froth flotation, and explain the physical principles on which it is based.
4. Distinguish between pyrometallurgy, hydrometallurgy, and electrometallurgy, and provide examples of each metallurgical process.
5. Describe the nature of slag, and indicate the means by which it is formed in pyrometallurgical operations.
6. Describe the pyrometallurgy of iron. You should know which ores are employed, the general design of a blast furnace, the ingredients in the blast furnace reactions, and the chemical reactions of major importance. You should know how a converter is used to refine the crude pig iron that is the product of the blast furnace operation.
7. Describe the hydrometallurgy of gold, including the chemical reactions of major importance.
8. Describe the Bayer process for purification of bauxite, including balanced chemical equations.
9. Describe how copper may be recovered from chalcopyrite by means of a hydrometallurgical process.
10. Describe the process by which sodium metal is obtained from NaCl, including balanced chemical equations for electrode processes.
11. Describe the Hall processes for obtaining aluminum, including balanced chemical equations for electrode processes.
12. Describe the electrometallurgical purification of copper, including balanced chemical equations for electrode processes.
13. Discuss the simple electron-sea model for metals, and indicate how it accounts for certain important properties of metals.
14. Describe how the extent of metallic bonding among the first transition elements is related to the number of valence electrons.
15. Describe the molecular-orbital model for metals, including the idea of bands of allowed energy levels. You should also be able to distinguish between metals and insulators in terms of this model.
16. Name the important types of alloys, and distinguish the different types.
17. Describe the nature of intermetallic compounds, and provide examples of such compounds and their uses.

## KEY TERMS

Among the more important terms and expressions used for the first time in this chapter are the following:
An **alloy** (Section 24.7) has the characteristic properties of a metal, and contains more than one element. Often there is one principal metallic component, with one or more other elements present in

smaller amounts. Alloys may be homogeneous or heterogeneous in nature.

The **Bayer process** (Section 24.4) is a hydrometallurgical procedure for purification of **bauxite** in recovery of aluminum from bauxite-containing ores.

**Calcination** (Section 24.3) is the heating of an ore to bring about its decomposition and the elimination of a volatile product. For example, a carbonate ore might be calcined to drive off $CO_2$.

A **converter** (Section 24.3) is a container in which crude iron from the blast furnace is refined.

The **Downs cell** (Section 24.5) is used to obtain sodium metal by electrolysis of molten NaCl.

**Electrometallurgy** (Section 24.5) refers to the use of electrolysis to obtain free metal from one of its compounds, or to purify a crude form of the metal.

In the **electron-sea model** (Section 24.6) for metals, the metallic structure is pictured as consisting of metal ions at the locations of the metallic atoms, with the electrons moving freely about in the spaces between the ions.

**Extractive metallurgy** (Section 24.2) is concerned with the steps involved in obtaining a metal in as pure a form as possible from raw ore as starting material.

**Flotation** (Section 24.2) is a technique for concentrating raw, finely ground ore by a process of physical separation of the desired ore from gangue through selective wetting with the use of flotation agents.

**Gangue** (Section 24.2) is material of little or no value that accompanies the desired mineral in most raw ores.

The **Hall process** (Section 24.5) is used to obtain aluminum by electrolysis of $Al_2O_3$ dissolved in molten cryolite, $Na_3AlF_6$.

**Hydrometallurgy** (Section 24.4) refers to aqueous chemical processes for recovery of a metal from an ore.

An **intermetallic compound** (Section 24.7) is a homogeneous alloy with definite properties and composition. Intermetallic compounds are stoichiometric compounds in the full sense, but their compositions are not readily understood in terms of ordinary chemical bonding theory.

**Leaching** (Section 24.7) refers to the selective dissolution of a desired mineral by passing an aqueous reagent solution through an ore.

The **lithosphere** (Section 24.1) is that portion of our environment consisting of the solid earth.

**Metallurgy** (Introduction) is the science of extracting metals from their natural sources, by a combination of chemical and physical processes. It is also concerned with the properties and structures of metals and alloys.

A **mineral** (Section 24.1) is a solid inorganic substance occurring in nature, such as calcium carbonate, which occurs as calcite.

**Pig iron** (Section 24.3) is the product of reduction of iron ore in a blast furnace.

In **pyrometallurgy** (Section 24.3) heat converts a mineral in an ore from one chemical form to another, and eventually to the free metal.

**Refining** (Section 24.3) is the process of converting an impure form of a metal into a more usable substance of well-defined composition. For example, crude pig iron from the blast furnace is refined in a converter to produce steels of desired compositions.

**Roasting** (Section 24.3) is thermal treatment of an ore to bring about chemical reactions involving the furnace atmosphere. For example, a sulfide ore might be roasted to form a metal oxide and $SO_2$.

**Slag** (Section 24.3) is a mixture of molten silicate minerals. Slags may be acidic or basic, according to the acidity or basicity of the oxide added to silica.

**Smelting** (Section 24.3) is a melting process in which the materials formed in the course of the chemical reactions that occur separate into two or more layers. For example, the layers might be slag and molten metal.

# EXERCISES

## Metallurgy

**24.1** Two of the most heavily utilized elements listed in Table 24.2 are Al and Fe. Indicate the most important natural sources of these elements, and the oxidation state of the metal in each case.

**24.2** One of the most important sources of copper is chalcopyrite. What is the chemical formula of this substance? What is the oxidation state of Cu in chalcopyrite?

**24.3** Provide a brief definition of each of the following terms: (a) calcination; (b) leaching; (c) gangue; (d) flotation.

**24.4** Provide a brief definition of each of the following terms: (a) slag; (b) smelting; (c) roasting; (d) refining.

**24.5** Complete and balance each of the following equations:

**(a)** $ZnCO_3(s) \xrightarrow{\Delta}$

**(b)** $Cu_2S(s) + O_2(g) \longrightarrow$

**(c)** $ZnS(s) + O_2(aq) \longrightarrow SO_4^{2-}(aq) +$

**(d)** $Al(OH)_3(s) + OH^-(aq) \longrightarrow$

**(e)** $UO_2(s) + CO_3^{2-}(aq) + O_2(g) \longrightarrow$
$$[UO_2(CO_3)_3]^{4-} + OH^-(aq)$$

**24.6** Complete and balance each of the following equations:

**(a)** $PbCO_3(s) \xrightarrow{\Delta}$

**(b)** $PbS(s) + O_2(g) \longrightarrow$

**(c)** $ZnO(s) + CO(g) \longrightarrow$

**(d)** $ZrC(s) + Cl_2(g) \longrightarrow$

**(e)** $Cu^{2+}(aq) + Cl^-(aq) + H_2SO_3(aq) \longrightarrow$
$$CuCl(s) + SO_4^{2-}(aq)$$

**24.7** Complete and balance each of the following equations:

**(a)** $H_2SeO_3(aq) + SO_2(aq) \longrightarrow Se(s) + SO_4^{2-}(aq)$

**(b)** $CaO(l) + P_2O_5(l) \longrightarrow$

**(c)** $K(l) + TiCl_4(g) \longrightarrow$

**(d)** $Pb(s) + NaOH(l) + NaNO_3(l) \longrightarrow$
$$Na_2PbO_3(l) + N_2(g) + H_2O(g)$$
(This is a fused salt reaction.)

**24.8** Complete and balance the following equations:

**(a)** $Na_2O(l) + As_2O_5(l) \longrightarrow$

**(b)** $MnO(s) + C(s) \longrightarrow$

**(c)** $Cu_2O(l) + CH_4(g) \longrightarrow Cu(l) +$

**(d)** $WO_3(s) + H_2(g) \longrightarrow$

**24.9** Write balanced chemical equations for molten state reactions between **(a)** CaO and $Al_2O_3$; **(b)** FeO and $SiO_2$; **(c)** MgO and $PbO_2$; **(d)** $Na_2O$ and $P_2O_5$.

**24.10** Place the following metal oxides in a list with the most basic oxide first and the most acidic oxide last: MgO, $Fe_2O_3$, $Re_2O_7$, $Li_2O$, $TiO_2$.

**24.11** Write balanced chemical equations for the reductions of FeO and $Fe_2O_3$ by $H_2$ or CO.

**24.12** What is the major reducing agent in the reduction of iron ore in a blast furnace? Write a balanced chemical equation for the reduction process.

**24.13** A charge of $2.0 \times 10^4$ kg of material containing 32 percent $Cu_2S$ and 7 percent FeS is added to a converter and oxidized. What mass of $SO_2(g)$ is formed?

**24.14** A charge of $1.2 \times 10^4$ kg of concentrate from a partial roast of a chalcopyrite ore, containing 26 percent FeO, is added to a furnace. $SiO_2$ is added to form a slag with FeO. Write the balanced equation for the reaction leading to slag formation, and calculate the mass of $SiO_2$ required to react with the FeO to form slag.

**24.15** What role does each of the following materials play in the chemical processes that occur in the blast furnace: **(a)** air; **(b)** limestone; **(c)** coke; **(d)** water? Write balanced chemical equations to illustrate your answers.

**24.16** In terms of the concepts discussed in Chapter 12, indicate why the molten metal and slag phases formed in the blast furnace shown in Figure 24.4 are immiscible.

**24.17** Describe how electrometallurgy could be employed to purify crude cobalt metal. Describe the compositions of the electrodes and electrolyte, and write out all electrode reactions.

**24.18** The element tin is generally recovered from deposits of the ore cassiterite, $SnO_2$. The oxide is reduced with carbon, and the crude metal is purified by electrolysis. Write balanced chemical equations for the reduction process and the electrode reactions in the electrolysis, assuming that an acidic solution of $SnSO_4$ is employed as electrolyte.

**24.19** In an electrolytic process nickel sulfide is oxidized in a two-step reaction:
$$Ni_3S_2(s) \longrightarrow Ni^{2+}(aq) + 2NiS(s) + 2e^-$$
$$NiS(s) \longrightarrow Ni^{2+}(aq) + S(s) + 2e^-$$

What mass of $Ni^{2+}$ is produced in solution by passage of a current of 66 A for a period of 8.0 h, assuming the cell to be 100 percent efficient?

**24.20** What mass of copper is deposited in an electrolytic refining cell by a passage of 240 A current for a period of 10 h, assuming 80 percent current efficiency?

**[24.21]** Using the data in Appendix D, estimate the free-energy change for each of the following reactions at $1200°C$:

**(a)** $PbO(s) + CO(g) \longrightarrow Pb(s) + CO_2(g)$

**(b)** $Si(s) + 2MnO(s) \longrightarrow SiO_2(s) + 2Mn(s)$

**(c)** $FeO(s) + H_2(g) \longrightarrow Fe(s) + H_2O(g)$

**[24.22]** In terms of thermodynamic functions, particularly free energies of formation, what conditions must be operative for refining iron in the converter shown in Figure 24.5 to be a workable process?

## Metals and Alloys

**24.23** Sodium is a highly malleable substance, whereas sodium chloride is not. Explain this difference in properties.

**24.24** Suppose that you have samples of magnesium, iron, sulfur, high-carbon steel, and calcium sulfate, all formed into solid, pencil-shaped rods. Compare these substances with respect to appearance, flexibility, melting point, and thermal conductivity. Account for the differences you would expect to observe.

**24.25** How does the electron-sea model account for the high electrical conductivity of copper metal?

**24.26** Discuss briefly the difference in bonding between metals and nonmetals.

**24.27** Moving across the fourth period of the periodic

table, hardness increases at first, reaching a maximum at Cr; the hardness then decreases as we move from Cr to Zn. Explain this trend.

**24.28** The densities of the elements K, Ca, Sc, and Ti are 0.86, 1.5, 3.2, and 4.5 g/cm$^3$, respectively. What factors are likely to be of major importance in determining this variation?

**24.29** Use the band theory to describe the differences among insulators, conductors, and semiconductors.

**[24.30]** Indicate whether each of the following is likely to be an insulator, a metallic conductor, an *n*-type semiconductor, or a *p*-type semiconductor: **(a)** germanium doped with Ga; **(b)** pure $CuAl_2$; **(c)** pure boron; **(d)** silicon doped with As; **(e)** pure Nb; **(f)** pure MgO.

**24.31** Tin exists in two allotropic forms; gray tin has a diamond structure and white tin has a close-packed structure. Which of these allotropic forms would you expect to be more metallic in character? Explain why the electrical conductivity of white tin is much greater than that of gray tin. Which form would you expect to have the longer Sn—Sn bond distance?

**24.32** The interatomic distance in silver metal, 2.88 Å, is much larger than the calculated Ag—Ag single-bond distance, 1.34 Å. Nevertheless, the melting point of silver, 906°C, and its high boiling point, 2164°C, suggest that there is strong bonding between silver atoms in the metal. Describe a model for the structure of silver metal that is consistent with these observed properties.

**24.33** Define the term *alloy*. Distinguish between solution alloys, heterogeneous alloys, and intermetallic compounds.

**24.34** Distinguish between substitutional and interstitial alloys. What conditions favor formation of substitutional alloys?

## Additional Exercises

**24.35** Copper exists in nature in both the free (elemental) and combined states, whereas aluminum is never found in the free state. Explain.

**24.36** Define *mineral*. Does every mineral have a definite, fixed composition? Explain.

**24.37** Three general categories of procedures needed to produce pure metals from ores are concentration, reduction, and refining. Using copper as an example, illustrate each of these processes.

**24.38** Using the concepts presented in Chapter 11, describe the forces that operate in the binding of potassium ethyl xanthate to $Cu_2S$ in froth flotation, and indicate how this bonding leads to flotation of the mineral.

**24.39** Three common oxide impurities in pig iron are Si, Mn, and P. If these are converted to oxides—$SiO_2$, MnO, and $P_4O_{10}$—what kind of oxides should be added to form slags with them?

**[24.40]** In steelmaking, when the slag present is "acidic" (containing $SiO_2$, FeO, and MnO, but little CaO), removal of phosphorus from iron through oxidation of the phosphorus does not occur. It does occur when the slag is "basic." Suggest an explanation.

**24.41** Complete and balance each of the following equations:

**(a)** $MoO_3(g) + MnO(l) \longrightarrow$

**(b)** $FeCl_2(s) + H_2(g) \xrightarrow{650°C}$

**(c)** $Mn_3O_4(s) + Al(l) \longrightarrow$

**(d)** $FeS(l) + Cu_2O(l) \longrightarrow FeO +$

**24.42** Write a balanced chemical equation to illustrate the use of each of the following reducing agents in metallurgy: **(a)** carbon; **(b)** $H_2$; **(c)** Na.

**24.43** Write balanced chemical equations to correspond to each of the following verbal descriptions. **(a)** NiO(s) can be solubilized by leaching with aqueous sulfuric acid. **(b)** After concentration, an ore containing the mineral carrolite, $CuCo_2S_4$, is leached with aqueous sulfuric acid to produce a solution containing copper and cobalt ions. **(c)** Titanium dioxide is treated with chlorine in the presence of carbon as reducing agent to form $TiCl_4$. **(d)** Under oxygen pressure, ZnS(s) reacts at 150°C with aqueous sulfuric acid to form soluble zinc sulfate, with deposition of elemental sulfur.

**24.44** In the early days of iron mining in the United States, the iron oxide mined in the Lake Superior region was reduced to pig iron in blast furnaces located near the mines. The furnaces used charcoal made from hardwood as the reducing agent (transportation savings prompted this procedure). What mass of pig iron containing 2.5 percent carbon was produced from each kilogram of iron ore, assuming that it is mined as magnetite, $Fe_3O_4$, of 70 percent purity?

**24.45** The crude copper that is subjected to electro-refining contains selenium and tellurium as impurities. Describe the probable fate of these elements during electrorefining, and relate your answer to the positions of these elements in the periodic table.

**24.46** Magnesium is obtained by electrolysis of molten $MgCl_2$. Several cells are connected in parallel by very large copper buses that convey current to the cells. Assuming that the cells are 96 percent efficient in producing the desired products in electrolysis, what mass of Mg is formed by passage of a current of 97,000 A for a period of 24 h?

**24.47** Zinc can be purified by electrolytic refining. A current of 75 A is used at an applied voltage of 3.5 V. Assuming that the electrolytic process is 94 percent efficient, how much electrical energy (kilowatt hours) is required to produce $1.0 \times 10^8$ kg of zinc metal?

**[24.48]** In the electrolytic purification of nickel, some iron may be present in the crude metal. It is found that as iron accumulates in the electrolyte solution the current efficiency of the electrolysis cell decreases when the contents of the cathode cell compartment are not isolated from the air. Write reactions involving iron species that could account for the loss in current efficiency. (Hint: Recall that iron can exist in solution in more than one oxidation state.)

**[24.49]** Suppose that a rich deposit of the mineral absolite, $CoO \cdot 2MnO_2 \cdot 4H_2O$, were found. Sketch out a scheme for recovery of cobalt metal from the ore containing this mineral.

**[24.50]** Write balanced chemical equations to correspond to the steps in the following brief account of the metallurgy of molybdenum. "Molybdenum occurs primarily as the sulfide, $MoS_2$. On boiling with concentrated nitric acid, a white residue of $MoO_3$ is obtained. This is an acidic oxide; from a solution of hot, excess concentrated ammonia, ammonium molybdate crystallizes on cooling. On heating ammonium molybdate, white $MoO_3$ is obtained. On heating to 1200°C in hydrogen, a gray powder of metallic molybdenum is obtained."

**24.51** Distinguish between **(a)** a substitutional and interstitial alloy; **(b)** an intermetallic compound and a het-erogeneous alloy; **(c)** a metal and an insulator; **(d)** the bonding in $Na_2(g)$ and in $Na(s)$.

**24.52** Treatment of iron surfaces at high temperatures with ammonia leads to "nitriding," in which nitrogen atoms are incorporated into the metal structure. What kind of alloy formation occurs in this case? How would you expect nitriding to alter the properties of the metal?

**24.53** Pure silicon is a very poor conductor of electricity. Titanium, which also possesses four valence-shell electrons, is a metallic conductor. Explain the difference.

**24.54** The enthalpies of sublimation of copper metal, copper(I) iodide, and iodine at 25°C are 338, 326, and 62 kJ/mol, respectively. Account for this variation in terms of the attractive forces that operate in each solid.

**[24.55]** The melting point of neodymium, Nd, is 1019°C. Considering its place in the periodic table and the melting points of other elements as shown in Figure 5.8, what number of electrons would you estimate to be involved in metallic bonding in this element? What do your conclusions suggest about the role of $4f$ electrons in the bonding?

# CHAPTER 25

# The Transition Metals

In Chapter 24 we considered the distribution of metallic elements in nature, the means by which they can be obtained from their ores, and the important characteristics of metals and alloys. In this chapter we will focus on the transition metals, that large block of elements in the central portion of the periodic table. These elements comprise three series of 10 elements each in the *d* block of the periodic table, as shown in Figure 25.1.

The transition elements include some of the most familiar and important elements, such as iron, copper, nickel, cobalt, and silver. They also include many elements that are likely to be unfamiliar to you but that are nonetheless important in modern technology. Even the least abundant transition metals have found uses as chemists and other scientists search for new materials to meet new performance requirements. To illustrate this point, Figure 25.2 shows the approximate composition of a high-performance jet engine. Notice that nearly the entire mass of the engine consists of transition metals. Notice also that iron, long the workhorse element of technology, is not even present to a significant extent.

The transition metals are numerous, and each has distinctive chemical properties. There is no point in our attempting to master the individual

# CONTENTS

**FIGURE 25.1**  The transition metals are those elements that occupy the *d* block of the periodic table.

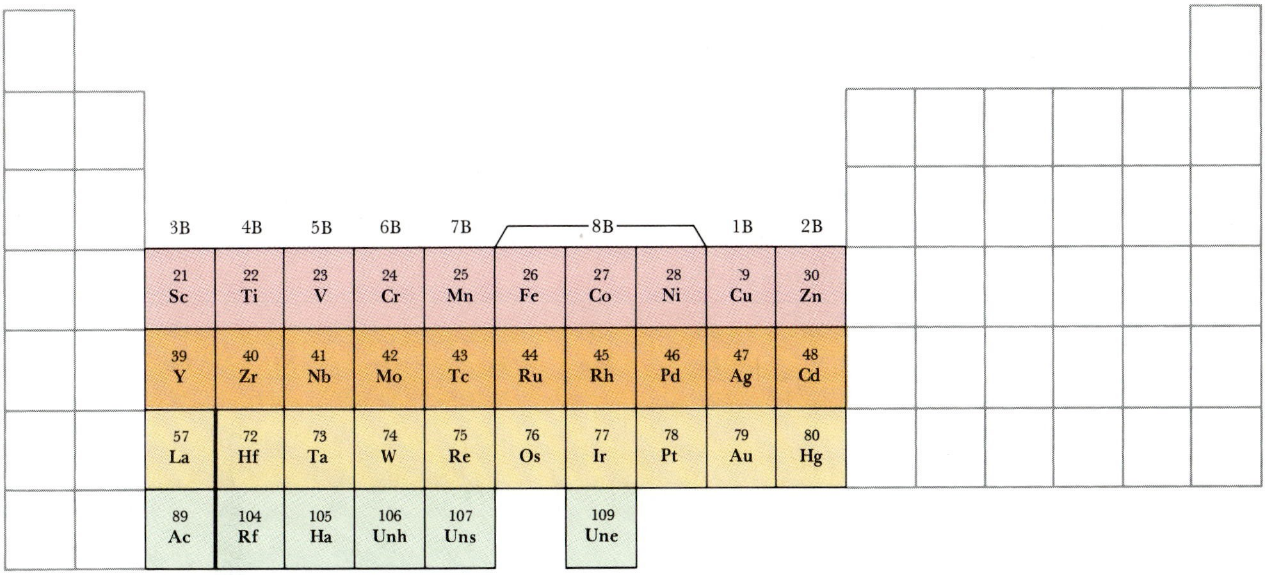

A ribbon of a nickel alloy. The ribbon was cast extremely rapidly from the molten alloy, producing a noncrystalline material referred to as a "metallic glass." Unlike the crystalline form of this alloy, this metallic glass is ductile and very flexible. Ribbons of this material are useful for brazing pieces of stainless steel together. (Allied Metglas Products)

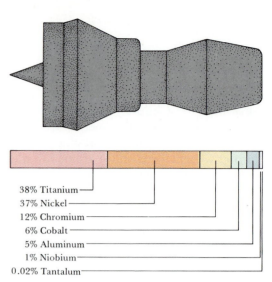

**FIGURE 25.2** Metallic elements employed in construction of a jet engine.

38% Titanium
37% Nickel
12% Chromium
6% Cobalt
5% Aluminum
1% Niobium
0.02% Tantalum

chemical properties of these elements. Rather, we will approach the subject by first considering some general aspects of their behavior and then briefly considering the aqueous solution behavior of a few of the more common metals. We will also examine the properties of transition metal oxides, and the magnetic properties of transition metals and their compounds. Finally, we will briefly discuss metals as strategic materials.

## 25.1 PHYSICAL PROPERTIES OF THE TRANSITION METALS

Some of the physical properties of the elements of the first transition series are listed in Table 25.1. It is important to distinguish those properties that are characteristic of the bulk element in the liquid or solid phase from those properties that are characteristic of isolated atoms, as encountered in the gas phase. Properties like melting point and density are properties of the solid metal. These properties are related to the extent of metallic bonding that holds the solid together (Section 24.6). Properties of this sort vary over a wide range, but often exhibit maxima near the middle of each transition series, somewhere in the vicinity of groups 6B to 8B.

**TABLE 25.1** Properties of the first-series transition elements

| Group: | 3B | 4B | 5B | 6B | 7B | 8B | | | 1B | 2B |
|---|---|---|---|---|---|---|---|---|---|---|
| Element: | Sc | Ti | V | Cr | Mn | Fe | Co | Ni | Cu | Zn |
| Electron configuration | $3d^14s^2$ | $3d^24s^2$ | $3d^34s^2$ | $3d^54s^1$ | $3d^54s^2$ | $3d^64s^2$ | $3d^74s^2$ | $3d^84s^2$ | $3d^{10}4s^1$ | $3d^{10}4s^2$ |
| First ionization energy (kJ/mol) | 631 | 658 | 650 | 653 | 717 | 759 | 758 | 737 | 745 | 906 |
| Atomic radius (Å) | 1.44 | 1.32 | 1.22 | 1.17 | 1.17 | 1.16 | 1.16 | 1.15 | 1.17 | 1.24 |
| Density (g/cm³) | 3.0 | 4.5 | 6.1 | 7.9 | 7.2 | 7.9 | 8.7 | 8.9 | 8.9 | 7.1 |
| $\Delta H_{atom}$ (kJ/mol) | 377 | 470 | 514 | 398 | 283 | 415 | 428 | 430 | 338 | 130 |
| Melting point (°C) | 1538 | 1660 | 1917 | 1857 | 1244 | 1537 | 1494 | 1455 | 1084 | 420 |

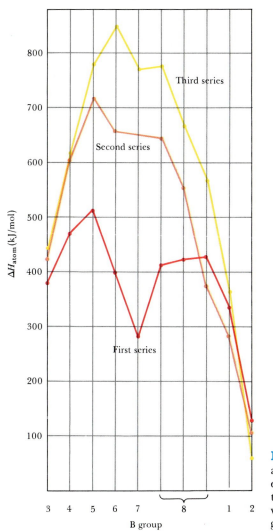

**FIGURE 25.3** Heats of atomization of the transition elements. These are the heats for the process, $M(s) \longrightarrow M(g)$, where $M(g)$ is the monatomic gas species.

For example, the heat of atomization, $\Delta H_{atom}$, listed in Table 25.1 and shown in Figure 25.3, reflects the extent of metallic bonding in the solid; in the atomization process $[M(s) \longrightarrow M(g)]$ the metal-metal bonds are broken.

Ionization energy and atomic radius are properties of isolated atoms. In contrast to those properties characteristic of the bulk metal, the properties of individual atoms show a relatively smooth and rather small variation across each series. For example, the atomic radii of the transition metals are shown in Figure 25.4. Notice that there is a slight decrease in size in moving across the first part of each series. This trend can be understood in terms of the concept of increasing effective nuclear charge, as described in Chapter 7. For example, as we move across the first series, electrons are being added to the $3d$ orbitals. The $3d$ electrons are quite effective in shielding the outer $4s$ electrons from the increasing

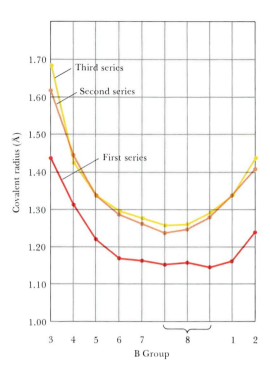

**FIGURE 25.4** Variation in covalent radius of the transition elements as a function of the periodic table group number.

nuclear charge. Thus the effective nuclear charge experienced by the outer $4s$ electrons increases only slowly, leading to a small decrease in radius across the series. The slight increase in radius at groups 1B and 2B is due to increasing electron-electron repulsions as the $3d$ subshell is completed.

The incomplete screening of the nuclear charge by added electrons produces a very interesting and important effect in the third transition metal series. In general, atomic size tends to increase going down a family because of the increasing principal quantum number of the outer-shell electrons (Section 7.6). Notice that the radii of the 3B elements follow the expected trend, Sc < Y < La. In group 4B we would similarly expect the radius to increase in the order Ti < Zr < Hf. However, Hf has virtually the same radius as Zr. This effect has its origin in the lanthanide series, the elements with atomic numbers 58 through 71, which occur between La and Hf. The filling of $4f$ orbitals through the lanthanide elements causes a steady increase in the effective nuclear charge, producing a contraction in size called the **lanthanide contraction**. This contraction just offsets the increase we would expect as we go from the second to the third series. You can see from Figure 25.4 that the second- and third-series transition metals in each group have about the same radii all the way across the series. As a consequence, the second- and third-series metals in a given group have great similarity in their chemical properties. For example, the chemical properties of zirconium and hafnium are remarkably similar. They always occur together in nature, and they are very difficult to separate.

## SAMPLE EXERCISE 25.1

The first ionization energies, $I_1$, of the group 3B and 6B elements are as follows:

| Group 3B | | Group 6B | |
|---|---|---|---|
| Sc | 631 kJ/mol | Cr | 653 kJ/mol |
| Y | 616 kJ/mol | Mo | 685 kJ/mol |
| La | 538 kJ/mol | W | 769 kJ/mol |

Account for the relative values among these elements in terms of their electronic structures.

**Solution:**  The decrease in $I_1$ with increasing atomic number in group 3B is what we expect. As the major quantum number increases, the outermost electrons are farther from the nucleus. If they experience about the same effective nuclear charge, as we might expect if the electron configurations are comparable in the series, the energy required for removal of the electron should decrease. The group 6B elements have higher ionization energies than do the corresponding group 3B elements. This again is expected; in moving from group 3B to group 6B the nuclear charge increases by 3; because the screening by the additional $d$ electrons is incomplete, the effective nuclear charge increases.

The variation in $I_1$ with increasing period in group 6B is not as easy to explain. The covalent atomic radius of molybdenum is larger than that of chromium. Thus we might expect a lower value for $I_1$ for Mo than for Cr, just as $I_1$ is lower for Y than for Sc.

The electronic configurations of Cr and Mo are entirely comparable (Table 25.1). The fact that $I_1$ is larger for Mo tells us that the effective nuclear charge experienced by the $5s$ electron in Mo is larger than that experienced by the $4s$ electron in Cr. The $5s$ electron in Mo must penetrate farther inside the electron distribution of the $4d$ orbitals than does the $4s$ electron inside the $3d$ electron distribution of Cr. The $5s$ electron in Mo therefore experiences a higher effective nuclear charge, with a resulting increase in $I_1$. Note that the electronic configuration for W is different from that for Cr or Mo (Table 25.1). The much higher ionization energy for W as compared with Mo, in spite of comparable covalent atomic radii, indicates that the outermost electrons in this element penetrate closer inside the other electrons, with a resultant higher effective nuclear charge.

The relative values of $I_1$ among the group 6B elements could not have been predicted on the basis of any of the ideas we have presented so far in this text. The example illustrates that in large atoms with many electrons the factors that influence energies and chemical behavior are complex and not easily predicted.

### PRACTICE EXERCISE
The elements cobalt, rhodium, and iridium are in the same column in group 8B. Which two of these would you expect to exhibit the greatest similarities in chemical properties?
*Answer:*  rhodium and iridium

---

The electron configurations of the transition metals and their ions were discussed in Sections 7.3 and 8.2. In the oxidation of transition metals to form cations, the outer $s$ electrons are removed first. Loss of additional electrons occurs from the $d$ subshell that lies just below the outer $s$ subshell in energy. For example, consider the formation of $Fe^{3+}$ from Fe. The electron configuration of Fe is $[Ar]3d^6 4s^2$. To form $Fe^{3+}$, we remove three electrons, starting with the two in the $4s$ subshell. Thus the electron configuration of $Fe^{3+}$ is $[Ar]3d^5$.

Transition-metal ions frequently possess partially occupied $d$ subshells. The existence of these $d$ electrons is partially responsible for several characteristics that the transition elements share with one another:

1. They often exhibit more than one stable oxidation state.
2. Many compounds of the transition elements are colored (see Figure 25.5). We will discuss the origin of these colors in Chapter 26.

## 25.2 ELECTRON CONFIGURATIONS AND OXIDATION STATES

**FIGURE 25.5** Sulfate salts and their solutions. From left to right: $Mn^{2+}$, $Fe^{2+}$, $Co^{2+}$, $Ni^{2+}$, $Cu^{2+}$, and $Zn^{2+}$. (Donald Clegg and Roxy Wilson)

**3.** Many compounds of the transition elements exhibit interesting magnetic properties. We will discuss this topic in Section 25.5.

## Oxidation States

Figure 25.6 summarizes the common oxidation states for the transition metals. The ones shown in red are those most frequently encountered either in solution or in solid compounds. The ones shown in blue are less common, but several well-known examples exist for each of these. For example, manganese is commonly found in solution in the $+2$ ($Mn^{2+}$) and $+7$ ($MnO_4^-$) oxidation states. In the solid state the $+4$ oxidation state (as in $MnO_2$) is common. The $+3$, $+5$, and $+6$ oxidation states are observed as relatively less stable species either in solution or in the solid state. Of course, the figure does not give a complete listing of all the oxidation states observed. First, there is a large class of compounds, called organometallic compounds, in which the metals are in low positive oxidation states (such as $+1$) or even in zero or negative oxidation states. We will not have the opportunity to discuss these compounds. Second, there are exotic compounds containing metals whose oxidation states are not listed in the figure.

Inspection of Figure 25.6 reveals some interesting trends:

**1.** Oxidation states of $+2$ and $+3$ are common, but they are more frequently observed in the first series than in the second and third. For example, in group 6B, the $+3$ oxidation state is common for Cr, occasionally observed for Mo, and only rarely found for W.

**2.** From Sc through Mn, the maximum oxidation state increases from $+3$ to $+7$, equaling in each case the total number of $4s$ plus $3d$ electrons of the atom. In the second and third series, the maximum is $+8$, an oxidation state found in $RuO_4$ and $OsO_4$. As we move to the right beyond Mn, Ru, or Os, the maximum observed oxidation state decreases. In general, the highest oxidation states are found only when the metals are combined with the most electronegative elements: O, F, and possibly Cl.

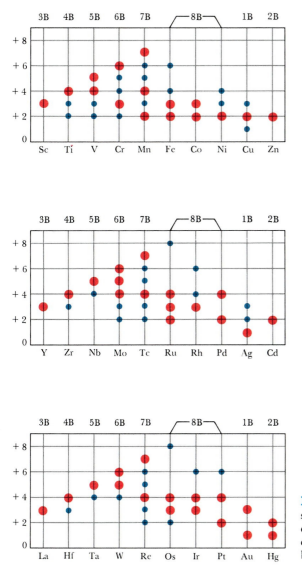

**FIGURE 25.6** Oxidation states of the transition elements. The most common oxidation states are indicated by red circles.

## SAMPLE EXERCISE 25.2

The manganate ion, $MnO_4^{2-}$, is stable only in strongly basic solution. In acidic solution it decomposes to form the permanganate ion and solid $MnO_2$. Write a balanced chemical equation to show this reaction.

**Solution:** In a disproportionation reaction the same species is both the oxidizing and the reducing agent. In this case we have manganese going from the $+6$ to the $+7$ and $+4$ oxidation states. Because there is a change in oxidation state of $+1$ for the oxidation and $-2$ for the reduction, there must be twice as much $MnO_4^-$ produced as $MnO_2$. The two half-reactions and overall reaction are:

$$2[MnO_4^{2-}(aq) \longrightarrow MnO_4^-(aq) + e^-]$$
$$MnO_4^{2-}(aq) + 2e^- + 4H^+(aq) \longrightarrow$$
$$MnO_2(s) + 2H_2O(l)$$
$$\overline{3MnO_4^{2-}(aq) + 4H^+(aq) \longrightarrow}$$
$$2MnO_4^-(aq) + MnO_2(s) + 2H_2O(l)$$

## PRACTICE EXERCISE

What is the oxidation state of Cr in each of the following compounds: **(a)** CrS; **(b)** $CrO_4^{2-}$; **(c)** $CrOCl_4^-$?
*Answers:* **(a)** $+2$; **(b)** $+6$; **(c)** $+5$

## Electrode Potentials and the Stability of Oxidation States

We often encounter the transition metals as ions in aqueous solution. This is especially true for the metals of the first transition series. Table 25.2 summarizes the common oxidation states and related electrode potentials for the first-series transition metals. This table summarizes a great deal of important chemical information about these metals. In considering these data it is important to keep several factors in mind and to note certain important trends:

1. Except for $Cu^{2+}$, all the 2+ ions of the first transition series exhibit negative standard electrode (reduction) potentials. Thus only $Cu^{2+}$ can be reduced by $H_2$ under standard conditions. Conversely, these metals generally oxidize readily. Thus it is not surprising that, with rare exceptions, all these metals are found in nature in the combined state.

2. The metal ions are present in water as the aquated ions (Section 17.8). The metal ion acts as a Lewis acid toward the surrounding water molecules. When ions or molecules other than water are also present in solution, the metal ion may form a metal complex (Section 18.5). We will have a great deal more to say about these complexes in Chapter 26. For the moment you should just keep in mind that the electrode potentials shown in Table 25.2 can be greatly modified by complex formation. In general, the formation of a very stable metal complex may be a key factor in stabilizing a particular oxidation state for a metal ion. As an example, the $Co^{3+}$ ion is not stable in water as the $Co(H_2O)_6{}^{3+}$ ion, but it is stable as the complex with ammonia, $Co(NH_3)_6{}^{3+}$:

**TABLE 25.2**  Oxidation states in aqueous solution for the first-series transition elements

| Element | Oxidation state | Reduction reaction | Standard electrode potential, $E^\circ_{298}$(V) |
|---------|-----------------|--------------------|--------------------------------------------------|
| Sc | +3 | $Sc^{3+}(aq) + 3e^- \longrightarrow Sc(s)$ | −2.1 |
| Ti | +2 | $Ti^{2+}(aq) + 2e^- \longrightarrow Ti(s)$ | −1.63 |
|    | +3 | $Ti^{3+}(aq) + 3e^- \longrightarrow Ti(s)$ | −0.37 |
| V  | +2 | $V^{2+}(aq) + 2e^- \longrightarrow V(s)$ | −1.2 |
|    | +3 | $V^{3+}(aq) + e^- \longrightarrow V^{2+}(aq)$ | −0.27 |
| Cr | +2 | $Cr^{2+}(aq) + 2e^- \longrightarrow Cr(s)$ | −0.91 |
|    | +3 | $Cr^{3+}(aq) + 3e^- \longrightarrow Cr(s)$ | −0.74 |
|    | +6 | $CrO_4{}^{2-}(aq) + 4H_2O(l) + 3e^- \longrightarrow Cr(OH)_3(s) + 5OH^-(aq)$ | −0.13 |
|    | +6 | $Cr_2O_7{}^{2-}(aq) + 14H^+(aq) + 6e^- \longrightarrow 2Cr^{3+}(aq) + 7H_2O(l)$ | +1.33 |
| Mn | +2 | $Mn^{2+}(aq) + 2e^- \longrightarrow Mn(s)$ | −1.18 |
|    | +7 | $MnO_4{}^-(aq) + 8H^+(aq) + 5e^- \longrightarrow Mn^{2+}(aq) + 4H_2O(l)$ | +1.51 |
|    | +7 | $MnO_4{}^-(aq) + 2H_2O(l) + 3e^- \longrightarrow MnO_2(s) + 4OH^-(aq)$ | +0.59 |
| Fe | +2 | $Fe^{2+}(aq) + 2e^- \longrightarrow Fe(s)$ | −0.440 |
|    | +3 | $Fe^{3+}(aq) + e^- \longrightarrow Fe^{2+}(aq)$ | +0.771 |
| Co | +2 | $Co^{2+}(aq) + 2e^- \longrightarrow Co(s)$ | −0.277 |
|    | +3 | $Co^{3+}(aq) + e^- \longrightarrow Co^{2+}(aq)$ | +1.842 |
| Ni | +2 | $Ni^{2+}(aq) + 2e^- \longrightarrow Ni(s)$ | −0.28 |
| Cu | +1 | $Cu^+(aq) + e^- \longrightarrow Cu(s)$ | +0.521 |
|    | +2 | $Cu^{2+}(aq) + 2e^- \longrightarrow Cu(s)$ | +0.337 |
| Zn | +2 | $Zn^{2+}(aq) + 2e^- \longrightarrow Zn(s)$ | −0.763 |

$$\text{Co(H}_2\text{O)}_6{}^{3+}(aq) + e^- \longrightarrow \text{Co(H}_2\text{O)}_6{}^{2+}(aq) \qquad E° = +1.95 \text{ V} \qquad [25.1]$$
$$\text{Co(NH}_3\text{)}_6{}^{3+}(aq) + e^- \longrightarrow \text{Co(NH}_3\text{)}_6{}^{2+}(aq) \qquad E° = +0.1 \text{ V} \qquad [25.2]$$

**3.** The simple aquated metal ions from the early part of the series tend to exhibit lower electrode (reduction) potentials than do those of the same charge at the other end of the series. For example, the standard electrode potential for reduction of $Cr^{3+}$ to $Cr^{2+}$ is $-0.41$ V, whereas that for reduction of $Co^{3+}$ to $Co^{2+}$ is $+1.8$ V. Thus the ions from the early part of the series are more difficult to reduce than those later in the series. Conversely, the ions early in the series are easier to oxidize than those later in the series. Indeed, *the higher oxidation states tend to become less stable relative to lower ones on going from left to right in a transition series.* This trend is consistent with an increasing effective nuclear charge, which makes it increasingly difficult to remove electrons.

Were we to consider electrode potentials for the ions in the second and third transition series, we would find one further general trend: *As we go down a group of transition metals, the higher oxidation states become increasingly more stable relative to the lower ones.*

---

**SAMPLE EXERCISE 25.3**

Which of the following would you expect to be a stronger oxidizing agent, $Fe^{3+}$ or $Ni^{3+}$?

**Solution:** Because an oxidizing agent is reduced in a redox reaction (Section 13.6), we want to know which ion gains electrons more readily. Locate Fe and Ni in the periodic table, and notice that Ni lies to the right of Fe. Because the higher oxidation states tend to become less stable relative to lower ones on going from left to right across a transition series, we might expect that $Ni^{3+}$ is reduced more readily than $Fe^{3+}$. Therefore, $Ni^{3+}$ should be the stronger oxidizing agent. In fact, $Ni^{3+}$ gains electrons so readily that it is very unstable. In its compounds and in solution, nickel is almost always found as $Ni^{2+}$.

**PRACTICE EXERCISE**
Which of the following do you expect to be the stronger oxidizing agent, $MnO_4{}^-$ or $ReO_4{}^-$?
*Answer:* $MnO_4{}^-$

---

Let's now consider briefly some of the chemistry of four common elements from the first transition series. As you read this material, you should look for evidence of the trends that illustrate the generalizations outlined above.

## 25.3 CHEMICAL PROPERTIES OF SELECTED TRANSITION METALS

### Chromium

Chromium is obtained mainly from the ore *chromite*, $FeCr_2O_4$. You can think of this substance as a mixed oxide of formula $FeO \cdot Cr_2O_3$. About 60 percent of the chromium produced goes into forming chromium-based alloys, about 20 percent into chemical use (including electroplating), and most of the rest into furnace bricks and other refractory (high-temperature) materials.

Chromium dissolves slowly in dilute hydrochloric or sulfuric acid, liberating hydrogen and forming a blue solution of chromous ion (see Figure 25.7, left);

$$\text{Cr}(s) + 2\text{H}^+(aq) \longrightarrow \text{Cr}^{2+}(aq) + \text{H}_2(g) \qquad [25.3]$$

Normally, however, you do not observe this blue color, because the

chromium(II) ion is rapidly oxidized in air to the violet-colored chromium(III) ion:

$$4Cr^{2+}(aq) + O_2(g) + 4H^+(aq) \longrightarrow 4Cr^{3+}(aq) + 2H_2O(l) \qquad [25.4]$$

When hydrochloric acid solution is used, the solution appears green due to the formation of complex ions containing chloride coordinated to chromium, for example, $Cr(H_2O)_4Cl_2^+$ (Figure 25.7, right).

Chromium is frequently encountered in aqueous solution in the $+6$ oxidation state. In basic solution the yellow chromate ion, $CrO_4^{2-}$, is the most stable. In acidic solution, the dichromate ion, $Cr_2O_7^{2-}$, is formed:

$$CrO_4^{2-}(aq) + H^+(aq) \rightleftharpoons HCrO_4^-(aq) \qquad [25.5]$$

$$2HCrO_4^-(aq) \rightleftharpoons Cr_2O_7^{2-}(aq) + H_2O(l) \qquad [25.6]$$

**FIGURE 25.7** The tube on the left shows the reaction of chromium metal with 6 $M$ sulfuric acid in the absence of air. If the air had not been excluded, the solution would have taken on a violet cast. The tube on the right shows the reaction of chromium metal with 6 $M$ hydrochloric acid in the presence of air. (Donald Clegg and Roxy Wilson)

You may recognize the second of these reactions as a condensation reaction, in which water is split out from between two $HCrO_4^-$ ions. Similar reactions occur among the oxyanions of the nonmetallic and other metallic elements (Section 23.4). The equilibrium between the dichromate and chromate ions is readily observable, because $CrO_4^{2-}$ is a bright yellow and $Cr_2O_7^{2-}$ is a deep orange, as seen in Figure 25.8. Note from the electrode potential listed in Table 25.2 that dichromate ion in acidic solution is a strong oxidizing agent, as evidenced by the large, positive reduction potential. By contrast, chromate ion in basic solution is not a particularly strong oxidizing agent.

Chromium in trace quantities is an essential element in human nutrition. However, at the concentrations employed in laboratory work the element is highly toxic.

**FIGURE 25.8** Lead chromate, $PbCrO_4$, on the left and potassium dichromate, $K_2Cr_2O_7$, on the right illustrate the difference in the colors of the chromate and dichromate ions. (Richard Megna/Fundamental Photographs)

We have already discussed the metallurgy of iron in considerable detail in Section 24.3. Let's consider here some of the important aqueous solution chemistry of the metal. Iron exists in aqueous solution in either the +2 (ferrous) or +3 (ferric) oxidation state. It often appears in natural waters because of contact of these waters with deposits of $FeCO_3$ ($K_{sp} = 3.2 \times 10^{-11}$). Dissolved $CO_2$ in the water can lead to dissolution of the mineral:

$$FeCO_3(s) + CO_2(aq) + H_2O(l) \longrightarrow Fe^{2+}(aq) + 2HCO_3^-(aq) \quad [25.7]$$

The dissolved $Fe^{2+}$, together with $Ca^{2+}$ and $Mg^{2+}$, contributes to water hardness (Section 14.6).

The standard reduction potentials given in Table 25.2 tell us much about the kind of chemical behavior we should expect to observe for iron. The potential for reduction from the +2 state to the metal is negative; conversely, the reduction from the +3 to the +2 state is positive. This tells us that iron should react with nonoxidizing acids such as dilute sulfuric acid or acetic acid to form $Fe^{2+}(aq)$, as indeed occurs. However, in the presence of air, $Fe^{2+}(aq)$ tends to oxidize to $Fe^{3+}(aq)$, as evidenced by the positive standard voltage for Equation 25.8:

$$4Fe^{2+}(aq) + O_2(g) + 4H^+(aq) \longrightarrow 4Fe^{3+}(aq) + 2H_2O(l) \quad E° = +0.46\ V \quad [25.8]$$

You may have seen instances in which water dripping from a faucet has left a brown stain in a sink (see Figure 25.9). The brown color is due to insoluble iron(III) oxide, formed by oxidation of iron(II) present in the water:

$$4Fe^{2+}(aq) + 8HCO_3^-(aq) + O_2(g) \longrightarrow 2Fe_2O_3(s) + 8CO_2(g) + 4H_2O(l) \quad [25.9]$$

When iron metal reacts with an oxidizing acid such as warm dilute nitric acid, $Fe^{3+}(aq)$ is formed directly:

$$Fe(s) + NO_3^-(aq) + 4H^+(aq) \longrightarrow Fe^{3+}(aq) + NO(g) + 2H_2O(l) \quad [25.10]$$

In the +3 oxidation state, iron is soluble in acidic solution as the hydrated ion, $Fe(H_2O)_6^{3+}$. However, this ion hydrolyzes readily (Section 17.8):

$$Fe(H_2O)_6^{3+}(aq) \rightleftharpoons Fe(H_2O)_5(OH)^{2+}(aq) + H^+(aq) \quad [25.11]$$

As an acidic solution of iron(III) is made more basic, a gelatinous red-brown precipitate, most accurately described as a hydrous oxide, $Fe_2O_3 \cdot nH_2O$, is formed (see Figure 25.10). In this formulation $n$ represents an indefinite number of water molecules, depending on the precise conditions of the precipitation. Usually, the precipitate that forms is represented merely as $Fe(OH)_3$. The solubility of $Fe(OH)_3$ is very low ($K_{sp} = 4 \times 10^{-38}$). It dissolves in strongly acidic solution, but not in basic solution. The fact that it does *not* dissolve in basic solution is the basis of the Bayer process, in which aluminum is separated from impurities, primarily iron(III) (Sections 18.5 and 24.4).

**Iron**

**FIGURE 25.9** Stain in a sink caused by iron in water, which is oxidized to $Fe^{3+}$ and precipitated as $Fe_2O_3$. (Page Poore)

**FIGURE 25.10** Addition of an NaOH solution to an aqueous solution of $Fe^{3+}$ causes $Fe(OH)_3$ to precipitate. (Donald Clegg and Roxy Wilson)

## Nickel

**FIGURE 25.11**
Anhydrous nickel(II) chloride, $NiCl_2$, is yellow; water solutions of this salt are green. (Donald Clegg and Roxy Wilson)

Nickel is not a particularly abundant element. It occurs chiefly in combination with arsenic, antimony, or sulfur. (You will note that both nickel and copper, which we will discuss next, are found mainly in combination with sulfur and related nonmetals, whereas the earlier transition elements, such as chromium or iron, are often found as oxide ores.) After separating nickel from other metallic elements that are usually present, the ore is roasted to NiO, which is then reduced with coke to form the crude metal. Crude nickel is normally refined by electrolysis, using nickel electrodes and an electrolyte containing both nickel sulfate and ammonium sulfate.

About 80 percent of the nickel produced goes into alloys. Among the more important of these is *stainless steel*, which also contains substantial amounts of chromium, and *Nichrome*, composed of 80 percent nickel and 20 percent chromium. Nichrome is used to make heating elements and in other applications that require a corrosion- and heat-resistant metal.

Nickel metal dissolves slowly in dilute mineral acids. Rather surprisingly, it does not dissolve in concentrated nitric acid. Attack of the strongly oxidizing acid on the metal surface results in formation of a protective oxide coat that stops further reaction.

In its compounds nickel is almost always found in the $+2$ oxidation state. Anhydrous nickel(II) chloride, $NiCl_2$, is a yellow solid. It dissolves in water to form a green solution that contains the hydrated nickel(II) cation, $Ni(H_2O)_6^{2+}$ (see Figure 25.11). The $Ni^{2+}$ ion forms complexes with many different anions and other neutral molecules, such as $NH_3$; these are discussed in Chapter 26. Among the less soluble compounds of nickel are the carbonate, $NiCO_3$, the oxalate, $NiC_2O_4$, and the sulfide, NiS.

## Copper

In Chapter 24 we discussed how copper can be produced from the ore chalcopyrite, $CuFeS_2$, and refined by electrolysis. In its aqueous solution chemistry, copper exhibits two oxidation states, $+1$ (cuprous) and $+2$ (cupric). Note that in the $+1$ oxidation state copper possesses a $d^{10}$ electron configuration. Salts of $Cu^+$ are often water insoluble and are mostly white in color. In solution the $Cu^+$ ion readily disproportionates:

$$2Cu^+(aq) \longrightarrow Cu^{2+}(aq) + Cu(s) \qquad K = 1.2 \times 10^6 \qquad [25.12]$$

Because of this reaction, and because copper(I) is readily oxidized to copper(II) under most solution conditions, the $+2$ oxidation state is by far the more common.

Many salts of $Cu^{2+}$, including $Cu(NO_3)_2$, $CuSO_4$, and $CuCl_2$, are water soluble. Copper sulfate pentahydrate, $CuSO_4 \cdot 5H_2O$, a widely used salt, has four water molecules bound to the copper ion, and a fifth held to the $SO_4^{2-}$ ion by hydrogen bonding. The salt is blue (it is called *blue vitriol*; see Figure 25.12). Aqueous solutions of $Cu^{2+}$, in which the copper ion is coordinated by water molecules, are also blue. Among the insoluble compounds of copper(II) are $Cu(OH)_2$, which is formed when NaOH is added to an aqueous $Cu^{2+}$ solution (see Figure 25.13). This blue compound readily loses water on heating to form black copper(II) oxide:

$$Cu(OH)_2(s) \longrightarrow CuO(s) + H_2O(l) \qquad [25.13]$$

**FIGURE 25.12** Crystals of copper(II) sulfate pentahydrate, $CuSO_4 \cdot 5H_2O$. (Paul Silverman/Fundamental Photographs)

**FIGURE 25.13** Addition of an NaOH solution to an aqueous solution of $Cu^{2+}$ causes $Cu(OH)_2$ to precipitate. (Donald Clegg and Roxy Wilson)

CuS is one of the least soluble copper(II) compounds ($K_{sp} = 6.3 \times 10^{-36}$). This black substance does not dissolve in NaOH, $NH_3$, or in nonoxidizing acids, such as HCl. However, it does dissolve in $HNO_3$, which oxidizes the sulfide to sulfur:

$$3CuS(s) + 8H^+(aq) + 2NO_3^-(aq) \longrightarrow 3Cu^{2+}(aq) + 3S(s) + 2NO(g) + 4H_2O(l) \qquad [25.14]$$

$CuSO_4$ is often added to water to stop algae or fungal growth, and other copper preparations are used to spray or dust plants to protect them from lower organisms and insects. Copper compounds are not generally toxic to human beings, except in massive quantities. Our daily diet normally includes from 2 to 5 mg of copper.

## 25.4 TRANSITION METAL OXIDES

The oxides of the transition elements are an interesting and very important group of compounds. That undesirable phenomenon known as corrosion is due primarily to formation of metal oxides. In the effort to inhibit corrosion, we are thus concerned with finding ways to inhibit oxide formation. Conversely, many metal oxides have very useful properties and play important roles in chemical technology and electronics.

Table 25.3 shows the oxides formed by metals of the first transition

**TABLE 25.3** Oxides of the first-series transition elements

| Element | | | | | | | | | |
|---------|---|---|---|---|---|---|---|---|---|
| **Sc** | **Ti** | **V** | **Cr** | **Mn** | **Fe** | **Co** | **Ni** | **Cu** | **Zn** |
| $Sc_2O_3$ | TiO | VO | $Cr_2O_3$ | MnO | FeO | CoO | NiO | $Cu_2O$ | ZnO |
| | $Ti_2O_3$ | $V_2O_3$ | $CrO_2$ | $Mn_2O_3$ | $Fe_3O_4$ | $Co_3O_4$ | $NiO_2$ | CuO | |
| | $TiO_2$ | $VO_2$ | $CrO_3$ | $MnO_2$ | $Fe_2O_3$ | $Co_2O_3$ | | | |
| | | $V_2O_5$ | | $Mn_2O_7$ | | $CoO_2$ | | | |

Many transition-metal oxides are used as paint pigments, because they are very stable compounds that are able to withstand exposure to sunlight and air without undergoing chemical change. We have noted that the colors in compounds of the transition elements are usually due to the presence of electrons in incompletely filled *d* orbitals. In Chapter 26 we will take up a discussion of the electronic transitions that give rise to these colors. As you might expect, the oxides that possess completely empty or completely filled 3*d* subshells (for example, $TiO_2$ and ZnO) are white. In fact, both of these oxides are used as white-paint pigments.

Very finely divided metal oxides are employed in high-grade paint finishes used in the automobile industry. When the particle size of the metal oxide pigment is on the order of 0.1 $\mu$m or smaller, the particles no longer efficiently scatter visible light rays. Thus, while they impart color to the lacquer in which they are mixed, they are transparent. A special form of $Fe_2O_3$ is widely used in this application. Small oxide particles are formed by precipitating $Fe(OH)_2$, then oxidizing it in air to form a substance of composition FeOOH. This is then dehydrated at high temperature to form small particles of $Fe_2O_3$. The iron oxide particles help prevent degradation of the lacquer by sunlight, because they strongly absorb the visible and ultraviolet rays. The brilliant metallic finish seen on many automobiles is achieved by mixing aluminum flakes about 10 to 50 $\mu$m in size into the lacquer.

series. By comparing the entries in this table with those in Figure 25.5, you can see that most of the oxidation states listed for the first-series transition elements are represented in the oxides. Notice in particular that for the elements through Mn, the highest oxidation state seen for each element is equal to the number of valence-shell electrons. The properties of the metal oxides reflect certain trends and regularities that should be kept in mind:

1. For any particular element the oxide of highest oxidation state will act in general as an oxidizing agent. However, the oxidizing power increases steadily as the oxidation state of the metal increases. Thus $Sc_2O_3$ and $TiO_2$ are not good oxidizing agents at all, but $CrO_3$ and $Mn_2O_7$ are powerful ones.

2. The acid-base behavior of a metal oxide depends on the oxidation state of the metal. In general, oxides in which the metal is in a low oxidation state are basic in character; those in which it is in a high oxidation state are acidic in character. Metal oxides are often insoluble in water, but their acidic or basic properties are revealed by whether they dissolve on reaction with acid or base. For example, FeO is insoluble in water or aqueous base, but it dissolves in acidic solution, thus exhibiting basic character.

3. The acid-base characteristics just described arise because there is a shift from ionic to covalent character in the metal-oxygen bond as the oxidation state of the metal increases. Thus the bonding in MnO is essentially ionic, but the bonding in $Mn_2O_7$ is polar covalent. This shift in bond character occurs because the metal's attraction for the electrons on the oxygen increases as the metal charge increases.

## SAMPLE EXERCISE 25.4

Predict the products that will occur upon reaction of **(a)** $V_2O_5$ with $H_2O$; **(b)** VO with a strong mineral acid. (Note: Assume that no oxidation-reduction reaction occurs.)

**Solution:** **(a)** $V_2O_5$ should be the anhydride of a strong acid. We note that the group 5B elements, like the group 5A elements, have five electrons in the orbitals beyond the nearest rare-gas element. By analogy with the behavior of group 5A elements, we would thus expect the acid formed by $V_2O_5$ to be $H_3VO_4$.

$$V_2O_5(s) + 3H_2O(l) \longrightarrow 2H_3VO_4(aq)$$

Here we have assumed that $H_3VO_4$ is a weak acid, in analogy with $H_3PO_4$. In fact, $K_{a1}$ for $H_3VO_4$ is $1.7 \times 10^{-4}$.

**(b)** The oxide VO should be basic; that is, its reaction with acid should be analogous to that of CaO or other basic oxides, leading to the +2 aquated metal ion:

$$VO(s) + 2H^+(aq) \longrightarrow V^{2+}(aq) + H_2O(l)$$

**PRACTICE EXERCISE** _____
Predict the product formed upon addition of $Mn_2O_7$ to water.
*Answer:* $HMnO_4$

---

The structures of the metal oxides can best be considered in terms of some extensions of the idea of close packing, discussed in Section 11.5. Anions generally have larger radii than do cations (Section 8.3). In metal oxides and other ionic compounds, the large anions in the lattice often assume a close-packed arrangement. A close-packed array of spheres, whether hexagonal or cubic close packed, occupies 74 percent of the total volume taken up by the structure as a whole. The remaining 26 percent consists of holes between the spheres. The smaller cations in the lattice occupy these holes. For example, $Fe_2O_3$ consists of a hexagonal close-packed arrangement of oxide ions, with $Fe^{3+}$ ions in the spaces between the oxide ions.

Examination of the close-packed structures reveals that they possess various types of holes. Two of these are shown in Figure 25.14. The first type is created by four large spheres arranged in a tetrahedral fashion. The second type is created by six large spheres arranged in an octahedral fashion. Trigonometry shows that a **tetrahedral hole** is smaller than an **octahedral hole**.

The size of the cation relative to that of the anion helps to determine the type of hole that the cation occupies. The most stable arrangement is one that maximizes the number of cation-anion interactions, because these lead to electrostatic attractive forces. At the same time, the arrangement should be one that prevents direct anion-anion contacts, because these represent repulsive electrostatic forces. A sphere whose radius is 0.225 times that of the larger sphere will fit perfectly into a tetrahedral hole. If the smaller sphere has a radius that is 0.414 times that of the larger spheres, it will just fit into an octahedral hole.

Consider the case where a cation occupies a tetrahedral hole. If the ratio of cation radius to anion radius, $r_c/r_a$, is 0.225, the cation just touches all four anions, and they just touch one another. If the cation were a little larger, so that $r_c/r_a$ were greater than 0.225, the cation would still be in touch with the four anions, but they would have been forced apart from one another, thus reducing the repulsive forces. In other words, this leads to a more stable arrangement. If we put that same cation in an octahedral hole, it would, in a sense, "rattle around" in the hole, because $r_c/r_a$ is less than 0.414. The anions would all be in contact, and a relatively less stable arrangement would result. For cations with radii such that $0.225 < r_c/r_a < 0.414$, therefore, a tetrahedral hole is normally the position of lowest energy.

**The Structure of Metal Oxides**

**FIGURE 25.14** Views of the tetrahedral and octahedral holes that exist between two layers of close-packed spheres.

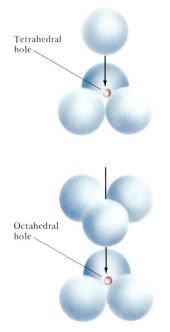

Tetrahedral hole

Octahedral hole

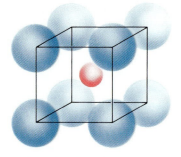

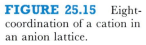

**FIGURE 25.15** Eight-coordination of a cation in an anion lattice.

**TABLE 25.4** Radius ratios and cation location in an anion lattice

| $r_c/r_a$ [a] | Coordination number of cation | Arrangement of anions about the cation |
|---|---|---|
| 0.225–0.414 | 4 | Tetrahedral |
| 0.414–0.732 | 6 | Octahedral |
| 0.732–1.000 | 8 | Cubic |

[a] If $r_c > r_a$, the ratio $r_a/r_c$ gives the coordination number of the anion and the arrangement of cations.

When the ratio of radii $r_c/r_a$ reaches a value of 0.414 or greater, the cation is no longer stable in the tetrahedral hole. Instead, the more stable location is the octahedral hole, which permits the cation to contact a greater number of anions. When $r_c/r_a > 0.414$, the cation forces the anions apart, preventing direct anion-anion contact while maintaining the stabilizing cation-anion contacts. For cation radii such that $0.414 < r_c/r_a < 0.732$, the most stable arrangement results when the cation is in an octahedral hole. When $r_c/r_a$ exceeds 0.732, the anions are no longer most stable in a close-packed array. Instead, they assume a simple cubic arrangement, which permits the cation to be in contact with eight anions, as illustrated in Figure 25.15. The number of stabilizing cation-anion contacts is thus further increased. These radius ratio guidelines, which indicate the most likely environment for cations in an anionic lattice, are summarized in Table 25.4. The radius ratio rules are not always followed, because other factors, such as the electronic configurations of the metal ions and covalency in the bonding, also play a role. Nevertheless, they are useful guidelines to the structures that might be expected.

**SAMPLE EXERCISE 25.5**

The mineral hematite, $Fe_2O_3$, consists of a hexagonal close-packed array of oxide ions with $Fe^{3+}$ ions occupying interstitial positions. Predict whether the $Fe^{3+}$ ions are in octahedral or tetrahedral holes. The radius of $Fe^{3+}$ is 0.65 Å; that of $O^{2-}$ ion is 1.45 Å.

**Solution:** The radius ratio $r_c/r_a$ is $0.65/1.45 = 0.45$. Based on the radius ratio, we would predict (Table 25.4) that the $Fe^{3+}$ ions would be located in octa-

hedral holes, as in fact they are.

**PRACTICE EXERCISE**

The mineral rutile, $TiO_2$, is approximately a close-packed array of $O^{2-}$ ions with $Ti^{4+}$ ions in interstitial positions. If the radius of $Ti^{4+}$ is 0.56 Å, and that of $O^{2-}$ is 1.45 Å, predict whether the $Ti^{4+}$ ions are in octahedral or tetrahedral holes.

*Answer:* tetrahedral

One of the most common oxide structures is that adopted by *corundum*, a form of $Al_2O_3$. In this structure, the oxide anions form a hexagonal close-packed lattice, with the cation in octahedral holes. The oxides $Ti_2O_3$, $V_2O_3$, $Cr_2O_3$, $Fe_2O_3$, and $Rh_2O_3$ all adopt the corundum structure. Many other oxides adopt the rutile ($TiO_2$) structure, illustrated in Figure 8.5. The oxide ions in the rutile structure are not close packed. The metal ions are surrounded by six anions, but there is some distortion, so that two of the anions are located at different distance than the other four. Among the transition metal oxides that adopt the rutile structure are $MnO_2$, $ZrO_2$, and $RuO_2$.

The magnetic properties of transition metals and their compounds are both interesting and important. Measurements of magnetic properties provide us with information about chemical bonding. In addition, many important uses are made of magnetic properties in modern technology.

In our discussion of chemical bonding in Section 9.6, we noted that substances that contain unpaired electrons are paramagnetic; that is, they are drawn into a magnetic field. The electron is the source of this magnetism. The electron possesses a "spin" that gives it a magnetic moment; that is, it behaves like a tiny magnet (Section 7.2). An atom or ion has a magnetic moment only if it possesses one or more unpaired electrons. For the transition metals, these unpaired electrons are in the $d$ orbitals.

When an atom or ion possessing one or more unpaired electrons becomes incorporated into a solid, those electrons may become involved in bonding and thus be effectively paired. If this occurs, the solid will not possess a magnetic moment. However, if unpaired electrons remain, the solid as a whole will exhibit some form of magnetism. Many different types of magnetic behavior are possible in solids. We will not consider all the possible situations, some of which are very complicated. However, the major classes of magnetic properties are readily understood on the basis of principles that we have already discussed in other contexts.

When all the electrons in a solid are paired, the substance exhibits **diamagnetic** behavior (Section 9.6). The magnetic moments of the paired electrons effectively cancel each other. However, when the substance is placed in a magnetic field, the motions of the electrons are perturbed in such a way that the substance is very weakly repelled by the magnet. A diamagnetic lattice is illustrated schematically in Figure 25.16($a$). Examples of diamagnetic solids include NaCl, $SiO_2$, copper metal, and most organic substances, such as methane, $CH_4$, and benzene, $C_6H_6$.

If the ions in a solid have unpaired electrons that are not influenced by electrons on other ions in the solid, the substance exhibits simple **paramagnetic** behavior. The magnetic moments on the individual ions are randomly oriented, as shown schematically in Figure 26.16($b$). When placed in a magnetic field, the magnetic moments become aligned roughly parallel to each other, producing a net attractive interaction with the magnet. Thus a paramagnetic substance is drawn into a magnetic field. The strength of this attraction is proportional to the strength of the magnetic field and to the number of unpaired electrons on the ions.

### Diamagnetism and Simple Paramagnetism

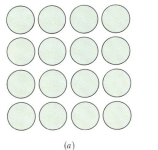
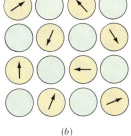
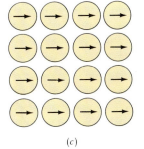
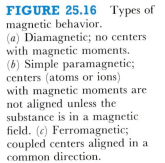

FIGURE 25.16   Types of magnetic behavior. ($a$) Diamagnetic; no centers with magnetic moments. ($b$) Simple paramagnetic; centers (atoms or ions) with magnetic moments are not aligned unless the substance is in a magnetic field. ($c$) Ferromagnetic; coupled centers aligned in a common direction.

$(a)$      $(b)$      $(c)$

**Ferromagnetism**

In a **ferromagnetic** solid the individual magnetic centers, which may be transition metal atoms or metal ions, interact with one another through the overlaps of atomic orbitals. By this means the electrons on a given center sense the orientations of electrons on neighboring centers. The most stable (lowest-energy) arrangement results when the spins of electrons on different centers are aligned in the same direction. When a ferromagnetic solid is placed in a magnetic field, the electrons tend to align strongly along the magnetic field. The attraction for the magnetic field that results may be as much as 1 million times larger than for a simple paramagnetic substance. When the external magnetic field is removed, the interactions between the electrons cause the solid as a whole to maintain a magnetic moment. Ferromagnetic behavior is illustrated schematically in Figure 25.16(c).

The most common examples of ferromagnetic materials are the elements Fe, Co, and Ni. You have perhaps done the experiment of inserting an iron nail into a coiled wire and passing a direct current through the wire for a time, as illustrated in Figure 25.17. When the current is turned off, the nail is seen to have developed a permanent magnetic moment, as evidenced by the manner in which it causes alignment of iron filings. We refer to it as a permanent magnet. Many alloy compositions exhibit ferromagnetism to a higher degree than do the pure metals themselves. Such alloys are employed in most applications of permanent magnets. Certain metal oxides (for example, $CrO_2$) also exhibit ferromagnetic behavior.

**FIGURE 25.17**  (a) Formation of a permanent magnet from an iron nail by the passage of direct current. (b) The presence of permanent magnetism is seen when the nail is laid on a sheet of paper and iron filings align in the field.

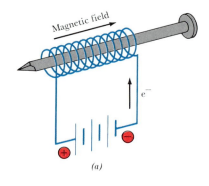

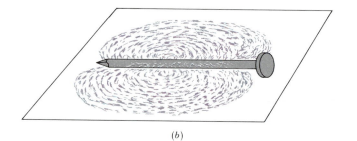

The Use of Ferromagnetic Oxides in
Recording Tapes

Several ferromagnetic metal oxides are used as recording media in magnetic recording tape. $CrO_2$,
which is one of the most widely used oxides, is
formed by reduction of $CrO_3$ or by a high-temperature reaction of two different chromium oxides with
one another:

$$2CrO_3(s) \longrightarrow 2CrO_2(s) + O_2(g) \qquad [25.15]$$

$$CrO_3(s) + Cr_2O_3(s) \longrightarrow 3CrO_2(s) \qquad [25.16]$$

Other oxide materials used include a special crys-
tallographic form of $Fe_2O_3$ and the mixed oxide
$CoFe_2O_4$. In magnetic recording it is important that
the ferromagnetic material be in a very finely divided, needle-shaped form. A diagram of the recording process is shown in Figure 25.18. When an
electrical current is passed through the coil, lines
of magnetic field are established in the core. The
field lines "leak" out and flow through the magnetic tape that passes in front of a gap. Variation
in the magnetic field, caused by modulation of the
signal applied to the core, causes a variation in the
orientation and strength of the magnetic moment
imparted to the ferromagnetic particles in the tape.
Played back, these variations recreate the recorded
sound.

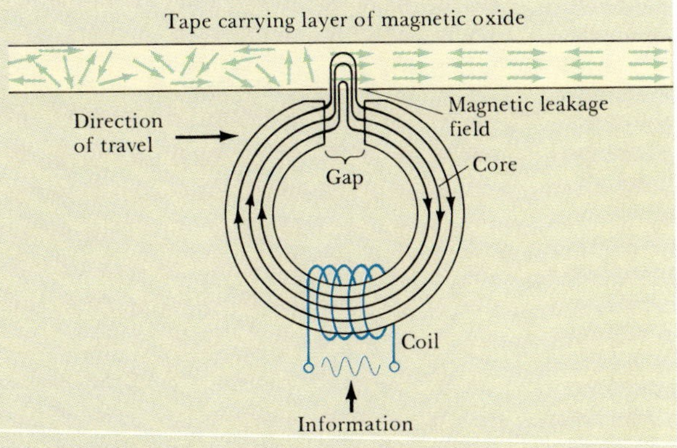

**FIGURE 25.18** Schematic illustration of magnetic
recording. As the tape is drawn past the gap in the core,
magnetic field lines extending into the magnetic oxide
in the tape cause an orientation of the magnetic
moments within the domains of the oxide. The
orientation and strength of the magnetic field varies
as the information is supplied to the core in the
form of a variable current.

From the material we have discussed in this and the preceding chapter
we see that a modern, high-technology society depends critically on the
availability of many metallic elements. It is important to the economic
and military security of a highly industralized nation that it have depend-
able access to the mineral sources from which these essential metals are
derived. In recent years this access has become a matter of deep concern
in the United States, because in fact this country has no domestic sources
of many of the most critical elements. Figure 25.19 shows the extent of
our dependence on imports in terms of percentages of annual consumption that are imported. Note that all of the elements listed in Figure 25.2
as important components of a modern jet engine are on this list. Furthermore, the most important sources of some of the metals are not highly
developed nations with which our diplomatic relations are secure.

    In response to this situation, the United States has developed a strate-
gic minerals policy. A part of this policy establishes stockpiles of the most
critical metals, so that if all foreign sources were cut off, the United
States could function for an extended time by drawing on the reserves.
Intensive efforts to develop new reserves within the United States, and to
recycle old metal, have been initiated.

## 25.6 METALS AS STRATEGIC MATERIALS

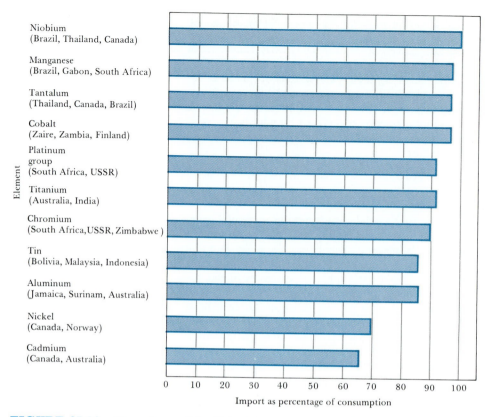

Element

Niobium
(Brazil, Thailand, Canada)

Manganese
(Brazil, Gabon, South Africa)

Tantalum
(Thailand, Canada, Brazil)

Cobalt
(Zaire, Zambia, Finland)

Platinum
group
(South Africa, USSR)

Titanium
(Australia, India)

Chromium
(South Africa, USSR, Zimbabwe)

Tin
(Bolivia, Malaysia, Indonesia)

Aluminum
(Jamaica, Surinam, Australia)

Nickel
(Canada, Norway)

Cadmium
(Canada, Australia)

0   10   20   30   40   50   60   70   80   90   100

Import as percentage of consumption

**FIGURE 25.19**   Dependence of the United States on imports of strategic metals.

# FOR REVIEW

## SUMMARY

In this chapter we have considered some of the more important general properties of the transition elements. We have also seen how these properties find useful application.

The transition elements are characterized by the incomplete filling of the $d$ orbitals. As we proceed through a given series of transition elements (that is, through a horizontal row), the effective nuclear charge, as seen by the valence electrons, increases slowly. The result is that the later transition elements in a given row tend to adopt lower oxidation states and have slightly smaller ionic radii for a given oxidation state. In comparing the elements of a given group—that is, elements located vertically with respect to one another in the table—we find that the atomic and ionic radii increase in the second series as compared with the first, but that the elements of the second and third series are very similar with respect to these properties. The effect is due to the **lanthanide contraction**. The lanthanide elements, atomic numbers 58 through 71, cause an increase in effective nuclear charge that compensates for the increase in major quantum number in the third series as compared with the second.

The presence of $d$ electrons in the transition elements leads to the existence of multiple oxidation states. In the transition elements of groups 3B through 7B the maximum oxidation state observed corresponds to the total number of *s* and *d* electrons available in the valence-shell orbitals. In the first transition series, the maximum oxidation state declines beyond group 7B. In the second and third series, the maximum is reached in group 8B with Ru and Os, respectively, and declines thereafter. The decline in each series can be ascribed to the increasing effective nuclear charge.

Metal oxides are a very important class of transition metal compounds. Because oxygen is a highly electronegative element, oxides of metals in high oxidation states are observed. Oxides of metals in high oxidation states, for example, $V_2O_5$ and $Mn_2O_7$, react with water to form acidic solutions. Oxides of metals in low oxidation states, for example, FeO, behave as bases in their reactions with water or with acidic solutions. The structures of solid metal oxides can be viewed as close-packed lattices of oxide ions, with the relatively smaller metal ions occupying the holes between oxide ions. The metal ion is most commonly found in either a tetrahedral hole, in which it has four oxide ion neighbors, or in an octahedral hole, in which it has six. When the ratio of cation radius to anion radius, $r_c/r_a$, is in the range $0.225 < r_c/r_a < 0.414$, a tetrahedral hole is preferred on electrostatic energy grounds. When the ratio is in the range $0.414 < r_c/r_a < 0.732$, an octahedral hole is preferred. When $r_c/r_a$ exceeds 0.732, an arrangement involving eight oxide ions about the metal ion is energetically preferred. This type of coordination requires a slight rearrangement of the oxide ions from the close-packed structure.

The presence of unpaired electrons in valence orbitals leads to interesting magnetic behavior in the transition elements and their compounds. Many substances can be classified in terms of one of the following magnetic types:

1. In a **diamagnetic** substance there are no unpaired electrons. The substance is repelled by a magnetic field.

2. In a simple **paramagnetic** substance, unpaired electrons on ions in a solid lattice or in solution are oriented by a magnetic field, so that the substance is drawn into the magnetic field. In a paramagnetic substance the unpaired spins on a given site do not interact significantly with those on other sites in the lattice or in solution.

3. In a **ferromagnetic** substance the unpaired spins on ions in the lattice are strongly affected by those on neighboring ions. In a magnetic field the orientations of spins in different domains become aligned along the magnetic field direction. When the magnetic field is removed, this common orientation remains, giving rise to a magnetic moment, as observed in permanent magnets.

Because so many uses are made of transition element in modern industry, these elements are important for national security. The United States has no significant deposits of some strategically important elements. To ensure an adequate supply of essential metals, the government has a strategic minerals policy that establishes stockpiles of critical metals and encourages the search for alternative sources, particularly the discovery of new reserves within the United States.

## LEARNING GOALS

Having read and studied this chapter, you should be able to:

1. Identify the transition elements on a periodic table.

2. Write the electronic configurations for the transition elements and their ions.

3. Describe the manner in which atomic properties such as ionization energy and atomic radius vary within each transition element series, and within each vertical group of transition elements.

4. Explain why the atomic radii and other atomic properties of second- and third-series transition elements of the same group are closely similar. To put it another way, you should be able to describe the lanthanide contraction and explain its origin.

5. Describe the general manner in which the maximum observed oxidation state varies as a function of group number among the transition elements, and account for the observed variation.

6. Describe the general manner in which the most stable oxidation state or states in aqueous solution vary as a function of group number among the first-series transition elements.

7. Write balanced chemical equations corresponding to the simple aqueous solution chemistry of chromium, iron, nickel, and copper, and account for the variations in chemical properties observed among these elements in terms of the characteristics of the elements themselves.

8. Describe the manner in which the highest oxidation state of the first-series transition metal oxides varies as a function of atomic number, and account for the observed variation.

9. Describe and explain the manner in which oxidizing strength, acid-base characteristics, and metal-oxygen bond character vary with the oxidation state of the metal.

10. Describe the factors that determine the most stable coordination environment for a metal ion in an oxide lattice.

11. Explain how the most stable coordination number for a metal ion in an oxide lattice varies with the ratio of cation to anion radii, $r_c/r_a$.

12. Name and describe the types of magnetic behavior discussed in this chapter.

13. Provide an explanation of each type of magnetic behavior in terms of the arrangements in the lattice of metal ions with unpaired spins, and their interactions with one another.

14. Explain what is meant by the term "strategic metal."

## KEY TERMS

Among the more important terms and expressions used for the first time in this chapter are the following:

A substance exhibits **ferromagnetism** (Section 25.5) when it possesses unpaired electrons on metal atoms or ions in the solid lattice, and when the magnetic moments on each metal ion are aligned with one another.

The **lanthanide contraction** (Section 25.1) refers to the gradual decrease in atomic and ionic radii with increasing atomic number among the lanthanide elements, atomic numbers 58 through 71. The decreases arise because of a gradual increase in effective nuclear charge through the lanthanide series.

An **octahedral hole** (Section 25.4) in a close-packed lattice is an opening in the lattice formed by six lattice units in contact with one another. The radius of the largest sphere that can be placed in this hole without forcing the larger spheres apart is 0.414 times as large as the radius of the larger spheres.

A **tetrahedral hole** (Section 25.4) in a close-packed lattice is an opening in the lattice formed by four lattice units in contact with one another. The radius of the largest sphere that can be placed in this hole without forcing the larger lattice spheres apart is 0.225 as large as the radius of the larger spheres.

## EXERCISES

### Physical Properties of Transition Elements

**25.1** Using only the periodic table for guidance, write the electron configurations for (**a**) Ti; (**b**) Ru; (**c**) Ni; (**d**) Cd.

**25.2** Using only the periodic table for guidance, write the electronic configurations for (**a**) Mn; (**b**) Zr; (**c**) Ag; (**d**) Re.

**25.3** How would you expect the heat of fusion for vanadium to compare with the heat of vaporization? Explain.

**25.4** How would you expect the heat of fusion of titanium to compare with that of vanadium?

**25.5** Which of the following properties is most closely related to bonding in the metal, and which is characteristic of individual atoms: (**a**) heat of sublimation; (**b**) boiling point; (**c**) first ionization energy; (**d**) malleability.

**25.6** Which of the following properties is most clearly related to the bonding in metals, and which is characteristic of individual atoms: (**a**) heat of fusion; (**b**) hardness; (**c**) atomic radius; (**d**) electrode potential for $M^{2+}(aq) + 2e^- \longrightarrow M(s)$?

**25.7** What is the lanthanide contraction? Why does it occur?

**25.8** What effect does the lanthanide contraction have on the properties of transition elements?

## Oxidation States

**25.9** Write the formula for the chloride corresponding to the highest expected oxidation state for each of the following elements: (**a**) Nb; (**b**) W; (**c**) Co; (**d**) Cd; (**e**) Re.

**25.10** Write the formula for the oxide corresponding to the highest expected oxidation state for each of the following elements: (**a**) Cr; (**b**) Zn; (**c**) Hf; (**d**) Os; (**e**) Sc.

**25.11** What accounts for the fact that chromium exhibits several oxidation states in its compounds, whereas aluminum exhibits only the $+3$ oxidation state?

**25.12** Account for the fact that zinc exhibits only the $+2$ oxidation state in its compounds, whereas copper exhibits both the $+1$ and $+2$ oxidation states.

**25.13** Write the expected electron configuration for each of the following ions: (**a**) $Ru^{4+}$; (**b**) $Hf^{3+}$; (**c**) $Ag^{3+}$; (**d**) $Co^{3+}$; (**e**) $V^{3+}$; (**f**) $Pt^{2+}$.

**25.14** Write the electron configurations for each of the following ions: (**a**) $Ti^{2+}$; (**b**) $Sc^{3+}$; (**c**) $W^{4+}$; (**d**) $Ir^{3+}$; (**e**) $Ni^{2+}$; (**f**) $Ag^+$.

**25.15** Which would you expect to be more easily oxidized, $Cr^{2+}$ or $Fe^{2+}$?

**25.16** Which would you expect to be more easily oxidized, $Ti^{2+}$ or $Ni^{2+}$?

**25.17** Which would you expect to be the better oxidizing agent, $CrO_4^{2-}$ or $WO_4^{2-}$?

**25.18** Which would you expect to be harder to reduce, $V_2O_5$ or $Ta_2O_5$?

**25.19** (**a**) What effect does hydration of the metal ion have on the electrode potential for the process $M^{2+}(aq) + 2e^- \longrightarrow M(s)$? (**b**) How would formation of a very sta-

ble metal complex in solution affect the value of the potential for reduction of the metal ion to the metal?

**25.20** The stability of the three complexes $Zn(H_2O)_4^{2+}$, $Zn(NH_3)_4^{2+}$, and $Zn(CN)_4^{2-}$ increases from the $H_2O$ to the $NH_3$ to the $CN^-$ complex. How do you expect the electrode potentials of these three complexes to compare to the processes in which Zn(II) is reduced to Zn metal?

## Chemical Properties

**25.21** How does the presence of air affect the relative stabilities of $Cr^{2+}$ and $Cr^{3+}$ in aqueous solution? Write a balanced chemical reaction as part of your response.

**25.22** Chromium(III) ion is amphoteric. Write equations to describe the reactions of $Cr_2O_3$ with sulfuric acid solution and with aqueous sodium hydroxide.

**25.23** What are the two most common oxidation states of chromium in aqueous solution?

**25.24** Give the chemical formulas and colors of the chromate and dichromate ions. Which of these ions is more stable in acidic solution?

**25.25** What are the two most common oxidation states of iron in aqueous solution?

**25.26** How does the presence of air affect the relative stabilities of the ferrous and ferric ions?

**25.27** Write balanced chemical equations for the reaction between iron and (**a**) hydrochloric acid; (**b**) nitric acid.

**25.28** Write balanced chemical equations for the reaction of (**a**) $Fe^{3+}(aq)$ and sodium hydroxide solution; (**b**) $FeCO_3(s)$ and hydrochloric acid solution.

**25.29** What is the common oxidation state of nickel in aqueous solution?

**25.30** What are the two most common oxidation states of copper in aqueous solution?

**25.31** What are the colors associated with the $Ni^{2+}(aq)$ and the $Cu^{2+}(aq)$ ions?

**25.32** Write a chemical equation for the disproportionation of $Cu^+$ in aqueous solution. What colors are usually associated with the $Cu^+(aq)$ and $Cu^{2+}(aq)$ ions?

## Metal Oxides

**25.33** Predict whether each of the following oxides will exhibit predominantly acidic or basic properties: (**a**) NiO; (**b**) $CrO_3$; (**c**) $Cu_2O$; (**d**) $Re_2O_7$.

**25.34** Write balanced chemical equations for each of the following descriptions. (**a**) $MnO_2$ is formed when $Mn(NO_3)_2$ is heated. (**b**) $Mn_3O_4$ is formed when Mn metal is heated in air. (**c**) Rhenium metal reacts with nitric acid to form an aqueous solution of the strong acid $HReO_4$. **(d)** Technetium(VII) oxide decomposes on heating to form technetium(IV) oxide. **(e)** Cobalt(II) oxide is converted to $Co_3O_4$ on heating in oxygen at 700 K. **(f)** Silver(I) oxide is oxidized by aqueous $S_2O_8^{2-}$ to AgO.

**25.35** Predict the type of hole in an oxide lattice that would be occupied by each of the following cations (radii in parentheses): **(a)** $Li^+$ (0.68 Å); **(b)** $Ti^{4+}$ (0.53 Å); **(c)** $Ni^{2+}$ (0.70 Å); **(d)** $Ca^{2+}$ (0.94 Å).

**25.36** From the ionic radii of each of the following metal ions, would you expect that the metal oxide would possess the corundum structure, as is found to be the case: $V^{3+}$, 0.74 Å; $Ti^{3+}$, 0.77 Å; $Rh^{3+}$, 0.75 Å? Explain.

## Magnetic Properties

**25.37** What, on an atomic level, distinguishes a paramagnetic material from a diamagnetic one? How does each behave in a magnetic field?

**25.38** What characteristics of a ferromagnetic material distinguish it from one that is paramagnetic? What type of interaction must occur in the solid to bring about ferromagnetic behavior?

**25.39** List three elements that are ferromagnetic.

**25.40** Predict whether the following solid substances are diamagnetic, paramagnetic, or ferromagnetic: **(a)** Ni; **(b)** Mg; **(c)** NaCl; **(d)** CoO.

## Additional Exercises

**25.41** List three general properties often associated with the transition elements.

**25.42** What is the order of energies of the following atomic orbitals in Re: **(a)** $4f$: **(b)** $6s$; **(c)** $5d$; **(d)** $5f$; **(e)** $5p$? List the lowest-energy orbital first, the highest-energy orbital last.

**25.43** Why do many transition elements exhibit multiple oxidation states?

**25.44** Explain why the heat of atomization of zinc is much smaller than the heat of atomization of vanadium (Table 25.1).

**25.45** The statement has been made that the chemistries of zirconium and hafnium are more nearly identical than that for any other two elements of a given group. What accounts for this similarity?

**25.46** How do the relative stabilities of high and low oxidation states vary within the $d$-block elements?

**25.47** Which compound in each of the following pairs is likely to be the stronger oxidizing agent: **(a)** $TiO_2$ or

$Mn_2O_7$; **(b)** $Cr_2O_3$ or $CrO_3$; **(c)** $FeCl_2$ or $FeCl_3$; **(d)** $NiO_2$ or $HfO_2$; **(e)** $Cu_2O$ or $Ag_2O$? Explain in each case.

**25.48** Write balanced chemical equations for each of the following verbal descriptions. **(a)** Vanadium(V) oxide is formed by thermal decomposition of ammonium metavanadate, $NH_4VO_3$. **(b)** Vanadium(V) oxide is reduced to vanadium(IV) oxide by thermal reduction with gaseous $SO_2$. **(c)** Vanadium(V) oxide is reduced by magnesium in hydrochloric acid to give a green solution of vanadium(III). **(d)** Niobium(V) chloride reacts with water to yield crystals of niobic acid, $HNbO_3$.

**25.49** Low-valent metal oxides generally dissolve in water only under conditions that lead to the aquated metal ion. High-valent metal oxides, however, generally dissolve to form oxygen-containing species, for example, $VO_4^{3-}$ or $CrO_4^{2-}$. Account for this difference in behavior.

**[25.50]** Graph the electrode potentials for the process $M^{2+}(aq) + 2e^- \longrightarrow M(s)$ for the first-series transition elements (Table 25.2). What trend is discernible in these data? How do you account for the observed variation?

**[25.51]** Using the data in Table 25.2, determine the standard electrode potentials for each of the following half-reactions:

**(a)** $Cr^{3+}(aq) + e^- \longrightarrow Cr^{2+}(aq)$
**(b)** $Cr_2O_7^{2-}(aq) + 12e^- + 14H^+(aq) \longrightarrow$
$$2Cr(s) + 7H_2O(l)$$
**(c)** $Co^{3+}(aq) + 3e^- \longrightarrow Co(s)$

**25.52** When a solution of $Cu^{2+}(aq)$ is reacted with aqueous sodium iodide, CuI and elemental iodine are formed. Write a balanced chemical equation for the reaction, and explain why it proceeds as it does.

**25.53** When $H_2S$ is bubbled through a solution containing $Fe^{3+}(aq)$, the precipitate that eventually forms is a mixture of iron(II) sulfide and sulfur. Write a balanced chemical equation for the reaction, and explain why it proceeds as it does.

**25.54** The vanadate ion, $VO_4^{3-}(aq)$, exists as such in strongly alkaline solution, pH > 13. As acid is added, a divanadate ion, and ions containing still more vanadium atoms, are formed. Write balanced chemical equations to describe the formation of the divanadate and trivanadate ions.

**25.55** Which of the following compounds would be expected to yield the most acidic aqueous solution: **(a)** $VCl_3$; **(b)** $FeSO_4$; **(c)** $CrO_3$; **(d)** $Co(NO_3)_3$?

**25.56** The radius of the sulfide ion is 1.90 Å. Assuming that the sulfides can be viewed as close-packed structures of $S^{2-}$ ions, predict the type of hole that each of the following cations will occupy: **(a)** $Zn^{2+}$ (0.74 Å); **(b)** $V^{5+}$ (0.54 Å); **(c)** $Tl^{3+}$ (0.88 Å).

**25.57** If cryolite, $Na_3AlF_6$, is viewed as a close-packed structure of fluoride ions, what types of interstitial positions do you predict the $Na^+$ and $Al^{3+}$ ions will occupy (radii: $F^-$, 1.33 Å; $Na^+$, 0.98 Å; $Al^{3+}$, 0.45 Å)?

**[25.58]** **(a)** What characteristics of $CrO_2$ render it useful as a magnetic tape recording medium? **(b)** Would a paramagnetic substance serve as a recording medium? Explain.

**[25.59]** The metals cobalt, chromium, manganese, platinum, and titanium receive most attention as strategic metals. By using references other than this text (for example, *Advanced Inorganic Chemistry*, F. A. Cotton and G. Wilkinson, 4th ed., John Wiley & Sons, Inc., New York, 1980; *Encyclopedia of Chemical Technology* (Kirk-Othmer), 3rd ed. (also published by Wiley); various issues of *Science* and *Chemical and Engineering News* during the period 1980 to present), describe for any three of these metals the principal mineral sources and their locations, annual U.S. consumption, and major uses.

## CHAPTER 26

# Chemistry of Coordination Compounds

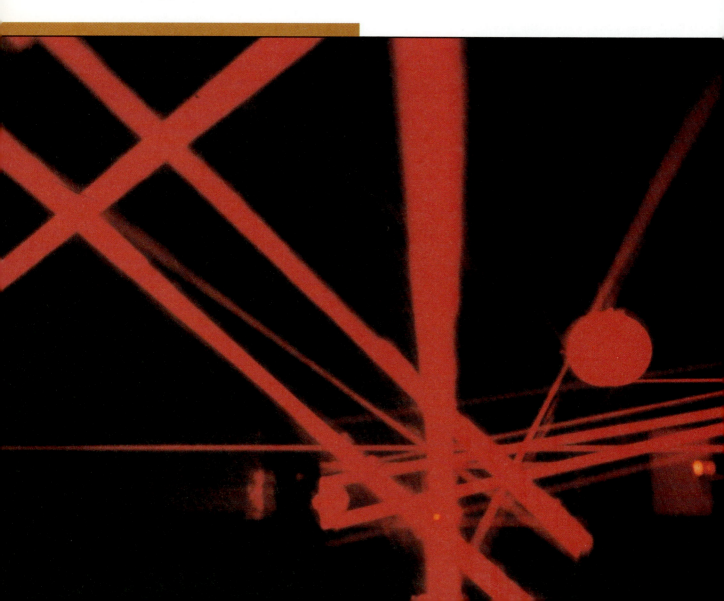

In earlier chapters we noted that metallic elements are characterized by a tendency to lose electrons in their chemical reactions. For this reason, positively charged metal ions play a primary role in the chemical behavior of metals. Of course, metal ions do not exist in isolation. In the first place they are accompanied by anions that serve to maintain charge balance. In addition, metal ions act as Lewis acids (Section 17.8). Neutral molecules or anions with unshared pairs of electrons may be bound to the metal center. On several occasions we have discussed compounds in which a metal ion is surrounded by a group of anions or neutral molecules. Examples include $[Au(CN)_2]^-$, discussed in connection with metallurgy in Section 24.4; hemoglobin, discussed in connection with the oxygen-carrying capacities of the blood in Section 14.4; $[Cu(CN)_4]^{2-}$ and $[Ag(NH_3)_2]^+$, encountered in our discussion of equilibria in Section 18.5. Such species are known as **complex ions** or merely **complexes**. Compounds containing them are called **coordination compounds**.

The molecules or ions that surround a metal ion in a complex are known as **ligands** (from the Latin word *ligare*, meaning "to bind"). Ligands are normally either anions or polar molecules. Furthermore, they have at least one unshared pair of valence electrons, as illustrated in the following examples:

## 26.1 THE STRUCTURE OF COMPLEXES

$$\overset{\displaystyle H}{\underset{\displaystyle H}{:\!\overset{..}{O}\!-\!H}} \qquad \overset{\displaystyle H}{\underset{\displaystyle H}{:\!N\!-\!H}} \qquad :\overset{..}{\underset{..}{Cl}}:^- \qquad :C\!\equiv\!N:^-$$

We will see that for most purposes it is adequate to think of the bonding between a metal ion and its ligands as an electrostatic interaction between the positive cation and the surrounding negative ions or dipoles oriented with their negative ends toward the metal ion. The ability of metal ions to form complexes normally increases as the positive charge of the cation increases and its size decreases. The weakest complexes are formed by the alkali metal ions, such as $Na^+$ and $K^+$. Conversely, the +2 and +3 ions of the transition elements generally excel in complex formation. In fact, many of the transition metal ions form complexes more readily than their charge and size would suggest. For example, on the basis of size alone we might expect that $Al^{3+}$ ($r = 0.45$ Å) would form complexes more readily than the larger $Cr^{3+}$ ion ($r = 0.62$ Å).

Light from a ruby laser. The heart of the laser is a single crystal ruby rod. Ruby consists of $Al_2O_3$ with a fraction of the $Al^{3+}$ ions replaced by $Cr^{3+}$. The 3$d$ electrons of the $Cr^{3+}$ ion undergo electronic transitions that form the basis of the laser action. (Jerry Mason, Science Photo Library/Photo Researchers)

However, with most ligands $Cr^{3+}$ forms much more stable complexes than does $Al^{3+}$. Thus the bonding in these complexes cannot be explained entirely on the basis of electrostatic attraction between metal ion and ligands. To account for some of the observed differences in complexes, we must assume that there is some degree of covalent character in the metal-ligand bond. Because metal ions have empty valence orbitals, they can act as Lewis acids (electron-pair acceptors). Because ligands have unshared pairs of electrons, they can function as Lewis bases (electron-pair donors). We can picture the bond between metal and ligand as the result of their sharing a pair of electrons that was initially on the ligand:

$$Ag^{+}(aq) + 2:N-H(aq) \longrightarrow \left[ H-N:Ag:N-H \right]^{+} (aq) \qquad [26.1]$$

We shall examine the bonding in complexes more closely in Section 26.8.

In forming a complex, the ligands are said to coordinate to the metal or to complex the metal. The central metal and the ligands bound to it constitute the **coordination sphere**. In writing the chemical formula for a coordination compound, we use square brackets to set off the groups within the coordination sphere from other parts of the compound. For example, the formula $[Cu(NH_3)_4]SO_4$ represents a coordination compound consisting of the $[Cu(NH_3)_4]^{2+}$ ion and the $SO_4^{2-}$ ion. The four ammonia groups in the compound are bound directly to the copper(II) ion.

As you might expect, a metal complex has different properties from either the metal ion or ligands from which it is derived. We are normally most interested in the effects of complex formation on the properties of the metal ion. The ease of oxidation or reduction of the metal ion may be drastically changed by complex formation. For example, $Ag^{+}$ is readily reduced in water:

$$Ag^{+}(aq) + e^{-} \longrightarrow Ag(s) \qquad E° = +0.799 \text{ V} \qquad [26.2]$$

By contrast, the $[Ag(CN)_2]^{-}$ ion is not at all readily reduced, because formation of the cyano complex stabilizes silver in the $+1$ oxidation state:

$$[Ag(CN)_2]^{-}(aq) + e^{-} \longrightarrow Ag(s) + 2CN^{-}(aq) \qquad E° = -0.31 \text{ V} \qquad [26.3]$$

Solubility properties may be dramatically different in the presence of complexing agents. For example, $CrBr_3$ is insoluble in water, but the complex $[Cr(H_2O)_6]Br_3$ is very soluble. Colors may be different. $CrBr_3$ is a deep green-black, but $[Cr(H_2O)_6]Br_3$ is violet. These differences in properties, illustrated in Figure 26.1, occur because the metal ion plus its surrounding ligands is a distinct species in its own right and so has characteristic physical and chemical properties.

The charge of a complex is the sum of the charges on the central metal and on its surrounding ligands. In $[Cu(NH_3)_4]SO_4$ we can deduce the charge on the complex if we first recognize $SO_4$ as being the sulfate ion

**FIGURE 26.1** The colors of $CrBr_3$ and $Cr(H_2O)_6Br_3$ are different because the immediate surroundings of the metal ion are different in the two compounds. (Donald Clegg and Roxy Wilson)

and therefore having a $-2$ charge. Because the compound is neutral, the complex ion must have a $+2$ charge, $[Cu(NH_3)_4]^{2+}$. The oxidation number of the copper must be $+2$ because the $NH_3$ groups are neutral:

$$+2 + 4(0) = +2$$
$$[Cu(NH_3)_4]^{2+}$$

## SAMPLE EXERCISE 26.1

What is the oxidation number of the central metal in $[Co(NH_3)_5Cl](NO_3)_2$?

**Solution:** The $NO_3$ group is the nitrate anion and has a $-1$ charge, $NO_3^-$. The $NH_3$ ligands are neutral; the Cl is a coordinated chloride and therefore has a $-1$ charge. The sum of all the charges must be zero:

$$x + 5(0) + (-1) + 2(-1) = 0$$
$$[Co(NH_3)_5Cl](NO_3)_2$$

The oxidation state of the cobalt, $x$, must therefore be $+3$.

### PRACTICE EXERCISE

What is the charge of the complex formed by a platinum(IV) metal ion surrounded by three ammonia molecules and three bromide ions?

*Answer:* $+1$

## SAMPLE EXERCISE 26.2

Given that a complex ion contains a chromium(III) bound to four water molecules and two chloride ions, write its formula.

**Solution:** The metal has a $+3$ oxidation number, water is neutral, and chloride has a $-1$ charge:

$$+3 + 4(0) + 2(-1) = +1$$
$$Cr(H_2O)_4Cl_2$$

Therefore, the charge on the ion is $+1$, $[Cr(H_2O)_4Cl_2]^+$.

### PRACTICE EXERCISE

Write the formula for the complex ion described in the Practice Exercise accompanying Sample Exercise 26.1.

*Answer:* $[Pt(NH_3)_3Br_3]^+$

The ligand atom that is bound directly to the metal is known as the **donor atom**. For example, nitrogen is the donor atom in the $[Ag(NH_3)_2]^+$ complex shown in Equation 26.1. The number of donor atoms attached to a metal is known as its **coordination number**. In $[Ag(NH_3)_2]^+$, silver has a coordination number of 2; in $[Cr(H_2O)_4Cl_2]^+$, chromium has a coordination number of 6.

Some metal ions exhibit constant coordination numbers. For example, the coordination number of chromium(III) and cobalt(III) is invariably 6, and that of platinum(II) is always 4. However, the coordination numbers of most metal ions vary with the ligand. The most common coordination numbers are 4 and 6.

The coordination number of a metal ion is often influenced by the relative sizes of the metal ion and the surrounding ligands. As the ligand gets larger, fewer can coordinate to the metal. This explains why iron is able to coordinate to six fluorides in $[FeF_6]^{3-}$, but to only four chlorides

**Coordination Numbers and Geometries**

**FIGURE 26.2** Structures of (a) $[Zn(NH_3)_4]^{2+}$ and (b) $[Pt(NH_3)_4]^{2+}$, illustrating the tetrahedral and square-planar geometries, respectively. These are the two common geometries for complexes in which the metal ion has a coordination number of four. The shaded surfaces shown in the figure are not bonds; they are included merely to assist in visualizing the shape of the metal complex.

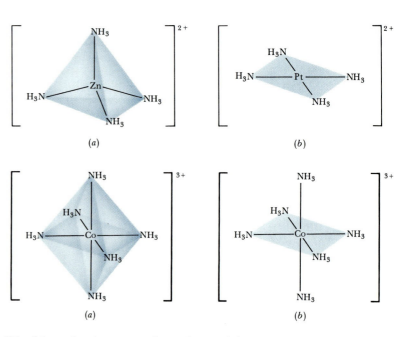

**FIGURE 26.3** Two representations of an octahedral coordination sphere, the common geometric arrangement for complexes in which the metal ion has a coordination number of 6.

in $[FeCl_4]^-$. Ligands that transfer substantial negative charge to the metal also produce reduced coordination numbers. For example, six neutral ammonia molecules can coordinate to nickel(II), forming $[Ni(NH_3)_6]^{2+}$; however, only four negatively charged cyanide ions coordinate, forming $[Ni(CN)_4]^{2-}$.

Four-coordinate complexes have two common geometries—tetrahedral and square planar—as shown in Figure 26.2. The tetrahedral geometry is the more common of the two and is especially common among the nontransition metals. The square-planar geometry is characteristic of transition metal ions with eight $d$ electrons in the valence shell, for example, platinum(II) and gold(III); it is also found in some copper(II) complexes.

Six-coordinate complexes have an octahedral geometry, as shown in Figure 26.3(a). The octahedron is often represented as a planar square with ligands above and below the plane, as in Figure 26.3(b). In this representation, the ligands along the vertical axis are geometrically equivalent to those in the plane.

## 26.2 CHELATES

The ligands that we have discussed so far, such as $NH_3$ and $Cl^-$, are known as **monodentate ligands** (from the Latin meaning "one-toothed"). These ligands possess a single donor atom. Some ligands have two or more donor atoms situated so that they can simultaneously coordinate to a metal ion. They are called **polydentate ligands** ("many-toothed"); because they appear to grasp the metal between two or more donor atoms, they are also known as **chelating agents** (from the Greek word *chele*, "claw"). One such ligand is ethylenediamine:

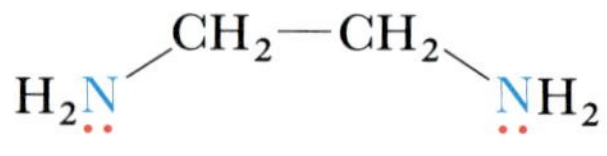

$$H_2\underset{\cdot\cdot}{N} \diagdown^{CH_2-CH_2}\diagdown \underset{\cdot\cdot}{N}H_2$$

This ligand, which is abbreviated en, has two nitrogen atoms (shown in color), which have unshared pairs of electrons. These donor atoms are sufficiently far apart that the ligand can wrap around a metal ion with the two nitrogen atoms simultaneously complexing to the metal in adjacent positions. The $[Co(en)_3]^{3+}$ ion, which contains three ethylenediamine ligands bound to the octahedral coordination sphere of cobalt(III), is shown in Figure 26.4. Notice that the ethylenediamine has been written in a shorthand notation as two nitrogen atoms connected by a line.

Ethylenediamine is an example of a **bidentate ligand**. The structures of several other bidentate ligands are shown in Figure 26.5.

The ethylenediaminetetraacetate ion is an important polydentate ligand:

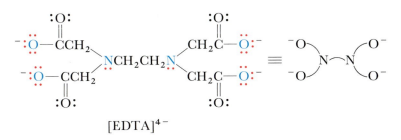

[EDTA]$^{4-}$

This ion, abbreviated EDTA$^{4-}$, has six donor atoms. It can wrap around a metal ion using all six of these donor atoms, as shown in Figure 26.6.

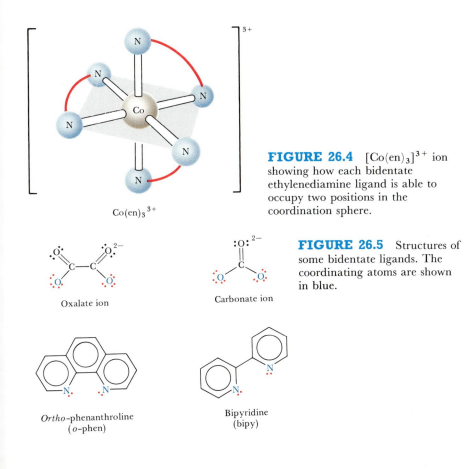

Co(en)$_3$$^{3+}$

**FIGURE 26.4** $[Co(en)_3]^{3+}$ ion showing how each bidentate ethylenediamine ligand is able to occupy two positions in the coordination sphere.

**FIGURE 26.5** Structures of some bidentate ligands. The coordinating atoms are shown in blue.

Oxalate ion

Carbonate ion

*Ortho*-phenanthroline (*o*-phen)

Bipyridine (bipy)

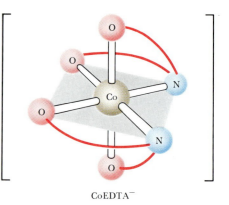

FIGURE 26.6
CoEDTA$^-$ ion, showing how the ethylenediaminetetraacetate ion is able to wrap around a metal ion, occupying six positions in the coordination sphere.

CoEDTA$^-$

In general, chelating agents form more stable complexes than do related monodentate ligands. This is illustrated by the formation constants for $[Ni(NH_3)_6]^{2+}$ and $[Ni(en)_3]^{2+}$, shown in Equations 26.4 and 26.5.

$$[Ni(H_2O)_6]^{2+}(aq) + 6NH_3(aq) \rightleftharpoons [Ni(NH_3)_6]^{2+}(aq) + 6H_2O(l) \qquad K_f = 4 \times 10^8 \qquad [26.4]$$

$$[Ni(H_2O)_6]^{2+}(aq) + 3en(aq) \rightleftharpoons [Ni(en)_3]^{2+}(aq) + 6H_2O(l) \qquad K_f = 2 \times 10^{18} \qquad [26.5]$$

Although the donor atom is nitrogen in both instances, $[Ni(en)_3]^{2+}$ has a stability constant nearly $10^{10}$ times larger than $[Ni(NH_3)_6]^{2+}$.

The generally larger formation constants for polydentate ligands as compared with the corresponding monodentate ligands is known as the **chelate effect**. Thermochemical studies of complex formation in aqueous solution show that in nearly all cases *the chelate effect is due to a more favorable entropy change for complex formation involving polydentate ligands*. As an example, let's compare the thermodynamic data at 25°C for formation of two closely related complexes of $Cd^{2+}$:

$$Cd^{2+}(aq) + 4CH_3NH_2(aq) \rightleftharpoons [Cd(CH_3NH_2)_4]^{2+}(aq)$$
$$\Delta G° = -37.2 \text{ kJ}; \quad \Delta H° = -57.3 \text{ kJ}; \quad \Delta S° = -67.3 \text{ J/K}$$

$$Cd^{2+}(aq) + 2H_2NCH_2CH_2NH_2(aq) \rightleftharpoons [Cd(H_2NCH_2CH_2NH_2)_2]^{2+}(aq)$$
$$\Delta G° = -60.7 \text{ kJ}; \quad \Delta H° = -56.5 \text{ kJ}; \quad \Delta S° = +14.1 \text{ J/K}$$

Recall from Section 19.6 that a large negative value for $\Delta G°$ corresponds to a large equilibrium constant for the forward reaction. The equilibrium constant for complex formation is thus much greater for the second reaction, with en as ligand. We see that the enthalpy changes in the two reactions are nearly the same. However, the entropy change is much more positive for the second reaction. It is indeed that more positive entropy change that accounts for the larger negative value for $\Delta G°$ for the second reaction.

The more positive entropy change associated with reactions involving polydentate ligands is related to some of the ideas regarding entropy discussed in Section 19.3. Ligands that are bound to the metal ion are constrained to remain with that metal ion; they are not free to move about the solution independently. Therefore, a negative entropy change is associated with binding to the metal ion. In the first reaction above, four

$CH_3NH_2$ molecules become bound to the metal ion, and four water molecules are freed to move about the solution. Although it might seem at first that this should lead to a net entropy change of zero, differences in the hydrogen-bonding ability of $H_2O$ as compared with the $CH_3NH_2$ molecules, and the tighter binding of $CH_3NH_2$ to the $Cd^{2+}$ ion as compared with water, lead to a net negative entropy change. However, when one ethylenediamine molecule binds to $Cd^{2+}$, two $H_2O$ molecules are released to move about freely in the solution. There is a net increase in the number of species that move about independently in the solution. The system has become more disordered, as it were, and the entropy change is accordingly more positive.

Living systems consist mainly of hydrogen, oxygen, carbon, and nitrogen. In fact, more than 99 percent of the atoms required by living cells are one of these four elements. Nevertheless, many other elements are known to be essential for life (see Figure 26.7). Nine of the essential elements are transition metals—vanadium, chromium, iron, copper, zinc, manganese, cobalt, nickel, and molybdenum. These elements owe their roles in living systems mainly to their ability to form complexes with a variety of donor groups present in biological systems. Many enzymes, the body's catalysts, require metal ions to function. We shall take a close took at enzymes in Chapter 28.

Although our bodies require only small quantities of certain metals, a deficiency of the element can lead to serious illness. For example, it was recently discovered that a deficiency of manganese in the diet can lead to convulsive disorders. This is an especially important consideration during pregnancy. Some epilepsy patients have been helped by addition of manganese to their diets.

Among the most important chelating agents in nature are those derived from the porphine molecule, shown in Figure 26.8. This molecule can coordinate to a metal using the four nitrogen atoms as donors. Upon coordination to a metal, the two $H^+$ shown bonded to nitrogen are displaced. Complexes derived from porphine are called **porphyrins**. Dif-

## Metals and Chelates in Living Systems

**FIGURE 26.7** The elements that are essential for life are indicated by colors. The violet color denotes the four most abundant elements in living systems (hydrogen, carbon, nitrogen, and oxygen). The orange color indicates the seven next-most abundant elements. The green color indicates the elements needed in only trace amounts.

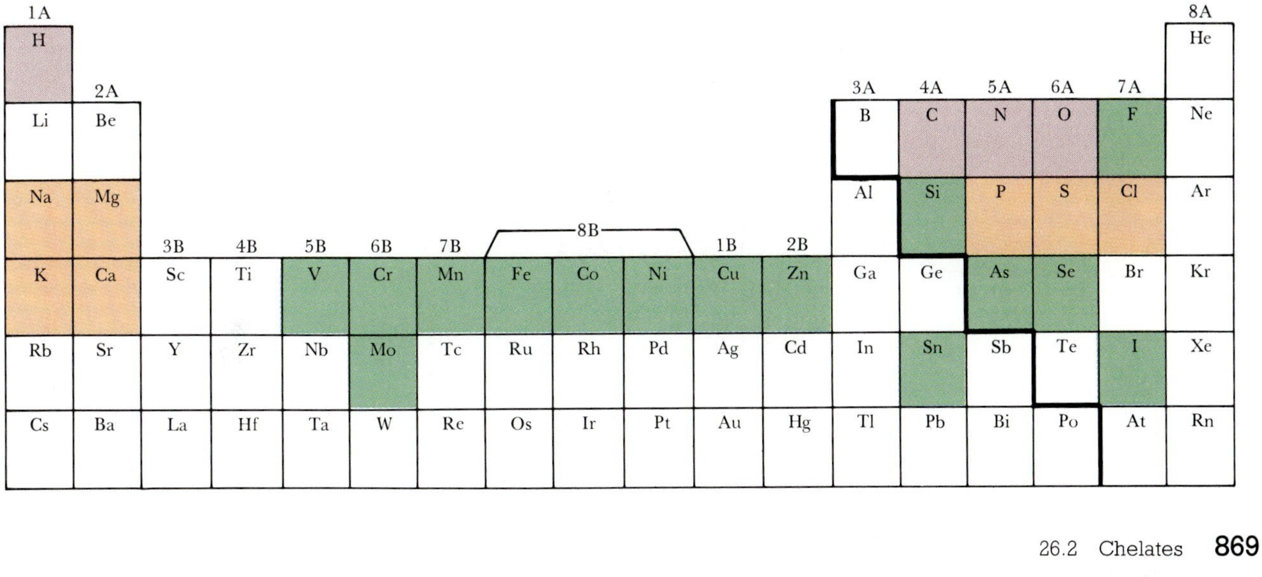

| 1A | | | | | | | | | | | | | | | | | 8A |
|----|----|----|----|----|----|----|----|----|----|----|----|----|----|----|----|----|----|
| H | 2A | | | | | | | | | | | 3A | 4A | 5A | 6A | 7A | He |
| Li | Be | | | | | | | | | | | B | C | N | O | F | Ne |
| Na | Mg | 3B | 4B | 5B | 6B | 7B | 8B | | | 1B | 2B | Al | Si | P | S | Cl | Ar |
| K | Ca | Sc | Ti | V | Cr | Mn | Fe | Co | Ni | Cu | Zn | Ga | Ge | As | Se | Br | Kr |
| Rb | Sr | Y | Zr | Nb | Mo | Tc | Ru | Rh | Pd | Ag | Cd | In | Sn | Sb | Te | I | Xe |
| Cs | Ba | La | Hf | Ta | W | Re | Os | Ir | Pt | Au | Hg | Tl | Pb | Bi | Po | At | Rn |

Complexing agents can often be added to a solution to prevent one or more of the customary reactions of a metal ion without actually removing it from the solution. For example, a metal ion that interferes with a chemical analysis can often be complexed and its interference thereby removed. In a sense the complexing agent hides the metal ion, and so we call it a sequestering agent. Because chelates perform this role more effectively than do monodentate ligands, the term *sequestering agent* is usually reserved for chelates.

One of the most common applications of sequestering agents is in complexing cations in natural waters to keep them from interfering with the action of soap or detergent molecules. As noted in Section 14.6, $Mg^{2+}$ and $Ca^{2+}$ ions react with soap to form a precipitate. Although these and other metal ions do not precipitate detergent molecules, they do complex with them, thereby interfering with their cleansing action.* Phosphates are effective and cheap sequestering agents for these ions. The most important phosphate used for this purpose is sodium tripolyphosphate:

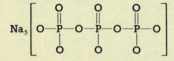

---

\* The action of soaps and detergents was discussed earlier, in Section 12.7; the formula of a typical detergent molecule is shown in Section 23.4.

Complexing agents also enhance the solubility of metal salts. For example, AgBr, the photosensitive material in photographic film, is insoluble in water, but it dissolves in the presence of thiosulfate ion, $S_2O_3^{2-}$:

$$AgBr(s) \rightleftharpoons Ag^+(aq) + Br^-(aq)$$
$$K_{sp} = 7.7 \times 10^{-13}$$

$$Ag^+(aq) + 2S_2O_3^{2-}(aq) \rightleftharpoons [Ag(S_2O_3)_2]^{3-}(aq)$$
$$K_f = 1.6 \times 10^{13}$$

Visualize the thiosulfate as shifting the first equilibrium to the right by complexing the $Ag^+$. Sodium thiosulfate decahydrate, $Na_2S_2O_3 \cdot 10H_2O$, known as hypo, is used in black-and-white photography to dissolve unexposed and undeveloped AgBr from the photographic film (Section 23.3).

Complex formation can be employed in the detection and analysis of metal ions in solution. For example, the $Fe^{3+}(aq)$ ion can be detected and its concentration measured in aqueous solution via formation of a complex with the thiocyanate ion, $SCN^-$:

$$Fe(H_2O)_6^{3+}(aq) + SCN^-(aq) \longrightarrow$$
$$Fe(H_2O)_5SCN^{2+}(aq) + H_2O(l) \quad [26.6]$$

Figure 26.8 illustrates this reaction. Addition of still more $SCN^-$ results in further displacements of water molecules by thiocyanate ions.

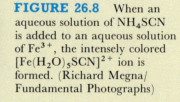

**FIGURE 26.8** When an aqueous solution of $NH_4SCN$ is added to an aqueous solution of $Fe^{3+}$, the intensely colored $[Fe(H_2O)_5SCN]^{2+}$ ion is formed. (Richard Megna/Fundamental Photographs)

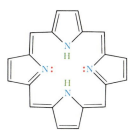

FIGURE 26.9 Structure of the porphine molecule. This molecule forms a tetradentate ligand with the loss of the two protons bound to nitrogen atoms. Porphine is the basic structure of porphyrins, compounds whose complexes play a variety of important roles in nature.

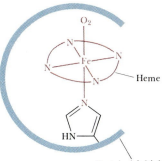

FIGURE 26.10 Schematic representation of oxyhemoglobin, showing one of the four heme units in the molecule. The iron is bound to four nitrogen atoms of the porphyrin, to a nitrogen from the surrounding protein, and to an $O_2$ molecule.

ferent porphyrins contain different metals and have different substituent groups attached to the carbon atoms at the ligand's periphery. Two of the most important porphyrins are heme, which contains iron(II), and chlorophyll, which contains magnesium(II). We discussed heme earlier, in Section 14.4. Hemoglobin contains four heme subunits; one of these subunits is shown in Figure 14.10. The iron is coordinated to the four nitrogen atoms of the porphyrin and also to a nitrogen atom from the protein that composes the bulk of the hemoglobin molecule. The sixth position around the iron is occupied either by $O_2$ (in oxyhemoglobin, the bright red form) or by water (in deoxyhemoglobin, the purplish-red form). Oxyhemoglobin is shown in Figure 26.10. As noted in Section 14.4, some groups, such as CO, act as poisons because of their ability to bind to iron more strongly than $O_2$ can.

When we have an insufficient quantity of iron in our diet, we suffer from iron-deficiency anemia. The lack of iron leads to a reduction in the amount of hemoglobin; we develop what advertisements have referred to as "iron-poor blood." Without hemoglobin to transport oxygen, our body's cells are unable to produce energy. Therefore, the symptoms of anemia include weakness and drowsiness. (See the Chemistry at Work feature on the next page.)

## 26.3 NOMENCLATURE

When complexes were first discovered and few were known, they were named after the chemist who originally prepared them. A few of these names persist; for example, $NH_4[Cr(NH_3)_2(NCS)_4]$ is known as Reinecke's salt. As the number of known complexes grew, they began to be named by color. For example, $[Co(NH_3)_5Cl]Cl_2$, whose formula was then written $CoCl_3 \cdot 5NH_3$, was known as purpureocobaltic chloride after its purple color. Once the structures of complexes were more fully understood, it became possible to name them in a more systematic manner. Let's consider two examples, which follow the Chemistry at Work feature.

Although iron is the fourth most abundant element in the earth's crust, living systems have difficulty in assimilating enough iron to satisfy their needs. Consequently, iron-deficiency anemia is a common problem in humans; in plants, chlorosis, an iron deficiency that results in yellowing of leaves, is also commonplace. Living systems have difficulty in assimilating iron because of changes that occurred in the earth's atmosphere in the course of geological time. The earliest living systems had a plentiful supply of soluble iron(II) in the oceans. However, when oxygen appeared in the atmosphere, vast deposits of iron(III) oxide formed. The amount of dissolved iron remaining was too small to support life. Microorganisms adapted to this problem by secreting an iron-binding compound, called a siderophore, that forms an extremely stable, water-soluble complex with iron(III). This complex is called ferrichrome; its structure is shown in Figure 26.11. The iron-binding strength of the siderophore is so great that it can extract iron from Pyrex glassware, and it readily solubilizes the iron in iron oxides.

The overall charge of ferrichrome is zero, which makes it possible for the complex to pass through the rather hydrophobic membrane walls of cells.

When a dilute solution of ferrichrome is added to a cell suspension, iron is found entirely within the cells in an hour. When ferrichrome enters the cell, the iron is removed through an enzyme-catalyzed reaction which reduces the iron to iron(II). Iron in the lower oxidation state is not strongly complexed by the siderophore. Microorganisms thus acquire iron by excreting siderophore into their immediate environment, then taking the ferrichrome complex into the cell. The overall process is illustrated in Figure 26.12.

In the human, iron is assimilated from food in the intestine. A protein called transferrin binds iron and transports it across the intestinal wall to distribute it to other tissue in the body. The normal adult human carries a total of about 4 g of iron. At any one time, about 3 g, or 75 percent, of the iron is in the blood, mostly in the form of hemoglobin. The remainder is carried by transferrin.

A bacterium that infects the blood requires a source of iron if it is to grow and reproduce. The bacterium excretes siderophore into the blood to compete with transferrin for the iron it holds. The formation constants for iron binding are about the same for transferrin and siderophore. The more iron

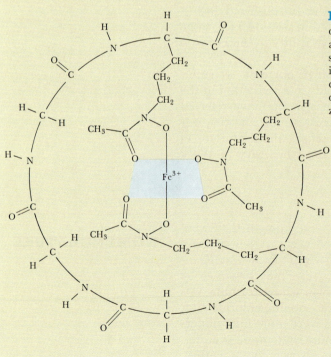

**FIGURE 26.11** The structure of ferrichrome. In this complex an $Fe^{3+}$ ion is coordinated by six oxygen atoms. The complex is very stable; it has a formation constant of about $10^{30}$. The overall charge of the complex is zero.

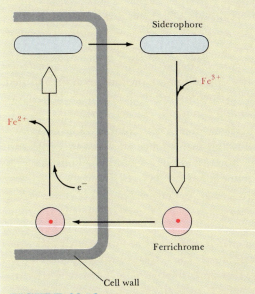

available to the bacterium, the more rapidly it can reproduce, and thus the more harm it can do. Several years ago New Zealand clinics regularly gave iron supplements to infants soon after birth. However, the incidence of certain bacterial infections was eight times higher in treated than in untreated infants. Presumably, the presence in the blood of more iron than absolutely necessary makes it easier for bacteria to obtain the iron necessary for growth and reproduction.

In the United States it is common medical practice to supplement infant formula with iron sometime during the first year of life. This practice is based on the fact that human milk is virtually devoid of iron. Given what is now known about iron metabolism by bacteria, many research workers in nutrition believe that iron supplementation is not generally justified or wise.

For bacteria to continue to multiply in the bloodstream they must synthesize new supplies of siderophore. It has been discovered that synthesis of siderophore in bacteria slows as the temperature is increased above the normal body temperature of 37°C, and it stops completely at 40°C. This suggests that fever in the presence of an invading microbe is a mechanism used by the body to deprive bacteria of iron.

**FIGURE 26.12** The iron-transport system of a bacterial cell. The iron-binding ligand, called the siderophore, is synthesized inside the cell and excreted into the surrounding medium. It reacts with $Fe^{3+}$ ion to form ferrichrome, which is then absorbed by the cell. Inside the cell the ferrichrome is reduced, forming $Fe^{2+}$, which is not tightly bound by the siderophore. Having released the iron for use in the cell, the siderophore may be recycled back into the medium.

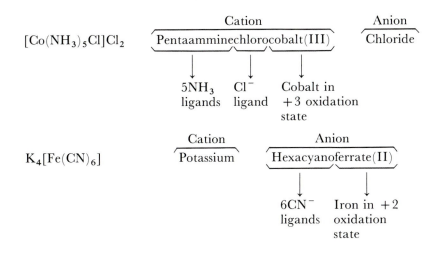

The rules of nomenclature are as follows:*

---

* The rules of nomenclature are approved by the International Union of Pure and Applied Chemistry and are subject to periodic revision. Some rules are clearly more important than others. For example, the order in which ligands are named (rule 2) is not as important as assigning each ligand its correct name.

1. *In naming salts, the name of the cation is given before the name of the anion.* Thus in $[Co(NH_3)_5Cl]Cl_2$ we name the $[Co(NH_3)_5Cl]^{2+}$ and then $Cl^-$.

2. *Within a complex ion or molecule, the ligands are named before the metal. Ligands are listed in alphabetical order, regardless of charge on the ligand. Prefixes that give the number of ligands are not considered part of the ligand name in determining alphabetical order.* Thus in the $[Co(NH_3)_5Cl]^{2+}$ ion, we name the ammonia ligands first, then the chloride, then the metal: pentaamminechlorocobalt(III). Note, however, that in writing the formula the metal is listed first.

3. *Anionic ligands end in the letter o, whereas neutral ones ordinarily bear the name of the molecule.* Some common ligands and their names are listed in Table 26.1. Special names are given to $H_2O$ (aqua) and $NH_3$ (ammine). Thus the terms *chloro* and *ammine* occur in the name for $[Co(NH_3)_5Cl]Cl_2$.

4. *A Greek prefix ( for example,* di-, tri-, tetra-, penta-, *and* hexa-) *is used to indicate the number of each kind of ligand when more than one is present.* Thus in the name for $[Co(NH_3)_5Cl]^{2+}$ we have pentaammine, indicating five $NH_3$ ligands. *If the name of the ligand itself contains a Greek prefix, such as* mono- *or* di-, *the ligand is enclosed in parentheses and alternate prefixes* (bis-, tris-, tetrakis-, pentakis-, *and* hexakis-) *are used.* For example, the name for $[Co(en)_3]Cl_3$ is tris(ethylenediamine)cobalt(III) chloride.

5. *If the complex is an anion, its name ends in* -ate. For example, in $K_4[Fe(CN)_6]$ the anion is called the hexacyanoferrate(II) ion. The suffix *-ate* is often added to the Latin stem, as in this example.

6. *The oxidation number of the metal is given in parentheses in Roman numerals following the name of the metal.* For example, the Roman numeral III is used to indicate the $+3$ oxidation state of cobalt in $[Co(NH_3)_5Cl]^{2+}$.

We apply these rules to the compounds listed below on the left to derive the names given in the second column:

| | |
|---|---|
| $K[Pt(NH_3)(N_3)_5]$ | potassium amminepentaazidoplatinate(IV) |
| $[Ni(C_5H_5N)_6]Br_2$ | hexapyridinenickel(II) bromide |
| $[Co(NH_3)_4(H_2O)CN]Cl_2$ | tetraammineaquacyanocobalt(III) chloride |
| $Na_2[MoOCl_4]$ | sodium tetrachlorooxomolybdate(IV) |
| $Na[Al(OH)_4]$ | sodium tetrahydroxoaluminate |

In the last example, the oxidation state of the metal is not mentioned in the name because aluminum is nearly always in the $+3$ oxidation state.

**TABLE 26.1**   Some common ligands

| Ligand | Ligand name |
|---|---|
| Azide, $N_3^-$ | Azido |
| Bromide, $Br^-$ | Bromo |
| Chloride, $Cl^-$ | Chloro |
| Cyanide, $CN^-$ | Cyano |
| Hydroxide, $OH^-$ | Hydroxo |
| Carbonate, $CO_3^{2-}$ | Carbonato |
| Oxalate, $C_2O_4^{2-}$ | Oxalato |
| Ammonia, $NH_3$ | Ammine |
| Ethylenediamine, en | Ethylenediamine |
| Water, $H_2O$ | Aqua |

## SAMPLE EXERCISE 26.3

Give the name of the following compounds: **(a)** $[Cr(H_2O)_4Cl_2]Cl$; **(b)** $K_4[Ni(CN)_4]$.

**Solution:** **(a)** We begin with the four waters, which are indicated as tetraaqua. Then there are two chloride ions, indicated as dichloro. The oxidation state of Cr is $+3$:

$$+3 + 4(0) + 2(-1) + (-1) = 0$$
$$[Cr(H_2O)_4Cl_2]Cl$$

Thus we have chromium(III). Finally, the anion is chloride. Putting these parts together, we have the compound's name: tetraaquadichlorochromium(III) chloride.

**(b)** The complex has four $CN^-$, which we indicate as tetracyano. The oxidation state of the nickel is zero:

$$4(+1) + 0 + 4(-1) = 0$$
$$K_4[Ni(CN)_4]$$

Because the complex is an anion, the metal is indicated as nickelate(0). Putting these parts together and naming the cation first, we have: potassium tetracyanonickelate(0).

### PRACTICE EXERCISE

Give the name of the following compounds: **(a)** $[Mo(NH_3)_3Br_3]NO_3$; **(b)** $(NH_4)_2[CuBr_4]$.
*Answers:* **(a)** triamminetribromomolybdenum (IV) nitrate; **(b)** ammonium tetrabromocuprate(II)

---

Given the name of a coordination compound, you should be able to write out the formula. Remember that cations are listed before anions and that the metal and ligands of the coordination sphere are enclosed within brackets.

## SAMPLE EXERCISE 26.4

Write the formula for bis(ethylenediamine)difluorocobalt(III) perchlorate.

**Solution:** The complex cation contains two fluorides, two ethylenediamines, and a cobalt with a $+3$ oxidation number. Knowing this, we can determine the charge on the complex:

$$+3 + 2(0) + 2(-1) = +1$$
$$[Co(en)_2F_2]$$

The perchlorate anion has a single negative charge, $ClO_4^-$. Therefore, only one is needed to balance the charge on the complex cation. The formula is thus $[Co(en)_2F_2]ClO_4$.

### PRACTICE EXERCISE

Write the formula for sodium diaquabis(oxalato)-ruthenate(III).
*Answer:* $Na[Ru(H_2O)_2(C_2O_4)_2]$

---

# 26.4 ISOMERISM

Sometimes two or more compounds have the same formula but differ in one or more physical or chemical properties, such as color, solubility, or rate of reaction with some reagent. Such compounds, which have the same collection of atoms arranged in different ways, are called **isomers**. Several types of isomerism are possible, as outlined in Figure 26.13.

## Structural Isomerism

**Structural isomers** differ in the bonding arrangements of the atoms; that is, they are chemically distinct. Many different types of structural isomerism are known in coordination chemistry. The two listed in Figure 26.13 are given as examples. **Linkage isomerism** is a relatively rare but interesting type that arises when a particular ligand is capable of

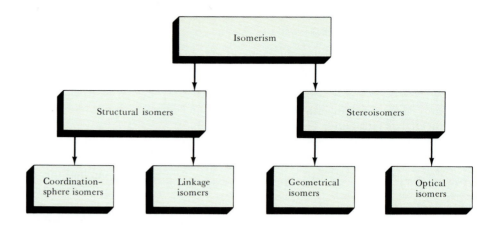

**FIGURE 26.13** Forms of isomerism in coordination compounds.

coordinating to a metal in two different ways. For example, the nitrite ion, $NO_2^-$, can coordinate through either a nitrogen or an oxygen atom, as shown in Figure 26.14. When it coordinates through the nitrogen atom the $NO_2^-$ ligand is called *nitro*; when it coordinates through the oxygen atom, it is called *nitrito* and generally written $ONO^-$. The isomers shown in Figure 26.14 differ in their chemical and physical properties. For example, the N-bonded isomer is yellow, while the O-bonded isomer is red. Other ligands that are capable of coordinating through either of two donor atoms include thiocyanate, $SCN^-$, whose potential donor atoms are N and S.

**Coordination-sphere isomers** differ in the ligands that are directly bonded to the metal, as opposed to being outside the coordination sphere in the solid lattice. For example, the compound $CrCl_3(H_2O)_6$ exists in three forms: $[Cr(H_2O)_6]Cl_3$ (a violet compound), $[Cr(H_2O)_5Cl]Cl_2 \cdot H_2O$ (a green compound), and $[Cr(H_2O)_4Cl_2]Cl \cdot 2H_2O$ (also a green compound). In the second and third compounds the water has been

**FIGURE 26.14** (*a*) Yellow N-bound and (*b*) red O-bound isomers of $[Co(NH_3)_5NO_2]^{2+}$.

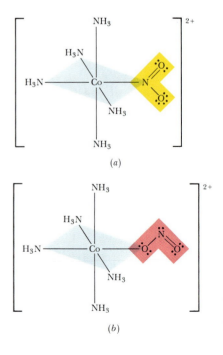

displaced from the coordination sphere by chloride ions and occupies a site in the solid lattice.

The most important form of isomerism is **stereoisomerism**. Stereoisomers have the same chemical bonds but different spatial arrangements. For example, in $[Pt(NH_3)_2Cl_2]$ the chloro ligands can be either adjacent or opposite to one another, as illustrated in Figure 26.15. This particular form of isomerism, in which the arrangement of the constituent atoms is different though the same bonds are present, is called **geometric isomerism**. Isomer (*a*), with like groups in adjacent positions, is called the **cis** isomer. Isomer (*b*), with like ligands across from one another, is called the **trans** isomer.

Geometric isomerism is possible also in octahedral complexes when two or more different ligands are present. The *cis* and *trans* isomers of tetraamminedichlorocobalt(III) ion are shown in Figure 26.16. Note that these two isomers possess different colors. Their salts also possess different solubilities in water. In general, geometric isomers possess distinct physical and chemical properties.

Because all of the corners of a tetrahedron are adjacent to one another, *cis-trans* isomerism is not observed in tetrahedral complexes.

**Stereoisomerism**

**FIGURE 26.15** (*a*) *Cis* and (*b*) *trans* geometric isomers of the square-planar $[Pt(NH_3)_2Cl_2]$.

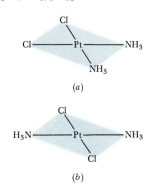

## SAMPLE EXERCISE 26.5

How many geometric isomers are there for $[Cr(H_2O)_2Br_4]^-$?

**Solution:** This complex has a coordination number of six and therefore presumably an octahedral geometry. Like $[Co(NH_3)_4Cl_2]^+$ (Figure 26.16), it has four ligands of one type and two of another. Consequently it possesses two isomers: one with $H_2O$ ligands across the metal from each other (the *trans* isomer) and one with $H_2O$ ligands adjacent (the *cis* isomer).

In general, the number of isomers of a complex can be determined by making a series of drawings of the structure with ligands in different locations. It is easy to overestimate the number of geometric isomers; sometimes different orientations of a single isomer are incorrectly thought to be different isomers. Therefore, you should keep in mind that if two structures can be rotated so that they are equivalent, they are not isomers of one another. The problem of identifying isomers is compounded by the difficulty we often have in visualizing three-dimensional molecules from their two-dimensional representations. It is easier to determine the number of isomers if we are working with three-dimensional models.

**PRACTICE EXERCISE**

How many isomers are there for square-planar $[Pt(NH_3)_2ClBr]$?
*Answer:* two

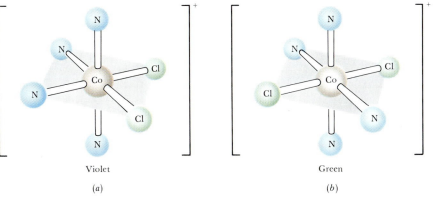

Violet

(*a*)

Green

(*b*)

**FIGURE 26.16** (*a*) *Cis* and (*b*) *trans* geometric isomers of the octahedral $[Co(NH_3)_4Cl_2]^+$ ion. (The symbol N represents the coordinated $NH_3$ group.)

**FIGURE 26.17** Our hands are nonsuperimposable mirror images of each other.

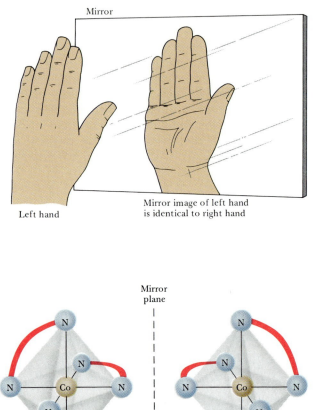

Mirror

Left hand

Mirror image of left hand is identical to right hand

**FIGURE 26.18** The two optical isomers of $[Co(en)_3]^{3+}$; notice that the ions are nonsuperimposable mirror images of each other.

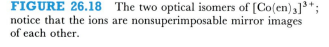

Mirror plane

A second type of stereoisomerism is known as **optical isomerism**. Optical isomers are nonsuperimposable mirror images of one another. They bear the same resemblance to one another that our left hand bears to our right hand. If you look at your left hand in a mirror, as shown in Figure 26.17, the image is identical to your right hand. Furthermore, your two hands are not superimposable on one another. A good example of a complex that exhibits this type of isomerism is the $[Co(en)_3]^{3+}$ ion. The two isomers of $[Co(en)_3]^{3+}$ and their mirror-image relationship to one another are shown in Figure 26.18. Just as there is no way that we can twist or turn our right hand to make it look identical to our left, so also there is no way to rotate one of these isomers to make it identical to the other. If we had models of each that we could handle, perhaps we could more easily satisfy ourselves of this fact. Molecules or ions that have nonsuperimposable mirror images are said to be **chiral** (pronounced KY-rul). Enzymes, the body's catalysts, are among the most highly chiral molecules known. As noted in Section 26.2, many enzymes contain complexed metal ions.

## SAMPLE EXERCISE 26.6

Does either *cis*- or *trans*-$[Co(en)_2Cl_2]^+$ have optical isomers?

**Solution:** To answer this question you should draw out both the *cis* and *trans* isomers of $[Co(en)_2Cl_2]^+$, then their mirror images. Note that the mirror image of the *trans* isomer is identical to the original. Consequently *trans*-$[Co(en)_2Cl_2]^+$ has no optical iso-mer. However, the mirror image of *cis*-$[Co(en)_2Cl_2]^+$ is not identical to the original. Consequently, there are optical isomers for this complex.

### PRACTICE EXERCISE

Does the square-planar complex ion $[Pt(NH_3)(N_3)ClBr]^-$ have optical isomers?
*Answer:* no

Most of the physical and chemical properties of optical isomers are identical. The properties of two optical isomers differ only if they are in a chiral environment—that is, one in which there is a sense of right- and left-handedness. For example, in the presence of a chiral enzyme, the reaction of one optical isomer might be catalyzed, whereas the other isomer would remain totally unreacted. Consequently, one optical isomer may produce a specific physiological effect within the body, whereas its mirror image produces a different effect or none at all.

Optical isomers are usually distinguished from each other by their inter-action with plane-polarized light. If light is polarized—for example, by passage through a sheet of Polaroid film—the light waves are vibrating in a single plane, as shown in Figure 26.19. If the polarized light is passed through a solution containing one optical isomer, the plane of polariza-tion is rotated either to the right (clockwise) or to the left (counterclock-wise). The isomer that rotates the plane of polarization to the right is said to be **dextrorotatory**; it is labeled the dextro, or *d*, isomer (Latin *dexter*, "right"). Its mirror image will rotate the plane of polarization to the left; it is said to be **levorotatory** and is labeled the levo, or *l*, isomer (Latin *laevus*, "left"). Because of their effect on plane-polarized light, chiral molecules are said to be **optically active**.

**FIGURE 26.19** Effect of an optically active solution on the plane of polarization of plane-polarized light. The unpolarized light is passed through a polarizer. The resultant polarized light thereafter passes through a solution containing a dextrorotatory optical isomer. As a result, the plane of polarization of the light is rotated to the right relative to an observer looking toward the light source.

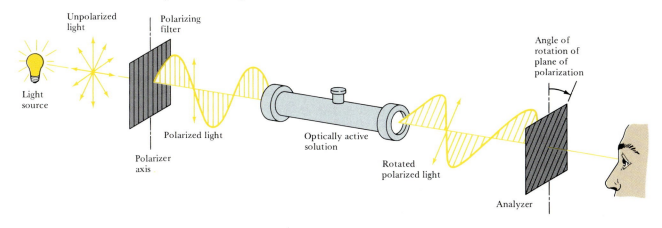

When a substance with optical isomers is prepared in the laboratory, the chemical environment during the synthesis is not usually chiral. Consequently, equal amounts of the two isomers are obtained; the mixture is said to be **racemic**. A racemic mixture will not rotate polarized light because the rotatory effects of the two isomers cancel each other. In order to separate the isomers from the racemic mixture, they must be placed in a chiral environment. For example, one optical isomer of the chiral tartrate anion,* $C_4H_4O_6^{2-}$, can be used to separate a racemic mixture of $[Co(en)_3]Cl_3$. If $d$-tartrate is added to an aqueous solution of the $[Co(en)_3]Cl_3$, $d$-$[Co(en)_3](d$-$C_4H_4O_6)Cl$ will precipitate, leaving $l$-$[Co(en)_3]^{3+}$ in solution.

## 26.5 LIGAND EXCHANGE RATES

Some complexes in solution exchange ligands at an extremely rapid rate, whereas others do so quite slowly. For example, addition of ammonia to an aqueous solution of $CuSO_4$ produces an essentially instantaneous color change as the pale blue $[Cu(H_2O)_4]^{2+}$ ion is converted to the deep blue $[Cu(NH_3)_4]^{2+}$ ion. When this solution is acidified, the pale blue color is regenerated again at a rapid rate:

$$[Cu(NH_3)_4]^{2+}(aq) + 4H_2O(l) + 4H^+(aq) \longrightarrow [Cu(H_2O)_4]^{2+}(aq) + 4NH_4^+(aq) \qquad [26.7]$$

In contrast, $[Co(NH_3)_6]^{3+}$ is more difficult to prepare than is $[Cu(NH_3)_4]^{2+}$. However, once it has been formed and placed in an acidic solution, reaction to form $NH_4^+$ takes several days. This tells us that the coordinated $NH_3$ groups are not readily removed from the metal. Complexes that like $[Cu(NH_3)_4]^{2+}$ undergo rapid ligand exchange are said to be **labile**; those that like $[Co(NH_3)_6]^{3+}$ undergo slow ligand exchange are said to be **inert**. The distinction between labile and inert complexes applies to how rapidly equilibrium is attained and not to the position of the equilibrium. For example, although $[Co(NH_3)_6]^{3+}$ is inert in acidified aqueous solutions, the equilibrium constant indicates that the complex is not thermodynamically stable under these conditions:

$$[Co(NH_3)_6]^{3+}(aq) + 6H_2O(l) + 6H^+(aq) \rightleftharpoons [Co(H_2O)_6]^{3+}(aq) + 6NH_4^+(aq) \qquad K_f \simeq 10^{20} \qquad [26.8]$$

The kinetic inertness of $[Co(NH_3)_6]^{3+}$ can be attributed to a high activation energy for the reaction.

Cobalt(III) is one of few metal ions to consistently form inert complexes; others include chromium(III), platinum(IV), and platinum(II). Complexes of these ions maintain their identity in solution long enough to permit study of their structures and properties. They were therefore among the first complexes studied. Much of our understanding of structure and isomerism comes from studies of these complexes.

---

* When sodium ammonium tartrate, $NaNH_4C_4H_4O_6$, is crystallized from solution, the two isomers form separate crystals whose shapes are mirror images of each other. In 1848, Louis Pasteur achieved the first separation of a racemic mixture into optical isomers; using a microscope he picked the "right-handed" crystals of this compound from the "left-handed" ones.

Many early systematic studies of coordination compounds involved complexes of cobalt(III), chromium(III), platinum(II), and platinum(IV), with ammonia as a ligand. One of the earliest reports of the preparation of an ammine complex dates from 1798, when a chemist by the name of Tassaert accidentally prepared an ammonia complex of cobalt. The compound was found to have the empirical formula $CoN_6H_{18}Cl_3$. Tassaert wrote the formula as $CoCl_3 \cdot 6NH_3$, suggesting that the compound was analogous to hydrated salts like $CoCl_2 \cdot 6H_2O$.

By 1890, many ammine complexes had been prepared, and a great deal of information about them had been gathered by a number of different investigators. By this time chemists had begun to wonder about how the atoms in these complexes were connected to each other and about the possible effect of these arrangements on the properties of the complexes. Among the observations that any successful theory would have to account for were the electrical conductivity of the complexes in solution and their behavior toward $AgNO_3$. Recall from our previous discussion (Section 12.4) that solutions of ionic substances conduct electrical current. The ease with which the solution conducts current is referred to as the conductivity. The conductivity of solutions of ionic substances increases with the total concentration of ions and is also greater for ions of higher charge. By comparing the conductivities of solutions of coordination compounds with those of simple salts, it was possible to determine the number and types of ions present in the solution. For example, the molar conductivity of an aqueous solution of $CoCl_3 \cdot 5NH_3$ is about the same as that of $CaCl_2$ and other 1:2 electrolytes. We can thus conclude that an aqueous solution of $CoCl_3 \cdot 5NH_3$ produces three ions, one that carries a $+2$ charge and two carrying negative charges. The number of free chloride ions present in the solution was determined by treating solutions with $AgNO_3$. When cold, freshly prepared solutions of $CoCl_3 \cdot 5NH_3$ were treated with $AgNO_3$, 2 mol of AgCl precipitated for each mole of complex; one chloride in the compound did not precipitate. Table 26.2 summarizes these results.

In 1893, a 26-year-old Swiss chemist, Alfred Werner, proposed a theory that successfully explained these facts and became the basis for our subsequent understanding of metal complexes. Werner's first basic postulate was that metals exhibit both primary and secondary valences. We now refer to these as the metal's oxidation state and coordination number, respectively. This postulate had no theoretical basis; it predated Lewis's

**TABLE 26.2**  Properties of some ammonia complexes of cobalt(III)

| Original formulation | Color | Ions per formula unit | Cl⁻ ions in solution per formula unit | Modern formulation |
|---|---|---|---|---|
| $CoCl_3 \cdot 6NH_3$ | Orange | 4 | 3 | $[Co(NH_3)_6]Cl_3$ |
| $CoCl_3 \cdot 5NH_3$ | Purple | 3 | 2 | $[Co(NH_3)_5Cl]Cl_2$ |
| $CoCl_3 \cdot 4NH_3$ | Green | 2 | 1 | trans-$[Co(NH_3)_4Cl_2]Cl$ |
| $CoCl_3 \cdot 4NH_3$ | Violet | 2 | 1 | cis-$[Co(NH_3)_4Cl_2]Cl$ |

theory of covalent bonding by 23 years. However, it allowed Werner to explain many experimental facts. Werner postulated a primary valence of three and a secondary valence of six for cobalt(III). He therefore wrote the formula for $CoCl_3 \cdot 5NH_3$ as $[Co(NH_3)_5Cl]Cl_2$. The ligands within the brackets satisfied cobalt's secondary valence of six; the three chlorides satisfied the primary valence of three. Werner proposed that the chlorides within the coordination sphere of cobalt(III) are bound so tightly that they are unavailable to contribute to the compound's conductivity or to react with $AgNO_3$. Thus $CoCl_3 \cdot 5NH_3$ consists of a $[Co(NH_3)_5Cl]^{2+}$ ion and two $Cl^-$ ions.

Werner also sought to deduce the arrangement of ligands around the central metal. He postulated that cobalt(III) complexes exhibit an octahedral geometry and sought to verify this postulate by comparing the number of observed isomers with the number expected for various geometries. For example, $[Co(NH_3)_4Cl_2]^+$ should exhibit two geometric isomers if it was octahedral. When he first postulated an octahedral geometry for cobalt(III), only one isomer of $[Co(NH_3)_4Cl_2]^+$ was known, the green *trans* isomer. In 1907, after considerable effort, Werner succeeded in isolating the violet *cis* isomer. Even before that, however, he had succeeded in isolating other *cis* and *trans* isomers of cobalt(III) complexes. The occurrence of two isomers was consistent with his postulate of octahedral geometry. Another result consistent with octahedral geometry was the demonstration that $[Co(en)_3]^{3+}$ and certain other complexes were optically active. In 1913, Werner was awarded the Nobel Prize in chemistry for his outstanding research work in the field of coordination chemistry.

---

**SAMPLE EXERCISE 26.7**

Suggest the structure for $CoCl_2 \cdot 6H_2O$.

**Solution:** By analogy to the ammonia complexes of cobalt(III), we might write the formula of this compound as $[Co(H_2O)_6]Cl_2$. Indeed, experimental evidence indicates that the water molecules are attached to the metal as ligands. Hydrated metal salts generally have water coordinated to the metal. However, water can also be hydrogen-bonded to the anion, particularly to oxyanions. For example, the familiar $CuSO_4 \cdot 5H_2O$ has four water molecules coordinated to $Cu^{2+}$ and one to $SO_4^{2-}$.

**PRACTICE EXERCISE**

Suggest the appropriate coordination compound formulation for the compound $PdCl_2 \cdot 3NH_3$. [Pd(II) forms square-planar complexes.]
*Answer:* $[Pd(NH_3)_3Cl]Cl$

---

## 26.7 COLOR AND MAGNETISM

Werner's theory helps us to understand many properties of complexes, including isomerism and conductivity. However, his theory must be extended before we can use it to explain other properties, such as the colors and magnetic properties of transition metal complexes. Studies of these two properties have played an important role in the development of more modern models for metal-ligand bonding. We have discussed the various types of magnetic behavior in Section 25.5; we also discussed the interaction of radiant energy with matter in Section 6.1. Let's briefly examine the significance of these two properties for transition metal complexes before we try to develop a model for metal-ligand bonding.

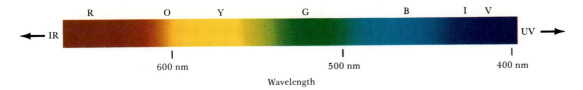

FIGURE 26.20 **FIGURE 26.20** Visible spectrum showing the relation between color and wavelength.

We have already mentioned that those oxides of the transition elements in which the valence level *d* orbitals are partially occupied tend to be colored (Section 25.4). Similarly, the complexes of transition metal ions exhibit a variety of colors. For example, the colors of several cobalt(III) complexes are listed in Table 26.2. From the list you can see that the color of the complex changes as the ligands surrounding the metal ion change. In general, the colors also depend on the particular metal, and on its oxidation state.

Before we can attempt to explain the origin of these colors, we need to review our earlier discussion of light and to introduce some new concepts. Recall first that visible light consists of electromagnetic radiation whose wavelength, $\lambda$, ranges from 400 nm to 700 nm (Figure 6.3). The energy of this radiation is inversely proportional to its wavelength, as discussed earlier, in Section 6.2:

$$E = h\nu = h(c/\lambda) \qquad [26.9]$$

The visible spectrum is shown in Figure 26.20. The colors of the spectrum, indicated by their first letters, spell out what appears to be a man's name: Roy G. Biv.

When a sample absorbs light, what we see is the sum of the remaining colors that strike our eyes. If a sample absorbs all wavelengths of visible light, none reaches our eyes from that sample. Consequently, it appears black. If the sample absorbs no light, it is white or colorless. If it absorbs all but orange, the sample appears orange. Each of these situations is shown in Figure 26.21. That figure shows one further situation; we also perceive an orange color when visible light of all colors except blue strikes our eyes. In a complementary fashion, if the sample absorbed only orange, it would appear blue; blue and orange are said to be **complementary colors**. Complementary colors can be determined with the aid of the artist's color wheel, shown in Figure 26.22. The colors that are complementary to one another, like orange and blue, are across the wheel from each other.

**Color**

## SAMPLE EXERCISE 26.8

The complex ion *trans*-$[Co(NH_3)_4Cl_2]^+$ absorbs light primarily in the red region of the visible spectrum (the most intense absorption is at 680 nm). What is the color of the complex?

**Solution:** Because the complex absorbs red light, its color will be the complementary color of red. From Figure 26.22, we see that this is green.

## PRACTICE EXERCISE

The $[Cr(H_2O)_6]^{2+}$ ion has an absorption band at about 630 nm. Which of the following colors—sky blue, yellow, green, or deep red—is most likely to describe this ion?

*Answer:* sky blue

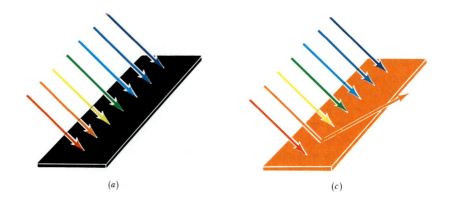

(a)                                              (c)

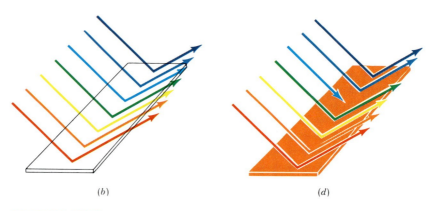

(b)                                              (d)

**FIGURE 26.21**    (a) An object is black if it absorbs all colors of light. (b) An object is white if it reflects all colors of light. (c) An object is orange if it reflects only this color and absorbs all others. (d) An object is also orange if it reflects all colors except blue, the complementary color of orange.

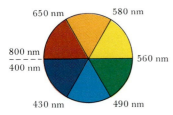

**FIGURE 26.22**  Artist's color wheel, showing the colors that are complementary to one another and the wavelength range of each color.

The amount of light absorbed by a sample as a function of wavelength is known as its **absorption spectrum**. The visible absorption spectrum of a transparent sample, such as a solution of *trans*-$[Co(NH_3)_4Cl_2]^+$, can be determined as shown in Figure 26.23. The spectrum of $[Ti(H_2O)_6]^{3+}$, which we shall discuss in the next section, is shown in Figure 26.24. The absorption maximum of $[Ti(H_2O)_6]^{3+}$ is at 510 nm. Because the sample absorbs most strongly in the green and yellow regions of the visible spectrum, it appears purple.

**Magnetism**

Many transition metal complexes exhibit simple paramagnetism, as described in Section 25.5. In such compounds the individual metal ions possess some number of unpaired electrons that interact strongly with one another. However, the metal ions are usually so far apart that there is not a strong interaction between the magnetic moment on one metal ion with that on the others, so the magnetic behavior observed is simple paramagnetism. It is possible to determine quantitatively the effective magnetic moment of the sample, and from that to deduce the number

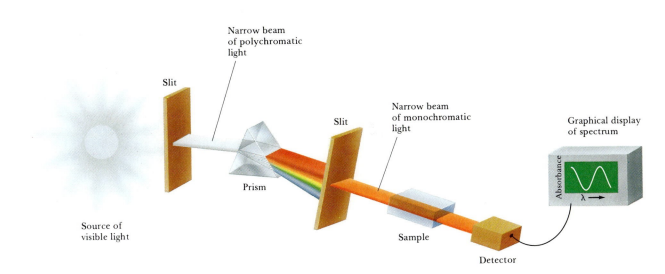

**FIGURE 26.23** Experimental determination of the absorption spectrum of a solution. The prism is rotated so that different wavelengths of light pass through the sample. The detector measures the amount of light reaching it, and this information can be displayed as the absorption at each wavelength.

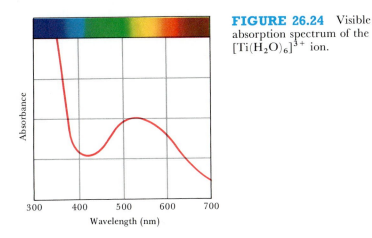

**FIGURE 26.24** Visible absorption spectrum of the $[Ti(H_2O)_6]^{3+}$ ion.

of unpaired electrons per metal ion. The experiments reveal some interesting comparisons. For example, in compounds of the complex ion $[Co(NH_3)_6]^{3+}$ there are no unpaired electrons, but in compounds of the $[CoF_6]^{3-}$ ion there are four per metal ion. Clearly, there is a major difference in the ways in which the electrons are arranged in the metal orbitals in these two cases, yet in both cases we are dealing with complexes of cobalt(III), with a $3d^6$, electron configuration. Any successful bonding theory must explain this and other related observations.

## 26.8 CRYSTAL-FIELD THEORY

Although the ability to form complexes is common to all metal ions, the most numerous and interesting complexes are formed by the transition elements. It has been recognized for a long time that the magnetic properties and colors of transition metal complexes are related to the presence of $d$ electrons in metal orbitals. In this section we will consider a model for bonding in transition metal complexes, called the **crystal-field theory**, that accounts very well for the observed properties of these interesting substances.*

We have already noted that the ability of a metal ion to attract ligands such as water around itself can be viewed as a Lewis acid-base interaction (Section 17.8). The base—that is, the ligand—can be considered to donate a pair of electrons into a suitable empty hybrid orbital on the metal, as shown in Figure 26.25. However, we can assume that much of the attractive interaction between the metal ion and the surrounding ligands is due to the electrostatic forces between the positive charge on the metal and negative charges on the ligands. If the ligand is ionic, as in the case of $Cl^-$ or $SCN^-$, the electrostatic interaction occurs between the positive charge on the metal center and the negative charge on the ligand. When the ligand is neutral, as in the case of $H_2O$ or $NH_3$, the negative ends of these polar molecules, containing an unshared electron pair, are directed toward the metal. In this case the attractive interaction is of the ion-dipole type (Section 11.4). In either case, the result is the same; the ligands are attracted strongly toward the metal center. At the same time, however, the ligands repel one another, because they possess the same charge, or because the negative ends of the molecular dipoles are directed toward the same center. In any metal complex, then, there is a balance between the attractive forces between metal and ligands and the ligand-ligand repulsive interactions. The most stable complex results when the geometrical arrangement of the ligands around the metal maximizes metal-ligand attractions while minimizing the ligand-ligand repulsions. The assembly of metal ion and ligands is lower in energy than the fully separated charges, as illustrated on the left in Figure 26.26.

In a six-coordinate complex the ligand-ligand repulsions cause the ligands to approach along the $x$, $y$, and $z$ axes, as shown in Figure 26.27. With the physical arrangement of ligands and metal ion shown in this figure as our starting point, let us now consider what happens to the energies of electrons in the metal $d$ orbitals as the ligands approach the metal ion. Keep in mind that the $d$ electrons are the outermost electrons of the metal ion. We know that the overall energy of the metal ion plus

*The name "crystal field" arose because the theory was first developed to explain the properties of solid, crystalline materials, such as ruby. The same theoretical model applies to complexes in solution.

**FIGURE 26.25**

Representation of the metal-ligand bond in a complex as a Lewis acid-base interaction. The ligand, which acts as a Lewis base, donates charge to the metal via a metal hybrid orbital. The bond that results is strongly polar, with some covalent character. For many purposes it is sufficient to assume that the metal-ligand interaction is entirely electrostatic in character, as is done in the crystal-field model.

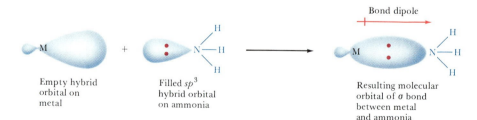

Empty hybrid orbital on metal

Filled $sp^3$ hybrid orbital on ammonia

Bond dipole

Resulting molecular orbital of $\sigma$ bond between metal and ammonia

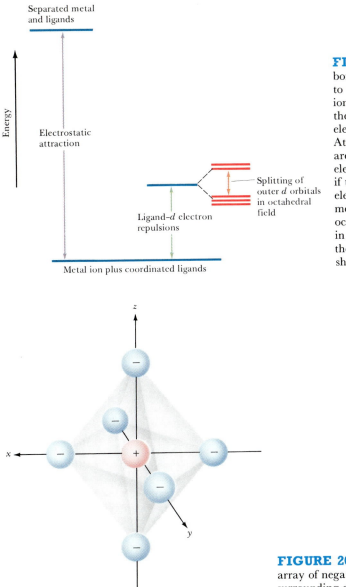

Separated metal
and ligands

Energy

Electrostatic
attraction

Ligand–*d* electron
repulsions

Splitting of
outer *d* orbitals
in octahedral
field

Metal ion plus coordinated ligands

**FIGURE 26.26** In the crystal-field model, the bonding between metal ion and donor atoms is considered to be largely electrostatic. The energy of the metal ion plus coordinated ligands is lower than that of the separated metal ion plus ligands because of the electrostatic attraction between metal ion and ligands. At the same time the energies of the metal *d* electrons are increased by the repulsive interaction between these electrons and the electrons of the ligands. However, if the arrangement of ligands is octahedral, the electrons that occupy the $d_{z^2}$ and $d_{x^2-y^2}$ orbitals are more strongly repelled by the ligands than the electrons occupying the $d_{xy}$, $d_{xz}$, and $d_{yz}$ orbitals. This difference in repulsive interactions gives rise to the splitting of the metal *d*-orbital energies, called crystal-field splitting, shown on the right.

**FIGURE 26.27** Octahedral array of negative charges surrounding a positive charge.

ligands will be lower (that is, more stable) when the ligands are drawn toward the metal center. At the same time, however, there is a repulsive interaction between the outermost electrons on the metal and the negative charges on the ligands. This interaction is called the crystal field. As a result, if we consider just the energies of the *d* electrons on the metal ion, we find that these have *increased*. This increase in energy is shown in the right part of Figure 26.26. But the *d* orbitals of the metal ion do not all behave in the same way under the influence of the ligand field. To see why this is so, recall the shapes of the five *d* orbitals, illustrated in Figure 26.28. In the isolated metal ion, these five orbitals are equivalent in

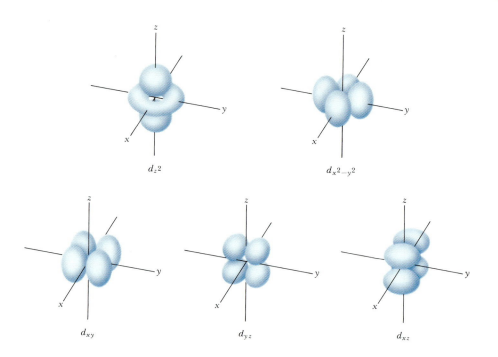

FIGURE 26.28 Shapes of the five $d$ orbitals. Remember that the lobes represent regions in which the electrons occupying an orbital are most likely to be found.

$d_{z^2}$

$d_{x^2-y^2}$

$d_{xy}$

$d_{yz}$

$d_{xz}$

energy. However, as the ligands approach the metal ion, the $d_{x^2-y^2}$ and $d_{z^2}$ orbitals, which are directed *along* the $x, y$, and $z$ axes, are more strongly repelled by the ligands than the $d_{xy}, d_{xz}$, and $d_{yz}$ orbitals. These latter are directed *between* the axes along which the ligands approach. Thus an energy separation, or splitting, occurs. The $d_{x^2-y^2}$ and $d_{z^2}$ orbitals are raised in energy, and the $d_{xz}, d_{yz}$, and $d_{xy}$ orbitals are lowered. This energy splitting is illustrated on the right side of Figure 26.26. In the material that follows we will concentrate on just the splitting of the $d$ orbital energies by the ligand field. This is illustrated in a slightly different form in Figure 26.29.

Let's examine how the crystal-field model accounts for the observed colors in transition metal complexes. The energy gap between the $d$ orbitals, labeled $\Delta$, is of the same order of magnitude as the energy of a photon of visible light. ($\Delta$ is sometimes referred to as the crystal-field splitting energy.) It is therefore possible for a transition-metal complex to absorb visible light, which thereby excites an electron from the lower energy $d$ orbitals into the higher-energy ones. The $[Ti(H_2O)_6]^{3+}$ ion

FIGURE 26.29 Energies of the $d$ orbitals in an octahedral crystal field.

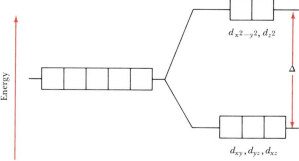

$d_{x^2-y^2}, d_{z^2}$

Energy

$\Delta$

$d_{xy}, d_{yz}, d_{xz}$

Gemstones, such as ruby and emerald, owe their color to the presence of trace amounts of transition metal ions. For example, replacement of a fraction of the aluminum in the colorless mineral corundum, $Al_2O_3$, produces several different gems: chromium forms ruby, manganese forms amethyst, and iron forms topaz. Sapphire, which occurs in a variety of colors but is most often blue, contains titanium and cobalt. Several other gems are produced by replacing a trace of aluminum in the colorless mineral beryl, $Be_3Al_2Si_6O_{18}$, with transition metal ions. For example, emerald contains chromium, and aquamarine contains iron. See Figure 26.30.

**FIGURE 26.30** A selection of gemstones, some mounted in rings: (a) aquamarine, (b) emerald, (c) amethyst, (d) blue sapphire, (e) ruby (an example of an uncut mineral sample), and (f) citrine quartz. In all these cases the colors are due to transition metal ions present as minor components in the parent mineral, such as quartz, alumina, or beryl. (Gem Media, a Division of the Gemological Institute of America)

provides a simple example, because titanium(III) has only one $3d$ electron. As shown in Figure 26.24, $[Ti(H_2O)_6]^{3+}$ has a single absorption peak in the visible region of the spectrum. The maximum absorption is at 510 nm ($3.9 \times 10^{-19}$ J/molecule). Light of this wavelength causes the $d$ electron to move from the lower-energy set of $d$ orbitals into the higher-energy set, as shown in Figure 26.31. The absorption of 510-nm radiation that produces this transition causes substances containing the $Ti(H_2O)_6^{3+}$ ion to appear purple.

The magnitude of the energy gap, $\Delta$—and consequently the color of a complex—depend on both the metal and the surrounding ligands. For

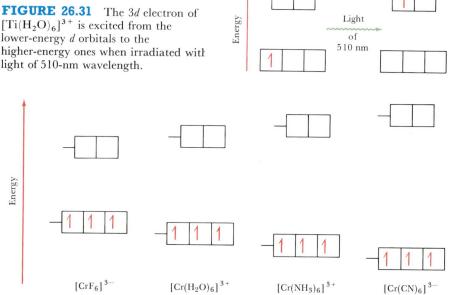

**FIGURE 26.31** The $3d$ electron of $[Ti(H_2O)_6]^{3+}$ is excited from the lower-energy $d$ orbitals to the higher-energy ones when irradiated with light of 510-nm wavelength.

**FIGURE 26.32**
Crystal-field splitting in a series of chromium(III) complexes.

| $[CrF_6]^{3-}$ | $[Cr(H_2O)_6]^{3+}$ | $[Cr(NH_3)_6]^{3+}$ | $[Cr(CN)_6]^{3-}$ |
| Green | Violet | Yellow | Yellow |

example, $[Fe(H_2O)_6]^{3+}$ is light violet, $[Cr(H_2O)_6]^{3+}$ is violet, and $[Cr(NH_3)_6]^{3+}$ is yellow. Ligands can be arranged in order of their abilities to increase the energy gap, $\Delta$. The following is an abbreviated list of common ligands arranged in order of increasing $\Delta$:

$$Cl^- < F^- < H_2O < NH_3 < en < NO_2^- \text{ (N-bonded)} < CN^-$$

This list is known as the **spectrochemical series**.

Ligands that lie on the low end of the spectrochemical series are termed weak-field ligands. Those that lie on the high end are termed strong-field ligands. Figure 26.32 shows schematically what happens to the crystal-field splitting when the ligand is varied in a series of chromium(III) complexes. (This is a good place to remind you that when a transition metal is ionized, the valence $s$ electrons are removed first. Thus the outer electron configuration for chromium is $[Ar]4s^1 3d^5$; that for $Cr^{3+}$ is $[Ar]3d^3$.) Notice that as the field exerted by the six surrounding ligands increases, the splitting of the metal $d$ orbitals increases. Because the absorption spectrum is related to this energy separation, these complexes vary in color.

---

## SAMPLE EXERCISE 26.9

Which of the following complexes of $Ti^{3+}$ exhibits the shortest wavelength absorption in the visible spectrum: $[Ti(H_2O)_6]^{3+}$; $[Ti(en)_3]^{3+}$; $[TiCl_6]^{3-}$?

**Solution:** The wavelength of the absorption is determined by the magnitude of the splitting between the $d$ orbital energies in the field of the surrounding ligands. The larger the splitting, the shorter the wavelength of the absorption corresponding to the transition of the electron from the lower- to the higher-energy orbital. The splitting will be largest for ethylenediamine, en, the ligand that is highest in the spectrochemical series. Thus the complex with the shortest wavelength absorption is $[Ti(en)_3]^{3+}$.

The absorption spectrum of $[Ti(NCS)_6]^{3-}$ shows a band that lies intermediate in wavelength between those for $[TiCl_6]^{3-}$ and $[TiF_6]^{3-}$. What can we conclude about the place of $NCS^-$ in the spectro-chemical series?

*Answer:* It lies between $Cl^-$ and $F^-$; that is, $Cl^- < NCS^- < F^-$

The crystal-field model helps us understand the magnetic properties and some important chemical properties of the transition metal ions. From our earlier discussion of electronic structure in atoms (Section 7.3) we expect that electrons will always occupy the lowest-energy vacant orbitals first and that they will occupy a set of degenerate orbitals one at a time with their spins parallel (Hund's rule). Thus, if we have one, two, or three electrons to add to the *d* orbitals in a complex ion, the electrons will go into the lower-energy set of orbitals, with their spins parallel, as shown in Figure 26.33. When we come to add a fourth electron, a problem arises. If the electron is added to the lower-energy orbital, an energy gain of magnitude $\Delta$ is realized, as compared with placing the electron in the higher-energy orbital. However, there is a penalty for doing this, because the electron must now be paired up with the electron already occupying the orbital. The energy required to do this, relative to putting it in another orbital with parallel spin, is called the **spin pairing energy**. The spin pairing energy arises from the greater electrostatic repulsion of electrons that share an orbital as compared with two that are in different orbitals.

The ligands that surround the metal ion, and the charge on the metal, often play major roles in determining which of the two electronic arrangements arises. Consider the $[CoF_6]^{3-}$ and $[Co(CN)_6]^{3-}$ ions. In both cases the ligands have a $-1$ charge. However, the $F^-$ ion, on the low end of the spectrochemical series, is a weak-field ligand. The $CN^-$ ion, on the high end of the spectrochemical series, is a strong-field ligand. It produces a larger energy gap than does the $F^-$ ion. The splitting of the *d* orbital energies in the complexes is compared in Figure 26.34.

A count of the electrons in cobalt(III) tells us that we have six electrons to place in the 3*d* orbitals. Let us imagine that we add these electrons one at a time to the *d* orbitals of the $CoF_6^{3-}$ ion. The first three will, of course, go into the lower-energy orbitals with spins parallel. The fourth electron could go into a lower-energy orbital, pairing up with

**Electron Configurations in Octahedral Complexes**

**FIGURE 26.33**  Electron configurations associated with one, two, and three electrons in the 3*d* orbitals in octahedral complexes.

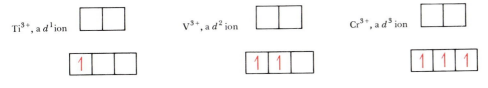

$Ti^{3+}$, a $d^1$ ion          $V^{3+}$, a $d^2$ ion          $Cr^{3+}$, a $d^3$ ion

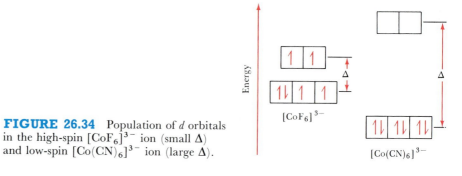

**FIGURE 26.34** Population of $d$ orbitals in the high-spin $[CoF_6]^{3-}$ ion (small $\Delta$) and low-spin $[Co(CN)_6]^{3-}$ ion (large $\Delta$).

one of those already present. This would result in an energy gain of $\Delta$ as compared with putting it in one of the higher-energy orbitals. However, it would cost energy in an amount equal to the spin pairing energy. Because $F^-$ is a weak-field ligand, $\Delta$ is not so large, and the more stable arrangement is the one in which the electron is placed in the higher-energy orbital. Similarly, the fifth electron we add goes into a higher-energy orbital. With all of the orbitals containing at least one electron, the sixth must be paired up, and it goes into a lower-energy orbital. In the case of the $[Co(CN)_6]^{3-}$ complex, the crystal-field splitting is much larger. The spin pairing energy is smaller than $\Delta$, so electrons are paired in the lower-energy orbitals, as illustrated in Figure 26.34.

The $[CoF_6]^{3-}$ complex is referred to as **high spin**; that is, the electrons are arranged so that they remain unpaired as much as possible. On the other hand, the electron arrangement found in the $[Co(CN)_6]^{3-}$ ion is referred to as **low spin**. These two different electronic arrangements can be readily distinguished by measuring the magnetic properties of the complex, as described earlier. The absorption spectrum also shows characteristic features that indicate the electronic arrangement.

**SAMPLE EXERCISE 26.10**

Predict the number of unpaired electrons in six-coordinate high-spin and low-spin complexes of $Fe^{3+}$.

High spin          Low spin

**Solution:** The $Fe^{3+}$ ion possesses five $3d$ electrons. In a high-spin complex these are all unpaired. In a low-spin complex the electrons are confined to the lower-energy set of $d$ orbitals, with the result that there is one unpaired electron:

**PRACTICE EXERCISE**

For which $d$ electron configurations does the possibility of a distinction between high-spin and low-spin arrangements exist in octahedral complexes?
*Answer:* $d^4$, $d^5$, $d^6$, $d^7$

**Tetrahedral and Square-Planar Complexes**

Thus far we have considered the crystal-field model only for complexes of octahedral geometry. When there are only four ligands about the metal, the geometry is tetrahedral, except for the special case of metal ions with a $d^8$ electron configuration, which we will discuss in a moment. The crystal-field splitting of the metal $d$ orbitals in tetrahedral complexes differs from that in octahedral complexes. Four equivalent ligands can interact with a central metal ion most effectively by approaching along the vertices of a tetrahedron. (Figure 25.14 offers a good comparison of the octahedral and tetrahedral geometries.) It turns out—and this is not easy to explain in just a few sentences—that the splitting of the metal $d$ orbitals in a tetrahedral crystal is just the opposite of that for the octahedral case. That is, three of the metal $d$ orbitals are raised in energy, and the other two are lowered, as illustrated in Figure 26.35. Because there are only four ligands instead of six, as in the octahedral case, the crystal-field splitting is much smaller for tetrahedral complexes. Calculations show that for the same metal ion and ligand set, the crystal-field splitting for a tetrahedral complex is only $\frac{4}{9}$ as large as for the octahedral complex. For this reason, all tetrahedral complexes are characterized by high spin; the crystal field is never large enough to overcome the spin-pairing energies.

Square-planar complexes, in which four ligands are arranged about the metal ion in a plane, represent a common geometrical form. One can think of the square-planar complex as formed by removing two ligands from along the vertical $z$ axis of the octahedral complex. As this happens the four ligands in the plane are drawn in more tightly. The changes that occur in the energy levels of the $d$ orbitals are illustrated in Figure 26.36.

Square-planar complexes are characteristic of metal ions with a $d^8$ electron configuration. They are nearly always low spin; that is, the eight $d$ electrons are spin-paired to form a diamagnetic complex. Such an electronic arrangement is particularly common among the heavier metals, such as Pd, Pt, Ir, and Au.

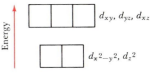

**FIGURE 26.35** Energies of the $d$ orbitals in a tetrahedral crystal field.

---

### SAMPLE EXERCISE 26.11

Four-coordinate nickel(II) complexes exhibit both square-planar and tetrahedral geometries. The tetrahedral ones, such as $[NiCl_4]^{2-}$, are paramagnetic; the square-planar ones, such as $[Ni(CN)_4]^{2-}$, are diamagnetic. Show how the $d$ electrons of nickel(II) populate the $d$ orbitals in the appropriate crystal-field-splitting diagram in each case.

**Solution:** Nickel(II) has an electron configuration of $[Ar]3d^8$. The population of the $d$ electrons in the two geometries is given at right:

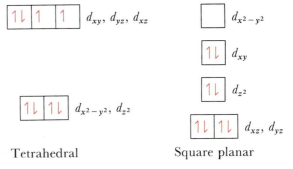

Tetrahedral        Square planar

### PRACTICE EXERCISE

How many unpaired electrons do you predict for the tetrahedral $[CoCl_4]^{2-}$ ion?

***Answer:*** three

---

We have seen that the crystal-field model provides a basis for explaining many features of transition metal complexes. In fact, it can be used to explain many observations in addition to those we have discussed. However, many lines of evidence show that the bonding between transition-metal ions and ligands must have some covalent character. Molecular-orbital theory (Sections 9.5 and 9.6) can also be used to describe the

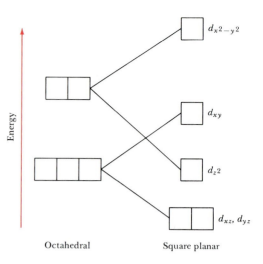

**FIGURE 26.36** Effect on the relative energies of the $d$ orbitals of removing the two negative charges from the $z$ axis of an octahedral complex. When the charges are completely removed, the square-planar geometry results.

$d_{x^2-y^2}$

$d_{xy}$

$d_{z^2}$

$d_{xz}, d_{yz}$

Octahedral          Square planar

Energy

bonding in complexes. However, the application of molecular-orbital theory to coordination compounds is beyond the scope of our discussion. The crystal-field model, although not entirely accurate in all details, provides an adequate and useful description.

## *FOR REVIEW*

### SUMMARY

**Coordination compounds** or **complexes** contain metal ions bonded to several surrounding anions or molecules known as **ligands**. The metal ion and its ligands comprise the **coordination sphere** of the complex. The atom of the ligand that bonds to the metal ion is known as the **donor atom**. The number of donor atoms attached to the metal ion is known as the **coordination number** of the metal ion. The common coordination numbers are four and six; the common coordination geometries are tetrahedral, square planar, and octahedral.

If a ligand has several donor atoms that can coordinate simultaneously to the metal, it is said to be **polydentate** and is referred to as a **chelating agent**. Two common examples are ethylenediamine (en), which is potentially bidentate, and ethylenediaminetetraacetate ($EDTA^{4-}$), which has six potential donor atoms. Many biologically important molecules, such as the porphyrins, are complexes of chelating agents.

Isomerism is common among coordination compounds. **Structural isomerism** involves differences in the bonding arrangements of the ligands. One simple form of structural isomerism, known as **linkage**

isomerism, occurs when a ligand is capable of coordinating to a metal through either of two donor atoms. **Coordination-sphere isomerism** occurs when two compounds with the same overall formula contain different ligands in the coordination sphere. **Stereoisomerism** involves complexes with the same chemical bonding arrangements but with differing spatial arrangements of ligands. The most common forms are **geometric** and **optical isomerism**. Geometric isomers differ from one another in the relative locations of donor atoms in the coordination sphere; the most common are *cis-trans* isomers. Optical isomers differ from one another in that they are nonsuperimposable mirror images of one another. Geometric isomers differ from one another in their chemical and physical properties; however, optical isomers differ only in the presence of a **chiral** environment. Most often optical isomers are distinguished from one another by their interactions with plane-polarized light; solutions of one isomer rotate the plane of polarization to the right (**dextrorotatory**), and solutions of its mirror image rotate the plane to the left (**levorotatory**). Chiral molecules are said to be **optically active**. A 50–50 mixture of two optical isomers does not rotate plane-polarized light and is said to be **racemic**.

Many of the early studies that served as a basis for our current understanding of complexes focused on complexes of chromium(III), cobalt(III), platinum(II), and platinum(IV). Complexes of these metal ions are **inert**, in that they undergo ligand exchange at a slow rate. Complexes undergoing rapid exchange are said to be **labile**.

Studies of the magnetic properties and colors of transition metal complexes have played an important role in formulation of bonding theories for these compounds. The **crystal-field theory** successfully accounts for many properties of coordination compounds. In this model the interaction between metal ion and ligand is viewed as electrostatic. The ligands produce an electric field that causes a splitting in the energies of the metal $d$ orbitals. In the **spectrochemical series** the ligands are listed in order of their ability to split the $d$ orbital energies in octahedral complexes.

In strong-field complexes the splitting of $d$ orbital energies is large enough to overcome spin-pairing energies, and $d$ electrons preferentially pair up in the lower-energy orbitals. Such complexes are **low spin**. When the ligands exert a relatively weak crystal field the electrons occupy the higher-energy $d$ orbitals in preference to pairing up in the lower-energy set, and the complexes are **high spin**.

The crystal-field model is applicable also to tetrahedral and square-planar complexes. However, the ordering of the $d$ orbital energies is different than in octahedral complexes.

## LEARNING GOALS

Having read and studied this chapter, you should be able to:

1. Determine either the charge of a complex ion, having been given the oxidation state of the metal, or the oxidation state, having been given the charge of the complex. (You will need to recognize the common ligands and their charges.)

2. Describe, with the aid of drawings, the common geometries of complexes. (You will need to recognize whether the common ligands are functioning as monodentate or polydentate ligands.)

3. Name coordination compounds, having been given their formulas, or write their formulas, having been given their names.

4. Describe the common types of isomerism, and distinguish between structural isomerism and stereoisomerism.

5. Determine the possible number of stereoisomers for a complex, having been given its composition.

6. Distinguish between inert and labile complexes.

7. Explain how the conductivity, precipitation reactions, and isomerism of complexes are used to infer their structures.

8. Explain how the magnetic properties of a compound can be measured and used to infer the number of unpaired electrons.

9. Explain how the colors of substances are related to their absorption and reflection of incident light.

10. Explain how the electrostatic interaction between ligands and metal $d$ orbitals in an octahedral complex results in a splitting of energy levels.

11. Explain the significance of the spectrochemical series.

12. Account for the tendency of electrons to pair in strong-field, low-spin complexes.

13. Sketch a representation of the $d$-orbital energy levels in a tetrahedral complex, and explain the reason for a smaller crystal-field splitting in this geometry as compared with octahedral complexes.

14. Sketch the $d$-orbital energy levels in a square-planar complex.

## KEY TERMS

Among the more important terms and expressions used for the first time in this chapter are the following:

A **chelating agent** (Section 26.2) is a polydentate ligand that is capable of occupying two or more sites in the coordination sphere.

**Chiral** (Section 26.4) means having a nonsuperimposable mirror image; for example, we might refer to a molecule as being chiral.

A *cis* geometrical arrangement (Section 26.4) has like groups adjacent to each other.

The **coordination number** (Section 26.1) of an atom is the number of adjacent atoms to which it is directly bonded; in a complex, the coordination number of the metal ion is the number of donor atoms to which it is bonded.

The **coordination sphere** (Section 26.1) of a complex consists of the metal ion and its surrounding ligands.

**Coordination-sphere isomers** (Section 26.4) are structural isomers of coordination compounds in which the ligands within the coordination sphere differ.

The **crystal-field theory** (Section 26.8) accounts for the colors and the magnetic and other properties of transition metal complexes in terms of the splitting of the energies of metal-ion $d$ orbitals by the electrostatic interaction with the ligands.

The term **dextrorotatory**, or merely **dextro** or $d$ (Section 26.4), is used to label a chiral molecule that rotates the plane of polarization of plane-polarized light to the right (clockwise).

The **donor atom** (Section 26.1) of a ligand is the one that bonds to the metal.

**Geometric isomers** (Section 26.4) have different spatial arrangements of donor atoms around the coordination sphere.

A **high-spin** complex (Section 26.8) has the same number of unpaired electrons as does the isolated metal ion.

An **inert** complex (Section 26.5) exchanges ligands at a slow rate.

**Isomers** (Section 26.4) are compounds whose molecules have the same overall composition but different structures.

A **labile** complex (Section 26.5) exchanges ligands at a rapid rate.

The term **levorotatory**, or merely **levo** or $l$ (Sec-

tion 26.4), is used to label a chiral molecule that rotates the plane of polarization of plane-polarized light to the left (counterclockwise).

A **ligand** (Section 26.1) is an ion or molecule that coordinates to a single central metal atom or to an ion to form a complex.

**Linkage isomers** (Section 26.4) are structural isomers of coordination compounds in which a ligand differs in its mode of attachment to a metal ion.

A **low-spin** complex (Section 26.8) has fewer unpaired electrons than does the isolated metal ion.

A **monodentate ligand** (Section 26.2) is one that binds to the metal ion via a single donor atom. It occupies one position in the coordination sphere.

**Optical isomers** (Section 26.4) are stereoisomers in which the two forms of the compound are nonsuperimposable mirror images.

A **polydentate ligand** (Section 26.2) is one in which two or more donor atoms can coordinate to the same metal ion. In a **bidentate ligand** the number of coordinating atoms bound to the metal is two.

The **spectrochemical series** (Section 26.8) is a list of ligands arranged in order of their abilities to split the $d$ orbital energies (using the terminology of the crystal-field model).

**Stereoisomers** (Section 26.4) are compounds possessing the same formula and bonding arrangement but differing in the spatial arrangements of the atoms.

**Structural isomers** (Section 26.4) are compounds possessing the same formula but differing in the bonding arrangements of the atoms.

The **trans** geometrical arrangement (Section 26.4) has like groups opposite each other.

# EXERCISES

### Structure and Nomenclature

**26.1** Indicate the coordination number about the metal and the oxidation number of the metal in each of the following complexes:
**(a)** $[Zn(en)_2]Br_2$
**(b)** $[Co(NH_3)_4Cl_2]Cl$
**(c)** $K[Co(C_2O_4)_2(NH_3)_2]$
**(d)** $K_2[MoOCl_4]$
**(e)** $K[Au(CN)_2]$.

**26.2** Indicate the coordination number about the metal and the oxidation number of the metal in each of the following complexes:
**(a)** $K_2[V(C_2O_4)_3]$
**(b)** $K_4[Fe(CN)_6]$
**(c)** $[Pd(NH_3)_2Cl_2]$
**(d)** $[Cr(en)_2F_2]NO_3$
**(e)** $K_2[HgCl_4]$.

**26.3** Sketch the structure of each of the following complexes:
**(a)** $[Zn(NH_3)_4]^{2+}$
**(b)** $[PtCl_2(en)]$
**(c)** $[Ag(CN)_2]^-$
**(d)** $trans$-$[PtH(Br)(ONO)_2]^{2-}$.

**26.4** Sketch the structure of each of the following complexes:
**(a)** $cis$-$[CrCl_2(en)_2]^+$
**(b)** $[FeBr_4]^-$
**(c)** $trans$-$[Pt(NH_3)_4Cl_2]^{2+}$
**(d)** $[Cd(en)(SCN)_2]$.

**26.5** Name each of the coordination compounds or plex ions listed in Exercises 26.3 and 26.4.

**26.6** Name each of the following complexes:
**(a)** $[Co(NO_2)_3(NH_3)_3]$
**(b)** $Cs_3[Cr(C_2O_4)_2Cl_2]$

(c)  $K_2[NiCl_4]$
(d)  $[Pd(en)_2][Cr(NH_3)_2Br_4]_2$
(e)  $K_3[IrCl_5(S_2O_3)]$
(f)  $[Co(en)(H_2O)F_3]$.

**26.7**  Using $Br^-$ and $NH_3$ as ligands, (a) give the formula of a six-coordinate palladium(IV) complex that would be a nonelectrolyte in aqueous solution; (b) give the coordination compound of Pt(II) that has about the same electrical conductivity as RbBr; (c) give an octahedral complex of Fe(III) containing two $NH_3$ groups.

**26.8**  Using $Cl^-$ and ethylenediamine, en, as ligands and $Rb^+$ as cation, (a) give the formulas of two six-coordinate cobalt(III) complexes that have approximately the same electrical conductivity in water as RbCl; (b) give the formula of a nickel(II) complex that has about the same electrical conductivity in water as $SrCl_2$; (c) give the formula for a tetrahedral complex of $Cd^{2+}$ that contains one en group.

**26.9**  Write the formulas for each of the following compounds, being sure to use brackets to indicate the coordination sphere:
(a)  hexaamminenickel(II) nitrate
(b)  tetraamminediazidocobalt(III) fluoride
(c)  potassium diaquatetranitritovanadate(III)
(d)  dichlorobis(ethylenediamine)cobalt(III) chloride
(e)  bis(ethylenediamine)zinc(II) tetrabromomercurate(II).

**26.10**  Write the formula for each of the following compounds, being sure to use brackets to indicate the coordination sphere:
(a)  pentaamminethiosulfatomanganese(III) sulfate
(b)  potassium trichloroethylenediaminenitritocobaltate(III)
(c)  tris(bipyridyl)ruthenium(III) nitrate
(d)  dichlorobis(*ortho*-phenanthroline)iron(III) chloride
(e)  sodium tris(oxalato)cobaltate(III).

**26.11**  Polydentate ligands can vary in the number of coordination positions they occupy. In each of the following indicate the most probable number of coordination positions occupied by the polydentate ligand present:
(a)  $[Co(NH_3)_4CO_3]Cl$
(b)  $[Cr(C_2O_4)_2(H_2O)_2]Br$
(c)  $[CrEDTA(H_2O)]^-$
(d)  $[Zn(en)_2](ClO_4)_2$.

**26.12**  Indicate the likely coordination number of the metal in each of the following complexes:
(a)  $[Zn(en)Cl_2]$
(b)  $[Co(o\text{-}phen)_2Cl_2]Cl$
(c)  $K_2[Hg(SCN)_4]$
(d)  $Na[FeEDTA]$.

## Isomerism

**26.13**  By writing formulas or drawing structures related to any one of the following complexes, illustrate (a) geometrical isomerism; (b) linkage isomerism; (c) optical isomerism; (d) coordination-sphere isomerism. The complexes are: $[Co(NH_3)_4(C_2O_4)]Cl_2$; $[Pd(NH_3)_2(ONO)_2]$; and *cis*-$[V(en)_2Cl_2]^+$.

**26.14**  What types of isomerism are present in each of the following pairs of compounds:
(a)  $[CoCl(H_2O)(en)_2]Cl_2$ and $[CoCl_2(en)_2]Cl\cdot H_2O$
(b)  $[PtCl_2(NH_3)_4]Br_2$ and $[PtBr_2(NH_3)_4]Cl_2$
(c)  $[Co(NH_3)_5SCN](NO_3)_2$ and $[Co(NH_3)_5NCS]$ $(NO_3)_2$?

**26.15**  Sketch all possible isomeric structures (there might be only one) for each of the following complex ions or coordination compounds:
(a)  $[Cd(en)Cl_2]$
(b)  $[Fe(C_2O_4)_2(NH_3)Cl]^{2-}$
(c)  $[PdCl_2(SCN)_2]^{2-}$
(d)  $[IrBr_3Cl(H_2O)_2]$.

**26.16**  Sketch all the possible stereoisomers for each of the following complex ions or compounds:
(a)  square-planar $[Pd(ox)(NH_3)Cl]^-$
(b)  tetrahedral $[Zn(NH_3)BrCl(SCN)]^-$
(c)  $[Fe(o\text{-}phen)(en)_2]^+$
(d)  $[Co(ox)_3]^{3-}$.

**26.17**  The compound $Co(NH_3)_5(SO_4)Br$ exists in two forms, one red and one violet. Both forms dissociate in solution to form two ions. Solutions of the red compound form a precipitate of AgBr on addition of $AgNO_3$ solution, but no precipitate of $BaSO_4$ on addition of $BaCl_2$ solution. For the violet compound, just the reverse occurs. From this evidence, indicate the structures of the complex ions in each case, and give the correct name of each compound.

**26.18**  Write the formulas for, and properly name, two possible coordination-sphere isomers with the formula $Cr(H_2O)_4(OH)Br_2$.

**26.19**  A palladium complex formed from a solution containing bromide ion and pyridine, $C_5H_5N$ ( a good donor toward metal ions), is found on elemental analysis to contain 37.6 percent bromine, 28.3 percent carbon, 6.60 percent nitrogen, and 2.37 percent hydrogen. The compound is slightly soluble in several organic solvents; its solutions in water or alcohol do not conduct electricity. It is found experimentally to have a zero dipole moment. Write the chemical formula and indicate its probable structure. Name the compound.

**26.20**  A manganese complex formed from a solution containing potassium bromide and oxalate ion is purified and analyzed. It contains 10.0 percent Mn, 28.6 percent potassium, 8.8 percent carbon, and 29.2 percent bromine. An aqueous solution of the complex has about the same electrical conductivity as an equimolar solution of $K_4[Fe(CN)_6]$. Write the formula of the compound, using brackets to denote the manganese and its coordination sphere. Name the compound.

## Color; Magnetism; Crystal-Field Theory

**26.21**  What is the observed color of a coordination compound that absorbs radiation of wavelength 580 nm?

**26.22**  A coordination compound is a bright blue in color. Assuming that this color is due to a single absorption band, what is the approximate wavelength of the absorption?

**26.23** Give the number of $d$ electrons associated with the central metal in each of the following complexes:
(a) $[Co(CN)_5]^{3-}$
(b) $[AuCl_4]^-$
(c) $[V(NH_3)_3Cl_3]^-$
(d) $[Ru(en)_3]^{2+}$
(e) $[Mn(CN)_6]^{3-}$.

**26.24** Give the number of $d$ electrons associated with the central metal ion in each of the following complexes:
(a) $[Ru(en)_3]Cl_3$
(b) $K[Cu(CN)_4]$
(c) $Na_3[Co(NO_2)_6]$
(d) $[MoEDTA]ClO_4$
(e) $K_3[ReCl_6]$.

**26.25** Explain why the $d_{xy}$, $d_{xz}$, and $d_{yz}$ orbitals lie lower in energy than the $d_{x^2-y^2}$ and $d_{z^2}$ orbitals in the presence of an octahedral arrangement of ligands about the central metal ion.

**26.26** What property of the ligand determines the size of the splitting of the $d$ orbital energies in the presence of an octahedral arrangement of ligands about a central transition metal ion? Explain.

**26.27** Explain why many cyano complexes of divalent transition metal ions are yellow, whereas many aqua complexes of these ions are blue or green.

**26.28** The ions $V^{3+}$ and $Ti^{2+}$ are isoelectronic; that is, both have a $3d^2$ electronic configuration. Assume you know the electronic spectrum of the $[V(H_2O)_6]^{3+}$ ion. How would you use this to predict the general character of the electronic spectrum for the as-yet unknown $[Ti(H_2O)_6]^{2+}$ ion? Explain.

**26.29** For each of the following metals, write the electronic configuration of the metal atom and the $+3$ ion: (a) Rh; (b) Mn; (c) Pd. Draw the crystal-field energy-level diagram for the $d$ orbitals of an octahedral complex, and show the placement of the $d$ electrons in each case, assuming a strong-field complex.

**26.30** For each of the following metals, write the electronic configuration of the metal atom and the $+2$ ion: (a) Ni; (b) Tc; (c) Cr. Draw the crystal-field energy-level diagram for the $d$ electrons of an octahedral complex, and show the placement of the $d$ electrons in each case, assuming a weak-field complex of the $+2$ ion.

**26.31** Draw the crystal-field energy-level diagram and show the placement of $d$ electrons for each of the following: (a) $[V(H_2O)_6]^{2+}$; (b) $[Mn(H_2O)_6]^{3+}$ (high spin); (c) $[Ru(CN)_6]^{3-}$ (one unpaired electron); (d) $[IrCl_6]^{3-}$ (low spin); (e) $[Cr(en)_3]^{2+}$ (four unpaired electrons); (f) $[NiF_6]^{4-}$.

**26.32** Draw the crystal-field energy-level diagrams and show the placement of electrons for the following complexes: (a) $[ZrCl_6]^{2-}$; (b) $[MnF_6]^{3-}$ (a high-spin complex); (c) $[Rh(NH_3)_6]^{3+}$ (a low-spin complex); (d) $[NiCl_4]^{2-}$; (e) $[PtBr_4]^{2-}$; (f) $[Cu(en)_3]^{2+}$.

**26.33** How would you go about determining the num-

ber of unpaired electrons in the compound $Na_2[CoCl_4]$? If the compound proved to have three unpaired electrons, how would you account for this in terms of crystal-field theory?

**26.34** The compound $[Fe(bipy)_3](NO_3)_3$ is paramagnetic. The magnitude of the paramagnetism indicates that there is one unpaired electron in the metal complex. Sketch the energy-level diagram for the complex and indicate the placement of electrons. Is the complex characterized by high spin or by low spin?

## Additional Exercises

**26.35** Distinguish between the following terms: (a) a chelate and a monodentate ligand; (b) coordination sphere and coordination number; (c) a labile and an inert complex; (d) a high-spin and a low-spin complex; (e) coordination-sphere isomerism and linkage isomerism; (f) optical isomerism and geometrical isomerism.

**26.36** Based on the molar conductance values listed below for the series of platinum(IV) complexes, write the formula for each complex so as to show which ligands are in the coordination sphere of the metal.

| Complex | Molar conductance $(\Omega^{-1})$[a] of 0.05 $M$ solution |
|---|---|
| $Pt(NH_3)_6Cl_4$ | 523 |
| $Pt(NH_3)_4Cl_4$ | 228 |
| $Pt(NH_3)_3Cl_4$ | 97 |
| $Pt(NH_3)_2Cl_4$ | 0 |
| $KPt(NH_3)Cl_5$ | 108 |

[a] The ohm $(\Omega)$ is a unit of resistance; conductance is the inverse of resistance.

**[26.37]** Both $Fe^{2+}$ and $Fe^{3+}$ complexes of $Cl^-$ and $Br^-$ are known, but only $Fe^{2+}$ complexes of $I^-$ exist. Suggest a reason for this.

**26.38** In Werner's early studies he observed that when the complex $[Co(NH_3)_4Br_2]Br$ was placed in water, the electrical conductivity of a 0.05 $M$ solution changed from an initial value of 191 $\Omega^{-1}$ to a final value of 374 $\Omega^{-1}$ over a period of an hour or so. Suggest an explanation of the observed results. Write a balanced chemical equation to describe the reaction.

**26.39** Solutions containing the $[Co(H_2O)_6]^{2+}$ ion absorb at about 520 nm; those containing the $[CoCl_4]^{2-}$ ion absorb at about 690 nm. What colors do you expect for the solutions? Why is the absorption maximum for the $[CoCl_4]^{2-}$ ion at a longer wavelength than for the $[Co(H_2O)_6]^{2+}$ ion?

**26.40** Draw out the $d$-orbital energy-level diagram for each of the following complexes, and indicate the most likely placement of $d$ electrons: (a) $[Au(CN)_4]^-$; (b) $[Ni(C_2O_4)_3]^{2-}$; (c) $[PtF_6]$; (d) $[Rh(CN)_6]^{2-}$.

**26.41** The complex $[Co(NH_3)_6]^{3+}$ contains no unpaired electrons, whereas the complex $[Mn(NH_3)_6]^{2+}$ contains five. Account for the difference in terms of the crystal-field model.

**26.42** Acetylacetone forms very stable complexes with many metallic ions. It acts as a bidentate ligand, coordinating to the metal at two adjacent positions. Suppose that one of the $CH_3$ groups of the ligand is replaced by a $CF_3$ group, as shown:

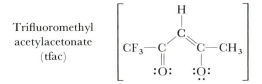

Trifluoromethyl acetylacetonate (tfac)

Sketch all possible isomers for the *tris* complex of tfac with cobalt(III). (You can use the symbol ●⌣○ to represent the ligand.)

**26.43** What changes would you expect in the absorption spectrum of complexes of V(III) as the ligand is varied in the order $F^-$, $NH_3$, $NO_2^-$?

**26.44** Indicate whether each of the following statements is true or false. If it is false, indicate how it would need to be modified to make it true. **(a)** Spin pairing energy is larger than $\Delta$ in low-spin complexes. **(b)** $\Delta$ is larger for complexes of $Mn^{3+}$ with a given ligand than for complexes of $Mn^{2+}$. **(c)** $[NiCl_4]^{2-}$ is more likely to be square-planar than is $[Ni(CN)_4]^{2-}$.

**26.45** The value of $\Delta$ for the $[CrF_6]^{3-}$ complex is 182 kJ/mol. Calculate the expected wavelength of the absorption corresponding to promotion of an electron from the lower-energy to the higher-energy $d$ orbital set in this complex. Should the complex absorb in the visible range? (You may need to review Sample Exercise 6.2; remember to divide by Avogadro's number.)

**26.46** Many trace metal ions exist in the bloodstream as complexes with amino acids (Figure 28.4) or small peptides. The anion of the amino acid glycine,

$$\underset{\substack{\\ H_2NCH_2\overset{\displaystyle O}{\overset{\|}{C}}-O^-}}{}$$

symbol gly, is capable of acting as a bidentate ligand, coordinating to the metal through nitrogen and $-O^-$ atoms. How many isomers are possible for **(a)** $[Zn(gly)_2]$ (tetrahedral); **(b)** $[Pt(gly)_2]$ (square planar);

**(c)** $[Co(gly)_3]$ octahedral). Sketch all possible isomers. Use N⌣O to represent the ligand.

**26.47** Write balanced chemical equations to represent the following observations. (In some instances, the complex involved has been discussed previously at some point in the text.) **(a)** Solid silver chloride dissolves in an excess of aqueous ammonia. **(b)** The green complex $[Cr(en)_2Cl_2]Cl$, on treatment with water over a long time, converts to a brown-orange complex. Reaction of $AgNO_3$ with a molar solution of the product results in precipitation of 3 mol of AgCl. (Write *two* reactions.) **(c)** Insoluble zinc hydroxide dissolves in excess aqueous ammonia. **(d)** A pink solution of $Co(NO_3)_2$ turns deep blue on addition of concentrated hydrochloric acid.

**26.48** In each of the following pairs of complexes, which would you expect to absorb at the longer wavelength: **(a)** $[CoF_6]^{4-}$, $[Co(CN)_6]^{4-}$; **(b)** $[V(H_2O)_6]^{2+}$, $[V(H_2O)_6]^{3+}$; **(c)** $[Mn(CN)_6]^{3-}$, $[MnCl_4]^-$? Explain your reasoning in each case.

**26.49** Draw the geometrical structure of each of the following complexes: **(a)** $[Pt(en)_2]^{2+}$; **(b)** *cis*-dichloroethylenediamineoxalatoferrate(III) ion; **(c)** *trans*-diamminedibromopalladium(II); **(d)** *trans*-$[Co(en)_2Br_2]^+$; **(e)** tetrabromocobaltate(II) ion: **(f)** *trans*-$[Mo(NCS)_2(en)_2]^+$; **(g)** pentaamminebromovanadium(II) ion.

**[26.50]** The red color of ruby is due to the presence of Cr(III) ions in octahedral sites in the close-packed oxide lattice. Draw the crystal-field splitting diagram for Cr(III) in this environment. Suppose that the ruby crystal is subjected to high pressure. What do you predict for the variation in the wavelength of absorption of the ruby as a function of pressure? Explain.

**[26.51]** The $d^3$ and $d^6$ electronic configurations are favorable for octahedral coordination, but not for tetrahedral. Explain why this is so in terms of crystal-field theory.

**[26.52]** Suppose that a transition metal ion were in a lattice in which it was in contact with just two nearby anions, located on a line along the $+$ and $-$ axis directions. Diagram the splitting of the metal $d$ orbitals that would result from such a crystal field. Assuming a strong field, how many unpaired electrons would you expect for a metal ion with six $d$ electrons?

## CHAPTER 27

# Organic Chemistry

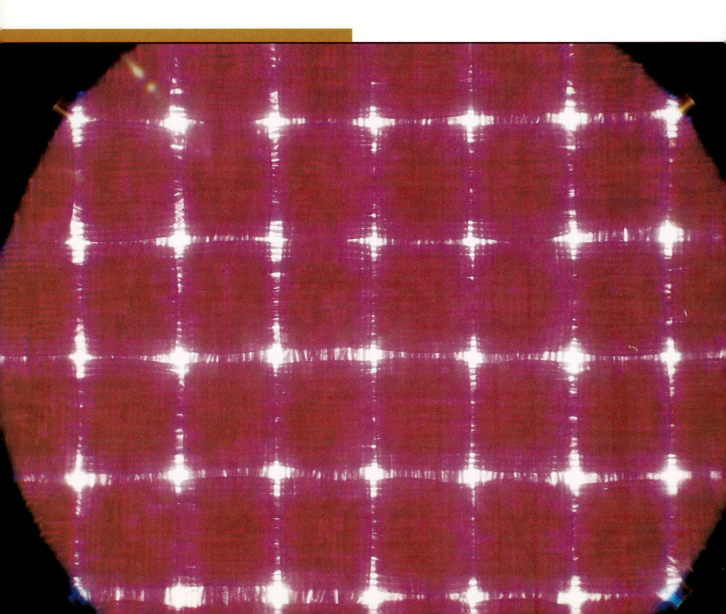

Organic chemistry deals mainly with compounds in which carbon is a principal element. There are no precise boundaries between organic chemistry and other areas of chemical science. The concept of a separate area of chemistry that could be thought of as organic developed out of the vitalist theory, which held that substances that make up living matter are fundamentally different from those that form inanimate matter. In 1828, Friedrich Wöhler, a German chemist, reacted potassium cyanate, KOCN, with ammonium chloride, $NH_4Cl$; much to his surprise he obtained urea, $H_2NCONH_2$, a well-known substance that had been isolated from the urine of mammals. Following Wöhler, many other organic substances were prepared from inorganic starting materials, and the vitalist theory gradually disappeared. Nevertheless, organic chemistry continued to develop as a distinct area of chemistry, partly because the raw materials for much of organic chemistry—oil, coal, wood, animal matter, and so forth—are of plant or animal origin.

In this chapter, we shall present a brief view of some of the elementary aspects of organic chemistry. We can do no more than hint at the magnitude of the subject. It has been estimated that there are now more than a million known organic substances. Each year several thousand new organic substances are discovered in nature or synthesized in the laboratory. These huge numbers might lead one to think that learning the subject of organic chemistry is hopelessly difficult. However, certain arrangements of atoms and groups of atoms, called **functional groups**, occur repeatedly in organic substances. These arrangements lead to particular chemical characteristics that are very similar among the compounds containing that functional group. By learning the characteristic chemical properties of these functional groups, you can understand the chemical characteristics of many organic substances.

**Hydrocarbons** contain only two elements, carbon and hydrogen. With such a limited range of composition, you might suppose that there would be little variety in the chemical properties of the hydrocarbons. However, such is not the case. The key structural feature of hydrocarbons, and for that matter of most other organic substances, is the presence of stable carbon-carbon bonds. Carbon alone among the elements is able to form stable, extended chains of atoms bonded through single, double, or triple bonds. No other element is capable of forming similar structures.

The hydrocarbons can be divided into four groups, the **alkanes**, **alkenes, alkynes**, and **aromatic hydrocarbons**. We have already encountered at least one member from each of these series. Figure 27.1

## 27.1  THE HYDROCARBONS

A 50-fold enlargement of a weave of nylon fibers. The first nylon was invented by Wallace Carrothers and co-workers in the 1930s at DuPont Central Research Laboratories, after ten years of research. Nylon is actually a family of related synthetic polyamide materials. Together they form the most generally important and widely useful of all synthetic polymers. (Runk, Schoenberger/Grant Heilman Photography)

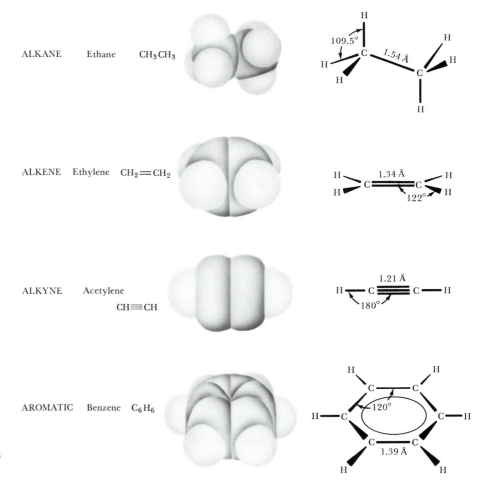

ALKANE    Ethane    $CH_3CH_3$

ALKENE    Ethylene    $CH_2 = CH_2$

ALKYNE    Acetylene

          $CH \equiv CH$

AROMATIC    Benzene    $C_6H_6$

**FIGURE 27.1** Names, geometrical structures, and molecular formulas of examples of each type of hydrocarbon.

shows the name, molecular formula, and geometrical structure of the simplest member that contains a carbon-carbon bond in each series.

The alkanes consist of carbon atoms bonded either to hydrogen or to other carbon atoms by four single bonds. Depending on its place in the alkane structure, a carbon atom may be bonded to three hydrogens and one carbon, two hydrogens and two carbons, a hydrogen and three carbons, or four carbons. The alkenes are hydrocarbons with one or more carbon-carbon double bonds. The simplest member of the alkene series is ethylene (see Figure 27.1); you might wish to review the electronic structure of ethylene, discussed in Section 9.4. In the alkynes, there is at least one carbon-carbon triple bond, as in the simplest member of the series, acetylene (Section 9.4). In the aromatic hydrocarbons the carbon atoms are connected in a planar ring structure, joined by both $\sigma$ and $\pi$ bonds between carbon atoms. Benzene (Figure 9.24) is the best-known example of an aromatic hydrocarbon. The nonaromatic hydrocarbons—that is, the alkanes, alkenes, and alkynes—are called **aliphatic** compounds to distinguish them from aromatic substances.

The members of the different series of hydrocarbons exhibit different chemical behaviors, as we shall see shortly. However, the hydrocarbons are very similar in many ways. Because carbon and hydrogen are not greatly different in electronegativity (2.5 for carbon, 2.2 for hydrogen), the C—H bond is not very polar. Hydrocarbons are formed entirely from C—H bonds and bonds between carbon atoms; this means that hydrocarbon molecules are relatively nonpolar. Thus they are very much unlike water and almost completely insoluble in water. Those hydrocarbons that are liquids are good solvents toward nonpolar molecules, but poor solvents toward ionic substances, such as sodium chloride, and polar substances, such as $NH_3$.

**The Alkanes**

Table 27.1 lists several of the simplest alkanes. Many of these substances are familiar because of their widespread use. Methane is a major component of natural gas and is used for home heating and in gas stoves and hot-water heaters. Propane is the major component of bottled, or LP, gas used for home heating, cooking, and so forth in areas where natural gas is not available. Butane is used in the disposable lighters sold in drug stores and in the fuel cannisters for gas camping stoves and lanterns. Alkanes with from 5 to 12 carbon atoms per molecule are found in gasoline.

The formulas for the alkanes given in Table 27.1 are written in a notation called the condensed structural formula. This notation reveals the way in which atoms are bonded to one another, but does not require drawing in all the bonds. For example, the Lewis structure and condensed structural formula for butane, $C_4H_{10}$, are:

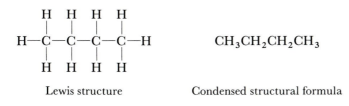

Lewis structure          Condensed structural formula

We shall frequently use either Lewis structures or condensed structural formulas to represent organic compounds. You should practice drawing the structural formulas from the condensed ones. Notice that each carbon atom in an alkane has four single bonds, while each hydrogen atom forms one single bond.

Each succeeding compound in the series listed in Table 27.1 has an additional $CH_2$ unit. A series such as that shown in Table 27.1 is known

**TABLE 27.1**   First several members of the straight-chain alkane series

| Molecular formula | Condensed structural formula | Name | Boiling point (°C) |
|---|---|---|---|
| $CH_4$ | $CH_4$ | Methane | −161 |
| $C_2H_6$ | $CH_3CH_3$ | Ethane | −89 |
| $C_3H_8$ | $CH_3CH_2CH_3$ | Propane | −44 |
| $C_4H_{10}$ | $CH_3CH_2CH_2CH_3$ | Butane | −0.5 |
| $C_5H_{12}$ | $CH_3CH_2CH_2CH_2CH_3$ | Pentane | 36 |
| $C_6H_{14}$ | $CH_3CH_2CH_2CH_2CH_2CH_3$ | Hexane | 68 |
| $C_7H_{16}$ | $CH_3CH_2CH_2CH_2CH_2CH_2CH_3$ | Heptane | 98 |
| $C_8H_{18}$ | $CH_3CH_2CH_2CH_2CH_2CH_2CH_2CH_3$ | Octane | 125 |
| $C_9H_{20}$ | $CH_3CH_2CH_2CH_2CH_2CH_2CH_2CH_2CH_3$ | Nonane | 151 |
| $C_{10}H_{22}$ | $CH_3CH_2CH_2CH_2CH_2CH_2CH_2CH_2CH_2CH_3$ | Decane | 174 |

as a **homologous series**. The general formula for all the compounds listed in the table is $C_nH_{2n+2}$, where $n$ is the number of carbon atoms. One of the characteristics of a homologous series is that all the compounds of the series can be described by the same general formula. We shall see several other examples of homologous series as we proceed.

The alkanes listed in Table 27.1 are called straight-chain alkanes, because all the carbon atoms are joined in a continuous chain. However, for alkanes consisting of four or more carbon atoms, other arrangements of the carbon atoms, consisting of branched chains, are possible. Figure 27.2 shows the Lewis structures, condensed structural formulas and space-

**FIGURE 27.2** Possible structures, names, and melting and boiling points of alkanes of formula $C_4H_{10}$ and $C_5H_{12}$.

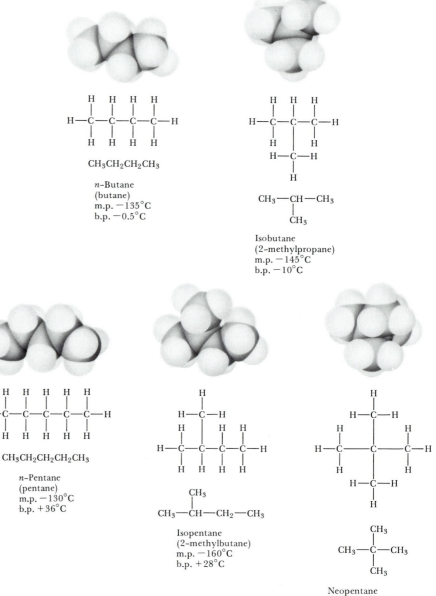

filling models for all the possible structures of alkanes containing four or five carbon atoms. Notice that the two possible forms of butane have the same molecular formula, $C_4H_{10}$. Similarly, the three possible forms of pentane have the same molecular formula, $C_5H_{12}$. Compounds with the same molecular formula, but with different structures, are called **isomers**. The isomers of a given alkane differ slightly from one another in physical properties. Note the melting and boiling points (°C) of the isomers of butane and pentane, given in Figure 27.2. The number of possible isomers increases rapidly with the number of carbon atoms in the alkane. For example, there are 18 possible isomers of octane, $C_8H_{18}$, and 75 possible isomers of decane, $C_{10}H_{22}$.

## Nomenclature of Alkanes

The first names given to the structural isomers shown in Figure 27.2 are the so-called common names. The straight-chain isomer is called the normal isomer, abbreviated by the prefix *n*-. The isomer in which one $CH_3$ group is branched off the major chain is labeled the iso-isomer (for example, isobutane). However, as the number of isomers grows, it becomes impossible to find a suitable prefix to denote each isomer. The need for a systematic means of naming organic compounds was recognized early in the history of organic chemistry. In 1892 an organization called the International Union of Chemistry met in Geneva, Switzerland, to formulate rules for systematic naming of organic substances. Since that time the task of keeping the rules for naming compounds up-to-date has fallen to the International Union of Pure and Applied Chemistry (IUPAC). It is interesting to note that through two devastating world wars and major social upheavals, the work of IUPAC has continued. Chemists everywhere, regardless of their nationality or political affiliation, subscribe to a common system for naming compounds.

The IUPAC names for the isomers of butane and pentane are the second ones given for each compound in Figure 27.2. The following rules summarize the procedures used to arrive at these names. We shall see that a similar approach is taken to write the names for other organic compounds.

1. Each compound is named for the longest continuous chain of carbon atoms present. For example, the longest chain of carbon atoms in isobutane is three (Figure 27.2). Consequently, this compound is named as a derivative of propane, which has three carbon atoms; in the IUPAC system it is called 2-methylpropane.

2. In general, a group that is formed by removing a hydrogen atom from an alkane is called an **alkyl group**. The names for alkyl groups are derived by dropping the *-ane* ending from the name of the parent alkane and adding *-yl*. For example, the methyl group, $CH_3$, is derived from methane, $CH_4$; likewise, the ethyl group, $C_2H_5$, is derived from ethane, $C_2H_6$. Table 27.2 lists several of the more common alkyl groups.

3. The location of an alkyl group along a carbon-atom chain is indicated by numbering the carbon atoms along the chain. Thus the name 2-methylpropane indicates the presence of a methyl ($CH_3$) group on the second carbon atom of a propane (three-carbon) chain. In general, the chain is numbered from the end that gives the lowest numbers for the alkyl positions.

4. If there is more than one substituent group of a certain type along the

**TABLE 27.2** Names and condensed structural formulas for several alkyl groups

| Group | Name |
|---|---|
| $CH_3$— | Methyl |
| $CH_3CH_2$— | Ethyl |
| $CH_3CH_2CH_2$— | *n*-Propyl |
| $CH_3CH_2CH_2CH_2$— | *n*-Butyl |
| $CH_3$<br>$HC$—<br>$CH_3$ | Isopropyl |
| $CH_3$<br>$CH_3$—$C$—<br>$CH_3$ | *t*-Butyl |

chain, the number of groups of that type is indicated by a prefix: *di-* (two), *tri-* (three), *tetra-* (four), *penta-* (five), and so forth. Thus the IUPAC name for neopentane (Figure 27.2) is 2,2-dimethylpropane. Dimethyl indicates the presence of two methyl groups; the 2,2- prefix indicates that both are on the second carbon atom of the propane chain.

## SAMPLE EXERCISE 27.1

Name the following alkane:

$$CH_3—CH—CH_3$$
$$|$$
$$CH_3—CH—CH_2$$
$$|$$
$$CH_3$$

**Solution:** To name this compound properly, you must first find the longest continuous chain of carbon atoms. This chain, extending from the upper left $CH_3$ group to the lower right $CH_3$ group, is five carbon atoms long:

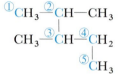

The compound is thus named as a derivative of

pentane. We could number the carbon atoms starting from either end. However, the IUPAC rules state that the numbering should be done so that the numbers of those carbons which bear side chains are as low as possible. This means that we should start numbering with the upper carbon. There is a methyl group on carbon number two, and one on carbon number three. The compound is thus called 2,3-dimethylpentane.

## PRACTICE EXERCISE

Name the following alkane:

$$CH_3—CH_2 \quad CH_3$$
$$| \qquad |$$
$$CH_3—CH—CH$$
$$|$$
$$CH_3$$

***Answer:*** 2,3-dimethylpentane

## SAMPLE EXERCISE 27.2

Write the condensed structural formula for 2-methyl-3-ethylpentane.

**Solution:** The longest continuous chain of carbon atoms in this compound is five. We can therefore begin by writing out a string of five C atoms:

$$C—C—C—C—C$$

We next place a methyl group on the second carbon, and an ethyl group on the middle carbon atom of the chain. Hydrogens are then added to all the other carbon atoms to make their covalences equal to four:

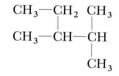

## PRACTICE EXERCISE

Write the condensed structural formula for 2,3-dimethylhexane.

***Answer:***

$$CH_3 \quad CH_3$$
$$| \qquad |$$
$$CH_3—CH—CHCH_2CH_2CH_3$$

## Structures of Alkanes

The Lewis structures and condensed structural formulas for alkanes do not tell us anything about the three-dimensional structures of these substances. As we would predict from the VSEPR model (Section 9.1), the geometry about each carbon atom in an alkane is tetrahedral; that is, the four groups attached to each carbon are located at the vertices of a tetrahedron. The three-dimensional structures can be represented as shown for methane in Figure 27.3. The bonding may be described as involving $sp^3$ hybridized orbitals on the carbon, as discussed earlier, in Section 9.3.

Rotation about the carbon-carbon single bond is relatively free. You

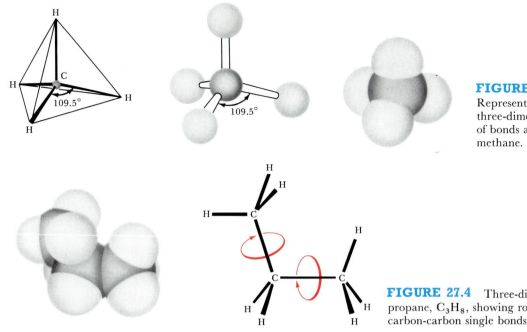

**FIGURE 27.3**
Representations of the three-dimensional arrangement of bonds about carbon in methane.

**FIGURE 27.4** Three-dimensional models for propane, $C_3H_8$, showing rotations about the carbon-carbon single bonds.

might imagine grasping the top left methyl group in Figure 27.4, which shows the structure of propane, and twisting it relative to the rest of the structure. Motion of this sort occurs very rapidly in alkanes at room temperature. Thus a long-chain alkane is constantly undergoing motions that cause it to change its shape, something like a length of chain that is being shaken.

One possible structural form for alkanes is that in which the carbon chain forms a ring, or cycle. Alkanes with this form of structure are called **cycloalkanes**. A few examples of cycloalkanes are shown in Figure 27.5. The cycloalkane structures are sometimes drawn as simple polygons, as illustrated in Figure 27.5. In this shorthand notation, each corner of the polygon represents a $CH_2$ group. This method of representation is similar to that used for aromatic rings as illustrated in Section 9.4. In the case of the aromatic structures, each corner represents a CH group.

Carbon rings containing fewer than six carbon atoms are strained, because the C—C—C bond angle in the smaller rings must be less than the 109.5° tetrahedral angle. The amount of strain increases as the rings get

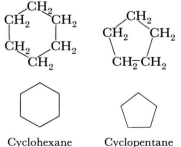

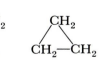

**FIGURE 27.5** Condensed structural formulas of three cycloalkanes.

Cyclohexane          Cyclopentane          Cyclopropane

smaller. In cyclopropane, which has the shape of an equilateral triangle, the angle is only $60°$; this molecule is therefore much more reactive than either propane, its straight-chain analog, or cyclohexane, which has no ring strain. Note that the empirical formula for the cycloalkanes is $C_nH_{2n}$, which differs from that for the straight-chain alkanes. The cycloalkanes thus form a separate homologous series.

**Alkenes**

The alkenes are close relatives of the alkanes. They differ in that alkenes have at least one carbon-carbon double bond in the molecule. Alkenes are sometimes referred to as **olefins**. The presence of a double bond results in two fewer hydrogens than would be present in an alkane. Because the alkenes possess fewer hydrogens than are needed to form the alkane, they are said to be **unsaturated**. As we shall see a little later, the presence of the double bond confers considerably more chemical reactivity on the alkenes than is found in the alkanes. The simplest alkene is $C_2H_4$, called ethene or ethylene. The next member of the series is $CH_3-CH{=}CH_2$, called propene or propylene. When there are more than three carbon atoms in the molecule, there are several possibilities for forming isomers. For example, the possible alkenes with four carbon atoms, and with molecular formula $C_4H_8$, are shown in Figure 27.6. The first compound shown has a branched chain; the other three have continuous chains of four carbon atoms. The compound is named for the length of the longest continuous chain of carbon atoms that contains the double bond. Of course, the ending of the name listed in Table 25.1 is changed from -*ane* to -*ene*. The location of the double bond is indicated by a prefix number that designates the number of the carbon atom that is part of the double bond and is nearest an end of the chain. The chain is always numbered from the end that brings us to the double bond sooner and hence gives the smallest number prefix. These rules are used to name the isomers of butene that are listed in Figure 27.6. For example, the compound on the left has a three-carbon chain; thus the parent alkene is considered to be propene. The only possible location of the double bond in propene is between the first and second carbons; thus a prefix indicating its location is unnecessary. Numbering the carbon chain from the end closer to the double bond places a methyl group on the second carbon. Thus the name of the isomer is 2-methylpropene.

In substances containing two alkene functional groups, each must be located by a number. The ending of the name is altered to identify the number of functional groups: diene (two double bonds), triene (three double bonds), and so forth. For example, 1,4-pentadiene is $CH_2{=}CH-CH_2-CH{=}CH_2$.

*Cis*- and *trans*-2-butene are **geometrical isomers**; that is, they are compounds that have the same molecular formula and the same groups bonded to one another, but that differ in the spatial arrangement of those groups. In the *cis* form, the two methyl groups are on the same side of

**FIGURE 27.6** Structures, names, and boiling points of the alkenes with molecular formula $C_4H_8$.

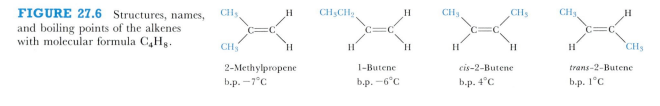

| 2-Methylpropene | 1-Butene | *cis*-2-Butene | *trans*-2-Butene |
| b.p. $-7°C$ | b.p. $-6°C$ | b.p. $4°C$ | b.p. $1°C$ |

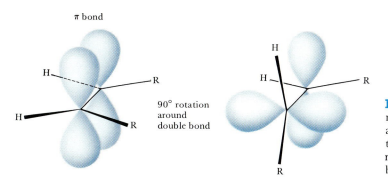

π bond

90° rotation
around
double bond

**FIGURE 27.7** Schematic illustration of rotation about a carbon-carbon double bond in an alkene. The overlap of the *p* orbitals that form the π bond is lost in the rotation. For this reason, rotation about carbon-carbon double bonds does not occur readily.

the double bond; in the *trans* form they are on opposite sides. We have already met examples of geometrical isomerism, in the chemistry of transition metal coordination compounds (Section 26.4). Geometrical isomers possess distinct physical properties and may even differ significantly in their chemical behavior in certain circumstances.

Recall from our earlier discussion of the geometry about carbon (Section 9.4) that the double bond between two carbon atoms consists of a σ and a π part. Figure 27.7 shows the geometrical arrangement as in a *cis* alkene. The bonding arrangement around each carbon is planar; that is, the carbon-carbon bond axis and the bonds to the other two groups, either hydrogen or carbon, are all in a plane. It is easy to see from Figure 27.7 that rotation about the carbon-carbon double bond will not be easy. The overlap of the *p* orbitals that form the π bond would be lost in rotation, and the π bond would be destroyed. Because rotation about the carbon-carbon bond is difficult, the *cis* and *trans* isomers can be separated and studied. If the interconversion were easy, we would always have an equilibrium mixture of the two forms rather than the pure isomers.

## SAMPLE EXERCISE 27.3

Name the following compound:

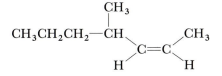

**Solution:** Because this compound possesses a double bond, it is an alkene. The longest continuous chain of carbons that contains the double bond is seven in length. Thus the parent compound is considered to be a heptene. The double bond begins at carbon number two (numbering from the end closest to the double bond); thus the parent hydrocarbon chain is named 2-heptene. Continuing the numbering

along the chain, a methyl group is bound at carbon atom number four. Thus the compound is 4-methyl-2-heptene. Finally, we note that the geometrical configuration at the double bond is *cis*; that is, the alkyl groups are bonded to the double bond on the same side. Thus the full name is 4-methyl-*cis*-2-heptene.

### PRACTICE EXERCISE

Draw the condensed structural formula for the compound *trans*-1,3-hexadiene.

*Answer:*   $CH_2{=}CH$
$\phantom{xxxxxx}\diagdown$
$\phantom{xxxxxxxx}C{=}CH$
$\phantom{xxxx}\diagup\phantom{xxxx}\diagdown$
$\phantom{xxx}H\phantom{xxxxxx}CH_2CH_3$

The alkynes are a series of unsaturated hydrocarbons that have one or more carbon-carbon triple bonds between carbon atoms. The empirical formula for simple alkynes is $C_nH_{2n-2}$. The simplest alkyne, acetylene,

**Alkynes**

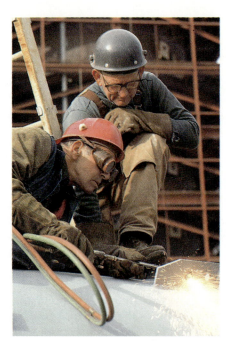

**FIGURE 27.8** Welding with an oxyacetylene torch. The heat of combustion of acetylene is exceptionally high, thus giving rise to a very high flame temperature when acetylene is combusted in oxygen. (© Robert Isear/ Photo Researchers, 1983)

is a highly reactive molecule. When acetylene is burned in a stream of oxygen in an oxyacetylene torch (Figure 27.8), the flame reaches a very high temperature, about 3200 K (Section 22.6). The oxyacetylene torch is widely used in welding, which requires high temperatures. Alkynes in general are highly reactive molecules. Because of their higher reactivity, they are not as widely distributed in nature as the alkenes; however, they are important intermediates in many industrial processes.

The alkynes are named by identifying the longest continuous chain in the molecule containing the triple bond and modifying the ending of the name as listed in Table 27.1 from *-ane* to *-yne*, as shown in the following sample exercise.

---

**SAMPLE EXERCISE 27.4**

Name the following compounds:

**(a)** $CH_3CH_2CH_2—C\equiv C—CH_3$

**(b)** $CH_3CH_2CH_2CH—C\equiv CH$
        $\quad\quad\quad\quad\quad | $
        $\quad\quad\quad\quad CH_2CH_2CH_3$

**Solution:** In (a) the longest chain of carbon atoms is six. There are no side chains. The triple bond begins at carbon atom number two (remember, we always arrange the numbering so that the smallest possible number is assigned to the carbon containing the multiple bond). Thus the name is 2-hexyne.

In (b) the longest continuous chain of carbon atoms is seven; but because this chain does not contain the triple bond we do not count it as derived from heptane. The longest chain containing the triple bond is six, so this compound is named as a derivative of hexyne, 3-propyl-1-hexyne.

**PRACTICE EXERCISE**

Draw the condensed structural formula for 4-methyl-2-pentyne.

*Answer:* $CH_3—C\equiv C—CH—CH_3$
$\quad\quad\quad\quad\quad\quad\quad\quad | $
$\quad\quad\quad\quad\quad\quad\quad CH_3$

---

We can obtain an estimate of the stabilization of the $\pi$ electrons in benzene by comparing the energy required to add hydrogen to benzene to form a saturated compound, with the energy required to hydrogenate simple alkenes. The hydrogenation of benzene to form cyclohexane can be represented as

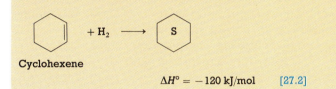

$$\Delta H° = -208 \text{ kJ/mol} \qquad [27.1]$$

(The s in the ring on the right indicates that it is a cycloalkane, with $CH_2$ groups at each corner.) The enthalpy change in this reaction is $-208$ kJ/mol. The heat of hydrogenation of the cyclic alkene cyclohexene, is $-120$ kJ/mol:

Cyclohexene

$$\Delta H° = -120 \text{ kJ/mol} \qquad [27.2]$$

Similarly, the heat released on hydrogenating 1,4-cyclohexadiene is $-232$ kJ/mol:

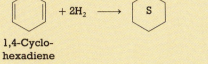

1,4-Cyclo-
hexadiene

$$\Delta H° = -232 \text{ kJ/mol} \qquad [27.3]$$

From these last two reactions it would appear that the heat of hydrogenating a double bond is about 116 kJ/mol for each bond. There is the equivalent of three double bonds in benzene. Thus we might expect that the heat of hydrogenating benzene would be about three times $-116$, or $-348$ kJ/mol, if benzene behaved as though it were "cyclohexatriene"; that is, if it behaved as though it were three double bonds in a ring. Instead, the heat released is much less than this, indicating that benzene is more stable than would be expected for three double bonds. The difference of 140 kJ/mol between $-348$ kJ/mol and the observed heat of hydrogenation, $-208$ kJ/mol, can be ascribed to stabilization of the $\pi$ electrons through delocalization in the $\pi$ orbitals that extend around the ring.

The aromatic hydrocarbons are a large and important class of hydrocarbons. The simplest member of the series is benzene (see Figure 27.1), with molecular formula $C_6H_6$. As we have already noted, benzene is a planar, highly symmetrical molecule. The molecular formula for benzene suggests a high degree of unsaturation. One might thus expect that benzene might resemble the unsaturated hydrocarbons and be highly reactive. In fact, however, benzene is not at all similar to alkenes or alkynes in chemical behavior. The great stability of benzene and the other aromatic hydrocarbons as compared with alkenes and alkynes is due to stabilization of the $\pi$ electrons through delocalization in the $\pi$ orbitals (Section 9.4).

There is no widely used systematic nomenclature for naming aromatic rings. Each ring system is given a common name; several aromatic compounds are shown in Figure 27.9. The aromatic rings are represented by hexagons with a circle inscribed inside to denote aromatic character. Each corner represents a carbon atom. Each carbon is bound to three other atoms—either three carbons or two carbons and a hydrogen. The hydrogen atoms are not shown. In naming derivatives of the aromatic hydrocarbons, it is often necessary to indicate the position in the aromatic ring

**Aromatic Hydrocarbons**

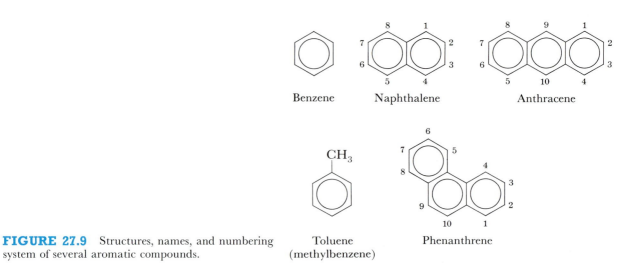

Benzene    Naphthalene    Anthracene

Toluene
(methylbenzene)

Phenanthrene

**FIGURE 27.9**  Structures, names, and numbering system of several aromatic compounds.

at which some side chain or other group is located. For this purpose the numbering shown in Figure 27.9 is used.

## 27.2  PETROLEUM

Petroleum is a complex mixture of organic compounds, mainly hydrocarbons, with smaller quantities of other organic compounds containing nitrogen, oxygen, or sulfur. Petroleum is formed over a period of millions of years by the decomposition of marine plants and animals.

The usual first step in the **refining**, or processing, of petroleum is separation of the crude oil into fractions on the basis of boiling points (Figure 27.10). The fractions commonly taken are shown in Table 27.3. As you might expect, the fractions that boil at higher temperatures are made up of molecules with larger numbers of carbon atoms per molecule. The fractions collected in the initial separation may require further processing to yield a usable product. For example, modifications must be

**FIGURE 27.10**  Petroleum is separated into fractions by distillation in a refinery, shown here. (Ken Biggs/Photo Researchers)

**TABLE 27.3**   Hydrocarbon fractions from petroleum

| Fraction | Size range of molecules | Boiling-point range (°C) | Uses |
|---|---|---|---|
| Gas | $C_1$–$C_5$ | −160 to 30 | Gaseous fuel, production of $H_2$ |
| Straight-run gasoline | $C_5$–$C_{12}$ | 30 to 200 | Motor fuel |
| Kerosene, fuel oil | $C_{12}$–$C_{18}$ | 180 to 400 | Diesel fuel, furnace fuel, cracking |
| Lubricants | $C_{16}$ and up | 350 and up | Lubricants |
| Paraffins | $C_{20}$ and up | Low-melting solids | Candles, matches |
| Asphalt | $C_{36}$ | Gummy residues | Surfacing roads, fuel |

made to the straight-run gasoline obtained from fractionation of petroleum to render it suitable for use as a fuel in automobile engines. Similarly, the fuel oil fraction may need additional processing to remove sulfur before it is suitable for use in an electrical power station or a home heating system. At present, the most commercially important single product from petroleum refining is gasoline.

**Gasoline**

Gasoline is a mixture of volatile hydrocarbons. Depending on the source of the crude oil, it may contain varying amounts of cyclic alkanes and aromatic hydrocarbons in addition to alkanes. Straight-run gasoline consists mainly of straight-chain hydrocarbons, which in general are not very suitable for use as fuel in an automobile engine. In an automobile engine, a mixture of air and gasoline vapor is ignited by the spark plug at the moment when the gas mixture inside the cylinder has been compressed by the piston. The burning of the gasoline should create a strong, smooth expansion of gas in the cylinder, forcing the piston outward and imparting force along the drive shaft of the engine. If the gas burns too rapidly, the piston receives a single hard slam rather than a strong, smooth push. The result is a "knocking" or pinging sound; the efficiency with which the energy of gasoline combustion is converted to power is reduced.

Gasolines are rated according to **octane number**. Gasolines with high octane numbers burn more slowly and smoothly and thus are more effective fuels, especially in engines in which the gas-air mixture is highly compressed. The more highly branched alkanes have higher octane numbers than the straight-chain compounds; some examples are shown in Table 27.4. The octane number of gasoline is obtained by comparing its knocking characteristics with those of "isooctane" (2,2,4-trimethylpentane) and heptane. Isooctane is assigned an octane number of 100, whereas heptane is assigned 0. Gasoline with the same knocking characteristics as a mixture of 90 percent isooctane and 10 percent heptane would be rated as 90 octane.

Because straight-run gasoline contains mostly straight-chain hydrocarbons, it has a low octane number. It is therefore subjected to a process called **cracking** to convert the straight-chain compounds into more desirable branched-chain molecules. Cracking is also used to convert some of the less volatile kerosene and fuel-oil fraction into compounds with lower molecular weights that are suitable for use as automobile fuel. In the cracking process, the hydrocarbons are mixed with a catalyst and heated

**TABLE 27.4** Octane numbers of some $C_7$ and $C_8$ hydrocarbons

| Name | Condensed structural formula | Octane number |
|------|------------------------------|:-------------:|
| *n*-Heptane | $CH_3CH_2CH_2CH_2CH_2CH_2CH_3$ | 0 |
| 2-Methylhexane | $CH_3CH_2CH_2CH_2{-}CH{-}CH_3$ <br> $\quad\quad\quad\quad\quad\quad\;\; \mid$ <br> $\quad\quad\quad\quad\quad\quad CH_3$ | 40 |
| Methylcyclohexane | (ring structure) | 75 |
| 2,3-Dimethylpentane | $\quad\quad\quad\quad\quad\quad CH_3$ <br> $\quad\quad\quad\quad\quad\quad \mid$ <br> $CH_3CH_2{-}CH{-}CH{-}CH_3$ <br> $\quad\quad\quad\quad \mid$ <br> $\quad\quad\quad\quad CH_3$ | 90 |
| 2,2,4-Trimethylpentane ("isooctane") | $\quad\quad\quad\quad\quad\quad CH_3$ <br> $\quad\quad\quad\quad\quad\quad \mid$ <br> $CH_3{-}CH{-}CH_2{-}C{-}CH_3$ <br> $\quad\quad\; \mid \quad\quad\quad \mid$ <br> $\quad\quad CH_3 \quad\quad CH_3$ | 100 |

to 400 to 500°C. The catalysts used are naturally occurring clay mineral or synthetic $Al_2O_3$–$SiO_2$ mixtures. In addition to forming molecules more suitable for gasoline, cracking results in the formation of hydrocarbons of lower molecular weight, such as ethylene and propene. These are used in a variety of processes to form plastics and other chemicals.

The octane number of a given blend of hydrocarbons can be improved by adding an **antiknock agent**, a substance that helps control the burning rate of the gasoline. Until recently, the most widely used substances for this purpose were tetraethyl lead, $(CH_3CH_2)_4Pb$, and tetramethyl lead, $(CH_3)_4Pb$. Although alkyl lead compounds were undoubtedly effective in improving gasoline performance, their use in gasolines has been drastically curtailed because of the environmental hazards associated with lead. This metal is highly toxic, and there is good evidence that the lead released from automobile exhausts is a general health hazard. The 1975 and later model cars are designed to operate with unleaded gasolines. The gasolines blended for these cars are made up of more highly branched components and more aromatic components, because these have relatively high octane ratings. In some areas of the United States octane rating is increased by addition of ethanol.

## 27.3 REACTIONS OF HYDROCARBONS

The hydrocarbons are capable of undergoing a variety of reactions with other substances. Many of these reactions are quite important in the chemical industry, because they lead to useful products or to substances that can in turn be converted to useful products.

### Oxidation

The most common oxidation reactions of hydrocarbons result in complete oxidation to form carbon dioxide and water:

$$CH_3CH_2CH_3 + 5O_2 \longrightarrow 3CO_2 + 4H_2O \qquad [27.4]$$

FIGURE 27.11 Drilling for oil on the North Slope in Alaska. (Joe Rychetnik/Photo Researchers)

$$CH_3CH{=}CH_2 + \tfrac{9}{2}O_2 \longrightarrow 3CO_2 + 3H_2O \qquad [27.5]$$

$$CH_3C{\equiv}CH + 4O_2 \longrightarrow 3CO_2 + 2H_2O \qquad [27.6]$$

$$C_6H_6 + \tfrac{15}{2}O_2 \longrightarrow 6CO_2 + 3H_2O \qquad [27.7]$$

These reactions are all highly exothermic. Combustion of the hydrocarbons results in release of energy that can be used to propel an auto or an airplane, to generate steam in the boiler of an electric power plant, or to heat a building. The tremendous demand for petroleum to meet the world's energy needs has led to the tapping of oil wells in such forbidding places as the North Sea and the North Slope of the continent in Alaska, (Figure 27.11).

Controlled oxidation of hydrocarbons yields organic substances that contain oxygen in addition to carbon and hydrogen. These products of controlled oxidations will be considered later when we discuss hydrocarbon derivatives.

**Addition Reactions**

Under appropriate conditions, it is possible to add atoms or groups of atoms to alkenes, alkynes, or aromatic hydrocarbons by disrupting the carbon-carbon multiple bonds. For example, ethylene reacts with bromine to form 1,2-dibromoethane, as in Equation 27.8:

$$H_2C{=}CH_2 + Br_2 \longrightarrow \begin{matrix} H_2C{-}CH_2 \\ \mid \quad\ \mid \\ Br \ \ \, Br \end{matrix} \qquad [27.8]$$

The pair of electrons that form the $\pi$ bonds in ethylene is uncoupled and is used to form two new bonds to the two bromine atoms. The $\sigma$ bond between the carbon atoms remains.

Addition of halogens to alkynes also occurs readily, as in the example shown in Equation 27.9:

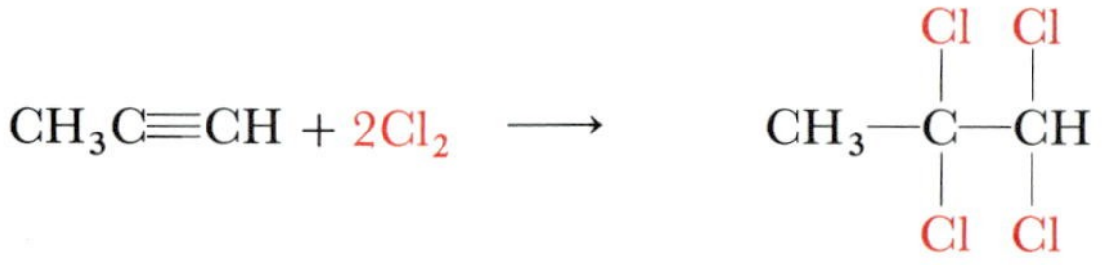

$$CH_3C\equiv CH + 2Cl_2 \longrightarrow \quad \underset{\underset{Cl\ \ Cl}{|}}{\overset{\overset{Cl\ \ Cl}{|}}{CH_3-C-CH}} \qquad [27.9]$$

1,1,2,2-Tetrachloropropane

Reaction of alkenes or alkynes with hydrogen halides also results in addition:

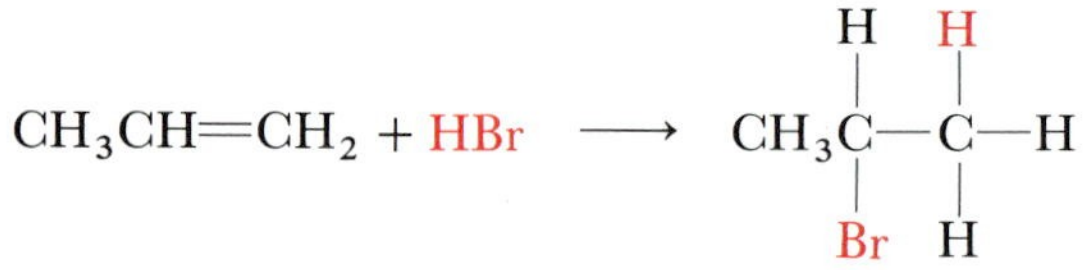

$$CH_3CH=CH_2 + HBr \longrightarrow \quad \underset{\underset{Br\ \ H}{|}}{\overset{\overset{H\ \ \ H}{|}}{CH_3C-C-H}} \qquad [27.10]$$

Notice that this reaction might have proceeded to give another product, in which the bromine atom is on the end carbon. When a hydrogen halide adds to an alkene, the more electronegative halogen atom always ends up on the carbon atom of the double bond that has the fewer hydrogen atoms. This rule, known as **Markovnikoff's rule**, was first formulated by the Russian chemist V. V. Markovnikoff.

---

**SAMPLE EXERCISE 27.5**

Predict the product of reaction of HCl with the following compound:

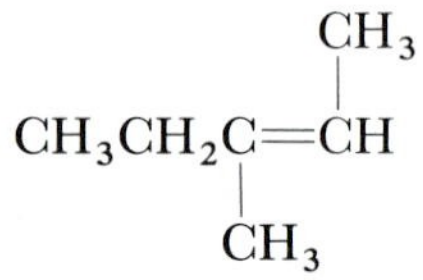

$$\underset{\underset{CH_3}{|}}{\overset{\overset{CH_3}{|}}{CH_3CH_2C=CH}}$$

**Solution:** The double-bonded carbon atom on the right has one hydrogen on it, the other has none. According to Markovnikoff's rule, the chloride of HCl will end up on the carbon atom on the left:

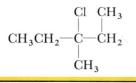

**PRACTICE EXERCISE**

The compound

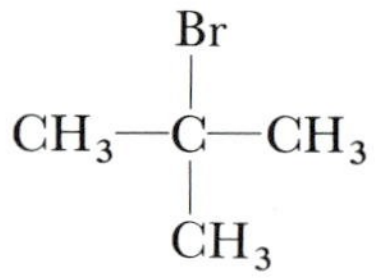

$$\underset{\underset{CH_3}{|}}{\overset{\overset{Br}{|}}{CH_3-C-CH_3}}$$

was formed by addition of HBr to an alkene. Draw the condensed structural formula of the alkene.

***Answer:*** $\underset{\underset{CH_3}{|}}{CH_3-C=CH_2}$

---

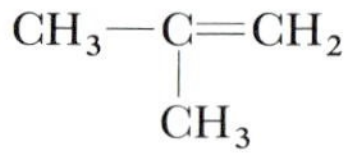

Markovnikoff's rule also applies to the addition of unsymmetrical reagents to alkynes. For example, reaction of HBr with 1-butyne yields 2,2-dibromobutane:

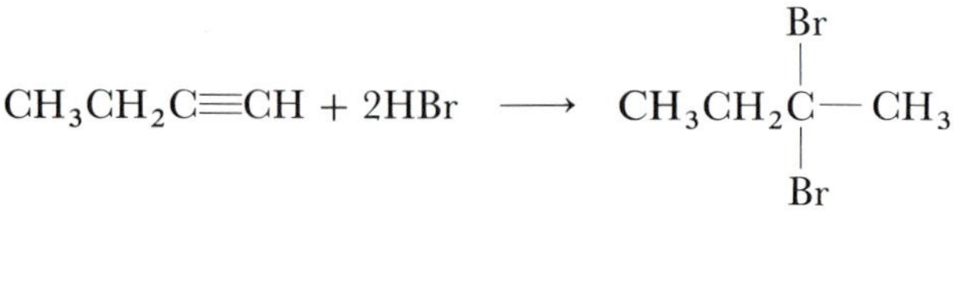

$$CH_3CH_2C\equiv CH + 2HBr \longrightarrow \quad \underset{\underset{Br}{|}}{\overset{\overset{Br}{|}}{CH_3CH_2C-CH_3}} \qquad [27.11]$$

Using an acid such as $H_2SO_4$ as catalyst, it is possible to add $H_2O$ to a double bond. The products of such reactions are **alcohols**; that is, compounds containing an OH group bonded to carbon. Markovnikoff's rule applies here as well. We consider the water molecule as being polarized as follows: $H^+\!-\!OH^-$. Thus the OH group adds to the carbon atom with fewer hydrogens:

$$CH_3CH\!=\!CH_2 + HOH \xrightarrow{H_2SO_4} CH_3\!-\!\overset{\displaystyle OH}{\overset{|}{C}H}\!-\!CH_3 \qquad [27.12]$$

Propene            2-Propanol

Addition of $H_2$ to an alkene converts it into an alkane. This reaction, referred to as hydrogenation, does not occur readily under ordinary temperature and pressure conditions. One of the reasons for the lack of reactivity of $H_2$ toward an alkene is the high bond energy of the $H_2$ bond. To promote the reaction it is necessary to use a catalyst that assists in the rupture of the H—H bond. The most widely used catalysts are heterogeneous and consist of finely divided metals on which $H_2$ is adsorbed. The action of these heterogeneous catalysts in reaction of $H_2$ with an alkene is described in detail in Section 15.6. Molecular hydrogen also reacts with alkynes in the presence of a catalyst to yield alkanes, as in Equation 27.13:

$$CH_3C\!\equiv\!CH + 2H_2 \xrightarrow{Ni} CH_3CH_2CH_3 \qquad [27.13]$$

Addition reactions of aromatic hydrocarbons proceed much less readily than such reactions of alkenes and alkynes. For example, benzene does not react at all with $Cl_2$ or $Br_2$ under ordinary conditions. However, if conditions are sufficiently rigorous, addition reactions may be forced to occur.

In a substitution reaction of a hydrocarbon, one or more hydrogen atoms are replaced by other atoms or groups. Substitution reactions are difficult to carry out with aliphatic compounds. One of the most important substitution reactions of alkanes is the replacement of hydrogen by a halogen atom. The chlorination of an alkane is a photo-initiated reaction. That is, it requires the use of light, which causes dissociation of the $Cl_2$ molecule, forming chlorine atoms:

**Substitution Reactions**

$$Cl_2 \longrightarrow 2Cl\cdot \qquad [27.14]$$

The chlorine atom has only seven valence-shell electrons. Species with an odd number of electrons are called **free radicals** (or simply radicals). They are frequently represented in chemical equations by showing a dot next to their chemical formula, representing their unpaired electron. Chlorine atoms are highly reactive and attack the alkane, removing a hydrogen and forming an alkyl radical. For example, when the alkane is ethane we have

$$Cl\cdot + CH_3CH_3 \longrightarrow HCl + CH_3CH_2\cdot \qquad [27.15]$$

The ethyl radical, $CH_3CH_2\cdot$, is another reactive species. It combines with molecular chlorine to give ethyl chloride and yet another chlorine atom:

$$CH_3CH_2\cdot + Cl_2 \longrightarrow CH_3CH_2Cl + Cl\cdot \qquad [27.16]$$

The resultant chlorine atom reacts with more ethane, continuing the cycle. Thus for each quantum of light absorbed by a chlorine molecule, many molecules of ethyl chloride may be formed. This reaction provides an example of a **radical chain** process. A disadvantage of radical chain reactions such as this is that they are not very selective. As the concentration of ethyl chloride builds up in the reaction, chlorine atoms may abstract hydrogen atoms from it, so that eventually dichloroethane and even more highly chlorinated molecules may be formed. Thus several products are formed in the reaction, and these must be carefully separated by distillation or some other separation procedure.

Substitution into alkenes or alkynes is difficult to carry out, because the presence of the reactive double or triple bond usually leads to addition reactions rather than substitution. By contrast, substitution into aromatic hydrocarbons is relatively easy. For example, when benzene is warmed in a mixture of nitric and sulfuric acid, hydrogen is replaced by the nitro group, $NO_2$:

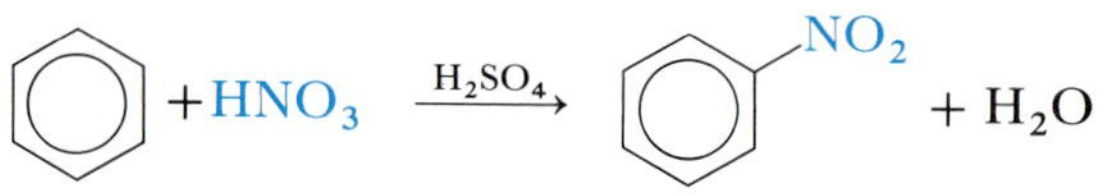

More vigorous treatment results in substitution of a second nitro group into the molecule:

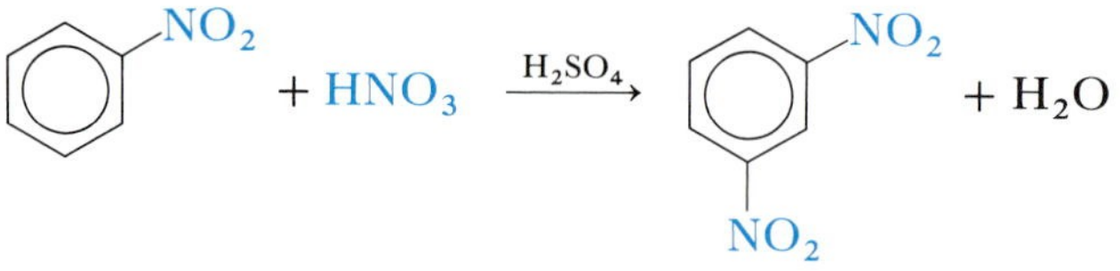

There are three possible isomers of benzene with two nitro groups attached. These three isomers are named *ortho-*, *meta-*, and *para*-dinitrobenzene:

*ortho*-Dinitrobenzene
m.p. 118°C

*meta*-Dinitrobenzene
m.p. 90°C

*para*-Dinitrobenzene
m.p. 174°C

Only the *meta* isomer is formed in the reaction of nitric acid with nitrobenzene.

Bromination of benzene is carried out using $FeBr_3$ as a catalyst:

$$\text{benzene} + Br_2 \xrightarrow{FeBr_3} \text{bromobenzene} + HBr \qquad [27.19]$$

In a similar reaction, called the **Friedel-Crafts reaction**, alkyl groups can be substituted onto an aromatic ring by reaction of an alkyl halide with an aromatic compound in the presence of $AlCl_3$:

$$\text{benzene} + CH_3CH_2Cl \xrightarrow{AlCl_3} \text{ethylbenzene} + HCl \qquad [27.20]$$

The substitution reactions of aromatic compounds occur via attack of a positively charged reagent on the ring. Substances such as $H_2SO_4$, $FeCl_3$, or $AlCl_3$ serve as catalysts by generating the positively charged species. For example, the bromination of benzene proceeds via the following steps:

$$FeBr_3 + Br_2 \rightleftharpoons FeBr_4^- + Br^+ \qquad [27.21]$$

$$\text{benzene} + Br^+ \longrightarrow \text{intermediate} \qquad [27.22]$$

$$\text{intermediate} + FeBr_4^- \longrightarrow \text{bromobenzene} + FeBr_3 + HBr \qquad [27.23]$$

The $Br^+$ ion has only six electrons in its valence shell. Using a pair of $\pi$ electrons from the benzene, it forms a single bond to a carbon atom. In doing so, it leaves the adjacent carbon atom with only six electrons in its valence shell. Such a species is not very stable. Loss of a proton from the carbon atom to which the bromine is attached converts the molecule to a more stable one, bromobenzene. The overall result is that bromine has replaced hydrogen on the benzene ring.

## 27.4 HYDROCARBON DERIVATIVES

We noted in the introduction to this chapter that certain groups or arrangements of atoms impart characteristic behavior to an organic molecule. These groups are called functional groups. Two of these groups have already been discussed—the double and triple carbon-carbon bonds—both of which impart considerable chemical reactivity to a hydrocarbon. The other functional groups contain elements other than carbon and hydrogen, most often oxygen, nitrogen, or a halogen. Compounds containing these elements are generally considered to be hydrocarbon derivatives, obtained from a hydrocarbon by replacing one or more of the hydrogens with functional groups. The compound can then be considered to consist of two parts: a hydrocarbon fragment such as an alkyl group (often designated R) and one or more functional groups.

Although a hydrocarbon derivative may consist largely of carbon and hydrogen, the hydrocarbon portion of the molecule may be essentially unchanged in various chemical reactions. The chemical properties are generally determined by the functional group. Now let's take a brief look at some of these functional groups and some of the properties they impart to organic compounds.

**Alcohols**

Alcohols are hydrocarbon derivatives in which one or more hydrogens of a parent hydrocarbon have been replaced by a hydroxyl or alcohol functional group, OH. Figure 27.12 shows the structural formulas and names of several alcohols. Note that the accepted name for an alcohol ends in *-ol*. The simple alcohols are named by changing the last letter in the name of the corresponding alkane to *-ol*—for example, ethan*e* becomes ethan*ol*. Where necessary, the location of the OH group is designated by an appropriate prefix numeral that indicates the number of the carbon atom bearing the OH group, as shown in the examples in Figure 27.12.

Aliphatic alcohols are classified according to the number of carbon groups bonded to the carbon that contains the OH group. If there is only one other carbon atom, as with ethanol or 1-propanol, the alcohol is termed a **primary alcohol**. If there are two other carbon groups attached, as in 2-propanol, the alcohol is **secondary**. If there are three other carbons, as in *t*-butanol, the alcohol is **tertiary**.

Because the OH group is quite polar, the presence of the hydroxyl group confers considerable polarity on the hydrocarbon molecule. Table 27.5 lists the solubilities of several alcohols in water at room temperature. In the alcohols having low molecular weights, the presence of the OH groups plays an important role in determining physical properties. However, as the length of the hydrocarbon chain increases, the OH group becomes less important in determining overall behavior (see also Section 12.3).

Figure 27.13 shows several familiar commercial products that consist

**FIGURE 27.12**  Structural formulas of several important alcohols.

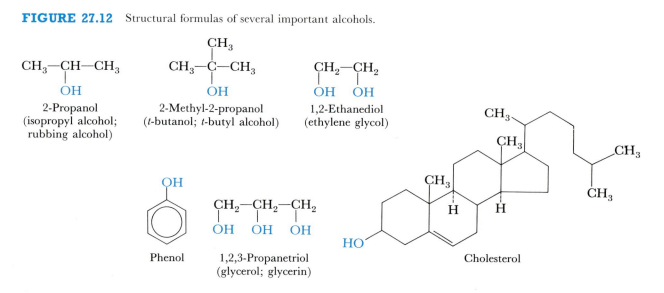

2-Propanol
(isopropyl alcohol;
rubbing alcohol)

2-Methyl-2-propanol
(*t*-butanol; *t*-butyl alcohol)

1,2-Ethanediol
(ethylene glycol)

Phenol

1,2,3-Propanetriol
(glycerol; glycerin)

Cholesterol

**TABLE 27.5** Solubilities of several straight-chain alcohols in water

| Alcohol | Boiling point (°C) | Solubility in water at 25°C (g/100 g $H_2O$) |
|---|---|---|
| Methanol | 65 | Miscible |
| Ethanol | 78 | Miscible |
| 1-Propanol | 97 | Miscible |
| 1-Butanol | 117 | 9 |
| Cyclohexanol | 161 | 5.6 |
| 1-Hexanol | 158 | 0.6 |

entirely or in major part of an organic alcohol. Let's consider how some of the more important alcohols are formed and used.

The simplest alcohol, methanol, has many important industrial uses, and it is produced on a large scale. Carbon monoxide and hydrogen are heated together under pressure in the presence of a metal oxide catalyst:

$$CO(g) + 2H_2(g) \xrightarrow[400°C]{200-300 \text{ atm}} CH_3OH(g) \qquad [27.24]$$

Because methanol has a very high octane rating as an automobile fuel, it has received increasing attention in recent years as a gasoline additive and as a fuel in its own right.

Ethanol, $C_2H_5OH$, is a product of the fermentation of carbohydrates such as sugar or starch. Bacterial cultures such as yeast work in the absence of air to convert carbohydrates into a mixture of ethanol and $CO_2$, as shown in Equation 27.25. In the process, they derive the energy necessary for growth:

$$C_6H_{12}O_6(aq) \xrightarrow{\text{yeast}} 2C_2H_5OH(aq) + 2CO_2(g) \qquad [27.25]$$

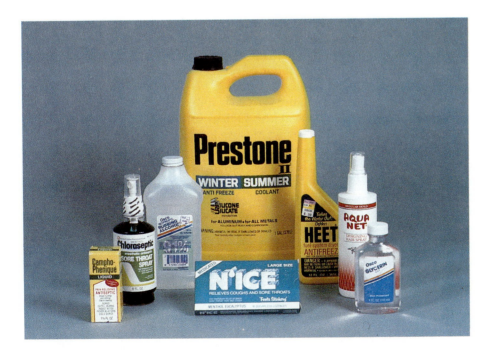

**FIGURE 27.13** Some commercial products that are composed entirely or mainly of alcohols. (Donald Clegg and Roxy Wilson)

This naturally occurring reaction is carried out under carefully controlled conditions to produce beer, wine, and other beverages in which ethanol is the active ingredient. It is often said that ethanol is the least toxic of the straight-chain alcohols. Although this is true in the strictest sense, the combination of ethanol and automobiles produces far more human fatalities each year than any other chemical agent.

Many alcohols are formed industrially by addition of the elements of water to an olefin. For example, isopropyl alcohol is formed by hydration of propene using sulfuric acid as a catalyst:

$$CH_3CH{=}CH_2 + H_2O \xrightarrow{H_2SO_4} CH_3{-}\underset{\underset{OH}{|}}{CH}{-}\underset{\underset{H}{|}}{\overset{\overset{H}{|}}{C}}{-}H \qquad [27.26]$$

Ethylene glycol, the major ingredient in automobile antifreeze, is formed in a two-stage reaction. First ethylene is reacted with hypochlorous acid:

$$CH_2{=}CH_2 + HOCl \longrightarrow \underset{\underset{OH}{|}}{CH_2}{-}\underset{\underset{Cl}{|}}{CH_2} \qquad [27.27]$$

Chlorine is removed from the product of this reaction by displacing the chloride ion with hydroxide ion:

$$\underset{\underset{OH}{|}\;\;\underset{Cl}{|}}{CH_2{-}CH_2} + OH^- \longrightarrow \underset{\underset{OH}{|}\;\;\underset{OH}{|}}{CH_2{-}CH_2} + Cl^- \qquad [27.28]$$

A reaction of this sort, in which hydroxide ion displaces another functional group, is called **base hydrolysis**.

Phenol is the simplest example of a compound with an OH group attached to an aromatic ring. One of the most striking effects of the aromatic group is the greatly increased acidity of the proton. Phenol is about 1 million times more acidic in water than a typical aliphatic alcohol such as ethanol. Even so, it is not a very strong acid ($K_a = 1.3 \times 10^{-10}$). Phenol is used industrially in the making of several kinds of plastics and in the preparation of dyes.

Cholesterol, shown in Figure 27.12, is an example of a biochemically important alcohol. Notice that the OH group forms only a small component of this rather large molecule. As a result, cholesterol is not very soluble in water (0.26 g per 100 mL of $H_2O$). Cholesterol is a normal component of our bodies. However, when present in excessive amounts it may precipitate from solution. It precipitates in the gallbladder to form crystalline lumps called gallstones. It may also precipitate against the walls of veins and arteries and thus contribute to high blood pressure and other cardiovascular problems. The amount of cholesterol in our blood is determined not only by how much cholesterol we eat, but by total dietary intake. There is evidence that excessive caloric intake leads the body to synthesize excessive cholesterol.

Compounds in which two hydrocarbon groups are bonded to one oxygen are called **ethers**. Ethers can be formed from two molecules of alcohol by splitting out a molecule of water. The reaction is thus a dehydration process; it is catalyzed by sulfuric acid, which takes up water to remove it from the system:

$$CH_3CH_2-OH + H-OCH_2CH_3 \xrightarrow{H_2SO_4} CH_3CH_2-O-CH_2CH_3 + H_2O \qquad [27.29]$$

A reaction in which water is split out from two substances is called a **condensation reaction**. An inorganic example of a condensation reaction was presented in our discussion of the chemistry of phosphates (Section 23.4). We shall encounter several additional examples in this chapter and in Chapter 28.

Ethers are used as solvents; both diethyl ether and tetrahydrofuran are common solvents for organic reactions.

$$CH_3CH_2-O-CH_2CH_3 + H_2O$$

Diethyl ether          Tetrahydrofuran (THF)

Several classes of organic compounds contain the **carbonyl group:**

$$>C=O$$

In **aldehydes** the carbonyl group has at least one hydrogen atom attached, as in the following examples:

$$\overset{O}{\overset{\|}{CH_3CH}} \qquad \overset{O}{\overset{\|}{HCH}}$$

Acetaldehyde          Formaldehyde

In **ketones** the carbonyl group occurs at the interior of a carbon chain and is therefore flanked by carbon atoms:

Acetone          Methyl ethyl ketone

Aldehydes and ketones can be prepared by careful oxidation of alcohols. It is fairly easy to oxidize alcohols. Complete oxidation results in formation of $CO_2$ and $H_2O$, as in the burning of methanol:

$$CH_3OH(g) + \tfrac{3}{2}O_2(g) \longrightarrow CO_2(g) + 2H_2O(g) \qquad [27.30]$$

Controlled oxidation to form other organic substances is carried out by using various oxidizing agents such as air, hydrogen peroxide $(H_2O_2)$, ozone $(O_3)$, or potassium dichromate $(K_2Cr_2O_7)$. We shall not concern ourselves with balancing most of these oxidation-reduction equations; instead, we shall simply show the source of oxygen as an O in parentheses.

Aldehydes are produced by oxidation of primary alcohols. For example, acetaldehyde is formed by oxidation of ethanol:

$$CH_3CH_2OH + (O) \longrightarrow CH_3\overset{\displaystyle O}{\overset{\|}{C}}H + H_2O \qquad [27.31]$$

Ethanol           Acetaldehyde

Ketones are produced by oxidation of secondary alcohols. For example, acetone is formed in large quantities by oxidation of isopropyl alcohol:

$$CH_3-\overset{\displaystyle OH}{\overset{|}{C}H}-CH_3 + (O) \longrightarrow CH_3-\overset{\displaystyle O}{\overset{\|}{C}}-CH_3 + H_2O \qquad [27.32]$$

Acetone

Ketones are less reactive than aldehydes because ketones do not possess the reactive C—H bond attached to the carbonyl carbon. Ketones are used extensively as solvents. Acetone, which boils at 56°C, is the most widely used. The carbonyl functional group imparts polarity to the solvent. Acetone is completely miscible with water, yet it dissolves a wide range of organic substances. Methyl ethyl ketone, $CH_3COCH_2CH_3$, which boils at 80°C, is also used industrially as a solvent.

**Carboxylic Acids**

Oxidation of a primary alcohol may yield either an aldehyde or a **carboxylic acid**, depending on reaction conditions. For example, mild oxidation of ethanol produces acetaldehyde (Equation 27.31), which under more vigorous conditions may be further oxidized to acetic acid:

$$CH_3\overset{\displaystyle O}{\overset{\|}{C}}H + (O) \longrightarrow CH_3\overset{\displaystyle O}{\overset{\|}{C}}OH \qquad [27.33]$$

The oxidation of ethanol to acetic acid is responsible for causing wine to turn sour, producing vinegar.

As seen in this example, carboxylic acids contain the following functional group:

$$-\overset{\displaystyle O}{\overset{\|}{C}}-OH$$

Carboxylic acids are widely distributed in nature and are important compounds in many industrial chemical processes. Figure 27.14 shows the

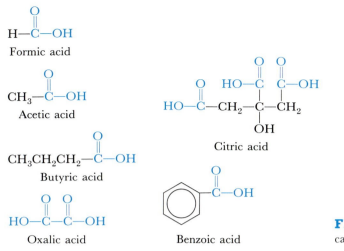

Formic acid

Acetic acid

Citric acid

Butyric acid

Oxalic acid

Benzoic acid

**FIGURE 27.14** Structural formulas of several familiar carboxylic acids.

structural formulas of several substances containing one or more carboxylic acid functional groups. Note that these substances are not named in a systematic way. The names of many acids are based on their historical origins. For example, formic acid was first prepared by extraction from ants; its name is derived from the Latin word *formica*, meaning "ant." Oxalic acid was first identified as a constituent of the *oxalis* family of plants.

Carboxylic acids are important in the manufacture of polymers used to make fibers, films, and paints. Among the compounds having low molecular weights, acetic acid is an industrially important compound. The most important method for manufacture of acetic acid is the reaction of methanol with carbon monoxide, in the presence of a rhodium catalyst:

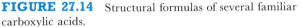

$$CH_3OH + CO \xrightarrow{\text{catalyst}} CH_3-\overset{\overset{\displaystyle O}{\|}}{C}-OH \qquad [27.34]$$

Notice that this reaction involves, in effect, the insertion of a carbon monoxide molecule between the $CH_3$ and OH groups. A reaction of this kind is called **carbonylation**.

Carboxylic acids may react with alcohols, with splitting out of water, to form **esters**:

**Esters**

$$CH_3-\overset{\overset{\displaystyle O}{\|}}{C}-OH + HO-CH_2CH_3 \longrightarrow CH_3-\overset{\overset{\displaystyle O}{\|}}{C}-O-CH_2CH_3 + H_2O \qquad [27.35]$$

Acetic acid          Ethanol          Ethyl acetate

As seen in this example, esters are compounds in which the H atom of a carboxylic acid is replaced by a hydrocarbon group:

The reaction between a carboxylic acid and an alcohol is another example of a condensation reaction. The reaction product, the ester, is named by using first the group from which the alcohol is derived and then the group from which the acid is derived.

## SAMPLE EXERCISE 27.6

Name the following esters:

(a)

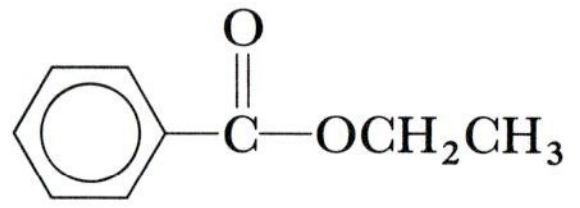

(b) 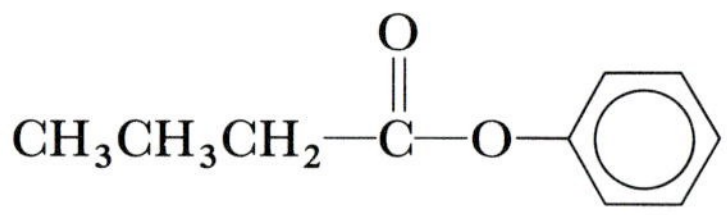 $CH_3CH_3CH_2$—

**Solution:** In (a) the ester is derived from ethanol and benzoic acid. Its name is therefore ethyl benzoate. In (b) the ester is derived from phenol and bu-

tyric acid. The residue from the phenol, $C_6H_5$, is called the phenyl group. The ester is therefore named phenyl butyrate.

## PRACTICE EXERCISE

Draw the structural formula for the compound propyl propionate.

*Answer:*

$$CH_3CH_2\overset{\overset{\displaystyle O}{\|}}{C}\!-\!O\!-\!CH_2CH_2CH_3$$

Esters generally have very pleasant odors; they are largely responsible for the pleasant aromas of fruit. Ethyl butyrate, for example, is responsible for the odor of apples. When esters are treated with acid or base in aqueous solution, they are hydrolyzed; that is, the molecule is split into its alcohol and acid components:

$$CH_3CH_2\overset{\overset{\displaystyle O}{\|}}{C}\!-\!O\!-\!CH_3 + Na^+ + OH^- \longrightarrow$$

Methyl propionate

$$CH_3CH_2\overset{\overset{\displaystyle O}{\|}}{C}\!-\!O^- + Na^+ + CH_3OH \qquad [27.36]$$

Sodium propionate          Methanol

In this example, the hydrolysis was carried out in basic medium. The products of the reaction are the sodium salt of the carboxylic acid and the alcohol.

## SAMPLE EXERCISE 27.7

Beginning with ethene as the only carbon-containing compound, suggest a series of reactions by which ethyl acetate might be prepared.

**Solution:** As shown in Equation 27.35, ethyl acetate can be prepared by a condensation reaction between acetic acid and ethanol. The ethanol can be prepared from ethene by $H_2SO_4$-catalyzed hydration:

$$CH_2{=}CH_2 + H_2O \xrightarrow{\;H_2SO_4\;} CH_3CH_2OH$$

This process is analogous to that given in Equation 27.26 for the hydration of propene. A portion of the ethanol thus produced can be oxidized to acetic acid, as shown in Equations 27.31 and 27.33.

## PRACTICE EXERCISE

What compounds are formed when propyl propionate is hydrolyzed in basic medium?
*Answer:* propyl alcohol and sodium propionate

**Amines** are organic bases (Section 17.4). They have the general formula $R_3N$, where R may be H or a hydrocarbon group, as in the following examples:

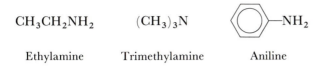

$$CH_3CH_2NH_2 \qquad (CH_3)_3N$$

Ethylamine      Trimethylamine      Aniline

Amines containing a proton can undergo condensation reactions with carboxylic acids to form **amides**:

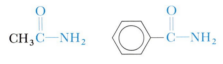

$$CH_3\overset{\displaystyle O}{\overset{\|}{C}}\!-\!OH + H\!-\!N(CH_3)_2 \longrightarrow CH_3\overset{\displaystyle O}{\overset{\|}{C}}\!-\!N(CH_3)_2 + H_2O \qquad [27.37]$$

We may consider the amide functional group to be derived from a carboxylic acid with an $NR_2$ group replacing the OH of the acid, as in these additional examples:

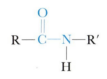

Acetamide        Benzamide

The amide linkage,

$$R\!-\!\overset{\displaystyle O}{\overset{\|}{C}}\!-\!\underset{\underset{\displaystyle H}{|}}{N}\!-\!R'$$

where R and R' are organic groups, is the key functional group in the structures of proteins, about which we shall have more to say in Chapter 28, Biochemistry.

## 27.5 POLYMERS

A **polymer** is a high-molecular-weight material formed from simple molecules called **monomers**. Polymers may be either synthetic or natural in origin. Natural polymers include proteins, starch, and cellulose, substances that we will discuss in Chapter 28. Our focus in the present discussion is on synthetic organic polymers. Synthetic polymers may be formed by either addition reactions (addition polymerization) or condensation reactions (condensation polymerization).

**Addition Polymerization**

In **addition polymerization**, unsaturated hydrocarbons or hydrocarbon derivatives are caused to react with one another. For example, addition polymerization of vinyl chloride produces polyvinyl chloride:

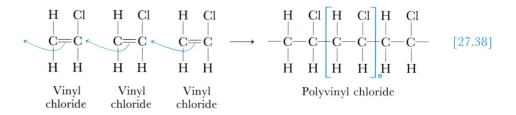

Vinyl chloride    Vinyl chloride    Vinyl chloride        Polyvinyl chloride

[27.38]

In such a polymerization reaction, $\pi$ bonds are broken and the electrons used to form new $\sigma$ bonds between the monomer units. In the preceding example, three vinyl chloride units are shown to combine. In practice, more than a thousand monomer units may combine to form a polymer (see Section 11.5).

Polymerization reactions are often written in a simplified form to stress the repeating unit of the polymer chain. For example, the addition polymerization of 2-methylpropene (whose common name is isobutylene) to produce polyisobutylene can be represented as follows:

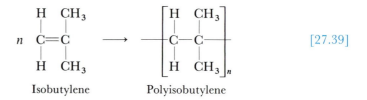

Isobutylene        Polyisobutylene

[27.39]

Polyisobutylene is a rubbery, rather gummy material that can be further treated to form a wide variety of useful products. In its finished form, it is known as butyl rubber. Many other alkenes can be similarly polymerized to form the variety of useful materials summarized in Table 27.6.

The polymerization reactions of alkenes may be catalyzed by a number of different types of substances. The choice of catalyst is important because it influences the structure and molecular weight of the resulting

**TABLE 27.6**    Alkenes that undergo addition polymerization

| Monomer formula | Monomer name | Polymer name | Uses |
|---|---|---|---|
| $CH_2{=}CH_2$ | Ethylene | Polyethylene | Coating for milk cartons, wire insulation, plastic bags |
| $CF_2{=}CF_2$ | Tetrafluoroethylene | Teflon | Insulation, bearings, frying-pan surfaces |
| $CH_2{=}CH$ <br>       $Cl$ | Vinyl chloride | Polyvinyl chloride (PVC) | Phonograph records, rainwear, piping |
| $CH_2{=}CH$ <br>       $CN$ | Acrylonitrile | Polyacrylonitrile (Orlon) | Rug fibers |
| $CH_2{=}CH$ <br>       $C_6H_5$ | Vinyl benzene (styrene) | Polystyrene | TV lead-in wire, combs, styrofoam insulation |
| $CH_2{=}CH$ <br>       $CH_3$ | Propylene | Polypropylene | Fibers for fabrics and carpets |

polymer and thus its properties. For example, polyethylene can be a pliable material with a low melting point or a much stiffer substance with a higher melting point, depending on the nature of the catalyst. One of the most important types of catalyst is the free radical (Section 27.3). Organic peroxides, compounds with O—O bonds, are exceptionally good sources of free radicals. These compounds decompose on heating, with the rupture of the O—O bond, to form a pair of free radicals:

$$R-\ddot{O}-\ddot{O}-R \; \rightleftharpoons \; 2R-\ddot{O}\cdot \qquad [27.40]$$

In this equation R could be any of a number of different organic groups. By placing a small amount of a peroxide in contact with an alkene and then heating it, radicals that can catalyze addition reactions are generated. For example, consider the polymerization of ethylene initiated (that is, started) by a radical formed from a peroxide:

$$R-\ddot{O}-\ddot{O}-R \; \rightleftharpoons \; 2R-\ddot{O}\cdot$$

$$R-\ddot{O}\cdot + H_2C{=}CH_2 \; \longrightarrow \; RO-\underset{\underset{H}{|}}{\overset{\overset{H}{|}}{C}}-\underset{\underset{H}{|}}{\overset{\overset{H}{|}}{C}}\cdot \qquad [27.41]$$

In Equation 27.41, the R—O radical reacts with a molecule of ethylene, forming a bond to one of the carbons. For this to occur, the pair of electrons forming the $\pi$ bond in ethylene must have uncoupled. One of them binds to the O—R group, the other to the second carbon. Thus, the product of this reaction is itself a free radical. This species goes on to react with a second molecule of ethylene:

$$RO-\underset{\underset{H}{|}}{\overset{\overset{H}{|}}{C}}-\underset{\underset{H}{|}}{\overset{\overset{H}{|}}{C}}\cdot + H_2C{=}CH_2 \; \longrightarrow \; RO-\underset{\underset{H}{|}}{\overset{\overset{H}{|}}{C}}-\underset{\underset{H}{|}}{\overset{\overset{H}{|}}{C}}-\underset{\underset{H}{|}}{\overset{\overset{H}{|}}{C}}-\underset{\underset{H}{|}}{\overset{\overset{H}{|}}{C}}\cdot \qquad [27.42]$$

By means of such successive reactions, the polymer grows in length. A reaction of this type is known as a chain reaction, because it is self-propagating. However, chain-termination reactions might occur. If two radicals come into contact, their unpaired spins might couple, and no further reaction would occur. In forming polymers by such free-radical reactions, we must control conditions carefully so that a polymer chain with the desired average length results.

In **condensation polymerization**, molecules are joined together through condensation reactions. The most important condensation polymers are polyamides and polyesters. The nylons are polyamides. They can be formed by reaction between a substance with two carboxylic acid functional groups and a substance with two amine functional groups. For example, nylon 6,6 is formed by heating a six-carbon diacid with a six-carbon diamine:

**Condensation Polymerization**

$$\text{---COOH} + \text{H}_2\text{N}-(\text{CH}_2)_6-\text{NH}_2 + \text{HOOC}-(\text{CH}_2)_4-\text{COOH}$$

$$+ \text{H}_2\text{N}-(\text{CH}_2)_6-\text{NH}_2 + \text{HOOC}-(\text{CH}_2)_4-\text{COOH} + \text{H}_2\text{N}\text{---}$$

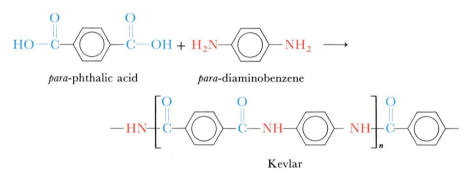

[27.43]

The polymer is formed by loss of water from between each pair of reacting functional groups. The resulting polymer is labeled 6,6 because there are six carbon atoms in the sections between each NH group along the chain.

Kevlar, a comparatively recent addition to the list of nylons, is formed by condensation between *para*-phthalic acid and *para*-diaminobenzene:

*para*-phthalic acid      *para*-diaminobenzene

Kevlar

Kevlar is notable for its exceptional strength and resistance to thermal decomposition. It is used as cording in heavy-duty truck tires and in making bulletproof vests (Figure 27.15).

**FIGURE 27.15**  This helicopter rotor blade is constructed from a composite of Kevlar and graphite fibers. The composite imparts great tensile strength and durability to a component that experiences high stresses during flight. (Courtesy of DuPont)

Condensation polymers can also result from formation of ester linkages, that is, by condensation of carboxylic acids with alcohols. In forming a polymer, it is important to use difunctional acids and alcohols. By varying the acid and alcohol functions, various kinds of polyester materials can be formed. The familiar polyester Dacron, from which so much of our clothing is produced, is a condensation polymer of ethylene glycol and *para*-phthalic acid:

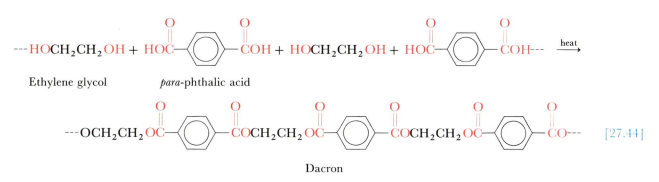

Ethylene glycol      *para*-phthalic acid

[27.44]

Dacron

The polymer formed under typical reaction conditions may have 80 to 100 units per molecule. This material can be melted without decomposition. The molten polymer is forced through tiny holes. It then cools and solidifies in the form of fibers. As the tiny fibers are stretched, the long molecules are forced to lie in a more or less parallel arrangement. The drawn or stretched fibers are then spun into a yarn.

# FOR REVIEW

## SUMMARY

In this chapter we have studied the chemical and physical characteristics of simple organic substances. **Hydrocarbons** are composed of only carbon and hydrogen. There are four major classes of hydrocarbons. The **alkanes** are composed of only carbon-hydrogen and carbon-carbon single bonds. The **alkenes** contain one or more carbon-carbon double bonds. **Alkynes** contain one or more carbon-carbon triple bonds. **Aromatic hydrocarbons** contain cyclic arrangements of carbon atoms bonded through both $\sigma$ and $\pi$ bonds. **Isomers** are substances that possess the same molecular formula but that differ in some other respect. Isomers may differ in the bonding arrangements of atoms within the molecule or in the geometrical arrangements of groups. **Geometrical isomerism** is possible in the alkenes because of restricted rotation about the C=C double bond.

The major sources of hydrocarbons are fossil fuels, such as coal and oil. Crude oil is separated by distillation into several fractions according to variations in volatility. The most important fraction is gasoline, used as automotive fuel. Less volatile fractions are **cracked**; that is, they are heated with catalysts to convert them to more volatile substances that have lower molecular weights and are suitable for gasoline. In the same process, straight-chain hydrocarbons are caused to rearrange to the more desirable branched-chain isomers, which have higher **octane numbers**.

Combustion of hydrocarbons is a highly exothermic process. The chief use of hydrocarbons is as a source of heat energy via combustion. The unsaturated hydrocarbons, the alkenes and alkynes, readily undergo **addition reactions** to the carbon-carbon multiple bonds. By contrast, addition reactions to aromatic hydrocarbons are difficult to carry out.

**Substitution reactions** are those in which one or more hydrogens of a hydrocarbon are replaced by some other atom or group. The substitution reactions of **aliphatic (nonaromatic) hydrocarbons** are not easily carried out. However, substitution reactions of aromatic hydrocarbons are easily accomplished in the presence of acid catalysts.

The chemistry of hydrocarbon derivatives is often dominated by the nature of their **functional groups**. The functional groups we have considered are summarized here:

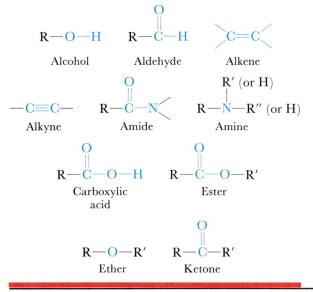

Alcohol    Aldehyde    Alkene

Alkyne    Amide    Amine

Carboxylic acid    Ester

Ether    Ketone

(Remember that R, R′, and R″ represent some hydrocarbon group—for example, methyl, $CH_3$, or phenyl, $C_6H_5$.)

**Alcohols** are hydrocarbon derivatives containing one or more OH groups. **Ethers** are related compounds that are formed by a **condensation reaction** of two molecules of alcohol, with water splitting out. Oxidation of primary alcohols can lead to formation of **aldehydes**; further oxidation of the aldehydes produces **carboxylic acids**. Oxidation of secondary alcohols leads to formation of **ketones**.

Carboxylic acids can form **esters** by a condensation reaction with alcohols, or they can form **amides** by condensation reaction with amines.

**Polymers** are large molecules formed by the combination of small, repeating units called **monomers**. **Addition polymerization** of alkenes is a source of many useful synthetic materials. These reactions can be catalyzed by many types of substances, including **free radicals**. **Condensation polymerization** of molecules with functional groups on both ends is also a source of important synthetic polymers, including polyamides and polyesters.

## LEARNING GOALS

Having read and studied this chapter, you should be able to:

1. List the four groups of hydrocarbons, and draw the structural formula of an example from each group.
2. List the names of the first 10 members of the alkane series of hydrocarbons.
3. Write the structural formula of an alkane, alkene, or alkyne, given its systematic (IUPAC) name.
4. Name an alkane, alkene, or alkyne, given its structural formula.
5. Give an example of geometrical isomerism in alkenes.
6. Explain why aromatic hydrocarbons do not readily undergo addition reactions as do the alkenes and alkynes.
7. List and describe the fractions obtained in petroleum refining.
8. Explain the general relationship between octane number and structure in alkanes, and describe the methods used to increase the octane numbers of straight-run gasolines.

9. Give examples of addition reactions of alkenes and alkynes, showing the structural formulas of reactants and products.
10. Describe Markovnikoff's rule and apply it to the addition reactions of alkenes and alkynes.
11. Write the steps in the photo-initiated chlorination of an alkane.
12. Give two or three examples of substitution of an aromatic compound.
13. Write structural formulas of molecules that are examples of an alcohol, an ether, an aldehyde, a ketone, a carboxylic acid, an ester, an amine, and an amide.
14. Describe the common characteristic of condensation reactions, and give an example of the condensation of alcohols to form ethers, condensations of alcohols and carboxylic acids to form esters, and condensation of amines and carboxylic acids to form amides.
15. Give an example of addition polymerization of an alkene and of condensation polymerization to form a polyamide or polyester.

## KEY TERMS

Among the more important terms and expressions used for the first time in this chapter are the following:

**Addition polymerization** (Section 27.5) is a reaction in which alkenes add together end to end by opening the carbon-carbon double bond to form long polymeric chain molecules.

In an **addition reaction** (Section 27.3) a reagent adds to the two carbon atoms of a carbon-carbon multiple bond.

**Alcohols** (Section 27.4) are hydrocarbons in which one or more hydrogens have been replaced by a hydroxyl group, OH.

**Aliphatic hydrocarbons** (Section 27.1) are those that are not aromatic; that is, alkanes, cycloalkanes, alkenes, and alkynes.

**Alkanes** (Section 27.1) are compounds of carbon and hydrogen containing only single carbon-carbon bonds.

**Alkenes** (Section 27.1) are hydrocarbons containing one or more carbon-carbon double bonds.

**Alkynes** (Section 27.1) are hydrocarbons containing one or more carbon-carbon triple bonds.

An **antiknock agent** (Section 27.2) is a substance added to automotive fuel to increase its octane number. This agent slows the rate of burning of fuel in the cylinder.

**Aromatic hydrocarbons** (Section 27.1) are hydrocarbon compounds that contain a planar, cyclic arrangement of carbon atoms linked by both $\sigma$ and $\pi$ bonds.

**Base hydrolysis** (Section 27.4) is a chemical reaction involving the replacement of a negatively charged functional group by aqueous hydroxide ion.

The **carbonyl functional group** (Section 27.4), $C=O$, is a characteristic feature of several organic functional groups, such as ketones and carboxylic acids.

**Carbonylation** (Section 27.4) is a reaction in which the carbonyl functional group is introduced into a molecule.

A **condensation polymer** (Section 27.5) is a substance having high molecular weight that is formed by elimination of water from between molecules.

**Cycloalkanes** (Section 27.1) are saturated hydrocarbons of general formula $C_nH_{2n}$ in which the carbon atoms form a closed ring.

A **free radical** (Section 27.3) is a species with an odd number of electrons.

A **functional group** (introduction; Section 27.4) is an atom or group of atoms that has characteristic chemical properties.

**Geometrical isomers** (Section 27.1) have the same molecular formulas and functional groups, but differ in the geometrical arrangements of atoms. For example, *cis*- and *trans*-2-butene are geometrical isomers.

In **homologous series** (Section 27.1), compounds contain common structural elements, but differ in the number of atoms making up the molecule. Members of a homologous series have the same general formula. For example, the aliphatic saturated alcohols have the general formula $C_nH_{2n+2}O$.

A **hydrocarbon** (Section 27.1) is a compound composed of carbon and hydrogen. Substituted hydrocarbons may contain functional groups composed of atoms other than carbon and hydrogen.

**Markovnikoff's rule** (Section 27.3) states that when a reagent of general formula XY adds to an alkene, the more negatively charged species, X or Y, ends up on the carbon with the fewer attached hydrogen atoms. For example, on addition of HCl to propene, Cl ends up on the middle carbon atom.

An **olefin** (Section 27.1) is a compound containing one or more carbon-carbon double bonds.

In a **substitution reaction** (Section 27.3), one atom or functional group bonded to a larger molecule is replaced by another.

**Unsaturated hydrocarbons** (Section 27.1) are nonaromatic compounds of carbon and hydrogen that contain one or more carbon-carbon multiple bonds.

## EXERCISES

### Hydrocarbon Structures and Nomenclature

**27.1** Write the molecular formulas of an alkane, alkene, alkyne, and aromatic hydrocarbon that in each case contains six carbon atoms.

**27.2** Write the Lewis structural formulas for all the alkanes with the molecular formula $C_6H_{14}$. Name each compound.

**27.3** Write the Lewis structural formula for each of the following hydrocarbons:
- **(a)** $CH_3CH(CH_3)CH_2CH_3$
- **(b)** $CH_3C{\equiv}CCH_3$
- **(c)** $CH_2{=}CHCH_3$
- **(d)** $CH_2{=}CHCH_2CH{=}CH_2$
- **(e)** $(CH_3)_2CHCH_3$.

**27.4** Write the condensed structural formulas of as many

aliphatic compounds as you can think of that have the molecular formula **(a)** $C_5H_8$; **(b)** $C_5H_{10}$.

**27.5** What are the characteristic bond angles **(a)** about carbon in an alkane; **(b)** about the carbon-carbon double bond in an alkene; **(c)** about the carbon-carbon triple bond in an alkyne?

**27.6** What are the characteristic hybrid orbitals employed by **(a)** carbon in an alkane; **(b)** carbon in a double bond in an alkene; **(c)** carbon in the benzene ring; **(d)** carbon in a triple bond in an alkyne.

**27.7** Write the condensed structural formula for each of the following compounds:
**(a)** 5-methyl-*trans*-2-heptene
**(b)** 3-chloropropyne
**(c)** *ortho*-dichlorobenzene
**(d)** 2,2,4,4-tetramethylpentane
**(e)** 2-methyl-3-ethylhexane
**(f)** 2-methyl-6-chloro-3-heptyne
**(g)** 1,5-dimethylnaphthalene.

**27.8** Write the condensed structural formula for each of the following compounds:
**(a)** 2,2-dimethylpentane
**(b)** 2,3-dimethylhexane
**(c)** *cis*-2-hexene
**(d)** methylcyclopentane
**(e)** 2-chlorobutane
**(f)** 1,2-dibromobenzene
**(g)** methylcyclobutane.

**27.9** Name the following compounds:

**(a)**
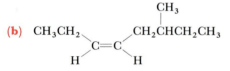

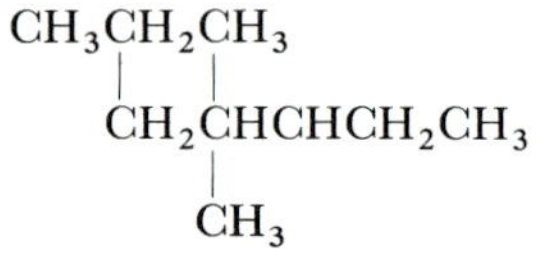

**(b)**

**(c)**
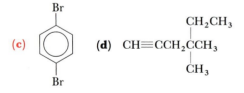

**(d)** $CH \equiv CCH_2\overset{\overset{\displaystyle CH_2CH_3}{|}}{\underset{\underset{\displaystyle CH_3}{|}}{C}}CH_3$

**27.10** Name the following compounds:

**(a)** 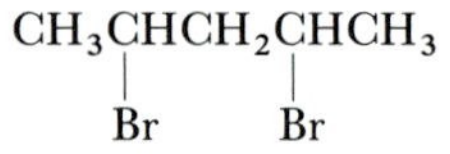 $CH_3\overset{}{\underset{\underset{\displaystyle Br}{|}}{C}}HCH_2\overset{}{\underset{\underset{\displaystyle Br}{|}}{C}}HCH_3$

**(b)** $CH_3CH\!=\!CHCH_2CH\!=\!CHCH_2CH_3$

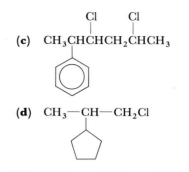

**(c)** $CH_3\overset{\overset{\displaystyle Cl}{|}}{C}HCH\overset{\overset{\displaystyle Cl}{|}}{C}HCH_2\overset{}{C}HCH_3$

**(d)** $CH_3 — CH — CH_2Cl$

**27.11** Indicate whether each of the following molecules is capable of *cis-trans* geometric isomerism; for those that are, draw the structures of the isomers: **(a)** 2,3-dichlorobutane; **(b)** 2,3-dichloro-2-butene; **(c)** 1,3-dimethylbenzene; **(d)** 4,4-dimethyl-2-pentyne.

**27.12** Draw the structural formulas for all possible geometric isomers for 2,6-octadiene.

## Petroleum

**27.13** Briefly describe the following terms: **(a)** cracking; **(b)** isomerization; **(c)** antiknock agent; **(d)** kerosene.

**27.14** Briefly describe the following terms: **(a)** branched-chain alkane; **(b)** paraffin; **(c)** octane number; **(d)** refining.

**27.15** What is the octane number of a mixture of 20 percent *n*-heptane, 30 percent methylcyclohexane, and 50 percent isooctane? (Refer to Table 27.4.)

**27.16** Describe two ways in which the octane number of a gasoline consisting of alkanes can be increased.

## Reactions of Hydrocarbons

**27.17** Give an example, in the form of a balanced equation, of each of the following chemical reactions: **(a)** substitution reaction of an alkane; **(b)** oxidation of an alkene; **(c)** addition reaction of an alkyne; **(d)** substitution reaction of an aromatic hydrocarbon.

**27.18** Using condensed structural formulas, write a balanced chemical equation for each of the following reactions: **(a)** hydrogenation of 1-butene; **(b)** addition of $H_2O$ to *cis*-2-butene using $H_2SO_4$ as a catalyst; **(c)** complete oxidation of cyclobutane; **(d)** reaction of benzene with 2-chloropropane in the presence of $AlCl_3$.

**27.19** Using Markovnikoff's rule, predict the product formed when HCl is reacted with each of the following compounds:
**(a)** $CH_3CH\!=\!CH_2$

**(b)** $\underset{\displaystyle CH_3}{\overset{\displaystyle CH_3}{}} C\!=\!C \underset{\displaystyle H}{\overset{\displaystyle CH_3}{}}$

**(c)** $CH_3C \equiv CH$

**27.20** Using Markovnikoff's rule, predict the product of each of the following reactions:
**(a)** addition of HBr to $CH_3CH_2CH\!=\!CHBr$

**(b)**   addition of $H_2O$ to

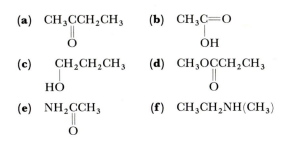

**(c)**   reaction of HBr with $(CH_3)_3CC\equiv CH$.

**27.21**   When cyclopropane is treated with HI, 1-iodopropane is formed. A similar type of reaction does not occur with cyclopentane or cyclohexane. How do you account for the activity of cyclopropane?

**27.22**   Why do addition reactions occur more readily with alkenes and alkynes than with aromatic hydrocarbons?

**27.23**   Predict the product or products formed in each of the following reactions:

**(a)**   $cis\text{-}CH_3CH{=}CHCH_3 + H_2 \xrightarrow{\text{Ni}}$

**(b)**   $CH_3CH_2C\equiv CH + HI \longrightarrow$

**(c)**   $CH_3C{=}CCH_2CH_3 + H_2O \xrightarrow{H_2SO_4}$
　　　　$\quad\ |\quad\ |$
　　　　$\quad CH_3\ CH_3$

**(d)**   $CH_2{=}CHCH_3 + Br_2 \longrightarrow$

**27.24**   Predict the product or products formed in each of the following reactions:

**(a)**   ⬡ $+ Br_2 \longrightarrow$

**(b)**   $(CH_3)_2C{=}CHCH_3 + HBr \longrightarrow$

**(c)**   $CH_3CH{=}CHCH_2CH_3 + O_2 \longrightarrow$

**(d)**   ⬡ $+ CH_3CHClCH_3 \xrightarrow{\text{AlCl}_3}$

**27.25**   Suggest a method of preparing ethylbenzene, starting with benzene and ethylene as the only organic reagents.

**27.26**   Write a series of reactions leading to *para*-bromoethylbenzene, beginning with benzene and using other reagents as needed.

**27.27**   The heat of combustion of decahydronaphthalene, $C_{10}H_{18}$, is $-6286$ kJ/mol. The heat of combustion of naphthalene, $C_{10}H_8$, is $-5157$ kJ/mol. (In each case $CO_2(g)$ and $H_2O(l)$ are the products.) Using these data and data in Appendix D, calculate the heat of hydrogenation of naphthalene. Does this value provide any evidence for aromatic character in naphthalene?

**27.28**   The molar heat of combustion of cyclopropane is 2089 kJ/mol; that for cyclopentane is 3317 kJ/mol. Calculate the heat of combustion per $CH_2$ group in the two cases; account for the difference.

### Hydrocarbon Derivatives

**27.29**   Identify the functional groups in each of the following compounds:

**(a)**   $CH_3CCH_2CH_3$
　　　　$\qquad\ \|$
　　　　$\qquad\ O$

**(b)**   $CH_3C{=}O$
　　　　$\quad\ \ |$
　　　　$\quad\ \ OH$

**(c)**   $CH_2CH_2CH_3$
　　　$|$
　　$HO$

**(d)**   $CH_3OCCH_2CH_3$
　　　　　$\ \|$
　　　　　$\ O$

**(e)**   $NH_2CCH_3$
　　　　$\quad\ \|$
　　　　$\quad\ O$

**(f)**   $CH_3CH_2NH(CH_3)$

**27.30**   Identify the functional groups in each of the following compounds:

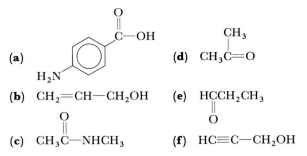

**(a)**

**(d)**   $CH_3\overset{CH_3}{\underset{|}{C}}{=}O$

**(b)**   $CH_2{=}CH{-}CH_2OH$

**(e)**   $HCCH_2CH_3$
　　　　$\quad\ \|$
　　　　$\quad\ O$

**(c)**   $CH_3\overset{O}{\overset{\|}{C}}{-}NHCH_3$

**(f)**   $HC\equiv C{-}CH_2OH$

**27.31**   What are the bond angles expected around each carbon atom in acetone, whose structural formula is shown below?

$$CH_3\overset{O}{\overset{\|}{C}}CH_3$$

**27.32**   Identify the carbon atom(s) in the following structure that has each of the following hybridizations: **(a)** $sp^3$; **(b)** $sp$; **(c)** $sp^2$.

$$HC\equiv C{-}\overset{O}{\overset{\|}{C}}{-}O{-}CH_3$$

**27.33**   What products, if any, occur when each of the following compounds is mildly oxidized:

**(a)**   $CH_3CHCH_2OH$
　　　　$\quad\ \ |$
　　　　$\quad\ \ CH_3$

**(b)**   $CH_3CH_2CH_2OH$

**(c)**   $CH_3\overset{O}{\overset{\|}{C}}CH_3$

**27.34**   What products, if any, occur when each of the following compounds is mildly oxidized?

**(a)**   $CH_3\overset{O}{\overset{\|}{C}}H$

**(b)**   $HOCH_2CH_2CH_2OH$

**(c)**   ⬡$\overset{CH_3}{\underset{H}{\overset{|}{\underset{|}{C}}OH}}$

**27.35** Draw the structures of the esters formed from **(a)** acetic acid and 2-propanol; **(b)** acetic acid and 1-propanol; **(c)** formic acid and ethanol. Name the compound in each case.

**27.36** Draw the structures of the compound formed by a condensation reaction between **(a)** benzoic acid and ethanol; **(b)** propionic acid and methyl amine; **(c)** acetic acid and phenol. Name the compound in each case.

**27.37** Write the structural formula for each of the following compounds: **(a)** 2-butanol; **(b)** 1,2-ethanediol; **(c)** methylformate; **(d)** diethyl ketone; **(e)** diethyl ether.

**27.38** Write the structural formula for each of the following compounds: **(a)** 3-chloropropionaldehyde; **(b)** methyl isopropyl ketone; **(c)** *meta*-chlorobenzaldehyde; **(d)** methyl-*cis*-2-butenyl ether; **(e)** *N*-diethylpropionamide.

**27.39** The IUPAC name for a carboxylic acid is based on the name of the hydrocarbon with the same number of carbon atoms. The ending *-oic* is appended, as in ethanoic acid, which is the IUPAC name for acetic acid,

Give the IUPAC name for each of the following acids:

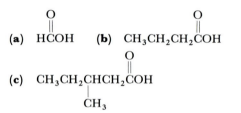

**27.40** Aldehydes and ketones can be named in a systematic way by counting the number of carbon atoms (including the carbonyl carbon) that they contain. The name of the aldehyde or ketone is based on the hydrocarbon with the same number of carbon atoms. The ending *-al*, for aldehyde, or *-one*, for ketone, is added as appropriate. Draw the structural formulas for the following aldehydes or ketones: **(a)** propanal; **(b)** 2-pentanone; **(c)** 3-methyl-2-butanone; **(d)** 2-methylbutanal.

**27.41** Write balanced chemical equations, with organic substances given by their condensed formulas, describing the preparation of **(a)** acetone from 1-propene; **(b)** propanoic acid (a carboxylic acid with three carbon atoms) from 1-propanol; **(c)** sodium acetate from ethene; **(d)** methyl acetate from methyl alcohol and acetic acid; **(e)** methyl ethyl ketone from 2-butanol.

**27.42** Using condensed formulas, write a balanced chemical equation for the condensation reactions that occur between the following substances: **(a)** methanol with itself; **(b)** methanol with ethanol; **(c)** acetic acid with 2-propanol; **(d)** formic acid with 1-propanol; **(e)** diethylamine with benzoic acid (see Figure 27.14); **(f)** methylethylamine with acetic acid.

**Polymers**

**27.43** Write a chemical equation for the addition polymerization of **(a)** acrylonitrile; **(b)** tetrafluoroethylene.

**27.44** If two different monomers are involved in addition polymerization, the resultant polymer is said to be a copolymer of the two monomers. Draw the formula for the copolymer of **(a)** styrene and propylene; **(b)** vinyl chloride ($CH_2\!=\!CHCl$) and 1,1-dichloroethylene ($CH_2\!=\!CCl_2$) to form Saran, the polymer used to make food wrapping.

**27.45** A rather simple condensation polymer can be made from ethylene glycol, $HOCH_2CH_2OH$, and oxalic acid,
$$HO\!-\!\overset{\displaystyle O}{\underset{\displaystyle \|}{C}}\!-\!\overset{\displaystyle O}{\underset{\displaystyle \|}{C}}\!-\!OH.$$ Sketch a portion of the condensation polymer chain obtained from these monomers.

**27.46** Draw the portion of the condensation polymer chain formed by the polymerization of 6-aminohexanoic acid,
$$H_2N\!-\!(CH_2)_5\!-\!\overset{\displaystyle O}{\overset{\displaystyle \|}{C}}\!-\!O\!-\!H$$
This material, called nylon 6 or Perlon, is used for making strong flexible fibers used in ropes and tire cords.

**Additional Exercises**

**27.47** Draw the structural formulas of two molecules with the formula $C_4H_6$.

**27.48** What is the molecular formula of **(a)** an alkane with 20 carbon atoms; **(b)** an alkene with 18 carbon atoms; **(c)** an alkyne with 12 carbon atoms?

**27.49** Draw the Lewis structures for the *cis* and *trans* isomers of 2-pentene. Does cyclopentane exhibit *cis-trans* isomerism? Explain.

**27.50** Classify each of the following substances as alkane, alkene, or alkyne (assuming that none are cyclic hydrocarbons): **(a)** $C_5H_{12}$; **(b)** $C_5H_8$; **(c)** $C_6H_{12}$; **(d)** $C_8H_{18}$.

**27.51** Give the IUPAC name for each of the following molecules:

$$\text{(a)} \quad CH_3\overset{\displaystyle OH}{\overset{\displaystyle |}{C}}HCH_3 \qquad \text{(b)} \quad CH_3OCH_3$$

$$\text{(c)} \quad CH_2\!=\!CHCH_2CH_2OH$$

$$\text{(d)} \quad CH_3\overset{\displaystyle CH_2CH_3}{\overset{\displaystyle |}{C}}\!=\!CHCH_2CH_3 \qquad \text{(e)} \quad CH_3C\!\equiv\!C\!-\!\overset{\displaystyle CH_3}{\underset{\displaystyle CH_3}{\overset{\displaystyle |}{\underset{\displaystyle |}{C}}}}H$$

**27.52** Describe how you would prepare **(a)** 2-bromo-2-chloropropane from propyne; **(b)** 1,2-dichloroethane from ethene; **(c)** a condensation polymer from $HOCH_2CH_2OH$ and
$$HO\overset{\displaystyle O}{\overset{\displaystyle \|}{C}}CH_2CH_2\overset{\displaystyle O}{\overset{\displaystyle \|}{C}}OH$$

**(d)** sodium benzoate from methylbenzoate; **(e)** polypropylene from propene; **(f)** 1-ethylnaphthalene from naphthalene and ethene.

**27.53** Write the formulas for all of the structural isomers of $C_3H_8O$.

**27.54** Explain why *trans*-1,2-dichloroethene has no dipole moment while *cis* 1,2-dichloroethene has a dipole moment.

**27.55** Would you expect cyclohexyne to be a stable compound? Explain.

**27.56** Identify all of the functional groups in each of the following molecules:

**(a)** $CH_2$=$CH$—$O$—$CH$=$CH_2$    (an anesthetic)

**(b)**
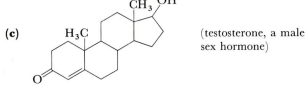
(acetylsalicyclic acid, aspirin)

**(c)**
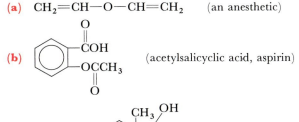
(testosterone, a male sex hormone)

**27.57** Write the structural formulas for each of the following: **(a)** an ether with the formula $C_3H_8O$; **(b)** an aldehyde with the formula $C_3H_6O$; **(c)** a ketone with the formula $C_3H_6O$; **(d)** a secondary alkyl fluoride with the formula $C_3H_7F$; **(e)** a primary alcohol with the formula $C_4H_{10}O$; **(f)** a carboxylic acid with the formula $C_3H_6O_2$; **(g)** an ester with the formula $C_3H_6O_2$.

**27.58** Give the condensed formulas for the carboxylic acid and the alcohol from which each of the following esters is formed:

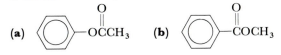

**27.59** In each of the following pairs, indicate which molecule is the more reactive, and give a reason for the greater reactivity: **(a)** butane and cyclobutane; **(b)** cyclohexane and cyclohexene; **(c)** benzene and 1-hexene; **(d)** 2-hexyne and 2-hexene.

**27.60** A 256-mg sample of an organic compound is combusted producing 512 mg of $CO_2$ and 209 mg of $H_2O$. At 127°C, 155 mg of this substance occupies a volume of 0.100 L with a pressure of 1.16 atm. What is the molecular formula of the compound? Describe a possible structure for the compound, and suggest some chemical test for the functional group.

**27.61** Bromination of butane requires irradiation. Write a series of reaction steps that can account for the bromination of butane under irradiation conditions.

**[27.62]** Suggest a process that would cause the breakup of a condensation polymer by essentially reversing the condensation reaction.

**[27.63]** An unknown organic compound is found on analysis to have the empirical formula $C_5H_{12}O$. It is slightly soluble in water. Upon careful oxidation it is converted into a compound of empirical formula $C_5H_{10}O$, which behaves chemically like a ketone. Indicate two or more reasonable structures for the unknown.

**[27.64]** Consider two hydrocarbons, A and B, whose molecular formulas are both $C_4H_8$. The first compound, A, is converted to $C_4H_{10}O$ when treated with water and sulfuric acid. Mild oxidation of this product produces $C_4H_8O$, which is an excellent solvent. Treatment with $K_2Cr_2O_7$ does not produce any further change in the $C_4H_8O$. In contrast, compound B gives a different $C_4H_{10}O$ substance upon treatment with sulfuric acid and water. Upon treating this $C_4H_{10}O$ substance with $K_2Cr_2O_7$, no oxidation occurs. Identify the two compounds, A and B, and trace all the reactions described that begin with these compounds, writing the names and structural formulas for all the organic compounds involved.

**[27.65]** An unknown substance is found to contain only carbon and hydrogen. It is a liquid that boils at 49°C at 1 atm pressure. Upon analysis it is found to contain 85.7 percent carbon and 14.3 percent hydrogen. At 100°C and 735 mm Hg, the vapor of the unknown has a density of 2.21 g/L. When it is dissolved in hexane solution and bromine water added, no reaction occurs. Suggest the identity of the unknown compound.

## CHAPTER 28

# Biochemistry

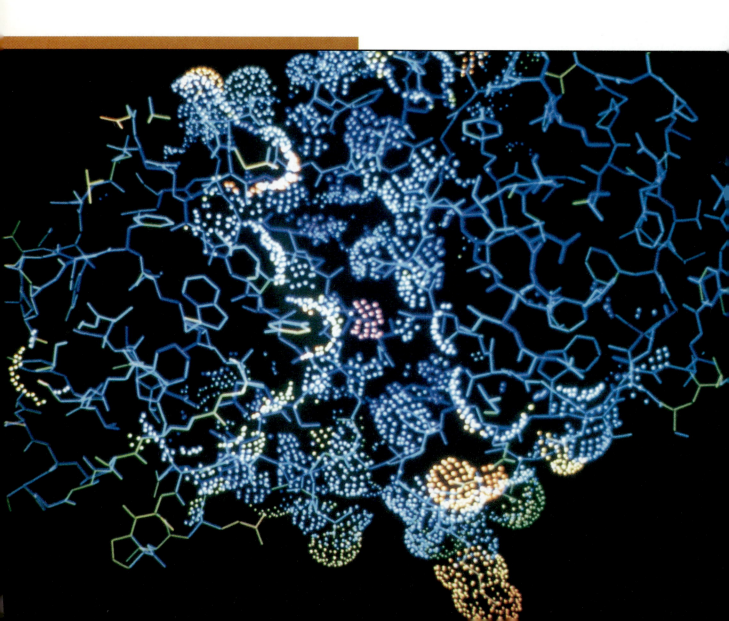

In earlier chapters of this text, we have considered various aspects of the physical world in which we live. We have discussed the chemistry of the atmosphere and of natural waters (Chapter 14) and of the solid earth (Chapter 24). In this final chapter we consider the **biosphere**, which is defined as that part of the earth in which living organisms are formed and live out their life cycles. The biosphere is not distinct from the atmosphere, natural waters, or the solid earth; rather, it is an integral part of it in those places where conditions permit life to exist. For living organisms to be sustained, there must be a supply of available energy, because the growth and maintenance of organisms require energy. Second, there must be an adequate supply of water, because organisms are composed largely of water and use it for exchange of materials with their environment.

## 28.1 ENERGY REQUIREMENTS OF ORGANISMS

Living organisms require energy for their maintenance, growth, and reproduction. The ultimate source of this energy is the sun. However, as living matter proliferated on earth, and as organisms became more and more specialized, many of them developed the capacity for obtaining energy indirectly, by using the energy stored in other organisms. For example, our bodies have essentially no capacity for directly using solar energy. We consume animal and plant materials to acquire substances that our bodies can use as energy sources.

There are two distinct reasons why living systems need energy. First, organisms rely on substances readily available in their surroundings for synthesis of compounds needed for their existence. Most of these reactions are endothermic. To make such reactions proceed, energy must be supplied from an outside source. Second, living organisms are highly organized. The complexity of all the substances that make up even the simplest of single-cell organisms and the relationships between all the many chemical processes occurring are truly amazing. In thermodynamic terms, living systems are very low in entropy as compared with the raw materials from which they are formed. The orderliness that is characteristic of living systems is attained by expenditure of energy.

Recall from the discussion in Section 19.5 that the free-energy change, $\Delta G$, is related to both the enthalpy and the entropy changes that occur in a process:

$$\Delta G = \Delta H - T \Delta S$$

If the entropy change in a process that builds up a living organism is

---

Computer graphic model of pepsin, a digestive enzyme found in the stomach. Part of the model shows the structure in the form of a skeleton of connected atoms; the other part shows it in the form of a space-filling model. The computer model is based upon X-ray diffraction data that show the general shape of the molecule and the locations of atoms other than hydrogen. (Professor T. L. Blundell, Science Photo Library/Photo Researchers)

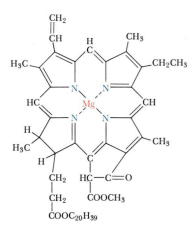

**FIGURE 28.1** Structure of chlorophyll *a*. All chlorophyll molecules are essentially alike; they differ only in details of the side chains.

negative (in other words, if a more highly ordered state results), the contribution to $\Delta G$ is positive. This means that the process becomes less spontaneous. Thus both the enthalpy and the entropy changes that result in the formation, maintenance, and reproduction of living systems are such that the overall process is nonspontaneous. To overcome the positive values for $\Delta G$ associated with the essential processes, living systems must be coupled to some outside source of energy that can be converted into a form useful for driving the biochemical processes. The ultimate source of this needed energy is the sun.

The major means for conversion of solar energy into forms that can be used by living organisms is **photosynthesis**. The photosynthetic reaction that occurs in the leaves of plants is conversion of carbon dioxide and water to carbohydrate, with release of oxygen:

$$6CO_2 + 6H_2O \xrightarrow{\ 48\ hv\ } C_6H_{12}O_6 + 6O_2 \qquad [28.1]$$

Note that formation of a mole of sugar, $C_6H_{12}O_6$, requires the absorption and utilization of 48 moles of photons. This needed radiant energy comes from the visible region of the spectrum (Figure 6.3). The photons are absorbed by photosynthetic pigments in the leaves of plants. The key pigments are the **chlorophylls**; the structure of the most abundant chlorophyll, called chlorophyll *a*, is shown in Figure 28.1.

Chlorophyll is a coordination compound; it contains a $Mg^{2+}$ ion bound to four nitrogen atoms arranged around the metal in a planar array. The nitrogen atoms are part of a porphyrin ring (Figure 26.8). Notice that there is a series of alternating double bonds in the ring that surrounds the metal ion. This system of alternating, or conjugated, double bonds gives rise to the strong absorptions of chlorophyll in the visible region of the spectrum. Figure 28.2 shows the absorption spectrum of chlorophyll as compared with the distribution of solar energy at the earth's surface. Chlorophyll is green in color because it absorbs red light (maximum absorption at 655 nm) and blue light (maximum absorption at 430 nm) and transmits green light.

The solar energy absorbed by chlorophyll is converted in a complex series of steps into chemical energy. This stored energy is then used to

**FIGURE 28.2** Absorption spectrum of chlorophyll (green curve), in comparison with the solar radiation at ground level (red curve).

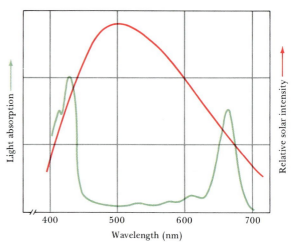

drive the reaction in Equation 28.1 to the right, a direction in which it is highly endothermic. Plant photosynthesis is nature's solar-energy-conversion machine; all living systems on earth are dependent on it for continued existence (Figure 28.3). A field of corn in the Midwest at the height of its growing season converts several percent of all the incident solar radiation into plant matter. It has been suggested that by cultivating about 6 percent of the land surface in the United States and using optimal growing conditions, we could obtain sufficient energy to satisfy all the energy needs of modern United States society.

Let us now turn our attention to the materials from which living organisms are formed. At some level of concentration, in some organism somewhere, almost every element seems to have a role to play in the biosphere. However, the elements of major importance in terms of their abundances in living systems are carbon, hydrogen, oxygen, nitrogen, phosphorus, and sulfur. Many other elements, including many metallic elements, are present in lesser quantities (refer back to Figure 26.7).

Much of the material present in living systems is in the form of polymers, molecules of high molecular weight (Section 27.5). These polymers in living systems, called **biopolymers**, can be classified into three broad groups: proteins, carbohydrates, and nucleic acids. The proteins and carbohydrates, along with a class of molecules called fats and oils, are the major sources of energy in the food supply of animals. Polymeric carbohydrates are the major construction material that gives form to plant systems; proteins perform a similar role in animal systems. The nucleic acids are the storehouse of information that determines the form of a living system and that regulates its reproduction and development.

**FIGURE 28.3** The absorption and conversion of solar energy that occurs in the maple leaf provides the energy necessary to drive all the living processes of the tree, including growth. (Calvin Larsen/Photo Researchers)

## 28.2 PROTEINS

**Proteins** are macromolecular substances present in all living cells. They serve as the major structural component in animal tissues; they are a key part of skin, cartilage, nails, and the skeletal muscles. Enzymes, the catalysts for the biochemical reactions occurring in all living systems, are proteins. Proteins transport vital substances within the body. For example, hemoglobin, which carries $O_2$ from the lungs to cells, is a protein. The antibodies that serve as a defense mechanism against undesirable substances within the organism are also composed of proteins.

The molecular weights of proteins vary from about 10,000 to over 50 million. Simple proteins are composed entirely of amino acids; other proteins, called **conjugated proteins**, are made up of simple proteins bonded to other kinds of biochemical structures. The most abundant elements present in proteins are carbon (50 to 55 percent), hydrogen (7 percent), oxygen (23 percent), and nitrogen (16 percent). Sulfur is present in most proteins to the extent of about 1 or 2 percent; phosphorus either is absent or is present to only a very slight extent.

Simple proteins are linear polymers of amino acids. The characteristic polymeric linkage in proteins is the amide linkage, noted in Section 27.4:

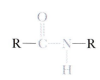

The particular functional group is called the **peptide** linkage when it is formed from amino acids. Proteins are sometimes referred to as **polypeptides**. Usually, however, the term "polypeptide" refers to a polymer of amino acids that has a molecular weight of less than 10,000. To see how the peptide linkage might serve to form a polymer we must investigate the structures of the amino acids, substances from which proteins are formed.

**Amino Acids**

Twenty amino acids are found to occur commonly in various proteins. All of these are α-amino acids; that is, the amino group is located on the carbon atom immediately adjacent to the carboxylic acid functional group. The general formula for an amino acid is as follows:

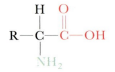

The amino acids differ in the nature of the R group. Figure 28.4 shows the structural formulas of the 20 amino acids found in most proteins. Although certain of the amino acids are more common than others, most large proteins contain most of the amino acids.

You can see from the condensed structural formula just shown that the α-carbon atom, which bears both the amino and carboxylic acid groups, has four different groups attached to it.* Any molecule containing a carbon with four different attached groups will be **chiral**. As we pointed out in Section 26.4, a chiral molecule is one that is not super-

* The sole exception is glycine, for which R = H. For this amino acid, there are two H atoms on the α-carbon atom.

**FIGURE 28.4** Condensed structural formulas of the amino acids, with the three-letter abbreviation for each acid.

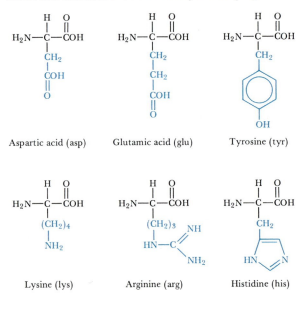

*Amino acids with hydrocarbon side chains*

**FIGURE 28.4** Continued

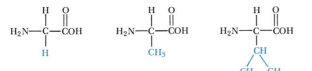

Glycine (gly)  Alanine (ala)  Valine (val)  Leucine (leu)

Isoleucine (ile)  Phenylalanine (phe)  Proline (pro)

*Amino acids with polar, neutral side chains*

Serine (ser)  Threonine (thr)  Methionine (met)  Cysteine (cys)

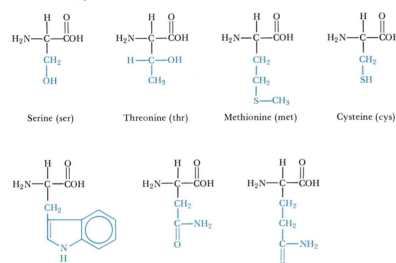

Tryptophan (trp)  Asparagine (asn)  Glutamine (gln)

imposable with its mirror image. Let us consider a particular amino acid, say alanine, and suppose that we have two such molecules that are mirror images of one another, as in Figure 28.5. This figure also shows two familiar objects that are mirror images of one another, a pair of hands. Your right and left hands are not superimposable. That is, if there were some way that you could hollow out your left hand and place it face down on a surface, there is no way that you could slide your right hand, also face down, into the left. (That is, of course, why you can't wear your right-hand glove on your left hand.) In the same way, one of the mirror-image forms of alanine cannot be superimposed on the other.

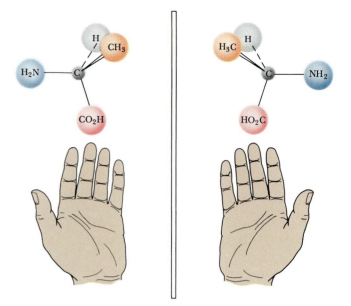

**FIGURE 28.5** Illustration of chiral character. The hands shown are mirror images of one another and are not superimposable. In the same way, the two alanine molecules that are mirror images are not superimposable.

The two molecules of a chiral substance that are mirror images of one another are called **enantiomers**. Because two enantiomers are not exactly the same, they are isomers of one another. This type of isomerism is called **configurational isomerism**, or **optical isomerism**. The two enantiomers of a pair are sometimes labeled *R*- (from the Latin *rectus*, "right") and *S*- (from the Latin *sinister*, "left") to distinguish them. Another widely used notation for distinguishing the enantiomers uses the prefix labels D- (from the Latin *dexter*, "right") and L- (from the Latin *laevus*, "left"). The enantiomers of a chiral substance possess the same physical properties, such as solubility, melting point, and so forth. Their chemical behavior toward ordinary chemical reagents is also the same. However, they differ in chemical reactivity toward other chiral molecules. It is a striking fact that all the amino acids in nature are of the *S*-, or L-, configuration at the carbon center (except glycine, which is not chiral). Only amino acids of this specific configuration at the chiral carbon center are biologically effective in forming polypeptides and proteins in most organisms; peptide linkages are formed in cells under such specific conditions that enantiomeric molecules can be distinguished.

Chiral substances differ from one another in their effect on the plane of **polarized light**. We saw an example of this in Section 26.4, which dealt with the isomerism of coordination compounds. Suppose that a beam of polarized light is passed through a solution containing a chiral substance such as alanine, using the arrangement shown in Figure 26.19. A solution of one of the enantiomers causes the plane of polarization to be rotated in one direction; a solution of the other enantiomer has the opposite effect. In each case the amount of rotation is proportional to the concentration of the solution and the length of the cell containing it. A solution containing equal concentrations of the two enantiomers, called a **racemic mixture**, causes no net rotation. Enantiomers of a chiral substance are often called **optical isomers** because of their effect on polarized light.

The relative amounts of the various amino acids in a protein vary with the nature and function of the protein material. Amino acids that have hydrocarbon side chains predominate in the insoluble, fiberlike proteins such as silk, wool, and collagen (a tissue-supporting protein). The amino acids with polar side chains are relatively more abundant in the water-soluble proteins. In addition, the polar groups often play an important role in determining the overall structure of the protein molecule, about which we shall have more to say later. Amino acids with acid or base side chains also help to increase water solubility. In addition, the functional groups on the side chains can act as sources of acid or base character in reactions catalyzed by acids or bases.

We have noted that the characteristic functional group in proteins is the amide, or peptide, linkage. This linkage is formed by a condensation reaction (Sections 27.4 and 27.5) between two amino acid molecules. As an example, alanine and glycine might form the dipeptide glycylalanine:

**Protein Structure**

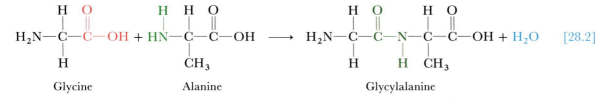

Glycine          Alanine                    Glycylalanine

Notice that the acid that furnishes the carboxyl group is named first, with a -*yl* ending; then the amino acid furnishing the amino group is named.

---

**SAMPLE EXERCISE 28.1**

Draw the structural formula for histidylglycine.

**Solution:**   The name for this dipeptide tells us that the amino group of glycine and the carboxylic acid function of the histidine are involved in formation of the peptide bond. The structure is therefore

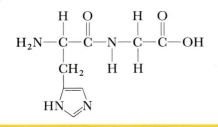

**PRACTICE EXERCISE**

Name the dipeptide that has this structure:

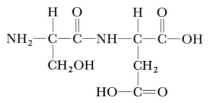

*Answer:*   serylaspartic acid

---

Because 20 different amino acids are commonly found in proteins and because the protein chain may consist of hundreds of these amino acids, the number of possible arrangements of amino acids is huge beyond imagination. Nature makes use of only certain of these combinations, but even so the number of different proteins is very large.

The arrangement, or sequence, of amino acids along the chain deter-

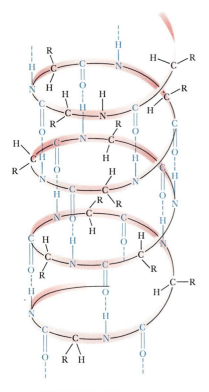

**FIGURE 28.6** α-Helix structure for a protein. The symbol R represents any one of the several side chains shown in Figure 28.4.

mines the **primary structure** of a protein. This primary structure gives the protein its unique identity. A change of even one amino acid can alter the biochemical characteristics of the protein. For example, sickle-cell anemia is a genetic disorder resulting from a single misplacement in a protein chain in hemoglobin. The chain that is affected contains 146 amino acids. The first seven in the chain are valine, histidine, leucine, threonine, proline, glutamic acid, and glutamic acid. In a person suffering from sickle-cell anemia, the sixth amino acid is valine instead of glutamic acid. This one substitution of an amino acid with a hydrocarbon side chain for one that has an acidic functional group in the side chain alters the solubility properties of the hemoglobin, and normal blood flow is impeded (see also Section 12.7).

Proteins in living organisms are not simply long, flexible chains with more or less random shape. Rather, the chain coils or stretches in particular ways that are essential to the proper functioning of the protein. This aspect of the protein structure is called the **secondary structure**. One of the most important and common secondary structure arrangements is the **α-helix**, first proposed by Linus Pauling and R. B. Corey. The helix arrangement is shown in schematic form in Figure 28.6. Imagine winding a long protein chain in a helical fashion around a long cylinder. The helix is held in position by hydrogen-bond interactions between an N—H bond and the oxygen of a carbonyl function located directly above or below. The pitch of the helix and the diameter of the cylinder must be such that (1) no bond angles are strained and (2) the N—H and C=O functional groups on adjacent turns are in proper position for hydrogen bonding. An arrangement of this kind is possible for some amino acids along the chain, but not for others. Large protein molecules may contain segments of chain that have the α-helical arrangement interspersed with sections in which the chain is in a random coil. The overall shape of the protein, determined by all the bends, kinks, and sections of a rodlike α-helical structure, is called the **tertiary structure**.

Figure 28.7 shows the structure of myoglobin, a protein of about 18,000 molecular weight, containing one heme group (Section 26.2). Myoglobin resides in the tissues; it stores oxygen until it is required for metabolic activities. Notice that the protein has sections of helical structure, separated by other segments of polypeptide.

You might suppose that a long, complex protein molecule could adopt almost any shape under a given set of conditions, but this is not the case. Certain foldings of the protein chain lead to lower-energy (more stable) arrangements than do other folding patterns. For example, a protein dissolved in aqueous solution folds in such a way that the nonpolar hydrocarbon portions are tucked within the molecule, whereas the more polar acidic and basic side-chain functional groups are projected into the solution.

When proteins are heated above the temperatures characteristic of living organisms, or when they are subjected to unusual acid or base conditions, they begin to lose their particular tertiary and secondary structure. As this happens the protein loses its biological activity; it is said to be **denatured**. When denaturation occurs under very mild conditions, it is often reversible. That is, a return to normal conditions results in a return of biological activity. However, denaturation may also be

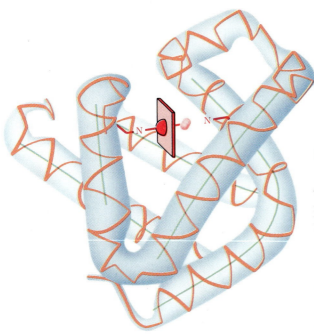

**FIGURE 28.7** The structure of myoglobin, a protein that acts to store oxygen in cells. This protein contains one heme unit, symbolized by the red sphere and surrounding square. The heme unit is bound to the protein through a nitrogen-containing ligand, represented by the red N on the left. In the oxygenated form an $O_2$ molecule is coordinated to the heme group, as shown. When the oxygen is not present, a nitrogen-containing ligand from the protein is bound to that site. The protein chain is represented by the continuous orange ribbon. The sections that possess a helical structure are denoted by the green line running through the center. The helical portions are separated by sections that have different arrangements of the amino acid chain. The protein wraps around to make a kind of pocket for the heme group.

irreversible; chemical changes that permanently alter the character of the protein may occur. As an example, the protein material in the white of an egg is irreversibly denatured when the egg is placed for a time in boiling water.

With this brief introduction to proteins, let us now consider the characteristics of a very important group of proteins, the enzymes.

## 28.3 ENZYMES

The human body is characterized by an extremely complex system of interrelated chemical reactions. All of these reactions must occur at carefully controlled rates, so that thousands of individual chemical components are maintained at proper concentration levels and the system is able to respond as needed to demands made upon it. Every one of the many thousands of chemical reactions occurring in a biochemical system can be represented by an ordinary chemical equation. Indeed, many of the same reactions can be carried out under ordinary laboratory conditions. What is extraordinary about biochemical systems is that so many reactions occur with great rapidity at moderate temperatures. We cited in Section 15.6 the example of sugar, which is quite rapidly oxidized in the body at 37°C to carbon dioxide and water. In the laboratory, sugar does not react with oxygen at room temperature at a measurable rate. It must be heated to rather high temperature—for example, in a burner flame—before it burns.

The oxidation of sugar in the body occurs by means of a series of more than two dozen biochemically catalyzed steps. The catalyst in each step is called an **enzyme**. Enzymes are formed in living cells and are protein in nature. The molecular weights of enzymes range from a low of perhaps 10,000 to a high of about 1 million. The enzyme may consist of just a single protein chain or may involve several chains that are loosely held together.

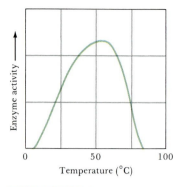

**FIGURE 28.8** A typical example of enzyme activity as a function of temperature. At the higher temperature, at which the activity of the enzyme drops to zero, the protein is denatured.

The reaction that the enzyme catalyzes occurs at a specific site on the protein; this is called the **active site**. The substances that undergo reaction at this site are called **substrates**. Besides the substrate, other substances, called **cofactors** or **coenzymes**, may be required in the enzyme-catalyzed reaction. For example, the enzyme may require the presence of $Mg^{2+}$ or some other metal ion, or it may require the presence and participation of a small organic molecule.

In many enzyme-catalyzed reactions, the coenzyme is one of the vitamins. An enzyme without its required coenzyme is called an **apoenzyme**. The combination of apoenzyme and coenzyme is called the **holoenzyme**:

$$\text{Apoenzyme} + \text{coenzyme} \longrightarrow \text{holoenzyme}$$

Enzymes possess certain characteristics not common to other types of catalysts. In the first place, they have a rather special sensitivity to temperature. Experimental studies have shown that the activity of a particular enzyme maximizes at around the normal temperature of the organism in which the enzyme is found. Figure 28.8 illustrates a typical activity versus temperature curve. When the temperature is raised above the usual operating temperature of the enzyme, the activity often temporarily increases, but then subsequently decreases. The secondary and tertiary structures of the protein chain, on which the activity of the active site depends, are the result of many weak forces that induce the chain to take up that particular arrangement. Heating causes the protein chain to come undone, as it were; the enzyme is denatured and loses its activity completely.

A second respect in which enzymes are special is that they often show a sharp change in activity with change in the acidity or basicity of the solution. This suggests that acid-base reactions are important in the catalyzed reaction. Third, enzymes may be very specific in catalyzing exclusively a particular kind of reaction and no other, or even in catalyzing a particular reaction for only one compound and no other. The degree of specificity of enzymes varies widely. For example, enzymes called carboxypeptidases are rather general in their action.* They catalyze the hydrolysis of polypeptides into amino acids:

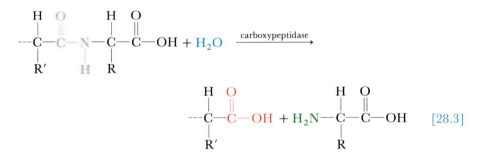

This equation shows only the end member of a peptide chain; the car-

---

* The names of most enzymes end in -*ase*. The -*ase* ending is attached to the name of the substrate on which the enzyme acts, or to the name of the type of reaction that the enzyme catalyzes. Thus *peptidases* are enzymes that act on peptides; *lipase* is an enzyme that acts on lipids; a *transmethylase* is an enzyme that catalyzes transfer of a methyl group.

**FIGURE 28.9** The active ingredient in meat tenderizers such as the product shown here is usually papain, an enzyme (a protease) present in the fruit of the papaya tree. Pineapple also contains a protease, called bromelin. Experienced cooks know that fresh pineapple should not be added to gelatin to make a molded fruit salad. The gelatin never hardens because the protein responsible for the setting action is digested by the bromelin present in the pineapple. (Richard Megna/Fundamental Photographs)

boxypeptidase attacks only the amide group at the end of the chain. However, it is active regardless of the nature of the particular side chains R and R′.

Another group of related enzymes, called *proteases*, catalyzes the hydrolysis of peptides into smaller lengths of polypeptide. These enzymes catalyze the hydrolysis reaction at any point in the polypeptide chain and not solely at the end member (Figure 28.9). Although all these peptidases catalyze the hydrolysis of peptides, they are not at all active in the hydrolysis of fats; an entirely separate set of enzymes is responsible for the latter reaction. The high degree of specificity that enzymes possess is necessary to maintain some degree of independence among all the reactions occurring in a complex organism.

Finally, many enzymes differ from ordinary nonbiochemical catalysts in that they are enormously more efficient. The number of individual, catalyzed-reaction events occurring per second at a particular active site, called the **turnover number**, is generally in the range $10^3$ to $10^7$ per second. Such large turnover numbers are indicative of reactions with very low activation energies.

## SAMPLE EXERCISE 28.2

Solutions of dilute (about 3 percent) hydrogen peroxide are stable for long periods of time, especially if a small amount of a stabilizer is added. However, when such a solution is poured onto an open cut or wound, the hydrogen peroxide decomposes very rapidly with evolution of oxygen and formation of water:

$$2H_2O_2(aq) \longrightarrow O_2(g) + 2H_2O(l)$$

How can you account for this result?

**Solution:** Because the reaction occurs so rapidly on contact with the open wound, some chemical species present at the wound must catalyze the decomposition of hydrogen peroxide. That substance is an enzyme called catalase. Catalase is present to ensure that hydrogen peroxide does not accumulate in cells, because it could interfere with many other cell reactions. Catalase is an efficient enzyme, with a very high turnover number.

**PRACTICE EXERCISE** _____

In 1926 James Sumner isolated the first pure, crystalline enzyme. This substance catalyzes the hydrolysis of urea to ammonia and carbon dioxide:

**PRACTICE EXERCISE** _____

In 1926 James Sumner isolated the first pure, crystalline enzyme. This substance catalyzes the hydrolysis of urea to ammonia and carbon dioxide:

$$H_2N-\overset{\overset{\displaystyle O}{\|}}{C}-NH_2(aq) + H_2O(l) \longrightarrow$$

$$2NH_3(aq) + CO_2(g)$$

**(a)** What is the name of this enzyme? **(b)** What type of biological substance is it?

*Answers:* **(a)** urease; **(b)** protein

---

Biochemists have been trying for a long time to discover how and why enzymes are so fantastically efficient and selective. During the past 20 years, the development of new experimental techniques for studying the structures of molecules and for following the rates of very rapid reactions has led to a much better understanding of how enzymes work.

The high turnover numbers observed for enzymes suggest that the substrate molecules cannot be very tightly bound to the enzyme; if they were, they might block the active site. Reaction would then be slow, because the active site is not quickly cleared out. Most enzyme systems that have been studied behave as though there were an equilibrium between the substrate ($S$) and the active site ($E$); we can write this equilibrium as an equation:

$$E + S \underset{k_{-1}}{\overset{k_1}{\rightleftharpoons}} ES \qquad [28.4]$$

The symbol $ES$ represents a species in which the substrate is attached in some way to the enzyme. This **enzyme-substrate complex**, as it is called, then reacts to give product ($P$) and free the active site:

$$ES \overset{k_2}{\longrightarrow} E + P \qquad [28.5]$$

In this picture, it is assumed that the substrate molecules may move off of and onto the active site very rapidly compared with the rate at which they undergo reaction to form products $P$. This means that the equilibrium described by Equation 28.4 is rapidly established between enzyme and substrate. It is also supposed that the equilibrium is such that most of the enzyme sites are not occupied by $S$ when the substrate is present at normal concentrations. Now suppose that we study and graph (Figure 28.10) the rate of the enzyme-catalyzed reaction as a function of increasing concentration of substrate $S$. When $S$ is present in low concentration, most of the enzyme active sites are not in use. Increasing the concentration of $S$ shifts the equilibrium in Equation 28.4 to the right, so that a larger number of enzyme-substrate complexes are formed. This in turn increases the overall rate of the reaction, because that rate depends on the concentration of $ES$, $[ES]$:

$$Rate = \frac{\Delta[P]}{\Delta t} = k_2[ES] \qquad [28.6]$$

However, with still further increases in the concentration of $S$, a sizable fraction of the active sites are occupied. Adding still more $S$ thus does not result in the same degree of increase in $[ES]$, and the rate does not increase as much. Eventually, with still higher concentration of $S$, *all* the

**FIGURE 28.10** Rate of product formation as a function of concentration in a typical enzyme-catalyzed reaction. The rate of product formation is proportional to substrate concentration in the region of low concentration. At high substrate concentrations, the active sites on the enzyme are all complexed; further addition of substrate does not affect reaction rate.

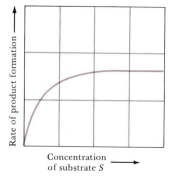

active sites are effectively occupied. Further increases in $S$ cannot result in faster reaction, because all available active sites are in use.

Many enzyme systems have been found to obey the sort of behavior illustrated in Figure 28.10. This strongly indicates that an enzyme-substrate complex is involved in the action of enzymes.

One of the earliest models for the enzyme-substrate interaction was the **lock-and-key model**, illustrated in Figure 28.11. The substrate is pictured as fitting neatly into a special place on the protein (the active site) tailored more or less specifically for that particular substrate. The catalyzed reaction occurs while the substrate is bound to the enzyme, and the reaction products then depart. Clearly, this picture of enzyme action has much in common with the models for heterogeneous catalysis that were discussed in Section 15.6. The difference is that greater specificity is brought into the enzyme picture.

The lock-and-key model for enzyme action goes a long way toward explaining many aspects of enzyme action, but it cannot tell the whole story. As the substrate molecules are drawn into the active site, they are somehow activated so that they are capable of extremely rapid reaction. This activation may result from the withdrawal or donation of electron density at a particular bond by the enzyme. In addition, in the process of fitting into the active site, the molecule may be distorted by the portions of the enzyme protein chain near the active site and thus made more reactive. If coenzymes are involved, the enzyme may promote reaction by binding both the coenzyme and substrate to the same site or to adjacent sites, thus keeping the reactants in close proximity. Enzymes, like all proteins, contain many acidic and basic sites along the chain, and these may be involved in acid-base reactions with the substrate. It is clear that the enzyme-coenzyme-substrate interactions can be very complex and are often difficult to unravel.

One of the more interesting corollaries of the lock-and-key model for enzyme action is that certain molecules should be capable of inhibiting the enzyme. Suppose that a certain molecule is capable of fitting the active site on the enzyme but is not reactive for one reason or another. If it is present in the solution along with substrate, this molecule competes with substrate for binding at the active sites. The substrate is thus prevented from forming the necessary enzyme-substrate complex, and the rate of product formation decreases. Metals that are highly toxic—for example, lead and mercury—probably operate as enzyme inhibitors. The heavy metal ions are especially strongly bound to sulfur-containing groups on protein side chains. By complexing very strongly to these sites on proteins, they interfere with the normal reactions of enzymes.

## The Lock-and-Key Model

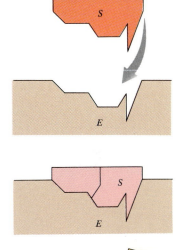

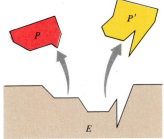

**FIGURE 28.11** Lock-and-key model for formation of the enzyme-substrate complex. Once the substrate ($S$) is bound to the enzyme ($E$), one or more products ($P$) may form and separate from the enzyme.

## SAMPLE EXERCISE 28.3

Many bacteria employ *para*-aminobenzoic acid (PABA) in a vital enzyme-catalyzed metabolic reaction. Sulfa drugs such as sulfanilamide (*para*-aminobenzenesulfonamide) kill bacteria. Explain this in terms of the concept of the enzyme-substrate complex.

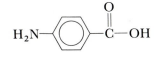

*para*-aminobenzoic acid (PABA)

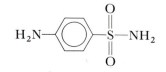

Sulfanilamide
(*para*-aminobenzenesulfonamide)

**Solution:** The structural formulas reveal that PABA and sulfanilamide are very similar in shape and in the locations of polar functional groups. The sulfa drug acts as an inhibitor by binding at the active site in the enzyme and blocking access by

PABA. Without this needed ingredient, the metabolic processes of the bacteria cease.

**PRACTICE EXERCISE**

Diisopropylphosphofluoridate (DFP) is a potent nerve poison. It has been found that DFP acts as an inhibitor of enzymatic reactions by binding to the active site. In one set of experiments it was determined that DFP had reacted with a serine side chain that is normally involved in the active site. From the structure of the product formed in the reaction of DFP with serine and the structure of DFP itself, suggest the nature of the reaction leading to binding of the inhibitor:

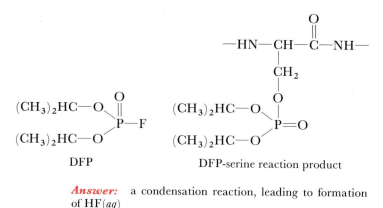

DFP                    DFP-serine reaction product

*Answer:* a condensation reaction, leading to formation of HF(*aq*)

---

## 28.4 CARBOHYDRATES

The carbohydrates are an important class of naturally occurring substances found in both plant and animal matter. The name **carbohydrate** (hydrate of carbon) comes from the empirical formulas for most substances in this class; they can be written as $C_x(H_2O)_y$. For example, **glucose**, the most abundant carbohydrate, has molecular formula $C_6H_{12}O_6$, or $C_6(H_2O)_6$. The carbohydrates are not really hydrates of carbon; rather, they are polyhydroxy aldehydes and ketones. The structural formula for glucose can be written as shown in Figure 28.12.

Notice that four of the carbon atoms in this molecule, numbered 2, 3, 4, and 5, are chiral. Thus there are many configurational isomers of glucose. Several naturally occurring sugars differ from glucose only in the configuration at one of these four chiral carbon atoms. The fact that these sugars have different biological properties is another example of the extreme specificity of biochemical systems. Many sugars are optically active substances; their solutions cause the plane of polarization of polarized light to be rotated as illustrated in Figure 26.19.

When glucose is placed in solution, the aldehyde functional group at one end of the molecule reacts with the hydroxy group at the other end to form a cyclic structure, called a **hemiacetal**:

**FIGURE 28.12** Linear structure of glucose.

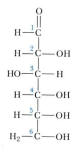

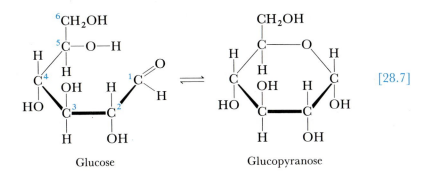

$$[28.7]$$

Glucose ⇌ Glucopyranose

Note that in forming the ring structure in Equation 28.7, the OH group on carbon 1 can be on the same side of the ring as the OH group of carbon 2 (the α form) or on the opposite side (the β form). This seemingly minor difference is very important in distinguishing starch from cellulose, as we shall see shortly.

Six-membered cyclic hemiacetal structures are called **pyranoses**. The name *glucopyranose* indicates that the sugar glucose forms a six-membered ring. Sugars may also form five-membered rings called **furanoses**.

Some sugars are ketones rather than aldehydes. One important example is **fructose**, shown here in its straight-chain (linear) form as well as the five-membered cyclic form called fructofuranose:

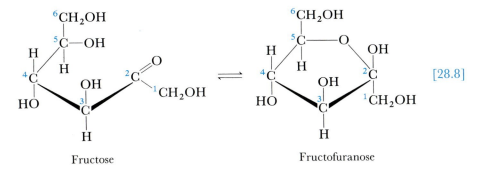

$$[28.8]$$

Fructose ⇌ Fructofuranose

Both glucose and fructose are examples of simple sugars, also called **monosaccharides**. Two or more such units can be joined together to form **polysaccharides**.

---

**SAMPLE EXERCISE 28.4**

Draw the structural formula for fructopyranose.

**Solution:** The name of the substance for which we are to draw the structure tells us that it is a six-membered ring. Thus beginning with fructose (its linear formula is given above), we complete a six-membered ring by closure with the OH group on the terminal carbon atom:

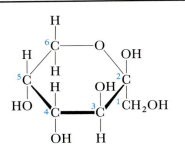

Notice that we might have drawn this structure so that the OH group of carbon atom 2 was on the opposite side of the ring as compared with the OH group of carbon 3. Thus there are two forms of fructopyranose.

## PRACTICE EXERCISE _____

The linear structure of the sugar D-mannose is as follows:

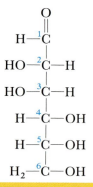

Draw the structural formula for mannopyranose.

*Answer:*

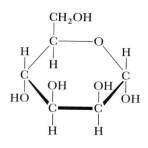

---

### Disaccharides

**Disaccharides** are formed by the joining together of two monosaccharide units. The monosaccharides are linked by a condensation reaction, in which a molecule of water is split out from between two hydroxyl groups, one on each sugar. Because there are several hydroxyl groups, there are several ways in which disaccharides can be formed. The structures of three common disaccharides—**sucrose** (table sugar), **maltose** (malt sugar), and **lactose** (milk sugar)—are shown in Figure 28.13. The word "sugar" makes us think of sweetness; all sugars are sweet, but they differ in the degree of sweetness we perceive when we taste them. Sucrose is about six times sweeter than lactose, about three times sweeter than maltose, slightly sweeter than glucose, but only about half as sweet as fructose. Disaccharides can be hydrolyzed; that is, they can be reacted with water in the presence of an acid catalyst to form the monosaccharides. When sucrose is hydrolyzed, the mixture of glucose and fructose that forms, called *invert sugar*,* is sweeter to the taste than the original sucrose. The sweet syrup present in canned fruits and candies is largely invert sugar formed from hydrolysis of added sucrose.

### Polysaccharides

**Polysaccharides** are made up of several monosaccharide units joined together by a bonding arrangement similar to those shown for disaccharides in Figure 28.13. The most important polysaccharides are starch, glycogen, and cellulose, which are formed from repeating glucose units.

**Starch** is not a pure substance; the term refers to a group of polysaccharides found in plants. Starches serve as a major method of food storage in plant seeds and tubers. Corn, potatoes, wheat, and rice all contain substantial amounts of starch. These plant products serve as major sources

---

* The term "invert sugar" comes from the fact that the rotation of the plane of polarized light by the glucose-fructose mixture is in the opposite direction, or inverted, from that of the sucrose solution.

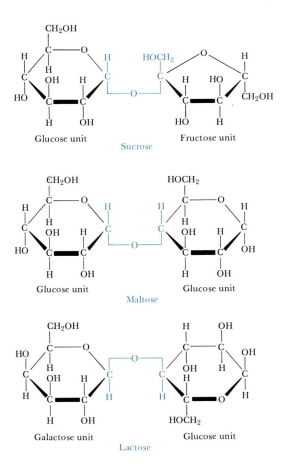

Glucose unit      Fructose unit

Sucrose

Glucose unit      Glucose unit

Maltose

Galactose unit      Glucose unit

Lactose

**FIGURE 28.13** The structures of three disaccharides. Note that in all three cases the two sugar units are joined by a linkage that is termed the α form.

of needed food energy for humans. Enzymes within the digestive system catalyze the hydrolysis of starch to glucose.

Some starch molecules are linear chains, whereas others are branched. All contain the type of chain structure illustrated in Figure 28.14. It is particularly important to note the geometrical arrangement in the joining of one ring to another; the glucose units are in the α form.

**Glycogen** is a starchlike substance synthesized in the body. Glycogen molecules vary in molecular weight from about 5000 to more than 5 million. Glycogen acts as a kind of energy bank in the body. It is concentrated in the muscles and liver. In muscles, it serves as an immediate source of energy; in the liver, it serves as a storage place for glucose and helps to maintain a constant glucose level in the blood.

**Cellulose** forms the major structural unit of plants. Wood is about 50 percent cellulose; cotton fibers are almost entirely cellulose. Cellulose

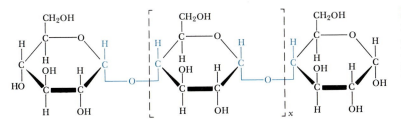

**FIGURE 28.14** Structure of the starch molecule. The molecule consists of many units of the kind enclosed in brackets, joined by linkages of the α form. (You can identify this form by noting the positions of the C—H bonds on the linking carbons in relation to the arrangements of groups on each ring.)

consists of a straight chain of glucose units, with molecular weights averaging more than 500,000. The structure of cellulose is shown in Figure 28.15. At first glance, this structure looks very similar to that of starch. However, the differences in the arrangements of bonds that join the glucose units are important. Notice that glucose in cellulose is in the $\beta$ form. The distinction is made clearer when we examine the structures of starch and cellulose in a more realistic three-dimensional representation, as shown in Figure 28.16. You can see that the individual glucose units have different relationships to one another in the two structures. Enzymes that readily hydrolyze starches do not hydrolyze cellulose. Thus you might chew up and swallow a pound of cellulose and receive no caloric value from it whatsoever, even though the heat of combustion per unit weight is essentially the same for both cellulose and starch. A pound of starch, on the other hand, would represent a substantial caloric intake. The difference is that the starch is hydrolyzed to glucose, which is eventually oxidized with release of energy. Cellulose, however, is not hydrolyzed by any enzymes present in the body, and so it passes through the body unchanged. Many bacteria contain enzymes, called cellulases, that hydro-

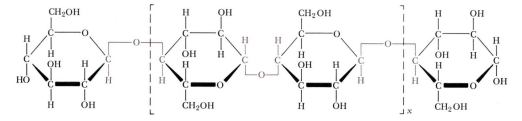

**FIGURE 28.15**   Structure of cellulose. Like starch, cellulose is a polymer. The repeating unit is shown between brackets. The linkage in cellulose is of the $\beta$ form, different from that in starch (see Figure 28.16).

**FIGURE 28.16**   Structures of starch (*a*) and cellulose (*b*). These representations show the geometrical arrangements of bonds about each carbon atom. It is easy to see that the glucose rings are oriented differently with respect to one another in the two structures.

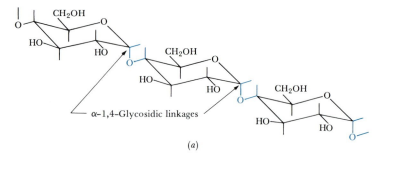

(*a*)

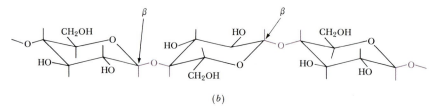

(*b*)

lyze cellulose. These bacteria are present in the digestive systems of grazing animals, such as cattle, that use cellulose for food.

Both plant and animal systems need means of storing energy in various chemical forms so that this energy supply can be tapped as needed later. In plant seeds, the stored energy is used to promote rapid growth after germination. In animals that hibernate during cold periods, the stored energy is needed when other sources of food are absent or scarce. One of the most important classes of compounds used for energy storage are the **fats** and **oils**. These substances are esters of long-chain carboxylic acids with 1,2,3-trihydroxypropane (glycerol) and are called **triglycerides**. Their general structural formula is shown in Figure 28.17. The groups $R_1$, $R_2$, and $R_3$ can be alike or different. The triglycerides are found as fat deposits in animals and as oils in nuts and seeds. Those that are liquid at room temperature are generally referred to as oils; those that are solid are called fats.

Hydrolysis of a fat or oil yields glycerol and the carboxylic acids of the $R_1$, $R_2$, and $R_3$ chains. The acids are commonly called fatty acids. Most commonly, the fatty acids contain from 12 to 22 carbons. It is an interesting fact that nearly all fatty acids contain an even number of carbon atoms, including the carboxylic acid carbon. Several of the more common fatty acids are listed in Table 28.1. The oils, which are obtained mainly from plant products (corn, peanuts, and soybeans), are formed primarily from unsaturated fatty acids. Animal fats (beef, butter, and pork) contain mainly saturated fatty acids.

Some of the oil extracted from cottonseed, corn, or soybean is used as liquid cooking oil. Other plant-derived oils are hydrogenated to convert the carbon–carbon double bonds in the acid chains to carbon–carbon single bonds. In the process, the oils are converted to solids; they are used to produce oleomargarine, peanut butter, vegetable shortening, and similar food products. As an example, trilinolein, a major constituent of

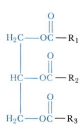

**FIGURE 28.17**  Structure of triglycerides. The groups $R_1$, $R_2$, and $R_3$ are hydrocarbon chains of from 3 to 21 carbons. The chains may be entirely saturated or may contain one or more *cis* olefin groups.

**TABLE 28.1**   Structure, name, and source of several fatty acids

| Formula or structure[a] | Name | Source |
|---|---|---|
| $CH_3(CH_2)_{10}COOH$ | Lauric acid | Coconuts |
| $CH_3(CH_2)_{12}COOH$ | Myristic acid | Butter |
| $CH_3(CH_2)_{16}COOH$ | Stearic acid | Animal fats |
| (structure: oleic acid) | Oleic acid | Corn oil |
| (structure: linoleic acid) | Linoleic acid | Vegetable oils |
| (structure: linolenic acid) | Linolenic acid | Linseed oil |

[a] Notice that the *cis* configuration is adopted at the double bonds in the unsaturated fatty acids.

cottonseed oil, is hydrogenated to form tristearin, which is solid at room temperature:

$$CH_3(CH_2)_4-CH=CH-CH_2-CH=CH-(CH_2)_7\overset{\overset{\displaystyle O}{\|}}{C}OCH_2$$

$$CH_3(CH_2)_4-CH=CH-CH_2-CH=CH-(CH_2)_7\overset{\overset{\displaystyle O}{\|}}{C}OCH \xrightarrow[\text{Ni}]{6H_2} CH_3(CH_2)_{16}\overset{\overset{\displaystyle O}{\|}}{C}OCH \quad [28.9]$$

$$CH_3(CH_2)_4-CH=CH-CH_2-CH=CH-(CH_2)_7\overset{\overset{\displaystyle O}{\|}}{C}OCH_2 \qquad CH_3(CH_2)_{16}\overset{\overset{\displaystyle O}{\|}}{C}OCH_2$$

Trilinolein (liquid)  Tristearin (mp 71°C)

Fats and oils are an important source of energy in our food supply. In the body they undergo hydrolysis to form glycerol and carboxylic acids. The hydrolysis is promoted by enzymes called **lipases:**

$$
\begin{array}{l}
H_2C-O\overset{\overset{\displaystyle O}{\|}}{C}-R \\[2mm]
HC-O\overset{\overset{\displaystyle O}{\|}}{C}-R + 3H_2O \xrightarrow{\text{lipase}} \quad
\begin{array}{l} H_2C-OH \\ HC-OH \\ H_2C-OH \end{array} + 3R-\overset{\overset{\displaystyle O}{\|}}{C}-OH \quad [28.10] \\[2mm]
H_2C-O\overset{\overset{\displaystyle O}{\|}}{C}-R
\end{array}
$$

This hydrolysis reaction, which takes place in aqueous solution, is hampered by the fact that fats and oils are essentially insoluble in water. As a result, not much hydrolysis occurs in the stomach. The gallbladder excretes compounds called bile salts into the small intestine to assist in the hydrolysis. The bile salts break up the larger droplets of fats into an emulsion (a suspension of very small droplets), so that hydrolysis can proceed more rapidly.

## 28.6 NUCLEIC ACIDS

The **nucleic acids** are a class of biopolymers present in nearly all cells. They are classified into two groups—**deoxyribonucleic acids (DNA)** and **ribonucleic acids (RNA)**. DNA molecules are very large; their molecular weights may range from 6 to 16 million. RNA molecules are much smaller, with molecular weights in the range 20,000 to 40,000. DNA is found primarily in the nucleus of the cell; RNA is found outside the nucleus, in the surrounding fluid called the cytoplasm.

Both DNA and RNA are polymers of a basic repeating unit called a **nucleotide**. To understand the structure of the polymer, we must first examine the structure of the nucleotide units. The nucleotide consists of three parts: (1) a phosphoric acid molecule, (2) a sugar in the furanose (five-membered ring) form, and (3) a nitrogen-containing organic base with a ring structure analogous to the ring structures of aromatic molecules. The sugar molecule found in RNA is ribose, shown in Figure 28.18. In DNA the sugar is deoxyribose, which differs from ribose only in the substitution of hydrogen for an —OH group on one carbon, as illustrated in Figure 28.18. In DNA the organic base may be any one of the four shown in Figure 28.19. An example of a complete nucleotide (deoxyadenylic acid) formed from these three components is illustrated in Figure 28.20. Substitution of one of the other bases for adenine in this structure produces one of the other three nucleotides that make up DNA.

Just as a polypeptide is formed by a condensation reaction between amino acids, creating a peptide bond, a polynucleotide is formed by a condensation reaction creating a phosphate ester bond. The condensation reaction results in elimination of water from between an OH group of the phosphoric acid and an OH group of the sugar. The formation of the polymeric chain is shown schematically in Figure 28.21. The organic bases in this illustration are simply represented as B; they might be the same or different bases.

DNA molecules consist of two linear strands of the sort shown in Figure 28.21, which are wound together in the form of a **double helix**. It is very confusing to look at a model of double-stranded DNA in which all the atoms are represented. However, we can see from a more schematic illustration, Figure 28.22, how the DNA molecule is constructed. Remember that the polymeric strand itself consists of alternating sugar and phosphate groups, represented as —S— and —P—, respectively. The various bases are attached to the polymeric strand at each sugar unit. The key to the double-helix structure for DNA is the formation of hydro-

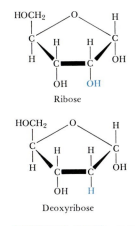

**FIGURE 28.18**  Ribose and deoxyribose, the two sugars found in RNA and DNA, respectively.

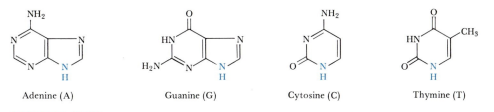

Adenine (A)  Guanine (G)  Cytosine (C)  Thymine (T)

**FIGURE 28.19**  The four organic bases present in DNA. In DNA each base is attached to a ribose molecule through, shown in color, a bond from the nitrogen.

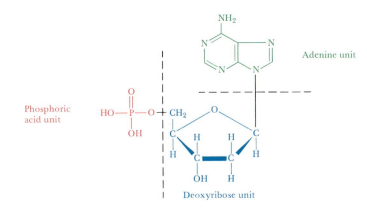

Phosphoric acid unit

Adenine unit

Deoxyribose unit

**FIGURE 28.20**  Structure of deoxyadenylic acid, a nucleotide formed from phosphoric acid, deoxyribose, and an organic base, adenine.

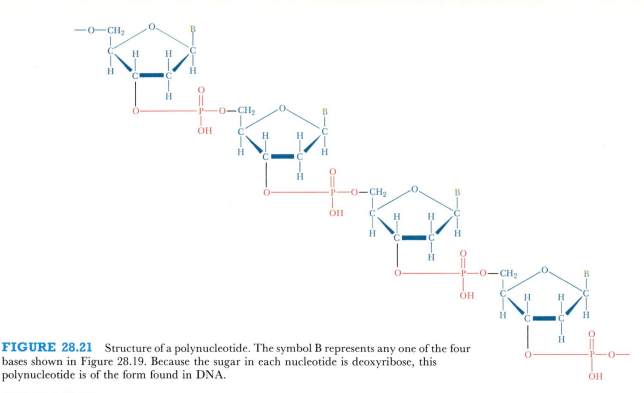

**FIGURE 28.21** Structure of a polynucleotide. The symbol B represents any one of the four bases shown in Figure 28.19. Because the sugar in each nucleotide is deoxyribose, this polynucleotide is of the form found in DNA.

**FIGURE 28.22** Schematic illustration of the double-stranded helical structure for DNA. The hydrogen-bond interactions between complementary base pairs is represented as illustrated in Figure 28.23.

gen bonds between bases on the two chains. When adenine and thymine are located opposite one another along the chain, they form a strong hydrogen bond, as shown in Figure 28.23. Similarly, guanine and cytosine form especially strong hydrogen bonds with one another. This means that the two strands of DNA are complementary; opposite each adenine on one chain there is always a thymine; opposite each guanine there is a cytosine. The double-helix structure for DNA (Figure 28.24), first proposed by James Watson and Francis Crick in 1953, has proved to be the key to understanding protein synthesis in cells, the means by which viral

**FIGURE 28.23** Hydrogen bonding between complementary base pairs. The hydrogen bonds shown here are responsible for formation of the double-stranded helical structure for DNA, as shown in Figure 28.22.

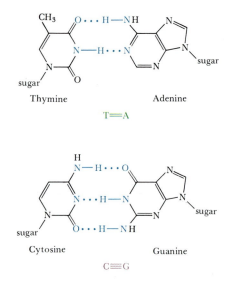

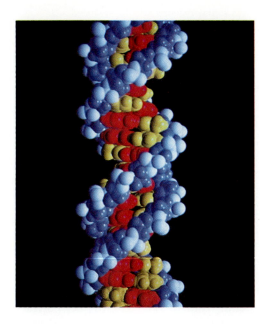

**FIGURE 28.24** A computer-generated model of a DNA double helix. The blue and white atoms represent the sugar-phosphate chains that wrap around the outside. Inside the chains are the bases, shown in red and yellow, formed by hydrogen-bonded pairs. The bases are stacked parallel to each other and perpendicular to the axis of the cylinder. (Richard Feldman/National Institutes of Health)

particles infect cells, the means by which genetic information is transmitted in reproduction of cells, and many other problems of central importance in molecular biology. Those themes are beyond the scope of this book; however, if you are interested in study in the area of life sciences, you will learn a good deal about such matters in other courses.

# *FOR REVIEW*

## SUMMARY

In this chapter, we have taken a brief look at some of the major constituents of the biosphere—that part of the physical world in which organisms live out their life cycles. The maintenance of life requires a source of energy as well as appropriate environmental conditions. The ultimate source of the needed energy is the sun. Plants convert solar energy to chemical energy in the process called **photosynthesis**. Solar energy is absorbed by a plant pigment called chlorophyll and then utilized to form carbohydrate and $O_2$ from $CO_2$ and $H_2O$.

**Proteins** are polymers of **amino acids**. They are the major structural materials in animal systems. **Enzymes**, the catalysts of biochemical reactions, are protein in nature. All naturally occurring proteins are formed from some 20 amino acids. The amino acids are **chiral** substances; that is, they are capable of existing as nonsuperimposable mirror-image isomers called **enantiomers**. Usually, only one of the enantiomers is found to be biologically active. Protein structure is determined by the sequence of amino acids in the chain, the coiling or stretching of the chain, and the overall shape of the complete molecule. All these aspects of protein structure are important in determining its biological activity. Heating or other forms of treatment may inactivate, or **denature**, the protein.

In this chapter, we also have considered the action of enzymes—their specificity, a model for how they may operate, and the ways in which they are sometimes inhibited.

**Carbohydrates**, which are formed from polyhydroxy aldehydes and ketones, are the major structural constituent of plants and a source of energy in both plants and animals. The three most important groups of carbohydrates are **starch**, which is found in plants, **glycogen**, which is found in mammals, and **cellulose**, which is also found in plants. All are **polysaccharides**; that is, they are polymers of a simple sugar, glucose.

**Fats** and **oils**, along with proteins and carbohydrates, form the important sources of energy in our food supply. The fats and oils are esters of long-chain acids with glycerol, 1,2,3-trihydroxypropane; they are often called **triglycerides**. The major sources of these substances are animal fats and the oils of plant seeds, such as corn, peanuts, cottonseeds, and soybeans.

The **nucleic acids** are biopolymers of high molecular weight that carry the genetic information necessary for cell reproduction. In addition, the nucleic acids control cell development through control of protein synthesis. The nucleic acids consist of a polymeric backbone of alternating phosphate and ribose sugar groups, with organic bases attached to the sugar molecules. The DNA polymer is a double-stranded helix that is held together by hydrogen bonding between matching organic bases situated across from one another on the two strands.

## LEARNING GOALS

Having read and studied this chapter, you should be able to:

1. Describe two distinct reasons why energy is required by living organisms.

2. List the several functions of proteins in living systems.

3. Define the terms "chiral" and "enantiomer," and draw the enantiomer of a given chiral molecule.

4. Write the reaction for formation of a peptide bond between two amino acids.

5. Explain the structures of proteins in terms of primary, secondary, and tertiary structure.

6. Explain the following characteristics of enzymes in terms of what is known about proteins: specificity; temperature dependence of enzyme activity, as shown in Figure 28.8; inhibition.

7. Explain the lock-and-key mechanism for enzyme action.

8. Distinguish the cyclic hemiacetal and linear forms of a sugar, and describe the difference between the pyranose and furanose forms of sugar.

9. Describe the manner in which monosaccharides are joined together to form polysaccharides.

10. Enumerate the major groups of polysaccharides, and indicate their sources and general functions.

11. Describe the structures of fats and oils, and list the sources of these substances.

12. Draw the structures of any of the nucleotides that make up the polynucleotide DNA.

13. Describe the nature of the polymeric unit of polynucleotides.

14. Describe the double-stranded structure of DNA, and explain the principle that determines the relationship between bases in the two strands.

## KEY TERMS

Among the more important terms and expressions used for the first time, or in a new context, in this chapter are the following:

The **active site** (Section 28.3) is the specific site in an enzyme at which the enzyme-catalyzed reaction occurs.

An **amino acid** (Section 28.2) is a carboxylic acid that contains an amino ($-NH_2$) group attached to the carbon atom adjacent to the carboxylic acid ($CO_2H$) functional group. Twenty different amino acids are important in living systems.

A **biopolymer** (Section 28.1) is a polymeric molecule of high molecular weight found in living systems. The three major classes of biopolymer are proteins, carbohydrates, and nucleic acids.

The **biosphere** (introduction) is that part of the earth in which living organisms can exist and live out their life cycles.

**Carbohydrates** (Section 28.4) are a class of substances formed from polyhydroxy aldehydes or ketones.

**Cellulose** (Section 28.4) is a polysaccharide of glucose; it is the major structural element in plant matter.

A **chiral** molecule (Section 28.2) is one that is not superimposable on its mirror image.

**Chlorophyll** (Section 28.1), found in plant leaves, is a plant pigment that plays a major role in conversion of solar energy to chemical energy in photosynthesis.

A **coenzyme**, or **cofactor** (Section 28.3), is a substance that is needed along with some enzyme if an enzyme-catalyzed reaction is to occur.

**Configurational isomers** (Section 28.2) are molecules that differ only by being nonsuperimposable mirror images of one another. Such isomers are often called **optical isomers** because their solutions affect the plane of polarized light differently.

**Denaturation** (Section 28.2) is the loss of biological activity in a protein because of disruption of its secondary and tertiary structure by heating, by the action of acids or bases, or by other influences.

**Deoxyribonucleic acid, DNA** (Section 28.6), is a polynucleotide in which the sugar component is deoxyribose (in the furanose-ring form).

The **double-helix** structure for DNA (Section 28.6) involves the winding of two DNA polynucleotide chains together in a helical arrangement. The two strands of the double helix are complementary, in that the organic bases on the two strands are paired for optimal hydrogen-bond interaction.

**Enantiomers** (Section 28.2) are two mirror-image molecules of a chiral substance. The enantiomers are nonsuperimposable.

**Enzymes** (Section 28.3) are proteins that act as catalysts in biochemical reactions.

**Fats** and **oils** (Section 28.5) are esters of long-chain carboxylic acids and the alcohol 1,2,3-trihydroxypropane (glycerol). These esters are also called **triglycerides**.

A **furanose** (Section 28.4) is a cyclic sugar molecule in the form of a five-membered ring.

**Glucose** (Section 28.4), a polyhydroxy aldehyde of formula $CH_2OH(CHOH)_4CHO$, is the most important of the monosaccharides.

**Glycogen** (Section 28.4) is the general name given to a group of polysaccharides of glucose that are synthesized in mammals and used as a means of carbohydrate energy storage.

In the **α-helix** structure (Section 28.2) for a protein, the protein is coiled in the form of a helix, with hydrogen bonds between $C{=}O$ and $N{-}H$ groups on adjacent turns.

A **hemiacetal** (Section 28.4) is a bonding arrangement formed by reaction of an aldehyde functional group with an alcohol. The hemiacetal linkage is formed when aldehyde sugars form the cyclic structure.

**Lipases** (Section 28.5) are enzymes that catalyze the hydrolysis of fats.

In the **lock-and-key model** for enzyme action (Section 28.3), the substrate molecule is pictured as fitting rather specifically into the active site on the enzyme. It is assumed that in being bound to the active site the substrate is somehow activated for reaction.

A **monosaccharide** (Section 28.4) is a simple sugar, most commonly containing six carbon atoms. The joining together of monosaccharide units by a condensation reaction results in formation of polysaccharides.

**Nucleic acids** (Section 28.6) are high-molecular-weight polymers of nucleotides.

A **nucleotide** (Section 28.6) is formed from a molecule of phosphoric acid, a sugar molecule, and an organic base. Nucleotides form linear polymers called DNA and RNA, which are involved in protein synthesis and cell reproduction.

**Photosynthesis** (Section 28.1) is the process occurring in plant leaves by which light energy is used to convert $CO_2$ and water to carbohydrates and oxygen.

**Polarized light** (Section 28.2) is electromagnetic radiation in which the waves that form the light move in a single plane.

The **primary structure** of a protein (Section 28.2) refers to the sequence of amino acids along the protein chain.

**Proteins** (Section 28.2) are biopolymers formed from amino acids.

A **pyranose** (Section 28.4) is a cyclic sugar molecule in the form of a six-membered ring.

**Ribonucleic acid, RNA** (Section 28.6), is a polynucleotide in which ribose (in the furanose-ring form) is the sugar component.

The **secondary structure** of a protein (Section 28.2) refers to the manner in which the protein is coiled or stretched.

**Starch** (Section 28.4) is the general name given to a group of polysaccharides that act as energy-storage substances in plants.

The **tertiary structure** of a protein (Section 28.2) refers to the overall shape of the molecule—specifically, the manner in which sections of the chain fold back upon themselves or intertwine.

**Turnover number** (Section 28.3) refers to the number of individual reaction events per unit time at a particular active site in an enzyme.

# EXERCISES

## Energy Requirements

**28.1** Explain the relationship between entropy and the energy needs of organisms.

**28.2** Name at least four processes occurring in an organism (such as yourself) that require energy. These may occur at the molecular level or may involve the complete organism.

**28.3** $\Delta G^\circ_{298}$ for oxidation of glucose in solution is $-2878$ kJ:

$$C_6H_{12}O_6(aq) + 6O_2(g) \longrightarrow$$
$$6CO_2(g) + 6H_2O(l) \qquad \Delta G^\circ_{298} = -2878 \text{ kJ}$$

**(a)** What is $\Delta G^\circ_{298}$ for photosynthesis? Is this reaction spontaneous under standard conditions? **(b)** What is $\Delta G^\circ_{298}$ for photosynthesis if $P_{O_2} = 0.02$ atm, $P_{CO_2} = 3.1 \times 10^{-4}$ atm, and $C_6H_{12}O_6 = 1.0 \times 10^{-3}$ $M$? (Assume that $H_2O$ can be omitted from the reaction quotient; refer to Section 19.6, if necessary.)

**28.4** Assume that $1 \times 10^{14}$ kg of carbon is photosynthetically fixed as glucose each year on the earth's surface. If the total solar energy falling on the earth's surface is $4 \times 10^{21}$ kJ/yr, and if 2878 kJ are required to produce a mole of glucose, calculate the percentage of the solar energy that is absorbed annually by photosynthetic processes.

**28.5** A tree might convert about 50 g of $CO_2$ per day into carbohydrate at its greatest rate of photosynthesis. How many grams of oxygen does the tree produce in 1 day at this rate? How many liters of $O_2$ (at STP) does this correspond to?

**28.6** Assuming that half the 48 mol of photons required in Equation 28.1 are of 430-nm wavelength and that half are of 650-nm wavelength, calculate the energy absorption represented by these 48 photons, on a molar basis. The free-energy change for Equation 28.1 is $+2878$ kJ. What is the approximate percentage efficiency of the plant system in converting the incoming radiant energy into glucose, assuming that all incoming photons are employed in the photosynthetic reaction? (See Sample Exercise 6.2.)

## Proteins

**28.7** **(a)** What is an $\alpha$-amino acid? **(b)** How do amino acids react to form proteins?

**28.8** How do the side chains (R groups) of amino acids affect their behavior?

**28.9** Draw the dipeptides formed by condensation reaction between glycine and valine.

**28.10** Write a chemical equation for the formation of aspartylcysteine from the constituent amino acids.

**28.11** Draw the structure of the tripeptide trp-gly-ser.

**28.12** What amino acids would be obtained by hydrolysis of the following tripeptide?

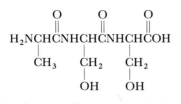

**28.13** In what form would you expect glycine to exist in basic solution? In acidic solution?

**28.14** Draw the structural formula for glutamic acid in **(a)** acidic solution; **(b)** basic solution.

**28.15** How many different tripeptides could one make from the amino acids glycine, valine, and alanine?

**28.16** Name the distinctly different tripeptides that could be formed from the two amino acids serine and proline.

**28.17** Draw the two enantiomeric forms of aspartic acid.

**28.18** Draw the two enantiomeric forms of cysteine.

**28.19** Describe the primary, secondary, and tertiary structures of proteins.

**28.20** Describe the role of hydrogen bonding in determining the $\alpha$-helix structure of a protein.

## Enzymes

**28.21** Provide a definition of each of the following terms: **(a)** enzyme; **(b)** denaturation; **(c)** holoenzyme; **(d)** peptidase.

**28.22** Provide a definition of each of the following terms: **(a)** apoenzyme; **(b)** specificity; **(c)** lipase; **(d)** turnover number.

**28.23** Normally, the rates of enzyme-catalyzed reactions increase linearly with increase in substrate concentration. What does this tell us about the position of equilibrium in Equation 28.4? Explain.

**28.24** In terms of the lock-and-key model, what characteristics must an enzyme inhibitor possess?

**28.25** One of the many remarkable enzymes in the human body is carbonic anhydrase, which catalyzes the release of dissolved carbon dioxide from the blood into the air of our lungs. If it were not for this enzyme, the body could not rid itself rapidly enough of the $CO_2$ accumulated by cell metabolism. The enzyme has a molecular weight of 30,000, contains one atom of zinc per protein molecule, and catalyzes the dehydration (the release to air) of up to $10^7$ $CO_2$ molecules per second. Which components of this description correspond to the terms "holoenzyme", "apoenzyme", "cofactor", and "turnover number"?

**28.26** Enzymes generally lose their effectiveness as biological catalysts when heated above a characteristic temperature. Explain the origin of this effect.

## Carbohydrates

**28.27** What is the difference between $\alpha$-glucose and $\beta$-glucose? Show the condensation of two glucose molecules to form a disaccharide with $\alpha$-linkages; with $\beta$-linkages.

**28.28** Identify each of the following as an $\alpha$ or a $\beta$ form of a hemiacetal:

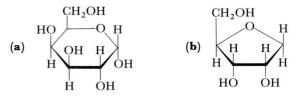

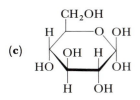

**(c)**

**28.29** The structural formula for the linear form of galactose is shown here. Draw the structure of the pyranose form of this sugar.

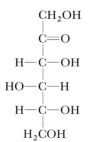

**28.30** The structural formula for the linear form of sorbose is shown here. Draw the structure of the furanose form of this sugar.

$$
\begin{array}{c}
CH_2OH \\
| \\
C{=}O \\
| \\
H{-}C{-}OH \\
| \\
HO{-}C{-}H \\
| \\
H{-}C{-}OH \\
| \\
H_2COH
\end{array}
$$

**28.31** Which carbon atoms in galactose (see Exercise 28.29) are chiral?

**28.32** Which carbon atoms in sorbose (see Exercise 28.30) are chiral?

## Fats and Oils

**28.33** Draw the structural formula for the triglyceride of oleic acid.

**28.34** Write a balanced chemical equation for complete hydrogenation of trilinolein. Assume partial hydrogenation occurs, with uptake of only 4 mol of $H_2$ per mole of fat. Draw the structural formulas of two possible products.

**28.35** The reaction between glycerol and a fatty acid is called esterification. Write the chemical equation for the esterification of a mole of glycerol with 3 mol of lauric acid.

**28.36** The hydrolysis of fats and oils is catalyzed by bases. Many household cleaners (such as aqueous ammonia so-lutions and Drāno, which contains NaOH) are basic. Write the chemical equation for the hydrolysis of the triglyceride of myristic acid.

## Nucleic Acids

**28.37** Describe a nucleotide. Draw the structural formula for deoxycytidine monophosphate in which cytosine is the organic base.

**28.38** A nucleoside consists of an organic base of the kind shown in Figure 28.18, bound to ribose or deoxyribose. Draw the structure for thymidine, formed from thymine and ribose.

**28.39** Write a balanced chemical equation for the condensation reaction between a mole of deoxyribose and a mole of phosphoric acid.

**28.40** An unknown substance undergoes hydrolysis under neutral conditions to yield 1 mol of phosphoric acid and an organic product. The same starting material undergoes hydrolysis under acidic conditions to yield guanine and ribose-monophosphate. Draw the structure of the unknown substance.

**28.41** Imagine a single DNA strand containing a section with the following base sequence: A, C, T, C, G, A. What is the base sequence of the complementary strand?

**28.42** When samples of double-stranded DNA are analyzed, the quantity of adenine present equals that of thymine. Similarly, the quantity of guanine equals that of cytosine. Explain the significance of these observations.

## Additional Exercises

**28.43** In a temperate climate 1.0 $m^2$ of leaf area absorbs about $2.0 \times 10^4$ kJ of energy per day. About 1.2 percent of this energy is used in photosynthesis. **(a)** Calculate the leaf area required to convert 10,000 kJ per day into plant matter. This is the approximate energy requirement of a person doing an average quantity of work per day (Section 4.8). **(b)** In terms of what you know about the composition of plants, explain why more area than this would be required to provide a 10,000-kJ daily diet for a person.

**28.44** In each of the following substances, locate the chiral carbon atoms, if any:

$$
\begin{array}{c}
\qquad\qquad O \\
\qquad\qquad \| \\
\textbf{(a)} \quad HOCH_2CH_2CCH_2OH \\
\\
\qquad\qquad OH \\
\qquad\qquad | \\
\textbf{(b)} \quad HOCH_2CHCCH_2OH \\
\qquad\qquad\qquad \| \\
\qquad\qquad\qquad O \\
\\
\qquad\quad O \quad CH_3 \\
\qquad\quad \| \quad\;\; | \\
\textbf{(c)} \quad HOCCHCHC_2H_5 \\
\qquad\qquad | \\
\qquad\qquad NH_2
\end{array}
$$

**28.45** Predict the products of the hydrolysis of each of the following compounds:

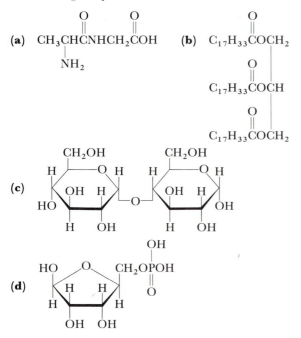

**(a)**

$$CH_3\overset{\displaystyle O}{\overset{\displaystyle \|}{C}}HC\overset{\displaystyle O}{\overset{\displaystyle \|}{N}}HCH_2\overset{\displaystyle O}{\overset{\displaystyle \|}{C}}OH$$
$$\underset{\displaystyle NH_2}{|}$$

**(b)** $C_{17}H_{33}\overset{\displaystyle O}{\overset{\displaystyle \|}{C}}OCH_2$

**28.46** Draw the condensed structural formula of each of the following tripeptides: **(a)** val-gly-thr; **(b)** pro-ser-ala.

**28.47** Glutathione is a tripeptide found in most living cells. Partial hydrolysis yields cys-gly and glu-cys. What structures are possible for glutathione?

**28.48** Phenylketonuria (PKU) is a disease caused by the lack in some individuals of the enzyme phenylalanine hydroxylase. This enzyme catalyzes the conversion of phenylalanine to tyrosine (both amino acids). The disease can lead to severe mental retardation. **(a)** Write the condensed structural formulas for phenylalanine and tyrosine. **(b)** Suggest the origin for the name given to the enzyme.

**28.49** The names of most enzymes end in *-ase*. The *-ase* ending is attached to the name of the substrate on which the enzyme acts, or to the type of reaction it catalyzes. Match the following enzyme names and reactions:

1. esterase
2. decarboxylase
3. urease
4. transmethylase
5. peptidase

**(a)** removal of carboxyl groups from compounds
**(b)** hydrolysis of peptide linkages
**(c)** transfers a methyl group
**(d)** formation of ester linkages
**(e)** hydrolysis of urea

**28.50** **(a)** Describe in qualitative terms how an enzyme works. **(b)** What is meant by "enzyme substrate"? By "enzyme inhibition"?

**28.51** The popular flavor enhancer MSG (monosodium glutamate) is the monosodium salt of glutamic acid. **(a)** What is its condensed structural formula? **(b)** Only the L-isomer is effective. Is this surprising?

**28.52** The standard free energy of formation of glycine($s$) is $-369$ kJ/mol, whereas that of glycylglycine($s$) is $-488$ kJ/mol. What is $\Delta G°$ for condensation of glycine to form glyclyglycine?

**28.53** Give the chain structural formulas for ribose and deoxyribose (see Figure 28.18).

**28.54** The enzyme invertase catalyzes the conversion of sucrose, a disaccharide, to invert sugar. When the concentration of invertase is $3 \times 10^{-7}$ $M$ and the concentration of sucrose is 0.01 $M$, invert sugar is formed at the rate of $2 \times 10^{-4}$ $M$/s. When the sucrose concentration is doubled, the rate of formation of invert sugar is doubled also. Assuming that the enzyme-substrate model is operative, is the fraction of enzyme tied up as complex large or small? Explain. Addition of innositol, another sugar, causes a decrease in rate of formation of invert sugar. Suggest a mechanism by which this occurs.

**28.55** Give a specific example of each of the following: **(a)** a disaccharide; **(b)** a sugar present in nucleic acids; **(c)** a sugar present in human blood serum; **(d)** a polysaccharide.

**28.56** The standard free energy of formation for aqueous solutions of glucose is $-917.2$ kJ/mol, whereas that of glycogen is $-662.3$ kJ/mol of glucose units. Derive a general expression for $\Delta G°$ for the formation of a glycogen molecule that contains $n$ units of glucose.

**28.57** Define, in your own words, the terms "condensation" and "hydrolysis". Why are such reactions so important in biochemistry?

**28.58** Write a complementary nucleic acid strand for the following strand, using the concept of complementary base pairing: TATGCA.

**28.59** The monoanion of adenosine monophosphate (AMP) is an intermediate in phosphate metabolism:

$$A-O-\overset{\displaystyle O^-}{\underset{\displaystyle O}{\overset{\displaystyle |}{\underset{\displaystyle \|}{P}}}}-OH = AMP-OH^-$$

where A = adenosine. If the $pK_a$ for this anion is 7.21, what is the ratio of $[AMP-OH^-]$ to $[AMP-O^{2-}]$ in blood at pH 7.40?

# Mathematical Operations

The numbers used in chemistry are often either extremely large or extremely small. Such numbers are conveniently expressed in the form

**A.1 Exponential Notation**

$$N \times 10^n$$

where $N$ is a number between 1 and 10, and $n$ is the exponent. Some examples of this exponential notation, which is also called scientific notation, follow:

1,200,000 is $1.2 \times 10^6$ (read "one point two times ten to the sixth power")

0.000604 is $6.04 \times 10^{-4}$ (read "six point oh four times ten to the negative fourth power")

A positive exponent, as in our first example, tells us how many times a number must be multiplied by 10 to give the long form of the number:

$$1.2 \times 10^6 = 1.2 \times 10 \times 10 \times 10 \times 10 \times 10 \times 10 \quad \text{(six tens)}$$
$$= 1,200,000$$

It is also convenient to think of the positive exponent as the number of places the decimal point must be moved to the *left* to give a number greater than 1 and less than 10: If we begin with 3450 and move the decimal point three places to left, we end up with $3.45 \times 10^3$.

In a related fashion, a negative exponent tells us how many times we must divide a number by 10 to give the long form of the number:

$$6.04 \times 10^{-4} = \frac{6.04}{10 \times 10 \times 10 \times 10}$$
$$= 0.000604$$

It is convenient to think of the negative exponent as the number of places the decimal point must be moved to the *right* to give a number greater than 1 but less than 10: If we begin with 0.0048 and move the decimal point three places to right, we end up with $4.8 \times 10^{-3}$.

In the system of exponential notation, with each shift of the decimal point one place to the right, the exponent *decreases* by 1:

$$4.8 \times 10^{-3} = 48 \times 10^{-4}$$

Similarly, with each shift of the decimal point one place to the left, the exponent *increases* by 1:

$$4.8 \times 10^{-3} = 0.48 \times 10^{-2}$$

Most scientific calculators have a key labeled **EXP** or **EE**, which is used to enter numbers in exponential notation. To enter the number $5.8 \times 10^3$, the key sequence is

On some calculators the display will show 5.8, then a space, followed by 03, the exponent. On other calculators, a small 10 is shown with an exponent 3.

To enter a negative exponent, use the key labeled $+/-$. For example, to enter the number $8.6 \times 10^{-5}$, the key sequence is

$$\boxed{8} \; \boxed{\cdot} \; \boxed{6} \; \boxed{\text{EXP}} \; \boxed{+/-} \; \boxed{5}$$

*When entering a number in exponential notation, do not key in the 10.*

In working with exponents, it is important to know that $10^0 = 1$. The following rules are useful for carrying exponents through calculations.

1. **Addition and Subtraction** In order to add or subtract numbers expressed in exponential notation, the powers of 10 must be the same:

$$(5.22 \times 10^4) + (3.21 \times 10^2) = (522 \times 10^2) + (3.21 \times 10^2)$$
$$= 525 \times 10^2 \text{ (3 significant figures)}$$
$$= 5.25 \times 10^4$$

$$(6.25 \times 10^{-2}) - (5.77 \times 10^{-3}) = (6.25 \times 10^{-2}) - (0.577 \times 10^{-2})$$
$$= 5.67 \times 10^{-2} \text{ (3 significant figures)}$$

When you use a calculator to add or subtract, you need not be concerned with having numbers with the same exponents, because the calculator automatically takes care of this matter.

2. **Multiplication and Division** When numbers expressed in exponential notation are multiplied, the exponents are added; when numbers expressed in exponential notation are divided, the exponent of the divisor is subtracted from the exponent of the dividend:

$$(5.4 \times 10^2)(2.1 \times 10^3) = (5.4)(2.1) \times 10^{2+3}$$
$$= 11 \times 10^5$$
$$= 1.1 \times 10^6$$

$$(1.2 \times 10^5)(3.22 \times 10^{-3}) = (1.2)(3.22) \times 10^{5-3}$$
$$= 3.9 \times 10^2$$

$$\frac{3.2 \times 10^5}{6.5 \times 10^2} = \frac{3.2}{6.5} \times 10^{5-2} = 0.49 \times 10^3$$
$$= 4.9 \times 10^2$$

$$\frac{5.7 \times 10^7}{8.5 \times 10^{-2}} = \frac{5.7}{8.5} \times 10^{7-(-2)} = 0.67 \times 10^9$$

$$= 6.7 \times 10^8$$

3. **Powers and Roots** When numbers expressed in exponential notation are raised to a power, the exponents are multiplied by the power; when the roots of numbers expressed in exponential notation are taken, the exponents are divided by the root:

$$(1.2 \times 10^5)^3 = 1.2^3 \times 10^{5 \times 3}$$
$$= 1.7 \times 10^{15}$$

$$\sqrt[3]{2.5 \times 10^6} = \sqrt[3]{2.5} \times 10^{6/3}$$
$$= 1.3 \times 10^2$$

Scientific calculators usually have keys labeled $x^2$ and $\sqrt{x}$ for squaring and taking the square root of a number, respectively. To take higher powers or roots, many calculators have $y^x$ and $\sqrt[x]{y}$ (or INV $y^x$) keys. For example, to perform the operation $\sqrt[3]{7.5 \times 10^{-4}}$ on such a calculator, you would key in $7.5 \times 10^{-4}$, press the $\sqrt[x]{y}$ key (or the INV and then the $y^x$ keys), enter the root, 3, and finally press =. The result is $9.1 \times 10^{-2}$.

---

### SAMPLE EXERCISE 1

Perform each of the following operations, using your calculator where possible: **(a)** write the number 0.0054 in standard exponential notation;
**(b)** $(5.0 \times 10^{-2}) + (4.7 \times 10^{-3})$;
**(c)** $(5.98 \times 10^{12})(2.77 \times 10^{-5})$;
**(d)** $\sqrt[4]{1.75 \times 10^{-12}}$.

**Solution:** **(a)** Because we move the decimal three places to the right to convert 0.0054 to 5.4, the exponent is $-3$:

$$5.4 \times 10^{-3}$$

Scientific calculators are generally able to convert numbers to exponential notation using one or two key strokes. Consult your instruction manual for how this operation is accomplished on your calculator.

**(b)** To add these numbers longhand, we must convert them to the same exponent:

$$(5.0 \times 10^{-2}) + (0.47 \times 10^{-2})$$
$$= (5.0 + 0.47) \times 10^{-2}$$
$$= 5.5 \times 10^{-2}$$

(Note that the result has only two significant figures.) To perform this operation on a calculator, we enter the first number, strike the + key, then enter the second number, and strike the = key.

**(c)** Performing this operation longhand, we have

$$(5.98 \times 2.77) \times 10^{12-5} = 16.6 \times 10^7 = 1.66 \times 10^8$$

On a scientific calculator, we enter $5.98 \times 10^{12}$, press the × key, enter $2.77 \times 10^{-3}$, and press the = key.

**(d)** To perform this operation on a calculator, we enter the number, press the $\sqrt[x]{y}$ key (or the INV and $y^x$ keys), enter 4, and press the = key. The result is $1.15 \times 10^{-3}$.

### PRACTICE EXERCISE

Perform the following operations: **(a)** write 67,000 in exponential notation, showing two significant figures; **(b)** $(3.378 \times 10^{-3}) - (4.97 \times 10^{-5})$; **(c)** $(1.84 \times 10^{15})/(7.45 \times 10^{-2})$; **(d)** $(6.67 \times 10^{-8})^3$.
*Answers:* **(a)** $6.7 \times 10^4$; **(b)** $3.328 \times 10^{-3}$; **(c)** $2.47 \times 10^{16}$; **(d)** $2.97 \times 10^{-22}$

---

## A.2 Logarithms

The common, or base 10, logarithm (abbreviated log) of any number is the power to which 10 must be raised to equal the number. For example, the common logarithm of 1000 (written log 1000) is 3, because raising 10 to the third power gives 1000: $10^3 = 1000$. Further examples are:

$$\log 10^5 = 5$$
$$\log 1 = 0$$
$$\log 10^{-2} = -2$$

In these examples, the logarithm can be obtained by simple inspection However, it is not possible to obtain the logarithm of a number like 3.25 by inspection. Many electronic calculators have a log key that can be used to obtain logs.

**Using Log Tables**   If you do not have a calculator with a log key, you can use a log table, such as that given in Appendix B. To find the logs of numbers from 1 to 10, we locate the first two digits in the first vertical column. (You should mentally insert a decimal point between the two-digit numbers in the first column.) We then move horizontally to the column headed by the third digit of the number. In this way we find that the log of 3.25 is listed as 5119. Because decimals are omitted from the table, the log of 3.25 is 0.5119. Further examples are:

$$\log 2.50 = 0.3979$$
$$\log 8.65 = 0.9370$$

To use a log table to obtain the logarithm of a number that is less than 1 or greater than 10, we must first write the number in standard exponential notation, as in the following examples:

$$\log 450 = \log (4.50 \times 10^2)$$
$$= \log 4.50 + \log 10^2$$
$$= 0.653 + 2 = 2.653$$
$$\log 0.0673 = \log (6.73 \times 10^{-2})$$
$$= \log 6.73 + \log 10^{-2}$$
$$= 0.828 - 2$$
$$= -1.172$$

Check these examples yourself, using either Appendix B or your calculator.

**Significant Figures and Logarithms**   The number of digits after the decimal point for the common logarithm of a measured quantity equals the number of significant figures in the original number. For example, if 23.5 is a measured quantity (three significant figures), then $\log 23.5 = 1.371$ (that is, there are three significant figures after the decimal point). If this rule had been followed in reporting $\log 2.50$ and $\log 8.65$ in the text above, the results would have been rounded off to 0.398 and 0.937, respectively.

**Obtaining Antilogarithms**   The process of finding a number given its logarithm is known as obtaining an antilogarithm. It is the reverse of taking a logarithm. Many electronic calculators employ a key labeled $\log^{-1}$ or INV log to obtain antilogs. Others employ the $10^x$ or $y^x$ key $(10^{\log x} = x)$. If you do not have a calculator that performs these operations, you can use a log table to obtain antilogs as the following examples illustrate:

1. Find the number whose logarithm is 5.322.

$$\text{antilog } 5.322 = \text{antilog } 0.322 \times \text{antilog } 5$$
$$= 2.10 \times 10^5$$

2. Find the number whose logarithm is $-2.133$.

$$\text{antilog } (-2.133) = \text{antilog } 0.867 \times \text{antilog } (-3)$$
$$= 7.37 \times 10^{-3}$$

**Mathematical Operations Using Logarithms**  Because logarithms are exponents, mathematical operations involving logarithms follow the rules for the use of exponents:

1. Multiplication and division:

$$\log ab = \log a + \log b$$
$$\log \frac{a}{b} = \log a - \log b$$

2. Powers and roots:

$$\log a^n = n(\log a)$$
$$\log a^{1/n} = \frac{1}{n}(\log a)$$

**Natural Logarithms**  Natural, or base $e$, logarithms (abbreviated ln) are the power to which $e$, which has the value $2.71828\ldots$, must be raised to equal a number. The relation between common and natural logarithms is as follows:

$$\ln a = 2.303 \log a$$

Many calculators have an ln key to obtain natural logarithms. To obtain the antilog of a natural logarithm, calculators often employ a $\ln^{-1}$ key or the combination of INV and ln keys. Other calculators use an $e^x$ key:

$$\text{if } \ln x = 3.4691, \text{ then } x = e^{3.4691} = 32.11$$

**pH Problems**  In general chemistry, logarithms are used most frequently in working pH problems. The pH is defined as $-\log [H^+]$, as discussed in Section 17.1. The following sample exercise illustrates this application.

## SAMPLE EXERCISE 2

**(a)** What is the pH of a solution whose hydrogen-ion concentration is 0.015 $M$? **(b)** If the pH of a solution is 3.80, what is its hydrogen-ion concentration?

**Solution:**

**(a)** $\text{pH} = -\log [\text{H}^+]$
$= -\log 0.015$
$= -\log (1.5 \times 10^{-2})$
$= -\log 1.5 - \log (10^{-2})$
$= -0.18 + 2 = 1.82$

**(b)** $\text{pH} = -\log [\text{H}^+] = 3.80$
$\log [\text{H}^+] = -3.80$
$[\text{H}^+] = \text{antilog} (-3.80)$
$= \text{antilog}\ 0.20$
$\times \text{antilog} (-4)$
$= 1.6 \times 10^{-4}\ M$

---

## A.3 Quadratic Equations

An algebraic equation of the form $ax^2 + bx + c = 0$ is called a quadratic equation. The two solutions to such an equation are given by the quadratic formula:

$$x = \frac{-b \pm \sqrt{b^2 - 4ac}}{2a}$$

---

## SAMPLE EXERCISE 3

Find $x$ if $2x^2 + 4x = 1$.

**Solution:** To solve the given equation for $x$, we must first put it in the form

$$ax^2 + bx + c = 0$$

and then use the quadratic formula. If

$$2x^2 + 4x = 1$$

then

$$2x^2 + 4x - 1 = 0$$

Using the quadratic formula, where $a = 2$, $b = 4$, and $c = -1$, we have

$$x = \frac{-4 \pm \sqrt{(4)(4) - 4(2)(-1)}}{2(2)}$$

$$= \frac{-4 \pm \sqrt{16 + 8}}{4} = \frac{-4 \pm \sqrt{24}}{4}$$

$$= \frac{-4 \pm 4.899}{4}$$

The two solutions are

$$x = \frac{0.899}{4} = 0.225$$

$$x = \frac{-8.899}{4} = -2.225$$

Often in chemical problems the negative solution has no physical meaning, and only the positive answer is used.

---

## A.4 Graphs

Often the clearest way to represent the interrelationship between two variables is to graph them. Usually, the variable that is being experimentally varied, called the independent variable, is shown along the horizontal axis ($x$-axis). The variable that responds to the change in the independent variable, called the dependent variable, is then shown along the vertical axis ($y$-axis). For example, consider an experiment in which we vary the temperature of an enclosed gas and measure its pressure. The independent variable is temperature, and the dependent variable is pressure. The data shown in Table 1 could be obtained by means of this experiment. These data are shown graphically in Figure 1. The relation-

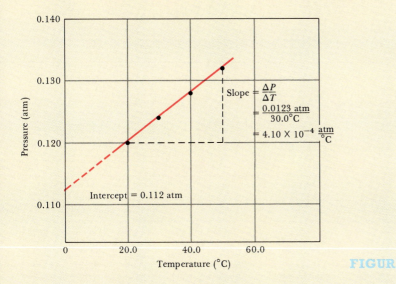

**TABLE 1** Interrelation between pressure and temperature

| Temperature (°C) | Pressure (atm) |
|---|---|
| 20.0 | 0.120 |
| 30.0 | 0.124 |
| 40.0 | 0.128 |
| 50.0 | 0.132 |

**FIGURE 1**

ship between temperature and pressure is linear. The equation for any straight-line graph has the form

$$y = ax + b$$

where $a$ is the slope of the line and $b$ is the intercept with the $y$-axis. In the case of Figure 1, we could say that the relationship between temperature and pressure takes the form

$$P = aT + b$$

where $P$ is pressure in atm and $T$ is temperature in °C. As shown on Figure 1, the slope is $4.10 \times 10^{-4}$ atm/°C, and the intercept—the point where the line crosses the $y$-axis—is 0.112 atm. Therefore, the equation for the line is

$$P = \left(4.10 \times 10^{-4} \frac{atm}{°C}\right) T + 0.112 \text{ atm}$$

# APPENDIX *B*

# Table of Four-Place Logarithms

|     | 0 | 1 | 2 | 3 | 4 | 5 | 6 | 7 | 8 | 9 |
|-----|------|------|------|------|------|------|------|------|------|------|
| 10  | 0000 | 0043 | 0086 | 0128 | 0170 | 0212 | 0253 | 0294 | 0334 | 0374 |
| 11  | 0414 | 0453 | 0492 | 0531 | 0569 | 0607 | 0645 | 0682 | 0719 | 0755 |
| 12  | 0792 | 0828 | 0864 | 0899 | 0934 | 0969 | 1004 | 1038 | 1072 | 1106 |
| 13  | 1139 | 1173 | 1206 | 1239 | 1271 | 1303 | 1335 | 1367 | 1399 | 1430 |
| 14  | 1461 | 1492 | 1523 | 1553 | 1584 | 1614 | 1644 | 1673 | 1703 | 1732 |
| 15  | 1761 | 1790 | 1818 | 1847 | 1875 | 1903 | 1931 | 1959 | 1987 | 2014 |
| 16  | 2041 | 2068 | 2095 | 2122 | 2148 | 2175 | 2201 | 2227 | 2253 | 2279 |
| 17  | 2304 | 2330 | 2355 | 2380 | 2405 | 2430 | 2455 | 2480 | 2504 | 2529 |
| 18  | 2553 | 2577 | 2601 | 2625 | 2648 | 2672 | 2695 | 2718 | 2742 | 2765 |
| 19  | 2788 | 2810 | 2833 | 2856 | 2878 | 2900 | 2923 | 2945 | 2967 | 2989 |
| 20  | 3010 | 3032 | 3054 | 3075 | 3096 | 3118 | 3139 | 3160 | 3181 | 3201 |
| 21  | 3222 | 3243 | 3263 | 3284 | 3304 | 3324 | 3345 | 3365 | 3385 | 3404 |
| 22  | 3424 | 3444 | 3464 | 3483 | 3502 | 3522 | 3541 | 3560 | 3579 | 3598 |
| 23  | 3617 | 3636 | 3655 | 3674 | 3692 | 3711 | 3729 | 3747 | 3766 | 3784 |
| 24  | 3802 | 3820 | 3838 | 3856 | 3874 | 3892 | 3909 | 3927 | 3945 | 3962 |
| 25  | 3979 | 3997 | 4014 | 4031 | 4048 | 4065 | 4082 | 4099 | 4116 | 4133 |
| 26  | 4150 | 4166 | 4183 | 4200 | 4216 | 4232 | 4249 | 4265 | 4281 | 4298 |
| 27  | 4314 | 4330 | 4346 | 4362 | 4378 | 4393 | 4409 | 4425 | 4440 | 4456 |
| 28  | 4472 | 4487 | 4502 | 4518 | 4533 | 4548 | 4564 | 4579 | 4594 | 4609 |
| 29  | 4624 | 4639 | 4654 | 4669 | 4683 | 4698 | 4713 | 4728 | 4742 | 4757 |
| 30  | 4771 | 4786 | 4800 | 4814 | 4829 | 4843 | 4857 | 4871 | 4886 | 4900 |
| 31  | 4914 | 4928 | 4942 | 4955 | 4969 | 4983 | 4997 | 5011 | 5024 | 5083 |
| 32  | 5051 | 5065 | 5079 | 5092 | 5105 | 5119 | 5132 | 5145 | 5159 | 5172 |
| 33  | 5185 | 5198 | 5211 | 5224 | 5237 | 5250 | 5263 | 5276 | 5289 | 5302 |
| 34  | 5315 | 5328 | 5340 | 5353 | 5366 | 5378 | 5391 | 5403 | 5416 | 5428 |
| 35  | 5441 | 5453 | 5465 | 5478 | 5490 | 5502 | 5514 | 5527 | 5539 | 5551 |
| 36  | 5563 | 5575 | 5587 | 5599 | 5611 | 5623 | 5635 | 5647 | 5658 | 5670 |
| 37  | 5682 | 5694 | 5705 | 5717 | 5729 | 5740 | 5752 | 5763 | 5775 | 5786 |
| 38  | 5798 | 5809 | 5821 | 5832 | 5843 | 5855 | 5866 | 5877 | 5888 | 5899 |
| 39  | 5911 | 5922 | 5933 | 5944 | 5955 | 5966 | 5977 | 5988 | 5999 | 6010 |
| 40  | 6021 | 6031 | 6042 | 6053 | 6064 | 6075 | 6085 | 6096 | 6107 | 6117 |
| 41  | 6128 | 6138 | 6149 | 6160 | 6170 | 6180 | 6191 | 6201 | 6212 | 6222 |
| 42  | 6232 | 6243 | 6253 | 6263 | 6274 | 6284 | 6294 | 6304 | 6314 | 6325 |
| 43  | 6335 | 6345 | 6355 | 6365 | 6375 | 6385 | 6395 | 6405 | 6415 | 6425 |
| 44  | 6435 | 6444 | 6454 | 6464 | 6474 | 6484 | 6493 | 6503 | 6513 | 6522 |
| 45  | 6532 | 6542 | 6551 | 6561 | 6571 | 6580 | 6590 | 6599 | 6609 | 6618 |
| 46  | 6628 | 6637 | 6646 | 6656 | 6665 | 6675 | 6684 | 6693 | 6702 | 6712 |
| 47  | 6721 | 6730 | 6739 | 6749 | 6758 | 6767 | 6776 | 6785 | 6794 | 6803 |
| 48  | 6812 | 6821 | 6830 | 6839 | 6838 | 6857 | 6866 | 6875 | 6884 | 6893 |
| 49  | 6902 | 6911 | 6920 | 6928 | 6937 | 6946 | 6955 | 6964 | 6972 | 6981 |

| | 0 | 1 | 2 | 3 | 4 | 5 | 6 | 7 | 8 | 9 |
|---|---|---|---|---|---|---|---|---|---|---|
| **50** | 6990 | 6998 | 7007 | 7016 | 7024 | 7033 | 7042 | 7050 | 7059 | 7067 |
| **51** | 7076 | 7084 | 7093 | 7101 | 7110 | 7118 | 7126 | 7135 | 7143 | 7152 |
| **52** | 7160 | 7168 | 7177 | 7185 | 7193 | 7202 | 7210 | 7218 | 7226 | 7235 |
| **53** | 7243 | 7251 | 7259 | 7267 | 7275 | 7284 | 7292 | 7300 | 7308 | 7316 |
| **54** | 7324 | 7332 | 7340 | 7348 | 7356 | 7364 | 7372 | 7380 | 7388 | 7396 |
| **55** | 7404 | 7412 | 7419 | 7427 | 7435 | 7443 | 7451 | 7459 | 7466 | 7474 |
| **56** | 7482 | 7490 | 7497 | 7505 | 7513 | 7520 | 7528 | 7536 | 7543 | 7551 |
| **57** | 7559 | 7566 | 7574 | 7582 | 7589 | 7597 | 7604 | 7612 | 7619 | 7627 |
| **58** | 7634 | 7642 | 7649 | 7657 | 7664 | 7672 | 7679 | 7686 | 7694 | 7701 |
| **59** | 7709 | 7716 | 7723 | 7731 | 7738 | 7745 | 7752 | 7760 | 7767 | 7774 |
| **60** | 7782 | 7789 | 7796 | 7803 | 7810 | 7818 | 7825 | 7832 | 7839 | 7846 |
| **61** | 7853 | 7860 | 7868 | 7875 | 7882 | 7889 | 7896 | 7903 | 7910 | 7917 |
| **62** | 7924 | 7931 | 7938 | 7945 | 7952 | 7959 | 7966 | 7973 | 7980 | 7987 |
| **63** | 7993 | 8000 | 8007 | 8014 | 8021 | 8028 | 8035 | 8041 | 8048 | 8055 |
| **64** | 8062 | 8069 | 8075 | 8082 | 8089 | 8096 | 8102 | 8109 | 8116 | 8122 |
| **65** | 8129 | 8136 | 8142 | 8149 | 8156 | 8162 | 8169 | 8176 | 8182 | 8189 |
| **66** | 8195 | 8202 | 8209 | 8215 | 8222 | 8228 | 8235 | 8241 | 8248 | 8254 |
| **67** | 8261 | 8267 | 8274 | 8280 | 8287 | 8293 | 8299 | 8306 | 8312 | 8319 |
| **68** | 8325 | 8331 | 8338 | 8344 | 8351 | 8357 | 8363 | 8370 | 8376 | 8382 |
| **69** | 8388 | 8395 | 8401 | 8407 | 8414 | 8420 | 8426 | 8432 | 8439 | 8445 |
| **70** | 8451 | 8457 | 8463 | 8470 | 8476 | 8482 | 8488 | 8494 | 8500 | 8506 |
| **71** | 8513 | 8519 | 8525 | 8531 | 8537 | 8543 | 8549 | 8555 | 8561 | 8567 |
| **72** | 8573 | 8579 | 8585 | 8591 | 8597 | 8603 | 8609 | 8615 | 8621 | 8627 |
| **73** | 8633 | 8639 | 8645 | 8651 | 8657 | 8663 | 8669 | 8675 | 8681 | 8686 |
| **74** | 8692 | 8698 | 8704 | 8710 | 8716 | 8722 | 8727 | 8733 | 8739 | 8745 |
| **75** | 8751 | 8756 | 8762 | 8768 | 8774 | 8779 | 8785 | 8791 | 8797 | 8802 |
| **76** | 8808 | 8814 | 8820 | 8825 | 8831 | 8837 | 8842 | 8848 | 8854 | 8859 |
| **77** | 8865 | 8871 | 8876 | 8882 | 8887 | 8893 | 8899 | 8904 | 8910 | 8915 |
| **78** | 8921 | 8927 | 8932 | 8938 | 8943 | 8949 | 8954 | 8960 | 8965 | 8971 |
| **79** | 8976 | 8982 | 8987 | 8993 | 8998 | 9004 | 9009 | 9015 | 9020 | 9025 |
| **80** | 9031 | 9036 | 9042 | 9047 | 9053 | 9058 | 9063 | 9069 | 9074 | 9079 |
| **81** | 9085 | 9090 | 9096 | 9101 | 9106 | 9112 | 9117 | 9122 | 9128 | 9133 |
| **82** | 9138 | 9143 | 9149 | 9154 | 9159 | 9165 | 9170 | 9175 | 9180 | 9186 |
| **83** | 9191 | 9196 | 9201 | 9206 | 9212 | 9217 | 9222 | 9227 | 9232 | 9238 |
| **84** | 9243 | 9248 | 9253 | 9258 | 9263 | 9269 | 9274 | 9279 | 9284 | 9289 |
| **85** | 9294 | 9299 | 9304 | 9309 | 9315 | 9320 | 9325 | 9330 | 9335 | 9340 |
| **86** | 9345 | 9350 | 9355 | 9360 | 9365 | 9370 | 9375 | 9380 | 9385 | 9390 |
| **87** | 9395 | 9400 | 9405 | 9410 | 9415 | 9420 | 9425 | 9430 | 9435 | 9440 |
| **88** | 9445 | 9450 | 9455 | 9460 | 9465 | 9469 | 9474 | 9479 | 9484 | 9489 |
| **89** | 9494 | 9499 | 9504 | 9509 | 9513 | 9518 | 9523 | 9628 | 9533 | 9538 |
| **90** | 9542 | 9547 | 9552 | 9557 | 9562 | 9566 | 9571 | 9576 | 9581 | 9586 |
| **91** | 9590 | 9595 | 9600 | 9605 | 9609 | 9614 | 9619 | 9624 | 9628 | 9633 |
| **92** | 9638 | 9643 | 9647 | 9652 | 9657 | 9661 | 9666 | 9671 | 9675 | 9680 |
| **93** | 9685 | 9689 | 9694 | 9699 | 9703 | 9708 | 9713 | 9717 | 9722 | 9727 |
| **94** | 9731 | 9736 | 9741 | 9745 | 9750 | 9754 | 9759 | 9763 | 9768 | 9773 |
| **95** | 9777 | 9782 | 9786 | 9791 | 9795 | 9800 | 9805 | 9809 | 9814 | 9818 |
| **96** | 9823 | 9827 | 9832 | 9836 | 9841 | 9845 | 9850 | 9854 | 9859 | 9863 |
| **97** | 9868 | 9872 | 9877 | 9881 | 9886 | 9890 | 9894 | 9899 | 9903 | 9908 |
| **98** | 9912 | 9917 | 9921 | 9926 | 9930 | 9934 | 9939 | 9943 | 9948 | 9952 |
| **99** | 9956 | 9961 | 9965 | 9969 | 9974 | 9978 | 9983 | 9987 | 9991 | 9996 |

# Properties
# of Water

|  | |
|---|---|
| Density: | 0.99987 g/cm$^3$ at 0°C |
|  | 1.00000 g/cm$^3$ at 4°C |
|  | 0.99707 g/cm$^3$ at 25°C |
|  | 0.95838 g/cm$^3$ at 100°C |
| Heat of fusion: | 6.008 kJ/mol at 0°C |
| Heat of vaporization: | 44.94 kJ/mol at 0°C |
|  | 44.02 kJ/mol at 25°C |
|  | 40.67 kJ/mol at 100°C |
| Ion-product constant, $K_w$: | $1.14 \times 10^{-15}$ at 0°C |
|  | $1.01 \times 10^{-14}$ at 25°C |
|  | $5.47 \times 10^{-14}$ at 50°C |
| Specific heat: | Ice ($-3$°C)—2.092 J/°C-g |
|  | Water at 14.5°C—4.184 J/°C-g |
|  | Steam (100°C)—1.841 J/°C-g |

Vapor pressure (mm Hg):

| $T$ (°C) | $P$ | $T$ (°C) | $P$ | $T$ (°C) | $P$ | $T$ (°C) | $P$ |
|---|---|---|---|---|---|---|---|
| 0 | 4.58 | 21 | 18.65 | 35 | 42.2 | 92 | 567.0 |
| 5 | 6.54 | 22 | 19.83 | 40 | 55.3 | 94 | 610.9 |
| 10 | 9.21 | 23 | 21.07 | 45 | 71.9 | 96 | 657.6 |
| 12 | 10.52 | 24 | 22.38 | 50 | 92.5 | 98 | 707.3 |
| 14 | 11.99 | 25 | 23.76 | 55 | 118.0 | 100 | 760.0 |
| 16 | 13.63 | 26 | 25.21 | 60 | 149.4 | 102 | 815.9 |
| 17 | 14.53 | 27 | 26.74 | 65 | 187.5 | 104 | 875.1 |
| 18 | 15.48 | 28 | 28.35 | 70 | 233.7 | 106 | 937.9 |
| 19 | 16.48 | 29 | 30.04 | 80 | 355.1 | 108 | 1004.4 |
| 20 | 17.54 | 30 | 31.82 | 90 | 525.8 | 110 | 1074.6 |

# Thermodynamic Quantities for Selected Substances at 25°C

| Substance | $\Delta H_f^\circ$ (kJ/mol) | $\Delta G_f^\circ$ (kJ/mol) | $S^\circ$ (J/mol-K) | Substance | $\Delta H_f^\circ$ (kJ/mol) | $\Delta G_f^\circ$ (kJ/mol) | $S^\circ$ (J/mol-K) |
|---|---|---|---|---|---|---|---|
| $Al(s)$ | 0 | 0 | 28.32 | $CaO(s)$ | −635.5 | −604.17 | 39.75 |
| $Al_2O_3(s)$ | −1669.8 | −1576.5 | 51.00 | $Ca(OH)_2(s)$ | −986.2 | −898.5 | 83.4 |
| $Ag^+(aq)$ | 105.90 | 77.11 | 73.93 | $CaSO_4(s)$ | −1434.0 | −1321.8 | 106.7 |
| $AgCl(s)$ | −127.0 | −109.70 | 96.11 | $Cl(g)$ | 127.7 | 105.7 | 165.2 |
| $Ba(s)$ | 0 | 0 | 63.2 | $Cl_2(g)$ | 0 | 0 | 222.96 |
| $BaCO_3(s)$ | −1216.3 | −1137.6 | 112.1 | $Co(s)$ | 0 | 0 | 28.4 |
| $BaO(s)$ | −553.5 | −525.1 | 70.42 | $Co(g)$ | 439 | 393 | 179 |
| $Br(g)$ | 111.8 | 82.38 | 174.9 | $Cr(s)$ | 0 | 0 | 23.6 |
| $Br^-(aq)$ | −120.9 | −102.8 | 80.71 | $Cr(g)$ | 397.5 | 352.6 | 174.2 |
| $Br_2(g)$ | 30.71 | 3.14 | 245.3 | $Cr_2O_3(s)$ | −1139.7 | −1058.1 | 81.2 |
| $Br_2(l)$ | 0 | 0 | 152.3 | $Cu(s)$ | 0 | 0 | 33.30 |
| $C(g)$ | 718.4 | 672.9 | 158.0 | $Cu(g)$ | 338.4 | 298.6 | 166.3 |
| $C(diamond)$ | 1.88 | 2.84 | 2.43 | $F(g)$ | 80.0 | 61.9 | 158.7 |
| $C(graphite)$ | 0 | 0 | 5.69 | $F_2(g)$ | 0 | 0 | 202.7 |
| $CCl_4(g)$ | −106.7 | −64.0 | 309.4 | $Fe(s)$ | 0 | 0 | 27.15 |
| $CCl_4(l)$ | −139.3 | −68.6 | 214.4 | $Fe^{2+}(aq)$ | −87.86 | −84.93 | 113.4 |
| $CF_4(g)$ | −679.9 | −635.1 | 262.3 | $Fe^{3+}(aq)$ | −47.69 | −10.54 | 293.3 |
| $CH_4(g)$ | −74.8 | −50.8 | 186.3 | $FeCl_3(s)$ | −400 | −334 | 142.3 |
| $C_2H_2(g)$ | 226.7 | 209.2 | 200.8 | $FeO(s)$ | −271.9 | −255.2 | 60.75 |
| $C_2H_4(g)$ | 52.30 | 68.11 | 219.4 | $Fe_2O_3(s)$ | −822.16 | −740.98 | 89.96 |
| $C_2H_6(g)$ | −84.68 | −32.89 | 229.5 | $Fe_3O_4(s)$ | −1117.1 | −1014.2 | 146.4 |
| $C_3H_8(g)$ | −103.85 | −23.47 | 269.9 | $H(g)$ | 217.94 | 203.26 | 114.60 |
| $CH_3OH(g)$ | −201.2 | −161.9 | 237.6 | $H^+(aq)$ | 0 | 0 | 0 |
| $CH_3OH(l)$ | −238.6 | −166.23 | 126.8 | $H_2(g)$ | 0 | 0 | 130.58 |
| $C_2H_5OH(l)$ | −277.7 | −174.76 | 160.7 | $HBr(g)$ | −36.23 | −53.22 | 198.49 |
| $CH_3COOH(l)$ | −487.0 | −392.4 | 159.8 | $HCl(g)$ | −92.30 | −95.27 | 186.69 |
| $C_6H_6(l)$ | 49.0 | 124.5 | 172.8 | $HF(g)$ | −268.61 | −270.70 | 173.51 |
| $C_6H_{12}O_6(s)$ | −1273.02 | −910.4 | 212.1 | $HI(g)$ | 25.94 | 1.30 | 206.3 |
| $C_6H_6(g)$ | 82.9 | 129.7 | 269.2 | $HNO_3(aq)$ | −206.6 | −110.5 | 146 |
| $CO(g)$ | −110.5 | −137.3 | 197.9 | $H_2O(g)$ | −241.8 | −228.61 | 188.7 |
| $CO_2(g)$ | −393.5 | −394.4 | 213.6 | $H_2O(l)$ | −285.85 | −236.81 | 69.96 |
| $Ca(s)$ | 0 | 0 | 41.4 | $H_2O_2(g)$ | −136.10 | −105.48 | 232.9 |
| $Ca(g)$ | 179.3 | 145.5 | 154.8 | $H_2O_2(l)$ | −187.8 | −120.4 | 109.6 |
| $CaCO_3(calcite)$ | −1207.1 | −1128.76 | 92.88 | $H_2S(g)$ | −20.17 | −33.01 | 205.6 |
| $CaF_2(s)$ | −1219.6 | −1167.3 | 68.87 | $H_2Se(g)$ | 29.7 | 15.9 | 219.0 |

| Substance | $\Delta H_f^\circ$ (kJ/mol) | $\Delta G_f^\circ$ (kJ/mol) | $S^\circ$ (J/mol-K) |
|---|---|---|---|
| $H_3PO_4(aq)$ | $-1288.3$ | $-1142.6$ | 158.2 |
| $Hg(g)$ | 60.83 | 31.76 | 174.89 |
| $Hg(l)$ | 0 | 0 | 77.40 |
| $I(g)$ | 106.60 | 70.16 | 180.66 |
| $I_2(s)$ | 0 | 0 | 116.73 |
| | | | |
| $I_2(g)$ | 62.25 | 19.37 | 260.57 |
| $K(g)$ | 89.99 | 61.17 | 160.2 |
| $KCl(s)$ | $-435.9$ | $-408.3$ | 82.7 |
| $KClO_3(s)$ | $-391.2$ | $-289.9$ | 143.0 |
| $KClO_3(aq)$ | $-349.5$ | $-284.9$ | 265.7 |
| | | | |
| $KNO_3(s)$ | $-492.70$ | $-393.13$ | 288.1 |
| $Mg(s)$ | 0 | 0 | 32.51 |
| $MgCl_2(s)$ | $-641.6$ | $-592.1$ | 89.6 |
| $MgO(s)$ | $-601.8$ | $-569.6$ | 26.8 |
| $Mn(s)$ | 0 | 0 | 32.0 |
| | | | |
| $Mn(g)$ | 280.7 | 238.5 | 173.6 |
| $MnO(s)$ | $-385.2$ | $-362.9$ | 59.7 |
| $MnO_2(s)$ | $-519.6$ | $-464.8$ | 53.14 |
| $N_2(g)$ | 0 | 0 | 191.50 |
| $NH_3(g)$ | $-46.19$ | $-16.66$ | 192.5 |
| | | | |
| $N_2H_4(g)$ | 95.40 | 159.4 | 238.5 |
| $NH_4CN(s)$ | 0.0 | — | — |
| $NH_4Cl(s)$ | $-314.4$ | $-203.0$ | 94.6 |
| $NH_4NO_3(s)$ | $-365.6$ | $-184.0$ | 151 |
| $NO(g)$ | 90.37 | 86.71 | 210.62 |
| | | | |
| $NO_2(g)$ | 33.84 | 51.84 | 240.45 |
| $NOCl(g)$ | 52.6 | 66.3 | 264 |
| $N_2O(g)$ | 81.6 | 103.59 | 220.0 |
| $N_2O_4(g)$ | 9.66 | 98.28 | 304.3 |
| $Na(g)$ | 107.7 | 77.3 | 51.5 |
| | | | |
| $NaBr(aq)$ | $-360.6$ | $-364.7$ | 141 |
| $NaCl(s)$ | $-410.9$ | $-384.0$ | 72.33 |
| $NaCl(aq)$ | $-407.1$ | $-393.0$ | 115.5 |
| $NaHCO_3(s)$ | $-947.7$ | $-851.8$ | 102.1 |
| $Na_2CO_3(s)$ | $-1130.9$ | $-1047.7$ | 136.0 |
| | | | |
| $NaNO_3(aq)$ | $-446.2$ | $-372.4$ | 207 |
| $NaOH(s)$ | $-425.6$ | $-379.5$ | 64.46 |
| $NaOH(aq)$ | $-469.6$ | $-419.2$ | 49.8 |
| $Ni(s)$ | 0 | 0 | 29.9 |

| Substance | $\Delta H_f^\circ$ (kJ/mol) | $\Delta G_f^\circ$ (kJ/mol) | $S^\circ$ (J/mol-K) |
|---|---|---|---|
| $Ni(g)$ | 429.7 | 384.5 | 182.1 |
| $O(g)$ | 247.5 | 230.1 | 161.0 |
| $O_2(g)$ | 0 | 0 | 205.0 |
| | | | |
| $O_3(g)$ | 142.3 | 163.4 | 237.6 |
| $OH^-(aq)$ | $-230.0$ | $-157.3$ | $-10.7$ |
| $P_2(g)$ | 144.3 | 103.7 | 218.1 |
| $P_4(g)$ | 58.9 | 24.4 | 280 |
| $PCl_3(l)$ | $-319.6$ | $-272.4$ | 217 |
| | | | |
| $PH_3(g)$ | 5.4 | 13.4 | 210.2 |
| $POCl_3(g)$ | $-542.2$ | $-502.5$ | 325 |
| $POCl_3(l)$ | $-597.0$ | $-520.9$ | 222 |
| $P_4O_6(s)$ | $-1640.1$ | — | — |
| $P_4O_{10}(s, \text{hexagonal})$ | $-2940.1$ | $-2675.2$ | 228.9 |
| | | | |
| $Pb(s)$ | 0 | 0 | 68.85 |
| $PbBr_2(s)$ | $-277.4$ | $-260.7$ | 161 |
| $PbCO_3(s)$ | $-699.1$ | $-625.5$ | 131.0 |
| $Pb(NO_3)_2(s)$ | $-451.9$ | — | — |
| $Pb(NO_3)_2(aq)$ | $-421.3$ | — | — |
| | | | |
| $PbO(s)$ | $-217.3$ | $-187.9$ | 68.70 |
| $Rb(g)$ | 85.8 | 55.8 | 170.0 |
| $RbCl(s)$ | $-430.5$ | $-412.0$ | 92 |
| $RbClO_3(s)$ | $-392.4$ | $-292.0$ | 152 |
| $S(s, \text{rhombic})$ | 0 | 0 | 31.88 |
| | | | |
| $SO_2(g)$ | $-296.9$ | $-300.4$ | 248.5 |
| $SO_3(g)$ | $-395.2$ | $-370.4$ | 256.2 |
| $SOCl_2(l)$ | $-245.6$ | — | — |
| $Sc(s)$ | 0 | 0 | 34.6 |
| $Sc(g)$ | 377.8 | 336.1 | 174.7 |
| | | | |
| $Si(g)$ | 368.2 | 323.9 | 167.8 |
| $Si(s)$ | 0 | 0 | 18.7 |
| $SiCl_4(l)$ | $-640.1$ | $-572.8$ | 239.3 |
| $SiO_2(s)$ (quartz) | $-910.9$ | $-856.5$ | 41.84 |
| $Ti(g)$ | 468 | 422 | 180.3 |
| | | | |
| $V(s)$ | 0 | 0 | 28.9 |
| $V(g)$ | 514.2 | 453.1 | 182.2 |
| $Zn(s)$ | 0 | 0 | 41.63 |
| $Zn(g)$ | 130.7 | 95.2 | 160.9 |
| $ZnO(s)$ | $-348.0$ | $-318.2$ | 43.9 |

# APPENDIX *E*

# Aqueous-Equilibrium Constants

**E.1**  Dissociation constants for acids at 25°C

| Name | Formula | $K_{a1}$ | $K_{a2}$ | $K_{a3}$ |
|------|---------|----------|----------|----------|
| Acetic | $HC_2H_3O_2$ | $1.8 \times 10^{-5}$ | | |
| Arsenic | $H_3AsO_4$ | $5.6 \times 10^{-3}$ | $1.0 \times 10^{-7}$ | $3.0 \times 10^{-12}$ |
| Arsenous | $H_3AsO_3$ | $6 \times 10^{-10}$ | | |
| Ascorbic | $HC_6H_7O_6$ | $8.0 \times 10^{-5}$ | $1.6 \times 10^{-12}$ | |
| Benzoic | $HC_7H_5O_2$ | $6.5 \times 10^{-5}$ | | |
| Boric | $H_3BO_3$ | $5.8 \times 10^{-10}$ | | |
| Carbonic | $H_2CO_3$ | $4.3 \times 10^{-7}$ | $5.6 \times 10^{-11}$ | |
| Chloroacetic | $HC_2H_2O_2Cl$ | $1.4 \times 10^{-3}$ | | |
| Citric | $H_3C_6H_5O_7$ | $3.5 \times 10^{-4}$ | $1.7 \times 10^{-5}$ | $4.0 \times 10^{-7}$ |
| Cyanic | $HCNO$ | $7.4 \times 10^{-4}$ | | |
| Formic | $HCHO_2$ | $1.8 \times 10^{-4}$ | | |
| Hydroazoic | $HN_3$ | $1.9 \times 10^{-5}$ | | |
| Hydrocyanic | $HCN$ | $4.9 \times 10^{-10}$ | | |
| Hydrofluoric | $HF$ | $6.8 \times 10^{-4}$ | | |
| Hydrogen chromate ion | $HCrO_4^-$ | $3.0 \times 10^{-7}$ | | |
| Hydrogen peroxide | $H_2O_2$ | $2.4 \times 10^{-12}$ | | |
| Hydrogen selenate ion | $HSeO_4^-$ | $2.2 \times 10^{-2}$ | | |
| Hydrogen sulfate ion | $HSO_4^-$ | $1.2 \times 10^{-2}$ | | |
| Hydrogen sulfide | $H_2S$ | $5.7 \times 10^{-8}$ | $1.3 \times 10^{-13}$ | |
| Hypobromous | $HBrO$ | $2 \times 10^{-9}$ | | |
| Hypochlorous | $HClO$ | $3.0 \times 10^{-8}$ | | |
| Hypoiodous | $HIO$ | $2 \times 10^{-11}$ | | |
| Iodic | $HIO_3$ | $1.7 \times 10^{-1}$ | | |
| Lactic | $HC_3H_5O_3$ | $1.4 \times 10^{-4}$ | | |
| Malonic | $H_2C_3H_2O_4$ | $1.5 \times 10^{-3}$ | $2.0 \times 10^{-6}$ | |
| Nitrous | $HNO_2$ | $4.5 \times 10^{-4}$ | | |
| Oxalic | $H_2C_2O_4$ | $5.9 \times 10^{-2}$ | $6.4 \times 10^{-5}$ | |
| Paraperiodic | $H_5IO_6$ | $2.8 \times 10^{-2}$ | $5.3 \times 10^{-9}$ | |
| Phenol | $HC_6H_5O$ | $1.3 \times 10^{-10}$ | | |
| Phosphoric | $H_3PO_4$ | $7.5 \times 10^{-3}$ | $6.2 \times 10^{-8}$ | $4.2 \times 10^{-13}$ |
| Propionic | $HC_3H_5O_2$ | $1.3 \times 10^{-5}$ | | |
| Pyrophosphoric | $H_4P_2O_7$ | $3.0 \times 10^{-2}$ | $4.4 \times 10^{-3}$ | |
| Selenous | $H_2SeO_3$ | $2.3 \times 10^{-3}$ | $5.3 \times 10^{-9}$ | |
| Sulfuric | $H_2SO_4$ | strong acid | $1.2 \times 10^{-2}$ | |
| Sulfurous | $H_2SO_3$ | $1.7 \times 10^{-2}$ | $6.4 \times 10^{-8}$ | |
| Tartaric | $H_2C_4H_4O_6$ | $1.0 \times 10^{-3}$ | $4.6 \times 10^{-5}$ | |

## E.2 Dissociation Constants for Bases at 25°C

| Name | Formula | $K_b$ |
|------|---------|-------|
| Ammonia | $NH_3$ | $1.8 \times 10^{-5}$ |
| Aniline | $C_6H_5NH_2$ | $4.3 \times 10^{-10}$ |
| Dimethylamine | $(CH_3)_2NH$ | $5.4 \times 10^{-4}$ |
| Ethylamine | $C_2H_5NH_2$ | $6.4 \times 10^{-4}$ |
| Hydrazine | $H_2NNH_2$ | $1.3 \times 10^{-6}$ |
| Hydroxylamine | $HONH_2$ | $1.1 \times 10^{-8}$ |
| Methylamine | $CH_3NH_2$ | $4.4 \times 10^{-4}$ |
| Pyridine | $C_5H_5N$ | $1.7 \times 10^{-9}$ |
| Trimethylamine | $(CH_3)_3N$ | $6.4 \times 10^{-5}$ |

## E.3 Solubility-Product Constants for Compounds at 25°C

| Name | Formula | $K_{sp}$ |
|------|---------|----------|
| Barium carbonate | $BaCO_3$ | $5.1 \times 10^{-9}$ |
| Barium chromate | $BaCrO_4$ | $1.2 \times 10^{-10}$ |
| Barium fluoride | $BaF_2$ | $1.0 \times 10^{-6}$ |
| Barium hydroxide | $Ba(OH)_2$ | $5 \times 10^{-3}$ |
| Barium oxalate | $BaC_2O_4$ | $1.6 \times 10^{-7}$ |
| Barium phosphate | $Ba_3(PO_4)_2$ | $3.4 \times 10^{-23}$ |
| Barium sulfate | $BaSO_4$ | $1.1 \times 10^{-10}$ |
| Cadmium carbonate | $CdCO_3$ | $5.2 \times 10^{-12}$ |
| Cadmium hydroxide | $Cd(OH)_2$ | $2.5 \times 10^{-14}$ |
| Cadmium sulfide | $CdS$ | $8.0 \times 10^{-27}$ |
| Calcium carbonate | $CaCO_3$ | $2.8 \times 10^{-9}$ |
| Calcium chromate | $CaCrO_4$ | $7.1 \times 10^{-4}$ |
| Calcium fluoride | $CaF_2$ | $3.9 \times 10^{-11}$ |
| Calcium hydroxide | $Ca(OH)_2$ | $5.5 \times 10^{-6}$ |
| Calcium phosphate | $Ca_3(PO_4)_2$ | $2.0 \times 10^{-29}$ |
| Calcium sulfate | $CaSO_4$ | $9.1 \times 10^{-6}$ |
| Cerium(III) fluoride | $CeF_3$ | $8 \times 10^{-16}$ |
| Chromium(III) fluoride | $CrF_3$ | $6.6 \times 10^{-11}$ |
| Chromium(III) hydroxide | $Cr(OH)_3$ | $6.3 \times 10^{-31}$ |
| Cobalt(II) carbonate | $CoCO_3$ | $1.4 \times 10^{-13}$ |
| Cobalt(II) hydroxide | $Co(OH)_2$ | $1.6 \times 10^{-15}$ |
| Cobalt(III) hydroxide | $Co(OH)_3$ | $1.6 \times 10^{-44}$ |
| α-Cobalt(II) sulfide[a] | $CoS$ | $4.0 \times 10^{-21}$ |
| Copper(I) bromide | $CuBr$ | $5.3 \times 10^{-9}$ |
| Copper(I) chloride | $CuCl$ | $1.2 \times 10^{-6}$ |
| Copper(I) sulfide | $Cu_2S$ | $2.5 \times 10^{-48}$ |
| Copper(II) carbonate | $CuCO_3$ | $1.4 \times 10^{-10}$ |
| Copper(II) chromate | $CuCrO_4$ | $3.6 \times 10^{-6}$ |
| Copper(II) hydroxide | $Cu(OH)_2$ | $2.2 \times 10^{-20}$ |
| Copper(II) phosphate | $Cu_3(PO_4)_2$ | $1.3 \times 10^{-37}$ |
| Copper(II) sulfide | $CuS$ | $6.3 \times 10^{-36}$ |
| Gold(I) chloride | $AuCl$ | $2.0 \times 10^{-13}$ |
| Gold(III) chloride | $AuCl_3$ | $3.2 \times 10^{-25}$ |
| Iron(II) carbonate | $FeCO_3$ | $3.2 \times 10^{-11}$ |
| Iron(II) hydroxide | $Fe(OH)_2$ | $8.0 \times 10^{-16}$ |
| Iron(II) sulfide | $FeS$ | $6.3 \times 10^{-18}$ |
| Iron(III) hydroxide | $Fe(OH)_3$ | $4 \times 10^{-38}$ |
| Lanthanum fluoride | $LaF_3$ | $7 \times 10^{-17}$ |
| Lanthanum iodate | $La(IO_3)_3$ | $6.1 \times 10^{-12}$ |
| Lead carbonate | $PbCO_3$ | $7.4 \times 10^{-14}$ |

| Name | Formula | $K_{sp}$ |
|------|---------|----------|
| Lead chloride | $PbCl_2$ | $1.6 \times 10^{-5}$ |
| Lead chromate | $PbCrO_4$ | $2.8 \times 10^{-13}$ |
| Lead fluoride | $PbF_2$ | $2.7 \times 10^{-8}$ |
| Lead hydroxide | $Pb(OH)_2$ | $1.2 \times 10^{-15}$ |
| Lead sulfate | $PbSO_4$ | $1.6 \times 10^{-8}$ |
| Lead sulfide | $PbS$ | $8.0 \times 10^{-28}$ |
| Magnesium hydroxide | $Mg(OH)_2$ | $1.8 \times 10^{-11}$ |
| Magnesium oxalate | $MgC_2O_4$ | $8.6 \times 10^{-5}$ |
| Manganese carbonate | $MnCO_3$ | $1.8 \times 10^{-11}$ |
| Manganese hydroxide | $Mn(OH)_2$ | $1.9 \times 10^{-13}$ |
| Manganese(II) sulfide | $MnS$ | $1.0 \times 10^{-13}$ |
| Mercury(I) chloride | $Hg_2Cl_2$ | $1.3 \times 10^{-18}$ |
| Mercury(I) oxalate | $Hg_2C_2O_4$ | $2.0 \times 10^{-13}$ |
| Mercury(I) sulfide | $Hg_2S$ | $1.0 \times 10^{-47}$ |
| Mercury(II) hydroxide | $Hg(OH)_2$ | $3.0 \times 10^{-26}$ |
| Mercury(II) sulfide | $HgS$ | $4 \times 10^{-53}$ |
| Nickel carbonate | $NiCO_3$ | $6.6 \times 10^{-9}$ |
| Nickel hydroxide | $Ni(OH)_2$ | $1.6 \times 10^{-14}$ |
| Nickel oxalate | $NiC_2O_4$ | $4 \times 10^{-10}$ |
| α-Nickel sulfide[a] | $NiS$ | $3.2 \times 10^{-19}$ |
| Silver arsenate | $Ag_3AsO_4$ | $1.0 \times 10^{-22}$ |
| Silver bromide | $AgBr$ | $5.0 \times 10^{-13}$ |
| Silver carbonate | $Ag_2CO_3$ | $8.1 \times 10^{-12}$ |
| Silver chloride | $AgCl$ | $1.8 \times 10^{-10}$ |
| Silver chromate | $Ag_2CrO_4$ | $1.1 \times 10^{-12}$ |
| Silver cyanide | $AgCN$ | $1.2 \times 10^{-16}$ |
| Silver iodide | $AgI$ | $8.3 \times 10^{-17}$ |
| Silver sulfate | $Ag_2SO_4$ | $1.4 \times 10^{-5}$ |
| Silver sulfide | $Ag_2S$ | $6.3 \times 10^{-50}$ |
| Strontium carbonate | $SrCO_3$ | $1.1 \times 10^{-10}$ |
| Tin(II) hydroxide | $Sn(OH)_2$ | $1.4 \times 10^{-28}$ |
| Tin(II) sulfide | $SnS$ | $1.0 \times 10^{-25}$ |
| Zinc carbonate | $ZnCO_3$ | $1.4 \times 10^{-11}$ |
| Zinc hydroxide | $Zn(OH)_2$ | $1.2 \times 10^{-17}$ |
| Zinc oxalate | $ZnC_2O_4$ | $2.7 \times 10^{-8}$ |
| α-Zinc sulfide[a] | $ZnS$ | $1.1 \times 10^{-21}$ |

[a] Some substances exist in more than one crystalline form; the prefix indicates the particular form for which $K_{sp}$ is listed.

# Standard Electrode Potentials at 25°C

| Half-reaction | $E°$ (V) |
|---|---|
| $Ag^+(aq) + e^- \rightleftharpoons Ag(s)$ | +0.799 |
| $AgBr(s) + e^- \rightleftharpoons Ag(s) + Br^-(aq)$ | +0.095 |
| $AgCl(s) + e^- \rightleftharpoons Ag(s) + Cl^-(aq)$ | +0.222 |
| $Ag(CN)_2^-(aq) + e^- \rightleftharpoons Ag(s) + 2CN^-(aq)$ | −0.31 |
| $Ag_2CrO_4(s) + 2e^- \rightleftharpoons 2Ag(s) + CrO_4^{2-}(aq)$ | +0.446 |
| | |
| $AgI(s) + e^- \rightleftharpoons Ag(s) + I^-(aq)$ | −0.151 |
| $Ag(S_2O_3)_2^{3-} + e^- \rightleftharpoons Ag(s) + 2S_2O_3^{2-}(aq)$ | +0.01 |
| $Al^{3+}(aq) + 3e^- \rightleftharpoons Al(s)$ | −1.66 |
| $H_3AsO_4(aq) + 2H^+(aq) + 2e^- \rightleftharpoons H_3AsO_3(aq) + H_2O(l)$ | +0.559 |
| $Ba^{2+}(aq) + 2e^- \rightleftharpoons Ba(s)$ | −2.90 |
| | |
| $BiO^+(aq) + 2H^+(aq) + 3e^- \rightleftharpoons Bi(s) + H_2O(l)$ | +0.32 |
| $Br_2(l) + 2e^- \rightleftharpoons 2Br^-(aq)$ | +1.065 |
| $BrO_3^-(aq) + 6H^+(aq) + 5e^- \rightleftharpoons \frac{1}{2}Br_2(l) + 3H_2O(l)$ | +1.52 |
| $Ca^{2+}(aq) + 2e^- \rightleftharpoons Ca(s)$ | −2.87 |
| $2CO_2(g) + 2H^+(aq) + 2e^- \rightleftharpoons H_2C_2O_4(aq)$ | −0.49 |
| | |
| $Cd^{2+}(aq) + 2e^- \rightleftharpoons Cd(s)$ | −0.403 |
| $Ce^{4+}(aq) + e^- \rightleftharpoons Ce^{3+}(aq)$ | +1.61 |
| $Cl_2(g) + 2e^- \rightleftharpoons 2Cl^-(aq)$ | +1.359 |
| $HClO(aq) + H^+(aq) + e^- \rightleftharpoons \frac{1}{2}Cl_2(g) + H_2O(l)$ | +1.63 |
| $ClO^-(aq) + H_2O(l) + 2e^- \rightleftharpoons Cl^-(aq) + 2OH^-(aq)$ | +0.89 |
| $ClO_3^-(aq) + 6H^+(aq) + 5e^- \rightleftharpoons \frac{1}{2}Cl_2(g) + 3H_2O(l)$ | +1.47 |
| | |
| $Co^{2+}(aq) + 2e^- \rightleftharpoons Co(s)$ | −0.277 |
| $Co^{3+}(aq) + e^- \rightleftharpoons Co^{2+}(aq)$ | +1.842 |
| $Cr^{3+}(aq) + 3e^- \rightleftharpoons Cr(s)$ | −0.74 |
| $Cr^{3+}(aq) + e^- \rightleftharpoons Cr^{2+}(aq)$ | −0.41 |
| $Cr_2O_7^{2-}(aq) + 14H^+(aq) + 6e^- \rightleftharpoons 2Cr^{3+}(aq) + 7H_2O(l)$ | +1.33 |
| $CrO_4^{2-}(aq) + 4H_2O(l) + 3e^- \rightleftharpoons Cr(OH)_3(s) + 5OH^-(aq)$ | −0.13 |
| | |
| $Cu^{2+}(aq) + 2e^- \rightleftharpoons Cu(s)$ | +0.337 |
| $Cu^{2+}(aq) + e^- \rightleftharpoons Cu^+(aq)$ | +0.153 |
| $Cu^+(aq) + e^- \rightleftharpoons Cu(s)$ | +0.521 |
| $CuI(s) + e^- \rightleftharpoons Cu(s) + I^-(aq)$ | −0.185 |
| $F_2(g) + 2e^- \rightleftharpoons 2F^-(aq)$ | +2.87 |
| | |
| $Fe^{2+}(aq) + 2e^- \rightleftharpoons Fe(s)$ | −0.440 |
| $Fe^{3+}(aq) + e^- \rightleftharpoons Fe^{2+}(aq)$ | +0.771 |
| $Fe(CN)_6^{3-}(aq) + e^- \rightleftharpoons Fe(CN)_6^{4-}(aq)$ | +0.36 |
| $2H^+(aq) + 2e^- \rightleftharpoons H_2(g)$ | 0.000 |
| $2H_2O(l) + 2e^- \rightleftharpoons H_2(g) + 2OH^-(aq)$ | −0.83 |

| Half-reaction | $E°$ (V) |
|---|---|
| $HO_2^-(aq) + H_2O(l) + 2e^- \rightleftharpoons 3OH^-(aq)$ | +0.88 |
| $H_2O_2(aq) + 2H^+(aq) + 2e^- \rightleftharpoons 2H_2O(l)$ | +1.776 |
| $Hg_2^{2+}(aq) + 2e^- \rightleftharpoons 2Hg(l)$ | +0.789 |
| $2Hg^{2+}(aq) + 2e^- \rightleftharpoons Hg_2^{2+}(aq)$ | +0.920 |
| $Hg^{2+}(aq) + 2e^- \rightleftharpoons Hg(l)$ | +0.854 |
| | |
| $I_2(s) + 2e^- \rightleftharpoons 2I^-(aq)$ | +0.536 |
| $IO_3^-(aq) + 6H^+(aq) + 5e^- \rightleftharpoons \frac{1}{2}I_2(s) + 3H_2O(l)$ | +1.195 |
| $K^+(aq) + e^- \rightleftharpoons K(s)$ | −2.925 |
| $Li^+(aq) + e^- \rightleftharpoons Li(s)$ | −3.05 |
| $Mg^{2+}(aq) + 2e^- \rightleftharpoons Mg(s)$ | −2.37 |
| | |
| $Mn^{2+}(aq) + 2e^- \rightleftharpoons Mn(s)$ | −1.18 |
| $MnO_2(s) + 4H^+(aq) + 2e^- \rightleftharpoons Mn^{2+}(aq) + 2H_2O(l)$ | +1.23 |
| $MnO_4^-(aq) + 8H^+(aq) + 5e^- \rightleftharpoons Mn^{2+}(aq) + 4H_2O(l)$ | +1.51 |
| $MnO_4^-(aq) + 2H_2O(l) + 3e^- \rightleftharpoons MnO_2 + 4OH^-(aq)$ | +0.59 |
| $HNO_2(aq) + H^+(aq) + e^- \rightleftharpoons NO(g) + H_2O(l)$ | +1.00 |
| | |
| $N_2(g) + 4H_2O(l) + 4e^- \rightleftharpoons 4OH^-(aq) + N_2H_4(aq)$ | −1.16 |
| $N_2(g) + 5H^+(aq) + 4e^- \rightleftharpoons N_2H_5^+(aq)$ | −0.23 |
| $NO_3^-(aq) + 4H^+(aq) + 3e^- \rightleftharpoons NO(g) + 2H_2O(l)$ | +0.96 |
| $Na^+(aq) + e^- \rightleftharpoons Na(s)$ | −2.71 |
| $Ni^{2+}(aq) + 2e^- \rightleftharpoons Ni(s)$ | −0.28 |
| $O_2(g) + 4H^+(aq) + 4e^- \rightleftharpoons 2H_2O(l)$ | +1.23 |
| | |
| $O_2(g) + 2H_2O(l) + 4e^- \rightleftharpoons 4OH^-(aq)$ | +0.40 |
| $O_2(g) + 2H^+(aq) + 2e^- \rightleftharpoons H_2O_2(aq)$ | +0.68 |
| $O_3(g) + 2H^+(aq) + 2e^- \rightleftharpoons O_2(g) + H_2O(l)$ | +2.07 |
| $Pb^{2+}(aq) + 2e^- \rightleftharpoons Pb(s)$ | −0.126 |
| $PbO_2(s) + HSO_4^-(aq) + 3H^+(aq) + 2e^- \rightleftharpoons PbSO_4(s) + 2H_2O(l)$ | +1.685 |
| | |
| $PbSO_4(s) + H^+(aq) + 2e^- \rightleftharpoons Pb(s) + HSO_4^-(aq)$ | −0.356 |
| $PtCl_4^{2-}(aq) + 2e^- \rightleftharpoons Pt(s) + 4Cl^-(aq)$ | +0.73 |
| $S(s) + 2H^+(aq) + 2e^- \rightleftharpoons H_2S(g)$ | +0.141 |
| $H_2SO_3(aq) + 4H^+(aq) + 4e^- \rightleftharpoons S(s) + 3H_2O(l)$ | +0.45 |
| $HSO_4^-(aq) + 3H^+(aq) + 2e^- \rightleftharpoons H_2SO_3(aq) + H_2O(l)$ | +0.17 |
| | |
| $Sn^{2+}(aq) + 2e^- \rightleftharpoons Sn(s)$ | −0.136 |
| $Sn^{4+}(aq) + 2e^- \rightleftharpoons Sn^{2+}(aq)$ | +0.154 |
| $VO_2^+(aq) + 2H^+(aq) + e^- \rightleftharpoons VO^{2+}(aq) + H_2O(l)$ | +1.00 |
| $Zn^{2+}(aq) + 2e^- \rightleftharpoons Zn(s)$ | −0.763 |

# Answers
# to Selected Exercises

## CHAPTER 1

**1.1** (a) Liquid; (b) solid; (c) gas; (d) liquid; (e) solid. **1.3** (a) Chemical; (b) physical; (c) chemical; (d) physical; (e) chemical; (f) chemical. **1.5** *Physical properties:* silver-white; soft; good conductor; boiling point = 883°C; violet-colored vapor. *Chemical properties:* prepared by passing electricity through molten salt; tarnishes rapidly in air; burns when heated in air; burns when heated in bromine vapor. **1.7** (a) Pure substance; (b) pure substance; (c) heterogeneous mixture; (d) heterogeneous mixture; (e) pure substance; (f) homogeneous mixture. **1.9** (a) C; (b) Na; (c) O; (d) Fe; (e) Mg; (f) Br; (g) Cu; (h) N. **1.11** To claim potassium as an element Davy argued that it derived from a more complex substance, potassium hydroxide, by decomposing that substance. To further substantiate his claim, Davy would have had to attempt converting potassium into still simpler substances. Since he could not readily do this, his claim was based mainly on the fact that the material he had formed had characteristic metallic properties and was very similar in character to sodium, a substance Davy also claimed to be an element. **1.13** (a) kg; (b) $m^3$; (c) m; (d) $m^2$; (e) s. **1.15** (a) milli; (b) nano; (c) centi; (d) micro. **1.17** (a) $4.53 \times 10^{-12}$ g; (b) $3.5 \times 10^4 \mu s$; (c) 13.27 m. **1.19** (a) 1.49 g/mL; (b) $1.17 \times 10^3$ g; (c) 1.60 mL. **1.21** (a) 20°C; (b) 378 K; (c) 5°F; (d) 32°C. **1.23** Exact: (a), (c), (d), (f). **1.25** (a) 3; (b) 2; (c) 4; (d) 4; (e) 3. **1.27** (a) $4.568 \times 10^6$; (b) $6.338 \times 10^3$; (c) $2.389 \times 10^{-3}$; (d) $9.876 \times 10^{-1}$; (e) $3.226 \times 10^{-4}$. **1.29** (a) 8.03; (b) 31.1; (c) 6.930; (d) $2.55 \times 10^{22}$. **1.31** (a) 91 cm; (c) 822 $cm^2$; (e) 25 m/s. **1.33** (b) 46 L; (d) 7.6 mi/hr; (f) 5.24 L. **1.35** 15 g CO. **1.37** Intensive: (b), (c), (e). **1.40** Density = 7.9 $g/cm^3$; the substance is iron. **1.43** (a) $1.245 \times 10^3$; (b) $6.50 \times 10^4$; (c) $5.975 \times 10^4$; (d) $4.56 \times 10^{-3}$. **1.47** Diameter = 0.830 cm. **1.49** Mass of toluene is 29.295 g; 33.8 mL in cylinder.

## CHAPTER 2

**2.1** Excess of one or another element can't alter the composition of the product, because the elements combine in a fixed ratio (law of constant composition or definite proportions). **2.3** (a) Gram O/gram N: A, 1.14; B, 2.28; C, 1.705. (b) Divide through by the smallest: A, 1.00; B, 2.00; C, 1.50. These data tell us that the ratios of O to N in the three compounds are related by small integers (2, 4, 3). **2.5** In both reactions oxygen from the air is the second reactant. The mass of rust produced equals the total mass of iron and oxygen reacted. When a match burns, there are several gaseous products in addition to the solid ash. The total mass of the un- noticed gases and the ash equals the mass of the match and oxygen reacted. **2.7** 12 excess electrons on the drop. **2.9** The Mg nuclei have a much smaller volume. There will be fewer direct hits. Also, positive charge in Mg nuclei is smaller than in Au nuclei; the charge repulsion between the alpha particles and the Mg nuclei will be less. Fewer alpha particles will be scattered in general, and fewer will be strongly back-scattered. **2.11** (a) $2.7 \times 10^2$ pm, 0.27 nm; (b) $3.7 \times 10^6$ Ir atoms. **2.13** (a) $^{80}$Br; 35 protons, 45 neutrons, 35 electrons; (b) $^{109}$Ag: 47 protons, 62 neutrons, 47 electrons; (c) $^{48}$Ti: 22 protons, 26 neutrons, 22 electrons. **2.15** (a) $^{39}_{19}$K; (b) $^{35}_{17}$Cl; (c) $^{29}_{14}$Si; (d) $^{32}_{16}$S, $^{34}_{16}$S.

**2.17**

| Symbol | $^{79}$Se | $^{80}$Br | $^{137}$Ba | $^{31}$P | $^{198}$Au |
|---|---|---|---|---|---|
| Protons | 34 | 35 | 56 | 15 | 79 |
| Neutrons | 45 | 45 | 81 | 16 | 119 |
| Electrons | 34 | 35 | 56 | 15 | 79 |
| Mass Number | 79 | 80 | 137 | 31 | 198 |

**2.19** (a) Mn (metal); (b) Br (nonmetal); (c) Cr (metal); (d) Ge (metalloid); (e) Hf (metal); (f) Se (nonmetal); (g) Ar (nonmetal). **2.21** (a) Ca and Sr are both in group 2A in adjacent rows and should have similar chemical and physical properties; (b) Al and Ga are in the same family and should be similar in many ways. **2.23** (a) $CH_2$; (b) $C_3H_8$; (c) $P_2O_5$; (d) $NO_2$; (e) CH; (f) $C_2HO_2$. **2.25** (a) $K^+$; (b) $F^-$; (c) $Ba^{2+}$; (d) $S^{2-}$; (e) $Al^{3+}$; (f) $Sc^{3+}$. **2.27** (a) $CaBr_2$; (b) $(NH_4)_2SO_4$; (c) $Mg(NO_3)_2$; (d) $Na_2S$; (e) KOH. **2.29** Molecular: $N_2O$, CO, $NF_3$. Ionic: $Na_2O$, CaO, $ScF_3$, KBr. **2.31** (a) Zinc chloride; (b) lead(II) chromate; (c) mercury(II) nitrate; (d) calcium cyanide, (e) iron(III) fluoride; (f) sodium carbonate. **2.33** (a) Hydroiodic acid; (b) carbonic acid; (c) nitrous acid; (d) $H_2CrO_4$; (e) HF; (f) HClO. **2.35** (a) Selenium dioxide; (b) carbon tetrachloride; (c) tetraphosphorus hexaoxide; (d) $IF_5$; (e) HCN; (f) $CS_2$. **2.37** (a) $ZnCO_3$, ZnO, $CO_2$; (b) HF, $SiO_2$, $SiF_4$, $H_2O$; (c) $SO_2$, $H_2O$, $H_2SO_3$; (d) $H_3P$ (or $PH_3$); (e) $HClO_4$, Cd, $Cd(ClO_4)_2$; (f) $VBr_3$.

**2.41**

| Symbol | $^{12}_{6}$C | $^{17}_{8}$O$^{2-}$ | $^{25}_{12}$Mg | $^{23}_{11}$Na$^+$ | $^{18}_{8}$O$^{2-}$ |
|---|---|---|---|---|---|
| Protons | 6 | 8 | 12 | 11 | 8 |
| Neutrons | 6 | 9 | 13 | 12 | 10 |
| Electrons | 6 | 10 | 12 | 10 | 10 |
| Net Charge | 0 | 2− | 0 | 1+ | 2− |

**2.44** (a) K; (b) Ca; (c) Ar; (d) Br; (e) Ge; (f) H; (g) Al; (h) O; (i) Ga. **2.46** (a) *Atomic number* is a measure of the number of protons in the atomic nucleus; the *mass number* gives the number of protons plus neutrons. (b) A *chemical property* is a characteristic of a substance as it undergoes a chemical change, that is, as it is converted into one or more other substances. A *physical property* (melting point, e.g.) is exhibited by a substance while it retains its chemical properties and composition. (c) The symbol Ca refers to the neutral calcium atom, with 20 electrons; the symbol $Ca^{2+}$ refers to the calcium ion, which has two electrons fewer than the neutral atom. (d) *Hydrochloric acid* is HCl; *chloric acid* is the oxyacid, $HClO_3$. (e) *Iron(II)* and *iron(III)* are two ionic forms of the element iron. In the iron(II) (or ferrous) form, the element has lost two electrons; in the iron(III) or ferric form, the element has lost three electrons. (f) *Sodium carbonate* is $Na_2CO_3$; *sodium bicarbonate* is $NaHCO_3$. (g) $H_2O$ is water; $H_2O_2$ is hydrogen peroxide. (h) A *metal* is an element with characteristic metallic properties, found mainly on the left side of the periodic table. *Nonmetals* are elements with characteristic properties different from those of metals; nonmetals are found on the upper right portion of the periodic table. (i) H represents the element hydrogen with 1 proton in the nucleus and one external electron per atom. He represents helium, which has 2 protons per nucleus and two electrons in each atom. (j) *Chloride* ion is $Cl^-$; *chlorate* ion is $ClO_3^-$. **2.48** (a) CaS, $Ca(HS)_2$; (b) HBr, $HBrO_3$; (c) AlN, $Al(NO_2)_3$; (d) FeO, $Fe_2O_3$; (e) $NH_3$, $NH_4^+$; (f) $K_2SO_3$, $KHSO_3$; (g) $Hg_2Cl_2$, $HgCl_2$; (h) $HClO_3$, $HClO_4$.

## CHAPTER 3

**3.1** (a) and (b) are inconsistent. **3.3** (a) 1, 1, 2; (c) 1, 4, 2, 1; (e) 1, 4, 1, 5. **3.5** (a) $2PH_3(g) + 4O_2(g) \longrightarrow 3H_2O(g) + P_2O_5(s)$; (c) $B_2S_3(s) + 6H_2O(l) \longrightarrow 2H_3BO_3(aq) + 3H_2S(g)$. **3.7** (a) $Zn(s) + H_2SO_4(aq) \longrightarrow H_2(g) + ZnSO_4(aq)$; displacement; (b) $2H_2O_2(aq) \longrightarrow 2H_2O(l) + O_2(g)$; decomposition; (c) $2Fe(s) + 3Cl_2(g) \longrightarrow 2FeCl_3(s)$; combination; (d) $Na_2CO_3(aq) + Ca(OH)_2(aq) \longrightarrow CaCO_3(s) + 2NaOH(aq)$; metathesis. **3.9** (a) $2C_4H_{10}(g) + 13O_2(g) \longrightarrow 8CO_2(g) + 10H_2O(l)$; (c) $Al(OH)_3(s) + 3HNO_3(aq) \longrightarrow Al(NO_3)_3(aq) + 3H_2O(l)$; (e) $2AgNO_3(aq) + H_2SO_4(aq) \longrightarrow Ag_2SO_4(s) + 2HNO_3(aq)$. **3.11** (a) $C_2H_5OH(l) + 3O_2(g) \longrightarrow 2CO_2(g) + 3H_2O(l)$; (b) $Pb(NO_3)_2(aq) + Na_2CrO_4(aq) \longrightarrow PbCrO_4(s) + 2NaNO_3(aq)$; (c) $2K(s) + H_2O(l) \longrightarrow 2KOH(aq) + H_2(g)$; (d) $H_2SO_4(aq) + 2KOH(aq) \longrightarrow K_2SO_4(aq) + 2H_2O(l)$ or $Ca(OH)_2(s) + H_2SO_4(aq) \longrightarrow CaSO_4(s) + 2H_2O(l)$; (e) $Ca(OH)_2(s) + 2HNO_3(aq) \longrightarrow Ca(NO_3)_2(aq) + 2H_2O(l)$ (Note: reaction of $Ca(OH)_2$ with $H_2SO_4$ leads to formation of insoluble $CaSO_4$); (f) $H_2SO_4(aq) + BaCl_2(aq) \longrightarrow BaSO_4(s) + 2HCl(aq)$. **3.13** Atomic weight = 24.32. **3.15** $^{107}Ag = 51.75\%$; $^{109}Ag = 48.25\%$. **3.17** Average atomic weight of Pb = 207.3 amu. **3.19** Since both are diatomic, each volume of chlorine or fluorine molecules reacting can produce two volumes of chlorine or fluorine atoms. "Two volumes of product" tells us that the atom ratio of chlorine to fluorine in the product must be the reacting volume ratio of 1:3. Thus the formula of the product is $ClF_3$. **3.21** Berzelius's value for the atomic weight of Zn was 129.0 g. This is high by a factor of two because he incorrectly assumed the formula of the oxide to be $ZnO_2$; the true formula is ZnO. Correcting this leads to a value of 64.50 g Zn/mol, in good agreement with the currently accepted value. **3.23** (a) 64.1; (d) 132.1; (g) 60.1. **3.25** (a) N = 30.45%, O = 69.55%; (e) Ca = 24.73%, C = 14.82%, O = 59.22%, H = 1.24%. **3.27** (a) 164.1 g; (b) 0.0152 mol $Ca(NO_3)_2$; (c) 53.3 g $Ca(NO_3)_2$; (d) $3.20 \times 10^{22}$ $Ca^{2+}$ ions. **3.29** (a) 132 g $CO_2$; (b) 0.238 g

$C_2H_4$; (c) 2.38 g Ar; (d) 4.83 g caffeine. **3.31** (a) 0.783 mol $Fe_2O_3$; (b) $7.22 \times 10^{-5}$ mol $Ca(OH)_2$; (c) $2.32 \times 10^{-4}$ mol $H_2O$; (d) 0.229 mol diamond. **3.33** $8.0 \times 10^{-8}$ mol; $4.8 \times 10^{16}$ molecules. **3.35** (a) $CH_4$; (b) 0.209 mol Fe, 0.313 mol O, thus $Fe_2O_3$; (c) $NH_2$. **3.37** (a) $CH_2O$; (c) $Na_3AlF_6$. **3.39** Empirical formula is $CH_3O_2$. **3.41** Empirical and molecular formulas are $C_9H_8O_4$. **3.43** $x = 8$; $Sr(OH)_2 \cdot 8H_2O$. **3.45** Molecular weight of fungal laccase is approximately $6.6 \times 10^4$ g. **3.47** (a) 2.63 g $NH_3$; (b) 6.52 g $CaCO_3$. **3.49** 4.37 kg $NH_4ClO_4$ per kg Al. **3.51** 5.249 g $Al_4C_3$; 19.44 g $AlCl_3$. **3.53** (a) 0.0515 mol $SiH_4$; (b) 3.71 g $H_2O$; (c) $5.54 \times 10^{-4}$ mol $SiH_4$ from $H_2O$. Thus, $H_2O$ is the limiting reagent; a maximum of 12.0 mg $SiH_4$ can be formed. **3.55** KBr is the limiting reagent; 5.45 g AgBr are formed in the precipitation reaction. **3.57** (a) Theoretical yield = 60.3 g $C_6H_5Br$; (b) 94.0% yield. **3.59** (a) 0.134 $M$; (b) 0.300 moles HCl; (c) 50.0 mL of 2.00 $M$ NaOH. **3.61** (a) 3.55 g $Na_2SO_4$; (b) 0.125 L solution. **3.63** (a) 41.7 mL; (b) 623 mL; (c) 0.465 $M$ $AgNO_3$ solution; (d) 0.235 g KOH. **3.65** 0.0912 L $Na_2CrO_4$. **3.67** (a) $Li_3N(s) + 3H_2O(l) \longrightarrow NH_3(g) + 3LiOH(aq)$; (c) $PBr_3(l) + 3H_2O(l) \longrightarrow H_3PO_3(aq) + 3HBr(aq)$; (e) $2CCl_4(l) + O_2(g) \longrightarrow 2CCl_2O(g) + 2Cl_2(g)$. **3.71** (a) 33.81% Sm; (c) 79.11% C. **3.75** Empirical and molecular formulas are $C_{10}H_{18}O$. **3.78** (a) $Al(OH)_3(s) + 3HCl(aq) \longrightarrow 3H_2O(l) + AlCl_3(aq)$; (b) 0.19 mol HCl. **3.81** $O_2$ is the limiting reagent for the reaction $2H_2(g) + O_2(g) \longrightarrow 2H_2O(g)$. Obtain 0.22 g $H_2$ and 29.2 g $H_2O$ upon completion of reaction. **3.84** (a) $3.47 \times 10^2$ kg $C_6H_{10}O_4$; (b) percent yield = 85.0%. **3.86** (a) 0.595 $M$; (b) 0.0231 mol $HNO_3$, 0.0462 $M$; (c) 0.382 $M$; (d) 0.041 mol KBr, 0.41 $M$. **3.88** (a) 0.133 $M$. **3.90** 2.8 L solution.

## CHAPTER 4

**4.1** (a) 2.1 kJ; (b) $5.1 \times 10^2$ cal; (c) This kinetic energy is transformed into work needed to deform the slug and into heat. **4.3** (a) Energy can be neither created nor destroyed, but it can be changed in form. (b) The well-defined part of the universe whose energy changes are being described. (c) The energy change of a system is equal in magnitude and opposite in sign to that of its surroundings. **4.5** (a) $\Delta E = -450$ J; (b) $\Delta E = +8360$ J; (c) $\Delta E = +154$ J. **4.7** (a) Independent; (b) dependent; (c) dependent. **4.9** Enthalpy change, $\Delta H$. **4.11** $\Delta H = +600$ J, $\Delta E = +460$ J. **4.13** (a) Exothermic; (c) 52.3 kJ produced; (d) $3.88 \times 10^{-2}$ g Al; (e) 78.4 kJ consumed. **4.15** (a) $-13.1$ kJ; (b) $-1.14$ kJ; (c) $+22.9$ kJ. **4.17** $H$ of $H_2O(s) < H_2O(l) < H_2O(g)$ for a given temperature and pressure. **4.19** If a reaction can be described as a series of steps, $\Delta H$ for the reaction is the sum of the enthalpy changes for each step. As long as we can describe a route where $\Delta H$ for each step is known, $\Delta H$ for any process can be calculated. **4.21** $\Delta H = +180.8$ kJ. **4.23** $\Delta H = -86$ kJ. **4.25** (a) $Ca(s) + C(s) + 1.5O_2(g) \longrightarrow CaCO_3(s)$; $\Delta H_f^\circ = -1207.1$ kJ. **4.27** $\Delta H_f^\circ = -1079$ kJ/mol $Ga_2O_3$. **4.29** $\Delta H^\circ = -847.6$ kJ. **4.31** (a) $\Delta H^\circ = -92.38$ kJ; (d) $\Delta H^\circ = -904.6$ kJ. **4.33** $\Delta H_f^\circ = -60.7$ kJ/mol. **4.35** (a) 3.58 kJ/°C; (b) 328 kJ. **4.37** $7.60 \times 10^3$ J. **4.39** $\Delta H = -87$ kJ/mol $CaCl_2$. **4.41** $2.75 \times 10^3$ kJ/mol of quinone. **4.43** (a) 8.67 kJ/°C; (b) 2.39 kJ/°C; (c) 4.82°C. **4.45** 94 Cal. **4.47** 463 kJ or 111 Cal. **4.49** (a) $C_3H_4$: $\Delta H = -1849.5$ kJ/mol or $4.61 \times 10^4$ kJ/kg; (b) $C_3H_6$: $\Delta H = -1926.3$ kJ/mol or $4.57 \times 10^4$ kJ/kg; (c) $C_3H_8$: $\Delta H = -2043.9$ kJ/mol or $4.63 \times 10^4$ kJ/kg. These three substances yield nearly identical quantities of heat per unit mass, but propane is marginally higher than the other two. **4.51** $2.05 \times 10^{14}$ kJ; $7.3 \times 10^6$ tons of coal. **4.53** The 1.0-kg object has a kinetic energy of 2.0 J, and the 2.0-kg object a kinetic energy of 1.0 J; the 1.0 kg object has the greater kinetic energy. **4.56** The system does 0.23 J of work on its surroundings. **4.59** $\Delta H = -319.5$ kJ. **4.63** The heat capacity of 1000 gal of water is $1.58 \times 10^4$ kJ/°C. $1.0 \times 10^4$ bricks also provide this heat capacity. **4.64** $\Delta H = -2.6$ kJ for the process as written; $-52$ kJ/mol $CuSO_4(aq)$ reacted. **4.68** $\Delta H = -1.5 \times 10^4$ kJ/L $NH_3(l)$; $\Delta H = -1.58 \times 10^4$ kJ/L $CH_3OH(l)$. **4.71** The sample is 55.0% boric oxide by weight.

## CHAPTER 5

**5.1** In Mendeleev's chart, the elements were arranged by increasing atomic weight, and in the modern chart, they are arranged by increasing atomic number. **5.3** (a) Representative elements: Rb, Al, Pb; (b) transition metal: Cu, Fe, Pt; (c) inner transition metals: U. **5.5** (a) Active metals: Ca, Li; (b) noble gases: Ar, Xe; (c) lanthanides: Ce. **5.7** Metals are good electrical conductors and nonmetals are poor ones, so the order of increasing metallic character and electrical conductivity is S < Si < Sn < Ca. **5.9** (a) Li; (b) Na; (c) Sn; (d) Al. **5.11** (a) $MgI_2$; (b) $Ga_2O_3$; (c) $K_2S$. **5.13** Ionic: $Na_2O$, $CaO$, $Fe_2O_3$; molecular: $N_2O$, $CO$, $P_2O_5$, $Cl_2O_7$. **5.15** (a) BaO; (b) MgO. **5.17** (a) $Na_2O(s) + H_2O(l) \longrightarrow 2NaOH(aq)$; (b) $CuO(s) + 2HNO_3(aq) \longrightarrow Cu(NO_3)_2(aq) + H_2O(l)$; (c) $SO_3(g) + H_2O(l) \longrightarrow H_2SO_4(aq)$; (d) $SeO_2(s) + 2NaOH(aq) \quad Na_2SeO_3(aq) + H_2O(l)$. **5.19** $2Mg(s) + O_2(g) \longrightarrow 2MgO(s)$; $3Mg(s) + N_2(g) \longrightarrow Mg_3N_2(s)$; $Mg(s) + H_2(g) \longrightarrow MgH_2(s)$; $Mg(s) + Cl_2(g) \longrightarrow MgCl_2(s)$. **5.21** The most easily oxidized metals are on the left side of the chart, particularly in groups 1A and 2A. The least easily oxidized are in the lower right of the transition metals, especially those in or near group 1B, the coinage metals. **5.23** (a) $2HCl(aq) + Ni(s) \longrightarrow NiCl_2(aq) + H_2(g)$; (d) $2HC_2H_3O_2(aq) + Mg(s) \longrightarrow Mg(C_2H_3O_2)_2(aq) + H_2(g)$. **5.25** (a) $2Al(s) + 3NiCl_2(aq) \longrightarrow 2AlCl_3(aq) + 3Ni(s)$; (b) no reaction; (c) $2Cr(s) + 3NiSO_4(aq) \longrightarrow Cr_2(SO_4)_3(aq) + 3Ni(s)$. **5.27** The activity of C > A > D > B. (C is most easily oxidized, B is least easily oxidized.) **5.29** Silver is less active than hydrogen, so it will not displace $H^+$ from HCl. In the reaction of nitric acid, $HNO_3$, with silver, it is the nitrate ion that oxidizes or removes electrons from the silver, not the $H^+$. **5.31** (a) $H^-$; (b) $N^{3-}$; (c) $Be^{2+}$; (d) $Ga^{3+}$. **5.33** (a) Bromine; (b) argon; (c) potassium; (d) aluminum; (e) hydrogen. **5.35** A: $CO_2$; B: $H_2$; C: $O_2$. **5.38** $H_2(g)$, $N_2(g)$, $O_2(g)$, $F_2(g)$, $Cl_2(g)$, $Br_2(l)$, $I_2(s)$. **5.40** The gases are $PH_3$, $N_2O$, and $O_3$. These are molecular compounds formed by nonmetals and are likely to be gases or have low boiling or melting points. The other choices are compounds formed from a metal and a nonmetal, which tend to be solids that melt at high temperatures. **5.42** (a) $F_2$; Al has a greater tendency to lose electrons than Si, but $F_2$ will gain electrons and form an anion. (b) Fe; Fe is above Cu on the activity series and will displace it. (c) $CrCl_3$; compounds formed from a metal and a nonmetal are usually solids that melt at high temperatures. (d) NiO; oxides of metals are basic and react with acids to form salts and water. Nonmetal oxides, such as $NO_2$, are themselves acidic in aqueous solution and wouldn't react with acids. **5.43** (b) $SnO(s) + 2HClO_4(aq) \longrightarrow Sn(ClO_4)_2(aq) + H_2O(l)$; (c) $SO_3(g) + Ca(OH)_2(aq) \longrightarrow CaSO_4(s) + H_2O(l)$.

## CHAPTER 6

**6.1** Wavelength of (d) cosmic rays < (e) X rays < (c) green light < (a) heat radiation < (b) FM radiation. **6.3** (a) $\lambda = 4.81 \times 10^{-6}$ m; (c) distance = $1.08 \times 10^{12}$ m. **6.5** $v = 6.59 \times 10^{14}/s$; the color is blue. **6.7** (a) $E = 2.98 \times 10^{-19}$ J; (b) $E = 3.0 \times 10^{-21}$ J; (c) $v = 3.24 \times 10^{15}/s$. **6.9** $E$ (ultraviolet) = $1.88 \times 10^{-18}$ J; $E$ (infrared) = $4.5 \times 10^{-21}$ J. The longer wavelength infrared photon clearly has the lower energy. **6.11** Blue to blue-violet; $E = 4.65 \times 10^{-19}$ J/photon; $E = 280$ kJ/mol. **6.13** A single photon of 540 nm wavelength has an energy $E = 3.68 \times 10^{-19}$ J. There are 11 photons in the signal. **6.15** $\lambda = 360$ nm. **6.17** (a) $E_{min} = 3.68 \times 10^{-19}$ J; (b) $v = 5.56 \times 10^{14}/s$; (c) $E$ of 440-nm photon = $4.51 \times 10^{-19}$ J; maximum kinetic energy of emitted electron = $8.4 \times 10^{-20}$ J. **6.19** (a) Absorbed; (b) emitted; (c) absorbed. **6.21** $\Delta E = 4.57 \times 10^{-19}$ J; $\lambda = 435$ nm. Light of this wavelength lies in the blue region of the spectrum. **6.23** (a) $r = 4.77 \times 10^{-10}$ m; (b) the larger the value of $Z$, the more strongly the electron is attracted to the nucleus and the smaller its orbit. **6.25** $\lambda = 1.1 \times 10^{-34}$ m. **6.27** $v = 4.51 \times 10^3$ m/s. **6.29** (a) $n = 5$; $l = 0, 1, 2, 3, 4$; (b) $l = 3$; $m_l = 3, 2, 1, 0, -1, -2, -3$. **6.31** (a) $n = 4$; $l = 2$; $m_l = 2, 1, 0, -1, -2$; (b) $n = 3$, $l = 0$, $m_l = 0$; $n = 3$, $l = 1$, $m_l = 1, 0, -1$; $n = 3$, $l = 2$, $m_l = 2, 1, 0, -1, -2$. **6.33** Incorrect atomic orbitals: $3f$, $2d$.

**6.35**

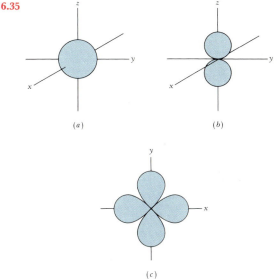

**6.37** The square of the wave function has the physical significance of an amplitude, or probability. The quantity $\psi^2$ at a given point in space is the probability of the electron's being located within a small volume around that point at any given instant. The total probability—that is, the sum for $\psi^2$ over all the space around the nucleus—must equal 1. **6.39** (a) $\lambda = 3.04$ m; (b) $\lambda = 405$ m. **6.42** (a) $v = 3.85 \times 10^{14}/s$; (b) $E = 2.55 \times 10^{-19}$ J. **6.45** (a) Work function = $3.46 \times 10^{-19}$ J/atom or 208 kJ/mol; (b) maximum kinetic energy of emitted photon = $1.28 \times 10^{-19}$ J. **6.48** The emission lines correspond to electronic transitions within the neon atoms. These transitions occur between electronic energy states of well-defined energies. Thus only certain energy *changes* can occur within the neon atom. These specific changes in energy give rise to emission at specific wavelengths. **6.49** An electron in a hydrogen atom has a higher energy in $n = 4$ than $n = 1$. The ground state is $n = 1$ and an excited state is $n = 4$. **6.52** $\Delta E = -1.06 \times 10^{-19}$ J; $n_i = 4$. **6.55** (a) $\lambda = 1.2 \times 10^{-11}$ m; (b) $\lambda = 8.8 \times 10^{-15}$ m. **6.58** (a) $2s$; (b) $4d$; (c) $5p$; (d) $3d$; (e) $4f$. **6.61** (a) 25; (b) 3; (c) 1; (d) 0 (forbidden designation); (e) 1; (f) 5; (g) 1. **6.64** (a) True, (b) False; the energy of the electron in the one-electron atom or ion depends on *both* $n$ and the nuclear charge, $Z$. (c) True, (d) True, (e) False; although $3p_z$ and $3d_{z^2}$ both have lobes pointing along the $z$ axis, the $3d_{z^2}$ has a donut-shaped lobe in the $x$-$y$ plane.

## CHAPTER 7

**7.1** (a) In hydrogen, orbitals with the same $n$ value are degenerate; (b) in many-electron atoms, orbitals with the same $n$ and $l$ values are degenerate. **7.3** The $3s$ electron in magnesium experiences a nuclear charge one higher than in sodium. This added unit of nuclear charge is partially, *but not completely*, cancelled by the presence of a second $3s$ electron. The overall effect is an increase in the apparent nuclear charge. **7.5** (a) $2s$; (b) $3d$; (c) $2p$. **7.7** (a) 10; (b) 14; (c) 14; (d) 6. **7.9** Li: $1s^2 2s^1$. $1s$ electrons: 1, 0, 0, $\frac{1}{2}$; 1, 0, 0, $-\frac{1}{2}$. $2s$ electron: 2, 0, 0, $\frac{1}{2}$, *or* 2, 0, 0, $-\frac{1}{2}$. **7.11** (a) False. It can have an $m_l$ value from 2 to $-2$; $l = 2$ for a $d$ orbital, regardless of the value for $n$, and $m_l$ ranges from $+l$ to $-l$. (b) True. **7.13** (b) Ge: $[\text{Ar}]4s^2 3d^{10}4p^2$; (f) Ti: $[\text{Ar}]4s^2 3d^2$.

**7.15** (b)

$4s$ $\qquad$ $3d$

(c) $5s$ $\qquad$ $4d$ $\qquad$ $5p$

**7.17** (a) Alkaline earth metals; (b) halogens; (c) alkali metals; (d) transition metals, 1B. **7.19** (a) 3; (b) 5; (c) 6; (d) 5. **7.21** Moving from left to right and ignoring discontinuities due to the special stability of half-filled shells, (a) effective nuclear charge increases, (b) atomic size decreases, (c) ionization energy increases, (d) electron affinity becomes more negative. **7.23** C < Cl < S < Co < Mg < Kr < K. (The position of Kr is uncertain because we cannot easily determine a *covalent* atomic radius for it.) **7.25** (a) Cl; (b) N; (c) Hf; (d) N (special stability of half-filled subshell); (e) Ge. **7.27** To form $\text{AlO}_2$, the aluminum would need to lose four electrons. This would mean removing a $2p$ electron. The energy cost of doing this is given by $I_4$ in Table 7.3. Note that it is much greater than $I_4$ for Si, because the effective nuclear charge experienced by the $2p$ electron in the $\text{Al}^{3+}$ ion is very high. The high energy cost of removing this electron rules out chemical behavior that corresponds to an $\text{Al}^{4+}$ species. **7.29** When an electron is added to the $3p$ orbital of Cl, it experiences electron-electron repulsions, but it also experiences some electron-nuclear attraction, because the other five $3p$ electrons do not completely shield it from the nucleus. Also, a stable octet of electrons is formed. In Ar the stable octet already exists. The added electron must go into a $4s$ orbital, in which it is nearly completely shielded from the nucleus. The arrangement is therefore not stable relative to Ar. **7.31** (a) K: $[\text{Ar}]4s^1$, Ca: $[\text{Ar}]4s^2$; (b) $\text{K}^+$, $\text{Ca}^{2+}$; (c) lower for K; (d) KOH, $\text{Ca(OH)}_2$; (e) 63°C for K, 851°C for Ca; (f) larger for K (Figure 7.7); (g) more negative for K (i.e., more exothermic.) Most of these differences can be accounted for in terms of the outer electron configurations and the tendency toward higher effective nuclear charge as we move from left to right in a row. The electron affinity of Ca is poor because it has a closed subshell of $1s$ electrons. (Compare Mg, Figure 7.10.) **7.33** (b) $\text{Ba}(s) + 2\text{H}_2\text{O}(l) \longrightarrow \text{Ba(OH)}_2(aq) + \text{H}_2(g)$; (d) $2\text{Na}(g) + \text{Br}_2(g) \longrightarrow 2\text{NaBr}(s)$.

**7.35** It is smaller for the $2s$ electron. We know that this is so, because the $2s$ energy level lies lower than the $2p$, and the energy level in turn is determined by the average electron-nuclear attraction. **7.38** The energy of electron (d) < (a) < (c) = (b). **7.41** (a) $1s^2 2s^2 2p^6 3s^2 3p^6 4s^2 3d^{10} 4p^6 5s^2 4d^{10} 5p^1$; (c) $1s^2 2s^2 2p^6 3s^2 3p^6 4s^2 3d^{10} 4p^6 5s^2 4d^{10} 5p^6 6s^2 5d^1$; (e) $1s^2 2s^2 2p^6 3s^2 3p^5$. **7.43** (a) Sr; (b) Cl; (c) As; (d) Fr; (e) Si, S. **7.46** The electron configuration of Na is $[\text{Ne}]3s^1$ and that of Mg is $[\text{Ne}]3s^2$. The first ionization energy for Mg is larger, because as the effective nuclear charge increases moving from left to right across a row, the ionization energy increases. After losing the first electron, $\text{Mg}^+$ has one more outer-shell electron to lose, but $\text{Na}^+$ has assumed the stable noble gas configuration of Ne; it requires a much larger amount of energy to remove an inner-shell electron from $\text{Na}^+$ than to remove an outer-shell electron from $\text{Mg}^+$. **7.49** O < F < Ne. **7.52** $\text{O}^-$ has a smaller effective nuclear charge than F and there are electrostatic repulsions when adding an electron to the negatively charged $\text{O}^-$, making $\Delta H$ for the second process more positive. **7.55** The $g$ orbitals correspond to $l = 4$. Thus, the lowest value that $n$ could have would be 5. There are $2l + 1$ orbitals of a given value of $l$ and $n$. Thus, there are 9 $g$ orbitals. There are no known elements with electrons occupying $g$ orbitals in the ground state. **7.59** When we move one place to the right in a horizontal row of the table—for example, from Li to Be—ionization energy increases. When we move downward in a given family—for example, from Be to Mg—ionization energy usually decreases. Atomic size decreases in moving one place to the right and increases in moving downward. Thus, two elements, such as Li and Mg, that are diagonally related tend to have similar ionization energies and atomic sizes. This in turn gives rise to some similarities in chemical behavior. Note, however, that the valences expected for the elements are not the same. That is, lithium still appears as $\text{Li}^+$, magnesium as $\text{Mg}^{2+}$.

## CHAPTER 8

**8.1** (a) $:\overset{\textbf{..}}{\underset{\textbf{.}}{\text{S}}}\cdot$ (c) $\cdot\overset{}{\text{Al}}\cdot$

**8.3** $\text{Na}\cdot + \text{Na}\cdot + \overset{\textbf{..}}{\underset{\textbf{..}}{\text{O}}}: \longrightarrow 2\text{Na}^+ + \left[:\overset{\textbf{..}}{\underset{\textbf{..}}{\text{O}}}:\right]^{2-}$

**8.5** (a) $\text{Sc}_2\text{O}_3$; (c) BaS. **8.7** (a) $\text{Ba}^{2+}$: [Xe] (noble-gas configuration); (c) $\text{Mn}^{2+}$: $[\text{Ar}]3d^5$. **8.9** The energy required to form the $\text{Ca}^{2+}$ and $\text{O}^{2-}$ ions is more than compensated by the stability associated with the CaO lattice. When a mole of $\text{Ca}^{2+}$ and $\text{O}^{2-}$ ions come together to form solid CaO, there is a large evolution of energy associated with the electrostatic attractions of the ions of opposite charge. The net energy in this process is called the *lattice energy*. **8.11** Mg has a larger charge $(2+)$ than Li $(1+)$, so the $\text{Mg}^{2+}$ ion is smaller. This in turn results in a much larger attractive interaction between the anions and cations. **8.13** (a) As $Z$ stays constant and the number of electrons increases, the electron-electron repulsions increase and the size increases. (b) Going *down* a family, the valence electron configuration of particles with like charge is the same, but the distance of these electrons from the nucleus increases and shielding by inner electrons increases. This outweighs the increase in $Z$ and causes the size of the particles to increase. **8.15** The $3p$ electron in $\text{K}^+$ experiences a larger effective nuclear charge, because the $4s$ electron removed in forming $\text{K}^+$ served to provide a *little* shielding of the nucleus from the $3p$ electrons (penetration effect). **8.17** (a) Kr; (b) Te.

**8.19** (b) $[:\ddot{O}—\ddot{C}l—\ddot{O}:]^-$ (c) $\ddot{O}=C=\ddot{O}$ There are two other possible resonance forms for $CO_2$, but based on formal charge arguments, these contribute very little to the overall structure. **8.21** (a) Odd number of electrons (19) in molecule, so one electron will be unpaired and one atom in the molecule cannot have an octet. (c) 34 electrons in the molecule; 10 electrons (expanded octet) in the valence shell of Te; (d) 24 electrons in the molecule, 6 electrons (incomplete octet) in the valence shell of B.

**8.23** (a)

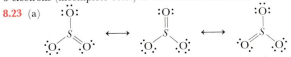

**8.25** The N—O bond orders in these three cases (that is, the average number of electron pairs in the N—O bond) is 2 for NO, $1\frac{1}{3}$ for $NO_3^-$, and $1\frac{1}{2}$ for $NO_2^-$. The N—O bond length is inversely related to N—O bond order (shorter NO bond length for higher NO bond order). Thus the N—O bond lengths vary in the order $NO < NO_2^- < NO_3^-$. **8.27** (a) $\Delta H = -46$ kJ. **8.29** (a) $\Delta H = -213$ kJ. **8.31** Average bond dissociation energy for Ti—Cl $= 430$ kJ/mol. **8.33** Electronegativity increases in moving from left to right in a row, because the effective nuclear charge is increasing. The atom thus has a greater capacity to attract electrons to itself in a bond. **8.35** (a) P < S < O; (b) Mg < Al < Si; (c) S < Br < Cl; (d) Si < C < N. **8.37** All except (b) and (c) are polar to some extent, because the two elements differ in electronegativity. The more electronegative in each case is (a) Cl; (d) F; (e) C. **8.39** (a) Mn, $+7$; (b) Fe, $+3$; (c) Xe, $+6$. **8.41** (a) Br: $+7, -1$; (c) Ba: $+2, 0$. **8.43** (a) $FeF_3$; (b) $MoO_3$; (d) vanadium(III) oxide. **8.45** (b) Copper is reduced from $+2$ to $+1$. Iodine is oxidized from $-1$ to 0. **8.47** (a) 0; (c) $+1$; (f) $-1$. **8.49** (a) MgO; (b) $SiO_2$; (c) $V_2O_5$; (d) $Cl_2O_7$. **8.51** (a) basic; (b) acidic; (c) basic; (d) acidic; (e) amphoteric; (f) basic. **8.53** (a) $ZnO(s) + 2HCl(aq) \longrightarrow ZnCl_2(aq) + H_2O(l)$; (b) $P_2O_5(s) + 6NaOH(aq) \longrightarrow 2Na_3PO_4(aq) + 3H_2O(l)$. **8.55** (a) $SO_3$; (b) $NO_2$; (c) $CrO_3$. **8.57** (a) For ionic oxides, the lower the lattice energy, the more soluble the oxide. Since $Al^{3+}$ has a higher positive charge than $Ca^{2+}$ and a correspondingly higher lattice energy, $Al_2O_3$ is less soluble than CaO. (b) For metal ions with equal charges the smaller cation will have the less soluble oxide, so CaO is less soluble than SrO. **8.59** (a) Group 3A or 3B; (b) group 2A; (c) group 5A or 5B. **8.61** $Cu^+$ in (a) and $Cd^{2+}$ in (b) do *not* have noble-gas electron configurations. **8.63** Each elemental transition metal has two $ns$ electrons in its outer shell, which will be lost before $(n-1)d$ electrons upon ionization. There is a small energy difference between the $ns$ and $(n-1)d$ subshells, so the $+2$ oxidation state will be stable, regardless of further ionization involving $(n-1)d$ electrons. **8.65** The steady decrease in lattice energies is due to the increasing size of the cation. The potential energy of interaction varies as $1/d$, where $d$ is the distance between ionic centers. Thus, the larger the cation, the larger $d$ and the lower the lattice energy. **8.68** (a) Cl; (b) $Ca^{2+}$; (c) $Ca^{2+}$; (d) $Ca^{2+}$. **8.70** (a) $SiH_4$; (b) $SF_2$; (c) $PCl_3$. **8.73** (a) $\ddot{N}=N=\ddot{O} \longleftrightarrow :N\equiv N—\ddot{O}:$ There is a third possible resonance form for $N_2O$, but its contribution to the overall structure is small, on the basis of formal charge arguments.

(c) $\overset{H}{\underset{H}{>}}C=\ddot{O}$. **8.76** (a) $\Delta H = +1547$ kJ. **8.79** (a) N, $+1$; O, $-2$; (b) K, $+1$; Bi, $+5$; O, $-2$. **8.82** The compound is most likely $GeO_2$ (m.p. $GeO_2 = 1115°C$).

## CHAPTER 9

**9.1** (a) Trigonal planar; (b) tetrahedral; (c) trigonal bipyramidal; (d) octahedral. **9.3** The electron-pair geometry describes the locations of electron pairs as predicted by VSEPR. The molecular geometry indicates the positions of the atoms in the molecule. In $H_2O$, there are four pairs of electrons around oxygen, so the electron-pair geometry is tetrahedral. Since there are two bonding and two nonbonding pairs, the molecular geometry (the arrangement of the three atoms) is bent. **9.5** (a) Tetrahedral, bent; (b) tetrahedral, trigonal pyramidal; (c) trigonal planar, trigonal planar. **9.7** $NF_3$—trigonal pyramidal; $BF_3$—trigonal planar; $ClF_3$—T-shaped. The molecular geometries or shapes differ because there is a different number of nonbonding (lone) pairs around each central atom. **9.9** Going from $NO_2^+$ to $NO_2^-$, the number of nonbonding electrons in the valence shell of N increases from 0 to 1 to 2 and the bond angle decreases correspondingly. **9.11** (a) 1—$109°$; 2—$120°$; (d) 7—$180°$, 8—$109°$. **9.13** (a) No; (b) yes; (c) no. **9.15** In $BeCl_2$ there are two pairs of electrons around Be, the structure is linear, and the bond dipoles cancel. In $SCl_2$ there are four pairs of electrons around S, and the two lone pairs do not cancel with the two S—Cl bonding pairs. **9.17** 11.7% ionic character. **9.19** (a) $sp^2$, $120°$ angles; (b) $dsp^3$, $90°$ and $120°$ angles; (c) $d^2sp^3$, $90°$ bond angles. **9.21** $H—\overset{..}{\underset{..}{S}}—H$. The electron-pair geometry is tetrahedral, and the molecular shape is bent. Each H—S $\sigma$ bond is formed by overlap of an $sp^3$ hybrid orbital on S with a $1s$ atomic orbital on H. **9.23** (a) $sp^2$; (b) $dsp^3$; (c) $sp^3$. **9.25** Two pure $p$ orbitals remain; two $\pi$ bonds can be formed. **9.27** (a) $\sim109°$ about the leftmost C, $sp^3$; $\sim120°$ about the right-hand C, $sp^2$. (b) The doubly bonded O can be viewed as $sp^2$, the other as $sp^3$; the nitrogen is $sp^3$ with approximately $109°$ bond angles. (c) 9 $\sigma$ bonds, one $\pi$ bond. **9.29** There are three equally correct Lewis structures (resonance forms) for nitrate ion, $NO_3^-$. The true model of the bonding is taken as an average or composite of the three forms. Each atom in $NO_3^-$ can be viewed as using $sp^2$ hybrid orbitals for $\sigma$ bonding, so each has a pure $2p$ orbital available and in the correct orientation to form a delocalized $\pi$ bond over the entire molecule. **9.31** (a) Bonding MO's are lower in energy than the starting atomic orbitals, and antibonding mo's are higher. (b) The electron density is concentrated between the nuclei in a bonding MO and away from the nuclei in an antibonding mo. **9.33** (a) *Bond order* is the net number of bonding electron pairs in a molecule [$\frac{1}{2}$(bonding $e^-$ − antibonding $e^-$)]. (b) *Paramagnetism* is the magnetic behavior of molecules with unpaired electrons. (c) An *antibonding molecular orbital* has electron density concentrated outside the region between the atoms and a higher energy than the isolated atomic orbitals that combined to form it **9.35** (a) BO = 2.5, paramagnetic; (c) BO = 3, diamagnetic. **9.37** In order of bond length, $O_2^+ < O_2 < O_2^- < O_2^{2-}$. **9.39** NO should have the lowest ionization energy. Both NO and $O_2^+$ have their highest energy electron in a $\pi^*$ orbital, but NO has a lower nuclear charge than $O_2^+$, so the electron in NO should be less tightly held. **9.41** The bond order of $N_2 > O_2 > C$ in diamond $= F_2$. As the bond order increases, the bond length decreases. For $F_2$ and diamond, which both have a bond order of one, F has a higher effective nuclear charge than C, attracts the valence electrons more tightly, and thus has a smaller bond length. **9.43** (a) Silicon does not readily form $\pi$ bonds, which are present in the graphite structure. (b) The interlocking network of four strong covalent C—C bonds to each carbon atom produces a rigid, hard structure. **9.45** (b) Trigonal pyramidal; (c) linear; (e) square planar.   **9.48**   (a) $3\sigma$, $1\pi$; (b) $1\sigma$, $2\pi$; (c) $2\sigma$, $1\pi$; (d) $3\sigma$, $0\pi$.   **9.52**   (b) The electron-pair geometry around each N is trigonal planar with bond angles of $\sim 120°$; the two N atoms must be coplanar to form the $\pi$ bond, so the entire molecule must be planar. (c) Each N atom will use $sp^2$ hybrid orbitals for $\sigma$ bonding and pure $p$ orbitals for the $\pi$ bond. (e) The two N—H bonds must be on opposite sides of the molecule to produce a dipole moment of zero. If the two N—H bonds were on one side of the molecule and the two lone pairs on the other side, the molecule would have a net dipole moment.   **9.55**   (a) $sp^3 \longrightarrow$ $sp^2$;   (b)   $dsp^3 \longrightarrow sp^3$;   (c)   $d^2sp^3 \longrightarrow dsp^3$.   **9.58**   $TeO_2(s)$ + $H_2O_2(aq) + 2H_2O(l) \longrightarrow Te(OH)_6(aq)$. Te goes from $+4$ to $+6$; O in $H_2O_2$ goes from $-1$ to $-2$. The Lewis structure for $Te(OH)_6$ has the six OH groups attached to the Te by $\sigma$ bonds in an octahedral arrangement; the Te uses $d^2sp^3$ hybrid orbitals. $TeO_6{}^{6-}$ has the same structure as $Te(OH)_6$, but the six protons attached to the oxygens have been removed.   **9.61**   $H_2 > Li_2 >$ $K_2 > Ca_2$. According to molecular-orbital theory, $Ca_2$ would have a bond order of zero and be unstable relative to isolated Ca atoms. The first three have a bond order of one, but the smaller the atoms, the stronger the bonds, so $H_2$ would have the largest bond-dissociation energy.

# CHAPTER 10

**10.1**   Most of the volume of a gas is empty space. It is easy to reduce this space by applying pressure. Molecules in liquids and solids are in close contact with each other, so compression has little effect.   **10.3**   (a) 92.2 kPa; (b) 97.3 kPa; (c) 0.998 atm; (d) 806 mm Hg.   **10.5**   (a) 52°F, 29.2 in Hg; (b) 32°C, 99.26 kPa.   **10.7**   (a) $2.7 \times 10^2$ Pa; (b) $9.1 \times 10^7$ Pa.   **10.9**   (a) 665 mm Hg; (b) 823 mm Hg.   **10.11**   Pressure, volume, temperature, number of moles of gas; three must be known to completely specify the state of a gas.   **10.13**   (a) $PV = c$; (b) $V \propto n$; (c) $P \propto T$; (d) $PV \propto T$.   **10.15**   (a) 8.38 atm;   (c)   0.244 mol.   **10.17**   9.83 kg   $O_2$;   $6.88 \times 10^3$ L.   **10.19**   Avogadro's law states that gas volume at constant temperature and pressure is proportional to the quantity of gas. According to the ideal gas equation, $V = n(RT/P)$. At constant temperature and pressure the quantity in parentheses is a constant. Thus $V$ is proportional to $n$.   **10.21**   (a) $1.78 \times 10^3$ mm Hg; (c) 2.84 L.   **10.23**   (a) 2.59 L.   **10.25**   (a) $2 \times 10^{-4}$ mol; (b) the cockroach will consume more than 20% of the available $O_2$.   **10.27**   (a) $7.5 \times 10^{-3}$ atm; (b) $2.45 \times 10^{-2}$ atm.   **10.29**   $4.2 \times 10^2$ mm Hg.   **10.31**   (a)   1.78 g/L.   **10.33**   (a)   $C_2N_2$.   **10.35**   Molecular weight = 89.4 g/mol.   **10.37**   10.2%.   **10.39**   $3.4 \times 10^{-9}$ g   Mg.   **10.41**   65 mL.   **10.43**   Yes; 2.69 atm remain.   **10.45**   (a) Same; (b) more collisions per unit time for $N_2$; (c) the $N_2$ distribution curve is spread out more toward higher speeds; (d) $N_2$ effuses 2.28 times faster.   **10.47**   (a) False–the average kinetic energy of a collection of gas molecules at a given temperature is the same for all gases. (b) True. (c) False—the molecules of a gas at a given temperature have *a distribution of* kinetic energies. (d) True.   **10.49**   (a) $2.22 \times 10^3$ m/s; (c) $2.50 \times 10^2$ m/s.   **10.51**   $NH_3$ will effuse 1.28 times faster than CO.   **10.53**   Molecular weight = $2.1 \times 10^2$ g/mol.   **10.55**   Real gases deviate from ideal behavior because of attractive forces among molecules and the finite volume of gas molecules. These were considered insignificant in the kinetic-molecular theory. Attractive forces lead to a smaller value of $PV$ than predicted by the ideal gas law, and the real volume of gas molecules leads to a larger value of $PV$ than predicted. That

gases can be liquefied at low temperature or high pressure is physical evidence of their departure from ideal behavior.   **10.57**   $SO_2$ is a polar molecular and will experience the largest intermolecular interactions, which are responsible for negative deviations. $SO_2$ should also have the largest volume correction.   **10.59**   (a) Ar atoms occupy 0.036% of the total volume at STP; (b) at 100 atm, 3.6%.   **10.61**   29.9 in Hg.   **10.64**   Any mass less than $5.1 \times 10^7$ g.   **10.66**   (a) 44 g/mol; (b) 64 g/L.   **10.70**   Molecular weight of vapor = 187 g/mol; it is a monomer.   **10.72**   5.92 g $CaH_2$.   **10.76**   The balloon will expand; $H_2$ ($M = 2$ g/mol) will effuse in through the walls of the balloon faster than He ($M = 4$ g/mol) will effuse out because the lower molecular weight gas effuses more rapidly.   **10.78**   $r(^{12}CO)/r(^{13}CO) = 1.018$; $r(^{12}CO_2)/r(^{13}CO_2) = 1.011$. Using CO will give a better separation.   **10.79**   $2Sc(s) + 6HCl(aq) \longrightarrow 2ScCl_3(aq) + 3H_2(g)$.

# CHAPTER 11

**11.1**   Gases are compressible because there are large distances between molecules; particles in the solid and liquid states are close together and are relatively incompressible.   **11.3**   Liquids and solids are incompressible; their particles are close together. For a pure substance, densities of the solid and liquid states are similar; density of the gas is much less.   **11.5**   Gas phase molecules have the highest average kinetic energy, liquid phase intermediate, and solid phase least. Lowering the temperature decreases the average kinetic energy of molecules, and the substance changes phases in order of decreasing average kinetic energy.   **11.7**   Condensing and freezing are the result of attractive forces that lead to a more ordered state; the excess energy is evolved as heat.   **11.9**   The molar heat of sublimation contains the heat of melting and the heat of vaporization, so it is always larger than the heat of vaporization alone, measured at the same temperature.   **11.11**   $\Delta H_{sub} = 50.95$ kJ/mol.   **11.13**   22.9 kJ.   **11.15**   The boiling point is the temperature at which the vapor pressure of the liquid equals the external pressure acting on the liquid; the lower the external pressure, the lower the boiling point. The melting point is the temperature at which the solid and liquid phases are in equilibrium; neither solids nor liquids are significantly affected by pressure because neither is very compressible.   **11.17**   (a) Surface tension draws the liquid into the shape with the smallest surface area, a sphere. (b) The gas pressure decreases from 1 atm to a much lower value, the vapor pressure of water at the cooler temperature. The external pressure of the atmosphere compresses the can. (c) Water boils at a lower temperature at the higher elevation because atmospheric pressure is lower. (d) 279°C is the critical temperature, the temperature above which the substance cannot be liquefied regardless of pressure.   **11.19**   The more carbon atoms, the higher the $\Delta H_{vap}$, and the stronger the intermolecular forces. The larger molecules are more readily polarized and therefore experience stronger London forces.   **11.21**   (a) $CH_3Cl$— dipole-dipole vs. ion-ion forces; (b) $Cl_2$— London dispersion forces in both, but $Cl_2$ has lower molecular weight; (c) $SO_2$—$SiO_2$ is a network solid with very high melting point; (d) HCl—dipole-dipole forces in both, no hydrogen bonding in HCl.   **11.23**   (a) London dispersion; (b) hydrogen bonding; (c) dipole-dipole; (d) ion-dipole.   **11.25**   $Br_2(l) \longrightarrow Br_2(g)$, $\Delta H = 30.71$ kJ; $Br_2(g) \longrightarrow 2Br(g)$, $\Delta H = 192.9$ kJ. The first process represents an overcoming of the intermolecular attractive energies. The second requires rupture of the Br—Br bond, an intramolecular interaction. As is usual, the *intermolecular* interactions are weaker than the *intra*-molecular ones. **11.27** (a) $PH_3 < AsH_3 < SbH_3$; (b) $CH_4 < CF_4 < CCl_4$; (c) $CH_4 < NH_3 < H_2O$. **11.29** Critical temperature increases with increasing strength of intermolecular forces. (a) $O_3$ is a bent molecule and has stronger intermolecular forces. (b) The heavier $SiCl_4$ has stronger dispersion forces. (c) Hydrogen bonding in $NH_3$ is a much stronger dipole-dipole force than in $PH_3$. **11.31** (a) $FeCl_2$; (b) $F_2$. **11.33** The boiling point, melting point, heat capacity, and heat of vaporization are all irregularly high for $H_2O$ with respect to similar compounds. **11.35** (a) Molecular; (b) atomic; (c) metallic; (d) ionic; (e) covalent (network). **11.37** (a) KBr, ionic vs. dispersion; (b) $SiO_2$, (network) covalent vs. dispersion; (c) Se, (network) covalent vs. dispersion; (d) $MgF_2$, higher charge, stronger electrostatic forces in solid. **11.39** The unit cell is the building block of the lattice. When repeated in three dimensions, it produces the crystal lattice. It is a parallelepiped with the characteristic distances $a$, $b$, and $c$, and angles $\alpha$, $\beta$, and $\gamma$. Unit cells can be primitive (lattice points only at the corners of the parallelepiped) or centered (lattice points at the corners and on faces or interior of the parallelepiped). **11.41** 1.28 Å; 8.97 g/cm³. **11.43** (a) $l = 5.387$ Å; (b) 5.55 Å. **11.45** (a) Face-centered cubic; (b) 4 Pt atoms/cell, face-centered cubic. **11.47** (a) 12; (b) 6; (c) 8. **11.49** Diffraction occurs when the wavelength of light is close to the separations between planes of atoms in the crystal. The 632 nm or 6320 Å wavelength is much longer than the $Na^+$—$Cl$ separation of 2.8 Å, so diffraction does not occur. **11.51** $d = 3.56$ Å. **11.53** (a) The water vapor would condense to form a solid at a pressure of around 4 mm Hg. At higher pressure, perhaps 5 atm or so, the solid would melt to form liquid water. This occurs because the melting point of ice, which occurs at 0°C at 1 atm, decreases with increasing pressure. (b) The water at $-1.0$°C and 0.30 atm is in the solid form. Upon heating, it melts to form liquid water at slightly above 0°C. The liquid converts to a vapor when the temperature reaches the point at which the vapor pressure of water reaches 0.3 atm (228 mm Hg). From Appendix C we see that this occurs just below 70°C.

**11.55**

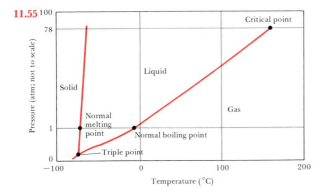

**11.57** $4.2 \times 10^2$ g. **11.61** Highest boiling point: RbCl, as the only ionic compound in the list; lowest boiling point: Ar, because it has the lowest mass and least complexity among all the nonpolar species listed. **11.64** Slope $= -3.15 \times 10^3$; $\Delta H_v = 60.3$ kJ/mol. **11.67** A most important attractive force would be that due to hydrogen bonding. In addition, there should be strong dipole-dipole interactions that bind a polar molecule such as $CH_3OH$ to the polar surface. **11.70** 1.248 Å. **11.73** $3.02 \times 10^{-2}$ g $H_2O$.

# CHAPTER 12

**12.1** (a) Weight percentage $= 4.76\%$. **12.3** (a) Mole fraction $CH_3OH = 6.98 \times 10^{-3}$. **12.5** (a) 0.0105 $M$; (d) 0.0125 $M$. **12.7** (a) 4.04 $m$; (b) 0.0962 $m$. **12.9** (a) Weight percentage $H_2O = 41.1\%$; (b) Mole fraction $H_2O = 0.692$; (c) 38.7 $m$; (d) 21.1 $M$. **12.11** (a) 12 $M$ HCl, 17 $m$ HCl, mole fraction HCl $= 0.77$. **12.13** 0.726 $M$ $Pb(NO_3)_2$, 0.755 $m$ $Pb(NO_3)_2$. **12.15** (a) 0.127 mol; (c) $3.80 \times 10^{-3}$ mol. **12.17** (a) 2.50 g KBr plus enough $H_2O$ to make 1.40 L of solution; (b) 11.4 g KBr plus 238.6 g $H_2O$; (c) 207 g KBr plus 1.52 kg $H_2O$; (d) 13.3 g KBr plus enough water to make 560 mL of solution. **12.19** (a) 0.38 $M = 0.76$ $N$ $Sn^{4+}$; (b) 0.111 $M$ $H_2SO_4$. **12.21** (a) ion-dipole; (b) London dispersion; (c) hydrogen bonding; (d) dipole-dipole. **12.23** $KCl < CH_3OH < C_6H_6$. **12.25** Enthalpy of solution is determined by the relative enthalpy required to separate the solute and solvent particles and the enthalpy released when the solute and solvent particles interact. In hydrated salts, the ions are already interacting with water molecules so that less enthalpy is released by new solute-solvent interactions, and the overall enthalpy of solution is more positive. **12.27** The energy that is required to separate the $K^+$ and $Br^-$ ions of KBr in forming a solution must be compensated for, or nearly compensated for, by the ion-solvent interactions. Because $CCl_4$ is a nonpolar molecule, with stable C—Cl bonds, there is no possibility for large, exothermic enthalpy terms corresponding to solvation of either $K^+$ or $Br^-$ in $CCl_4$. **12.29** $C_{N_2} = 1.5 \times 10^{-3}$ $M$; $C_{He} = 9.2 \times 10^{-4}$ $M$. **12.31** The solubility of a gas is determined in part by the attractive forces between the gas and solvent molecules. At higher temperatures the increased kinetic energies of the molecules counter the effects of these attractive interactions. **12.33** (a) Weak; (b) strong; (c) nonelectrolyte; (d) strong. **12.35** $NH_3$. **12.37** (a) Entirely ionic; (b) entirely molecular; (c) entirely molecular; (d) mixture of ions and molecules. **12.39** (a) 0.14 $M$ $Na^+$, 0.14 $M$ $OH^-$; (b) 0.25 $M$ $Ca^{2+}$, 0.50 $M$ $Br^-$; (c) 0.25 $M$ $CH_3OH$; (d) 0.13 $M$ $K^+$, $Na^+$, $ClO_3^-$; 0.067 $M$ $SO_4^{2-}$. **12.41** The presence of non-volatile solute reduce the capacity of volatile solute molecules to escape to the gas phase. Fewer gaseous solvent molecules leads to a lower vapor pressure above the solution. **12.43** 0.15 $M$ KCl. **12.45** (a) 186 mm Hg; (b) 184 mm Hg. **12.47** (a) $T_f = -115.0$°C, $T_b = 78.6$°C; (c) $T_f = -1.5$°C, $T_b = 100.4$°C. **12.49** 0.03 $m$ glycerine $< 0.03$ $m$ benzoic acid $< 0.02$ $m$ KBr. **12.51** Molecular weight $= 267$ g/mol. **12.53** Arctic has 1.06 $m$ ionic solutes; Atlantic has 1.12 $m$. **12.55** The outline of a light beam passing through a colloid is visible, while a beam passing through a true solution is invisible unless collected on a screen. The suspended particles in a colloid are large enough to scatter light, but those in a solution are not. **12.57** Hydrophobic: a and c; hydrophilic: b. **12.59** To remain dispersed the gold particles must be extremely small, so that their kinetic energies of motion can be influenced by collisions with solvent molecules. Under these conditions, the gravitational force will not be large relative to kinetic energies, and the particles will remain in suspension. Also, the particles must acquire a common electrical charge, by adsorption of ions, so that they are repelled by one another. This keeps them from aggregating to form larger particles that could settle out. **12.64** 2.72 kg $CuSO_4 \cdot 5H_2O$; 1.36 $M$ $CuSO_4$. **12.67** Increased disorder: a and d. **12.68** The outer periphery of the BHT molecule is mostly hydrocarbonlike groups, such as —$CH_3$. The one OH group is rather buried inside and so probably does little to enhance solubility in water. Thus, BHT accumulates in fats and has a fairly long residence time in the body. **12.70** Osmotic pressure = 0.953 mm Hg. **12.73** Vapor pressure = 89.0 mm Hg. **12.76** Molecular formula is $C_9H_{13}O_3N$; molecular weight = 183 g/mol.

## CHAPTER 13

**13.1** (a) $SO_2(g) + 2OH^-(aq) \longrightarrow SO_3{}^{2-}(aq) + H_2O(l)$; (c) $Zn(s) + Cu^{2+}(aq) \longrightarrow Cu(s) + Zn^{2+}(aq)$. **13.3** (b) and (c) are soluble. **13.5** $K^+$ or $Ba^{2+}$. **13.7** (a) $ZnS(s) + 2H^+(aq) \longrightarrow H_2S(g) + Zn^{2+}(aq)$; (b) $Ba^{2+}(aq) + CO_3{}^{2-}(aq) \longrightarrow BaCO_3(s)$; (c) $3H^+(aq) + PO_4{}^{3-}(aq) \rightleftharpoons H_3PO_4(aq)$ ($H_3PO_4$ is a relatively weak acid), (d) $H^+(aq) + OH^-(aq) \longrightarrow H_2O(l)$. **13.9** The driving force of a reaction (for example, formation of a gas, a precipitate, or a nonelectrolyte) leads to net chemical change, often involving the removal of a species from solution. **13.11** (a) Soluble: $SO_2(g)$ and $H_2O(l)$, a nonelectrolyte, are formed; (b) soluble; $SrI_2$ and $SrCl_2$ are soluble salts. **13.13** (a) Gas, $H_2S$; (b) precipitate, $BaCO_3$; (c) weak electrolyte, $H_3PO_4$; (d) nonelectrolyte, $H_2O$. **13.15** (a) $Cu(OH)_2(s) + 2HNO_3(aq) \longrightarrow Cu(NO_3)_2(aq) + 2H_2O(l)$; (b) $MnCl_2(aq) + H_2S(g) \longrightarrow MnS(s) + 2HCl(aq)$. **13.17** Arrhenius postulated that all acids produce hydrogen ions in aqueous solution. The presence of $H^+$ in both $HCl(aq)$ and $H_2SO_4(aq)$ dominates the properties of the two solutions. **13.19** In solution, water molecules surround HBr molecules, and the H-Br polar covalent bond is broken to form $H^+$ and $Br^-$, each hydrated. These mobile charged particles enable the solution to conduct electricity. **13.21** When polar $HCl(g)$ dissolves in toluene, a nonpolar solvent, the solvent-solute interactions are not strong enough to separate the molecule into $H^+$ and $Cl^-$. Since no $H^+$ are present, the solution doesn't react with zinc or conduct electricity. In water the dipole-dipole forces cause HCl to dissociate into ions, the solution conducts electricity, and the $H^+$ can react with Zn. **13.23** (a) $NH_4{}^+$; (c) $NH_3$; (d) $H_3PO_4$. **13.25** (b) Acid: $H_2C_2O_4$; conjugate base: $HC_2O_4{}^-$; base: $H_2O$; conjugate acid: $H_3O^+$. **13.27** (a) Yes; (b) no. **13.29** (a) S, $-2$ to 0; N, $+5$ to $+2$; (c) Cr, $+3$ to $+6$; Cl, $+1$ to $-1$. **13.31** (a) $2N_2H_4(g) + N_2O_4(g) \longrightarrow 3N_2(g) + 4H_2O(g)$; (b) $N_2O_4$, oxidizing agent; $N_2H_4$, reducing agent. **13.33** (b) $ClO^-(aq) + 4OH^-(aq) \longrightarrow ClO_3{}^-(aq) + 2H_2O(l) + 4e^-$; oxidation; (d) $SO_4{}^{2-}(aq) + 4H^+(aq) + 2e^- \longrightarrow SO_2(g) + 2H_2O(l)$; reduction. **13.35** (a) $Cr_2O_7{}^{2-}(aq) + I^-(aq) + 8H^+(aq) \longrightarrow 2Cr^{3+}(aq) + IO_3{}^-(aq) + 4H_2O(l)$; (e) $2MnO_4{}^-(aq) + Br^-(aq) + H_2O(l) \longrightarrow 2MnO_2(s) + BrO_3{}^-(aq) + 2OH^-(aq)$. **13.37** (a) $3MnO_4{}^{2-}(aq) + 4H^+(aq) \longrightarrow 2MnO_4{}^-(aq) + MnO_2(s) + 2H_2O(l)$; (b) $4Au(s) + 8CN^-(aq) + O_2(g) + 2H_2O(l) \longrightarrow 4[Au(CN)_2]^-(aq) + 4OH^-(aq)$. **13.39** Present: $Ag^+$, $Sr^{2+}$; absent: $Ni^{2+}$, $Mn^{2+}$. **13.42** An aqueous solution of HCl conducts an electrical current because the HCl has reacted with water to form aquated hydrogen ions, $H(aq)^+$, and aquated chloride ions, $Cl(aq)^-$. HCl does *not* react with benzene. It is slightly soluble in benzene, but exists in that solvent as nonionized HCl molecules. Because there are neutral, they do not move when an electrical potential is applied. **13.45** (c) Acid: $H_2O$; base: $ClO_2{}^-$; (d) acid: $HCO_2H$; base: $CN^-$. **13.49** (a) $2MnO_4{}^-(aq) + 5H_2O_2(aq) \longrightarrow 2Mn^{2+}(aq) + 5O_2(g) + 10H^+(aq)$; reducing agent. (b) $2I^-(aq) + H_2O_2(aq) + 2H^+(aq) \longrightarrow I_2(s) + 2H_2O(l)$; oxidizing agent.

## CHAPTER 14

**14.1** *Troposphere* (0–11 km) temperature varies from 290 K to 220 K; *stratosphere* (11–50 km) temperature varies from 220 K to 270 K; *mesosphere* (50–85 km) temperature varies from 270 K to 180 K; *thermosphere* (85 km–upwards) temperature increases from 180 K beyond 1200 K. **14.3** For Ar, $P = 7.00$ mm Hg; for Ne, $P = 1.36 \times 10^{-2}$ mm Hg. **14.5** 569 nm. **14.7** The photon must possess sufficient energy to produce rupture of the NO bond. Also, the NO molecule must be able to absorb a photon having at least the minimum required energy. **14.9** For two oxygen atoms to form $O_2$, a third body must be present to carry off the excess energy, as described in Section 14.3. This is much more likely to occur at 50 km than at 120 km because the atmosphere is much denser at the lower elevation; there are many more atoms or molecules per unit volume. **14.11** First step: $\Delta H = -198.9$ kJ; second step: $\Delta H = -190.9$ kJ. **14.13** $P = 2.0 \times 10^{-4}$ mm Hg; $6.7 \times 10^{18}$ $O_3$ molecules/m$^3$. **14.15** CO in unpolluted air is typically 0.05 ppm, about 10 ppm in urban air. A major source is automobile exhaust. $SO_2$ is less than 0.01 ppm in unpolluted air, about 0.08 ppm in urban air. A major source is coal and oil-burning power plants, but there is also some $SO_2$ in auto exhausts. NO is about 0.01 ppm in unpolluted air, about 0.05 ppm in urban air. It comes mainly from auto exhausts. **14.17** $5.66 \times 10^3$ mol $SO_4{}^{2-}$/mi$^2$ per year. **14.19** 0.09 $M$. **14.21** $1.5 \times 10^{12}$ L. **14.23** A high concentration of dissolved $O_2$ supports biodegradation of organic matter by aerobic bacteria. Lack of $O_2$ indicates pollution by organic and other materials requiring $O_2$ for decomposition. **14.25** 2.5 g $O_2$. **14.27** $Ca^{2+}(aq) + 2HCO_3{}^-(aq) \longrightarrow CaCO_3(s) + CO_2(g) + H_2O(l)$. **14.29** $4.8 \times 10^5$ g $Ca(OH)_2$ and $4.2 \times 10^5$ g $Na_2CO_3$. **14.31** The formation of ozone from reaction of O and $O_2$ is favored by a high concentration of $O_2$. Thus, as altitude decreases the formation reaction rate increases. (There is also a generally higher concentration of molecules to act as third bodies, and this helps as well.) However, the intensity of radiation of sufficient energy to cause photodissociation of $O_2$ decreases as altitude decreases. The two effects, working in opposite directions, lead to a maximum in the $O_3$ level at about 22 km. **14.34** The ionization energies of the metal atoms are much lower than those of any of the atomic or molecular ions present at 120 km ($O_2{}^+$, $O^+$, $NO^+$). Thus, these ions react with the metal atoms as illustrated for $NO^+$: $M + NO^+ \longrightarrow M^+ + NO$. **14.37** The answer lies in Figure 14.2. Atmospheric pressure drops rapidly with increasing elevation. Thus at 11 km the pressure is only about 250 mm Hg, and it declines to perhaps 1 mm Hg at 50 km. The number of molecules per unit volume is thus much lower on average in the stratosphere. **14.39** $CF_2Cl_2(g) + h\nu \longrightarrow CF_2Cl(g) + Cl(g)$; $CF_2Cl(g) + h\nu \longrightarrow CF_2(g) + Cl(g)$; $Cl(g) + O_3(g) \longrightarrow ClO(g) + O_2(g)$; $ClO(g) + O(g) \longrightarrow Cl(g) + O_2(g)$. **14.40** 7.9% of the blood is in the form of carboxyhemoglobin. **14.44** (a) $NO(g) + h\nu \longrightarrow N(g) + O(g)$; (b) $NO(g) + h\nu \longrightarrow NO^+(g) + e^-$; (e) $Cl_2(g) + H_2O(l) \longrightarrow HClO(aq) + H^+(aq) + Cl^-(aq)$. **14.46** Primary treatment removes large and small suspended particles by filtration and sedimentation, respectively. Secondary treatment decomposes organic matter by the action of aerobic bacteria and sedimentation. Neither mode removes detergent phosphates, which are soluble (not effected by filtration or sedimentation) and contain phosphorus in its most highly oxidized state (not susceptible to aerobic bacteria).

# CHAPTER 15

**15.1**

| Time (min) | Time interval (min) | Concentration ($M$) | $\Delta M$ | Rate ($M$/min) |
|---|---|---|---|---|
| 0 | | 1.85 | | |
| | 79 | | 0.18 | $2.3 \times 10^{-3}$ |
| 79 | | 1.67 | | |
| | 79 | | 0.15 | $1.9 \times 10^{-3}$ |
| 158 | | 1.52 | | |
| | 158 | | 0.22 | $1.4 \times 10^{-3}$ |
| 316 | | 1.30 | | |
| | 316 | | 0.30 | $0.95 \times 10^{-3}$ |
| 632 | | 1.00 | | |

**15.3** From the slopes of the tangents to the graph, the rates are $t = 100$ min, $1.9 \times 10^{-3}$ $M$/min and $t = 500$ min, $9.4 \times 10^{-4}$ $M$/min. **15.5** (a) $-\Delta[H_2O_2]/\Delta t = 2\Delta[O_2]/\Delta t$; (c) $-\Delta[Br^-]/\Delta t = \frac{5}{3}\Delta[Br_2]/\Delta t$. **15.7** (a) NO appears at the rate of 6 mol/s; $H_2O$ appears at the rate of 9 mol/s. **15.9** (a) $6.08 \times 10^{-5}$ $M$/s. **15.11** (a) First order in A, first order in B, second order overall; (b) second order in A, zero order in B, second order overall; (c) first order in A, second order in B, third order overall.

**15.13**

| Rate law | Units of $k$ | Effect of doubling [A] | Effect of doubling [B] |
|---|---|---|---|
| $k[A][B]$ | $1/M\text{-s}$ | double rate | double rate |
| $k[A]^2$ | $1/M\text{-s}$ | four-fold increase | no effect |
| $k[A][B]^2$ | $1/M^2\text{-s}$ | double rate | four-fold increase |

**15.15** (a) rate $= k[S_2O_8^{2-}][I^-]$; (b) $k = 6.1 \times 10^{-3}/M\text{-s}$; (c) $-\Delta[S_2O_8^{2-}]/\Delta t = 1.5 \times 10^{-5}$ $M$/s; (d) $\Delta[SO_4^{2-}]/\Delta t = 1.5 \times 10^{-5}$ $M$/s. **15.17** $k = 1.6 \times 10^{-5}$/s **15.19** Plot $\log P_{SO_2Cl_2}$ vs. $t$, slope $= -9.53 \times 10^{-6}$/s; $k = 2.19 \times 10^{-5}$/s **15.21** A plot of $\log [C_{12}H_{22}O_{11}]$ vs. time is linear; the reaction is first order. $k = -2.30(\text{slope}) = -2.30(-1.60 \times 10^{-3}/\text{min}) = 3.68 \times 10^{-3}/\text{min} = 6.13 \times 10^{-5}$/s. **15.23** (a) 0.146 mol $N_2O_5$ remain; (b) $t = 236$ s. **15.25** $k = 3.53 \times 10^{-5}/M\text{-s}$ **15.27** (b) 123 kJ/mol. **15.29** No. The value of A, related to frequency and effectiveness of collisions, is different for each reaction, and $k$ is proportional to A. **15.31** A plot of $\log k$ vs. $1/T$ has a slope of $-2.47 \times 10^3$; slope $= -E_a/2.3R$; $E_a = 47$ kJ/mol. **15.33** (a) $k_{873} = 1.19 \times 10^{-4}$ $M$/s. **15.35** (b) First step: $-\Delta[NO]/\Delta t = k[NO][Br_2]$; second step: $-\Delta[NOBr_2]/\Delta t = k[NOBr_2][NO]$. (c) $NOBr_2$ is an intermediate. (d) The second step is slow relative to the first. If the second step were fast, the reaction would be first order in NO. **15.37** Both (b) and (d) are consistent with the observed rate law. **15.39** A catalyst is present at the beginning of the reaction, whereas an intermediate is formed during the reaction. They are *alike* in that neither appears in the balanced chemical reaction for the process. **15.41** Two factors which determine the activity of a heterogeneous catalyst are total surface area and the number of active sites per unit amount of catalyst (absence of impurities); these are strongly dependent on the preparation and handling of the individual catalyst. **15.43** As illustrated in Figure 15.16, the two C—H bonds that exist on each carbon of the ethylene molecule before adsorption are retained in the process in which a D atom is added to each C (assuming we use $D_2$ rather than $H_2$). To put two deuteriums on a single carbon, one of the already existing C—H

bonds in ethylene must be broken while the molecule is adsorbed. The H atom moves off as an adsorbed atom and is replaced by a D. This requires a larger activation energy than simply adsorbing $C_2H_4$ and adding one D atom to each carbon. **15.45** (a) Concentrations of reactants, temperature, addition of a catalyst. (b) Increases in concentrations of reactants result in more frequent collisions per unit time, so a greatr probability of reaction arises. An increase in temperature results in an increase in average kinetic energy of the reactants, so that a larger fraction may possess sufficient energy to form product. A catalyst lowers the activation energy for reaction, so that a larger fraction of the encounters between reactants can lead to products. **15.48** (a) [A] and rate double; (b) $k$ and rate change; (c) the concentration changes cancel each other, and $k$ and rate are unchanged; (d) $k$ and rate increase. **15.51** Rate $= k[(CH_3)_3CBr]$; $k = 1.0 \times 10^{-2}$/s. **15.55** The slope of a plot of $\log k$ vs. $1/T$ is $-3.4 \times 10^3$; $E_a = (-\text{slope})2.3R = 65$ kJ/mol. **15.57** Better control of reactions; lower energy costs due to lower temperatures; fewer side reactions, thus higher yields of desired products; possibly a more favorable equilibrium constant at the lower temperature (as in $NH_3$ synthesis from $N_2$ plus $H_2$). **15.61** The data can be tested for first order reaction behavior by graphing log (percentage oxidation of organic matter) vs. time. Such a graph is linear with slope $= -0.10$, and thus $k = 0.23$/day.

# CHAPTER 16

**16.1** (a) $K = [SO_2]^2[O_2]/[SO_3]^2$; homogeneous; (b) $K = [H_2O]^2[N_2]$; heterogeneous; (c) $K = [H_2]^4/[H_2O]^4$; heterogeneous. **16.3** Equilibrium concentrations for a certain reaction can be reached by starting with either pure reactants or pure products, which suggests that the forward and reverse processes are occurring simultaneously at the same rate. **16.5** $K_p = 8.82 \times 10^{-4}$. **16.7** (a) $K_c = 4.90 \times 10^{-2}$; (b) $K_c = 416$; (c) $K_c = 5.21$. **16.9** $K_c = 0.16$; $K_p = 14$. **16.11** 0.038 mol NO react, 0.062 mol remain; 0.012 mol $H_2$ remain; 0.138 mol $H_2O$ form; 0.019 mol $N_2$ form; (b) $K_c = (0.019)(0.138)^2/(0.062)^2(0.012)^2 = 6.5 \times 10^2$. **16.13** (a) $K_c = 6.42 \times 10^{-2}$; (b) $K_p = 1.98$; (c) total pressure $= 0.968$ atm. **16.15** In each case compare the reaction quotient with $K_p = 4.51 \times 10^{-5}$. (d) $Q = 9 \times 10^{-3}$; $Q > K_p$; reaction proceeds to the left to attain equilibrium. **16.17** [HBr] $= 4.2$ $M$. **16.19** $P_{CO_2} = 1.49 \times 10^{-3}$ atm. **16.21** $P_{I_2} = P_{B_2} = 1.2 \times 10^{-2}$ atm. **16.23** $[H_2S] = [NH_3] = 1.1 \times 10^{-2}$ $M$. **16.25** (a) Shift equilibrium to right, more CO will be formed. (b) No effect on equilibrium; so long as *any* CO is present, the amount does not influence the equilibrium. (c) Equilibrium is shifted to the right, direction in which reaction is endothermic. (d) An increase in pressure causes shift in equilibrium to left. (e) No effect. (f) Removal of CO causes a shift to the right. **16.27** (a) No effect; (b) no effect; (c) increase equilibrium constant; (d) no effect. **16.30** (a) At equilibrium, the forward and back reactions occur at the same rate. (b) A catalyst would lower the activation energy barrier. (c) A catalyst would increase the rates of both forward and reverse reactions, leaving the ratio a constant. (d) Since the reaction is endothermic, $K$ should increase with increasing temperature. **16.31** $k_f/k_r = [NO_2][O_2]/[NO][O_3] = K_c$. **16.33** (a) $K = [Ag(NH_3)_2^+]/[Ag^+][NH_3]^2$; (b) $K = [Ag^+]^2[CrO_4^{2-}]$; (c) $K = [H^+][NO_2^-]/[HNO_2]$; (d) $K = $

[$Zn^{2+}$]/[$Cu^{2+}$]; (e) $K = $[$NH_4^+$][$OH^-$]/[$NH_3$]. **16.35** (a) [$SO_2$] = 0.50 $M$; [$NO_2$] = 1.0 $M$; [$SO_3$] = 1.5 $M$; [$NO$] = 3.0 $M$; (b) $K_c$ = 9.0. **16.37** $K_c = 3.61 \times 10^{-3}$. **16.42** $K_p$ = 1.8. **16.44** (a) [$NO$] = $1.78 \times 10^{-2}$ $M$; [$O_2$] = $2.89 \times 10^{-2}$ $M$; (b) $K_c$ = 0.53. **16.47** (a) Left; (b) right; (c) left (a volume increase would produce a lower pressure, and the system will respond by a shift in equilibrium toward the side that produces more moles of gas); (d) right; (e) left. **16.50** $CO(g) + 2H_2(g) \rightleftharpoons CH_3OH(l)$; $\Delta H = -238.6 - (-110.5) = -128.1$ kJ. The process is exothermic; therefore, the equilibrium constant decreases with increasing temperature. All the gas components of the reaction are on the left. The extent of reaction would therefore increase with an increase in total pressure. **16.53** $Q = 8.3 \times 10^{-6}$. Since $Q > K_p$, the system will shift to the left to attain equilibrium. A catalyst that promoted the attainment of equilibrium would *decrease* the [$CO$] in the exhaust. **16.56** (a) [$Br$] = $3.20 \times 10^{-2}$ $M$; (b) 0.0160 or 1.60%.

# CHAPTER 17

**17.1** Water has a very small but measurable electrical conductivity at room temperature, indicating the presence of a small number of ions. The current is carried by $H^+(aq)$ and $OH^-(aq)$ formed in the autoionization. **17.3** (a) Acidic; (b) basic; (c) neutral; (d) basic. **17.5** (a) pH = 3.70; (c) pH = 11.52. **17.7** (a) [$H^+$] = $3.98 \times 10^{-4}$ $M$. **17.9** (a) pOH = 3.28; pH = 10.72. (b) [$H^+$] = $1.0 \times 10^3$[$OH^-$]; pOH = 8.50; pH = 5.50. **17.11** (a) 10; (b) $10^3$. **17.13** (b) [$OH^-$] = $7.0 \times 10^{-3}$; pH = 11.85. (c) [$OH^-$] = 0.250 $M$; pH = 13.398. **17.15** (a) $HNO_2$. **17.17** (a) $HClO_3(aq) \rightleftharpoons H^+(aq) + ClO_3^-(aq)$; $K_a = $[$H^+$][$ClO_3^-$]/[$HClO_3$]. **17.19** [$H^+$] = [$C_3H_5O_3^-$] = $3.6 \times 10^{-3}$ $M$; $K_a = 1.4 \times 10^{-4}$. **17.21** (a) [$H^+$] = $2.7 \times 10^{-5}$ $M$. **17.23** (a) [$H^+$] = $2.8 \times 10^{-3}$ $M$; 0.70% dissociation. **17.25** [$H^+$] = $4.6 \times 10^{-3}$ $M$; 23% dissociation (note that the approximation cannot be used in this case.) **17.27** $HX(aq) \rightleftharpoons H^+(aq) + X^-(aq)$; $K_a = $[$H^+$][$X^-$]/[$HX$]. Assume that the percent of acid that dissociates is small. Let [$H^+$] = [$X^-$] = $y$. $K_a = y^2/$[$HX$]; $y = K_a^{1/2}$[$HX$]$^{1/2}$. Percent dissociation = $y/$[$HX$] × 100. Substituting for $y$, percent dissociation = $100K_a^{1/2}$[$HX$]$^{1/2}/$[$HX$] or $100K_a^{1/2}/$[$HX$]$^{1/2}$. That is, percent ionization varies inversely as the square root of the concentration of HX. **17.29** [$H^+$] = [$X^-$] = $1.9 \times 10^{-2}$ $M$; [$HX$] = 0.18 $M$; $K_a = 2.0 \times 10^{-3}$. **17.31** [$H^+$] = $5.7 \times 10^{-3}$ $M$; pH = 2.24. The approximation that the first dissociation is less than 5% of the total acid concentration is not valid; the quadratic equation must be solved rigorously. The [$H^+$] produced from the second and third dissociations is small with respect to that present from the first step; the second and third ionizations can be neglected when calculating the [$H^+$] and pH. **17.33** (a) $C_3H_7NH_2(aq) + H_2O(l) \rightleftharpoons C_3H_7NH_3^+(aq) + OH^-(aq)$; $K_b = $[$C_3H_7NH_3^+$][$OH^-$]/[$C_3H_7NH_2$]. (b) $CN^-(aq) + H_2O(l) \rightleftharpoons HCN(aq) + OH^-(aq)$; $K_b = $[$HCN$][$OH^-$]/[$CN^-$]. **17.35** (a) [$OH^-$] = $9.2 \times 10^{-6}$ $M$; pH = 8.96. **17.37** (a) $K_b = 2.2 \times 10^{-11}$; (c) $K_b = 1.6 \times 10^{-7}$. **17.39** $SeO_3^{2-}(aq) + H_2O(l) \rightleftharpoons HSeO_3^-(aq) + OH^-$; $K_{b_1} = 1.9 \times 10^{-6}$; $HSeO_3^-(aq) + H_2O(l) \rightleftharpoons H_2SeO_3(aq) + OH^-(aq)$; $K_{b_2} = 4.4 \times 10^{-12}$. **17.41** $CO_3^{2-}$ is the strongest base ($H_2CO_3$ is the weakest conjugate acid), so $Na_2CO_3$ will produce the most basic solution.

**17.43** (a) Basic; (b) basic; (c) acidic; (d) basic. **17.45** $H_3O^+ \ll NO_3^- \ll N_3^- < HPO_4^{2-}$. **17.47** (a) Acidic: $CH_3NH_3Br$, $Zn(NO_3)_2$; (b) basic: KCNO, $Ba(C_2H_3O_2)_2$, $Na_2HPO_4$; (c) $Na_2HPO_4$ is the most basic. **17.49** $MgH_2 < PH_3 < H_2Se < HBr$. **17.51** (a) True. (b) In a series of acids that have the same central atom, acid strength increases with the number of *oxygen* atoms bonded to the central atom. (c) $H_2Te$ is a stronger acid than $H_2S$ because the H—Te bond is longer, weaker, and more easily dissociated than the H—S bond. **17.53** Replacement of H by the more electronegative chlorine atoms causes the central carbon to become more positively charged, thus withdrawing more electrons from the attached COOH group. This, in turn, causes the O—H bond to be more polar, so that $H^+$ is more readily ionized. pH of 0.10 $M$ $CH_3COOH$ = 2.87; 0.10 $M$ $CH_2ClCOOH$ = 1.95.

**17.55**

| Theory | Acid | Base |
|---|---|---|
| Lewis | electron-pair acceptor | electron-pair donor |
| Arrhenius | forms $H^+$ ions in water | produces $OH^-$ in water |
| Brønsted-Lowry | proton ($H^+$) donor | proton acceptor |

The Brønsted-Lowry theory is more general than the Arrhenius theory, because it is based on a unified model for what processes are responsible for acidic or basic character, and it shows the relationship between these processes. The Lewis theory is more general still, because it is not restricted to $H^+$ as the acidic species. Other substances that can be viewed as electron-pair acceptors are encompassed under the definition of "acid." **17.57** (a) Acid: $Fe(ClO_4)_3$; base: $H_2O$. (b) Acid: $H_2O$; base: $CN^-$. **17.59** (a) $CdI_2$; (b) $Fe(NO_3)_3$; (c) $CoCl_3$. **17.61** (a) [$H^+$] = $8 \times 10^{-7}$ $M$, acidic; (d) [$H^+$] = $1 \times 10^{-12}$ $M$, basic. **17.64** pH of a "neutral" solution = 6.63. **17.67** (a) 9.0%; (b) 9.0%. **17.70** (a) $KNO_2$ contains $NO_2^-$, which is the conjugate base of the weak acid $HNO_2$. $NO_2^-$ produces $OH^-$ in solution by hydrolysis, which raises the pH. (b) Lowers pH. **17.72** (a) $NH_3$ cannot be a proton donor in water. Suppose there were a base strong enough to remove a proton from $NH_3$. That base would preferentially react with $H_2O$, because $H_2O$ is a stronger proton donor (acid) than $NH_3$. There cannot be a stronger base in a solvent than the conjugate base characteristic of that solvent. In water the characteristic conjugate base is $OH^-$. (b) pH = 12.18. **17.75** [$HSO_4^-$] = $1.8 \times 10^{-2}$; [$SO_4^{2-}$] = $6.8 \times 10^{-3}$; [$H^+$] = $3.2 \times 10^{-2}$ $M$. **17.78** (a) pH = 10.30; (b) $pK_a$ = 7.90, pH = 4.60. **17.81** (a) pH = 5.39. (b) In a strongly acid solution the —COO$^-$ group would be protonated, so glycine would exist as $^+H_3NCH_2COOH$. In strongly basic solution the —NH$_3^+$ group would be deprotonated, so glycine would be in the form $H_2NCH_2COO^-$. (c) At pH 10, the ratio of the deprotonated form $H_2NCH_2COO^-$ to form II $^+H_3NCH_2COO^-$ is 1.67 to 1. There is more of the deprotonated form, but not twice as much. At pH = 2, the ratio of form II to the fully protonated form $^+H_3NCH_2COOH$ is 0.43. About two-thirds of the glycine molecules are in the protonated form and one-third exist as form II. **17.84** We normally ignore the autoionization of water in considering the [$H^+(aq)$] in an acidic solution, because the [$H^+$] produced by autoionization is very small compared to [$H^+$] produced by the acid. However, in this case, the concentration of HBr, $1.0 \times 10^{-9}$ $M$, is small compared to [$H^+$] produced by autoionization of water, $1.0 \times 10^{-7}$ $M$. Thus pH = 7.0.

# CHAPTER 18

**18.1** (b) Decrease: $NH_4^+$ is a weak acid, and $ClO_4^-$ is a *very* weak base; (d) decrease. **18.3** (a) pH = 3.44. **18.5** $pK_a$ = 7.80. **18.7** (a) $\Delta pH$ = +0.77; (b) $\Delta pH$ = +0.05; (c) $\Delta pH$ = +2.98. **18.9** (a) $H^+(aq) + C_7H_5O_2^-(aq) \rightleftharpoons HC_7H_5O_2(aq)$. (b) This reaction can be written as the sum of two reactions: I. $C_7H_5O_2^-(aq) + H_2O(l) \rightleftharpoons HC_7H_5O_2(aq) + OH^-(aq)$, $K_b$ = $1.5 \times 10^{-10}$; II. $H^+(aq) + OH^-(aq) \rightleftharpoons H_2O(l)$, $K = 1.0 \times 10^{14}$. $K$ for the overall reaction $= K_I \times K_{II} = 1.5 \times 10^4$. (c) $[Na^+]$ = $[ClO_4^-]$ = 0.10 M; $[HC_7H_5O_2]$ = 0.10 M; $[H^+]$ = $[C_7H_5O_2^-]$ = $2.5 \times 10^{-3}$ M. **18.11** Add a few drops of 1 M HCl to a sample from each, and test the pH of the resulting solutions using pH paper, indicator solutions, or a pH meter. The pH of the buffered solution will change very little from 9.0; that of the unbuffered solution will decrease significantly. **18.13** pH = 8.18. **18.15** (a) $[HC_7H_5O_2]$ = 0 067 M; $[C_7H_5O_2^-]$ = 0.033 M; pH = 3.88. (b) pH = 1.30. **18.17** (a) pH = 4.57. (b) $[HC_2H_3O_2]$ = 0.16 M; $[C_2H_3O_2^-]$ = 0.090 M; pH = 4.49. (c) pH = 4.64. **18.19** (a) $[HCO_3^-]/[H_2CO_3]$ = 11; (b) $[HCO_3^-]/[H_2CO_3]$ = 5.4. **18.21** (a) The quantity of base required to reach the equivalence point is the same in the two titrations. (c) The pH at the equivalence point is higher in the titration of a weak acid. **18.23** (b) 21.0 mL NaOH solution. **18.25** (a) pH = 1.54; (b) pH = 3.30. **18.27** (a) pH = 7.00; (b) pH = 3.52. **18.29** (a) $K_{sp} = [Cd^{2+}][S^{2-}]$; (c) $K_{sp} = [Ce^{3+}][F^-]^3$. **18.31** $K_{sp} = 7.63 \times 10^{-9}$. **18.33** (a) $[Ag^+] = [IO_3^-] = 1.00 \times 10^{-4}$; $K_{sp} = 1.08 \times 10^{-8}$; (c) $K_{sp} = 8.2 \times 10^{-12}$. **18.35** (a) $9.1 \times 10^{-9}$ M; (b) $2.8 \times 10^{-14}$ M. **18.37** $Q = (0.050)(1.0 \times 10^{-6})^2 = 5.0 \times 10^{-14}$; $Q < K_{sp}$. Therefore, $Mn(OH)_2$ will not precipitate. **18.39** $[Cr^{3+}]/[Ce^{3+}] = 6.6 \times 10^{-5}/8.0 \times 10^{-10} = 8.2 \times 10^4$. **18.41** More soluble in acid: $Ag_2CO_3$, $CeF_3$, $Cd(OH)_2$. $Ag_2SO_4$ will be slightly more soluble in more concentrated acid solutions because of the equilibrium $SO_4^{2-}(aq) + H^+(aq) \rightleftharpoons HSO_4^-(aq)$. **18.43** (a) The solubility is 2.5 M. **18.45** The solubility is $3 \times 10^{-5}$ M. **18.47** $[KMnO_4]$ = 0.11 M. **18.49** (a) $Zn(OH)_2(s) + 2H^+(aq) \longrightarrow Zn^{2+}(aq) + 2H_2O(l)$; Lewis base: $OH^-$; Lewis acid: $H^+$. (b) $Cd(CN)_2(s) + 2CN^-(aq) \longrightarrow [Cd(CN)_4]^{2-}(aq)$; Lewis base: $CN^-$; Lewis acid: $Cd^{2+}$. **18.51** $[Cu^{2+}] = 2 \times 10^{-12}$ M. **18.53** $K = K_{sp} \times K_f = 2 \times 10^{11}$. **18.55** The first two experiments eliminate group 1 and 2 ions (Figure 18.12). Since there is no insoluble carbonate precipitate in the filtrate from the third experiment, group 4 ions can be ruled out. The ions which might be in the sample are those from group 3—$Al^{3+}$, $Fe^{2+}$, $Zn^{2+}$, $Cr^{3+}$, $Ni^{2+}$, $Co^{2+}$, and $Mn^{2+}$—and those of group 5—$NH_4^+$, $Na^+$, and $K^+$. **18.57** (a) Make the solution acidic using 0.5 M HCl; saturate with $H_2S$. CdS will precipitate, ZnS will not. (b) Add excess NaOH; $Fe(OH)_3(s)$ does not dissolve, but $Cr^{3+}$ forms the soluble complex $[Cr(OH)_4]^-$. **18.59** (b) $K_{sp}$ for those in group 3 is much larger, so to exceed $K_{sp}$, a higher $[S^{2-}]$ is required. This is achieved by making the solution more basic. **18.61** The greatest buffer capacity is possessed by solution (a), because the concentrations of the acid and conjugate base available to react with added base or acid is largest. Solution (b) has the smallest buffer capacity, because there is no acid-conjugate base equilibrium present. **18.64** $K_a = 8.4 \times 10^{-6}$. **18.66** $pK_a$ = 4.68. **18.69** (a) True. The presence of the strong base inhibits the dissociation reaction of the weak base, so $[OH^-]$ is determined essentially by the concentration of the strong base. (c) False. The *capacity* to resist changes in pH is related to the concentrations of the acid and base pair. When the concentra-

tions are high, there is more reagent present to react with added acid or base. **18.72** 10 g $NaC_2H_3O_2$ and 13 mL glacial acetic acid. **18.75** Solubility = $[Cu^{2+}]$ = $2.8 \times 10^{-4}$ M. **18.76** $[Pb^{2+}]$ in solution $1.6 \times 10^{-3}$ M. At pH = 1, $Q = 1.1 \times 10^{-22}$ and $Q > K_{sp}$. PbS will precipitate from the solution. **18.79** (a) $[HS^-] = 6 \times 10^{-6}$ M. **18.82** HgS and PbS will precipitate, but NiS will not. **18.86** $[Ag^+] = 1.5 \times 10^{-8}$ M.

# CHAPTER 19

**19.1** Spontaneous: a, d, e; nonspontaneous: b, c. **19.3** Berthelot's suggestion is incorrect. Some spontaneous processes that are non-exothermic are the expansion of certain pressurized gases, the dissolving of one liquid in another, and the dissolving of many salts in water. **19.5** (a) Increases; (d) decreases. **19.7** (a) $N_2$ at 0.1 atm; (b) $Br_2(g)$. **19.9** (a) $S°$ for $Hg(g)$ is 175 J/mol-K, and for $Hg(l)$ it is 77 J/mol-K. $S°$ is higher for $Hg(g)$ because it occupies a much larger volume in space. (b) $S°$ for $F_2(g)$ is 203 J/mol-K, and for $F(g)$ it is 159 J/mol-K. $S°$ is higher for $F_2$ ($F_2$ has more ways to store energy). **19.11** (a) $-\Delta S$; (b) $+\Delta S$. **19.13** (a) $\Delta S° = -7.4$ J/mol-K. This value is near zero because the number of moles (and the kinds of molecules) are the same on both sides of the equation. The reason for the slightly negative value of $\Delta S°$ cannot be determined from the equation alone. (c) $\Delta S° = +91.8$ J/mol-K. It is positive because there are six moles of gas among the products and five among the reactants. **19.15** (a) The free-energy change, $\Delta G$, is given by $\Delta G = \Delta H - T\Delta S$, when $\Delta H$ and $\Delta S$ represent the enthalpy change and entropy change, respectively. (b) There is no necessary relationship between $\Delta G$ and the rate of a process. Even when $\Delta G$ is large and negative, a process may not occur at a measurable rate, because the energy barrier to reaction is too high. **19.17** $\Delta G° = -1.23 \times 10^5$ J. The process is highly spontaneous, so one does not expect that $H_2O_2(g)$ would be stable in the gas phase at 298 K. **19.19** (a) $\Delta G° = -49.4$ kJ, spontaneous; (b) $\Delta G° = -109.6$ kJ, spontaneous. **19.21** (a) $\Delta G$ is negative at low temperatures, positive at high temperatures. The reaction proceeds in the forward direction spontaneously at lower temperatures but spontaneously reverses at higher temperatures. (b) $\Delta G$ is positive at all temperatures, and the reaction is nonspontaneous in the forward direction at all temperatures. **19.23** (a) $\Delta H° = +229.7$ kJ; $\Delta S° = 156$ J/K. The reaction is nonspontaneous at low temperatures and will become spontaneous at some very high temperature. (b) $\Delta G°(900 K)$ = 89 kJ. **19.25** (a) $\Delta G$ becomes more positive. **19.27** (a) $\Delta G° = -5.40$ kJ; (b) $\Delta G = -14.54$ kJ. **19.29** (a) $\Delta G° = 2.60$ kJ; $K_P = 0.350$. **19.31** (a) $P_{CO_2}$ at 298 K = $1.6 \times 10^{-23}$ atm. **19.33** (a) $\Delta G° = +23.9$ kJ. (b) By definition, $\Delta G = 0$ when the system is at equilibrium. (c) $\Delta G = -11.6$ kJ. **19.35** (a) $-890.4$ kJ of heat produced/mol $CH_4$ burned; (b) $w_{max} = -817.2$ kJ/mol $CH_4$. **19.37** (a) False. The feasibility of manufacturing $NH_3$ from $N_2$ and $H_2$ depends on whether the reaction proceeds far to the right before arriving at equilibrium. The position of the equilibrium depends on $\Delta H$ *and* the entropy change for the process. (b) True. (c) False. Spontaneity relates to the position of equilibrium in a process, not to the rate at which that equilibrium is approached. (d) True. (e) False. Such a process may or may not be spontaneous. Spontaneous processes are exothermic and/or lead to increased *disorder* in the system. **19.40** There is no inconsistency. The second law states that in any spontaneous process there is an increase in the entropy of the *universe*. In the present case, there is a decrease in entropy of the system, but it is more than offset by an increase in entropy of the surroundings. **19.43** Spontaneous at 25°C: (a), (b); spontaneous at higher temperatures: (a), (c). **19.46** (a) $\Delta G°(718\ K) = -23.4\ kJ$. **19.48** (ii) (a) $\Delta H°$ is positive (the uncombined elements are in a higher energy state than in $KI(l)$; $\Delta S°$ is positive (production of a gas from a liquid increases disorder in the system). Positive $\Delta H°$ leads to a smaller value of $K$, while positive $\Delta S°$ increases the value of $K$, particularly at high temperatures. (b) At standard conditions, $K < 1$ for this reaction. (c) $K$ will increase at higher temperatures due to the positive value for $\Delta S°$. **19.50** (a) For the aqueous equilibrium $K^+(\text{plasma}) \rightleftharpoons K^+(\text{muscle})$, $K = 1$ and $\Delta G° = 0$. Therefore, $\Delta G = 8.75\ kJ$. (b) The minimum amount of work required to transfer one mole of $K^+$ is 8.75 kJ, although in practice more than minimum work is required.

## CHAPTER 20

**20.1** The salt bridge provides a mechanism by which ions not directly involved in the redox reaction can migrate into the anode and cathode compartments to maintain charge neutrality of the solutions. **20.3** (a) Cathode: $NO_3^-(aq) + 4H^+(aq) + 3e^- \rightleftharpoons NO(g) + 2H_2O(l)$; anode: $Sn^{2+}(aq) \rightleftharpoons Sn^{4+}(aq) + 2e^-$. (b) $E° = 0.81\ V$. **20.5** (a) Anode, Sn; cathode, Cu; (b) The copper electrode gains mass as Cu is plated out, and the tin electrode loses mass as Sn is oxidized; (c) Cu $(+)$, Sn $(-)$; (d) $E° = 0.473\ V$. (e) $Cu^{2+}(aq) + Sn(s) \longrightarrow Cu(s) + Sn^{2+}(aq)$. **20.7** (a) $E° = 0.78\ V$. (b) "Under standard conditions" refers to solutions that are 1 $M$ in concentration, at 25°C, with all gases at 1 atm pressure.

**20.9** (a)–(c).

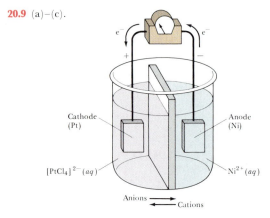

Cathode (Pt)  Anode (Ni)

$[PtCl_4]^{2-}(aq)$   $Ni^{2+}(aq)$

Anions ⟶ ⟵ Cations

(d) Anions flow from the cathode compartment to the anode compartment; cations flow in the opposite direction; (e) $E° = 1.01\ V$. **20.11** (a) Oxidizing strength: $Cu^{2+}(aq) < O_2(g) < Cr_2O_7^{2-}(aq) < Cl_2(aq) < H_2O_2(aq)$; (b) reducing strength: $H_2O_2(aq) < I^-(aq) < Sn^{2+} < Zn(s) < Al(s)$. **20.13** Any *reduced* species in Table 20.1 for which the oxidation potential is more positive than $-0.43\ V$; these include $Zn(s)$, $H_2(g)$, etc. ($Fe(s)$ is questionable). **20.15** Spontaneous: (a) and (b). **20.17** (a) True; (b) false. The free-energy change is an *extensive* property; it depends on the number of moles of substance reacting. (c) False. $\Delta G°$ has units of joules or kJ, and $E°$ is expressed in volts. **20.19** (a) $\Delta G° = -235\ kJ$. **20.21** (c) $E° = +0.44\ V$; $K = 1 \times 10^{74}$. **20.23** (a) $K = 7 \times 10^{33}$; (b) $K = 2 \times 10^{-7}$. **20.25** (a) $E$ is lowered by increasing $[Br^-]$ and shifting the equilibrium to the left. (d) $E$ is unchanged because

increasing the quantity of a pure substance does not change its "concentration." **20.27** $E = 1.299\ V$. **20.29** pH = 1.23; $[H^+] = 0.059\ M$. **20.31** 319 g $MnO_2$. **20.33** (a) Discharge: $Cd(s) + NiO_2(s) + 2H_2O(l) \longrightarrow Cd(OH)_2(s) + Ni(OH)_2(s)$. In charging, the reverse reaction happens. (b) $E° = +1.25\ V$. **20.35** $E°$ for the half-cell reaction $Cd(s) \longrightarrow Cd^{2+}(aq) + 2e^-$ is $+0.40\ V$, smaller than the value of $+0.76\ V$ for the corresponding oxidation of Zn. Thus, the overall cell emf will be reduced. **20.37** The products are different because in aqueous electrolysis water is reduced in preference to $Mg^{2+}$. Thus, the overall reactions in the two cases are: $MgCl_2(l) \longrightarrow Mg(l) + Cl_2(g)$ and $2Cl^-(aq) + 2H_2O(l) \longrightarrow Cl_2(g) + H_2(g) + 2OH^-(aq)$. **20.39** Chloride is oxidized in preference to water because of the large overvoltage for the formation of $O_2(g)$.

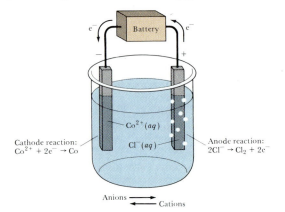

Cathode reaction: $Co^{2+} + 2e^- \rightarrow Co$    Anode reaction: $2Cl^- \rightarrow Cl_2 + 2e^-$

Anions ⟶ ⟵ Cations

**20.41** 3.66 g Zn. **20.43** (a) 3.83 L $Cl_2(g)$; (b) 0.342 mol $NaOH(aq)$. **20.45** (a) True; (b) true; (c) false. Overvoltage requires a higher applied potential than calculated theoretically. Thus, more energy than the minimum is expended in producing the cell reaction. **20.47** $w_{max} = 1.51 \times 10^5\ J$. **20.49** (a) $3.2 \times 10^5$ g Mg; (b) $6.1 \times 10^3$ kWh. **20.51** No. The potential for oxidation of cobalt to $Co^{2+}$, $+0.28\ V$, is lower than that for oxidation of iron to $Fe^{2+}$, $+0.44\ V$. To afford cathodic protection a metal must be *more* readily oxidized than iron. **20.53** The added amines, acting as Brønsted bases, keep the $[H^+]$ in the antifreeze solution low. This keeps the equilibrium in Equation 20.42 shifted to the left, and the corrosion slows down. **20.55** (a) Reaction at anode: $5Fe^{2+}(aq) \longrightarrow 5Fe^{3+}(aq) + 5e^-$; reaction at cathode: $8H^+(aq) + MnO_4^-(aq) + 5e^- \longrightarrow Mn^{2+}(aq) + 4H_2O(l)$. Electrons move from the Pt electrode in the iron solution to the Pt electrode in the $MnO_4^-(aq)$ solution. Anions migrate through the salt bridge from the cathode beaker to the anode beaker; cations migrate in the opposite direction. The electrode in the iron-containing beaker has a negative sign, that in the $MnO_4^-$ beaker has a positive sign. (b) $E° = -0.77\ V + 1.51\ V = 0.74\ V$. **20.57** (a) $Zn(s)$ is oxidized to $Zn^{2+}(aq)$; $Ag_2O(s)$ is reduced to $Ag(s)$. (b) Zn is the negative electrode (anode), and Ag is the positive electrode (cathode). (c) $E° = +1.107\ V$. **20.60** (a) $\Delta G = -232\ kJ$; (b) approximately 6 mol ATP per mol $O_2$. **20.63** The cell reaction is: $2HPO_3^{2-}(aq) + IO_3^-(aq) + 2OH^-(aq) \longrightarrow 2PO_4^{3-}(aq) + IO^-(aq) + 2H_2O(l)$; $E° = 1.61\ V$, $K = 1 \times 10^{109}$. **20.66** 0.079 mol $e^-$ used; 0.50 mol In produced per mol $e^-$. Therefore indium is in the $+2$ oxidation state in the molten halide. **20.69** (a) $E_{min} = 1.34\ V$. (b) A larger voltage than this calculated $E_{min}$ is required to overcome cell resistance, if the reaction is to proceed at a reasonable rate. There may also be an overvoltage for reduction of $Br^-(aq)$ at Pt, although this should not be large. **20.73** Mg acts as a sacrificial anode: $Mg(s) \longrightarrow Mg^{2+}(aq) + 2e^-$. The corresponding cathode reaction is reduction of water: $2H_2O(l) + 2e^- \longrightarrow H_2(g) + 2OH^-(aq)$. Mg replaces the iron of the ship's hull as the anode material. **20.76** Titanium forms a very stable oxide, as indicated by the $E°$ values given. This oxide coat is like the coat of $Al_2O_3$ that forms on aluminum; it completely blankets the surface of the metal and prevents further reaction. **20.78** (a) $1.4 \times 10^{14}$ C; (b) for the process $2H_2O(l) \longrightarrow O_2(g) + 2H_2(g)$, $E° = -1.23$ V; at $P_{H_2} = P_{O_2} = 300$ atm, $E_{min} = -1.34$ V; (c) for 2 mol $e^-$/mol $H_2$, energy $= -n\mathscr{F}E = 2 \times 10^{14}$ J; (d) cost $= \$1.3 \times 10^7$ (13 million dollars!).

## CHAPTER 21

**21.1** (a) 8 protons, 9 neutrons; (b) 42 protons, 57 neutrons. **21.3** (a) $^{181}_{72}Hf \longrightarrow {}^{0}_{-1}e + {}^{181}_{73}Ta$; (b) $^{226}_{88}Ra \longrightarrow {}^{222}_{86}Rn + {}^{4}_{2}He$. **21.5** (b) $^{7}_{4}Be + {}^{0}_{-1}e$ (orbital electron) $\longrightarrow {}^{7}_{3}Li$; (d) $^{98}_{42}Mo + {}^{2}_{1}H \longrightarrow {}^{1}_{0}n + {}^{99}_{43}Tc$. **21.7** $7\alpha$ emissions, $4\beta$ emissions. **21.9** (b) $^{14}_{7}N + {}^{1}_{1}H \longrightarrow {}^{11}_{6}C + {}^{4}_{2}He$. **21.11** Radioactive: (b), which has a low neutron/proton ratio, (c), which has a low neutron/proton ratio, and (e), which has an atomic number greater than 83. Stable: (a) and (d). **21.13** Positron emission (in which a proton is converted to a neutron) is most likely to occur from $^{51}Mn$, the nuclide with the lowest neutron/proton ratio. **21.15** The suggestion is not reasonable. The energies of nuclear states are very large relative to ordinary temperatures. Merely changing the temperature by less than 100 K would not significantly affect the behavior of nuclei with regard to nuclear decay rates. **21.17** (a) No (low neutron/proton ratio, should be a positron emitter) (b) no (low neutron/proton ratio, should be a positron emitter or possibly undergo orbital electron capture); (c) no (high neutron/proton ratio, should be a beta emitter); (d) no (high atomic number, should be an alpha emitter). **21.19** $^{66}_{32}Ge \longrightarrow {}^{66}_{31}Ga + {}^{0}_{-1}e$; 1.56 mg $^{66}Ge$ remains after 10.0 h. **21.21** $k = 0.0306$ yr$^{-1}$, $t_{1/2} = 22.6$ yr. **21.23** $k = 8.15 \times 10^{-3}$ min$^{-1}$, $9.6 \times 10^{-9}$ g $^{139}Ba$ remain. **21.25** Taking only the spontaneous radioactive decay of $^{131}I$ into account, the activity after 32 days should be 5.7 counts per minute, which is the observed activity. The plants do not absorb iodide. **21.27** $k = 1.5 \times 10^{-10}$ yr$^{-1}$; the original rock contained 16.2 mg $^{238}U$ and is $1.9 \times 10^9$ yr old. **21.29** $1.70 \times 10^3$ yr. **21.31** (a) $8.25 \times 10^{-3}$ amu or $1.23 \times 10^{-12}$ J per nucleon. **21.33** $5.14 \times 10^{-12}$ J/nucleus; $3.09 \times 10^{12}$ J/mol. **21.35** $1.71 \times 10^5$ kg/day; $2.09 \times 10^8$ g $^{235}U$. **21.37** (d) $^{59}Co$, it has the largest binding energy per nucleon, and binding energy gives rise to mass defect. **21.39** (a) $^{235}_{92}U + {}^{1}_{0}n \longrightarrow {}^{160}_{62}Sm + {}^{72}_{30}Zn + 4{}^{1}_{0}n$; (b) $^{239}_{94}Pu + {}^{1}_{0}n \longrightarrow {}^{144}_{58}Ce + {}^{94}_{36}Kr + 2{}^{1}_{0}n$. **21.41** The major problem has to do with the very high neutron fluxes. These neutrons cause nuclear reactions that alter the core materials in a fundamental way. For example, an (n, p)-type nuclear reaction would convert an atom of the core material into an atom of the element with one lower atomic number. Also, radiation can lead to other kinds of damage in the solid, causing structural defects. **21.43** The $^{59}Fe$ is incorporated into the diet component, which in turn is fed to the rabbits. After a time, blood samples are removed from the animals, the red blood cells separated, and the radioactivity of the sample measured. If the iron in the dietary compounds has been incorporated into blood hemoglobin, the blood cell sample should show beta emission. Samples can be taken at various times to determine the rate of iron uptake, rate of loss of the iron from the blood, and so

forth. **21.45** The extremely high temperature is required to overcome the electrostatic charge repulsions between the nuclei in forcing them together to react. **21.47** (a) The *electron* and *positron* possess the same mass but opposite electrical charge; (c) a *curie* is a measure of the number of nuclear disintegrations that occur in a sample per second, and a *rem* is a measure of the effective dose of radiation received by a human. **21.50** $\mathscr{N}_t/\mathscr{N}_0 = 0.055$, $5.5\%$. **21.53** Of the three types of radiation, alpha rays are least penetrating and gamma rays most. Thus, biological damage from *external* sources is least for alpha rays. However, when radioactive materials are ingested or inhaled, the damage which is caused by alpha emissions for a given level of energy deposition is ten times greater than that caused by beta or gamma rays. **21.56** The most massive radionuclides have the highest neutron/proton ratios and are likely to decay by a process that lowers this ratio, such as beta decay. The least massive nuclides will decay by a process that increases the neutron/proton ratio; positron emission or orbital electron capture. **21.59** $4.6 \times 10^{13}$ $^{90}Sr$ nuclei, or $6.9 \times 10^{-9}$ g $^{90}Sr$. **21.62** $7.4 \times 10^{-2}$ disintegration/s; $2.5 \times 10^{-6}$ rad/yr; $2.5 \times 10^{-5}$ rem/yr.

## CHAPTER 22

**22.1** Metals: Sc, Ga, Sr; nonmetals: Ar, I; semimetals: As. **22.3** Bismuth should be most metallic. It has a metallic luster and a relatively low ionization energy. $Bi_2O_3$ is soluble in acid but not in base, characteristic of the basic properties of the oxide of a metal rather than of the oxide of a nonmetal. **22.5** (a) N; (b) K; (c) K in the gas phase (lowest ionization energy), Li in aqueous solution (most positive $E°$ value); (d) N has the smallest *covalent* radius; Ne is difficult to compare because it doesn't form compounds; (e) N. **22.7** (a) $IF_3$, which has a greater difference in electronegativity. **22.9** (a) N is too small a central atom to fit five fluorine atoms, and it does not have available $d$ orbitals, which can help accommodate more than eight electrons. (b) Si does not readily form $\pi$ bonds, which are necessary to satisfy the octet rule for both atoms in the molecule. (c) As has a lower electronegativity than N; that is, it more readily gives up electrons to an acceptor and is more easily oxidized. **22.11** (a) $NaNH_2(s) + H_2O(l) \longrightarrow NH_3(aq) + Na^+(aq) + OH^-(aq)$; (b) $2C_3H_7OH(l) + 9O_2(g) \longrightarrow 6CO_2(g) + 8H_2O(l)$. **22.13** $^{1}_{1}H$—protium; $^{2}_{1}H$—deuterium; $^{3}_{1}H$—tritium. **22.15** Like other elements in group IA, hydrogen has only one valence electron. Like other elements in group 7A, hydrogen needs only one electron to complete its valence shell. **22.17** (b) $C(s) + H_2O(g) \longrightarrow CO(g) + H_2(g)$. **22.19** (a) $NaH(s) + H_2O(l) \longrightarrow Na^+(aq) + OH^-(aq) + H_2(g)$; (b) $Fe(s) + H_2SO_4(aq) \longrightarrow Fe^{2+}(aq) + SO_4^{2-}(aq) + H_2(g)$. **22.21** (a) Ionic; (b) molecular; (c) metallic. **22.23** $7.9 \times 10^8$ kg $H_2$. **22.25** As an oxidizing agent in steel-making; to bleach pulp and paper; in oxyacetylene torches; in medicine to assist in breathing.

**22.27** 

Ozone has two resonance forms; the molecular structure (similar to that of $SO_2$, Sample Exercise 9.1) is bent, with an O—O—O bond angle of approximately $120°$. The $\pi$ bond in ozone is delocalized over the entire molecule; neither individual O—O bond is a full double bond, so the observed O—O distance of 1.28 Å is greater than the 1.21 Å distance in $O_2$, which has a full O—O double bond. **22.29** (a) $CaO(s) + H_2O(l) \longrightarrow Ca^{2+}(aq) + 2OH^-(aq)$; (b) $Al_2O_3(s) + 6H^+(aq) \longrightarrow 2Al^{3+}(aq) + 3H_2O(l)$; (c) $Na_2O_2(s) + 2H_2O(l) \longrightarrow 2Na^+(aq) + 2OH^-(aq) + H_2O_2(aq)$; (d) $N_2O_3(g) + H_2O(l) \longrightarrow 2HNO_2(aq)$; (e) $2KO_2(s) + 2H_2O(l) \longrightarrow 2K^+(aq) + 2OH^-(aq) + O_2(g) + H_2O_2(aq)$; (f) $NO(g) + O_3(g) \longrightarrow NO_2(g) + O_2(g)$. **22.31** (a) Neutral; (b) acidic; (c) basic; (d) amphoteric. **22.33** Assume that all reactions occur in basic solution; the half-reaction for the reduction of $H_2O_2$ is $H_2O_2(aq) + 2e^- \longrightarrow 2OH^-(aq)$. (a) $H_2O_2(aq) + S^{2-}(aq) \longrightarrow 2OH^-(aq) + S(s)$; (b) $SO_2(g) + 2OH^-(aq) + H_2O_2(aq) \longrightarrow SO_4^{2-}(aq) + 2H_2O(l)$. **22.35** Formation of ammonia; as an inert gas in manufacturing processes; as a coolant in liquid form. **22.37** (a) $HNO_2$, $+3$; (b) $N_2H_4$, $-2$; (c) $KCN$, $-3$; (f) $Li_3N$, $-3$.

**22.39** (b)

The molecular geometry around nitrogen is trigonal planar, but the hydrogen atom is not required to lie in this plane. (There is a third resonance form, which makes a smaller contribution to the true structure.) (c) $\ddot{N}{=}N{=}\ddot{O}$; the molecular geometry is linear. (There are two other possible resonance forms involving triple bonds.) **22.41** (a) $Mg_3N_2(s) + 6H_2O(l) \longrightarrow 3Mg(OH)_2(s) + 2NH_3(aq)$; (c) $4NH_3(g) + 3O_2(g) \xrightarrow{\Delta} 2N_2(g) + 6H_2O(g)$. **22.43** (a) $4Zn(s) + 2NO_3^-(aq) + 10H^+(aq) \longrightarrow 4Zn^{2+}(aq) + N_2O(g) + 5H_2O(l)$; (b) $4NO_3^-(aq) + S(s) + 4H^+(aq) \longrightarrow 4NO_2(g) + SO_2(g) + 2H_2O(l)$, or $6NO_3^-(aq) + S(s) + 4H^+(aq) \longrightarrow 6NO_2(g) + 2H_2O(l) + SO_4^{2-}(aq)$; **22.45** (a) $2NO_3^-(aq) + 10e^- + 12H^+(aq) \longrightarrow N_2(g) + 6H_2O(l)$; $E^\circ = +1.25$ V; (b) $2NH_4^+(aq) \longrightarrow N_2(g) + 8H^+(aq) + 6e^-$; $E^\circ = -0.27$ V. **22.47** $1.1 \times 10^{10}$ kg $CH_4$ are required to form the $NH_3$. **22.49** (a) Fire extinguishers; as a coolant (dry ice); carbonation of beverages; manufacture of $Na_2CO_3 \cdot 10H_2O$. **22.51** (a) $HCN$; (b) $SiC$; (c) $CaCO_3$; (d) $CaC_2$. **22.53** (a) $[:C{\equiv}N:]^-$; (d) $\ddot{S}{=}C{=}\ddot{S}$. **22.55** (a) $ZnCO_3(s) \xrightarrow{\Delta} ZnO(s) + CO_2(g)$; (b) $BaC_2(s) + 2H_2O(l) \longrightarrow Ba^{2+}(aq) + 2OH^-(aq) + C_2H_2(g)$; (c) $C_2H_4(g) + 3O_2(g) \longrightarrow 2CO_2(g) + 2H_2O(g)$. **22.57** (a) $2Mg(s) + CO_2(g) \longrightarrow 2MgO(s) + C(s)$; (b) $6CO_2(g) + 6H_2O(l) \xrightarrow{h\nu} C_6H_{12}O_6(aq) + 6O_2(g)$; (c) $CO_3^{2-}(aq) + H_2O(l) \longrightarrow HCO_3^-(aq) + OH^-(aq)$. **22.59** 4.06 L of $C_2H_2(g)$. **22.61** Physical properties: luster, electrical and thermal conductivity, malleability; chemical properties: bonding characteristics of oxides (metal oxides are ionic solids, nonmetal oxides are molecular compounds), acid/base properties of aqueous solutions of the oxides, type of ionic species present in aqueous solution (metals exists mainly as cations, nonmetals as anions or oxyanions). **22.64** 1 ft$^3$ = 28.3 L; at STP, 1 ft$^3$ = 1.26 mol of gas; (a) $7.9 \times 10^{10}$ ft$^3$ of $H_2$ are produced. **22.67** (a) $SO_2$; (b) $N_2O_5$. **22.70** $Ba(s) + O_2(g) \longrightarrow BaO_2(s)$; $BaO_2(s) + 2H^+(aq) + SO_4^{2-}(aq) \longrightarrow BaSO_4(s) + H_2O_2(aq)$. **22.73** Urea is 46.6% N by mass. **22.76** For this reaction, the temperature dependence of the equilibrium constant and the reaction rate are opposing effects. At ambient temperatures, the rate of the reaction is extremely slow, and equilibrium is not established in an economically viable length of time. To establish a reasonable rate, the reaction must be carried out at high temperature. **22.78** Lewis basicity is dependent on the presence of a lone pair of electrons that can be donated to a Lewis acid. All three species

meet this requirement; in fact, they are isoelectronic with the general electronic structure: $[:X{\equiv}Y:]^n$. Given the same number of electrons in all species, the one with the highest nuclear charge on both atoms, $N_2$, will be the poorest donor. The species with the lowest charge on one atom and the lowest *total* nuclear charge, $CN^-$, will be the strongest base. **22.80** (a) $2MnO_4^-(aq) + 5H_2O_2(aq) + 6H^+(aq) \longrightarrow 2Mn^{2+}(aq) + 5O_2(g) + 8H_2O(l)$; (c) $2I^-(aq) + H_2O_2(aq) + 2H^+(aq) \longrightarrow I_2(aq) + 2H_2O(l)$.

## CHAPTER 23

**23.1** (a) $IO_3^-$; $+5$; (d) $NaClO$; $+1$; (f) $XeO_3$; $+6$. **23.3** (a) Potassium chlorate; (c) aluminum chloride; (e) paraperiodic acid. **23.5** (a) Linear; (b) square planar; (d) octahedral around the central iodine. **23.7** Xenon has a lower ionization energy than argon; because the valence electrons are not as strongly attracted to the nucleus, they are more readily promoted to a state in which the atom can form bonds with fluorine. Also, Xe is larger and can more easily accommodate an expanded octet of electrons. **23.9** (a) Van der Waals intermolecular attractive forces increase with increasing numbers of electrons in the atoms. (b) $F_2$ reacts with water: $F_2(g) + H_2O(l) \longrightarrow 2HF(g) + O_2(g)$. That is, fluorine is too strong an oxidizing agent to exist in water. (c) HF has extensive hydrogen bonding. (d) Oxidizing power is related to electronegativity. Electronegativity decreases in the order given. **23.11** *Fluorine* is used to prepare fluorocarbons such as Freon 12 ($CF_2Cl_2$), used as a refrigerant and propellant, and teflon, used as a lubricant. *Chlorine* is used to prepare vinyl chloride ($C_2H_3Cl$) for plastics, and in bleach containing $ClO^-$. *Bromine* is used to produce silver bromide, $AgBr$, used in photographic film. *Iodine* is used in the form of KI as an additive to table salt and as $I_2$ in the antiseptic tincture of iodine. **23.13** (a) $PBr_5(l) + 4H_2O(l) \longrightarrow H_3PO_4(aq) + 5H^+(aq) + 5Br^-(aq)$; (b) $IF_5(l) + 3H_2O(l) \longrightarrow H^+(aq) + IO_3^-(aq) + 5HF(aq)$; (c) $SiBr_4(l) + 4H_2O(l) \longrightarrow Si(OH)_4(s) + 4H^+(aq) + 4Br^-(aq)$. **23.15** (a) $Br_2(l) + 2OH^-(aq) \longrightarrow BrO^-(aq) + Br^-(aq) + H_2O(l)$; (b) $Cl_2(g) + 2Br^-(aq) \longrightarrow Br_2(l) + 2Cl^-(aq)$; (c) $Br_2(l) + H_2O_2(aq) \longrightarrow 2Br^-(aq) + O_2(g) + 2H^+(aq)$. **23.17** (a) $2ClO_3^-(aq) + 10Fe^{2+}(aq) + 12H^+(aq) \longrightarrow Cl_2(g) + 10Fe^{3+}(aq) + 6H_2O(l)$; (b) $E^\circ = +0.70$ V. **23.19** The average Xe-F bond enthalpies in the three compounds: $XeF_2$, 134 kJ; $XeF_4$, 134 kJ; $XeF_6$, 130 kJ. They are remarkably constant in the series. **23.21** (a) $H_2SeO_3$, $+4$; (c) $H_2Te$, $-2$; (e) $CaSO_4$, $+6$. **23.23** (a) Potassium thiosulfate; (b) aluminum sulfide; (c) sodium hydrogen selenite; (d) selenium hexafluoride (or selenium(VI) fluoride).

**23.25** (a)

Tetrahedral

(b)

Octahedral

**23.27** (d) $Se(s) + 2H_2SO_4(l) \xrightarrow{\Delta} SeO_2(g) + 2SO_2(g) + 2H_2O(l)$; (e) $SO_3(g) + H_2SO_4(l) \longrightarrow H_2S_2O_7(l)$; (f) $H_2SeO_3(aq) + H_2O_2(aq) \longrightarrow 2H^+(aq) + SeO_4^{2-}(aq) + H_2O(l)$. **23.29** (b) $Cr_2O_7^{2-}(aq) + 3H_2SO_3(aq) + 2H^+(aq) \longrightarrow 2Cr^{3+}(aq) + 3SO_4^{2-}(aq) + 4H_2O(l)$. **23.31** Te is less electronegative than Se or S, so it withdraws less charge from the oxygens, making the O—H bonds less polar and less likely to ionize as $H^+$.

Also, Te is larger than Se or S and in aqueous solution is likely to be coordinated by solvent water molecules. This interaction increases the electron density on the central tellurium and further reduces the withdrawal of charge from the O—H bonds in $H_2TeO_4$. **23.33** (a) $H_3PO_4$, $+5$; (b) $H_3AsO_3$, $+3$; (e) $K_3P$, $-3$. **23.35** (b) Arsenic acid; (c) tetraphosphorus decaoxide. **23.37** (a) $PCl_4^+$ is tetrahedral, $PCl_6^-$ is octahedral. (b) $PCl_4^+$, $sp^3$; $PCl_6^-$, $d^2sp^3$. (c) The ionic form is stabilized in the solid state by the lattice energy that is gained by forming the ions. **23.39** (a) Only two of the hydrogens in $H_3PO_3$ are bound to oxygen. The third is attached directly to phosphorus and not readily ionized, because the H—P bond is not very polar. (b) The smaller, more electronegative nitrogen withdraws more electron density from the O—H bond, making it more polar and more likely to ionize. (c) Phosphate rock consists of $Ca_3(PO_4)_2$, which is only slightly soluble in water. The phosphorus is unavailable for plant use. (d) $N_2$ can form stable $\pi$ bonds to complete the octet of both N atoms. Because phosphorus atoms are larger than nitrogen atoms, they do not form stable $\pi$ bonds with themselves and must form $\sigma$ bonds with several other phosphorus atoms (producing $P_4$ tetrahedra or sheet structures) to complete their octets. **23.41** (a) $2Ca_3(PO_4)_2(s) + 6SiO_2(s) + 10C(s) \longrightarrow P_4(g) + 6CaSiO_3(l) + 10CO_2(g)$; (b) $3H_2O(l) + PCl_3(l) \longrightarrow H_3PO_3(aq) + 3H^+(aq) + 3Cl^-(aq)$. **23.43** (a) $SiO_2$, $+4$; (b) $GeCl_4$, $+4$; (c) $NaBH_4$, $+3$; (d) $SnCl_2$, $+2$. **23.45** In $C_2H_4$ and $C_2H_2$ the carbon atoms are joined by $\pi$ bonds as well as $\sigma$ bonds. Silicon does not form Si—Si $\pi$ bonds of great strength, so compounds involving such bonds are unstable relative to other bonding configurations. **23.47** (a) Carbon; (b) lead; (c) silicon. **23.49** $GeCl_4(g) + 2H_2O(l) \longrightarrow GeO_2(s) + 4HCl(g)$; $SiCl_4(g) + 2H_2O(l) \longrightarrow SiO_2(s) + 4HCl(g)$. **23.51** $KSi_3AlO_8$. **23.53** (a) $SiO_4^{4-}$; (b) $SiO_3^{2-}$; (c) $SiO_3^{2-}$. **23.56** (a) Sulfur is a less electronegative element than oxygen and can be expected to more readily give up its electrons in the course of being oxidized. (c) Astatine is radioactive and not present in nature to a significant extent. (g) In silicate sheets the overall charge is less than in aluminosilicates because of replacement of $Si^{4+}$ by $Al^{3+}$. Cations are present between layers. Electrostatic interactions between the cations and the sheets determine the rigidity of the structure. **23.58** 0.533 atm. **23.61** 28.9 $M$. **23.64** $8Fe(s) + S_8(s) \longrightarrow 8FeS(s)$; $S_8(s) + 16F_2(g) \longrightarrow 8SF_4(g)$ or $S_8(s) + 24F_2(g) \longrightarrow 8SF_6(g)$; $S_8(s) + 8O_2(g) \longrightarrow 8SO_2(g)$; $S_8(s) + 8H_2(g) \longrightarrow 8H_2S(g)$. **23.66** The valence-shell electron arrangement controls the upper and lower limits of oxidation state. For group 6A the configuration is $ns^2np^4$. Addition of 2 electrons or loss of 6 electrons produces a closed shell. Thus, the oxidation state limits are $-2$ to $+6$. **23.69** (a) $2H_2Se(g) + O_2(g) \longrightarrow 2H_2O(g) + 2Se(s)$; (b) $\Delta G° = -489.0$ kJ; $\Delta G° = -2.303RT \log K$, $K = 5 \times 10^{85}$. **23.71** $GeO_2(s) + C(s) \xrightarrow{\Delta} Ge(l) + CO_2(g)$; $Ge(l) + 2Cl_2(g) \longrightarrow GeCl_4(l)$; $GeCl_4(l) + 2H_2O(l) \longrightarrow GeO_2(s) + 4HCl(g)$; $GeO_2(s) + 2H_2(g) \longrightarrow Ge(s) + 2H_2O(l)$. **23.74** $2.0 \times 10^3$ mol air in the room (average molecular weight of air $= 29.0$ g/mol); $4.0 \times 10^{-2}$ mol $H_2S$ allowable; 3.5 g FeS required. **23.77** The atomic radii calculated for N in the three compounds are: NOF $- 0.80$ Å, NOCl $- 0.98$ Å, NOBr $- 0.99$ Å. The possible resonance forms are $X—N{=}O \longleftrightarrow X{=}N—O$. Because fluorine is a second row element, we expect that $\pi$ bond formation is more important than for Cl or Br. The $\pi$ character of the X—N bond in the second structure suggests that this form will be more important in NOF than NOCl or NOBr and that the N—F bond should be shorter than otherwise expected.

## CHAPTER 24

**24.1** Iron: hematite, $Fe_2O_3$; magnetite, $Fe_3O_4$; and taconite, mainly $Fe_3O_4$. Aluminum: bauxite, $Al_2O_3 \cdot xH_2O$; corundum, $Al_2O_3$, and gibbsite, $Al_2O_3 \cdot 3H_2O$. In ores, iron is present as the $+3$ ion, or as both the $+2$ and $+3$ ions, as in magnetite. Aluminum is always present in the $+3$ oxidation state. **24.3** (a) Heating of an ore to cause decomposition with loss of a volatile product. (b) A process in which a desired mineral component of an ore is selectively dissolved. (c) The undesirable or waste portion of a raw ore. (d) A physical concentration process in which a desired ore is wetted by an agent so that it is "floated" away from gangue in a frothing process. **24.5** (b) $Cu_2S(s) + 2O_2(g) \longrightarrow 2CuO(s) + SO_2(g)$; (d) $Al(OH)_3(s) + OH^-(aq) \longrightarrow [Al(OH)_4]^-(aq)$; (e) $2UO_2(s) + 6CO_3^{2-}(aq) + O_2(g) + 2H_2O(l) \longrightarrow [UO_2(CO_3)_3]^{4-}(aq) + 4OH^-(aq)$. **24.7** (a) $H_2SeO_3(aq) + 2SO_2(aq) + H_2O(l) \longrightarrow Se(s) + 2SO_4^{2-}(aq) + 4H^+(aq)$; (c) $K(l) + TiCl_4(g) \longrightarrow Ti(s) + 4KCl(s)$. **24.9** (a) $CaO(l) + Al_2O_3(l) \longrightarrow Ca(AlO_2)_2(l)$; (b) $FeO(l) + SiO_2(l) \longrightarrow FeSiO_3(l)$. 24.11 $FeO(s) + H_2(g) \longrightarrow Fe(l) + H_2O(g)$; $Fe_2O_3(s) + 3H_2(g) \longrightarrow 2Fe(l) + 2H_2O(g)$; $FeO(s) + CO(g) \longrightarrow Fe(l) + CO_2(g)$; $Fe_2O_3(s) + 3CO(g) \longrightarrow 2Fe(l) + 3CO_2(g)$. **24.13** $3.6 \times 10^3$ kg $SO_2$ (or, to one significant figure, $4 \times 10^3$ kg $SO_2$). **24.15** (a) Air serves mainly to oxidize coke to CO; this exothermic reaction also provides heat for the furnace: $2C(s) + O_2(g) \longrightarrow 2CO(g)$ $\Delta H = -110$ kJ. (b) Limestone, $CaCO_3$, is the source of basic oxide for slag formation: $CaCO_3(s) \xrightarrow{\Delta} CaO(s) + CO_2(g)$; $CaO(l) + SiO_2(l) \longrightarrow CaSiO_3(l)$. **24.17** To purify electrochemically, use a soluble cobalt salt such as $CoSO_4 \cdot 7H_2O$ as electrolyte. Reduction of $H_2O$ does not occur because of overvoltage. Anode reaction: $Co(s) \longrightarrow Co^{2+}(aq) + 2e^-$; cathode reaction: $Co^{2+}(aq) + 2e^- \longrightarrow Co(s)$. **24.19** $5.8 \times 10^2$ g $Ni^{2+}(aq)$. **24.21** (c) $\Delta G = \Delta H - T\Delta S$; $\Delta H° = +30.1$ kJ; $\Delta S = +24.5$ J/K; $\Delta G = -6.0$ kJ. **24.23** Sodium is metallic; each atom is bonded to many others. When the metal lattice is distorted, many bonds remain intact. In NaCl the ionic forces are strong, and the ions are arranged in very regular arrays. The ionic forces tend to be broken along certain cleavage planes in the solid, and the substance does not tolerate much distortion before cleaving. **24.25** In the electron-sea model the electrons move about in the metallic lattice while the atoms remain more or less fixed in position. Under the influence of an applied potential the electrons are free to move throughout the structure, giving rise to conductivity. **24.27** According to the molecular-orbital or band theory of metallic bonding, the maximum bond order and highest bond strength occurs in metals with six to eight valence electrons. Assuming that hardness is directly related to the bond strength between metal atoms, Cr, with six valence electrons, should have the highest bond strength and greatest hardness in the fourth row. **24.29** A *conductor* has partially filled energy bands; an *insulator* has energy bands that are either completely filled or completely empty, with a large energy gap between full and empty bands. A *semiconductor* has a filled energy band separated by a small energy gap from an empty band. **24.31** White tin is more metallic in character; it has a higher conductivity and larger Sn-Sn distance (3.02 Å) than gray tin (2.81 Å). **24.33** An *alloy* contains atoms of more than one element and has the properties of a metal. In a *solution alloy* the components are randomly dispersed. In a *heterogeneous alloy* the components are not evenly dispersed and can be distinguished at a macroscopic level. In an *intermetallic compound* the components have interacted to form a compound substance, as in $Cu_3As$. **24.35** Al ($E°_{ox} = +1.66$ V) is a much more active metal than Cu ($E°_{ox} = -0.337$ V), and any of the free element is certain to be oxidized by atmospheric agents such as $O_2$. **24.38** The potassium ethyl xanthate molecule has a polar, ionic end and a relatively nonpolar organic end. The polar end interacts with the surface of the mineral through ion-ion electrostatic interactions. The nonpolar end protrudes into the solvent. These nonpolar ends tend to form a layer on bubble surfaces. **24.41** (a) A simple acid-base reaction is most likely: $MoO_3(g) + MnO(l) \longrightarrow MnMoO_4(l)$; (c) $3Mn_3O_4(s) + 8Al(l) \longrightarrow 4Al_2O_3(s) + 9Mn(l)$. **24.44** 493 g pig iron. **24.47** $3.1 \times 10^5$ kWh. **24.50** $5MoS_2(s) + 14H^+(aq) + 14NO_3^-(aq) \longrightarrow 5MoO_3(s) + 10SO_2(g) + 7N_2(g) + 7H_2O(l)$; $MoO_3(s) + 2NH_3(aq) + H_2O(l) \longrightarrow (NH_4)_2MoO_4(s)$; $(NH_4)_2MoO_4(s) \longrightarrow 2NH_3(g) + H_2O(g) + MoO_3(s)$; $MoO_3(s) + 3H_2(g) \longrightarrow Mo(s) + 3H_2O(g)$. **24.53** Si(s) adopts the diamond structure; its 4 valence electrons are involved in 4 localized bonds to its nearest neighbors. Ti(s) exists in a close-packed lattice with 12 nearest neighbors; its valence electrons cannot be localized between pairs of atoms but are delocalized and free to move throughout the structure. In terms of the band model (Figure 24.13), Ti has a partially filled valence band and Si a filled one.

# CHAPTER 25

**25.1** (a) $[Ar]3d^24s^2$; (b) $[Kr]4d^65s^2$. **25.3** The heat of fusion of a metal should be much smaller than the heat of vaporization, because fusion does not involve substantial breaking of metal-metal bonds, which hold the atoms together in both the solid and liquid states. **25.5** Isolated atoms: (c); metallic bonding: (a), (b), (d). **25.7** The lanthanide contraction is the name given to the decrease in atomic size due to the buildup in effective nuclear charge as we move through the lanthanides (elements 58–71) and beyond them. **25.9** See Figure 25.6 (a) $NbCl_5$; (b) $WCl_6$. **25.11** Chromium, $[Ar]3d^54s^1$, has 6 valence electrons, some or all of which can be involved in bonding, leading to multiple stable oxidation states. Al, $[Ne]3s^23p^1$, has only three valence electrons, which are all lost or shared during bonding, producing the +3 state exclusively. **25.13** (a) $[Kr]4d^4$; (b) $[Xe]4f^{14}5d^1$. **25.15** $Cr^{2+}$. **25.17** $CrO_4^{2-}$. **25.19** (a) Hydration causes $E°$ for reduction to be more negative; (b) formation of a stable metal complex would cause $E°$ for reduction to be more negative. **25.21** $Cr^{2+}(aq)$ is immediately oxidized to $Cr^{3+}(aq)$ in the presence of $O_2$ from air, according to the reaction: $4Cr^{2+}(aq) + O_2(g) + 4H^+(aq) \longrightarrow 4Cr^{3+}(aq) + H_2O(l)$. **25.23** Cr(III) and Cr(VI). **25.25** Fe(II) and Fe(III). **25.27** (a) $Fe(s) + 2HCl(aq) \longrightarrow FeCl_2(aq) + H_2(g)$; (b) $Fe(s) + 4HNO_3(aq) \longrightarrow Fe(NO_3)_3(aq) + NO(g) + 2H_2O(l)$. **25.29** Ni(II). **25.31** $Ni^{2+}(aq)$ is green; $Cu^{2+}(aq)$ is blue. **25.33** Acidic: (b), (d); basic: (a), (c). **25.35** Tetrahedral: (b); octahedral: (a), (c), (d). **25.37** The unpaired electrons in a paramagnetic material cause it to be weakly attracted into a magnetic field. A diamagnetic material, where all electrons are paired, is very weakly repelled by a magnetic field. **25.39** Fe, Co, Ni. **25.41** Color, magnetism, and variable oxidation state. **25.44** Heat of atomization is related to strength of metal-metal bonding. In Zn the $3d$ electrons are paired in nonbonding orbitals and held tightly to the Zn nucleus by its increased effective nuclear charge; they are essentially unavailable for metal-metal bonding. Vanadium has a lower effective nuclear charge and unpaired valence electrons,

which are involved in metal-metal bonding. **25.47** (a) $Mn_7O_7$—higher oxidation state of metal; (d) $NiO_2$—metals near the right of a transition series tend to be less stable in high oxidation states due to increasing effective nuclear charge across the series. **25.49** When the metal center is highly charged, the metal-oxygen bond is polar covalent and not easily dissociated. When the metal is in a lower oxidation state, the metal-oxygen bond has more ionic character. The ionic metal oxide dissociates in water to form the aquated metal ion and $O^{2-}$, a strong Lewis base, which reacts with water to form $OH^-(aq)$. **25.51** (a) $E° = -0.40$ V; (b) $E° = +0.30$ V; (c) $E° = +0.429$ V. **25.54** $VO_4^{3-}(aq) + H^+(aq) \longrightarrow HVO_4^{2-}(aq)$; $2HVO_4^{2-}(aq) \longrightarrow V_2O_7^{4-}(aq) + H_2O(l)$; $V_2O_7^{4-}(aq) + H^+(aq) \longrightarrow HV_2O_7^{3-}(aq)$; $2HV_2O_7^{3-}(aq) + HVO_4^{2-}(aq) \longrightarrow V_3O_{10}^{5-}(aq) + H_2O(l)$. **25.57** $Na^+$: cubic (if the structure is a bit distorted) or octahedral; $Al^{3+}$: tetrahedral.

# CHAPTER 26

**26.1** (a) Coordination number = 4, oxidation number = +2; (c) 6, +3. **26.3** (a)

$$\left[ \begin{array}{c} N \\ | \\ N - Zn - N \\ | \\ N \end{array} \right]^{2+}$$
(b)
Cl, N–Pt–Cl, N

**26.5** Tetraamminezinc(II); dichloroethylenediamineplatinum(II); dicyanoargentate(I); *trans*-bromohydridodinitritoplatinate(II). **26.7** (a) $[Pd(NH_3)_2Br_4]$; (b) $K[Pt(NH_3)Br_3]$ or $[Pt(NH_3)_3Br]Br$; (c) $K[Fe(NH_3)_2Br_4]$. **26.9** (b) $[Co(NH_3)_4(N_3)_2]F$; (e) $[Zn(en)_2][HgBr_4]$. **26.11** (a) $CO_3^{2-}$ is bidentate; (b) $C_2O_4^{2-}$ is bidentate; (c) EDTA is pentadentate; (d) ethylenediamine (en) is bidentate.

**26.13** (a)

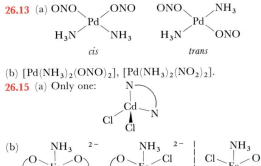

(b) $[Pd(NH_3)_2(ONO)_2]$, $[Pd(NH_3)_2(NO_2)_2]$.
**26.15** (a) Only one:

(b)

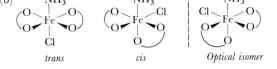

**26.17** Cobalt(III) complexes are generally inert. Thus the ions that form precipitates are outside the coordination sphere. The red complex is $[Co(NH_3)_5SO_4]Br$, pentaamminesulfatocobalt(III) bromide; the violet complex is $[Co(NH_3)_5Br]SO_4$, pentaamminebromocobalt(III) sulfate. **26.19** $[Pd(NC_5H_5)_2Br_2]$; square planar, *trans* isomer; *trans*-dibromodipyridinepalladium(II). **26.21** Blue to blue-violet. **26.23** (a) Seven; (c) three. **26.25** The $d_{x^2-y^2}$ and $d_{z^2}$ metal orbitals point directly toward the six ligands in an octahedron. Electrons in these metal orbitals experience greater electrostatic repulsion with the negative charge or dipole of the ligands than electrons in the $d_{xy}$, $d_{xz}$, and $d_{yz}$ orbitals. Thus, the $d_{xy}$, $d_{xz}$, and $d_{yz}$ orbitals are lower in energy. **26.27** A yellow color is due to absorption of light around 400–

430 nm, a blue color to absorption near 620 nm. The shorter wavelength corresponds to a higher energy electron transition and larger $\Delta$ value. Cyanide is a stronger field ligand, and its complexes are expected to have larger $\Delta$ values than aqua complexes.

**26.29** (a) $[Kr]4d^85s^1$, $[Kr]4d^6$  (b) $[Ar]3d^54s^2$, $[Ar]3d^4$

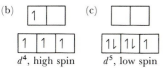

**26.31** All complexes in this exercise are six-coordinate octahedral.

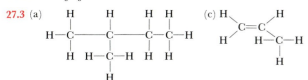

**26.33** Use the experimental procedure in Figure 9.31, employing a weighed sample of known magnetism for calibration purposes. Complex is tetrahedral, $d^7$ (see Figure 26.34). **26.36** $[Pt(NH_3)_6]Cl_4$; $[Pt(NH_3)_4Cl_2]Cl_2$; $[Pt(NH_3)_3Cl_3]Cl$; $[Pt(NH_3)_2Cl_4]$; $K[Pt(NH_3)Cl_5]$. **26.39** $[Co(H_2O)_6]^{2+}$ should be pink and $[CoCl_4]^{2-}$ blue-green. $[CoCl_4]^{2-}$ is tetrahedral; the $\Delta$ values of tetrahedral complexes are only 4/9 as large as for the same ligand in octahedral complexes. Thus the $d$-$d$ transition energy is lower in the tetrahedral complex, and the absorption occurs at a longer wavelength. **26.41** Cobalt is present as $Co^{3+}$, so the crystal-field splitting is greater than for $Mn^{2+}$. Cobalt is $d^6$ and low spin, manganese is $d^5$ and high spin. **26.43** The absorption spectra of $[VF_6]^{3-}$, $[V(NH_3)_6]^{3+}$, and $[V(NO_2)_6]^{3-}$ should exhibit a trend toward absorption at shorter wavelength (higher energy). **26.45** The complex will absorb in the visible, around 660 nm, and appear green. **26.48** Longer wavelength absorption corresponds to a smaller splitting of the $d$ orbital energies. (a) $[CoF_6]^{4-}$, because $F^-$ is a weaker field ligand than $CN^-$. **26.51** For orbital splitting in an octahedral field, the $d^3$ and $d^6$ configurations have electrons only in the lower energy $d$ orbitals, but for tetrahedral splitting, they have one or more electrons in the higher-energy $d$ orbitals. Thus, tetrahedral arrangements are disfavored.

## CHAPTER 27

**27.1** Alkane—$C_6H_{14}$;   alkene—$C_6H_{12}$;   alkyne—$C_6H_{10}$; aromatic—$C_6H_6$.

**27.3** (a)

**27.5** (a) 109°; (b) 120°; (c) 180°.

**27.7** (a)

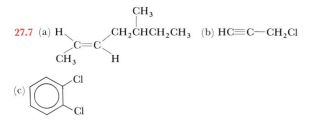

**27.9** (a) *Cis*-6-methyl-3-octene; (c) *para*-dibromobenzene.

**27.11** (a) No;

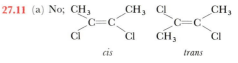

(c) no; (d) no.

**27.13** (a) *Cracking* is a high-temperature, catalyzed process in which straight-chain hydrocarbons are converted to branched-chain molecules. (b) An *isomerization* reaction is any process in which a molecule is converted into another structural isomer. (c) An *antiknock* agent is a chemical added to gasolines to improve their burning characteristics in internal combustion engines.

**27.15** 72. **27.17** (a) $CH_3CH_2CH_2CH_2CH_3(g) + Cl_2(g) \xrightarrow{h\nu}$ $CH_3CH_2CH_2CHClCH_3(g) + HCl(g)$; (b) $2CH_3CH{=}CH_2(g) + 9O_2(g) \longrightarrow 6CO_2(g) + 6H_2O(g)$. **27.19** (a) $CH_3CHCH_3$; with Cl below

(b) $(CH_3)_2CCH_2CH_3$ with Cl below; (c) $CH_3CCH_3$ with Cl below or $CH_3C{=}CH_2$ with Cl below.

**27.21** The 60° C—C—C angles in the cyclopropane ring cause strain that provides a driving force for reactions that result in ring opening. There is no comparable strain in the five- or six-membered rings. **27.23** (b) $CH_3CH_2CI{=}CH_2$ or $CH_3CH_2CI_2CH_3$; (c) $(CH_3)_2C(H)(OH){-}C(CH_3)(C_2H_5)$ or $(CH_3)_2C(HO)(H){-}C(CH_3)(C_2H_5)$

**27.25** $C_2H_4(g) + HBr(g) \longrightarrow CH_3CH_2Br(l)$; $C_6H_6(l) + CH_3CH_2Br(l) \xrightarrow{AlCl_3} C_6H_5CH_2CH_3(l) + HBr(g)$. **27.27** Heat of hydrogenation of naphthalene = $-300$ kJ/mol; for $C_2H_4 = -137$ kJ/mol. The second value applies to one double bond, so for five double bonds we would expect about $-685$ kJ. Since hydrogenation of naphthalene produces only $-300$ kJ, we can conclude that there is some special stability associated with the aromatic system in this molecule. **27.29** (a) Ketone; (b) carboxylic acid; (c) alcohol; (d) ester; (e) amide; (f) amine. **27.31** About each methyl ($CH_3$) carbon, 109°; about the carbonyl carbon, 120°, planar. **27.33** (a) $CH_3CHC(O)H$ with $CH_3$ below; (b) $CH_3CH_2C(O)H$ or

$CH_3CH_2C(O)OH$; (c) no reaction.

**27.35** (a) $CH_3C(H)O{-}C(O)CH_3$ with $CH_3$ below (2-propylacetate) (b) $CH_3CH_2CH_2O{-}C(O)CH_3$ (1-propylacetate)

(c) $CH_3CH_2O{-}C(O)H$ (ethylformate)

**27.37** (b) $HOCH_2CH_2OH$; (d) $CH_3CH_2C(O)CH_2CH_3$; (e) $CH_3CH_2OCH_2CH_3$. **27.39** (a) Methanoic acid; (b) butanoic acid; (c) 3-methylpentanoic acid.

**27.41** (a) $CH_2{=}CHCH_3 + H_2O \xrightarrow{H_2SO_4} CH_3\underset{\underset{OH}{|}}{C}HCH_3$;

$CH_3\underset{\underset{OH}{|}}{C}HCH_3 + (O) \longrightarrow CH_3\underset{\underset{O}{\|}}{C}CH_3 + H_2O$;

(b) $CH_3CH_2CH_2OH + 2(O) \longrightarrow CH_3CH_2\overset{\overset{O}{\|}}{C}{-}OH + H_2O$

**27.43** (a) $nCH_2{=}\underset{\underset{CN}{|}}{C}H \longrightarrow \left[\begin{array}{cc} \overset{\overset{H}{|}}{C} & \overset{\overset{H}{|}}{C} \\ \underset{\underset{H}{|}}{} & \underset{\underset{CN}{|}}{} \end{array}\right]_n$

(b) $nCF_2{=}CF_2 \longrightarrow \left[\begin{array}{cc} \overset{\overset{F}{|}}{C} & \overset{\overset{F}{|}}{C} \\ \underset{\underset{F}{|}}{} & \underset{\underset{F}{|}}{} \end{array}\right]_n$

**27.45**

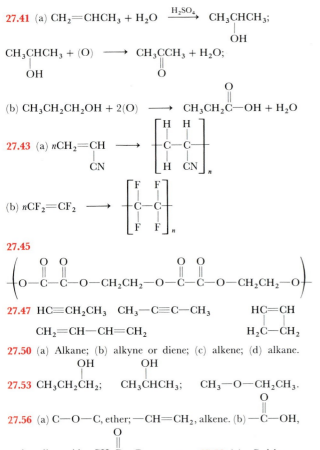

**27.47** $HC{\equiv}CH_2CH_3$  $CH_3{-}C{\equiv}C{-}CH_3$  $\begin{array}{c} HC{=}CH \\ H_2C{-}CH_2 \end{array}$

$CH_2{=}CH{-}CH{=}CH_2$

**27.50** (a) Alkane; (b) alkyne or diene; (c) alkene; (d) alkane.

**27.53** $CH_3CH_2\underset{\underset{OH}{|}}{C}H_2$;  $CH_3\underset{\underset{OH}{|}}{C}HCH_3$;  $CH_3{-}O{-}CH_2CH_3$.

**27.56** (a) C—O—C, ether; —CH=CH$_2$, alkene. (b) $\overset{\overset{O}{\|}}{C}{-}OH$, carboxylic acid; $CH_3\overset{\overset{O}{\|}}{C}{-}O{-}$, ester.  **27.59** (a) *Cyclobutane*. There is strain in the four-membered ring because the C—C—C angles must be less than the desired 109°. (c) *1-hexene*. The alkene readily undergoes addition, whereas the aromatic hydrocarbon is extremely stable, due to resonance.  **27.62** Hydrolysis of the ester or amide linkages under strong base conditions. **27.65** Cyclopentane.

# CHAPTER 28

**28.1** Organisms require energy for maintainance of all living functions. The conversion of external energy to biochemical energy is a spontaneous process, so the accompanying entropy change is positive. The entropy change within the plant may be negative, but the overall entropy of the universe (organism + surroundings) increases.  **28.3** (a) $\Delta G^\circ_{298} = 2878$ kJ/mol glucose, nonspontaneous; (b) $\Delta G^\circ_{298} = 2957$ kJ.  **28.5** 36 g $O_2(g)$, or 25.4 L.  **28.7** (a) An alpha amino acid contains an $NH_2$ function on the carbon adjacent to the carboxylic acid function. (b) In protein formation, amino acids undergo a condensation reaction between the amino group of one molecule and the carboxylic acid group of another to form the amide linkage.  **28.9** Two peptides are possible: $H_2NCH_2CONHCH(CH(CH_3)_2)COOH$ (glycylvaline) and $H_2NCH(CH(CH_3)_2)CONHCH_2COOH$ (valylglycine).

**28.11**

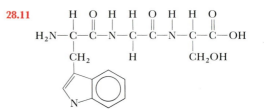

**28.13** In base: $H_2NCH_2CO_2^-$; in acid: $H_3NCH_2COOH^+$.
**28.15** Six: gly-val-ala, gly-ala-val, val-gly-ala, val-ala-gly, ala-gly-val, ala-val-gly.  **28.17** The drawing should resemble Figure 28.5, with the —CH$_3$ group of alanine replaced by the —CH$_2$COOH group of aspartic acid.  **28.19** The *primary structure* of a protein refers to the sequence of amino acids in the chain. The *secondary structure* is the configuration (helical, folded, open, etc.) of the protein chain. The *tertiary structure* is the overall shape of the protein, determined by the way the segments come together or pack.  **28.21** (a) An *enzyme* is a protein that acts as a catalyst in a biochemical reaction. (b) *Denaturation* of a protein is loss of its chemical activity due to disruption of its secondary or tertiary structure. (c) A *holoenzyme* is an enzyme protein plus other essential components called coenzymes or cofactors. (d) A *peptidase* is an enzyme that catalyzes the hydrolysis of a protein or polypeptide chain.  **28.23** That an increase in substrate concentration produces a roughly proportional increase in the rate indicates that most of the active sites are unoccupied at a particular instant and that the equilibrium in Equation 28.4 lies to the left.  **28.25** Holoenzyme: zinc plus protein; apoenzyme: protein alone; cofactor: zinc (present as $Zn^{2+}$); turnover number: $1 \times 10^7$ $CO_2$ molecules/second.  **28.27** When the glucose chain closes to form the hemiacetal (Equation 28.7), α-glucose has the "new" OH group on carbon 1 on the same side of the six-membered ring as the OH group on the adjacent carbon 2, while the β form has the OH groups on carbon 1 and carbon 2 on opposite sides of the ring.

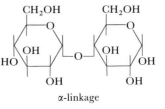

α-linkage

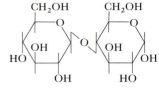

β-linkage

**28.29** Both the β form (shown here) and the α form (OH on $C^1$ on same side of ring as OH on $C^2$) are possible.

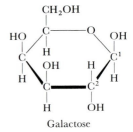

Galactose

**28.31** Numbering the carbon atoms from the top down, atoms 2, 3, 4, and 5 are chiral, because they carry four different groups on each.

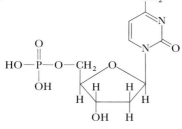

$cis$-CH$_3$(CH$_2$)$_7$CH=CH(CH$_2$)$_7$—$\overset{\displaystyle O}{\overset{\|}{C}}$—O—CH$_2$

$cis$-CH$_3$(CH$_2$)$_7$CH=CH(CH$_2$)$_7$—$\overset{\displaystyle O}{\overset{\|}{C}}$—O—CH

$cis$-CH$_3$(CH$_2$)$_7$CH=CH(CH$_2$)$_7$—$\overset{\displaystyle O}{\overset{\|}{C}}$—O—CH$_2$

**28.35** C$_3$H$_5$(OH)$_3$ + 3C$_{11}$H$_{23}\overset{\displaystyle O}{\overset{\|}{C}}$—OH $\longrightarrow$

C$_3$H$_5$(O$\overset{\displaystyle O}{\overset{\|}{C}}$C$_{11}$H$_{23}$)$_3$ + 3H$_2$O.

**28.37** A nucleotide consists of a nitrogen-containing aromatic compound, a sugar in the furanose (5-membered) ring form, and a phosphoric acid molecule. The structure of deoxycytidine monophosphate is:

**28.39** C$_4$H$_7$CH$_2$OH + H$_3$PO$_4$ $\longrightarrow$

C$_4$H$_7$CH$_2$—O—PO$_3$H$_2$ + H$_2$O.

**28.41** —A—C—T—C—G—A—
         |   |   |   |   |   |
         T—G—A—G—C—T—  $\longleftarrow$  complementary strand

**28.43** (a) 42 m$^2$. (b) This leaf area would result in conversion of about 10,000 kJ per day to plant matter. However, not all this energy is available to the human ingesting the plant. Some energy will be in the form of cellulose, which is not digestible, some goes into root formation, and some is required for various functions within the living plant.

**28.45** (a) CH$_3$CH$\overset{\displaystyle O}{\overset{\|}{C}}$—OH + H$_2$NCH$_2\overset{\displaystyle O}{\overset{\|}{C}}$—OH;
        |
        NH$_2$

Alanine          Glycine

(b) 3 mol C$_{17}$H$_{33}\overset{\displaystyle O}{\overset{\|}{C}}$—OH + C$_3$H$_5$(OH)$_3$.

(c) 2 mol α-glucose     (d) 

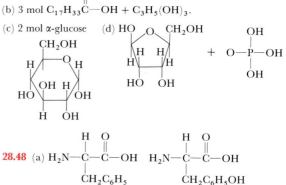

**28.48** (a) H$_2$N—$\overset{\displaystyle H}{\underset{\displaystyle CH_2C_6H_5}{C}}$—$\overset{\displaystyle O}{\overset{\|}{C}}$—OH   H$_2$N—$\overset{\displaystyle H}{\underset{\displaystyle CH_2C_6H_5OH}{C}}$—$\overset{\displaystyle O}{\overset{\|}{C}}$—OH

(b) The only difference between the two amino acids is an OH group on the phenyl (C$_6$H$_5$) ring. The enzyme catalyzes the placement of this hydroxyl group, so it is called a hydroxylase.

**28.51** (a) NH$_2$—CH$\overset{\displaystyle O}{\overset{\|}{C}}$—O$^-$  Na$^+$
              |
              CH$_2$
              |
              CH$_2$
              |
              COOH

(b) The flavor-enhancing properties of MSG probably relate to the metabolic functions of glutamic acid and its sodium salt. Since the L-isomer is the naturally occurring form of glutamic acid, it is not surprising that only the L-isomer has flavor-enhancing properties.   **28.54** That the rate doubles with a doubling of concentration of the sugar tells us that the fraction of enzyme tied up in an enzyme-substrate complex is small. Innositol is behaving like a competitor with sucrose for binding in the active site.   **28.57** *Condensation* is the joining of two neutral molecules by a link formed when a small molecule, usually water, is eliminated. *Hydrolysis* is the reverse process: adding water to split up two molecules. These reactions are central to biochemistry because all biopolymers are formed through condensation reactions, and many metabolic processes involve either condensation or hydrolysis.

# Index

mean free path and thermal conductivity, *326*
molecular weights and densities of, 315-16
natural, *122*
noble. *See* Noble gases
nonideal, 326-30
parts per million, *454*
phase diagrams of, *371-72*
pressure, *302*, 319, *326f*, 327-28, *328f*
  concept of, 302-5, *303f*
  correction to, 329
  Dalton's Law of partial, 313-15, 454
  variation with temperature, *311f*, 311-12
separations, *325f*, 325
van der Waals equation, *329*
vaporization or evaporation, *341*, *342f*
volume of, 301, 306-8, 327-28, *328f*
  correction to, 329
  measuring, 317
  quantity of gas and, 306-8
  temperature and, 306, 320
water, *731*
Gasification of coal, 123-24
Gasoline, 913-14
Gas (vapor), *3*, 301
Gay-Lussac, Joseph Louis, 71, 306-8
Geiger, Hans, 36
Geiger counter, *703f*, *703*
Gemstones, *889f*, 889
Generon system, 325
Genetic damage from radiation, *714*
Geometrical isomers, *908*-9
Geometric isomerism, *877f*, *877*
Geometries of complexes, *866f*, 866, 882
Geometry, molecular. *See* Molecular geometries
Gerlach, Walter, 196
Germanium (Ge), 136, 790
  energy gaps in, *825t*
Gibbs, J. Willard, 635
Glass, 796-97, *797t*
Glucopyranose, 953
Glucose, 120, *393f*, 393, *952*
  combustion of, *115f*, 115-16
  oxidation, 642-43, *643f*
Glycerin, viscosity and surface tension of, *349t*

Glycine, 945
Glycogen, *955*
Glycylalanine, 945
Gold (Au), 42
  hydrometallurgy of, 815
*g* orbital, 181
Graduated cylinder, 14
Graham, Thomas, 322
Graham's law of effusion, *322-23*
Graphite, 292-94, 724, *749-51*, *751f*
  structure of, *293f*
Graphs, 972-73
Gravitational force, 302-3
Gravity, 38
Greeks, ancient, 29
"Greenhouse effect," 469
Ground state, *175*, 197
  of hydrogen atom, 183
Group 1A elements. *See* Alkali metals (Group 1A)
Group 2A elements. *See* Alkaline earth metals
Group 4A elements
  electron configuration of, 790
  general characteristics of, 790-91
  properties of, 790-91, *791t*
  *See also* specific elements
Group 5A elements
  electron configuration, 783
  general characteristics of, 783-84
  properties of, 784, *784t*
  *See also* specific elements
Group 6A elements, 155-57, *156t*
  electron configuration, 775
  general characteristics of, 775-76, *776t*
  occurrences and preparation of, 776-77
  properties and uses of, 777-79
  *See also* specific elements
Group 7A elements. *See* halogens
Group 8A elements. *See* noble gases
Group (family) in periodic table, 42, *138*, *142*
Guldberg, Cato Maximilian, 521

# H

Haber, Fritz, 520, 524, 532, 538-40, 625
Haber process, *519-20*, 533, *742*

Hahn, Otto, 710
Half-life, *491-92*, 493, *699-708*, *700t*
  calculations based on, 701
Half-reaction, *246*
  oxidation-reduction equation balancing and, 438-41
Halide(s)
  hydrogen, *769t*, 769-72, *770f*
    properties of, *272t*
  hydrolysis of nonmetal, 772
  metal, 810
  phosphorus, *785t*, 785-86
  salts, 157-58
Halite. *See* Sodium chloride
Hall, Charles M., 819
Hall process, *818f*, *819*, *820f*, 820
Halogens (group 7A), 42, *139*, *139t*, *157t*, 157-59
  addition to alkynes, 915-16
  electron affinities of, 212, *213t*
  halides, 769-72
  interhalogen compounds, 772-74, *772t*
  molecular forms, 44
  occurrences of, 765-66, *766t*
  oxyacids and oxyanions of, *774t*, 774-75
  properties and preparation of, *766t*, 766-68
  uses of, 768-69
  valence-electron configuration of, 765
Hardness, water, *474*, 847
Heat(s), 101, *102*
  of atomization of transition elements, *839f*, 839
  capacity, *116-17*
    for water, molar, *343t*, 343-44, *344f*
  content, 17
  $\Delta E$ and, 104-5
  $\Delta$, 107
  flow, 117
  of formation, *113-16*, *114t*
    calculating heats of reactions from, 114-15
    standard, *114*, *114t*, *977-78t*
  of fusion, 113
  specific, *117*, *117t*
  of vaporization, 113
Heavy water, 727-28, *728t*
Heisenberg, Werner, 179-80
Helium (He), 41, 42, 159, 763
  discovery of, 173
  electron configuration of, 198, *204f*, 204-6

Polysaccharides, *953*, *954*-57
Polyunsaturated hydrocarbon, 731-32
Polyvinyl chloride, 928
*p* orbitals, 181, 184*f*, 185*f*, 185-86, 279
  interactions between, 287*f*, 287
Porphyrin ring, 940
Porphyrins, *869*-71, 871*f*
Positron, *692*
  emission, 694, 695*f*
Positron emission transaxial tomography (PET), 688, 689, 705*f*, 705
Potassium bromide, 4
Potassium chlorate, 156, 774
  decomposition of, 316-17
  manganese dioxide as catalyst for decomposition of, 504
Potassium chloride, dissolution in water, 628-29
Potassium iodide (KI), 69
Potassium (K), 41, 42, 150
  electron configuration of, 200
  reaction in water, 65*f*
Potential energy, *102*
Potential
  oxidation ($E_{ox}$), 655-56
  reduction ($E_{red}$), 655-56
  standard electrode, *655*-57, 656*t*, 981-82*t*
Power plant, nuclear, 711*f*, 711-13
Precipitate, *69*
Precipitation
  criteria for, 605-14
  effect of pH, 606-7
  reaction, 66, *69*, 422
  and solubility rules, 422-24
  of sulfides, 607-9
Precision, *18*
Prefixes, SI unit, 13
Pressure, *302*
  atmospheric, 107, 303-4, 305, 452*f*, 452-53
    boiling point and, 348
  concept of, 302-5, 303*f*
  critical, *348*, 349*t*, *371*
  Dalton's law of partial, 313-15, 454
  deviations from ideal gas behavior and, 326*f*, 327-28, 328*f*
  equilibrium and changes in, 534-35
  gas, *302*, 319, 326*f*, 327-28, 328*f*

van der Waals' correction to, 329
  variation with temperature, 311*f*, 311-12
osmotic, 405-7, 406*f*, *406*
phase diagrams of, 371*f*, 371-72, 372*f*
solubility and, 394*f*, 394-97, 396*f*
standard temperature and (STP), *309*
vapor, 341*f*, *341*, 346-47
  of solutions, 400-402, 403*f*, 403
Pressure-volume work, 106-9
Priestley, Joseph, 734*f*, 734
Primary alcohol, *920*, 924
Primary structure of protein, *946*
Primitive cubic, *362*
Principal quantum number (*n*), *174*, 181, 182*t*
Product concentrations, equilibrium shifts and, 532-34
Products, *62*
Propane, 65, 903
  combustion of, 70
  structure of, 907*f*
Properties of matter, *4*
  extensive, 17
  intensive, 17
Propionaldehyde, 466
Proteases, 949
Proteins, 121
  denatured, *946*-47
  amino acids, 941-45, 942*f*
  conjugated, *941*
  fuel values of, 121
  structure of, 945-47
Protium ($^1_1H$), *727*, 728
Proton, *689*
  charge of, *38*
  hydrated, 428-30
  mass, 39, 704
Proton-transfer reactions. *See* Aqueous equilibria
*p*-type semiconductor, 826
Pure compound, 10
Pure substances, *4*
*P-V* (pressure-volume) work, 106-9
pX scales, 551
Pyranoses, *953*
Pyridine, 563
Pyrites, 782

Pyrometallurgy, 809-15
  calcination, *809*
  formation of steel, 813-15
  of iron, 811-13
  refining, *811*
  roasting, *809*-10
  slag and, *810*-11
  smelting, *810*

## Q

Quadratic equations, 972
Qualitative analyses for metallic elements, *614*-16
Quantitative analysis, *614*
Quantities, stoichiometrically equivalent, 81
Quantum, *169*
Quantum number
  azimuthal (*l*), 181, 181*t*, 182*t*
  electron spin ($m_s$), *196*
  magnetic ($m_l$), 181, 182*t*
  principal (*n*), *174*, 181, 182*t*
Quantum theory, 168-73
  continuous and line spectra, 171-73
  photoelectric effect and, 170*f*, 170-71
Quantum (wave) mechanics, *177*
  of atom, 180-83
Quartz, 338, 339, 369, 796, 807*f*
Quicklime (calcium oxide), 755

## R

Racemic mixture, *880*, *944*
Radiant energy, 102
  dual nature of, 177
  wave characteristics of, 165-66, 166*f*, 167*f*
  wavelength units for, 167*t*
Radiation
  alpha, 35-36
  beta, 35-36
  biological effects of, 714-16, 716*t*
  doses, 715-16, 716*t*
  from earth, 468-69
  electromagnetic. *See* Radiant energy
  gamma, 35-36
  ozone layer absorption of, 458
Radical chain process, *918*

Radicals, free, *466, 714, 917,* 929
Radii, atomic, 206-8, *207f*
Radioactive decay, *690*
  half-life, 699-703
  types of, 691-92, *692t*
Radioactive series, *695, 696f*
Radioactive wastes, 712
Radioactivity, 34-36, *35, 690,*
    691-96
  detection of, 703-4
Radiocarbon dating, 700-701
Radioimmunoassay (RIA), 705
Radioisotopes, *690,* 704
  half-lives of, *700t*
Radiotracer, *704,* 705
Radium (Ra), 35, 42
Radon (Rn), 42, 159, 763
  electron configuration of, 201
Rad (Radiation absorbed dose),
    *715*
Rain, acid, 156, *462-64, 463f*
Rainbow, 172
Ramsay, Sir William, *158f,* 158-59
Ramsen, Ira, 5
"Random walk," diffusion of gases
    in, *324f,* 324-26
Raoult's law, *401-2*
Rare-earth (lanthanide) elements,
    138, *201*
Rare gases, *42*
Rate constant, *487,* 488
Rate-determining step, *501-2,* 504
Rate laws, *487,* 487-88, 489, 492
  of elementary reactions,
      500-501
  of multistep mechanisms, 501-4
Rate of reaction. *See* Reaction rate
Ray(s)
  cathode, *32*
  X, 34, 365-68, *367f, 367t*
  alpha, *690, 715*
  beta, 715
  gamma, *691*-92, *714,* 715
Rayleigh, Lord, 158-59
RBE, *715*
Reactant concentrations
  equilibrium shifts and, 532-34
  reactant or product
      concentration change and,
      *533f,* 533
  time and, 489-93
Reactants, *62*
  limiting, 83-85, *84f, 84*
Reaction(s), *4-6,* 64-70, *66t,* 725-27
  addition, 915-17

branching chain, *709,* 710
chain fission, *709f,* 709
combustion, 65, 725
critical, *709*
  subcritical, *709*
  supercritical, *709f,* 709
Brønsted acid-base (proton
    transfer), 726
condensation, *787*
double displacement
    (metathesis), *66, 66t,* 725
endothermic and exothermic,
    105, 109
  energy expenditure and, 101
  enthalpies of, 109-11
Friedel-Crafts, 919
half-life of, *491*-92, *493*
  of hydrocarbons, 914-19
Lewis acid-base, 726, 810
molecularity of, 499, *500*
nonspontaneous, 642-43
nuclear, 690-91
precipitation, 66, *69,* 422
prediction of direction of, to
    achieve equilibrium, *529t,*
    529-30
quantities of gases involved in
    chemical, 316-18
single displacement, *66t, 148,*
    725-26
spontaneous, 642
substitution, 917-19
thermite, *419f*
thermonuclear, *499, 713-14*
unimolecular, *49,* 500
*See also* Aqueous solutions;
    Spontaneous reactions
Reaction mechanism, *498-504*
  evidence of, 503
Reaction orders, *487,* 487-88
Reaction quotient, *530, 530t*
Reaction rate, 483-89
  dependence on concentrations,
      486-89
  effect of catalysts, 504-10
  temperature dependence of,
      494-98
Reactors, nuclear, 711-13
Reagent, limiting, 83-85, *84f, 84*
Recharging of lead storage
    battery, 665
Recording tapes, use of
    ferromagnetic oxides in,
    *855f,* 855
Redox reactions. *See* Oxidation-

reduction reactions
Reducing agents. *See* Oxidizing
    (oxidant) and reducing
    (reductant) agents
Reduction, defined, 147, 246
Reduction potential ($E_{red}$), 655-56
Reference (zero-energy) state, 175
Refining, *811, 912*-13
Reinecke's salt, 871
Rems (roentgen equivalent for
    man), *715*
Representative elements, 138,
    *139f, 202*
  electron configuration of, 202
Resonance forms, 236-39, *238*
Reverse osmosis, *471, 472f*
Ribonucleic acids (RNA), *958*
Ribose, *959f*
Roasting, *809-10*
Rock, phosphate, 784
Rods, control, *711f, 711*
Roentgen, Wilhelm, 34, 365
Roosevelt, F.D., 710
Root-mean-square (rms) speed,
    *319,* 320, 322
Rotational motion, *632f, 632*
Rows (periods) of periodic table,
    *138*
Rubber, butyl, 928
Rubidium (Rb), 42, 150
  electron configuration of, 201
Rust, 4, 146
  iron, *676f,* 676-79, *677f, 678f*
Rutherford, Daniel, 696, 740
Rutherford, Ernest, 35-37, 197
Rydberg constant ($R_H$), 174

## S

Salinity of seawater, *470*
Salt(s), *68,* 399
  chlorite, 774
  enhanced corrosion caused by,
      677
  halide, 157-58
  hypochlorite, 774
  nomenclature for, 874
  Reinecke's, 871
  solubility equilibria for slightly
      soluble, 602-5
  solutions, acid-base properties
      of, 568-70
Salt bridge, voltaic cell using, *653f,*
    653, 664

# PHYSICAL AND CHEMICAL CONSTANTS

| | |
|---|---|
| Atomic mass unit | $1 \text{ amu} = 1.6605402 \times 10^{-27}$ kg |
| | $6.0221367 \times 10^{23} \text{ amu} = 1$ g |
| Avogadro's number | $N = 6.0221367 \times 10^{23}/\text{mol}$ |
| Boltzmann's constant | $k = 1.38066 \times 10^{-23}$ J/K |
| Electron rest mass | $m_e = 5.485799 \times 10^{-4}$ amu |
| | $= 9.1093897 \times 10^{-28}$ g |
| Electronic charge | $e = 1.60217733 \times 10^{-19}$ C |
| Faraday's constant | $\mathfrak{F} = Ne = 9.6485309 \times 10^4$ C/mol |
| Gas constant | $R = Nk = 8.31452$ J/K-mol |
| | $= 0.0820579$ L-atm/K-mol |
| Neutron rest mass | $m_n = 1.00866$ amu |
| | $= 1.67495 \times 10^{-24}$ g |
| Pi | $\pi = 3.1415926536$ |
| Planck's constant | $h = 6.6260755 \times 10^{-34}$ J-s |
| Proton rest mass | $m_p = 1.0072765$ amu |
| | $= 1.672623 \times 10^{-24}$ g |
| Speed of light (in vacuum) | $c = 2.997925 \times 10^8$ m/s |

# COMMON IONS

## Positive Ions (Cations)

### 1+
Ammonium ($NH_4^+$)
Copper (I) or cuprous ($Cu^+$)
Hydrogen ($H^+$)
Potassium ($K^+$)
Silver ($Ag^+$)
Sodium ($Na^+$)

### 2+
Barium ($Ba^{2+}$)
Calcium ($Ca^{2+}$)
Chromium (II) or chromous ($Cr^{2+}$)
Cobalt (II) or cobaltous ($Co^{2+}$)
Copper (II) or cupric ($Cu^{2+}$)
Iron (II) or ferrous ($Fe^{2+}$)
Lead (II) or plumbous ($Pb^{2+}$)
Magnesium ($Mg^{2+}$)
Manganese (II) or manganous ($Mn^{2+}$)
Mercury (II) or mercuric ($Hg^{2+}$)
Tin (II) or stannous ($Sn^{2+}$)
Zinc ($Zn^{2+}$)

### 3+
Aluminum ($Al^{3+}$)
Chromium (III) or chromic ($Cr^{3+}$)
Iron (III) or ferric ($Fe^{3+}$)

## Negative Ions (Anions)

### 1−
Acetate ($C_2H_3O_2^-$)
Bromide ($Br^-$)
Chlorate ($ClO_3^-$)
Chloride ($Cl^-$)
Cyanide ($CN^-$)
Fluoride ($F^-$)
Hydride ($H^-$)
Hydrogen carbonate or bicarbonate ($HCO_3^-$)
Hydrogen sulfite or bisulfite ($HSO_3^-$)
Hydroxide ($OH^-$)
Iodide ($I^-$)
Nitrate ($NO_3^-$)
Nitrite ($NO_2^-$)
Perchlorate ($ClO_4^-$)
Permanganate ($MnO_4^-$)

### 2−
Carbonate ($CO_3^{2-}$)
Chromate ($CrO_4^{2-}$)
Oxide ($O^{2-}$)
Peroxide ($O_2^{2-}$)
Sulfate ($SO_4^{2-}$)
Sulfide ($S^{2-}$)
Sulfite ($SO_3^{2-}$)

### 3−
Arsenate ($AsO_4^{3-}$)
Phosphate ($PO_4^{3-}$)